# Understanding Bioinformatics

In memory of Arno Siegmund Baum ז"ל

# Understanding Bioinformatics

Marketa Zvelebil & Jeremy O. Baum

NEW YORK AND LONDON

**Senior Publisher:** Jackie Harbor
**Editor:** Dom Holdsworth
**Development Editor:** Eleanor Lawrence
**Illustrations:** Nigel Orme
**Typesetting:** Georgina Lucas
**Cover design:** Matthew McClements, Blink Studio Limited
**Production Manager:** Tracey Scarlett
**Copyeditor:** Jo Clayton
**Proofreader:** Sally Livitt
**Accuracy Checking:** Eleni Rapsomaniki
**Indexer:** Lisa Furnival
**Vice President:** Denise Schanck



10-digit ISBN 0-8153-4024-9 (paperback)
13-digit ISBN 978-0-8153-4024-9 (paperback)

**Library of Congress Cataloging-in-Publication Data**

Zvelebil, Marketa J.
Understanding bioinformatics / Marketa Zvelebil & Jeremy O. Baum.
p. ; cm.
Includes bibliographical references and index.
ISBN-13: 978-0-8153-4024-9 (pbk.)
ISBN-10: 0-8153-4024-9 (pbk.)
1. Bioinformatics.
[DNLM: 1. Computational Biology--methods. QU 26.5 Z96u 2008] I. Baum, Jeremy O. II. Title.
QH324.2.Z84 2008
572.80285--dc22

2007027514

Published by Garland Science, Taylor & Francis Group, LLC, an informa business
270 Madison Avenue, New York, NY 10016, USA, and
2 Park Square, Milton Park, Abingdon, OX14 4RN, UK.

Printed in India.

15 14 13 12 11 10 9 8 7 6 5 4

Taylor & Francis Group, an informa business

Visit our Web site at http://www.garlandscience.com

# PREFACE

The analysis of data arising from biomedical research has undergone a revolution over the last 15 years, brought about by the combined impact of the Internet and the development of increasingly sophisticated and accurate bioinformatics techniques. All research workers in the areas of biomolecular science and biomedicine are now expected to be competent in several areas of sequence analysis and often, additionally, in protein structure analysis and other more advanced bioinformatics techniques.

When we began our research careers in the early 1980s all of the techniques that now comprise bioinformatics were restricted to specialists, as databases and user-friendly applications were not readily available and had to be installed on laboratory computers. By the mid-1990s many datasets and analysis programs had become available on the Internet, and the scientists who produced sequences began to take on tasks such as sequence alignment themselves. However, there was a delay in providing comprehensive training in these techniques. At the end of the 1990s we started to expand our teaching of bioinformatics at both undergraduate and postgraduate level. We soon realized that there was a need for a textbook that bridged the gap between the simplistic introductions available, which concentrated on results almost to the exclusion of the underlying science, and the very detailed monographs, which presented the theoretical underpinnings of a restricted set of techniques. This textbook is our attempt to fill that gap.

Therefore on the one hand we wanted to include material explaining the program methods, because we believe that to perform a proper analysis it is not sufficient to understand how to use a program and the kind of results (and errors!) it can produce. It is also necessary to have some understanding of the technique used by the program and the science on which it is based. But on the other hand, we wanted this book to be accessible to the bioinformatics beginner, and we recognized that even the more advanced students occasionally just want a quick reminder of what an application does, without having to read through the theory behind it.

From this apparent dilemma was born the division into Applications and Theory Chapters. Throughout the book, we wrote dedicated Applications Chapters to provide a working knowledge of bioinformatics applications, quick and easy to grasp. In most places, an Applications Chapter is then followed by a Theory Chapter, which explains the program methods and the science behind them. Inevitably, we found this created a small amount of duplication between some chapters, but to us this was a small sacrifice if it left the reader free to choose at what level they could engage with the subject of bioinformatics.

We have created a book that will serve as a comfortable introduction to any new student of bioinformatics, but which they can continue to use into their postgraduate studies. The book assumes a certain level of understanding of the background biology, for example gene and protein structure, where it is important to appreciate the variety that exists and not only know the canonical examples of first-year textbooks. In addition, to describe the techniques in detail a level of mathematics is required which is more appropriate for more advanced students. We are aware that many postgraduate students of bioinformatics have a background in areas such as computer science and mathematics. They will find many familiar algorithmic approaches presented, but will see their application in unfamiliar territory. As they read the book they will also appreciate that to become truly competent at bioinformatics they will require knowledge of biomedical science.

There is a certain amount of frustration inherent in producing any book, as the writing process seems often to be as much about what cannot be included as what can. Bioinformatics as a subject has already expanded to such an extent, and we had to be careful not to diminish the book's teaching value by trying to squeeze every possible topic into it. We have tried to include as broad a range of subjects as possible, but some have been omitted. For example, we do not deal with the methods of constructing a nucleotide sequence from the individual reads, nor with a number of more specialized aspects of genome annotation.

The final chapter is an introduction to the even-faster-moving subject of systems biology. Again, we had to balance the desire to say more against the practical constraints of space. But we hope this chapter gives readers a flavor of what the subject covers and the questions it is trying to answer. The chapter will not answer every reader's every query about systems biology, but if it prompts more of them to inquire further, that is already an achievement.

We wish to acknowledge many people who have helped us with this project. We would almost certainly not have got here without the enthusiasm and support of Matthew Day who guided us through the process of getting a first draft. Getting from there to the finished book was made possible by the invaluable advice and encouragement from Chris Dixon, Dom Holdsworth, Jackie Harbor, and others from Garland Science. We also wish to thank Eleanor Lawrence for her skills in massaging our text into shape, and Nigel Orme for producing the wonderful illustrations. We received inspiration and encouragement from many others, too many to name here, but including our students and those who read our draft chapters.

Finally, we wish to thank the many friends and family members who have had to suffer while we wrote this book. In particular JB wishes to thank his wife Hilary for her encouragement and perseverance. MZ wishes to specially thank her parents, Martin Scurr, Nick Lee, and her colleagues at work.

Marketa Zvelebil

Jeremy O. Baum

May 2007

# A NOTE TO THE READER

## Organization of this Book

### Applications and Theory Chapters

Careful thought has gone into the organization of this book. The chapters are grouped in two ways. Firstly, the chapters are organized into seven parts according to topic. Within the parts, there is a second, less traditional, level of organization: most chapters are designated as either Applications or Theory Chapters. This book is designed to be accessible both to students who wish to obtain a working knowledge of the bioinformatics applications, as well as to students who want to know how the applications work and maybe write their own. So at the start of most parts, there are dedicated Applications Chapters, which deal with the more practical aspects of the particular research area, and are intended to act as a useful hands-on introduction. Following this are Theory Chapters, which explain the science, theory, and techniques employed in generally available applications. These are more demanding and should preferably be read after having gained a little experience of running the programs. In order to become truly proficient in the techniques you need to read and understand these more technical aspects. On the opening page of each chapter, and in the Table of Contents, it is clearly indicated whether it is an Applications or a Theory Chapter.

### Part 1: Background Basics

*Background Basics* provides three introductory chapters to key knowledge that will be assumed throughout the remainder of the book. The first two chapters contain material that should be well-known to readers with a background in biomedical science. The first chapter describes the structure of nucleic acids and some of the roles played by them in living systems, including a brief description of how the genomic DNA is transcribed into mRNA and then translated into protein. The second chapter describes the structure and organization of proteins. Both of these chapters present only the most basic information required, and should not in any way be regarded as an adequate grounding in these topics for serious work. The intention is to provide enough information to make this book self-sufficient. The third chapter in this part describes databases, again at a very introductory level. Many biomedical research workers have large datasets to analyze, and these need to be stored in a convenient and practical way. Databases can provide a complete solution to this problem.

### Part 2: Sequence Alignments

*Sequence Alignments* contains three chapters that deal with a variety of analyses of sequences, all relating to identifying similarities. Chapter 4 is a practical introduction to the area, following some examples through different analyses and showing some potential problems as well as successful results. Chapters 5 and 6 deal with several of the many different techniques used in sequence analysis. Chapter 5 focuses on the general aspects of aligning two sequences and the specific methods employed in database searches. A number of techniques are described in detail, including dynamic programming, suffix trees, hashing, and chaining. Chapter 6 deals with methods involving many sequences, defining commonly occurring patterns, defining the profile of a family of related proteins, and constructing a multiple alignment. A key technique presented in this chapter is that of hidden Markov models (HMMs).

## Part 3: Evolutionary Processes

*Evolutionary Processes* presents the methods used to obtain phylogenetic trees from a sequence dataset. These trees are reconstructions of the evolutionary history of the sequences, assuming that they share a common ancestor. Chapter 7 explains some of the basic concepts involved, and then shows how the different methods can be applied to two different scientific problems. In Chapter 8 details are given of the techniques involved and how they relate to the assumptions made about the evolutionary processes.

## Part 4: Genome Characteristics

*Genome Characteristics* deals with the analysis required to interpret raw genome sequence data. Although by the time a genome sequence is published in the research journals some preliminary analysis will have been carried out, often the unanalyzed sequence is available before then. This part describes some of the techniques that can be used to try to locate genes in the sequence. Chapter 9 describes some of the range of programs available, and shows how complex their output can be and illustrates some of the possible pitfalls. Chapter 10 presents a survey of the techniques used, especially different Markov models and how models of whole genes can be built up from models of individual components such as ribosome-binding sites.

## Part 5: Secondary Structures

*Secondary Structures* provides two chapters on methods of predicting secondary structures based on sequence (or primary structure). Chapter 11 introduces the methods of secondary structure prediction and discusses the various techniques and ways to interpret the results. Later sections of the chapter deal with prediction of more specialized secondary structure such as protein transmembrane regions, coiled coil and leucine zipper structures, and RNA secondary structures. Chapter 12 presents the underlying principles and details of the prediction methods from basic concepts to in-depth understanding of techniques such as neural networks and Markov models applied to this problem.

## Part 6: Tertiary Structures

*Tertiary Structures* extends the material in Part 5 to enable the prediction and modeling of protein tertiary and quaternary structure. Chapter 13 introduces the reader to the concepts of energy functions, minimization, and *ab initio* prediction. It deals in more detail with the method of threading and focuses on homology modeling of protein structures, taking the student in a stepwise fashion through the process. The chapter ends with example studies to illustrate the techniques. Chapter 14 contains methods and techniques for further analysis of structural information and describes the importance of structure and function relationships. This chapter deals with how fold prediction can help to identify function, as well as giving an introduction to ligand docking and drug design.

## Part 7: Cells and Organisms

*Cells and Organisms* consists of two chapters that deal in some detail with expression analysis and an introductory chapter on systems biology. Chapter 15 introduces the techniques available to analyze protein and gene expression data. It shows the reader the information that can be learned from these experimental techniques as well as how the information could be used for further analysis. Chapter 16 presents some of the clustering techniques and statistics that are touched upon in Chapter 15 and are commonly used in gene and protein expression analysis. Chapter 17 is a standalone chapter dealing with the modeling of systems processes. It introduces the reader to the basic concepts of systems biology, and shows what this exciting and rapidly growing field may achieve in the future.

## Appendices

Three appendices are provided that expand on some of the concepts mentioned in the main part of this book. These are useful for the more inquisitive and advanced reader. Appendix A deals with probability and Bayesian analysis, Appendix B is mainly associated with Part 6 and deals with molecular energy functions, while Appendix C describes function optimization techniques.

# Organization of the Chapters

## Learning Outcomes

Each chapter opens with a list of learning outcomes which summarize the topics to be covered and act as a revision checklist.

## Flow Diagrams

Within each chapter every section is introduced with a flow diagram to help the student to visualize and remember the topics covered in that section. A flow diagram from Chapter 5 is given below, as an example. Those concepts which will be described in the current section are shown in yellow boxes with arrows to show how they are connected to each other. For example two main types of optimal alignments will be described in this section of the chapter: local and global. Those concepts which were described in previous sections of the chapter are shown in grey boxes, so that the links can easily be seen between the topics of the current section and what has already been presented. For example, creating alignments requires methods for scoring gaps and for scoring substitutions, both of which have already been described in the chapter. In this way the major concepts and their inter-relationships are gradually built up throughout the chapter.

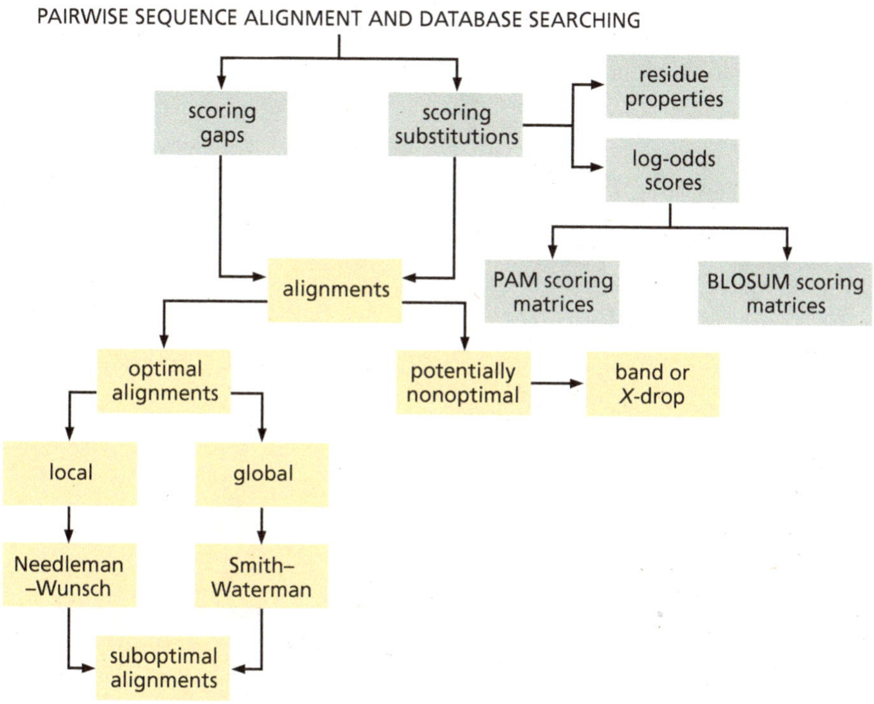

PAIRWISE SEQUENCE ALIGNMENT AND DATABASE SEARCHING

scoring gaps

scoring substitutions

residue properties

log-odds scores

PAM scoring matrices

BLOSUM scoring matrices

alignments

optimal alignments

potentially nonoptimal

band or *X*-drop

local

global

Needleman–Wunsch

Smith–Waterman

suboptimal alignments

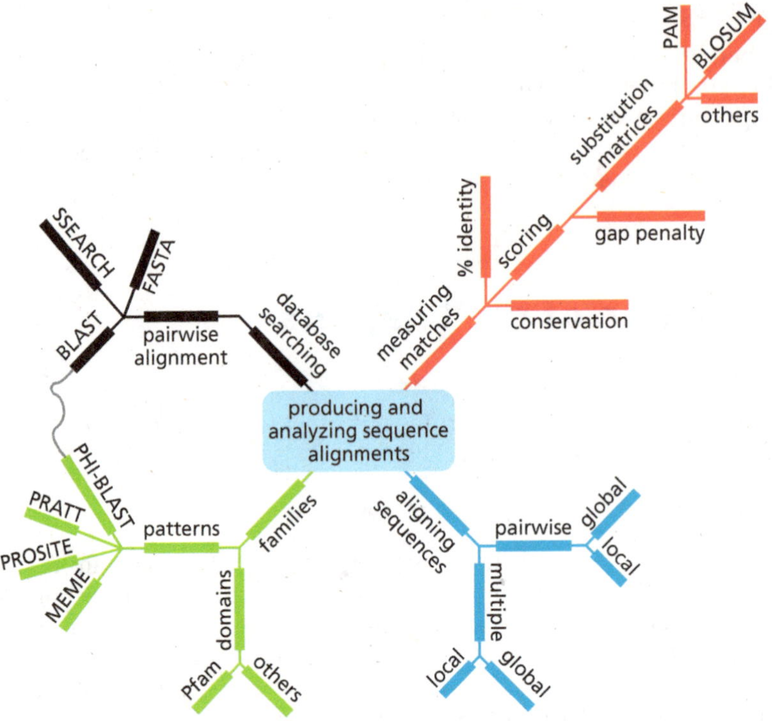


## Mind Maps

Each chapter has a mind map, which is a specialized pedagogical feature, enabling the student to visualize and remember the steps that are necessary for specific applications. The mind map for Chapter 4 is given above, as an example. In this example, four main areas of the topic 'producing and analyzing sequence alignments' have been identified: measuring matches, database searching, aligning sequences, and families. Each of these areas, colored for clarity, is developed to identify the key concepts involved, creating a visual aid to help the reader see at a glance the range of the material covered in discussing this area. Occasionally there are important connections between distinct areas of the mind map, as here in linking BLAST and PHI-BLAST, with the latter method being derived directly from the former, but having a quite different function, and thus being in a different area of the mind map.

## Illustrations

Each chapter is illustrated with four-color figures. Considerable care has been put into ensuring simplicity as well as consistency of representation across the book. Figure 4.16 is given below, as an example.

(A)

p110δ YCVATYVLGIGDRHSDNIMIRESGQLFHIDFGHFLGNFKTKFGINRERVP
p110β YCVASYVLGIGDRHSDNIMVKKTGQLFHIDFGHILGNFKSKFGIKRERVP
p110γ YCVATFVLGIGDRHNDNIMITETGNLFHIDFGHILGNYKSFLGINKERVP
p110α YCVATFILGIGDRHNSNIMVKDDGQLFHIDFGHFLDHKKKKFGYKRERVP

(B)

| name | combined p-value | motifs |
|---|---|---|
| p110α | 5.03e-127 | 2 1 3 |
| p110β | 1.22e-142 | 2 1 3 |
| p110δ | 7.09e-139 | 2 1 3 |
| p110γ | 2.13e-119 | 2 1 3 |

(C)

P11G pig 5.9e-161 2 1 6 3 4
PRKD human 0.34 2 5 6 3 2 1

### Further Reading

It is not possible to summarize all current knowledge in the confines of this book, let alone anticipate future developments in this rapidly developing subject. Therefore at the end of each chapter there are references to research literature and specialist monographs to help readers continue to develop their knowledge and skills. We have grouped the books and articles according to topic, such that the sections within the Further Reading correspond to the sections in the chapter itself: we hope this will help the reader target their attention more quickly onto the appropriate extension material.

### List of Symbols

Bioinformatics makes use of numerous symbols, many of which will be unfamiliar to those who do not already know the subject well. To help the reader navigate the symbols used in this book, a comprehensive list is given at the back which quotes each symbol, its definition, and where its most significant occurrences in the book are located.

### Glossary

All technical terms are highlighted in bold where they first appear in the text and are then listed and explained in the Glossary. Further, each term in the Glossary also appears in the Index, so the reader can quickly gain access to the relevant pages where the term is covered in more detail. The book has been designed to cross-reference in as thorough and helpful a way as possible.

## Garland Science Website

Garland Science has made available a number of supplementary resources on its website, which are freely available and do not require a password. For more details, go to www.garlandscience.com/gs_textbooks.asp and follow the link to *Understanding Bioinformatics*.

### Artwork

All the figures in *Understanding Bioinformatics* are available to download from the Garland Science website. The artwork files are saved in zip format, with a single zip file for each chapter. Individual figures can then be extracted as jpg files.

### Additional Material

The Garland Science website has some additional material relating to the topics in this book. For each of the seven parts a pdf is available, which provides a set of useful weblinks relevant to those chapters. These include weblinks to relevant and important databases and to file format definitions, as well as to free programs and to servers which permit data analysis on-line. In addition to these, the sets of data which were used to illustrate the methods of analysis are also provided. These will allow the reader to reanalyze the same data, reproducing the results shown here and trying out other techniques.

# LIST OF REVIEWERS

The Authors and Publishers of *Understanding Bioinformatics* gratefully acknowledge the contribution of the following reviewers in the development of this book:

| | |
|---|---|
| Stephen Altschul | National Center for Biotechnology Information, Bethesda, Maryland, USA |
| Petri Auvinen | Institute of Biotechnology, University of Helsinki, Finland |
| Joel Bader | Johns Hopkins University, Baltimore, USA |
| Tim Bailey | University of Queensland, Brisbane, Australia |
| Alex Bateman | Wellcome Trust Sanger Institute, Cambridge, UK |
| Meredith Betterton | University of Colorado at Boulder, USA |
| Andy Brass | University of Manchester, UK |
| Chris Bystroff | Rensselaer Polytechnic University, Troy, USA |
| Charlotte Deane | University of Oxford, UK |
| John Hancock | MRC Mammalian Genetics Unit, Harwell, Oxfordshire, UK |
| Steve Harris | University of Oxford, UK |
| Steve Henikoff | Fred Hutchinson Cancer Research Center, Seattle, USA |
| Jaap Heringa | Free University, Amsterdam, Netherlands |
| Sudha Iyengar | Case Western Reserve University, Cleveland, USA |
| Sun Kim | Indiana University Bloomington, USA |
| Patrice Koehl | University of California Davis, USA |
| Frank Lebeda | US Army Medical Research Institute of Infectious Diseases, Fort Detrick, Maryland, USA |
| David Liberles | University of Bergen, Norway |
| Peter Lockhart | Massey University, Palmerston North, New Zealand |
| James McInerney | National University of Ireland, Maynooth, Ireland |
| Nicholas Morris | University of Newcastle, UK |
| William Pearson | University of Virginia, Charlottesville, USA |
| Marialuisa Pellegrini-Calace | European Bioinformatics Institute, Cambridge, UK |
| Mihaela Pertea | University of Maryland, College Park, Maryland, USA |
| David Robertson | University of Manchester, UK |
| Rob Russell | EMBL, Heidelberg, Germany |
| Ravinder Singh | University of Colorado, USA |
| Deanne Taylor | Brandeis University, Waltham, Massachusetts, USA |
| Jen Taylor | University of Oxford, UK |
| Iosif Vaisman | University of North Carolina at Chapel Hill, USA |

# CONTENTS IN BRIEF

# CONTENTS

## Part 2 Sequence Alignments

APPLICATIONS CHAPTER

### Chapter 4 Producing and Analyzing Sequence Alignments

# Part 4 Genome Characteristics

APPLICATIONS CHAPTER

## Chapter 9 Revealing Genome Features



THEORY CHAPTER

## Chapter 10 Gene Detection and Genome Annotation

## Part 7 Cells and Organisms

### Chapter 15 Proteome and Gene Expression Analysis



### Chapter 16 Clustering Methods and Statistics

# PART 1

# BACKGROUND BASICS

The understanding of a number of biological concepts is very important in order to enable any type of bioinformatics research. As bioinformatics deals in large part with the analysis of DNA and proteins, the first two chapters of this book give an overview of some of the most important aspects, essential for bioinformatics analysis, of DNA, genes and the way genes code for proteins, and protein structures.

Another vital aspect of bioinformatics is the use of, and sometimes the creation of, databases—places used to store, retrieve, and analyze data. Therefore the third chapter in this part of the book gives an overview of some of the more common database aspects that the reader should be familiar with before proceeding with the rest of the book.

All three chapters are introductory and in no way attempt to deal with any of the subjects in detail. Many books have been written on all the topics of these three chapters, and the reader should consult these books (examples given in the Further Reading sections) for additional information.

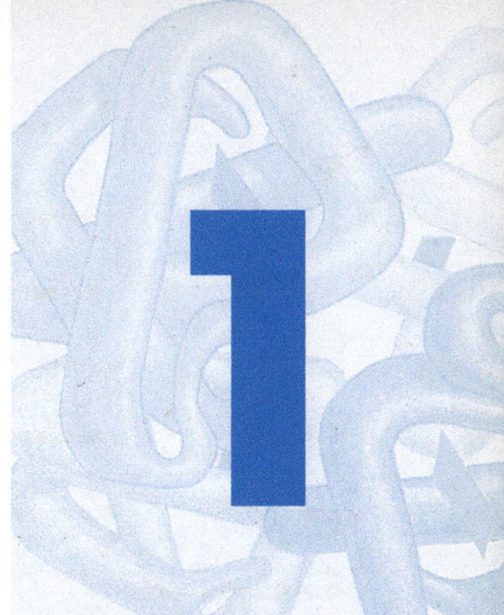

# THE NUCLEIC ACID WORLD

## When you have read Chapter 1, you should be able to:

State the chemical structures of nucleic acids.

Explain base-pairing and the double helix.

Explain how DNA stores genetic information.

Summarize the intermediate role of mRNA between DNA and proteins.

Outline how mRNA is translated into protein by ribosomes.

Outline how gene control is exercised by binding to short nucleotide sequences.

Show that eukaryotic mRNA often has segments (introns) removed before translation.

Discuss how all life probably evolved from a single common ancestor.

Summarize how evolution occurs by changes to the sequence of genomic DNA.

It is amazing to realize that the full diversity of life on this planet—from the simplest bacterium to the largest mammal—is captured in a linear code inside all living cells. In almost exactly the way the vivid detail of a musical symphony or a movie can be digitally recorded in a binary code, so the four base units of the DNA molecule capture and control all the complexity of life. The crucially important discovery of the link between DNA, proteins, and the diversity of life came during the twentieth century and brought about a revolution in the understanding of genetics. Since that time we have amassed increasing amounts of information on the sequences of DNA, RNA, and proteins and the great variety of these molecules in cells under differing conditions. The growth of this information threatens to outstrip our ability to analyze it. It is one of the key challenges facing biologists today to organize, study, and draw conclusions from all this information: the patterns within the sequences and experimental data, the structure and the function of the various types of molecules, and how everything interacts to produce a correctly functioning organism. Bioinformatics is the name we give to this study and our aim is to use it to obtain greater understanding of living systems.

Information about nucleic acids and proteins is the raw material of bioinformatics, and in the first two chapters of this book we will briefly review these two types of biomolecules and their complementary roles in reproducing and maintaining life. This review is not a comprehensive introduction to cell and molecular biology; for that you should consult one of the excellent introductory textbooks of cell and molecular biology listed in Further Reading (p. 24). Rather, it is intended simply as a reminder of those aspects of nucleic acids and proteins that you will need as a background to the bioinformatics methods described in the rest of the book. More

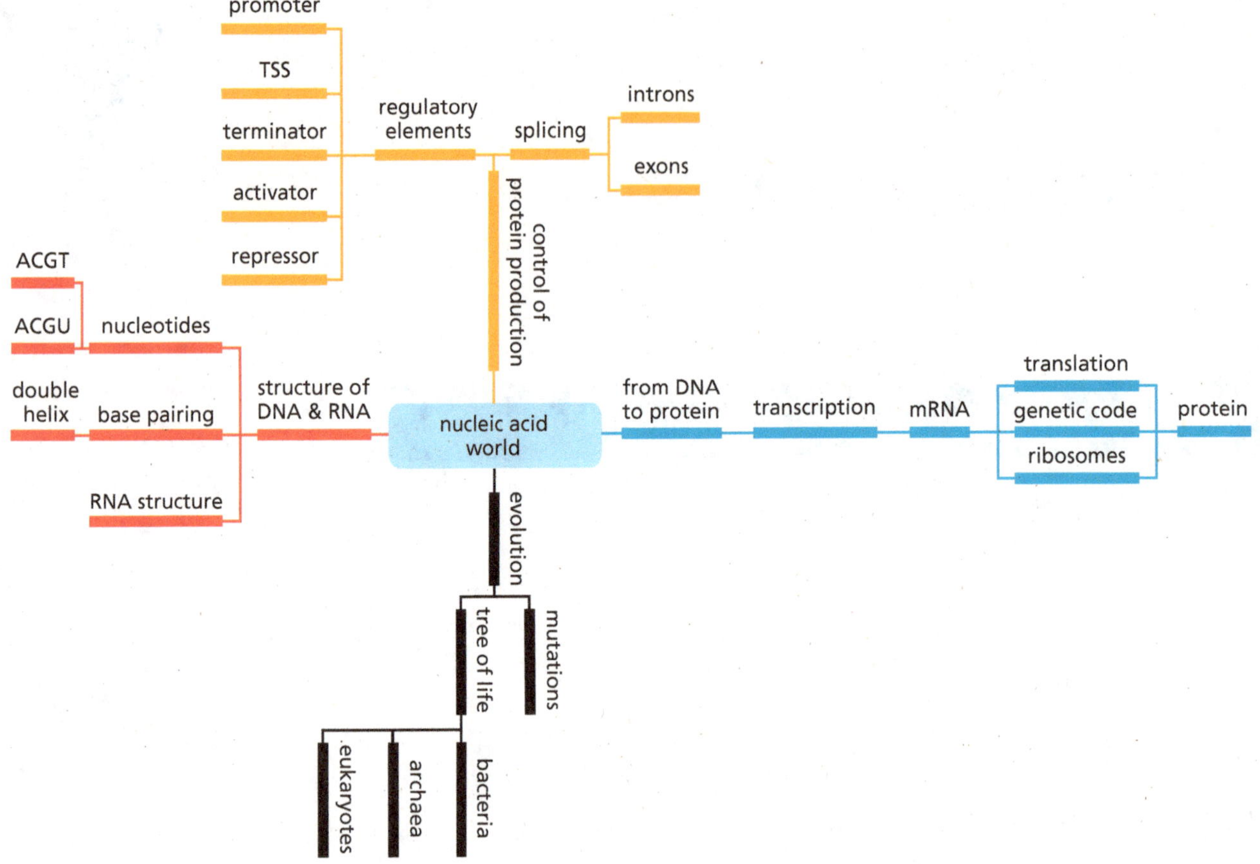


**Mind Map 1.1**
**A mind map schematic of the topics covered in this chapter and divided, in general, according to the topic sections.** This is to help you visualize the topics, understand the structure of the chapter, and memorize the important elements.

information about the biological context of the bioinformatics problems we discuss is also given in the biology boxes and glossary items throughout the book.

This chapter will deal with the nucleic acids—**deoxyribonucleic acid** (**DNA**) and **ribonucleic acid** (**RNA**)—and how they encode proteins, while the structure and functions of proteins themselves will be discussed in Chapter 2. In these two chapters we shall also discuss how DNA changes its information-coding and functional properties over time as a result of the processes of **mutation**, giving rise to the enormous diversity of life, and the need for bioinformatics to understand it.

The main role of DNA is information storage. In all living cells, from unicellular bacteria to multicellular plants and animals, DNA is the material in which genetic instructions are stored and is the chemical structure in which these instructions are transmitted from generation to generation; all the information required to make and maintain a new organism is stored in its DNA. Incredibly, the information required to reproduce even very complex organisms is stored on a relatively small number of DNA molecules. This set of molecules is called the organism's **genome**. In humans there are just 46 DNA molecules in most cells, one in each chromosome. Each DNA molecule is copied before cell division, and the copies are distributed such that each daughter cell receives a full set of genetic information. The basic set of 46 DNA molecules together encode everything needed to make a human being. (We will skip over the important influence of the environment and the nature–nurture debate, as they are not relevant to this book.)

Proteins are manufactured using the information encoded in DNA and are the molecules that direct the actual processes on which life depends. Processes essential to life, such as energy metabolism, biosynthesis, and intercellular communication, are all carried out through the agency of proteins. A few key processes such as the synthesis of proteins also involve molecules of RNA. Ignoring for a moment some of the complexity that can occur, a gene is the information in DNA that directs the manufacture of a specific protein or RNA molecular form. As we shall see, however, the organization of genes within the genome and the structure of individual genes are somewhat different in different groups of organisms, although the basic principles by which genes encode information are the same in all living things.

Organisms are linked together in evolutionary history, all having evolved from one or a very few ancient ancestral life forms. This process of evolution, still in action, involves changes in the genome that are passed to subsequent generations. These changes can alter the protein and RNA molecules encoded, and thus change the organism, making its survival more, or less, likely in the circumstances in which it lives. In this way the forces of evolution are inextricably linked to the genomic DNA molecules.

## 1.1 The Structure of DNA and RNA

Considering their role as the carrier of genetic information, DNA molecules have a relatively simple chemical structure. They are linear polymers of just four different **nucleotide** building blocks whose differences are restricted to a substructure called the **base** (see Flow Diagram 1.1). For efficient encoding of information, one might have expected there to be numerous different bases, but in fact there are only four. This was one of the reasons why the true role of DNA as the carrier of genetic information was not appreciated until the 1940s, long after the role of the chromosomes in heredity was apparent. But although chemically simple, genomic DNA molecules are immensely long, containing millions of bases each, and it is the order of these bases, the **nucleotide sequence** or **base sequence** of DNA, which encodes the information for making proteins.

The three-dimensional structure of DNA is also relatively simple, involving regular double helices. There are also larger-scale regular structures, but it has been clearly established that the information content of DNA is held at the level of the base sequence itself.

RNA molecules are also linear polymers, but are much smaller than genomic DNA. Most RNA molecules also contain just four different base types. However, several classes of RNA molecules are known, some of which have a small proportion of other bases. RNA molecules tend to have a less-regular three-dimensional structure than DNA.

### DNA is a linear polymer of only four different bases

The building blocks of DNA and RNA are nucleotides, which occur in different but chemically related forms. A nucleotide consists of a nitrogen-containing base that is linked to a five-carbon sugar ring at the 1′ position on the ring, which also carries a phosphate group at the 5′ position (see Figure 1.1A). In DNA the sugar is deoxyribose, in RNA it is ribose, the difference between the two being the absence or presence,

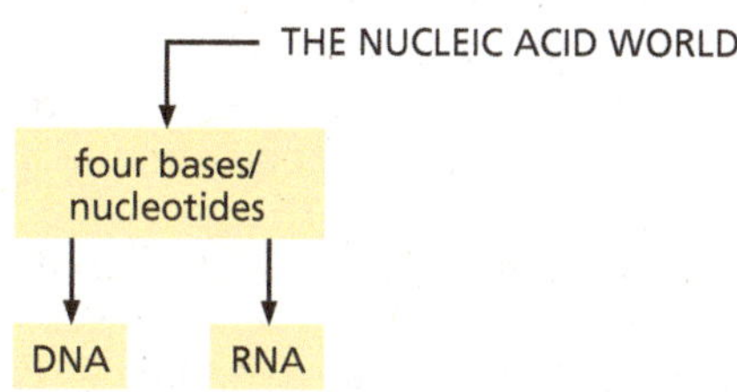


**Flow Diagram 1.1**
**The key concept introduced in this section is that DNA and RNA are composed of subunits called nucleotides, with only four different nucleotide types in a molecule, but a different set of four nucleotides in each of the two types of nucleic acid.**

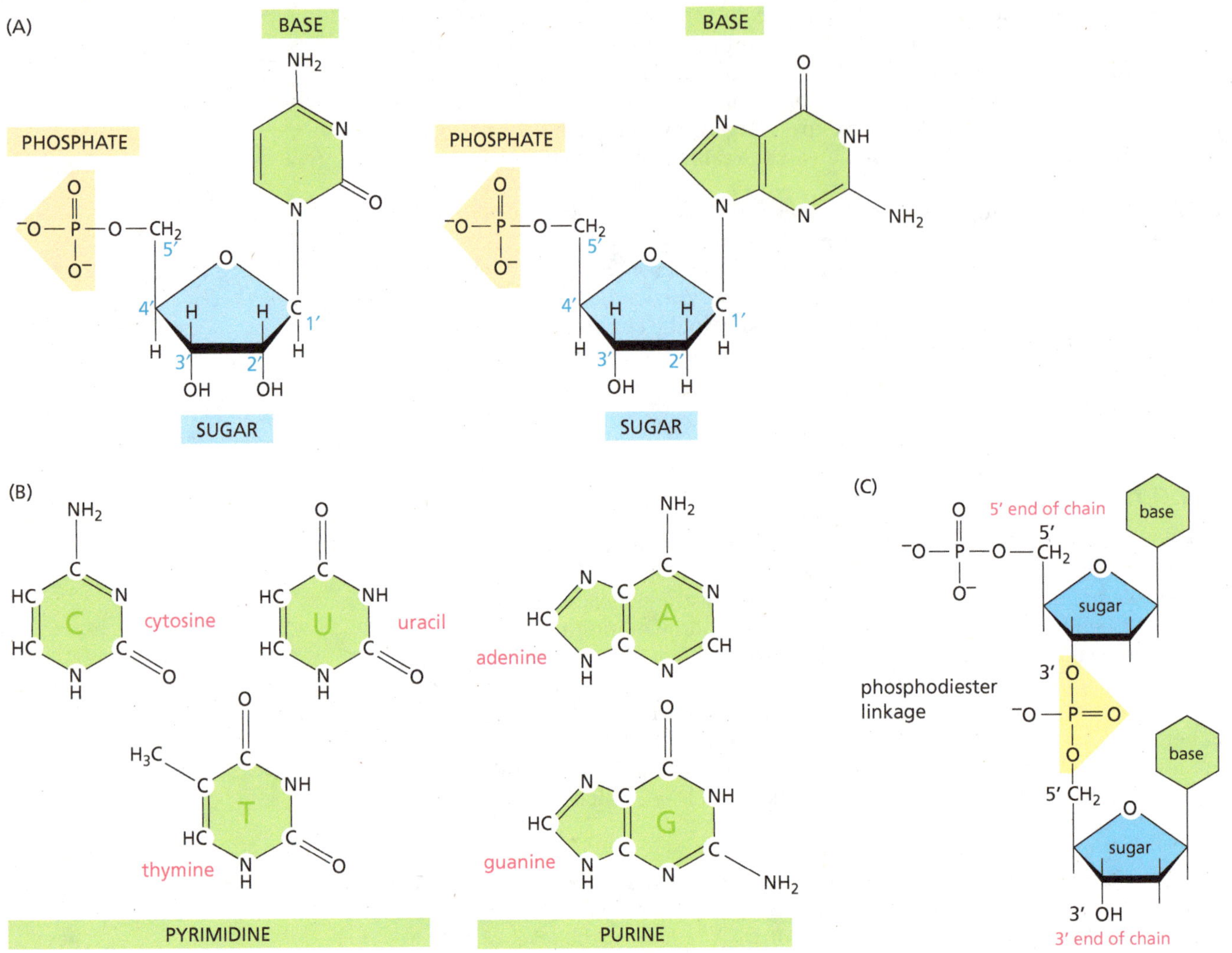

(A)

BASE NH2 PHOSPHATE O O− P O CH2 5′ N N O O 4′ H H C 1′ H 3′ 2′ H OH OH SUGAR

BASE O N NH PHOSPHATE O O− P O CH2 5′ N N NH2 O− O 4′ H H C 1′ H 3′ 2′ H OH H SUGAR

(B)

cytosine uracil thymine adenine guanine

PYRIMIDINE PURINE

(C)

5′ end of chain base 5′ sugar 3′ phosphodiester linkage base 5′ CH2 sugar 3′ OH 3′ end of chain

**Figure 1.1**
**The building blocks of DNA and RNA.** (A) Left, cytidylate (ribo-CMP); right, deoxyguanylate (dGMP). Each consists of three main parts: a phosphate group, a pentose sugar, and a base. It is the base part that is responsible for the differences in the nucleotides. (B) The bases fall into two groups: the purines and the pyrimidines. The purines consist of a 6- plus a 5-membered nitrogen-containing ring while the pyrimidines have only one 6-membered nitrogen-containing ring. (C) It is the phosphate group that is involved in linking the building blocks together by a phosphodiester linkage in DNA. (From B. Alberts et al., Molecular Biology of the Cell, 4th ed. New York: Garland Science, 2002.)

respectively, of a hydroxyl group at the 2′ position. These two types of nucleotide are referred to as **deoxyribonucleotides** and **ribonucleotides,** respectively. Apart from this, the only difference between nucleotides is the base, which is a planar ring structure, either a pyrimidine or a purine (see Figure 1.1B). In DNA only four different bases occur: the purines guanine (G) and adenosine (A) and the pyrimidines cytosine (C) and thymine (T). In most forms of RNA molecule there are also just four bases, three being the same as in DNA, but thymine is replaced by the pyrimidine uracil (U).

Each nucleic acid chain, or strand, is a linear polymer of nucleotides linked together by phosphodiester linkages through the phosphate on one nucleotide and the hydroxyl group on the 3′ carbon on the sugar of another (see Figure 1.1C). This process is carefully controlled in living systems so that a chain is exclusively made with either deoxyribonucleotides (DNA) or ribonucleotides (RNA). The resulting chain has one end with a free phosphate group, which is known as the **5′ end**, and one end with a free 3′ hydroxyl group, which is known as the **3′ end**. The base sequence or nucleotide sequence is defined as the order of the nucleotides along the chain from the 5′ to the 3′ end. It is normal to use the one-letter code given above to identify the bases, starting at the 5′ end, for example AGCTTAG.

There are instances of bases within a nucleic acid chain being modified in a living cell. Although relatively rare, they can be of great interest. In the case of DNA, in vertebrates cytosine bases can be methylated (addition of a $-CH_3$ group). This can

**Figure 1.2**
**Two scientists whose work was influential on James Watson and Francis Crick when they elucidated the structure of DNA.** (A) Maurice Wilkins. (B) Rosalind Franklin. ( A and B courtesy of Science Photo Library.)

result in certain genes being rendered inactive, and is involved in the newly discovered phenomenon known as **genomic imprinting**, where the change that may occur in the offspring depends on whether the gene is maternally or paternally inherited. The class of RNA molecules called tRNA (see Section 1.2) have modifications to approximately 10% of their bases, and many different modifications have been seen involving all base types. These changes are related to the function of the tRNA.

## Two complementary DNA strands interact by base-pairing to form a double helix

The key discovery for which the Nobel Prize was awarded to James Watson and Francis Crick, who drew on the work of Rosalind Franklin and others (see Figure 1.2), was the elucidation in 1953 of the double-helical structure of the DNA molecule, in which two strands of DNA are wound around each other and are held together by hydrogen bonding between the bases of each strand. Their structure was an early example of model building of large molecules, and was based on knowledge of the chemical structure of the constituent nucleotides and experimental X-ray diffraction data on DNA provided by, among others, Maurice Wilkins, who was also awarded a share in the Nobel Prize. (Current methods of model building as applied to proteins are discussed in Chapter 13.)

All the bases are on the inside of the double helix, with the phosphate-linked sugars forming a backbone on the outside (see Figure 1.3A). Crucial to Watson and Crick's success was their realization that the DNA molecules contained two strands and that the base-pairing follows a certain set of rules, now called **Watson–Crick base-pairing**, in which a specific purine pairs only with a specific pyrimidine: A with T, and C with G. Each strand of a DNA double helix therefore has a base sequence that is **complementary** to the base sequence of its partner strand. The bases interact through hydrogen bonding; two hydrogen bonds are formed in a T–A base pair, and three in a G–C base pair (see Figure 1.3B). This complementarity means that if you know the base sequence of one strand of DNA, you can deduce the base sequence of the other. Note, however, that the two strands are antiparallel, running in opposite directions, so that the complementary sequence to AAG is not TTC but CTT (see Figure 1.3A).

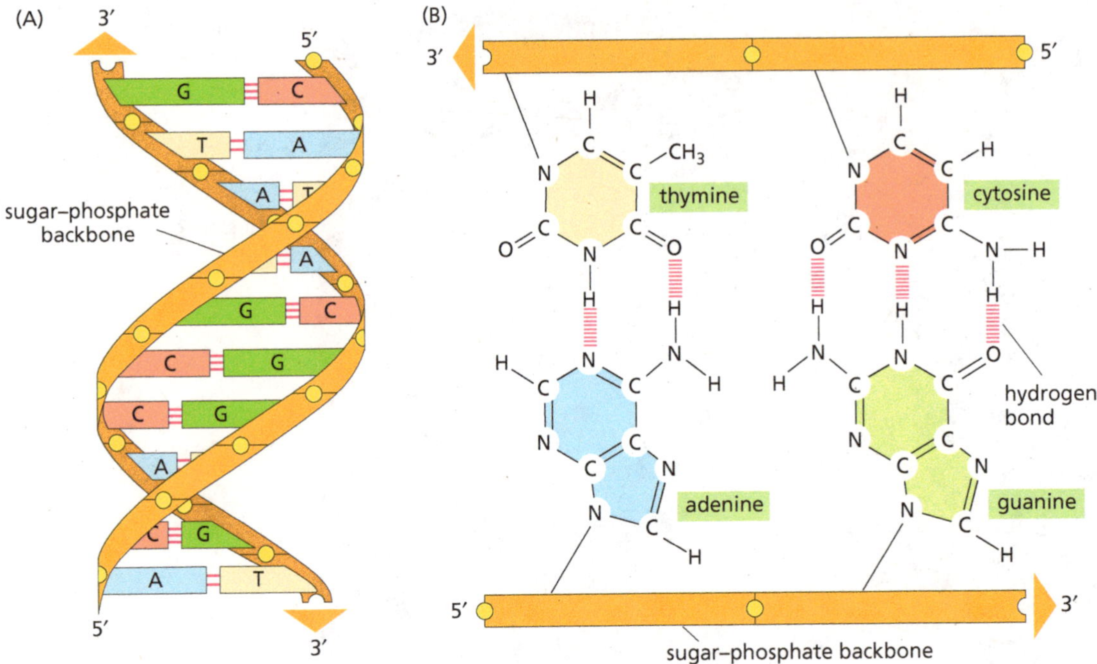


**Figure 1.3**
**The double helical structure of DNA.** (A) DNA exists in cells mainly as a two-stranded coiled structure called the double helix. (B) The two strands of the helix are held together by hydrogen bonds (shown as red lines) between the bases; these bonds are referred to as base-pairing. (From B. Alberts et al., Molecular Biology of the Cell, 4th ed. New York: Garland Science, 2002.)

Hydrogen bonds are noncovalent bonds, which in biomolecules are much weaker than covalent bonds, often involving a binding energy of only 5–20 kJ $mol^{-1}$. As a consequence the two strands of the DNA double helix can be relatively easily separated. There are a number of processes in which this strand separation is required, for example when the molecules of DNA are copied as a necessary prelude to cell division, a process called **DNA replication**. The separated strands each serve as a template on which a new complementary strand is synthesized by an enzyme called DNA polymerase, which moves along the template successively matching each base in the template to the correct incoming nucleotide and joining the nucleotide to the growing new DNA strand. Thus each double helix gives rise to two new identical DNA molecules (see Figure 1.4). Genetic information is thus preserved intact through generations of dividing cells. The actual biochemical process of DNA replication is complex, involving a multitude of proteins, and will not concern us here, but at its heart is this simple structural property of complementarity.

One of the truly remarkable features of DNA in living systems is how few errors there are in the complementarity of the two strands. The error rate is approximately 1 base in $10^9$. This is very important, as it is vital to transmit the genome as accurately as possible to subsequent generations. Many alternative base-pairings are possible, although these are less favorable than the Watson–Crick base-pairing. These energies can be used to predict the expected rate of incorrect base-pairing. However, genomic DNA shows a much lower error rate than expected. This is a

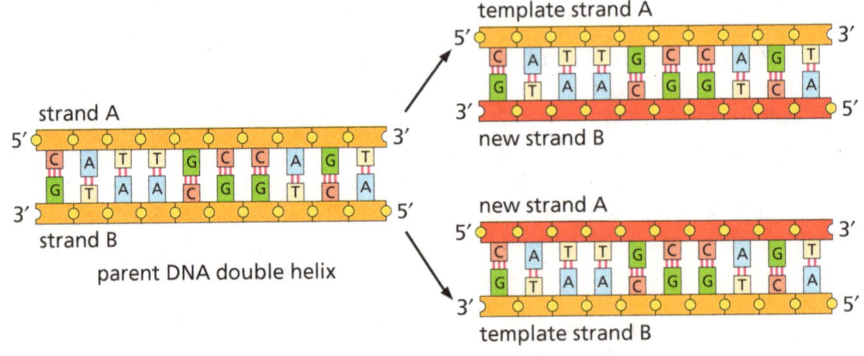


**Figure 1.4**
**DNA replication.** DNA duplicates itself by first separating the two strands, after which enzymes come and build complementary strands base-paired to each of the old strands. (From B. Alberts et al., Molecular Biology of the Cell, 4th ed. New York: Garland Science, 2002.)

result of the controlled way in which the DNA polymerase builds the second strand, including mechanisms for checking and correcting erroneous base-pairings.

DNA strands can pair with a DNA or RNA strand of complementary sequence to make a double-stranded DNA or DNA/RNA hybrid. This property forms the basis of a set of powerful experimental molecular biology techniques. Known generally as nucleic acid **hybridization**, it is exploited in applications such as DNA **microarrays** (described in Chapter 15), *in situ* hybridization to detect the activity of specific genes in cells and tissues, and fluorescence *in situ* hybridization (FISH) for visually locating genes on chromosomes.

## RNA molecules are mostly single stranded but can also have base-pair structures

In contrast to DNA, almost all RNA molecules in living systems are single stranded. Because of this, RNA has much more structural flexibility than DNA, and some RNAs can even act as enzymes, catalyzing a particular chemical reaction. The large number of hydrogen bonds that can form if the RNA molecule can double back on itself and create base-pairing makes such interactions highly energetically favorable. Often, short stretches of an RNA strand are almost or exactly complementary to other stretches nearby. Two such stretches can interact with each other to form a double-helix structure similar to that of DNA, with loops of unpaired nucleotides at the ends of the helices. The interactions stabilizing such helices are often not standard Watson–Crick pairing. These structures are often referred to as **RNA secondary structure** by analogy with protein secondary structure described in Section 2.1.

It is likely that all RNA molecules in a cell have some regions of stable three-dimensional structure due to limited base-pairing, the unpaired regions being very flexible and not taking up any specific structure. In some cases RNA folds up even further, packing helices against each other to produce a molecule with a rigid structure. The three-dimensional structures that result have a functional role for many RNA molecules. An example of such structures is shown in Figure 1.5, and is fairly typical in that a significant fraction of the sequence is involved in base-pairing interactions. The prediction of ribosomal secondary structures and three-dimensional structure is not covered in this book, but introductory references are given in Further Reading. Even more complex interactions are possible, one example of

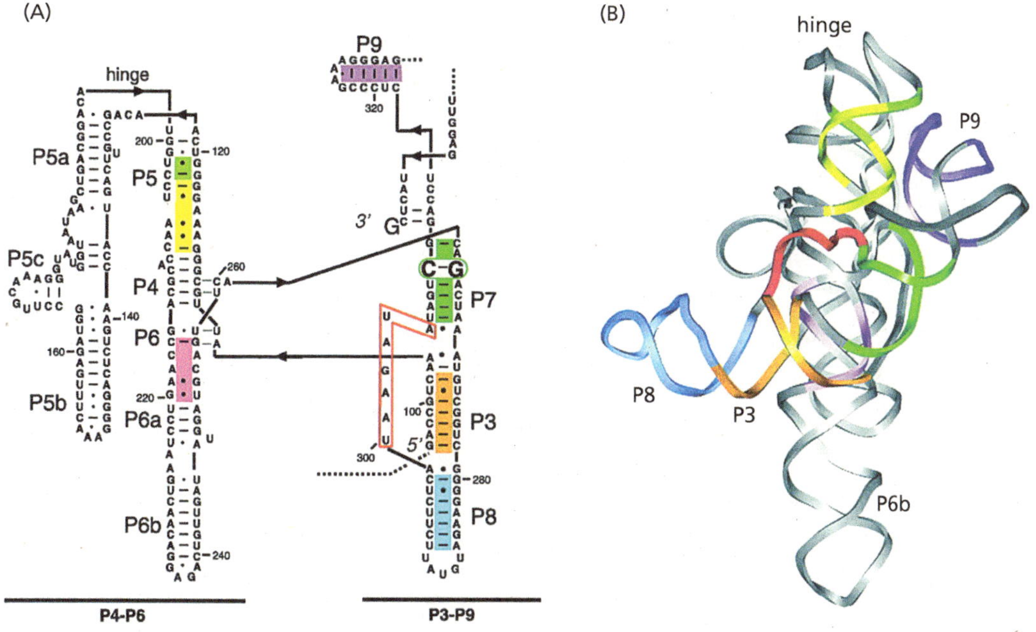


**Figure 1.5**
**Example three-dimensional structure of RNA.** The structure shown is the Tetrahymena ribozyme. (A) Nucleotide sequence showing base-pairing and loops. (B) Three-dimensional structure as determined by x-ray crystallography. Entry 1GRZ of MSD database. (From B.L. Golden et al., A preorganized active site in the crystal structure of Tetrahymena ribozyme, *Science* 282:259–264, 1998. Reprinted with permission from AAAS.)

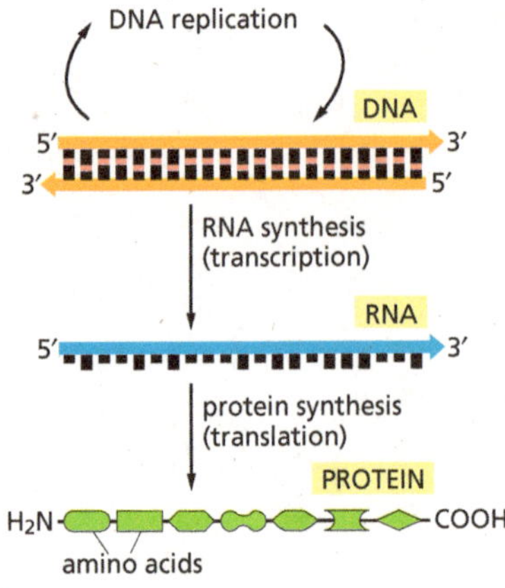


**Figure 1.6**
**Central dogma.** This is the basic scheme of the transcription of DNA to RNA (mRNA) which is then translated to proteins. Many enzymes are involved in these processes. (From B. Alberts et al., Molecular Biology of the Cell, 4th ed. New York: Garland Science, 2002.)

which is illustrated in Figure 1.5A. To the right of the P4 and P6 labels are four bases that are involved in two separate base-pairings, in each case forming a base triplet. Such interactions have been observed on several occasions, as have interactions involving four bases.

## 1.2 DNA, RNA, and Protein: The Central Dogma

There is a key relationship between DNA, RNA, and the synthesis of proteins, which is often referred to as the central dogma of molecular biology. According to this concept, there is essentially a single direction of flow of genetic information from the DNA, which acts as the information store, through RNA molecules from which the information is translated into proteins (see Figure 1.6 and Flow Diagram 1.2). This basic scheme holds for all known forms of life, although there is some variation in the details of the processes involved. The proteins are the main working components of organisms, playing the major role in almost all the key processes of life. However, not all the genetic information in the DNA encodes proteins. Molecules such as RNA can also be the end product, and other regions of genomes have as yet no known function or product. The genomic DNA encodes all molecules that are necessary for life, whether they are the proteins (such as enzymes) involved in nearly all biological activities or RNA important for translation and transcription.

For example, all the information needed to build and maintain a human being is contained in just 23 pairs of DNA molecules, comprising the chromosomes of the human genome. These molecules are amongst the largest and longest known, the smallest having 47 million bases and the largest 247 million bases, with the entire human genome being composed of approximately 3 billion bases. Even bacterial genomes, which are much smaller than this, tend to have several million bases. The DNA of each chromosome encodes hundreds to thousands of proteins, depending on the chromosome, each protein being specified by a distinct segment of DNA. In simple terms, this segment is the gene for that protein. In practice, a **gene** is considered also to include surrounding regions of noncoding DNA that act as control regions, as will be described in Section 1.3. These are involved in determining whether the gene is active—in which case the protein is produced—or is inactive.

**Flow Diagram 1.2**
**The key concept introduced in this section is that DNA is transcribed into mRNA which is then translated to make protein molecules.** The shaded boxes show the concepts introduced in the previous section, the yellow-colored boxes refer to concepts in this section. This color coding is used throughout the book.

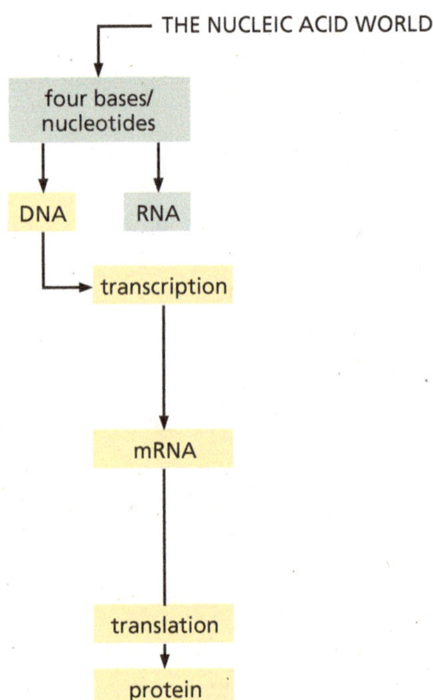

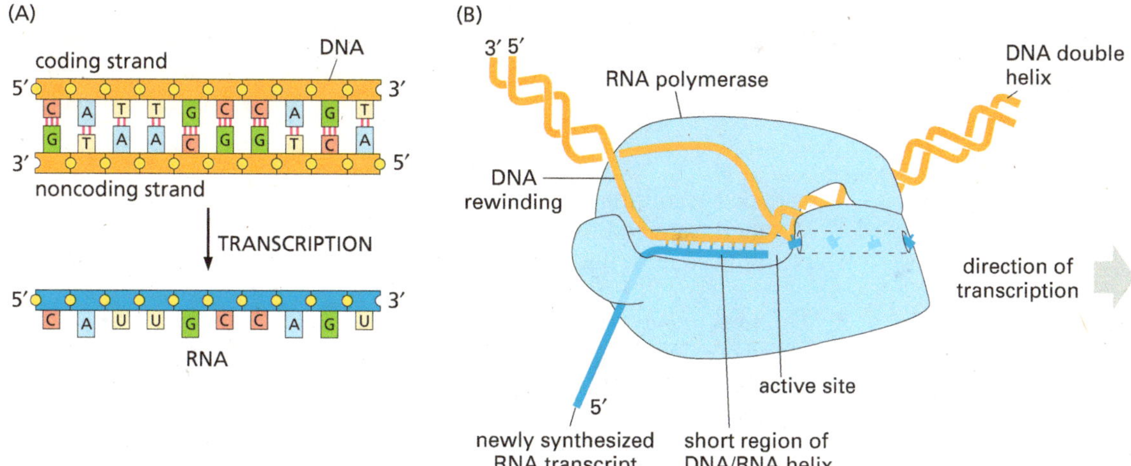


**Figure 1.7**
**Transcription.** (A) One strand of the DNA is involved in the synthesis of an RNA strand complementary to the strand of the DNA. (B) The enzyme RNA polymerase is involved in the transcription process. It reads the DNA and recruits the correct building blocks of RNA to string them together based on the DNA code. (From B. Alberts et al., Molecular Biology of the Cell, 4th ed. New York: Garland Science, 2002.)

The sequence of the protein-coding region of a gene carries information about the protein sequence in a coded form called the genetic code. This DNA sequence is decoded in a two-stage process (see Figure 1.6), the stages being called transcription and translation. Both stages involve RNA molecules and will now be described.

## DNA is the information store, but RNA is the messenger

The information encoded in DNA is expressed through the synthesis of other molecules; it directs the formation of RNA and proteins with specific sequences. As is described in detail in Chapter 2, proteins are linear polymers composed of another set of chemical building blocks, the **amino acids**. There are some 20 different amino acids in proteins, and their varied chemistry (see Figure 2.3) makes proteins much more chemically complex and biochemically versatile molecules than nucleic acids.

The sequence of bases in the DNA of a gene specifies the sequence of amino acids in a protein chain. The conversion does not occur directly, however. After a signal to switch on a gene is received, a single-stranded RNA copy of the gene is first made in a process called **transcription**. Transcription is essentially similar to the process of DNA replication, except that only one of the DNA strands acts as a template in this case, and the product is RNA not DNA (see Figure 1.7). RNA synthesis is catalyzed by enzymes called RNA polymerases, which, like DNA polymerases, move along the template, matching incoming ribonucleotides to the bases in the template strand and joining them together to make an RNA chain. Only the relevant region of DNA is transcribed into RNA, therefore the RNA is a much smaller molecule than the DNA it comes from. So while the DNA carries information about many proteins, the RNA carries information from just one part of the DNA, usually information for a single protein. RNA transcribed from a protein-coding gene is called **messenger RNA (mRNA)** and it is this molecule that directs the synthesis of the protein chain, in the process called translation, which will be described in more detail below. When a gene is being transcribed into RNA, which is in turn directing protein synthesis, the gene is said to be **expressed**. Expression of many genes in a cell or a set of cells can be measured using DNA or RNA expression microarrays (see Chapter 15).

Only one of the DNA strands in any given gene is transcribed into RNA. As the RNA must have a **coding** sequence that can be correctly translated into protein, the DNA strand that acts as the physical template for RNA synthesis does not carry the coding sequence itself, but its complement. It is therefore known as the **noncoding strand** or **anticoding** or **antisense strand**. The sequence of the other, non-template DNA strand is identical to that of the messenger RNA (with T replacing U), and this strand is called the coding or **sense strand**. This is the DNA sequence that is written out to represent a gene, and from which the protein sequence can be

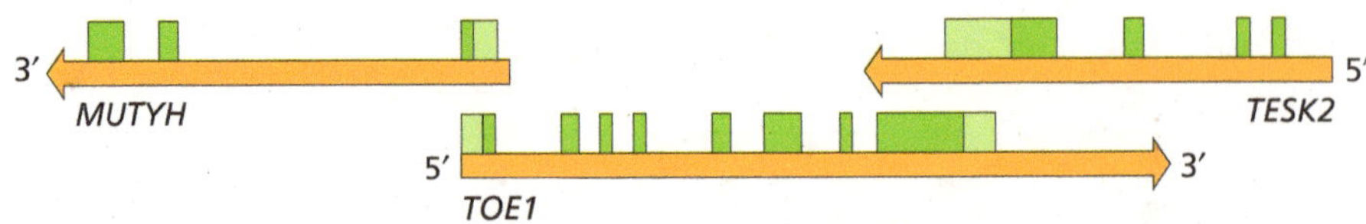


**Figure 1.8**
**Overlapping genes.** A schematic showing the overlap of three human genes. The dark green boxes show protein-coding regions of the DNA (exons) while the light green boxes show regions that are untranslated. (Adapted from V. Veeramachaneni et al., Mammalian overlapping genes: the comparative perspective, *Genome Res.* 14:280–286, 2004.)

deduced according to the rules of the genetic code. Note that the polymerase transcribes the anticoding strand in the direction from 3′ to 5′, so that the mRNA strand is produced from the 5′ to the 3′ end.

Although only one segment of the DNA strand is transcribed for any given gene, it is also possible for genes to overlap so that one or both strands at the same location encode parts of different proteins. This most commonly occurs in **viruses** as a means of packing as much information as possible into a very small genome. However, overlapping genes can also occur in mammals; recently 774 pairs of overlapping genes were identified in the human genome (see Figure 1.8).

The genomic DNA sequence contains more information than just the protein sequences. The transcriptional apparatus has to locate the sites where gene transcription should begin, and when to transcribe a given gene. At any one time, a cell is only expressing a few thousand of the genes in its genome. To accomplish this regulated gene expression, the DNA contains control sequences in addition to coding regions. We shall return to these regulatory regions later, after first discussing the details of the coding of protein sequences.

## Messenger RNA is translated into protein according to the genetic code

The **genetic code** refers to the rules governing the correspondence of the base sequence in DNA or RNA to the amino acid sequence of a protein. The essential

**Table 1.1**
**Standard genetic code.** The corresponding amino acid is given next to each codon. The three-letter amino acid code defined in Table 2.1 is used.

| First letter of the codon (5′ end) | Second letter of the codon: U | | C | | A | | G | | Third letter of the codon (3′ end) |
|---|---|---|---|---|---|---|---|---|---|
| U | UUU | Phe | UCU | Ser | UAU | Tyr | UGU | Cys | U |
| | UUC | Phe | UCC | Ser | UAC | Tyr | UGC | Cys | C |
| | UUA | Leu | UCA | Ser | UAA | Stop | UGA | Stop | A |
| | UUG | Leu | UCG | Ser | UAG | Stop | UGG | Trp | G |
| C | CUU | Leu | CCU | Pro | CAU | His | CGU | Arg | U |
| | CUC | Leu | CCC | Pro | CAC | His | CGC | Arg | C |
| | CUA | Leu | CCA | Pro | CAA | Gln | CGA | Arg | A |
| | CUG | Leu | CCG | Pro | CAG | Gln | CGG | Arg | G |
| A | AUU | Ile | ACU | Thr | AAU | Asn | AGU | Ser | U |
| | AUC | Ile | ACC | Thr | AAC | Asn | AGC | Ser | C |
| | AUA | Ile | ACA | Thr | AAA | Lys | AGA | Arg | A |
| | AUG | Met | ACG | Thr | AAG | Lys | AGG | Arg | G |
| G | GUU | Val | GCU | Ala | GAU | Asp | GGU | Gly | U |
| | GUC | Val | GCC | Ala | GAC | Asp | GGC | Gly | C |
| | GUA | Val | GCA | Ala | GAA | Glu | GGA | Gly | A |
| | GUG | Val | GCG | Ala | GAG | Glu | GGG | Gly | G |

problem is how a code of four different bases in nucleic acids can specify proteins made up of 20 different types of amino acids. The solution is that each amino acid is encoded by a set of three consecutive bases. The three-base sets in RNA are called **codons**, and genetic code tables conventionally give the genetic code in terms of RNA codons. The standard genetic code is given in Table 1.1. From this table you can see that the genetic code is **degenerate**, in that most amino acids can be specified by more than one codon. The degeneracy of the genetic code means that you can deduce the protein sequence from a DNA or RNA sequence, but you cannot unambiguously deduce a nucleic acid sequence from a protein sequence.

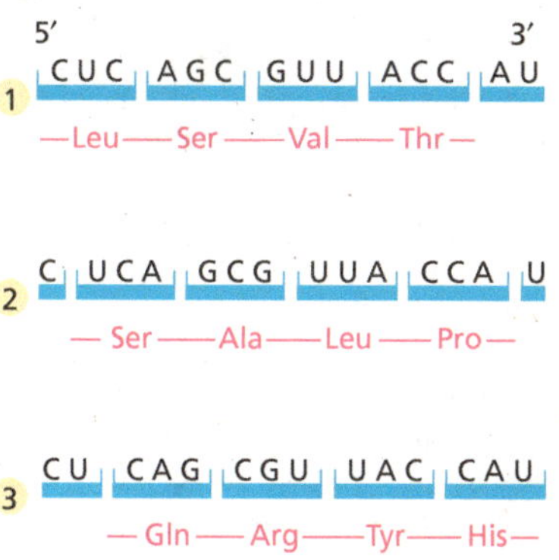


**Figure 1.9**
**The three reading frames of a strand of mRNA.** Each reading frame starts one nucleotide further giving rise to a different protein sequence. (From B. Alberts et al., Molecular Biology of the Cell, 4th ed. New York: Garland Science, 2002.)

There are three codons that do not encode an amino acid but instead signal the end of a protein-coding sequence. These are generally called the stop codons. The signal to start translating is more complex than a single codon, but in most cases translation starts at an AUG codon, which codes for the amino acid methionine. This initial methionine residue is often subsequently removed from the newly synthesized protein. In general, all life uses the same genetic code, but there are some exceptions and care should be taken to use the appropriate code when deducing amino acid sequences from DNA sequences. The code tables can be accessed through many of the sequence databases and through the National Center for Biotechnology Information (NCBI) Taxonomy section.

The translation of bases to amino acids occurs in nonoverlapping sets of three bases. There are thus three possible ways to translate any nucleotide sequence, depending on which base is chosen as the start. These three **reading frames** give three different protein sequences (see Figure 1.9). In the actual translation process the detailed control signals ensure that only the appropriate reading frame is translated into protein. When trying to predict protein-coding sequences in DNA sequences, information about the control signals is often lacking so that one needs to try six possible reading frames, three for each DNA strand. Usually, only one of these reading frames will produce a functional protein. Proteins tend (with notable exceptions) to be at least 100 amino acids in length. The incorrect reading frames often have a stop codon, which results in a much shorter translated sequence. When analyzing bacterial genome sequences, for example, to predict protein-coding sequences, reading frames are identified that give a reasonable length of uninterrupted protein code, flanked by appropriate start and stop signals, called **open reading frames** or **ORFs**. Gene prediction is discussed in Chapters 9 and 10.

## Translation involves transfer RNAs and RNA-containing ribosomes

RNAs have a variety of roles in cells but are mainly involved in the transfer of information from DNA and the use of this information to manufacture proteins. There are three main classes of RNA in all cells—messenger RNA (mRNA), **ribosomal RNA (rRNA)**, and **transfer RNA (tRNA)**—as well as numerous smaller RNAs with a variety of roles, some of which we will encounter in this book. The role of mRNA has been described above. rRNAs and tRNAs are involved in the process of mRNA translation and protein synthesis.

The mRNA produced by transcription is translated into protein by ribosomes, large multimolecular complexes formed of rRNA and proteins, in a process called **translation**. Ribosomes consist of two main subunit complexes, one of which binds the mRNA (see Figure 1.10A). Amino acids do not recognize the codons in mRNA directly and their addition in the correct order to a new protein chain is mediated by the tRNA molecules, which transfer the amino acid to the growing protein chain when bound to the ribosome. These small tRNA molecules have a three-base **anticodon** at one end that recognizes a codon in mRNA, and at the other end a site to which the corresponding amino acid becomes attached by a specific enzyme. This system is the physical basis for the genetic code. There are different tRNAs corresponding to every amino acid, but some tRNA anticodons can bind to several similar codons. One common mechanism for this flexibility, which is called **wobble base-pairing**, involves the occurrence of modified bases in the anticodon. The human tRNA set has 48 different anticodons, but some species have less than 40. tRNA is a good example of an RNA molecule with a specific three-dimensional structure that is crucial for its function, and an example structure is shown in Figure 1.11. The identification of tRNA genes is described in Chapter 10, and the generalized sequence and secondary structure of tRNA molecules are shown in Figure 10.2.

A simplified outline of the process of translation is given in Figure 1.10B and we shall not go into details of it here. The decoding of mRNA starts at its 5′ end, which is the end of the mRNA that is first synthesized at transcription. This is the key justification for the DNA sequence of a gene being conventionally written as the sequence of the sense or coding strand starting at its 5′ end. The site of binding of the ribosome gradually moves along the mRNA toward the 3′ end. The tRNA molecules bind to the ribosome and are the physical link between the mRNA and the growing protein chain. The enzymatic activity in the ribosome that joins amino acids together to form a protein chain is due to the rRNA, not to the ribosomal proteins.

**Figure 1.10**
**Translation.** (A) Schematic structure of the ribosome showing the binding site for the mRNA and the three tRNA binding sites. (B) A simplified view of the three steps of translation during which residue 4 is added to the C-terminal end of the growing protein chain. These steps are repeated for every residue in the protein. (From B. Alberts et al., Molecular Biology of the Cell, 4th ed. New York: Garland Science, 2002.)

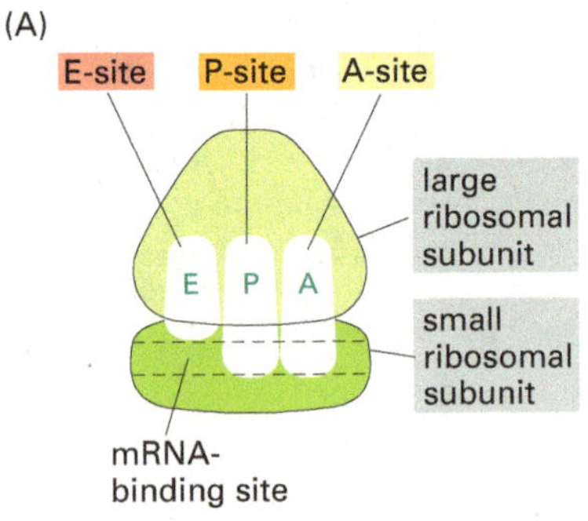


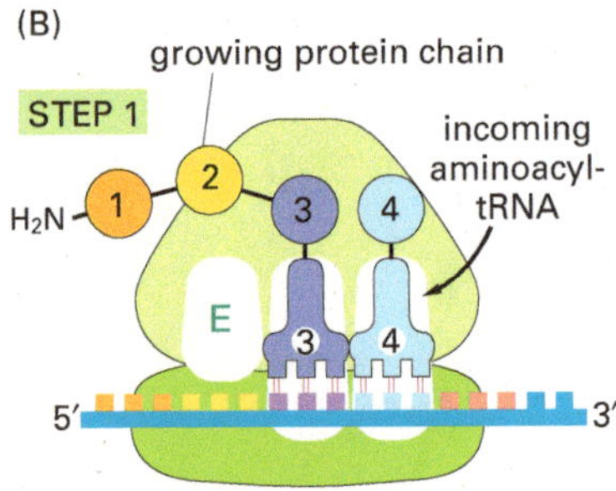


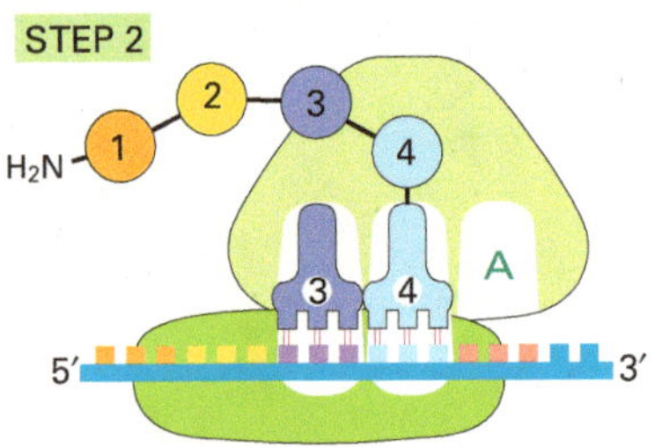


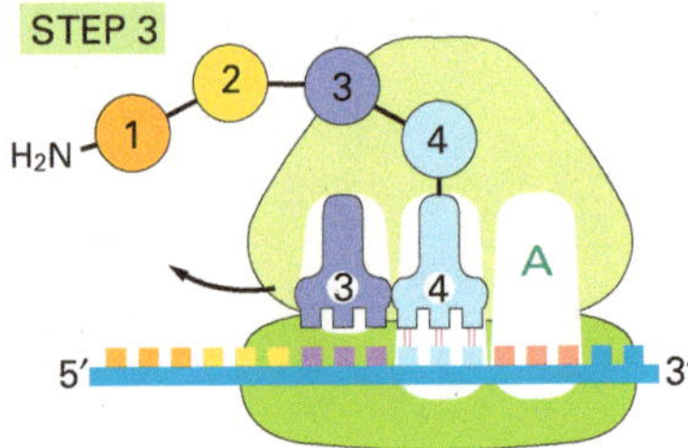


## 1.3 Gene Structure and Control

The description in the previous section of the details of the central dogma focused almost exclusively on the way the protein sequence information is stored in the genome and its conversion from nucleotides to amino acids. Little attention was paid to the ways in which these processes are controlled. Additionally, there are further complications, especially in **eukaryotes**, whose genes often have a more complicated structure including noncoding regions called introns between protein-coding regions. Expressing such genes involves an extra step in converting the DNA information to proteins, called RNA splicing, in which the mRNA produced initially is modified to remove the introns.

**Figure 1.11**
**The structure of a tRNA molecule.** (A) A schematic of the Phe-tRNA molecule showing the arrangement of the base-pairing and loops. (B) The actual three-dimensional structure of the same tRNA molecule colored to show the equivalent regions as in the schematic. (From B. Alberts et al., Molecular Biology of the Cell, 4th ed. New York: Garland Science, 2002.)

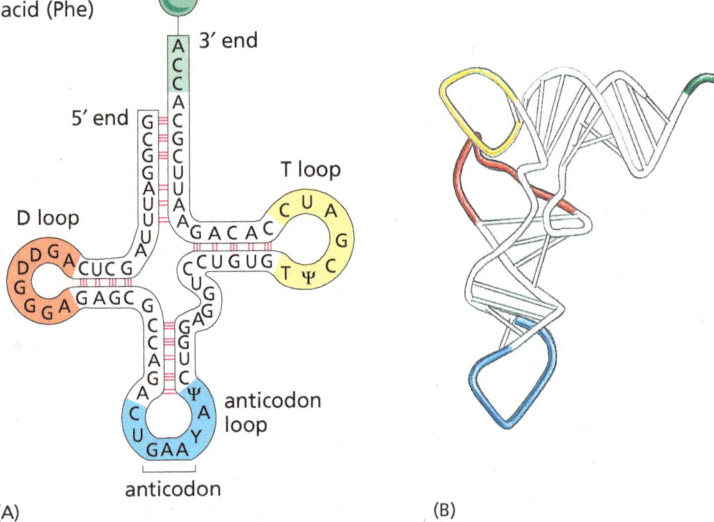

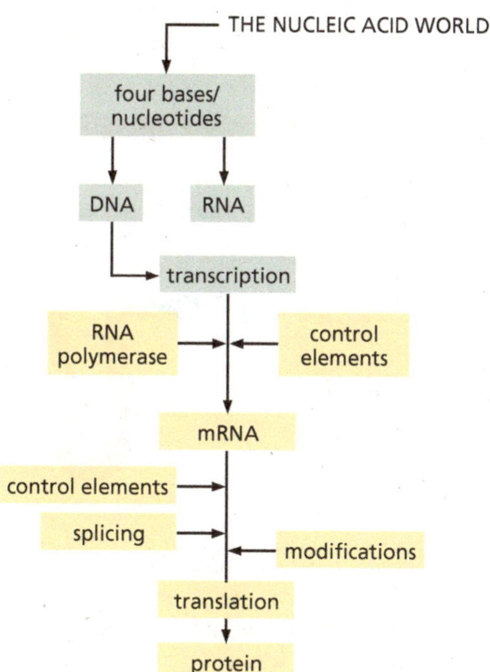


**Flow Diagram 1.3**
**The key concept introduced in this section is that the processes of transcription and translation are subject to several distinct and sophisticated control mechanisms.**

The regulation of many processes that interpret the information contained in a DNA sequence relies on the presence of short signal sequences in the DNA (see Flow Diagram 1.3). There are many different signal sequences, the general term for which is a **regulatory element**. For example, the molecules involved in transcription and translation require signals to identify where they should start and stop. In addition there are signals that are involved in the control of whether transcription occurs or not. The majority of these regulatory sequences are binding sites for specialized regulatory proteins that interact with the DNA to regulate processes such as transcription, RNA splicing, and translation. In order to interpret a DNA sequence correctly, it is important to have some understanding of the nature of these signals and their roles, although the precise mechanisms by which gene regulatory proteins act are not relevant to this book and are not discussed further. In this section we will survey the main aspects of the control of genes and how control structures occur in gene structure.

There are two distinct classes of organism—**prokaryotes** and eukaryotes—whose translation and transcription processes need to be considered separately. For a description of the general characteristics of prokaryotic and eukaryotic organisms see Section 1.4. We shall briefly review the general types of noncoding regulatory sequences found in prokaryotic and eukaryotic genes and introduce their roles. As prokaryotic control regions are, in general, less complicated than those of eukaryotic genes, we shall use them first to describe the basic signals that direct RNA polymerase to bind and start transcription in the appropriate place.

## RNA polymerase binds to specific sequences that position it and identify where to begin transcription

The control regions in DNA at which RNA polymerase binds to initiate transcription are called **promoters**. RNA polymerase binds more tightly to these regions than to the rest of the DNA and this triggers the start of transcription. Additional proteins binding with the polymerase are also required in order to activate transcription. Bacterial promoters typically occur immediately before the position of the **transcription start site (TSS)** and contain two characteristic short sequences, or motifs, that are the same or very similar in the promoters for different genes.

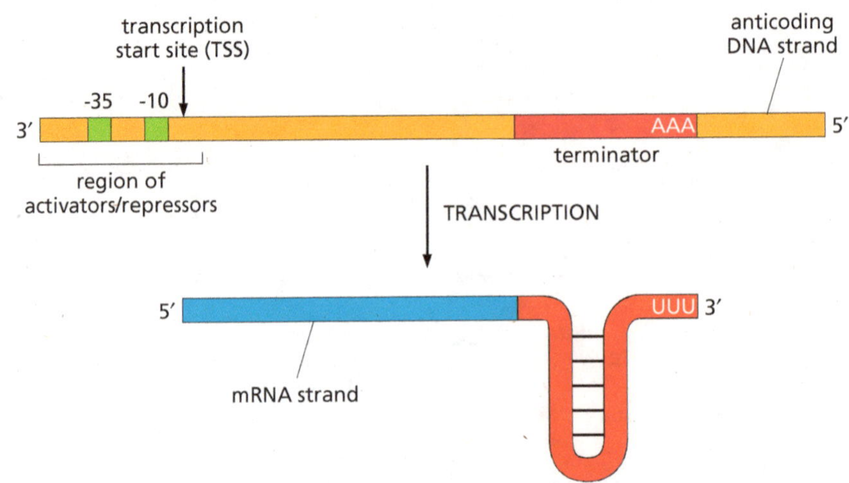


**Figure 1.12**
**The start and stop signals for prokaryotic transcription.** The signals to start transcription are short nucleotide sequences that bind transcription enzymes. The signal to stop transcription is a short nucleotide sequence that forms a loop structure preventing the transcription apparatus from continuing.

One of these motifs is centered at approximately 10 bases before the start, conventionally written as –10, and the other at approximately 35 bases before, written as –35 (see Figure 1.12). These two sequences are essential for the tight binding of the RNA polymerase, and they position it at the appropriate location and in the correct orientation to start transcription on the right strand and in the right direction. Sequences located before the start point of transcription are commonly called **upstream sequences**. Sequences located after the start point of transcription are called **downstream sequences**. This terminology is easy to remember if one thinks of the direction of transcription as the flow of a river.

One of the problems in finding promoters in DNA sequences is that the sequence outside the particular conserved motifs varies considerably, and even the motifs vary somewhat from gene to gene. Motifs like these are often described in terms of their **consensus sequence,** which is made up of the bases that are most frequently found at each position. In *Escherichia coli,* for example, the consensus sequence for the –10 motif is TATAAT, and that of the –35 motif is TTGACA. Furthermore, the separation between these two motifs can be between 15 and 19 bases. Note that by convention these are the sequences found on the coding DNA strand, whereas in reality the polymerase binds to double-stranded DNA, which also contains the complementary anticoding strand sequence. RNA polymerase binds to a variety of sequences but binds more strongly the closer the promoter is to a consensus sequence. The tighter the binding, the more frequently the region will be transcribed. Such variation in sequence is a common feature of many control sites, and makes it harder to locate them.

The termination of transcription is also controlled by sequence signals. In bacteria the **terminator signal** is distinct from the promoter in that it is active when transcribed to form the end of the mRNA strand. In the mRNA the terminator sequence produces two short stretches of complementary sequence that can base-pair to form an RNA double helix, usually involving at least four to five CG base pairs; this is followed, usually within five bases, by at least three consecutive U nucleotides (see Figure 1.12). The prokaryotic terminator sequence is more variable than the promoter sequence, and is usually not included in genome annotations, as described in Chapter 9.

In addition to the promoter sequences, many bacterial genes have additional controls, including binding sites for proteins other than the RNA polymerase. Some of these proteins improve the efficiency of correct binding of the RNA polymerase and are called **activators**, while others called **repressors** block the promoter sites and thus inhibit the expression of the gene. These additional repressor and activator proteins have a profound influence on whether, and when, transcription actually

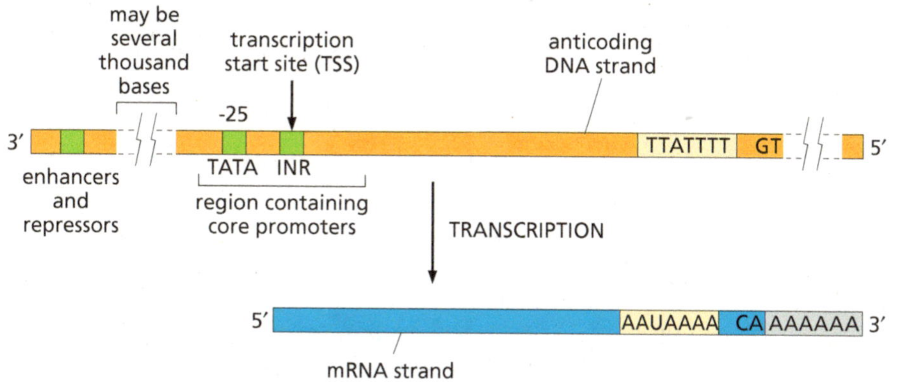


**Figure 1.13**
**The start and stop signals for eukaryotic transcription.** All the signals are short sequences that bind enzymes involved in this complex process.

occurs, and so are of crucial biological importance. In this book, however, we shall only be concerned with locating genes, not in trying to dissect their higher-level control, so this aspect of gene control will not be discussed in any detail.

## The signals initiating transcription in eukaryotes are generally more complex than those in bacteria

In bacteria all genes are transcribed by a single type of RNA polymerase, whereas in eukaryotes there are three different RNA polymerases, each of which transcribes a different type of gene. All the mRNA that codes for proteins is produced by RNA polymerase II, and in this and later chapters we shall focus mainly on this class of genes. The other RNA polymerases are concerned with transcribing genes for tRNAs, rRNAs, and other types of RNA and these have different types of promoters from genes transcribed by RNA polymerase II.

There are differences in the mechanisms for initiating transcription in eukaryotes compared with bacteria, but the principles are the same. Perhaps the most important difference is the much greater degree of variation in the set of promoters present to initiate transcription. There is a set of **core promoter** DNA sequence signals located in the region of the transcription start site (see Figure 1.13) that are initially bound by a complex of proteins known as **general transcription initiation factors**, which in turn recruit the RNA polymerase to the DNA in the correct position to start transcription. The most important core promoter sequence in genes transcribed by RNA polymerase II is about 25 nucleotides upstream from the start of transcription; this is called the **TATA box** and is characterized by a TATA sequence motif. Details of the sequence variation of this promoter are given in Figure 10.12A, but it should also be noted that this signal is not present in all eukaryotic genes. The transcription factor that binds this motif is called the TATA-binding protein (TBP). Many other protein components are involved in initiating and regulating RNA polymerase activity, but any given gene will only require a small subset for activation, and thus only have a subset of the promoter signals. There appears to be no promoter ubiquitous in eukaryotic genes. In further contrast to the situation in prokaryotes, some of the eukaryotic protein-binding sites that modify RNA polymerase activity can be more than a thousand bases away from the promoter. It is thought that the intervening DNA loops round so that all the gene regulatory proteins come together in a complex that regulates the activity of the RNA polymerase and determines whether or not it starts transcription.

Although termination signals have been identified for both RNA polymerases I and III, no specific signal has been identified for RNA polymerase II. However, it may yet remain to be identified because following transcription the mRNA transcript contains a AAUAAA sequence signal that results in cleavage of the 3′ end of the transcript at a site some 10–30 bases after the signal and addition of a series of adenosine nucleotides to form a new 3′ end. This occurs very quickly and removes information about any signal which may be present to terminate transcription. In eukaryotes, this initial mRNA transcript is further modified before translation, as will be described below.

## Eukaryotic mRNA transcripts undergo several modifications prior to their use in translation

The major difference between eukaryotes and prokaryotes in terms of their transcription and translation processes is that the eukaryotic mRNA transcripts are substantially modified before translation. Two of these modifications have no effect on the final protein product. The first modification, which occurs whilst transcription is still in progress, involves the addition of a modified guanosine nucleotide (7-methylguanosine) to the 5′ end of the transcript, a process called **RNA capping**. The last modification, which also occurs while transcription continues, is that which produces the 3′ end of the transcript as mentioned previously; this modification consists of two separate steps. The first step is the cleavage of the mRNA transcript after a CA sequence. The second step, called **polyadenylation**, results in approximately 200 adenosine nucleotides being added to the 3′ end.

The other mRNA modification that occurs in eukaryotes has a significant effect on the final protein products. The major structural difference between the protein-coding genes of prokaryotes and those of eukaryotes is that the protein-coding DNA of most plant and animal genes is interrupted by stretches of noncoding DNA called **introns**; the blocks of protein-coding sequence between the introns are known as **exons** (see Figure 1.14). Introns have lengths which vary from 10 to over 100,000 nucleotides, whereas exons tend to an average length of 100–200 necleotides and rarely exceed 1000 nucleotides. Most bacterial protein-coding genes, on the other hand, have an uninterrupted coding sequence. Introns are found in the genes of most eukaryotes but are less frequent in some genomes, for example that of the yeast *Saccharomyces cerevisiae*. They occur very rarely in prokaryotes.

The existence of introns necessitates an extra step between transcription and translation in eukaryotes, which is known as **RNA splicing**. The complete gene is initially transcribed into RNA; the introns are then excised and the exons joined, or spliced, together to provide a functional mRNA that will give the correct protein sequence when translated (Figure 1.14). In most protein-coding genes, RNA splicing is carried out by a complex called a spliceosome, which consists of small nuclear RNA molecules (snRNAs) and proteins. This complex contains the enzymatic activity that cleaves and rejoins the RNA transcript. The excised intron forms a circular structure called a lariat with a branching point usually at an adenine base (Figure 1.14). The lariat RNA is subsequently degraded. There are particular sequence motifs present at the sites at which the RNA transcript is spliced, as well as the position which will become the lariat branch point. However, these

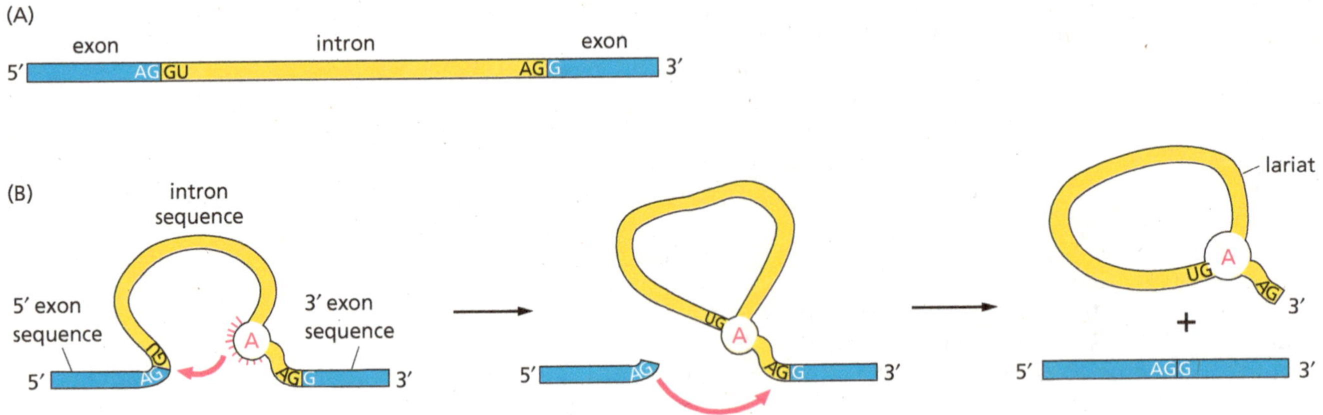


**Figure 1.14**
**The simple schematic of the splicing of an intron.** (A) A linear schematic that shows a segment of pre-mRNA with an intron in yellow. The donor splice site has the conserved dinucleotide GU at the start of the intron. The acceptor splice site has the conserved dinucleotide AG at the end of the intron. (B) The intron is spliced out in a two-stage process: firstly creating a loop structure, the lariat, involving the adenine base (colored red), followed by joining the two exons together releasing the intron. In this case the intron occurred within the codon AGG, which is formed when the exons are spliced together. (B, from B. Alberts et al., Molecular Biology of the Cell, 4th ed. New York: Garland Science, 2002.)

sequence motifs show considerable variability, with the exception of the first and last two bases of each intron. In most cases these are GU and AG as shown in Figure 1.14, but a few instances of another signal, namely AU and AC, have been found in some complex eukaryotes. These are known as AT–AC or U12 introns, after the DNA sequences or one of the components of the spliceosome, respectively. In some even rarer cases the intron RNA itself has splicing activity.

It should be noted that although protein-coding sequences require a whole number of codons to encode the protein sequence, and thus are a multiple of three bases in length, individual exons do not have this requirement. Codons can be derived from the spliced ends of two consecutive exons, as shown in Figure 1.14. This can lead to further complications in gene prediction, as it adds to the difficulty of correctly identifying the exons and from them the protein sequence. For this and other reasons the possible existence of introns in eukaryotic genes significantly complicates the process of gene prediction in eukaryotic compared with prokaryotic genomes, as is discussed in more detail in Chapters 9 and 10.

Although usually all the exons are retained and joined together during RNA splicing, there are cases where this does not happen and **alternative splicing** occurs, excluding some exons or parts of exons, and thus producing different versions of a protein from the same gene. Alternative splicing is quite common in the genes of humans and other mammals, and is thought to be one means by which the relatively small number of genes present in the genome can specify a much greater number of proteins.

## The control of translation

There are various sequence motifs in the mRNA transcripts that indicate the beginning and end of a protein-coding sequence and the site at which the mRNA initially binds to a ribosome. Most protein-coding sequences start with a methionine codon, typically AUG, and invariably end with one of the stop codons of the genetic code (see Table 1.1). In bacterial DNA there is also a distinct short sequence at the 5′ end of the mRNA known as the **Shine–Dalgarno sequence** that indicates the ribosome-binding site. This has a typical consensus sequence of AGGAGGU and occurs a few bases upstream of the AUG translation start codon.

Eukaryotes do not use a Shine–Dalgarno sequence to position the ribosome, but instead have components that bind specifically to the 7-methylguanosine nucleotide at the 5′ end of all eukaryotic mRNA transcripts. The ribosome binds to this and then starts to search for an AUG codon. Occasionally, the first AUG is missed and another downstream codon is used instead. Termination of translation occurs on encountering a stop codon as in prokaryotes.

There is one feature of gene organization in bacteria that is rarely found in eukaryotes, and which relates to the way the ribosome binds near the translation start site. Functionally related protein-coding sequences are often clustered together into **operons** (see Figure 1.15). Each operon is transcribed as a single mRNA transcript and the proteins are then separately translated from this one long molecule. This has the advantage that only one control region is required to activate the simultaneous

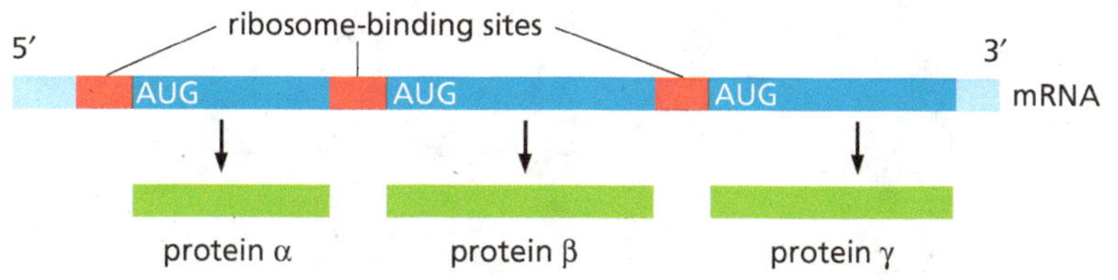


**Figure 1.15**
**A schematic of operon structure.** This single mRNA molecule contains three separate protein-coding regions that produce proteins α, β, and γ on translation. The mRNA contains three separate ribosome-binding sites (red).

**Flow Diagram 1.4**
**The key concept introduced in this section is that the process of evolution is based on mutations of the DNA that result in different life forms, which may then be subjected to different evolutionary selective pressures.**

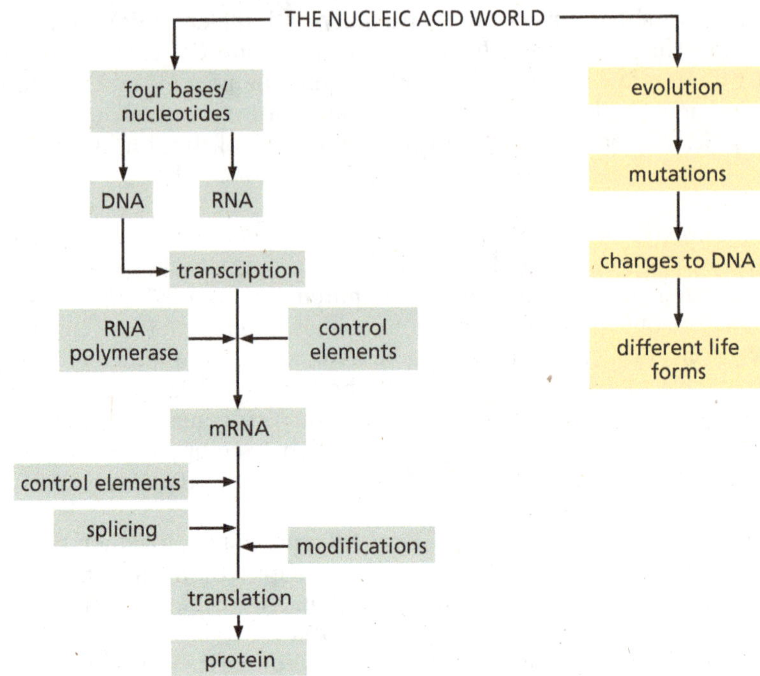


expression of, say, all the enzymes required for a particular metabolic pathway. Not all bacterial genes are contained in operons; many are transcribed individually and have their own control regions.

## 1.4 The Tree of Life and Evolution

The integrity of the genetic information in DNA is carefully protected, and without this protection life would soon disintegrate. Nevertheless, errors inevitably arise in the genome, and these are extremely important as they also provide the genetic variation on which natural selection and evolution can act. Over very long periods of time some of those changes that do not prove to cause their carriers a disadvantage are likely to spread and eventually occur in all genomes of the species (see Flow Diagram 1.4). In this way species can evolve, and ultimately they can evolve into entirely new species. It is generally thought that all existing life has evolved from a single common, very distant ancestor. The evolutionary relationship of known species to each other is commonly described as the tree of life (see Figure 1.16). In this section we will briefly describe the most basic divisions of the tree of life, and some of the modifications which are seen in genomes and their consequences.

**Figure 1.16**
**Tree of life.** Evolution branches out like a real tree where the roots are the origin and the branches are the different groups of life form.

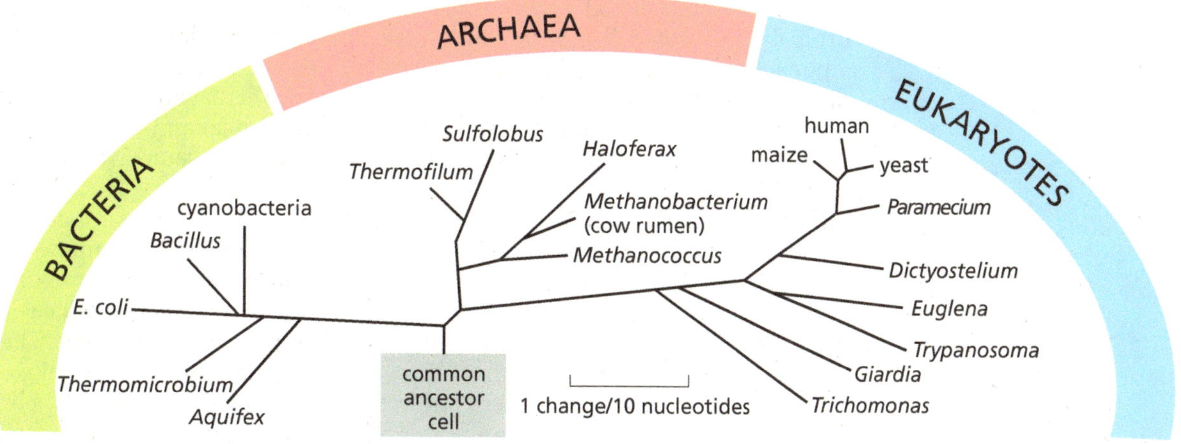


**Figure 1.17**
**The tree of life based on analysis of molecular sequences.** Note that multicellular organisms are all in a small region of the tree. (From B. Alberts et al., Molecular Biology of the Cell, 4th ed. New York: Garland Science, 2002.)

## A brief survey of the basic characteristics of the major forms of life

In this context, all living organisms can be divided into two vast groups: the prokaryotes, which are further divided into the unicellular bacteria and archaea, and the eukaryotes, which include all other living organisms (see Figure 1.17). Another class of objects that contain nucleic acid instructions for their reproduction is the viruses. These have very small genomes that encode the proteins that make up the virus structure, but viruses can only replicate inside a living cell of another organism, as they hijack the cell's biochemical machinery for replicating DNA and synthesizing proteins. Viruses may have either DNA or RNA genomes. Although viral genes follow the basic rules by which DNA encodes RNAs and proteins, it is worth noting that some viral genomes have unusual features not commonly present in cellular genomes, such as overlapping genes, which need careful interpretation.

The prokaryotes are a vast group of unicellular microorganisms. Their cells are simple in structure, lacking a nucleus and other intracellular organelles such as mitochondria and chloroplasts. Taxonomically, the prokaryotes comprise two **superkingdoms**, or **domains**, called the Bacteria and the Archaea, which in evolutionary terms are as distinct from each other as both are from the rest of the living world. Their DNA is usually in the form of a single circular chromosome (although linear chromosomes are known in prokaryotes), containing a single circular DNA molecule, and is not enclosed in a nucleus. In favorable growing conditions, prokaryotes reproduce rapidly by simple cell division, replicating their DNA at each division so that each new cell receives a complete set of genetic instructions. Many bacteria also contain extrachromosomal DNAs in the form of **plasmids**, small circular DNAs that carry additional genes and which can often be transmitted from bacterium to bacterium. Genes for drug resistance, the ability to utilize unusual compounds, or the ability to cause disease are often carried on plasmids. The best-studied bacterium, and the one that for many years provided virtually all our knowledge about the processes of transcription and translation, is the gut bacterium *Escherichia coli*, abbreviated to *E. coli*.

All other living organisms are eukaryotes and belong to the domain **Eukarya**. All animals, plants, fungi, algae, and protozoa are eukaryotes. The eukaryotes include both multicellular and unicellular organisms. Unicellular eukaryotes widely used as model organisms for genetic and genomic studies are the yeasts and unicellular algae such as *Chlamydomonas*. Eukaryotic cells are larger and more complicated than those of prokaryotes. The DNA is contained inside a nucleus, and is highly packaged with histones and other proteins into a number of linear chromosomes, ranging from two to hundreds depending on the organism. Humans have 46 chromosomes in their body cells (made up of two sets of 23 chromosomes inherited from each parent), the fruit fly *Drosophila* has 8, petunias have 14, while the king crab has 208. There appears to be no particular reason why the DNA is divided up into such different numbers of chromosomes in different organisms; the actual numbers of genes in the genomes of multicellular organisms are much more similar and vary between 20,000 and 30,000 for those organisms whose genomes have been sequenced to date.

Eukaryotic cells are highly compartmentalized, with different functions being carried out in specialized organelles. Two of these are of particular interest here, as they contain their own small genomes. **Mitochondria** contain the components for the process of energy generation by aerobic respiration, and **chloroplasts** contain the molecular components for photosynthesis in plant and algal cells. These two organelles are believed to be the relics of prokaryotic organisms engulfed by the ancestors of eukaryotic cells and still retain small DNA genomes of their own—mitochondrial DNA (mitDNA) and chloroplast DNA—and the protein machinery to transcribe and translate them. These genomes encode some of the proteins specific to mitochondria and chloroplasts, but most of their proteins are now encoded by genes in the eukaryotic cell nucleus.

## Nucleic acid sequences can change as a result of mutation

There are a number of occasions, such as DNA replication, when the genomic DNA is actively involved in processes that leave it vulnerable to damage. Sometimes this damage will be on a large scale, such as the duplication of whole genes, but often it involves just a single base being incorrectly replicated. The general term used to describe such damage is mutation. Depending on which part of the DNA sequence is affected, mutations can have a drastic effect on the information encoded, leading to changes in the sequence of encoded proteins, or the loss of control signals. Genes can be rendered inactive or proteins dysfunctional, although mutations can also have beneficial effects (see Box 1.1). In organisms that use sexual reproduction, unless the DNA affected is in a germ cell, the DNA will not be used to generate the genomes of future generations, and so will only affect the organism in which the damage occurred. In such cases, the organism might suffer, especially if the mutation causes uncontrolled cell growth and division, as happens in tumors. The alternative is that the mutation is transmitted through to the next generation, in which case it has a chance to eventually become part of the normal genome of the species.

### Box 1.1 There is more to genomes than protein and RNA genes

Not all the DNA sequence in a genome contains a meaningful message or a known function. Regions without a message are sometimes referred to as junk DNA, although this term should not be taken too literally as we have much still to learn, and these regions may yet come to be seen as functional in some way. Mammalian genomes contain large amounts of this type of DNA, both in the form of introns and between genes. Simpler eukaryotic organisms have less, and bacteria have very little.

Much of the so-called junk DNA is in the form of highly repeated DNA sequences, which form significant percentages of the genomes of many higher organisms. Many of these **DNA repeats** are due to DNA elements known as transposons, which can copy themselves and insert themselves into other parts of the genome. Transposons are present in both bacteria and eukaryotes, but in mammalian genomes they have multiplied to a much greater extent and appear to have largely lost their function, existing now simply as apparently functionless sequence repeats.

On a final note, changes in DNA sequences that occur during evolution occasionally destroy some of the control sequences needed for a gene to be expressed. When this happens, the resultant inactive gene is called a **pseudogene**. Soon after the initial inactivation, pseudogenes can be hard to distinguish from active genes, but over time they accumulate many more mutations than active genes, and so diverge to (ultimately) random sequence.

The fate of the mutation, to be lost or to be retained, depends on the process of natural selection that is the cornerstone of the theory of evolution.

The existence of similar DNA and protein sequences in different organisms is a consequence of the process of evolution that has generated the multitude of living organisms from an ancient ancestor held in common. The sequence similarities reveal details of the ways that mutations arise and of the balance of forces which will result in only a small group of mutations being preserved through evolutionary selection. Therefore some details of the mechanism of mutation are of relevance in a number of areas of bioinformatics, including sequence alignment (see especially Sections 4.3 and 5.1) and phylogenetic analysis (see especially Sections 7.2 and 8.1). In the phylogenetic analysis described in Chapters 7 and 8 an attempt is made to reconstruct the evolutionary history of a set of sequences. This requires a detailed model of evolution, which requires a comprehensive understanding of the kinds of mutations that occur, their effects, and the process of natural selection by which they are either accepted or lost.

## Summary

We have tried in this chapter to give a brief introduction to the nucleic acids and their role in living systems. We have focused exclusively on the role of nucleic acids in genomes, although a look at any introductory textbook on molecular biology will show that, in addition, single nucleotides play a part in many other processes. We have described how the sequence of DNA can encode proteins, and how simple sequence signals are used to control the interpretation of the genomic DNA. The material is sufficient to allow a novice to appreciate the techniques discussed in this book, particularly those for gene detection. You should be aware, however, that there are probably exceptions to every general statement we have made in this chapter! (See Box 1.2.) For example, under certain circumstances, the codon UGA can code for the unusual amino acid selenocysteine instead of being understood as a stop signal. Many organisms have their own peculiarities, and one should preferably seek out an expert for advice.

### Box 1.2 Things are usually not that simple!

This chapter has given a brief introduction to the key role of nucleic acids in the storage, interpretation, and transmission of genetic information. The many genome sequencing projects are producing results at a phenomenal rate, and the techniques of bioinformatics, such as are described in Chapters 9 and 10, are required to identify and characterize the functional components of the genomes. When working on such projects, it must be remembered that the descriptions given in this and the other chapters are general, and wherever possible care must be taken to discover the specific details applicable to the organism of interest.

A further warning is required in that even some of the fundamental concepts described in this chapter are much less well defined than might be supposed. Two brief illustrations of this will be given, dealing with the definition of a gene and of the human genome. Processes such as alternative splicing are making it ever harder to agree on a definition of a gene. The human genome is another concept whose definition is proving harder to agree on than expected. Recent studies have shown that there is much greater variation in the genome in humans than was expected. As well as many small point mutations that were anticipated, it was found that there were a surprising number of large-scale differences between humans. As a result it is no longer clear how to define a fully representative human genome. There are reviews in Further Reading that explain these points in more detail.

# Further Reading


## General References

Alberts B, Johnson A, Lewis J, Raff M, Roberts K & Walter P (2008) Molecular Biology of the Cell, 5th ed. New York: Garland Science.

Strachan T & Read A (2004) Human Molecular Genetics, 3rd ed. London: Garland Science.

## 1.1 The Structure of DNA and RNA

Dahm R (2005) Friedrich Miescher and the discovery of DNA. *Dev. Biol.* 278, 274–288.

Watson JD & Crick FHC (1953) Molecular structure of nucleic acids. *Nature* 171, 737–738.

## 1.2 DNA, RNA and Protein: The Central Dogma

Veeramachaneni V, Makalowski W, Galdzicki M et al. (2004) Mammalian overlapping genes: the comparative perspective. *Genome Res.* 14, 280–286.

## 1.3 Gene Structure and Control

Portin P (1993) The concept of the gene: short history and present status. *Q. Rev. Biol.* 68, 173–223.

## 1.4 The Tree of Life and Evolution

Kaessmann H & Paabo S (2002) The genetical history of humans and the great apes. *J. Intern. Med.* 251, 1–18.

## Box 1.2

Check E (2005) Patchwork people. *Nature* 437, 1084–1086.

Pearson H (2006) What is a gene? *Nature* 441, 399–401.

# PROTEIN STRUCTURE

2

## When you have read Chapter 2, you should be able to:

Discuss how proteins are important for function.

Describe how amino acids are the building blocks of proteins.

Show how amino acids are joined together to form a protein chain.

Explain that secondary structures are repeating structures found in most proteins.

Describe how the whole protein chain folds up into a fold.

Discuss how the structure of the fold is important for its function.

Outline how proteins often exist as a complex of more than one fold.

If there is one class of molecules which could be said to live life it would be the proteins. They are responsible for catalyzing almost all the chemical reactions in the cell (RNA has a more limited but important role, as we saw in Chapter 1), they regulate all gene activity, and they provide much of the cellular structure. There is speculation that life may have started with nucleic acid chemistry only, but it is the extraordinary functional versatility of proteins that has enabled life to reach its current complex state. Proteins can function as enzymes catalyzing a wide variety of reactions necessary for life, and they can be important for the structure of living systems, such as those proteins involved in the cytoskeleton. The size of a protein can vary from relatively small to quite large macromolecules.

As we saw in Chapter 1, the DNA sequence of a gene can be analyzed to give the **amino acid sequence** of the protein product. In that aspect alone, the ready availability of DNA sequences of genes and whole genomes from the 1980s onward revolutionized biology, as it opened up this vital shortcut to determining the amino acid sequence of virtually any protein. Bioinformatics uses this sequence information to find related proteins and thus gather together knowledge that can help deduce the likely properties of unknown proteins, plus their structures and functions. Knowing the relationship between a protein's structure and its function provides a greater understanding of how the protein works, and thus often enables the researcher to propose experiments to explore how modifying the structure will affect the function. As the vast majority of currently marketed pharmaceuticals act by interacting with proteins, structure–function studies are vital to the design of new drugs, and bioinformatics has an important role in speeding up this process and enabling computer modeling of these interactions, as we shall see in Chapter 14.

**Mind Map 2.1**
A mind map visualization of aspects of protein structure.

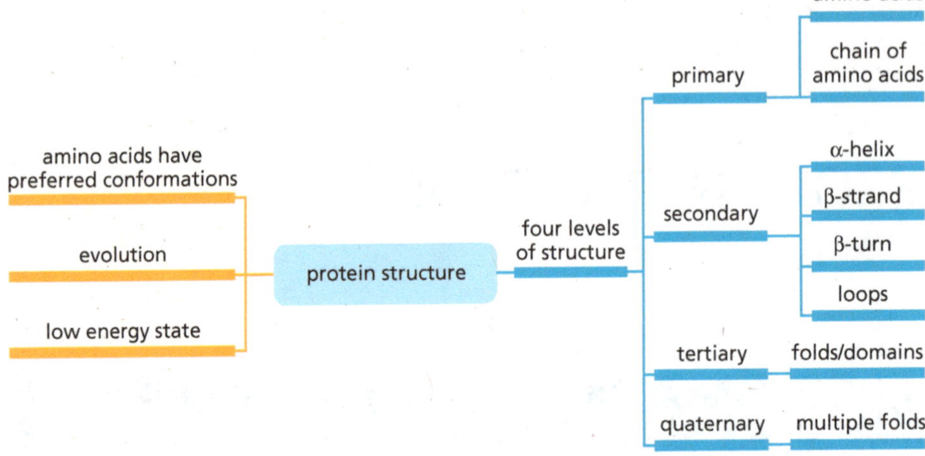


There are many excellent books that describe protein structures, form, and classification in detail (see Further Reading). In this chapter we will introduce proteins and outline some concepts that are important for bioinformatics, providing a minimum amount of information required to understand the techniques and algorithms described in later chapters. Additional information is given where needed in those chapters.

## 2.1 Primary and Secondary Structure

A protein folds into a three-dimensional structure, which is determined by its protein sequence. The fold of the protein consists of repeating structural units called secondary structures, that will be discussed in this section (see Flow Diagram 2.1). The fold of the protein is very important for the way the protein will function, and whether it will function correctly. Therefore the study of the ways in which proteins fold and understanding how they fold is an important area of bioinformatics, as well as predicting the fold of a protein from its sequence.

### Protein structure can be considered on several different levels

The analysis of protein structure by experimental techniques such as X-ray crystallography and nuclear magnetic resonance (NMR) has shown that proteins adopt distinct structural elements. In general there are four levels of protein structure to consider (see Figure 2.1). The **primary structure** is the protein sequence, the types

**Flow Diagram 2.1**
The key concepts introduced in this section are that proteins are linear polymers of amino acids whose order is often called the primary structure or sequence, and that proteins can fold into a limited number of regular three-dimensional forms called secondary structures.

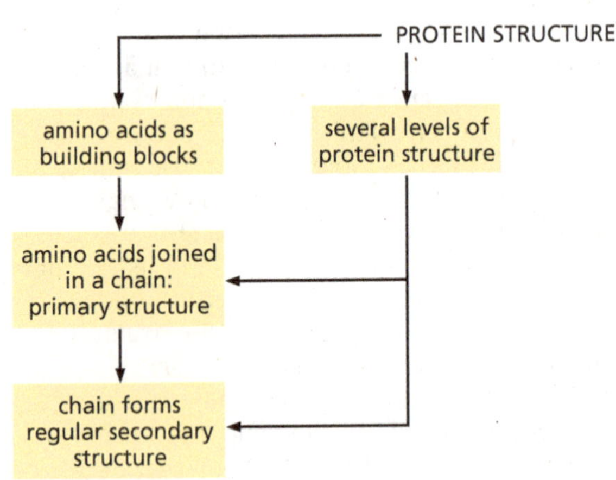

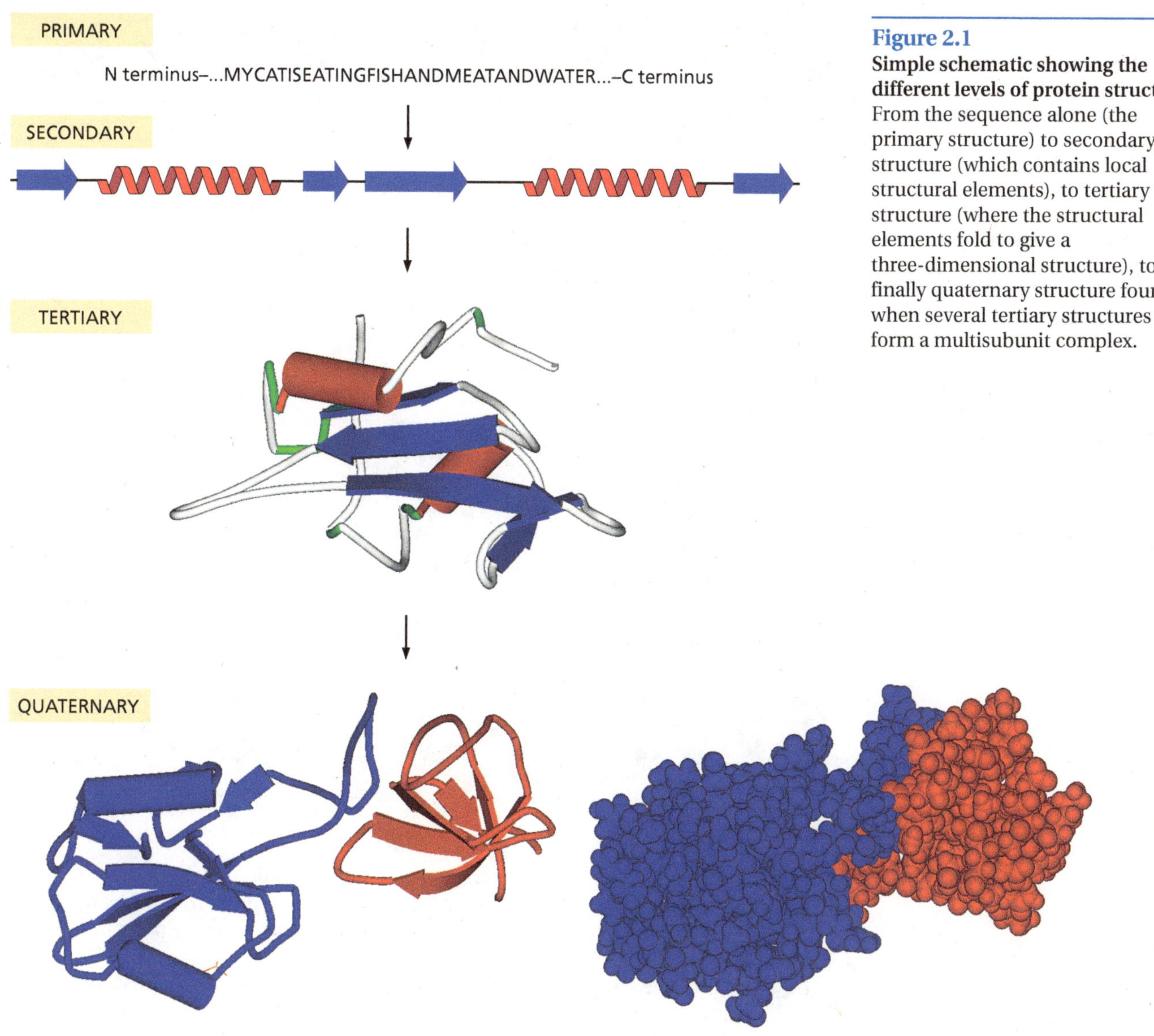


**Figure 2.1**
**Simple schematic showing the different levels of protein structure.** From the sequence alone (the primary structure) to secondary structure (which contains local structural elements), to tertiary structure (where the structural elements fold to give a three-dimensional structure), to finally quaternary structure found when several tertiary structures form a multisubunit complex.

and order of the amino acids in the protein chain; the **secondary structure** is the first level of protein folding, in which parts of the chain fold to form generic structures that are found in all proteins. The **tertiary structure** is formed by the further folding and packing together of these elements to give the final three-dimensional **conformation** unique to the protein. Many functional proteins are formed of more than one protein chain, in which case the individual chains are called protein subunits. The subunit composition and arrangement in such multisubunit proteins is called the **quaternary conformation**. The structure adopted by a protein chain, and thus its function, is determined entirely by its amino acid sequence, but the rules that govern how a protein chain of a given sequence folds up are not yet understood and it is impossible to predict the folded structure of a protein de novo from its amino acid sequence alone. Helping to solve this problem is one of the challenges facing bioinformatics.

## Amino acids are the building blocks of proteins

Proteins are made up of 20 types of naturally occurring amino acids (see Table 2.1), with a few other amino acids occurring infrequently. These 20 amino acids consist solely of the elements carbon (C), nitrogen (N), oxygen (O), and hydrogen (H), with the exception of cysteine and methionine, which also contain sulfur (S). The structure

**Table 2.1**
**The 20 amino acids colored according to their properties.** The amino acid name, the three-letter code, and the one-letter code are given. Note that the one-letter code for amino acids can have the same letter as found in the one-letter codes that denote nucleotides, but they mean different things. So for example the letters C, G, A, and T in a DNA sequence would stand for cytosine, guanine, adenine, and thymine, respectively. In protein sequences these same letters would refer to cysteine, glycine, alanine, and threonine, respectively.

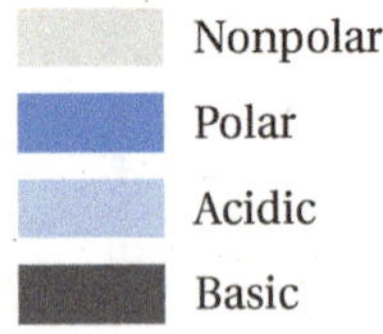


| Amino acid | Three-letter code | One-letter code | Comment |
|---|---|---|---|
| Glycine | Gly | G | Only -H as side chain |
| Alanine | Ala | A | |
| Valine | Val | V | |
| Leucine | Leu | L | |
| Isoleucine | Ile | I | |
| Proline | Pro | P | Side chain to N bond |
| Phenylalanine | Phe | F | |
| Methionine | Met | M | |
| Tryptophan | Trp | W | |
| Cysteine | Cys | C | Forms disulfide bonds |
| Asparagine | Asn | N | Amide N polar |
| Glutamine | Gln | Q | Amide N polar |
| Serine | Ser | S | -OH group polar |
| Threonine | Thr | T | -OH group polar |
| Tyrosine | Tyr | Y | -OH group polar |
| Aspartic acid | Asp | D | |
| Glutamic acid | Glu | E | |
| Histidine | His | H | |
| Lysine | Lys | K | |
| Arginine | Arg | R | |

of an amino acid can be divided into a common **main chain** part (see Figure 2.2) and a side chain that differs in chemical structure among the different amino acids. The side chain is attached to the main chain carbon atom known as the α-carbon ($C_\alpha$).

The 20 amino acids are given in Table 2.1 along with the abbreviations that will be used in this book. When the amino acid sequence of a protein is written out, it is usually given in the one-letter code. We will first look at the side-chain properties that distinguish one amino acid from another, and then explore in some detail the main chain component, whose properties are responsible for most of the general features of protein structure.

## The differing chemical and physical properties of amino acids are due to their side chains

The functional properties of proteins are almost entirely due to the side chains of the amino acids. Each type of amino acid has specific chemical physical properties

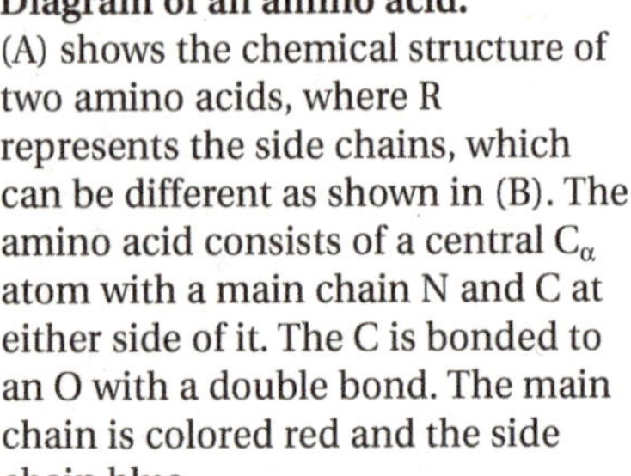

**Figure 2.2**
**Diagram of an amino acid.** (A) shows the chemical structure of two amino acids, where R represents the side chains, which can be different as shown in (B). The amino acid consists of a central $C_\alpha$ atom with a main chain N and C at either side of it. The C is bonded to an O with a double bond. The main chain is colored red and the side chain blue.

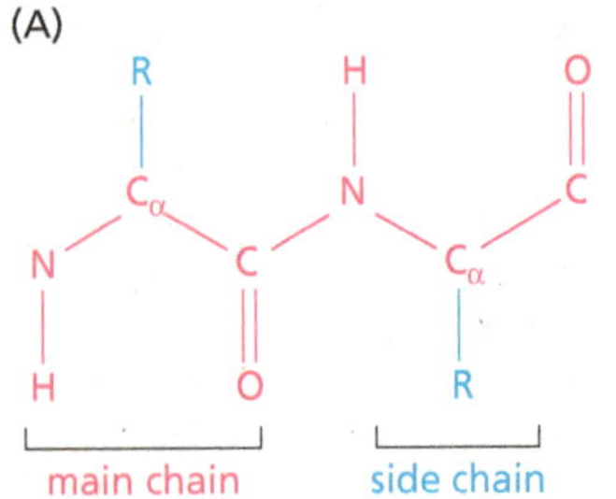


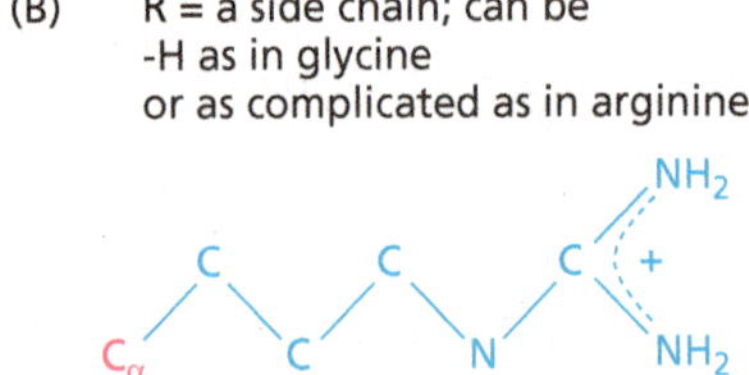

that are conferred on it by the structure and chemical properties of its side chain. They can, however, be classified into overlapping groups that share some common physical and chemical properties, such as size and electrical charge. The smallest amino acid is glycine, which has only a hydrogen atom as its side chain. This endows it with particular properties such as great flexibility. The other extreme of side-chain flexibility is represented by proline, an amino acid that has a side chain bonded to the main-chain nitrogen atom, resulting in a rigid structure. Some amino acids have uncharged side chains and these are generally **hydrophobic** (not liking water, therefore tend to be buried within the protein surrounded by other hydrophobic amino acids) while others are positively or negatively charged. The charged or polar amino acids are **hydrophilic**; they like to be surrounded by water molecules with which they can form interactions. Buried inside the protein they will often have another oppositely charged hydrophilic residue to form an interaction with, for example, the long side chain of the amino acid arginine, with its $-NH_3^+$ groups is (relatively) large and basic (positively charged). The amino acid side chains are shown in Figure 2.3.

As there are 20 distinct amino acids that occur in proteins, there can be $20^n$ different polypeptide chains of length $n$. For example, a polypeptide chain 250 amino acids in length will be one of more than $10^{325}$ alternative different sequences. Clearly, the sequences that do occur are only a tiny fraction of those possible. Often only a few sequence modifications are needed to destabilize the three-dimensional conformation of a protein, and so it is probable that the majority of these alternative sequences will not adopt a stable conformation.

## Amino acids are covalently linked together in the protein chain by peptide bonds

The primary structure of a protein is the sequence of amino acids in the linear protein chain, which consists of covalently linked amino acids. This linear chain is often called a polypeptide chain. The amino acids are linked by peptide bonds, which are formed by a **condensation reaction** (the loss of a water molecule) between the backbone carboxyl group of one amino acid and the amino group of another (see Figure 2.4). When linked together in this way, the individual amino acids are conventionally called amino acid residues. There are two key properties of peptide bonds that are important in determining protein structure, and will be discussed in this chapter. First, the bond is essentially flat, or planar; that is to say that the carbon, nitrogen, and carbonyl oxygen atoms involved in the bond all essentially lie in the same plane. This limits rotation around the bond. Second, both the -NH and -C=O groups are capable of noncovalent hydrogen-bonding interactions; this type of interaction is key to the folding up of the protein chain into a three-dimensional structure.

The peptide bonds and connecting $C_\alpha$ atoms are referred to as the protein backbone or main chain. A polypeptide will have a reactive amino group at one end and a reactive carboxyl group at the other end. These are referred to as the **amino terminus** (or **N terminus**) and **carboxy terminus** (or **C terminus**), respectively. When a protein is synthesized by translation of an mRNA (see Chapter 1), the amino terminus is synthesized first. Protein sequences are thus conventionally written starting with the N terminus on the left. The relation of DNA coding-strand sequence to mRNA sequence to protein sequence is shown in Figure 2.5

In principle, the atoms at either end of a bond are free to rotate in any direction but due to atomic constraints this is not the case. **Torsion angles** are used to define the conformation around bonds that can rotate and are defined for four atoms (see Figure 2.6A). Torsion angles are usually described in the range of $-180°$ to $+180°$ and are the number of degrees that atoms 1 and 2 have to move to eclipse atoms 3 and 4. The groups on either side of the peptide bond can rotate to some extent around the

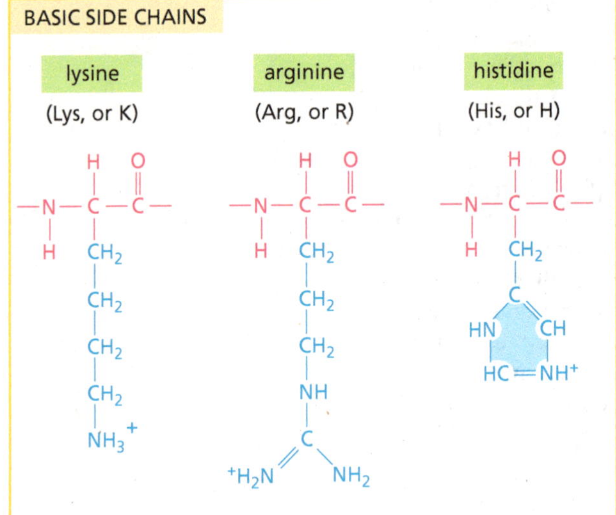
BASIC SIDE CHAINS
lysine
(Lys, or K)
arginine
(Arg, or R)
histidine
(His, or H)

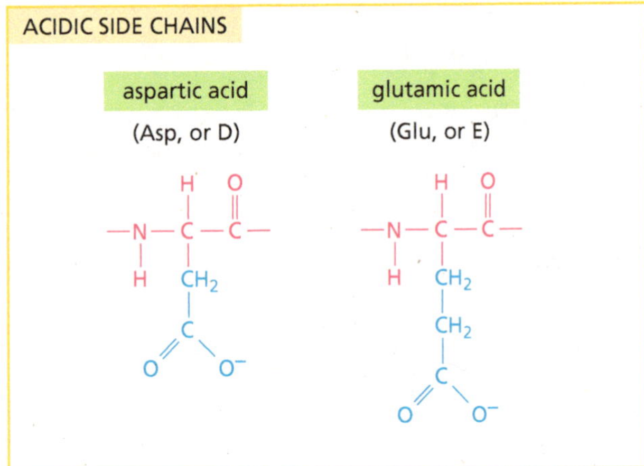
ACIDIC SIDE CHAINS
aspartic acid
(Asp, or D)
glutamic acid
(Glu, or E)

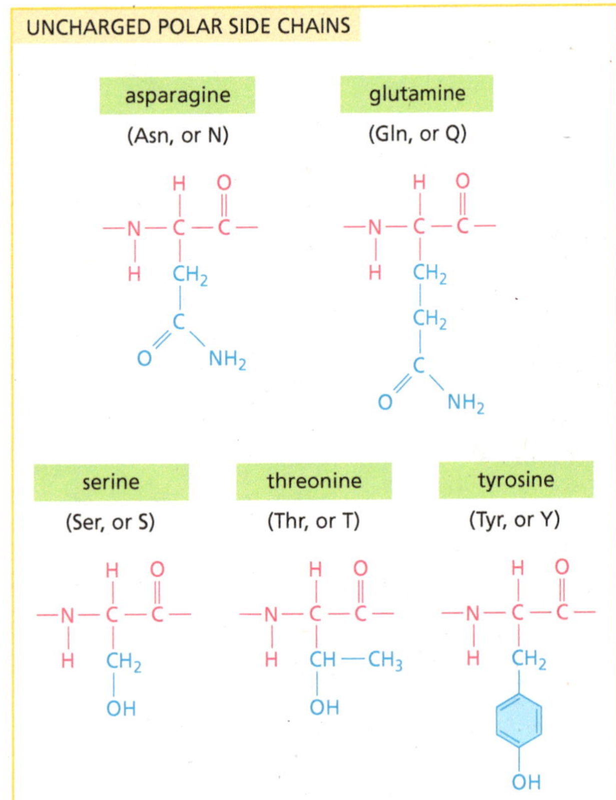
UNCHARGED POLAR SIDE CHAINS
asparagine
(Asn, or N)
glutamine
(Gln, or Q)
serine
(Ser, or S)
threonine
(Thr, or T)
tyrosine
(Tyr, or Y)

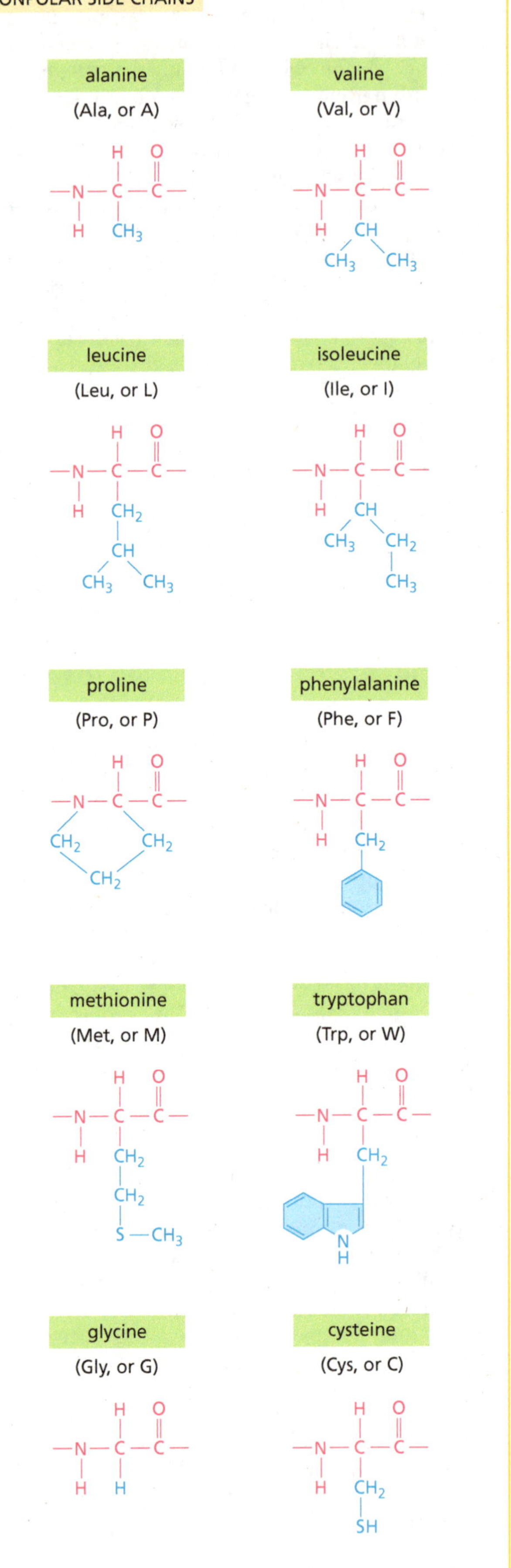
NONPOLAR SIDE CHAINS
alanine
(Ala, or A)
valine
(Val, or V)
leucine
(Leu, or L)
isoleucine
(Ile, or I)
proline
(Pro, or P)
phenylalanine
(Phe, or F)
methionine
(Met, or M)
tryptophan
(Trp, or W)
glycine
(Gly, or G)
cysteine
(Cys, or C)

**Figure 2.3**
**The side chains of the 20 commonly occurring amino acids.** They are arranged according to their physicochemical properties: acidic, basic, uncharged polar, and nonpolar (hydrophobic). Some have additional properties, such as aromaticity (for example Phe). (Adapted from B. Alberts et al., Molecular Biology of the Cell, 4th ed. New York: Garland Science, 2002.)

bond. But the peptide bond itself is planar due to the phenomenon known as electronic resonance (the movement of delocalized electrons within the bond) that gives it a partial double-bond character, and leads to a slightly shorter C-N bond than might otherwise be expected. The torsion angle within the peptide bond (usually called $\omega$) is rarely more than 10° from planarity (see Figure 2.6B). The polypeptide

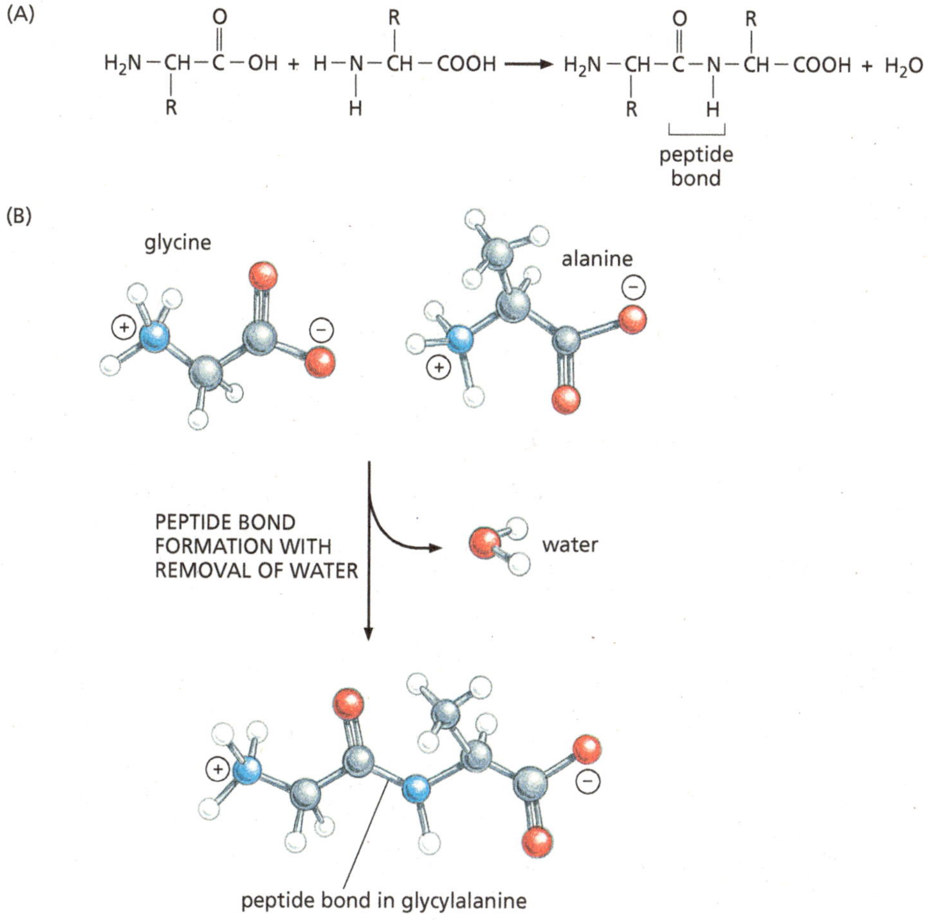

(A)

$$H_2N-CH(R)-C(=O)-OH + H-N(H)-CH(R)-COOH \longrightarrow H_2N-CH(R)-C(=O)-N(H)-CH(R)-COOH + H_2O$$

peptide bond

(B)

glycine

alanine

PEPTIDE BOND FORMATION WITH REMOVAL OF WATER

water

peptide bond in glycylalanine

**Figure 2.4**
**Peptide bonds.** (A) gives the chemical formulae of the peptide bond that is formed between amino acids to make a polypeptide chain. (B) illustrates the above in a diagrammatic form. (B, from B. Alberts et al., Molecular Biology of the Cell, 4th ed. New York: Garland Science, 2002.)

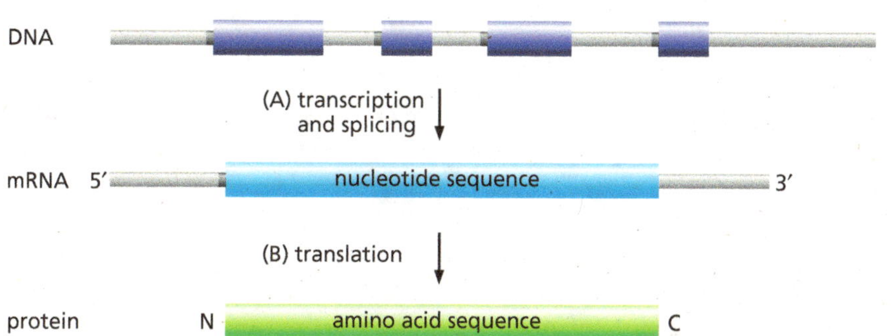


**Figure 2.5**
**Transcription and translation.** The relation of DNA coding-strand sequence to mRNA sequence to protein sequence. The exons (purple boxes) of the DNA are transcribed into mRNA which, using other molecules (see Chapter 1), directs the protein sequence.

## Box 2.1 Interactions between protein atoms

Interactions between protein atoms can be either covalent or noncovalent. The covalent bond, which occurs when an electron pair is shared between two atoms, is a stronger bond than the noncovalent interaction. Noncovalent interactions can take place by different means, such as hydrogen bonding or van der Waals or electrostatic interactions.

Atoms can be thought of as a nucleus surrounded by a cloud of constantly moving electrons. Sometimes there are more electrons at one end of the atom than at the other; this forms a dipole. A van der Waals interaction occurs when atoms that have oppositely oriented dipoles are near each other (see Figure B2.1). These interactions are weak.

Less weak are the hydrogen bonds, where there is a weak sharing of electrons between a hydrogen atom (donor) attached to a nitrogen or an oxygen and an acceptor atom, which in proteins is usually a nitrogen or an oxygen, although other atoms can act as acceptors as well.

Electrostatic interactions occur between groups that are of opposite charge.

**Figure B2.1**
**Two atoms with opposing dipoles.**

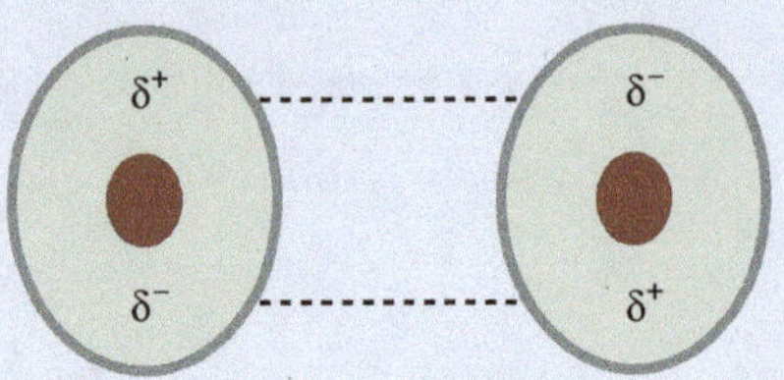


backbone can therefore be thought of as a chain of flat peptide units connected by $C_\alpha$ atoms. Each $C_\alpha$ atom makes two single bonds, within this backbone, one to the N and one to the carbonyl C, giving rise to two torsion angles per residue. These angles are known as $\phi$ (phi) and $\psi$ (psi) (see Figure 2.6C) and are the main source of flexibility of the polypeptide chain, allowing it to fold up. Although, in principle, $\phi$ and $\psi$ could take any value between –180° and +180°, they usually fall between certain limits, and residues with a given side chain have a strong preference for a particular value giving rise to repeating regular structural units in a protein. These preferences arise in two ways. First, **steric hindrance** prevents certain combinations of angles from occurring. Second, the hydrogen-bonding ability of the peptide groups can lead to selective stabilization for other angle choices. If several consecutive residues have the same ($\phi$,$\psi$) values, the resulting structure will be regular and repetitive. Some regular structures will be stabilized by hydrogen bonding, and we will describe the most commonly occurring structures below.

Additionally, the peptide bond is usually in the atomic conformation called the **trans conformation**, as this is thermodynamically more favorable than the alternative **cis conformation** (see Figure 2.7). This preference is because of clashes between two adjacent side chains (steric hindrance) in the cis conformation. Because of the particular nature of its side chain, proline is found in the cis conformation more frequently than other amino acids.

**Figure 2.6**
**Torsion angles.** (A) Angles between atoms can be calculated and are referred to as dihedral or torsion angles. (B) The angle of the peptide bond (O-C-N-H atoms) has a partial double-bond character and is therefore more rigid. (C) The angles of the bonds flanking the peptide bond can rotate more freely and are called phi ($\phi$) and psi ($\psi$); these are the main source of conformational flexibility of the polypeptide chain. Preferences for a limited number of values for these angles arise from steric hindrance and hydrogen bonding.

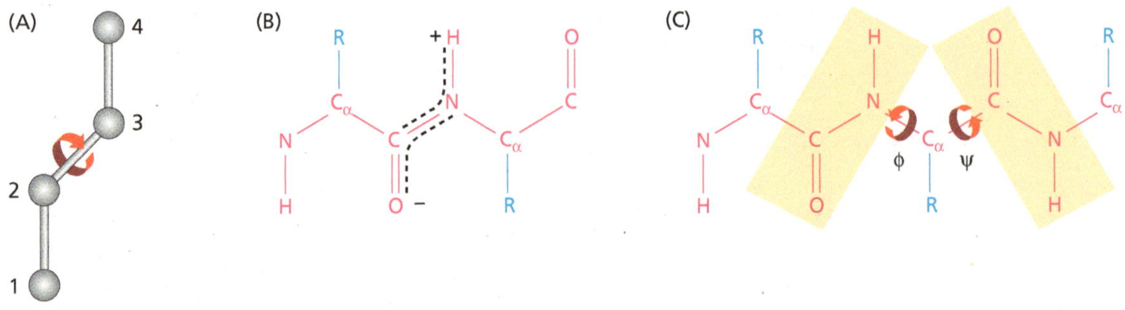

## Box 2.2 Covalent bonds

Covalent bonds are formed between two atoms by sharing electrons. The number of bonds an atom can make depends on how many electrons are available for this type of sharing. Most atoms form single bonds sharing one electron each between them. But there are some atoms that can form multiple bonds, such as double bonds. Covalent bonds can be classified according to their bond order. Single bonds have a bond order of 1; double bonds a bond order of 2. Non-integral bond orders are also possible, explained by the theory of resonance. In proteins many bonds with orders intermediate between 1 and 2 occur (usually shown as a dashed line beside a solid line), such as in peptide bonds and in carbon rings where the resonance makes them planar (Figure B2.2).

The arrangement of covalent bonds in a molecule describes the **molecular configuration**, but this does not necessarily define a single three-dimensional structure. Single covalent bonds allow for free rotation about the torsion angle of the bond, subject to any steric hindrance. Multiple bonds create constraints to the movement around the bond, resulting in some cases in distinct conformations such as the trans and cis forms of peptide bonds.

**Figure B2.2**
**A six-membered ring with single bonds and resonant bonds.** The ring with single bonds has two hydrogens per carbon atom, while the ring with resonant bonds has only one hydrogen per carbon atom and is planar.

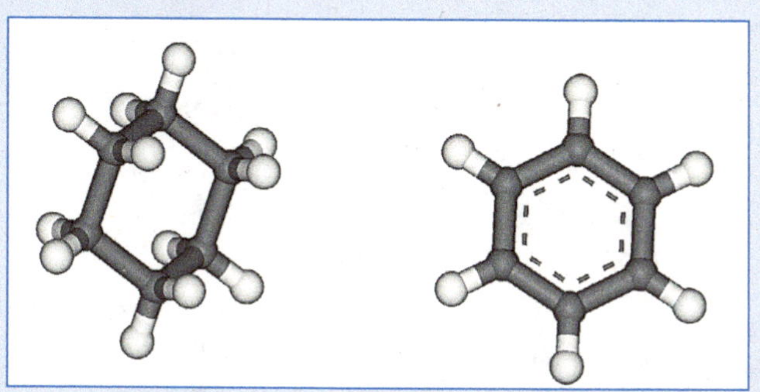

Perhaps the most important conformational property of a protein residue is its $(\phi,\psi)$ value. To reflect this, Ramachandran proposed plotting these values in a two-dimensional plot for all the residues of protein structures, and calculating the regions of $(\phi,\psi)$ values where most of the amino acids were found (these are referred to as favorable regions) and those $(\phi,\psi)$ value areas of the plot where it is rare to find an amino acid (these regions are referred to as unfavorable). Such plots are called the Ramachandran plots and are very useful tools to estimate how plausible a predicted protein structure is, as described in Chapter 13. The regions that were highly populated correspond to the regular structures (α-helix, β-strand) that are formed by the amino acid sequence in a protein structure (see Figure 2.8).

## Secondary structure of proteins is made up of α-helices and β-strands

In nature in general, regular structures are often helical, and a helix is found as a secondary structure in proteins. Helices arise as a result of energetically favorable hydrogen bonding between atoms of the backbone of the protein chain. In stretches of the chain with repetitive $(\phi,\psi)$ values of approximately (–60°, –60°), an

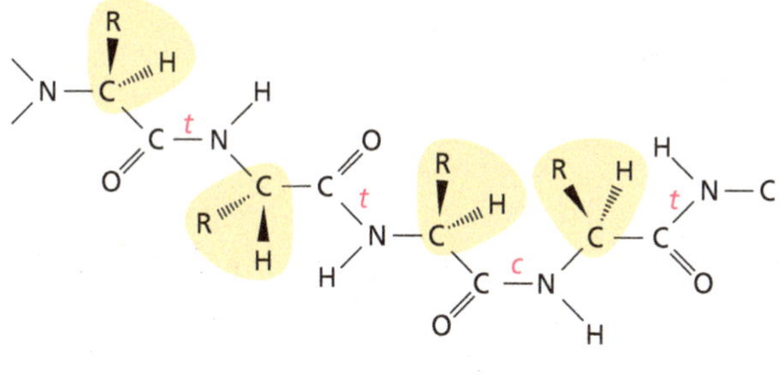

**Figure 2.7**
**Trans versus cis conformation of a peptide bond.** The figure illustrates the reasons for a preference for the trans (t). A cis (c) conformation results in a considerable increase in steric hindrance (atom clashes) between the two side-chain atoms.

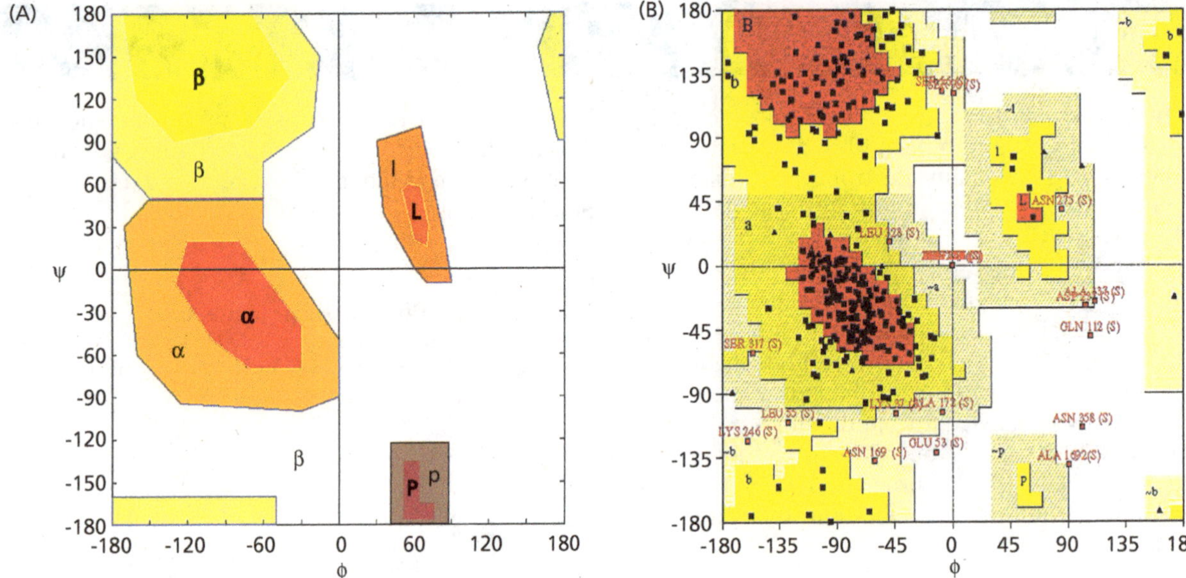


**Figure 2.8**
**Ramachandran plot.** (A) An ideal Ramachandran plot with the angle ϕ plotted on the x-axis and the angle ψ plotted on the y-axis. The yellow areas indicate the preferred conformation for β-sheet residues, the regular α-helix, the left-handed helix, and P (p) the epsilon structure. The darker colors show more favorable conformations. (B) The actual Ramachandran plot as calculated for a model of a kinase enzyme. Some residues fall within the disallowed areas (non-colored) and are identified by the residue type and sequence number.

especially stable right-handed helix is formed called the **α-helix** (see Figure 2.9). Left-handed helices can exist in proteins, but side-chain interactions favor the right-handed helix, and this is overwhelmingly predominant.

The proteins we will deal with in this book are composed of one or more globular domains, with most α-helices limited to a single domain, which restricts them usually to 2 nm in length. This is useful to bear in mind when predicting secondary structure from protein sequences, as described in Chapters 11 and 12. However, there are some instances of globular proteins with long α-helices extending across domains, of which an extreme example is hemagglutinin of the influenza virus, which contains an α-helix over 100 μm long. Many α-helices in globular proteins are only one or two turns long.

In an α-helix, the planes of the peptide bonds are nearly parallel with the helix axis and the dipoles within the helix are aligned such that all C=O groups point in the **C-terminal** direction and all N-H groups point in the **N-terminal** direction. The side chains point outward from the helix axis and are generally oriented toward the N terminus. The main chain -C=O and -NH groups in an α-helix are hydrogen bonded to a peptide bond four residues away [O($i$) to N($i$+4)]. Within the helix, all the backbone groups that can be are involved in hydrogen bonding. This gives a very regular and stable arrangement (see Figure 2.9).

It should be noted that the structures found in globular proteins are not perfectly regular, so it is frequently difficult to define the precise ends of the helices, and in some cases the hydrogen-bonding patterns are intermediate between these idealized forms. Therefore prediction of these structures using bioinformatics programs is made more difficult. A detailed discussion of this issue is given in Section 12.1.

In the 1960s another common regular repeating structure was found in globular proteins that consisted of extended strands aligned with each other to permit favorable hydrogen bonding. A single extended chain of this type is called a **β-strand**, and a set of β-strands hydrogen bonded together side by side forms a β-sheet (Figure 2.10). A β-strand has two sides that can participate in hydrogen bonding. The -NH groups occupy all of one side, and the -C=O groups the other; one side is proton accepting and the other proton donating. This dictates how β-strands are added to the β-sheet. Often, β-sheets are displayed by drawing out the pattern of

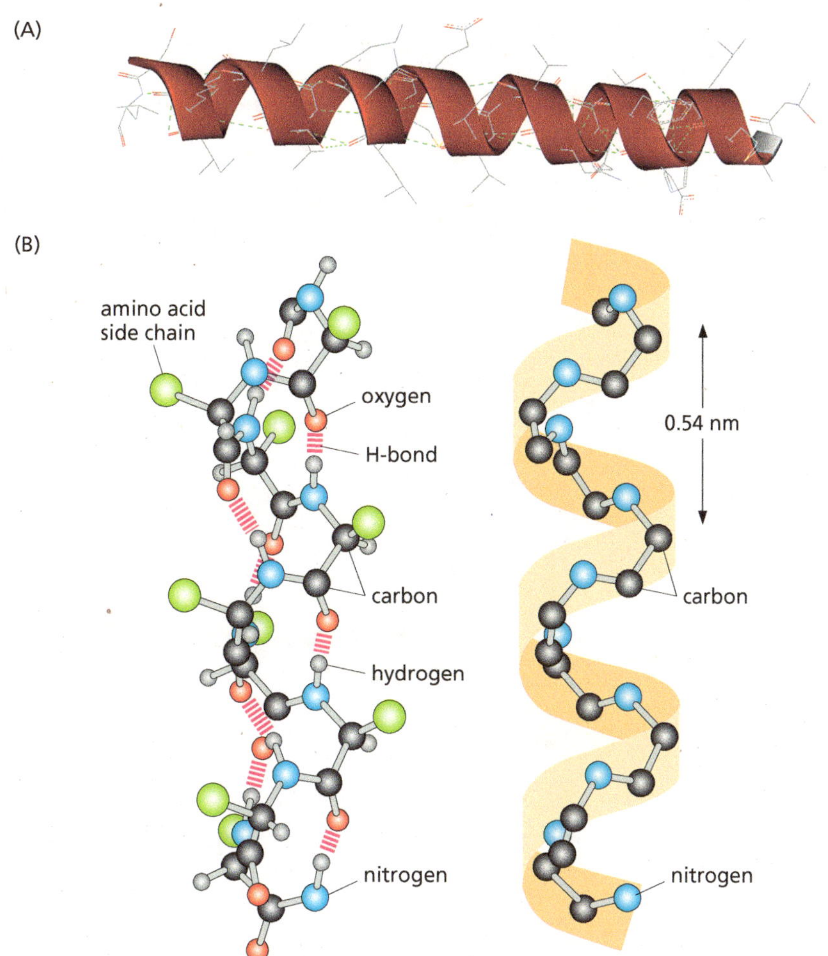


**Figure 2.9**
**Hydrogen bonding in the α-helix.** In the α-helix all hydrogen (H-) bonds involve the same element of secondary structure. (A) A representation of the α-helix. (B) The helical structure repeats itself every 0.54 nm (5.4 Å) along the helix axis, therefore we say that the α-helix has a pitch of 0.54 nm. α-Helices have 3.6 amino acid residues per helical turn. The separation of residues along the helix axis is 0.54/3.6 or 0.15 nm (1.5 Å); i.e., the α-helix has a rise per residue of 0.15 nm (1.5 Å).

hydrogen bonding involved. The backbone in a β-strand is not perfectly flat, but undulates. It is possible for the backbone to be flat, with the chain fully extended, but this is not the energetically favored conformation. One key difference between the β-strand and α-helices is that there are no short-range hydrogen-bonding interactions within an individual β-strand. The strands are stabilized by hydrogen bonding between different strands.

## Several different types of β-sheet are found in protein structures

As noted above, β-strands must face each other with correct orientation for hydrogen bonding to occur. Even when this criterion is met, there are still two alternative orientations: the adjoining strands can run parallel (see Figure 2.10C) to each other, or in opposite directions (antiparallel) as is seen in Figure 2.10B. The difference between them is most easily seen by locating the N and C termini of the strands. β-Strands are often drawn in schematics of protein structure as arrows pointing in the N-terminal to C-terminal direction. In their ideal form, β-sheets consist of an all-parallel or all-antiparallel arrangement of β-strands. These ideal β-sheets rarely occur in real proteins, and often there is a mixture of parallel and antiparallel β-strands in the same sheet. The order of the β-strands in the β-sheet is often apparently random when the β-strands are numbered from the N terminus onward. β-Sheets are usually not flat, being rather twisted instead, and a continuous β-sheet can even form a complete cylinder. The protein pores, the porins, that allow substances to pass through the outer membrane of some bacteria are formed of such cylinders. Like α-helices, β-strands are usually restricted to a single globular

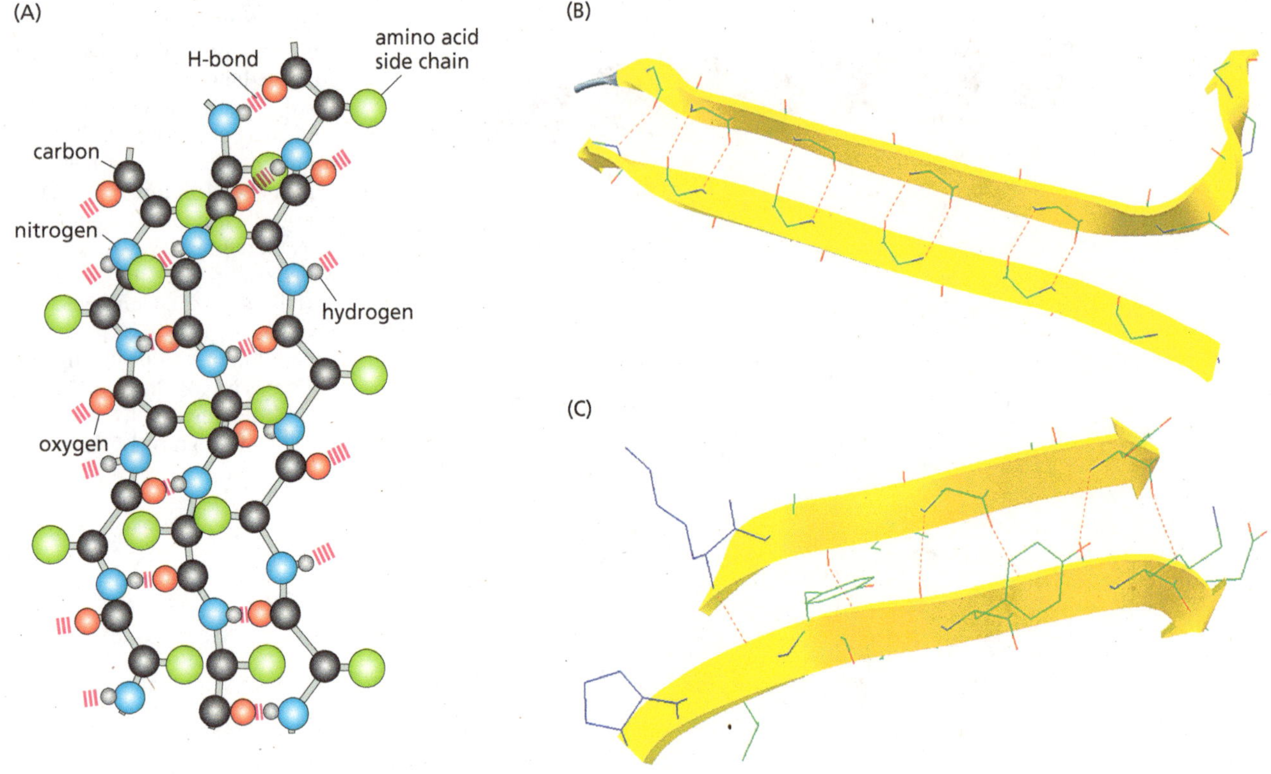


**Figure 2.10**
**β-Sheets are composed of sets of β-strands hydrogen bonded together.** A β-sheet may be formed from the folding of a continuous stretch of the protein chain, or can be formed by β-strands from different parts of the chain coming together. As with other regular protein structures, the hydrogen bonds are formed mainly between backbone groups. (A) A β-strand has two sides that can participate in hydrogen bonding. The -NH groups occupy all of one side, and the -C=O groups the other; one side is proton-accepting and the other proton-donating. This property dictates how β-strands are added to the β-sheet. Often β-sheets are displayed by drawing out the pattern of hydrogen bonding involved. In parallel β-sheets the strands all run in one direction, whereas in antiparallel sheets they all run in opposite directions. (B) An antiparallel β-sheet with the hydrogen bonds in red dots and (C) a parallel β-sheet. Note the different but specific hydrogen-bonding patterns in the antiparallel and parallel β-sheets. From an examination of many β-sheets seen in globular proteins, it appears that the conformation of the β-strands is less fixed in terms of the $(\phi,\psi)$ angles than is the case with α-helices. The $(\phi,\psi)$ angles for idealized (i.e., perfectly straight) β-strands is (–120°, +120°), but almost all β-strands are significantly curved and thus have some deviation from these values. As mentioned above in the case of helices, real structures in globular proteins are considerably distorted from the ideal, and deciding which residues are actually in a β-strand is nontrivial (see Section 12.1).

domain, and so rarely exceed 2 nm in length. However, β-sheets have been found that extend over more than one domain, comprising many β-strands.

## Turns, hairpins, and loops connect helices and strands

Usually somewhere in the range of 50–80% of residues in globular proteins can be classified into one of the regular structures described above. The third type of secondary structure element is the **β-turn**. These are short regions where the protein chain takes a 180° change in direction, doubling back on itself (see Figure 2.11). Turns are found, for example, between two adjacent β-strands in β-sheets formed from a continuous stretch of protein chain. The remainder of the protein structure has much less order, and can be viewed as simply the connecting pieces (**loops**) that allow the α-helices and β-strands to pack, creating the protein structure. Whilst this view may not be too unrealistic from the viewpoint of structure alone, it is not at all the case when considering function. Most of the surface of proteins (and hence the accessible regions) is not regular structure. These surface parts often hold substrate-binding

**Box 2.3 Chameleon sequences**

Some short sequences of amino acids can adopt either an α-helical or a β-strand conformation, depending on the protein in which they occur. These sequences are known as chameleon sequences. Recently it has been found that this ability of a protein sequence played a part in bovine spongiform encephalopathy. The prion protein associated with this disease can adopt different stable conformations. The conformation found in the diseased state is a mixture of helices and sheets, while in the normal state the prion protein consists of a bundle of helices. Consequently, predictions based on local sequence information will not be able to resolve such conflicts as found in the prion protein. For a further discussion of the consequences for secondary structure prediction, see Figure 12.12 and Section 12.3.

residues and catalytic residues, and are therefore extremely important. For example, the ligand-binding site of the phosphoinositol kinases (which will be used later in this book) is in part formed by a relatively large loop on the surface. This loop is so mobile that its structure could not be seen in a crystallographic analysis.

Any chain between two regular structures is referred to as a loop. In many cases a loop will contain a turn (or even several). In general there are no classifications for loops, but there is an important exception. In antibody recognition, immunoglobulins employ loops at the edge of a β-sheet to recognize the antigen. There are vast numbers of different immunoglobulin structures, all with the same overall chain fold, but it is the difference at these loops that results in different affinities. With many structures known, it has been observed that the loops take up one of a limited number of structures (called canonical forms), so that in this particular case the loops have been classified. This type of classification is important when trying to predict both the structure and function of the protein.

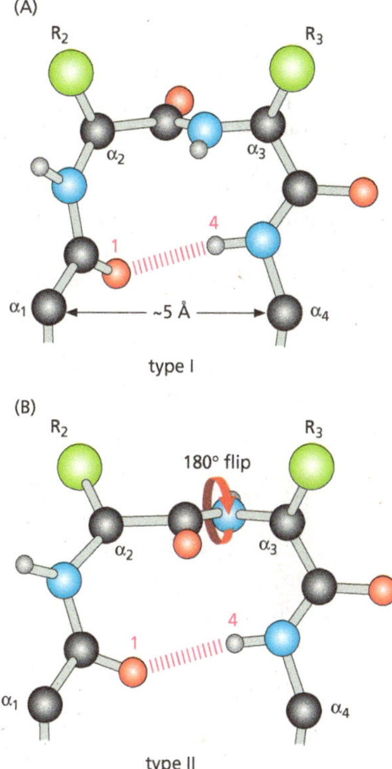


**Figure 2.11**
**Turns.** The two common types of tight turn, called β-turns type I and type II, respectively. The side chain of R3 is usually hydrogen (glycine).

## 2.2 Implication for Bioinformatics

In part, bioinformatics concerns itself with the analysis of protein sequence to predict the secondary structure, the tertiary structure, and the function of the protein, as well as its relationship to other proteins. Different secondary structures tend to have subtle differences in chemical environments, resulting in amino acid preferences. In addition, amino acid preferences are seen at particular locations in proteins due to the functional role they play, for example as catalytic residues or stabilizing the overall protein structure. These aspects of proteins, which influence structure prediction and analysis, are described in more detail below.

### Certain amino acids prefer a particular structural unit

Due to the various properties of the amino acid side chains, certain residues or certain types are found more often in one or the other of the structural units. Some residues have been classified, for example, as α-helix breakers (poor formers) or formers. Proline, for example, is a poor helix former due to the fact that its backbone nitrogen atom is already bonded to its own side chain and cannot form hydrogen bonds within the helix, and its rigid structure causes some steric hindrance. Other poor α-helix formers are Gly and Tyr, while Ala, Glu, Leu, and Met are good α-helix formers, whereas Val, Ile, Tyr, Cys, Trp, Phe, and Thr are more frequently found in β-strands than in α-helices. In addition, Gly, Asn, Pro, Ser, and Asp are more likely to be found in turn segments than in any other structure. These types of preferences have been the driving forces in secondary structure prediction programs (especially in the early development of the field) and are discussed further in Chapters 11 and 12.

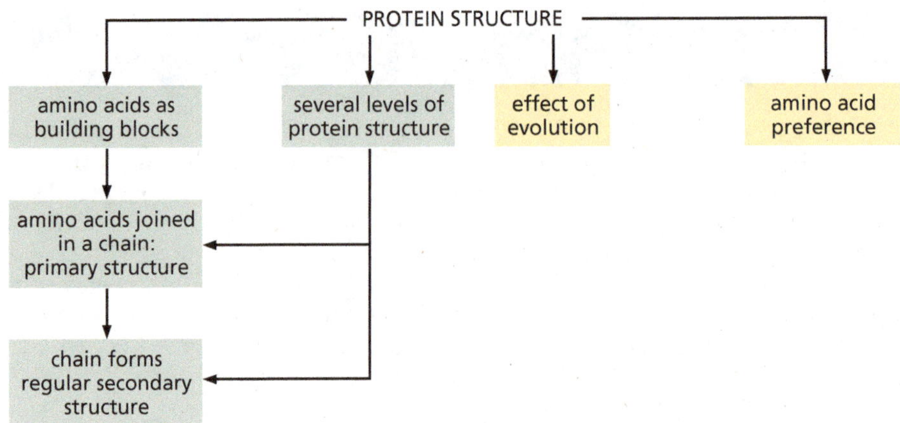


**Flow Diagram 2.2**
**The key concept introduced in this section is that, for a variety of reasons related to protein structure and function, each position in a protein sequence has an associated amino acid preference that is a reflection of the evolutionary selection pressure on the organism.**

## Evolution has aided sequence analysis

Protein sequence similarity is a powerful tool for characterizing protein function and structure since an enormous amount of information is conserved throughout the evolutionary process (Chapters 7 and 8). Proteins that have a common ancestor are referred to as being **homologous** (see Figure 2.12). Sequence alignment and database search techniques (Chapters 4 and 5) can identify homologous proteins. Homologous proteins usually have a similar three-dimensional structure with related active sites and binding domains. Therefore homologous proteins will also often have related functions, although this is not always the case. Most amino acids that change during evolution are found in regions that are not structurally or functionally important, such as many of the loops (or variable) regions. If the homologous protein is also functionally related then the amino acids involved in function are often conserved during evolution, which helps in identifying the function of a new protein (see Flow Diagram 2.2).

## Visualization and computer manipulation of protein structures

There are a number of programs available that read the coordinate file and convert it to a visible three-dimensional representation of the protein. The protein can be rotated, specific regions highlighted, and some measurements can be calculated. Some of these programs are very powerful and can be of great use in analyzing the structural properties and molecular function, as well as allowing for the manual modification of the molecule. Some of the programs are free or low cost, such as Chimera, Yasara, and DeepView. Others are extremely powerful programs that allow

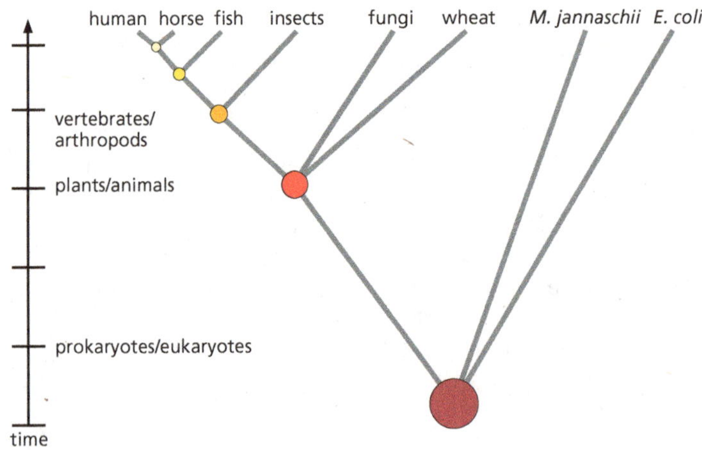


**Figure 2.12**
**Tree of life.** This tree shows how life may have evolved from a common ancestor giving the diversity of life forms that inhabit Earth.

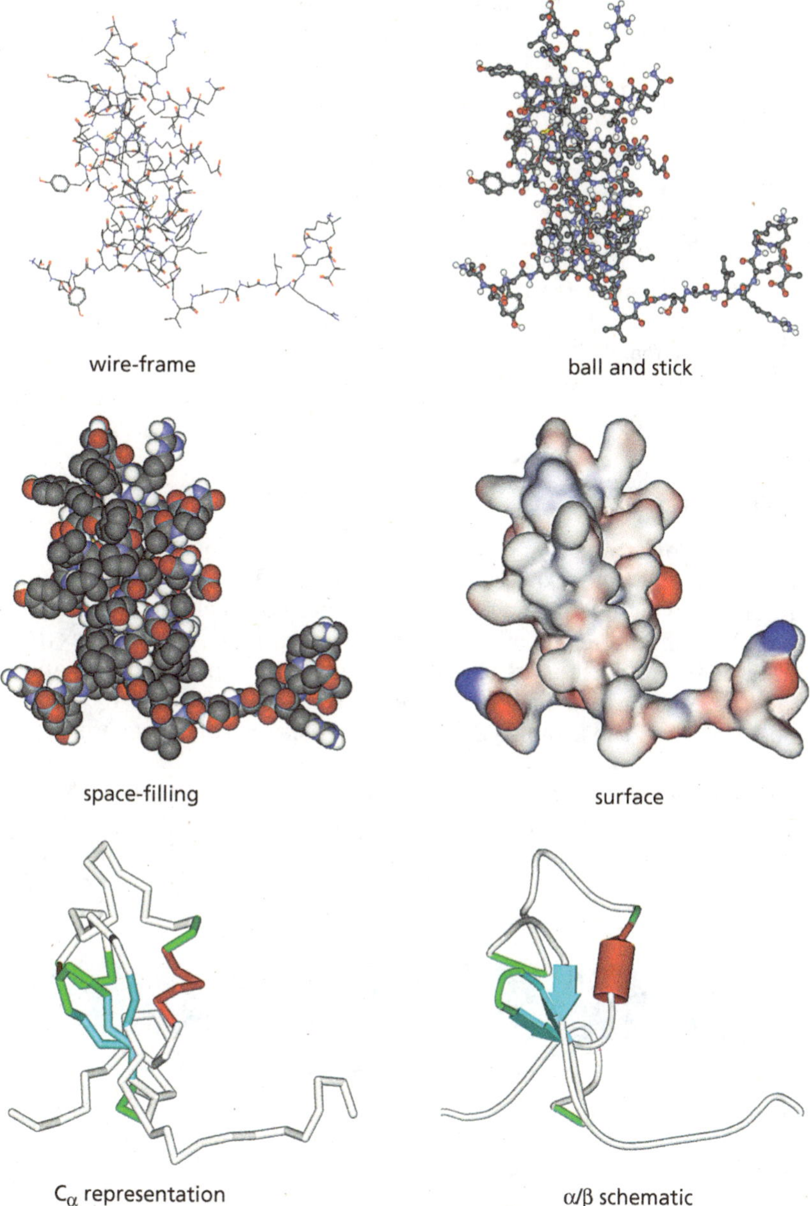


**Figure 2.13**
**Molecular representations.** The different representations that can be used to illustrate molecules, from very simple ones that only use the $C_\alpha$ or backbone atoms to space-filling models of all atoms in the structure.

the user to carry out computationally intensive modifications to the molecule, but are expensive. It would be helpful for the reader to gain some experience of using these programs before reading the secondary and tertiary structure prediction chapters (see Chapters 11 to 14).

There are many styles for viewing molecular structures, including those with atomic-level detail such as space-filling models, ball and stick models, and wire-frame models (also called stick models or skeletal models), as well as surface models. However, it is often desirable to have a simplified model of the protein, such as **backbone** or **$C_\alpha$ models** and schematic (cartoon) models. Such models can be represented on a computer screen and can be represented in different styles and colors. Molecular models are usually based upon an atomic coordinate file, which in general give the $(x,y,z)$ coordinates of each atom. Figure 2.13 shows some of the more common structural representations.

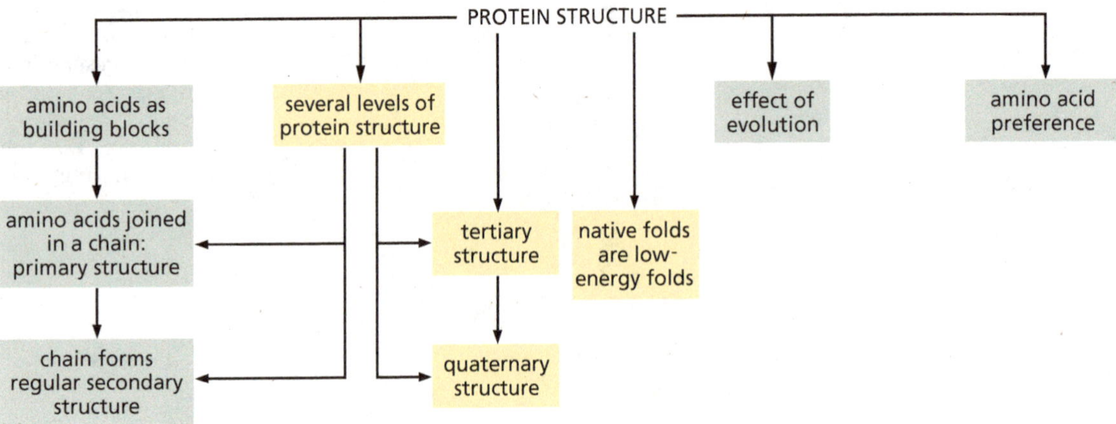


**Flow Diagram 2.3**
**The key concepts introduced in this section are that most proteins exist in a well-defined three-dimensional structure that is thermodynamically favorable, and that proteins often occur as complexes of several protein chains.**

## 2.3 Proteins Fold to Form Compact Structures

Most **protein folds** seem to be dominated by the appearance of regular structures such as α-helices. It is now known that these are the first elements to form during protein folding, which confirms that they are the most stable components of the protein fold. Furthermore, the overall structure of a protein is so complex that these smaller, more ordered elements are useful as an aid to understanding and classifying the fold. Most proteins are composed of some combination of α-helices and β-strands. There are also proteins that are all α-helix and some that are all β-sheet.

Protein chains themselves rarely have any biological function. Only when the chain has folded up into a three-dimensional structure (however small) does the protein have functional activity. Some proteins are enzymes that bind other molecules (ligands) and catalyze their biochemical reactions (see Figure 2.14), others act by

### Box 2.4 Supersecondary structure

Supersecondary structures are fold-elements composed of specific combinations of α-helices and/or β-strands that are often repeated throughout protein structures.

The figure illustrates schematically the most commonly occurring supersecondary structures: (A) a βαβ repeat; (B) a β-meander; and (C) a special β-unit called the Greek Key structure, which occurs when an antiparallel sheet doubles back on itself. Part D in the figure is a cartoon representation of a Greek Key form gamma β-crystallin [Protein Data Bank (PDB) code 1AMM].

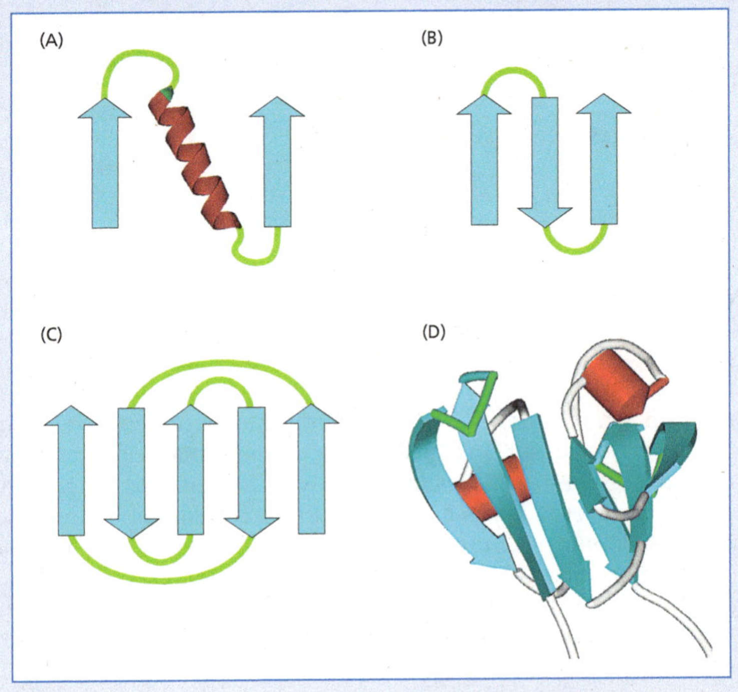


**Figure B2.3**
**Some examples of supersecondary structure.**

binding other proteins and influencing their activity, and yet others bind to DNA and regulate gene expression. Some proteins have a purely structural function, making up the fabric of the cell. Large numbers of proteins are released, or secreted, from cells and act as chemical messengers, influencing the behavior of other cells by acting on yet another large functional class of proteins, known as receptors, on cell surfaces.

The proteins that will concern us most in this book are globular proteins, proteins that are roughly spherical when folded up. The other main class of proteins are fibrous proteins, such as the keratin of wool and hair, and the silk protein. As globular proteins, or proteins composed of multiple globular units, are the proteins that carry out most cellular functions, they are the most studied. Some globular proteins are composed of a single, folded, globular mass; in others, the protein chain folds up into several such discrete structural units, each of which is termed a protein **domain**. Many proteins are composed of a multiplicity of different domains, each with a particular biochemical or binding function. The possibility of multidomain structure poses challenges when comparing protein sequences or interpreting the amino acid sequence of an unknown protein, as we shall see in Chapter 4.

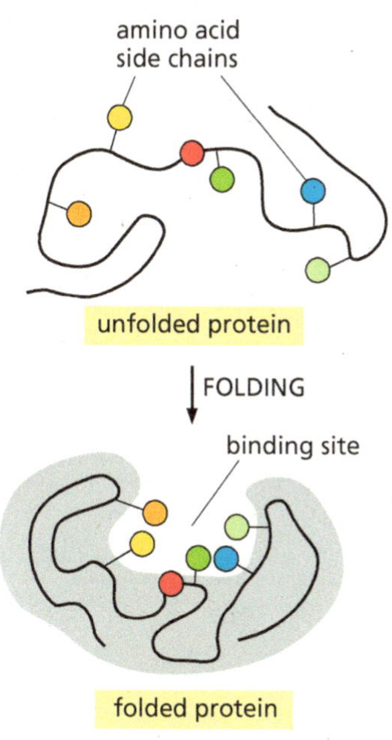


**Figure 2.14**
**Distant residues can come close in the folded structure.** A schematic which shows that when a polypeptide chain (primary structure) folds into a tertiary structure, residues that are far apart from each other in the sequence can come close together to form a functional unit, in this case a binding/catalytic site. (From B. Alberts et al., Molecular Biology of the Cell, 4th ed. New York: Garland Science, 2002.)

## The tertiary structure of a protein is defined by the path of the polypeptide chain

In the tertiary structure of a protein, various combinations of secondary structure pack together to form a compactly folded mass. For simplicity we will discuss tertiary structure here as if a protein were folding up into just one globular domain, but many proteins are composed of multiple domains joined by stretches of polypeptide chain. In a multidomain protein it is thought that each domain folds independently of the others. As will soon become clear in later chapters, bioinformatics questions are often concerned with comparing the sequences and structures of different domains rather than whole proteins. A domain can be anything from 50 to around 350 amino acids in length. The core of each domain is mainly composed of tightly packed α-helices, β-sheets, or a mixture of both. The three-dimensional structure of a protein is known as its conformation. More specifically, the spatial path of any given folded polypeptide chain is known as its fold.

As proteins exist in an aqueous environment, folding tends to result in hydrophobic regions of the protein ending up in the interior, while more hydrophilic regions are on the outside. A variety of noncovalent interactions stabilize the fold, dominated by hydrogen bonding and the clustering of hydrophobic groups. Secondary structures pack together in a variety of ways, such as the formation of β-sheets from either parallel or antiparallel β-strands. The atoms pack together very efficiently in most natural proteins.

There appears to be a limited number of ways in which secondary structures fold into domains. There are several instances where proteins that seem to be completely unrelated in terms of sequence are found to have the same fold and some researchers estimate that there may be only a few thousand different folds in nature. Currently, there are more than 35,000 known protein structures, and these are classified into approximately 2000 fold families. The fact that so many proteins fold into a similar structure even if their sequences are not very similar means that we can use bioinformatics tools to model structures of various proteins on similar folds as will be described in Section 13.2.

## The stable folded state of a protein represents a state of low energy

A protein chain starts to fold as soon as it has been synthesized and thermodynamic considerations mean that the final fold it adopts is a state of low **free energy**. This is discussed in more detail in Section 13.1 and Appendix B. Folded proteins are

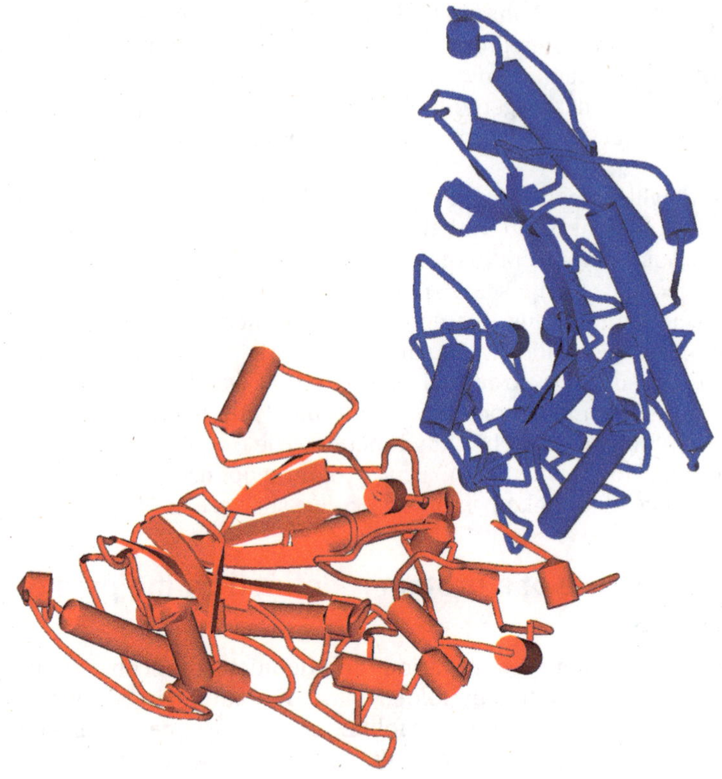

**Figure 2.15**
**The quaternary structure of muscle creatine kinase.** This structure is formed by the complex of two identical chains and structures. Therefore it is called a homodimer.

generally stable in the conditions in which they have to operate, but in a wider sense, proteins are unstable thermodynamically. Most proteins start to unfold above about 60°C, as the noncovalent bonds that hold them together are broken by thermal energy; unfolded proteins are said to be **denatured**. As it becomes denatured, a protein loses its function. Only specialized microorganisms are capable of living at temperatures this high.

The stability of a protein is a fine balance between a number of effects. Predominant among these is the hydrophobic nature of many side chains, and so structure that depends on hydrophobic interactions is more stable away from contact with water. As a result there is a tendency for large hydrophobic groups to cluster inside the protein, with most of the polar groups on the protein surface. This is an oversimplification, however, as many nonpolar groups do reside on the surface, occasionally even having an important role in binding nonpolar ligands.

Hydrogen bonds also play an important role in the structure of the protein, although it is not as simple as the hydrophobic effect. The folded structure will form internal hydrogen bonds, both between the main chain atoms and some of the side chain atoms. However, in an unfolded state the atoms form hydrogen bonds with water. Therefore all these water-mediated hydrogen bonds are lost when the protein folds, which is not conducive to folding, but the release of all the otherwise hydrogen-bonded water creates more entropy in the solvent and this contributes to the folding effect.

Additionally, the residues in a protein pack together very effectively to occupy space, and there is some evidence that those proteins that have unoccupied space in the interior are less stable.

## Many proteins are formed of multiple subunits

Individual folded polypeptides can interact with each other to form protein complexes or quaternary structures. Oligomeric proteins contain more than one

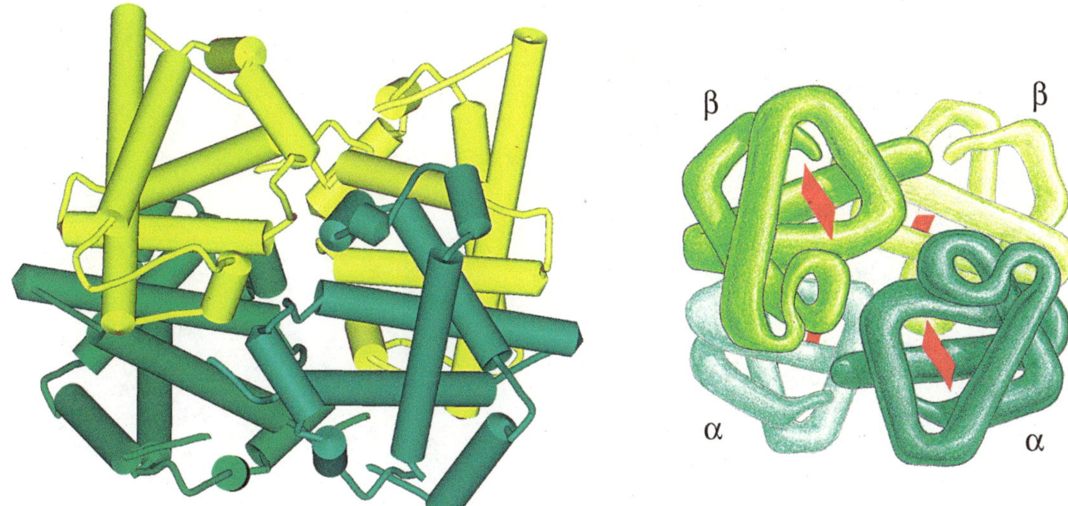


polypeptide chain. The individual chains are referred to as subunits or monomers of the oligomer. The assembly of these monomers is the quaternary structure. A protein consisting of two subunits is called a dimer, three a trimer, four a tetramer, and so on. The subunits can be the same protein sequence, such as in the dimer of muscle creatine kinase (see Figure 2.15), or made up of different polypeptide chains, such as the tetrameric hemoglobin, which has two α and two β subunits (see Figure 2.16). A quaternary structure associates in a similar way as the secondary or tertiary structure using noncovalent interactions. However, when quaternary structures assemble they often have hydrophobic groups that interact and exclude water. Often quaternary structure forms only when the complex is active, and frequently there is a sophisticated control of the activity involving several subunits. It is much more difficult to predict or model the structure of a quaternary structure compared to a single domain protein.

**Figure 2.16**
**The quaternary structure of bovine deoxyhemoglobin which is a heterotetramer.** In other words it is made up of different chains and structures from four folds. Two folds are the same therefore it consists of two homodimers. The heme groups are shown in red. (From B. Alberts et al., Molecular Biology of the Cell, 4th ed. New York: Garland Science, 2002.)

## Summary

This chapter introduced the general principles associated with protein structure. Proteins are linear polymers composed of amino acids. They have a complex structure that includes some more regular features. Large proteins consist of several domains, each of which will have some regular secondary structures. The structures form a scaffold on which the active groups are located, ready to bind ligands, or carry information, and other roles in order for the protein to function properly.

One of the main aims of bioinformatics is to predict and analyze the structure of proteins and the relationship of the structure to the function. Many programs have been designed to predict protein secondary structure from the primary sequence information. These types of studies are described in Chapters 11 and 12 of this book. Further methods of modeling have been proposed that predict the three-dimensional structure of proteins as well as the quaternary complexes and these are described in Chapter 13. Some aspects of structure function analysis are dealt with in Chapter 14. To perform accurate predictions and to obtain more information about the protein sequence under study, it has to be aligned to other proteins to find homologs. This type of research is dealt with in detail, both for protein and nucleotide sequences, in Chapters 4 to 8 of this book.

## Further Reading

Alberts B, Johnson A, Lewis J, Raff M, Roberts K & Walter P (2008) Molecular Biology of the Cell, 5th ed. New York: Garland Science.

Branden C & Tooze J (1999) Introduction to Protein Structure, 2nd ed. New York: Garland Science.

Lesk AM (2000) Introduction to Protein Architecture. Oxford: Oxford University Press.

Mezei M (1998) Chameleon sequences in the PDB. *Protein Eng* 11, 411–414.

Schulz GE & Schirmer RH (1984) Principles of Protein Structure. New York: Springer Verlag.

3

# DEALING WITH DATABASES

## When you have read Chapter 3, you should be able to:

Explain why databases are the backbone of bioinformatics research.

Discuss how flat files were the first type of database, and why they are still used today.

Show that relational databases are better for searching across tables.

Outline the many other types of database structure that exist.

Explain why databases contain both data and annotations of data.

Discuss how many different types of database exist.

Explain why the quality of data is important.

Explain why checking the data and human curation are necessary.

The first databases were simply collections of data. For example old cards-in-a-box catalogs were noncomputerized databases. However, a database is more than just a collection of data. The modern database indeed stores data, but it also contains quite complex technology (such as ORACLE) to store the data in a structured manner and it is a complex model of tables and accompanying connections. A database is a sophisticated arrangement of storage, methods of storing, and architecture. It can be likened to a tax collection office, which not only contains the data about people's tax payments, but has a physical means of storing it, and methods by which the data can be **input**, accessed, and analyzed. In this context, the term architecture refers not only to the organized structure in which the data are kept but also to the building itself which provides the resources needed by the staff (also part of the architecture!) to process the data.

It is just over 50 years ago that the first protein sequence, that of bovine insulin, was determined by Frederick Sanger. Ten years later there were already attempts to collect all known sequences in a single database as an aid to the analysis of relationships between similar sequences. At the same time, programs for extracting and analyzing these sequences were written and the field of bioinformatics began, although it did not receive this name for some years.

The number of documented nucleotide sequences now numbers in the hundreds of thousands, and there are over a hundred thousand protein sequences too. This explosion in the number of sequences has made the use of electronic databases for storage and analysis essential. There has been a parallel increase in the quantity of data in other areas of biomedical research, such as molecular structures, and through the use

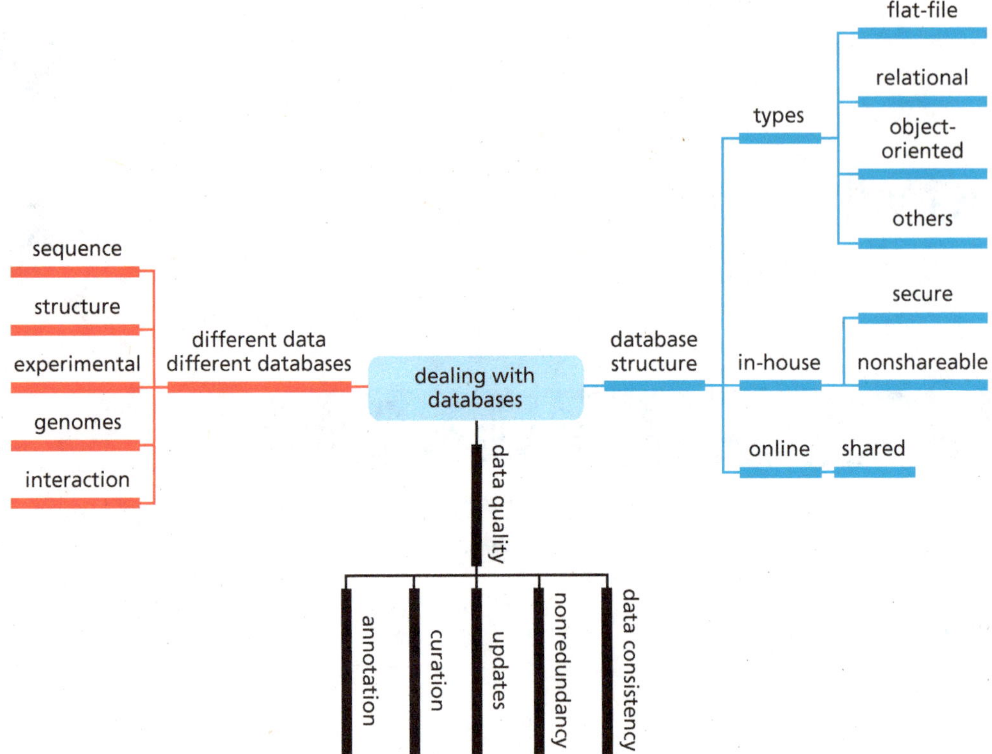


**Mind Map 3.1**
**The schematic representation of topics important in understanding general aspects of databases.** The mind map highlights the points that there are many different databases used by bioinformaticians, that the database structure can vary, and also that data quality is very important.

of new experimental techniques such as microarrays and gene expression measurements. The need for databases has similarly increased in these areas. The existence of many different databases in closely related areas makes it useful to include cross references between related entries in different databases. As a result, today many of these databases can be regarded as linked together into a large network of information covering a broad range of biomedical and chemical research.

There are many different ways in which databases can be designed, both in terms of the ways the information is stored and the ways it can be retrieved and analyzed. There is no need to have a detailed technical understanding of these aspects in order to use databases, but we will describe some of the basic concepts as they can help in making effective use of these data sources.

Although there are many different types of databases, we will give an overview of the types most commonly used in bioinformatics research. Only a small fraction of the complete set of databases will be mentioned here, but the reader interested in discovering more about those not covered will be directed to an extensive list.

It is important to be able to have confidence in the accuracy of the data extracted from these sources. For this, certain aspects of database maintenance need to be understood before accessing any type of data for further analysis. Data quality issues are described in the last section of this chapter.

## 3.1 The Structure of Databases

A database is a repository of information that has a specific structure that enables the entering and extraction of data and in many cases also aids analysis of the data (see Flow Diagram 3.1). In general this database structure consists of files or tables, each containing numerous **records** and **fields**. Figure 3.1A shows an example of a very simple database table, in this case a single page with a contact list, with three

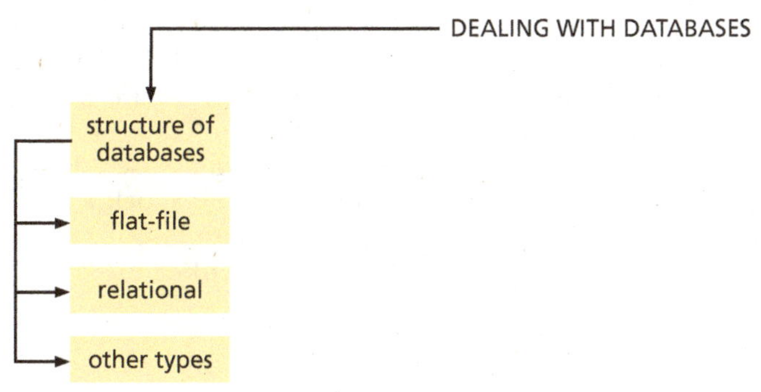


**Flow Diagram 3.1**
**The key concept introduced in this section is that there are several distinct forms of electronic databases, each with particular advantages and disadvantages.**

records each storing the details of one individual. There are three fields—Name, Telephone and Address—for each record.

A more complicated example would be a database of gene sequences stored in paper form in a filing cabinet, with gene data for each species stored in a separate file. Each file would contain many pages, each holding the information about a single gene. The information given about each gene will be in several distinct parts, such as the name of the gene, the gene sequence, or the name of the protein encoded by the gene. Each of these different pieces of information can occur in all genes, so that often the page used is printed with a standard form, with each section of the form, called a field, used to record one of the types of information. When databases are stored in electronic form their structure has many similarities to the paper form. Often a single computer file stores the entire database, and is the equivalent of the filing cabinet. Electronic database files consist of tables, which are the equivalent of the individual files in the cabinet. Thus, a gene sequence database might contain a separate table for each organism. Each gene would be listed in a separate row of the table, called a record, the electronic equivalent of the page. Each record will consist of several different pieces of information given in different columns of the table, and called fields. An example of the beginning of a record for gene TCP1-beta of *Saccharomyces cerevisiae* is shown in Figure 3.1B. This illustrates that the GenBank flat-file format is readable by humans as well as computers, with the field names shown here in blue. The complete record is very long, and so only the top section is shown.

(A)

| CONTACT LIST | | |
|---|---|---|
| NAME | TELEPHONE | ADDRESS |
| S. Claus | 0203 450 | The North Pole, Lapland |
| M. Mouse | 0202 453 | Disneyworld, Florida |
| A. Moonman | 0104 459 | Craterland, The Moon |

(B) GenBank Flat-File Format

```
LOCUS       SCU49845     5028 bp    DNA
DEFINITION  Saccharomyces cerevisiae TCP1-beta gene, partial cds, and
             Axl2p
            (AXL2) and Rev7p (REV7) genes, complete cds.
ACCESSION   U49845
VERSION     U49845.1  GI:1293613
KEYWORDS    .
SOURCE      Saccharomyces cerevisiae (baker's yeast)
  ORGANISM  Saccharomyces cerevisiae
            Eukaryota; Fungi; Ascomycota; Saccharomycotina;
           Saccharomycetes;
            Saccharomycetales; Saccharomycetaceae; Saccharomyces.
```

**Figure 3.1**
**Two examples of flat-file database structures.** (A) shows a contact list as a flat-file database in which a record holds the contact information for an individual, and consists of a number of fields (in this case three), such as name, telephone number, and address. (B) shows that a flat-file format can be very useful and is still used today, especially with text-handling computer languages. This is an example of a flat-file format obtained from the GenBank sequence databank. It is a very small part of the complete record, and the words in blue are the field names.

There are various types of electronic databases which differ in their structure. A structurally simple example of a database is the flat-file format, while a much more complex and therefore more versatile database structure is the relational form. Both of these will be discussed below in more detail. More modern database management structures include **object-oriented databases**, **data warehouses**, and **distributed databases**. These are also briefly described below. Note that a computerized database needs software that is used to control the database; this software is referred to as a database management system or DBMS for short.

## Flat-file databases store data as text files

A flat-file database is the simplest form of a database, where collections of data, such as nucleotide and amino acid sequence, are stored as either a large single text file or as a collection of different text files. The file contains records, generally one record per line. The data files are flat, as in a sheet of paper, in contrast to more complex and versatile models such as a **relational database** discussed next.

A common example of a flat-file database is a simple name and address list. In this example the record will consist of a small and fixed number of fields such as a name, address, and contact number. Flat-file databases can be created by hand, for example when you write on a piece of paper the name, address, and phone number of your friends you will have created a flat-file database. However, although useful, this has its limitations; as the number of friends grows and as friends change addresses the database has to be updated, which would end up with the piece of paper being filled up and looking very messy with many crossed-out fields. Therefore using computers and computer programs to manipulate the data in a flat-file format is useful and was first implemented a long time ago. In the nineteenth century Herman Hollerith decided that any resident of the United States could be encoded by a string of exactly 80 digits and letters consisting of name, age, and other information, filled with blank spaces to make everyone's entry the same length. This led to the first-ever computerized census, in 1890, encoded on punch cards that were read by machines (see Figure 3.2).

Nowadays, electronic flat-file databases can be organized and maintained by sophisticated management software that provide a mechanism by which data can

**Figure 3.2**
**First computer databases.** This computer was designed in the late 19th century, and first used in the 1890 United States census. Hollerith developed an integrating tabulator housing separate adding machines—the upright units—that could simultaneously add totals recorded in separate areas, or fields, of a punched card. (Courtesy of Science Photo Library.)

be easily input and can be retrieved using detailed search queries. However, in general, the data available to biologists involves so many different aspects that it can be more effectively stored using more sophisticated database structures. Flat files are still used, especially to distribute the data (see Figure 3.1B), because many of the more complex database structures depend on specific, often expensive software, whereas the flat files can be read and analyzed by many alternative programs according to the user's preferences.

## Relational databases are widely used for storing biological information

Probably the commonest type of database used for biological information is the relational database. A relational database stores the data within a number of tables. Each table consists of records and fields (rows and columns) as described above and in general these will be different for each table. However, each table will be linked to at least one other by a shared field called a **key**. It is these keys that distinguish a relational database from a flat-file database. It should be noted that not all fields can be keys, as the data for a key must be unique in each record. A relational database can have several different fields acting as keys, even in the same table, and the structure of the database is often drawn to show the keys linking different tables together.

For example, in Figure 3.3 a table called *protab1* is shown where each row contains a code for a protein, the name of the protein, the length of its sequence, and which species it comes from. In this case *protab1* is linked to other tables in a relational database by the protein code field, which is the key. This enables the extraction of related data from many tables so that, for example, using the protein code we can extract the name of a protein from *protab1* and the sequence of that protein from *protab2*, a table that only stores the protein code and protein sequence. In this way, the information held in different tables can be combined.

In relational database applications a set of operators is provided that allows for easy manipulation and analysis of the data. Some of these operators are mathematical in nature (addition, multiplication, and so on) and some are concerned with data handling, such as selecting specific subtypes of data from different tables. Frequently, the operations performed on the data in the tables produce results that are displayed in new tables. In the protein sequence example just described, the results table would contain the protein code, protein name, and protein sequence. In relational database management systems, the operators are written in query-specific languages that are based on relational algebra. **Structured Query Language** (**SQL**) is

**protab1**

| Protein-code | Protein-name | Length | Species-origin |
|---|---|---|---|
| P1001 | Hemoglobin | 145 | Bovine |
| P1002 | Hemoglobin | 136 | Ovine |
| P1003 | Eye Lens Protein | 234 | Human |
| ..... | | | |

**protab2**

| Protein-code | Protein-sequence |
|---|---|
| P1001 | MDRTTHGFDLKLLSPRTVNQWLMLALFFGHS... |
| P1002 | MDKTSHGFEIKLLTPKKLQQWLMIAIYFGHT... |
| P1003 | SRTHEEEGKLMQWPPRPLYIALFTEPPYP... |
| ..... | |

**Figure 3.3**
**Two example tables representing the relational database model.** In this structure the records are connected by an identifier, often referred to as the primary key. There can be a number of subidentifiers or secondary keys. These connect various types of data from as many tables as necessary. Any field can be a key as long as it is unique.

commonly used and the examples below show how data could be extracted from the tables in Figure 3.3. The first query is a simple extraction of data from a single table.

*Query 1*

```
SELECT protein-code, protein-name
FROM protab1
WHERE species-origin = 'Bovine';
```

The SQL commands are in capital letters (this is a common standard when writing SQL), and all this simple bit of SQL code means is: select all protein codes and protein names from the table called "protab1" where the species of origin is bovine. The second query is slightly more complex and extracts data from two different tables using the key as the common factor.

*Query 2*

```
SELECT protab1.protein-name, protab2.protein-sequence
FROM protab1, protab2
WHERE protab1.protein-code = protab2.protein-code
AND protab1.protein-code = 'P1002';
```

In this example we are telling the program to search and extract all protein names from protab1 and all sequences from protab2 where protein-code is p1002 in both tables.

These are just simple examples of basic queries. The queries can become extremely complicated when one deals with complex database structures containing many connections between many different tables. A detailed explanation of how relational databases and query languages work is outside the scope of this book; the interested reader should consult Further Reading on p. 67.

## XML has the flexibility to define bespoke data classifications

Recent years have seen a significant effort to develop **eXtensible Markup Language (XML)** as a general tool for the storage of data and information. XML, a very powerful system for marking up (annotating) data, is one of many markup languages, including hypertext markup language (HTML) and XHTML, which are commonly used to write Web pages. In fact HTML and XHTML are subsets of XML, although they are considerably less powerful.

The key feature of these languages is the use of identifiers called tabs, which can enclose sections of data, for example <title>Understanding Bioinformatics</title>. While languages such as HTML can only use a restricted, small set of tabs, most of which are concerned with the presentation format (for example <bold> which indicates use of bold font), XML has mechanisms that allow arbitrary tags to be used. It is possible to construct and disseminate sets of tags appropriate for specialized data and for particular applications. For example, a <publisher> tag could be defined and used to identify book publishers. Recent versions of many word processing applications also use special sets of tags, allowing them to generate XML files. The ability to design appropriate specialized ways to classify data is one of the features that has made XML an attractive alternative to the databases described above. A particularly useful feature of XML is that it uses a plain file format, which makes XML files very portable and accessible. An increasing number of bioinformatics databases are being made available in XML format, although often their master copy is maintained using other solutions such as relational databases.

Having data marked up in XML or any other markup language is only useful if there are methods available to access and interpret the data. HTML and XHTML are interpreted by Web browsers, which identify the tags and use them as instructions for formatting and displaying the data. Interpreting programs have been written for some specialized XML tag sets, for example the word processing applications already mentioned. In other cases, especially those where the tags identify the nature of the data, it is more appropriate to use general programs to extract the data, and to create bespoke applications if further analysis is required.

The extraction of data from an XML file is very similar to database querying. For example, although the underlying technical details might be very different, if an XML file and a database both held lists of the same books, the result of extracting/querying all book titles would be the same. There is an equivalent of SQL for XML files, called Xquery, so that extracting information is equally easy from either source. The XML language does not automatically have the complexity of database applications, so that XML files may not necessarily be the method of choice. However, they have the advantage of great flexibility, so that in certain situations they are likely to be preferred.

## Many other database structures are used for biological data

There are other database models that are also used for biological systems, such as the object-oriented model. Object-oriented databases do not necessarily need SQL or any such special type of database programming language, and can use languages such as Java or C++. The key feature of object-oriented databases is their user interface. They have an apparent structure that consists of objects that correspond closely to the objects and concepts which occur in the field of interest. For example, the database structure might have cell objects with components and properties reflecting those of biological cells. The user of such a database can frame their queries in more familiar terms than is often possible using relational databases as described above. However, the actual structure used to store the information in an object-oriented database may still involve tables and have little, if any, real differences when compared to a relational database.

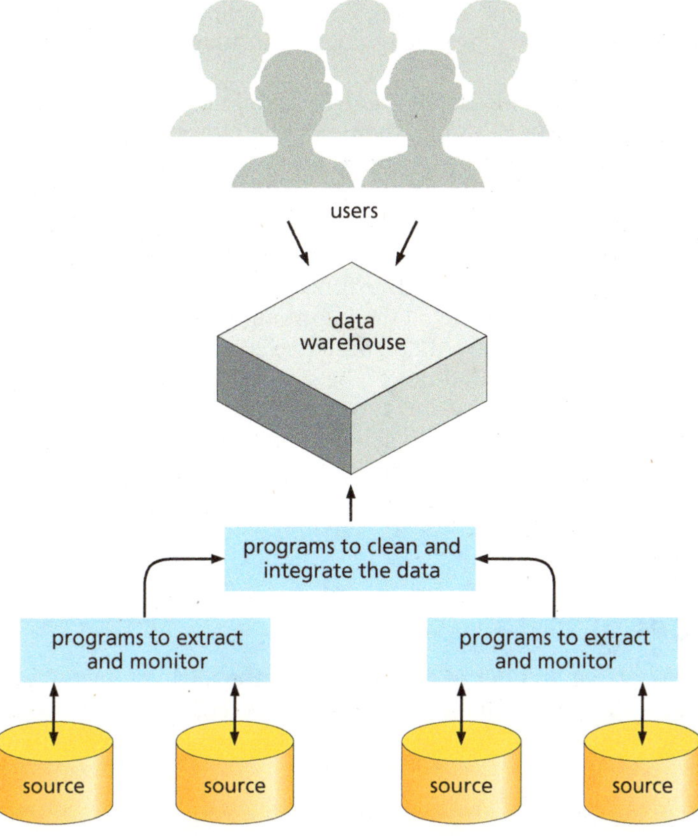


**Figure 3.4**
**A schematic of data warehousing.** The data come from different sources and are extracted automatically by special programs that also usually monitor the original source for updates. The data have to then be cleaned, checked, and put into a similar format so that integration can take place. Finally, the data are stored in a single database and are accessible to users.

Another form is the distributed database, in which although under the control of a central database management system, the actual database part (tables) may be stored in more than one computer located either in the same room or located anywhere but connected through the network. An example of a distributed database is the Reciprocal Net, which is a distributed database for crystallographers to store information about molecular structures. The central database management is at Indiana University but there are 19 other participating sites, ranging from those in the United States to those in the United Kingdom as well as Australia.

Data warehousing has been developed to overcome the problem of data heterogeneity (see next section) and data management between databases. The data warehouse basically stores information that has been integrated into one database (see Figure 3.4). It is, in effect, the opposite of the distributed database model.

Most large biological database sites, such as the Macromolecular Structure Database (MSD) at the European Bioinformatics Institute (EMBL-EBI), are made up of a mixture of the above models. MSD uses the relational model as well as having the data warehouse model for its structural information, integrating information about secondary structure, active sites, ligands, as well as information from external databases.

### Databases can be accessed locally or online and often link to each other

Access to a database can be local, usually through user interface programs written on-site, for example in Java, or external through Web-based interfaces. A database stored locally has the advantages of faster access, more flexibility in designing specific queries, and security. However, it is necessary not only to have enough disk space to keep a database but also to invest time and money for the management and maintenance of the database. Therefore the preferred choice for most users is to access external databases via the Internet. A lot of effort has gone into providing Web-based forms and interfaces to biological and related databases and database resources, as can be seen by exploring the major bioinformatics database serving sites at the National Center for Biotechnology Information (NCBI) and EMBL-EBI.

Often many biological and related databases contain records that are relevant to the topic of interest. For example, there may be relevant entries for a protein in many different databases including those for protein sequence, protein structure, and gene sequence data. If each of these databases had to be separately queried it could take a long time to learn all that is known about the system, and some information may be missed if all databases are not searched. However, almost all databases in the biomedical area have some **links** to relevant entries in other databases, making it much easier to collate all the information. No database can be assumed to have links to all other relevant databases, so some care is still needed to obtain all the information, but much time can be saved. Note that some major sites such as NCBI and EBI augment these links so that not all accessible versions of a particular database necessarily have the same set of links. These links have made the collection of publicly available databases one of the most powerful resources in biomedical research.

## 3.2 Types of Database

There is a very large number of databases covering a wide range of scientific data with much duplication, both real and apparent. Just discovering what exists can be a difficult task, because many of the databases are created by small, independent teams and hosted on their own Web servers. In this section we will give a brief survey of the existing databases (see Flow Diagram 3.2).

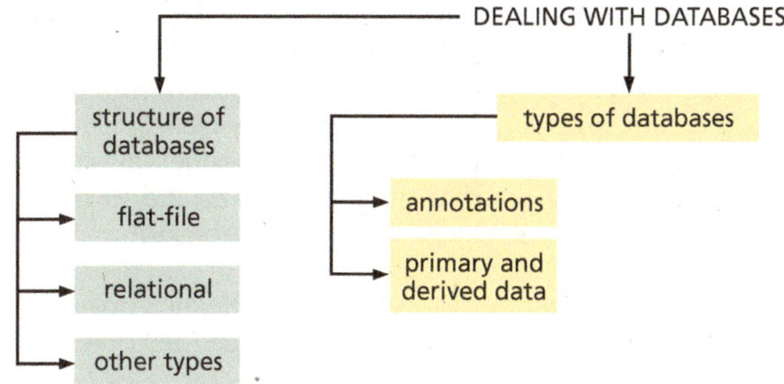


**Flow Diagram 3.2**
The key concept introduced in this section is that most bioinformatics databases contain some results from analysis of the primary data, greatly increasing their usefulness.

An important feature of many databases is that they do not only store the data that are supposedly their reason for being created. For example, entries in the major protein and nucleotide sequence database often have a large amount of relevant non-sequence information. For clarity, in the remainder of this chapter we will use the term **data** to refer to the key information in the database entry: for example, the sequence, structure, or expression levels together with the minimal information required to identify the data, for example the name of the gene (and the organism) whose sequence is contained in the entry. All of the additional information in the database entry will be referred to as the **annotation**, and can include links to related entries in other databases, interpretation of the data, and relevant research citations. We will begin by briefly looking at some aspects of database annotation.

## There's more to databases than just data

The minimal content of a database entry would include just the data and the data's identity (for example, protein name and source organism) and the author/submitter responsible for the entry. However, usually much more information is given, referred to as annotation, including published papers reporting the data, other known facts or interpretations (for example, enzyme reactions and substrates, or gene structures), and lists of highly relevant entries in other databases (for example, entries about a sequence motif, if one is present). Thus, even though the majority of databases concentrate almost wholly on one particular aspect (for example, DNA sequence, protein structure, and human genetic disorders) they contain information about a wide range of related aspects.

When the annotation lists related entries in other databases these are usually given as **URL** pointers, called links, with which one can very quickly surf through many databases. This makes the set of databases an extremely useful information resource often far more powerful than standard literature reviews. A successful database search will reveal many different aspects of the area of interest, and provide raw data and its interpretation as well as important research papers.

In addition to simply providing information, some of the databases present their data in an interactive graphical display. This enables the data to be browsed akin to thumbing the pages of a book. Sometimes this is the best way to become familiar with general features of the data, as well as being a more intuitive way to search through some types of data. In addition, some databases also provide programs for the online analysis of their data.

## Primary and derived data

A distinction is sometimes made between databases of so-called primary data and those that contain secondary data derived from these primary sources. In some cases the primary data include the raw experimental results, for example scans of gene expression arrays and two-dimensional (2-D) proteomic gels. In many cases the primary data are the initial experimental interpretation, for example nucleotide sequences. Although many protein sequences have been deduced from nucleotide sequences, the main protein sequence databases such as SWISS-PROT are often regarded as primary data sources. In general, except for experimental errors, these data are regarded as very reliable. (Note that this does not extend to the further annotation that might be present, especially in sequence databases.)

Examples of secondary databases are those that contain collections of conserved protein sequence motifs, or comparisons of multiple sequences that give measures of sequence similarity and relatedness. Because they can only be based on the data existing at the time, some of the entries in this kind of database may be proved incorrect once more primary data have been collected. For example, new protein sequence information may mean that a sequence hitherto regarded as a complete conserved motif turns out not to be conserved at all, or to be only part of a motif. If such secondary databases are not rederived at regular intervals, their data must be treated with a degree of suspicion.

## How we define and connect things is important: Ontologies

The organization of the complex data found in databases, as well as merging data from various places, is difficult when data are described using different terminology for each database. Therefore formal and explicit specifications of the terms used and the relationships between them have been defined; these are called **ontologies**. The specific terms defined in ontologies can be reused between different biological fields; this simplifies the sharing of data and the sharing of knowledge about the data, and facilitates analysis by computer programs, such as the GOTM program that is used to analyze and visualize sets of genes using the Gene Ontology. This widely used Gene Ontology (GO) is a collaborative project across many laboratories to provide a controlled vocabulary that describes gene and gene-associated information (but not gene products) for all organisms. The building blocks of any ontology are terms. For example in GO, each GO term has a number and a name, such as *signal transduction*. Each term is also assigned to one of three subontologies in GO: molecular function, cellular component, and biological process. Sharing a vocabulary such as GO between databases is a step toward unification of databases. Similarly there is MGED (Microarray Gene Expression Data), whose terms are developed to describe gene

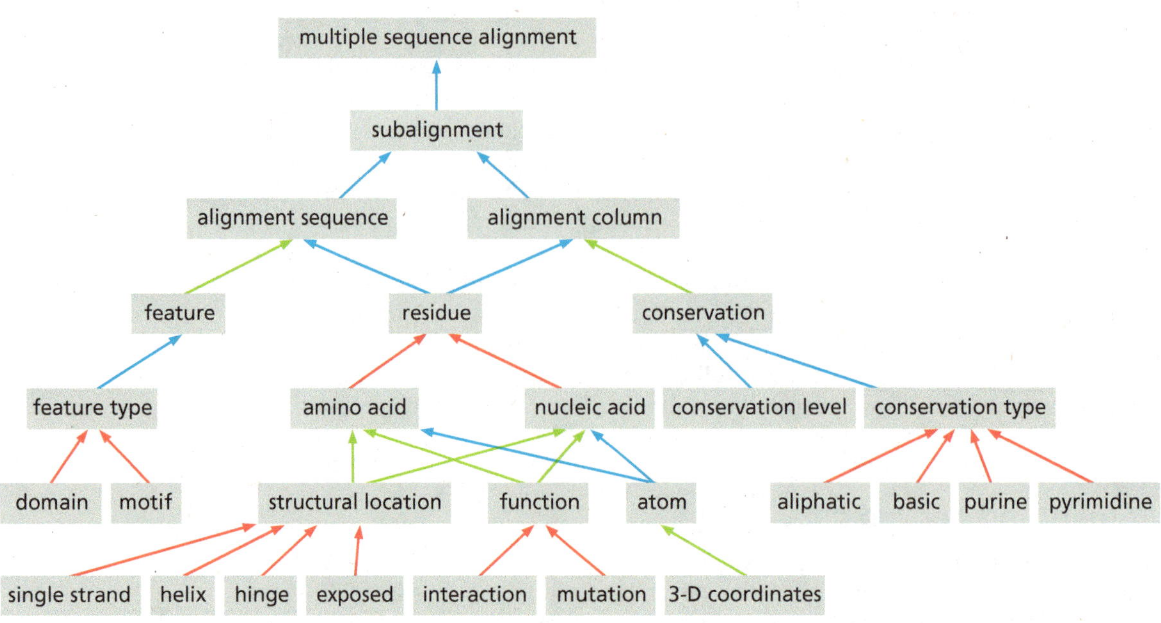


**Figure 3.5**
**Schematic diagram of a section of the MAO ontology architecture.** The gray boxes show terms and the arrows represent relationships between them. Red arrows indicate "is a" relationship; for example, a domain is a feature type. Green arrows indicate "is attribute;" for example, function is an attribute of amino acid. Blue arrows indicate "part of," thus an atom is part of an amino acid. (From J.D. Thompson et al., MAO: a multiple alignment ontology for nucleic acid and protein sequences, N.A.R. 33 (13):4164–4171, 2005, by permission of Oxford University Press.)

expression experiments, sequence ontology project (SOP), which aims to define terms that describe features of a nucleotide or a protein sequence, and a multiple alignment ontology (MAO) for sequences. Figure 3.5 shows a graphical representation of a part of the MAO ontology structure. In MAO the sequence is placed into a context of the evolutionary family. It provides standard descriptions of most multiple sequence alignments and their methods, as well as integration of other relevant data such as structural or functional information. This facilitates inclusion of information based on conservation and evolution in the final analysis.

## 3.3 Looking for Databases

Beyond a thorough literature and Internet search, there is a key site to use when looking for relevant databases. Every year the first new issue in the journal *Nucleic Acids Research* (NAR) is devoted to papers reporting new and updated databases. In addition to the articles presented, there is a list of the URLs of databases that have been reported in these annual issues of NAR. This list, called the Molecular Biology Database Collection and numbering 858 entries in 2006, is available from the NAR home page. To give a feel for the range of material available in the biomedical databases, in 2006 the Molecular Biology Database Collection was divided into 14 categories (see Figure 3.6). Some of these categories will be described in more detail below as they are closely related to the bioinformatics analyses that are dealt with in the rest of this book (see Flow Diagram 3.3). Because of their importance as primary data sources, we will discuss sequence databases in some more detail.

There are a number of centers that have been funded to provide access to a large number of major databases in an integrated environment, to facilitate their use by the research community. The major database centers include the NCBI (http://www.ncbi.nlm.nih.gov/), the EBI (http://www.ebi.ac.uk), and the Sanger Institute (http://www.sanger.ac.uk). Each has links to over 100 different databases and is a very useful starting point for sequence, structure, and genome analysis. However, although these include most of the major databases, they are only a small fraction of the total available, and often a more specialist database can provide better information in the case of specific queries. In particular, a number of high-quality genome-specific databases have been produced, which would probably be the best starting point for learning about a gene or protein of that species.

### Sequence databases

Nucleotide sequence related databases (8% of the 858 listed in the Molecular Biology Database Collection) include major international collaborations, such as GenBank and the EMBL-EBI Nucleotide Sequence database, as well as resources

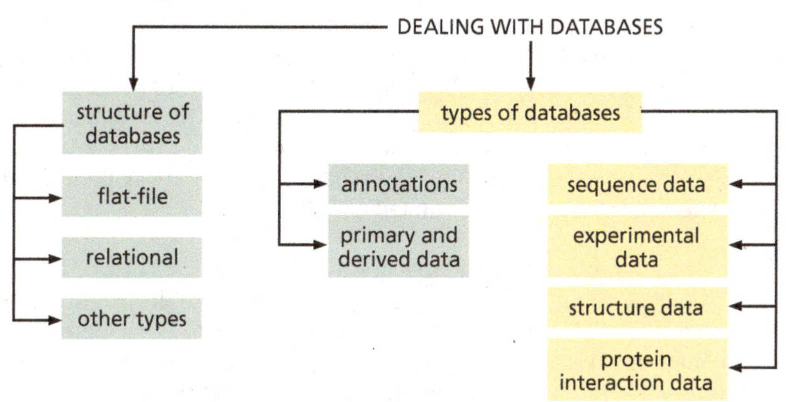


**Flow Diagram 3.3**
**The key concept introduced in this section is that bioinformatics databases can be classified according to the type of data they contain.**

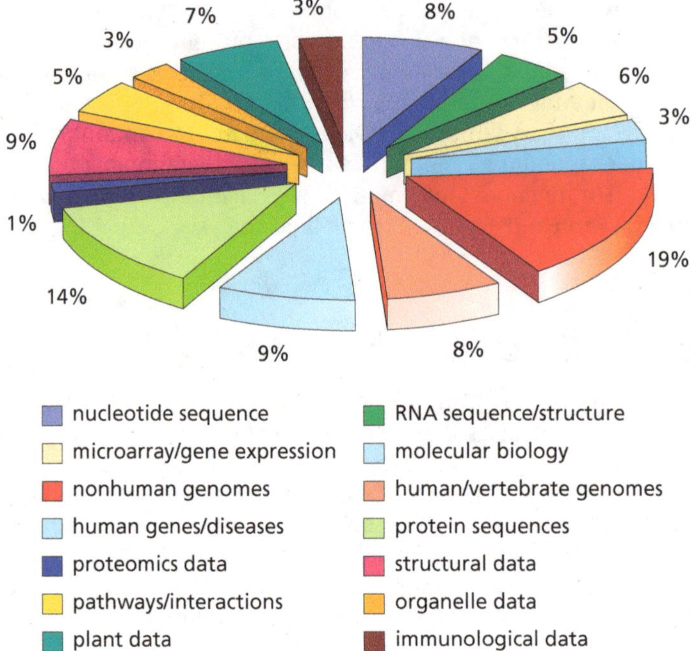


**Figure 3.6**
**Distribution of the type of databases as classified at the Nucleic Acids Research (NAR) Molecular Biology Database Collection Web site.** In 2006 there were 858 databases listed in total, classified into 14 major categories, of which the genome (27%) and sequence (26%) databases form the largest sections.

that are more gene specific with information on introns, exons, and splice sites, as well as motifs and transcriptional regulators and sites. RNA-specific databases comprise 5% of the total and most have data on secondary structure and other aspects in addition to sequence. Genomic databases form a large part of the database list, with 19% nonhuman and 8% human and other vertebrate genomes.

There are a number of different types of DNA sequences stored in the databases containing information about nucleic acids. These differ in the way they have been obtained, and each type provides different biological information and must be treated differently in terms of their analysis. The first type is raw genomic sequence, representing the sequence of chromosomal DNA. This is the type of DNA sequence derived from genome sequencing projects and is deposited in GenBank and the organism-specific DNA sequence databases. These sequences include all the elements present in genomic DNA, including noncoding regions, introns, and control regions, as well as the sequences that code for proteins and RNAs. A second type of DNA sequence that can be encountered, known as **cDNAs**, are the sequences of DNA molecules that have been synthesized by reverse transcription (copying of RNA into DNA) of mRNA molecules. The mRNA present at the time of the experiment will depend on the nature of the sample, for example the type of cell or tissue, the particular stage of development, or the particular disease. The set of cDNA database entries for that sample represents the genes actually being expressed in that sample. Because they are synthesized using the mRNA as a template, these DNA sequences lack any introns that might exist in the gene and any control sequences that lie outside the region transcribed into RNA (see Chapter 1). The third type of DNA sequence held in databases is known as **expressed sequence tags** (**ESTs**). An EST is a partial cDNA sequence. As with cDNAs, a library of ESTs indicates the range of genes being expressed in a sample, and they can be used to scan genome sequences to help identify genes. The NCBI site hosts a special database of ESTs called dbEST.

Protein sequence databases form a large group of the NAR list (14%). These include the major sequence databases such as UniProtKB, with its highly annotated component Swiss-Prot, and the NCBI Protein Database, both being efforts to collect information on all protein sequences. These protein databases are often compiled and annotated from raw nucleotide sequence data. For example,

**Figure 3.7**

**An extract from a GenBank DNA sequence file with the DNA sequence of human LIM domain 7.** The type of information on each line is preceded by the field name. A number of accession and version numbers are given, including cross-references (xref) which are links to related entries in other databases. Many lines have been omitted from the file after the "SOURCE" line identifying the sequence as of human origin. The section from "gene" to the line after the "CDS" line identifies the protein-coding sequence (CDS) as only bases 1261 to 5310 of the sequence. The final section of the file, from the "ORIGIN" line, is the formatted nucleotide sequence. Note that GeneID is the same as in the microarray data (see Figure 3.9).

```
LOCUS       NM_005358                7235 bp    mRNA    linear   PRI 02-AUG-2006
DEFINITION  Homo sapiens LIM domain 7 (LMO7), mRNA.
ACCESSION   NM_005358
VERSION     NM_005358.4  GI:111119012
KEYWORDS    .
SOURCE      Homo sapiens (human)
…
gene             1..7235
                 /gene="LMO7"
                 /note="synonyms: LOMP, FBX20, FBXO20, KIAA0858"
                 /db_xref="GeneID:4008"
                 /db_xref="HGNC:6646"
                 /db_xref="HPRD:05078"
                 /db_xref="MIM:604362"
CDS              1261..5310
                 /gene="LMO7"
ORIGIN
        1 ggaaagaagt ggaataatta ggaacctagg gtggggtagg gtagcaggac atttcaaaca
       61 ttaatgagca tatgagattc caggtcttgt taaaatgcaa attctgattc agctggtagg
      121 tgaggtctga gattgtgcat ttctaacaag cactcagata atcttaaggc tgttggcccc
      181 agggtcacac ttatagtgat tttctagaac ccagttgggg aagtgaatct tgggcaggag
      241 aaatacacac ctcttgcatt gagtttggag atctcatctg atataacttt ttaagaaaga
      301 aaaataattt tccaaatatc caattgataa gctttcccac taagtggctt tcccactaag
      361 tggctgcgtt atgaaaattg cttcactttg aaacttctgg tcttggtaat atagaatttc
      421 tgtgttctca cagtgcttga ttgagaatat gatattgaga ttatggcata aaatatagtg
      481 gctgtacaaa aaaaaataca ttattaggat ctctaacaat tatgtaaaag tcattgcttc
      541 atgggtagag ctcaaacttt ggtgtgagac ctggttttat tcttggcact tactctgagt
      601 tgtcttaggc aaattaatac cttaagcaaa aatattctca tgtacatttt acatgagaat
      661 tataaatgaa gtacataaag tccagcagtc acaaatgtta tctattatta ccatcgtcct
      721 aagactgcaa tcagctatag tgaaagtagt ctcaaagatt gtttcataaa tcatcagatt
      781 cacctaattt tctaaagaat ttaaataagg agatggaatg aatagattgc attttgtttc
      841 catgcacagg ggaactgtgc atatttcttc tgtgactcgg aaatggttta acttttaaaa
      901 atcccaaaat agctgaagtt agcagacatg caatttacca aggatgattg gaatttttat
      961 ctttcctgta ataatactat acccaagcac actgctcatg aggaaaacat ttttatgtga
     1021 atcttttact cttgggggca aagaatgctg tttttctttt tgataactat gtttatagaa
     1081 tctaaatcac cctgagcaat tatttcaaca tctaaagtta ttattaccat tcatgtttca
     1141 tttatagcta tttgaatttt gatgaatttc aatatggtgc tacagtgata gggcaagtgc
     1201 aaataagttc aatatatggg tacggtctaa agctatttta atttttttat tacaactgct
     1261 atgaagaaaa ttaggatatg ccatattttc acgttttaca gttggatgtc ctatgatgtt
     1321 ctcttccaga gaacagagct cggagctctg gaaatttgga ggcaactgat atgtgctcat
     1381 gtctgcatct gtgtgggttg gctgtatctc agggacagag tctgcagcaa aaaagatata
     1441 attttgagga ctgaacaaaa ttcaggaagg actattctca ttaaggcagt aacagagaag
     1501 aattttgaaa caaaagattt tcgagcctct ctagaaaatg gtgttctgct gtgtgatttg
     1561 attaataagc ttaaacctgg cgtcattaag aagatcaata gactgtctac accaatagca
     1621 ggattggata atataaacgt tttcttgaaa gcttgtgaac agattggatt gaaagaagcc
     1681 cagcttttcc atcctggaga tctacaggat ttatcaaatc gagtcactgt caagcaagaa
     1741 gagactgaca ggagagtgaa aaatgttttg ataacattgt actggctggg aagaaaagca
     1801 caaagcaacc cgtactataa tggtccccat cttaatttga aagcgtttga gaatctttta
     1861 ggacaagcac tgacgaaggc actcgaagac tccagcttcc tgaaaagaag tggcagggac
     1921 agtggctacg gtgacatctg gtgtcctgaa cgtggagaat ttcttgctcc tccaaggcac
     1981 cataagagag aagattcctt tgaaagcttg gactctttgg gctcgaggtc attgacaagc
     2041 tgctcctctg atatcacgtt gagagggggg cgtgaaggtt ttgaaagtga cacagattcg
     2101 gaatttacat ttaagatgca ggattataat aaagatgata tgtcgtatcg aaggatttcg
     2161 gctgttgagc caaagactgc gttacccttc aatcgttttt tacccaacaa aagtagacag
     2221 ccatcctatg taccagcacc tctgagaaag aaaaagccag acaaacatga ggataacaga
     2281 agaagttggg caagcccggt ttatacagaa gcagatggaa cattttcaag actctttcaa
     2341 aagatttatg gtgagaatgg gagtaagtcc atgagtgatg tcagcgcaga agatgttcaa
     2401 aacttgcgtc agctgcgtta cgaggagatg cagaaaataa aatcacaatt aaaagaacaa
     2461 gatcagaaat ggcaggatga ccttgcaaaa tggaaagatc gtcgaaaaag ttacacttca
     2521 gatctgcaga agaaaaaaga agagagagaa gaaattgaaa agcaggcact tgagaagtct
     2581 aagagaagct ctaagacgtt taaggaaatg ctgcaggaca gggaatccca aaatcaaaag
     2641 tctacagttc cgtcaagaag gagaatgtat tcttttgatg atgtgctgga ggaaggaaag
     2701 cgaccccct a caatgactgt gtcagaagca agttaccaga gtgagagagt agaagagaag
     2761 ggagcaactt atccttcaga aattcccaaa gaagattcta ccacttttgc aaaaagagag
     2821 gaccgtgtaa caactgaaat tcagcttcct tctcaaagtc ctgtggaaga acaaagccca
     2881 gcctctttgt cttctctgcg ttcacggagc acacaaatgg aatcaactcg tgtttcagct
     2941 tctctcccca gaagttaccg gaaaactgat acagtcaggt taacatctgt ggtcacacca
     3001 agaccctttg gctctcagac aaggggaatc tcatcactcc ccagatctta cacgatggat
     3061 gatgcttgga agtataatgg agatgttgaa gacattaaga gaactccaaa caatgtggtc
     3121 agcacccctg caccaagccc ggacgcaagc caactggctt caagcttatc tagccagaaa
     3181 gaggtagcag caacagaaga agatgtgaca aggctgccct ctcctacatc cccttctca
     3241 tctctttccc aagaccaggc tgccacttct aaagccacat tgtcttccac atctggtctt
     3301 gatttaatgt ctgaatctgg agaaggggaa atctccccac aaagagaagt ctcaagatcc
     3361 caggatcagt tcagtgatat gagaatcagc ataaaccaga cgcctgggaa gagtcttgac
     3421 tttgggttta caataaaatg ggatattcct gggatcttcg tagcatcagt tgaagcaggt
     3481 agcccagcag aattttctca gctacaagta gatgatgaaa ttattgctat taacaacacc
     3541 aagttttcat ataacgattc aaaagagtgg gaggaagcca tggctaaggc tcaagaaact
     3601 ggacacctag tgatggatgt gaggcgctat ggaaaggctg gttcacctga aacaaagtgg
     3661 attgatgcaa cttctggaat ttacaactca gaaaaatctt caaatctatc tgtaacaact
     3721 gatttctccg aaagccttca gagttctaat attgaatcca aagaaatcaa tggaattcat
     3781 gatgaaagca atgcttttga atcaaaagca tctgaatcca tttctttgaa aaacttaaaa
     3841 aggcgatcac aatttttga acaaggaagc tctgattcgg tggttcctga tcttccagtt
     3901 ccaaccatca gtgccccgag tcgctgggtg tgggatcaag aggaggagcg gaagcggcag
     3961 gagaggtggc agaaggagca ggaccgccta ctgcaggaaa aatatcaacg tgagcaggag
     4021 aaactgaggg aagagtggca aagggccaaa caggaggcag agagagagaa ttccaagtac
     4081 ttggatgagg aactgatggt cctaagctca aacagcatgt ctctgaccac acgggagccc
     4141 tctcttgcca cctgggaagc tacctggagt gaagggtcca agtcttcaga cagagaagga
     4201 acccgagcag gagaagagga gaggagacag ccacaagagg aagttgttca tgaggaccaa
     4261 ggaaagaagc cgcaggatca gcttgttatt gagagagaga ggaaatggga gcaacagctt
     4321 caggaagagc aagagcaaaa gcggcttcag gctgaggctg aggagcagaa gcgtcctgcg
     4381 gaggagcaga agcgccaggc agagatagag cgggaaacat cagtcagaat ataccagtac
     4441 aggaggcctg ttgattccta tgatatacca aagacagaag aagcatcttc aggttttctt
     4501 cctggtgaca ggaataaatc cagatctact actgaactgg atgattactc cacaaataaa
     4561 aatggaaaca ataaatattt agaccaaatt gggaacatga cctcttcaca gaggagatcc
     4621 aagaaagaac aagtaccatc aggagcagaa ttggagaggc aacaaatcct tcaggaaatg
     4681 aggaagagaa caccccttca caatgacaac agctggatcc gacagcgcag tgccagtgtc
     4741 aacaaagagc ctgttagtct tcctgggatc atgagaagag gcgaatcttt agataacctg
     4801 gactcccccc gatccaattc ttggagacag cctccttggc tcaatcagcc cacaggattc
     4861 tatgcttctt cctctgtgca agactttagt cgcccaccac ctcagctggt gtccacatca
     4921 aaccgtgcct acatgcggaa cccctcctcc agcgtgcccc caccttcagc tggctccgtg
     4981 aagacctcca ccacaggtgt ggccaccaca cagtccccca ccccgagaag ccattcccct
     5041 tcagcttcac agtcaggctc tcagctgcgt aacaggtcag tcagtgggaa gcgcatatgc
     5101 tcctactgca ataacattct gggcaaagga gccgccatga tcatcgagtc cctgggtctt
     5161 tgttatcatt tgcattgttt taagtgtgtt gcctgtgagt gtgacctcgg aggctcttcc
     5221 tcaggagctg aagtcaggat cagaaaccac caactgtact gcaacgactg ctatctcaga
     5281 ttcaaatctg gacggccaac cgccatgtga tgtaagcctc catacgaaag cactgttgca
     5341 gatagaagaa gaggtggttg ctgctcatgt agatctataa atatgtgttg tatgtctttt
     5401 ttgctttttt tttaaaaaaa agaataactt tttttgcctc tttagattac atagaagcat
     5461 tgtagtcttg gtagaaccag tatttttgtt gtttatttat aaggtaattg tgtgtgggga
     5521 aaagtgcagt atttacctgt tgaattcagc atcttgagag cacaagggaa aaaataagaa
     5581 cctacgaata tttttgaggc agataatgat ctagtttgac tttctagtta gtggtgtttt
     5641 gaagagggta ttttattgtt ttttaaaaaa aggttcttaa acattatttg aaatagttaa
     5701 tataaataca taattgcatt tgctctgttt attgtaatgt attctaaatt aatgcagaac
     5761 catatggaaa atttcattaa aatctatccc caaatgtgct ttctgtatcc ttccttctac
     5821 ctattattct gatttttaaa aatgcagtta atgtaccatt tatttgcttg atgaagggag
     5881 ctctattttc tttaccagaa atgttgctaa gtaattccca atagaaagct gcttattttc
     5941 attaatgaaa aataaccatg gtttgtatac tagaagtctt cttcagaaac tggtgagcct
     6001 ttctgttcaa ttgcatttgt aaataaactt gctgatgcat ttaacgagtg ggtcgtcttt
     6061 ttcttaggtg tatgtgtctg acctcaggcc ttttagccat atttcagtat gtggcctttt
     6121 ttgatgttat gttttatcca gtagctttac taaggtataa ttgatgtaat aaactgcata
     6181 tatttaaagt gtatactttg acaaattttg acatggtgta taccttcgaa actatgccac
     6241 agtctggatg tgtttactga aacattttaa taaggaagtt tatttttgat aaagttatgt
     6301 ttttggatac aatatatttg tatggtgaga gtgatgaatt gttggatcat ttgatctttt
     6361 actaacccca tgataaaagg agaagacaac agtgagctta gaatatctat aaagcaaaaa
     6421 atgtagtctc ttgtttaaaa aatctggagc gggaatgcaa ggatacaaaa ctttagcatg
     6481 ctttgagcaa aaatttaaac ttactggaat cttttataat aatgtaagtg gaatggagga
     6541 ttctaggaac tgagaactgt attggaatag gttcaaaata tgtaagaaat gctaatgtgg
     6601 gagataaaaa ttttatttag tacttattct gattattatt aaagtaataa tgtgttcctt
     6661 gaggataact tgtcaaatgc cccaaagcat aaagaatata attctgaatc ccaaattcca
     6721 aagacaagaa ctctgtgttt gaattcattc tgcatataat tatttataag tatagattgt
     6781 gaatttttcc atgttcttaa aattattttt atcttttttc atggttgcat agtgctccat
     6841 tgtttggcct tggtaatatt tagttgataa ttccattact gtgtattttt cacttgtttc
     6901 taagatcaaa cattttaata tgtgcatgtt atatataaat atgtaaattc tgtgatactc
     6961 tatgatcatc tctttcttta tattattttc atagacatga aatagttgct cagagattat
     7021 gcattttaag acactcatag tatatattgc caaagtggtt tccagaaagg cactgctggc
     7081 ttcgactcct ataagcagca cgtgggcttg ttcatctcac tgcatgttta tgaagataca
     7141 gttcttttgc cttgttctct gcctgatgtg tatgcagagg cagccctcaa tatgcagtgg
     7201 ttgaataaat gaatgaagaa accactaaaa aaaaa
```

UniProtKB is produced by analysis of all the translations of the EMBL database nucleotide sequences. It has two components: Swiss-Prot mentioned above, which has manual annotations incorporated, and the TrEMBL component, which is only annotated using a computer. The latter has more entries, but the annotation is not as accurate as Swiss-Prot.

The sequence databases tend to contain very extensive annotation, and large teams are needed to help with this intensive task. In the case of nucleotide sequences, the typical features identified include the presence of open reading frames, introns, and promoter sites (see Chapter 1), as well as translated protein sequence. The protein sequence databases have equally detailed annotation with more emphasis on the protein, with various properties including their localization, their biological targets, sequence motifs, active sites, and domains (see Chapter 2). A partial entry for a DNA sequence of the LM07 gene is illustrated in Figure 3.7 and for its protein sequence in Figure 3.8.

Secondary databases such as Blocks and Prodom supply information regarding sequence or structural patterns found in proteins. This information is very useful when searching for proteins that are related by evolution (see Chapters 4 and 6). The proteins are grouped together according to sequence or structural similarities such as analogous active sites or substructures. In addition to the general DNA and protein sequence databases, there are many databases that specialize in specific groups of sequences. There are specialist databases covering most areas, including repetitive DNA elements, protein motifs, and particular classes of proteins or RNA molecules.

## Microarray databases

Microarray data and **gene expression** databases make up 6% of the list. In contrast, only 1% of databases deal with protein expression. These have data on 2-D acrylamide gels with their protein expression levels, as well as posttranslational modifications and mass spectroscopy identifications for some or all the spots on a 2-D gel.

Microarray databases are a repository of data from microarray experiments, often accompanied by data analysis and tools to visualize the raw image. Gene expression databases also contain expression data collected by other experimental methods such as SAGE (serial analysis of gene expression) and EST sequencing. The databases contain expression data and often extensive annotation as well as techniques to visualize the numerical data and statistical analysis programs. One such resource is the Stanford Microarray Database (SMD) where the data from more than 7000 microarray experiments are available and can be viewed, downloaded, and analyzed using the tools provided by the SMD site (see Figure 3.9). Another important microarray database is ArrayExpress, which is also a repository for microarray data. In addition there is an ArrayExpress Data Warehouse that stores gene-indexed expression profiles from a curated subset of experiments from the database.

## Protein interaction databases

For most proteins to carry out their function they have to interact with other molecules, including other proteins. Therefore protein interaction databases are important resources for understanding function and building up biological networks that then can be used in systems biology (see Chapter 17). There are a number of protein interaction databases available, each with different advantages and also limitations. The Database of Interacting Proteins (DIP) contains information only on protein–protein interactions but employs rigorous criteria for evaluating the reliability of each interaction. The Molecular INTeraction database (MINT) contains additional information on protein, nucleic acid, and lipid interactions. The Biomolecular Interaction Network (BIND) describes interactions at the atomic level for protein, DNA, and RNA. pSTIING (protein Signalling, Transcriptional Interaction and Inflammation Networks Gateway) is a Web-based application as well as an interaction database. pSTIING integrates protein–protein, protein–anything else interactions as well as transcriptional associations. The search engine is very powerful, allowing for integrated searching, for example linking disease types with protein interaction networks. It also allows for the integration of experimental data such as from gene expression experiments with the protein networks. Figure 3.10 shows a snapshot of a pSTIING pathway and dynamically generated further interactions.

## Structural databases

Structural databases make up 9% of the 2006 NAR list, and include those containing information on the structure of small molecules, carbohydrates, nucleic acids

*Note: most headings are clickable, even if they don't appear as links. They link to the user manual or other documents.*

| Entry information | |
|---|---|
| Entry name | **LMO7_HUMAN** |
| Primary accession number | **Q8WWI1** |
| Secondary accession numbers | O15462 O95346 Q9UKC1 Q9UQM5 Q9Y6A7 |
| Integrated into Swiss-Prot on | March 15, 2004 |
| Sequence was last modified on | March 15, 2004 (Sequence version 2) |
| Annotations were last modified on | July 25, 2006 (Entry version 39) |
| **Name and origin of the protein** | |
| Protein name | **LIM domain only protein 7** |
| Synonyms | **LOMP**<br>**F-box only protein 20** |
| Gene name | **Name:** LMO7<br>Synonyms: FBX20, FBXO20, KIAA0858 |
| From | Homo sapiens (Human) [TaxID: 9606] |
| Taxonomy | Eukaryota; Metazoa; Chordata; Craniata; Vertebrata; Euteleostomi; Mammalia; Eutheria; Euarchontoglires; Primates; Haplorrhini; Catarrhini; Hominidae; Homo. |

**References**

[1] NUCLEOTIDE SEQUENCE [MRNA] (ISOFORM 3), AND TISSUE SPECIFICITY.
**TISSUE**=Brain, and Peripheral blood leukocyte;
DOI=10.1007/s00439-001-0646-6; PubMed=11935316 [NCBI, ExPASy, EBI, Israel, Japan]
Rozenblum E., Vahteristo P., Sandberg T., Bergthorsson J.T., Syrjakoski K., Weaver D., Haraldsson K., Johannsdottir H.K., Vehmanen P., Nigam S., Golberger N., Robbins C., Pak E., Dutra A., Gillander E., Stephan D.A., Bailey-Wilson J., Juo S.-H.H., Kainu T., Kallioniemi O.-P.;
"A genomic map of a 6-Mb region at 13q21-q22 implicated in cancer development: identification and characterization of candidate genes.";
Hum. Genet. 110:111-121(2002).

| Key | From | To | Length | Description | FTId |
|---|---|---|---|---|---|
| CHAIN | 1 | 1683 | 1683 | LIM domain only protein 7. | PRO_0000075824 |
| DOMAIN | 54 | 168 | 115 | CH. | |
| DOMAIN | 1042 | 1128 | 87 | PDZ. | |
| DOMAIN | 1612 | 1678 | 67 | LIM zinc-binding. | |

```
        10         20         30         40         50         60
MKKIRICHIF TFYSWMSYDV LFQRTELGAL EIWRQLICAH VCICVGWLYL RDRVCSKKDI

        70         80         90        100        110        120
ILRTEQNSGR TILIKAVTEK NFETKDFRAS LENGVLLCDL INKLKPGVIK KINRLSTPIA

       130        140        150        160        170        180
GLDNINVFLK ACEQIGLKEA QLFHPGDLQD LSNRVTVKQE ETDRRVKNVL ITLYWLGRKA
```

**Figure 3.8**
**An LIM domain 7 entry from a protein point of view from the Swiss-Prot database.** As in Figure 3.7 the sequence is given at the bottom of the entry (shown in part here). A major section of the database entry consists of annotations that are linked to other tables, files, and databases. For example, in the References section there are links to the PubMed research literature database, with links to the equivalent entry in five worldwide locations of this database, including the United States, Europe, and Japan.

(DNA and RNA), and proteins. These are the results of study using various experimental techniques, usually X-ray crystallography or nuclear magnetic resonance (NMR). The most widely used databases are the Structural Bioinformatics Protein Databank (RCSB, PDB) and the Macromolecular Structure Database (MSD) at EBI. Figure 3.11 shows the results of a text search through the RCSB server with the keywords "LIM domain." These types of databases are very important for structural analysis and homology modeling (see Chapters 13 and 14). Protein secondary

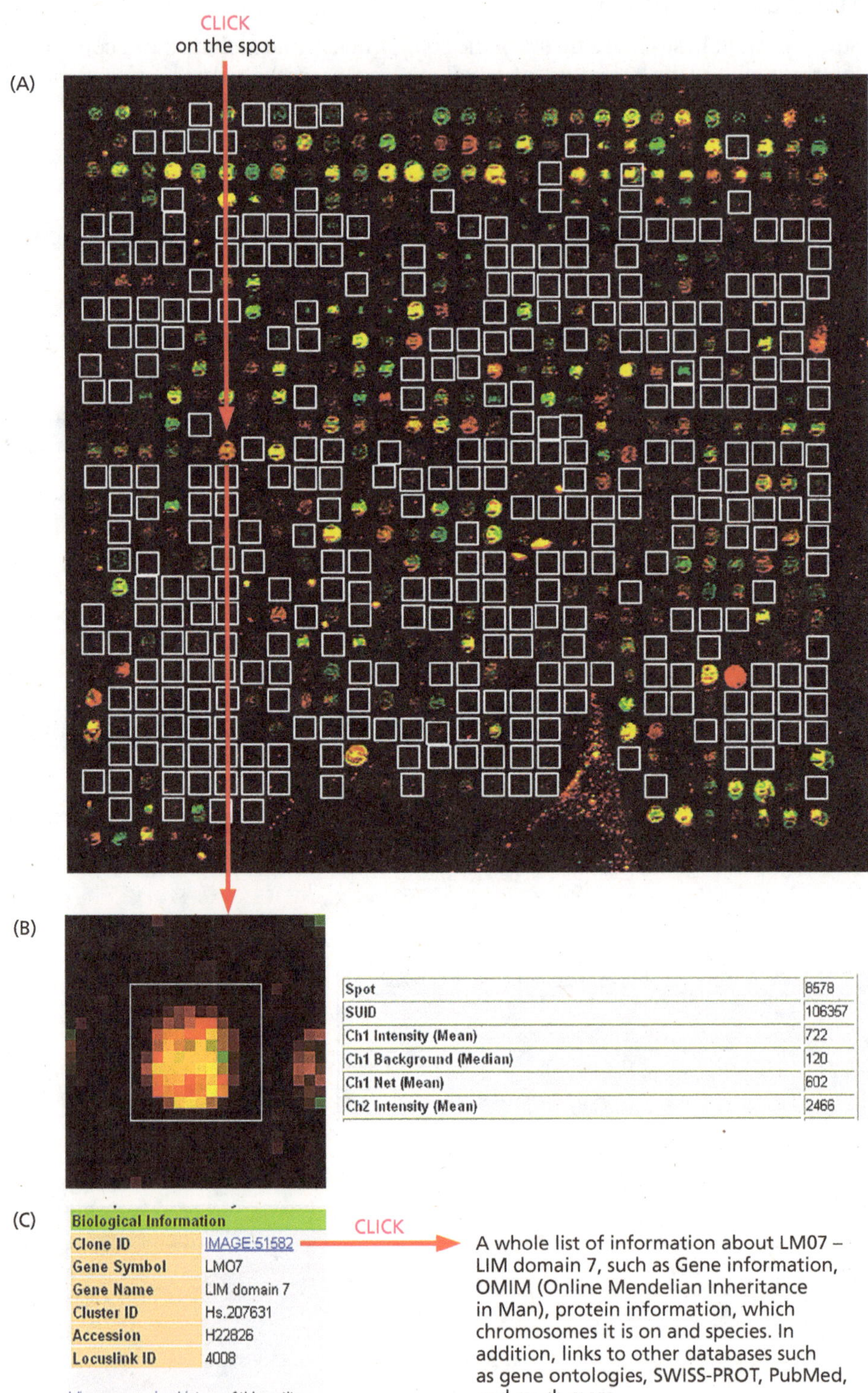
CLICK
on the spot
(A)
(B)
(C)
CLICK
A whole list of information about LM07 – LIM domain 7, such as Gene information, OMIM (Online Mendelian Inheritance in Man), protein information, which chromosomes it is on and species. In addition, links to other databases such as gene ontologies, SWISS-PROT, PubMed, and much more.
View expression history of this entity

| Spot | 8578 |
|---|---|
| SUID | 106357 |
| Ch1 Intensity (Mean) | 722 |
| Ch1 Background (Median) | 120 |
| Ch1 Net (Mean) | 602 |
| Ch2 Intensity (Mean) | 2466 |

| Biological Information | |
|---|---|
| Clone ID | IMAGE:51582 |
| Gene Symbol | LMO7 |
| Gene Name | LIM domain 7 |
| Cluster ID | Hs.207631 |
| Accession | H22826 |
| Locuslink ID | 4008 |

**Figure 3.9**
**Searching through the SMD microarray database returns lots of information.** It also gives you the possibility to see the whole microarray image, which you can enlarge (A) and choose a spot of interest and click on it (B) to give you a large view of the spot with additional data such as numerical spot information and biological annotation of the spot, the gene whose expression is measured (LIM domain 7), and linkable IDs. Clicking on the clone ID number gives (C) further annotation and links to other databases.

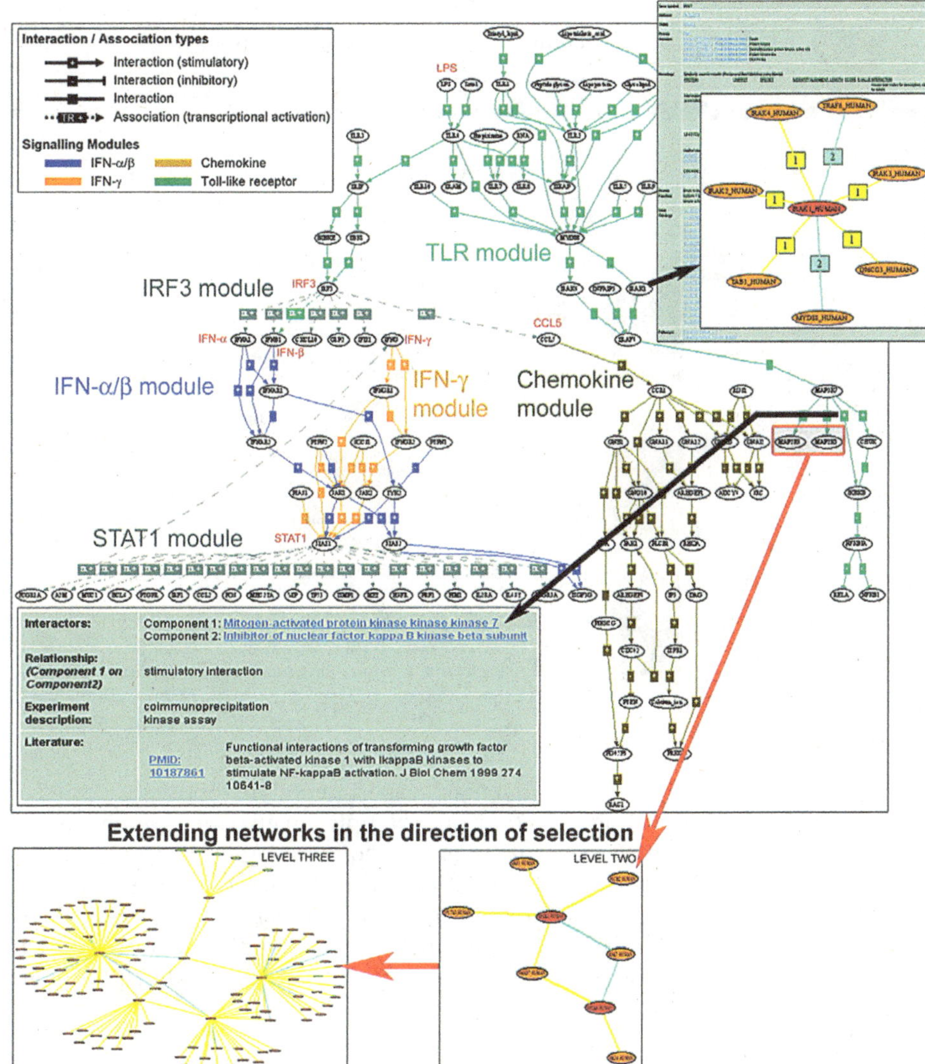


**Figure 3.10**
**Interaction maps generated with pSTIING.** The pathway is divided into functional modules, for example the Chemokine module. Each component node (protein or gene) is linked to more extensive information about the component including interaction partners, domains, gene ontology, other homologous proteins, and links to other databases. By clicking on the node, some or all the information can be displayed and further interactions can be viewed, extending the interaction network.

structure databases have been produced by analysis of the structures for protein folds and classifying them according to the conservation of the fold. These include CATH and SCOP and are discussed in Chapter 14.

## 3.4 Data Quality

One of the great strengths of bioinformatics databases is that as well as the basic data they often contain further analysis and interpretation and links to relevant entries in other databases. However, analysis that is based upon incorrect data is likely to lead to erroneous conclusions. Hence it is extremely important that the data provided by databases be as accurate as possible, ideally even to the point of including details of possible errors and levels of uncertainty (see Flow Diagram 3.4). Furthermore, some results of analysis are more accurate than others, so that they also require some measure of reliability. These accuracy issues are often noted in the annotation of each entry, making these bioinformatics databases very useful. In this section we will start by looking at data accuracy and how to identify uncertainties. Following this, we will look at accuracy issues related to data analysis. There are two distinct ways in which this can be done: computer-based analysis and manual curating. Both methods make an important contribution to ensuring the accuracy of database information.

**Figure 3.11**
**An example of the results of a search through the RCSB structural database with the keywords "LIM domain."** Ninety-one structures were found, of general LIM domain proteins and associated proteins. Links are provided to further details of each structure.

A further potential complication is that many experimental methods are applied to only small parts of the complete system. For example, an enzyme is often studied purified from the cell, or protein structures are frequently reported as only a fragment of the complete molecule. In these cases, while the data might be accurate, caution is required when extrapolating back to the complete cellular system. The database information must include sufficient experimental details, or references to details in published papers, that the information can be understood in its proper context.

## Nonredundancy is especially important for some applications of sequence databases

There are numerous occasions on which the same work has been subsequently repeated, either by the same researchers or others. A typical example occurs when the complete genome of an organism is sequenced, as often some of the genes will have been individually sequenced previously. Another often-encountered situation is that a confirmatory sequencing of a clone reveals a small sequence difference

**Flow Diagram 3.4**
**The key concept introduced in this section is that it is important to consider the quality of the data within a database, how good quality can be achieved, and how it can be maintained.**

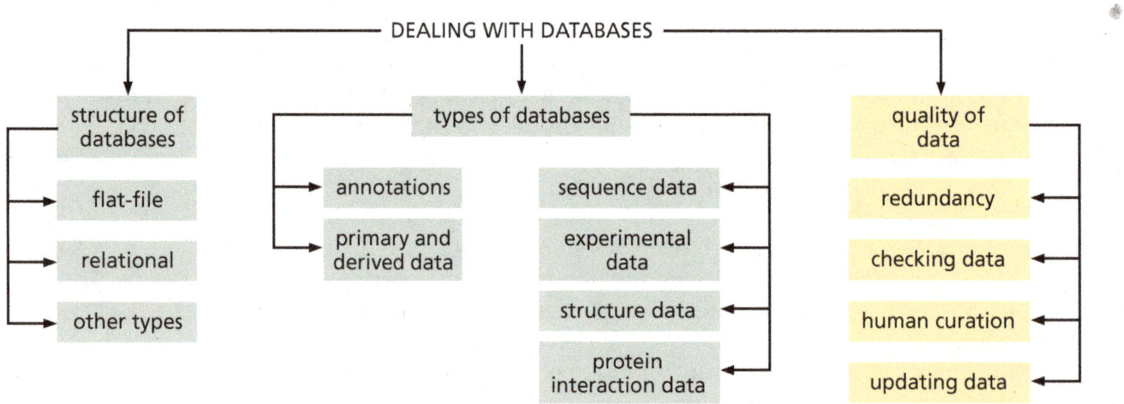

with respect to the database entry, which may be due to errors in the original sequencing or to the presence of mutations.

The database should ideally contain the results from all published work, but this can lead to many entries with identical or almost identical data. In the case of identical data, the entries are said to be **redundant**, in that for many applications they both provide the same information. A typical application for which this is the case is identifying similar database sequences to a query sequence. In most cases the data are more usefully held in a single entry in the database, referring to all the independent experiments, and summarizing all the separate database entries. This allows researchers to discover all the information by reading the single entry. In sequence databases it is normal practice to include in this single entry data that are almost but not exactly identical due to mutations or possible experimental errors. A database constructed in this fashion is known as a **nonredundant database**, and is the usual form in which the database is queried. Note that the redundant database with the original separate entries must be maintained, so that if in future errors come to light, such as attribution to the wrong species, the corrected files can be used to generate a corrected nonredundant database.

Although this issue has been presented in terms of several entries in the same database, it can also arise when several databases have been constructed that hold overlapping sets of data. For example, there used to be several protein sequence databases, all of which had some unique data as well as a large amount of common data. Simply combining these databases resulted in a large degree of redundancy that had to be removed for efficient use.

## Automated methods can be used to check for data consistency

Usually the data present in database entries should have certain properties that can be checked automatically. For example, a fully defined DNA sequence should consist only of the four bases A, C, G, and T. However, sometimes there is experimental uncertainty about the base identity at certain positions, leading to an extended character set that describes all the possible uncertainty (see Table 3.1). A similar situation occurs with protein sequences, leading to the one-letter codes B, Z, and X to describe uncertainty in the amino acid present at a sequence position. Other aspects of sequences can also be checked; for example, that the sequence length and molecular weight, if reported, agree with the sequence.

Some data can be checked even more rigorously. The bonding geometry found in protein structures is known to show only limited variation, for example the length of the main chain double bond linking the C and O atoms (see Figure 2.2) averages

**Table 3.1**
**The nomenclature used to represent base uncertainty in DNA sequences.**

| Letter | Base uncertainty |
|---|---|
| M | A or C |
| R | A or G |
| W | A or T |
| S | C or G |
| Y | C or T |
| K | G or T |
| V | A or C or G |
| H | A or C or T |
| D | A or G or T |
| B | C or G or T |
| X or N | any nucleotide |

0.123 nm (1.23 Å) with a standard deviation of 0.002 nm (0.02 Å). Protein structures are defined in the databases by atomic coordinates relative to an origin, so that the geometry is not explicitly given. However, it is easy to calculate the bond lengths of a structure in the database and to check them against their known ranges. This can be done for all bond lengths, bond angles, and torsion angles, as well as chirality (see Chapter 2). Sometimes, when an error is detected, it is possible to identify and correct the error in the atomic coordinates. In other cases the database entry can be annotated to describe the error. Another typical feature of protein structure entries in the databases is that they often lack certain atoms or even entire residues. This is due to the nature of the experimental techniques used, and frequently relates to the degree of freedom of movement of regions of the molecule. Again, these missing atoms should be identified in the annotation.

Other forms of data present in the database entries may also be amenable to automated checking. For example, in the entry for a gene expression experiment, a check can be made that expression levels are given for every gene under every condition. Wherever cross-references are given to entries in other databases a check can be made that the databases and entries exist. These sorts of automated checking are usually only able to detect errors and highlight them for manual resolution. In addition, they may often fail to identify other more subtle errors. In some cases, such as when submitting microarray data to certain databases, the data must conform to a specific standard that is intended to include sufficient experimental information in order for the work to be accurately reproduced. One such data standard for microarray experiments is called MIAME. MIAME describes the Minimum Information About a Microarray Experiment that is needed to enable the interpretation of the results of the experiment unambiguously and, potentially, to reproduce the experiment.

As the number and size of databases grew, and annotations became more extensive, some potential problems were recognized in electronic text searching. Often alternative spellings of the same word were encountered, as well as alternative names. For example, it is not uncommon for an enzyme to have three or four alternative names. As a consequence it can be difficult to locate all relevant database entries without searching for all names or spelling variants. To circumvent this problem, ontologies have been proposed covering specific disciplines. By restricting the words used in databases to those in the ontologies, text searches can be rendered more effective. Automated methods can be used to identify alternative terms and replace them with approved terms, as well as to find misspellings.

## Initial analysis and annotation is usually automated

The data submitted in new database entries are increasingly generated in electronic form by the experiments. In such cases it is relatively easy to include details of important parameters used in obtaining the data. These can be extremely useful in subsequent interpretation. An example of this occurs in the MSD molecular structure database, especially for those structures derived from crystallographic methods. As well as details that may only be of interest to other experimentalists, crystallographers report measures called the resolution and *R* factor, which are a very useful guide to the overall accuracy of the structure. The resolution indicates how much measured diffraction data were included in the work, while the *R* factor measures the correlation between the structure and the experimental measurements. Both give strong indications of the effective limit of accuracy with which atomic coordinates can be determined. It is common practice when analyzing general protein molecular geometry only to include structures whose resolution is better than some threshold, such as 0.2 nm (2.0 Å).

As regards analysis of data already incorporated in the database, in many cases (and certainly with sequences and molecular structure data) many of the forms of analysis applied in bioinformatics are fully automatable. Many of these techniques are described in detail in this book in the following chapters. An example of such a method is the identification of similar sequences by alignment of all sequences in a database. For example, by using such techniques on a protein sequence it is possible to identify the protein family to which it belongs and its likely function. The methods usually include statistical analysis that can assess the likely significance of the result, information which should ideally be presented in the database entry. One way of increasing the reliability of the analysis is to reduce the level of detail, for example identifying a more general class of enzyme function instead of trying to predict exactly which compounds are involved. However, this can render the information too vague to be useful.

There is a particularly high potential for misleading analysis when genes are identified in nucleotide sequences by applying purely computational methods, a situation which is common in the early stages of genome sequence annotation. Such methods are described in detail in Chapters 9 and 10. Sometimes there are no experimental data available to support the predicted genes. In such cases the genes and the proteins they encode are often labeled as "hypothetical" in the databases. The gene prediction methods available are still relatively inaccurate, especially for eukaryotes, so that considerable caution is required when encountering these entries in the databases. In some cases the gene may be correctly predicted, but it is also quite possible that no such gene exists, or that errors have been made in the details of the prediction, resulting in a different protein sequence for example. Often considerable experimental work occurs after genome sequencing is complete to try to obtain experimental evidence for these hypothetical genes and proteins.

## Human intervention is often required to produce the highest quality annotation

While computer-based analysis has the benefit of being easily carried out in an objective way for all database entries, it cannot produce annotation of the highest quality because the methods used are not perfect and rely on data being present in electronically accessible forms. In contrast, human database annotation can use data in any form, such as information buried in the text of a research paper or the knowledge of an expert in the field.

An example of the use of manual curation to produce high-quality annotation is the Swiss-Prot protein sequence database, now a component of the UniProtKB universal protein resource. However, production of such annotated entries is very time consuming, and needs input from specialists. For example, the UniProtKB database involves numerous researchers worldwide who have volunteered to assist in the manual curation of entries relating to one or more specific **protein families**. So far volunteers have been found for over 130 protein families. However, this is still only a small proportion of all the entries in that database.

Manual annotation usually is most effective for entries whose data have been well characterized experimentally, or are related to other entries for which this is the case. As mentioned above, this relationship is usually one of similarity. If no experimental evidence is available even for related entries, manual annotation is likely to provide only limited benefit.

## The importance of updating databases and entry identifier and version numbers

Databases need to be regularly updated to reflect new information, not just including new entries but also updating and correcting existing ones. In fields that are actively evolving, major and minor releases may be required, with a few major updates each year but minor updates on a weekly or even daily basis. Individual database entries will change less often, but it is important that it is easy to recognize if the current entry differs from a copy made earlier. This can be achieved by using version numbers for entries, or alternatively by reporting the most recent date on which changes were made, as can be seen in the "Entry information" section of the database shown in Figure 3.8. However, this is only of use if there is also a unique identifier for the entry that is fixed and can be used to ensure the two versions are indeed of the same entry. In Figure 3.8 this is called the "Primary accession number."

There are occasions when a decision is taken to make an existing database entry obsolete. In most cases the data from this entry will still be in the database, but now in a different entry with a different unique identifier. It is important that the current database maintains some record of the obsolete entry identifier, the reasons for the decision, and the fate of the data. This information and version records will sometimes be of great value in understanding the reasons why a repeated study produces different results.

The eukaryotic genome sequencing projects are an extreme example of the importance of the issues just discussed. The experimental methods used involve breaking the genome into many small overlapping pieces, which can be individually sequenced, and then using the overlaps to assemble these into the complete chromosomes. Many of these projects were, and are, funded subject to intermediate data being made publicly available very soon after it is obtained. As a consequence, a database of the sequence data has to be constructed before the complete chromosomes have been assembled. As the project progresses, the assembly progresses, resulting in ever-longer sequences formed by merging the smaller sequences. This results in some database entries becoming obsolete. Until the assembly has been completed, features such as genes cannot be identified by reliable sequence base numbers, as the true start of the chromosome will not be known. Every time sequences are merged into a larger assembly there is a possibility of the sequence numbers changing. Great care is needed to use suitable methods to identify features in the database such that they can be traced over the development of the assembly to its final state.

## Summary

In this chapter we have introduced the reader to the concept of a database and looked at the wide variety of databases publicly available and easily accessible via the Internet. There are many Web sites that serve locally created, highly specialized databases, as well as large resource centers that integrate many key databases into a unified network that facilitates identifying connections in the data, which is one of the main aspects of bioinformatics analysis.

We have highlighted the importance of accuracy, both in data and in the annotations. Equally important is that the database is kept up to date, as analysis based on outdated or incorrect data will also be outdated and, quite possibly, incorrect.

Databases are often the starting point of many types of bioinformatics research that will be described in the following chapters. They are a powerful tool for storing, sharing, and describing data, as well as for extracting information for further understanding and analysis. They can be regarded both as data repositories and online libraries.

Finally, to appreciate the range of data available in the public databases and the numerous ways in which it can be presented, the reader is recommended to go to the NAR database list and click on the links, exploring all the possibilities, and let their curiosity lead them on.

# Further Reading


## 3.1 The Structure of Databases

Bressan S & Catania B (2006) Introduction to Database Systems. New York: McGraw Hill Higher Education.

Date CJ (1995) An Introduction to Database Systems, 6th ed. Boston: Addison-Wesley.

Kim W (1990) Introduction to Object-Oriented Databases. Cambridge MA: MIT Press.

Riccardi G (2001) Principles of Database Systems with Internet and Java Applications. Boston: Addison-Wesley.

Stein LD (2003) Integrating biological databases. *Nat. Rev. Genet.* 4, 337–345.

## 3.2 Types of Database

**How we define and connect things is important: Ontologies**

Ashburner M, Ball CA, Blake JA et al. (2000) The Gene Ontology Consortium. Gene ontology: tool for the unification of biology. *Nat. Genet.* 25, 25–29.

Bard J (2003) Ontologies: Formalising biological knowledge for bioinformatics. *Bioessays* 25, 501–506.

Bard JB & Rhee SY (2004) Ontologies in biology: design, applications and future challenges. *Nat. Rev. Genet.* 5, 213–222.

Gruber TR (1993) A translation approach to portable ontology specification. *Knowledge Acquisition* 5, 199–220.

Thompson JD, Holbrook SR, Katoh K et al. (2005) MAO: a multiple alignment ontology for nucleic acid and protein sequences. *Nucleic Acids Res.* 33, 4164–4171.

Zhang B, Schmoyer D, Kirov S & Snoddy J (2004) GOTree Machine (GOTM): a web-based platform for interpreting sets of interesting genes using Gene Ontology hierarchies. *BMC Bioinformatics* 5, 5:16.

## 3.3 Looking for Databases

**Sequence databases**

Apweiler R, Bairoch A & Wu CH (2004) Protein sequence databases. *Curr. Opin. Chem. Biol.* 8, 76–80.

Bairoch A, Apweiler R, Wu CH et al. (2005) The Universal Protein Resource (UniProt). *Nucleic Acids Res.* 33, D154–D159.

Benson DA, Karsch-Mizrachi I, Lipman DJ et al. (2005) GenBank. *Nucleic Acids Res.* 33, D34–D38.

Cochrane G, Aldebert P, Althorpe N et al. (2006) EMBL Nucleotide Sequence Database: developments in 2005. *Nucleic Acids Res.* 34, D10–D15.

Wheeler DL, Barrett T, Benson DA et al. (2006) Database resources of the National Center for Biotechnology Information. *Nucleic Acids Res.* 34, D173–D180.

**Microarray databases**

Brazma A, Sarkans U, Robinson A et al. (2002) Microarray data representation, annotation and storage. *Adv. Biochem. Eng. Biotechnol.* 77, 113–139.

Gollub J, Ball CA, Binkley G et al. (2003) The Stanford Microarray Database: data access and quality assessment tools. *Nucleic Acids Res.* 31, 94–96.

Gollub J, Ball CA & Sherlock G (2006) The Stanford Microarray Database: a user's guide. *Methods Mol. Biol.* 338, 191–208.

Parkinson H, Sarkans U, Shojatalab M et al. (2005) ArrayExpress—a public repository for microarray gene expression data at the EBI. *Nucleic Acids Res.* 33, D553–D555.

**Protein interaction databases**

Bader GD, Betel D & Hogue CW (2003) BIND: the Biomolecular Interaction Network Database. *Nucleic Acids Res.* 31, 248–250.

Ng A, Bursteinas B, Gao Q et al. (2006) pSTIING: a 'systems' approach towards integrating signalling pathways, interaction and transcriptional regulatory networks in inflammation and cancer. *Nucleic Acids Res.* 34, D527–D534.

Salwinski L, Miller CS, Smith AJ et al. (2004) The Database of Interacting Proteins: 2004 update. *Nucleic Acids Res.* 32, D449–D451.

Zanzoni A, Montecchi-Palazzi L, Quondam M et al. (2002) MINT: a Molecular INTeraction database. *FEBS Lett.* 513, 135–140.

**Structural databases**

Berman HM, Westbrook J, Feng Z et al. (2000) The Protein Data Bank. *Nucleic Acids Res.* 28, 235–242.

Berman HM, Henrick K & Nakamura H (2003) Announcing the worldwide Protein Data Bank. *Nat. Struct. Biol.* 10, 980.

Lo Conte L, Brenner SE, Hubbard TJP et al. (2002) SCOP database in 2002: refinements accommodate structural genomics. *Nucleic Acids Res.* 30, 264–267.

Murzin AG, Brenner SE, Hubbard T & Chothia C (1995) SCOP: a structural classification of proteins database for the investigation of sequences and structures. *J. Mol. Biol.* 247, 536–540.

Orengo CA & Thornton JM (2005) Protein families and their evolution—A structural perspective. *Annu. Rev. Biochem.* 74, 867–900.

Pearl FM, Bennett CF, Bray JE et al. (2003) The CATH database: an extended protein family resource for structural and functional genomics. *Nucleic Acids Res.* 31, 452–455.

The Reciprocal Net. http://www.reciprocalnet.org/

Velankar S, McNeil P, Mittard-Runte V et al. (2005) E-MSD: an integrated data resource for bioinformatics. *Nucleic Acids Res.* 33, D262–D265.

## 3.4 Data Quality

Ashelford KE, Chuzhanowa NA, Fry JC et al. (2005) At least 1 in 20 16S rRNA sequence records currently held in public repositories is estimated to contain substantial anomalies. *Appl. Environ. Microbiol.* 71, 7724–7736.

Gilks WR, Audit B, de Angelis D et al. (2005) Percolation of annotation errors through hierarchically structured protein sequence databases. *Math. Biosci.* 193, 223–234.

Weichenberger CX & Sippl MJ (2006) NQ-Flipper: validation and correction of asparagine/glutamine amide rotamers in protein crystal structures. *Bioinformatics* 22, 1397–1398.

**MIAME**

Brazma A, Hingamp P, Quackenbush J et al. (2001) Minimum information about a microarray experiment (MIAME)—toward standards for microarray data. *Nat. Genet.* 29, 365–371.

# PART 2

# SEQUENCE ALIGNMENTS

DNA and proteins are coded for by chemicals that can be represented by one-letter codes. This makes the comparison of different DNA or protein sequences possible and aids in identifying regions of similarity that may show functional, structural, or evolutionary relationships between the sequences. Therefore this is a very important part of bioinformatics analysis and is also often the first step in the analysis of newly found DNA or protein sequences.

The first chapter introduces some of the concepts of sequence alignment programs and illustrates how these programs work. The rest of this chapter then concentrates on the practical aspects of alignments and how these methods can also be used to search databases or sequence patterns to find homologs and functionally related proteins. Examples are used to illustrate the effectiveness of sequence alignments.

The subsequent two chapters focus on the methods used for pairwise and multiple sequence alignment, pattern searching, and database searching in detail, giving the algorithmic steps used in the models. Study of these two chapters should enable students to start designing their own algorithms.

# PRODUCING AND ANALYZING SEQUENCE ALIGNMENTS

4

APPLICATIONS CHAPTER

## When you have read Chapter 4, you should be able to:

Determine homology by sequence alignment.

Describe different uses of protein and DNA sequence alignments.

Define scoring alignments.

Make alignments between two sequences.

Make multiple alignments between many sequences.

Compare local alignment techniques for finding limited areas of similarity.

Explain global alignment techniques for matching whole sequences.

Search databases for homologous sequences.

Look for patterns and motifs in a protein sequence.

Use patterns and motifs to locate proteins of similar function.

The revolution in genetic analysis that began with recombinant DNA technology and the invention of DNA sequencing techniques in the 1970s has, 30 years later, filled vast databases with nucleotide and protein sequences from a wide variety of organisms. Genomes that have now been completely sequenced include human, mouse, chimpanzee, the fruit fly *Drosophila*, the nematode *Caenorhabditis*, and the yeast *Saccharomyces*, as well as numerous bacteria, archaea, and viruses. Although entries for nucleotide and protein sequences in databases such as GenBank, dbEST, and UniProt KB now number many millions, nothing is known about the structure or function of the proteins specified by many of them. Converting this sequence information into useful biological knowledge is now the main challenge.

To find out more about a newly determined sequence, it is subjected to the process of sequence analysis. There are many aspects to this, depending on the source of the sequence and what you ultimately want to find out about it. In this chapter, we will focus on one of the key stages in most sequence analyses: the alignment of different sequences to detect homology and the comparison of a novel sequence with those in the databases to see whether there is any similarity between them. The practical use of techniques and programs for general alignment, database searching, and pattern searching will be described in this chapter, with the main focus on the alignment and analysis of protein sequences. The theory underlying programs for pairwise alignment is described in Chapter 5 and that dealing with multiple alignments in Chapter 6, for both nucleic acid and protein sequences. Techniques and programs for detecting genes and other sequence features in genomic DNA are dealt with in Chapters 9 and 10.

**Mind Map 4.1**
**A mind map of the four major sections relating to sequence analysis and alignment: aligning sequences, searching through databases, measuring how well sequences match, and looking for families of proteins.**

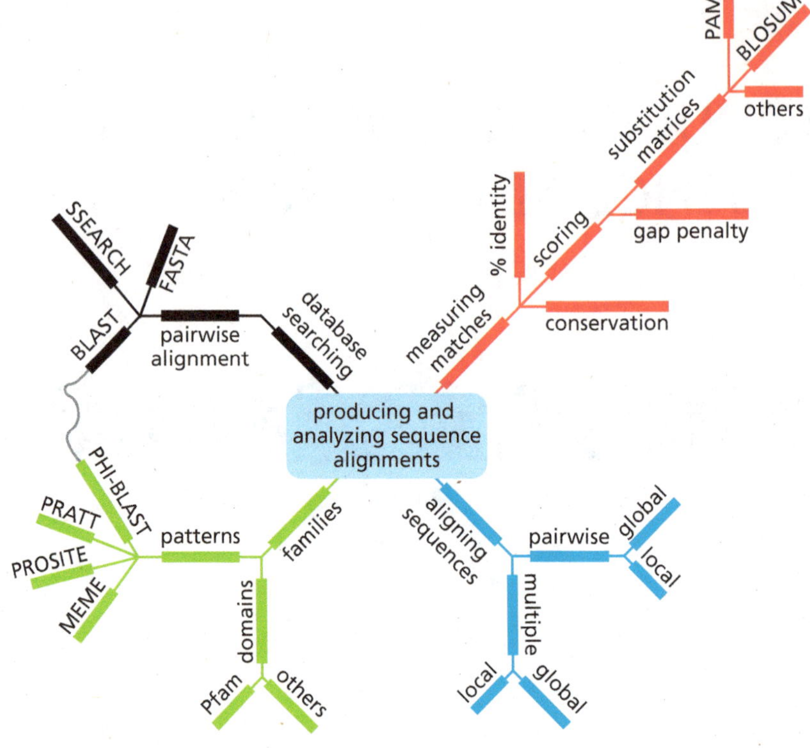


The identification of similar sequences has a multitude of applications. For raw, uncharacterized genomic DNA sequences, comparison with sequences in a database can often tell you whether the sequence is likely to contain, or be part of, a protein-coding gene. The similarity search may retrieve a known gene or family of genes with a strong similarity to the new sequence. This will provide the first clues to the type of protein the new gene encodes and its possible function. Similarities in sequence can also help in making predictions about a protein's structure (see Chapters 11–14). Sequences of proteins or DNAs from different organisms can also be compared in order to construct phylogenetic trees, which trace the evolutionary relationships between species or within a family of proteins (see Chapters 7 and 8).

As well as many general and specialized databases of DNA and protein sequences, the fully sequenced genomes of various organisms are now available (see Chapter 3), providing vast amounts of information for comparison. It is, however, important to remember that although many newly discovered sequences will share some or considerable similarity to sequences in the databases, there will still be many that are unique.

## 4.1 Principles of Sequence Alignment

Devising ways of comparing sequences has never been straightforward, not just because of the vast amounts of information now available for searching. The difficulties arise because of the many ways DNA and protein sequences can change during evolution. Mutation and selection over millions of years can result in considerable divergence between present-day sequences derived from the same ancestral gene. Bases at originally corresponding positions, and the amino acids they encode, can change as a result of point mutation, and the sequence lengths can be quite different as a result of insertions and deletions. Even more dramatic changes may have occurred; for example, the fusion of sequences from two different genes. Gene

## Box 4.1 Genes and pseudogenes

Pseudogenes are sequences in genomic DNA that have a similar sequence to known protein-coding genes but do not produce a functional protein. They are assumed to arise after gene duplication, when one of the gene copies undergoes mutation that either prevents its transcription or disrupts its protein-coding sequence. The human genome is estimated to contain up to 20,000 pseudogenes. As the pseudogene sequence is no longer under selection to retain protein function, it will generally accumulate further mutations at a higher rate than the functional gene. Despite this, many pseudogenes retain considerable sequence similarity to their active counterparts. One case has even been found in which the RNA from a transcribed pseudogene regulates the expression of the corresponding functional gene.

duplications are common in eukaryotic genomes, and in many cases mutation has disabled one copy of a gene so that it is either no longer expressed or, if transcribed, does not produce a functional protein. Such genes are called pseudogenes (see Box 4.1) and can be found in homology searches.

On superficial inspection, such changes in gene sequence and length can effectively mask any underlying sequence similarity. To reveal it, the sequences have to be aligned with each other to maximize their similarities. This crucial step in sequence comparison is the main topic of the first half of this chapter (Sections 4.1 to 4.5). Alignment methods are at the core of many of the software tools used to search the databases, and in the second half of the chapter we will describe some of these tools and how they can be used to retrieve similar sequences from the databases (Sections 4.6 to 4.10). The first steps to consider are shown in Flow Diagram 4.1.

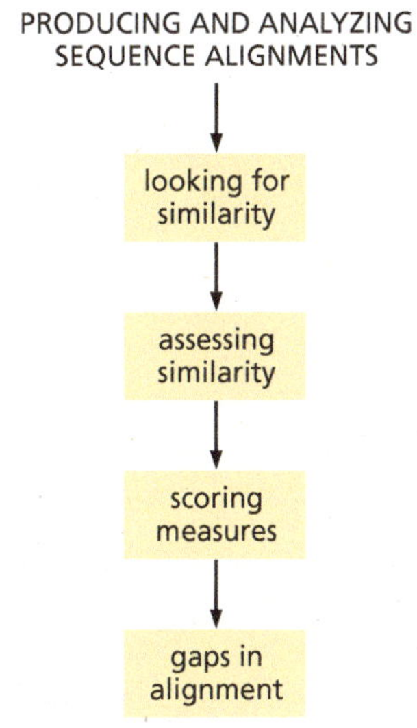


**Flow Diagram 4.1**
**The key concept introduced in these first four sections is that in order to assess the similarity of two sequences it is necessary to have a quantitative measure of their alignment, which includes the degree of similarity of two aligned residues as well as accounting for insertions and deletions.**

## Alignment is the task of locating equivalent regions of two or more sequences to maximize their similarity

As the result of mutation, even the sequences of the same protein or gene from two closely related species are rarely identical. Ideally, what we want to achieve when comparing sequences is to line them up in such a way that, when they do derive from a common ancestor, bases or amino acids derived from the same ancestral base are aligned. Without information to the contrary, this is best achieved by maximizing the similarity of aligned regions.

To illustrate the general principle, take the two hypothetical amino acid sequences THISSEQUENCE and THATSEQUENCE. If we align them so that as many identical letters as possible pair up we get

where the letters in red type are identical. As we can easily see with such short and similar sequences, this alignment clearly identifies their strong similarity to each other.

So far so good, but when sequences become more different from each other, they become more difficult to compare. How would we go about comparing the two sequences THATSEQUENCE and THISISASEQUENCE, in which a mutation has led to the insertion of the three amino acids I, S, A into one of the original sequences? Simply lining them up from the beginning loses much of the similarity we can see exists. More subtly, because of the difference in length, it also creates false matches between noncorresponding positions.

```
T H A T S E Q U E N C E
| |             |       |
T H I S I S A S E Q U E N C E
```

To get round this problem, **gaps** are introduced into one or both of the sequences so that maximum similarity is preserved.

There is never just one possible alignment between any two sequences, and the best one is not always obvious, especially when the sequences are not very similar to each other. At the heart of sequence-comparison and database-searching methods are algorithms for testing the fit of each alignment generated, giving it a quantitative score, and filtering out the unsatisfactory ones according to preset criteria.

## Alignment can reveal homology between sequences

In all methods of sequence comparison, the fundamental question is whether the similarities perceived between two sequences are due to chance, and are thus of little biological significance, or whether they are due to the derivation of the sequences from a common ancestral sequence, and are thus homologous. The terms "homology" and "similarity" are sometimes used interchangeably, but each has a distinct meaning. **Similarity** is simply a descriptive term telling you that the sequences in question show some degree of match. Homology, in contrast, has distinct evolutionary and biological implications. In the molecular biological context, it is generally defined as referring specifically to similarity in sequence or structure due to descent from a common ancestor. Homologous genes are therefore genes derived from the same ancestral gene. During their evolutionary history they will have diverged in sequence as a result of accumulating different mutations.

Because homology implies a common ancestor, it can also imply a common function or structure for two homologous proteins, which can be a useful pointer to function if one of the proteins is known only from its sequence. The operation of natural selection tends to result in the acceptance of mutations that preserve the folding and function of a protein, whereas those that destroy folding or function will be eliminated. However, similar or identical aligned residues may simply be due to relatively recent divergence of the two sequences, and so care must be taken not to overestimate their functional importance. Moreover, mutation and selection can generate proteins with new functions but relatively little change in sequence. Therefore, sequence similarity does not always imply a common function.

Conversely, there are proteins with very little sequence similarity to each other but in which a common protein fold and function are preserved. Consequently, low sequence similarity does not necessarily rule out common function or homology. Such cases require extra information, such as structural or biochemical knowledge, to demonstrate their true relationship.

Sequences can also be significantly similar to each other, and yet not be evolutionarily homologous, as a result of **convergent evolution** for similar function (see Box 4.2). In this case, identical or very similar aligned residues can be argued to have an important functional role. Convergent evolution does not, however, usually produce highly similar sequences of any great length.

All these considerations have to be taken into account when analyzing the results of a database search. An alignment of two sequences is, in effect, a hypothesis about which pairs of residues have evolved from the same ancestral residue. But an alignment in itself does not imply an evolutionary order of events, so that the two

## Box 4.2 Convergent and divergent evolution

Convergent evolution is the evolutionary process in which organs, proteins, or DNA sequences that are unrelated in their evolutionary origin independently acquire the same structure or function. This usually reflects a response to similar environmental and selective pressures. Convergent evolution is contrasted with the process of **divergent evolution**, which produces different structures or sequences from a common ancestor. An example of convergent evolution for function can be seen in the wings of insects and bats. Although adapted to the same function—that of flight—insect wings and bat wings do not derive from the same ancestral structure.

**Figure B4.1**
**(A) Bat wings and (B) butterfly wings.** (A, courtesy of Ron Austing/Science Photo Library.)

alternatives of homology and convergent evolution cannot usually be distinguished without additional information.

Sequence comparison methods have to take account of such factors as the types of mutation that occur over evolutionary time, differences in the physicochemical properties of amino acids and their role in determining protein structure and function, and the selective pressures that result in some mutations being accepted and others being eliminated. One has to consider the evolutionary processes that are responsible for sequence divergence and find a way to include the salient features in practicable schemes for testing the goodness of fit of the alignment. These must be quantitative and hence involve a score. Such **scoring schemes** can then be incorporated in algorithms designed to generate the best possible alignments. Finally, ways must be found to discriminate between fortuitously good alignments and those due to a real evolutionary relationship.

As we shall see in this chapter, all computational methods of sequence comparison take account of these factors in some way.

### It is easier to detect homology when comparing protein sequences than when comparing nucleic acid sequences

For most purposes, comparisons of protein sequences show up homology more easily than comparisons of the corresponding DNA sequences. There are many reasons for this greater sensitivity. First, there are only four letters in the DNA alphabet compared to the 20 letters in the protein alphabet, and so a DNA sequence, of necessity, provides less information at each sequence position than does a protein sequence. In other words, there is a much greater probability that a match at any one position between two DNA sequences will have occurred by chance. Therefore, the degree of similarity, as judged by some appropriate quantitative score, needs to be greater between DNA sequences than between protein sequences for the alignment to be of importance. As we shall see later in this chapter, ways have been devised of determining the likelihood that one amino acid can be substituted for another during evolution, and this provides additional information beyond simple identity for scoring an alignment and determining homology.

Second, as we saw in Chapter 1, the genetic code is redundant; that is, there are two or more different codons for most amino acids (see Table 1.1). This means that identical amino acid sequences can be encoded by different nucleotide sequences. Finally, the complex three-dimensional structure of a protein, and hence its function, is determined by the amino acid sequence. The importance of maintaining protein function usually leads to amino acid sequences changing less over evolutionary time than homologous DNA sequences. In this chapter we will concentrate for the most part on protein sequence analysis.

There are many circumstances, however, in which it is necessary to compare DNA sequences: when searching for promoters and other regulatory sequences, for example, or in whole-genome comparisons. DNA alignment is also performed, to some extent, as part of gene identification (see Chapters 9 and 10).

## 4.2 Scoring Alignments

### The quality of an alignment is measured by giving it a quantitative score

Two homologous sequences are often so different that a correct or best alignment is not obvious by visual inspection. Furthermore, the large numbers of sequences that can be examined for similarity nowadays oblige us to use automated computational methods to judge the quality of an alignment, at least as an initial filter.

Because it is possible for two sequences to be aligned in a variety of different ways, including the insertion of gaps to improve the number of matched positions, how does one objectively determine which is the best possible alignment for any given pair of sequences? In practice, this is done by calculating a numerical value or **score** for the overall similarity of each possible alignment so that the alignments can be ranked in some order.

We can then work on the basis that alignments of related sequences will give good scores compared with alignments of randomly chosen sequences, and that the correct alignment of two related sequences will ideally be the one that gives the best score. The alignment giving the best score is referred to as the **optimal alignment**, while others with only slightly worse scores are often called **suboptimal alignments**. No one has yet devised a scoring scheme that perfectly models the evolutionary process, which is so complex that it defies any practical method of modeling. The implication of this is that the best-scoring alignment will not necessarily be the correct one, and conversely, that the correct alignment will not necessarily have the best score. However, the scoring schemes now in common use, and which are described in this chapter, are generally reliable and useful in most circumstances, as long as the results are treated with due caution and regard for biological plausibility. In principle, a scoring scheme can either measure similarity or difference, the best score being a maximum in the former case and a minimum in the latter.

### The simplest way of quantifying similarity between two sequences is percentage identity

**Identity** describes the degree to which two or more sequences are actually identical at each position, and is simply measured by counting the number of identical bases or amino acids matched between the aligned sequences. Identity is an objective measure and can be precisely defined. **Percentage** or **percent identity** is obtained by dividing the number of identical matches by the total length of the aligned region and multiplying by 100. For the THATSEQUENCE/THISISASEQUENCE comparison, for example, the alignment given on page 74 is the best that can be achieved, and has a percentage identity score of 68.75% (11 matches over a total length of 16 positions, including the gaps).

One might think that an alignment of completely unrelated sequences would have a percentage identity of zero. However, as there are only four different nucleotides in nucleic acid sequences, and only 20 different amino acids in protein sequences, there is always a small but finite probability for any aligned sequences that identical residues will be matched at some positions. Because there are often hundreds of residues in a protein sequence and thousands in a nucleotide sequence, unrelated sequences are expected to align matches at several positions. The length of the sequence matters: a 30% identity over a long alignment is less likely to have arisen by chance than a 30% identity over a very short alignment. Statistically rigorous methods have been devised to measure the significance of an alignment, which will be discussed later in connection with database searches and in Section 5.4.

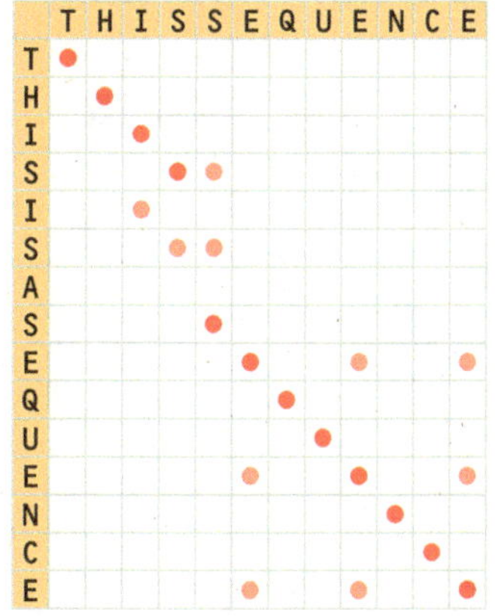


**Figure 4.1**
**Dot-plot representations.** A dot-plot matrix of the THISSEQUENCE/ THISISASEQUENCE example where red dots represent identities that are due to true matching of identical residue-pairs and pink dots represent identities that are due to noise; that is, matching of random identical residue-pairs.

## The dot-plot gives a visual assessment of similarity based on identity

A dot matrix or **dot-plot** is one of the simplest ways to compare sequence similarity graphically, and can be used for both nucleotide and protein sequences. To compare two sequences X and Y, one sequence is written out vertically, with each residue in the sequence represented by a row, while the other is written horizontally, with each residue represented by a column. Each residue of X is compared to each residue of Y (row to column comparison) and a dot is placed where the residues are identical. In the simplest scoring system, identical residues are scored as 1 and nonidentical residues as 0, and dots are placed at all positions that contain a 1. For example, if we take the pair THISSEQUENCE/THISISASEQUENCE pair, then a simple dot-plot will look like that illustrated in Figure 4.1. The dots in red, which form diagonal lines, represent runs of matched residues. The pink dots scattered either side of the diagonals are the same residues found elsewhere in the sequence. The diagonals are interrupted by a few cells, where a gap has been inserted.

Dot-plots can be useful for identifying intrasequence repeats in either proteins or nucleic acids. However, dot-plots suffer from background noise. To distinguish dot-patterns arising from background noise from significant dot-patterns it is usually necessary to apply a filter. The most widely used filtering procedure uses

**Figure 4.2**
**Two views of dot-plot representations of an SH2 sequence compared with itself.** (A) Unfiltered dot-plot (window length = 1 residue). The identity between the two sequences is shown by the unbroken identity diagonal. Nevertheless, there is still background noise. (B) Dot-plot of the same sequence comparison with a window of 10 residues and a minimum identity score within each window set to 3. The background noise has all been removed, leaving only the identity diagonal.

(A)

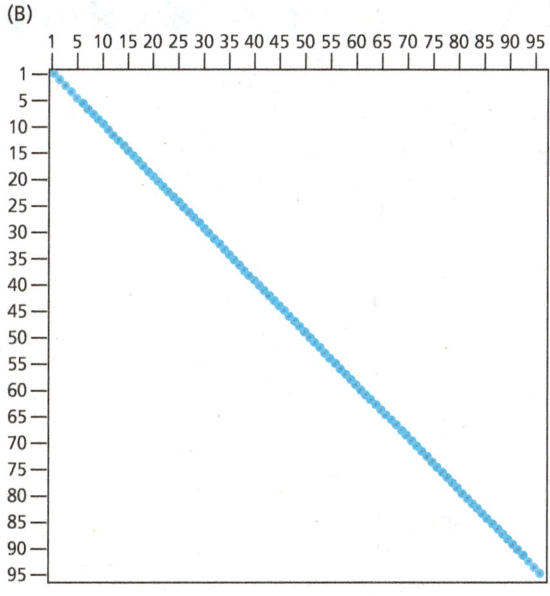

overlapping fixed-length windows and requires that the comparison achieve some minimum identity score summed over that window before being considered; that is, only diagonals of a certain length will survive the filter. Figure 4.2 shows a dot-plot between two identical SH2 sequences (see Box 4.3).

Figure 4.2A has a window length of 1; in other words, every residue is considered individually. Although the diagonal line indicating matched identical residues is clear and unbroken, as one would expect from a comparison of two identical sequences, there is still a certain amount of background noise detracting from the result, as most types of amino acid occur more than once in the sequence. Figure 4.2B shows the same comparison with a window of 10 residues and a minimum score for each window set to 3. Only the main diagonal is now seen, representing the one-to-one matching of the identical sequences.

Most dot-plot software provides a default window length and this is sufficient for an initial analysis. But one can use the window length to greater effect by varying it depending on what one is searching for. Window length can be set, for example, to the length of an exon when comparing coding sequences, or to the size of an average secondary structure within a protein when looking for structural motifs. When searching for internal repeats, the length of the repeat can be used to cut out background noise. In addition, rather than using 0 and 1 as the scores for nonidentical and identical residues, other values can be used and the score can be varied depending on the type of residues involved.

Figure 4.3 illustrates how a dot-plot can be used to identify repeats within a sequence. It shows two dot-plot calculations on the protein BRCA2 encoded by the breast cancer susceptibility gene *BRCA2*. This protein contains eight repeats of a short sequence of around 39 amino acids, called the BRC repeat (see Box 4.4). Figure 4.3A shows an unfiltered version of a self-comparison dot-plot of a region of BRCA2 containing two BRC repeats. The background noise is so strong that it is very difficult to pick out the repeats. Figure 4.3B shows a highly filtered dot-plot of the same comparison in which a diagonal line is now visible. This is the identity diagonal, where the one-to-one alignment of the sequence with itself is highlighted. But two other runs of dots are now also visible; these represent the internal BRC repeats.

## Box 4.3 The SH2 protein-interaction domain

The SH2 or Src-homology 2 domain is a small domain of about 100 residues found in many proteins involved in intracellular signaling in mammalian cells. It gets its name from the protein tyrosine kinase Src, where it was first found. It is one of numerous protein-interaction domains found in signaling proteins, which recognize and bind to particular features on other proteins to help pass the signal onward. SH2 domains bind specifically to phosphotyrosines on proteins; these are formed by the phosphorylation—the modification by covalent addition of a phosphate group—of tyrosine residues in specific peptide motifs by protein tyrosine kinases. This type of kinase is often part of, or associated with, cell-surface receptors, and is activated in response to an extracellular signal. The phosphotyrosine-binding site on SH2 domains consists of two pockets. One is conserved and binds the phosphotyrosine residue (pY); the other is more variable in sequence between different SH2 domains and binds residues located downstream from the pY, thereby conferring specificity on the protein–protein interaction. Because of its role in intracellular signaling, the SH2 domain is an important potential drug target for a number of diseases, including cancer and osteoporosis.

**Figure B4.2**
**A ribbon representation of an SH2 domain.**

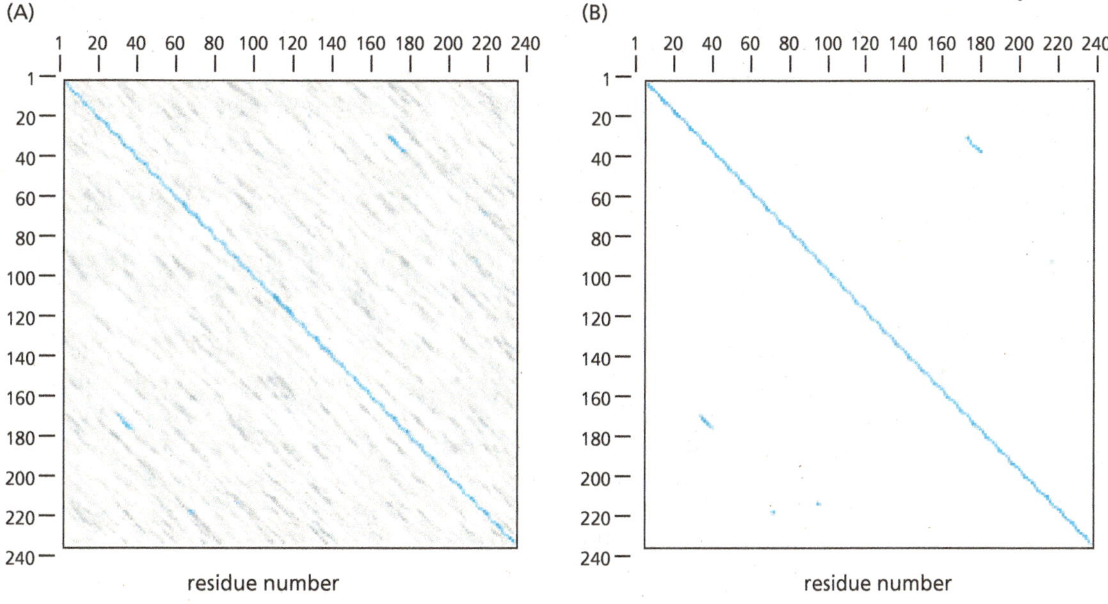


**Figure 4.3**
**Two dot-plots involving the breast cancer susceptibility gene protein BRCA2, which contains the small BRCA2 repeat domain.** (A) An unfiltered self-comparison dot-plot of part of the human BRCA2 sequence containing two BRCA2 repeats (the first and second BRCA repeat in the sequence). The background noise is so strong that it is very difficult to pick out the repeats. (B) The same dot-plot with a window length of 30 and a minimum score of 5. In addition to the identity diagonal there are two other clear diagonal runs of dots that represent the two internal BRCA2 repeats.

## Genuine matches do not have to be identical

Although it is the simplest alignment score to obtain, and can be very useful as a quick test of the quality of an alignment, percentage identity is a relatively crude measure and does not give a complete picture of the degree of similarity of two sequences to each other, especially in regard to protein sequences. For example,

### Box 4.4 The breast cancer susceptibility genes *BRCA1* and *BRCA2*

Two genes that confer increased susceptibility to breast cancer have been identified: the *BRCA1* gene on chromosome 17 in 1994 and the *BRCA2* gene on chromosome 13 in 1995. Women with a mutation in either *BRCA1* or *BRCA2* are at increased risk of developing breast, ovarian, and some other cancers by a given age than those without a mutation. The normal role of the BRCA1 and BRCA2 proteins, which are not structurally related, is to associate with the protein RAD51, a protein essential for the repair of double-strand breaks in DNA. Mutations in *BRCA1* or *BRCA2* can thus partly disable this repair mechanism, leading to more errors in DNA repair than usual, an increased mutation rate, and, ultimately, a greater risk of tumorigenesis. The BRCA2 protein has a number of repeats of 39 amino acids, the BRC repeats. Eight BRC repeats in BRCA2 are defined in the Pfam database, of which six are highly conserved and are involved in binding RAD51.

**Figure B4.3**
**BRC repeats of the BRCA2 protein as defined by the Pfam database.**

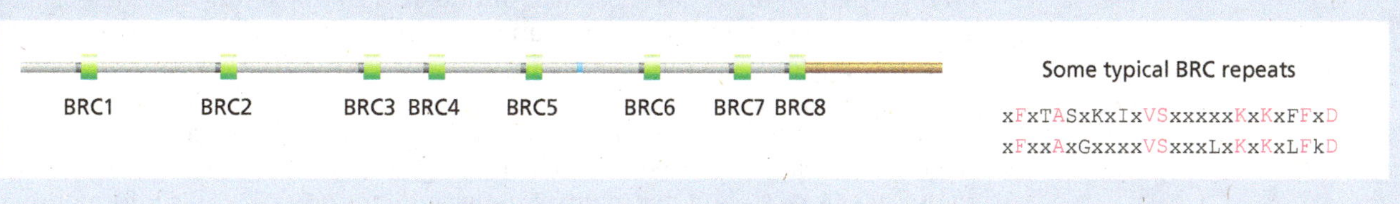

simply scoring identical matches as 1 and mismatches as 0 ignores the fact that the type of amino acid involved is highly significant. In particular, certain nonidentical amino acids are very likely to be present in the same functional position in two related sequences, and thus are likely to represent genuine matches. This is chiefly because certain amino acids resemble each other closely in their physical and/or chemical properties (see Figure 2.3) and can thus substitute functionally for each other. Mutational changes that replace one amino acid with another having similar physicochemical properties are therefore more likely to have been accepted during evolution. So pairs of amino acids with similar properties will often represent genuine matches rather than matches occurring randomly.

The simplest way of taking this into account is simply to count such similar pairs of amino acids as matches, and to refer to the score as **percent similarity**. In the now familiar example sequences below, red is used to indicate residues that are similar but not identical. Here the sequences have been realigned to take into account similarity as well as identity. Isoleucine (I) and alanine (A) are similar as they are both hydrophobic, whereas serine (S) and threonine (T) both have an -OH group in their side chain and are polar.

Not all similar amino acid pairs are equally likely to occur, however, and more sophisticated measures of assessing similarity are more commonly used. In these, each aligned pair of amino acids is given a numerical score based on the probability of the relevant change occurring during evolution. In such scoring schemes, pairs of identical amino acids are assigned the highest score; then, pairs of amino acids with similar properties (such as isoleucine and leucine) score more highly than those with quite different properties (such as isoleucine and lysine), which are rarely found in corresponding positions in known homologous protein sequences.

Other properties of amino acids can be added into scoring schemes for greater accuracy. For example, the type of residue involved should be taken into account. Many cysteine residues are highly conserved because of their important structural role in forming disulfide bonds, and tryptophan residues are usually key components of the hydrophobic cores of proteins. To mimic this, the scores for matching residues can be varied according to the type, with pairs of cysteines and tryptophans, for example, being assigned particularly high values. When aligned amino acid pairs are given varying scores in this way, summing the values at all positions gives the **overall alignment score**.

Most currently used alignment-scoring schemes for protein sequences measure the relative likelihood of an evolutionary relationship compared to chance. The theory behind such assessments is explained further in Section 5.1. With such schemes, the higher the alignment score, the more likely it is that the aligned sequences are homologous.

Ideally, it would be possible to decide unequivocally whether two sequences are homologous by simply looking at their best alignment score. This turns out to be more difficult than might be imagined, as the significance of the score will depend on the length of the sequences, their amino acid composition, and the number of sequences being compared—for example when searching a large database. We shall return to this topic later in the chapter.

The concept of similarity, rather than identity, has little relevance to comparisons of nucleotide sequences, especially in generating alignments. Purines tend to mutate to purines (A ↔ G) and pyrimidines to pyrimidines (C ↔ T). This information can be used to help construct phylogenetic trees (see Sections 7.2 and 8.1), but is not helpful for sequence alignment. In the case of an alignment of nucleotide sequences, the scoring scheme is almost always very simple. For example, in the database-searching program FASTA, which is discussed later and in Section 5.3, a score of +5 for matching bases and –4 for mismatches has been found to be effective for DNA database searches. This simpler scoring scheme is sufficiently sensitive to be useful in part because of the much higher percentage identity expected if there is significant homology between the sequences, since there are only four types of bases as compared to 20 amino acids.

### There is a minimum percentage identity that can be accepted as significant

What is the minimum percentage identity that can reasonably be accepted as significant? Burkhard Rost analyzed more than a million alignments of pairs of protein sequences for which structural information was available to find a cut-off for the level of sequence identity below which alignment becomes unreliable as a measure of homology. He found that 90% of sequence pairs with identity at or greater than 30% over their whole length were pairs of structurally similar proteins. Given both sequence and structural similarity, one can usually be confident that two sequences are homologous, so 30% sequence identity is generally taken as the threshold for an initial presumption of homology. Below about 25% sequence identity, however, Rost found that only 10% of the aligned pairs represented structural similarity. The region between 30% and 20% sequence identity has been called the **twilight zone,** where homology may exist but cannot be reliably assumed in the absence of other evidence. Even lower sequence identity (<20%) is referred to as the **midnight zone.**

### There are many different ways of scoring an alignment

The function of an alignment score is to provide a single numerical value for the degree of similarity or difference between two sequences. Most current applications measure similarity, and in this case the highest scores are best. A few applications, particularly those used for generating phylogenetic trees (see Chapters 7 and 8), use a score related to sequence difference, usually known as a **distance**, in which case the most closely related sequences give alignments with the lowest scores. The measure of difference between two homologous sequences from different species is sometimes called the **genetic** or **evolutionary distance**.

There is no a priori reason why residue pair alignment scores cannot be negative, for example to represent especially unlikely alignments. In fact, some of the popular techniques require scores that can be negative, and most commonly used schemes have both positive and negative scores for pairs of residues.

Scoring schemes have to represent two salient features of an alignment. On the one hand, they must reflect the degree of similarity of each pair of residues; that is, the likelihood that both are derived from the same residue in the presumed common ancestral sequence. On the other hand, they must assess the validity of inserted gaps. Ways of quantifying these two features will be described separately here, although in fact they are used together to arrive at the final score. We will first go through the ways of assessing the degree of similarity for pairs of aligned residues.

## 4.3 Substitution Matrices

### Substitution matrices are used to assign individual scores to aligned sequence positions

For alignments of protein sequences, the score is assigned to each aligned pair of amino acids is generally determined by reference to a **substitution matrix**, which defines values for all possible pairs of residues. Various types of substitution matrices have been used over the years. Some were based on theoretical considerations, such as the number of mutations that are needed to convert one amino acid into another, or similarities in physicochemical properties. The most successful, however, use actual evidence of what has happened during evolution, and are based on analysis of alignments of numerous homologs of well-studied proteins from many different species.

The choice of which substitution matrix to use is not trivial because there is no one correct scoring scheme for all circumstances. There is a wide range of variation in the similarity of sequences, from almost complete identity to a few percent. On one occasion we may need to align and score closely related sequences, whereas on another we may want to identify very distant relationships reliably. In the first case, the scoring scheme should be strongly biased toward giving high values to perfect matches and highly conserved substitutions. In the second case, a wider range of substitutions should be treated favorably.

Most scoring schemes for amino acid sequences use as reference a 20 × 20 substitution matrix, representing the 20 amino acids found in proteins. Each cell of the matrix is occupied by a score representing the likelihood that that particular pair of amino acids will occupy the same position through true homology, compared to the likelihood of their occurring as a random match. The most important scoring matrices will be described below, with general guidance as to which one to use when. A more comprehensive description of the theory underlying the scoring schemes discussed here is given in Section 5.1.

When an alignment is made, each aligned amino acid pair is given a score from the substitution matrix. These scores are then summed to give the overall score (*S*) of the alignment. For example, using the BLOSUM-62 matrix (see Figure 4.4A) we would score our example alignment as follows (in this case "U" represents an unknown residue; that is, a residue that could not be identified by sequencing techniques and is thus not given a score).

| | | | | | | | | | | | | |
|---|---|---|---|---|---|---|---|---|---|---|---|---|
| Seq1: | T | H | I | S | S | E | Q | U | E | N | C | E |
| Seq2: | T | H | A | T | S | E | Q | U | E | N | C | E |
| Score: | 5 | 8 | –1 | 1 | 4 | 5 | 5 | 0 | 5 | 6 | 9 | 5 |

Therefore the overall score *S* for this alignment equals 52. The BLOSUM matrices are described in more detail below.

## The PAM substitution matrices use substitution frequencies derived from sets of closely related protein sequences

A commonly used set of substitution matrices is based on the observed amino acid substitution frequencies in alignments of homologous protein sequences. These matrices were first developed by Margaret Dayhoff and her co-workers in the 1960s and 1970s, and have been found to be superior to substitution schemes that use only the physicochemical similarities of amino acids, as they use real data to model the evolutionary process. The sequences used to generate these matrices were all very similar, allowing the alignment to be made with confidence. In addition, the high similarity meant that there was a high probability that amino acid differences at an alignment position were due to just a single mutation event, over a short period of time, since it is unlikely that more than one mutation would occur at the same site. A phylogenetic tree (see Section 7.1) was constructed for the protein sequences, from which the individual mutations that had occurred could be deduced. From this tree, the researchers calculated the ratio of the number of changes undergone by each type of amino acid to the total number of occurrences of that amino acid in the sequence set.

From these ratios it was possible to calculate the probabilities that any one amino acid would mutate into any other over a given period of evolutionary time. The final matrix of substitution scores is a logarithmic matrix of the mutation probabilities. Probabilities are converted to logarithms so that the final alignment score can be calculated by summation of the individual scores from aligned pairs of amino acids, rather than by multiplication of probabilities.

(A)

| | C | S | T | P | A | G | N | D | E | Q | H | R | K | M | I | L | V | F | Y | W |
|---|---|---|---|---|---|---|---|---|---|---|---|---|---|---|---|---|---|---|---|---|
| C | 9 | | | | | | | | | | | | | | | | | | | |
| S | −1 | 4 | | | | | | | | | | | | | | | | | | |
| T | −1 | 1 | 5 | | | | | | | | | | | | | | | | | |
| P | −3 | −1 | −1 | 7 | | | | | | | | | | | | | | | | |
| A | 0 | 1 | 0 | −1 | 4 | | | | | | | | | | | | | | | |
| G | −3 | 0 | −2 | −2 | 0 | 6 | | | | | | | | | | | | | | |
| N | −3 | 1 | 0 | −2 | −2 | 0 | 6 | | | | | | | | | | | | | |
| D | −3 | 0 | −1 | −1 | −2 | −1 | 1 | 6 | | | | | | | | | | | | |
| E | −4 | 0 | −1 | −1 | −1 | −2 | 0 | 2 | 5 | | | | | | | | | | | |
| Q | −3 | 0 | −1 | −1 | −1 | −2 | 0 | 0 | 2 | 5 | | | | | | | | | | |
| H | −3 | −1 | −2 | −2 | −2 | −2 | 1 | −1 | 0 | 0 | 8 | | | | | | | | | |
| R | −3 | −1 | −1 | −2 | −1 | −2 | 0 | −2 | 0 | 1 | 0 | 5 | | | | | | | | |
| K | −3 | 0 | −1 | −1 | −1 | −2 | 0 | −1 | 1 | 1 | −1 | 2 | 5 | | | | | | | |
| M | −1 | −1 | −1 | −2 | −1 | −3 | −2 | −3 | −2 | 0 | −2 | −1 | −1 | 5 | | | | | | |
| I | −1 | −2 | −1 | −3 | −1 | −4 | −3 | −3 | −3 | −3 | −3 | −3 | −3 | 1 | 4 | | | | | |
| L | −1 | −2 | −1 | −3 | −1 | −4 | −3 | −4 | −3 | −2 | −3 | −2 | −2 | 2 | 2 | 4 | | | | |
| V | −1 | −2 | 0 | −2 | 0 | −3 | −3 | −3 | −2 | −2 | −3 | 3 | 2 | 1 | 3 | 1 | 4 | | | |
| F | −2 | −2 | −2 | −4 | −2 | −3 | −3 | −3 | −3 | −3 | −1 | −3 | −3 | 0 | 0 | 0 | −1 | 6 | | |
| Y | −2 | −2 | −2 | −3 | −2 | −3 | −2 | −3 | −2 | −1 | 2 | −2 | −2 | −1 | −1 | −1 | −1 | 3 | 7 | |
| W | −2 | −3 | −2 | −4 | −3 | −2 | −4 | −4 | −3 | −2 | −2 | −3 | −3 | −1 | −3 | −2 | −3 | 1 | 2 | 11 |

(B)

| | C | S | T | P | A | G | N | D | E | Q | H | R | K | M | I | L | V | F | Y | W |
|---|---|---|---|---|---|---|---|---|---|---|---|---|---|---|---|---|---|---|---|---|
| C | 9 | | | | | | | | | | | | | | | | | | | |
| S | −1 | 3 | | | | | | | | | | | | | | | | | | |
| T | −3 | 2 | 4 | | | | | | | | | | | | | | | | | |
| P | −3 | 1 | −1 | 6 | | | | | | | | | | | | | | | | |
| A | −3 | 1 | 1 | 1 | 3 | | | | | | | | | | | | | | | |
| G | −5 | 1 | −1 | −2 | 1 | 5 | | | | | | | | | | | | | | |
| N | −5 | 1 | 0 | −2 | 0 | 0 | 4 | | | | | | | | | | | | | |
| D | −7 | 0 | −1 | −2 | 0 | 0 | 2 | 5 | | | | | | | | | | | | |
| E | −7 | −1 | −2 | −1 | 0 | −1 | 1 | 3 | 5 | | | | | | | | | | | |
| Q | −7 | −2 | −2 | 0 | −1 | −3 | 0 | 1 | 2 | 6 | | | | | | | | | | |
| H | −4 | −2 | −3 | −1 | −3 | −4 | 2 | 0 | −1 | 3 | 7 | | | | | | | | | |
| R | −4 | −1 | −2 | −1 | −3 | −4 | −1 | −3 | −3 | 1 | 1 | 6 | | | | | | | | |
| K | −7 | −1 | −1 | −2 | −2 | −3 | 1 | −1 | −1 | 0 | −2 | 2 | 5 | | | | | | | |
| M | −6 | −2 | −1 | −3 | −2 | −4 | −3 | −4 | −4 | −1 | −4 | −1 | 0 | 8 | | | | | | |
| I | −3 | −2 | 0 | −3 | −1 | −4 | −2 | −3 | −3 | −3 | −4 | −2 | −2 | 1 | 6 | | | | | |
| L | −7 | −4 | −3 | −3 | −3 | −5 | −4 | −5 | −4 | −2 | −3 | −4 | −4 | 3 | 1 | 5 | | | | |
| V | −2 | −2 | 0 | −2 | 0 | −2 | −3 | −3 | −3 | −3 | −3 | −3 | −4 | 1 | 3 | 1 | 5 | | | |
| F | −6 | −3 | −4 | −5 | −4 | −5 | −4 | −7 | −6 | −6 | −2 | −4 | −6 | −1 | 0 | 0 | −3 | 8 | | |
| Y | −1 | −3 | −3 | −6 | −4 | −6 | −2 | −5 | −4 | −5 | −1 | −6 | −6 | −4 | −2 | −3 | −3 | 4 | 8 | |
| W | −8 | −2 | −6 | −7 | −7 | −8 | −5 | −8 | −8 | −6 | −5 | 1 | −5 | −7 | −7 | −5 | −8 | −1 | −1 | 12 |

**Figure 4.4**
**Amino acid substitution scoring matrices.** (A) The BLOSUM-62 matrix and (B) the PAM120 substitution matrix. Each cell represents the score given to a residue paired with another residue (row × column). The values are given in half-bits, as discussed in Section 5.1. The colored shading indicates different physicochemical properties of the residues (see Figure 2.3): small and polar, yellow; small and nonpolar, white; polar or acidic, red; basic, blue; large and hydrophobic, green; aromatic, orange.

There is more than one such matrix and each matrix corresponds to a particular quantity of **accepted mutations** — mutations that have been retained in the sequence. This quantity is measured in PAM units, where PAM stands for Point Accepted Mutations (accepted point mutations per 100 residues), and these matrices are generally called **PAM matrices**. One of the more frequently used substitution matrices corresponds to 250 PAM, which means that 250 mutations have been fixed on average per 100 residues; that is, many residues have been subject to more than one mutation. The matrix itself is called PAM250. This amount of change is near the limit of detection of distant relationships. Other matrices, such as PAM120, correspond to a smaller amount of mutation (see Figure 4.4B)

The currently used PAM matrices, also known as Dayhoff mutation data matrices (MDMs), were originally created in 1978. More recent matrices have also been constructed using newer and larger data sets. The PET91 matrix, for example, represents a new generation of Dayhoff-type matrices.

## The BLOSUM substitution matrices use mutation data from highly conserved local regions of sequence

The **BLOSUM matrix** is another very commonly used amino acid substitution matrix that depends on data from actual substitutions. It was derived much more recently than the Dayhoff matrices, in the early 1990s, using local multiple alignments rather than global alignments. First, a large set of aligned highly conserved short regions was generated from analysis of the protein-sequence database SWISS-PROT. The sequences were then clustered into groups according to similarity, so that sequences were grouped together if they exceeded a specified threshold for percentage identity. Substitution frequencies for all possible pairs of amino acids were then calculated between the clustered groups (without the construction of phylogenetic trees) and used to compute BLOSUM (BLOck SUbstitution Matrix) scores. Various BLOSUM matrices are obtained by varying the percentage cut-off for clustering into similarity groups. For example, the commonly used BLOSUM-62 matrix was derived using a threshold of 62% identity (see Figure 4.4).

## The choice of substitution matrix depends on the problem to be solved

With many scoring matrices available, it is hard to know which one to use. Within a group of matrices such as the PAM or BLOSUM series, different ones, for example PAM250 versus PAM120 or BLOSUM-50 versus BLOSUM-80, are more suitable for different types of problem. The PAM matrix number indicates evolutionary distance whereas the BLOSUM matrix number refers to percentage identity. When aligning sequences that are anticipated to be very distantly related, matrices such as PAM250 and BLOSUM-50 may therefore be preferable. PAM120 and BLOSUM-80 may perform better for more closely related sequences.

Some matrices have been derived using additional information; the STR matrix, for example, includes information from known protein structures. Because protein structure is more conserved than sequence, more distantly related proteins can be compared using such methods, even when sequence alignment alone would not pick up any significant relationship.

Some scoring matrices have been designed to work well in special situations. For example, the matrices SLIM (ScoreMatrix Leading to Intra-Membrane) and PHAT (Predicted Hydrophobic And Transmembrane matrix) are especially designed for membrane proteins, where the characteristic amino acid composition and the selective forces for acceptable mutations are different from those for soluble proteins. In 2006, there were 94 matrices collected in a database list called AAINDEX and searchable at GenomeNet.

As well as the degree of evolutionary distance, the length of the sequences to be aligned must be taken into account when choosing a suitable matrix. This is especially relevant when searching databases against a query sequence, as the length of the sequence is taken into account when assessing the significance of the score: the shorter the sequence, the higher the score needs to be in order to be judged significant. Short sequences need to use matrices designed for short evolutionary time scales, such as PAM40 or BLOSUM-80. Longer sequences of 100 residues or more can use matrices intended for use with longer evolutionary time scales (such as PAM250 and BLOSUM-50). The reasons why the significance of a score depends on the length of the sequences to be aligned are discussed in more detail in Section 5.4.

## 4.4 Inserting Gaps

### Gaps inserted in a sequence to maximize similarity with another require a scoring penalty

Homologous sequences are often of different lengths as the result of insertions and deletions (**indels**) that have occurred in the sequences as they diverged from the ancestral sequence. Their alignment is generally dealt with by inserting **gaps** in the sequences to achieve as correct a match as possible. To signify that an insertion or deletion has occurred, a letter or stretch of letters in one sequence is paired up with blank spaces (usually indicated by hyphens) inserted into the other sequence to achieve a better match.

Gaps must be introduced judiciously: forcing two sequences to match up simply by inserting large numbers of gaps will not reflect reality and will produce a meaningless alignment. To place limits on the introduction of gaps, alignment programs use a **gap penalty**: each time a gap is introduced, the penalty is subtracted from the score, decreasing the overall score of the alignment. Structural analysis has shown that fewer insertions and deletions occur in sequences of structural importance, and that insertions tend to be several residues long rather than just a single residue long. This information can be included in the scoring scheme by placing a smaller penalty on lengthening an existing gap (**gap extension penalty**) than on introducing a new gap, thus penalizing single-residue gaps relatively more. The best alignment is thus the one that returns the maximum score for the smallest number of introduced gaps.

Gap penalties can usually be varied in an alignment program, so the user has to decide what gap penalty to use. It should be kept in mind that the insertion of a gap must improve the quality of the alignment and therefore the maximum-match value. If a gap penalty is set high, then fewer gaps will be inserted into the alignment, as their inclusion will radically decrease the maximum-match value. If a low gap penalty is chosen, then more and larger gaps will be inserted. Therefore, if you are searching for sequences that are a strict match for your query sequence, the gap penalty should be set high. This will often retrieve a region, or regions, of very closely related sequence. If you are searching for similarity between distantly related sequences, the gap penalty should be set low. Note that suitable gap-penalty values may be different with different substitution matrices. It is advisable to start, when possible, with a combination of matrix and gap penalties that have been reported to give optimal performance.

In some alignment programs, a gap score depends on the type of residue with which the gap is aligned. Some types of residues are more likely to be conserved than others because their side chains tend to be more important in determining structure or function. An example is tryptophan, and so a gap aligned with a tryptophan will exact a larger gap penalty than a gap aligned with a glycine, for example.

(A)

```
Bovine PI-3Kinase p110a           LNWENPDIMSELLFQNNEIIFKNGDDLRQDMLTLQIIRIMENIWQNQGLDLRMLPYGCLSIGDCVGLIEVVRNSHTIMQIQCKGGLKGAL
cAMP-dependent protein kinase     --WENPAQNTAHLDQFERIKTLGTGSFGRVMLVKHMETGNHYAMKILDKQKVVKLKQIEHTLNEKRILQAVNFPFLVKLEFSFKDNSNLY

Bovine PI-3Kinase p110a           QFNSHTLHQWLKDKNKGEIYDAAIDLFTRSCAGYCVATFILGIGDRHNSNIMVKDDGQLFHIDFGHFLDHKKKFGYKRERVPFVLTQDF
cAMP-dependent protein kinase     MVMEYVPGGEMFSHLRRIGRFSEPHARFYAAQIVLTFEYLHSLDLIYRDLKPENLLIDQQGYIQVTDFGFAKRVKGRTWXLCGTPEYLAP

Bovine PI-3Kinase p110a           LIVISKGAQECTKTREFERFQEMCYKAYLAIRQHANLFINLFSMMLGSGMPELQSFDDIAYIRKTLALDKTEQEALEYFMKQMNDAHHGG
cAMP-dependent protein kinase     EIILSKGYNKAVDWWALGVLIYEMAAGYPPFFADQPIQIYEKIVSGKVRFPSHFSSDLKDLLRNLLQVDLTKRFGNLKNGVNDIKNHKWF

Bovine PI-3Kinase p110a           WTTKMDWIFHTIKQHALN-----------------------------------
cAMP-dependent protein kinase     ATTDWIAIYQRKVEAPFIPKFKGPGDTSNFDDYEEEEIRVXINEKCGKEFSEF
```

(B)

```
Bovine PI-3Kinase p110a           LNWENPDIMSELLFQNNEIIFKNGDDLRQDMLTLQIIRIMENIWQNQGLDLRMLPYGCLSIGDCVGLIEVVRNSHTIMQIQCKGGLKGAL
cAMP-dependent protein kinase     ?-WENPAQNTAHLDQFERIKTLGTGSFGRVMLVKHM--ETGNHYAMKILDKQKV-VKLKQIEHTLNEKRILQAVNFPFLVKLEFSFKDN-

Bovine PI-3Kinase p110a           QFNSHTLHQWLKDKNKGEIYDAAIDLFTRSCAGYCVATFILGIGDRHNSNIMVKD-DGQLFHIDFGHFLDHKKKKFGYKRERVPFVL--T
cAMP-dependent protein kinase     -SNLYMVMEYVPGGEMFSHLRR-IGRFSEPHARFYAAQIVLTFEYLHSLDLIYRDLKPENLLIDQQGYIQVTDFGFAKRVKGRTWXLCGT

Bovine PI-3Kinase p110a           QDFL---IVISKGAQECTKTREFERF-QEMC--YKAYLAIRQHANLFINLFSMMLGSGMPELQSFDDIAYIRKTLALDKTEQEALEYFMK
cAMP-dependent protein kinase     PEYLAPEIILSKGYNKAVDWWALGVLIYEMAAGYPPFFA-DQPIQIYEKIVSGKVRF--PSHFSSDLKDLLRNLLQVDLTKR--FGNLKN

Bovine PI-3Kinase p110a           QMNDAHHGGWTTKMDWI-----------------------FHTIKQHAL----N----------
cAMP-dependent protein kinase     GVNDIKNHKWFATTDWIAIYQRKVEAPFIPKFKGPGDTSNFDDYEEEEIRVXINEKCGKEFSEF
```

**Figure 4.5**

**Pairwise alignments of the PI3-kinase p110α and a cAMP-dependent protein kinase.** Note that the protein kinase sequence is considerably longer than the p110α sequence. (A) An alignment where the gap penalty has been set very high. Gaps have therefore only been inserted at the beginning and end of the sequences. The percentage identity of this alignment is 10%. (B) An alignment with a very low gap penalty. Many more gaps have been inserted to maximize the number of matched residues. Especially apparent is the lone matched pair of asparagine (N) residues in the carboxy-terminal region. The percentage identity of this alignment is 18%. Green shading, identical amino acids.

It is best to start with the default values given by the program you are using and then raise or lower the penalty to obtain a desired alignment. However, the number of gaps should always be kept to the minimum possible. Figure 4.5 shows two pairwise alignments of a **phosphatidylinositol-3-OH** kinase sequence (from bovine PI3-kinase p110α) and a **protein kinase** sequence from a cyclic AMP (cAMP)-dependent protein kinase (see Box 4.5), which have only limited similarity to each other.

In the first alignment (see Figure 4.5A) the gap penalty was set very high; therefore the program inserts as few gaps as possible. Any inserted gaps are found at the ends of the sequence, as often, unless there is an obvious relationship between the terminal amino acids, end gaps are not penalized. In the second alignment (see Figure 4.5B) the gap penalty was set very low; the effect is that many more gaps are inserted and the number of matched amino acids is increased (identities are shown in green). Although there are more matched residues in the alignment with low gap penalties, this does not necessarily mean that it is more accurate. In sequences that share such low homology as these, expert knowledge, such as the location of active-site residues, has to be used to decide if the alignment is accurate.

## Dynamic programming algorithms can determine the optimal introduction of gaps

In practice, it is nearly always necessary to insert gaps into sequences when aligning them. The most obvious way of finding the best alignment with gaps would be to generate all possible gapped alignments, find the score for each, and select the highest-scoring alignment. This would be enormously time consuming,

## Box 4.5 Protein kinases and phospholipid kinases

**Phosphorylation** is one of the commonest ways of rapidly altering a protein's activity. The enzymes that phosphorylate proteins are known as protein kinases and add phosphate groups to specific amino acid residues in the protein. Most, such as the cAMP-dependent protein kinases, phosphorylate serine or threonine residues, whereas others phosphorylate tyrosine residues. The effect of protein phosphorylation can be reversed by phosphoprotein phosphatases, which specifically remove the phosphate group. Because of their important roles as regulators of cellular activity and behavior, the activity of protein kinases is, in general, tightly controlled. The cAMP-dependent protein kinases, for example, are activated by binding the intracellular second messenger cAMP, which is specifically generated in response to a variety of extracellular signals acting at cell-surface receptors.

The phosphatidylinositol-3-OH kinases (PI3-kinases) phosphorylate inositol phospholipids in the cytoplasmic surface of the cell membrane, adding a phosphate group to position 3 on the inositol ring. Other members of this family, the PI4-kinases, phosphorylate the inositol ring on position 4. The phosphorylated lipids then specifically bind and activate other proteins, such as protein kinases, to initiate intracellular signal transduction cascades. PI3-kinases are involved in initiating the pathway by which the hormone insulin controls carbohydrate metabolism. PI3-kinases and protein kinases have very little sequence similarity to each other except in the enzymatic kinase domain.

however. For example, approximately $10^{75}$ alignments would need to be generated for a sequence of only 100 residues. It only became practicable to incorporate gaps into an alignment with the development of **dynamic programming algorithms**. These avoid unnecessary exploration of the bulk of alignments that can be shown to be nonoptimal. The name "dynamic programming" reflects the fact that the precise behavior of the algorithm is established only when it runs (in other words, dynamically) because it depends on the sequences being aligned.

The first algorithm to use dynamic programming for sequence comparison was that of S. B. Needleman and C. D. Wunsch, published in 1970. Their technique is still the core of many present-day alignment and sequence-searching methods. In their method, gaps, regardless of length, have an associated penalty score; newer methods use more complicated gap penalties. The actual values of the gap scores can be varied depending on the type of scoring matrix being used. One rule always followed is that gaps can never be aligned with each other.

The basic concept of a Needleman–Wunsch-type algorithm is that comparisons are made on the basis of all possible pairs of amino acids that could be made between the two sequences. All possible pairs are represented as a two-dimensional matrix, in which one of the sequences to be aligned runs down the vertical axis and the other along the horizontal axis. All possible comparisons between any number of pairs are given by pathways through the array, each of which can be scored. The principles and method of the algorithm are dealt with in detail in Section 5.2. The general idea is to grow the alignment from the amino or carboxy terminus, at each step rejecting all possible alignments except that with the best score.

# 4.5 Types of Alignment

## Different kinds of alignments are useful in different circumstances

The general principles outlined in the previous sections can be used to make different types of alignment (see Flow Diagram 4.2). Two closely related homologous sequences will generally be of approximately the same length, so that their alignment

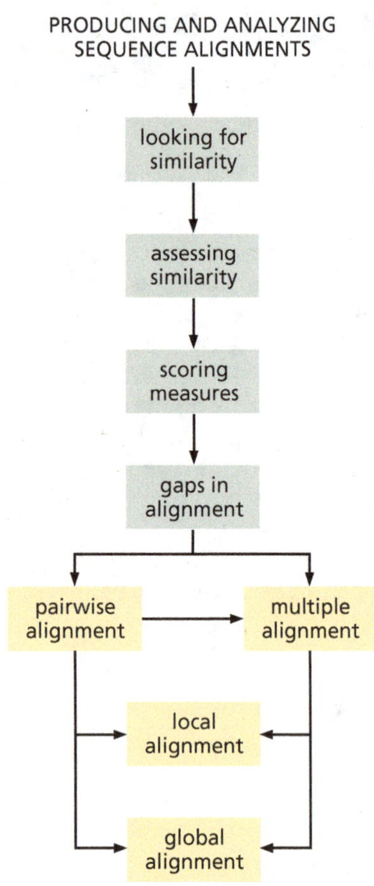


**Flow Diagram 4.2**
**The key concept introduced in this section is that there are several different types of sequence alignment, one of which will be the most appropriate for a particular problem.**

will cover the full range of each sequence. This is referred to as a **global alignment**, and is generally the appropriate one to use when you want to compare or find closely related sequences that are similar over their whole length.

On the other hand, there are many cases where only parts of sequences are related. A simple example is the amino acid sequences of two proteins each consisting of two domains, with only one domain common to both proteins and the other domains completely unrelated. In this case, the only meaningful alignment will be a **local alignment** of the shared domain. Looking only at global alignments may not reveal the limited but important similarity between the sequences. This is particularly the case for comparisons between multidomain proteins, such as PI3-kinases, which consist of a number of small protein domains strung together (see Figure 4.6). Local alignment programs are therefore useful for detecting shared domains in such proteins.

When searching through a sequence database with a query sequence from an unknown protein, local alignment is a very useful tool to use initially. Once sequences with regions of high similarity are found using local alignment, global alignment can be used to align the rest of the sequence that is not so similar. Local alignment is also a good tool for identifying particular functional sites from which sequence patterns and motifs can be derived.

A widely used local alignment algorithm is the Smith–Waterman algorithm, which is a modification of the Needleman–Wunsch algorithm. Instead of looking at each sequence in its entirety, which is what the Needleman–Wunsch algorithm does, the Smith–Waterman method compares segments of all possible lengths and chooses the segment that optimizes the similarity measure. The scoring matrix used must include both positive and negative scores, and only alignments with a positive total score are considered. Therefore, if on extending the alignment at a particular step none of the possible alignments has a positive score, all previous alignments are

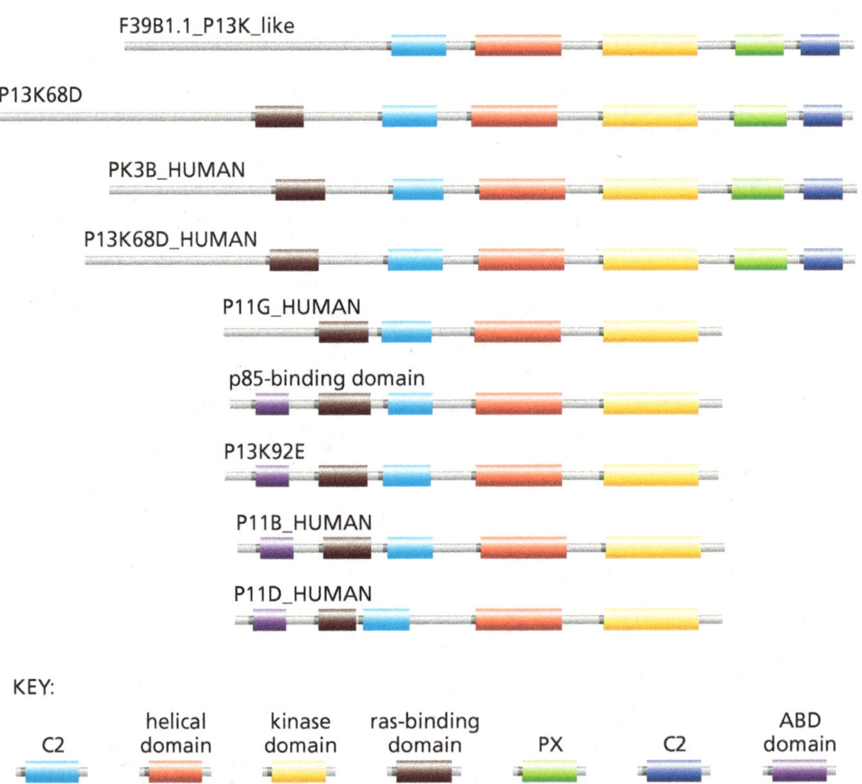


**Figure 4.6**
**PI3-kinase is a multidomain protein.** One possible output from a search of the Pfam database with the p100α PI3-kinase catalytic domain (yellow bar) is shown here. The figure also shows the complete domain structure of the protein family comprising the PI3-kinases and the related PI4-kinases, which catalyze phosphorylation of position 4 of the inositol ring of inositol phospholipids. The other domains and their arrangement are represented by the other colored bars.

rejected, and new ones are considered starting from that point. This makes the calculation sensitive to the precise match and mismatch scores and gap penalties. Section 5.2 describes the algorithm in detail.

Figure 4.7 shows an example of local versus global alignment of the complete protein sequences of the bovine PI3-kinase p110α and the cAMP-dependent protein kinase shown in Figure 4.5, using the Web-based programs ALIGN (global) and LALIGN (local). Although these proteins share structural homology within the core kinase catalytic domain, there is very little sequence homology. Figure 4.7A shows that local alignment of the catalytic domains has identified one important conserved region, out of five regions that were aligned. This region is involved in catalysis and also contains the three-residue motif DFG, which is conserved between many kinases. Figure 4.7B shows that, in this case, a global alignment fails to identify this region. The percentage sequence identity for these two sequences is very low (17.8%), well into the midnight zone of sequence alignment.

For both global and local alignments, methods exist for making **pairwise alignments**, that is, the alignment of just two sequences, and for making **multiple alignments**, in which more than two sequences are aligned with each other. In this part of the chapter, we have mainly used examples of pairwise alignments to illustrate the general principles of alignment scoring and quality assessment. Multiple alignment introduces yet another dimension to the computational problems of alignment. The theory is dealt with in detail in Chapter 6, but a few general points are described here.

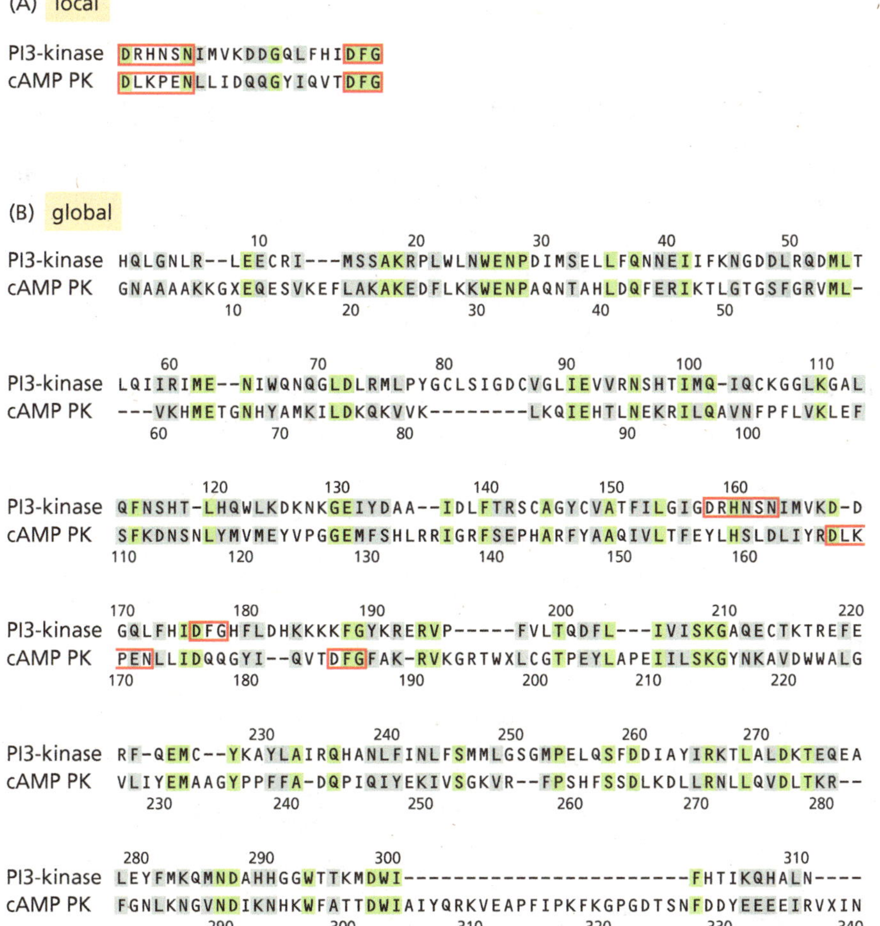


**Figure 4.7**
**Local and global alignments.** The complete sequences of PI3-kinase p110α and the cAMP-dependent protein kinase (cAMP PK) shown in Figure 4.5 were compared. (A) Local alignment using the program LALIGN (a subset of the FASTA package) has matched a short conserved region in the kinase domains that contains the functionally important residues D and N in the DLKPEN sequence and the DFG repeat common to nearly all kinases. (B) Because of the low overall sequence similarity, a standard global alignment of these two sequences using the program ClustalW has not matched these functionally important residues (boxed in each sequence). Green shading, identical amino acids; gray shading, similar amino acids.

## Multiple sequence alignments enable the simultaneous comparison of a set of similar sequences

Multiple alignments can be used to find interesting patterns characteristic of specific protein families, to build phylogenetic trees, to detect homology between new sequences and existing families, and to help predict the secondary and tertiary structures of new sequences, as we shall see in more detail in Chapters 11 to 14.

In general, the alignment of multiple sequences will give a more reliable assessment of similarity than a pairwise alignment. The reason for this is that ambiguities in a pairwise comparison can often be resolved when further sequences are compared. Multiple alignment provides more information than pairwise alignment on the individual amino acid positions, such as the overall similarity and evolutionary relationships. This is especially important when using sequence-comparison methods to construct taxonomic phylogenetic trees. Multiple alignment is especially useful for illustrating sequence conservation throughout the aligned sequences. Such conservation over many sequences can identify amino acids that are important for function or for the structural integrity of the protein fold.

## Multiple alignments can be constructed by several different techniques

A number of methods are available for generating multiple alignments. One of these is an extension of the dynamic programming method, so that instead of a two-dimensional matrix for a pair of sequences, an alignment of $n$ protein sequences uses an $n$-dimensional matrix. However, this is limited by the prohibitively large computational requirement of the algorithm, and none of the examples discussed below uses this technique.

Other methods, while often using dynamic programming to align pairs of sequences, use other techniques to combine these together into one multiple alignment. Tree or hierarchical methods of multiple alignment are widely used, for example in the multiple alignment program ClustalW. This method first compares all the sequences in a pairwise fashion, then performs a **cluster analysis** on the pairwise data to generate a hierarchy of sequences in order of their similarity (see Figure 4.8A). The hierarchy is a simple phylogenetic tree and is often referred to as the **guide tree**. A multiple alignment is then built based on the guide tree by first aligning the most similar pairs, then aligning the other sequences with these pairs until all the sequences have been aligned (Figure 4.8B). However accurate this method is, there are problems with it in that any errors in the initial alignments cannot be corrected later as new information from other sequences is added. This

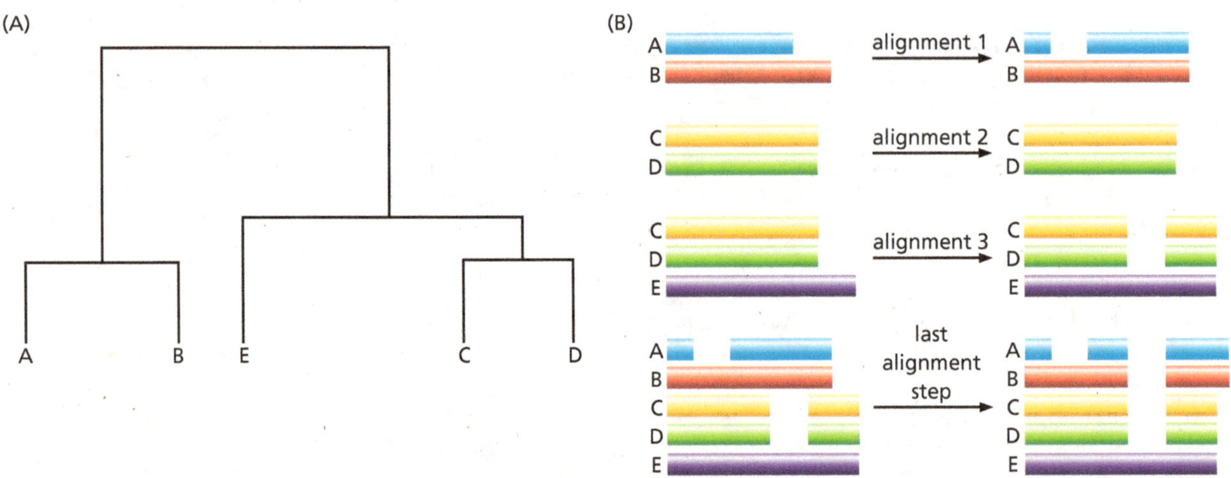


**Figure 4.8**
**The tree method for the multiple alignment of sequences A, B, C, D, and E.** Pairwise alignments are first made between all possible pairs of sequences—that is, AB, AC, AD, and so on—to determine their relative similarity to each other (not shown). (A) A cluster analysis is performed on this preliminary round of alignments, and the individual sequences are ranked in a tree according to their similarity to each other. (B) In the next step, the most similar sequences are aligned in pairs as far as possible. These are then aligned to the next closest sequence. This is repeated until all sequences or groups of sequences are aligned.

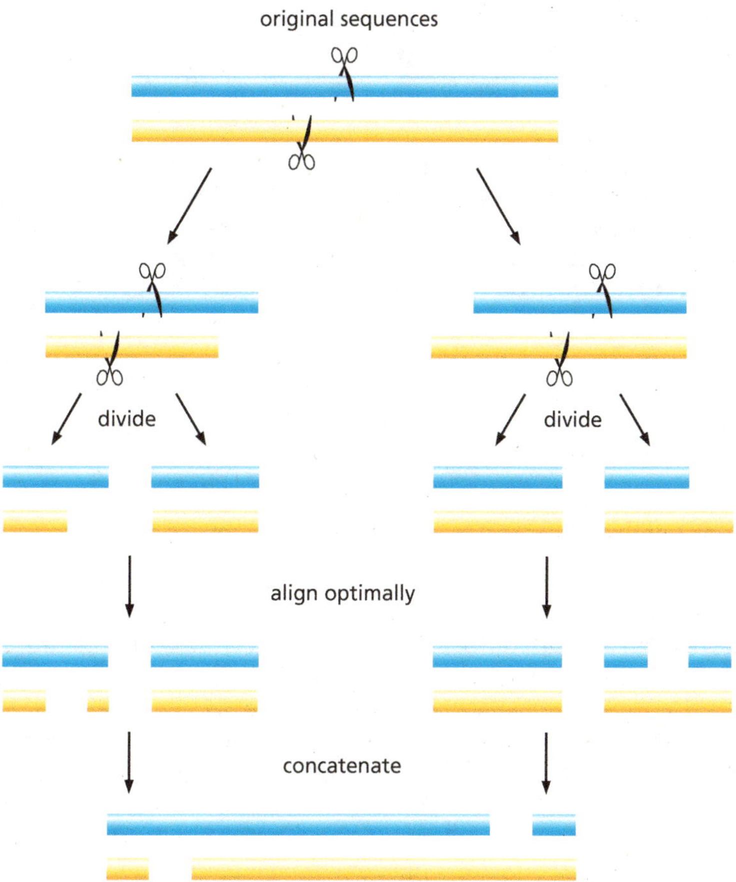


**Figure 4.9**
**The divide-and-conquer method of multiple alignment.** The sequences to be aligned are divided into two regions, then into four, and so on until the segments are considered small enough for accurate optimal alignment. The segments are then aligned and in the last step the alignments are concatenated to form the final complete multiple alignment.

difficulty has been avoided in iterative or stochastic sampling procedures as in the Barton and Sternberg program (see Chapter 6).

Other methods for building multiple alignments include the segment method, the consensus method, and the **divide-and-conquer method**. In the divide-and-conquer alignment, the sequences are first cut several times to reduce the length of the sequences to be aligned, the cut sequences are then aligned, and they are finally concatenated into a multiple alignment (see Figure 4.9). Initially, each sequence is divided into two segments at a suitable cut-position somewhere close to the midpoint of the sequences. This procedure is repeated until the sequences are shorter than a predetermined size, which is set as a parameter of the divide-and-conquer algorithm. Therefore the problem of aligning one family of long sequences is divided into several smaller alignment tasks. The segments are then aligned. The last step concatenates the short alignments, giving a multiple alignment of the original sequences.

## Multiple alignments can improve the accuracy of alignment for sequences of low similarity

The same proteins with which we illustrated local versus global alignment—a cAMP-dependent protein kinase and a PI3-kinase—will be used to illustrate the improvement multiple alignment can make to the alignment of sequences of low similarity. Figure 4.10A shows part of a pairwise alignment between the protein kinase and the PI3-kinase. The active-site region and the DFG pattern are not aligned. Figure 4.10B shows the result of a multiple alignment between five different PI3-kinases and the protein kinase made using the program ClustalW with the default settings. The effect of the multiple alignment is to give added weight to

```
(A) p110α        TFILGIGDRHNSNIMVKDDG-QLFHIDFGHFLDHKKKKFGYKRERVPFVLT--QDFLIVI 142
    cAMP-kinase  QIVLTFEYLHSLDLIYRDLKPENLLIDQQGYIQVTDFGFAKRVKGRTWXLCGTPEYLAPE 179

(B) p110β        SYVLGIG----------DRHSDNINVKKTGQLFHIDFGHILGNFKSKFGIKRERVPFILT 136
    p110δ        TYVLGIG----------DRHSDNIMIRESGQLFHIDFGHFLGNFKTKFGINRERVPFILT 136
    p110α        TFILGIG----------DRHNSNIMVKDDGQLFHIDFGHFLDHKKKKFGYKRERVPFVLT 135
    p110γ        TFVLGIG----------DRHNDNIMITETGNLFHIDFGHILGNYKSFLGINKERVPFVLT 135
    p110_dicti   TYVLGIG----------DRHNDNLMVTKGGRLFHIDFGHFLGNYKKKFGFKRERAPFVFT 135
    cAMP-kinase  QIVLTFEYLHSLDLIYRDLKPENLLIDQQGYIQVTDFGFAKRVKGRTWXLCG--TPEYLA 177
```

**Figure 4.10**
**Pairwise and multiple alignments of part of the catalytic domains of five PI3-kinases and a cAMP-dependent protein kinase.** (A) Pairwise alignment of PI3-kinase p110α and the protein kinase does not align the important active-site residues and the DFG motif (in green). (B) Multiple alignment of the protein kinase with a set of five PI3-kinases (which have considerable overall homology to each other) has the effect of forcing the best-conserved regions to be matched. Here the DFG motif and the important N and D (green) residues are aligned correctly in all the sequences. In addition it is apparent that a G (green) is also totally conserved (identical) and that three more residues are conserved in their physicochemical properties (blue).

the conserved residues within the PI3-kinases, resulting in a better alignment for that region of the kinase domain.

## ClustalW can make global multiple alignments of both DNA and protein sequences

ClustalW uses a tree method of multiple alignment as described briefly above. The program is easy to use with the default settings and can be accessed from a number of Web sites. To use it, one must have collected a set of sequences, perhaps from a database search. Either protein or DNA sequences can be used. The sequences are cut and pasted into a dialog box; you can then run the program immediately with the default settings (for gap penalties and type of scoring matrix, for example). All the settings can be changed if required.

## Multiple alignments can be made by combining a series of local alignments

DIALIGN is a relatively recent method for multiple alignment developed by Burkhard Morgenstern and colleagues. Whereas standard alignment programs such as ClustalW compare residues one pair at a time and impose gap penalties, DIALIGN constructs pairwise and multiple alignments by comparing whole ungapped segments several residues long. The alignment is then constructed from pairs of equal-length gap-free segments, which are termed diagonals because they would show up as diagonal lines in the respective pairwise comparison matrices. The segment length varies between diagonals. Many diagonals overlap, and the program has to find a set that can be combined into one consistent alignment (see Section 6.5). As the segments are gap-free there is no need to use a gap-penalty parameter. Every diagonal is given a weight reflecting the degree of similarity between the two segments involved. The overall score of an alignment is the sum of the weights of all the diagonals, and the program finds the alignment with the maximum score. A threshold can be set so that diagonals are considered only if their weights exceed this threshold, so that regions of lower similarity are ignored. As DIALIGN is a local alignment method it may not align the whole sequence, and may align several blocks of residues with unaligned regions between them.

Figure 4.11 illustrates the alignment of five SH2 domain sequences using ClustalW, DIALIGN, and the divide-and-conquer algorithm (DCA) methods compared with the structural/functional alignment from BAliBase, which can be considered accurate. All three methods fail to some extent to align the residues of the first helix correctly, inserting a gap. ClustalW does slightly worse in this region by splitting the helix, but is better in conserving the integrity of the second core block around the FLVR region important for binding. DCA does not align the last helix as well as ClustalW or DIALIGN. However, all the alignment programs are generally good and useful in that they often produce alignments very close to the correct ones based on extra information, such as those found in BALiBase.

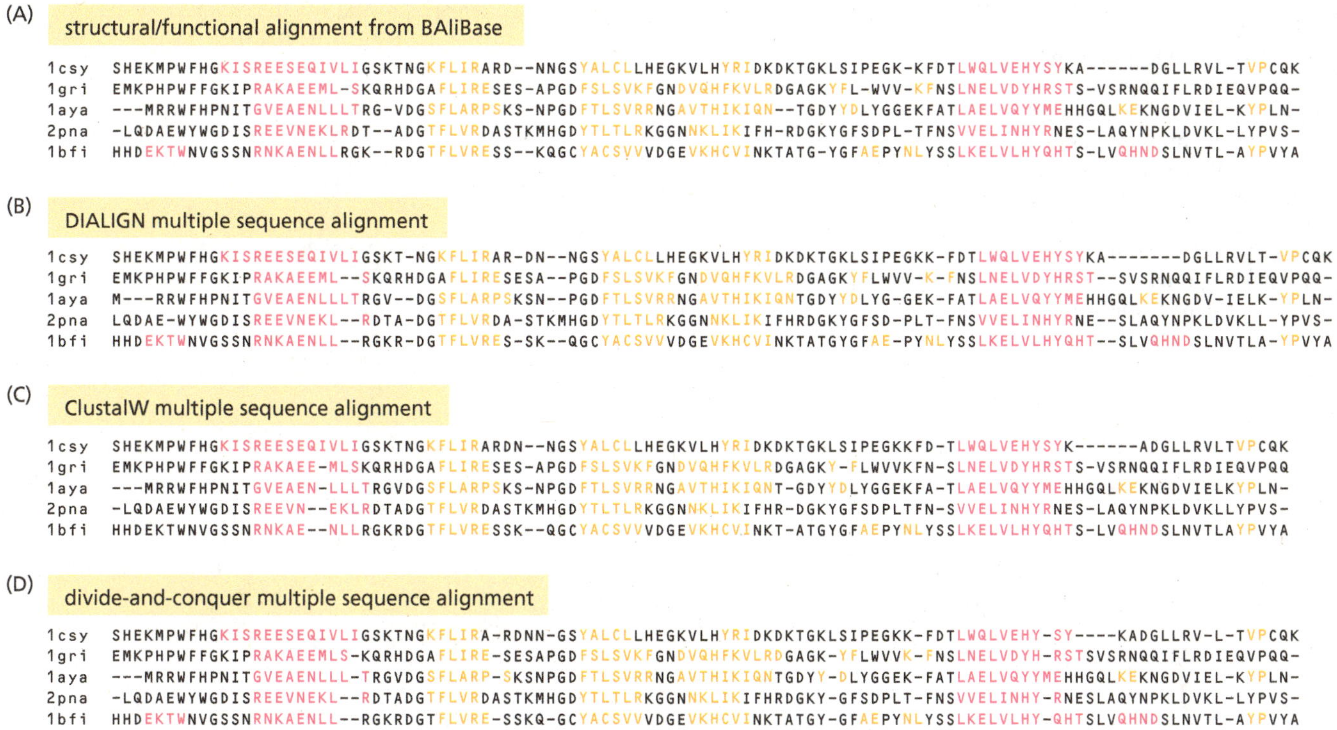

(A) structural/functional alignment from BAliBase

```
1csy  SHEKMPWFHGKISREESEQIVLIGSKTNGKFLIRARD--NNGSYALCLLHEGKVLHYRIDKDKTGKLSIPEGK-KFDTLWQLVEHYSYKA------DGLLRVL-TVPCQK
1gri  EMKPHPWFFGKIPRAKAEEML-SKQRHDGAFLIRESES-APGDFSLSVKFGNDVQHFKVLRDGAGKYFL-WVV-KFNSLNELVDYHRSTS-VSRNQQIFLRDIEQVPQQ-
1aya  ---MRRWFHPNITGVEAENLLLTRG-VDGSFLARPSKS-NPGDFTLSVRRNGAVTHIKIQN--TGDYYDLYGGEKFATLAELVQYYMEHHGQLKEKNGDVIEL-KYPLN-
2pna  -LQDAEWYWGDISREEVNEKLRDT--ADGTFLVRDASTKMHGDYTLTLRKGGNNKLIKIFH-RDGKYGFSDPL-TFNSVVELINHYRNES-LAQYNPKLDVKL-LYPVS-
1bfi  HHDEKTWNVGSSNRNKAENLLRGK--RDGTFLVRESS--KQGCYACSVVVDGEVKHCVINKTATG-YGFAEPYNLYSSLKELVLHYQHTS-LVQHNDSLNVTL-AYPVYA
```

(B) DIALIGN multiple sequence alignment

```
1csy  SHEKMPWFHGKISREESEQIVLIGSKT-NGKFLIRAR-DN--NGSYALCLLHEGKVLHYRIDKDKTGKLSIPEGKK-FDTLWQLVEHYSYKA-------DGLLRVLT-VPCQK
1gri  EMKPHPWFFGKIPRAKAEEML--SKQRHDGAFLIRESESA--PGDFSLSVKFGNDVQHFKVLRDGAGKYFLWVV-K-FNSLNELVDYHRST--SVSRNQQIFLRDIEQVPQQ-
1aya  M---RRWFHPNITGVEAENLLLTRGV--DGSFLARPSKSN--PGDFTLSVRRNGAVTHIKIQNTGDYYDLYG-GEK-FATLAELVQYYMEHHGQLKEKNGDV-IELK-YPLN-
2pna  LQDAE-WYWGDISREEVNEKL--RDTA-DGTFLVRDA-STKMHGDYTLTLRKGGNNKLIKIFHRDGKYGFSD-PLT-FNSVVELINHYRNE--SLAQYNPKLDVKLL-YPVS-
1bfi  HHDEKTWNVGSSNRNKAENLL--RGKR-DGTFLVRES-SK--QGCYACSVVVDGEVKHCVINKTATGYGFAE-PYNLYSSLKELVLHYQHT--SLVQHNDSLNVTLA-YPVYA
```

(C) ClustalW multiple sequence alignment

```
1csy  SHEKMPWFHGKISREESEQIVLIGSKTNGKFLIRARDN--NGSYALCLLHEGKVLHYRIDKDKTGKLSIPEGKKFD-TLWQLVEHYSYK------ADGLLRVLTVPCQK
1gri  EMKPHPWFFGKIPRAKAEE-MLSKQRHDGAFLIRESES-APGDFSLSVKFGNDVQHFKVLRDGAGKY-FLWVVKFN-SLNELVDYHRSTS-VSRNQQIFLRDIEQVPQQ
1aya  ---MRRWFHPNITGVEAEN-LLLTRGVDGSFLARPSKS-NPGDFTLSVRRNGAVTHIKIQNT-GDYYDLYGGEKFA-TLAELVQYYMEHHGQLKEKNGDVIELKYPLN-
2pna  -LQDAEWYWGDISREEVN--EKLRDTADGTFLVRDASTKMHGDYTLTLRKGGNNKLIKIFHR-DGKYGFSDPLTFN-SVVELINHYRNES-LAQYNPKLDVKLLYPVS-
1bfi  HHDEKTWNVGSSNRNKAE--NLLRGKRDGTFLVRESSK--QGCYACSVVVDGEVKHCVINKT-ATGYGFAEPYNLYSSLKELVLHYQHTS-LVQHNDSLNVTLAYPVYA
```

(D) divide-and-conquer multiple sequence alignment

```
1csy  SHEKMPWFHGKISREESEQIVLIGSKTNGKFLIRA-RDNN-GSYALCLLHEGKVLHYRIDKDKTGKLSIPEGKK-FDTLWQLVEHY-SY----KADGLLRV-L-TVPCQK
1gri  EMKPHPWFFGKIPRAKAEEMLS-KQRHDGAFLIRE-SESAPGDFSLSVKFGNDVQHFKVLRDGAGK-YFLWVVK-FNSLNELVDYH-RSTSVSRNQQIFLRDIEQVPQQ-
1aya  ---MRRWFHPNITGVEAENLLL-TRGVDGSFLARP-SKSNPGDFTLSVRRNGAVTHIKIQNTGDYY-DLYGGEK-FATLAELVQYYMEHHGQLKEKNGDVIEL-KYPLN-
2pna  -LQDAEWYWGDISREEVNEKL--RDTADGTFLVRDASTKMHGDYTLTLRKGGNNKLIKIFHRDGKY-GFSDPLT-FNSVVELINHY-RNESLAQYNPKLDVKL-LYPVS-
1bfi  HHDEKTWNVGSSNRNKAENLL--RGKRDGTFLVRE-SSKQ-GCYACSVVVDGEVKHCVINKTATGY-GFAEPYNLYSSLKELVLHY-QHTSLVQHNDSLNVTL-AYPVYA
```

**Figure 4.11**
**Known structural alignments can be useful in checking sequence alignments.** (A) Multiple alignment of the sequences of five SH2 domains according to their sequence/structure alignment in BAliBase. α-Helices are shown in red and β-strands in yellow. (B) Multiple alignment for the same set of sequences obtained by DIALIGN. (C) Alignment obtained by ClustalW. (D) Alignment obtained by the divide-and-conquer method. There is not much difference in performance between the algorithms (all were run with the default settings), although some alignment programs break the secondary structure element indicated by dashes. The coded names of the domains on the left are their identification numbers in the Protein Data Bank (PDB).

Once a satisfactory alignment has been obtained, there are now numerous programs available through the Web that allow you to view, analyze, and even edit alignments. AMAS (Analyze Multiply Aligned Sequences), CINEMA (Colour Interactive Editor for Multiple Alignments), and ESPript (Easy Sequencing in Postscript) are but a few.

### Alignment can be improved by incorporating additional information

The alignment of two or more sequences can be improved by incorporating expert knowledge such as known structural properties of one or more sequences. For example, if the structure of one of the proteins to be aligned is known, then the gap penalty can be increased for regions of known secondary structures such as α-helices or β-strands, as these regions are less likely to suffer insertions or deletions. This will mean that few or no gaps are introduced into these regions. On the other hand, gap penalties can be decreased for loop regions, in which insertions and deletions are better tolerated.

Often the results of an automatic alignment program benefit from manual final adjustment. For example, if specific residues are known to be important for structure, function, or ligand binding, then manual realignment may be necessary to match these residues.

## 4.6 Searching Databases

Searching sequence databases now has a part to play in nearly every branch of molecular biology, and is crucial for making sense of the sequence data becoming available from the genome projects. For example, one may wish to search the database with a DNA sequence to locate and identify a gene in a new genome. When a protein sequence is available, then searching through the database can be used to identify the potential function. Sometimes one wishes to find the gene for

**Flow Diagram 4.3**
**The key concept introduced in this and the following section is that applications have been designed to overcome the problems associated with searching a database for sequences that are similar to a query sequence, including the need to pay special attention to the statistical significance of the alignment scores obtained.**

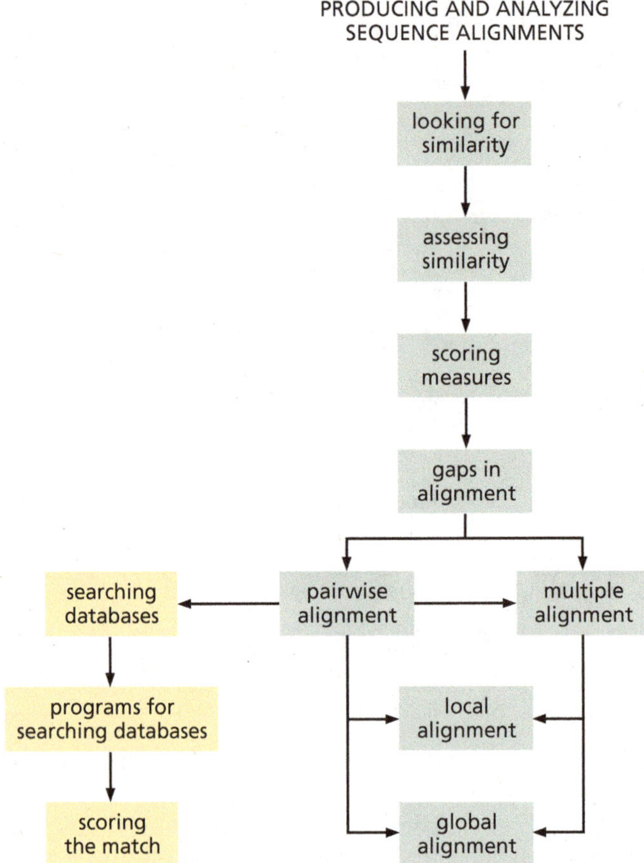


a particular protein in a genome, which can be done by searching with a homologous protein or DNA sequence.

We will now discuss the practical task of searching sequence databases to find sequences that are similar to the query sequence or search sequence that we submit to them (Flow Diagram 4.3). When searching a database with a newly determined DNA or protein query sequence, one does not usually know whether an expected similarity might span the entire query sequence or just part of it; similarly, one does not know if the match will extend along the full length of a database sequence or only part of it. Therefore, one initially needs to look for local alignments between the query sequence and any sequence in the database. The top-scoring database sequences are then candidates for further analysis.

Database searching needs to be both sensitive, in order to detect distantly related homologs and avoid **false-negative** searches, and also specific, in order to reject unrelated sequences with fortuitous similarity (**false-positive** hits). This is not an easy balance to achieve, and search results should be scrutinized with care.

In general, it is not possible to decide from a visual inspection of the alignment whether the database and query sequences are truly homologous. However, analysis of the score statistics has provided us with useful measures to estimate the validity of a hit. This important aspect of database searching, which is required to interpret any database search correctly, is discussed later in this chapter and in more detail in Section 5.4.

## Fast yet accurate search algorithms have been developed

The sequence databases are now extremely large and growing daily. This means that aligning a query sequence with sequences in a database requires considerable computer resources. In the past, this exceeded the available computing power and so great effort was put into developing fast yet accurate alignment methods. Almost all database search programs currently in use are modifications of the rigorous methods discussed earlier. The Needleman–Wunsch and Smith–Waterman methods are rigorous in the sense that given a scoring scheme they are guaranteed to find the best-scoring alignments between two sequences. Two suites of programs are in common use for database searching: FASTA and BLAST. These use dynamic programming, but only for database entries that have a segment sufficiently similar to the query sequences. The methods used to find these entries are purely heuristic; that is, not rigorous.

## FASTA is a fast database-search method based on matching short identical segments

FASTA is a popular database-searching program that increases the speed of a search at the expense of some sensitivity. It speeds up the searching process by using **k-tuples**, short stretches of *k* contiguous residues. In protein searches *k* can equal 1 or 2, while 6 is a typical value for DNA. The program makes up a dictionary of all possible k-tuples within the query sequence. Each entry contains a list of numbers that describe the location of the k-tuple in the query sequence. This is called **hashing**, and the theory behind it is described in Section 5.3. Therefore, for each k-tuple in the searched sequences, FASTA only has to consult the dictionary to find out if it occurs in the query sequence. However, sensitivity is reduced because a partial match of a k-tuple (for example, AC to AG in DNA) is ignored. Therefore, although speed increases with the length of a k-tuple, sensitivity will decrease.

In the first step of the FASTA method all possible pairwise k-tuples are identified: these can be considered as diagonals in a set of dot-plots. In the second step, alignments of these diagonals are rescored using a scoring matrix such as one of those described above. In this step, the k-tuple regions are also extended without including gaps, and only those that score above a given threshold are retained. In the third step, the program checks to see if some of the highest-scoring diagonals can be joined together. Finally, the search sequences with the highest scores are aligned to the query sequence using dynamic programming. The final alignment score ranks the database entries and the highest-scoring set is reported.

## BLAST is based on finding very similar short segments

BLAST (Basic Local Alignment Search Tool) or Wu-BLAST (a version of BLAST developed at Washington University, St Louis) is one of the most widely used database-search program suites. It relies on finding core similarity, which is defined by a window of preset size (called a "word") with a certain minimum density of matches (for DNA) or with an amino-acid similarity score above a given threshold (for proteins). Note that these amino acid word-matches do not only include identities and that they are scored with a standard substitution matrix. In the first step, all suitable matches are located in each database sequence. Subsequently, matches are extended without including gaps, and on this basis the database sequences are ranked. The highest-scoring sequences are then subjected to dynamic programming to obtain the final alignments and scores. BLAST and Wu-BLAST can be run with or without the use of gaps. The gapped setting of BLAST, which is usually the default setting, reports the best local alignments and is suitable for most applications. Both the FASTA and BLAST methods are described in detail in Section 5.3.

## Different versions of BLAST and FASTA are used for different problems

Many of the search algorithms can be used to search either nucleic acid or protein sequences, or even to search a protein-sequence database using a nucleic acid

**Table 4.1**
**The various algorithms within the FASTA package are given with descriptions of their uses.** Equivalent programs in the BLAST package are highlighted. The ktup parameter of fasta defines the length of a k-tuple, as explained in Section 5.3.

| Program | Description | BLAST equivalent |
|---|---|---|
| fasta | Protein compared to protein database or DNA to DNA database. For protein, ktup = 2 by default (ktup = 1 is more sensitive); default for DNA is 6; 4 or 3 is more sensitive. 1 should be used for short DNA stretches. | blastp/blastn |
| ssearch | Uses Smith–Waterman algorithm. Can search protein to protein or DNA to DNA. Can be more sensitive than fasta with protein sequences. | |
| fastx/fasty | DNA compared to protein database. DNA translated into all three frames. fasty slower than fastx but better. Used to see if DNA encodes a protein. | blastx |
| tfastx/tfasta | Protein compared to DNA database. Mainly used to identify EST sequences. This is preferred over fastx as protein comparison is more sensitive than DNA. | tblastn (tblastx compares translated DNA to translated DNA database) |
| fastf | Mixed peptide sequence (such as obtained by Edman degradation) compared to protein database. | |
| tfastf | Mixed peptide sequence compared to DNA database. | |

sequence and vice versa. However, you need to choose the correct program for the required type of search. In BLAST, for example, one can choose among blastp, which compares an amino acid query sequence against a protein-sequence database; blastn, which compares a nucleotide query sequence against a nucleic acid sequence database; blastx, which compares a nucleotide query sequence translated in all reading frames against a protein-sequence database; tblastn, which compares a protein query sequence against a nucleotide-sequence database dynamically translated in all reading frames; and finally, tblastx, which compares the six possible translations of a nucleotide query sequence against the six frame translations of a nucleotide-sequence database. The FASTA suite has similar versions of these search programs (see Table 4.1).

## PSI-BLAST enables profile-based database searches

Variations of BLAST such as PSI-BLAST (Position-Specific Iterative BLAST) have been devised. This suite of programs makes use of features characteristic of a particular protein family to identify related sequences in a protein database, and can identify related sequences that are too dissimilar to be found in a straightforward BLAST search. In PSI-BLAST, a **profile**, or **position-specific scoring matrix (PSSM)**, of a set of sequences is constructed from a multiple alignment of the highest-scoring hits returned in an initial BLAST search (see Section 6.1). The PSSM is created by calculating new scores for each position in this alignment. A highly conserved residue at a particular position is assigned a high positive score, while other residues at that position are assigned high negative scores. At positions that are weakly conserved throughout the alignment, all residues are given scores near zero. The profile generated is used to replace the substitution matrix in a subsequent BLAST search. This process can be repeated many times; each time, the results from the search are used to refine the profile. This type of iterative searching results in increased sensitivity and has been used to good effect in protein-fold recognition programs such as 3D-PSSM (see Chapter 13).

Ways of extracting more distantly related homologous sequences and finding links between known families are now being explored. Such methods include, for example, the use of **intermediate sequences**; that is, sequences that are found in more than one family. Suppose we submit an unknown sequence A to a database search and among the significant hits there is a protein called, for example, mediator protein. We then submit an unknown sequence B to the same database search, and this also returns mediator protein with a significant score. Then, especially if more than one such intermediate sequence is found, we can deduce that sequences A and B are homologous, as their families are related. Such ideas can be automated for ease of application.

### SSEARCH is a rigorous alignment method

Despite the computational requirements, some programs have been written that use rigorous methods to search databases. SSEARCH is a search program based on the Smith–Waterman algorithm and is therefore slower than either BLAST or FASTA. SSEARCH performs a rigorous search for similarity between a query sequence and the database. Other search algorithms based on the Smith–Waterman method have been written and are gaining in popularity as computer power increases.

## 4.7 Searching with Nucleic Acid or Protein Sequences

### DNA or RNA sequences can be used either directly or after translation

In general, nucleic acid sequence searches are more difficult to handle and analyze than protein sequence searches. However, most primary data will be in the form of nucleic acid sequences. If you have an untranslated DNA or RNA sequence and you want to know if the DNA codes for a protein, you can use fasta, ssearch, or blastn (see Table 4.1) to search the EST (expressed sequence tag), EMBL, or nr (nonredundant) databases, or one of the species-specific genome EST databases, such as EST-Rodent. The results may well be confusing, in that a lot of partial sequence matches will be found. Many retrieved sequences will also be unknown sequences. An easier search can be made using fastx/fasty (or blastx), which will translate the DNA in all three reading frames on both strands—six translations in all—and search a protein database of choice. More details and examples of dealing with DNA sequences can be found in Chapter 9.

### The quality of a database match has to be tested to ensure that it could not have arisen by chance

How good is an alignment and how believable are the results of a database search? These vital questions must be answered before any further use can be made of the results. Every alignment reported will have been selected on the basis of its score. What we need to know is whether the score is greater than we would expect from the alignment of the sequence with a random (unrelated) sequence. However, there is a complex relationship between the score and the significance of the sequence similarity. For one thing, as each pair of aligned residues contributes to the score, longer sequences are expected to give higher alignment scores, assuming the same degree of similarity.

If a large number of random sequences are generated and aligned with the query sequence, the resulting alignment scores will follow a particular distribution. Because we always choose the best-scoring alignment, the distribution will be related to the extreme-value distribution (see Section 5.4). Through application of

**Figure 4.12**

**The results of a search of the SWISS-PROT protein sequence database using BLAST with PI3-kinase p100α as the query sequence.** (A) Output from a standard BLAST search. Each line reports a separate database sequence. The penultimate column gives the alignment score, and the last column the *E*-value. Hits before the arrow are significant, while below the arrow the hit does not have a significant score. (B) A BLAST search on one month's new sequences, using the same query sequence as in (A), finds only two matches. One is a PI4-kinase, which has most of its sequence aligned to the query sequence (magenta line). The other has only a small region aligned (black line) and a borderline score. (C) Output from a Conserved Domain Database (CDD) search.

this distribution it is possible to estimate the probability of two random sequences aligning with a score greater than or equal to *S*. This is usually reported as an **expectation value** or ***E*-value**, and is used to order the database search results.

The programs BLAST and FASTA calculate an *E*-value, which is the number of alignments with a score of at least *S* that would be expected by chance alone in searching a complete database of *n* sequences. These *E*-values can vary from 0 to *n*. For example, by chance alone, you would expect to find three sequence alignments with an *E*-value of 3.0 or less in a database search, so an *E*-value of 3.0 suggests that the database sequence is not related to the query sequence. Quite closely related sequences often give very small *E*-values of $10^{-20}$ or less, and such scores clearly indicate a significant similarity of the database and query sequences. However, we really need to know how large an *E*-value can be while still reliably indicating a significant sequence similarity. It is important to remember that the *E*-value depends on the sequence length and the number of sequences in the database as well as on the alignment score.

In general, the smaller the *E*-value the better the alignment, and the higher the percentage identity the more secure the assessment of the significance of the similarity between the database sequence and the query sequence. The default *E*-value threshold in many search packages is set to either 0.01 or 0.001. However, most programs permit the user to set the *E*-value threshold, and matches above that threshold will not be included in the output.

To test new or existing sequence-alignment programs and their scoring schemes one can compare the alignment obtained by the program against carefully constructed alignments that are based on known structural features or biological function. There are databases of such accurately aligned sequences, such as BAliBase.

## Choosing an appropriate *E*-value threshold helps to limit a database search

To illustrate some of the possible sequence searches, alignments, and analyses that can be carried out via the Web, we will use two examples: the catalytic domain from a PI3-kinase and the protein-interaction domain SH2. Structural information and an accurate alignment in the BAliBase database are available for the family of SH2 domains.

The human *Syk* tyrosine kinase carboxy-terminal SH2 domain is the first query sequence. Protein searches with BLAST through the SWISS-PROT database gave 149 sequences below the default *E*-value cut-off. All these were SH2-related domains. That is a lot of information to cope with. All the *E*-values were very low, indicating that all the hits were significant. This is a case of result overload. Decreasing the *E*-value cut-off would have no effect in this case, as all the hits were far below the threshold used.

(A)

```
sp|P32871|P11A BOVIN   PHOSPHATIDYLINOSITOL 3-KINASE CATALYTI...  680  0.0
sp|P42336|P11A HUMAN    PHOSPHATIDYLINOSITOL 3-KINASE CATALYT...  676  0.0
sp|P42337|P11A MOUSE    PHOSPHATIDYLINOSITOL 3-KINASE CATALYT...  674  0.0
sp|P42338|P11B HUMAN    PHOSPHATIDYLINOSITOL 3-KINASE CATALYT...  338  9e-93
sp|O35904|P11D MOUSE    PHOSPHATIDYLINOSITOL 3-KINASE CATALYT...  332  7e-91
sp|O00329|P11D HUMAN    PHOSPHATIDYLINOSITOL 3-KINASE CATALYT...  331  2e-90

sp|P47473|RIR1 MYCGE   RIBONUCLEOSIDE-DIPHOSPHATE REDUCTASE A...   34  0.59
```

(B)

Distribution of 2 Blast Hits on the Query Sequence

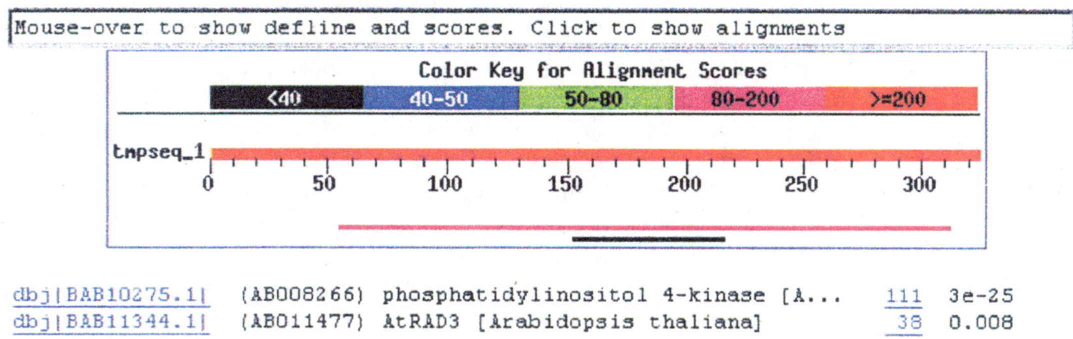


```
dbj|BAB10275.1|  (AB008266) phosphatidylinositol 4-kinase [A...  111  3e-25
dbj|BAB11344.1|  (AB011477) AtRAD3 [Arabidopsis thaliana]          38  0.008
```

(C)

.. This CD alignment includes 3D structure. To display structure, download Cn3D v3.00!

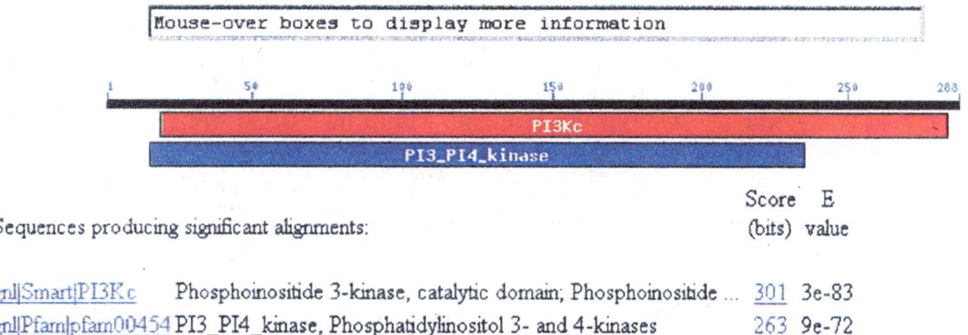


| Sequences producing significant alignments: | | Score (bits) | E value |
|---|---|---|---|
| gnl\|Smart\|PI3Kc | Phosphoinositide 3-kinase, catalytic domain; Phosphoinositide ... | 301 | 3e-83 |
| gnl\|Pfam\|pfam00454 | PI3_PI4_kinase, Phosphatidylinositol 3- and 4-kinases | 263 | 9e-72 |

gnl|Smart|PI3Kc, Phosphoinositide 3-kinase, catalytic domain; Phosphoinositide 3-kinase isoforms participate in a variety of processes, including cell motility, the Ras pathway, vesicle trafficking and secretion, and apoptosis. These homologues may be either lipid kinases and/or protein kinases: the former phosphorylate the 3-position in the inositol ring of inositol phospholipids. The ataxia telangiectesia-mutated gene produced, the targets of rapamycin (TOR) and the DNA-dependent kinase have not been found to possess lipid kinase activity. Some of this family possess PI-4 kinase activities.

Add query to multiple alignment, display up to 10 sequences most similar to the query

```
         Length = 265
         Score =  301 bits (763), Expect = 3e-83

Query:  19    IIFKNGDDLRQDMLTLQIIRIMENIWQNQGLDLRMLPYGCLSIGDCVGLIEVVRNSHTIM  78
Sbjct:  2     IIFKHGDDLRQDMLILQILRIMESIWETESLDLCLLPYGCISTGDKIGMIEIVKDATTIA  61
```

To reduce the large number of hits one could search a subset of the data, for example only the newest sequences in the database (the "month" option; that is, those deposited in the last month) or a specific genome database. A search through sequences released in the past month detected eight sequences, all with significant scores, of which three had not been identified. A search through the *Drosophila* genome data yielded three hits, all of which are unknown. Taking one of the regions that matched our SH2 (a section of the *Drosophila* 3R chromosome arm) and searching with this sequence through SWISS-PROT yielded a highly significant hit to an SH2 domain of a rat protein. So we may have identified a previously unknown *Drosophila* sequence as containing an SH2 domain.

This example illustrates a search through the database with a family that is very well represented and shows the problems that can arise. We will now look at a family that is not so well represented—the PI3-kinases.

First the SWISS-PROT database was searched using the catalytic domain protein sequence from the PI3-kinase p110α using BLAST with the *E*-value cut-off set to 1. Thirty-two hits were found. In this list there are three near-identical isoforms of p110α which have an *E*-value of 0.0; that is, the chance of obtaining such a match with random sequence is taken to be zero. There is one match that is not significant: ribonucleoside-diphosphate reductase, with an *E*-value of 0.59 (see Figure 4.12A). From the assigned function this is clearly a different enzyme, but the enzymatic reactions of both this reductase and the kinases involve a nucleotide, which might have led to some small degree of similarity between the sequences. Any such speculation would need further and more thorough investigation. If we rerun the search with the *E*-value cut-off set to 0.01 (the advised setting) only the significant matches are retrieved.

Searching with BLAST through a subset of sequences such as those that have only been released in one month found two hits: one is a homolog of PI3-kinase, a PI4-kinase with a significant *E*-value, and the other is a segment of an *Arabidopsis thaliana* protein, atRad3, with an *E*-value score of borderline significance. From the length of the matched sequence illustrated in the search output (see Figure 4.12B) the segment seems far too short to be of interest; compare the length of the matched PI4-kinase, in magenta. For this reason the hit can now be reclassified as not significant.

Another useful option available within the BLAST search server is a concurrent search of the Conserved Domain Database (CDD) entries. Figure 4.12C shows the results of using this option on the PI3-kinase sequence. Proteins often contain several domains, and the program CD-Search can potentially identify domains present in a protein sequence. CDD contains domains derived mainly from the SMART and Pfam protein-family databases. To identify conserved domains in a protein sequence, the CD method uses the BLAST algorithm where the query sequence is matched with a PSSM designed from the conserved domain alignments. Matches are shown as either a pairwise alignment of the query sequence to a representative domain sequence or as a multiple alignment.

A FASTA search with the p110α sequence through SWISS-PROT with default settings (k-tuple = 2) yielded 36 hits, of which eight had a nonsignificant score (see Figure 4.13). Of these eight, ribonucleoside-diphosphate reductase was also found by BLAST. Although both FASTA and BLAST report an *E*-value, the actual values are different, which reflects subtle differences in the methods used. An SSEARCH search of the SWISS-PROT database with default settings found 29 significant hits. SSEARCH, a more rigorous method, found fewer hits than BLAST or FASTA.

## Low-complexity regions can complicate homology searches

Among the many features that can complicate a sequence-similarity search is the occurrence of **low-complexity regions** in protein sequences. These are regions with a highly biased amino acid composition, often runs of prolines or acidic amino

```
the best scores are:                                                           E(86391)
SW:P11A BOVIN P32871 PHOSPHATIDYLINOSITOL 3-KINAS   (1068) 2228 493 1.2e-138
SW:P11A HUMAN P42336 PHOSPHATIDYLINOSITOL 3-KINAS   (1068) 2216 490 7.4e-138
SW:P11A MOUSE P42337 PHOSPHATIDYLINOSITOL 3-KINAS   (1068) 2204 488 4.5e-137
SW:P11B HUMAN P42338 PHOSPHATIDYLINOSITOL 3-KINAS   (1070) 1126 254 1.1e-66

                          ↓ other sequences

SW:ESR1 YEAST P38111 ESR1 PROTEIN.                  (2368)  144  41    0.028
SW:PRA2 USTMA P31303 PHEROMONE RECEPTOR 2.          ( 346)  116  35    0.35
SW:TEL1 YEAST P38110 TELOMER LENGTH REGULATION PR   (2787)  127  37    0.42
SW:YA51 METJA Q58451 HYPOTHETICAL PROTEIN MJ1051.   ( 513)  112  34    0.91
SW:RIR1 MYCGE P47473 RIBONUCLEOSIDE-DIPHOSPHATE R   ( 721)  106  33       3
SW:YAY1 SCHPO Q10209 HYPOTHETICAL 44.8 KDA PROTEI   ( 392)   99  31     5.1
SW:PAFA CAVPO P70683 PLATELET-ACTIVATING FACTOR A   ( 436)   96  30     8.8
SW:KC47 ORYSA P29620 CDC2+/CDC28-RELATED PROTEIN    ( 424)   95  30     9.9
```

**Figure 4.13**
**Output from a search of the SWISS-PROT protein sequence database using FASTA with PI3-kinase p110α as the query sequence.** Thirty-six hits were obtained. Eight of these have a nonsignificant score (below the arrow). One of these, ribonucleoside-diphosphate reductase, was also found by BLAST. The *E*-values in FASTA are different from those in BLAST.

acids. In some cases, self-comparison dot-plots (see page 77) can identify low-complexity regions in a protein sequence. Alignments of such regions in different proteins can achieve high scores, but these can be misleading and can obscure the biologically significant hits. It is better to exclude low-complexity regions when constructing the alignment. By default, the BLAST program filters the query sequence for low-complexity regions. In the BLAST output file, an X marks regions that have been filtered out (using SEG for proteins and DUST for DNA) (see Box 5.2).

Figure 4.14A shows a self-comparison dot-plot of human prion protein precursor (PrP), an abnormal form of which is found in large amounts in the brains of people with neurodegenerative diseases such as Creutzfeldt–Jakob disease (CJD) and kuru (see Box 4.6). It has several low-complexity regions, which are seen as dark diagonal lines (apart from the main identity diagonal) and the ordered dark-shaded regions. Figure 4.14B shows a search for homologs of the human PrP. The extensive low-complexity regions have been filtered out in the query sequence (as indicated by the strings of Xs). A BLAST search of SWISS-PROT with human PrP with the low-complexity filter turned on gave approximately 40 hits, all prion proteins. One of

## Box 4.6 Prions: Proteins that can exist in different conformations

Scrapie in sheep, bovine spongiform encephalopathy (BSE) in cattle, and Creutzfeldt–Jakob disease (CJD), fatal familial insomnia, and kuru in humans are rare, fatal, transmissible, neurodegenerative diseases known generally as the transmissible spongiform encephalopathies, after the characteristic damage they do to the brain. They can arise sporadically, or as a result of the inheritance of a faulty gene, or can be transmitted by ingestion of infected material. Kuru, which was relatively common in people in the Eastern Highlands of Papua New Guinea in the 1950s and 1960s, was found to be caused by the ritual custom of eating the brains of dead relatives, while a variant form of CJD (vCJD), which has appeared only recently, is thought to be caused by people having eaten BSE-infected meat products.

The causal agent in the spongiform encephalopathies is believed to be an infectious protein, a prion, rather than a DNA or RNA virus. Prions are normal proteins that have the property of being able to convert into an alternative stable conformation, which is associated with disease, although the mechanism by which prions cause cell death and neurodegeneration is not yet fully understood. The normal form of the prion protein ($PrP^{c}$) is a monomer with a structure consisting mainly of α-helices, and is mainly found at the cell surface, whereas the abnormal form ($PrP^{Sc}$), is mainly β-sheet and has a tendency to aggregate into clumps. $PrP^{Sc}$ itself appears to be able to induce the conversion of $PrP^{c}$ into $PrP^{Sc}$. The prion protein is an example of a metastable protein, where the same or similar sequences can exist in different stable structural forms.

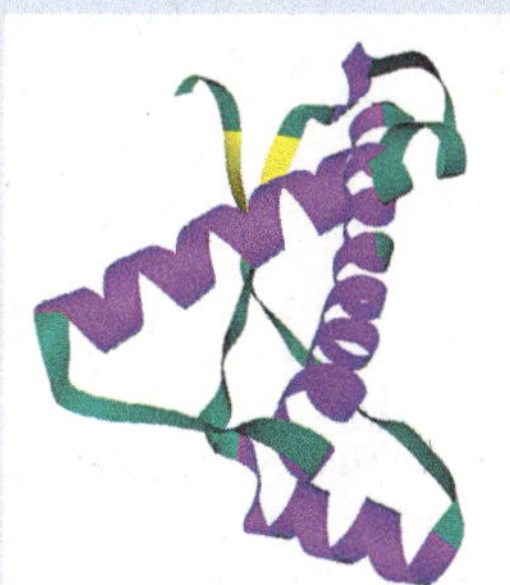

**Figure B4.4**
**A ribbon representation of the normal form of prion protein, $PrP^{c}$.**

(A)

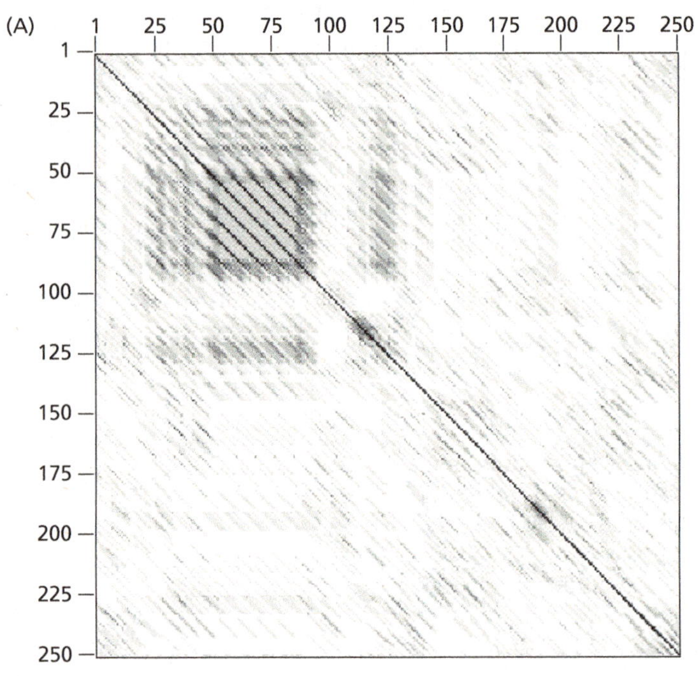


(B)

```
>sp[P04156]PRIO HUMAN  MAJOR PRION PROTEIN PRECURSOR (PRP) (PRP27-30) (PRP33-35C) (ASCR)
          Length = 253

 Score =  312 bits (792), Expect = 5e-85
 Identities = 154/236 (65%), Positives = 154/236 (65%)

Query: 64   MANLGCWMLVLFVATWSDLGLCKKRPKPGGWNTGGSRYPGQGSPGGNRYXXXXXXXXXXX 123
             MANLGCWMLVLFVATWSDLGLCKKRPKPGGWNTGGSRYPGQGSPGGNRY
Sbjct: 1    MANLGCWMLVLFVATWSDLGLCKKRPKPGGWNTGGSRYPGQGSPGGNRYPPQGGGGWGQP 60

Query: 124  XXXXXXXXXXXXXXXXXXXXXXXXXXXXXXXXXXTHSQWNKPSKPKTNMKHMXXXXXXXX 183
             XXXXXXXXXXXXXXXXXXXXXXXXXXXXXXXXXXTHSQWNKPSKPKTNMKHM
Sbjct: 1    HGGGWGQPHGGGWGQPHGGGWGQPHGGGWGQGGGTHSQWNKPSKPKTNMKHMAGAAAAGA 120

Query: 184  XXXXXXXXXXXXXXXRPIIHFGSDYEDRYYRENMHRYPNQVYYRPMDEYSNQNNFVHDCV 243
                            RPIIHFGSDYEDRYYRENMHRYPNQVYYRPMDEYSNQNNFVHDCV
Sbjct: 121  VVGGLGGYMLGSAMSRPIIHFGSDYEDRYYRENMHRYPNQVYYRPMDEYSNQNNFVHDCV 180

Query: 244  NITIKQHXXXXXXXXXXXXXXDVKMMERVVEQMCITQYERESQAYYQRGSSMVLFS 299
             NITIKQH              DVKMMERVVEQMCITQYERESQAYYQRGSSMVLFS
Sbjct: 181  NITIKQHTVTTTTKGENFTETDVKMMERVVEQMCITQYERESQAYYQRGSSMVLFS 236
```

**Figure 4.14**
**Dealing with low-complexity regions of sequence.** (A) The low-complexity regions are clearly visible on a self-comparison dot-plot of a human prion precursor protein (PrP). They are indicated by the black diagonal lines on either side of the identity diagonal and by the ordered dark-shaded regions. (B) Results of a database search with PrP from which sequences of low complexity have been filtered out by application of the program SEG, which marks them with Xs (top row of the alignment).

these is shown aligned with the query sequence in the figure. When the filter was turned off, the number of hits rose to 220, most of which were not homologous.

Sometimes one may wish to study low-complexity regions in particular. For example, in the case of the tubulin and actin gene clusters it is thought that amplification of the protein-coding genes may be related to these regions. There are options in BLAST that allow you to select these regions for study.

## Different databases can be used to solve particular problems

To some extent, the choice of which database to search will depend on which databases are provided by the site that runs the search algorithms. Most sites contain a selection of the most popular databases, such as GenBank for DNA sequences, SWISS-PROT for annotated protein sequences, TrEMBL, a translated EMBL DNA-sequence database, and PDB, a database of protein structures with sequences (see Chapter 3). Some sites also provide access to expressed sequence tag (EST) databases, such as dbEST, and genome-sequence databases from some of the fully sequenced genomes

In general, a first pass should be run on a generic protein- or nucleic acid sequence database. You can also carry out a search on the PDB to see if your query sequence has a homolog with known structure. More specific searches can be performed to answer particular questions. For example, if it is suspected that the query sequence belongs to a family of immune-system proteins, the search could be carried out on the Kabat database, which contains sequences of immunological interest. If the sequence originates from a mouse, you may want to know if a homolog exists in the rat, *Drosophila*, or human genomes, and should therefore search the databases containing sequences from the appropriate species. You also need to check that you are searching a database that is up to date; sites such as those at NCBI and EBI are regularly updated.

If no match is found to the query sequence, it does not necessarily mean that there is no homolog in the databases, just that the similarity is too weak to be picked up by existing techniques. Techniques are continually being improved and the amount of sequence data continues to increase; you should therefore periodically resubmit your sequence.

Many other sequence-related databases can usefully be searched and provide additional information. For example the Sequences Annotated by Structure (SAS) server is a collection of programs and data that can help identify a protein sequence by using structural features that are the result of sequence searches of annotated PDB sequences. Residues in the sequences of known structures are colored according to selected structural properties, such as residue similarity, and are displayed using a Web browser. SAS will perform a FASTA alignment of the query sequence against sequences in the PDB database and return a multiple alignment of all hits. Each of the hits is annotated with structural and functional features. That information can be used to annotate the unknown protein sequence. Further links are provided to the separate PDB files. Databases such as Clusters of Orthologous Groups (COGs) and UniGene can help in gene discovery, gene-mapping projects, and large-scale expression analysis. Sites such as Ensembl provide convenient access for gene searches in many different annotated eukaryotic genomes and useful associated information.

## 4.8 Protein Sequence Motifs or Patterns

If the similarity between an unknown sequence and a sequence of known function is limited to a few critical residues, then standard alignment searches using BLAST or FASTA against general sequence databases such as GenBank, dbEST, or SWISS-PROT will fail to pick up this relationship. What is required is a method of searching for the occurrence of short sequence patterns, or **motifs** (see Flow Diagram 4.4). A motif, in general, is any conserved element of a sequence alignment, whether composed of a short sequence of contiguous residues or a more distributed pattern. Functionally related sequences will share similar distribution patterns of critical functional residues that are not necessarily contiguous. For example, conserved amino acid residues comprising an enzyme's active site may be distant from each other in the protein sequence but will still occur in a recognizable pattern because of the constraints imposed by the requirement for them to come together in a particular spatial configuration to form the active site in the three-dimensional structure.

There are three different types of activity associated with pattern searching. A query sequence can be searched for patterns (from a patterns database) that could help suggest functional activity. A sequence database can be searched with a specific pattern, for example to determine how many gene products in a genome have a

**Flow Diagram 4.4**
The key concepts introduced in this and the following two sections are that sequence patterns can be very useful in identifying protein function and that special pattern databases and search programs have been designed to assist in identifying patterns in a query sequence.

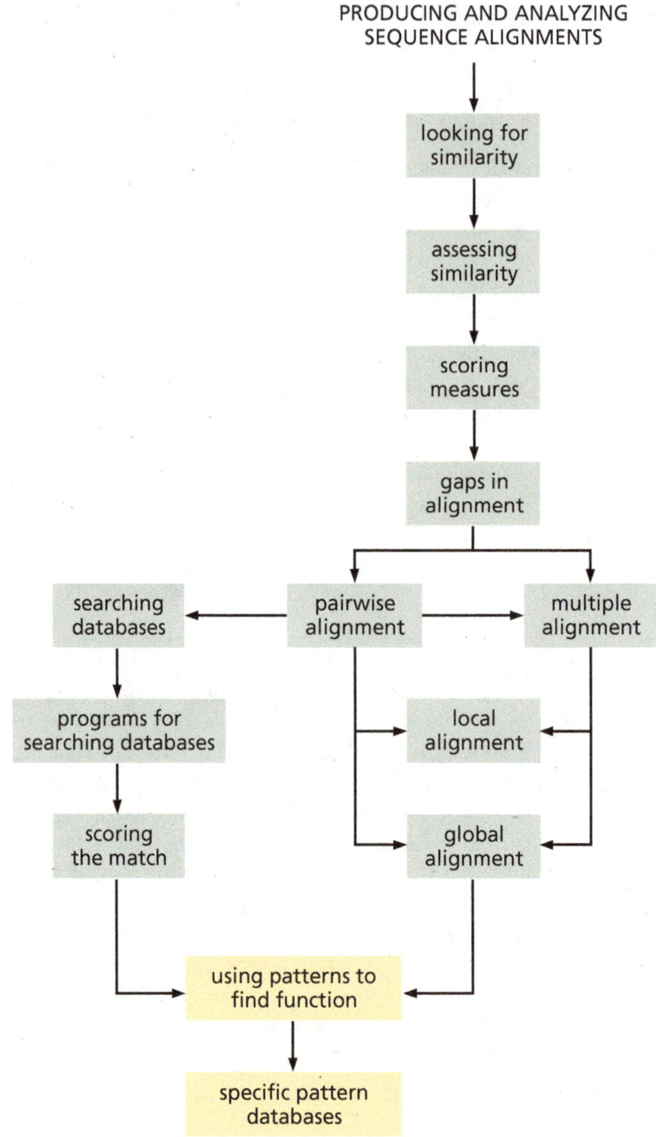


specific function. Lastly, we may want to define a new pattern specific for a selected set of sequences.

In searches with new sequences, the whole database is searched and expert knowledge, such as the known function of a homologous protein, is then used to extrapolate the function of the new sequence. In contrast, when new patterns and motifs are used to search a database, the expert knowledge is needed right at the beginning to construct the motifs that are intended to identify the specific function or any other physicochemical property.

Pattern and motif searches are mostly used with protein sequences rather than nucleotide sequences, as the greater number of different amino acids makes protein patterns more efficient in discriminating truly significant hits. In addition, many of the patterns identify biological function, which is mediated at the protein level. There are, however, particular problems in DNA- and genome-sequence analysis where searching for motifs is useful (see Chapters 9 and 10).

## Creation of pattern databases requires expert knowledge

The construction of patterns or motifs is of prime importance in characterizing a protein family, and much time and energy has gone into constructing pattern and motif databases that one can search with an unknown sequence. One of the most important steps is careful selection of the sequences used to define the pattern. If these do not all share the same biological properties for which you wish to define a pattern, you will almost certainly encounter problems later. Thus, experimental evidence of function or clear homology is necessary for all the sequences used.

Some pattern databases have been constructed by hand by inspection of large amounts of data. This is very time consuming, but necessary, as the task of extracting a pattern is a complex one, depending on expert knowledge of the structures and/or functions of the sequences involved. For example, analysis of the X-ray structure of a protein can delineate the functional residues involved in an enzyme active site or a regulatory binding site, and an initial pattern can be generated. This pattern can then be refined by multiple alignment of sequences of other members of the same structural or functional protein family. If no structural data are available, multiple sequence alignment of short regions of similarity, assessed alone or in conjunction with experimental data on biochemical and cellular function, can be used to extract a pattern.

The simplest method of constructing a pattern or motif is the consensus method, in which the most similar regions in a global multiple sequence alignment are used to construct a pattern. Those positions in the alignment that are all occupied by the same residue (or a limited subset of residues) are used to define the pattern at these positions, by specifying just the allowed residues at each position. More sophisticated patterns can be generated using scoring tables to assess the similarity of the matched amino acids. In this case, instead of just defining the pattern as requiring, for example, a glutamic or aspartic acid at a given position, different residues at this position have associated scores.

## The BLOCKS database contains automatically compiled short blocks of conserved multiply aligned protein sequences

Sequence motifs can also be defined automatically from the multiple alignment of a specified set of sequences. Blocks are multiple alignments of ungapped segments of protein sequence corresponding to the most highly conserved regions of the proteins. The blocks for the BLOCKS database are compiled automatically by looking for the most highly conserved regions in groups of proteins documented in the PROSITE database. The blocks are then calibrated against the SWISS-PROT database to obtain a measure of the chance distribution of matches. The calibrated blocks make up the BLOCKS database, against which a new sequence can be searched. Both protein and DNA sequences (automatically translated into a protein sequence) can be submitted to search the BLOCKS database. The BLOCKS Web site, in addition to providing the BLOCKS database, will align your set of sequences and automatically design a block with which you can search SWISS-PROT. Generating blocks from your set of sequences and searching with them can find sequences that have very weak sequence similarity but are, nevertheless, functionally related. Generating patterns and/or blocks is also a useful method to search for hints to function within an unknown sequence.

Another program that will analyze a set of sequences for similarities and produce a motif for each pattern it discovers is MEME (Multiple Expectation maximization for Motif Elicitation). MEME characterizes motifs as position-dependent probability matrices. The probability of each possible letter occurring at each possible position in the pattern is given in the matrices. Single MEME motifs do not contain gaps and therefore patterns with gaps will be divided by the program into two or more separate motifs.

The program takes the group of DNA or protein sequences provided by the user and creates a number of motifs. The user can choose the number of motifs that MEME will produce. MEME uses statistical techniques to choose the best width,

**Figure 4.15**
**Residues that contribute to one of the blocks returned by the BLOCKS database after submission of the PI3-kinase p100α sequence.** (A) A block for four homologous sequences, and (B) for 31 homologous sequences. These representations are called logos, and are computed using a position-specific scoring matrix. This block contains the active-site amino acids and the DFG kinase motif. The size of the letters indicates the level of conservation and the colors indicate physicochemical properties of the residues: acidic, red; basic, blue; small and polar, white; asparagine and glutamine, green; sulfur-containing amino acids, yellow; hydrophobic, black; proline, purple; glycine, gray; aromatic, orange.

number of occurrences, and description for each motif (see Section 6.6). The motifs found by MEME can also be given in BLOCKS format to allow further analysis as described below.

If we submit the PI3-kinase p110α sequence and four homologs to the BLOCKS program it creates six separate blocks of high similarity. Figure 4.15A illustrates the block that contains residues important in PI3-kinase catalysis and ATP binding. Letters that are large and occupy the whole position represent identities in the multiple sequence alignment (see Section 6.1 for further details on this sequence representation). The SCAGY, DRH, and DFG motifs that are the markers for PI3-kinases are identified by the BLOCKS program and form part of the conserved regions. If more distant sequences are submitted, fewer residues will form the highly conserved regions with the largest residues, as shown in Figure 4.15B where 31 sequences were aligned.

The six blocks can now be submitted to a database search using the program LAMA (Local Alignment of Multiple Alignments), which compares multiple alignments of protein sequences with each other. The program searches the BLOCKS database, the PRINTS database (see below), or your own target data, to see if similar blocks or patterns already exist. This is a sensitive search technique, detecting weak sequence relationships between protein families. The LAMA search of the BLOCKS database has identified seven blocks, of which three are significant: these are PI3/4-kinase signatures.

The blocks can also be submitted to a MAST (Motif Alignment and Search Tool) search of one of the online nucleotide- or protein-sequence databases. MAST is a program that searches for motifs—highly conserved regions or blocks. Here we submit the six PI3-kinase blocks to a MAST search of SWISS-PROT (to use this program you need to have an e-mail address to receive the results). Twenty-four sequences were found with significant scores, with the PI3-kinase sequences all scoring more highly than the PI4-kinases.

The same set of PI3-kinase p110α sequences was submitted to the MEME motif-generating program. The number of motifs to be generated was set to six (the same number found by BLOCKS). The top-scoring motif (see Figure 4.16A) describes similar residues as the BLOCK motif described above (see Figure 4.15). The MEME motif starts at the end of the SCAGY motif (Y), contains the active site D, N, and DFG residues, and extends a bit further than the BLOCKS motif. A nice feature of the MEME program is that it generates a figure containing a summary of all motifs (see Figure 4.16B). This illustrates where the motifs are located with respect to each other within all the sequences (only three are shown for clarity). Submitting the MEME motifs to a search through SWISS-PROT finds 21 matches. Matches with significant

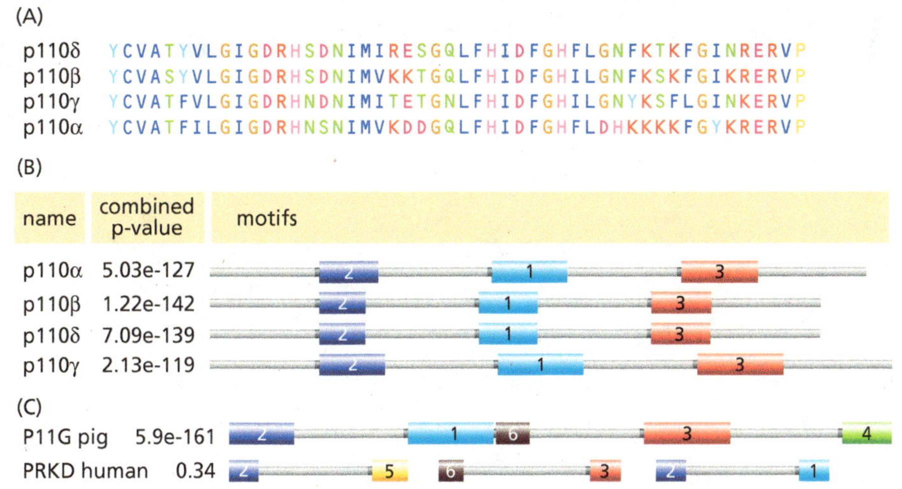


**Figure 4.16**
**MEME generates motifs.** (A) The top-scoring patterns are color coded according to the physicochemical properties of the amino acid side chains: dark blue is used for the residues ACFILVM; green for NQST; magenta to indicate DE; red color is used for residues KR; pink for H; orange for G; yellow for P; and light blue shows Y. (B) Summary motif information where each motif is represented by a colored block. The number in a block gives the scored position of the motif. The light blue block, number 1, contains the motif described in (A). The combined p-value of a sequence measures the strength of the match of the sequence to all the motifs. (C) Illustration of how lower-scoring motif matches can still find interesting and true homologs. The distances between the motif blocks are not representative.

scores are all PI3-kinases and PI4-kinases. The significant scores usually match most, if not all, of the motifs submitted. However, lower scores can be informative as well; distant relationships can be found if only a subset of the motifs matches.

For example, a search using the motifs of PI3-kinases finds the DNA-dependent protein kinase catalytic subunit (PRKD), which has shared kinase activity with the PI3-kinases. Four of the six motifs are matched (Figure 4.16C) and some are repeated within the DNA-dependent kinase. Simple sequence-alignment searches through the sequence databases may not pick up this type of relationship, although in this case a blastp search with p110α through SWISS-PROT matches PRKD with a score that would be considered significant, and a search with FASTA gives a borderline score.

## 4.9 Searching Using Motifs and Patterns

### The PROSITE database can be searched for protein motifs and patterns

The PROSITE database is a compilation of motifs and patterns extracted from protein sequences and compiled by inspection of protein families. This database can be searched with an unknown protein sequence to obtain a list of hits to possible patterns or protein signatures. It is also possible to create your own pattern in the manner of a PROSITE pattern to search another sequence database. The syntax of a PROSITE pattern consists of amino acid residues interspersed with characters that denote the rules for the pattern, such as distances between residues, and so on (see Table 4.2).

For example, a pattern for the kinase active site, starting from the conserved DRH and making use of the very conserved DFG region, can be created manually from the 31 sequences used in the BLOCKS example.

D-R-[KH]-X-[DE]-N-[IL]-[MILV](2)-X(3)-G-X-[LI]-X(3)-D-F-G

Inputting this pattern into the ScanProsite Web page and running it against the SWISS-PROT database of protein sequences obtained 92 hits; all were PI3 (PI4)-kinases or protein kinases. If, on the other hand, we submit the catalytic domain of the PI3-kinase p110α sequence to be scanned through the PROSITE database to see if there are any existing patterns, the search retrieves two signature

| Code | Explanation | Example explanation | Examples |
|---|---|---|---|
| One-letter codes | Standard amino acids | | G-L-L-M-S-A-D-F-F-F |
| - | All positions must be separated by - | | G-L-L-M-S-A-D-F-F-F |
| X | Any amino acid | Any amino acid allowed in second place | G-X-L-M-S-A-D-F-F-F |
| [] | Two or more possible amino acids | L or I allowed in second place | G-[LI]-L-M-S-A-D-F-F-F |
| {} | Disallowed amino acids | R or K not allowed in sixth place | G-[LI]-L-M-S-A-{RK}-F-F-F |
| (*n*) *n* = number | Repetition can be indicated by a number in brackets after the amino acid | F repeated three times | G-[LI]-L-M-S-A-{RK}-F(3) |
| (*n*,*m*) | A range: only allowed with X | One to three positions with any amino acids (X) allowed | G-[LI]-L-M-S-A-{RK}-X(1,3) |
| < | Pattern at amino-terminal of sequence | | |
| > | Pattern at carboxy-terminal of sequence | | |

**Table 4.2**
**The various codes used to define a PROSITE protein pattern for a search through a sequence database.**

sequences for PI3- and PI4-kinases. These give us a much more specific search signature for the PI3/4-kinase family, but do not tell us, for example, that this kinase family is also similar to the protein kinase family. The patterns for the signatures are:

(1) [LIVMFAC]-K-X(1,3)-[DEA]-[DE]-[LIVMC]-R-Q-[DE]-X(4)-Q

(2) [GS]-X-[AV]-X(3)-[LIVM]-X(2)-[FYH]-[LIVM](2)-X-[LIVMF]-X- D-R-H-X(2)-N

The second signature pattern contains at its right-hand end (underlined) the start of the kinase pattern we created above to scan SWISS-PROT. Pattern 1 and the rest of pattern 2 contain conserved regions within the PI3/4-kinase families that are amino-terminal to our created pattern.

## The pattern-based program PHI-BLAST searches for both homology and matching motifs

The BLAST set of programs also has a version that uses motifs in the query sequence as a pattern. PHI-BLAST (Pattern Hit Initiated BLAST) uses the PROSITE pattern syntax shown in Table 4.2 to describe the query protein motif. The specified pattern need not be in the PROSITE database and can be user generated. PHI-BLAST looks for sequences that not only contain the query-specific pattern but are also homologous to the query sequence near the designated pattern. Because PHI-BLAST uses homology as well as motif matching, it generally filters out those sequences where the pattern may have occurred at random. On the NCBI Web server, PHI-BLAST is integrated with PSI-BLAST, enabling one or more subsequent PSI-BLAST database searches using the PHI-BLAST results.

## Patterns can be generated from multiple sequences using PRATT

The program PRATT can be used to extract patterns conserved in sets of unaligned protein sequences. The patterns are described using the PROSITE syntax. The power of PRATT is that it requires no knowledge of possible existing patterns in a set of sequences. Figure 4.17 shows the results for the PI3-kinase p110α family. The pattern illustrated in the figure contains the DFG motif which is highlighted in the second PROSITE pattern.

```
PRATT output :

   p110-a:    qlfhi DFGHFLDhkKkkFGykRERVPFVLTqDFLiViskGaQE ctktr
   p110-b:    qlfhi DFGHILGnfKskFGikRERVPFILTyDFIhViqqGkTG ntekf
   p110-d:    qlfhi DFGHFLGnfKtkFGinRERVPFILTyDFVhViqqGkTN nsekf
   p110-g:    nlfhi DFGHILGnyKsfLGinKERVPFVLTpDFLfVm--GtSG kktsp

D-F-G-H-[FI]-L-[DG]-x(2)-K-x(2)-[FL]-G-x(2)-[KR]-E-R-V-P-F-[IV]-L-T-x-D-F-[ILV]-x-V-x(1,3)-G-x-[QST]-[EGN]
```

**Figure 4.17**
**PRATT pattern search.** Sequences of the four types of PI3-kinase sequences (α, β, γ, and δ) have been submitted to PRATT to automatically create PROSITE-like patterns from a multiple alignment. This figure shows the alignment block and a PRATT-generated PROSITE pattern of the region that contains the DFG motif (shaded in green and boxed in red).

## The PRINTS database consists of fingerprints representing sets of conserved motifs that describe a protein family

The PRINTS database is a next-generation pattern database consisting of fingerprints representing sets of conserved motifs that describe a protein family. The fingerprint is used to predict the occurrence of similar motifs, either in an individual sequence or in a database. The fingerprints were refined by iterative scanning of the OWL composite sequence database: a composite, nonredundant database assembled from sources including SWISS-PROT, sequences extracted from NBRF/PIR protein sequence database, translated sequences from GenBank, and the PDB structural database. A composite, or multiple-motif, fingerprint contains a number of aligned motifs taken from different parts of a multiple alignment. True family members are then easy to identify by virtue of possessing all the elements of the fingerprint; possession of only part of the fingerprint may identify subfamily members. A search of the PRINTS database with our PI3-kinase sequence found no statistically significant results.

## The Pfam database defines profiles of protein families

Pfam is a collection of protein families described in a more complex way than is allowed by PROSITE's pattern syntax. It contains a large collection of multiple sequence alignments of protein domains or conserved regions. Hidden Markov model (HMM)-based profiles (see Section 6.2) are used to represent these Pfam families and to construct their multiple alignments. Searching the Pfam database involves scanning the query sequence against each of these HMM profiles. Using these methods, a new protein can often be assigned to a protein family even if the sequence homology is weak. Pfam includes a high proportion of extracellular protein domains. In contrast, the PROSITE collection emphasizes domains in intracellular proteins—proteins involved in signal transduction, DNA repair, cell-cycle regulation, and apoptosis—although there is some overlap. A search of the Pfam database allows you to look at multiple alignments of the matched family, view protein domain organization (see Figure 4.6), follow links to other databases by clicking on the boxed areas, and view known protein structures.

A search in Pfam using the sequence of the PI3-kinase p110α catalytic domain will find the PI3/4-kinase family. You can then retrieve the multiple alignment that has been used to define the family and obtain a diagram of the domain structure of the whole family. (Clicking on a domain will call up another Web page of detailed information.) Figure 4.6 shows a snapshot of the interactive diagram; the yellow boxed area is the catalytic domain upon which the search was based.

Only the most commonly used pattern and profile databases have been described here; links to others are given on the Publisher's Web page.

# 4.10 Patterns and Protein Function

## Searches can be made for particular functional sites in proteins

There are techniques other than simple sequence comparison that can identify functional sites in protein sequences. In contrast to the methods discussed above, which tend to cover a very wide range of biological functions, these techniques are usually made available in programs which predict only one specific functional site.

For example, signals from the environment are transmitted to the inside of the cell where they induce biochemical reaction cascades called signal transduction pathways. These result in responses such as cell division, proliferation, and differentiation and, if not properly regulated, cancer. During signal transduction, cellular components are chemically modified, often transiently. One of the key modifications used in these pathways is the addition and removal of phosphate groups. Sites susceptible to such modification can be predicted by the NetPhos server, which uses neural network methods to predict serine, threonine, and tyrosine phosphorylation sites in eukaryotic proteins. PROSITE also has patterns describing sites for phosphorylation and other posttranslational modifications, but specific programs such as NetPhos are expected to be more accurate.

## Sequence comparison is not the only way of analyzing protein sequences

Apart from sequence comparison and alignment methods, there are various other ways of analyzing protein sequences to detect possible functional features. These techniques can be useful either when you have found a homolog in a database search and want to analyze it further, or when you have failed to find any similar sequence homolog and have no other avenue open. The physicochemical properties of amino acids, such as polarity, can be useful indicators of structural and functional features (see Chapter 2). There are programs available on the Web that plot hydrophobicity profiles, the percentage of residues predicted to be buried, some secondary-structure prediction (see Chapters 11 and 12), and percentage accessibility. ProtScale is one easy-to-use Web site that allows many of the above protein properties to be plotted.

Hydrophobic cluster analysis (HCA) is a protein-sequence comparison method based on α-helical representations of the sequences, where the size, shape, and orientation of the clusters of hydrophobic residues are compared. Hydrophobic cluster analysis can be useful for comparing possible functions of proteins of very low sequence similarity. It can also be used to align protein sequences. The patterns generated by HCA via the online tool drawhca can be compared with any other sequences one is interested in. It has been suggested that the effectiveness of HCA

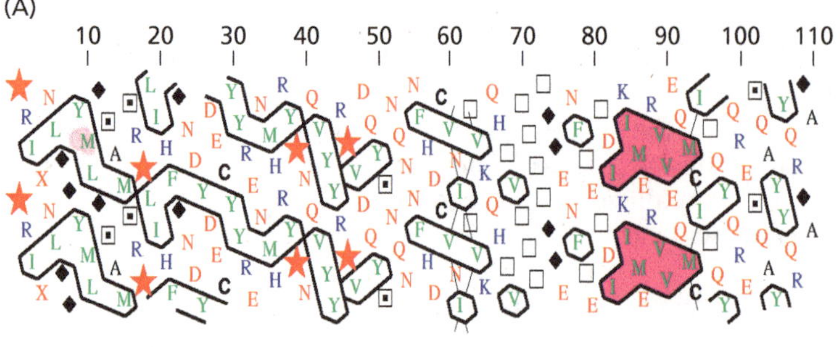


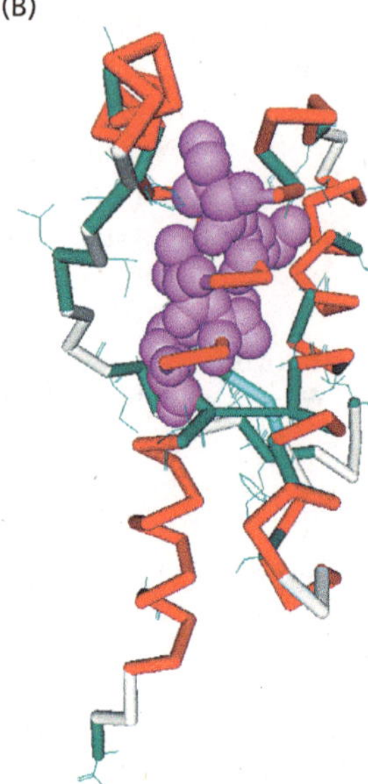


**Figure 4.18**
**(A) Hydrophobic cluster analysis (HCA) of the prion protein using drawhca.** Hydrophobic residues are in green, acidic in red, and basic in blue. A star indicates proline, a diamond glycine, an open box threonine, and a box with a dot serine. The same types of residues tend to cluster together, forming hydrophobic or charged patches. One such patch is highlighted in magenta. (B) X-ray structure of the same protein, with the same residues highlighted in magenta. As shown by the X-ray structure, the patch found by HCA forms a hydrophobic core in the interior of the protein.

### Box 4.7 **Protein localization signals**

Proteins are all synthesized on ribosomes in the cytosol but, in eukaryotic cells in particular, have numerous final destinations: the cell membrane, particular organelles, or secretion from the cell. Intrinsic localization signals in the protein itself help to direct it to its destination and these can often be detected by their sequence characteristics. Proteins sorted through the endoplasmic reticulum (ER) for secretion or delivery to the cell membrane and some other organelles usually have a characteristic signal sequence at the amino-terminal end. This interacts with transport machinery in the ER membrane, which delivers the protein into the ER membrane or into the lumen. The signal sequence is often subsequently removed. Signal sequences are characterized by an amino-terminal basic region and a central hydrophobic region, and these features are used to predict their presence.

for comparison originates from its ability to focus on the residues forming the hydrophobic cores of globular proteins. Figure 4.18 shows the prion protein patterns that were generated using the program drawhca.

Information about the possible location of proteins in the cell can sometimes be obtained by sequence analysis. Membrane proteins and proteins destined for organelles such as the endoplasmic reticulum and nucleus contain intrinsic sequence motifs that are involved in their localization. Most secreted proteins, for example and other proteins that enter the endoplasmic reticulum protein-sorting pathway, contain sequences known as signal sequences when they are newly synthesized (see Box 4.7). The PSORT group of programs predicts the presence of signal sequences by looking for a basic region at the amino-terminal end of the protein sequence followed by a hydrophobic region. A score is calculated on the basis of the length of the hydrophobic region, its peak value, and the net charge of the amino-terminal region. A large positive score means that there is a high possibility that the protein contains a signal sequence. More methods of analyzing protein sequences to deduce structure and function are described in Chapters 11 to 14.

## Summary

The comparison of different DNA or protein sequences to detect sequence similarity and evolutionary homology is carried out by a process known as sequence alignment. This involves lining up two or more sequences in such a way that the similarities between them are optimized, and then measuring the degree of matching. Alignment is used to find known genes or proteins with sequence similarity to a novel uncharacterized sequence, and forms the basis of programs such as BLAST and FASTA that are used to search sequence databases. Similarities in sequence can help make predictions about a protein's structure and function. Sequences of proteins or DNAs from different organisms can be compared to construct phylogenetic trees, which trace the evolutionary relationships between species or within a family of proteins.

The degree of matching in an alignment is measured by giving the alignment a numerical score, which can be arrived at in several different ways. The simplest scoring method is percentage identity, which counts only the number of matched identical residues, but this relatively crude score will not pick up sequences that are only distantly related. Other scoring methods for protein sequences take into account the likelihood that a given type of amino acid will be substituted for another during evolution, and these methods give pairs of aligned amino acids numerical scores which are summed to obtain a score for the alignment. The probabilities are obtained from reference substitution matrices, which have been compiled from the analysis of alignments of known homologous proteins. Because insertions or deletions often occur as two sequences diverge during evolution, gaps must usually be inserted in either sequence to maximize matching, and scoring schemes exact a penalty for each gap. As there are 20 amino acids, compared to only four different nucleotides, it is easier to detect homology in alignments of protein sequences than in nucleic acid sequences since chance matches are less likely.

There are several different types of alignment. Global alignments estimate the similarity over the whole sequence, whereas local alignments look for short regions of similarity. Local alignments are particularly useful when comparing multi-domain proteins, which may have only one domain in common. Multiple alignments compare a set of similar sequences simultaneously and are more accurate and more powerful than pairwise alignments in detecting proteins with only distant homology to each other.

Algorithms that automate the alignment and scoring process have been devised and are incorporated into various programs. Once an alignment has been scored, the significance of the score has to be tested to determine the likelihood of its arising by chance. Factors such as the length of the alignment and the total number of sequences in the database are taken into account. In database search programs such as BLAST and FASTA, potential matches are evaluated automatically and given a significance score, the *E*-value.

Databases may also be searched to find proteins of similar structure or function by looking for conserved short sequence motifs or discontinuous patterns of residues. These are likely to relate to a functional feature, such as an active site, a binding site, or to structural features. When sufficient members of a protein family have been sequenced, a characteristic profile of the family, summarizing the most typical sequence features, can be derived and can be used to search for additional family members. Database searches can also be widened to include structural information, where available. This is useful for finding homologs which have diverged so much in sequence that their sequence similarity can no longer be detected, but which retain the same overall structure.

# Further Reading

## 4.1 Principles of Sequence Alignment

Bignall G, Micklem G, Stratton MR et al. (1997) The BRC repeats are conserved in mammalian BRCA2 proteins. *Hum. Mol. Genet.* 6, 53–58.

Brenner SE, Chothia C & Hubbard TJP (1998) Assessing sequence comparison methods with reliable structurally identified distant evolutionary relationships. *Proc. Natl Acad. Sci. USA* 95, 6073–6078.

Durbin R, Eddy S, Krogh A & Mitchison G (1998) Biological Sequence Analysis. Cambridge: Cambridge University Press.

Higgins D & Taylor W (eds) (2000) Bioinformatics. Sequence, Structure and Databanks, chapters 3–5, 7. Oxford: Oxford University Press.

Shivji MKK, Davies OR, Savill JM et al. (2006) A region of human BRCA2 containing multiple BRC repeats promotes RAD51-mediated strand exchange. *Nucleic Acids Res.* 34, 4000–4011.

## 4.2 Scoring Alignments

**Twilight zone and midnight zone**

Doolittle RF (1986) Of URFs and ORFs: A Primer on How to Analyze Derived Amino Acid Sequences. Mill Valley, CA: University Science Books.

Doolittle RF (1994) Convergent evolution: the need to be explicit. *Trends Biochem. Sci.* 19, 15–18.

Rost B (1999) Twilight zone of protein sequence alignments. *Protein Eng.* 12, 85–94.

## 4.3 Substitution Matrices

Dayhoff MO, Schwartz RM & Orcutt BC (1978) A model of evolutionary change in proteins. In Atlas of Protein Sequence and Structure (MO Dayhoff ed.), vol. 5, suppl. 3, pp. 345–352. Washington, DC: National Biomedical Research Foundation.

Henikoff S & Henikoff JG (1992) Amino acid substitution matrices from protein blocks. *Proc. Natl Acad. Sci. USA* 89, 10915–10919.

## 4.4 Inserting Gaps

Goonesekere NCW & Lee B (2004) Frequency of gaps observed in a structurally aligned protein pair database suggests a simple gap penalty function. *Nucleic Acids Res.* 32, 2838–2843.

## 4.5 Types of Alignment

Gotoh O (1982) An improved algorithm for matching biological sequences. *J. Mol. Biol.* 162, 705–708.

Needleman SB & Wunsch CD (1970) A general method applicable to the search for similarities in the amino acid sequence of two proteins. *J. Mol. Biol.* 48, 443–453.

Smith TF & Waterman MS (1981) Identification of common molecular subsequences. *J. Mol. Biol.* 147, 195–197.

**ClustalW**

Higgins DW, Thompson JD & Gibson TJ (1996) Using CLUSTAL for multiple sequence alignments. *Methods Enzymol.* 266, 383–402.

**DIALIGN**

Morgenstern B (1999) DIALIGN 2: improvement of the segment-to-segment approach to multiple sequence alignment. *Bioinformatics* 15, 211–218.

Morgenstern B, Frech K, Dress A & Werner T (1998) DIALIGN: finding local similarities by multiple sequence alignment. *Bioinformatics* 14, 290–294.

## 4.6 Searching Databases

**BLAST**

Altschul SF, Gish W, Miller W et al. (1990) Basic Local Alignment Search Tool. *J. Mol. Biol.* 215, 403–410.

**FASTA**

Pearson WR & Lipman DJ (1988) Improved tools for biological sequence comparison. *Proc. Natl. Acad. Sci. USA* 85, 2444–2448.

**PSI-BLAST**

Altschul SF, Madden TL, Schäffer AA et al. (1997) Gapped BLAST and PSI-BLAST: a new generation of protein database search programs. *Nucleic Acids Res.* 25, 3389–3402.

## 4.8 Protein Sequence Motifs or Patterns

**MEME**

Bailey TL & Elkan C (1995) Unsupervised learning of multiple motifs in biopolymers using expectation maximization. *Mach. Learn.* 21, 51–80.

## 4.9 Searching Using Motifs and Patterns

**MOTIF**

Smith HO, Annau TM & Chandrasegaran S (1990) Finding sequence motifs in groups of functionally related proteins. *Proc. Natl Acad. Sci. USA* 87, 826–830.

**PRATT**

Jonassen I (1997) Efficient discovery of conserved patterns using a pattern graph. *Comput. Appl. Biosci.* 13, 509–522.

Jonassen I, Collins JF & Higgins DG (1995) Finding flexible patterns in unaligned protein sequences. *Protein Sci.* 4, 1587–1595.

## 4.10 Patterns and Protein Function

**HCA**

Lemesle-Varloot L, Henrissat B, Gaboriaud C et al. (1990) Hydrophobic cluster analysis: procedures to derive structural and functional information from 2-D-representation of protein sequences. *Biochimie* 72, 555–574.

# PAIRWISE SEQUENCE ALIGNMENT AND DATABASE SEARCHING

THEORY CHAPTER

**When you have read Chapter 5, you should be able to:**

Compare and contrast different scoring schemes.

Summarize the techniques for obtaining the best-scoring alignments of a given type.

Describe ways to reduce the computational resources required.

Speed up database searches using index techniques.

Evaluate approximations used in common database search programs.

Summarize techniques for aligning DNA and protein sequences together.

Identify sequences of low complexity.

Identify significant alignments on the basis of their score.

Summarize the techniques for alignments involving complete genome sequences.

The identification of homologous sequences and their optimal alignment is one of the most fundamental tasks in bioinformatics. As will be seen in many other chapters of this book, there are very few topics in bioinformatics which do not at some stage involve these techniques. A large part of Chapter 4 was devoted to an introduction to some practical aspects of sequence comparison and alignment. In this chapter we will focus in considerable detail on these methods and the science that lies behind them, but will restrict our attention solely to methods of aligning two sequences. The problems of obtaining multiple alignments and profiles and of identifying patterns will be discussed in Chapter 6.

Given two sequences, and allowing gaps to be inserted as was described in Section 4.4, it is possible to construct a very large number of alignments. Of these, there will be an optimal alignment, which in the ideal case perfectly identifies the true equivalences between the sequences. However, there will be many alternative alignments with varying degrees of error that could potentially be seriously misleading. Furthermore, the fact that an alignment can be constructed for any two sequences, even ones with no meaningful equivalences, has the potential to be even more misleading. Therefore, all useful methods of sequence alignment must not only generate alignments but also be able to compare them in a meaningful way and to provide an assessment of their significance.

Both the comparison of alignments and the assessment of their significance require a method of scoring. As discussed in Chapter 4, by considering the evolutionary processes that are responsible for sequence divergence it is possible to find ways of including their salient features in an alignment scoring system. If the scoring

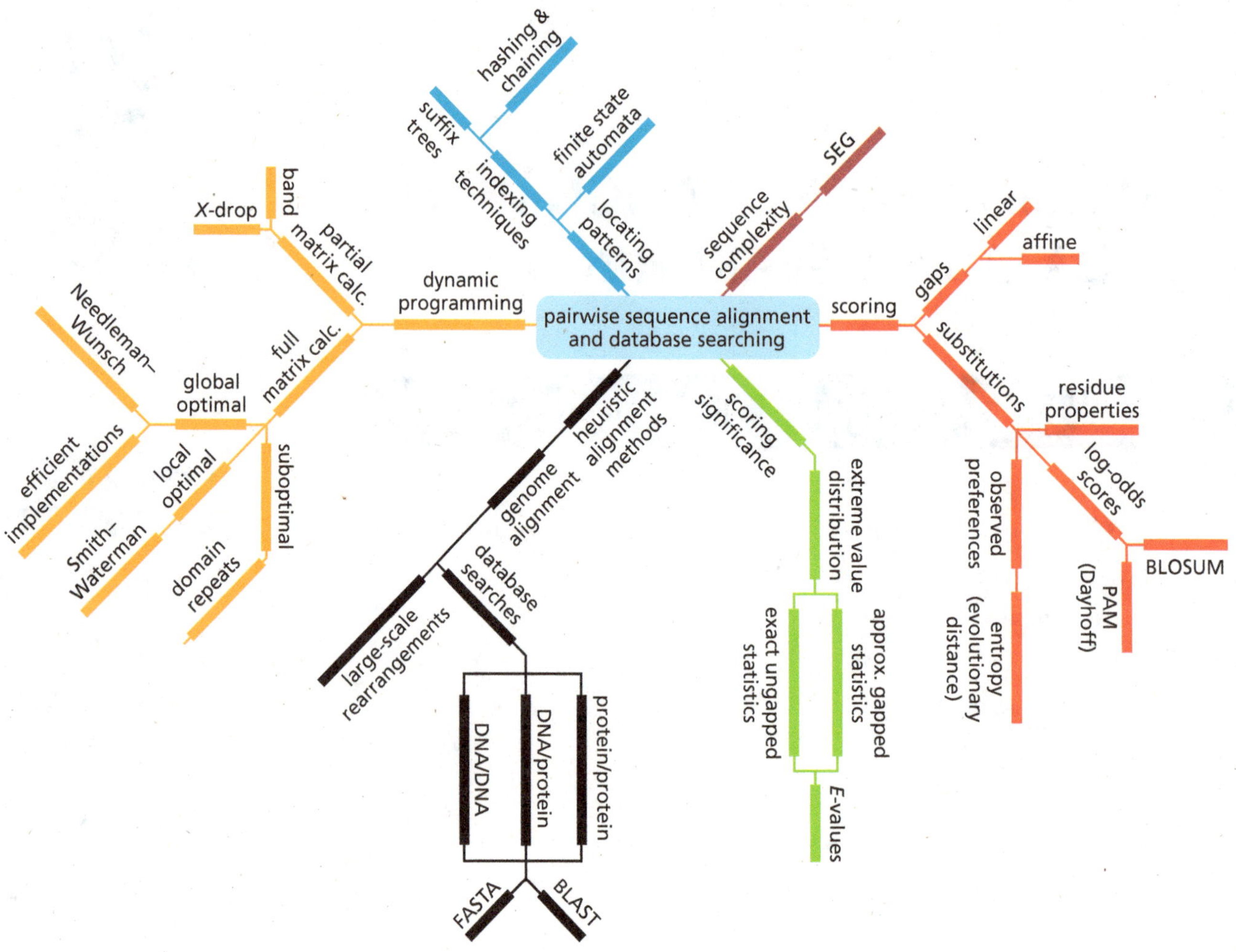


**Mind Map 5.1**
**The theory of pairwise alignment and searching the database depicted in a mind map.**

scheme is accurate and appropriate, all meaningful alignments should have optimal scores, i.e., they will have a better score when compared to any alternative. In Section 5.1 some of the best scoring schemes will be presented.

The number of alternative alignments is so great, however, that efficient methods are required to determine those with optimal scores. Fortunately, algorithms have been derived that can be guaranteed to identify the optimal alignment between two sequences for a given scoring scheme. These methods are described in detail in Section 5.2, although it should be noted that the emphasis is mostly on the scientific rather than the computer science aspects. As long as only single proteins or genes, or small segments of genomes, are aligned, these methods can be applied with ease on today's computers. When searching for alignments of a query sequence with a whole database of sequences it is usual practice to use more approximate methods that speed up the search. These are described in Section 5.3.

Finding the best-scoring alignment between two sequences does not guarantee the alignment has any scientific validity. Ways must be found to discriminate between fortuitously good alignments and those due to a real evolutionary relationship. Section 5.4 presents some of the concepts behind the theory of assessing the statistical significance of alignment scores.

The large number of complete genome sequences has led to increased interest in aligning very long sequences such as whole genomes and chromosomes. As described in Section 5.5, these applications require a number of approximations and techniques to increase the speed and reduce the storage requirements. In

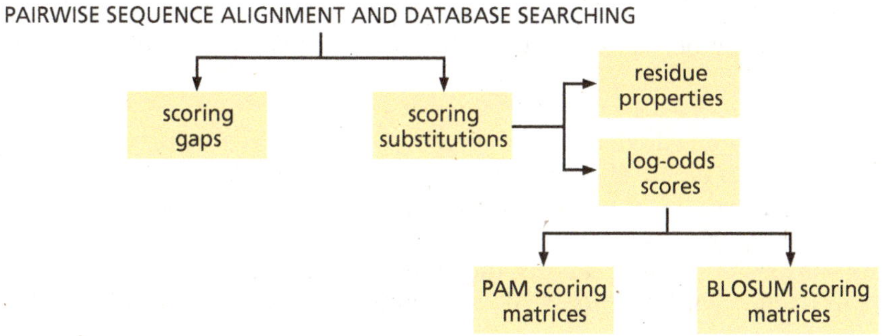

PAIRWISE SEQUENCE ALIGNMENT AND DATABASE SEARCHING

scoring gaps

scoring substitutions

residue properties

log-odds scores

PAM scoring matrices

BLOSUM scoring matrices

**Flow Diagram 5.1**
**The key concept introduced in this section is that if alignments of two sequences are assigned a quantitative score based on evolutionary principles then meaningful comparisons can be made.** Several alternative approaches have been suggested, resulting in a number of different scoring schemes including those which account for insertions and deletions.

addition, the presence of large-scale rearrangements in these sequences has required the development of new algorithms.

## 5.1 Substitution Matrices and Scoring

As discussed in Chapter 4, the aim of an alignment score is to provide a scale to measure the degree of similarity (or difference) between two sequences and thus make it possible to quickly distinguish among the many subtly different alignments that can be generated for any two sequences. Scoring schemes contain two separate elements: the first assigns a value to a pair of aligned residues, while the second deals with the presence of insertions or deletions (indels) in one sequence relative to the other. For protein sequence alignments, reference substitution matrices (see Section 4.3) are used to give a score to each pair of aligned residues. Indels necessitate the introduction of gaps in the alignment, which also have to be scored. The total score $S$ of an alignment is given by summing the scores for individual alignment positions. Special scoring techniques that are applicable only to multiple alignments will be dealt with in Sections 6.1 and 6.4.

One might think that a relatively straightforward way of assessing the probability of the replacement of one amino acid by another would be to use the minimum number of nucleotide base changes required to convert between the codons for the two residues. However, most evolutionary selection occurs at the level of protein function and thus gives rise to significant bias in the mutations that are accepted. Therefore the number of base changes required cannot be expected to be a good measure of the likelihood of substitution. Currently used reference substitution matrices are based on the frequency of particular amino acid substitutions observed in multiple alignments that represent the evolutionary divergence of a given protein. The substitution frequencies obtained thus automatically take account of evolutionary bias.

Two methods that have been used in deriving substitution matrices from multiple sequence alignments will be described here. These have provided two sets of matrices in common use: the PAM and the BLOSUM series. Both these matrices can be related to a probabilistic model, which will be covered first. The key concepts involved in deriving scoring schemes for sequence alignments are outlined in Flow Diagram 5.1.

### Alignment scores attempt to measure the likelihood of a common evolutionary ancestor

The theoretical background of alignment scoring is based on a simple probabilistic approach. Two alternative mechanisms could give rise to differences in DNA or protein sequences: a random model and a nonrandom (evolutionary) model. By generating the probability of occurrence of a particular alignment for each model, an assessment can be made about which mechanism is more likely to have given rise to that alignment.

In the random model, there is no such process as evolution; nor are there any structural or functional processes that place constraints on the sequence and thus cause similarities. All sequences can be assumed to be random selections from a given pool of residues, with every position in the sequence totally independent of every other. Thus for a protein sequence, if the proportion of amino acid type $a$ in the pool is $p_a$, this fraction will be reproduced in the amino acid composition of the protein.

The nonrandom model, on the other hand, proposes that sequences are related—in other words, that some kind of evolutionary process has been at work—and hence that there is a high correlation between aligned residues. The probability of occurrence of particular residues thus depends not on the pool of available residues, but on the residue at the equivalent position in the sequence of the common ancestor; that is, the sequence from which both of the sequences being aligned have evolved. In this model the probability of finding a particular pair of residues of types $a$ and $b$ aligned is written $q_{a,b}$. The actual values of $q_{a,b}$ will depend on the properties of the evolutionary process.

Suppose that in one position in a sequence alignment, two residues, one type $a$ and the other type $b$, are aligned. The random model would give the probability of this occurrence as $p_a p_b$, a product, as the two residues are seen as independent. The equivalent value for the nonrandom model would be $q_{a,b}$. These two models can be compared by taking the ratios of the probabilities, called the **odds ratio**, that is $q_{a,b}/p_a p_b$. If this ratio is greater than 1 (that is, $q_{a,b} > p_a p_b$) the nonrandom model is more likely to have produced the alignment of these residues. However, a single model is required that will explain the complete sequence alignment, so all the individual terms for the pairs of aligned residues must be combined.

In practice, there is often a correlation between adjacent residues in a sequence; for example, when they are in a hydrophobic stretch of a protein such as a transmembrane helix (see Chapter 2). This type of correlation is ignored in both of these models, so that each position in an alignment will be regarded as independent. In that case, the odds ratios for the different positions can be multiplied together to give the odds ratio for the entire alignment:

$$\prod_u \left( \frac{q_{a,b}}{p_a p_b} \right)_u \qquad \text{(EQ5.1)}$$

where the product is over all positions $u$ of the alignment.

It is frequently more practical to deal with sums rather than products, especially when small numbers are involved. This can easily be arranged by taking logarithms of the odds ratio to give the **log-odds ratio**. This ratio can be summed over all positions of the alignment to give $S$, the score for the alignment:

$$S = \sum_u \log \left( \frac{q_{a,b}}{p_a p_b} \right)_u = \sum_u \left( s_{a,b} \right)_u \qquad \text{(EQ5.2)}$$

where $s_{a,b}$ is the score (that is, the **substitution matrix** element) associated with the alignment of residue types $a$ and $b$. A positive value of $s_{a,b}$ means that the probability of those two residues being aligned is greater in the nonrandom than in the random model. The converse is true for negative $s_{a,b}$ values. $S$ is a measure of the relative likelihood of the whole alignment arising due to the nonrandom model as compared with the random model. However, as discussed later in this chapter, a

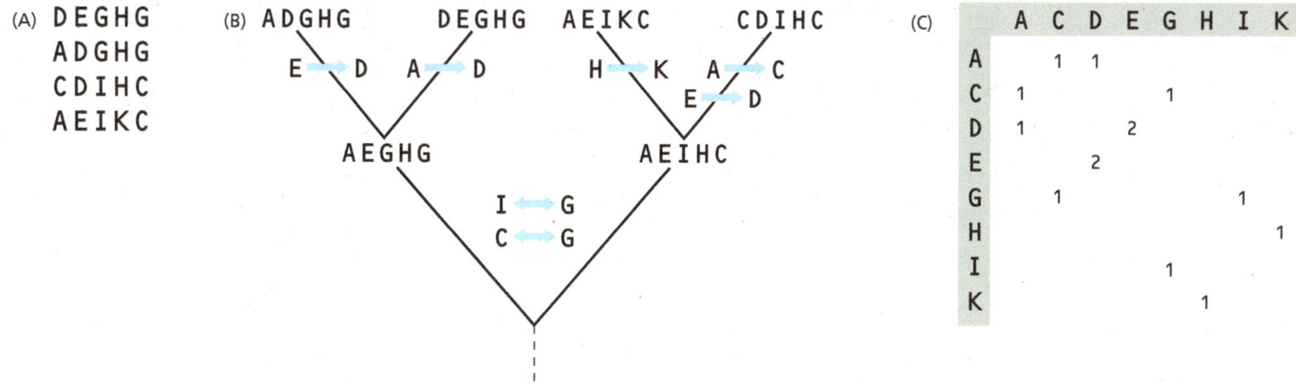


| | A | C | D | E | G | H | I | K |
|---|---|---|---|---|---|---|---|---|
| A | | 1 | 1 | | | | | |
| C | 1 | | | | 1 | | | |
| D | 1 | | | 2 | | | | |
| E | | | 2 | | | | | |
| G | | 1 | | | | | 1 | |
| H | | | | | | | | 1 |
| I | | | | | 1 | | | |
| K | | | | | | 1 | | |

**Figure 5.1**
**The identification of accepted point mutations in the derivation of the PAM/MDM amino acid substitution scoring matrices.** (A) Part of an alignment that might have occurred in the protein sequence data used to obtain the Dayhoff PAM matrices. (B) The phylogenetic tree derived from this alignment. Note that this tree hypothesizes the smallest number of mutations required to explain the observed sequences, but does not necessarily represent the actual evolutionary history. See Chapters 7 and 8 for further details on phylogenetic trees. The direction of some mutations cannot be resolved due to lack of knowledge of the oldest ancestral sequence. This is also the case for the other ancestral sequences shown, which are specified here for clarity only. (C) The matrix of accepted substitutions derived from this tree. Note that all observed mutations have been assumed to be equally likely in either direction, and hence are counted in both directions.

positive $S$ is not a sufficient test of the alignment's significance. There will be a distribution of values of $S$ for a given set of sequences, which can be used to determine significant scores.

From this discussion one would expect there to be both positive and negative $s_{a,b}$ values. In practice this is not always the case, because each $q_{a,b}/p_a p_b$ term can be multiplied by a constant. Multiplication by a constant $X$ will result in a term $\log X$ being added to each $s_{a,b}$. This could result in all $s_{a,b}$ values being positive, for example. The alignment score $S$ is shifted by $L_{aln} \log X$ for an alignment of length $L_{aln}$. Similarly, all the $s_{a,b}$ values can be multiplied by a constant. In both cases, scores of alternative alignments of the same length retain the same relative relationship to each other. However, local alignments discussed below involve comparing alignments of different lengths, in which case adding a constant $\log X$ to each $s_{a,b}$ will have an effect on the relative scores.

Note that the link between substitution matrices and the log-odds ratio described by Equation EQ 5.2 may exist even if the matrix was derived without any reference to this theory. Firstly, note that the **expected score** for aligning two residues can be written

$$E\left(s_{a,b}\right) = \sum_{a,b} p_a p_b s_{a,b} \tag{EQ5.3}$$

If this is negative, and there is at least one positive score, one can in principle perform the reverse procedure to obtain the $q_{a,b}$ given the $s_{a,b}$ and the $p_a$. The procedure is not entirely straightforward, and interested readers should refer to the Further Reading at the end of the chapter.

## The PAM (MDM) substitution scoring matrices were designed to trace the evolutionary origins of proteins

As we saw in Section 4.3, one commonly used type of matrix for protein sequences is the point accepted mutations (PAM) matrix, also known as the mutation data matrix (MDM), derived by Margaret Dayhoff and colleagues from the analysis of multiple alignments of families of proteins, including cytochrome *c*, α- and β-globin (the protein chains of which hemoglobin is composed), insulin A and B chains, and ferredoxin.

These raw data are biased by the amino acid composition of the sequences, the differing rates of mutation in different protein families, and sequence length. The first step in an attempt to remove this bias is to calculate, for each alignment, the exposure of each type of amino acid residue to mutation. This is defined as the compositional fraction of this residue type in the alignment multiplied by the total number of mutations (of any kind) that have occurred per 100 alignment positions.

See below for a description of how the number of mutations is calculated. The value for each residue type is summed over all the alignments in the data set to give the total exposure of that residue type. The mutability of a specific residue type *a* is defined as the ratio of the total number of mutations in the data that involve residue *a* divided by the total exposure of residue *a*. Usually this is reported relative to alanine as a standard, and is referred to as the **relative mutability**, $m_a$. A few residues have higher mutability than alanine, but more notable are those that are less likely to change, especially cysteine and tryptophan. The mutability of these residues is approximately one-fifth that of alanine.

Phylogenetic trees are constructed for each alignment by a method that infers the most likely ancestral sequence at each internal node and hence postulates all the mutations that have occurred. The observed substitutions are tabulated in a matrix, ***A*** (**accepted point mutation matrix**) according to the residue types involved (see Figure 5.1). It is assumed that a mutation from residue type *a* to type *b* was as likely as a mutation from *b* to *a*, so all observed mutations are included in the count for both directions. Where there is uncertainty as to which mutations have occurred, all possibilities are treated as equally likely, and fractional mutations are added to the matrix of accepted substitutions. The matrix element values are written $A_{a,b}$. Dayhoff's 1978 dataset contained 1572 observed substitutions, but even so, 35 types of mutations had not been observed, for example tryptophan (W) → alanine (A). This was due to the relatively small dataset of highly similar sequences that was used.

From this information a **mutation probability matrix**, ***M***, can be defined. Each element $M_{a,b}$ gives the probability that a residue of type *b* will be replaced by one of type *a* after a given period of evolutionary time. The residue *b* has a likelihood of mutating that is proportional to $m_b$. The expected fraction of mutations of *b* into residue *a* can be obtained from the accepted point mutation matrix (see Figure 5.1C) with elements $A_{a,b}$. Thus the off-diagonal ($a \neq b$) terms of *M* are given by the formula

$$M_{a,b} = \frac{\Lambda m_b A_{a,b}}{\sum_a A_{a,b}} \qquad \text{(EQ5.4)}$$

where $\Lambda$ is a constant that accounts in part for the proportionality constant of $m_b$ and in part for the unspecified evolutionary time period. Note that matrix ***M*** is not symmetrical because of the residue relative mutability $m_b$. The diagonal terms ($a$ = $b$) of this matrix, corresponding to no residue change, are

$$M_{b,b} = 1 - \Lambda m_b \qquad \text{(EQ5.5)}$$

Note that the sum of all off-diagonal elements $M_{a,b}$ involving mutation from a given residue *b* to another type and the element $M_{b,b}$ equals 1, as required for ***M*** to be a probability matrix.

The percentage of amino acids unchanged in a sequence of average composition after the evolutionary change represented by the matrix ***M*** can be calculated as

$$100\sum_b f_b M_{b,b} = 100\sum_b f_b\left(1 - \Lambda m_b\right) \qquad \text{(EQ5.6)}$$

where $f_b$ is approximately the frequency of residue type *b* in the average composition. In fact $f_b$ is the total exposure of residue *b* normalized so that the sum of all values is 1, and is the residue composition weighted by the sequence mutation rate. The value of $\Lambda$ is selected to determine this percentage of unchanged residues. If $\Lambda$ is chosen to make this sum 99, the matrix represents an evolutionary change of 1 PAM (that is, one accepted mutation per 100 residues).

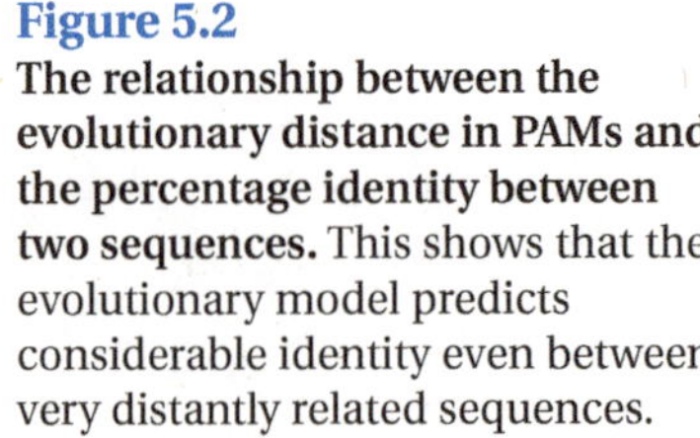

**Figure 5.2**
**The relationship between the evolutionary distance in PAMs and the percentage identity between two sequences.** This shows that the evolutionary model predicts considerable identity even between very distantly related sequences.

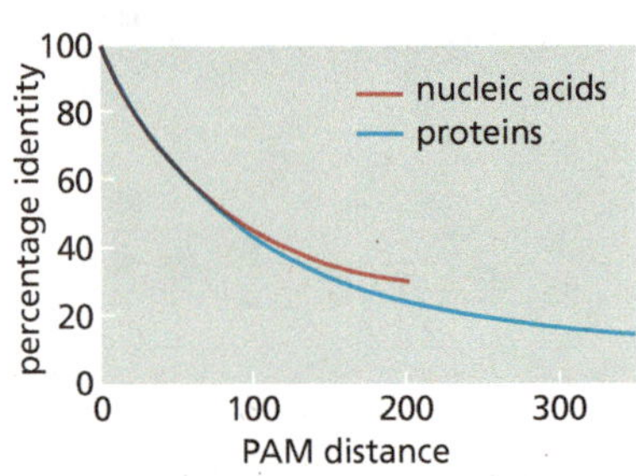

| | | | | | | | | | | | | | | | | | | | | |
|---|---|---|---|---|---|---|---|---|---|---|---|---|---|---|---|---|---|---|---|---|
| **C** | 11 | | | | | | | | | | | | | | | | | | | |
| **S** | 1 | 2 | | | | | | | | | | | | | | | | | | |
| **T** | -1 | 1 | 2 | | | | | | | | | | | | | | | | | |
| **P** | -2 | 1 | 1 | 6 | | | | | | | | | | | | | | | | |
| **A** | -1 | 1 | 2 | 1 | 2 | | | | | | | | | | | | | | | |
| **G** | -1 | 1 | -1 | -1 | 1 | 5 | | | | | | | | | | | | | | |
| **N** | -1 | 1 | 1 | -1 | 0 | 0 | 3 | | | | | | | | | | | | | |
| **D** | -3 | 0 | -1 | -2 | 0 | 1 | 2 | 5 | | | | | | | | | | | | |
| **E** | -4 | -1 | -1 | -2 | -1 | 0 | 1 | 4 | 5 | | | | | | | | | | | |
| **Q** | -3 | -1 | -1 | 0 | -1 | -1 | 0 | 1 | 2 | 5 | | | | | | | | | | |
| **H** | 0 | -1 | -1 | 0 | -2 | -2 | 1 | 0 | 0 | 2 | 6 | | | | | | | | | |
| **R** | -1 | -1 | -1 | -1 | -1 | 0 | 0 | -1 | 0 | 2 | 2 | 5 | | | | | | | | |
| **K** | -3 | -1 | -1 | -2 | -1 | -1 | 1 | 0 | 1 | 2 | 1 | 4 | 5 | | | | | | | |
| **M** | -2 | -1 | 0 | -2 | -1 | -3 | -2 | -3 | -3 | -2 | -2 | -2 | -2 | 6 | | | | | | |
| **I** | -2 | -1 | 1 | -2 | 0 | -3 | -2 | -3 | -3 | -3 | -3 | -3 | -3 | 3 | 4 | | | | | |
| **L** | -3 | -2 | -1 | 0 | -1 | -4 | -3 | -4 | -4 | -2 | -2 | -3 | -3 | 3 | 2 | 5 | | | | |
| **V** | -2 | -1 | 0 | -1 | 1 | -2 | -2 | -2 | -2 | -3 | -3 | -3 | -3 | 2 | 4 | 2 | 4 | | | |
| **F** | 0 | -2 | -2 | -3 | -3 | -5 | -3 | -5 | -5 | -4 | 0 | -4 | -5 | 0 | 0 | 2 | 0 | 8 | | |
| **Y** | 2 | -1 | -3 | -3 | -3 | -4 | -1 | -2 | -4 | -2 | 4 | -2 | -3 | -2 | -2 | -1 | -3 | 5 | 9 | |
| **W** | 1 | -3 | -4 | -4 | -4 | -2 | -5 | -5 | -5 | -3 | -3 | 0 | -3 | -3 | -4 | -2 | -3 | -1 | 0 | 15 |
| | **C** | **S** | **T** | **P** | **A** | **G** | **N** | **D** | **E** | **Q** | **H** | **R** | **K** | **M** | **I** | **L** | **V** | **F** | **Y** | **W** |

**Figure 5.3**
**The PET91 version of the PAM250 substitution matrix.** Scores that would be given to identical matched residues are in blue; positive scores for nonidentical matched residues are in red. The latter represent pairs of residues for which substitutions were observed relatively often in the aligned reference sequences.

According to the Dayhoff model of evolution, to obtain the probability matrices for higher percentages of accepted mutations, the 1-PAM matrix is multiplied by itself. This is because the model of evolution proposed was a Markov process (Markov models are described in detail in Section 10.3). If the 1-PAM matrix is squared, it gives a 2-PAM matrix; if cubed, a 3-PAM matrix; and so on. These correspond to evolutionary periods twice and three times as long as the period used to derive the 1-PAM matrix. Similarly, a 250-PAM matrix is obtained by raising the 1-PAM matrix to the 250th power.

These matrices tell us how many mutations have been accepted, but not the percentage of residues that have mutated: some may have mutated more than once, others not at all. For each of these matrices, evaluation of Equation EQ5.6 gives the actual percentage identity to the starting sequence expected after the period of evolution represented by the matrix. The percentage identity does not decrease linearly with time in this model (see Figure 5.2).

So far, a probability matrix has been obtained, but not a scoring matrix. As discussed earlier, such a score should involve the ratio of probabilities derived from nonrandom and random models. The matrix $\boldsymbol{M}$ gives the probability of residue $b$ mutating into residue $a$ if the two sequences are related, that is, if substitution is nonrandom. It already includes a term for the probability of occurrence of residue $b$ in the total exposure term used to calculate $m_b$. Thus the only term needed for the random-model probability is $f_a$, the likelihood of residue $a$ occurring by chance. Hence, the scoring matrix was originally derived from $\boldsymbol{M}$ by the formula

$$s_{a,b} = 10\log_{10}\left(\frac{M_{a,b}}{f_a}\right) \qquad \text{(EQ5.7)}$$

and is in fact a symmetrical matrix. The resultant scoring matrices *s* are usually named PAM*n*, where *n* is the number of accepted point mutations per 100 residues in the probability matrix ***M*** from which they are derived. The exact scaling factors and logarithm base are to some degree arbitrary, and are usually chosen to provide integer scores with a suitable number of significant figures. In Figure 4.4B the PAM120 matrix is shown, but a scaling factor of $2\log_2$ has been used instead of $10\log_{10}$, so that the values are in units of half bits (a bit is a measure of information) (see Appendix A).

There was a lack of sequence data available when the original work was done to derive the PAM matrices, and in 1991 the method was applied to a larger dataset, producing the PET91 matrix that is an updated version of the original PAM250 matrix (see Figure 5.3). This matrix shows considerable differences for aligning two tryptophan (W) residues, with a score of 15, as opposed to two alanine residues (A), which score only 2. About one-fifth of the scores for nonidentical residues have positive scores, considerably more than occurs in PAM120 (see Figure 4.4B), reflecting the longer evolutionary period represented by the matrix.

## The BLOSUM matrices were designed to find conserved regions of proteins

The BLOCKS database containing large numbers of ungapped multiple local alignments of conserved regions of proteins, compiled by Steven and Jorja Henikoff, became available in 1991 (see Section 4.8). These alignments included distantly related sequences, in which multiple base substitutions at the same position could be observed. The BLOCKS database was soon recognized as a resource from which substitution preferences could be determined, leading to the BLOSUM substitution score matrices.

There are two contrasts with the data analysis used to obtain the PAM matrices. First, the BLOCKS alignments used to derive the BLOSUM matrices include sequences that are much less similar to each other than those used by Dayhoff, but whose evolutionary homology can be confirmed through intermediate sequences. In addition, these alignments are analyzed without creating a phylogenetic tree and are simply compared with each other.

A direct comparison of aligned residues does not model real substitutions, because in reality the sequences have evolved from a common ancestor and not from each other. Nevertheless, as the large sequence variation prevents accurate construction of a tree, there is no alternative. However, if the alignment is correct, then aligned residues will be related by their evolutionary history and therefore their alignment will contain useful information about substitution preferences. Another argument in favor of direct analysis of aligned sequence differences is that often the aim is not to recreate the evolutionary history, but simply to try to align sequences to test them for significant similarity. The intention in producing the BLOSUM matrices was to find scoring schemes that would identify conserved regions of protein sequences.

One of the key aspects of the analysis of the alignment blocks is to weight the sequences to try to reduce any bias in the data. This is necessary because the sequence databases are highly biased toward certain species and types of proteins, which means there are many very similar sequences present. The weighting involves clustering the most similar sequences together, and different matrices are produced according to the threshold *C* used for this clustering. Sequences are clustered together if they have ≥ *C*% identity, and the substitution statistics are calculated only between clusters, not within them. Weights are determined according to the number of sequences in the cluster. For a cluster of $N_{seq}$ sequences, each sequence is assigned a weight of $1/N_{seq}$. The weighting scheme was used to obtain a series of substitution matrices by varying the value of *C*, with the matrices named

(A)

| | 1 2 3 4 5 |
|---|---|
| 1 | ATCKQ |
| 2 | ATCRN |
| 3 | ASCKN |
| 4 | SSCRN |
| 5 | SDCEQ |
| 6 | VDCEN |
| 7 | TECRQ |

(B)

| | $q_{QN}$ | $q_{NN}$ | $q_{QQ}$ | $p_N$ | $p_Q$ |
|---|---|---|---|---|---|
| C=62% | 0.114 | 0.057 | 0.029 | 0.171 | 0.143 |
| C=50% | 0.117 | 0.025 | 0.058 | 0.142 | 0.175 |
| C=40% | – | – | – | – | – |

**Figure 5.4**
**Derivation of the BLOSUM amino acid substitution scoring matrices.** (A) An example ungapped alignment block with one fully conserved position (cysteine) colored red. In this case, sequences have been clustered if they are at least 50% identical to any other within the cluster. Thus for example, although sequences 1 and 4 are only 20% identical to each other, they are in the same cluster, as they are both 60% identical to sequence 2. Similar clustering of sequence data was used to derive the BLOSUM-50 matrix. If the same data were used to derive the BLOSUM-62 matrix (that is, $C = 62\%$) none of these sequences would cluster together, as no two of them share more than 60% identical residues, leading to very different sequence weights. (B) Values of $q_{a,b}$ and $p_a$ for asparagine (N) and glutamine (Q) residues for three different values of $C$. For $C = 40\%$ all the sequences belong in the same cluster, and since $q_{a,b}$ and $p_a$ measure intercluster alignment statistics, they cannot be calculated in this case.

as BLOSUM-62, for example, in the case where $C = 62\%$. The sequences of all the alignments used to obtain the Dayhoff matrices fall in a single cluster for $C \leq 85\%$, indicating that they were much more similar to each other than those used for BLOSUM.

The derivation of substitution data will be illustrated using the example alignment in Figure 5.4A and the case of $C = 50\%$, which would lead to the BLOSUM-50 matrix. There are three sequence clusters, with four, two, and one sequences, giving weights to their constituent sequences of $^1/_4$, $^1/_2$, and 1, respectively. These weights are applied to the counts of observed aligned residues to produce the weighted frequencies $f_{a,b}$. The observed probability ($q_{a,b}$) of aligning residues of types $a$ and $b$ is given by

$$q_{a,b} = \frac{f_{a,b}}{\sum\limits_{1 \leq b \leq a}^{20} f_{a,b}} \tag{EQ5.8}$$

Note that this ignores which sequence the residue $a$ or $b$ has come from, so that $f_{a,b}$ and $f_{b,a}$, and hence $q_{a,b}$ and $q_{b,a}$, are equal.

Consider the calculation of $q_{a,b}$ for asparagine (N) and glutamine (Q) residues, which occur only in column 5 in Figure 5.4A. If $C = 62\%$, then all sequence clusters will contain just one sequence and each sequence will have a weight of 1. (No pair of sequences share more than 60% identical residues.) In this case there are 21 ($7 \times 6/2$) distinct cluster pairs and thus 21 pairs of aligned residues in any single alignment column. Counting these, there are 12 QN pairs ($= f_{Q,N}$), 3 QQ ($= f_{Q,Q}$), and 6 NN ($= f_{N,N}$), making a total of 21 pairs. As all sequence weights are 1, they play no real part in this calculation. As there are five alignment columns, the total number of aligned pairs (the denominator of Equation EQ5.8) is 105. From these data the $q_{a,b}$ can be obtained, as listed in Figure 5.4B. Note that if other columns had contained N and Q, they would also have needed to be included in the calculation.

Considering the case of $C = 50\%$, the sequences separate into three clusters, so the total number of cluster pairs at position 5 will be 3 ($3 \times 2/2$). The top, middle, and bottom clusters can be regarded as being {$^1/_4$Q, $^3/_4$N}, {$^1/_2$Q, $^1/_2$N}, and {Q}, respectively, at this position. Remembering to consider only pairs between clusters and not within them, the weighted number of QN aligned pairs is calculated as

$$f_{Q,N} = \left(\frac{1}{4} \times \frac{1}{2}\right) + \left(\frac{3}{4} \times \frac{1}{2}\right) + \left(\frac{3}{4} \times 1\right) + \left(\frac{1}{2} \times 1\right) = \frac{14}{8} \tag{EQ5.9}$$

where the first term is for Q residues of the top cluster and N residues of the second cluster; the second term is for N residues of the top cluster and Q residues of the

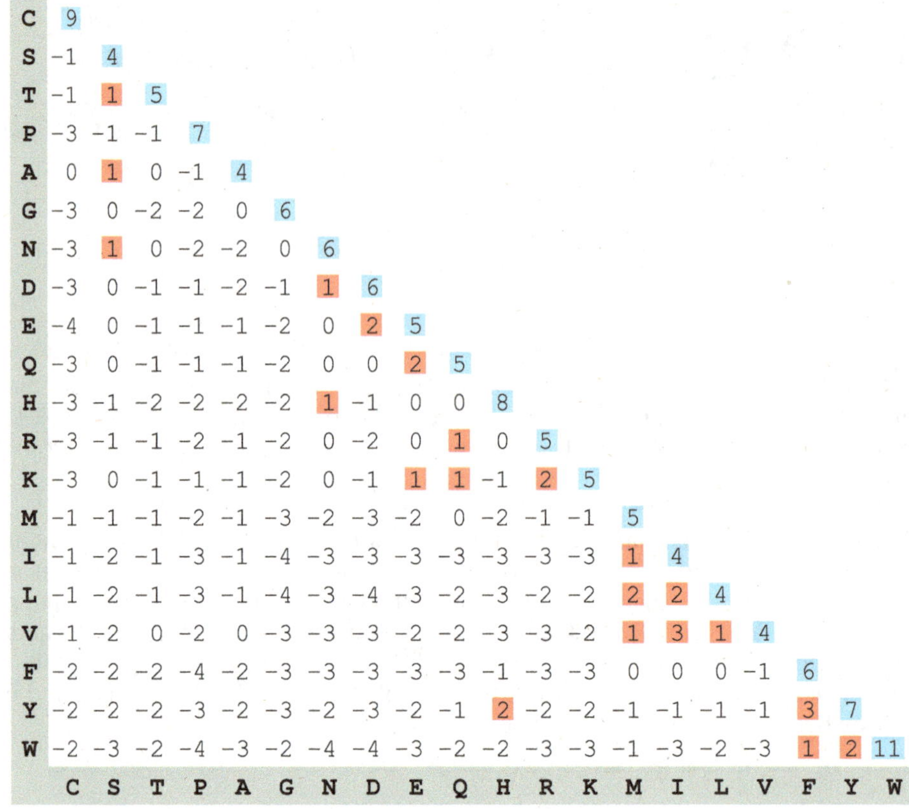

| | C | S | T | P | A | G | N | D | E | Q | H | R | K | M | I | L | V | F | Y | W |
|---|---|---|---|---|---|---|---|---|---|---|---|---|---|---|---|---|---|---|---|---|
| **C** | 9 | | | | | | | | | | | | | | | | | | | |
| **S** | -1 | 4 | | | | | | | | | | | | | | | | | | |
| **T** | -1 | 1 | 5 | | | | | | | | | | | | | | | | | |
| **P** | -3 | -1 | -1 | 7 | | | | | | | | | | | | | | | | |
| **A** | 0 | 1 | 0 | -1 | 4 | | | | | | | | | | | | | | | |
| **G** | -3 | 0 | -2 | -2 | 0 | 6 | | | | | | | | | | | | | | |
| **N** | -3 | 1 | 0 | -2 | -2 | 0 | 6 | | | | | | | | | | | | | |
| **D** | -3 | 0 | -1 | -1 | -2 | -1 | 1 | 6 | | | | | | | | | | | | |
| **E** | -4 | 0 | -1 | -1 | -1 | -2 | 0 | 2 | 5 | | | | | | | | | | | |
| **Q** | -3 | 0 | -1 | -1 | -1 | -2 | 0 | 0 | 2 | 5 | | | | | | | | | | |
| **H** | -3 | -1 | -2 | -2 | -2 | -2 | 1 | -1 | 0 | 0 | 8 | | | | | | | | | |
| **R** | -3 | -1 | -1 | -2 | -1 | -2 | 0 | -2 | 0 | 1 | 0 | 5 | | | | | | | | |
| **K** | -3 | 0 | -1 | -1 | -1 | -2 | 0 | -1 | 1 | 1 | -1 | 2 | 5 | | | | | | | |
| **M** | -1 | -1 | -1 | -2 | -1 | -3 | -2 | -3 | -2 | 0 | -2 | -1 | -1 | 5 | | | | | | |
| **I** | -1 | -2 | -1 | -3 | -1 | -4 | -3 | -3 | -3 | -3 | -3 | -3 | -3 | 1 | 4 | | | | | |
| **L** | -1 | -2 | -1 | -3 | -1 | -4 | -3 | -4 | -3 | -2 | -3 | -2 | -2 | 2 | 2 | 4 | | | | |
| **V** | -1 | -2 | 0 | -2 | 0 | -3 | -3 | -3 | -2 | -2 | -3 | -3 | -2 | 1 | 3 | 1 | 4 | | | |
| **F** | -2 | -2 | -2 | -4 | -2 | -3 | -3 | -3 | -3 | -3 | -1 | -3 | -3 | 0 | 0 | 0 | -1 | 6 | | |
| **Y** | -2 | -2 | -2 | -3 | -2 | -3 | -2 | -3 | -2 | -1 | 2 | -2 | -2 | -1 | -1 | -1 | -1 | 3 | 7 | |
| **W** | -2 | -3 | -2 | -4 | -3 | -2 | -4 | -4 | -3 | -2 | -2 | -3 | -3 | -1 | -3 | -2 | -3 | 1 | 2 | 11 |

**Figure 5.5**
**The BLOSUM-62 substitution matrix scores in half bits.** Scores that would be given to identical matched residues are in blue; positive scores for nonidentical matched residues are in red. The latter represent pairs of residues for which substitutions were observed relatively often in the aligned reference sequences.

second cluster; the third term is for the Q residue of the third cluster; and the fourth term is for the N residues of each of the other clusters. The equivalent values for $f_{\mathrm{N,N}}$ and $f_{\mathrm{Q,Q}}$ are 3/8 and 7/8, respectively. Dividing these $f_{a,b}$ values by 15 (the weighted total number of aligned pairs in the data, three for each alignment position) gives the $q_{a,b}$ values shown in Figure 5.4B. The values for $f_{\mathrm{Q,Q}}$ and $f_{\mathrm{N,N}}$ differ between $C$ = 62% and $C$ = 50%, because in the latter case most of the N residues are in one cluster. Note that in the case of $C$ = 40% there is only one cluster, as each sequence is at least 40% identical to at least one other, so that no intercluster residue pairing exists to be counted! This example shows how the value of $C$ can affect the derived substitution matrix in a complicated manner.

The scores for the residue pairs are obtained using the log-odds approach described earlier. The estimate $e_{a,b}$ of the probability of observing two residues of type $a$ and $b$ aligned by chance is given (as in deriving PAM matrices) by a weighted composition of the sequences, with $p_a$ being approximately the fraction of all residues that is type $a$. The background residue frequencies $p_a$ are defined by

$$p_a = q_{a,a} + \sum_{a \neq b} \frac{q_{a,b}}{2} \qquad \text{(EQ5.10)}$$

(A)

| | A | C | G | T |
|---|---|---|---|---|
| **A** | 67 | -96 | -20 | -117 |
| **C** | -96 | 100 | -79 | -20 |
| **G** | -20 | -79 | 100 | -96 |
| **T** | -117 | -20 | -96 | 67 |

(B)

| | A | C | G | T |
|---|---|---|---|---|
| **A** | 91 | -114 | -31 | -123 |
| **C** | -114 | 100 | -125 | -31 |
| **G** | -31 | -125 | 100 | -114 |
| **T** | -123 | -31 | -114 | 91 |

(C)

| | A | C | G | T |
|---|---|---|---|---|
| **A** | 100 | -123 | -28 | -109 |
| **C** | -123 | 91 | -140 | -28 |
| **G** | -28 | -140 | 91 | -123 |
| **T** | -109 | -28 | -123 | 100 |

**Figure 5.6**
**Three nucleotide substitution scoring matrices, derived by Chiaromonte and co-workers.** Each matrix was obtained by analysis of alignments of distinct regions of the human and mouse genomes with different G+C content. (A) was derived from the CFTR region which is 37% G+C; (B) was derived from the HOXD region which is 47% G+C; (C) was derived from the hum16pter region which is 53% G+C. Each matrix was scaled to give a maximum score of 100.

Figure 5.4B shows how these are affected by the choice of *C*. For identical residues, $e_{a,a}$ is $p_a^2$, whereas for different residues, $e_{a,b}$ is $2p_ap_b$, because there are two ways in which one can select two different residues. The BLOSUM matrices are defined using information measured in bits, so the formula for a general term is

$$s_{a,b} = \log_2\left(\frac{q_{a,b}}{e_{a,b}}\right) \qquad \text{(EQ5.11)}$$

One of the most commonly used of these matrices is BLOSUM-62, which is shown in Figure 5.5 in units of half bits, obtained by multiplying the $s_{a,b}$ in Equation EQ5.11 by 2.

## Scoring matrices for nucleotide sequence alignment can be derived in similar ways

The methods described above for deriving scoring matrices have been illustrated with examples from protein sequences. The same techniques can be applied to nucleotides, although often simple scoring schemes such as +5 for a match and –4 for a mismatch are used. With certain exceptions, such as 16S rRNA and repeat sequences, until quite recently almost all sequences studied were for protein-coding regions, in which case it is usually advantageous to align the protein sequences. This has changed with the sequencing of many genomes and the alignment of long multigene segments or even whole genomes.

Matrices have been reported that are derived from alignments of human and mouse genomic segments (see Figure 5.6). The different matrices were derived from sequence regions with different G+C content, as it is thought that this influences the substitution preferences. This is to be contrasted with the different PAM and BLOSUM matrices, which are based on evolutionary distance. The matrices were derived using a similar approach to that described for the BLOSUM series. However, there is a difference worth noting when dealing with DNA sequences. Any alignment of DNA sequences implies a second alignment of the complementary strand. Thus, every observation of a T/C aligned pair implies in addition an aligned A/G pair.

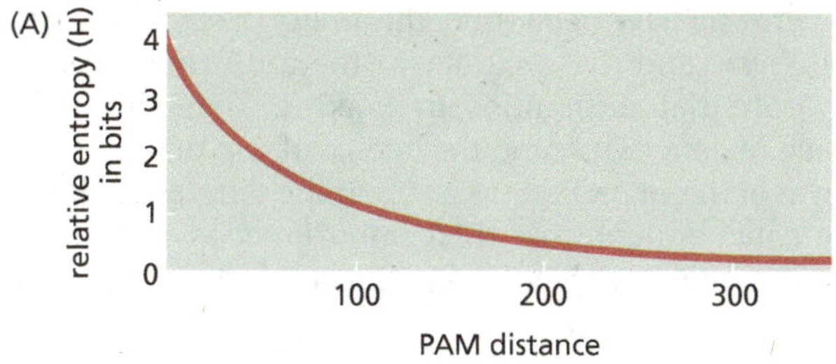


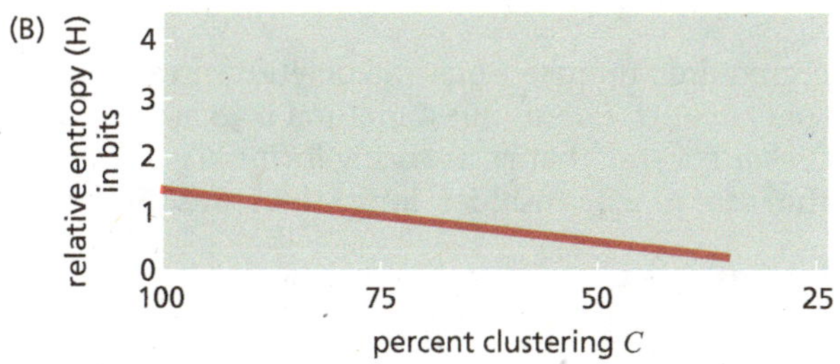


**Figure 5.7**
**Plots of the relative entropy *H* in bits for two different series of substitution matrices.** (A) Plot for the PAM matrices according to PAM distance (Data from Altschul, 1991.) (B) Plot for the BLOSUM matrices according to the percentage cut-off *C* used in clustering alignment blocks. (Data from Henikoff and Henikoff, 1992.)

## The substitution scoring matrix used must be appropriate to the specific alignment problem

Many other substitution scoring matrices have been derived. Some are based on alternative ways of analyzing the alignment data, while others differ in the dataset used. Some are intended for specialized use, such as aligning transmembrane sequences. The matrices can be compared in three ways, focusing on (1) the relative patterns of scores for different residue types, (2) the actual score values, or (3) the practical application of the matrices.

Cluster analysis of the scores can be used to see if matrices distinguish between the amino acids in different ways. For example, one matrix may be strongly dominated by residue size, another by polarity. This may improve our understanding of the evolutionary driving forces or, alternatively, highlight shortcomings in the sequence data used to derive the matrix, but probably will not assist in determining the relative usefulness of matrices for constructing alignments.

In common with most proposed amino acid substitution matrices, the PAM and BLOSUM matrix series include both positive and negative scores, with the average score being negative. The actual score values can be summarized in two measures that have some practical use. The **relative entropy** ($H$) of the nonrandom model with respect to the random model is defined as

$$H = \sum_{a,b} q_{a,b} s_{a,b} = \sum_{a,b} q_{a,b} \log\left(\frac{q_{a,b}}{p_a p_b}\right) \qquad \text{(EQ5.12)}$$

The scores $s_{a,b}$ are summed, weighted by $q_{a,b}$, and $H$ is a measure of the average information available at each alignment position to distinguish between the nonrandom and random models, and is always positive. (See Appendix A for further discussion of this measure.) Figure 5.7 shows the variation of $H$ for different PAM and BLOSUM matrices. The shortest local alignment that can have a significant score is in part dependent on the relative entropy of the scoring matrix used, as discussed later in the chapter. The other measure of score values is the expected score, defined in Equation EQ5.3, which is usually—but not necessarily—negative, for example –0.52 for BLOSUM-62. This measure has been found to influence the variation of alignment scores with alignment length.

Perhaps the best way to compare matrices is to see how well they perform with real data. Two different criteria have been used: the ability to discover related sequences in searching a database, and the accuracy of the individual alignments derived. There are many potential difficulties in making a meaningful comparison, including the choice of data to use and the choice of gap penalties. Although some matrices do perform better at certain tasks, often the differences for the commonly used matrices are small enough that their importance is unclear, especially as a poor choice of gap penalties can have a significant effect.

## Gaps are scored in a much more heuristic way than substitutions

A scoring scheme is required for insertions and deletions in alignments, as they are common evolutionary events. The simplest method is to assign a gap penalty $g$ on aligning any residue with a gap; that is, a scoring formula $g = -E$, where $E$ is a positive number. If the gap is $n_{\text{gap}}$ residues long, then this **linear gap penalty** is defined as

$$g\left(n_{\text{gap}}\right) = -n_{\text{gap}}E \tag{EQ5.13}$$

Usually, no account is taken of the type of residue aligned with the gap, although making the value of $E$ vary with residue type would easily do this. The observed preference for fewer and longer gaps can be modeled by using a higher penalty to initiate a gap [the **gap opening penalty** (GOP), designated $I$] and then a lower penalty to extend an existing gap [the **gap extension penalty** (GEP), designated $E$]. This leads to the **affine gap penalty** formula

$$g\left(n_{\text{gap}}\right) = -I - \left(n_{\text{gap}} - 1\right)E \tag{EQ5.14}$$

for a gap of $n_{\text{gap}}$ residues. Note that an alternative definition can give rise to the formula

$$g\left(n_{\text{gap}}\right) = -I - n_{\text{gap}}E \tag{EQ5.15}$$

Again, residue preferences can easily be added to this scheme by varying the value of $E$.

The values of the gap parameters need to be carefully chosen for the specific substitution scoring matrix used. Failure to optimize these parameters can significantly degrade the performance of the overall scoring scheme. This is illustrated by the worked example later in the chapter. Some matrices seem less sensitive to gap parameterization than others. In practice, as optimization is a lengthy process, most workers use previously reported optimal combinations. Typical ranges of the parameters for protein alignment are 7–15 for $I$ and 0.5–2 for $E$.

## 5.2 Dynamic Programming Algorithms

For any given pair of sequences, if gaps are allowed there is a large number of possibilities to consider in determining the best-scoring alignment. For example, two sequences of length 1000 have approximately $10^{600}$ different alignments, vastly more than there are particles in the universe! Given the number and length of known sequences it would seem impossible to explore all these possibilities. Nevertheless, a class of algorithms has been introduced that is able to efficiently explore the full range of alignments under a variety of different constraints. They are known as dynamic programming algorithms, and efficiently avoid needless exploration of the majority of alignments that can be shown to be nonoptimal. There are several variants that produce different kinds of alignments, as outlined in Flow Diagram 5.2.

The key property of dynamic programming is that the problem can be divided into many smaller parts. Consider the following alignment:

$$X_1 \cdots X_u\ X_{u+1} \cdots X_v\ X_{v+1} \cdots X_L$$

$$Y_1 \cdots Y_u\ Y_{u+1} \cdots Y_v\ Y_{v+1} \cdots Y_L$$

in which the subscripts $u$, $v$, etc. refer to alignment positions rather than residue types, so that $X_u$, $Y_v$, and so on each correspond to a residue or to a gap. The alignment has been divided into three parts, with positions labeled $1 \rightarrow u$, $u + 1 \rightarrow v$, and

**Flow Diagram 5.2**
**The key concept introduced in this section is that the method of dynamic programming can be applied with minor modifications to solve several related problems of determining optimal and near-optimal pairwise sequence alignments.**

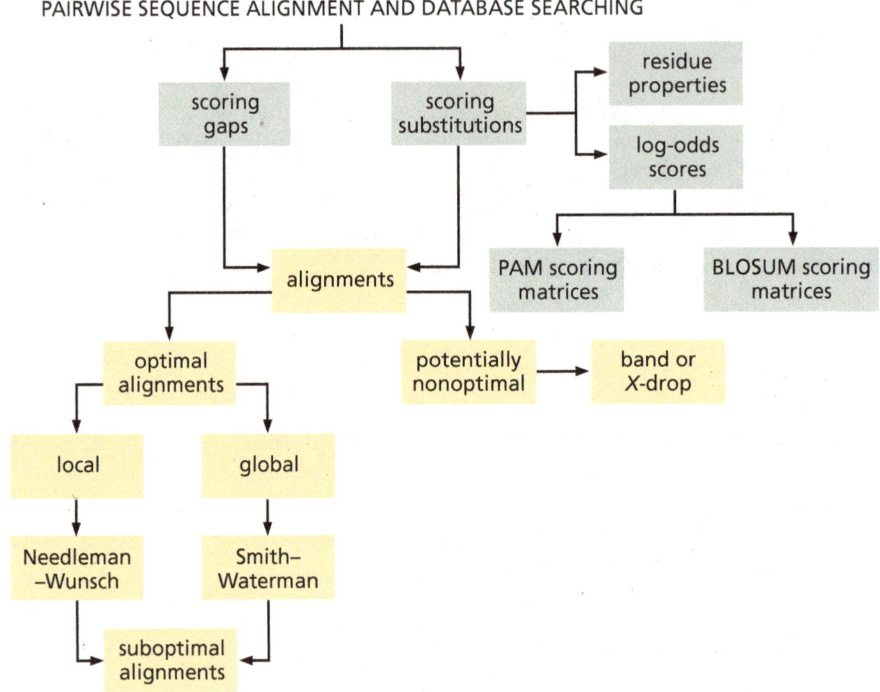


$v + 1 \rightarrow L$. Because scores for the individual positions are added together, the score of the whole alignment is the sum of the scores of the three parts; that is, their contributions to the score are independent. Thus, the optimal global alignment can be reduced to the problem of determining the optimal alignments of smaller sections. A corollary to this is that the global optimal alignment will not contain parts that are not themselves optimal. While affine gap penalties require a slightly more sophisticated argument, essentially the same property holds true for them as well.

Starting with sufficiently short sub-sequences, for example the first residue of each sequence, the optimal alignment can easily be determined, allowing for all possible gaps. Subsequently, further residues can be added to this, at most one from each sequence at any step. At each stage, the previously determined optimal subsequence alignment can be assumed to persist, so only the score for adding the next residue needs to be investigated. A worked example later in the section will make this clear. In this way the optimal global alignment can be grown from one end of the sequence. As an alignment of two sequences will consist of pairs of aligned residues, a rectangular matrix can conveniently represent these, with rows corresponding to the residues of one sequence, and columns to those of the other.

Until the global optimal alignment has been obtained, it is not known which actual residues are aligned. All possibilities must be considered or the optimal alignment could be missed. This is not as impossible as it might seem.

Saul Needleman and Christian Wunsch published the original dynamic programming application in this field in 1970, since then many variations and improvements have been made, some of which will be described here. There have been three different motivations for developing these modifications. Firstly, global and local alignments require slightly different algorithms. Secondly, but less commonly, certain gap-penalty functions and the desire to optimize scoring parameters have resulted in further new schemes. Lastly, especially in the past, the computational requirements of the algorithms prevented some general applications. For example, the basic technique in a standard implementation requires computer memory proportional to the product $mn$ for two sequences of length $m$ and $n$. Some algorithms have been proposed that reduce this demand considerably.

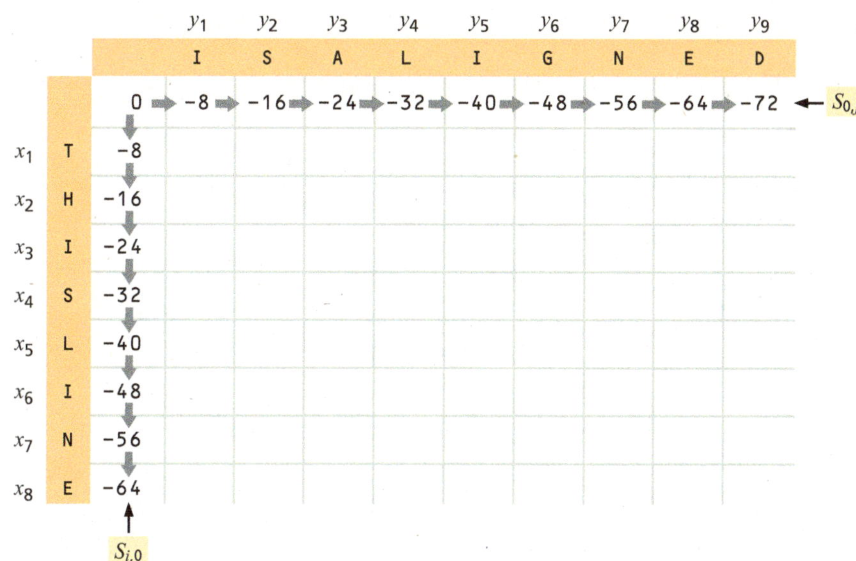


**Figure 5.8**
**The initial stage of filling in the dynamic programming matrix to find the optimal global alignment of the two sequences THISLINE and ISALIGNED.** The initial stage of filling in the matrix depends only on the linear gap penalty, defined in Equation EQ5.13 with $E$ set to –8. The arrows indicate the cell(s) to which each cell value contributes.

## Optimal global alignments are produced using efficient variations of the Needleman–Wunsch algorithm

We will introduce dynamic programming methods by describing their use to find optimal global alignments. Needleman and Wunsch were the first to propose this method, but the algorithm given here is not their original one, because significantly faster methods of achieving the same goal have since been developed. The problem is to align sequence $x$ ($x_1x_2x_3...x_m$) and sequence $y$ ($y_1y_2y_3...y_n$), finding the best-scoring alignment in which all residues of both sequences are included. The score will be assumed to be a measure of similarity, so that the highest score is desired. Alternatively, the score could be an evolutionary distance (see Section 4.2), in which case the smallest score would be sought, and all uses of "max" in the algorithm would be replaced by "min."

The key concept in all these algorithms is the matrix $\boldsymbol{S}$ of optimal scores of sub-sequence alignments. The matrix has $(m + 1)$ rows labeled $0 \rightarrow m$ and $(n + 1)$ columns labeled $0 \rightarrow n$. The rows correspond to the residues of sequence $\boldsymbol{x}$, and the columns to those of sequence $\boldsymbol{y}$ (see Figure 5.8). We shall use as a working example the alignment of the sequences $\boldsymbol{x}$ = THISLINE and $\boldsymbol{y}$ = ISALIGNED, with the BLOSUM-62 substitution matrix as the scoring matrix (see Figure 5.5). Because the sequences are small they can be aligned manually, and so we can see that the optimal alignment is:

```
TH I S - L I - N E -
   | |   | |   | |
- - I S A L I G N E D
```

This alignment might not produce the optimal score if the gap penalty were set very high relative to the substitution matrix values, but in this case it could be argued that the scoring parameters would then not be appropriate for the problem. In the matrix in Figure 5.8, the element $S_{i,j}$ is used to store the score for the optimal alignment of all residues up to $x_i$ of sequence $\boldsymbol{x}$ with all residues up to $y_j$ of sequence $\boldsymbol{y}$. The sequences ($x_1x_2x_3...x_i$) and ($y_1y_2y_3...y_j$) with $i < m$ and $j < n$ are called sub-sequences. Column $S_{i,0}$ and row $S_{0,j}$ correspond to the alignment of the first $i$ or $j$ residues with the same number of gaps. Thus, element $S_{0,3}$ is the score for aligning sub-sequence $y_1y_2y_3$ with a gap of length 3.

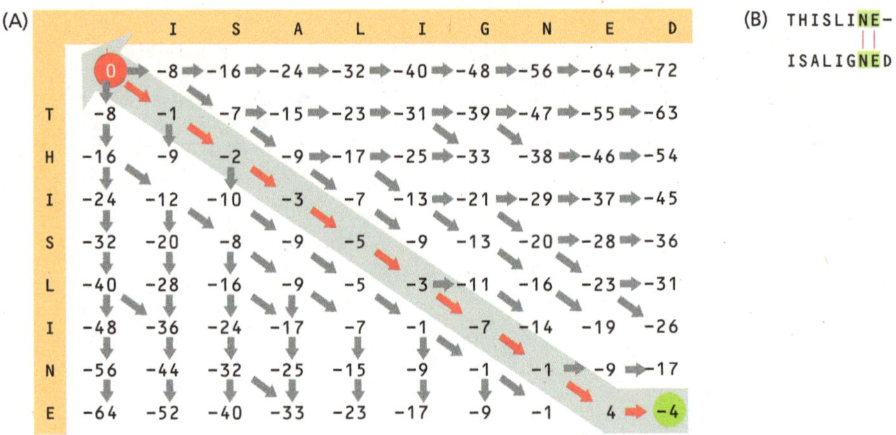


**Figure 5.9**
**The dynamic programming matrix (started in Figure 5.8) used to find the optimal global alignment of the two sequences THISLINE and ISALIGNED.** (A) The completed matrix using the BLOSUM-62 scoring matrix and a linear gap penalty, defined in Equation EQ5.13 with *E* set to –8. (See text for details of how this was done.) The red arrows indicate steps used in the traceback of the optimal alignment. (B) The optimal alignment returned by these calculations, which has a score of –4.

To fill in this matrix, one starts by aligning the beginning of each sequence; that is, in the extreme upper left-hand corner. We could equally well start at the end of the sequences (extreme bottom right-hand corner), but then the matrix should be labeled $1 \rightarrow m+1$ and $1 \rightarrow n+1$. The elements $S_{i,0}$ and $S_{0,j}$ are easy to fill in, because there is only one possible alignment available. $S_{i,0}$ represents the alignment

$$
\begin{array}{ccccc}
a_1 & a_2 & a_3 & \cdots & a_i \\
- & - & - & \cdots & -
\end{array}
$$

We will start by considering a linear gap penalty $g$ of $-8n_{gap}$ for a gap of $n_{gap}$ residues, giving the scores of $S_{i,0}$ and $S_{0,j}$ as $-8i$ and $-8j$, respectively. This starting point with numerical values inserted into the matrix is illustrated in Figure 5.8.

The other matrix elements are filled in according to simple rules that can be understood by considering a process of adding one position at a time to the alignment. There are only three options for any given position, namely, a pairing of residues from both sequences, and the two possibilities of a residue from one sequence aligning with a gap in the other. These three options can be written as:

$$
\begin{array}{ccc}
\cdots x_i & \cdots - & \cdots x_i \\
\cdots y_j & \cdots y_j & \cdots -
\end{array}
$$

The scores associated with these are $s(x_i,y_j)$, $g$, and $g$, respectively. The value of $s(x_i,y_j)$ is given by the element $s_{a,b}$ of the substitution score matrix, where $a$ is the residue type of $x_i$ and $b$ is the residue type of $y_j$. The change in notation is solely to improve the clarity of the following equations.

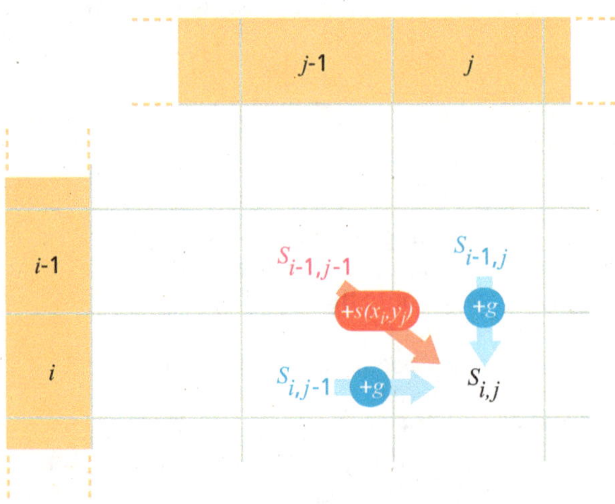


**Figure 5.10**
**Illustration of the application of Equation EQ5.17 to calculate an element of the dynamic programming matrix.** Only a small part of the matrix is shown, as only this part contributes directly to the calculation of the element.

Consider the evaluation of element $S_{1,1}$, so that the only residues that appear in the alignment are $x_1$ and $y_1$. The left-hand possibility of the three possibilities could only occur starting from $S_{0,0}$, as all other alignments will already contain at least one of these two residues. The middle possibility can only occur from $S_{1,0}$ because it requires an alignment that contains $x_1$ but not $y_1$. Similar reasoning shows that the right-hand possibility can only occur from $S_{0,1}$. The three possible alignments have the following scores:

$$\begin{aligned} S_{0,0}+s(x_1,y_1) &= 0+s(\mathrm{I,T}) = 0-1 = -1 \\ S_{1,0}+g &= -8-8 = -16 \\ S_{0,1}+g &= -8-8 = -16 \end{aligned} \qquad \text{(EQ5.16)}$$

where $s$(I,T) has been obtained from Figure 5.5. Of these alternatives, the optimal one is clearly the first. Hence in Figure 5.9, $S_{1,1} = -1$. Because $S_{1,1}$ has been derived from $S_{0,0}$ an arrow has been drawn linking them in the figure.

An identical argument can be made to construct any element of the matrix from three others, using the formula

$$S_{i,j} = \max\begin{cases} S_{i-1,j-1} + s(x_i,y_j) \\ S_{i-1,j} + g \\ S_{i,j-1} + g \end{cases} \qquad \text{(EQ5.17)}$$

The maximum ("max") implies that we are using a similarity score. Figure 5.10 illustrates this formula in the layout of the matrix. Note that it is possible for more than one of the three alternatives to give the same optimal score, in which case arrows are drawn for all optimal alternatives. The completed matrix for the example sequences is given in Figure 5.9A. Note that the number of steps in this algorithm is proportional to the product of the sequence lengths.

We now have a matrix of scores for optimal alignments of many sub-sequences, together with the global sequence alignment score. This is given by the value of $S_{m,n}$, which in this case is $S_{8,9} = -4$. Note that this is not necessarily the highest score in the matrix, which in this case is $S_{8,8} = 4$, but only $S_{m,n}$ includes the information from both complete sequences. For each matrix element we know the element(s) from which it was directly derived. In the figures in this chapter, arrows are used to indicate this information.

(A)

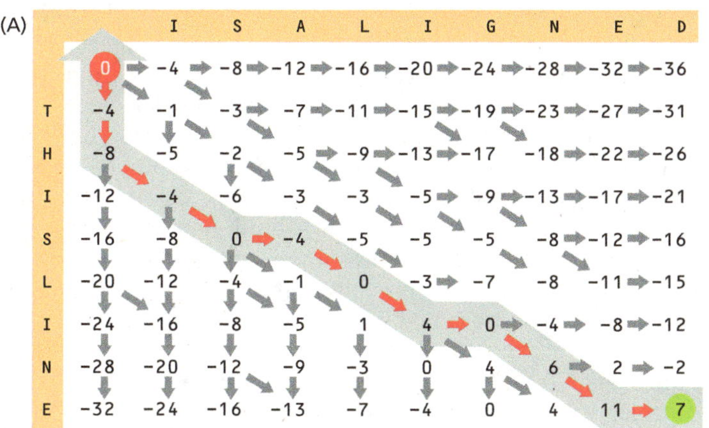


(B)

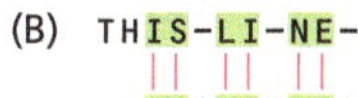


**Figure 5.11**
**Optimal global alignment of two sequences, identical to Figure 5.9, except for a change in gap scoring.** The linear gap penalty, defined in Equation EQ5.13 using a value of –4 for the parameter *E*. (A) The completed matrix using the BLOSUM-62 scoring matrix. (B) The optimal alignment, which has a score of 7.

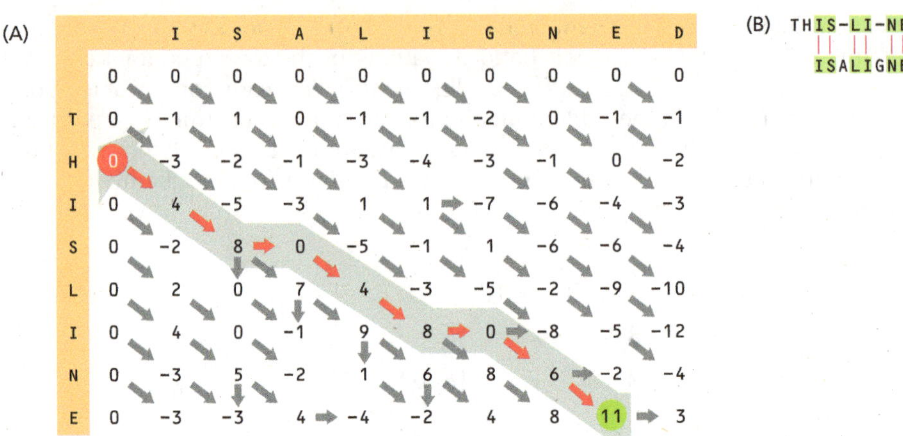


**Figure 5.12**
**Dynamic programming matrix for semiglobal alignment of the same sequences as in Figure 5.9.** In this case, end gaps are not penalized. (A) The completed matrix for determining the optimal global alignment of THISLINE and ISALIGNED using the BLOSUM-62 scoring matrix with a linear gap penalty, defined in Equation EQ5.13 with $E$ set to –8 and with end gaps not penalized. (B) The optimal alignment, which has a score of 11.

We can use the information on the derivation of each element to obtain the actual global alignment that produced this optimal score by a process called **traceback**. Beginning at $S_{m,n}$ we follow the arrows back through the matrix to the start ($S_{0,0}$). Thus, having filled the matrix elements from the beginning of the sequences, we determine the alignment from the end of the sequences. At each step we can determine which of the three alternatives given in Equation EQ5.17 has been applied, and add it to our alignment. If $S_{i,j}$ has a diagonal arrow from $S_{i-1,j-1}$, that implies the alignment will contain $x_i$ aligned with $y_j$. Vertical arrows imply a gap in sequence $x$ aligning with a residue in sequence $y$, and vice versa for horizontal arrows. The traceback arrows involved in the optimal global alignment in Figure 5.9A are shown in red. When tracing back by hand, care must be taken, as it is easy to make mistakes, especially by applying the results to residues $x_{i-1}$ and $y_{j-1}$ instead of $x_i$ and $y_j$.

The traceback information is often stored efficiently in computer programs, for example using three bits to represent the possible origins of each matrix element. If a bit is set to zero, that path was not used, with a value of one indicating the direction. Such schemes allow all this information to be easily stored and analyzed to obtain the alignment paths.

Note that there may be more than one optimal alignment if at some point along the path during traceback an element is encountered that was derived from more than one of the three possible alternatives. The algorithm does not distinguish between these possible alignments, although there may be reasons for preferring one to the others. Such preference would normally be justified by knowledge of the molecular structure or function. Most programs will arbitrarily report just one single alignment.

The alignment given by the traceback is shown in Figure 5.9B. It is not the one we expected, in that it contains no gaps. The carboxy-terminal aspartic acid residue (D) in sequence y is aligned with a gap only because the two sequences are not the same length. We can readily understand this outcome if we consider our chosen gap penalty of 8 in the light of the BLOSUM-62 substitution matrix. The worst substitution score in this matrix is –4, significantly less than the gap penalty. Also, many of the scores for aligning identical residues are only 4 or 5. This means that if we set such a high gap penalty, a gap is unlikely to be present in an optimal alignment using this scoring matrix. In these circumstances, gaps will occur if the sequences are of different length and also possibly in the presence of particular residues such as tryptophan or cysteine which have higher scores.

If instead we use a linear gap penalty $g(n_{gap}) = -4n_{gap}$, the situation changes, as shown in Figure 5.11, which gives the optimal alignment we expected. Because the gap penalty is less severe, gaps are more likely to be introduced, resulting in a different alignment and a different score. In this particular case, four additional gaps occur, two of which occur within the sequences. The overall alignment score is 7, but this alignment would have scored –13 with the original gap penalty of 8.

This example illustrates the need to match the gap penalty to the substitution matrix used. However, care must be taken in matching these parameters, as the performance also depends on the properties of the sequences being aligned. Different parameters may be optimal when looking at long or short sequences, and depending on the expected sequence similarity (see Figure 4.5 for a practical example).

A simple modification of the algorithm allows the sequences to overlap each other at both ends of the alignment without penalty, often called **semiglobal alignment**. In this way, better global alignments can be obtained for sequences that are not the same length. Instead of applying the gap penalty scores to matrix elements $S_{i,0}$ and $S_{0,j}$, we set these to zero. The remaining elements are calculated as before (Equation EQ5.17). However, instead of traceback beginning at $S_{m,n}$, it starts at the highest-scoring element in the bottom row or right-most column. This is illustrated in Figure 5.12 for the gap penalty $g(n_{gap}) = -8n_{gap}$. Note that now the expected alignment is obtained, despite the gap penalty being so high. This modified algorithm gives the same optimal alignment with a gap penalty of $g(n_{gap}) = -4n_{gap}$.

The methods presented above are for use when scoring gaps with a linear penalty of the form $g(n_{gap}) = -n_{gap}E$. If we wish to differentiate between penalties for starting and extending a gap, using a scoring scheme such as $g(n_{gap}) = -I - (n_{gap} - 1)E$, a slightly different algorithm is required. The problem is not simply one of ensuring that we know if a gap is just being started or is being extended. In the previous algorithm, the decision for $S_{i,j}$ could be made without knowing the details of the alignment for any of $S_{i-1,j}$, $S_{i-1,j-1}$, or $S_{i,j-1}$. Now we need to know if these alignments ended with gaps, which means knowing details of their derivation, particularly involving $S_{i-2,j}$ and $S_{i,j-2}$.

Consider the general case of using a gap penalty, which is simply written as a function of the gap length $n_{gap}$; that is, $g(n_{gap})$. Now, for any matrix element $S_{i,j}$, one must consider the possibility of arriving at that element directly via insertion of a gap of length up to $i$ in sequence $\boldsymbol{x}$ or $j$ in sequence $\boldsymbol{y}$. Some of these routes are illustrated in Figure 5.13. The algorithm now has to be modified to

$$S_{i,j} = \max\begin{cases} S_{i-1,j-1} & + & s(x_i, y_j) \\ [S_{i-n_{gap1},j} & + & g(n_{gap1})]_{1 \le n_{gap1} \le i} \\ [S_{i,j-n_{gap2}} & + & g(n_{gap2})]_{1 \le n_{gap2} \le j} \end{cases} \qquad \text{(EQ5.18)}$$

The algorithm now has a number of steps proportional to $mn^2$, where $m$ and $n$ are the sequence lengths with $n > m$. This is a significant increase in requirements over the original, because there are approximately $n$ terms involving gaps used to evaluate each matrix element. However, when we use the specific affine gap penalty formula of Equation EQ5.14 it is possible to reformulate things and obtain the full matrix in $mn$ steps again. Let us define $V_{i,j}$ as

$$V_{i,j} = \max\left\{S_{i-n_{gap1},j} + g(n_{gap1})\right\}_{1 \le n_{gap1} \le i} \qquad \text{(EQ5.19)}$$

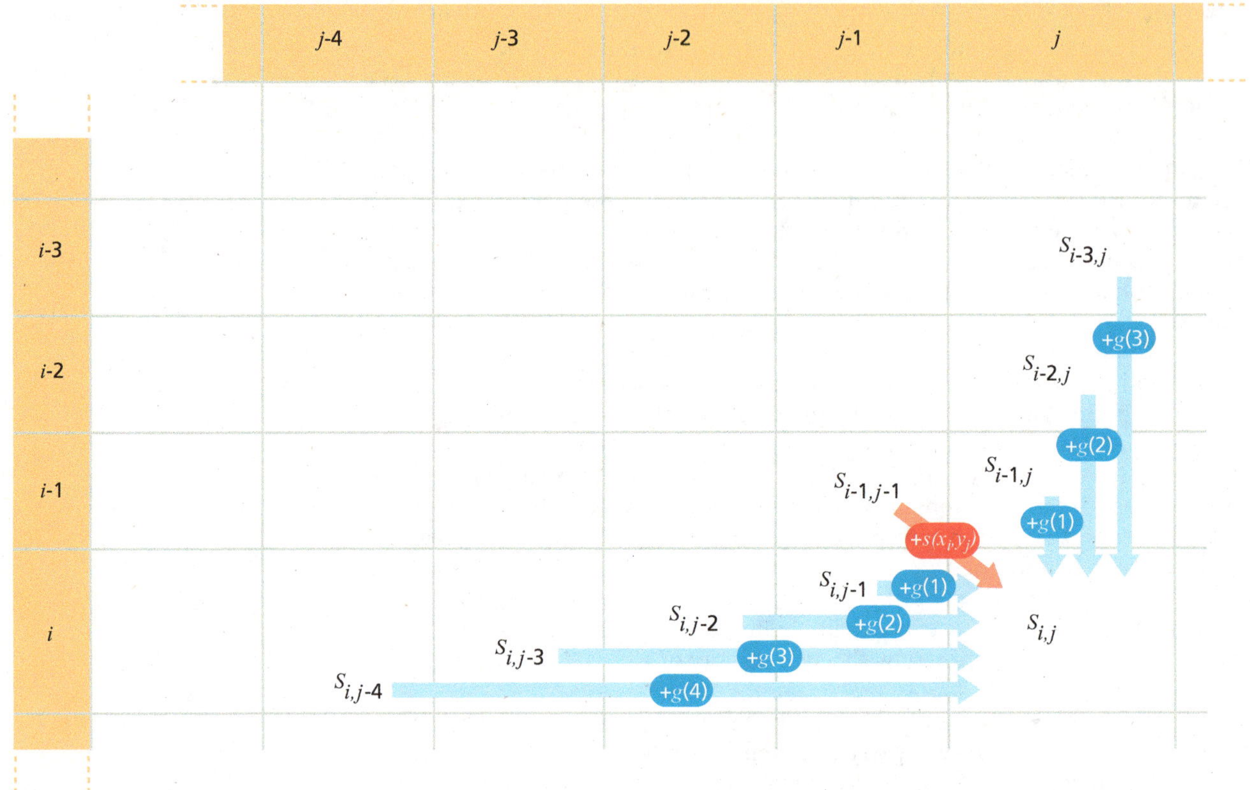


**Figure 5.13**
**Illustration of some of the possible paths that could be involved in calculating the alignment score matrix for a general gap penalty g($n_{gap}$).** All possible gap sizes must be explored, with $S_{i,j}$ being chosen as the maximum of all these possibilities.

from which

$$V_{i,j} = \max\begin{cases} S_{i-1,j} + g(1) \\ [S_{i-n_{\text{gap}1},j} + g(n_{\text{gap}1})]_{2\le n_{\text{gap}1}\le i} \end{cases} \quad \text{(EQ5.20)}$$

$$= \max\begin{cases} S_{i-1,j} + g(1) \\ [S_{i-n_{\text{gap}1}-1,j} + g(n_{\text{gap}1}+1)]_{1\le n_{\text{gap}1}\le i-1} \end{cases} \quad \text{(EQ5.21)}$$

$$= \max\begin{cases} S_{i-1,j} - 1 \\ [S_{i-n_{\text{gap}1}-1,j} + g(n_{\text{gap}1}) - E]_{1\le n_{\text{gap}1}\le i-1} \end{cases} \quad \text{(EQ5.22)}$$

But from Equation EQ5.19 substituting $i-1$ for $i$, we can see that

$$V_{i-1,j} = \max\left\{S_{i-n_{\text{gap}1}-1,j} + g(n_{\text{gap}1})\right\}_{1\le n_{\text{gap}1}\le i-1} \quad \text{(EQ5.23)}$$

so that

$$V_{i,j} = \max\begin{cases} S_{i-1,j} - I \\ V_{i-1,j} - E \end{cases} \quad \text{(EQ5.24)}$$

Thus we have a recursive equation for the elements $V_{i,j}$, involving aligning residues in sequence $x$ with gaps in $y$. This can readily be evaluated from a starting point of

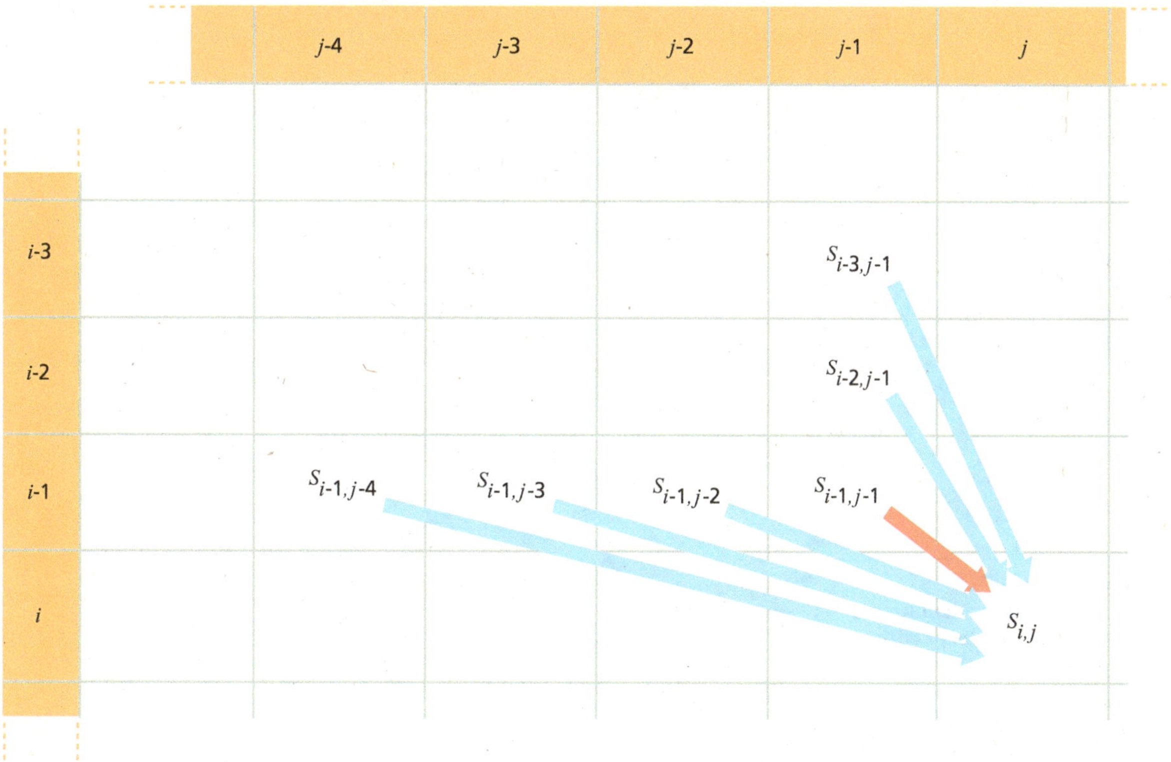

**Figure 5.14**
**Illustration of some of the possible paths that could be involved in calculating the alignment score matrix with the original Needleman and Wunsch algorithm.** All possible gap sizes must be explored, with $S_{i,j}$ being chosen as the maximum of all these possibilities. The interpretation differs from the previous matrices in that if a path stops at an element $S_{i,j}$ it indicates that residues $x_i$ and $y_j$ are aligned with each other. Thus the path $S_{i-1,j-2} \rightarrow S_{i,j}$ represents the alignment of $x_{i-1}$–$x_i$ with $y_{j-2}y_{j-1}y_j$; that is, an insertion in sequence $x$ aligned with $y_{j-1}$.

$V_{1,j} = S_{0,j} - I$ (Equation EQ5.19 with $i = 1$). In a similar manner, we can define $W_{i,j}$, involving aligning residues in sequence $y$ with gaps in $x$ as

$$W_{i,j} = \max \begin{cases} S_{i,j-1} & - \quad I \\ W_{i,j-1} & - \quad E \end{cases} \qquad \text{(EQ5.25)}$$

This can readily be evaluated from a starting point of $W_{i,1} = S_{i,0} - I$. These two recursive formulae can be substituted into Equation EQ5.18 to give the faster ($nm$ steps) algorithm

$$S_{i,j} = \max \begin{cases} S_{i-1,j-1} + s(x_i, y_i) \\ V_{i,j} \\ W_{i,j} \end{cases} \qquad \text{(EQ5.26)}$$

It should be noted that the original algorithm of Needleman and Wunsch differs in some key details from those described here. Their method is slower (requiring more computing steps) and is rarely used today. The types of path involved in calculating the matrix elements with their algorithm are shown in Figure 5.14. Note that the interpretation of the paths through the matrix differs from that presented above (see the figure legend for details).

## Local and suboptimal alignments can be produced by making small modifications to the dynamic programming algorithm

Often we do not expect the whole of one sequence to align well with the other. For example, the proteins may have just one domain in common, in which case we

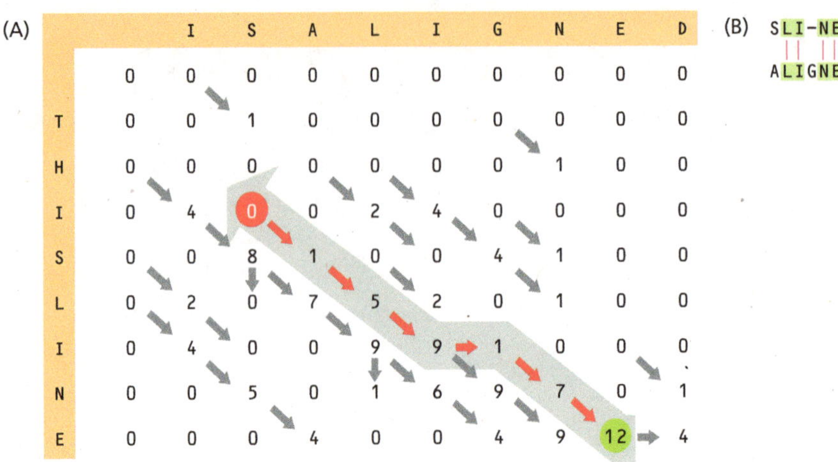


**Figure 5.15**
**The dynamic programming calculation for determining the optimal local alignment of the two sequences THISLINE and ISALIGNED.** (A) The completed matrix using the BLOSUM-62 scoring matrix with a linear gap penalty, defined in Equation EQ5.13 with *E* set to –8. (B) The optimal alignment, determined by the highest-scoring element, which has a score of 12.

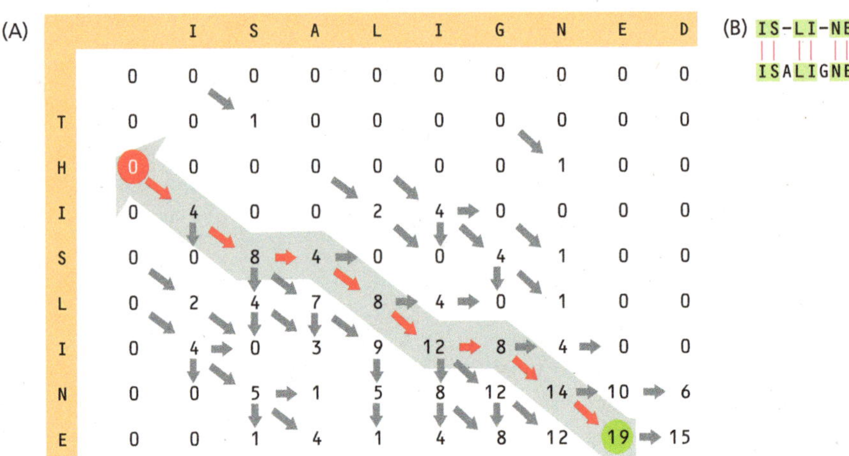


**Figure 5.16**
**Optimal local alignment calculation identical to Figure 5.15, except with a linear gap penalty with *E* set to –4.** (A) The completed matrix for determining the optimal local alignment of THISLINE and ISALIGNED using the BLOSUM-62 scoring matrix. (B) The optimal alignment, identified by the highest-scoring element in the entire matrix, which has a score of 19.

want to find this high-scoring zone, referred to as a local alignment (see Section 4.5). In a global alignment, those regions of the sequences that differ substantially will often obscure the good agreement over a limited stretch. The local alignment will identify these stretches while ignoring the weaker alignment scores elsewhere.

It turns out that a very similar dynamic programming algorithm to that described above for global alignments can obtain a local alignment. Smith and Waterman first proposed this method. However, it should be noted that the method presented here requires a similarity-scoring scheme that has an expected negative value for random alignments and positive value for highly similar sequences. Most of the commonly used substitution matrices fulfill this condition. Note that the global alignment schemes have no such restriction, and can have all substitution matrix scores positive. Under such a scheme, scores will grow steadily larger as the alignment gets larger, regardless of the degree of similarity, so that long random alignments will ultimately be indistinguishable by score alone from short significant ones.

The key difference in the local alignment algorithm from the global alignment algorithm set out above is that whenever the score of the optimal sub-sequence alignment is less than zero it is rejected, and that matrix element is set to zero. The scoring scheme must give a positive score for aligning (at least some) identical residues. We would expect to be able to find at least one such match in any alignment worth considering, so that we can be sure that there should be some positive alignment scores. Another algorithmic difference is that we now start traceback from the highest-scoring matrix element wherever it occurs.

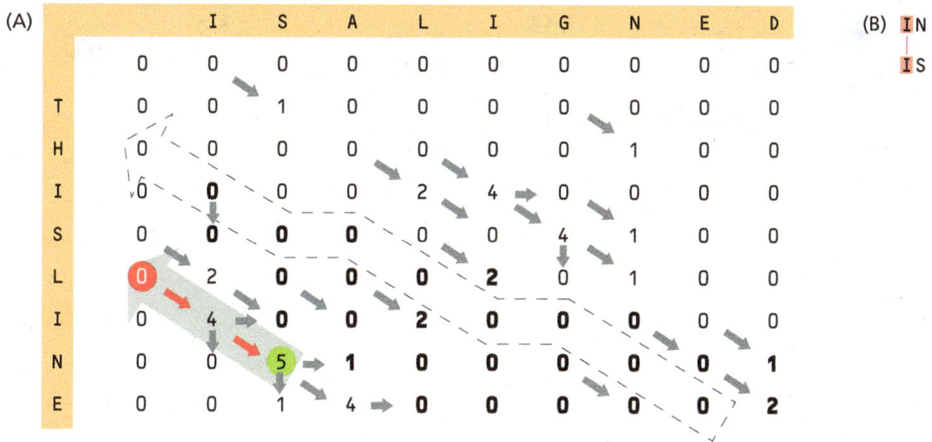


**Figure 5.17**
**The dynamic programming calculation that follows on from the calculation shown in Figure 5.16 to find the best suboptimal local alignment.** (A) The completed matrix for determining the best suboptimal local alignment of THISLINE and ISALIGNED using the BLOSUM-62 scoring matrix with a linear gap penalty with $E$ set to –4. The matrix elements that were involved in the optimal local alignment have been set to zero and are shown with bold font. The matrix elements that have changed value from Figure 5.16(A) are also shown with bold font, and extend below and to the right of the optimal alignment. (B) The best suboptimal alignment, identified by the highest-scoring element in the entire matrix, has a score of 5.

The extra condition on the matrix elements means that the values of $S_{i,0}$ and $S_{0,j}$ are set to zero, as was the case for global alignments without end gap penalties. The formula for the general matrix element $S_{i,j}$ with a general gap penalty function $g(n_{gap})$ is

$$S_{i,j} = \max \begin{cases} S_{i-1,j-1} & + \quad s(x_i, y_j) \\ [S_{i-n_{gap1},j} & + \quad g(n_{gap1})]_{1 \le n_{gap1} \le i} \\ [S_{i,j-n_{gap2}} & + \quad g(n_{gap2})]_{1 \le n_{gap2} \le j} \\ 0 & \end{cases} \qquad \text{(EQ5.27)}$$

which only differs from Equation EQ5.18 by the inclusion of the zero. The same modifications as above can be applied for the cases of linear gap penalty given in Equation EQ5.13 and affine gap penalty given in Equation EQ5.14.

Figures 5.15 and 5.16 show the optimal local alignments for our usual example in the two cases of linear gap penalties $g(n_{gap}) = -8n_{gap}$ and $-4n_{gap}$, respectively. Both result in removal of the differing ends of the sequences. In the first case, the higher gap penalty forces an alignment of serine (S) and alanine (A) in preference to adding a gap to reach the identical IS sub-sequence. Lowering the gap penalty in this instance improves the result to give the local alignment we would expect.

Sometimes it is of interest to find other high-scoring local alignments. A common instance would be the presence of repeats in a sequence. There will usually be a number of alternative local alignments in the vicinity of the optimal one, with only slightly lower scores. These will have a high degree of overlap with the optimal alignment, however, and contain little, if any, extra information beyond that given by the optimal local alignment. Of more interest are those suboptimal local alignments that are quite distinct from the optimal one. Usually their distinctness is defined as not sharing any aligned residue pairs.

An efficient method has been proposed for finding distinct suboptimal local alignments. These are alignments in which no aligned pair of residues is also found aligned in the optimal or other suboptimal alignments. They can be very useful in a variety of situations such as aligning multidomain proteins. Sometimes a pair of proteins has two or more domains in common but other regions with no similarity. In such cases it is useful to obtain separate local alignments for each domain, but only one of these will give the optimal score, the others being suboptimal alignments. The method starts as before by calculating the optimal local alignment. Then, to ensure that any new alignment found does not share any aligned residues

## Box 5.1 Saving space by throwing away the intermediate calculations

In this chapter the steps required to identify optimal sequence alignments have been specified, but there has been no discussion of the precise coding used in a program. Such details are in general beyond the scope of this book, although important for the production of efficient and practical tools. The computer science and other specialist texts cited in Further Reading at the end of the chapter should be consulted for details of efficient coding techniques. However, we will briefly examine one algorithmic trick that can substantially increase the range of alignment problems feasible with limited computer resources.

All the methods presented so far calculate the whole of matrix **S**, and by default it might be assumed that the entire matrix was stored for traceback analysis. If the sequences to be compared have lengths of up to a few thousand residues, the matrix can be stored quite easily on current computers. However, particularly in the case of nucleotide sequences, we might wish to align much longer sequences, for example mitochondrial DNA and even whole genomes. Storage of the whole matrix **S** is often not possible for such long sequences, sometimes requiring gigabytes of memory. Without using an alternative, this memory problem could prevent the use of these dynamic programming methods on such data. However, a neat solution is available by modifying the algorithm to store just two rows of the matrix. Note that further calculation is then needed to recover the actual alignment, as the basic method presented here only provides the optimal alignment score.

The key to this algorithm is to notice that the calculation of any element only requires knowledge of the results for the current and previous row (see Figure 5.10 and Equations EQ5.17 and EQ5.26); we could have chosen to work with columns instead. The two rows will be labeled $R_{0,j}$ and $R_{1,j}$. The scoring scheme used here will involve the linear gap penalty of Equation EQ5.13, but the affine gap penalty scheme can also be used. Similarly, the different variations in the treatment of end gaps can be incorporated.

The steps can be summarized as follows. Firstly, initialize row 0 according to the specific algorithm. For standard Needleman–Wunsch with a linear gap penalty, this means setting $R_{0,j} = -jE$ for $j = 0 \rightarrow$ m. For the 0th column, $R_{1,0}$ is then set to $-E$ or whatever other boundary conditions are required. This is equivalent to the initialization of the full matrix method above. We now step through each residue $x_i$ of sequence $x$, from $x_1$ to $x_n$, calculating all the elements of row $R_{1,j}$, which at each step is the equivalent of the $i$th row in the full matrix. These elements are labeled with the letter $j$, and correspond to residue $y_j$ of sequence $y$. Thus the elements of the $R_{0,j}$ row correspond to the matrix elements $S_{i-1,j}$, and those of the $R_{1,j}$ row to $S_{i,j}$. The other elements of $R_{1,j}$ are assigned a value according to the equation (equivalent to Equation EQ15.17)

$$R_{1,j} = \max \begin{cases} R_{0,j-1} + s(x_i, y_j) \\ R_{0,j} - E \\ R_{1,j-1} - E \end{cases} \quad \text{(BEQ5.1)}$$

Once all the elements of the $R_{1,j}$ row have been calculated, the value of each matrix element of $R_{1,j}$ is transferred to $R_{0,j}$, which now represents matrix row 1. In practical programming, pointers would be used, so that this step would take virtually no time. For the 0th column, $R_{1,0}$ is then set to $-2E$ or whatever other boundary conditions are required, with the $R_{1,j}$ elements now representing matrix row 2. For the other columns, $R_{1,j}$ is assigned a value as before. In this way results keep being overwritten, but sufficient are kept to continue the calculation.

Certain scores must be saved, the details differing according to the precise type of alignment sought. Thus for the basic Needleman–Wunsch scheme, only $R_{m,n}$ need be stored, while Smith–Waterman requires the highest-scoring element and the associated values of $i$ and $j$. The storage requirements are twice the length of the shorter sequence. Because this technique only stores two rows of the matrix, it makes it possible to use dynamic programming to align complete bacterial genomes.

However, the saving in memory required comes at a price, and the traceback procedure is much more complicated than that which can be used if the full matrix calculation has been stored. The traceback now involves considerably more calculation and so production of the overall alignment will require longer computing times. Versions of this technique exist for all the types of alignment mentioned in this chapter. Over the past few years much effort has been devoted to optimizing alignment programs for use with whole genome sequences. See Section 5.5 for more practical methods concerning whole genome sequence alignment. In contrast to the methods above, the methods discussed in that section all involve approximations, but a good argument can be made that such techniques are more appropriate for that type of problem.

with the optimal alignment, it is necessary to set all the matrix elements on this initial path to zero (see Figure 5.17). Subsequently, the matrix needs to be recalculated to include the effect of these zeroed elements. If we recalculate the matrix, only a relatively small number of elements near the optimal local alignment will have new values. The key to the efficiency of the technique is the recognition that these modified elements can only affect elements to their right and below them (see Figure 5.13). Working along a matrix row from the zeroed elements, the new values are monitored until an element is found whose value is unchanged from the original calculation. Further elements on this row will also be unchanged, so calculation can now move to the next row down. In this way the new matrix is obtained with a minimum of work. The suboptimal local alignment is identified by locating the highest matrix element. Further suboptimal alignments can be found by repeating the procedure, zeroing every element involved in a previous alignment. For affine gap penalties the alignment elements in the matrices $\boldsymbol{V}$ and $\boldsymbol{W}$ are also forced to zero and must also be unchanged before a row calculation can stop.

In the example shown in Figure 5.17, the first suboptimal local alignment is found for the linear gap penalty $g(n_{gap}) = -4n_{gap}$; that is, following on from Figure 5.16. All the matrix elements of the optimal alignment are in bold font, as are those elements that have been recalculated. Only 27 elements needed to be recalculated, all below or to the left of the elements of the optimal alignment. If we had not monitored for unchanged elements we would have had to recalculate 54 elements.

## Time can be saved with a loss of rigor by not calculating the whole matrix

High-scoring local alignments indicating significant sequence similarity usually do not have many insertions and deletions. Global alignments may contain large insertions, for example if there is a whole domain inserted in one sequence, but such situations are only rarely encountered. In terms of the matrix, this means that these alignments generally follow a diagonal quite closely. This has led people to try to save time in deriving alignments by only calculating matrix elements within a

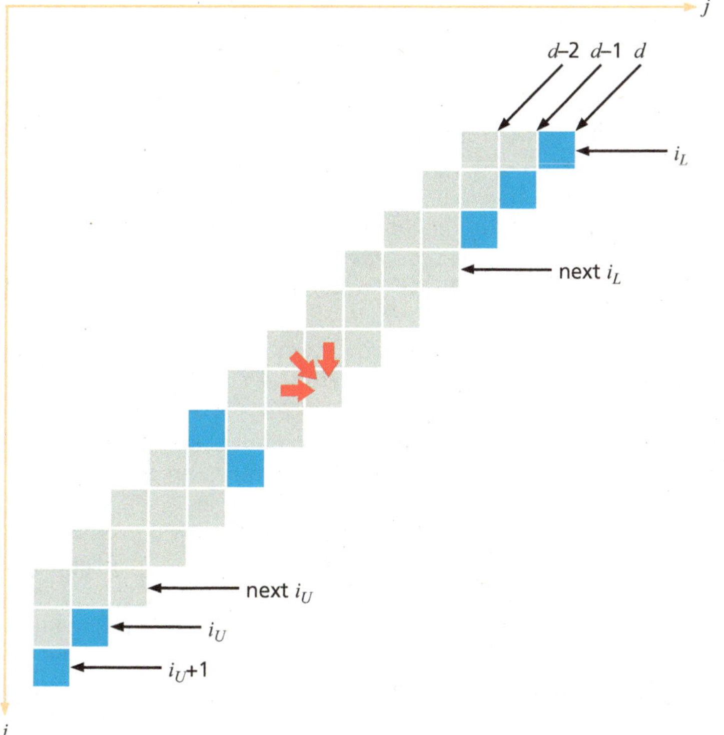


**Figure 5.18**
**The *X*-drop method, proceeding by antidiagonals indexed by $d = i + j$.** The boxes shaded in blue are matrix elements which have a score that falls *X* below the current best score, and therefore have been assigned a value of $-\infty$. The normal three possible paths used to determine the score are shown in one case by red arrows. (Adapted from Z. Zhang et al., A greedy algorithm for aligning DNA sequences, *J. Comput. Biol.* 7 (1–2):203–214, 2000.)

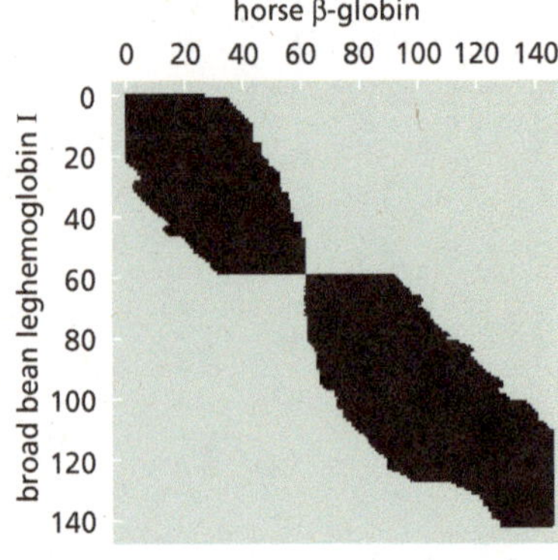


**Figure 5.19**
**The matrix elements calculated in the BLAST program using the *X*-drop method for the example of aligning broad bean leghemoglobin I and horse β-globin.** The value of $X$ used was 40, with BLOSUM-62 substitution matrix and an affine gap penalty of $g(n_{gap}) = -10 - n_{gap}$. The alignment starts at $S_{60,62}$, alanine in both sequences, which is the location determined by the process illustrated in Figure 5.24. Calculated elements are shown black. (From S.F. Altschul et al., Gapped BLAST and PSI-BLAST: a new generation of protein database search programs, *N.A.R.* 25 (17):3389–3402, 1997, by permission of Oxford University Press.)

short range of a specified diagonal. Note that such techniques are not guaranteed to find the optimal alignment!

There are two ways of assigning the diagonal around which the alignment will be sought. One can assume that both sequences are similar at a global level, and use the diagonal $S_{i,i}$ of the matrix. Alternatively, especially in the case of database searches, one can use the diagonal of a high-scoring ungapped local alignment found in an initial alignment search, as discussed later.

There are two ways of restricting the coverage of the matrix away from the central diagonal. If a limit is imposed for the maximum difference $M_{ins}$ between the number of insertions in each of the two sequences, the algorithm can simply be set to include only $M_{ins}$ diagonals on either side of the central one; that is $(2M_{ins} + 1)$ diagonals in total. The database search program FASTA (see Section 4.6 and below) uses this technique, especially for nucleotide sequences, when $M_{ins}$ is frequently set to 15. A second method of restricting the matrix elements examined is to limit the search according to how far the scores fall below the current best score. In this ***X*-drop method**, used in the database searching program BLAST (see Section 4.6 and below), the current best score is monitored. When calculating elements along a row, calculation stops when an element scores a preset value $X$ below this best score, and calculation restarts on the next row. A large value of $X$ results in more of the matrix being evaluated, in which case the true optimal alignment is more likely to be determined.

The $X$-drop algorithm that follows aligns two sequences $\boldsymbol{x}$ and $\boldsymbol{y}$ of length $m$ and $n$, respectively. The matrix elements are processed for each antidiagonal in turn. The $d$th antidiagonal is defined by $d = i + j$. Only a restricted region of the $d$th antidiagonal is evaluated, defined by variables $i_L$ and $i_U$, as illustrated in Figure 5.18. The current best alignment score is stored in the variable $S_{best}$.

In some cases, such as occur in the database search methods described in Section 5.3, the algorithm is started at element $S_{i_0,j_0}$, which has been identified by a preliminary local alignment step. In this instance $S_{best}$ is initially set to the value of element $S_{i_0,j_0}$, and the other variables are initialized to $d = i_0 + j_0$ and $i_L = i_U = i_0$. Alternatively, the algorithm can be started at element $S_{0,0}$, which has the value zero, so that $S_{best}$ is initially set to zero, as are $d$, $i_L$, and $i_U$. With these initial values, the algorithm proceeds as follows.

Each antidiagonal is evaluated in turn. We will describe the method as proceeding from smaller to larger values of $d$. The changes required to proceed to decreasing values of $d$ will be discussed afterwards. If there are elements of the antidiagonal to calculate, residues $x_i$ are evaluated in order of increasing $i$, from $i_L$ to $i_U + 1$. For each value of $i$ the relevant $j$ for this antidiagonal can be obtained using $j = d - i$. This identifies the matrix element $S_{i,d-i}$ under consideration. Initial evaluation is as described previously, for example, Equation EQ5.17 when a linear gap penalty is used.

After this initial evaluation of the antidiagonal elements, a check is first made to see if the score of any of these elements improves on the best score so far, in which case $S_{best}$ is set to this value. The antidiagonal elements are then evaluated to identify any that fall more than $X$ below this current best score, $S_{best}$. All elements that have such low scores are assigned the value $-\infty$, so that they play no further part in later calculations.

In the next step, $i_L$ and $i_U$ are redetermined. $i_L$ is chosen as the lowest $i$ in the selected region of the current antidiagonal such that $S_{i,j} \neq -\infty$. $i_U$ is set to the highest $i$ in the selected region of the current antidiagonal such that $S_{i,j} \neq -\infty$. Sometimes this will result in a smaller region being defined for the next antidiagonal, the situation illustrated in Figure 5.18. When the elements at the edge of the antidiagonal region do not have the value $-\infty$ the region must be extended. This extension might be based on the calculation of scores for further elements of the antidiagonal, or may be limited to a prespecified number of extra elements. If $i_U + 1 \le i_L$ there are no matrix elements in the selected region, and the calculation is finished. Otherwise, the next antidiagonal, that is, the $(d + 1)$th, is now evaluated.

If the calculation started at an element $S_{i_0,j_0}$, that is not $S_{0,0}$, it must be carried out in reverse as well to trace the alignment toward the start of the sequences. Note that there is no difference in principle between using dynamic programming to find optimal alignments forward or backward. Some indices need to be changed, however, as for example, element $S_{i,j}$ now depends on the values of $S_{i+1,j+1}$, $S_{i+1,j}$, and $S_{i,j+1}$. The score of such an element $S_{i,j}$ now relates to an alignment starting with residues $x_i$ and $y_j$ and ending at $x_u$ and $y_v$. The full alignment is found by two trace-backs, one from the forward and one from the backward region of the matrix, both of which end at $S_{i_0,j_0}$. Figure 5.19 shows an example of the matrix elements calculated for a real alignment by an algorithm like this.

## 5.3 Indexing Techniques and Algorithmic Approximations

The huge increase in the size of the sequence databases and their daily updating has obliged the general research community to access these databases through central facilities. This makes it easier for people to be confident of searching all known data, but serves to concentrate the demands for similarity searches on a few machines. Even though the power of computers has greatly increased over the years, the methods of full-matrix dynamic programming described above are too demanding for general use in database searches.

A number of alternative procedures have been developed that are considerably faster, although there is a penalty to pay in that they do not guarantee to find the highest-scoring alignment. The key to their success is the use of **indexing techniques** to locate possible high-scoring short local alignments. These initial local alignments are then extended, subject to certain constraints, to provide scores that are used to rank the database sequences by similarity. In most implementations, modified dynamic programming algorithms are used to examine the best-scoring sequences and to produce final scores and alignments.

In the following sections we will examine in detail the two major methods implemented in the two program suites in widest use today: BLAST and FASTA. These both start by considering very short segments of sequence that they call "words," "k-tuples," or "k-mers." A k-tuple (as used in FASTA) is simply a stretch of *k* residues in the query sequence. A k-mer (as used in BLAST) is a stretch of *k* residues, which when aligned with all *k* residue stretches of the query sequence will score above some threshold value (*T*) at least once. The term "word" is used more generally, meaning simply any short sub-sequence. The initial steps of both methods find high-scoring ungapped local alignments, referred to in BLAST as **high-scoring segment pairs** (**HSPs**), of which the highest-scoring one for a given pair of sequences is the **maximal segment pair** (**MSP**). Flow Diagram 5.3 gives an outline of the steps covered in this section.

### Suffix trees locate the positions of repeats and unique sequences

One method of indexing uses a device known as the **suffix tree**. A variant of this technique is used in the BLAST programs. The example given here will be for a nucleotide sequence because an example using a realistic protein sequence would be too complex to illustrate the method. Consider a short segment of a nucleotide sequence ATCCGAGGATATCGA$, where the $ is used to identify the end of the sequence. This has a number of short repeats, such as AT. A suffix tree is a way of

**Flow Diagram 5.3**
**The key concept introduced in this section is that database sequence similarity searches require refinements in the pairwise alignment methods to make them more efficient, although at the cost of decreasing the chance of finding the best-scoring alignment.**

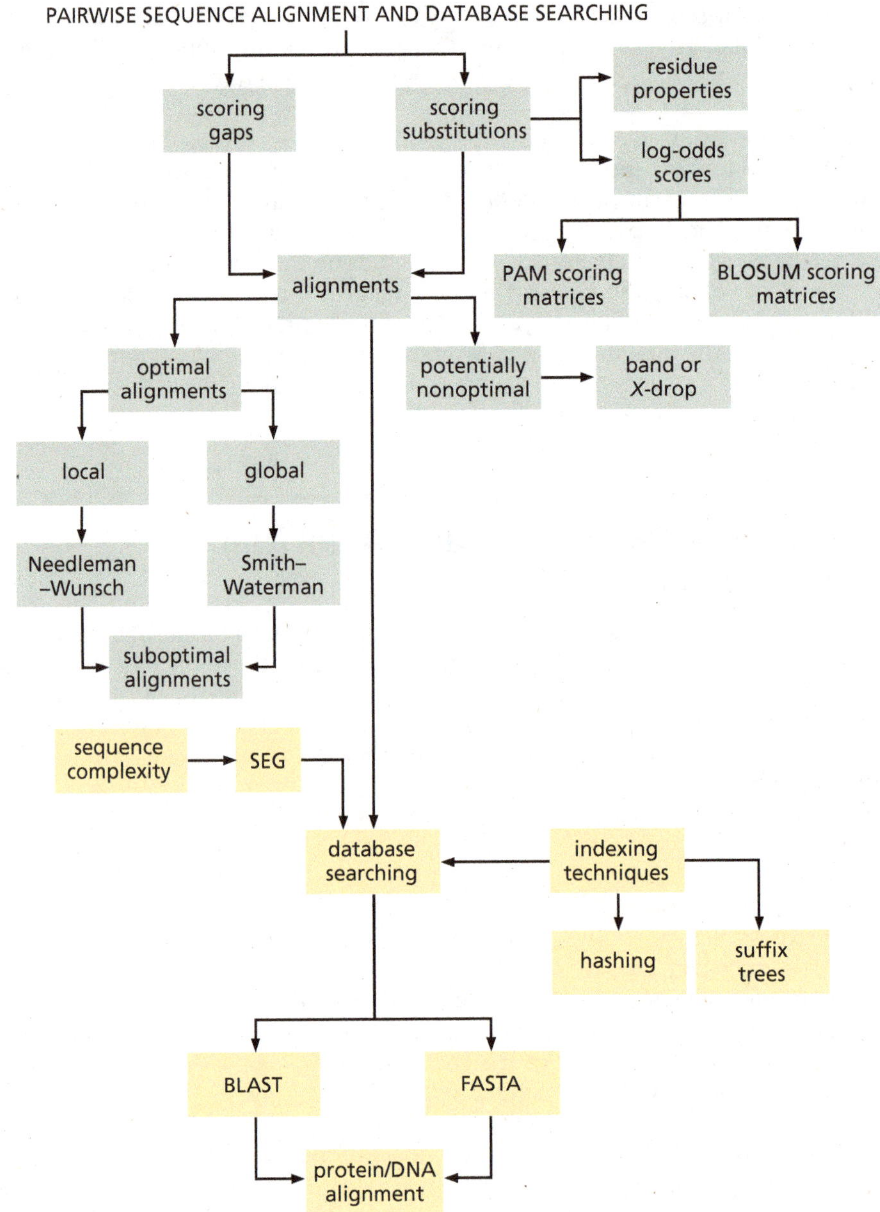


representing the sequence that uses these repeats to reduce the space needed for a full description. In addition, the form of the tree makes it very easy to find specific sequences and sequence repeats.

A **suffix** is defined as the shortest sub-sequence starting at a particular position that is unique in the complete sequence and can therefore be used to clearly identify that position. For example, the third base in the sequence above is C. As there are several Cs in the whole sequence, the sub-sequence C is not sufficiently unique to label position 3. However, by including the next base, as in CC, we have found a unique sub-sequence, and the suffix at position 3 is CC.

The suffix tree for the example sequence is given in Figure 5.20. Looking at it you can easily see that the longest repeats are the triplets ATC and CGA, which occur at positions (1,11) and (4,13), respectively. The efficiency of this technique may not be so apparent with this example because the sequence is rather short. Although longer sequences will tend to have longer suffixes, they will also tend to have more repeats, resulting in a more efficient tree.

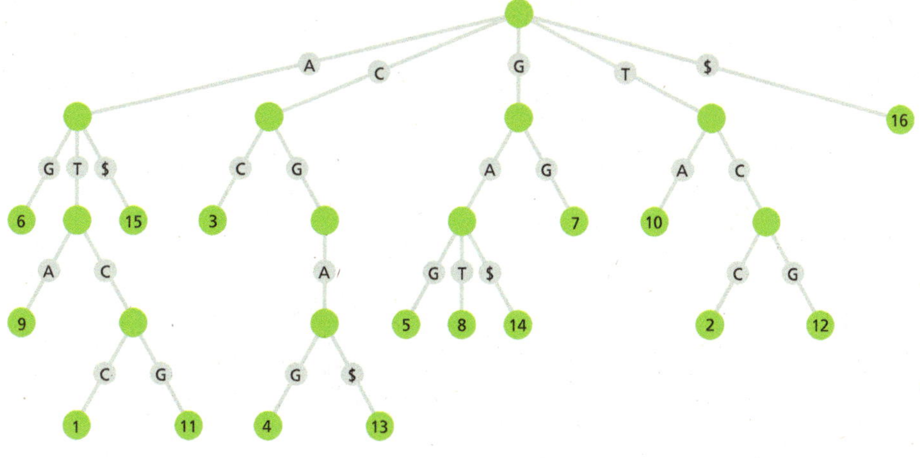


**Figure 5.20**
**The suffix tree for the sequence ATCCGAGGATATCGA$.** There are alternative ways of showing and labeling suffix trees, and often they are labeled with the whole suffix, but this form has been used because parallels can then be seen readily with finite-state automata (Figure 5.23). The numbered terminal nodes (leaves) correspond with the numbered positions in the sequence. Thus 6 refers to the sixth position, at which the base is A. The suffix for position 6 is AG. All suffix positions are clustered according to the 5′-terminal part of their sequence (or amino terminal if dealing with a protein sequence), grouping together all suffixes starting with a G for example.

Constructing the tree is straightforward. First, all positions in the sequence are grouped according to their base type; for a protein sequence this would be amino acid residue type. These groups correspond to the first row of nodes from the root. Each of these groups is then regrouped according to the following base to give the second row of nodes. This procedure is continued, stopping for a group when it only contains one sequence position. For a sequence of length $L$, this method requires a number of steps proportional to $L \log L$. Faster methods, requiring a number of steps proportional to $L$, are known but are more involved.

## Hashing is an indexing technique that lists the starting positions of all k-tuples

The basic aim of hashing is to construct a list of the starting positions of all k-tuples that occur in a query sequence. If we subsequently want to find where a particular k-tuple occurs in the sequence we simply look up the list. This procedure is used in FASTA.

Before construction of the list can begin we need to create a code for each k-tuple. Suppose we are dealing with nucleotide sequences, and k-tuples of length $k = 3$. Because there are four possible bases, there are $4^3$ (= 64) possible k-tuples, trinucleotides in this case. We can easily create a number code for these. First, assign each base a number from 0 to 3, for example A = 0, C = 1, G = 2, and T = 3. If we label this variable $e$, any 3-tuple $x_i x_{i+1} x_{i+2}$ can be assigned a number ($c_i$) according to the formula

$$c_i = e(x_i)4^2 + e(x_{i+1})4^1 + e(x_{i+2})4^0 \qquad \text{(EQ5.28)}$$

For example, the trinucleotides AAA, CAA, ACA, and AAC would be assigned the numbers 0, 16, 4, and 1, respectively. The $c_i$ will vary from 0 to 63 (= $4^3 - 1$). As an example, the sequence TAAAACTCTAAC has 10 trinucleotides with $c_1$ to $c_{10}$ given by 48, 0, 0, 1, 7, 31, 55, 28, 48, and 1, respectively. In the case of protein sequences, the amino acids would be numbered from 0 to 19, and instead of using powers of 4, powers of 20 would be used.

If the sequence in question is of length $L$, there will be $(L - k + 1)$ k-tuples that will be coded into the values of $c_1$ to $c_{n-k+1}$. Because we ultimately want to search for particular k-tuples, which means finding particular values of the $c_i$, we need to sort the $c_i$ into numerical order. For the example given above, the ordering will be $c_2$, $c_3$, $c_4$, $c_{10}$, $c_5$, $c_8$, $c_6$, $c_1$, $c_9$, $c_7$. There are many textbooks on numerical algorithms that give details of efficient sorting methods, so they will not be discussed here (see Further Reading).

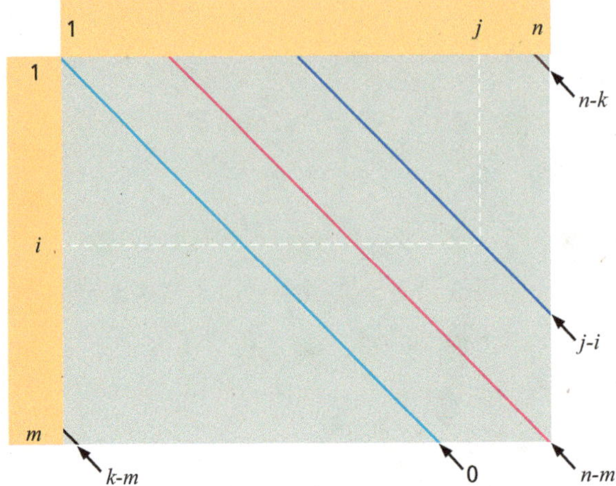


**Figure 5.21**
**Definition of the labeling of matrix diagonals $d_{j-i}$.** A word length of $k$ is used, so the last $(k-1)$ elements of each row and column are not filled. This is why the diagonal labels range from $k-m$ to $n-k$.

For each different k-tuple we need to know whether it occurs, and if so at which base(s) it starts. There are several ways of doing this, of which a particularly efficient one is **chaining**. For sequence of length $L$, two arrays are required: one, $\boldsymbol{a}$, as long as the number of different possible k-tuples ($4^k$ for nucleotides) and one, $\boldsymbol{b}$, of length $(L-k+1)$. The element $a_i$ contains a pointer to the first base of the first occurrence of the $i$th k-tuple if it exists in the sequence, or else a value that signifies its absence. Suppose that this pointer is to base $x_j$. In that case, $b_j$ contains a pointer to $x_k$, the first base of the second instance of the $i$th k-tuple, or else a value that signifies there is no second instance. If there is an $x_k$ in $b_j$, then $b_k$ will contain information about the presence of a further instance of the $i$th k-tuple.

The first eight elements of $\boldsymbol{a}$ for the example sequence TAAAACTCTAAC will be (2, 4, 0, 0, 0, 0, 0, 5), where a value 0 indicates the absence of that particular 3-tuple. The elements of $\boldsymbol{b}$ are (9, 3, 0, 10, 0, 0, 0, 0, 0, 0). This contains three pointers because three 3-tuples are present more than once in the sequence. It can be seen that no 3-tuple occurs more than twice, because these pointers in $\boldsymbol{b}$ are to elements that contain 0. Hashing and chaining techniques are used to great effect in the FASTA programs, as we discuss next.

## The FASTA algorithm uses hashing and chaining for fast database searching

A series of programs have been written by William Pearson and colleagues that allow fast and accurate searching of both protein and nucleic acid databases with both protein and nucleotide query sequences. They are based on a very fast heuristic algorithm that has four distinct steps, which are applied to each database sequence independently. In the first step, local ungapped alignments of k-tuples are located. These are then scored using a standard scoring scheme, and only the highest-scoring aligned regions are retained. Still retaining the ungapped regions, an attempt is then made to join these into a single crude alignment for the pair of sequences, resulting in an initial alignment score. This score is used to rank the database sequences. In the final step, the highest-scoring sequences are aligned using dynamic programming. Depending on the program parameters selected, this may use only a band of the full matrix including the region containing the crude alignment. We will now describe some of these steps in more detail.

FASTA starts by hashing and chaining the query sequence. A parameter, ktup, gives the size of the k-tuples to be hashed, and is usually set to 2 amino acids for proteins and 6 bases for nucleotides. Note that these values produce 400 and 4096 different possible k-tuples, respectively. As the program will search for k-tuples shared by both sequences, using smaller values of the ktup parameter makes the search more

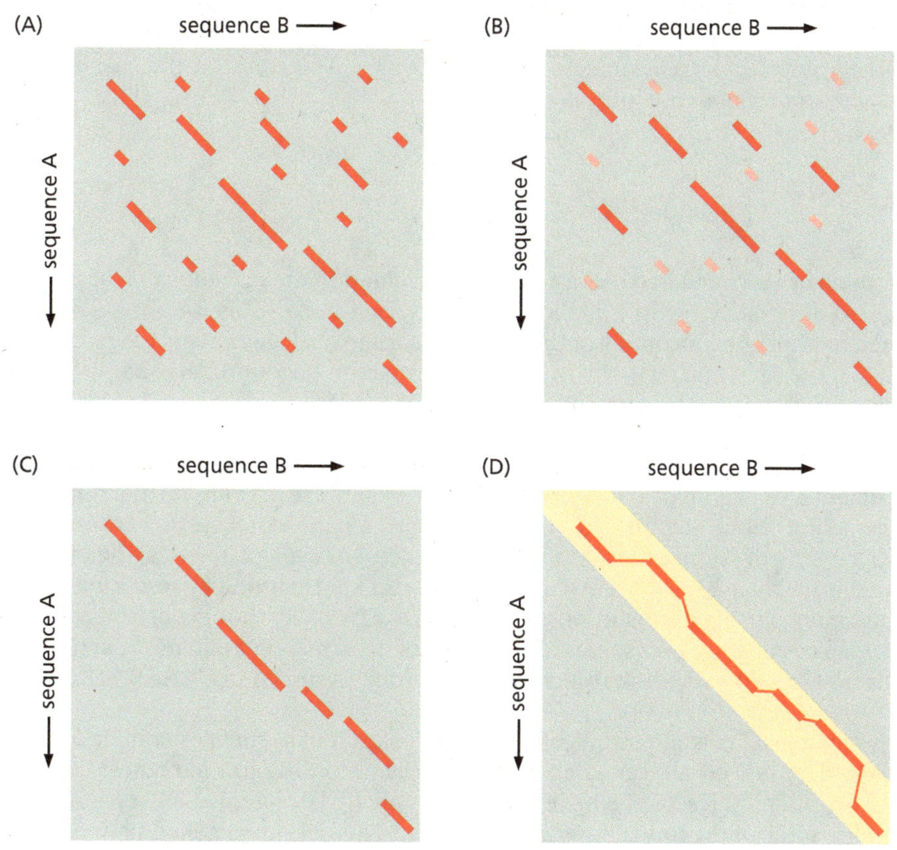


**Figure 5.22**
**The four steps in the FASTA method for database searching.** (A) For a given database sequence, all ungapped local alignments (against the query sequence) of suitably high score are shown. (B) The ten highest-scoring alignments are rescored with the PAM250 matrix, the highest having score init1. (C) Attempts are made to join together some of these ungapped alignments, allowing some gaps, to obtain score initn. (D) Dynamic programming is used to extend the alignment and give the final score opt.

sensitive. However, this is usually only beneficial for distantly related sequences, and searches based on family profiles (see Sections 4.6 and 6.1) would be expected to be even more sensitive.

In what follows, we will assume that the chaining arrays $a$ and $b$ have been calculated as described in the previous section. For each database entry, successive k-tuples of the sequence are searched against these arrays to identify where identical k-tuples align in the two sequences. As these alignments are ungapped, they lie on diagonals in an alignment matrix (such as the one in Figure 5.11). We want to find those diagonals that have many such k-tuples aligned along them.

Before giving the details of the search algorithm, a few comments about labeling diagonals in an alignment matrix will be helpful. As in the alignment matrices shown in Section 5.2, the alignment diagonals descend to the right. Suppose the two sequences have $m$ and $n$ residues, respectively. Remembering that there are $(m - k)$ k-tuples in an $m$ residue sequence, there will be $(n + m - 2k + 1)$ diagonals, which can be labeled $d_{k-m}$ to $\mathrm{d}_{n-k}$ (see Figure 5.21).

With this diagonal labeling, if the same k-tuple is located at position $i$ of the query sequence and at position $j$ of the database sequence, the alignment of these k-tuples will lie on the diagonal $d_{j-i}$. The chaining arrays make it simple to determine all the identical k-tuples, and on which diagonals they lie. All such aligned k-tuples are given the same (positive) score $s$. We just need to account for residues between these aligned k-tuples to complete the scoring of diagonals. These can be scored with a penalty $g(l)$ proportional to their length $l$. Using these scores, a straightforward algorithm finds the highest-scoring local regions of the diagonals. This is related to the Smith–Waterman algorithm, although as there are only matching residues it is much simpler (and quicker). A score is maintained for each diagonal of the alignment matrix. Let this score be $s_{j-i}$ for diagonal $d_{j-i}$, and suppose the last k-tuple match on this diagonal was at location $j_0$. As the database sequence is scanned, another k-tuple match is found on $d_{j-i}$. The score of the diagonal is updated according to the formula

$$s_{j-i} = s + \max \begin{cases} s_{j-i} & + \quad g(l) \\ & 0 \end{cases} \tag{EQ5.29}$$

The value of $l$ is calculated using $j_0$, and $j_0$ is updated to the current value of $j$. Note the zero alternative, as in Equation EQ5.27, which makes this a local alignment search method. The locations of the highest-scoring local alignments are recorded, so that more than one region of the same diagonal can be found. An example of the result of this step can be seen in Figure 5.22A.

Current implementations of this scheme identify the 10 highest-scoring ungapped alignments for each particular pair of sequences and use them in further analysis. In assessing these alignments, no account has yet been taken of conservative replacements (if comparing amino acid sequences) or even of identical matches in runs shorter than $k$ residues. We now address this shortcoming by rescoring these 10 alignments using a substitution matrix such as PAM250, thus generating a more reasonable score. In the case of nucleotide sequences, matches and mismatches are by default scored +5 and –4, respectively. This stage is shown in Figure 5.22B.

Early versions of this program (called FASTP for protein sequences and FASTN for nucleotides) ranked all the database sequences according to the highest-scoring local alignment, the score being referred to as "init1." The later FASTA versions first combine some of the top-scoring alignments into a single longer alignment using a simple dynamic programming technique. The scoring scheme for this technique is based on the scores of the individual alignments and a joining penalty to score the regions between them. The resultant alignment score is called "initn," and is used to make a preliminary ranking of the database sequences. This stage is shown in Figure 5.22C.

In the last step, suitably high-scoring database sequences are further investigated with a Smith–Waterman local alignment procedure to produce the final alignment and score. This will introduce gaps to give the best alignment. In most cases the older versions of the program used a banded version of the algorithm, centered on the initial approximate alignment. In general, a band of 16-residue width was used, except in the case of protein sequence alignments with 1-tuple indexing, when a 32-residue band was employed. This stage is shown in Figure 5.22D. In later versions, such as FASTA3, all alignments of protein sequences use the full-matrix Smith–Waterman method by default. The default for nucleotide sequences is still to use the banded version, as the full-matrix method is very time consuming for the longer sequences.

The resultant alignment score, opt, is used for the final sequence ranking, but the ranking itself is according to the estimated significance of the score. This is based on theories such as the extreme-value distribution discussed in the final part of the chapter. The significance estimates involve the sequence lengths as well as the score, and thus the ranking reported can differ from that based directly on opt. By default, the alignments reported by FASTA are those given by the Smith–Waterman method, and therefore may contain gaps.

By restricting the initial search to ungapped perfect matches, joining these together in a simple way, and using the resulting score to filter out very dissimilar database sequences, FASTA can achieve a considerable speed-up in database searches over a straightforward application of the Smith–Waterman method. Furthermore this has been achieved for a very small loss of sensitivity.

## The BLAST algorithm makes use of finite-state automata

To evaluate the alignment of a database sequence with the query sequence, one needs to know the significance of its score relative to that expected for a random sequence. It proved very hard to derive a theory from which the significance could be calculated. The inclusion of gaps in alignments proved to be one of the major complications. In 1990, Samuel Karlin and co-workers derived a theory for ungapped local alignment scores. The BLAST programs were written to take advantage of the rigorous scoring significance estimates that could now be derived for ungapped local alignments.

For this reason the original program did not consider gaps at all, and reported one or more local ungapped alignments per pair of sequences. Subsequent versions of BLAST allow gaps in alignments after an initial search without them, so that currently BLAST will produce a final local gapped alignment. In this respect the latest versions of BLAST are similar to FASTA.

The initial stage of BLAST uses short words to search for identities in the database sequence. It differs from FASTA in two respects. First, FASTA only looks for k-tuples that are identical to query-sequence k-tuples. BLAST, on the other hand, searches for k-mers that would score above a given threshold (*T*) when aligned with a query k-mer. These aligned k-mers need not be identical. Second, FASTA uses hashing and chaining to aid rapid identification of k-tuple matches. BLAST uses a scheme based on **finite-state automata** (**FSA**) to achieve the same goal. The input for such an automaton is a linear string of symbols taken one at a time. The automaton is designed to be able to identify particular patterns in the input string. These patterns may be more complex than a specific sequence and can cover a range of variation. Such automata usually report the existence of one or more of these patterns, possibly including location information, by emitting data under specific circumstances.

Each state of an automaton has well-defined responses to any possible input, responses that can include both the transition to a new state and the emission of symbols. A key response to a new symbol is not to accept it, meaning that the input string is rejected. Some automata always start reading a new string from a particular state, in which case rejection can also be written as transition back to this state. Figure 5.23 shows an example of such a finite-state automaton, in which rejection (denoted by $\P_1$ and $\P_2$) results in transition back to state 0.

As **hidden Markov models** (**HMMs**) are discussed in detail in further chapters (especially Sections 6.2 and 9.3) it is useful to note here that in contrast to HMMs, the transitions and emissions in FSA models are not probabilistic. Each input symbol leads to a deterministic outcome.

Unlike FASTA, at all stages BLAST calculates scores using a realistic substitution matrix such as BLOSUM-62 (that is, it does not use a simple sum of identities such

| | INPUT | OUTPUT |
|---|---|---|
| $\P_1$: | ~CHY | none |
| | H | CHH |
| | Y | CHY |
| $\P_2$: | ~CH | none |
| | H | CYH |

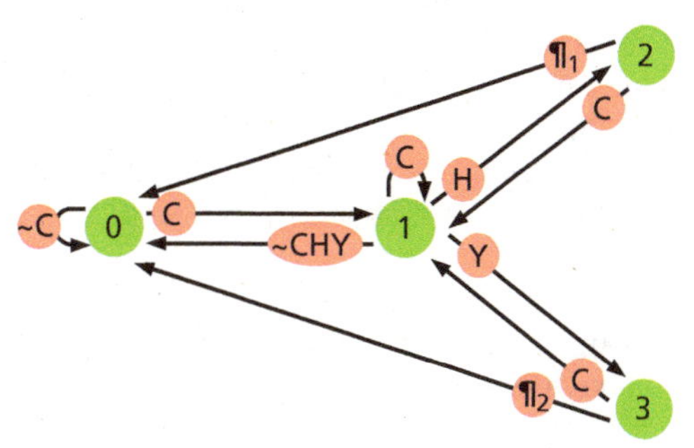


**Figure 5.23**
**The deterministic finite-state automaton that can be used to find instances of the 3-mers CHH, CHY, and CYH in an input sequence.** Input starts at state 0. The input that causes a particular transition is given near the start of the relevant arrow. Thus the transition from state 2 to state 1 is triggered by input C. The symbol ~ means "not," so ~CHY means any input except C, H or Y. The transitions $2 \rightarrow 0$ and $3 \rightarrow 0$ are triggered by several different inputs as listed, some of which also result in an output from the automaton. No other transitions produce an output. The output in this example is simply the 3-mer matched in the input, but it could be the residue number of the start of the matching 3-mers, or any other useful information.

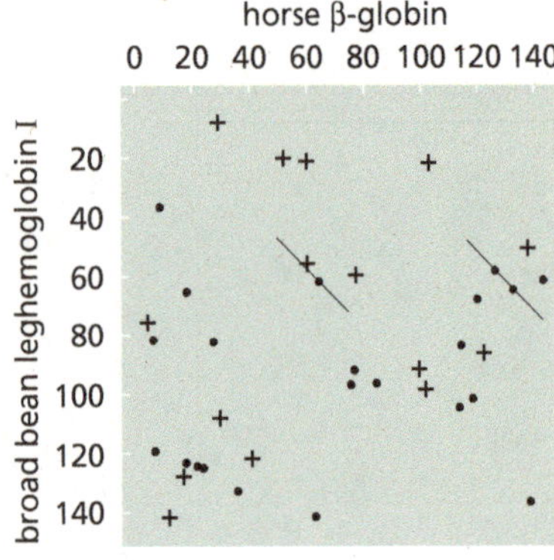


**Figure 5.24**
**Illustration of BLAST word hits for a comparison of horse β-globin and broad bean leghemoglobin I.** The + symbols indicate the 15 hits with $T = 13$, as used in the original BLAST algorithm. All 15 would be extended to give ungapped HSPs. The • symbols show a further 22 hits with $T = 11$, the setting of the more recent gapped BLAST program. From this total of 37 hits, there are two pairs on the same diagonal within 40 residues. Only these two are extended, as shown by the lines. The left-hand one gives the higher ungapped HSP score, and is subsequently extended into a gapped alignment as shown in Figure 5.19. (Redrawn from S.F. Altschul et al., Gapped BLAST and PSI-BLAST: a new generation of protein database search programs, *N.A.R.* 25 (17):3389–3402, 1997, by permission of Oxford University Press.)

as Equation EQ5.29); as in FASTA, nucleotide sequences are by default scored +5 and –4 for matches and mismatches, respectively. Typically, a k-mer of length 3 is used for protein sequences, and of length 11 for nucleotide sequences. In the following discussion, unless explicitly stated, we will assume we are dealing with protein sequences.

For each 3-mer in the query sequence all possible 3-mers that have an alignment score greater than $T$ are generated. This list is then used to scan the database sequence to find all possible similar regions. A typical value for $T$ was 14 in early versions of BLAST when used with the BLOSUM-62 matrix, meaning an average alignment score per residue of over 4. This meant that 3-mers composed only of alanine, isoleucine, leucine, serine, or valine could not reach the threshold (see Figure 5.5), as they all score a maximum of 4. (Some implementations of the algorithm allow the exact 3-mer, i.e., the identical tripeptide, in such a case.) In fact, the large number of negative scores in the BLOSUM-62 matrix, and the maximum possible score of 11 (for aligning two tryptophan residues), mean that all the possible 3-mers are highly similar, if not identical, to the query 3-mer. If $T$ is raised, the number of possible 3-mers is reduced, and they will be even closer to the query 3-mer. This will be the case even for 3-mers including residues that have high scores, such as cysteine and histidine. For example, if $T = 19$, only CHH, CHY, and CYH will score sufficiently highly against the query sequence 3-mer CHH with BLOSUM-62.

For protein sequences there are a total of $20^3 = 8000$ possible different 3-mers, and an $n$-residue query protein will have $(n - 2)$ 3-mers. Each possible 3-mer must be associated with the position of the query sequence 3-mer. The default choice of parameters can result in about 50 words, making a long list of potentially tens of thousands of k-mers for even an average-length query sequence.

For nucleotide sequences, a word size of 11 is often used, and instead of allowing nonidentical matches, only exact matches to these query words are allowed. In this case, construction of the word list used for searching resembles that in FASTA.

Returning to the case where we are matching nonidentical k-mers, given the list of k-mers, a deterministic finite-state automaton can be constructed such that if the database sequences are input to the automaton, all possible matches will be output in an efficient way. Finite-state automata are best explained with an example, and Figure 5.23 shows one for the CHH 3-mer mentioned above. The figure assumes that only the three 3-mers CHH, CHY, and CYH are to be searched for.

The automaton is started in state 0. The database sequence is input to the automaton one residue at a time. From state 0, only state 1 can be reached, requiring a C, because all desired 3-mers start with a C. Note, however, that all inputs to state 0 that are not C (~C) are regarded as causing a transition from 0 to itself (shown by the curled arrow on the left in Figure 5,23). In fact the results of any input to any state are defined by an unambiguous transition, which is why this is called a **deterministic finite-state automaton.**

From state 1, an H input leads to state 2, which can only be reached this way, requiring an input sequence CH. Similarly, state 3 can only be reached with the input sequence CY. Thus we are gradually building up the desired 3-mers, in a way analogous to the suffix tree (see Figure 5.20). If the input at state 1 is C there is a transition back to state 1, because this second C can be regarded as a second attempt to start one of the 3-mers. All other inputs give a transition back to state 0, from where the search starts all over again with the next input.

From states 2 and 3, unless the input is C, the transition is to state 0. Certain inputs result in the output of a 3-mer name, and these transitions can only occur if the input sequence contains the relevant 3-mer at that point. Thus, if the input sequence is CHCYHC, the states visited will be 0121301 and the transition to state 0 will be accompanied by the output CYH. Thus the 3-mer has been identified. In the BLAST algorithm, instead of storing the 3-mer sequence, as in the first stage of FASTA the positions of the 3-mers in the two sequences are kept for further analysis in the next stage of the algorithm. An example of the results at this stage is shown in Figure 5.24.

In a real case, with several thousand k-mers to find, a diagram of the automaton is extremely large, with many crossing transitions. Whenever input is such as to require restarting the search with a new initial residue, transition is made to the state that can reuse as much as possible of the failed k-mer. For example, if a state was reached via input ABCD but no desired k-mer has sequence ABCDE, then input E will cause a transition to a state that can only be reached via input of BCDE, CDE, DE, or E, or else will return to state 0. These transitions will be considered in the order given, to try to retain as much information as possible. These transition choices depend only on the k-mer list, and so can be constructed with fixed transitions as soon as the list is available. Using this approach, all suitably high-scoring k-mer matches between the query and database sequences can be listed.

In the original version of BLAST, all k-mer matches scoring above $T$ are extended in both directions without using gaps. Such extended ungapped local alignments are called high-scoring segment pairs (HSPs). The score is monitored and the extension is stopped when the score falls by some set amount $X_u$ from the maximum found so far for this match. This procedure tries to allow an HSP to contain a region of lower similarity. For protein sequences, a typical value $X_u$ for the permitted drop in score is 20. It is possible that the best-scoring HSP has a sufficiently poor-scoring internal region that the extension will stop prematurely, but the parameters are set to try to minimize the likelihood of this occurring. The highest-scoring region (MSP, maximal segment pair) for this pair of sequences is used with some statistical measures discussed in the final part of this chapter to estimate the significance of the alignment. In some cases, more than one HSP may score above some threshold value, in which case two or more HSP scores may be used in determining the significance. The alignment reported by BLAST can be a relatively short stretch of the whole sequence, and no attempt is made to extend it further using gaps.

Newer versions of BLAST take a different approach to generating alignments from the initial hits, called the two-hit method. This starts from the premise that any significant alignment is likely to have at least two high-scoring k-tuple matches on the same diagonal of the alignment matrix. Thus the initial matches are searched to find pairs on the same diagonal within a given distance (typically 40 residues) of each other. The scoring threshold $T$ for k-mers is then lowered, for example to 11 for protein sequences scored with BLOSUM-62, producing more k-mers, and thus more initial hits (see Figure 5.24). However, only a few of these will pair up suitably close to each other on the same diagonal. Only the second hit of such pairs is extended, initially ungapped, as for the older BLAST version. The $X_u$ parameter is used in obtaining the extension, as described above. This ungapped extension must produce an HSP with a score greater than a threshold value $S_g$ if it is not to be discarded. $S_g$ is set such that approximately 2% of database sequences will have an HSP of greater score.

Alignments with scores exceeding $S_g$ are used to seed a dynamic programming calculation of a gapped alignment. This is started from the center of the highest-scoring 11-mer of the HSP. The matrix is filled both forward and backward, as described in Section 5.2, with elements calculated until the score falls a set amount $X_g$ below the current highest-scoring alignment. In this way the amount of calculation is minimized without restricting the alignment to a predetermined band of the matrix. Only one gapped alignment is generated for a database sequence, and its score is used to determine the significance, as discussed in Section 5.4.

The parameters of the gapped BLAST program are set to make it approximately three times as fast as the ungapped version, yet more sensitive. This is achieved by severely reducing the number of extensions attempted. A gapped extension takes approximately 500 times as long as an ungapped one.

## Comparing a nucleotide sequence directly with a protein sequence requires special modifications to the BLAST and FASTA algorithms

There are several situations where a comparison between a nucleotide sequence and a protein sequence is necessary. Most protein databases tend to be nonredundant and only contain highly reliable sequences (that is, minor variants and many hypothetical genes are excluded). When analyzing new sequences, a much stronger signal for significant similarity can be obtained by comparing against protein sequences. Thus, new nucleotide sequences are often compared with the protein sequence databases. Conversely, there are occasions when one wants to search for homologous proteins in large genomic sequences.

Two new problems arise in these circumstances. First, a nucleotide sequence can be translated into protein in six different reading frames (three on each strand), so that there are six different potential protein sequences to be examined for each nucleotide sequence. Second, insertion or deletion errors can be present in the nucleotide sequence. This can result in the actual protein sequence being in different reading frames for different parts of the sequence, a situation called **frameshift**. The algorithms described earlier for comparing protein with protein or nucleotide with nucleotide sequences require modifications to allow for these new factors.

In the BLAST program suite the two programs blastx and tblastx allow searches with different reading frames, but neither allows for frameshifts. In addition to the 20 possible amino acids, there may be some stop codons present, and by default, the score for aligning a stop codon with an amino acid is taken as the most negative score in the substitution matrix, although other scoring schemes are possible. Both these programs convert the nucleotide sequence into protein sequence and then work exactly as a standard protein BLAST search. The only difference is that the query-sequence reading frame used in each alignment is reported.

The FASTA program suite contains four programs relevant to this problem. Two of these (tfastx and tfasty) are for searching nucleotide databases with protein sequences, and the others (fastx and fasty) are for nucleotide query-sequence searches of protein databases. All these programs can account for frameshifts to some degree, as well as the alternative reading-frames, and thus are, in principle, more powerful than blastx at generating suitable alignments. However, in practice blastx is still very useful, and all these programs have their place in database searching.

The fastx and tfastx programs use an algorithm that only allows for nucleotide insertions and deletions. The possibility of a base being incorrectly given, for example an A where there should have been a T, is not considered. The nucleotide sequence is translated in all six frames, each set of three being considered in a single run of the alignment program.

Each of the three reading frames is analyzed in a separate matrix. However, at each step, the possibility of moving between matrices is considered. Thus, the alignment could arrive at matrix element $S_{i,j}$ from $S_{i-1,j-1}$ in the same frame, or from $T_{i-1,j-1}$ or $U_{i-1,j-1}$ in the alternative frames. The other ($i-1,j$ and $i,j-1$) elements are also possible sources of the new $(i,j)$ element. Any move between matrices incurs an extra penalty — the frameshift penalty. This can be set higher when the sequences are expected to be more accurate, to prevent excessive numbers of frameshifts in the alignments, and is often set a little higher than the gap-opening penalty. The full details of the algorithm will not be given here, but a sequence example is given to illustrate the effects of frameshifts.

## Box 5.2 Sometimes things just aren't complex enough

Many protein and nucleotide sequences contain regions that can be described as being of low compositional complexity: low-complexity regions or **simple sequences**. Examples include stretches of identical amino acids in proteins, repeated short sequences, and longer DNA repeats (see Box 1.1). It is estimated that roughly half of all database sequences contain at least one such region. If alignments are made with these regions present, many spurious similarities will be reported because many unrelated sequences contain similar low-complexity regions. These stretches also cause problems in database searches because they are nonrandom, violating the assumptions on which the calculations of statistical significance are based (Section 5.4). It is important to be able to define these stretches and mask them to prevent their biasing the database search results. Many databases have such sequences masked when they are made available for sequence searches, so that often there is only a need to find these regions in the query sequence.

Three different properties of these simple sequences can be distinguished. Precisely repeating sequences are referred to as patterns. These can be very short, for example ATT, and are not necessarily in a single contiguous block, for example ATTCATTGCATTATT. If there is a clear period of repeat—for example, ATTATTATTATT has a repeat of three—this is called a periodicity. The periodicity need not necessarily be an integer, as can occur, for example, in an amino acid repeat relating to an α-helix. Furthermore, patterns and periodicities can both involve some degree of error, in that the repeat need not be exact. The final property that we will discuss is the compositional complexity, which is a measure of bias in the sequence composition.

In general terms, a sequence of length $L$ is made up from $N_{\text{type}}$ possible different components (i.e., $N_{\text{type}}$ is 20 for proteins, 4 for nucleic acids) in a composition defined by the $a$th component occurring $n_a$ times. The general complexity-state vector is defined as a list of these integers $n_a$ in numerical order, disregarding the specific components. Thus both nucleotide sequences ATA and CAC are represented by the same vector {2, 1, 0, 0}. The number of distinct sequences of length $L$ with this composition is given by

$$\frac{L!}{\prod_{a=1}^{N_{\text{type}}} n_a!} \qquad \text{(BEQ5.2)}$$

(The term $L!$ is called "$L$ factorial", and means the product $L \times (L-1) \times (L-2) \ldots 3 \times 2 \times 1$, so $4! = 4 \times 3 \times 2 \times 1 = 24$. By definition, $0! = 1$.) The $n_a$ can vary in value from 0 to $L$, and will always sum to $L$. The number of these $n_a$ that have the value $c$ is written $r_c$, and the $r_c$ will sum to $N_{\text{type}}$. The number of different compositions that will give rise to the same complexity-state vector is given by

$$\frac{N_{\text{type}}!}{\prod_{c=0}^{L} r_c!} \qquad \text{(BEQ5.3)}$$

Thus the total number of distinct sequences corresponding to a particular complexity-state vector is

$$\frac{L!}{\prod_{a=1}^{N_{\text{type}}} n_a!} \times \frac{N_{\text{type}}!}{\prod_{c=0}^{L} r_c!} \qquad \text{(BEQ5.4)}$$

The more distinct sequences available to a complexity state, the more complex the sequence.

As an example consider the five-nucleotide sequence ATTAT. The composition of this sequence can be represented as $\{T_3, A_2, C_0, G_0\}$ where the bases have been ordered according to their abundance. There are 10 possible sequences with the same composition: AATTT, ATATT, ATTAT, ATTTA, TAATT, TATAT, TATTA, TTAAT, TTATA, and TTTAA. Note that this is 5!/(3!2!). Now consider how many different compositions are possible by switching the bases around but maintaining the proportion of bases at 3:2:0:0. There are 12 of these: $A_3C_2$, $A_3G_2$, $A_3T_2$, $C_3A_2$, $C_3G_2$, $C_3T_2$, $G_3A_2$, $G_3C_2$, $G_3T_2$, $T_3A_2$, $T_3C_2$, and $T_3G_2$. Note that this is 4!/(2!1!1!). Therefore for the general complexity state represented by the vector {3, 2, 0, 0} there are a total of 10 × 12 = 120 distinct DNA sequences. For comparison, the complexity state represented by vector {5, 0, 0, 0} has only four unique DNA sequences [(5!/5!) × 4!/(3!1!)], compared to 360 for the state {2, 2, 1, 0} [5!/(2!2!1!) × 4!/(2!1!1!)].

Several programs are available that attempt to distinguish between simple and other regions of a sequence. We will only discuss one, the program SEG, which uses compositional complexity as a measure to determine regions of simple sequences. SEG works in two steps to determine regions that satisfy certain complexity constraints. The compositional complexity $I_{\text{SEG}}$, which is a measure of the information required per sequence position to specify a particular sequence, given the composition, is defined by

$$I_{\text{SEG}} = \frac{1}{L} \log_{N_{\text{type}}} \left[ \frac{L!}{\prod_{a=1}^{N_{\text{type}}} n_a!} \right] \qquad \text{(BEQ5.5)}$$

## Box 5.2 Sometimes things just aren't complex enough (continued)

For example, for a five-nucleotide sequence, $I_{SEG}$ can vary from 0 for {5, 0, 0, 0} to ~1.92 for {2, 1, 1, 1}. In the first step of the SEG method, a more computationally efficient approximation is used to search the sequence. Windows of length $L$ are identified for which the value of

$$I'_{SEG} = -\sum_{a=1}^{N_{type}} \frac{n_a}{L}\left(\log_2 \frac{n_a}{L}\right) \qquad \text{(BEQ5.6)}$$

is less than a given threshold $I_{SEG1}$. Note that $I'_{SEG}$ approximates $I_{SEG}$ for large $L$. These initial low-complexity regions are augmented by any overlapping windows that have a value of $I'_{SEG}$ that is exceeded by a less-strict threshold $I_{SEG2}$ (that is, $I_{SEG2} > I_{SEG1}$).

In the second step, SEG determines the sub-sequence of each initial low-complexity region whose composition has the least probability of occurrence, based on a model with all residues equally likely. The probability is calculated using the formula

$$P_{SEG} = \frac{1}{N_{type}^{L}}\left(\frac{L!}{\prod_{a=1}^{N_{type}} n_a!}\right)\left(\frac{N_{type}!}{\prod_{c=0}^{L} r_c!}\right) \qquad \text{(BEQ5.7)}$$

By inclusion of the first term, this value can be compared for different window sizes.

Versions of SEG are available that are designed to search for specific periodicities, which can be useful in some instances. For example, a number of proteins have regions of low complexity, often short repeats, which can indicate nonglobular structure. Since these sequences tend not to be exact repeats, SEG can be a powerful tool for identifying these regions. DUST is an equivalent program for DNA sequences. There are other programs designed to search for specific known repeat sequences such as the DNA repeats found in many genomes, usually defined in a small repeat database.

Consider the following short stretch of sequence and three forward translations:

```
                ACCAGAGCCAACT

Frame 1          T  R  A  N
Frame 2           P  E  P  T
Frame 3            Q  S  Q
```

The translations have been placed under the central base of each codon, so that T is placed under the first C of ACC. Consider all the possible moves, including frameshifts, from a matrix element at position 2 in the translated sequence to one at position 3. This means that the alignment of TR, PE, and QS has already been considered, and we are now considering adding A, P, or Q, referring to frame 1, 2, or 3, respectively. When translated back into nucleotides, the possible interpretations of the sequence are:

1 → 2 (AGA)*G*(CCA)

1 → 3 (AGA)*GC*(CAA)

2 → 1 (GA(**G**)CC) or (GAG)*CC*(AAC)

2 → 3 (GAG)*C*(CAA)

3 → 1 (A(**GC**)C) or (AGC)*C*(AAC)

3 → 2 (AG(**C**)CA) or (AGC)*CA*(ACT)

where codons are in parentheses. Bases that are ignored in translation (deletions) are in italic, while those used in two successive codons (insertions) are in bold. Thus the sequence written (A(**GC**)C) is interpreted as (AGC)(GCC). Similarly (AGC)*CA*(ACT) is interpreted as (AGC)(ACT). Note that deletions and insertions only occur at the boundaries of codons; that is, the sequence ACGT is only interpreted as ACG or CGT, never ACT or AGT.

A greater variety of errors in the nucleotide sequence can be allowed for by fasty and tfasty. Any two, three, or four consecutive nucleotides can be interpreted as a codon, of which only the middle alternative does not involve an indel. In each case, base errors are considered. A penalty scheme for modifying the codons is combined with a BLOSUM-50 scoring of the aligned amino acids to find the best codon for that alignment; if indels are involved they also incur a penalty. Note that in this case, as well as allowing all four interpretations of ACGT mentioned above, base-pair changes are also considered. If the base change and frameshift penalties are not sufficiently punitive, almost any pair of sequences could be aligned with a high score. Both of these are commonly set to up to twice the gap opening penalty.

## 5.4 Alignment Score Significance

In this section we will examine how to determine whether the score of an optimal alignment is significantly higher than would be expected for two unrelated sequences (see Flow Diagram 5.4). This is not the only way of trying to assess significance, but it is probably the most sensitive currently available. The simpler technique of observing the percentage of identical or similar residues in the alignment (see Section 4.2), although useful, is far less precise.

If the alignment scores were normally distributed—that is, Gaussian—then knowledge of their mean and standard deviation would allow us to calculate the probability of observing any given score using standard tables. The situation is not so simple, however, because the score that is reported in a database search is that of the optimal alignment; that is, it is the best possible score for that particular pair of sequences. This means that the scores are always from the extreme end of the distribution of all alignment scores.

The statistics of optimal alignment scores have only been rigorously derived in the case of ungapped local alignments, for which the scores follow an extreme-value distribution. Using this theory, precise evaluation of score significance for ungapped local alignments is readily calculable. For gapped local alignments, only approximations are available for the statistical score distribution. Both gapped and ungapped local alignments have been found to have very similar distributions, but differ in their parameterization.

Before discussing the details of the practical application of these score distributions, two general points about scores have to be made. First, one can ask how much information is required to define the position of an alignment in two sequences. Information is usually measured as bits, which can be regarded as yes or no answers. For example, distinguishing the numbers 0–255 requires eight digits in a binary (i.e., 0 or 1) number representation, which is $\log_2 256$. For a sequence of length $m$, we have to distinguish among the $m$ possible alignment starting positions, which in general requires $\log_2 m$ bits of information. If this sequence is aligned to an $n$-residue sequence, positioning the start of the alignment on this second sequence will require a further $\log_2 n$ bits of information. Therefore the alignment requires a total of $\log_2 m + \log_2 n = \log_2(mn)$ bits of information to define its start position on both sequences.

**Flow Diagram 5.4**
**The key concept introduced in this section is that when comparing optimal alignments between a query sequence and a database of sequences, care must be taken to evaluate the true significance of the alignment score, since even the best scoring database sequence cannot be assumed to show significant similarity.**

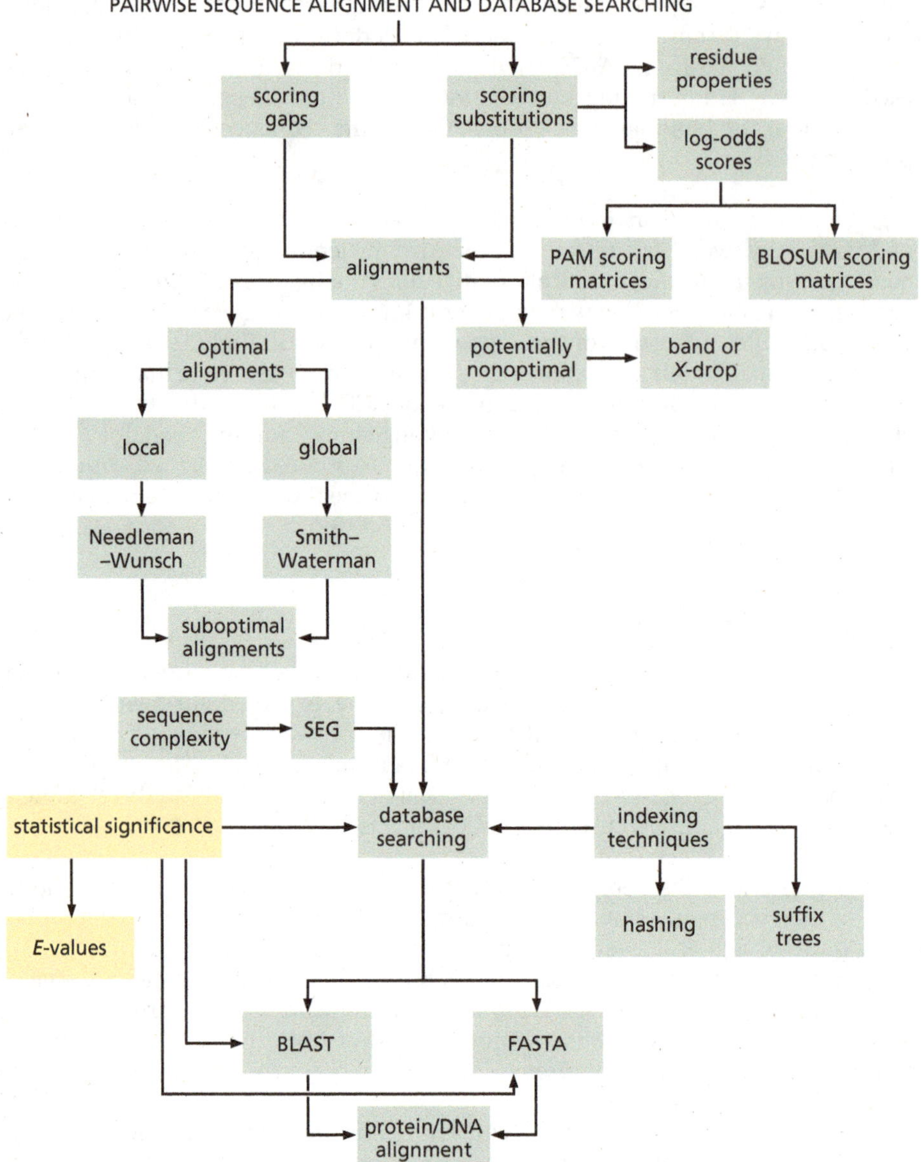


In performing a database search, if $m$ is the length of the query sequence, then $n$ is the total length of the database entries. Considering a protein sequence database of 100 million residues and an average-length query sequence of 250 residues, a score of approximately 35 bits would be required to define the location of the best alignment between the sequences.

The second general aspect of optimal alignment scores relates to how they vary with the length of the alignment. Waterman has shown that for global alignments with gaps, the score grows linearly with sequence length. For local alignments the situation is more complex, and depends on a property of the substitution matrix used. If the expected score of a random sequence with a given substitution matrix, given by Equation EQ5.3 is positive, the local alignment score grows linearly with sequence length regardless of the gap penalties. However, if the expected score is negative, which is the case for most amino acid substitution matrices in common use, then unless gap penalties are very low, the optimal local alignment score will grow logarithmically; that is, as log $n$ where $n$ is the number of residues. In what follows, it is assumed we are using scoring parameters such that the scores grow logarithmically with sequence length. The alignment scores need to be corrected for length as part of the process of determining their significance.

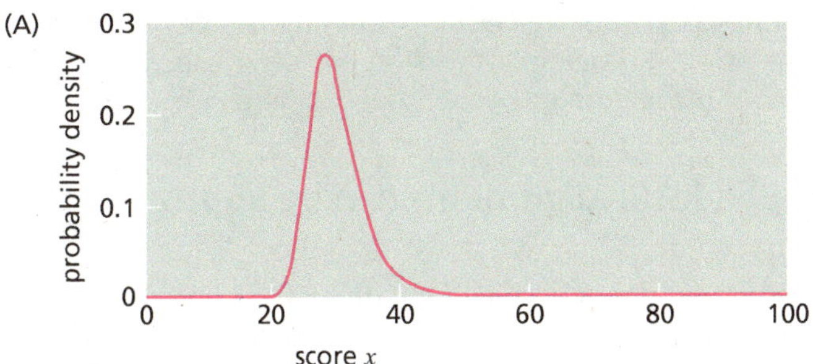


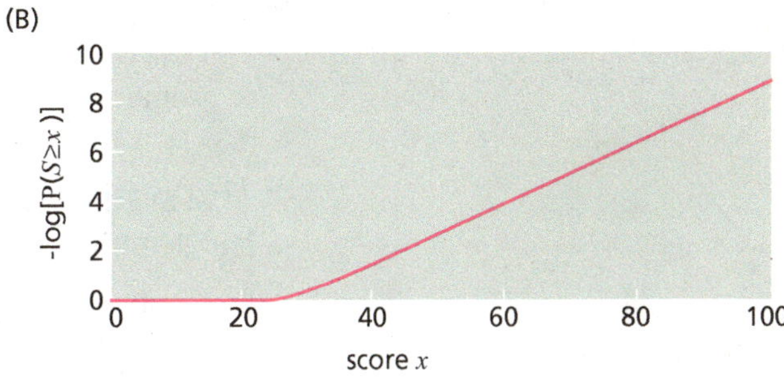


**Figure 5.25**
**The extreme-value distribution of optimal alignment scores.** (A) The distribution for λ and $K$ of 0.286 and 0.055, respectively, for two sequences both of length 245 residues. (These values are from Altschul and Gish (1996) for a BLOSUM-62 substitution matrix, with affine gap opening and extension penalties –12 and –1, respectively.) (B) The probability, given the distribution in part A of an observed score $S$ being at least $x$, plotted so that a value $y$ on the vertical axis represents a probability $10^{-y}$.

The distribution of alignment scores for optimal ungapped local alignments has been rigorously derived from first principles. A database search only carries out further analysis of the highest-scoring alignment of each database entry with the query sequence. Omitting the considerable amount of analysis required, it can be proved that the optimal ungapped local alignment score follows the Gumbel extreme-value distribution. With $m$ and $n$ defined as above, this distribution peaks at a value

$$U = \frac{\ln(Kmn)}{\lambda} \qquad \text{(EQ5.30)}$$

where $K$ and λ are constants that depend on the scoring matrix used and the sequence composition. λ is a scaling parameter of the substitution matrix, and is the unique positive solution of the equation

$$\sum_{a,b} p_a p_b e^{s_{a,b}\lambda} = 1 \qquad \text{(EQ5.31)}$$

where the summation is over all residue types $a$ and $b$, the $p_a$ and $p_b$ are frequencies of occurrence of the residues as defined in Section 5.1, and $s_{a,b}$ are the substitution scores.

The probability of the alignment score $S$ being less than $x$ is given by the cumulative distribution function

$$P(S < x) = \exp\left(-e^{-\lambda(x-U)}\right) \qquad \text{(EQ5.32)}$$

from which the complementary probability of the score being at least $x$ is

$$P(S \geq x) = 1 - \exp\left(-e^{-\lambda(x-U)}\right) \qquad \text{(EQ5.33)}$$

Substituting for $U$ we arrive at the formula for the probability of obtaining an alignment of score $S$ greater than a value $x$:

$$P(S \geq x) = 1 - \exp\left(-Kmne^{-\lambda x}\right) \qquad \text{(EQ5.34)}$$

The extreme-value distribution is shown in Figure 5.25. The key feature of interest is the tail for high values. Notice that this decays much more slowly than the low-value tail; that is, the distribution is asymmetric with a bias to high values. It is important that we have the correct distribution if we are to estimate the significance of any given score accurately. In general, if $P(S \geq x)$ is less than 0.01, the alignment is significant at the 1% level. The level chosen as cut-off depends on the particular problem, and is discussed in more detail in Section 4.7.

Formulae exist for calculating $K$ and $\lambda$ for a given substitution matrix and sequence database composition. The original (ungapped) version of BLAST used this theory to estimate the significance of the alignments generated in a database search.

### The statistics of gapped local alignments can be approximated by the same theory

For gapped local alignments, examination of actual database searches has shown that the scores of optimal alignments also fit an extreme-value distribution. However, in this case there is no rigorous theory to provide the parameters $\lambda$ and $K$. These can be estimated by studying sample database searches for particular values of the scoring schemes (including gap penalties). Note, however, that they will depend on the composition of the database sequences, so that the parameterization should really be done according to the actual database to be searched. Both BLAST and FASTA use such methods to derive score significance.

The programs in the BLAST and FASTA suites report $E$-values, which are related to the probability $P(S \geq x)$. The value of $P$ is calculated for the particular lengths of the query and database sequences, and varies from 0 to 1. The $E$-values are obtained from this by multiplying by the number of sequences in the database. Thus, if there are $D$ database sequences, the $E$-values range from 0 to $D$.

The $E$-value is the number of database sequences not related to the query sequence that are expected to have alignment scores greater than the observed score. Thus an $E$-value of 1 is not significant, as one unrelated database sequence would be expected to have such a score. The $E$-value deemed to indicate a significant similarity as shown by the alignment score is a practical problem discussed in Section 4.7.

## 5.5 Aligning Complete Genome Sequences

As more genome sequences have become available, of different species and also of different bacterial strains, there are several reasons for wanting to align entire genomes to each other. Such alignments can assist the prediction of genes because if they are conserved between species this will show up clearly in the alignment. In addition, other regions of high conservation between the genomes can indicate other functional sequences. The study of genome evolution needs global alignments to identify the large-scale rearrangements that may have occurred. Although in principle the dynamic programming methods presented in Section 5.2 could be used, the lengths of the sequences involved makes the demands on computer time and disk space prohibitive. As in the case of database searches, other techniques must be used that are not guaranteed to find the optimal scoring alignment.

When aligning two complete genome sequences, problems arise that are quite distinct from those discussed above in relation to aligning two related proteins or making simple database searches. In the case of complete genomes, the alignment problem is more complex because the intergene regions may be subject to higher mutation rates; in addition, large-scale rearrangements may have occurred. Many discrete, locally similar segments may exist, separated by dissimilar regions. Unlike in the BLAST and FASTA methods, therefore, it is insufficient to determine a single location from which to extend the alignment. The solution is to modify the indexing techniques for genome alignments to locate a series of anchors. The relationship of this subject to the other topics discussed in this chapter is shown in Flow Diagram 5.5.

A second problem is the linking together of the anchors to form a scaffold for the alignment, and doing this in such a way as to identify the large-scale rearrangements.

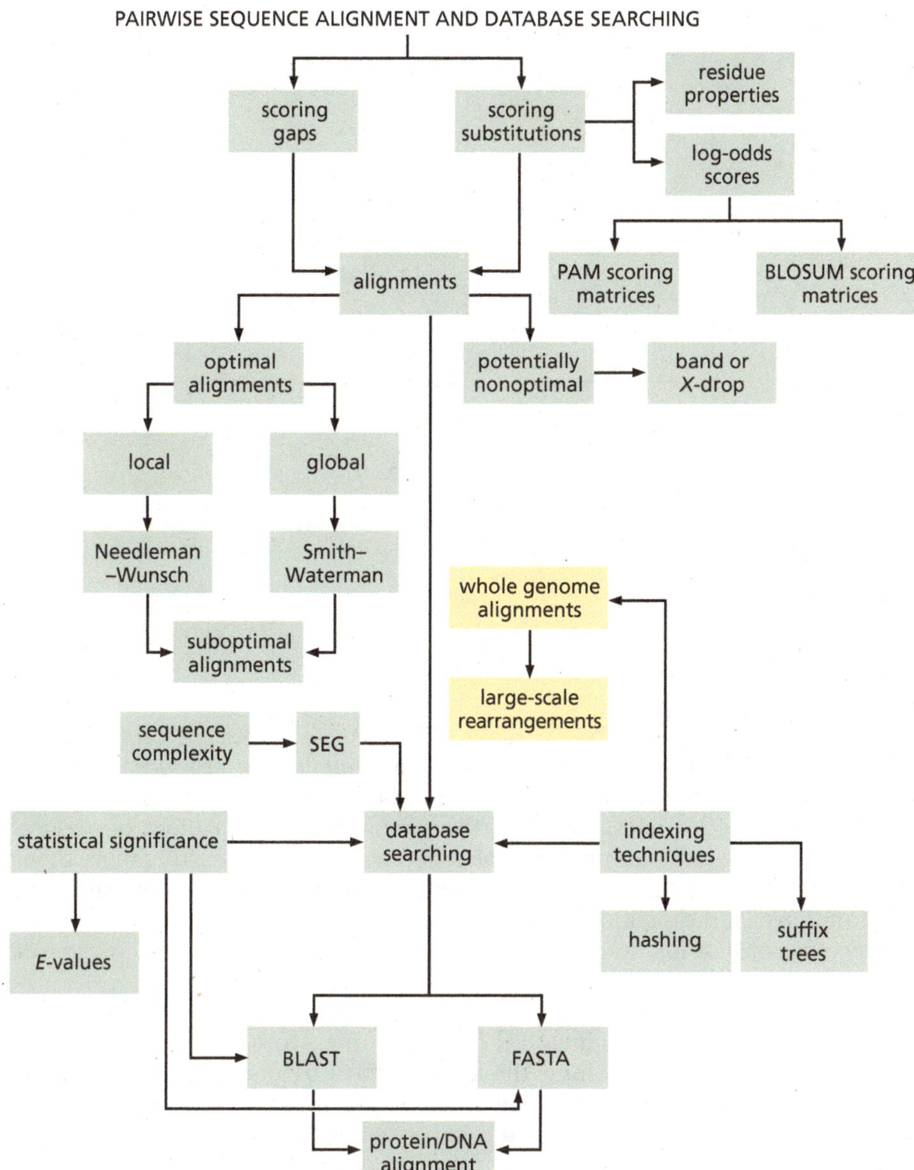


**Flow Diagram 5.5**
**The key concept introduced in this section is that when aligning extremely long sequences, refinements are required in order to make the alignment search more efficient.** The section also deals with the observation that long sequences often show large-scale rearrangements, a feature that needs to be included in the alignment methods.

Both of these problems require complex solutions that are only briefly outlined here. This area is still undergoing rapid development, and new techniques may yet emerge that are a significant advance in the field.

Another application involving a complete genome sequence is the alignment to it of many smaller nucleotide sequences. These could be from a variety of sources, including sequence data from related species at various stages of assembly. The latter task, in particular, can give rise to the problems noted above.

## Indexing and scanning whole genome sequences efficiently is crucial for the sequence alignment of higher organisms

Both FASTA and BLAST use indexing techniques to speed up database searches, and both index the query sequence. For genome alignments it is now practical to index the complete genome sequence, as long as the computer used has sufficient memory to store the whole index. In contrast to usual database searches with single gene sequences, this requires careful planning of the data storage. If the index is larger than the available computer memory, the time required for the alignment

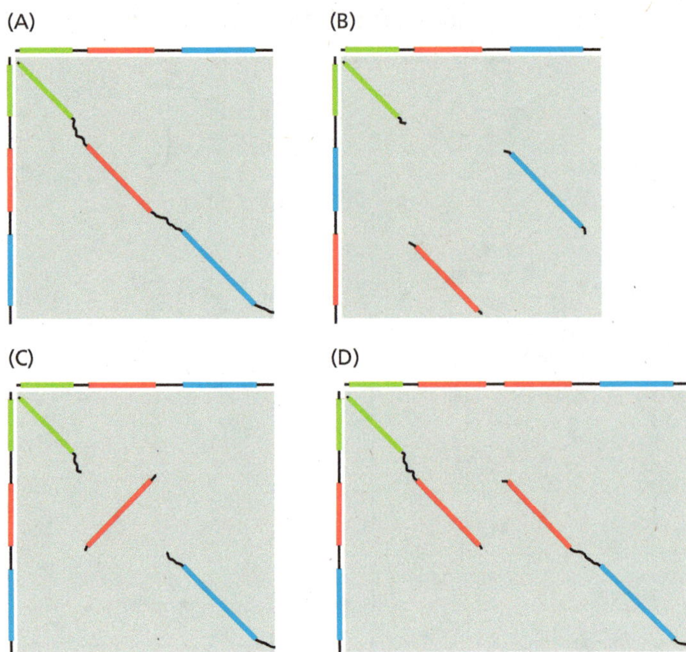


**Figure 5.26**
**Four examples of possible relationships between two long nucleotide sequences.** (A) Both sequences are similar along their whole length, with three particularly similar segments identified. No rearrangements during evolution from the most recent common ancestor need be proposed. This is the only case in which a global alignment can be given. (B) A translocation has changed the order of the red and blue segments for one of the sequences. (C) The red segment has been inverted in the horizontal sequence. (D) The red segment has been duplicated in the horizontal sequence. The black lines indicate regions of alignment outside the main segments. (Adapted from M. Brudno et al., Global alignment: finding rearrangement during alignment, *Bioinformatics* 19, Supplement 1:i54–i62, 2003.)

can increase by a factor of four or more. If suffix trees are used for the indexing, then with efficient techniques the space requirements are proportional to the sequence length. Hashing methods require storage according to the lengths of the sequence and the k-tuples. However, the space required can often be reduced significantly if the most common k-tuples, defined as those occurring more than a threshold number of times, are omitted on the grounds that they are likely to be uninformative for the alignment. The hashing methods often also reduce the index size by only considering nonoverlapping k-tuples; that is, bases $1 \rightarrow k$, $k + 1 \rightarrow 2k$, and so on.

The index is used to identify identical k-tuples in the two sequences. Because the sequences involved are very long, the k-tuples must be much longer than in the example of database searching discussed above, or many matches will be found with random sequence. The expected number of matches depends on the lengths of the k-tuples and the percentage identity of the aligned sequences, and was thoroughly analyzed during development of the BLAT (BLAST-like Alignment Tool) package. k-tuples of 10–15 bases are used by the Sequence Search and Alignment by Hashing Algorithms (SSAHA) program, and lengths of 20–50 bases have been used in suffix tree methods.

One technique that has proved to be particularly effective is the spaced seed method. When searching using the long k-tuples mentioned above, all $k$ positions of the k-tuples are compared. It has been found that these are not the most efficient

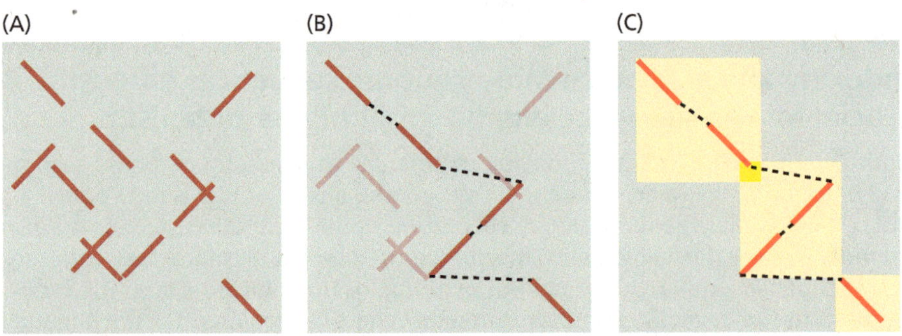

**Figure 5.27**
**An example of the application of the SLAGAN method.** (A) The set of identified potential anchors for the alignment of two long nucleotide sequences. (B) The path of anchors identified. The anchors are ordered progressively along the vertical sequence but not the horizontal one because of large-scale rearrangements. (C) The regions identified as locally consistent. Individual alignments are determined for each of the three regions. (Adapted from M. Brudno et al., Global alignment: finding rearrangement during alignment, *Bioinformatics* 19, Supplement 1:i54–i62, 2003.)

probes for identifying matches that are useful seeds for genome alignment. In the development of the PatternHunter program it was found that searching for a specific pattern of 12 bases in a stretch of 19 was more effective. The 19mer can be represented as 1110100110010101111, where 1 indicates a position whose match is sought, and 0 indicates a position whose base is disregarded. The later versions of the BLASTZ program made a further improvement by allowing any one of the 12 positions to align bases A and G or bases C and T, so long as the other 11 sites were perfect matches.

### The complex evolutionary relationships between the genomes of even closely related organisms require novel alignment algorithms

Many anchor points must be identified between the two genome sequences, each initially as matched k-tuples which are usually extended in a gapless way as in FASTA and BLAST. The original MUMmer approach was to include only unique k-tuples common between the sequences; that is, k-tuples present just once on each sequence. Such matches can be expected with confidence to identify truly homologous segments. Most methods (including MUMmer 2) are not so restrictive, and include k-tuples that occur several times on a sequence. As a consequence, many matches are recorded, some of which are expected to be incorrect.

In contrast to the database search methods, which only need to select a single anchor for the alignment that is then extended using dynamic programming, for genome alignment a set of anchors is needed that spans the lengths of the genomes, identifying the true homologous segments. This step is usually similar to the third step of FASTA (see Figure 5.22C) and involves dynamic programming where each anchor and potential gap is given a score. Anchors are not allowed to overlap and must be arranged sequentially along the sequences.

In addition, a number of large-scale sequence rearrangements may have occurred since the last common ancestor of the species whose sequences are being aligned. As a consequence, the correct alignment will not have the straightforward character of the alignments discussed so far. Some examples of common rearrangements are shown in Figure 5.26, and a real example is shown for aligning equivalent chromosomes in mouse and rat in Figure 10.21. Most existing programs cannot automatically recognize large-scale rearrangements and thus cannot correctly assign segments to the overall alignment. An exception is the Shuffle-Limited Area Global Alignment of Nucleotides (SLAGAN) program, which attempts to identify rearrangements. The FASTA-type step referred to above requires each successive anchor to occur in succession along both sequences. However, any rearrangements that produce sequence reversal of a segment (such as in Figure 5.26C) will result in the successive anchors occurring progressively but in opposite directions along the two sequences. SLAGAN only requires the anchors to occur in succession along one of the sequences. A further step then identifies those regions that contain anchors that are locally consistent with a sequence reversal segment. An alignment is obtained for each of these regions. An example of a successful application of this method is shown in Figure 5.27.

In all the methods, an attempt is made to extend identified homologous segments, often using standard alignment techniques but trying to identify the limits beyond which the sequences are not recognizably similar. The alignments derived by these methods are often fragments separated by unaligned regions.

## Summary

In the first part of this chapter, we looked at the derivation of the specialized algorithms that are used in automated methods to determine the optimal alignment between two sequences. Such algorithms are needed because if the insertion of gaps is allowed there are a very large number of possible different alignments of two sequences. Before an algorithm can be applied, a scoring scheme has to be devised so that the resulting alignments can be ranked in order of quality of matching. Scoring schemes involve two distinct components: a score for each pair of aligned residues, which assesses the likelihood of such a substitution having occurred during evolution, and a penalty score for adding gaps to account for the insertions or deletions that have also occurred during evolution. The first score has its foundations in the concept of log-odds, and is assigned on the basis of reference substitution matrices that have been derived from the analysis of real data in the form of multiple alignments. The penalty scores given to gaps have a more ad hoc basis, and have been assigned on the basis of the ability of combined (substitution and gap) scoring schemes to reproduce reference sequence alignments. (Confidence in these reference alignments is usually based on a combination of structural information and regions of sufficiently high similarity.)

The automated construction and assessment of gapped alignments only became possible with the development of dynamic programming techniques, which provide a rigorous means of exploring all the possible alignments of two sequences. The essence of dynamic programming is that optimal alignments are built up from optimal partial alignments, residue by residue, such that nonoptimal partial alignments are efficiently identified and discarded. The technique ensures that the optimal alignment(s) will be identified. A number of different schemes have been developed, which treat gaps in different ways. Only minor variations in the dynamic programming algorithms are required to alter the type of alignment produced from global to local, or to identify repeat or suboptimal alignments. Thus the same technique can be used to answer several different problems of sequence similarity and relatedness.

The problem with dynamic programming methods is that despite their efficiency they can place heavy demands on computer memory and take a long time to run. The speed of calculation is no longer as serious a barrier as it has been in the past, but the problem of insufficient computer memory persists, particularly as there are now many very long sequences, including those of whole genomes, available for comparison and analysis. Some modifications of the basic dynamic programming algorithm have been made that reduce the memory and time demands. One way of reducing memory requirements is by storing not the complete matrix but only the two rows required for calculations. However, to recover the alignment from such a calculation takes longer than if all the traceback information has been saved. By only calculating a limited region of the matrix, commonly a diagonal band, both time and space saving can be made, although at the risk of not identifying the correct optimal alignment.

Often the first step in a sequence analysis is to search databases to retrieve all related sequences. Such searches depend on making pairwise alignments of the query sequence against all the sequences in the databases, but because of the scale of this task, fast approximate methods are usually used to make such searches more practicable. The algorithms for two commonly used search programs—BLAST and FASTA—make use of indexing techniques such as suffix trees and hashing to locate short stretches of database sequences highly similar or identical to parts of the query sequence. Attempts are then made to extend these to longer, ungapped local alignments which are scored, the scores being used to identify database sequences that are likely to be significantly similar. This process is considerably faster than applying full-matrix dynamic programming to each database sequence. At this point, both techniques revert to the more accurate methods to examine the highest-scoring sequences, in order to determine the optimal local alignment and score, but this is only done for a tiny fraction of the database entries.

There are instances where it is necessary to align a protein sequence with a nucleotide sequence. In such cases one solution is simply to translate the nucleotide sequence in all possible reading frames and then use protein–protein alignments. However, this approach is rather bad at dealing with errors in the sequence that give rise to frameshifts. The dynamic programming method can be modified to overcome this problem by using three matrices, one for each possible reading frame in the given strand direction, with a set of rules for moving from one matrix to another. In this way, an optimal alignment can be generated that uses more than one reading frame, simultaneously proposing sequence errors.

Another problem that arises in database searches is the occurrence of regions of low sequence complexity, which can cause spuriously high alignment scores for unrelated sequences. Such regions can be defined either by identifying a repeating pattern or on the basis of their composition, and can then be omitted from the comparison.

To be sure that two sequences are indeed homologous, it is important to know when the alignment score reported is statistically significant. In the case of sequence alignments, the statistical analysis is very difficult; in fact the theory has not yet been fully developed for alignments including gaps. Part of the reason for this difficulty is that the alignments reported are always those with the best scores. These will not be expected to obey a normal distribution, but rather an extreme-value distribution. Furthermore, the scores are also dependent on the scoring scheme used, the sequence composition and length, and the size of the database searched. For alignments without gaps, formulae have been derived that allow the score to be converted to a probability from which the significance can be gauged.

The last section of the chapter deals with the alignment of very long sequences, for example those of whole genomes. The straightforward application of dynamic programming is often not feasible because of the lack of computer resources. This difficulty can be overcome by using similar indexing methods to those in the database search programs. However, there are significant differences, which lead to the indexing being applied to the genome sequence rather than the query sequence, and to searching for much longer identical segments than in database searches. Nevertheless, there is often an additional problem because frequently there are large-scale genome rearrangements even over relatively short periods of evolutionary time. To overcome this, local alignments must be identified that can then be joined together into a global alignment by a specific dynamic programming technique that allows for translations and inversions of segments of the sequence.

# Further Reading

## 5.1 Substitution Matrices and Scoring

**The PAM (MDM) substitution scoring matrices were designed to trace the evolutionary origins of proteins**

Altschul SF (1991) Amino acid substitution matrices from an information theoretic perspective. *J. Mol. Biol.* 219, 555–565.

Dayhoff MO, Schwartz RM & Orcutt BC (1978) A model of evolutionary change in proteins. In Atlas of Protein Sequence and Structure, vol 5 suppl 3 (MO Dayhoff ed.), pp 345–352. Washington, DC: National Biomedical Research Foundation.

Jones DT, Taylor WR & Thornton JM (1992) The rapid generation of mutation data matrices from protein sequences. *Comput. Appl. Biosci.* 8, 275–282. *(The PET91 version of PAM matrices.)*

Yu Y-K, Wootton JC & Altshul SF (2003) The compositional adjustment of amino acid substitution matrices. *Proc. Natl Acad. Sci. USA* 100, 15688–15693.

See appendix for deriving target frequencies.

**The BLOSUM matrices were designed to find conserved regions of proteins**

Henikoff S & Henikoff JG (1992) Amino acid substitution matrices from protein blocks. *Proc. Natl Acad. Sci. USA* 89, 10915–10919.

**Scoring matrices for nucleotide sequence alignment can be derived in similar ways**

Chiaromonte F, Yap VB & Miller W (2002) Scoring pairwise genomic sequence alignments. *Pac. Symp. Biocomput.* 7, 115–126.

**The substitution scoring matrix used must be appropriate to the specific alignment problem**

Altschul SF (1991) Amino acid substitution matrices from an information theoretic perspective. *J. Mol. Biol.* 219, 555–565.

May ACW (1999) Towards more meaningful hierarchical classification of amino acid scoring matrices. *Protein. Eng.* 12, 707–712. *(A comparative study of substitution matrices in terms of the way they group amino acids.)*

Yu Y-K & Altschul SF (2005) The construction of amino acid substitution matrices for the comparison of proteins with non-standard compositions. *Bioinformatics* 21, 902–911.

Yu Y-K, Wootton JC & Atschul SF (2003) The compositional adjustments of amnio acid substitution matrices. *Proc. Natl Acad. Sci USA* 100, 15688–15693.

**Gaps are scored in a much more heuristic way than substitutions**

Goonesekere NCW & Lee B (2004) Frequency of gaps observed in a structurally aligned protein pair database suggests a simple gap penalty function. *Nucleic Acids Res.* 32, 2838–2843. *(This recent paper on scoring gaps for protein sequences includes a good listing of older work.)*

## 5.2 Dynamic Programming Algorithms

Waterman MS (1995) Introduction to Computational Biology: Maps, Sequences and Genomes, chapter 9. London: Chapman & Hall. (*A more computational presentation of dynamic programming alignment algorithms.)*

**Optimal global alignments are produced using efficient variations of the Needleman–Wunsch algorithm**

Gotoh O (1982) An improved algorithm for matching biological sequences. *J. Mol. Biol.* 162, 705–708.

Needleman SB & Wunsch CD (1970) A general method applicable to the search for similarities in the amino acid sequence of two proteins. *J. Mol. Biol.* 48, 443–453.

**Local and suboptimal alignments can be produced by making small modifications to the dynamic programming algorithm**

Smith TF & Waterman MS (1981) Identification of common molecular subsequences. *J. Mol. Biol.* 147, 195–197.

Waterman MS & Eggert M (1987) A new algorithm for best subsequence alignments with application to tRNA–tRNA comparisons. *J. Mol. Biol.* 197, 723–728.

**Time can be saved with a loss of rigor by not calculating the whole matrix**

Zhang Z, Berman P & Miller W (1998) Alignments without low-scoring regions. *J. Comput. Biol.* 5, 197–210.

Zhang Z, Scwartz S, Wagner L & Miller W (2000) A greedy algorithm for aligning DNA sequences, *J. Comput. Biol.* 7 (1–2), 203–214.

## 5.3 Indexing Techniques and Algorithmic Approximations

**Suffix trees locate the positions of repeats and unique sequences; Hashing is an indexing technique that lists the starting positions of all k-tuples**

Gusfield D (1997) Algorithms on Strings, Trees and Sequences: Computer Science and Computational Biology. Cambridge: Cambridge University Press.

Waterman MS (1995) Introduction to Computational Biology: Maps, Sequences and Genomes, chapter 8. London: Chapman & Hall.

**The FASTA algorithm uses hashing and chaining for fast database searching; The BLAST algorithm makes use of finite-state automata**

Altschul SF, Gish W, Miller W et al. (1990) Basic Local Alignment Search Tool. *J. Mol. Biol.* 215, 403–410. (*The original BLAST paper.*)

Altschul SF, Madden TL, Schäffer AA et al. (1997) Gapped BLAST and PSI-BLAST: a new generation of protein database search programs. *Nucleic Acids Res.* 25, 3389–3402.

Pearson WR & Lipman DJ (1988) Improved tools for biological sequence comparison. *Proc. Natl Acad. Sci.USA* 85, 2444–2448. *(The original FASTA paper.)*

Pearson WR, Wood T, Zhang Z & Miller W (1997) Comparison of DNA sequences with protein sequences. *Genomics* 46, 24–36. (*fastx, fasty.*)

**Box 5.2: Sometimes things just aren't complex enough**

Morgulis A, Gertz EM, Schäffer AA & Agarwala R (2006) A fast and symmetric DUST implementation to mask low-complexity DNA sequences. *J. Comput. Biol.* 13, 1028–1040.

Wootton JC & Federhen S (1996) Analysis of compositionally biased regions in sequence databases. *Methods Enzymol.* 266, 554–571 (*SEG.*)

## 5.4 Alignment Score Significance

Altshul SF & Gish W (1996) Local alignment statistics. *Methods Enzymol.* 266, 460–480.

Mott R (2000) Accurate formula for P-values of gapped local sequence and profile alignments. *J. Mol. Biol.* 300, 649–659.

Waterman MS (1995) Introduction to Computational Biology: Maps, Sequences and Genomes, chapter 11. London: Chapman & Hall.

## 5.5 Alignments Involving Complete Genome Sequences

The field of genome sequence alignment is moving very quickly. Some useful references in this area are:

Brudno M, Malde S, Poliakov A et al. (2003) Glocal alignment: finding rearrangements during alignment. *Bioinformatics* 19(suppl. 1), i54–i62. *SLAGAN.*

Delcher AL, Kasif S, Fleischmann RD et al. (1999) Alignment of whole genomes. *Nucleic Acids Res.* 27, 2369–2376. *MUMmer.*

Delcher AL, Phillippy A, Carlton J & Salzberg SL (2002) Fast algorithms for large-scale genome alignment and comparison. *Nucleic Acids Res.* 30, 2478–2483. *MUMmer 2.*

Kent WJ (2002) BLAT—the BLAST-like alignment tool. *Genome Res.* 12, 656–664.

Ma B, Tromp J & Li M (2002) PatternHunter: faster and more sensitive homology search. *Bioinformatics* 18, 440–445.

Ning Z, Cox AJ & Mullikin JC (2001) SSAHA: A fast search method for large DNA databases. *Genome Res.* 11, 1725–1729.

Schwartz S, Kent WJ, Smit A et al. (2003) Human–mouse alignments with BLASTZ. *Genome Res.* 13, 103–107.

# PATTERNS, PROFILES, AND MULTIPLE ALIGNMENTS

THEORY CHAPTER

**When you have read Chapter 6, you should be able to:**

Explain how to produce position-specific scoring matrices (PSSMs).

Describe the methods to overcome a lack of data.

Compare and contrast scoring methods for alignments and profiles.

Explain the graphical illustration of sequence profiles and patterns.

Explain how to produce profile HMMs.

Describe the alignment of profiles with sequences and other profiles.

Obtain multiple alignments.

Discover sequence patterns.

Discuss the statistical scoring of patterns.

The great majority of protein sequences share significant similarity with others in the databases, as can be determined by the techniques described in Chapters 4 and 5. When these sequences are carefully compared they can reveal important information about the special role played by specific residues at particular sequence locations. The degree and kind of residue conservation found can improve our understanding of the interplay of protein function and structure. Furthermore, sequences that are all related by a common ancestral sequence hold the key to uncovering the evolutionary history since that ancestor. Techniques that are able to recover this information are presented in Chapters 7 and 8. All of these facets of protein sequence analysis are dependent at some stage on having a multiple alignment or profile constructed from all the available sequences. The main focus of this chapter is the description of the scientific bases of the many different techniques that have been proposed to achieve this crucial task.

In addition, many cases have been identified of short sequences, usually showing high levels of conservation, that have been found to correlate with specific functional properties, such as serine protease activity. These patterns are now used for function prediction purely on the basis of sequence. The last part of this chapter will explore the ways in which such patterns can be identified, including the use of methods that do not require alignments.

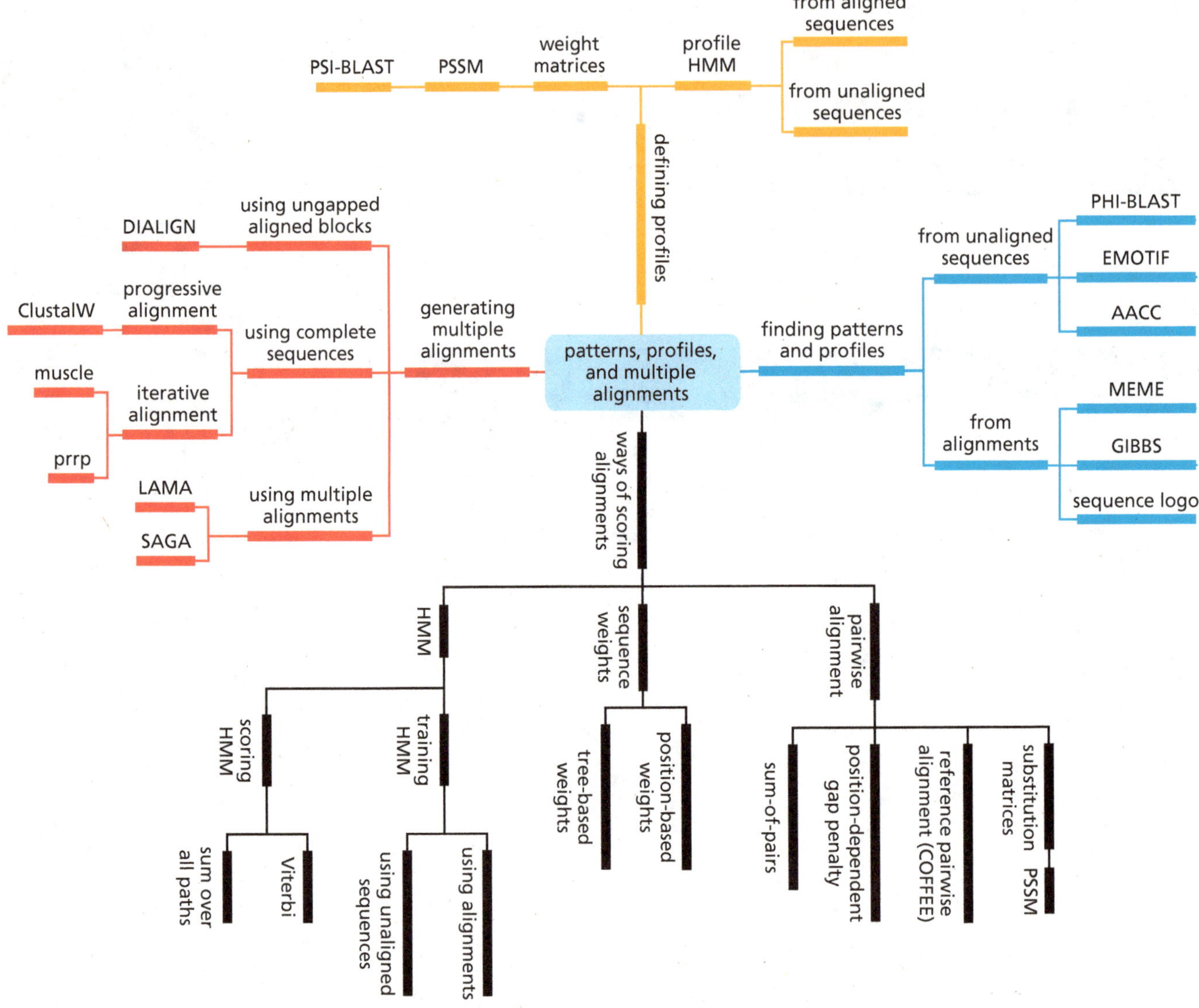


**Mind Map 6.1**
**A mind map illustrating the topics covered in patterns, profiles, and multiple alignments.** A large part is devoted to the all-important subject of scoring schemes.

Because of their general use in multiple alignment methods, in the first part of the chapter we describe methods for producing profiles, which at one level can be regarded as representations of alignments. Multiple alignments can be used in various ways to generate special scoring schemes for searching for other similar sequences. Position-specific scoring matrices (PSSMs), which take into account the position in the alignment when scoring matches, have been very successful in this context. Programs that use PSSMs include PSI-BLAST, for searching sequence databases, and LAMA, for searching a database of alignments with a query alignment.

We then discuss the use of hidden Markov models (HMMs), especially profile HMMs to define a sequence profile of a protein family. Sequence profiles can then be used to search for other family members. HMMs are based on a sound probabilistic theory, and are used in many multiple alignment and sequence-profile programs. Unlike PSSMs, HMMs do not require an alignment, and can produce a description of a family of related sequences without any prior alignment. Relationships between sequence families can be discovered by aligning profiles, which can be the most sensitive way to detect homology.

Most methods of multiple alignment are based on modifications of the pairwise dynamic programming techniques described in Section 5.2. In these methods the alignment is built up by adding sequences one at a time. As with pairwise alignments, the goodness-of-fit of the sequence to the alignment is tested by giving it a score, and the overall quality of the multiple alignment can also be tested quantitatively. The scoring schemes used have many similarities to those used for pairwise alignments but, as we shall see, additional scoring schemes are available, and different approaches can be taken to the practical task of building up the alignment.

There are several other ways of constructing multiple alignments that differ in important respects from the pairwise dynamic programming approach. In a later part of the chapter we look at methods that construct alignments using all the sequences simultaneously.

The final part of the chapter deals with methods that can detect common sequence patterns in a multiple alignment, and those that can identify patterns in unaligned sequences. These have been used to create pattern databases such as PROSITE, which can be used to help predict a protein's function on the basis of the presence of specific sequences.

## 6.1 Profiles and Sequence Logos

In Chapter 5 we looked at some techniques for aligning one sequence to another. Very often many similar sequences are known, all presumably descendants of a common ancestral sequence. From experience these sequences will be expected to have similar properties at equivalent regions, leading for example to protein sequences sharing a common protein fold and patterns of residue conservation. A frequently encountered problem is the alignment of a sequence to such a set of similar sequences. Rather than align the new sequence to a single member of the set, it would be better to find a way to represent the general properties of the set of sequences such that the new sequence can be aligned to this representation. Such a representation is called a profile, and in this section we will consider the ways in which profiles can be constructed and used in pairwise alignment of new sequences (see Flow Diagram 6.1).

Profiles can be constructed in a form that makes it possible to use the dynamic programming techniques of Section 5.2 to align sequences to the profile. An important aspect of generating these profiles is the frequent lack of sufficient data for parameterization. A number of techniques are presented to overcome this problem. Following this we will briefly look at how the PSI-BLAST program uses these techniques. Finally, a method will be presented that gives a graphical display of the residue preferences of profile positions.

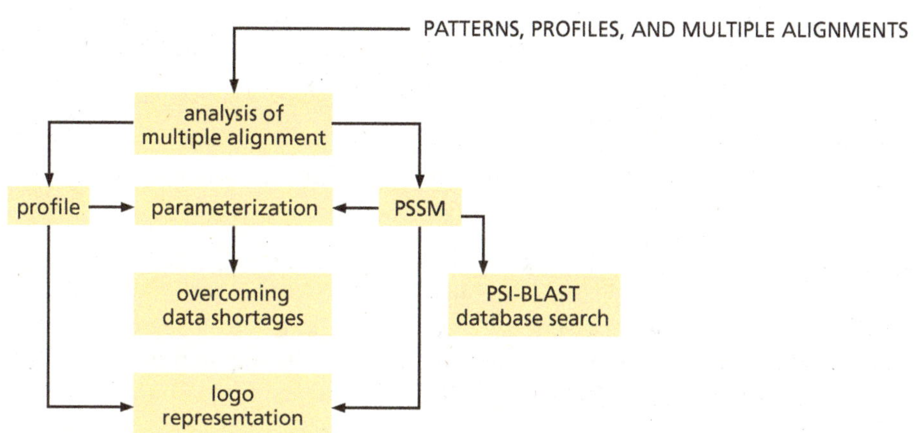


**Flow Diagram 6.1**
**The key concept introduced in this section is that analysis of multiple alignments can identify the specific preferences of each alignment position, which can be used to define a profile.** The profile can be used to define a scoring scheme such as a position-specific scoring matrix (PSSM), which can be used to search for further examples of the profile, and can also be illustrated graphically as a logo.

## Position-specific scoring matrices are an extension of substitution scoring matrices

All the alignment methods discussed in Chapter 5 apply a substitution score matrix to an alignment of residue type *a* with residue type *b* without regard for their environment; that is, the score $s_{a,b}$ for aligning these two residue types is always the same. Note that in Section 5.2 this was written $s(x_i,y_j)$ where $x_i$ is the residue at the *i*th position of sequence $\mathbf{x}$. Similarly, the gap penalties discussed in Section 5.1 are the same regardless of where the gaps occur along the sequence. One can view database searches with these general scoring schemes as using the query sequence without any added information beyond the general features of evolution used to generate the substitution score matrix.

One of the common uses of database searches is to discover all known sequences that belong to the same sequence family as the query sequence, namely those sequences that align well over the whole of a specific region (often the entire length) of the query. Sequence variability within the family will usually prevent all members from being detected with a search based on a single query sequence. One method to try to identify potentially missed sequences is to perform several searches with different family members in the hope that all members will be identified by at least one of the searches. When the family examples used for further searches have been detected in the initial query-sequence search, and each newly discovered sequence is used in a further query, this method is referred to as an **iterated sequence search (ISS)**.

An alternative and usually more efficient way of finding all family members takes account of known residue preferences at each alignment position. This information is obtained from the alignment of an initial set of family members, such as those discovered from the first database search. Inclusion of these position-specific preferences in the scoring scheme is achieved with the use of a scoring profile in which each alignment position has its own substitution scores. Such constructions are often referred to as position-specific scoring matrices (PSSMs), although usually this terminology is reserved for cases without the position-dependent gap penalties that will also be described. The alignment of a sequence to a PSSM can proceed using dynamic programming in exactly the same way as described in Section 5.2 for aligning two sequences, except that the scores differ for each column. We will now discuss some of the ways of obtaining profiles, including methods that are designed to overcome the problems that may arise as a result of a lack of data.

In order to generate a PSSM, a set of sequences is required, all of which are aligned to a common reference. Two sources of such alignments exist: the results of a database search or a multiple alignment. In the first case, the common reference is the query sequence with which each database entry has been aligned. The common reference for multiple alignments is harder to define, but all the sequences are aligned with reference to all the other sequences. Either source can be used to define the preferences for each alignment position. The scores assigned must represent the residue preference found at a given position. If the alignment only contains a few sequences it will not be possible to determine an accurate residue preference at all alignment positions. This problem can be tackled in several related ways. A further aspect of determining PSSMs is that the data usually require weighting.

Suppose we have an alignment of $N_{seq}$ sequences with $L_{aln}$ positions; that is, $L_{aln}$ alignment columns. The PSSM for this alignment will also have $L_{aln}$ columns, each of which will (for protein sequences) have 20 rows. Each row corresponds to an amino acid type. An example of a PSSM can be seen in the top part of Figure 11.24, where the alignment is used to improve secondary structure prediction. However, note that in that example the PSSM has been transposed, so that the 20 rows have become 20 columns. It is possible to use one or two extra rows with parameters that relate to position-dependent gap penalties. Such matrices are usually referred to as profiles rather than PSSMs. (A PSSM, with only 20 rows, will use identical gap penalties at all sequence positions.) The values assigned to the PSSM are a weighted function of the values of a standard substitution matrix of the form discussed in Section 5.1, for example BLOSUM-45. We will write a substitution score matrix element $s_{a,b}$ for the alignment of residue types $a$ and $b$, and will label the elements of the PSSM $m_{u,a}$ for column $u$ and row (residue type) $a$.

One possible derivation of PSSM values uses the average of the scores of the residue types found at each alignment position. If a particular alignment column contains a perfectly conserved tyrosine residue, for example, the score on aligning a residue from another sequence to that position is taken to be the same as if we were dealing with just a single tyrosine residue. The elements of that column in the PSSM are the $s_{a,b}$ elements that relate to tyrosine; that is, for row (residue type) $a$ they are $s_{a,\mathrm{Y}}$, using the one-letter amino acid code for the subscript. If instead the residue preference of that column was exactly shared by tyrosine and tryptophan, then each row $a$ of the PSSM column will have the score $(s_{a,\mathrm{Y}} + s_{a,\mathrm{W}})/2$. Generalizing this, if there are $n_{u,b}$ residues of type $b$ at column $u$, comprising a fraction $f_{u,b}$ of the column residues, i.e.,

$$f_{u,b} = \frac{n_{u,b}}{N_{seq}} \qquad \text{(EQ6.1)}$$

then the score associated with row $a$ and column $u$ will be

$$m_{u,a} = \sum_{\substack{\text{residue}\\ \text{types } b}} f_{u,b} s_{a,b} \qquad \text{(EQ6.2)}$$

If the residue type in an alignment column is found to be highly conserved, it seems sensible to give the preferred residues extra support, because the residue types rarely found are probably highly disfavored at that position. Rather than using the fraction $f_{u,b}$, the following logarithmic form of weighting the substitution scores has been proposed:

$$m_{u,a} = \sum_{\substack{\text{residue}\\ \text{types } b}} \frac{\ln\left(1 - f'_{u,b}\right)}{\ln\left(1/\left(N_{seq} + 1\right)\right)} s_{a,b} \qquad \text{(EQ6.3)}$$

where $f'_{u,b}$ differs from $f_{u,b}$ as defined by Equation EQ6.1, in that the denominator is $(N_{seq} + 1)$ instead of $N_{seq}$. (This is necessary to avoid the numerator becoming $-\infty$ for alignment columns only containing a single residue type.) The value of the ratio of the logs varies between 0 and 1 as does $f_{u,b}$, but residues present in a smaller fraction of the sequences are relatively under-weighted.

Although the two methods just described have been applied successfully to create useful PSSMs, neither is of the log-odds ratio form that was shown in Section 5.1 to be particularly appropriate for alignment scoring. We will define the probability of residue type $a$ occurring in column $u$ of the PSSM as $q_{u,a}$, and the probability of residue type $a$ occurring at any position in any sequence, including those not related to the PSSM sequence family (i.e., the background frequency), as $p_a$. The $q_{u,a}$ are the PSSM equivalent of the $q_{a,b}$ of Section 5.1, which are the probability of aligning two residues of types $a$ and $b$ when they are part of a meaningful alignment. The log-odds form for a PSSM element can then be written

$$m_{u,a} = \log \frac{q_{u,a}}{p_a} \qquad \text{(EQ6.4)}$$

If there are sufficient sequence data available, $q_{u,a}$ can be identified with $f_{u,a}$ as given by Equation EQ6.1. The $p_a$ are readily obtained from the analysis of database composition.

Because PSSMs are often used in database searches for other members of the sequence family, it is important to understand the scoring statistics, as was the case for BLAST and FASTA searches as discussed in Section 5.4. As was explained then, each substitution matrix has a value $\lambda$ associated with it that has a strong influence on the statistics. The following general formula applies to log-odds substitution matrices relating the alignment of two residues of types $a$ and $b$:

$$q_{a,b} = p_a p_b e^{\lambda s_{a,b}} \qquad \text{(EQ6.5)}$$

(Compare this equation with Equation EQ5.31, which was encountered in the discussion of the significance of alignment scores.) Equation EQ6.5 can be rearranged to give

$$s_{a,b} = \frac{\ln\left(q_{a,b} / p_a p_b\right)}{\lambda} \qquad \text{(EQ6.6)}$$

By analogy, an alternative to Equation EQ6.4 is

$$m_{u,a} = \frac{\log\left(q_{u,a} / p_a\right)}{\lambda} \qquad \text{(EQ6.7)}$$

where the value of $\lambda$ can be specified to control the scoring statistics. It is a simple scaling factor for the scores, and so could be omitted. However, this method of obtaining PSSM values has been applied in the PSI-BLAST method described below, where the $\lambda$ parameter is used to selectively scale the scores as desired.

If position-dependent gap penalties are included, these can be assigned manually on the basis of the location of secondary structural elements to give smaller penalties for creating alignment gaps between these elements. An alternative approach is to have a position-specific multiplier of the gap penalty at column $u$, $g'_u$, such that an affine gap penalty for a gap of length $n_{\text{gap}}$ extending across this column has a penalty

$$g_u\left(n_{\text{gap}}\right) = g'_u\left[-I - \left(n_{\text{gap}} - 1\right)E\right] \qquad \text{(EQ6.8)}$$

where $I$ and $E$ are the gap opening and extension penalties, respectively (see Equation EQ5.14). One proposed assignment of values to $g'_u$ starts by identifying the length of the longest gap that occurs that includes column $u$ and the highest possible score in the substitution matrix used. These values are used to scale $g'_u$ so that the penalty applied to the longest observed gap is of the same magnitude as the highest possible score in the substitution matrix.

Thus far, we have treated each sequence in the alignment equally. The best PSSM will represent the full range of diversity within the sequence family in an unbiased fashion. However, a partial set of family sequences will most probably be biased toward a certain subgroup. Hence, we must weight the different sequences, and the weighting should be reduced for very similar sequences. We will look at two sequence weighting schemes, one of which applies the weight to all residues of a sequence, the other specifying different weights for each alignment column.

Some PSSMs have been derived using a sequence weighting scheme proposed by Peter Sibbald and Patrick Argos. In this scheme, weights are assigned to the sequences on the basis of an iterative procedure. The sequence weights are first initialized to zero. Random aligned sequences are generated where the residue at each position is chosen at random from the residues (including any gap) that occur at that particular position in the aligned sequences. The closest sequence or equally close sequences to each random sequence are identified, and a weight of 1 is evenly distributed between them. The sequence weights are normalized to sum to 1. Further random sequences are generated until the sequence weights are seen to have converged.

The position-based sequence-weight method assigns weights based on a multiple alignment. The basic unit used to assign weights is not the whole sequence, but the individual alignment columns. For each column, the number of different residues present is counted. If there are $m$ different residues, each is assigned a weight of $1/m$. Then for each residue type, the number of sequences that have this residue at this position is counted. If there are $n$ sequences with this residue, the weight is equally divided amongst them; that is, the weight becomes $1/mn$. These weights can be used as sequence weights for the individual column, and the total for all aligned residues is always 1. If desired, an overall weight for the whole sequence can be defined by adding the individual column weights and then normalizing by the number of columns, so that the sequence weights also sum to 1. This weighting scheme is illustrated in Figure 6.1 for a simple example alignment.

If the weights are labeled $w_u^x$ for column $u$ of sequence $\boldsymbol{x}$ (including the possibility that there may be no variation across the alignment columns) the above formulae can be modified by replacing $f_{u,b}$ with

$$f''_{u,b} = \frac{\sum_{\text{sequences } x} w_u^x \delta_{u,b}^x}{\sum_{\text{sequences } x} w_u^x} \qquad \text{(EQ6.9)}$$

where $\delta_{u,b}^x$ is 1 if sequence $\boldsymbol{x}$ has residue $b$ in alignment column $u$, and 0 otherwise. Note that both weighting methods discussed above do not require the denominator sum of Equation EQ6.9, as they both produce weights that sum to 1 for a column.

| 12345 | Weight |
|---|---|
| HSAPL | 0.280 |
| HTADV | 0.171 |
| HTAEV | 0.171 |
| HTGLI | 0.188 |
| HTGVI | 0.188 |

**Figure 6.1**
**Illustration of the position-based sequence weight scheme of Henikoff and Henikoff.** There are five sequences, each of them is five residues long. Column 1 contains only one residue type (H) but there are five occurrences, so each histidine has a weight of ⅕. Column 4 contains five different residue types, each of which will be weighted as ⅕. Note that this scheme does not distinguish between these two situations, which can be understood by observing that neither can be used to prefer any sequence over any other. In column 2, there are two residue types (S and T), each of which will be assigned an initial weight of ½. As there is only one instance of S, it is assigned a weight of ½. The four Ts will each have a weight of ⅛. After assigning weights to each residue, the sequence weights are obtained by adding the residue weights together and dividing by the number of columns. The resulting sequence weights are given in the right-hand column.

## Methods for overcoming a lack of data in deriving the values for a PSSM

The log-odds schemes for calculating the PSSM elements $m_{u,a}$ using the formulae given above all have the property that if any residue type is not observed in the column $u$, the score for aligning that residue type in that column will always have the value $-\infty$. As a consequence, no sequence being aligned to the PSSM will be able to align that residue type at that location. While this might be appropriate for perfectly conserved and functionally vital positions, even in extreme cases it is almost certainly too restrictive. (We most probably want to be able to use the PSSM to align any sequences that code for related but dysfunctional proteins.) Furthermore, the absence of a particular residue type in a particular column of the alignment used to derive the PSSM is more likely to indicate a lack of sequence alignment data rather than the true residue preferences. Such situations are common in fitting profile models, and a variety of possible solutions have been proposed that will now be described.

The $p_a$ of Equations EQ6.4 and EQ6.7 does not cause the $-\infty$ values mentioned above, nor does their estimation suffer from a lack of data. The problem rests entirely with the estimation of the $q_{u,a}$ and is almost always due to a lack of data. (The exception to this would be a rare case of a truly perfectly conserved residue type—an extremely rare occurrence.) All the methods described to overcome the problem can include sequence weights but they make the formulae more complex, so for simplicity we will not use them here. The starting point for estimating $q_{u,a}$ is to use $f_{u,a}$ as given by Equation EQ6.1.

A simple way of trying to overcome the lack of data would be to assume at least one occurrence of each residue type at each alignment position. It is preferable to treat all residues and all columns the same way, so in this case Equation EQ6.1 is modified to

$$q_{u,a} = \frac{n_{u,a}+1}{N_{seq}+20} \tag{EQ6.10}$$

in the case of a PSSM with 20 rows (i.e., no position-dependent gap parameters). Note that the denominator becomes ($N_{seq}$ + 20) as this is the sum of the numerators for the 20 different terms. The sum of the $q_{u,a}$ for all possible $a$ must always equal 1 as each sequence must be represented at column $u$. The inclusion of the extra observation means that $q_{u,a}$ will never be 0 nor ever reach 1. It is as if we have increased the amount of data available by 20 residues in each column, and such additional data are usually called **pseudocounts.**

It is easy to see that there are more sophisticated ways of adding pseudocounts that take advantage of the knowledge we have of the properties of sequences. We know that the amino acid composition of proteins is not uniform and, as discussed in Section 5.1, the frequency of occurrence $p_a$ of residue type $a$. This knowledge can readily be incorporated into the formula to obtain

$$q_{u,a} = \frac{n_{u,a}+\beta p_a}{N_{seq}+\beta} \tag{EQ6.11}$$

where the parameter $\beta$ is a simple scaling parameter that determines the total number of pseudocounts in an alignment column. The advantage of introducing the $\beta$ parameter is that we can easily adjust the relative weighting of the pseudocounts and real data. When there are a lot of data (i.e., $N_{seq}$ is large) there is little if any need for pseudocounts, and $\beta$ should be much smaller than $N_{seq}$, whereas when there are less data, $\beta$ should be larger relative to $N_{seq}$. A simple formula that has been found useful is to make $\beta$ equal to $\sqrt{N_{seq}}$, although this can result in $\beta$ being too small for small values of $N_{seq}$. At large values of $N_{seq}$ this formula approaches $q_{u,a} = f_{u,a}$ as desired. Often an additional parameter $\alpha$ is used to weight the observed data, giving a formula more easily expressed in terms of the $f_{u,a}$ than the $n_{u,a}$:

$$q_{u,a} = \frac{\alpha f_{u,a}+\beta p_a}{\alpha+\beta} \tag{EQ6.12}$$

In the absence of any data, the pseudocounts would completely determine the PSSM values. In Bayesian analysis terms, the pseudocounts represent the **prior distribution,** which expresses our prior knowledge of the system before we introduce the data. (They can also be seen as an expression of our prejudices and bias!) See Appendix A for a discussion of Bayesian analysis.

We can improve on the pseudocount distribution suggested by Equations EQ6.11 and EQ6.12, because the substitution matrices contain a more accurate distribution based on the data from which they were derived. By definition, log-odds substitution matrices have embedded in them the information about the relative probabilities of residues aligning because their sequences are related as opposed to purely random alignment. This is expressed by the terms $q_{a,b}/p_a p_b$ for residue types $a$ and $b$. Rearranging Equation EQ6.5 we find

$$\frac{q_{a,b}}{p_a p_b} = e^{\lambda s_{a,b}} \qquad \text{(EQ6.13)}$$

If a column $u$ contains a fraction $f_{u,b}$ of type $b$ residues, the probability of finding a residue of type $a$ aligned with these is proportional to $f_{u,b}q_{a,b}/p_a p_b$. Adding these terms for all residue types $b$ gives the overall probability of finding a residue of type $a$ aligned at column $u$ on the basis of existing residues in that column. Because residue type $a$ occurs at a background frequency $p_a$, this sum needs to be multiplied by $p_a$ to obtain a suitable pseudocount for residue type $a$. Thus the formula for the number of pseudocounts of residue type $a$ is

$$g_{u,a} = p_a \sum_{\substack{\text{residue} \\ \text{types } b}} \frac{f_{u,b} q_{a,b}}{p_a p_b} = \sum_{\substack{\text{residue} \\ \text{types } b}} \frac{f_{u,b}}{p_b} q_{a,b} \qquad \text{(EQ6.14)}$$

Note that because of the summation, the $g_{u,a}$ can never be 0. If the intermediate results are available that were used to derive the substitution matrix, the values of the terms $q_{a,b}/p_a p_b$ may be available. If this is not the case, they can be recovered from the substitution scores $s_{a,b}$ by applying Equation EQ6.13, although the data are likely to be less accurate. Replacing $p_a$ with this in Equation EQ6.12 we obtain the improved estimate of $q_{u,a}$:

$$q_{u,a} = \frac{\alpha f_{u,a} + \beta g_{u,a}}{\alpha + \beta} \qquad \text{(EQ6.15)}$$

One possible value for $\alpha$ is $N_{seq} - 1$. If the PSSM elements $m_{u,a}$ are calculated using Equation EQ6.7, then with only one sequence this particular $\alpha$ will result in the $m_{u,a}$ being the substitution matrix values $s_{a,b}$ (see Equation EQ6.13). The value of $\beta$ controls how much the PSSM is biased to the substitution matrix when there are few data available. In PSI-BLAST (see below) a value of 10 is used by default.

The derivation of pseudocounts from substitution scoring matrices goes some way toward using a more realistic prior distribution. However, that method as well as all the others described so far have qualitatively inaccurate features. Only the relative frequency of different amino acids in an alignment column affects the PSSM values. All PSSM columns based on the observation of a single fully conserved amino acid will have the same values regardless of whether the alignment contained just three or a thousand sequences. Clearly the latter case, with no exceptions in a thousand observations, should be treated differently, with far greater penalties for the presence of a different residue than when the PSSM has been derived from only three sequences. The second incorrect feature is that no distinction is made between the different possible environments of a residue. The probabilities associated in these methods with any two aligned residue types are the same in all circumstances. However, it is well known that alignment columns often show a clear preference for a small set of residues with similar properties, such as charge, size, or polarity. In reality the alignment probabilities will differ for a given pair of residue types according to these preferences. The probability of aligning a leucine to a column will differ according to whether that column has a perfectly conserved leucine, conserved residues that are small and hydrophobic, or no apparent residue preference.

It is possible to avoid the shortcomings of the methods described so far by using a **Dirichlet mixture**, more accurately described as a mixture of **Dirichlet distribution densities**. Dirichlet distributions are often used as the prior distribution in Bayesian analysis of data where each observation is one of a limited set of possibilities. In this case, the 20 amino acids constitute the set, and each Dirichlet density describes a particular residue composition that might occur at an alignment column. The Dirichlet mixture is a linear combination of these densities with multiplying coefficients that are the probability of their occurrence. Therefore these coefficients sum to 1. A set of densities and mixture coefficients has been fitted by Kimmen Sjölander and co-workers to the BLOCKS database of protein multiple alignments. The details of the fitting are beyond the scope of this book, but involve maximum likelihood and the EM (expectation maximization) method. For further information on these methods see Further Reading.

Figure 6.2 gives the values for a nine-component Dirichlet mixture derived from the BLOCKS database. The *i*th Dirichlet density component is labeled Blocks9.*i* and has a linear mixture coefficient (probability of occurrence in the BLOCKS data) $c_i$. The amino acid distribution of each density component is defined by the numbers in each column, labeled $\beta_{i,a}$ for residue type *a*. Note that these do not add to 1, their sum $\beta_i$ being given for each density, so the frequency for residue *a* in density *i* is given by $\beta_{i,a}/\beta_i$. The density components can be analyzed to identify their residue preferences. As the BLOCKS database like all real protein sequences has a nonuniform residue composition—equivalent to the background composition $p_a$ of the previous formulae—the component residue preferences are best measured as the component frequency relative to the background frequency. These preferences are shown on the bottom row of Figure 6.2, where all residues present at greater than twice the background frequency are listed. For example, the first component favors

| | Blocks9.1 | Blocks9.2 | Blocks9.3 | Blocks9.4 | Blocks9.5 | Blocks9.6 | Blocks9.7 | Blocks9.8 | Blocks9.9 |
|---|---|---|---|---|---|---|---|---|---|
| $c_i$ | 0.182962 | 0.057607 | 0.089823 | 0.079297 | 0.083183 | 0.091122 | 0.115962 | 0.066040 | 0.234006 |
| A | 0.270671 | 0.021465 | 0.561459 | 0.070143 | 0.041103 | 0.115607 | 0.093461 | 0.452171 | 0.005193 |
| C | 0.039848 | 0.010300 | 0.045448 | 0.011140 | 0.014794 | 0.037381 | 0.004737 | 0.114613 | 0.004039 |
| D | 0.017576 | 0.011741 | 0.438366 | 0.019479 | 0.005610 | 0.012414 | 0.387252 | 0.062460 | 0.006722 |
| E | 0.016415 | 0.010883 | 0.764167 | 0.094657 | 0.010216 | 0.018179 | 0.347841 | 0.115702 | 0.006121 |
| F | 0.014268 | 0.385651 | 0.087364 | 0.013162 | 0.153602 | 0.051778 | 0.010822 | 0.284246 | 0.003468 |
| G | 0.131916 | 0.016416 | 0.259114 | 0.048038 | 0.007797 | 0.017255 | 0.105877 | 0.140204 | 0.016931 |
| H | 0.012391 | 0.076196 | 0.214940 | 0.077000 | 0.007175 | 0.004911 | 0.049776 | 0.100358 | 0.003647 |
| I | 0.022599 | 0.035329 | 0.145928 | 0.032939 | 0.299635 | 0.796882 | 0.014963 | 0.550230 | 0.002184 |
| K | 0.020358 | 0.013921 | 0.762204 | 0.576639 | 0.010849 | 0.017074 | 0.094276 | 0.143995 | 0.005019 |
| L | 0.030727 | 0.093517 | 0.247320 | 0.072293 | 0.999446 | 0.285858 | 0.027761 | 0.700649 | 0.005990 |
| M | 0.015315 | 0.022034 | 0.118662 | 0.028240 | 0.210189 | 0.075811 | 0.010040 | 0.276580 | 0.001473 |
| N | 0.048298 | 0.028593 | 0.441564 | 0.080372 | 0.006127 | 0.014548 | 0.187869 | 0.118569 | 0.004158 |
| P | 0.053803 | 0.013086 | 0.174822 | 0.037661 | 0.013021 | 0.015092 | 0.050018 | 0.097470 | 0.009055 |
| Q | 0.020662 | 0.023011 | 0.530840 | 0.185037 | 0.019798 | 0.011382 | 0.110039 | 0.126673 | 0.003630 |
| R | 0.023612 | 0.018866 | 0.465529 | 0.506783 | 0.014509 | 0.012696 | 0.038668 | 0.143634 | 0.006583 |
| S | 0.216147 | 0.029156 | 0.583402 | 0.073732 | 0.012049 | 0.027535 | 0.119471 | 0.278983 | 0.003172 |
| T | 0.147226 | 0.018153 | 0.445586 | 0.071587 | 0.035799 | 0.088333 | 0.065802 | 0.358482 | 0.003690 |
| V | 0.065438 | 0.036100 | 0.227050 | 0.042532 | 0.180085 | 0.944340 | 0.025430 | 0.661750 | 0.002967 |
| W | 0.003758 | 0.071770 | 0.029510 | 0.011254 | 0.012744 | 0.004373 | 0.003215 | 0.061533 | 0.002772 |
| Y | 0.009621 | 0.419641 | 0.121090 | 0.028723 | 0.026466 | 0.016741 | 0.018742 | 0.199373 | 0.002686 |
| $\beta_i$ | 1.180650 | 1.355830 | 6.664360 | 2.081410 | 2.081010 | 2.568190 | 1.766060 | 4.987680 | 0.099500 |
| | AST | FHWY | EQ | KQR | ILM | IV | DEN | M | GPW |

**Figure 6.2**
**The nine-component Dirichlet mixture derived from the BLOCKS alignment database.** The components are labeled Blocks9.1 to Blocks9.9, and their mixture coefficients $c_i$ are given in the first row. Subsequent rows give the distributions (unnormalized) for the individual amino acids in each component. Those amino acids of a component that are present at more than double the background frequency are highlighted in yellow, and summarized in the last row. Thus positively and negatively charged residues dominate components Blocks9.4 and Blocks9.7, respectively. The overall amino acid composition predicted by this mixture is obtained by weighting these component distributions by their coefficients $c_i$. (Data K. Sjölander, K. Karplus, M. Brown et al. Dirichlet mixtures: a method for improved detection of weak but significant protein sequence homology. *Comput. Appl. Biosci.* 12:327–345, 1996.)

small neutral residues; the second component favors aromatic residues. The ninth component has the largest mixture coefficient, and is the most common component in the BLOCKS database. It is the component with the closest composition to the background frequency, and favors highly conserved individual residues such as tryptophan and proline. Other Dirichlet mixtures have been derived to represent amino acid alignments, including one with twenty components that is regarded as an improvement over the Blocks9 mixture.

The application of Dirichlet mixtures is more complex than using pseudocounts, because for each alignment column the weights have to be assigned to each of the components. The equations for residue type $a$ in alignment column $u$ is given by

$$q_{u,a} = \sum_{\substack{\text{Dirichlet}\\ \text{components } i}} P\left(\vec{\beta}_i \middle| \vec{n}_u, \theta\right) \frac{n_{u,a} + \beta_{i,a}}{N_{seq} + \beta_i} \qquad \text{(EQ6.16)}$$

where the term $P\left(\vec{\beta}_i \middle| \vec{n}_u, \theta\right)$ means the posterior probability (see Appendix A for a definition of this term) of the $i$th Dirichlet component with a vector of residue coefficients $\vec{\beta}_i$, given the vector of observed residue occurrences $\vec{n}_u$ in alignment column $u$. The symbol $\theta$ stands for all the Dirichlet mixture parameters. This term is the weight of the $i$th Dirichlet component. This equation should be compared with Equation EQ6.11.

After considerable algebraic manipulation, it is possible to derive a useful expression for terms $q'_{u,a}$, which can then be normalized to sum to 1 to obtain the $q_{u,a}$. The formula is

$$q'_{u,a} = \sum_{\substack{\text{Dirichlet}\\ \text{components } i}} c_i \frac{B\left(\vec{n}_u + \vec{\beta}_i\right)}{B\left(\vec{\beta}_i\right)} \left(\frac{n_{u,a} + \beta_{i,a}}{N_{seq} + \beta_i}\right) \qquad \text{(EQ6.17)}$$

where the mixture coefficients $c_i$ are now clearly involved, and the function $B$ is the Beta function, which is defined in terms of the more familiar mathematical Gamma function ($\Gamma$) by

$$B\left(\vec{n}_u + \vec{\beta}_i\right) = \frac{\prod_{\substack{\text{residue}\\ \text{type } b}} \Gamma\left(n_{u,b} + \beta_{i,b}\right)}{\Gamma\left(N_{seq} + \beta_i\right)} \qquad \text{(EQ6.18)}$$

Thus the observed residue counts in a column can readily be combined with the coefficients of the Dirichlet mixture components to obtain an improved estimate of the $q_{u,a}$.

Figure 6.3 compares some of the alternative methods discussed to overcome a lack of data, using as an example two cases. In one case, only a single sequence is present in the data, with a leucine residue at the column of interest. In the other case, 10 sequences all have a leucine at that position. The column composition is identical in the two cases, but clearly there is more information when the data consist of 10 sequences instead of just one. The figure shows how different methods can incorporate that information, with the Dirichlet mixture method clearly superior to the others.

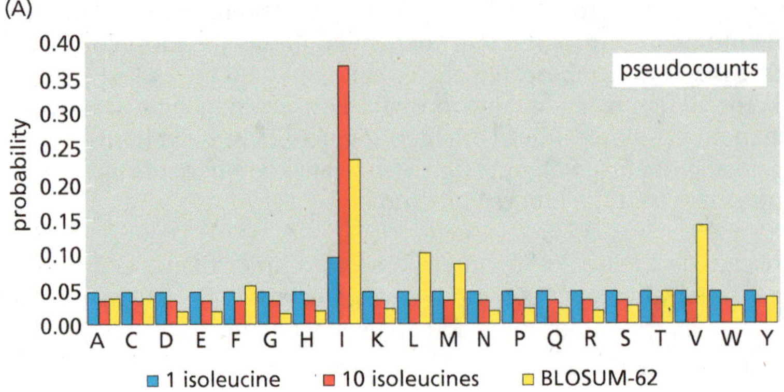


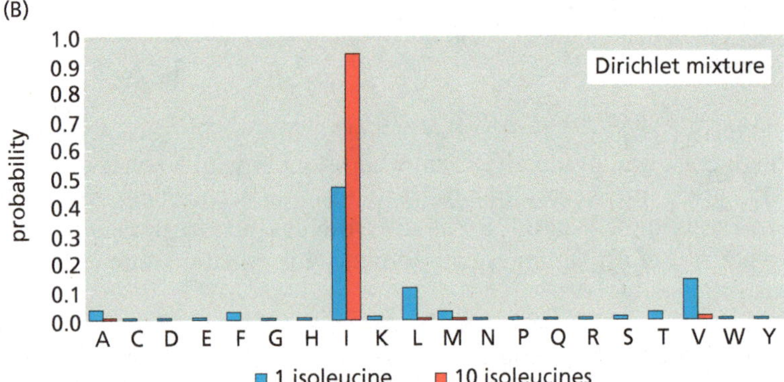


**Figure 6.3**
**The estimated probabilities for amino acids according to some methods for dealing with a lack of data.** The results shown are those given when the data indicate a conserved site with only isoleucine residues. Two cases are considered, observing either 1 or 10 isoleucine residues at the site. (A) The estimates given by using the BLOSUM-62 matrix with Equation EQ6.3 compared with those obtained using a pseudocount method of adding 1 to each observed amino acid frequency (Equation EQ6.10). Note that Equation EQ6.3 is only sensitive to observed composition, so that the probabilities are the same. (B) The estimates given by the nine-component Dirichlet mixture of Sjölander et al. (see Figure 6.2).

## PSI-BLAST is a sequence database searching program

When a PSSM is carefully constructed it can be a very sensitive tool in database searches, finding many distant members of a protein sequence family not easily found by a standard sequence search. For this reason Steven Altschul and co-workers decided to enhance their BLAST database searching method to incorporate PSSMs. They created the program PSI-BLAST (Position-Specific Iterated BLAST) which has proved extremely successful.

A PSI-BLAST database search incorporates a number of steps. The first step usually involves performing a standard BLAST search of a database using the BLOSUM-62 substitution matrix with a single query sequence. This results in an initial set of related sequences. These sequences will be determined as those whose BLAST score gives an *E*-value smaller than a predetermined cut-off, often quite stringent to ensure that only truly significant matches are identified (see Section 5.4). The alignments of these significant matches with the query sequence are used to create a PSSM. The PSSM is then scanned against the database using a variant of the BLAST program to identify new sequences with suitably small *E*-values. If this second search finds some newly identified related sequences, these are used to update the PSSM. Successive cycles of PSSM refinement and database searching can be carried out until no new sequences are found.

Clearly, the cut-off threshold of the *E*-value is a critical parameter. If it is too large, as well as homologs, false positive sequences (i.e., unrelated to the initial query sequence) will be collected. These will result in PSSM corruption, which in further cycles will most likely give rise to a substantial fraction of false positive sequences. If the cut-off is too small, the run may fail to identify some of the more distantly related sequences, or will require many search cycles to locate them. A commonly used *E*-value cut-off is 0.001, although this may still result in a corrupted PSSM in some cases.

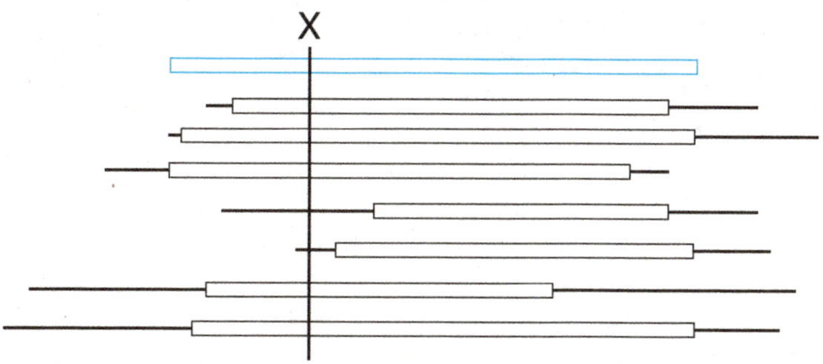


**Figure 6.4**
**Illustration of how PSI-BLAST uses the pairwise sequence alignments from a database search in PSSM construction.** The BLAST local alignments to the query sequence (the top blue sequence) are shown as rectangles. At each residue position of the query, a PSSM is constructed using only those sequences whose BLAST alignments involve that position. Thus at position X only six residues are considered in order to derive the PSSM, as only six of the sequences (including the query sequence) have local alignments including this position.

We will concentrate here on how the PSSM is constructed in PSI-BLAST. The PSSM is restricted to those residues that have been aligned to a residue in the query sequence. This removes from consideration any residues of database sequences that align with insertions in the query sequence. Thus the PSSM will have the same length as the query sequence. The PSSM constructed does not explicitly consider gaps, the usual gap penalties being used even when searching with the PSSM. Thus the gaps are treated exactly as in BLAST, i.e., not position-specific.

For any given alignment column, only those sequences that actually have a residue aligned are considered, so that the number of sequences in the alignment changes from column to column (see Figure 6.4). Each sequence is weighted using the position-based sequence weight scheme (see Figure 6.1), slightly modified to include gaps as another residue type, and to ignore fully conserved residues. The resulting weighted frequencies are used in Equations EQ6.14 and EQ6.15, which are then substituted into Equation EQ6.7 to obtain the PSSM parameters. The value of $\alpha$ used is not $N_{seq} - 1$, where $N_{seq}$ is the number of sequences, but $N' - 1$, where $N'$ is the number of different residue types observed in the column, including gaps, and thus varies from 1 to 21. The initial version of PSI-BLAST used standard substitution score matrices and Equation EQ6.13, but recent versions use the values of the terms $q_{a,b}/p_a p_b$.

The scaling $\lambda$ used is that given by the chosen substitution matrix. The gap penalties used are those applied in a standard BLAST run in combination with this substitution matrix. This results in the scoring statistics for the PSSM being the same as for standard BLAST with the same matrix. This was confirmed using alignment data for real sequences, as the theory has not been derived for these alignments. Consequently, the measure of significance for PSI-BLAST scores is readily available, and one can readily ascertain which new sequences should be included in the recalculation of the PSSM. The technique has proved very successful, as shown by a comparison with some other sequence-search methods (see Table 6.1). A further refinement of $\lambda$, resulting in improved accuracy, has been applied in recent versions of PSI-BLAST. The value is modified to take account of deviations from the scoring statistics that result from compositional bias in the sequences.

## Representing a profile as a logo

The score parameters of a PSSM are useful for obtaining alignments, but do not easily show the residue preferences or conservation at particular positions. This residue information is of interest because it is suggestive of the key functional sites of the protein family. A suitable graphical representation would make the identification of these key residues easier. One solution to this problem uses information theory, and produces diagrams that are called **logos**.

In any PSSM column $u$ residue type $a$ will occur with a frequency $f_{u,a}$. The entropy (uncertainty—see Appendix A) in that position is defined by

$$H_u = -\sum_a f_{u,a} \log_2 f_{u,a} \qquad \text{(EQ6.19)}$$

**Table 6.1**
**An illustration of the greater effectiveness of the PSI-BLAST method as compared with some other algorithms for detecting significantly similar sequences.**

| Protein family | SWISS-PROT id | Smith–Waterman | Gapped BLAST | PSI-BLAST |
|---|---|---|---|---|
| Interferon α | P05013 | 53 | 53 | 53 |
| Serine protease | P00762 | 275 | 275 | 286 |
| Serine protease inhibitor | P01008 | 108 | 108 | 111 |
| Glutathione transferase | P14942 | 83 | 81 | 142 |
| Ras | P01111 | 255 | 252 | 375 |
| Globin | P02232 | 28 | 28 | 623 |

A protein sequence from each of the example families (designated by a protein i.d. in the SWISS-PROT sequence database) was submitted to a search using one of the three methods. For each search algorithm the number of sequences found for that family is reported. The Smith–Waterman method uses full matrix dynamic programming as described in Section 5.2 to determine the local alignment of optimal score. Gapped BLAST uses a method described in Section 5.3 that is faster but in principle cannot guarantee to find the optimal alignment. In all cases except the Ras family it is as successful as Smith–Waterman. The PSI-BLAST method is notably more successful than the other two, especially in three cases shown here, indicating that it is considerably more sensitive in homology searches. In the first step a gapped BLAST search is run, so it always reports at least as many similar sequences. Only in the case of interferon α does it fail to find any more sequences. (Data from Table 3 of Altschul *et al.*, 1997.)

where the summation is over all the residue types, and the entropy units are bits of information. If only one residue is found at that position, all terms are zero and $H_u$ is zero; that is, there is no uncertainty. The maximum value of $H_u$ occurs if all residues are present with equal frequency, in which case $H_u$ takes the value $\log_2 20$ for proteins. The information present in the pattern at that position $I_u$ is given by

$$I_u = \log_2 20 - H_u \qquad \text{(EQ6.20)}$$

Thus a position with a perfectly conserved residue will have the maximum amount of information.

In practice this equation must be slightly modified. Firstly, the average amino acid composition will not be uniform, and the entropy of that composition can be used instead of $\log_2 20$. Also, account must be taken of the small sample sizes, which potentially underestimate the information content. (If there are only two sequences, for example, $H_u$ can never exceed $\log_2 2$.) For details of this correction see Further Reading.

Thomas Schneider and Michael Stephens used the information measure to produce a simple graphic that shows not only the amount of information present at each position but also the residues and their contributions. The contribution of a residue is simply defined as $f_{u,a}I_u$. At every position the residues are represented by their one-letter code, with each letter having a height proportional to their contribution. An example is given in Figure 6.5. These logos are a good visual summary of the profile, although they cannot easily cope with variable insertions. The same method can be equally successfully applied to nucleotide sequences, as will be seen in Chapter 10.

**Figure 6.5**
**Example of a protein sequence alignment logo, taken from the BLOCKS database.** This is block IPB000399E, constructed from an alignment of 186 sequences of TDP (thymine diphosphate)- binding enzymes. This region of sequence is also present in PROSITE (entry PS00187), which reports a TPP (thiamine pyrophosphate)- binding pattern [LIVMF]-[GSA]-x(5)-P-x(4)-[LIVMFYW]-x-[LIVMF]-x-G-D-[GSA]-[GSAC]. (See Table 4.2 for the notation used in PROSITE patterns.) Note that this pattern is only found in a subset of 44 of the 186 sequences whose alignment is shown here. The conserved proline of this pattern is in column 18, and the G-D-[GSA] at columns 27–29. Produced at http://weblogo.berkeley.edu

## 6.2 Profile Hidden Markov Models

In Section 5.3 we saw how finite-state automata can be used to find a specified subsequence or string within a sequence (see Figure 5.23). The automaton model has to be specially constructed for the desired string. The model consists of several states and it can only be in one type of state at any one time. There is a limited set of transitions allowed between states, and the output of the model (the sequence) is referred to as the emission of, in this case, one residue at a time. This idea can be extended to make models that can be used in many other sequence-recognition applications. An important aspect of some of these more complex models is the introduction of probabilities for transitions between states and emissions from states. The class of models that will be presented here are generally called hidden Markov models (HMMs) (see Flow Diagram 6.2). HMMs can be designed for many different tasks, of which sequence alignment is just one. In this section we will study HMMs designed to align a sequence to a profile. Other applications of HMMs will be encountered elsewhere in this book, for example for gene identification in Chapter 10 and for transmembrane helix prediction in Section 12.5, as it is a versatile way of constructing a variety of models of relevance to bioinformatics.

An HMM is defined by having a set of **states**, each of which has a limited number of **transitions** to other states and a limited number of **emissions** from a given state. Each transition between states will have an assigned probability, the value of which is independent of the history of previous states encountered. It is this property that makes these a class of **Markov models**. Each of the models considered here has a **start state** and an **end state**, and any **path** through the model from the start to the end state will produce a sequence. In these HMMs the state emission is of a residue of the sequence. As will be shown below, there are many alternative paths through the model that can produce the same sequence, and the $i$th residue of the sequence may have been emitted from any of a number of alternative states. The sequence alone has no information about the state from which each residue arose. This is the hidden nature of these models.

It is possible to define an HMM that contains the information present in a multiple alignment so that it becomes an alternative representation of a PSSM. The resultant model is usually referred to as a **profile HMM**. Profile HMMs can be seen as a more sophisticated version of a PSSM, especially because of the position-dependent way in which insertions and deletions are treated. Many HMM practitioners emphasize the fact that HMMs have a sound theoretical foundation,

**Flow Diagram 6.2**
**The key concept introduced in this section is that a class of hidden Markov models (HMMs) can be constructed to represent a sequence profile, and methods are defined for generating such models for use in searching for the profile in sequences.**

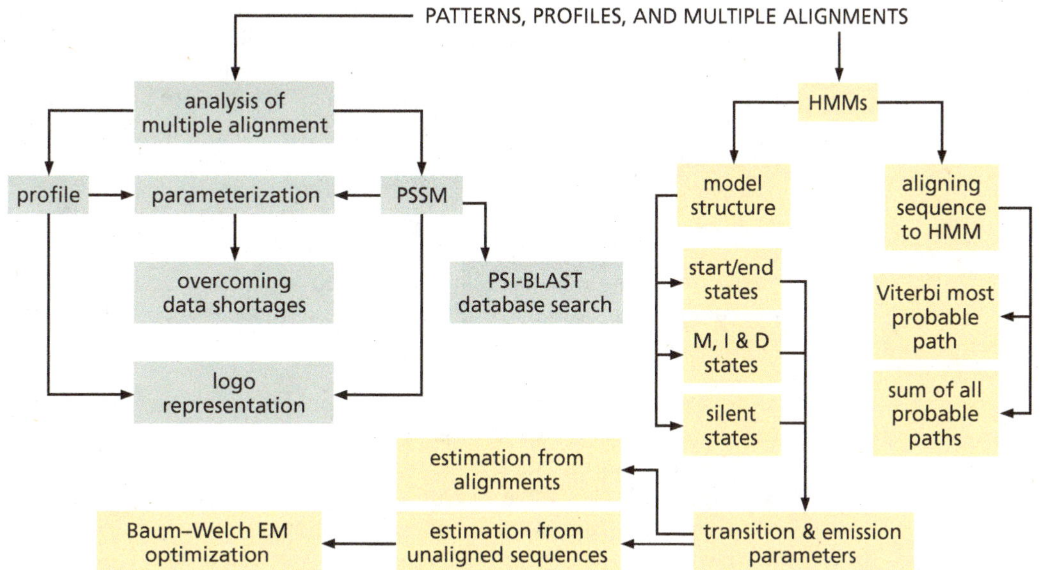

permitting accurate estimation of the likelihood of observing the alignment produced. This can be especially useful when searching for new members of a sequence family. It is always preferable to be able to take decisions about the inclusion of a distantly related member based on precise statistical analysis. (This is why the PSI-BLAST PSSM parameters were scaled so carefully.)

This section starts with a look at the structure of profile HMMs. Each alignment position has a standard structure, consisting of several different states with a limited set of allowed transitions. Some additional model architectures are required to permit the alignment of sequences that either do not fully or do not only cover the profile. After this, the model parameters will be defined and a method presented for deriving their values when a multiple alignment is available. (This is the exact analogy of fitting the PSSM values and, as will be seen, involves a similar concern for the potential lack of sufficient data.) Once a profile HMM has been defined, sequences can be aligned to it. Any path through the model and the associated emitted sequence can be assigned a probability based on the transition and emission probabilities. As is the case for pairwise sequence alignment, there are many possible alignments, paths, of which we are most likely only to be interested in the optimal alignments. These will be shown to be obtainable for profile HMMs in ways related to the dynamic programming techniques shown in Section 5.2. However, profile HMMs can produce the same alignment by many different paths through different states. This is not the case with the dynamic programming methods mentioned, for which a particular alignment is associated with only one path. A way to identify the most probable sequence will be discussed, as will be the ability to estimate the most likely state from which a given residue was emitted. The final part of this section will explore how a profile HMM can be parameterized using unaligned sequences.

## The basic structure of HMMs used in sequence alignment to profiles

All the HMMs discussed here consist of a number of states that are linked together by transitions. An example of a state is one that matches (i.e., aligns) residues at a specific position (column) of an alignment. This state may be linked to the matching state for the next alignment position by a transition that specifies the direction of the connection. All transitions have an associated probability, such that for every state that has transitions to other states, the sum of all such probabilities must sum to exactly 1. Thus, it is certain that a transition will occur from any such state, although it is possible for the transition to be directly back to the same state. The only state present in the HMMs discussed here that does not have this property is the end state, because on reaching the end state the HMM terminates, so that state has no transitions. The sum of probabilities for all transitions to a given state need not add to 1, resulting in certain states being more likely to be visited.

A state of a profile HMM may emit a residue, so that a path through the model via a series of states will define the alignment of a sequence. The order of the residues emitted along the path must correspond to that of the query sequence. In general, any state that emits residues can emit any type and has defined probabilities for each possible type, with the total emission probability summing to 1. Not all states in a model are required to emit residues, and those that do not are sometimes called **silent states**.

It should be noted that there are some significant differences between the HMM and the finite-state automaton of Figure 5.23. In the latter, the sequence is emitted on transition from one state to another, and the transitions are labeled with these residues. In HMMs, the emission occurs when the state has been entered, and the transitions themselves are not directly connected to the query sequence. In addition, the costs associated with transitions in finite-state automata are not necessarily

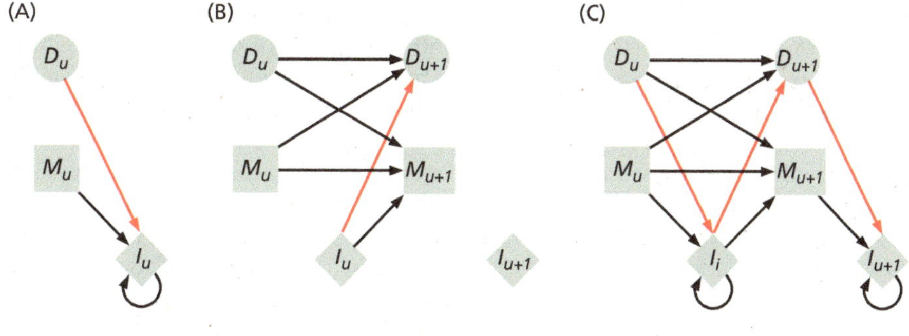


**Figure 6.6**
**The repeating architecture of a profile HMM.** (A) The transitions between the three states of a profile HMM associated with position $u$. The match, insert, and delete states are labeled $M_u$, $I_u$, and $D_u$, respectively. Note the red transition, which is often not included in models. (B) The transitions from the three states of a profile HMM associated with position $u$ and those states associated with position $u + 1$. Again, a red transition is often omitted from models. (C) The full model, being the two sets of transitions combined.

probabilities, and paths through such models are unlikely to have a probabilistic interpretation.

There are three possible outcomes when considering aligning the next residue of a query sequence to a reference sequence alignment. First, the query sequence residue may align (or match) with the next residue of the reference. Second, it may correspond to an insertion relative to the reference. Third, the query sequence residue may correspond to a later position in the reference, indicating a gap (or deletion) in the query. Profile HMMs contain a set of three states associated with each alignment position to model these alternatives. For the $u$th alignment position the match, delete, and insert states will be referred to as $M_u$, $D_u$, and $I_u$, respectively, and will be represented on diagrams by a square, circle, and diamond, respectively (see Figure 6.6). The delete state is an example of a silent state, as it does not emit any residues of the sequence.

A few transitions are allowed between $M_u$, $D_u$, and $I_u$ in all profile HMMs (see Figure 6.6A). As the insert state emissions correspond to residues inserted after the $u$th position, there is a transition from $M_u$ to $I_u$. In many models there is also a transition from $D_u$ to $I_u$, although a number of models omit this because when such transitions are present they tend to have low probability, often resulting in low-scoring paths. Furthermore, the alignments they represent can be obtained by alternative paths through the HMM involving the match state. In addition, there is a transition from the $I_u$ state to itself to allow insertions of more than one residue before alignment position $u + 1$. (The alignment is strictly only of those residues emitted by the match states, so that sequence residues emitted by the insert states are not regarded as occupying an alignment position.)

Several transitions connect the $u$th position states to states at position $u + 1$, all in the forwards direction from $u$ to $u + 1$ (see Figure 6.6B). The $M_{u+1}$ and $D_{u+1}$ states can be reached from all three of the $u$th position states. This means that there is a path connecting successive delete states, corresponding to a large deletion. The insert states are only accessed via transitions from the match and delete states at the same position. Transitions between insert states $I_u$ and $I_{u+1}$ are not allowed because they would imply the absence of a match state between them, and hence delete state $D_{u+1}$ should be involved. However, note that the transition from $I_u$ to $D_{u+1}$ is often omitted, for the same reasons given above in the case of $D_u$ to $I_u$. The full set of states for positions $u$ and $u + 1$ are shown in Figure 6.6C, together with all the transitions involving just this set of states.

As discussed above, in addition to transitions between states, HMMs also have emissions from states. Match and insert states emit a residue, while delete states do not. Recalling the discussion of substitution matrices in Section 5.1, there are two basic models of sequence alignment to consider: the random and nonrandom (that is, related) models. If there is an insertion in the query sequence relative to the HMM, the residues involved are, by definition, not related to residue information in the HMM. The emission probabilities for insert states are often assigned according to the basic residue frequencies $p_a$ as described in Section 5.1. In contrast, the residues emitted by match states are related, and so the emission probabilities depend on the multiple alignment. They are most closely related to the PSSMs discussed above, as they vary with position in the alignment.

Unlike the PSSMs discussed above, which were constructed with as many positions as in the query sequence, the structure of the equivalent profile HMM is not quite as straightforward. Firstly, in contrast to PSSMs, profile HMMs can be obtained from a set of unaligned sequences. In such cases the number of match states is often set to the average length of the sequences that define the profile, since this will hopefully reduce the need for insert and delete states. However, it is often useful to try some alternative models, as they can lead to better results. The HMM program SAM examines the existing HMM to identify match states that are used by relatively few sequences, or insert states used by many sequences, and then respectively removes or inserts positions in the HMM. The process of exploring alternative model structures is sometimes called **model surgery**. When fitting a profile HMM to a multiple alignment a different approach is often taken, in which those alignment columns that have more gaps than some specified fraction are assumed to be insertions, other columns being taken as indicating a match state.

Each transition and emission in the model is a parameter that needs to be defined. The basic HMM structure for an alignment position normally has either nine or seven transitions (see Figure 6.6C). If all insert states are assigned identical emission probabilities, these can be precalculated from the residue composition or refined as 20 parameters. The match-state emission probabilities will be different for each alignment position, each requiring 20 parameters. Ignoring the insert-state emission, this adds up to as many as 29 parameters for each alignment position, so that a profile HMM with 100 match positions and associated states, transitions, and emissions will have several thousand parameters. This has important consequences, as almost certainly a lot of sequence data will be required to derive a good parameter set for an HMM of this size.

The path taken through the model, resulting in the emission of a sequence, can be interpreted as assigning each of the residues to be either aligned to a particular HMM position (corresponding to the particular match state) or else inserted at a particular position (corresponding to the particular insert state). Note that when comparing the assignments for several sequences, those residues from different sequences that were emitted by the same match state are regarded as aligned with each other. However, those emitted by the same insert state are not regarded as aligned, even though they may be shown in the same column of a summary alignment.

So far we have only discussed the states required in the section of the profile HMM that aligns the relevant region of the query sequence (for example, the domain that the profile HMM represents). The HMM must also contain special sections for starting and ending the alignment. In addition, some HMMs contain features designed to cope with query sequences containing unrelated regions of sequence as well as the relevant region, for example other domains not modeled in this HMM. Such HMMs model local as opposed to global alignments. Other HMMs are designed to allow for repetition of segments of sequence, even a complete domain. We will now examine these other features of profile HMMs as used in some of the commonly available packages.

All the profile HMMs described here require states at the start and end of the model. Figure 6.7A shows a profile HMM configuration including such start and end states. The start state does not emit any residues and has no transitions to itself, but has three transitions to other states. All paths through the HMM start at this state. The transition to $M_1$ will occur when the query sequence amino-terminal residue matches the first position in the profile HMM. If the query sequence has extra residues at its amino terminus, these will be emitted by the $I_0$ state. Finally, if the

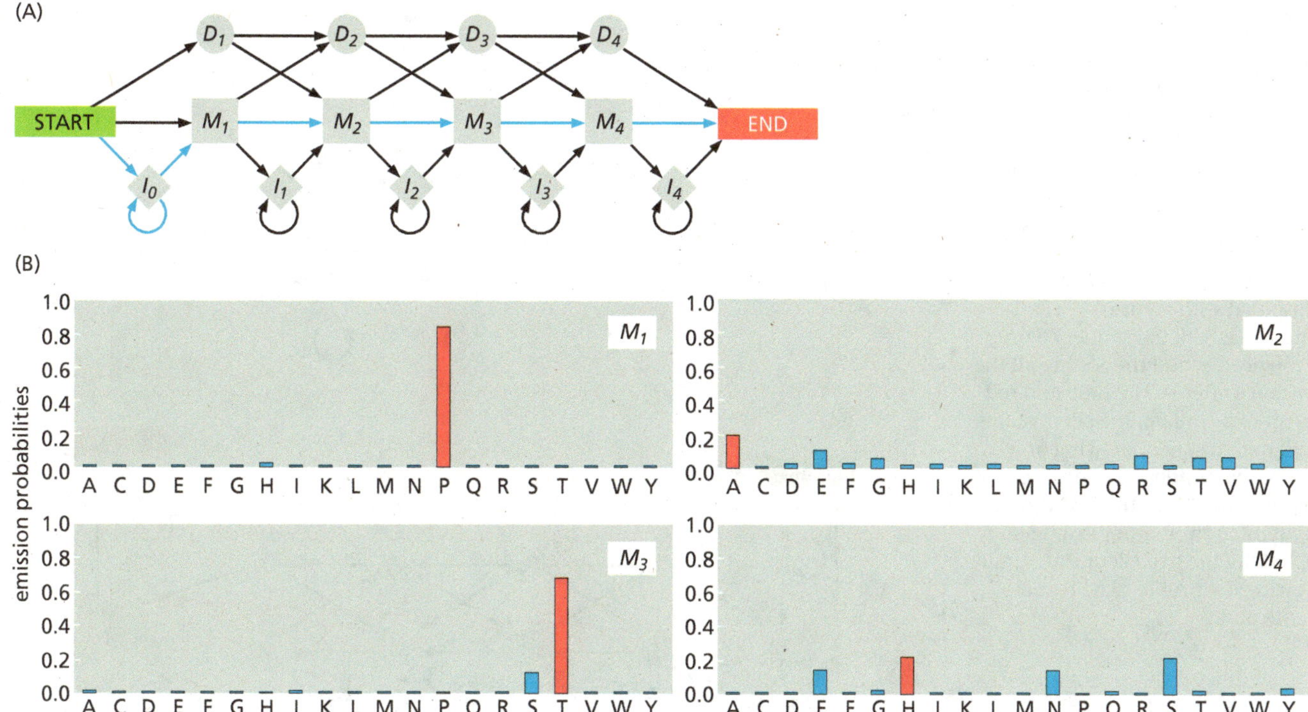


**Figure 6.7**
**A complete profile HMM model, with a start and end state and four match states.** The existence of states $I_0$ and $I_4$ allows the profile to occur anywhere within a larger sequence. (A) The organization of the states and allowed transitions. Note that each of the transitions in the model will in general have a different value for the transition probabilities. One of the possible paths through the model is marked by blue transition arrows, and produces a sequence that has two or more residues at the amino terminus, followed by the profile with no insertions, immediately followed by the carboxyl terminus. (B) The emission probabilities from the four match states. Ignoring the residues emitted from state $I_0$, the sequence PATH has the greatest probability of being emitted, but many other sequences are almost as likely to be emitted, for example, PETS. Note that the emissions for each state sum to 1.

first-position residue is missing from the query the transition used will be from start to $D_1$. The end state shown in Figure 6.7A has transitions to it from the three states associated with the final profile position and does not emit a residue, nor have any transitions from itself. All paths through the HMM terminate at this state. In the model shown, any extra residues in the query sequence after the profile will have to be emitted by the last insert state. Figure 6.7B shows example emission probabilities from the four match states. These sum to 1 for each state, and show that each alignment position has particular residue preferences. For example, state $M_1$ represents a highly conserved proline, and $M_3$ a conserved small polar residue. The insert state emission probabilities are not shown, but would normally be the same for all insertion states and related to the overall amino acid composition.

All the models considered here represent a profile against which a query sequence is to be aligned. We must allow for the query sequence to be wholly unrelated to the profile. In the model discussed so far, such an unrelated sequence would be emitted by paths that visited many delete and insert states. It should be recalled that only the sequence emitted by the match states can be regarded as aligned to the profile HMM, and any residues emitted by insert states are unaligned by definition. In addition, an unrelated sequence will align to a profile HMM with a very low probability or log-odds score. As in the case of database searches, unrelated sequences will usually be identified on these quantitative scores rather than a qualitative assessment of the alignment itself.

Another possible situation arises if the query sequence is longer than the profile, so that there are segments of sequence that are unrelated. The initial and final insert states can represent any unrelated flanking sequences, allowing for a local alignment of the query sequence. However, the model of Figure 6.7 is not the full equivalent of Smith–Waterman local alignment (see Section 5.2) because the path taken through the HMM by the sequence will still include all profile positions. Alignment of only a part of the profile will involve a path through many delete states, which will almost certainly occur with very low probability if the model is parameterized for sequences with the full-length profile. A more realistic local alignment model is shown in Figure 6.8A. This involves two new silent states that are directly connected

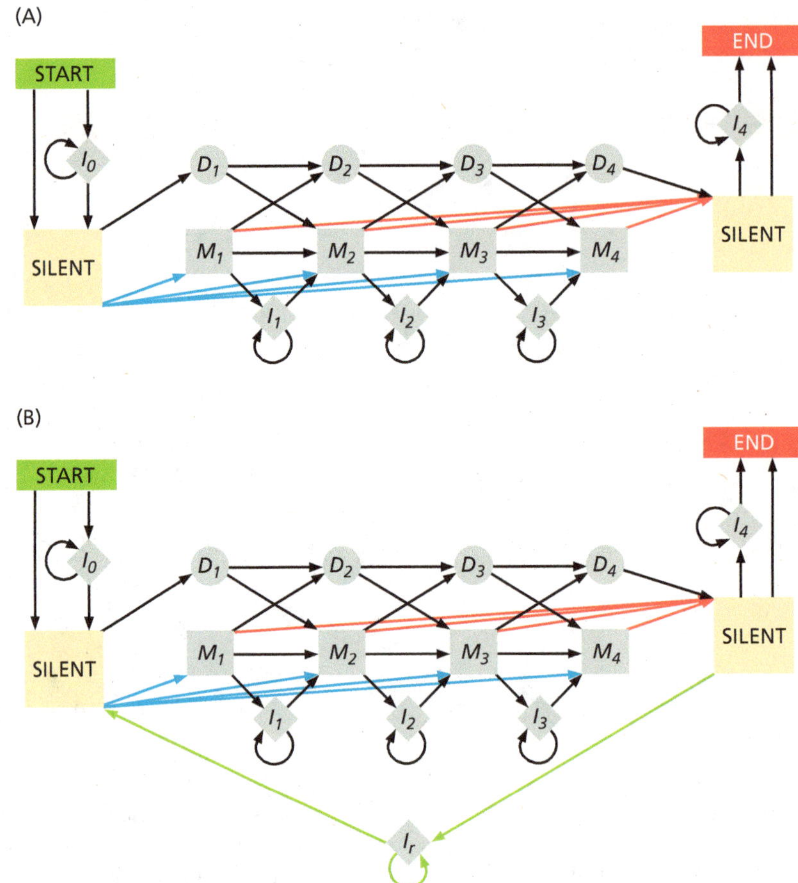


**Figure 6.8**
**Complete profile HMM models for local and repeat local profile alignment.** (A) Complete profile HMM model allowing for local alignment, including the presence of only part of the profile in the query sequence. Note the two silent states, and the change in architecture in the region of the $I_0$ and $I_4$ states, as compared with Figure 6.7. The $I_0$ state models all the sequence prior to the profile, and the $I_4$ state all the sequence after it. The blue and red transitions connect the silent states to all match states, allowing only a part of the profile to be included in the path. (B) The same HMM augmented by another insert state ($I_r$) linking the two silent states, and allowing the profile to be repeated on the path.

to every match state of the model, as well as to two insert states that model the flanking sequences of the query. These silent states are not necessary, but they reduce the number of transitions required in the model, making parameterization easier. Without them, the start and end states would also need to be connected to all match states to allow for the absence of flanking regions. Note that the red transitions of Figure 6.6C are missing in this model.

Although the insert states $I_0$ and $I_4$ of Figures 6.7 and 6.8 can model any flanking sequence segments that are not part of the profile, further design is needed to appropriately model such regions. For example, if simple insert states are used, at least one residue must be emitted at each end of the sequence, and these terminal residues will not be part of the profile alignment. This can be avoided by a more complex model structure. In the SAM program such structures are referred to as **free insertion modules** (FIMs) and, in addition to the insert state, include a delete state. The FIM is always entered via this delete state and can be left via either state. This allows for the possibility of no residues in a flanking sequence. The delete state has a transition to the FIM insert state, the insert state has a transition back to itself, and both have transitions that leave the FIM.

As described so far, with probabilities constrained to sum to 1 for every state, all transitions and emissions in a profile HMM will affect the overall probability or score for the sequence. This is undesirable in the case of flanking sequences, as it is only the region related to the profile that is of interest. The contribution of these flanking regions can be controlled by relaxing the probability constraints to set the transition from the FIM delete to the FIM insert state and all the transitions leaving a FIM a probability of 1. The only remaining influence on scoring will be from the emitted residues, but these are usually assigned the probabilities of the null, i.e., random model, and as a result have no influence.

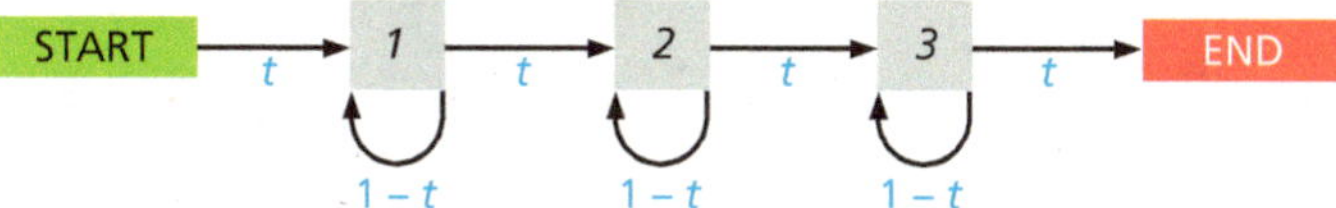


**Figure 6.9**
**A simple HMM used in the discussion of the length dependence of HMM models.** There are three match states between the start and end states. As there are no insert or delete states, all the match states must be included at least once in each path.

The HMM can be further enhanced to permit repeats of the profile by connecting the silent states of Figure 6.8A together. The connecting transition goes via an insert state (shown in Figure 6.8B) or more correctly via a FIM to model any intervening sequence, and must lead back to the state that has transitions to all the profile match states. This is the model used in the program HMMER2 but is closely related to those used in several other HMM packages.

The final issue relating to the basic HMM structure that we will consider is that the number of model states and the transition structure have an important influence on the distribution of path lengths through the HMM. This is easiest to understand by looking at some simpler HMMs, such as the one shown in Figure 6.9. All paths through the model shown in Figure 6.9 must pass at least once through each of the emitting states 1, 2, and 3, each of which emits a residue. Therefore, the minimal sequence length this HMM can model is three residues long. Furthermore, states 1, 2, and 3 have identical transition probabilities. Thus, any path corresponding to a sequence of $L$ residues will involve three transitions of probability $t$ (to move from start to end states) and $(L-3)$ transitions of probability $(1-t)$. Thus the probability of any path emitting $L$ residues is $t^3(1-t)^{L-3}$. However, there are many different equally probable paths that emit $L$ residues. We need to know how many in order to calculate the probability of this model producing an $L$-residue sequence. First, note that every path must begin with a transfer from the start state into match state 1. Thus, the paths can only differ in their remaining $(L-1)$ transitions, of which there will be $(L-3)$ transitions returning to the same match state (1 to 1, 2 to 2, or 3 to 3). The paths can only differ in the ordering of these $(L-3)$ transitions within the remaining $(L-1)$. Combinatorial analysis tells us that there are

$$ {}^{L-1}C_{L-3} = \frac{(L-1)!}{(L-3)!2!} \qquad \text{(EQ6.21)} $$

different orderings and hence different paths. As these are all of equal probability, the probability of a path in this model having length $L$ is ${}^{L-1}C_{L-3}\, t^3(1-t)^{L-3}$. For a general model with $N$ match states, the probability of a path of length $L$ is

$$ {}^{L-1}C_{L-N}\, t^N(1-t)^{L-N} = \frac{(L-1)!}{(L-N)!(N-1)!} t^N(1-t)^{L-N} \qquad \text{(EQ6.22)} $$

This distribution is shown in Figure 6.10 for several different values of $t$ and numbers of match states $N$. It can have an important influence on the success of the HMM, because the model should ideally have the same distribution of lengths as the set of sequences it is trying to represent. If $t$ is made sufficiently small, the most likely path lengths can become very long, but such models are not encountered in realistic protein-sequence problems.

## Estimating HMM parameters using aligned sequences

The structure and parameters of a profile HMM can be derived from a set of protein sequences. These sequences may already be in a multiple alignment or they may as yet be unaligned, waiting for the HMM to do the work. Here we will consider parameterization using alignments, leaving the unaligned case until later.

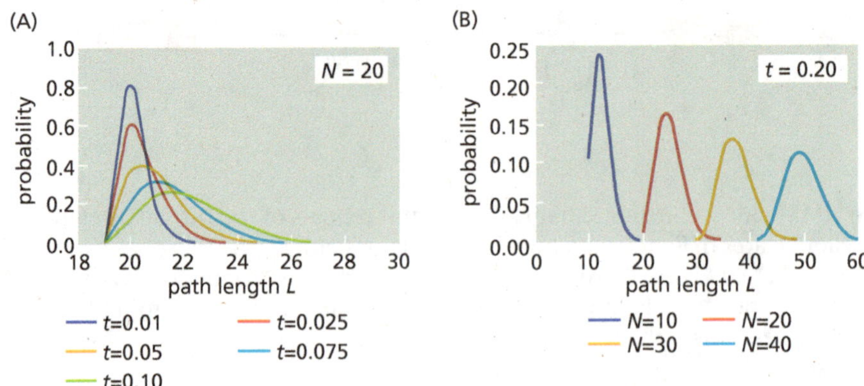


**Figure 6.10**
**Length distributions for the simple HMMs shown in Figure 6.9.** (A) For a model with 20 match states, the probability of paths of length $L$ for different values of the probability of a transition back to the same match state $t$. (B) The probability of paths of length $L$ for models with different numbers of match states when the transition probability $t$ is 0.20.

To use a multiple sequence alignment for profile HMM parameterization we must first decide which alignment positions correspond to which HMM match states, and which are to be modeled as insert states. An alignment column that contains no gaps should be assigned to a match state, and one with a majority of gaps to an insert state, but intermediate situations need careful consideration. Usually, a threshold proportion of gaps is selected to determine the assignments. Once these assignments have been made, the path that each sequence follows through the HMM can be deduced. In this way, the alignment data can be translated into the frequencies of transitions between particular states along the path and the frequencies of emission of individual residues from particular states. Note that more sophisticated parameterization methods have been proposed that simultaneously try to find the best column assignment during the parameterization (see Further Reading).

Given a set of sequences with known paths through the profile HMM, parameterization is in principle straightforward. All states in the HMM, except the end state, have some transitions from that state to others (in some cases including looping transitions back to itself). The total probability for these transitions must sum to 1 for each state, so that there is certainty that one of these transitions will be taken. (Exceptions such as states in FIMs will be ignored here.) For the state $u$ the transition to another state $v$ will be written $t(u,v)$. The summation over all states $w$ that are connected to state $u$ by transitions (including possibly state $u$ itself) gives

$$\sum_{w} t(u,w) = 1 \tag{EQ6.23}$$

If the paths taken by the sequences used for parameterization contain a total of $m_{u,v}$ transitions from $u$ to state $v$, then the model transition probabilities can be estimated by

$$t(u,v) = \frac{m_{u,v}}{\sum_{w} m_{u,w}} \tag{EQ6.24}$$

Emission probabilities for match states can be derived in an analogous way. The emission probabilities $e_{M_u}(a)$ for a residue $a$ from the $u$th match state $M_u$ add up to 1 where all possible residue types are considered. If the parameterization data contain $n_{M_u,a}$ emissions of residue $a$ from match state $M_u$ (which would result from observing that number of occurrences of residue $a$ in the alignment column corresponding to match state $M_u$), then the model emission probability can be estimated by

$$e_{M_u}(a) = \frac{n_{M_u,a}}{\sum_b n_{M_u,b}} \tag{EQ6.25}$$

where the summation is over all the residue types $b$. The insert state emission probabilities $e_{I_u}(a)$ are not calculated in this way. The model for insert states is the random model, so that the probabilities are usually taken from the overall amino acid composition of a selected data set. Thus

$$e_{I_u}(a) = p_a \tag{EQ6.26}$$

Unfortunately, parameterization is more difficult in practice because there is almost always a lack of data. If there are insufficient sequences for parameterization, many of the counts $m_{u,v}$ and especially $n_{M_u,a}$ will be zero or very small, resulting in very poor estimates of the probabilities. This is especially the case for emission parameters. For protein sequence HMMs a minimum of 20 sequences is required simply to observe each possible match state emission, but many more are required to obtain realistic estimates of the emission probabilities. If no transitions or emissions have been observed, this will lead to $t(u,v)$ or $e_{M_u}(a)$ being set to zero for these events. Using zero probabilities will result in the HMM being unable to accurately align sequences that require these events as they will be impossible in the model. An HMM that has zero probability for emission of a particular residue from a given match state cannot align a sequence with that residue at that position. It can only make the residue an insertion before or after the position, in which case it does not count as aligned. Even when a relatively large quantity of sequence data is available, parameterization problems often occur for positions with very highly conserved residues or sequence segments.

This problem is identical to that discussed above in the context of PSSM parameterization. In the case of emission probabilities the solution is very similar. The formulae presented earlier for $q_{u,a}$ can be applied immediately, so that for example the equivalent of Equation EQ6.10 is

$$e_{M_u}(a) = \frac{n_{M_u,a} + 1}{\sum_b n_{M_u,b} + 20} \tag{EQ6.27}$$

Other pseudocount methods can be applied, deriving expressions by reference to the previous formulae for PSSM parameters.

## Scoring a sequence against a profile HMM: The most probable path and the sum over all paths

Given a parameterized profile HMM, any given path through the model will emit a sequence with an associated probability. This path probability will be the product of all the transition and emission probabilities along the path. In addition, the path defines how the emitted sequence is aligned to the model. Residues emitted from match states will be aligned to that position in the profile. Residues emitted from insert states should not be regarded as aligned, but the position of the insertion is clear. Normally, the sequence is specified and we wish to find the alignment and score against the profile HMM.

There can be many paths through the HMM that result in the same emitted sequence, each with a different alignment to the profile. This is analogous to the many possible alignments of one sequence to another. Of these many paths, some or one will be the most probable, analogous to the best-scoring alignment in dynamic programming. However, the probability of the sequence being emitted by the HMM is given by the sum of the probabilities for all possible paths. This has no equivalent in dynamic programming.

Methods exist for calculating the most probable path and the total probability, both of which are quite similar to the dynamic programming methods discussed in Section 5.2. Using probabilities is computationally impractical, because in realistic situations it leads to very small numbers. This technical problem can be avoided by taking logarithms. However, log-odds scores (the log of the ratio of probabilities of alternative models) are preferable, using a random model as in Section 5.1, as this is easier to analyze to provide an estimate of the significance of the score. The random model of an $L$ residue sequence is the product of all the $L$ terms $p_a$ (the residue composition frequency). If each of the $L$ emission probabilities, each for a particular state $u$ emitting $e_u(a)$, is divided by its associated $p_a$ then the emission terms will include all the terms of the random model and the transition terms $t(u,v)$ can remain unchanged. Therefore the equations that follow will contain terms such as $\ln(e_u(a)/p_a)$ and $\ln(t(u,v))$ where $u$ and $v$ are model states and $a$ is the emitted residue type.

We will deal first with calculating the most probable path, which can be found using the **Viterbi algorithm**. This will be applied to a profile HMM with a structure very similar to that shown in Figure 6.7A, except that it will have an arbitrary number of sets of match $M$, delete $D$, and insert $I$ states. The states are labeled as was described earlier when presenting the basic profile HMM structure. At a profile position $u$, these states will be called $M_u$, $I_u$, and $D_u$. Hence the transition probability from $M_u$ to $M_{u+1}$ will be written $t(M_u,M_{u+1})$. The start and end states will here be labeled "*Start*" and "*End*," respectively. However, the formulae will only be shown for the regular repeating structure. The formulae involving the *Start* and *End* states are a modification of those presented here. If the query or emitted sequence $\boldsymbol{x}$ is $L$ residues long and is labeled $x_1x_2x_3...x_L$, then the emission probability for residue $x_i$ from insert state $I_u$ is written $e_{I_u}(x_i)$.

During the Viterbi algorithm, a record must be kept of the highest probability or best log-odds score up to that point in the model and for a given amount of emitted sequence. For example, at the state $D_u$, when the sequence up to and including residue $x_i$ has been emitted, the highest probability will be written $V_{D_u}(x_i)$ and the best log-odds score will be written $V_{D_u}(x_i)$. Note that this calculation is equivalent to a dynamic programming matrix with one side of length equal to the number of states in the HMM, and the other side being the length of the query or emitted sequence

The Viterbi algorithm, like dynamic programming, starts with very short emitted sequences involving paths of only very few HMM states, so that the best path probabilities and log-odds scores are readily calculated, and then builds on these. The calculation usually uses initialization values that start the path at a particular state with no sequence emitted. For example, in probability calculations the initialization for the model in Figure 6.7A might be $v_{Start}(0) = 1$, $v_u(0) = 0$ for any or other state $u$. The log-odds equivalents are $V_{Start}(0) = 0$, $V_u(0) = -\infty$ for any or other state $u$. This results in all paths beginning at the *Start* state.

The key stage in the algorithm is to derive values for the states in the $u$th profile position based on values at states which have transitions to them, wholly or mostly from the $u$–1th profile position. The following equations deal only with the basic profile HMM model shown in the central part of Figure 6.7A. We will also assume that the insert state emission probabilities are given by the amino acid composition $p_a$ for residue type $a$, written $p_{x_i}$ in the following equations where they represent emission of the residue type present at position $x_i$.

The formulae for the probability calculation are

$$
\begin{aligned}
v_{M_u}(x_i) &= e_{M_u}(x_i)\max\begin{cases} v_{M_{u-1}}(x_{i-1})t(M_{u-1},M_u) \\ v_{I_{u-1}}(x_{i-1})t(I_{u-1},M_u) \\ v_{D_{u-1}}(x_{i-1})t(D_{u-1},M_u) \end{cases} \\
v_{I_u}(x_i) &= p_{x_i}\max\begin{cases} v_{M_u}(x_{i-1})t(M_u,I_u) \\ v_{I_u}(x_{i-1})t(I_u,I_u) \end{cases} \\
v_{D_u}(x_i) &= \max\begin{cases} v_{M_{u-1}}(x_i)t(M_{u-1},D_u) \\ v_{D_{u-1}}(x_i)t(D_{u-1},D_u) \end{cases}
\end{aligned}
\qquad \text{(EQ6.28)}
$$

Note that other transitions, such as those shown in red in Figure 6.6C, may also occur in some other model profile HMMs. These would lead to extra terms corresponding to these extra transitions. For traceback to determine the path, a record must be kept of which term was selected as the maximum in each formula. For computational reasons the logarithms of probabilities are always used, which replaces the multiplication with addition. For example,

$$
\begin{aligned}
\log v_{M_u}(x_i) &= \log e_{M_u}(x_i) + \max\begin{cases} \log v_{M_{u-1}}(x_{i-1}) + \log t(M_{u-1},M_u) \\ \log v_{I_{u-1}}(x_{i-1}) + \log t(I_{u-1},M_u) \\ \log v_{D_{u-1}}(x_{i-1}) + \log t(D_{u-1},M_u) \end{cases} \\
\log v_{I_u}(x_i) &= \log p_{x_i} + \max\begin{cases} \log v_{M_u}(x_{i-1}) + \log t(M_u,I_u) \\ \log v_{I_u}(x_{i-1}) + \log t(I_u,I_u) \end{cases} \\
\log v_{D_u}(x_i) &= \max\begin{cases} \log v_{M_{u-1}}(x_i) + \log t(M_{u-1},D_u) \\ \log v_{D_{u-1}}(x_i) + \log t(D_{u-1},D_u) \end{cases}
\end{aligned}
\qquad \text{(EQ6.29)}
$$

Using a **null model** the same as the model used for insert emissions (Equation EQ6.26), we can obtain log-odds scores. Using the HMM model of Figure 6.6C, we can derive log-odds scores as

$$
\begin{aligned}
V_{M_u}(x_i) &= \log\left(\frac{e_{M_u}(x_i)}{p_{x_i}}\right) + \max\begin{cases} V_{M_{u-1}}(x_{i-1}) + \log t(M_{u-1},M_u) \\ V_{I_{u-1}}(x_{i-1}) + \log t(I_{u-1},M_u) \\ V_{D_{u-1}}(x_{i-1}) + \log t(D_{u-1},M_u) \end{cases} \\
V_{I_u}(x_i) &= \max\begin{cases} V_{M_u}(x_{i-1}) + \log t(M_u,I_u) \\ V_{I_u}(x_{i-1}) + \log t(I_u,I_u) \\ V_{D_u}(x_{i-1}) + \log t(D_u,I_u) \end{cases} \\
V_{D_u}(x_i) &= \max\begin{cases} V_{M_{u-1}}(x_i) + \log t(M_{u-1},D_u) \\ V_{I_{u-1}}(x_i) + \log t(I_{u-1},D_u) \\ V_{D_{u-1}}(x_i) + \log t(D_{u-1},D_u) \end{cases}
\end{aligned}
\qquad \text{(EQ6.30)}
$$

For comparison with the previous equations, these are for the basic profile HMM unit as shown in Figure 6.6C, with all nine possible transitions given in these equations. Only the match state $M_u$ equations include an emission term because the delete state $D_u$ does not emit, and the insert state $I_u$ emission term becomes 0 on dividing by the null model. Note that the emitted residue is dictated by the query sequence, and must be the the next residue in the sequence $\boldsymbol{x}$.

The termination of this calculation will depend on the precise details of the model in the region of the end state. If the end state is called *End* the value of $V_{End}(x_L)$ will be the log-odds score of this best path. Because this state does not emit a residue, this last formula will lack emission terms.

Algorithms quite similar to these can calculate the total probability or total log-odds score for all paths that emit the query sequence. The main difference is that instead of determining the maximum of three alternatives, the three probabilities are summed. However, there is a significant practical difference, because actual probabilities must be calculated, not log-odds scores. The method is known as the **forward algorithm** because it progresses from the start state to the end state of the HMM. At the state $M_u$, when the sequence up to and including residue $x_i$ has been emitted, the total probability is written $f_{M_u}(x_i)$ and the total log-odds score is written as $F_{M_u}(x_i)$. The initialization is performed as was discussed previously. By analogy with Equation EQ6.28 we have

$$
\begin{aligned}
f_{M_u}(x_i) &= e_{M_u}(x_i)\left[f_{M_{u-1}}(x_{i-1})t(M_{u-1},M_u) + f_{I_{u-1}}(x_{i-1})t(I_{u-1},M_u) + f_{D_{u-1}}(x_{i-1})t(D_{u-1},M_u)\right] \\
f_{I_u}(x_i) &= p_{x_i}\left[f_{M_u}(x_{i-1})t(M_u,I_u) + f_{I_{u-l}}(x_{i-1})t(I_{u-1},I_u)\right] \\
f_{D_u}(x_i) &= \left[f_{M_{u-1}}(x_i)t(M_{u-1},D_u) + f_{D_{u-1}}(x_i)t(D_{u-1},D_u)\right]
\end{aligned}
\qquad \text{(EQ6.31)}
$$

The log-odds score version of this, by analogy to Equation EQ6.30 is

$$
\begin{aligned}
F_{M_u}(x_i) &= \log\left(\frac{e_{M_u}(x_i)}{p_{x_i}}\right) + \log\left[t(M_{u-1},M_u)e^{F_{M_{u-1}}(x_{i-1})} + t(I_{u-1},M_u)e^{F_{I_{u-1}}(x_{i-1})} + t(D_{u-1},M_u)e^{F_{D_{u-1}}(x_{i-1})}\right] \\
F_{I_u}(x_i) &= \log\left[t(M_u,I_u)e^{F_{M_u}(x_{i-1})} + t(I_u,I_u)e^{F_{I_u}(x_{i-1})} + t(D_u,I_u)e^{F_{D_u}(x_{i-1})}\right] \\
F_{D_u}(x_i) &= \log\left[t(M_{u-1},D_u)e^{F_{M_{u-1}}(x_i)} + t(I_{u-1},D_u)e^{F_{I_{u-1}}(x_i)} + t(D_{u-1},D_u)e^{F_{D_{u-1}}(x_i)}\right]
\end{aligned}
\qquad \text{(EQ6.32)}
$$

As for the Viterbi algorithm, $F_{End}(x_L)$ will be the total log-odds score for obtaining the query sequence via all possible paths with this HMM. Again the *End* state does not emit a residue, so the formula to calculate it will lack this emission term.

An equivalent algorithm has also been proposed going backwards from the *End* state to the *Start* state. This is called the **backward algorithm**, and defines the terms $b$ and $B$ that are exactly analogous to the terms $f$ and $F$. We are only interested here in the probabilities $b$ as they will be used later. The probability of the sequence $x_{i+1}$ to $x_L$ being emitted, starting from state $M_u$ when it emitted residue $x_i$, is written $b_{M_u}(x_i)$. The initialization is $b_u(x_L) = t(u,End)$ for all states $u$, which is only non-zero for those states with a direct transition to the *End* state. By analogy with Equations EQ6.28 and EQ6.31, using the HMM model of Figure 6.7A we have

$$
\begin{aligned}
b_{M_u}(x_i) &= t(M_u,M_{u+1})e_{M_{u+1}}(x_{i+1})b_{M_{u+1}}(x_{i+1}) + t(M_u,I_u)e_{I_u}(x_{i+1})b_{I_u}(x_{i+1}) + t(M_u,D_{u+1})b_{D_{u+1}}(x_i) \\
b_{I_u}(x_i) &= t(I_u,M_{u+1})e_{M_{u+1}}(x_{i+1})b_{M_{u+1}}(x_{i+1}) + t(I_u,I_u)e_{I_u}(x_{i+1})b_{I_u}(x_{i+1}) \\
b_{D_u}(x_i) &= t(D_u,M_{u+1})e_{M_{u+1}}(x_{i+1})b_{M_{u+1}}(x_{i+1}) + t(D_u,D_{u+1})b_{D_{u+1}}(x_i)
\end{aligned}
\qquad \text{(EQ6.33)}
$$

Note that $f_u(x_i)$ is the probability of obtaining the sequence $\boldsymbol{x}$ from the start up to the $i$th residue in a path that ends at the $u$th model state. Similarly, $b_u(x_i)$ is the probability of obtaining the sequence $\boldsymbol{x}$ from the $(i + 1)$th residue to the end having emitted the $i$th residue by the time state $u$ was left. Also note that $f_{End}(x_L) = b_{Start}(0) = P(\boldsymbol{x})$, the probability of observing the whole sequence $\boldsymbol{x}$ from the HMM.

## Estimating HMM parameters using unaligned sequences

Returning now to the case where the sequences to be used for parameterization are not aligned, we must begin by deciding on the number of match states. These sequences (called the training sequences) should have been selected on the basis of having a common sequence region, possibly only a small part of the complete sequence. The number of match states should be fixed according to the anticipated features of the common region; that is, how many alignment columns are expected to contain residues as opposed to gaps in a majority of the sequences.

A starting set of parameters is required, which will be improved by iteration as described below. This set should be as good as possible to increase the chances of determining a good parameter set. For this reason, Dirichlet priors are often used to obtain a first estimate of the parameters. Furthermore, because there are general problems in locating the optimal parameter set, it is often advisable to generate several starting parameter sets and try each of them.

Several algorithms can be used to obtain HMM parameters. Pierre Baldi has proposed a technique involving maximum likelihood estimation methods where the optimization occurs via gradient descent. For more detail see Further Reading, although some of the key concepts are discussed in Appendix C.

The method we will discuss here is the **Baum–Welch expectation maximization** algorithm. The essential idea is to try to estimate the number of times each transition and emission occurs in the model when using the training sequences. These values are obtained by summation over all possible paths that produce the sequences, using the HMM with the current parameter set. The method uses the forward and backward algorithms. (A faster but less accurate variant uses only the highest-scoring paths; that is, Viterbi paths.) The estimated transition and emission frequencies are used to derive an improved parameter set. The training sequences are then rerun with the HMM parameters now set at their new values. Several cycles are run until the parameters have converged. This method is an example of the expectation maximization (EM) method (see Further Reading), and involves iteratively improving the values of the parameters, such that the likelihood of observing the sequences with the HMM should increase with each iteration.

First, note that $f_u(x_i)t(u,v)e_v(x_{i+1})b_v(x_{i+1})$ is the probability of emitting the sequence $\boldsymbol{x}$ summed over all paths that pass through states $u$ and $v$, emitting $x_{i+1}$ at state $v$. From this, the probability that the transition from state $u$ to state $v$ is taken with the emission of $x_{i+1}$, is given by

$$\frac{f_u(x_i)t(u,v)e_v(x_{i+1})b_v(x_{i+1})}{P(\boldsymbol{x})} \qquad \text{(EQ6.34)}$$

where $P(\boldsymbol{x})$ is the probability of observing sequence $\boldsymbol{x}$ by any path. (As noted above, the value of $P(\boldsymbol{x})$ can be obtained by the forward or backward algorithm.) By summing this formula over all values of sequence position $i$ and over all training sequences, one can obtain the expected number of transitions $m_{u,v}$ for the training set. For the model illustrated in Figure 6.6C, with extra transitions allowed, we obtain the nine equations

$$m_{M_u,M_{u+1}} = \sum_{seqs\ x}\left[\frac{1}{P(x)}\sum_i f_{M_u}(x_i)\,t(M_u,M_{u+1})\,e_{M_{u+1}}(x_{i+1})\,b_{M_{u+1}}(x_{i+1})\right]$$

$$m_{I_u,M_{u+1}} = \sum_{seqs\ x}\left[\frac{1}{P(x)}\sum_i f_{I_u}(x_i)\,t(I_u,M_{u+1})\,e_{M_{u+1}}(x_{i+1})\,b_{M_{u+1}}(x_{i+1})\right]$$

$$m_{D_u,M_{u+1}} = \sum_{seqs\ x}\left[\frac{1}{P(x)}\sum_i f_{D_u}(x_i)\,t(D_u,M_{u+1})\,e_{M_{u+1}}(x_{i+1})\,b_{M_{u+1}}(x_{i+1})\right]$$

$$m_{M_u,I_u} = \sum_{seqs\ x}\left[\frac{1}{P(x)}\sum_i f_{M_u}(x_i)\,t(M_u,I_u)\,q_{x_{i+1}}\,b_{I_u}(x_{i+1})\right]$$

$$m_{I_u,I_u} = \sum_{seqs\ x}\left[\frac{1}{P(x)}\sum_i f_{I_u}(x_i)\,t(I_u,I_u)\,q_{x_{i+1}}\,b_{I_u}(x_{i+1})\right]$$

$$m_{D_u,I_u} = \sum_{seqs\ x}\left[\frac{1}{P(x)}\sum_i f_{D_u}(x_i)\,t(D_u,I_u)\,q_{x_{i+1}}\,b_{I_u}(x_{i+1})\right]$$

$$m_{M_u,D_{u+1}} = \sum_{seqs\ x}\left[\frac{1}{P(x)}\sum_i f_{M_u}(x_{i+1})\,t(M_u,D_{u+1})\,b_{D_{u+1}}(x_{i+1})\right]$$

$$m_{I_u,D_{u+1}} = \sum_{seqs\ x}\left[\frac{1}{P(x)}\sum_i f_{I_u}(x_{i+1})\,t(I_u,D_{u+1})\,b_{D_{u+1}}(x_{i+1})\right]$$

$$m_{D_u,D_{u+1}} = \sum_{seqs\ x}\left[\frac{1}{P(x)}\sum_i f_{D_u}(x_{i+1})\,t(D_u,D_{u+1})\,b_{D_{u+1}}(x_{i+1})\right]$$

(EQ6.35)

Using similar logic, and this time restricting paths to those that emit a residue of type *a* at state $M_u$ (that is, the residue emitted can be any residue *a* in the sequence ***x***), we obtain

$$n_{M_u,a} = \sum_{seqs\ x}\left[\frac{1}{P(x)}\sum_{i|x_i=a} f_{M_u}(x_i)\,b_{M_u}(x_i)\right]$$

(EQ6.36)

where the summation is over all residues of type *a* in the whole sequence. The insert state emission probabilities can also be fitted with an equivalent equation for $I_u$.

The process starts with rough estimates of the transition and emission parameters. These values are used to obtain a new set of parameters, using Equations EQ6.35 and EQ6.36 with EQ6.27 or other methods to take account of a lack of data as described in Section 6.1. An improved parameter set should make the emission of the training sequences more likely. The likelihood of observing all the training sequences is given by the product of the $P(\mathbf{x})$ for all sequences ***x***. Normally the log likelihood is used, which is the sum of the $F_{End}(x_L)$ for each sequence. The method is repeatedly used to (hopefully) improve the HMM parameters, until the log likelihood of the training data does not improve any further. After repeating this procedure from several alternative starting points, the best parameters are kept, and the HMM is regarded as parameterized.

Thus far we have not considered weighting the training sequences, even though we might expect some to be more helpful in the parameterization than others. The sequence weighting methods discussed in Section 6.1 are only applicable to aligned sequences and so do not help here, unless the poorly parameterized HMM is used for their "alignment". (Aligning each sequence to the profile HMM cannot be said to create a true multiple alignment, despite the results appearing in a similar form. However, the output could be treated in the same way to obtain sequence weights.) Most profile HMM programs weight training sequences according to their log-odds score. Typically the worst-scoring sequence is assigned a weight of 1, and the others are weighted on a scale that assigns a weight of 0.01–0.001 to the best-scoring sequence. These weights are typically updated during each iteration of the parameterization.

## 6.3 Aligning Profiles

The methods of generating PSSMs and profile HMMs from sequences are now well established, and several databases exist that provide such models for different protein families and subfamilies. It has become common practice to use these models in searching for remote homologs, during which sequences are aligned to them. More recently it has been appreciated that aligning two PSSMs or profile HMMs can be even more effective at identifying remote homologs and evolutionary links between protein families (see Flow Diagram 6.3). (Generally families are thought to have no evolutionary relationship to each other.) As a result several methods have been proposed that can perform such analysis.

### Comparing two PSSMs by alignment

The alignment of two PSSMs cannot proceed by a standard alignment technique. Consider the alignment of two columns, one from each PSSM. As neither represents a residue or even a mixture of residues, but both in fact are scores, there is no straightforward way of using them together to generate a score for use in an alignment algorithm. The solution to this problem is to use measures of the similarity between the scores in the two columns.

The program LAMA (Local Alignment of Multiple Alignments) solves one of the easiest formulations of this problem, not allowing any gaps in the alignment of the PSSMs. Suppose the two PSSMs $A$ and $B$ consist of elements $m^{A}_{u,a}$ and $m^{B}_{v,a}$ for residue type $a$ in columns $u$ and $v$, respectively. Various methods can be used to try

**Flow Diagram 6.3**
**The key concept introduced in this section is that methods have been designed that align two profiles.** There is a hierarchy of related protein sequences, so that often two profiles can be aligned to identify features of their common ancestral sequence.

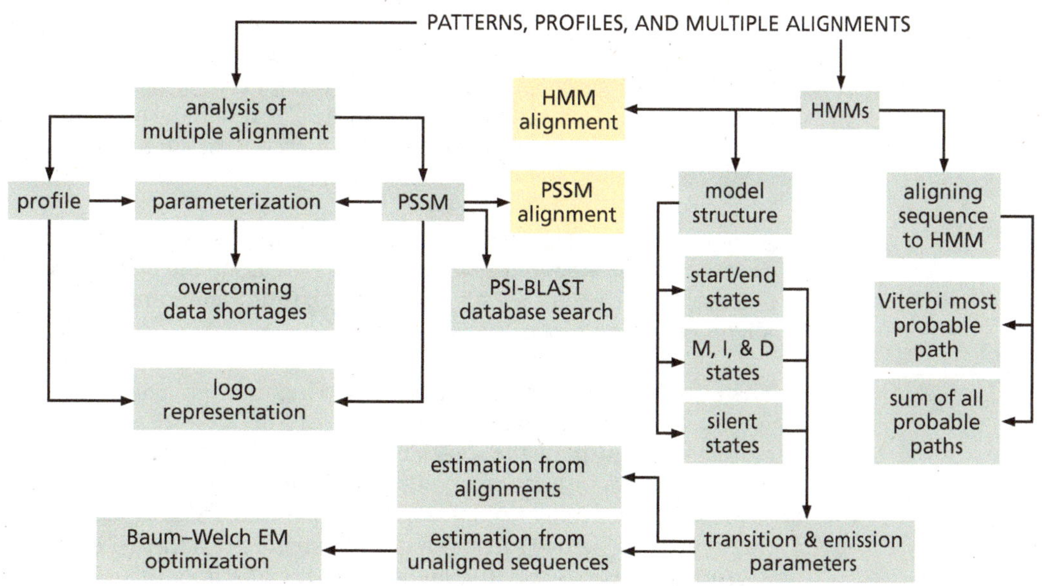

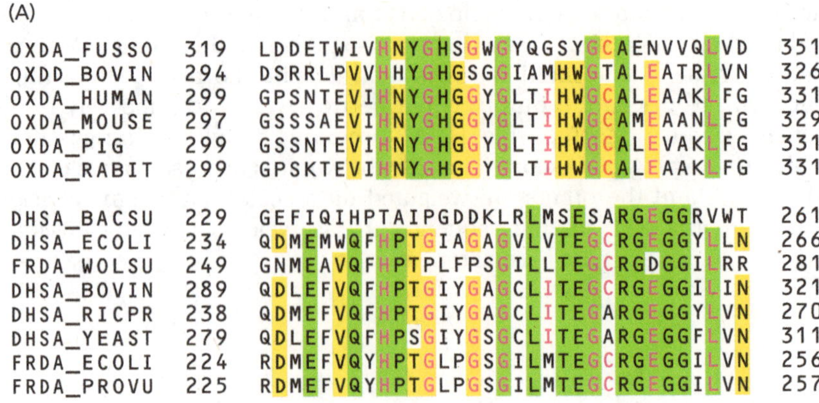


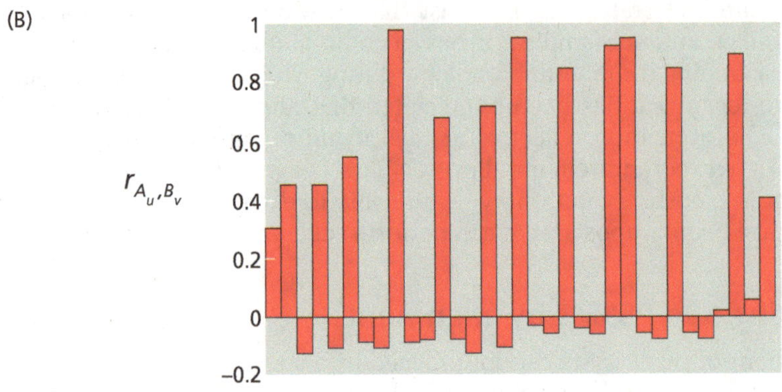


**Figure 6.11**
**Illustration of LAMA alignment of two protein families using PSSMs.** (A) The top protein family sequence alignment is of a set of D-amino oxidase flavoenzymes, entry BL00677D of the BLOCKS database. The lower alignment is of a set of succinate dehydrogenase and fumarate reductase enzymes, from entry BL00504D of the BLOCKS database. The sequences are identified by their SWISS-PROT identifiers. Highly conserved positions in each alignment have a colored background, green indicating identity and yellow highly conserved alignment columns. Highly conserved residues identified after aligning these two alignments are shown by red letters. (B) A plot of the LAMA column scores $r_{A_u,B_v}$ (Equation EQ6.37) for the aligned alignments. This plot has been scaled so that the scores are vertically below the relevant columns in the alignments of part A. (Adapted from S. Pietrokovski, Searching databases of conserved sequence regions by aligning protein multiple-alignments, *N.A.R.* 24:3836–3845, 1996.)

to measure the similarity of these two columns. LAMA uses the **Pearson correlation coefficient** $r_{A_u,B_v}$ defined as

$$r_{A_u,B_v} = \frac{\sum_a \left(m^A_{u,a} - \overline{m}^A_u\right)\left(m^B_{v,a} - \overline{m}^B_v\right)}{\sqrt{\sum_a \left(m^A_{u,a} - \overline{m}^A_u\right)^2 \sum_a \left(m^B_{v,a} - \overline{m}^B_v\right)^2}} \tag{EQ6.37}$$

where $\overline{m}^A_a$ and $\overline{m}^B_v$ are the means of the respective PSSM columns and the summations are over all residue types $a$. The value of $r_{A_u,B_v}$ ranges from 1 for identical columns to –1. In a random alignment test it was found that the average column score was just slightly negative (–0.05). The score for aligning two PSSMs is defined as the sum of $r_{A_u,B_v}$ for all aligned columns.

As no gaps are permitted in aligning the two PSSMs, all possible alignments can readily be scored by simply sliding one PSSM along the other, allowing for overlaps at either end of each PSSM. The highest-scoring alignment is then taken as the best alignment of the two families.

There still remains the difficulty of assessing the significance of a given score. The scores are proportional to the length of the shorter PSSM, but this is easily corrected. To estimate the score distribution, the columns of the PSSMs are shuffled many times, recording the possible alignment scores each time. (Shuffling is used to preserve the overall residue composition of the PSSMs.) Assuming normal statistics, these can be used to estimate the mean and standard deviation of the score distribution. In this way the highest-scoring alignment can be assigned a $z$-value (number of standard deviations from the mean) and hence a statistical significance. There is no sound basis for this analysis, but it has proved useful, and all alignments with $z$-scores greater than 5.6 are taken as significant. Once a significant alignment has been detected, a plot of the $r_{A_u,B_v}$ values can help to identify the columns for which the families have similar residues (see Figure 6.11).

Golan Yona and Michael Levitt formulated an alternative way of aligning PSSMs called prof_sim. This allows for gap insertion in the PSSMs and uses the Smith–Waterman local alignment technique. Instead of using the PSSM score values, this method uses the original sequence alignments to define the amino acid distribution in each column. These distributions are compared for one column at a time from each PSSM using a score that measures the similarity of the two distributions and their deviation from the average distribution.

Yet another method has been proposed, called COMPASS (COmparison of Multiple Protein alignments with Assessment of Statistical Significance), which also performs local alignments including gaps. The main difference when compared with prof_sim is that the column residue distributions are calculated using sequence weights and pseudocounts in a similar way to that used by PSI-BLAST (see Equations EQ6.14 and EQ6.15) except that gaps are treated as another residue type. Column comparisons are scored using a method related to log-odds. By carefully following the BLAST and PSI-BLAST methodology, COMPASS scores can be readily converted into $E$-values, making assessment of the score significance easier and more rigorous.

## Aligning profile HMMs

The first method we will describe is called COACH (COmparison of Alignments by Constructing HMMs), and its intended application is the alignment of two alignments by firstly reducing one of them to a profile HMM. This HMM defines the equivalence of alignment columns and the HMM states. A modification of the Viterbi method is then used to find the most probable set of paths, which together emit the other alignment. (The probabilities for each sequence path must be multiplied together to get the overall probability for the alignment.) For details of the algorithm, see Further Reading. The paths can be used to construct the alignment of the two alignments. The advantage of this method over the PSSM-based ones presented above is that the gap scoring is position dependent.

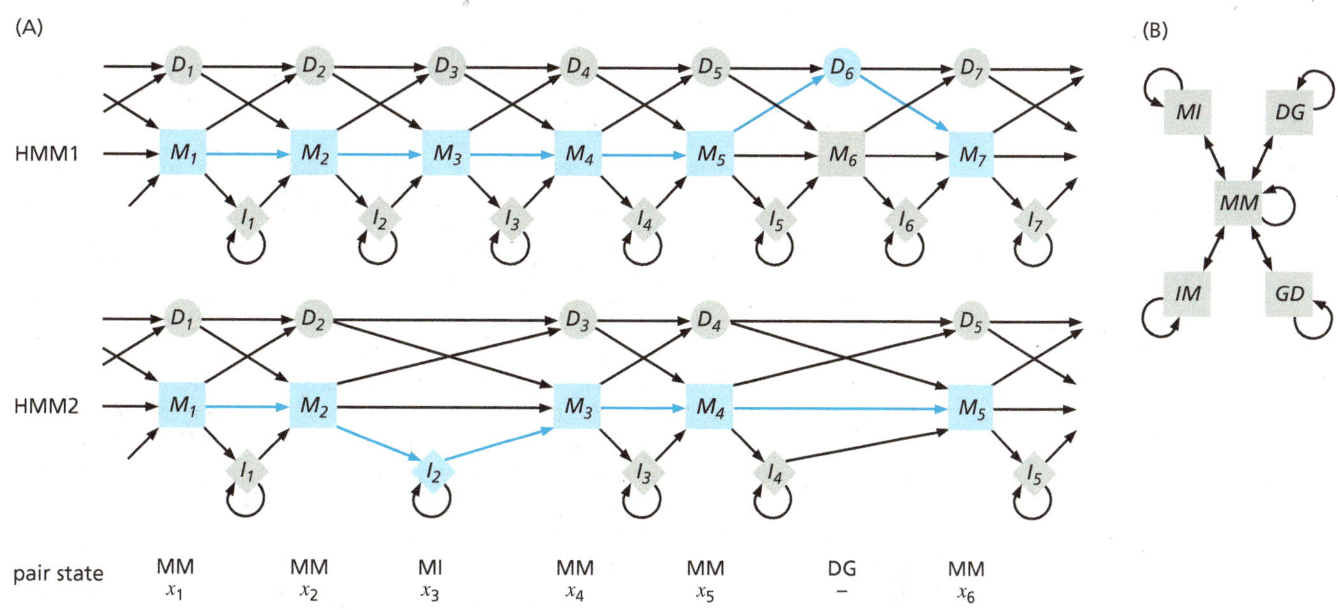


**Figure 6.12**
**An example of the alignment of two HMMs using HHsearch.** (A) Two HMMs are aligned over a region that results in a six-residue sequence segment being emitted by both models. HMM2 has an inserted residue (state $I_2$) that is aligned with the residue emitted by the $M_3$ state of HMM1. Note that position 6 of HMM1 has no equivalent in HMM2, so that its delete state $D_6$ aligns with a gap inserted in HMM2. The alignment corresponds to a path through a pair state model shown in (B). The pair states of this path are shown at the bottom of (A), as is the corresponding emitted sequence. (Adapted from J. Söding, Protein homology detection by HMM-HMM comparison, *Bioinformatics* 21:951–960, 2005.)

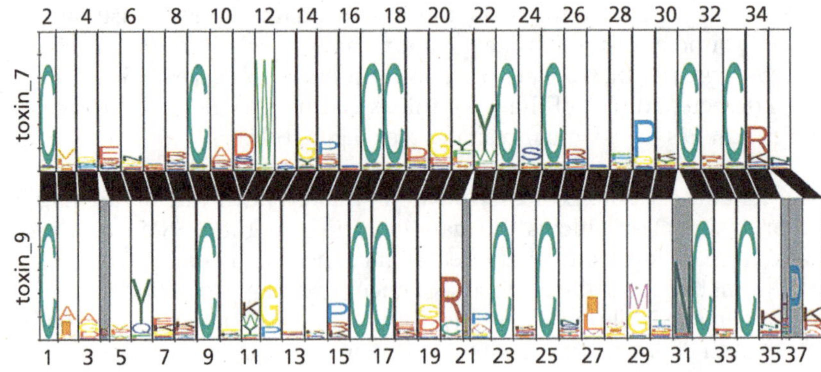


**Figure 6.13**
**An example of the visualization of the alignment of two HMMs using logos to illustrate the emission probabilities for different residue types at each match state.** The top HMM is the Toxin_7 family, while the lower HMM is the Toxin_9 family, both HMMs are from the Pfam database. The similarities in the pattern of cysteine residues are striking, as are some differences between the HMMs, such as the preferred tyrosine at position 6 in Toxin_9 and the preferred tryptophan at position 12 in Toxin_7. (From B. Schuster-Böckler and A. Bateman, Visualizing profile-profile alignment: pairwise HMM logos, *Bioinformatics* 21 (12):2912–2913, 2005, by permission of Oxford University Press.)

The HHsearch method of Johannes Söding aligns two profile HMMs and is designed to identify very remote homologs. It uses a variant of the Viterbi method to find the alignment of the two HMMs that has the best log-odds score. For details of the algorithm, see Further Reading. An example of an alignment of two HMMs is shown in Figure 6.12A. Both HMMs emit the same sequence $x_1x_2x_3x_4x_5x_6$, but using (in this case) different paths in the two HMMs. The states in each HMM are aligned and are labeled in the figure. For example, two aligned match states are labeled *MM*. In addition to the usual match, insert, and delete states, gap states *G* can also occur. The emitting states (*M* and *I*) can only be aligned with each other; the same restriction holds for the silent states *D* and *G*. Both *II* and *DD* aligned pair states could occur in principle, but ignoring them makes the model and computations more tractable, and has not been found to lead to any difficulty in obtaining good alignments. Figure 6.12B shows the possible aligned pair states and allowed transitions between them. Other transitions are also possible, but were ignored for reasons similar to those employed in ignoring the red transitions shown in Figure 6.6. For truly related HMMs the ignored pair states and transitions would be expected only rarely, if ever. Some account is also taken of predicted secondary structure for the two HMMs, justified by the observation that protein folds are much better conserved than sequences.

The representation of Figure 6.12 is not especially informative because the sequence details are omitted. A solution to this problem has been proposed that shows both HMMs aligned but includes a logo representation of the emission probabilities at each match state (see Figure 6.13).

## 6.4 Multiple Sequence Alignments by Gradual Sequence Addition

When a number of related sequences have been identified, for example as a result of database searches such as those described in Section 4.6 or using PSI-BLAST, it is usually informative to construct a multiple alignment that includes all the sequences (see Flow Diagram 6.4). Multiple alignments are more powerful for comparing similar sequences than profiles because they align all the sequences together, rather than using a generalized representation of the sequence family. If there are subgroups of sequences with extra features in common that do not occur in the complete sequences set, these features may be lost on generating a profile. The multiple alignment methods described here should be able to identify and analyze such features. The profile methods can do so only after the subgroup has been identified and separated. Identifying such a subgroup can lead to creating a new profile with which to search for other sequences. Using profiles in database searches is the most powerful current technique for identifying distant homologs. In addition, databases exist with collections of profiles for specific subfamilies. These have been found to be extremely useful in sequence analysis.

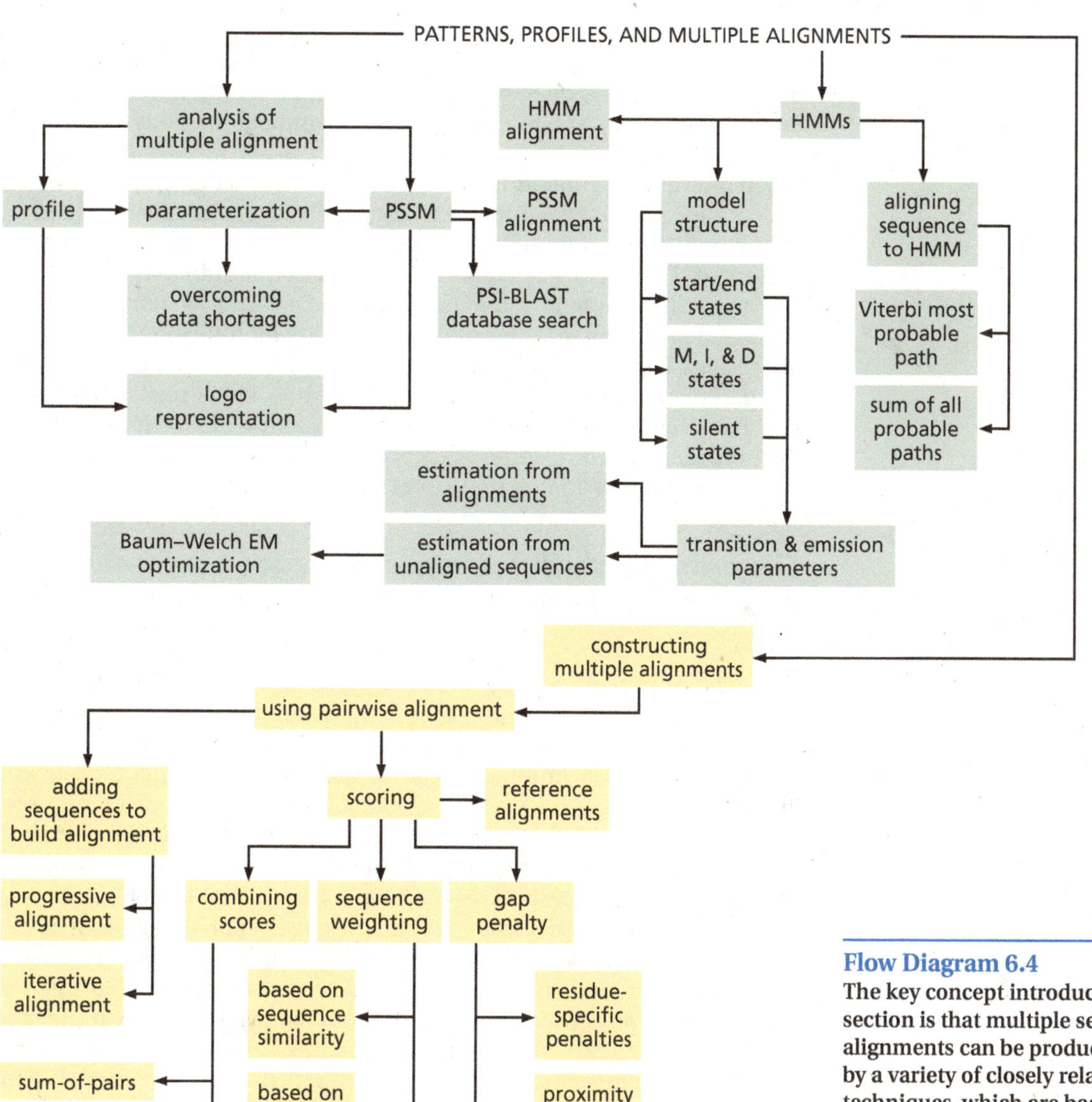


**Flow Diagram 6.4**
**The key concept introduced in this section is that multiple sequence alignments can be produced by a variety of closely related techniques, which are based on the pairwise sequence alignment methods.**

One way of building a multiple alignment is simply to superpose each of the pairwise alignments from a database search using the query sequence as the reference. In favorable cases this method can reveal conserved regions. However, this method of construction does not take account of the direct alignment of pairs of sequences not including the query, and is therefore unlikely to give the optimal multiple alignment. In practice, the pairwise alignments can often be significantly improved by multiple alignment with other similar sequences. An example of this is shown in Figure 4.10A, which shows that a pairwise alignment of PI3-kinase p110α and a protein kinase does not align the important active-site residues. Figure 4.10B shows a multiple alignment of the same protein kinase with a set of five PI3-kinases: the multiple alignment has successfully aligned the functionally important residues.

The pairwise dynamic programming algorithms described in Section 5.2 can produce the highest-scoring alignment of two sequences given a substitution scoring matrix and a gap scoring method. It is possible to modify the algorithms to find the optimal alignment of more than two sequences. This will be discussed briefly in the next section but is actually too demanding of computer memory and time to be very practical. As a result many alternative multiple alignment methods have been proposed, which are not guaranteed to find the optimal alignment but can nevertheless find good alignments with the limited resources available. The majority of these heuristic methods create a multiple alignment by gradually building it up, adding sequences one at a time. This is often referred to as **progressive alignment**, although the same method also occurs at the start of iterative alignment methods discussed later.

When multiple alignments are built up by gradually adding sequences, the order in which they are added can be crucial to the successful generation of an accurate alignment. A number of alternative ways of determining this order will be presented. Once this has been done, the actual multiple alignment itself can be constructed. Although some methods use scoring schemes related to those discussed in Section 5.1, they often incorporate modifications designed to improve the results. For example, many methods use weights for the different sequences, and gap scoring can be much more sophisticated. Other methods use very different scoring schemes, and the range of schemes used will be reviewed. The final part of this section will look at some aspects of building the multiple alignment using the scores and order of adding sequences.

## The order in which sequences are added is chosen based on the estimated likelihood of incorporating errors in the alignment

The dynamic programming methods such as Smith–Waterman or Needleman–Wunsch, presented in Section 5.2 in the context of pairwise alignment, can be extended to simultaneously align more than two sequences. The resulting algorithms are, however, impractical with current computational capabilities unless limited to about 6–10 sequences of moderate (200–300 residue) length. This is because the matrices that have to be constructed will have as many dimensions as there are sequences, so that to store a full matrix for even six sequences would require almost 60,000 GB of memory (at 1 byte per matrix element). Even using a linear gap penalty (see Section 5.1) the calculation of a single matrix element will require consideration of all possible combinations of gaps and residue matches among the sequences; that is, $2^{N_{seq}} - 1$ alternatives for $N_{seq}$ sequences. In fact, even the practical limits given above are for modified dynamic programming algorithms that attempt to limit the regions of the matrix explored to those most likely to contain the optimum alignment. See Further Reading for references to methods that use such modified algorithms. Even with ever-growing computing resources, these methods will not be practical when hundreds of sequences are involved, and more heuristic algorithms must be used. Most of these are based on building up a multiple alignment by gradually adding sequences, as will now be discussed.

In a group of related sequences, some will be more similar to the average while others will be relatively different. Those sequences that are more similar can usually be aligned with high confidence, meaning that a high proportion of the residues will be expected to align with their true equivalents in the other sequences. In contrast, often less confidence can be placed in the alignment of more divergent sequences, which can be expected to give more alignment errors. Where errors occur in an **intermediate alignment** (an alignment to which further sequences remain to be added) they can cause further errors during the continuing construction of the multiple alignment. This is largely because the intermediate alignments are kept fixed as further sequences are added. Therefore it is desirable to delay the addition of sequences whose alignment is more likely to contain errors until as late as possible.

Note that some multiple alignment programs report the alignment as soon as all sequences have been added, and are therefore especially susceptible to these errors. A number of methods try to circumvent this potential problem, some by removing sequences from the multiple alignment or dividing it into two subsets of aligned sequences and realigning them. These and other related techniques are called **iterative alignment**, and are described in detail later in this section.

An illustration of the potential importance of the order of adding sequences can be seen in the following example. If the two sequences GG and DGG are to be aligned to the reference sequence DGD, and GG is the first one to be aligned, both

```
D – G – D
    |
    G – G
```

and

```
D – G – D
    |
G – G
```

would score equally in most scoring schemes. However, given the third sequence, DGG, aligning the second sequence as -GG would clearly be preferable

```
D – G – D
    |
    G – G
    |   |
D – G – G
```

But if the first two sequences have already been aligned according to the second scheme, then this better multiple alignment is not available in the progressive alignment method. Such methods are often referred to as greedy because once the method has determined an intermediate alignment, it will continue to use it even though, subsequently, there may be better alternatives.

Most multiple alignment methods deal with this issue by analyzing every pair of sequences to get a measure of their similarity. Of course, to obtain a very accurate similarity measure between two sequences requires the correct alignment as in the multiple alignment, which will not be available at this point. The majority use this measure to derive a phylogenetic tree, usually referred to in this context as the guide tree, which is then used to guide the order of constructing the multiple alignment. In the example above, most measures used would result in a guide tree that indicates that DGD and DGG should be aligned first, followed by GG. As already shown, this leads to a better final alignment. The full explanation of how the guide tree is used will be given later.

We will not discuss the algorithms used to generate such trees here, as this topic is covered in detail in Chapters 7 and 8. The methods used are amongst the simplest and fastest, mostly the clustering methods UPGMA or neighbor-joining (see Section 8.2). The important feature for discussion here is the use of a simple measure of the similarity of each pair of sequences.

Several multiple alignment programs, including ClustalW and T-Coffee, perform Needleman–Wunsch global alignment for every pair of sequences, and from these alignments obtain the measure used in constructing the guide tree. To be precise, this measure is usually the evolutionary distance, a measure of dissimilarity, which is discussed in detail in Section 8.1. The simplest evolutionary distance measure is the percentage of alignment sites at which different residues have been aligned. However, obtaining a precise estimate of the evolutionary distance between two sequences is not trivial, and numerous formulae have been proposed, as discussed in Chapter 8.

When the number of sequences to be aligned is very large, the Needleman–Wunsch alignment of all sequence pairs can take a very long time. Therefore some faster although more approximate methods of obtaining a distance measure have been proposed. The MUSCLE and MAFFT methods both have the option of using measures based on the presence of $k$-mers (stretches of $k$ residues) in common for two sequences. The advantage of this approach is that the two sequences do not need to be aligned. Typically, hexamers are used, and instead of distinguishing all 20 amino acids, they are grouped according to their physicochemical properties. Measures used include such properties as the fraction of hexamers not in common to the two sequences.

The ProbCons method differs from those discussed above, in that every pair of sequences is aligned using a pair HMM (the HMM equivalent of Needleman–Wunsch alignment). Because of the probabilistic nature of HMMs it is possible to define an estimate of the accuracy of an alignment. This estimate is then used in a procedure similar to UPGMA to generate the guide tree.

Another method involving HMMs is used in SATCHMO. Every sequence is used to generate its own profile HMM, and the similarity measure used is the average score of one sequence in the profile HMM of the other, and vice versa. The most similar sequences according to this measure are aligned first. However, SATCHMO does not use a guide tree, and the way a multiple alignment is constructed will be presented later.

The computer time and space requirements of generating similarity measures and guide trees can easily dominate multiple alignment programs unless care is taken to choose the most efficient algorithms. The paper in which MUSCLE is presented discussed this point, and compares the less accurate but more efficient methods such as *k*-mers and UPGMA with the results from Needleman–Wunsch and neighbor-joining. In that case the faster methods were fortuitously found to ultimately produce the more accurate multiple alignments, but this result may not hold for other work.

The information such as pair alignments used to generate the guide tree is often not used further in generating the multiple alignment. There are exceptions to this as shown below, including those methods that use the COFFEE (Consistency based Objective Function For alignmEnt Evaluation) scoring system and some schemes for weighting sequence that use the guide tree branch lengths. The ProbCons method uses the pair HMM alignments to derive terms referred to as quality match scores, discussed in more detail later, which are then used as the scoring scheme for the multiple alignment.

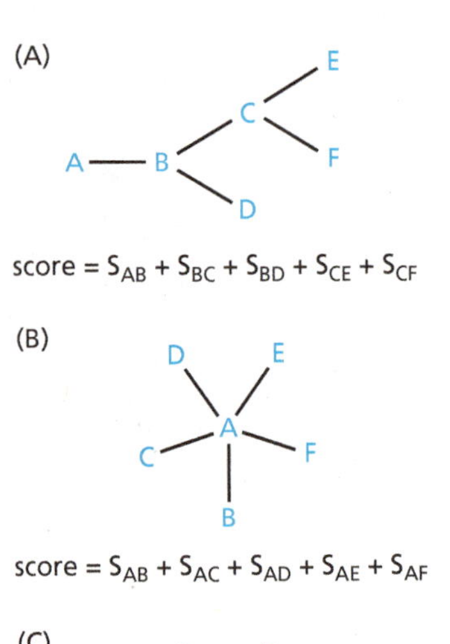


**Figure 6.14**
**Some possible general scoring schemes based on pairwise alignment scores for use in scoring multiple alignments.** Three alternative methods are shown for the case of six sequences (A–F). Part (A) shows a phylogenetic tree; (B) star; and (C) sum-of-pairs. In the formulae, $S_{AB}$ is the score for the alignment of sequences A and B, and so on.

## Many different scoring schemes have been used in constructing multiple alignments

Most of the early sequence alignment programs used scoring schemes closely related to those described in Chapters 4 and 5 for pairwise alignment and database searching. Of these the most complex is probably that used by ClustalW, which will be described below. More recently a number of other methods for scoring have been used, some of which will also be presented.

By analogy to the pairwise alignment problem as discussed in Section 5.2 we need a quantitative score of the quality of a multiple alignment. In pairwise alignment, the scores used have been based on measures of evolutionary distance and a similar basis proves useful here. Assuming that all the sequences being aligned have evolved from a common ancestor (as is often the case), their evolutionary relationship can be described in the form of a **phylogenetic tree**, as is shown in Figure 6.14A. However, we do not know these relationships at the time of making the alignment. As an interim measure, we could assume relationships between pairs of sequences as shown in Figures 6.14B and 6.14C. Based on these we need to propose a method for scoring an alignment. However, although the scoring of a multiple alignment must take account of all the sequences, we are only going to deal with those scoring methods that involve combinations of pairwise alignment scores.

This means that the substitution matrices such as PAM and BLOSUM presented in Sections 4.3 and 5.1 for scoring pairwise alignments can be used here.

A **star tree** (Figure 6.14B) would be obtained if one of the sequences were arbitrarily chosen to be the central node and scored against every other sequence, but no other pairs are directly scored against each other. This is the scheme that is, in effect, being used when combining database search results using the query sequence as a template. For highly similar sequences this method can generate a reasonable alignment (not necessarily the best), but when the percentage identity between sequences is low, multiple alignments obtained by this method can be very poor.

In most multiple alignments there is no rationale to choose one sequence to be the central node (unlike in a database search) and therefore the star tree method is not appropriate. The most commonly used alternative general scoring scheme is known as **sum-of-pairs** or **SP** (Figure 6.14C). It involves scoring the alignment of every possible pair of sequences and adding all the scores together.

When the sum-of-pairs method is used to score alignments, all sequences should not be regarded as equally independent or useful, and should be weighted to take account of this. At one extreme, two identical sequences will not provide any more information than one sequence alone, and so they should be given lower weights. Conversely, two very different sequences will provide a lot of information about acceptable alternative residues at particular positions and so should be weighted more highly.

We will now discuss some simple weighting schemes that are based on the branch lengths of a phylogenetic tree constructed from the sequences. The multiple alignment program ClustalW uses such a scheme, and also makes a distinction between the weights applied when generating a new multiple alignment from individual sequences and those applied when adding a sequence to an existing alignment. When generating a new alignment, weights are determined from the guide tree. To obtain a sequence weight, the path from each sequence to the tree root is analyzed. Each branch on the path contributes to the sequence weight an amount proportional to its length divided by the number of sequences that share the ancestral nodes of that branch (see Figure 6.15). When two very similar sequences occur, their common ancestor will be very recent, so most of their paths to the root of the tree will be shared, and their weights will be effectively halved, as in B and C in the example shown in Figure 6.15.

When sequences are added to an existing alignment, the weighting required is different. Now, an almost identical sequence in the alignment should be weighted more highly, as the new sequence should have almost the same alignment. To achieve this, pairwise alignments are used to calculate the distance of the new sequence(s) from those already in the alignment. The new weights are simply (1 – distance). In ClustalW, these are then multiplied by the original weights from the tree (see Figure 6.15) to give final weights for adding the sequence. Note that there

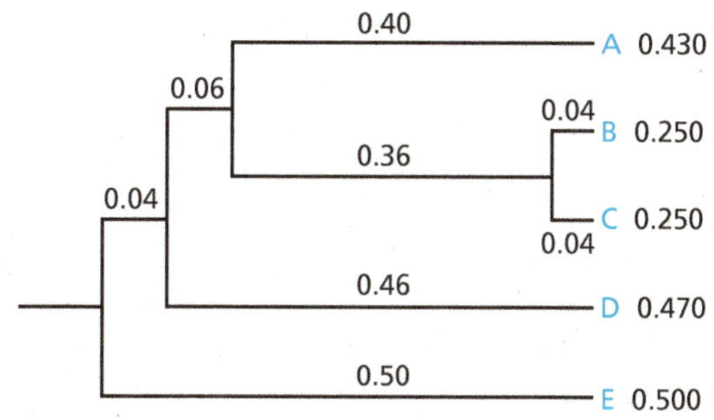


**Figure 6.15**
**Illustration of the sequence weighting scheme applied in ClustalW when constructing a multiple alignment.** The five sequences A–E have been related by a tree based on their pairwise alignment scores, with branch lengths as indicated. The sequence weights, listed against each sequence on the right of the tree, are the sum of branch lengths from each sequence to the tree root (left-hand end of tree), each branch length being divided by the number of sequences whose paths to the root share that branch. For example, the weight of sequence A is obtained as 0.40 + (0.06/3) + (0.04/4) = 0.430.

are a number of alternative schemes for obtaining sequence weights, many of which do not use trees. Some of these were described in Section 6.1.

If the alignment of two protein sequences is biologically meaningful, there will also be a high degree of similarity between the two protein structures specified by these sequences. For two structures with the same protein fold, most of the secondary structural elements should be preserved as a constant core. There can, of course, be some additional secondary structural elements, as well as minor rearrangements. However, the essential difference to be expected between related structures is that the polypeptide chain loops connecting elements of secondary structure will vary in size. This will show up in the sequence alignment as gaps of different lengths within the loop sequences, i.e., gaps tend to occur in specific regions that correspond to parts of the protein structure where there are residues exposed to the solvent.

The many known protein structures have been analyzed to examine if there are any preferences for particular residues in the loop regions. Attempts have been made to distinguish between the physicochemical type of residues between which gaps are inserted and the residues of an insertion. These have the same residue preferences, but their preferences are significantly different from those found in secondary structural elements. For soluble globular proteins, the dominant preference is for hydrophobic residues not to be associated with regions containing indels (insertions or deletions) as these loop regions generally represent regions of the protein structure exposed to the solvent. For membrane proteins, the preferences may be different, but as yet, far fewer structures of membrane proteins are known.

The program ClustalW uses a scale of residue-specific gap opening penalty factors (see Section 5.1). The actual calculation of the affine gap penalties in this program entails a complicated modification of a set of initial default values. In deriving the gap opening penalties, the manipulations take account of sequence lengths, sequence similarity, and average mismatch scores, as well as the positions of gaps in intermediate alignments (partial alignments produced during the generation of the complete multiple alignment). One of the aims of this scheme is to concentrate gaps into the same alignment column, which is achieved in part by raising the gap opening penalties to either side of an existing gap. The method is ad hoc, but seems to work quite effectively. An example alignment showing the effects of the variation in gap opening penalty is given in Figure 6.16.

The discussions above have presented ways of scoring a multiple alignment that are based closely on the scoring schemes used for pairwise alignment. An alternative approach has been proposed that uses the pairwise alignments of all sequence-pairs

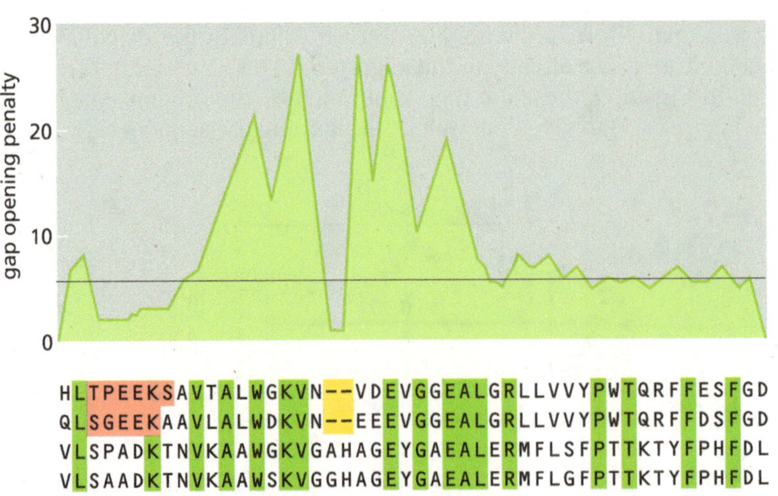


**Figure 6.16**
**The variation of the gap opening penalty along an example alignment as applied in the ClustalW program.** The initial gap opening penalty (GOP) is shown by the horizontal line. Note the very low GOP at the location of a gap in the alignment, and the very high GOP to either side of the gap. The other region of low GOP is due to the two hydrophilic stretches shown in red. (Adapted from D.G. Higgins et al., *Methods in Enzymology* 288, 1996.)

as the basis of a scoring scheme. The set of pairwise alignments is regarded as a reference library. When scoring a multiple alignment, the alignment of two individual sequences within it is compared with the equivalent pairwise alignment in the library. The score is based on the consistency found between these two pairwise alignments, in other words the frequency of occurrence of identical alignment columns. The scoring function based on this has been called COFFEE.

The alignment of sequences $\boldsymbol{x}$ and $\boldsymbol{y}$ can be extracted from the multiple alignment to give an alignment $A(\boldsymbol{x},\boldsymbol{y})$. The length of this alignment is the number of positions that contain at least one residue from either sequence and is given by $|A(\boldsymbol{x},\boldsymbol{y})|$. When compared at every position with a given pairwise alignment, $R(\boldsymbol{x},\boldsymbol{y})$ of the same sequences, taken from the reference library, a score can be assigned. The number of identical alignment columns $n_{ident}(R(\boldsymbol{x},\boldsymbol{y}), A(\boldsymbol{x},\boldsymbol{y}))$ is counted between the aligned sequence-pairs $R(\boldsymbol{x},\boldsymbol{y})$ and $A(\boldsymbol{x},\boldsymbol{y})$. The pairwise alignment $R(\boldsymbol{x},\boldsymbol{y})$ in the reference library is assigned a weight $w_{R(\boldsymbol{x},\boldsymbol{y})}$ that is intended to be a measure of its quality. The weight assigned to each library alignment is the percentage identity between the two sequences according to this specific alignment. The total score $S$ for an alignment of $N_{seq}$ sequences is given by

$$S = \frac{\sum_{x<y}^{N_{seq}} \sum_{\text{Lib}} w_{R(x,y)} n_{\text{ident}} \left[ R(x,y), A(x,y) \right]}{\sum_{x<y}^{N_{seq}} \sum_{\text{Lib}} w_{R(x,y)} \left| A(x,y) \right|} \qquad \text{(EQ6.38)}$$

where the $\sum_{\text{Lib}}$ is over the library of alignments for those alignments that contain sequences $\boldsymbol{x}$ and $\boldsymbol{y}$.

The T-Coffee alignment program uses a score based on the COFFEE scheme. Each pair of residues that are aligned in one of the reference library alignments is assigned a score that is to be used in scoring the multiple alignment. Initially, the score is the weight assigned to that particular pairwise alignment (the percentage identity). This score is subsequently modified to favor regions of the alignment that are consistent with the equivalent aligned region in other library alignments. The method is referred to as library extension, and is done by considering all possible intermediate sequences as illustrated in Figure 6.17. When calculating the library extension for the alignment of sequences $\boldsymbol{x}$ and $\boldsymbol{y}$, the alignments of $\boldsymbol{x}$ with $\boldsymbol{z}$, and $\boldsymbol{y}$ with $\boldsymbol{z}$, need to be considered, where $\boldsymbol{z}$ is all other sequences present in the library. Residues that are aligned identically in all three pairs receive extra weight, and in this way each residue can be assigned a unique weight in each library alignment. These weights are then used by T-Coffee in the construction of the multiple alignment.

The library of reference pairwise alignments may contain more than one alignment for a given pair of sequences. These alternatives may have been generated by different alignment techniques (for example global or local alignments), by using different substitution matrices or gap scores. It is also possible to include sequence alignments based on structural alignments, a method known as 3D-Coffee.

The SATCHMO program uses the COACH method to align two alignments. (These two alignments can include single sequences as well as intermediate alignments.) This was presented in the previous section, and uses the HMM scores that result from the profile HMM parameterization of one of the alignments. The scores are added for the path taken by each sequence of the other alignment through the profile HMM. As mentioned above, ProbCons uses match-quality scores to score the multiple alignment. These are a measure of the probability that two given residues in two sequences are aligned, taking into account the other alignments involving

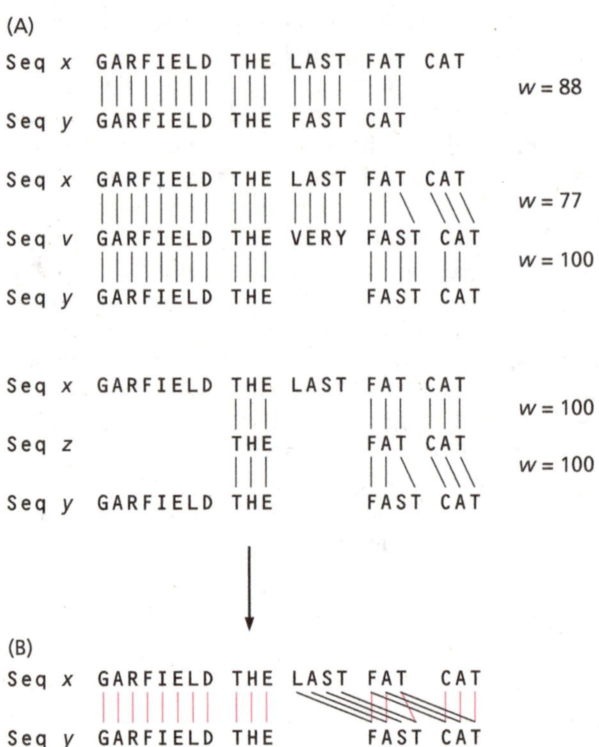


**Figure 6.17**
**The library extension process used in the COFFEE scoring method to assign weights to individual aligned residues of reference pairwise alignments.** Three possible alignments for sequence ***x*** and ***y*** are illustrated in (A): that resulting from direct alignment of the sequences ***x*** and ***y***; that resulting from aligning ***x*** with ***v***, and then ***v*** with ***y***; and that resulting from aligning ***x*** with ***z***, and then ***z*** with ***y***. The initial weights $w$ given to the right of each pairwise alignment are related to the percentage identity of the shorter sequence. (B) The alignments are combined to produce a single alignment of sequences ***x*** and ***y*** in the extended library. In the direct alignment of ***x*** and ***y*** the first residues $x_1$, $y_1$, are glycine 'G' in both sequences, with a corresponding weight of 88. When ***x*** and ***y*** are aligned through ***v***, the G is matched between all three sequences. The weight added to the extended library is the sum of the previous weight and the lowest weight of the two weights in the two new alignments (i.e., 77 + 88 for the G in sequence-pair ***x*** and ***y***). No further information is gained about this first residue G from the alignment of ***x*** and ***y*** through ***z***, and thus it has no influence on the weight associated with aligning $x_1$ and $y_1$. (Adapted from C. Notredame, D.G. Higgins and J.Heringa, T-Coffee: a novel method for fast and accurate multiple sequence alignment, *J.Mol.Biol.* 302:205–217, 2000.)

these sequences, and are derived from the pair HMM alignment calculations used to generate the guide tree. The scores are affected by how consistently the two residues are aligned in the pairwise alignments. To illustrate the concept of consistency in this context, if for example residues $x_i$ and $y_j$ are aligned in the (***x***,***y***) pairwise alignment, then if $x_i$ is also found aligned to residue $z_k$ in the (***x***,***z***) alignment, $y_j$ will also be aligned to $z_k$ in the (***y***,***z***) alignment if the alignments are consistent.

## The multiple alignment is built using the guide tree and profile methods and may be further refined

There are a number of alternative methods available to construct a multiple alignment given a guide tree and a scoring scheme. There is a choice of using dynamic programming or HMM alignment techniques, and in addition there are some ways to try to speed up the creation of good alignments. A number of techniques have also been proposed that try to refine the multiple alignment produced by iteratively realigning sequences to it.

The essential idea of progressive alignment is to add the sequences one at a time, producing intermediate alignments that are subsequently used to build the final multiple alignment. In this step-by-step construction process three different types of alignment may occur: aligning a sequence with a sequence, aligning a sequence with an intermediate alignment, and aligning two intermediate alignments. When using an intermediate alignment in a subsequent step, the only modification of the intermediate alignment that is allowed is the addition of gaps. These gaps must be added at the same site within all sequences in the intermediate alignment (see Figure 6.18). Clearly, with this limitation for intermediate alignments the multiple alignment produced will depend on the order in which the sequences are added, defined usually by the guide tree as was discussed earlier.

Each internal node of the guide tree corresponds to an intermediate alignment of all the sequences descended from that node (see Figure 6.19). These intermediate alignments are built up gradually so that the first steps involve pairs of sequences, but as the alignment progresses, intermediate multiple alignments will be aligned with each other or with another sequence.

The alignment of a pair of sequences is usually performed exactly as described in Section 5.2. When generating a combined alignment that involves intermediate alignments, consideration needs to be given to the scoring method. The simplest, but least accurate, technique involves finding the best-scoring pair of sequences, one from each intermediate alignment, and then using them as a template to construct the combined alignment. However, most currently used techniques take advantage of the methods of profile alignment discussed earlier in this chapter.

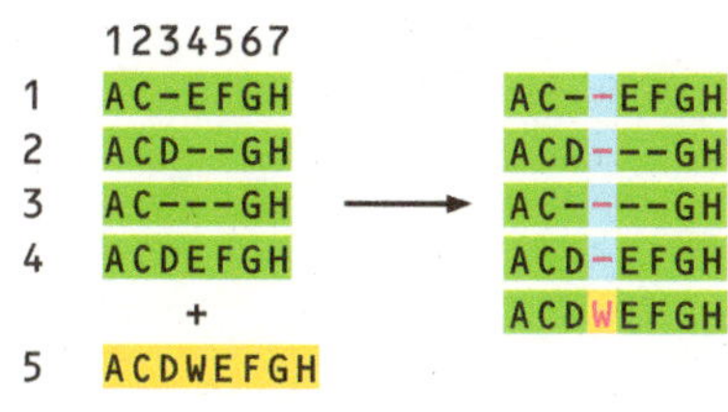


**Figure 6.18**
**Illustration of profile alignment and the scoring problems caused by gaps.** The addition of sequence 5 to the intermediate alignment of sequences 1–4 requires a gap to be inserted after position 3. As a consequence, sequence 4 has a new gap, scored with a gap opening penalty, while the other three sequences all require a gap extension penalty to be added.

Programs such as ClustalW use dynamic programming with the sum-of-pairs scoring method that takes account of all the sequences. At each step, the scores of all pairs of sequences must be taken into account, remembering that any gap introduced must be introduced at the same point in all sequences in the relevant intermediate alignment (see Figure 6.18). This would require a lot of computational effort if the calculation were done in a naive fashion. This can be overcome because it is possible to rearrange the terms for much more efficient calculation by using profiles.

We will consider the alignment of two intermediate alignments, $A$ and $B$, with $n_A$ and $n_B$ sequences, respectively. The residue in the $u$th column of the $x$th sequence of $A$ will be denoted $x^A_u$. Suppose that we are using a linear gap penalty with no distinction between gap opening and extension. In that case, each intermediate alignment can be treated as a single alignment that is the average of its constituent sequences. For any given alignment column $u$ of intermediate alignment $A$ we can obtain the frequency of occurrence of residue $a$, $f^A_{u,a}$. These frequencies can then be used to obtain the score for aligning a residue $b$ with this column, which is simply given by

$$m^A_{u,b} = \sum_{\substack{\text{residue}\\ \text{types } a}} f^A_{u,a}\, s_{a,b} \qquad \text{(EQ6.39)}$$

where $s_{a,b}$ is an element of the substitution matrix, and the summation is over all possible residue types. By comparison with Equation EQ6.2 it can be seen that the set of all $m^A_{u,b}$ is the alignment profile.

The total score $S_{A_u,B_v}$ or aligning column $u$ from intermediate alignment $A$ and column $v$ from intermediate alignment $B$ can then be written in several alternative ways:

$$\begin{aligned} S_{A_u,B_v} &= \sum_{x\in A \text{ seqs}}^{n_A} \sum_{y\in B \text{ seqs}}^{n_B} s\left(x^A_u, y^B_v\right) \\ &= \sum_{x\in A \text{ seqs}}^{n_A} m^B_{v,x^A_u} = \sum_{y\in B \text{ seqs}}^{n_B} m^A_{u,y^B_v} = \sum_{\substack{\text{residue}\\ \text{types } a}} m^A_{u,a} f^B_{v,a} = \sum_{\substack{\text{residue}\\ \text{types } a}} m^B_{v,a} f^A_{u,a} \end{aligned} \qquad \text{(EQ6.40)}$$

Depending on the number of sequences in each intermediate alignment relative to the number of possible residue types, one of these formulae can be chosen for

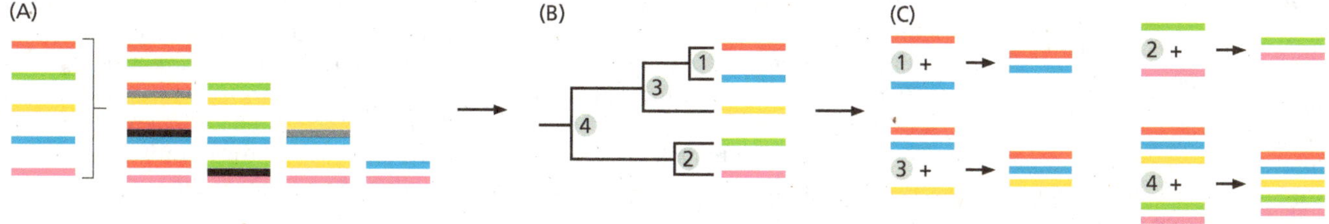


**Figure 6.19**
**Schematic illustrating the steps involved in a typical progressive alignment technique.** (A) Initial pairwise alignments are made and scored in order to find sequences most similar to each other. The darkness of the space between the aligned sequences indicates the degree of similarity, going from white, indicating low similarity, to black, indicating high similarity. (B) Construction of the guide tree using the similarity scores determined in (A). The numbers 1 to 4 identify the nodes of the tree, and correspond to intermediate and final alignments. (C) Each of the labeled tree nodes corresponds to a step in building up the complete alignment by creating intermediate alignments composed of the sequences associated with the nodes.

optimal efficiency. ProbCons uses a very similar method of progressive alignment. The terms $s\left(x_u^A, y_v^B\right)$ in Equation EQ6.40 are the usual elements of the substitution matrix. The combined alignment of A and B is calculated by dynamic programming using the scores $S_{A_u,B_v}$.

These formulae must be adjusted if affine gap scores are used. The alignment in Figure 6.18 illustrates the consequences of distinguishing between gap opening and extension. The intermediate alignment shown (sequences 1 to 4) requires an insertion after the residue D to allow for the extra residue W in the new sequence, 5. For sequence 4 this corresponds to a gap opening, and for the others to a gap extension. Care must be taken to account correctly for gaps in these circumstances.

Rather than use the standard pairwise or profile alignment algorithms, which can still take a long time for large numbers of sequences, some programs use fast methods that attempt to identify diagonals (i.e., gapless segments) of the optimal alignment. Once identified, the dynamic programming can be restricted to a small region around these diagonals, rather than performing calculations over the complete dynamic programming matrix. One of the current, most accurate methods, developed by Kazutaka Katoh and colleagues and named MAFFT (Multiple Alignment using Fast Fourier Transform), uses such a technique. In order to speed up the alignment steps, a fast Fourier transform (FFT) method is used to quickly identify key regions of sequence similarity. For the Fourier transform, the sequence is represented by the general physicochemical properties of the residues. The method is significantly faster than ClustalW or T-Coffee (which is no doubt also due to the different methods used to obtain the guide tree as discussed above) and has been used to successfully align several thousand sequences. MUSCLE can also identify diagonals, using a technique based on $k$-mers, but in that case it was found that this resulted in a reduction in the accuracy of the final alignment for minimal improvements in speed, and so is not the default method. MAFFT also gives more accurate alignments with the full dynamic programming matrix calculation.

(A)

(B) For the tree:

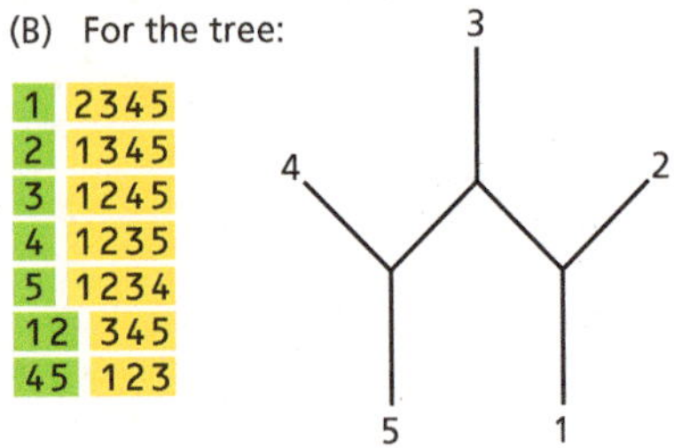


**Figure 6.20**
**Lists of the realignments to be made and scored for an alignment of five sequences (1–5) in an iterative alignment method.** (A) Some iterative alignment methods involve all possible realignments; that is, all possible unique combinations of subsets of the sequences. Each line represents a realignment of the aligned sequences in green with the aligned sequences in yellow to form a new complete alignment, keeping the green and yellow alignments fixed (as in Figure 6.18). (B) An iterative alignment method involving only the splits that exist in the guide tree. (See Section 7.1 for a definition of splits.) In both cases the best-scoring new alignment can be used as the starting point for a repeat round of realignment. The process stops when there has been no improvement in the score during the last cycle or when a given number of cycles is exceeded.

The progressive methods discussed above use a guide tree to define an order of addition of sequences to create a multiple alignment in which each sequence is fixed once it has been added. Several attempts have been made to improve on this by allowing some element of realignment of sequences to the existing multiple alignment. Geoff Barton and Michael Sternberg proposed the most straightforward of these. Once the initial multiple alignment has been produced, each sequence in turn is removed and realigned to the remaining alignment. When the sequence is realigned, all details of how it had previously been aligned to the other sequences are removed. This procedure is continued until no further change is observed.

When only a few sequences are being aligned, some comprehensive searching is possible. For $N_{seq}$ sequences there are $2^{N_{seq}-1} - 1$ different ways of splitting the sequences into two groups (see Figure 6.20A). Improvements can be sought from the initial multiple alignment by realigning these groups and selecting the final alignment with the best score. This can be repeated until no further improvement is obtained. When there are too many sequences the same approach can be modified to look at some of the smaller intermediate alignments, or to choose a random subset of the possible splits. One way of limiting the search in a rational way is to restrict the splits to those present in the guide tree (see Figure 6.20B).

In the program prrp, Osamu Gotoh has added a further level of iteration for the case of weighted-sequence sum-of-pairs scoring. Since the weights are derived from an initial alignment of sequences, it is more consistent to reevaluate the weights with the new multiple alignment, and then proceed to a new alignment. This program also tries to identify regions of the alignment that are more reliable than others and are thus expected to be locally optimal. Leaving these out of the iterative procedure effectively reduces the sequence length to be aligned and makes the program more efficient.

## 6.5 Other Ways of Obtaining Multiple Alignments

The previous section presented methods of constructing multiple sequence alignments based on gradual addition of sequences to a growing multiple alignment. Here we discuss two alternative techniques (see Flow Diagram 6.5). The first of these is the program DIALIGN, which focuses on short ungapped alignments, from which the complete alignment is built. The second, SAGA, uses a genetic programming technique to optimize the complete alignment while always considering all sequences.

### The multiple sequence alignment program DIALIGN aligns ungapped blocks

We saw in Section 5.3 that exploring ungapped local alignments, a technique used in FASTA and BLAST database searches, can generate useful pairwise alignments. The basic concept of the DIALIGN multiple alignment method is that the complete alignment can be constructed from such local alignments between pairs of sequences. The name DIALIGN derives from the fact that these ungapped alignments show up as diagonals in a dynamic programming matrix (see Figure 4.3 where diagonals are

**Flow Diagram 6.5**
**The key concept introduced in this section is that it is also possible to obtain multiple sequence alignments by methods that do not involve constructing pairwise sequence alignments at any stage.**

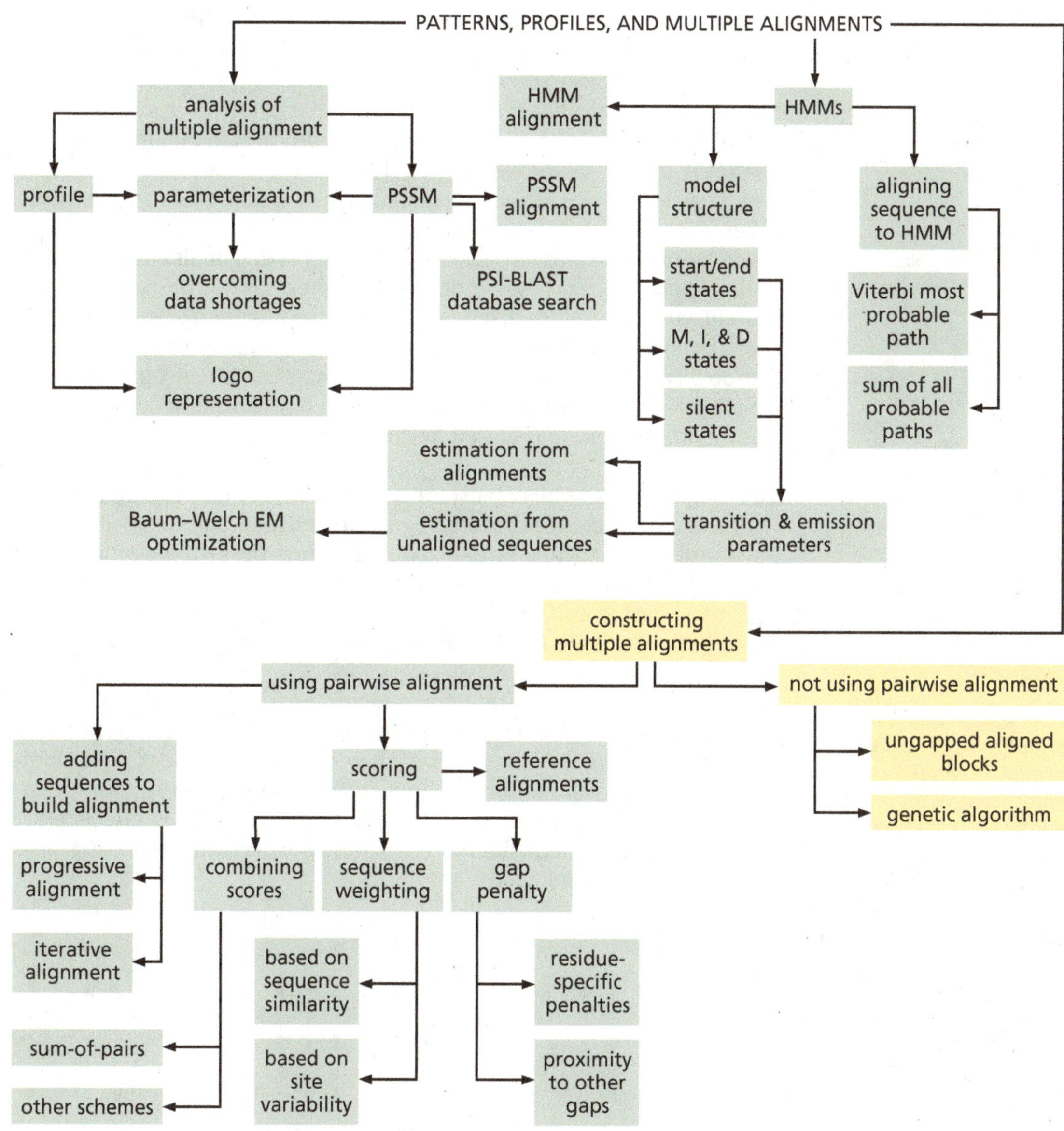

shown), and we will refer to such alignments as **diagonals** in this context. The method is unusual in that regions of a sequence that do not form part of a diagonal are not considered part of the final multiple alignment. For comparison, recall that residues emitted by the insert states of profile HMMs are also not regarded as aligned. Two stages are involved in producing the final alignment. First, all possible pairs of sequences are aligned to determine the set of all diagonals, and to assign weights to diagonals. These are then built up into a multiple alignment, adding consistent diagonals in order of decreasing weight. As low-scoring diagonals are often not included in the initially constructed alignment, the program iteratively explores all remaining unaligned regions in an attempt to include them whenever possible.

All possible diagonals between each pair of sequences are considered. For a diagonal $D$ of length $l$ residues, a score $s$ is obtained by adding the individual scores for each position in the alignment obtained from a standard substitution matrix. Two of the key requirements of this method are that each diagonal has an associated weight, and that diagonals of different lengths can be compared in a meaningful way. To obtain the weighting of a diagonal, we consider two random sequences of the same length as the sequences under consideration. One can calculate the probability that any diagonal of length $l$ in these random sequences has a score at least as large as $s$, $P(l,s)$. The weight of the diagonal is then given by $w(D) = -\log(P(l,s))$. A diagonal with an average score will have a weight of approximately 0. By including the lengths in the weighting scheme, diagonals of different lengths can be compared. This is necessary for the next stage of the alignment.

The pairwise weighting scheme described so far ignores a crucial aspect of biologically meaningful multiple alignments. These should contain regions that are conserved in several if not all of the sequences. From the separate pairwise alignments, such regions will produce a set of diagonals that overlap in a consistent way with each other. Such sets of diagonals are almost certainly an important component of the correct multiple alignment, and the weighting of diagonals can be biased towards this group. To do this, every pair of diagonals in the set is examined to find overlaps. These overlaps are themselves diagonals that can be assigned a weight. The weight of the overlap diagonal is then added to the existing weights of the two original diagonals. The details of weight calculation are different, but the concept is quite similar to that described in the library extension of the COFFEE scoring scheme (see Figure 6.17). In this way the weighting is biased toward overlapping diagonals.

The diagonals must be consistent with each other (see Figure 6.21). To clarify this concept, if the residues that are aligned in two diagonals are listed together and the pairs ordered according to one sequence, then the two diagonals are consistent if the residues of the other sequence are simultaneously in order. One further necessary condition is that no residue occurs more than once in the list, as this would mean it had been aligned to two residues on the other sequence.

Once all possible pairs of sequences have been examined, the large set of diagonals accumulated is used to construct the multiple alignment. We start including diagonals by choosing them in order of their weights. In general, not all diagonals can be used, as they will not form a single consistent set (see Figures 6.21A and 6.21C). It is necessary to find a subset of the diagonals that is consistent. Every time another diagonal is considered it must be checked to ensure that it is consistent with those already selected.

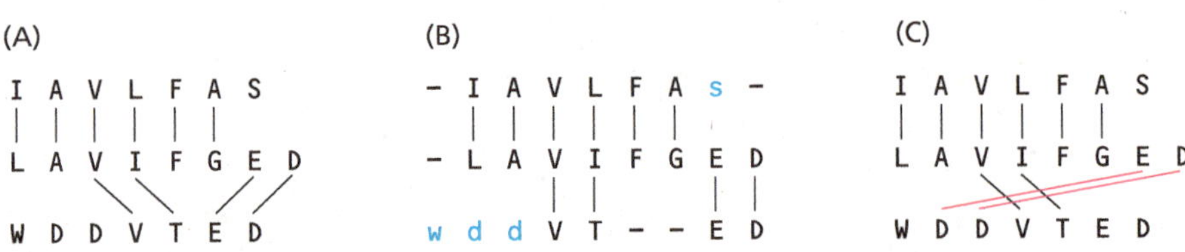


**Figure 6.21**
**The use of diagonals in the DIALIGN method, and the concept of a consistent set.** (A) All three diagonals shown are consistent with each other, leading to a possible alignment shown in (B) with those residues that are not included in the alignment (and thus are unaligned) shown in blue. (C) Another set of diagonals for the same three sequences, but the diagonal shown in red is inconsistent with the two other diagonals. The three diagonals cannot be present together in an alignment.

Once all diagonals have been considered for addition to the growing alignment, some sequence regions may remain unaligned. These regions are now realigned in pairs to generate a new set of diagonals that can be considered for addition to the existing alignment. Usually only two such iterations are needed to produce the final alignment. Any residues that have not been involved in any of the diagonals used to construct the alignment are regarded as unaligned (see Figure 6.21B).

The availability of a growing number of genome sequences has led to a need for multiple alignments of very large sequences. Some techniques that can deal with these problems have been discussed in Section 5.5. Michael Brudno and colleagues, finding DIALIGN too slow for such tasks, developed the CHAOS (CHains Of Seeds) algorithm to rapidly identify some unambiguous short local alignments (seeds) that allow the sequences to be divided into smaller regions. DIALIGN can align the remaining smaller regions sufficiently quickly that the combination can align three 1 Mb sequences in a few hours.

## The SAGA method of multiple alignment uses a genetic algorithm

Cedric Notredame and Desmond Higgins took a completely different approach to the multiple alignment problem and proposed using a **genetic algorithm** to find the optimal alignment. In their method, called SAGA (Sequence Alignment by Genetic Algorithm), a collection of possible alignments is modified according to a set of rules to form a new collection. The set of alignments is referred to as a **generation**, and in its normal implementation there are 100 alignments in each generation. Several hundred successive generations may be produced en route to the final alignment. Subsequent generations are designed to gradually improve the alignment scores until an optimum is found.

Three basic procedures are used to create the next generation. First, a subset of the current generation of alignments is selected to be passed on to the next generation. Subsequently, this subset of alignments is used to create new members of the next generation. An important last step is to ensure that all the alignments are different, as SAGA does not allow duplication of an alignment in any generation. These procedures will now be considered in detail.

Once a new generation has been formed it must be scored. Each multiple alignment can be scored using scoring methods similar to those discussed previously. However, SAGA is less restricted in the scoring schemes allowed than most of the other methods considered, and can also use the COFFEE scoring function described above.

After each multiple alignment in the generation has been scored, the next step is to select the subset that will survive to the next generation. Alignments are chosen for survival directly on the basis of their scores: the highest-scoring alignments survive. Typically, half the new generation is formed by survival from the preceding one and the other half is derived by selected modification (breeding). This modification process uses either one or two of the surviving alignments to create one new alignment. Not all the surviving alignments are used in the breeding process, however, while some alignments are used more than once. The surviving alignments are assigned an expected number of offspring (EO), typically between 0 and 2. This number is based on the alignment scores, where better scores give more offspring. Alignments are selected for breeding at random with a probability proportional to their EO. If an alignment is selected for breeding, its EO is reduced by 1 in subsequent random selections.

Breeding involves taking one (sometimes two) parent alignments and modifying them according to a randomly chosen method referred to as an operator. SAGA has 22 different operators that, initially, are equally likely to be selected. The operators

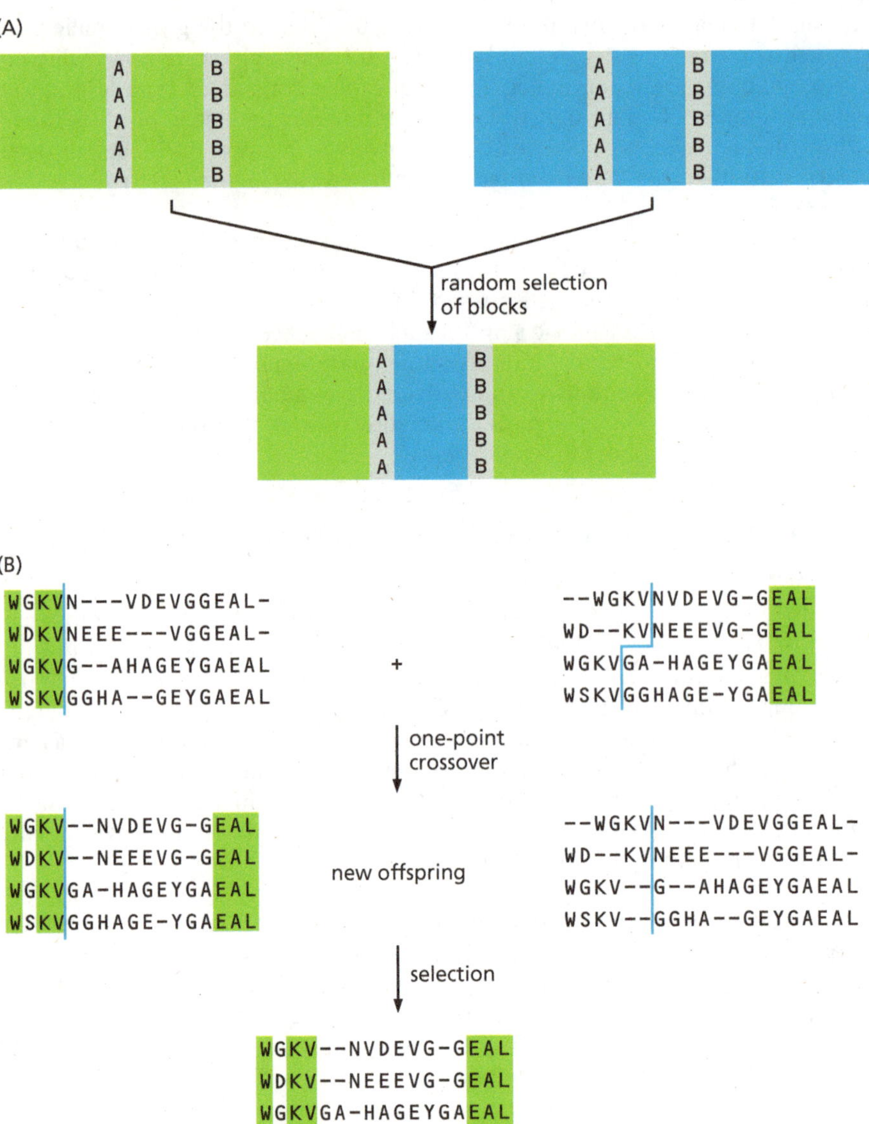


**Figure 6.22**
**Illustration of the crossover operator in SAGA and the two ways in which it can create the offspring of the next generation.** (A) The uniform crossover operator. The two parent alignments in green and blue are consistent between columns A and B. That means that columns A and B are the same in the two alignments, so that the central regions correspond to the same residues of each sequence. An offspring alignment is produced by randomly choosing three colored blocks from the parents. (B) The one-point crossover operator. One parent alignment, in this case the one on the left, is divided into two parts by a straight cut between two alignment columns. The other parent is divided into two parts in such a way that the sequences are separated at the equivalent positions. Two alternative offspring can be constructed from these, of which only the higher-scoring one is kept. (B, adapted from C. Notredame and D.G. Higgins, SAGA: sequence alignment by genetic algorithm, *N.A.R.* 24:1515–1524, 1996.)

used in SAGA can be divided into those that require two parents as input, producing two offspring of which the better-scoring one is kept, and those that take one parent and produce a single offspring.

An example of this process is shown in Figure 6.22. The "crossover operator" takes two alignments and combines parts from each to create two offspring. It can combine alignments in two different ways, referred to as uniform (see Figure 6.22A) and one-point (see Figure 6.22B). The uniform method searches for two alignment columns that are identical in the two parents and then swaps the region between them. In the one-point method a point is randomly selected in one parent at which the alignment is cut into two, preserving all the alignment columns. The other parent is cut in such a way that each sequence is split as in the first, and gaps are inserted to repair any jagged edges. The halves from different parents can now be combined, producing two offspring, of which the better-scoring one is kept. Of these two methods, the uniform method has been found to be most successful at deriving good offspring.

The "gap insertion operator" is an example of a single parent operator, shown in Figure 6.23. In this case a crude phylogenetic tree is constructed from the parent,

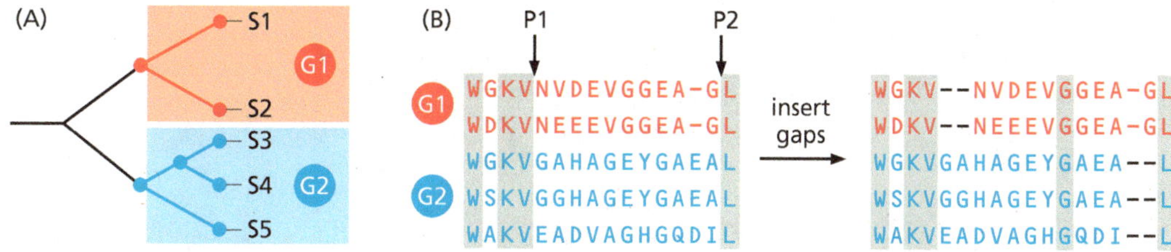


**Figure 6.23**
**Illustration of the gap insertion operator as defined in the SAGA alignment method.** A crude phylogenetic tree is estimated for a parent generation alignment (A) and is randomly split into two subtrees, giving two sets of sequences (G1 and G2). (B) Two positions are randomly chosen in the alignment (P1 and P2) and a gap of random length (length 2 in this case) is inserted at position P1 in the G1 set of sequences, while the same length gap is inserted at position P2 in the G2 set of sequences.

and a split is chosen from this tree. A random-length gap is then inserted at the same, randomly chosen place in all the sequences of one split. The sequences of the other split receive a gap of the same length at a nearby location.

Other operators move small blocks of residues or gaps in a random or partially optimal way, and another group of operators performs a limited local optimal alignment. Some of these operators make considerable efforts to obtain good offspring by local optimization, but without the random operators the process would get stuck in poor local minima. Both forms of operator are necessary for SAGA to work efficiently and effectively.

The resultant offspring alignment is accepted into the new generation as long as it is not a duplicate of one already present. As different operators are found to vary in their efficiency at producing new offspring, the weightings used in randomly selecting operators are modified to favor success. Once the new generation is complete it can be scored and compared to the previous generation. The process is terminated when no improvement in score has occurred for a given number of generations.

All this requires initial construction of the first generation of sequence alignments. For this, a simple procedure suffices in which each sequence is offset a random number of residues from the amino terminus. The offsets are commonly up to a quarter of the sequence length. This procedure is followed for each member of the first generation making sure that there are no alignment duplicates.

Once it has been decided that the run has converged, the best-scoring multiple alignment can be reported as the final result. However, there may be many other alignments in the last generation that have potentially useful information. Other high-scoring alignments may suggest useful alternatives in some regions, and where they are in agreement can also be used as a measure of confidence in the common features.

## 6.6 Sequence Pattern Discovery

So far we have considered alignments and profiles that involve either the complete sequence of a protein or (in the case of local alignments) a substantial region such as a complete protein domain. Shorter segments will probably not score sufficiently highly to be regarded as significant. (See the discussion in Section 5.4 on score significance.) However, examination of multiple alignments of large numbers of related sequences shows that high conservation tends to occur only for short stretches of sequence. These sequences often have a significant role in stabilizing important protein structural features or are actively involved in the protein's function. These shorter regions are often very useful for finding or confirming putative members of protein families that are highly diverged from other members, as in such cases the global alignment score is often too low to be confident of family assignment.

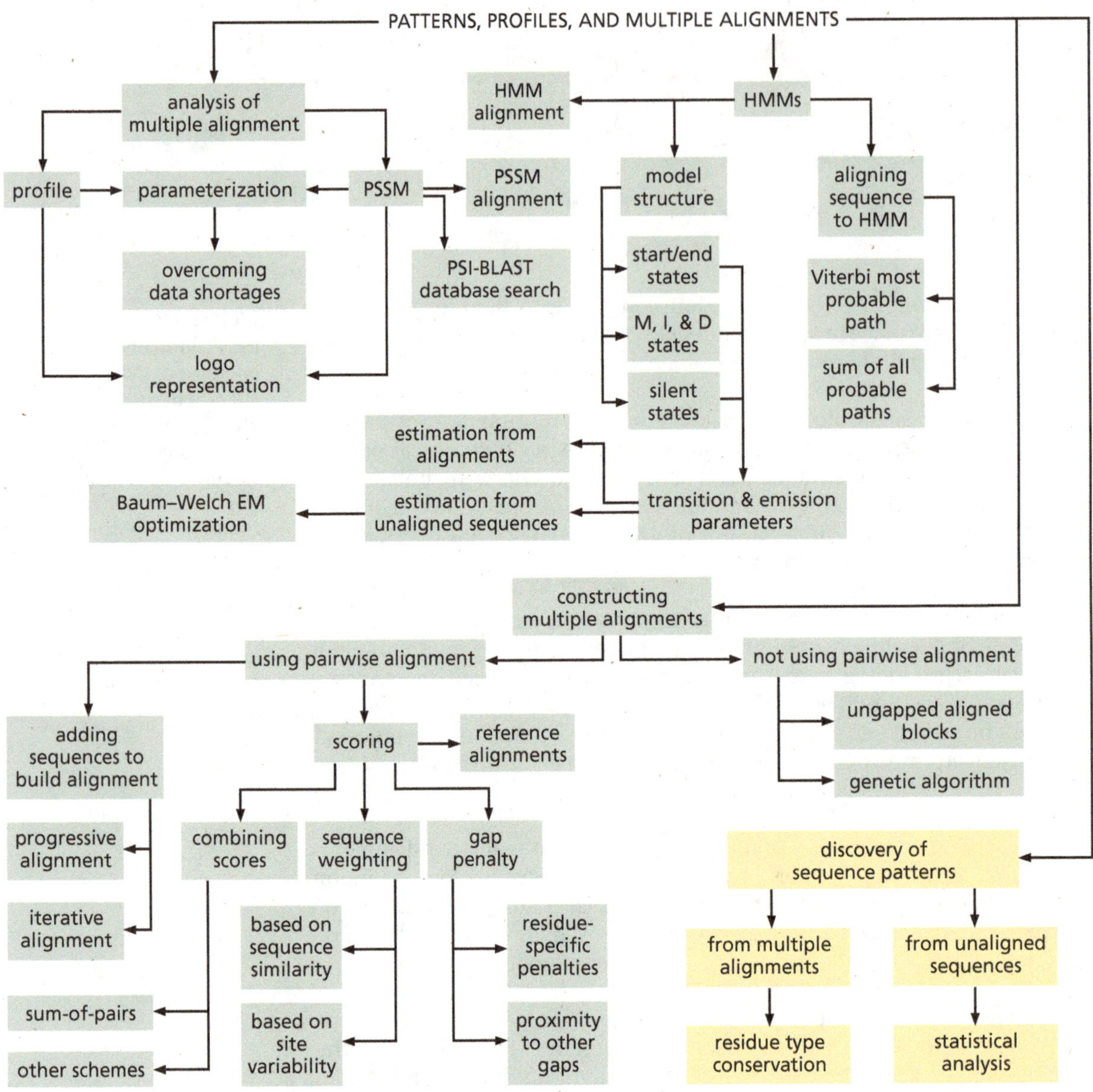


**Flow Diagram 6.6**
**The key concept introduced in this section is that a variety of methods exist for identifying frequently occurring sequence patterns within a set of sequences, even when the sequences have not been previously aligned.**

The ideal sequence pattern for such predictive purposes will occur only in a specific protein family or subfamily, or only in sequences with a particular function. Such a pattern often has high predictive value for the protein function. If a pattern is too highly specified, many sequences of the family or with the function of interest will be missed because their equivalent region will differ slightly from the defined pattern. Such sequences are often referred to as **false negatives** for that pattern. At the other extreme are patterns that detect all the desired sequences but at the expense of obtaining many **false positives** (sequences with that pattern that do not belong to the family). Very few patterns are perfect in detecting all the correct sequences and no others.

Here we will cover some of the ways in which short, highly conserved sequence patterns can be discovered from a set of sequences (see Flow Diagram 6.6). The techniques described fall into two distinct groups. The first of these takes a multiple alignment previously generated, and analyzes it for conserved patterns. The alternative is to search all sequences for a particular pattern, leading if desired to a partial multiple alignment based on the positions of the patterns found. The first group of methods should find any pattern that happens to be present, regardless of its form. The second group of methods requires searching with a comprehensive set of patterns, and therefore requires more computer resources, but has the advantage of not relying on the accuracy of the alignment. Furthermore, some of the second group of methods can locate repeated patterns in the same sequence. However, the set of patterns searched for will of necessity be incomplete, creating the possibility of not discovering a pattern in the sequence.

Before describing these techniques in detail it is useful to look briefly at how patterns can be described. In Section 6.1 the logo representation based on information theory was presented. This is just as useful for the analyses of short sequence patterns as it is for profiles. An example logo is shown in Figure 6.5. The other description of patterns that we will explore gives a representation that is extremely helpful in defining patterns in text, and is therefore the form used by most programs.

As useful patterns are expected to have highly conserved positions with a very limited number of allowed residues, a representation not too dissimilar to that of a basic sequence is often possible. This can, moreover, be used to represent insertion variability, and so is a useful way of writing sequence patterns. Three sequential fully conserved residues are represented as, for example, GYT, sometimes written G-Y-T to distinguish the three positions. If several residues can occur at a given position, they are represented as, for example [YWF] as in G-[YWF]-T. Note that there is no further detail about relative preferences for each residue type. Thus positions 27–29 of the logo of Figure 6.5 can be written as [G]-[DE]-[GSA]. This scheme, with a few extensions, is used to define many of the PROSITE database entries (see Table 4.2).

Finally, it should be noted that there are two separate tasks in this area, of which only one will be described in this section. Here we will survey some of the methods used for identifying sequence patterns. Often, as well as producing a pattern-finding program using the algorithm, a complementary program is provided to locate discovered patterns in other sequences. For example, the MAST program will search for patterns discovered by MEME, and scan does the same job for Gibbs. Practical examples of pattern searching are described in Sections 4.8 to 4.10.

## Discovering patterns in a multiple alignment: eMOTIF and AACC

From an analysis of two databases of multiple alignments, Craig Nevill-Manning and co-workers derived a number of sets of amino acids that they refer to as **substitution groups**. The sets contain residues that are often found to substitute for one another, but in addition, the residues of one set were found to substitute for those of another set only rarely. A total of 20 such sets were found in common to both databases and are used in the program eMOTIF. These 20 substitution groups are shown in Figure 6.24A. Note that, in addition, there are the 20 individual residues to consider, making a total of 40 groups.

eMOTIF analyzes every position in a given multiple alignment, initially simply determining the single substitution group that covers all the observed residues. If none of the 20 groups gives full coverage, the position is assigned the group of all residues (written "." in the discussion of MOTIF). At each position one can define allowable groups, namely those that define subsets of the initial group.

eMOTIF examines the patterns that can be defined by choosing all combinations of allowable groups at the different positions. For each such pattern, the number of sequences that contain it is readily obtained, giving an estimate of the coverage of the family by this pattern. (In some cases a lack of full coverage may still lead to useful patterns that distinguish between recognized subsets of the protein family; that is, subfamilies.) A lower limit is set to the percentage of sequences that contain the

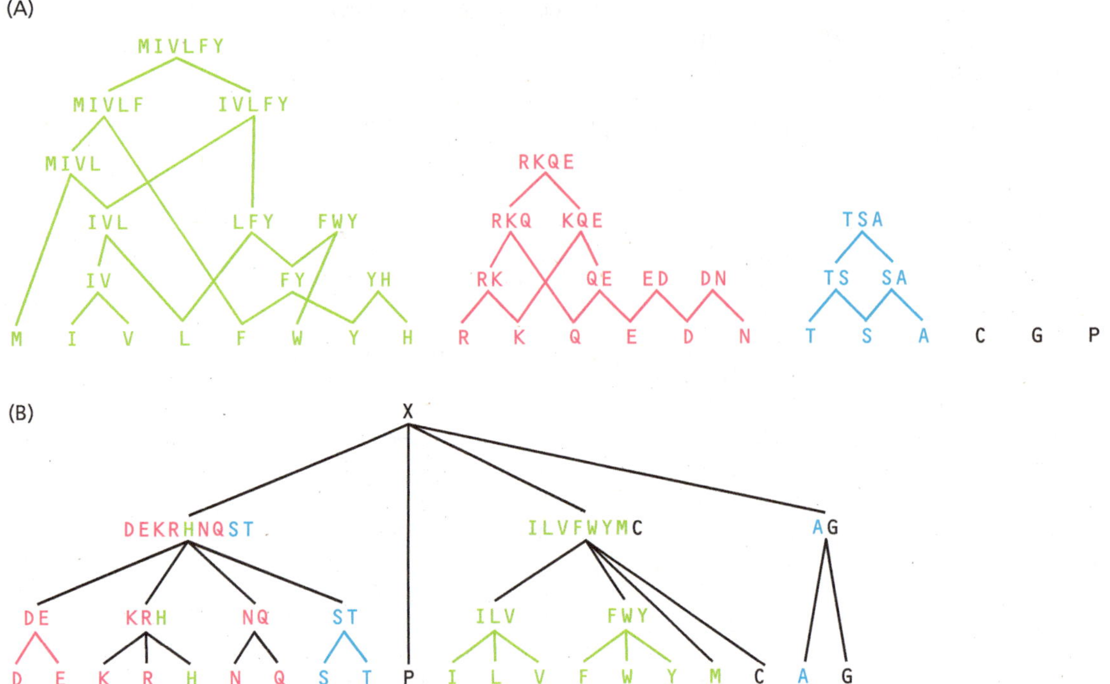


**Figure 6.24**
**Amino acid residue groupings used to determine conserved residue types in alignments during searches for sequence patterns.** (A) The groupings used in eMOTIF. (B) The groupings used in AACC. The residue letters are colored as in A to aid comparison of these alternative groupings.

pattern, the default being 30%. This reduces the number of patterns to be explored, speeding up the program and also limiting the number of false negatives. The probability of each pattern occurring by random is estimated using the observed residue frequencies in a large sequence database. This leads directly to a prediction of the number of false positives in a database search. These two calculations permit some discrimination between potentially useful patterns. Those that are predicted to give rise to fewer false positives for the same level of cover are always favored.

One way of improving the ability to identify the desired sequences from sequence patterns involves the use of several patterns in conjunction. First, one determines a pattern that is highly specific for a subset of the sequences. These are then removed, and another pattern obtained that fits a further subset of sequences with high specificity. In this way a set of patterns is found, which together can distinguish the sequence members with a low error rate. Often these patterns are found to separate the sequences into subfamilies that have some biological significance.

Randal Smith and Temple Smith developed an alternative pattern-derivation method that also uses aligned sequences. The technique, called AACC (Amino Acid Class Covering), uses an alternative set of residue groupings, as shown in Figure 6.24B, and is a modification of the progressive alignment techniques described earlier in this chapter. The sequences are arranged using a guide tree and pairwise alignment is initiated. At each stage, once the alignment has been obtained by dynamic programming, the two sequences are condensed to one. If a position has identical residues, or residues from the same grouping, in each sequence it is simply represented by that residue or residue group. If the residues or groups differ, the smallest group that contains both residues is used to represent the position. In this way, internal nodes of the guide tree are assigned a single sequence that represents those of the lower leaves. When the root node has been reached, its sequence will contain the pattern common to all the sequences.

This method is very straightforward, but has two main disadvantages compared with eMOTIF. First, unless all the sequences contain very similar patterns, none will be found. A single distantly related sequence could prevent AACC reporting a pattern shared by almost all the other sequences. Second, because no analysis is made of the likelihood of the pattern occurring by random chance, it is possible that the reported pattern will result in many false positives in a database search.

## Probabilistic searching for common patterns in sequences: Gibbs and MEME

Finding patterns in an already existing multiple alignment is relatively fast and easy, but if there are errors in the alignment, there is no possibility of correction. So although searching unaligned sequences for a given pattern is more computationally intensive, techniques based on this approach are preferred. The difficulty lies in the large number of different patterns that need to be explored. One way to restrict the number of possible patterns is to consider only patterns that occur in a short contiguous stretch of sequence. This reduces the problem sufficiently so that quite sophisticated methods can be applied by comparison with those used when searching for variable-width patterns. Both methods involve estimating the probability of obtaining observed patterns given a pattern model.

The Gibbs program searches for a pattern of length $W$ present in a set of $N_{seq}$ sequences. It uses a probabilistic formalism, distinguishing between background residues, that is, those not in the pattern, and residues at specific positions in the pattern. The model has $(W + 1)$ different preferred residue distributions (that is, residue probabilities), one for each pattern position ($q_{u,a}$ for residue type $a$ to be at the $u$th pattern position) and one for the background ($\rho_a$). The latter is the set of probabilities applicable for any positions not designated in the pattern. The final element of the Gibbs model is the position in each sequence at which the pattern starts: $x_s$ for the $\boldsymbol{x}$th sequence.

In its basic form this method assumes a single occurrence of the pattern in each sequence. If some of the sequences do not contain the pattern, this can reduce the likelihood of finding it, but in practice the method is quite robust. Modifications can be made to allow for a number of negative examples. The model can be further enhanced to find several copies of the same pattern in each sequence, and to find several patterns simultaneously. These more complex models will not be discussed further here, except to say that the number of patterns must be stated in advance.

The search starts with a fixed pattern length of $W$ and a random set of locations $x_s$. This means that the initial pattern is completely random. One sequence $\boldsymbol{y}$ is selected at random and an attempt is made to improve its pattern position as defined by $y_s$. First, new values of $q_{u,a}$ are obtained from the $(N_{seq} - 1)$ other sequences using

$$q_{u,a} = \frac{n_{u,a} + \beta p_a}{N_{\text{seq}} - 1 + \beta} \qquad \text{(EQ6.41)}$$

where $n_{u,a}$ is the number of times residue $a$ is found at position $u$ in the pattern located in the other $(N_{seq} - 1)$ sequences by the parameters $x_s$. The pseudocounts are proportional to the residue frequencies $p_a$ in the whole dataset, and weighted to sum to $\beta$, which is taken as $\sqrt{N}$ in the standard implementation of this method. (Compare this with Equation EQ6.11.) The background frequencies $\rho_a$ are obtained in an analogous fashion with the same pseudocounts. Using these values, the probability can be calculated that any stretch of $W$ residues starting at the $i$th residue $y_i$ in sequence $\boldsymbol{y}$ is or is not the pattern. The likelihood that it is the pattern is obtained using the $q_{u,a}$ to get $Q_{y_i}$, the likelihood that it is not by using the $\rho_a$ to get $P_{y_i}$. Weights are then assigned to each position $y_i$ using $w_{y_i} = Q_{y_i} / P_{y_i}$, and normalizing these values so that they sum to 1 over the whole sequence. The new $y_s$ is chosen by random selection with probabilities given by the normalized $w_{y_i}$. This will be biased toward a region of sequence similar to the patterns defined by the other sequences.

This process is iterated many times (approximately $100N_{seq}$) until convergence has occurred. When several of the $x_s$ correctly locate a pattern, the $w_{y_i}$ will be more likely to favor the correct starting residue. Otherwise, the $x_s$ will explore the sequence essentially at random until eventually by chance the pattern is picked up, hence the large number of iterations.

Using only the step above, the model can become stuck with the sequences correctly aligned for the pattern, but the window of $W$ residues shifted from the correct pattern so as to cover only part of it. To get round this problem, after several iterations an alternative method of finding new $x_s$ is tried. All the $x_s$ are moved $t$ residues from their current positions, where $t$ takes all values positive and negative up to some threshold $T$. At each value of $t$ an equivalent of $w_{y_i}$ is calculated, normalized over the $(2T + 1)$ alternatives, and a random selection made to obtain the new set of $x_s$.

So far we have taken the pattern width $W$ as a given constant. We could simply run the program several times with different values. The Gibbs program can try to estimate $W$ by calculating a quantity that is proportional to the quantity of information per parameter of the model. This allows an objective estimation of the pattern width as that which has the most information per model parameter.

The program MEME uses many similar concepts to Gibbs, such as the residue probabilities, but uses the EM (expectation maximization) method (see Further Reading). In the program Gibbs, the parameter $x_s$ points at the position of the pattern in sequence $\boldsymbol{x}$ of length $L_{\boldsymbol{x}}$. In MEME, each sequence position $i$ that can be the start of a pattern of length $W$ (that is, residues 1 to $L_{\boldsymbol{x}} - W + 1$) has a probability proportional to $w'_{x_i}$ actually being the start of the pattern. This is not identical to the $w_{y_i}$ in Gibbs as will become clear below.

One of the parameters in MEME is the expected number of appearances of a pattern in the sequences, $N_{\text{exp}}$. This allows MEME to work with many negative examples, when $N_{\text{exp}}$ will be less than the number of sequences $N_{seq}$, and will simply define the pattern on the basis of the $N_{\text{exp}}$ estimated occurrences. The same parameter can also be used when there is more than one pattern instance expected in each sequence.

MEME systematically uses each sub-sequence of length $W$ in the set of sequences to provide an initial estimate of the pattern residue probabilities $q_{u,a}$. A modified form of Equation EQ6.41 is used, the difference being that the pseudocounts are often derived using a Dirichlet distribution. Using these, the probability of obtaining the sequence of length $W$ starting at position $i$ in sequence $\boldsymbol{x}$ as a pattern $w'_{x_i}$ is obtained by simple multiplication of the $q_{u,a}$. By analogy with the $w_{y_i}$, the $w'_{x_i}$ must be normalized, but in this case the normalization is such that the sum of the $w'_{x_i}$ over all sequences is $N_{\text{exp}}$.

Some additional modification of the $w'_{x_i}$ is required. First, as they represent probability, they must have values between 0 and 1. Second, long repeating patterns such as AAAAAA could lead to many consecutive $w'_{x_i}$ having values close to 1, when in fact they all refer to the same instance of the pattern. To circumvent this situation, the $w'_{x_i}$ are also constrained to sum to a maximum of 1 over any $W$ consecutive sites. From these $w'_{x_i}$ the $q_{u,a}$ can be reestimated, and these can be used to calculate the log likelihood of the model given the data. This is an example of the expectation maximization (EM) optimization method, and can be iterated to convergence, defined in this case to be a negligible change in the $q_{u,a}$.

MEME carries out a single EM iteration for every length $W$ sub-sequence in the sequences, and finds the model with the highest log likelihood. This model is then iterated to convergence, at which point the largest $w'_{x_i}$ will predict the presence of instances of the pattern, now defined by the converged $q_{u,a}$. These patterns can now be probabilistically erased by weighting those sequence positions by $(1 - w'_{x_i})$ after which the method can be restarted to locate other patterns. In this way, MEME can find several patterns in the same dataset. MEME searches for patterns of varying widths by restarting the algorithm with different values of *W* and by a pattern expansion step. The expansion step attempts to add (or remove) columns at the edges of the motif. An objective function that approximates the *E*-value of the motif is used to choose the best motif. This function considers the width of the motif, number of occurrences of the motif, and information content of the motif, as well as the size of the training set.

## Searching for more general sequence patterns

The methods described in the previous section will find a pattern that has several highly conserved residues in a short sequence segment. In many cases, patterns have been found that have a few highly conserved residues but these are separated by positions that have little or no residue preference, and the highly conserved positions are a small minority of the whole pattern segment. This can cause problems for methods such as Gibbs and MEME by reducing the distinction between the pattern and the background. The methods described next have been designed to search specifically for such patterns.

The MOTIF program searches for patterns of the form X–*x*1–Y–*x*2–Z where X, Y, and Z are particular residues and *x*1, *x*2 are unspecified intermediate sequence segments of length between zero and some specified maximum (often around 20). Such patterns are often called 3-patterns or, in general, *d*-patterns, where there are *d* positions at which a single residue occurs, including the first and last residues of the segment. Searches are limited in practice owing to the very large number of different patterns possible. For example, given that there are 20 different amino acids there are 1.8 million possible different 3-patterns with intervening stretches of up to 15 residues.

MOTIF scans the sequences for 3-patterns, scoring them in a very large array. Attention is focused on those patterns found more often than expected on the basis of a statistical model. Motifs with more than three conserved positions will result in several 3-patterns; for example, the pattern A–B–C–D will be observed as A–B–C, A–.–C–D, A–B–.–D, and B–C–D. Some of these are filtered out before analyzing the data according to the number of observed instances of each 3-pattern. A 3-pattern that passes this test is used to align the sequences in the region of the potential pattern, aligned according to the observed 3-patterns. A region of 30–55 residues centered on the pattern is realigned crudely (not allowing insertions). The final definition of the pattern is given by finding all columns with fully conserved residues, all columns with more than half the sequences having the same residue at that position, and all positions whose mean sum-of-pairs score with the PAM250 matrix (see Figure 5.3) is positive.

MOTIF uses a very inefficient technique, because the sequences from which the pattern is to be found contain far fewer patterns than are possible. If the method were being used to locate patterns in 12 sequences each 250 residues long, each sequence would have (250 – 10 + 1), that is, 241, different sub-sequences of length 10. In total, there will be 2892 sub-sequences. For longer patterns this number decreases slowly. For 2-patterns with up to eight intervening residues, there are 3600 different 2-patterns up to 10 residues long. As seen above, the number of 3-patterns is very large by comparison. From a computational perspective, therefore, rather than searching the sequences for a set of patterns it is more efficient to discover common patterns from the sub-sequences.

Given the range of possible patterns, this is still difficult to do efficiently. The PRATT method is perhaps the most advanced yet. It is efficient enough to allow searches of patterns of up to 50 residues long, consisting of up to 10 specified positions. Other positions are completely unspecified, and the length between two specified positions can be variable (within limitations). Patterns must occur in the sequences a specified minimum number of times. All these patterns are scored according to a measure of their information content. This ensures the patterns with more specified residues are preferred. Otherwise, more general patterns may be found more frequently and then given greater significance.

Attempting to identify residue preferences in those positions previously undefined can further refine the highest-scoring patterns. In addition, positions after the carboxy-terminal end of the pattern are analyzed to try to extend it. The allowed residue combinations are restricted as for eMOTIF and AACC, but they are fully under user control; they are a user-defined list.

## Summary

This chapter has dealt with the many ways in which a set of related sequences can be analyzed to create a multiple alignment or to identify the key common features, even short conserved patterns. It has also dealt with ways that sets of sequences can be compared by aligning their alignments or profiles.

A multiple alignment can be analyzed to determine the degree of conservation at any position. This leads to the concepts of sequence profiles and position-specific scoring matrices. These tools are useful for discovering related sequences in database searches, because they can represent the general characteristics of a sequence family better than a single example. The PSI-BLAST database search method is one of the most powerful of these methods available.

A more sophisticated way of representing the properties of a sequence family is to use a profile hidden Markov model. These models are based on sound probabilistic principles, and can represent insertions and deletions in a consistent way. This comes at a cost, however, as HMMs require numerous parameters to be specified. Nevertheless, HMMs are perhaps the most sensitive way available to represent a protein family.

Once a profile representation is available for a set of sequences, either as a PSSM or a profile HMM, it can be used to find further related sequences, or to compare with other profiles to explore the relationship between protein families.

Many techniques have been proposed for constructing multiple alignments. Most are closely related to the pairwise alignment techniques described in Chapters 4 and 5, using dynamic programming as the basic step. Other approaches have been proposed, including the use of HMMs. Some more sophisticated scoring schemes are possible, because the protein structures are assumed to be related, leading to predictions about regions of sequence insertion and conservation. The majority of multiple alignment techniques have to choose the order in which the sequences are added. Some make only one attempt to construct the alignment using this order while others make more extensive searches to find the optimal alignment. A few methods have been proposed that are not based on pairwise alignments, of which two were given here: one based on sequence blocks and the other on genetic programming.

If there are very highly conserved regions in a multiple alignment, it can be relatively easy to propose a sequence pattern that is a useful indicator of membership of the family. Often, however, the pattern is subtle, and methods have been proposed to search for these patterns. Some analyze multiple alignments, but in difficult cases these may contain errors that obscure the pattern. The best pattern-finding methods do not require alignments and can search for quite complex patterns involving variable-length insertions and partially conserved positions.

# Further Reading

The literature in this field is vast, necessitating careful selection of methods for presentation here. We have attempted to cover the range of applications and theoretical techniques without exhaustively listing all variations and implementations. More complete presentations in specific areas are given by:

Durbin R, Eddy S, Krogh A & Mitchison G (1998) Biological Sequence Analysis. Cambridge: Cambridge University Press. (*This is an excellent introduction to the use of HMM techniques in this area.*)

Gotoh O (1999) Multiple Sequence Alignment: Algorithms and Applications. *Adv. Biophys.* 36, 159–206.

Higgins D & Taylor WR (eds) (2000) Bioinformatics. Sequence, Structure and Databanks. Oxford: Oxford University Press. (*Especially Chapters 3, 4, 5, & 7.*)

The references for those methods specifically referred to in this chapter are as follows:

## 6.1 Profiles and Sequence Logos


Altschul SF, Carroll RJ & Lipman DJ (1989) Weights for data related by a tree. *J. Mol. Biol.* 207, 647–653.

Altschul SF, Madden TL, Schäffer AA et al. (1997) Gapped BLAST and PSI-BLAST: a new generation of protein database search programs. *Nucleic Acids Res.* 25, 3389–3402.

Gribskov M, McLachlan AD & Eisenberg D (1987) Profile analysis: Detection of distantly related proteins. *Proc. Natl Acad. Sci.USA* 84, 4355–4358.

Henikoff S & Henikoff JG (1994) Position-based sequence weights. *J. Mol. Biol.* 243, 574–578.

Henikoff S & Henikoff JG (1997) Embedding strategies for effective use of information from multiple sequence alignments. *Protein Sci.* 6, 698–705.

Luthy R, Xenarios I & Bucher P (1994) Improving the sensitivity of the sequence profile method. *Protein Sci.* 3, 139–146.

Schäffer AA, Aravind L, Madden TL et al. (2001) Improving the accuracy of PSI-BLAST protein database searches with compositions-based statistics and other refinements. *Nucleic Acids Res.* 29, 2994–3005.

Schneider TD & Stephens RM (1990) Sequence logos: a new way to display consensus sequences. *Nucleic Acids Res.* 18, 6097–6100.

Schneider TD, Stormo GD, Gold L & Ehrenfeucht A (1986) Information content of binding sites on nucleotide sequences. *J. Mol. Biol.* 188, 415–431.

Sibbald PR & Argos P (1990) Weighting aligned protein or nucleic acid sequences to correct for unequal representation. *J. Mol. Biol.* 216, 813–818.

Sjölander K, Karplus K, Brown M et al. (1996) Dirichlet mixtures: a method for improved detection of weak but significant protein sequence homology. *Comput. Appl. Biosci.* 12, 327–345.

Tatusov RL, Altschul SF & Koonin EV (1994) Detection of conserved segments in proteins: Iterative scanning of sequence databases with alignment blocks. *Proc. Natl Acad. Sci. USA* 91, 12091–12095.


## 6.2 Profile Hidden Markov Models


Baldi P, Chauvin Y, Hunkapiller T & McClure MA (1994) Hidden Markov models of biological primary sequence information. *Proc. Natl Acad. Sci. USA* 91, 1059–1063.

Barrett C, Hughey R & Karplus K (1997) Scoring hidden Markov models. *Comput. Appl. Biosci.* 13, 191–199.

Eddy SR (1998) Profile hidden Markov models. *Bioinformatics* 14, 755–763.

Karchin R & Hughey R (1998) Weighting hidden Markov models for maximum discrimination. *Bioinformatics* 14, 772–782.


Two more advanced HMM models that include the phylogeny explicitly are:


Holmes I & Bruno WJ (2001) Evolutionary HMMs: a Bayesian approach to multiple alignment. *Bioinformatics* 17, 803–820.

Qian B & Goldstein RA (2003) Detecting distant homologs using phylogenetic tree-based HMMs. *Proteins* 52, 446–453.


## 6.3 Aligning Profiles


Edgar RC & Sjölander K (2004) COACH: profile–profile alignment of protein families using hidden Markov models. *Bioinformatics* 20, 1309–1318.

Pietrokovski S (1996) Searching databases of conserved sequence regions by aligning protein multiple-alignments. *Nucleic Acids Res.* 24, 3836–3845. (*LAMA*)

Sadreyev R & Grishin N (2003) COMPASS: A tool for comparison of multiple protein alignments with assessment of statistical significance. *J. Mol. Biol.* 326, 317–336.

Schuster-Bockler B & Bateman A (2005) Visualizing profile–profile alignment: pairwise HMM logos. *Bioinformatics* 21, 2912–2913.

Söding J (2005) Protein homology detection by HMM–HMM comparison. *Bioinformatics* 21, 951–960. (*HHsearch*)

Wang G & Dunbrack RL (2004) Scoring profile-to-profile sequence alignments. *Protein Sci.* 13, 1612–1626.

Yona G & Levitt M (2002) Within the twilight zone: A sensitive profile-profile comparison tool based on information theory. *J. Mol. Biol.* 315, 1257–1275. (*prof_sim*)


## 6.4 Multiple Sequence Alignments by Gradual Sequence Additions


Brudno M, Chapman M, Göttgens B et al. (2003) Fast and sensitive multiple alignment of large genomic sequences. *BMC Bioinformatics* 4, 66–77. (*CHAOS and DIALIGN*)

Chenna R, Sugawara H, Koike T et al. (2003) Multiple sequence alignment with the Clustal series of programs. *Nucleic Acids Res.* 31, 3497–3500.

Do CB, Mahabhashyam MSP, Brudno M & Batzoglou S (2005) ProbCons: Probabilistic consistency-based multiple sequence alignment. *Genome Res.* 15, 330–340.

Edgar RC (2004) MUSCLE: multiple sequence alignment with high accuracy and high throughput. *Nucleic Acids Res.* 32, 1792–1797.

Edgar RC (2004) MUSCLE: a multiple sequence alignment with reduced time and space complexity. *BMC Bioinformatics* 5, 113.

Gupta SK, Kececioglu J & Schäffer AA (1995) Improving the practical space and time efficiency of the shortest-paths approach to sum-of-pairs multiple sequence alignment. *J. Comput. Biol.* 2, 459–472. (*MSA*)

Higgins DW, Thompson JD & Gibson TJ (1996) Using CLUSTAL for multiple sequence alignments. *Methods Enzymol.* 266, 383–402. (*ClustalW*)

Katoh K, Misawa K, Kuma K & Miyata T (2002) MAFFT: a novel method for rapid multiple sequence alignment based on fast Fourier transform. *Nucleic Acids Res.* 30, 3059–3066.

Katoh K, Kuma K, Toh H & Miyata T (2005) MAFFT version 5: improvement in accuracy of multiple sequence alignment. *Nucleic Acids Res.* 33, 511–518.

Notredame C, Holm L & Higgins DG (1998) COFFEE: an objective function for multiple sequence alignments. *Bioinformatics* 14, 407–422.

Notredame C, Higgins DG & Herringa J (2000) T-Coffee: A novel method for fast and accurate multiple sequence alignment. *J. Mol. Biol.* 302, 205–217.

O'Sullivan O, Suhre K, Abergel C et al. (2004) 3DCoffee: Combining protein sequences and structures within multiple sequence alignments. *J. Mol. Biol.* 340, 385–395.

Thompson JD, Higgins DW & Gibson TJ (1994) CLUSTAL W: improving the sensitivity of progressive multiple sequence alignment through sequence weighting, position-specific gap penalties and weight matrix choice. *Nucleic Acids Res.* 22, 4673–4680.

## 6.5 Other Ways of Obtaining Multiple Alignments

Morgenstern B (1999) DIALIGN 2: Improvement of the segment-to-segment approach to multiple sequence alignment. *Bioinformatics* 15, 211–218.

Morgenstern B, Frech K, Dress A & Werner T (1998) DIALIGN: Finding local similarities by multiple sequence alignment. *Bioinformatics* 14, 290–294.

Notredame C &, Higgins DG (1996) SAGA: sequence alignment by genetic algorithm. *Nucleic Acids Res.* 24, 1515–1524.

## 6.6 Sequence Pattern Discovery

Bailey TL & Elkan C (1995) Unsupervised learning of multiple motifs in biopolymers using expectation maximization. *Machine Learning* 21, 51–83. (*MEME*)

Jonassen I (1997) Efficient discovery of conserved patterns using a pattern graph. *Comput. Appl. Biosci.* 13, 509–522.

Jonassen I, Collins JF & Higgins DH (1995) Finding flexible patterns in unaligned protein sequences. *Protein Sci.* 4, 1587–1595. (*PRATT*)

Lawrence CE, Altschul SF, Boguski MS et al. (1993) Detecting subtle sequence signals: a Gibbs sampling strategy for multiple alignment. *Science* 262, 208–214. (*Gibbs*)

Nevill-Manning CG, Wu TD &, Brutlag DL (1998) Highly specific protein sequence motifs for genome analysis. *Proc. Natl Acad. Sci. USA* 95, 5865–5871. (*EMOTIF*)

Smith RF & Smith TF (1990) Automatic generation of primary sequence patterns from sets of related protein sequences. *Proc. Natl Acad. Sci. USA* 87, 118–122. (*AACC*)

Smith HO, Annau TM & Chandrasegaran S (1990) Finding sequence motifs in groups of functionally related proteins. *Proc. Natl Acad. Sci. USA* 87, 826–830. (*MOTIF*)

One area not covered at all in this chapter is the evaluation of the different methods. Several workers have proposed test datasets. Readers interested in this area could start by examining:

Mizuguchi K, Deane CM, Blundell TL & Overington JP (1998) HOMSTRAD: a database of protein structure alignments for homologous families. *Protein Sci.* 7, 2469–2471.

Raghava GPS, Searle SMJ, Audley PC et al. (2003) OXBench: A benchmark for evaluation of protein multiple sequence aligment accuracy. *BMC Bioinformatics* 4, 47–70.

Thompson JD, Plewniak F & Poch O (1999) A comprehensive comparison of multiple sequence alignment programs. *Nucleic Acids Res.* 27, 2682–2690. (*BAliBASE*)

Thompson JD, Plewniak F & Poch O (1999) BAliBASE: a benchmark alignment database for the evaluation of multiple alignment programs. *Bioinformatics* 15, 87–88. (*BAliBASE*)

# PART 3

# EVOLUTIONARY PROCESSES

Evolution is a vital part of the process of life. In general, the changes that give rise to evolution occur in the form of mutations at the level of the genetic sequence but are often observed at the protein level, when the effect of the change is noticeable. This is not always the case, as some mutations at the genetic level will not change the code for the protein and some important mutations may not occur in the regions that code for proteins.

This part of the book consists of two chapters: the first is an Applications Chapter, while the second is a Theory Chapter. The first chapter in this part explains the basic ideas involved in reconstructing the evolutionary history of the gene or protein sequences and illustrates how the methods can be applied to various scientific problems. The second chapter gives details of the techniques involved in evolutionary analysis.

# RECOVERING EVOLUTIONARY HISTORY

7

APPLICATIONS CHAPTER

**When you have read Chapter 7, you should be able to:**

- Discuss the evolutionary basis of sequence variation.
- Show how phylogenetic trees represent evolutionary history.
- Outline basic phylogenetic tree structure.
- Show the consequences of gene duplication for phylogenetic tree construction.
- Summarize the methods of constructing trees from a sequence alignment.
- Select an evolutionary model for use in phylogenetic analysis of sequences.
- Recount practical examples of tree construction.
- Test the support for tree features.
- Propose function from phylogenetic analysis.

When a group of aligned sequences shows significant similarity to each other, as described in Chapters 4 to 6, this can usually be taken as evidence that they are the result of divergent evolution from a common ancestral sequence. In this case, the sequence alignment will contain traces of the evolutionary history of these sequences. It is possible to infer this history by complex analysis of a multiple sequence alignment. In this and the following chapter we will describe the ways in which the evolutionary history can be recovered and used to investigate the evolutionary relationships between present-day sequences.

This type of analysis can be applied to a number of distinct problems. By studying sequences that have both a common ancestor and common function—known as **orthologous sequences** or **orthologs**—from different species, one can investigate the evolutionary relationships between species. These results can usefully be compared with an accepted taxonomic classification, which has probably been derived using different data. Alternatively, by studying a set of sequences from a single protein class we can examine the evolution of functions such as enzymatic activity. Techniques have also been developed that attempt to reconstruct ancestral sequences in order to examine the molecular properties of these hypothetical ancient genes and their products. That aspect of phylogenetic analysis is beyond the scope of this book and those interested should consult the Further Reading at the end of the chapter.

In this chapter we will consider the reconstruction of the evolutionary history of a set of sequences from a multiple alignment, and how this history can be represented as a graphical structure called a **phylogenetic tree**. We start by describing

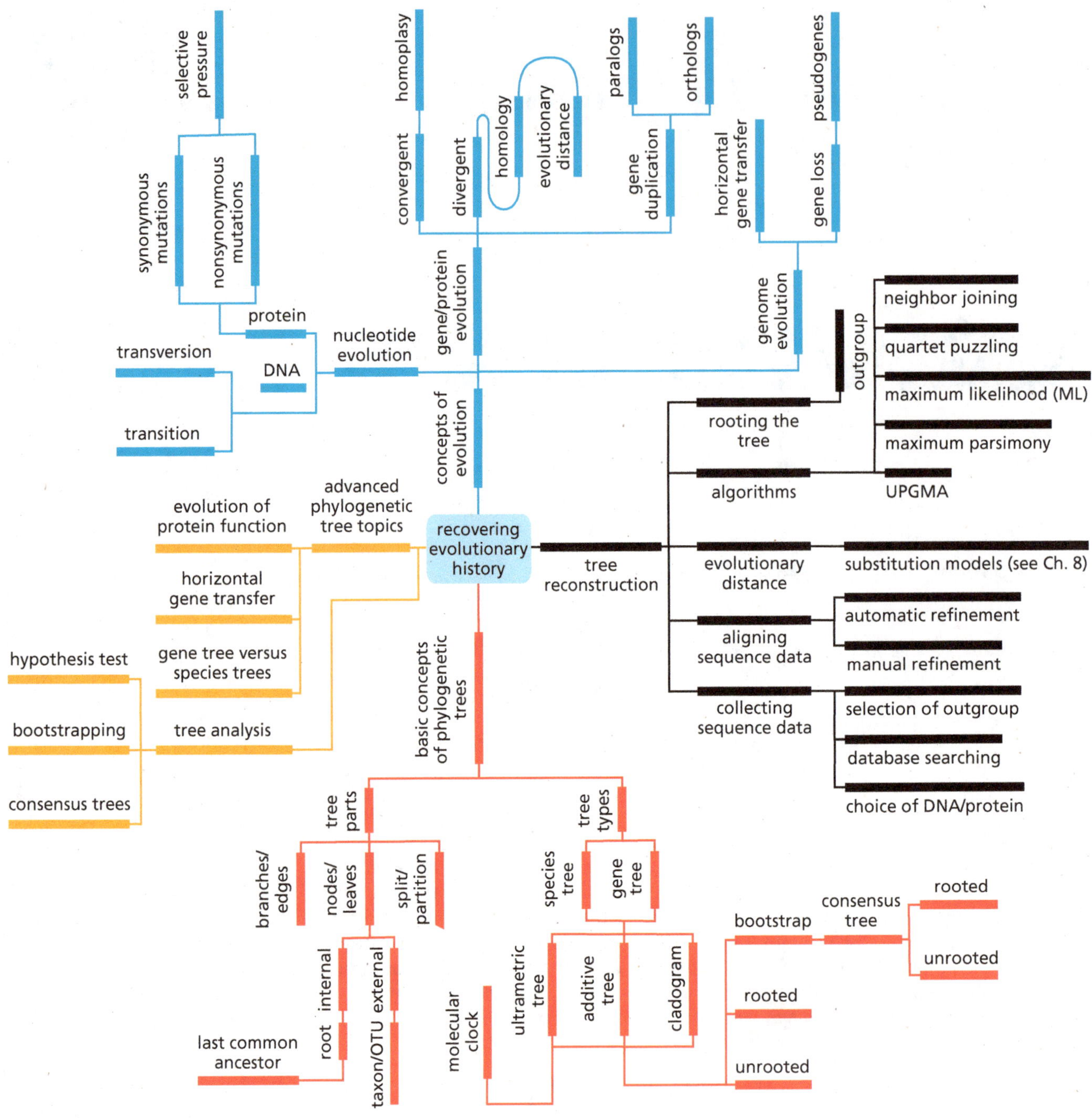


**Mind Map 7.1**
**A mind map showing the major steps that are taken when reconstructing phylogenetic trees.**

the types of trees and their features and how they can be interpreted. The evolutionary history we hope to recover from the data will encompass many different types of mutational events, and we will discuss the relative evolutionary importance of different types of mutation in sequence change. Often, simplifications must be introduced into the analysis, restricting the possible evolutionary history. Understanding these limitations clarifies the kinds of issues that can be explored and the way that answers may be derived.

In Chapters 4 to 6, we examined the relationships between sequences in terms of the degree of similarity of the sequences to each other. When analyzing sequences with evolution in mind, however, it is the differences between them—often summarized as the evolutionary distance or genetic distance—that are of interest and which need to be quantified and scored. Before one can quantify the changes that have occurred during evolution and generate an evolutionary distance, a simplified model of evolution as a framework is needed. Numerous models have been proposed, and a number of them are presented with the emphasis on the assumptions made and the parameters used. As mutation events occur at the nucleotide level, most of the models reflect this. Some models of protein evolution have been proposed similar to those discussed in connection with the derivation of the PAM matrices (see Section 5.1), which consider evolution at the level of amino acid changes.

Many different ways of reconstructing the evolutionary history of a set of sequences from a multiple alignment have been proposed. These can be divided into methods that use the sequence alignment directly and those that use the evolutionary distances. In this chapter we look at the practicalities of selecting an evolutionary model and tree construction method, and how to justify the choice; the detailed algorithmic description and theoretical justifications are considered in Chapter 8.

As the actual relationship of one sequence or species to another is a matter of historical fact, fixed by the course of past evolution, one would hope to recover this true tree whatever the type of data used. But, as we shall see, differences can occur depending on the data available, the particular model of molecular evolution used, and the tree-building method.

## 7.1 The Structure and Interpretation of Phylogenetic Trees

The purpose of the phylogenetic tree representation is to summarize the key aspects of a reconstructed evolutionary history. Even if a multiple alignment contained an unambiguous record of every mutation that had occurred (which is rarely the case, as we shall see), a shorthand method of representing this in an easily understandable way is still needed. In this section we will present the different types of phylogenetic tree and explain how they should be interpreted. We will also define many of the terms used in this area. The section will end with a description of ways in which the reliability and uncertainty of some of the tree features can be indicated (see Flow Diagram 7.1).

### Phylogenetic trees reconstruct evolutionary relationships

A phylogenetic tree is a diagram that proposes an hypothesis for the reconstructed evolutionary relationships between a set of objects—which provide the data from which the tree is constructed. These objects are referred to as the **taxa** (singular **taxon**) or **operational taxonomic units** (**OTUs**) and in phylogenies based on sequence data they are the individual genes or proteins. When orthologous sequences from different species are being used with the aim of determining relationships between species, the taxa are labeled with the species name. Such trees are called **species trees** to distinguish them from those trees that are intended to show the relationships between the genes or proteins in a large gene family. Species trees can also be constructed from data other than sequences, such as the morphological features used in traditional taxonomy, the presence of certain restriction sites in the DNA, or the order of a particular set of genes in the genome. Such data can often be treated, in principle, in much the same way as sequences; with the proviso that differences may be qualitative rather than quantitative. When examining a phylogenetic tree, care should be taken to note exactly what information was used to produce it because, as will be discussed in detail later in this chapter, the evolutionary history of a set of related genes is often not the same as that of the species from which the genes were selected.

The basic features of phylogenetic trees are illustrated in Figure 7.1, using imaginary bird species as the objects or taxa. The species are connected via a set of lines,

**Flow Diagram 7.1**
The key concept introduced in this section is that there are several different types of phylogenetic tree that can be used to represent the evolutionary history of a group of sequences, some of which convey much more quantitative information than others.

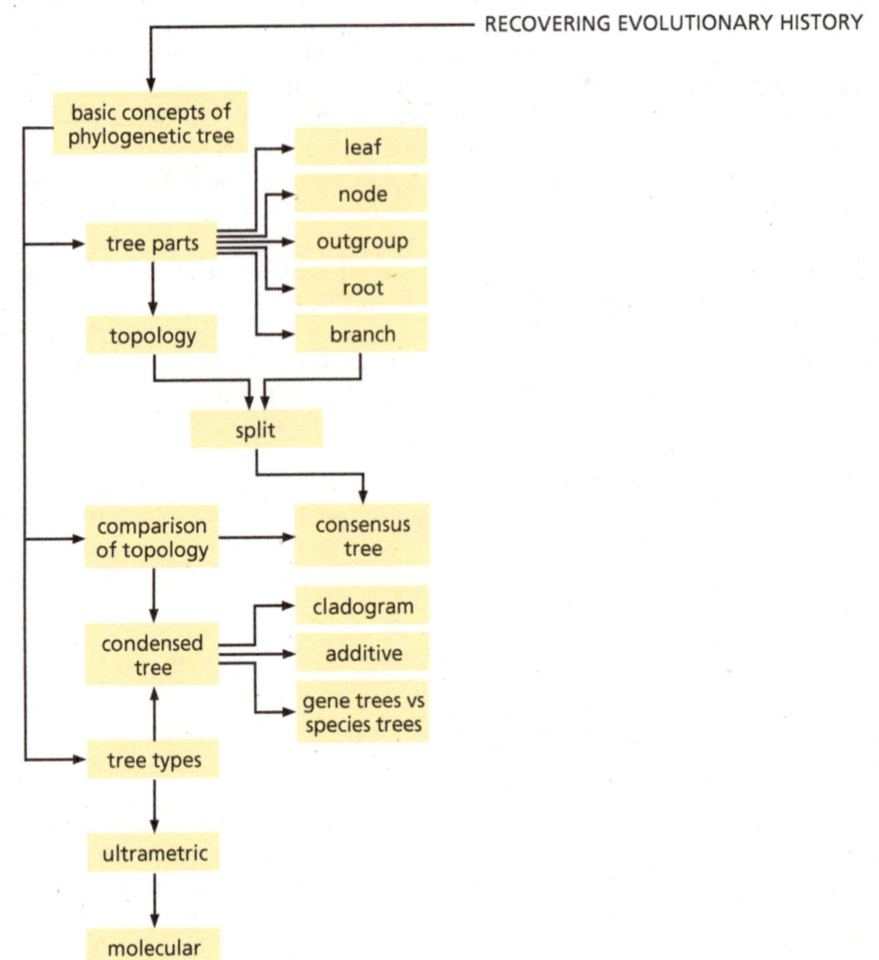


called **branches** or **edges**, which represent the evolutionary relationships between them. Those species that occur at the **external nodes** or **leaves** of the tree are either existing species that have not as yet evolved into new species, or else are extinct species whose lineage died out without leaving any descendants. To be precise, we should say that the dataset used to produce the tree lacks such descendant species, as it is quite possible for the dataset not to contain data from all related species. The branches do not join any of the external nodes together directly, but via **internal nodes,** which represent **ancestral states** that are hypothesized to have occurred during evolution. In our bird species tree, for example, the internal branch points represent **speciation events** that produced two descendant divergent species. Likewise, when sequence data have been used to build a species tree, with each species represented by one sequence, the internal nodes represent the ancestral sequences from which the present-day sequences have diverged after speciation. However, in trees representing gene or protein families, which may contain several sequences from a single species, a branch point can also represent the point at which a gene duplication event has occurred within a genome, followed by the divergence of the two copies within the same species. We shall return to this important point later.

Under normal evolution, a given taxon will have evolved from a single recent ancestor and if it does not become extinct first, may diverge into two separate descendant taxa. As a consequence, any internal node of a phylogenetic tree is expected to have three branches; one to an ancestor and the other two to descendants. This branching pattern is called **bifurcating** or **dichotomous**. The external nodes, which have not yet produced descendants, have a single connection to their

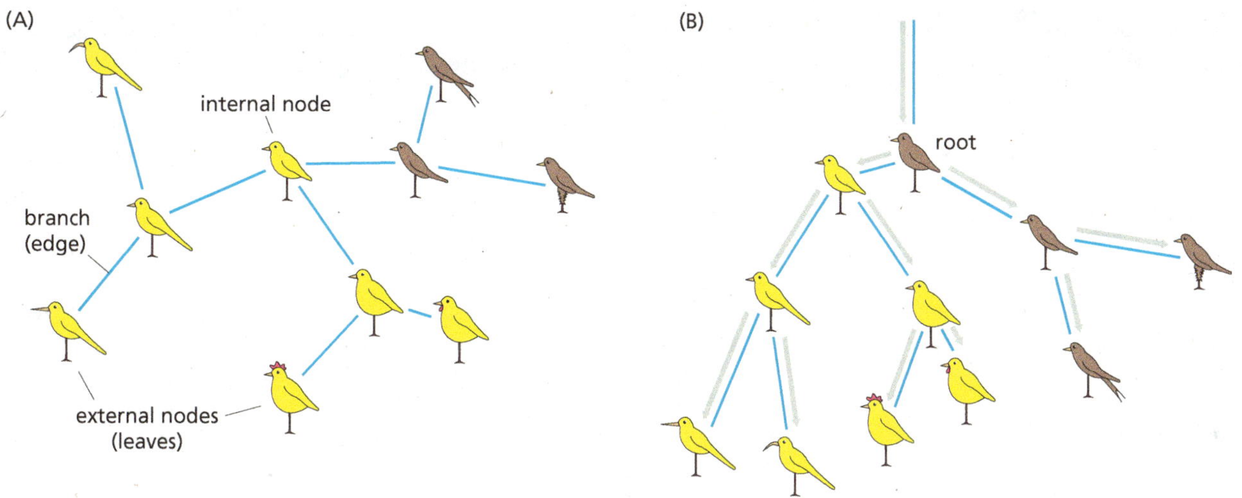


**Figure 7.1**
**Unrooted and rooted phylogenetic trees.** These trees reconstruct the evolutionary history of a set of six imaginary extant bird species, which are shown at the outermost tips of the branches—the external nodes or leaves. These are the species from which the data to construct the tree have been taken. These data could be morphological data or sequence data. The birds shown at the internal nodes are the predicted (reconstructed) extinct ancestor species. (A) A fully resolved unrooted tree. This tree is fully resolved in that each internal node has three branches leading from it, one connecting to the ancestor and two to descendants. However, the direction of evolution along the internal branches—that is, which ancestral species has evolved from which—remains undetermined. Thus we cannot distinguish from this tree alone whether the yellow birds evolved from a brown bird, or vice versa. (B) A fully resolved rooted tree for the same set of existing species. The brown bird marked "root" can now be distinguished as the last common ancestor of all the yellow and brown birds. The line upward from the root bird indicates where the ancestors of the root bird would be. In a rooted tree, there is a clear timeline (shown as a gray arrow) from the root to the leaves, and it is clear which species has evolved from which. Thus, the yellow birds did evolve from a brown bird. Apart from the region around the root in (B), the two trees are identical in the relationship between the taxa and give the same information. The arrangement of the branches in space is different, but the information in a phylogenetic tree is contained solely in the branch connections and branch lengths.

ancestral taxon. There are very rare evolutionary events that do not obey this rule. Two species can merge into one, as is thought to have occurred when eukaryotic cell ancestors entered into symbiosis with the prokaryotes that eventually became mitochondria and chloroplasts (see Section 1.4). Alternatively, a single population might simultaneously give rise to three or more distinct new species. If all the internal nodes have three branches the tree is said to be **fully resolved**, meaning that it hypothesizes a location for all the expected speciation or gene duplication events. A **partially resolved tree** will have at least one internal node with four or more branches, sometimes described as **multifurcating** or **polytomous**.

Trees can be either unrooted (see Figure 7.1A) or rooted (see Figure 7.1B). **Rooted trees** represent the divergence of a group of related species from their last common ancestor, the **root**, by successive branching events over time. **Unrooted trees**, on the other hand, show the evolutionary relationship between taxa but do not identify the last common ancestor. In a rooted tree the direction of evolution along each of the branches is unambiguous. In an unrooted tree, by contrast, which ancestral species evolved from which is not clear once one gets to the internal branches.

There are two further components to the description of a phylogenetic tree—the **tree topology** or the way it branches, and the branch lengths. In some types of tree there are no defined branch lengths, and in general it is the topology that is the main interest (see Box 7.1). Even for a small number of sequences there are many possible trees with different topologies. Each of these tree topologies represents a possible evolutionary history that differs from any tree with alternative topology in

## Box 7.1 Tree Topology

Tree topology is the organization of the branches that make the tree structure. An individual tree topology is defined by the specific connectivity between the nodes in the tree. At this rather abstract level there are two distinct types of tree topology to be considered: one involves the identification of particular taxa with specific places on the tree (that is, the leaves are labeled with the taxa); the other is solely concerned with the branch structure, the leaves remaining unlabeled. A tree constructed from $N$ currently existing taxa (an $N$-taxa tree) will have $N$ leaves and $N$ external branches. An unrooted, fully resolved $N$-taxa tree will have $N-3$ internal branches and $N-2$ internal nodes. The equivalent numbers for a rooted, fully resolved $N$-taxa tree are $N-2$ and $N-1$, respectively. For a given number of taxa there are many alternative unlabeled tree topologies, each of which could be labeled in many different ways. Some of the methods for reconstructing trees need to sample amongst all of the possible trees and require efficient ways to do this. With the same number of taxa, there are more topologies for rooted trees than unrooted trees. The number of possible topologies increases very quickly with the number of taxa (see Equation EQ8.33); for 10 taxa they exceed 1 million. It is likely that very few of these possible topologies will be a good representation of the data.

the way that major events such as speciation and gene duplication are hypothesized to have occurred. The differences are usually in both event order and ancestry. The major task of phylogenetic tree reconstruction is to identify from the numerous alternatives the topology that best describes the evolution of the data.

Apart from the presence or absence of a root, there are three basic types of tree representation. Figure 7.2A shows a rooted **cladogram**, which shows us the genealogy of the taxa but says nothing about the timing or extent of their divergence. In this type of tree the branch lengths have no meaning, and only the tree topology is defined. The cladogram simply tells us that the four taxa—the four birds—have a common ancestor that initially evolved into two descendants, one the common ancestor of the yellow birds and the other the common ancestor of the brown birds. Although the tree has been drawn with different branch lengths, so that the brown birds appear to have diverged more recently than the yellow birds, it is important to note that this is only for artistic effect, and the branch lengths have no scientific basis. Unlike in Figure 7.1, the ancestors are not shown on the tree, only the branches. This is the normal way of presenting trees, with the ancestors only implied by the internal nodes. Most tree-reconstruction methods do not explicitly predict any properties for the ancestral states, and thus would have no useful representation to put at internal nodes. It is also easier to see the connectivity in this form of tree representation. Because the techniques used to reconstruct trees using sequences almost always give more quantitative information, cladograms are rarely used in sequence analysis.

Figure 7.2B shows an **additive tree** constructed from the same data as Figure 7.2A, in which branch lengths represent a quantitative measure of evolution, such as being proportional to the number of mutations that have occurred. The tree shown here is rooted, but unrooted additive trees can also be made. The evolutionary distance between any two taxa is given by the sum of the lengths of the branches connecting them. This tree contains all the information given by Figure 7.2A, but also tells us that the common ancestor of the yellow birds is a smaller evolutionary distance (3) from the root than is the common ancestor of the brown birds (4). From this representation alone, however, we cannot say for certain that the brown birds diverged from their ancestor at a later time than the yellow birds from their ancestor because we have no information about the rates of mutation along the branches. Evolution from the root to the brown birds' ancestor could have occurred at a much faster rate than to the yellow birds' ancestor. If that were the case, the brown birds could actually have diverged before the yellow birds evolved from their

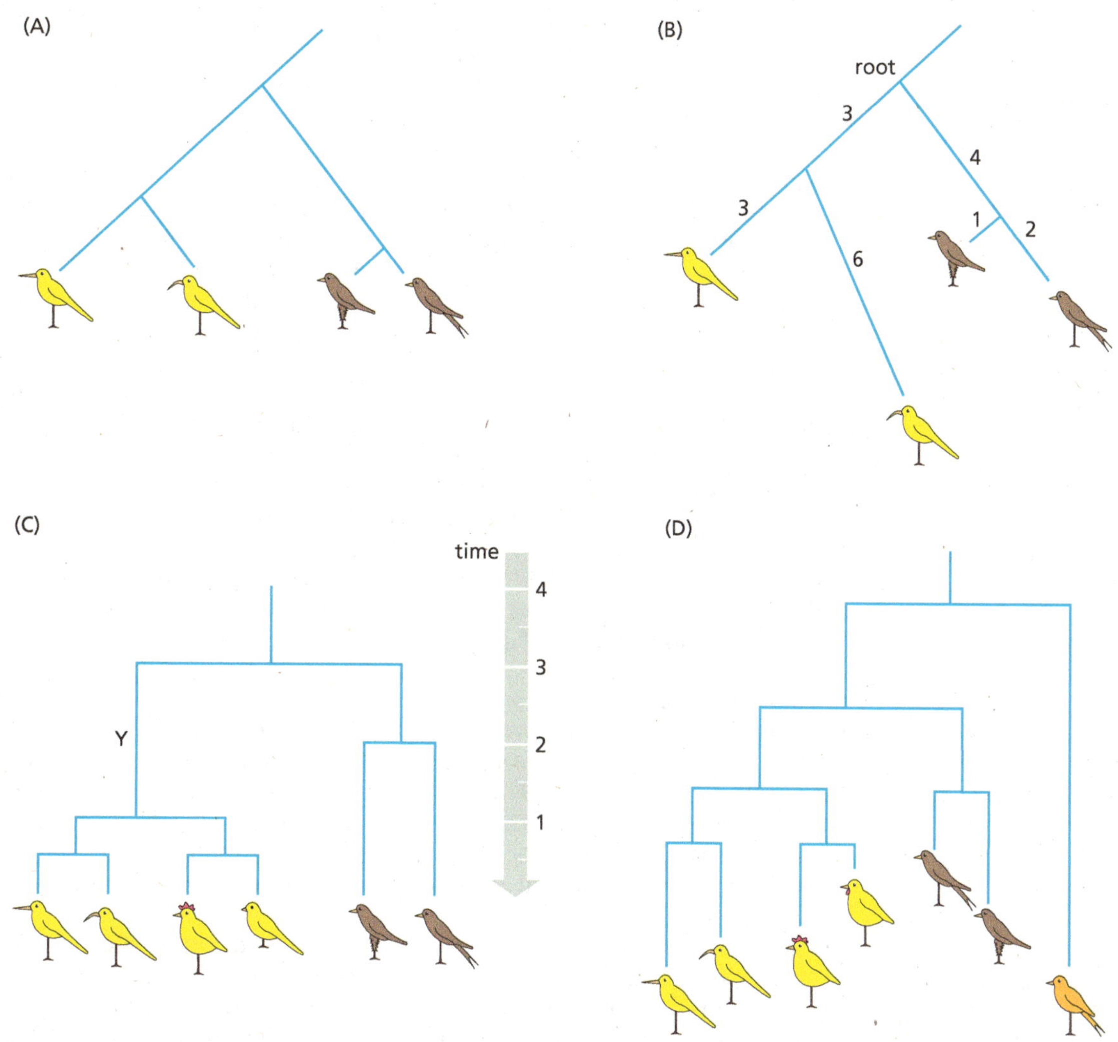


**Figure 7.2**
**Four different rooted phylogenetic trees, illustrating the variety of types of tree.** (A) A cladogram in which branch lengths have no meaning. (B) An additive tree, in which branch lengths are a measure of evolutionary divergence. The branch lengths here are given in arbitrary units proportional to the number of mutations per site, from which we can see that the evolutionary distance between the two yellow birds (3 + 6 = 9) is three times that between the two brown birds (1 + 2 = 3). (C) An ultrametric tree, which in addition to the properties of the additive tree has the same constant rate of mutation assumed along all branches. The scale on the right of the tree is in this case proportional to time as well as to the number of mutations per site. This tree is an ultrametric version of the tree shown in Figure 7.1B. (D) An additive tree for the same set of species as in Figure 7.1B, which has been rooted by the addition of data for a distantly related outgroup (orange bird). Note that in this tree the external nodes are at different distances from the root.

most recent ancestor. But although we cannot make confident assertions about the relative times of events on an additive tree, by adding up the lengths of branches it is possible to make meaningful comparisons of the relative amounts of evolutionary divergence.

Figure 7.2C shows an **ultrametric tree,** which in addition to the properties of the additive tree, has the same constant rate of mutation assumed along all branches. This last property is often referred to as a **molecular clock,** because one can, in principle, measure the actual times of evolutionary events from such trees. All ultrametric trees have a root, and one axis of the tree is directly proportional to time. In our depiction this is the vertical axis, with the present day at the bottom and the last common ancestor at the top. The horizontal lines in Figure 7.2C have no meaning, and are simply used to display the tree in a visually pleasing way on the page. Note that the evolutionary distance from a common ancestor to all its descendants is the same, a condition not usually observed in an additive tree.

Sequence data often do not conform to a molecular clock, however. Non-clock-like sequence evolution probably results from a variety of causes, including changes in mutation rate as a result of changes in evolutionary pressure and increasing constraints on change in an organism's morphological structure and metabolism. If an ultrametric tree is used to represent such data it can lead to an incorrect tree topology as well as incorrect branch lengths. In such cases an additive tree will be more accurate. Unlike ultrametric trees, which are always rooted, additive trees often lack a root. When a rooted tree is required, the most accurate method of placing the root is to include in the dataset a group of homologous sequences from species or genes that are only distantly related to the main set of species or genes under study. This is known as the **outgroup**, while the group of more closely related species are known as the ingroup. Figure 7.2D shows an additive tree constructed using the same species as Figure 7.2C with the addition of the orange bird as an outgroup. The root can then be located between the outgroup and the remainder of the taxa. There are other ways of placing a root if the dataset does not include an outgroup, and these are mentioned later. They are, however, less reliable.

There are many different ways in which a tree can be drawn to illustrate a given set of relationships between taxa. For example, as shown in Figure 7.2D, the additive tree can be drawn in a style similar to that of an ultrametric tree. As is the case for the ultrametric tree, in this representation only one axis is a measure of evolutionary events, now proportional to accepted mutation events instead of time; the other axis is for visual clarity. If as is usually the case there is no constant molecular clock, the leaves will not all occur at the same distance from the root.

There are a number of ways in which a tree can be drawn to show the same information. For example, it is possible to reflect a part of the tree (a **subtree**) about the internal branch connecting it to the rest of the tree. Reflecting the yellow birds' subtree in Figure 7.2C about the branch Y swaps the pair of yellow birds that will be drawn next to the brown birds. However, as long as branch lengths and connectivity are maintained, the trees will have the same meaning. Such changes are often made to bring particular taxa into proximity to illustrate a point more clearly.

It is important to be aware of the distinction between sequence-based phylogenetic trees intended primarily to show evolutionary relationships between species (see Figure 7.3A), and those intended to show relationships between members of a family of homologous genes or proteins from different species (see Figure 7.3B). A gene family tree charts the way in which an ancestral gene has become duplicated and reduplicated within genomes, giving rise to distinct gene subfamilies, and how the copies have diverged within the same species during evolution as well as between species. In gene family trees, some branch points represent gene duplication events within the same species while others represent speciation events.

## Tree topology can be described in several ways

Some tree-construction methods can produce several alternative trees that seem equally good at representing the data. Another occasion when alternative trees can arise is when different methods, models, or parameters are used to analyze the same data. Finally, there are occasions when it is interesting to compare the reconstructed evolutionary history produced by two or more sets of data, such as different genes from the same set of species. We therefore need ways of describing tree topology in a form that makes it possible to compare the topologies of different trees.

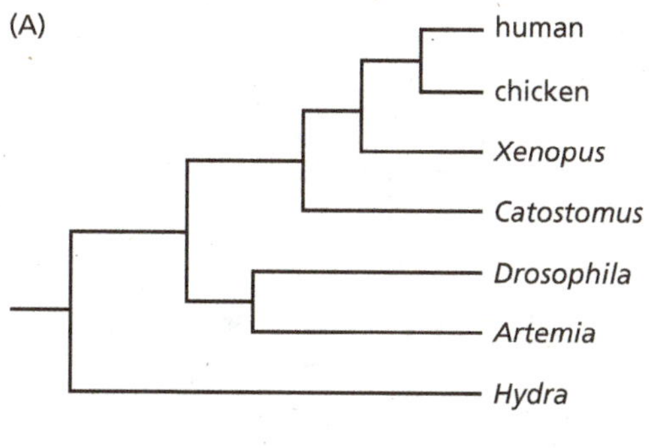


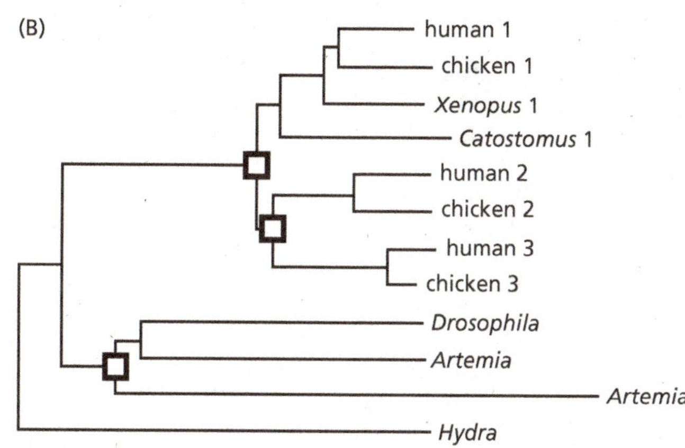


**Figure 7.3**
**Examples of a species tree and a gene tree.** (A) A species tree showing the evolutionary relationships between seven eukaryotes, with one more distantly related to the others (*Hydra*) used as an outgroup to root the tree. *Xenopus* is a frog, *Catostomus* a fish, *Drosophila* a fruit fly, and *Artemia* the brine shrimp. (B) The gene tree for the $Na^+$–$K^+$ ion pump membrane protein family members found in the species shown in (A). In some species, e.g., *Hydra*, and *Xenopus*, only one member of the family is known, whereas other species, such as humans and chickens, have three members. The small squares at nodes indicate gene duplications, discussed in detail in Section 7.2. (Adapted from N. Iwabe et al. Evolution of gene families and relationship with organismal evolution: rapid divergence of tissue-specific genes in the early evolution of chordates. *Mol. Biol. Evol.* 13:483–493, 1996.)

The graphical views of trees are convenient for human visual interpretation, but not for other tasks such as comparison. One way of summarizing basic information about a tree in computer-readable format is to subdivide or split it into a collection of subgroups. Every branch in a tree connects two nodes, and if that branch is removed, the tree is divided into two parts. Such a division is called a **split** or **partition**, and a tree contains as many splits as there are branches. Note that since every split contains two groups, which together make up the entire tree, only one group need be given to define the other. The set of splits formed by removing an external branch are inherently uninteresting because every possible tree for the same set of taxa will produce this set. Splits involving internal branches are more interesting as they can confirm a common origin of a set of taxa. Figure 7.4A shows a tree connecting eight mammalian taxa. By removing the branch with the label X, it can be seen that the sea lion and seal form a group of their own, distinct from the other animals. Figure 7.4B shows the set of internal branch splits for this tree. A fully resolved unrooted tree with $N$ taxa has $N-3$ splits; the equivalent number in the case of a rooted tree is $N-2$.

When splits are calculated for a rooted tree, one of the two groups of taxa will always be **monophyletic**; that is, the group will contain all the descendant taxa from the ancestor represented by the internal node at the end of the cut branch more distant from the root. On occasions, it has been found that some of the groups of organisms traditionally classified together are not monophyletic, suggesting that the classification scheme needs revision. We shall see this in the species tree produced using 16S RNA discussed later in this chapter.

The complete list of splits of a tree can be written in computer-readable form using the **Newick** or **New Hampshire format**. Each split is written as a bracketed list of the taxa as in (sea_lion, seal). (Note the use of the underline character, "sea_lion" not "sea lion," so that a computer does not misunderstand this as two taxa called "sea" and "lion.") A complete tree can be described in a similar fashion that identifies every split. Figure 7.4A can be written as

((raccoon, bear), ((sea_lion, seal), ((monkey, cat), weasel)), dog);

from which all possible splits are identifiable by all groups enclosed by matching parentheses. The semicolon (;) at the end completes the Newick format, indicating that this is the end of the tree.

Given the many graphical ways in which the same tree can be represented through reflection about internal branches, it can be difficult to be certain that two trees do in fact differ only in their aesthetics. In fact, most trees can be represented in many different ways using the Newick format by listing the splits in different orders.

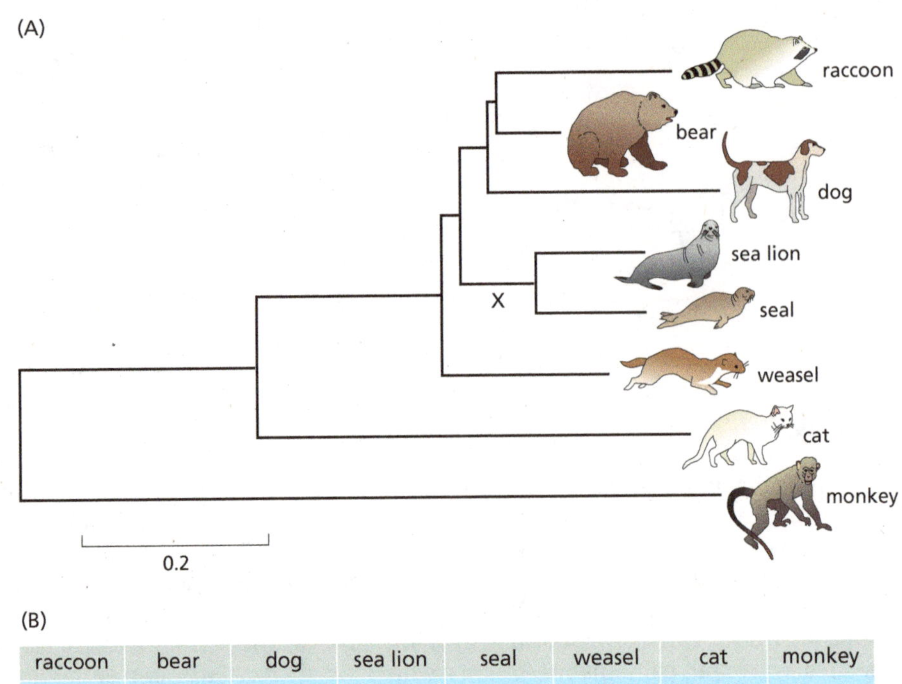


| raccoon | bear | dog | sea lion | seal | weasel | cat | monkey |
|---|---|---|---|---|---|---|---|
| * | * | | | | | | |
| * | * | * | | | | | |
| | | | * | * | | | |
| * | * | * | * | * | | | |
| * | * | * | * | * | * | | |

**Figure 7.4**
**A tree can be represented as a set of splits.** (A) An unrooted additive tree using fictitious data for eight mammalian taxa. The horizontal lines carry the information about evolutionary change; the vertical lines are purely for visual clarity. This common depiction of an unrooted tree is drawn as if the tree were rooted at the midpoint of the distance between the most widely separated taxa; that is, the two taxa connected by the longest total line length. Note that because the tree is unrooted, the branch connecting the monkey to the rest of the tree is not an internal branch. Thus, the evolutionary distance from the monkey to its nearest internal node is represented by the sum of the lengths of the two horizontal lines connected by the leftmost vertical line in the figure. The scale bar refers to branch length and in this case represents a genetic distance of 0.2 mutations per site. (B) A table representing all the possible internal branch splits of the tree shown in (A). The columns correspond to the taxa and the rows to the split. The two groups of each split are shown by labeling the taxa in one group with an asterisk, and leaving the others blank. As every split contains two groups that together make up the entire tree, only one group need be given to define the other.

However, the format is computer-readable making it easier for trees to be compared. More information, such as branch lengths, can be added to this format, so that the tree could be written

((raccoon:0.20, bear:0.07):0.01, ((sea_lion:0.12, seal:0.12):0.08,

((monkey:1.00, cat:0.47):0.20, weasel:0.18):0.02):0.03, dog:0.25);

The branch distances follow colons (:). If you are not used to this format, it is not always easy to see which branch the distances refer to, but looking at Figure 7.4A should make it clearer.

## Consensus and condensed trees report the results of comparing tree topologies

As mentioned above, there are occasions when a set of trees might be produced that are regarded as equally representative of the data. These trees may differ in their topology and branch lengths. Differences in topology imply a disagreement about the speciation and/or gene duplication events, and it is important to quantify these uncertainties. The topology comparison methods are based on the concept of the frequency of occurrence of particular splits in the set of trees, which renders the problem relatively straightforward.

There are two different circumstances where this analysis may prove useful, which differ in the treatment of the trees. When the trees have been produced by several methods of tree construction applied to a single set of data, or by the same method applied to several different datasets, all trees are treated equally in the analysis. In these cases the analysis can identify support across a range of techniques or data

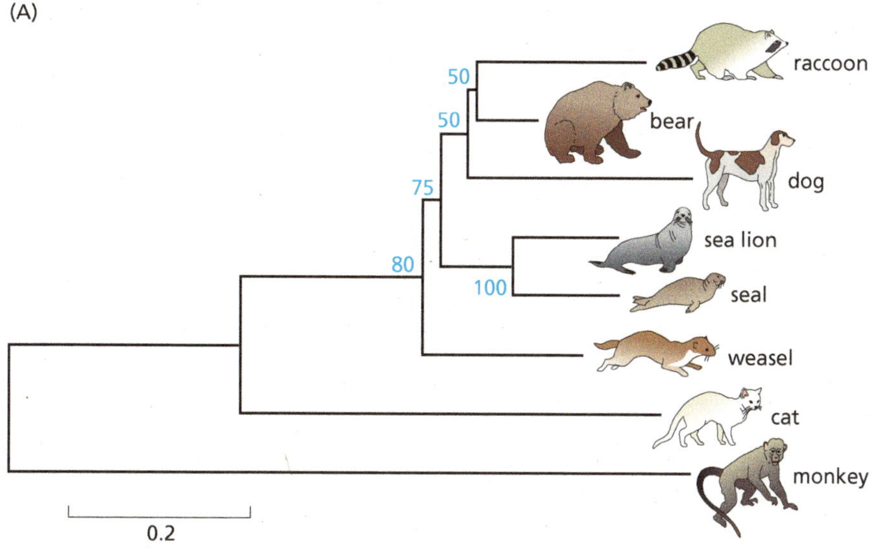


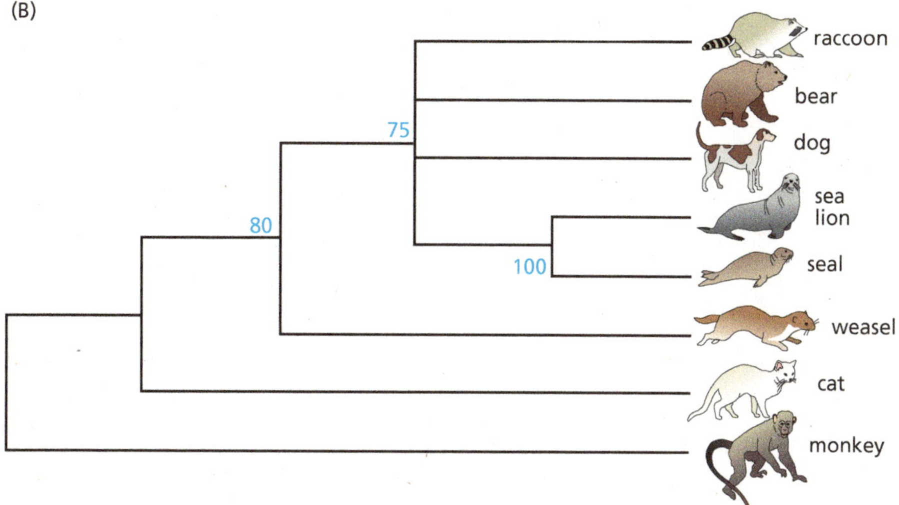


**Figure 7.5**
**A condensed tree showing well-supported features is derived by applying the bootstrap procedure.** The bootstrap procedure assigns values to individual branches that indicate whether their associated splits are well supported by the data. (A) Each internal branch of the tree shown in Figure 7.4 has been given a number that indicates the percentage occurrence of these branches in a bootstrap test. (B) A condensed tree is produced by removing internal branches that are supported by less than 60% of the bootstrap trees. This procedure results in multifurcating nodes. Note that the branch lengths no longer have meaning, so all line lengths are for aesthetic purposes only.

for a given topology. However, the analysis will not indicate the degree of support present in a dataset for particular topological features produced on applying a given tree construction method. **Bootstrap analysis** is designed to estimate this support, based on repeating the tree construction for different samplings of the same dataset. This technique is discussed in more detail in Box 8.4. In this case, during tree topology comparison the tree constructed from the actual data is treated as the standard against which all the other trees are compared.

When bootstrapping is used, all the observed splits of the tree produced from the original data are listed, and each bootstrap tree produced from sampled data is examined to see if it contains the same splits. The percentage of the bootstrap trees that contain each split is either reported in a splits list or displayed on the tree as a number. The tree of Figure 7.4A with bootstrap percentages added to all internal branches is shown in Figure 7.5A. Sometimes, as a visual aid, all internal branches that are not highly supported by the bootstrap are removed. Such a tree is called a **condensed tree**, and if poorly supported branches do occur, when they are removed such a tree will have multifurcating internal nodes. In the case of the tree shown in Figure 7.5A, if all branches with bootstrap support less than 60% are removed, then the raccoon, bear, dog, and common ancestor of the sea lion and seal all diverge from the same internal node (Figure 7.5B). This shows that the data

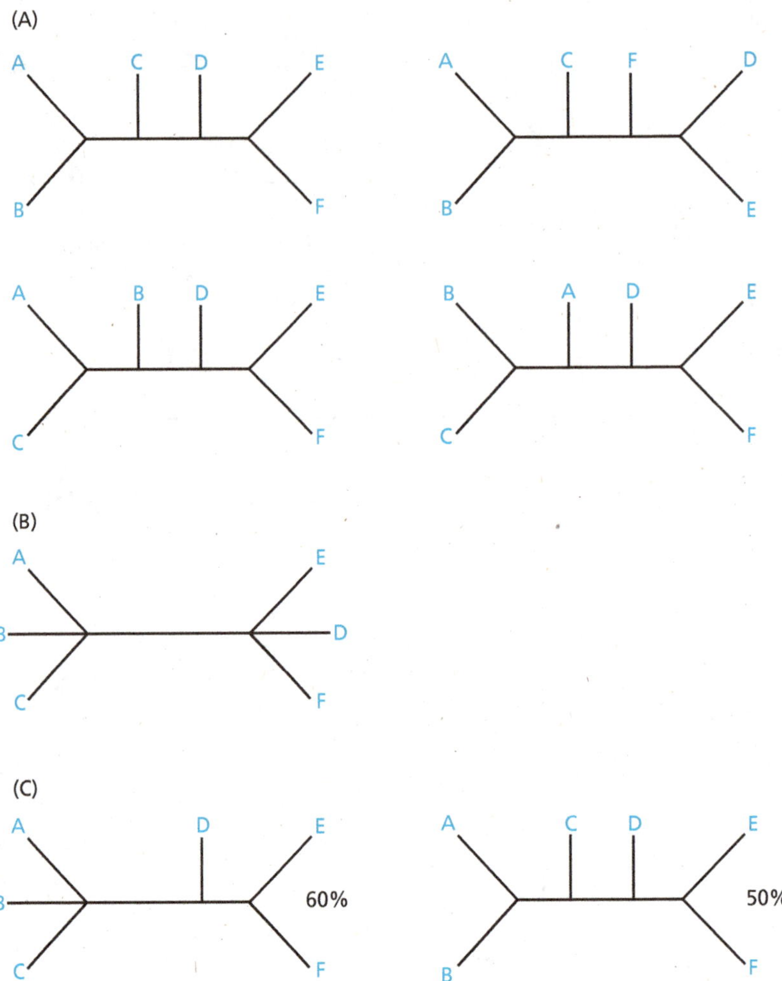


**Figure 7.6**
**Consensus trees show features that are consistent between trees.** Assuming that the four trees in (A) are all equally strongly supported they can be represented by the strict consensus tree shown in (B), in which only splits that occur in all the trees are represented: that is, (A,B,C) and (D,E,F). (C) Majority-rule consensus trees (60% and 50%) for the four trees in (A). The (A,B) split occurs in only 50% of the trees, and thus is not included separately in the 60% consensus tree, whereas the (E,F) split occurs in 75%. The (A,B) split can be included in the 50% tree.

do not show high support for any particular order for the speciation events that gave rise to the raccoon, bear, dog, and seal/sea lion lineages.

When several equally well-supported trees are obtained from the same data, it can be useful to identify in a single tree diagram those features that are always (or frequently) observed. These are known as the **consensus features**. Such a procedure can also be useful when trees obtained using different data are expected to reveal essentially the same evolutionary history. This would be the case when several different genes are used to infer the relationships between a set of species. Even though the trees constructed with the different genes may differ, one can at least identify those features that are shared by all or some trees. For example, certain sequences may always group together, despite the details of their interrelationships differing in different trees, and the grouping itself may be the most informative feature. A useful way of summarizing this information is to calculate a **consensus tree** in which only sufficiently commonly occurring topological features are retained (see Figure 7.6). Trees used to generate a consensus tree do not need to have been generated using the same evolutionary model or tree construction method. Despite looking like a condensed tree, consensus trees are different, in principle, in that the features omitted may well have strong support in some of the individual trees. However, the strong support for the omitted features will be limited to the minority of combinations of data, model, and method that produced those particular features.

There are a number of approaches to generating consensus trees, two of which are particularly commonly used. **Strict consensus trees** only show those topological features that occur in all trees, and are thus fully supported. Thus, any branching-pattern conflicts between trees are resolved by making the relevant sequences all emanate from the same node; that is, the tree has multifurcating nodes (see Figure 7.6B). **Majority-rule consensus trees** allow a degree of disagreement between trees before the branch pattern is removed. An X% majority-rule consensus tree will contain any branch pattern that occurs in at least X% of the trees and is the majority pattern (see Figure 7.6C). In this way, some idea of the strongly supported features of the tree is obtained. In the trees shown in Figure 7.6A, for example, the split (A,B,C) is always observed, and therefore can be considered relatively reliable. There is no established way of determining branch lengths for consensus trees.

## 7.2 Molecular Evolution and its Consequences

To reconstruct the evolutionary history of DNA or protein sequence data, we need to understand some of the principles of molecular evolution and appreciate how molecular evolution is constrained by biological considerations. The Darwinian concept of evolution by natural selection concentrates on the consequences of evolutionary changes for the fitness of the organism: its ability to survive and transmit its genes to the next generation by producing offspring. Fitness depends on the properties of the organism as a whole, and thus change at the DNA sequence level will be constrained by considerations of how it affects protein expression and function, and how these affect cellular properties and whole-organism physiology and behavior. Since the sequences we work with are, by definition, from lineages

**Flow Diagram 7.2**
**The key concept introduced in this section is that there are many different processes that can occur during the evolution of a gene, so that evolutionary history can become extremely complex.**

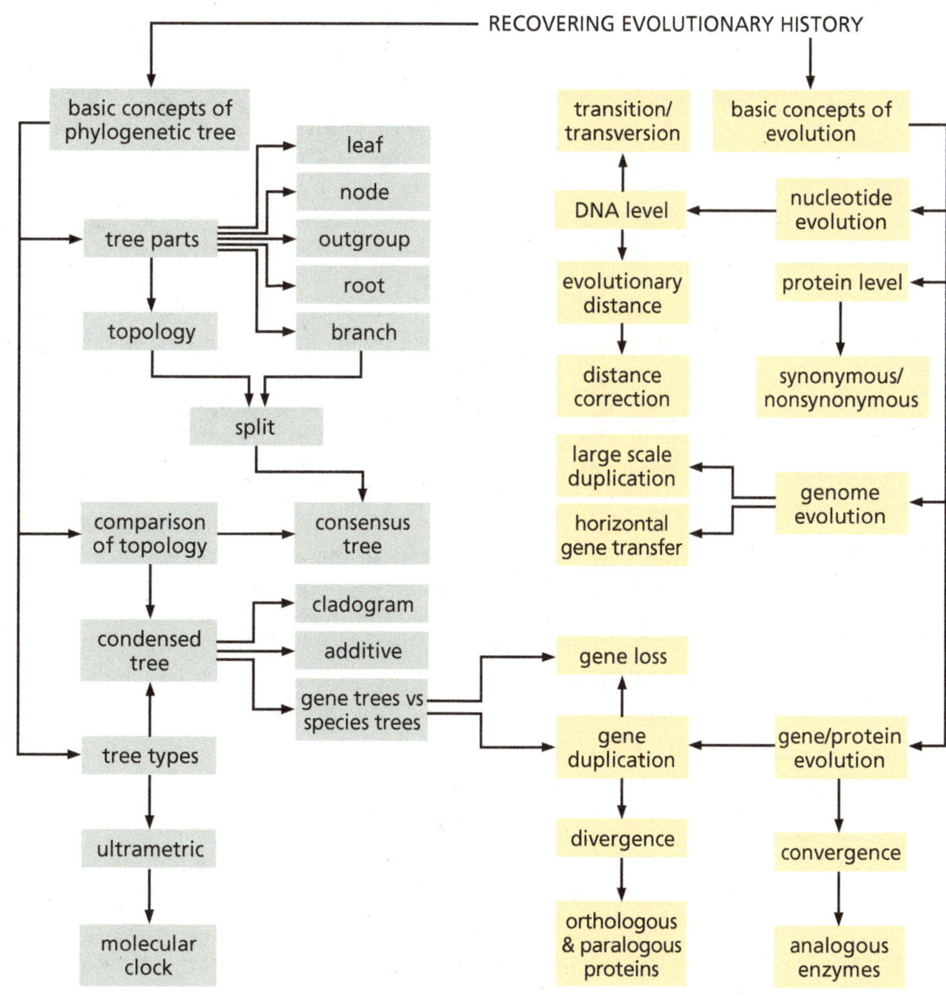

that have survived the evolutionary process, these aspects of evolution will be inherent in the data. In the case of sequence-based phylogenetic reconstruction, the changes that occur in genomic DNA are the main focus and, where relevant, their effects on amino acid sequence. We shall look first at changes that affect just one or a few nucleotides, and then at events such as gene duplication and gene loss. Finally, we will briefly explore evolutionary changes that occur at the scale of chromosomes or whole genomes (see Flow Diagram 7.2).

## Most related sequences have many positions that have mutated several times

Phylogenetic reconstruction from a set of homologous sequences would be considerably easier than it is if two conditions had held during sequence evolution; first, that all the sequences evolved at a constant mutation rate for all mutations at all times; and second, that the sequences have only diverged to a moderate degree such that no position has been subjected to more than one mutation. If the latter condition were true, once the sequences have been accurately aligned, all the mutational events could be observed as nonidentical aligned bases and the mutation could be assumed to be from one base to the other. If, in addition, the first condition held, then the number of observed differences between any two aligned sequences would be directly proportional to the time elapsed since they diverged from their most recent common ancestor. Reconstruction of a phylogenetic tree in such cases would present no problem.

In reality, neither condition usually holds. As we saw in Section 5.1, over the long periods of time separating most present-day sequences from each other, many sequence positions will have mutated several times. Even apparently conserved bases may in the past have mutated to a base that subsequently mutated back to the original base; any such pairs of mutations that have occurred are undetectable from the sequence alignment. The simplest way of estimating the evolutionary distance from an alignment is to count the fraction of nonidentical alignment positions, to obtain a measure called the ***p*-distance**. However, because of the possibility of overlapping mutations, this *p*-distance is almost always an underestimate of the number of mutations that actually occurred (see Figure 7.7) and corrections are needed to improve the estimated evolutionary distance.

A way of correcting this error is essential to obtain accurate results in phylogenetic tree reconstruction, which depends on estimations of the actual numbers of mutations that have occurred. Various models of evolution have been proposed for use in phylogenetic tree reconstruction that attempt to predict the amount of multiple mutation and thus the actual mutation rate at each position. The features of the evolutionary processes that affect the relative mutation rate of a base in a particular position in a sequence can be incorporated into models. From an evolutionary model an equation can be formulated to make a **distance correction** to the number of mutations observed directly in the alignment, which will hopefully convert the observed distance between the aligned sequences into a value nearer to the correct evolutionary distance. Some tree-reconstruction methods rely on the distance correction being directly applied to the data, while others can use the model directly without the need for the correction to be explicitly defined. Details of such models and their application are given later in this chapter, with the more technical aspects described in Section 8.1.

## The rate of accepted mutation is usually not the same for all types of base substitution

The models of evolution used for phylogenetic analysis define base mutation rates and substitution preferences for each position in the alignment. The simplest models assume all rates to be identical and time-invariant with no substitution

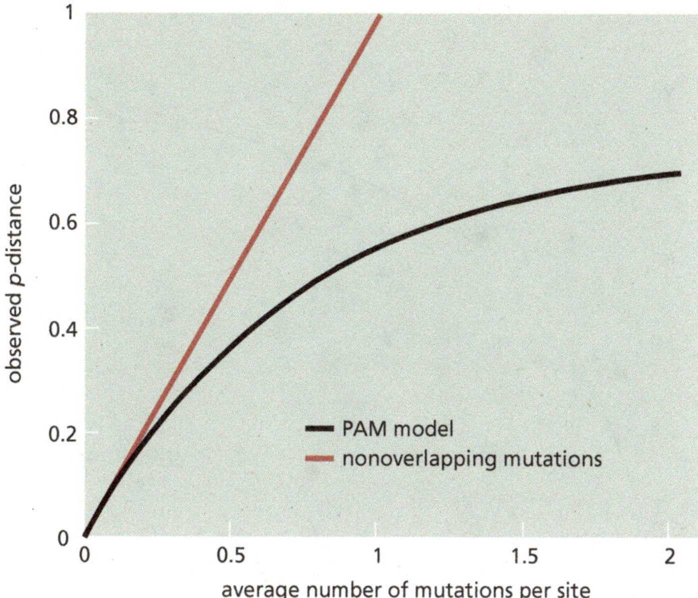


**Figure 7.7**
**The number of observed mutations is often significantly less than the actual number of mutations because of overlapping mutations.** The straight red line represents the *p*-distance—the fraction of nonidentical sites in an alignment—that would be observed if each site only received one mutation at most. The observed *p*-distance in an alignment is plotted (black line) against the average number of mutations at each site as calculated by the PAM model described in Section 5.1. This can be compared with Figure 5.2, which shows the fraction of identical alignment sites for different PAM distances. It can be seen that the observed fraction of nonidentical sites in an alignment is always an underestimate of the actual number of mutations that have occurred, except when the number of mutations is very small.

preferences, but more sophisticated models have been proposed that relax these assumptions. To use the models properly, it is important to understand their limitations and how accurate their assumptions are. Most mutations that are retained in DNA will have occurred during DNA replication as the result of uncorrected errors in the replication process. Whether a mutation is retained or whether it is immediately lost from the population's gene pool will depend on many factors. In the case of protein-coding sequences, these factors include whether it alters the amino acid sequence or not, and what effect amino acid changes at the various positions in a protein-coding sequence have on protein function. If the altered protein has altered function or is nonfunctional, this can affect cellular functions and thus the organism as a whole. The consequence of all these influences is that the rate of mutation and the substitution preference can vary at each position along the genome. To add to the complications, there is also likely to be a variation in mutational preferences with time, as organisms probably had to evolve under different evolutionary pressures at different times.

Models of molecular evolution have been formulated that take account of at least some of these considerations, although the more complex effects of mutation at the level of the organism cannot be sufficiently defined to be modeled in this way. Because of the structure of DNA (see Figure 1.3), if a purine base is replaced by another purine on mutation, or a pyrimidine by a pyrimidine, the structure will suffer little if any distortion. Such mutations are called **transition** mutations, as opposed to **transversions** in which purines become pyrimidines or vice versa (see Figure 7.8A). One factor that is included in some evolutionary models is the fact that transition mutations are much more commonly observed than transversions (see Figure 7.8B). This is despite the fact that there are twice as many ways of generating a transversion than of generating a transition. The observed relative preference for transition mutations over transversion mutations is described as the **transition/transversion ratio**, and will be written $R$, defined as the number of transitions per transversion during the evolution of the sequences being studied. The value of $R$ relates to mutations that have been accepted during the evolutionary process, and may be very different from the relative rates for the chemical and physical mechanisms that lead to their initial occurrence. If all the mutations shown in Figure 7.8A were equally likely, then $R$ would be ½. However, values of 4 and above are often seen, indicating that in those cases there is some evolutionary pressure significantly favoring transition mutations. In practice, $R$ varies significantly between different sequences, so that if the evolutionary model requires a value for $R$ this should be obtained from the data being

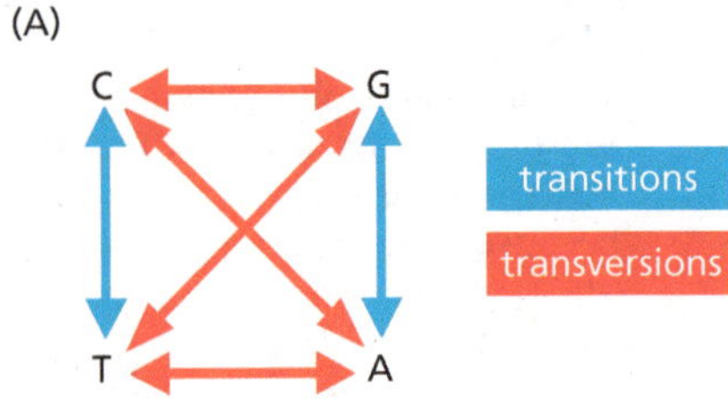


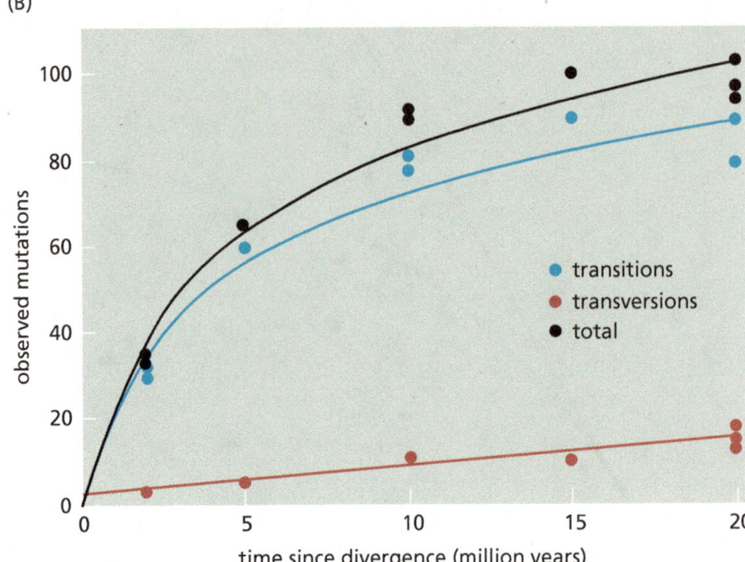


**Figure 7.8**
**Transition and transversion mutations.** (A) The possible transitions (blue) and transversions (red) between the four bases in DNA. Note that there are twice as many ways of generating a transversion than a transition. (B) The observed numbers of transitions, transversions, and total mutations in an aligned set of cytochrome c oxidase subunit 2 (COII) mitochondrial gene sequences from the mammalian subfamily Bovinae. (Data from L.L. Janacek et al. Mitochondrial gene sequences and the molecular systematics of the Arteriodactyl subfamily Bovinae. *Mol. Phylogenet. Evol.* 6:107–119, 1996.)

used. The calculation of *R* from the alignment is not as straightforward as one might think and is explained in Box 8.1.

## Different codon positions have different mutation rates

Another factor that affects the acceptance rates of mutations in protein-coding sequences is the effect of the mutation on the amino acid sequence, and thus, potentially, on the function of the protein. Nucleotide mutations that do not change the encoded amino acid are called **synonymous mutations**, and it is apparent from the standard genetic code (see Table 1.1) that most changes at the third codon position will be synonymous. These mutations can generally be considered to be neutral, that is, to have no effect. One exception might be when tRNAs that recognize the new codon are present at significantly different levels from those for the

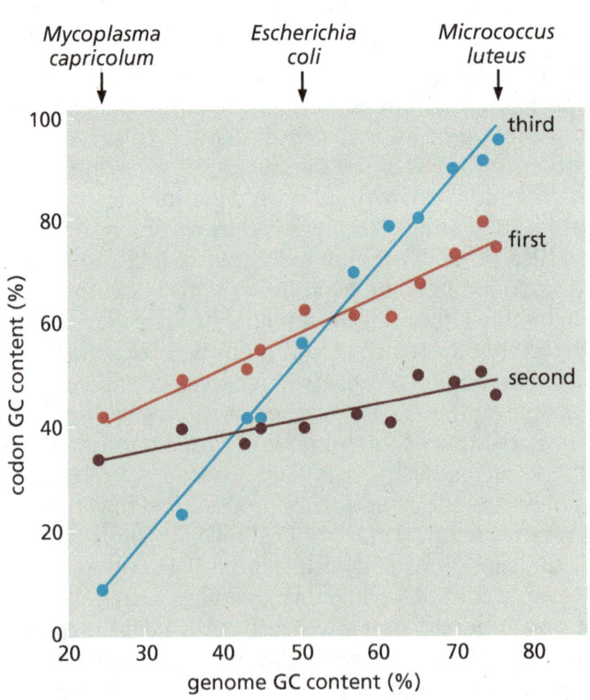


**Figure 7.9**
**A comparison of the average percentage GC content of codons in different bacteria.** The points on each line represent percentage GC values for each of 11 bacteria at the codon position indicated, plotted against the overall genome percentage GC content. While all three codon positions adapt to some extent to the compositional bias of the genome, the third position adapts most. In some bacteria this even results in a more extreme percentage GC value at the third codon position than for the overall genome. The data were taken from a large sample of genes in each organism, but the analysis is expected to carry over to the entire genome. (Adapted from A. Muto and S. Osawa, The guanine and cytosine content of genomic DNA and bacterial evolution, *Proc. Natl Acad. Sci. USA* 84:116–119, 1987.)

original codon, thus potentially affecting the rate of translation. When a nucleotide change alters the encoded amino acid, the change is said to be **nonsynonymous**. The protein product will have a different sequence, and may thus have altered properties: it might, for example, be nonfunctional or less efficient, or be less, or more, stable. Whether the function is affected will depend on where the change occurs in the protein chain and what type of amino acid is substituted. While complete nonfunction is likely to be unfavorable, whether other, more subtle changes turn out to be unfavorable, beneficial or neutral will also depend on the evolutionary pressures being exerted on the organism at the time. The relative likelihood of synonymous and nonsynonymous mutations being retained in the genome is discussed in Box 7.2.

Because nucleotide substitutions at the third codon position are almost always synonymous, the accepted mutation rate at these sites would be expected to be higher than at the first and second positions, and this has been verified from sequence data. The fact that almost all third-position mutations are synonymous is also shown by the phenomenon of **biased mutation pressure**. Most of a bacterial genome (in contrast to those of most eukaryotes) consists of protein-coding sequences, and so when such a genome has a very high or very low GC content, this bias must be accommodated within the codon usage. An examination of the GC content of the three codon positions in a number of bacterial genomes demonstrates the considerably greater flexibility of the third position to respond to this pressure (see Figure 7.9).

Evolutionary models exist that can accommodate variation in mutation rates at different sites, usually using the Gamma correction (see Section 8.1). If the dataset involves long evolutionary timescales, many of the third codon positions may have experienced multiple mutations and show almost random base content. In such cases it has been found useful to remove the third codon sites from the data before further analysis.

## Only orthologous genes should be used to construct species phylogenetic trees

The key assumption made when constructing a phylogenetic tree from a set of sequences is that they are all derived from a single ancestral sequence, i.e., they are homologous. As we will now see, there are different forms of homology, and some homologs are more suitable than others for particular kinds of phylogenetic analysis. Homologous genes can arise through a variety of different biological processes. One is by speciation, which results in two homologous genes diverging in different lineages. Pairs of homologous genes derived this way are described as orthologous and called orthologs. Orthologs can be formally defined as pairs of genes whose last common ancestor occurred immediately before a speciation event (see Figure 7.10). Because the function of the gene is at least initially required in each new species, and will thus likely be conserved, pairs of orthologous genes are expected to have the same or almost identical biochemical functions in their respective species. In fact, the term ortholog is often employed in this more restrictive manner to mean those genes in different species that are both homologous and carry out the same function. Note that the first definition, although expecting that orthologs will most likely have the same function, does not demand it.

Another way in which homologous genes arise is by **gene duplication**, the process by which a gene becomes copied, within the same genome. Gene duplication is believed to be a key process in evolution and has occurred to some extent in all organisms whose genomes have been explored. A pair of genes arising from a gene duplication event is described as **paralogous**, and are called **paralogs**, which can be more formally defined as a pair of genes whose most recent common ancestor occurred immediately before a gene duplication event (see Figure 7.10B). These

## Box 7.2 The Influence of Selective Pressure on the Observed Frequency of Synonymous and Nonsynonymous Mutations

A mutation occurring in an individual organism will either be maintained and become ubiquitous in the species gene pool or will be quickly lost. The fate of the mutation will depend on the selective pressure on the species at the time and on whether the mutation makes a significant difference in the fitness of the organism. In the absence of selective pressure, the mutation will be kept or lost through a series of random events: random genetic drift. If a mutation confers an advantage on the organism, then **positive selection** will increase the likelihood of it being kept. Conversely, a mutation that causes the organism to be disadvantaged will most likely be lost due to **negative** (or **purifying**) **selection**. It has been suggested in the neutral theory of evolution that in practice most mutations do not encounter strong selection.

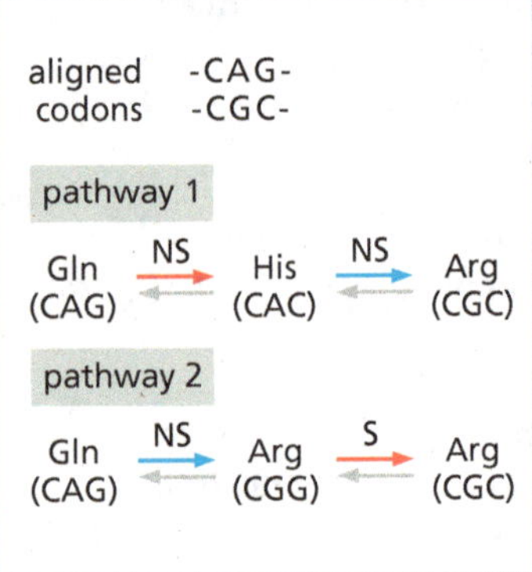

Given two aligned protein-coding sequences, it is, in principle, possible to identify the type of selection operating during the evolution of each from their common ancestor. If we ignore any possible effects on translation, synonymous mutations will not affect the fitness of the organism and thus are not selected for or against. Thus, the proportion of synonymous mutations in a protein-coding gene will form a baseline against which the proportion of nonsynonymous mutations can be measured, and this ratio can be used to determine whether that gene has been subject to positive selection, negative selection, or no selection. If the resulting protein is more effective as a result of a nonsynonymous mutation, under selective pressure the mutation is likely to be retained. Thus the observation of a greater number of nonsynonymous mutations than expected under neutral evolution implies that the gene was being subject to positive selection. If a mutation decreases the effective role of a protein, under selective pressure it is likely to be lost. Negative selection will result in fewer nonsynonymous mutations than expected, implying that change is being strongly selected against; that is, the sequence is being conserved.

There are a number of ways of counting the numbers of synonymous and nonsynonymous mutations. The following is due to Masatoshi Nei and Takashi Gojobori. Suppose that a sequence alignment aligns the two codons CAG (Gln) and CGC (Arg). If the two observed base changes are assumed to have occurred as two separate events, and assuming no other mutations occurred, the evolution of CAG (Gln) to CGC (Arg), or vice versa, could have occurred by either of two pathways (see Figure B7.1). The first pathway has two nonsynonymous mutations, whereas the second pathway only has one, as the second change is synonymous. Since we cannot distinguish between the two possibilities, these two aligned codons are counted as 1.5 nonsynonymous and 0.5 synonymous mutations (by adding up the number of mutations in each category and dividing by the number of pathways). If a proposed pathway contains a stop codon at some stage, the entire pathway is ignored in the calculation. Counting mutations is trivial for aligned codons that are identical or that differ in just one base, and those that differ in all three bases are counted by considering all six possible pathways. Adding these up over the whole alignment produces the total number of synonymous differences ($S_d$) and nonsynonymous differences ($N_d$).



**Figure B7.1**
**Two possible mutation pathways that connect the aligned codons CAG and CGC.** The pathways consist of steps, each of which is a single base change. Transitions and transversions are colored as in Figure 7.8, and classified as synonymous (S) or nonsynonymous (NS).

If one wants to identify the type of selection operating during evolution of a particular gene, as well as counting the numbers of mutations, it is necessary to assess the number of possible sites of synonymous and nonsynonymous mutations. This is done by considering in turn each of the codons in the two sequences. As each of the three codon positions can mutate into any of three other bases, each codon can mutate nine possible codons with a single mutation. In the example given in Figure B7.2, one of the nine possible mutations results in a stop codon, so this mutation is ignored. Each codon position is considered separately. The first position gives rise to three nonsynonymous mutations out of three, and counts overall as one possible nonsynonymous mutation site. By similar arguments, the third position counts as one possible synonymous mutation site. At the second site only the two mutations avoiding a stop codon are considered, and they also give rise (as two out of two) to one possible nonsynonymous mutation site. The total number of possible mutation sites for a codon is three, in this case two nonsynonymous and one synonymous. In this way each codon can be assigned a number of possible synonymous ($S$) and nonsynonymous ($N$) mutation sites, and the total number of sites is calculated as the average for the two sequences.

## Box 7.2 The Influence of Selective Pressure on the Observed Frequency of Synonymous and Nonsynonymous Mutations (continued)

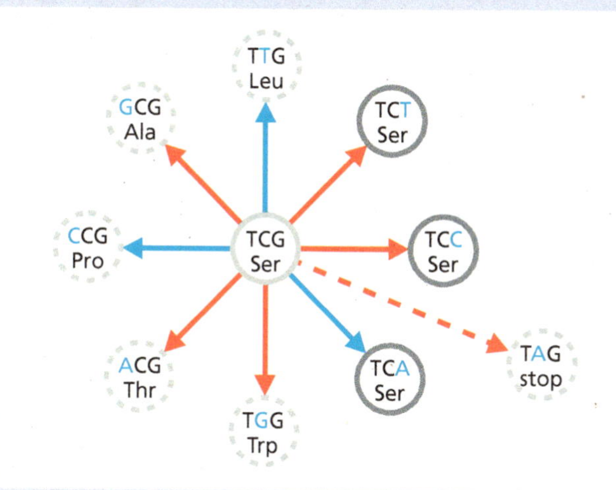


**Figure B7.2**
**The nine possible single-base mutations of the TCG triplet. Mutations are labeled according to whether they involve a transition (blue) or transversion (red), and whether they are synonymous (bold circle) or nonsynonymous (dashed circle).** One possible mutation (dashed) leads to the stop triplet TAG. (Adapted from Figure 1 of Goldman & Yang, 1994.)

From this analysis, the simplest estimate of the proportion of synonymous mutations observed from the possible synonymous sites is $p_S = S_d/S$, with the nonsynonymous equivalent being $p_N = N_d/N$. However, these values ignore the fact that there will almost certainly have been multiple mutations at the same site, so that these are underestimates. In the same way, a distance correction can be applied to these values. In the case of the Nei–Gojobori method, the Jukes–Cantor model, which is described in Section 8.1, is used which gives the synonymous substitutions per synonymous site as

$$d_S = \frac{3}{4}\ln\left(1-\frac{4}{3}p_S\right) \tag{BEQ7.1}$$

and the nonsynonymous equivalent as

$$d_N = -\frac{3}{4}\ln\left(1-\frac{4}{3}p_N\right) \tag{BEQ7.2}$$

$d_S$ and $d_N$ are sometimes also called $K_s$ and $K_a$, respectively.

To determine whether positive selection has occurred during the evolution of the two sequences, it is necessary to apply a statistical test to determine if the value of $d_N$ is significantly greater than that of $d_S$. This involves estimating the uncertainty (variance) of the values and is beyond the scope of this book (see Further Reading). There are other methods of estimating these values: some take account of transition/transversion mutational preferences while others depend on more complicated techniques such as likelihood (see Further Reading).

An interesting application of the $d_S$ measure has been proposed that arises from the assumed minimal involvement of selection in determining the fate of synonymous mutations. The application involves distinguishing between sets of duplicated genes within a genome when more than one large-scale duplication event has occurred. In this case the difference between the $d_S$ measure within the two sets should be indicative of the time since divergence, and the technique has indeed identified two major duplication events at different times in *Arabidopsis* (see Figure B7.3).

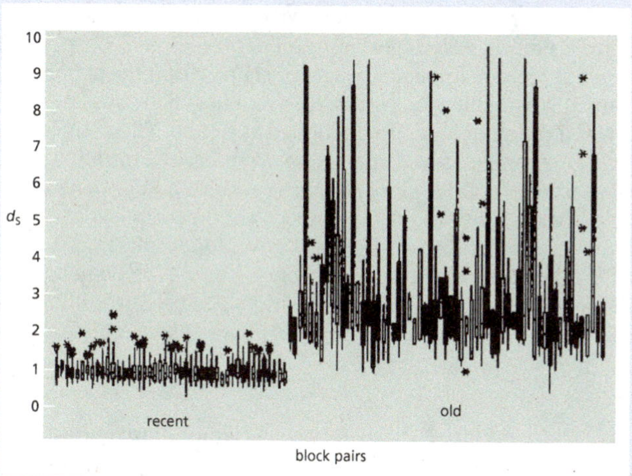


**Figure B7.3**
**At least two distinct periods of large-scale duplication events can be distinguished in *Arabidopsis* from examination of synonymous mutations.** Each column represents the $d_S$ value for a pair of sequence segments (blocks) resulting from duplication. Each pair of blocks has in common at least six duplicated genes. The spread of $d_S$ measurements (one for each gene pair) is shown for each block, with more extreme values shown by asterisks. There is one set of block pairs with a similar $d_S$ measurement of about 1, which can be postulated to have arisen due to a major gene duplication event of relatively recent origin. The other blocks have larger $d_S$ values, indicating much greater sequence divergence, and thus are likely to have arisen from one or more earlier duplication events. (Adapted from G. Blanc, K. Hokamp and K.H. Wolfe, A recent polyploidy superimposed on older large-scale duplications in the Arabidopsis genome, *Genome Res.* 13:137–144, 2003.)

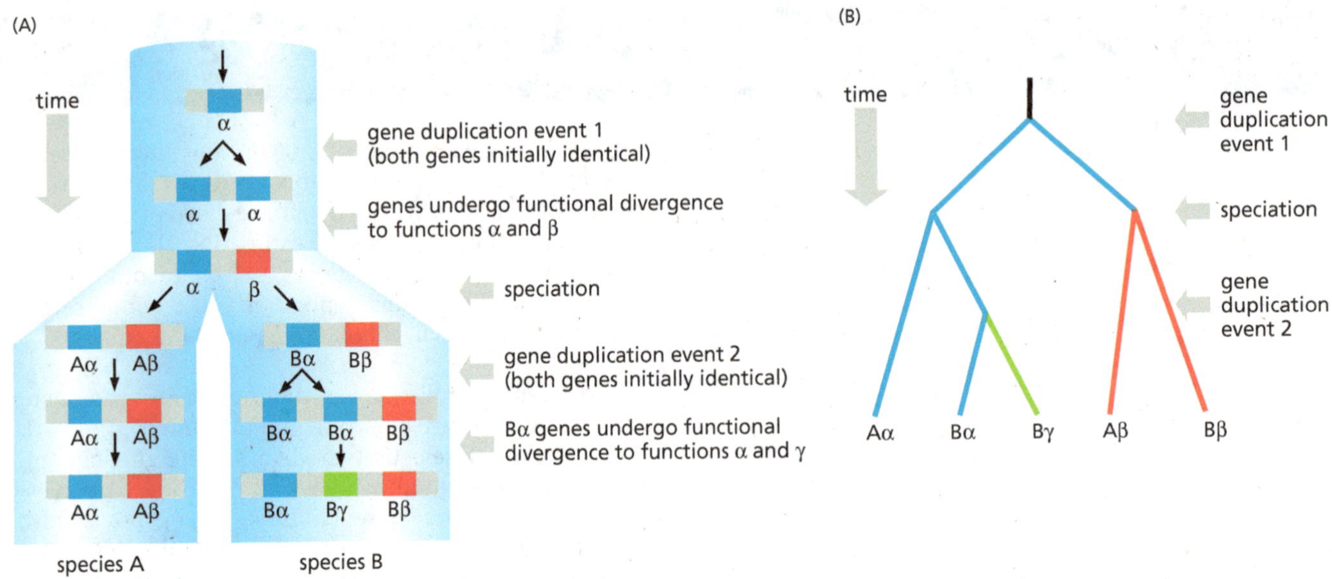


**Figure 7.10**
**The evolutionary history of a gene that has undergone two separate duplication events.** (A) A species tree is depicted by the pale blue cylinders, with the branch points (nodes) in the cylinders representing speciation events. In the ancestral species a gene is present as a single copy and has function α (blue). At some time a gene duplication event occurs within the genome, producing two identical gene copies, one of which subsequently evolves a different function, identified as β (red). These are paralogous genes. Later a speciation event occurs, resulting in two species (A and B) both containing genes α and β. Gene Bα (in species B) subsequently undergoes another duplication event, which after further divergent evolution results in genes Bα and Bγ, the latter with a new function γ (green). The Bα gene is still functionally very similar to the original gene. At the end of this period of evolution, all five genes in both species are homologous, with three orthologous pairs: Aβ/Bβ, Aα/Bα, and Aα/Bγ. The Bα and Bγ genes are paralogous, as are any other combinations except the orthologous pairs. Note that Aα and Bγ are orthologs despite their different functions, and so if the intention is to study the evolution of a particular functional product, such as the α function, we need to be able to distinguish the Aα/Bα pair from the Aα/Bγ pair. This can be done using sequence similarity, which would be expected to be greater for the Aα/Bα pair as they will be evolving under almost identical evolutionary pressures. Errors in functional orthology assignment can easily occur, depending on sequence and functional similarity and whether all related genes have been discovered. (B) The phylogenetic tree that would be drawn for these genes, here drawn as a cladogram.

duplicate genes can then diverge through evolution, developing new functions. Occasionally, the extra gene will evolve to have a new function of use to the organism, and selective pressure will preserve this new gene. Alternatively, both genes may evolve to different functions that together accomplish the required functional aspects of the original gene as well as possible new functions. The detail of the mechanisms whereby this divergence produces new protein functions is hotly debated and beyond the scope of this book, but see Further Reading. Figure 7.10 illustrates a typical sequence of events that can lead to a set of homologous genes in two species, some of which are orthologs and some paralogs that have acquired new functions.

Because initially after duplication there will only be a requirement for one of the genes, often instead of evolving to an alternative function one of the genes becomes nonfunctional through mutation. Nonfunctional genes can arise through the loss of control sequences, resulting in a failure to generate the protein product, or alternatively by modification of the protein sequence rendering the protein inactive. Genes that have mutated so as to no longer give rise to protein products are called **pseudogenes**. Usually these pseudogenes will then steadily accumulate mutations until they are no longer detectable. This entire process is known as **gene loss**.

Note that gene loss can also occur without gene duplication. The occurrence of gene loss can make the gene and species trees appear very different (see Figure 7.11).

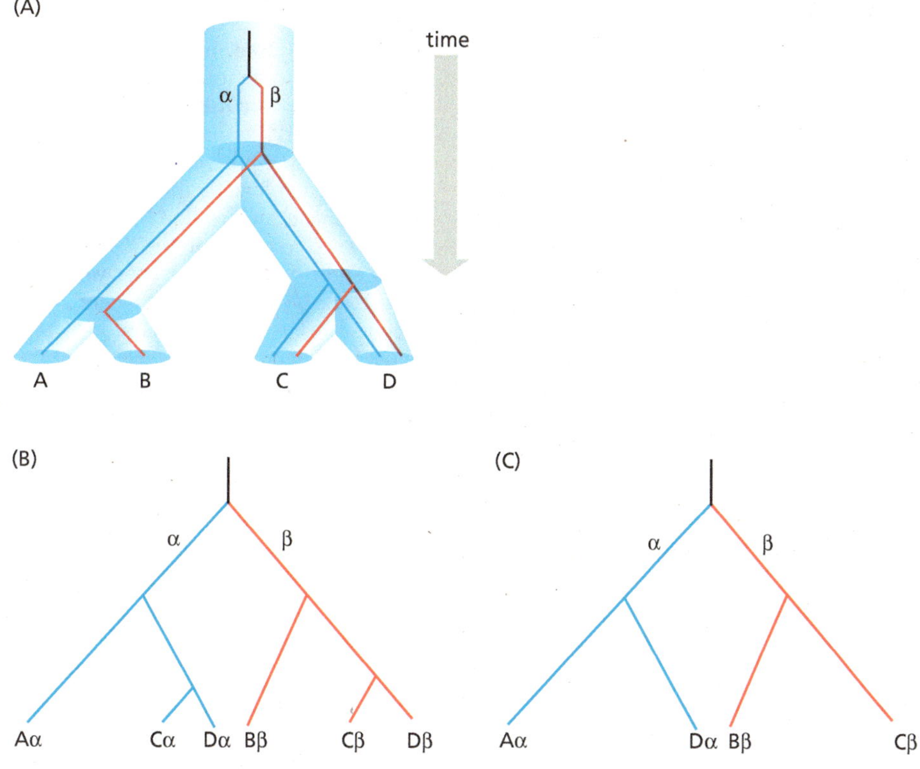


**Figure 7.11**
**The effects of gene loss and missing gene data on phylogenetic trees.** (A) The phylogenetic tree for the species A, B, C, and D is depicted using pale blue cylinders. The branch points (nodes) in the cylinders represent speciation events. The evolutionary history of a gene family is shown by the lines in the cylinders. A gene duplication of the common ancestral gene gives the paralogs α (blue) and β (red). The genes α in all species are orthologs, as are all the genes β. As a result of gene loss, in species A the β gene has been lost, and the same has happened for gene α in species B. Both the α and β genes are present in species C and D. (B) The gene tree for α and β corresponding to their history as shown in (A). (C) The gene tree that would result if only the four genes shown were included, as might happen if the Cα and Dβ genes were not known. From this tree the erroneous conclusion would be drawn that species A was more closely related to D than to B. On the basis of this tree alone the gene duplication could not be spotted, and all four genes would be thought to be orthologous.

Attempts have been made to combine gene and species trees so as to clearly identify the speciation and duplication events and the gene losses. Such trees are called **reconciled trees**, an example of which, reconciling the trees shown in Figure 7.3, is shown in Figure 7.12.

As can be seen in Figure 7.10B, only orthologous sequences will identify the speciation times, paralogous sequences producing the gene duplication events. Therefore, species phylogenetic trees should ideally be constructed using only orthologous sequences. But distinguishing orthologs from paralogs is not always easy. The best indication of orthologs apart from similarity of sequence is identity of function. But as experimental proof of gene function is frequently not available, we usually have to fall back on sequence similarity. One approach to identifying orthologs in this way is to construct a phylogenetic tree using a large set of potential functional homologs. As we saw in Figures 7.3 and 7.12, such a tree can help distinguish duplication events (paralog formation) from speciation events (ortholog formation). But phylogenetic tree reconstruction is sometimes not practical for this purpose. Some protein families and superfamilies comprise thousands of homologous sequences. Constructing a tree using all these sequences, if feasible at all, would only be possible with the least sophisticated methods, and not a great deal of confidence could be placed in the results. The process would also be very demanding of computer resources. The easiest and quickest current method of identifying paralogs and orthologs, without the need to generate phylogenetic trees, is by using the COG (Clusters of Orthologous Groups) and KOG (Eukaryotic Orthologous Groups) methodology (see Box 7.3).

Although orthologs often have the same function, not all proteins with identical or similar function are orthologs, or even homologs. As discussed in Chapter 4, unrelated nonhomologous genes can sometimes develop equivalent functions as the result of convergent evolution, although sequence identity between such genes is usually very limited. Convergence is best characterized in enzymes. The nonhomologous serine proteases chymotrypsin and subtilisin, for example, have independently evolved an identical catalytic mechanism (see Figure 7.13). Such functionally

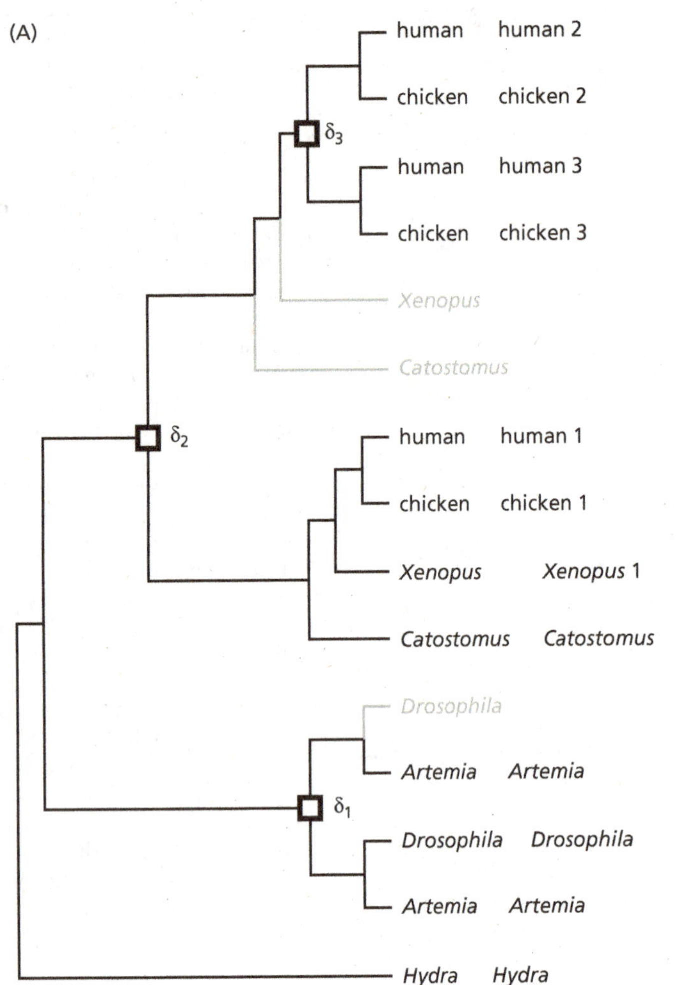


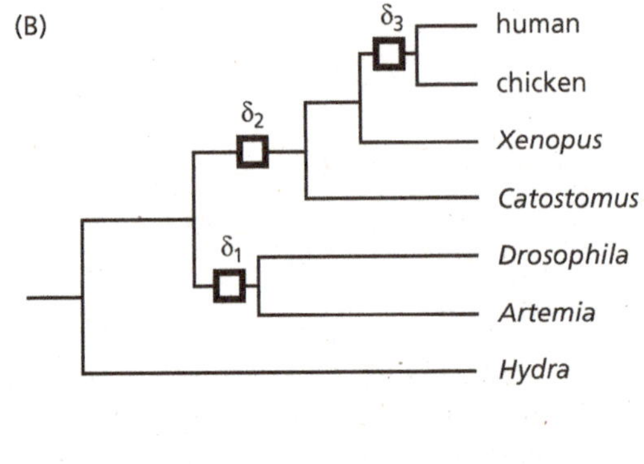


**Figure 7.12**
**An example of a reconciled tree and the equivalent species tree.** (A) shows the reconciled tree that combines the information from the two trees of Figure 7.3, which are for a selected set of eukaryotes and the $Na^+$–$K^+$ ion pump membrane protein family. Three gene losses are identified, drawn on the tree in gray lines. Each external node is labeled first by the species and then by the protein, according to the labels on the trees of Figure 7.3. Three duplication events were identified, shown as the boxes labeled $\delta_1$, $\delta_2$, and $\delta_3$. (B) shows the related species tree, on which the duplication events have been superposed along the branches where they occurred. (Part A is taken from Figure 8 of Page and Charleston, 1997.)

similar yet unrelated enzymes are referred to as **analogous enzymes**. Two proteins with similar three-dimensional folds but no common evolutionary ancestor are also referred to as analogous. Two sequences that are similar by convergent evolution do not share a common ancestor, and thus should not be used together in the same phylogenetic tree. Sequence similarities that are not due to homology are known generally as **homoplasy**. Convergent evolution is just one cause of homoplasy; others are parallel evolution and evolutionary reversal (see Further Reading).

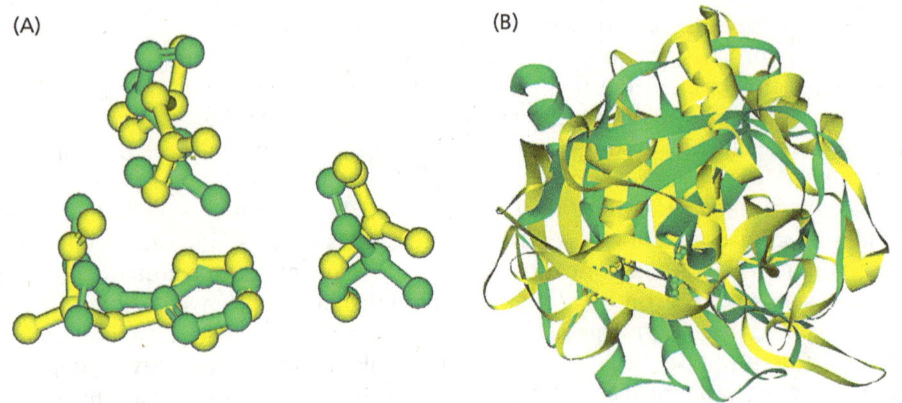


**Figure 7.13**
**Chymotrypsin and subtilisin have independently evolved to an identical catalytic triad of residues.** (A) The superposed catalytic triad of histidine, aspartic acid, and serine residues in the active sites of chymotrypsin (yellow) and subtilisin (green). The almost exact equivalence makes it extremely likely that the catalytic mechanism is identical in the two enzymes. (B) The complete folds of these two proteins with the triad still superposed, showing that the two protein folds are completely unrelated, and that therefore the two proteins do not share a common ancestor. (PDB database entries 1AB9 and 1CSE.)

## Box 7.3 Identifying Paralogs and Orthologs via COGs and KOGs

The COG (Clusters of Orthologous Groups) and KOG (euKaryotic Orthologous Groups) databases have been constructed using a careful analysis of BLAST hits. First, low-complexity sequence regions and commonly occurring domains are masked to prevent spurious hits and also to improve the likelihood that the statistical score analysis (*E*-values) will be correct (see Chapters 4 and 5 for details). All gene sequences from one genome are then scanned against all from another genome, noting the best-scoring BLAST hits (BeTs) for each gene, and this is repeated for all possible pairs.

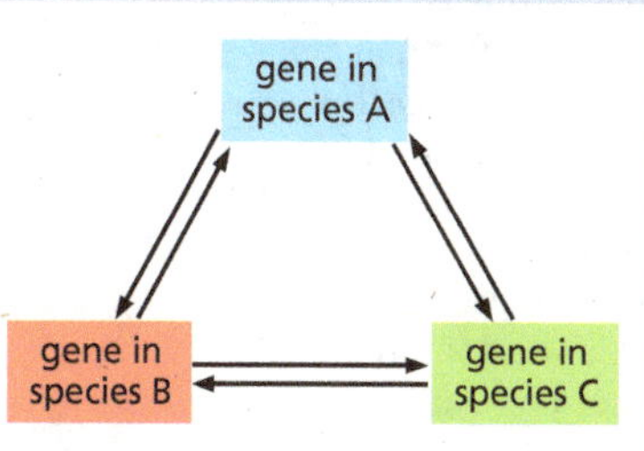


**Figure B7.4**
**The simplest type of cluster of orthologous genes (COG).** Each arrow represents the best-scoring BLAST hit (BeT) found on searching for the gene at the start of the arrow in the genome of the gene at the end of the arrow. If all three genes in the three species have each other as BeTs, then the three genes form a COG. More complex arrangements of BeTs can occur in COGs, especially when there are more species. See Further Reading for more details.

Paralogous genes within a genome that result from gene duplication since divergence of two species are identified as those that give a better-scoring BLAST hit with each other than their BeTs with the other genome. Orthologous genes are found as groups of genes from different genomes that are reciprocal BeTs of each other, the simplest COG making a triangle of genes from three genomes (see Figure B7.4). This approach would have correctly identified orthologous and paralogous genes in the example shown in Figure 7.11B. By careful analysis of such BeT relationships, clusters of orthologous genes can be constructed for large numbers of genomes. In the case of Figure 7.11B there should be two COGs, associated with the α and β parts of the tree.

Gene loss can cause problems, as the remaining paralogous gene may provide a misleading BeT relationship, resulting in two COGs being combined. An example of this would occur on comparing species A and B of Figure 7.11. Examining the initial COGs for sequence-similarity patterns can sometimes identify the two correct COGs. All sequences in a COG or a KOG are assumed to have a related function, and thus the method can be used to predict gene and protein function.

Figure 7.14 shows the orthologous and homologous relationships between the genes of the chicken genome and the human and puffer fish genomes. The method used to define these relationships is based on the scores of BLAST searches within genomes. This approach identified a core set of orthologs in all three genomes that are also likely to have conserved functions. The use of orthologous genes to predict function depends on the function having been conserved over the evolutionary period since divergence. This assumption is, in principle, testable using the statistics of synonymous versus nonsynonymous mutations as discussed in Box 7.2. If the proportion of nonsynonymous mutations ($d_N$) is observed to be greater than that observed for synonymous mutations ($d_S$), this is indicative of selection for retention of function.

Although duplication events usually lead to the existence of two closely related genes they can also result in a longer protein sequence if the duplication occurs within a single gene. This should be detectable through the use of dot-plots or by aligning the sequence to itself (see Section 4.2), which will reveal significant similarity between two sections. Note that the set of sequences used in phylogenetic analysis must be of the equivalent regions from the ancestral sequence. If one is studying the evolution of the organisms as opposed to the genes, gene duplication and gene loss can cause considerable difficulties. One way around this problem is to examine trees for many different genes, looking for the evolutionary history supported by a majority of trees. An alternative approach to this is to try to restrict the analysis to those genes that appear not to have duplication events in their evolutionary past.

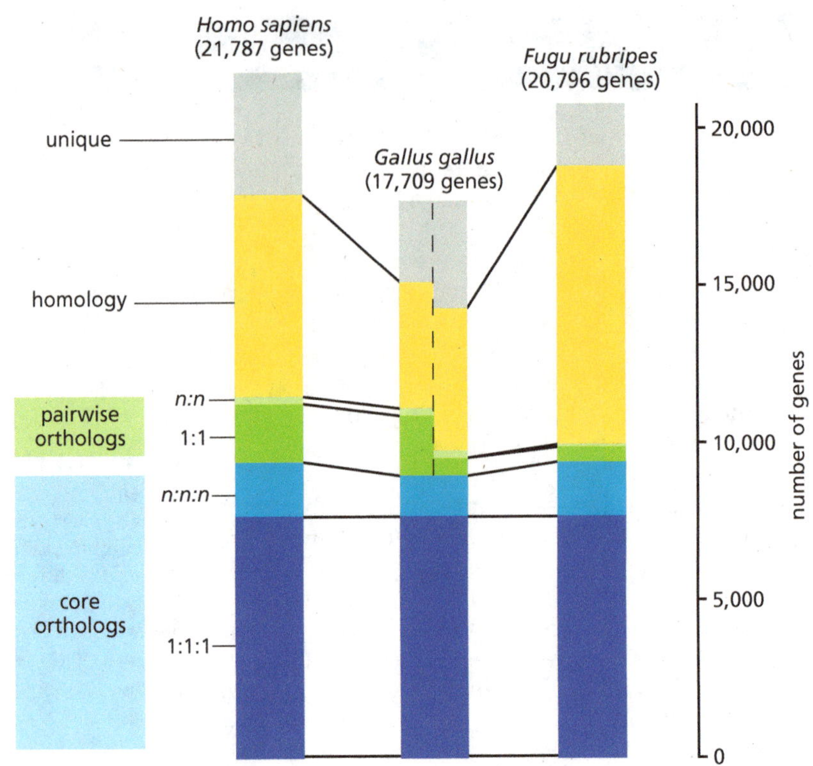


**Figure 7.14**
**The numbers of orthologous, homologous, and unique genes in the human (*Homo sapiens*), chicken (*Gallus gallus*), and puffer fish (*Fugu rubripes*) genomes.** Orthologs present as one copy in each of the three species are represented in dark blue on the histogram and are labeled 1:1:1; genes represented in lighter blue and labeled n:n:n are orthologous genes present in all three species, which have been duplicated in at least one of the species. These two groups are considered as core orthologs for the three species. Orthologs (and duplicated orthologs) found in only two species are labeled pairwise orthologs, represented as 1:1 (or n:n) and colored green. Homologous genes for which no clear orthologous relationships can be determined in the pairwise comparison are in yellow, and unique genes are in gray. (Reprinted by permission from Macmillan Publishers Ltd: *Nature* 432:695–716, International Chicken Genome Sequencing Consortium, Sequence and comparative analysis of the chicken genome provide unique perspectives on vertebrate evolution, 2004.)

Another evolutionary event that can confuse phylogenetic analysis is the process of **horizontal gene transfer** (**HGT**), also known as **lateral gene transfer** (**LGT**). This event is significantly different from all the others discussed so far, in that it involves a gene from one species being transferred into another species: the term "horizontal" is applied to contrast it with the vertical transmission from parent to offspring. For a long time HGT was thought not to occur in eukaryotes, but viral genes have been identified in the human genome. However, HGT is far more

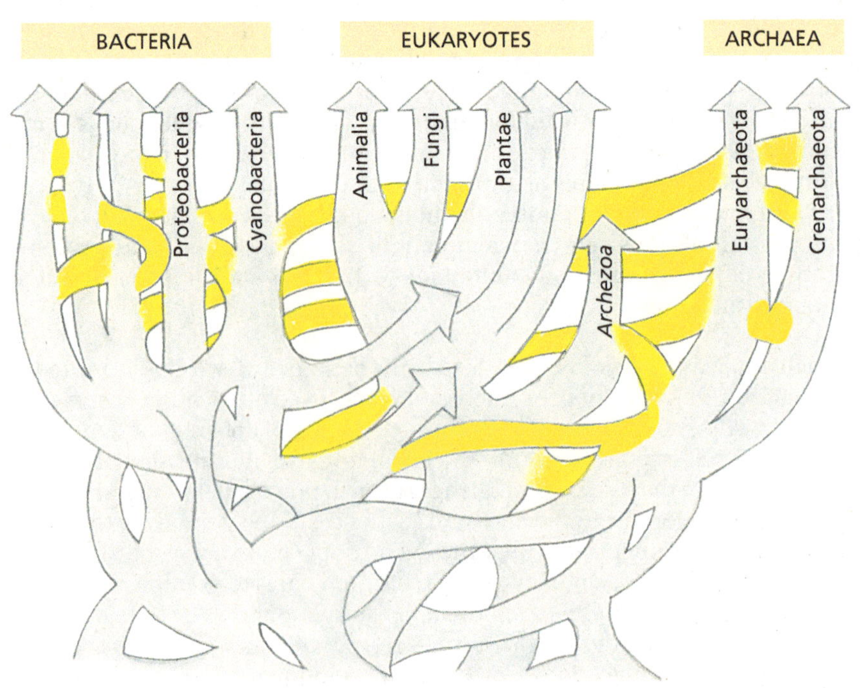


**Figure 7.15**
**A sketch of the tree of life including a considerable amount of horizontal gene transfer (HGT).** HGT is distinguished from normal vertical gene transmission by yellow shading. The tree should not be interpreted too precisely, but does show that Eukaryotes are largely unaffected by HGT except in their earliest evolutionary period, whereas the Archaea and Bacteria continue to experience these events. (Adapted from W.F. Doolittle, Phylogenetic classification and the universal tree, *Science* 284:2124–2128, 1999.)

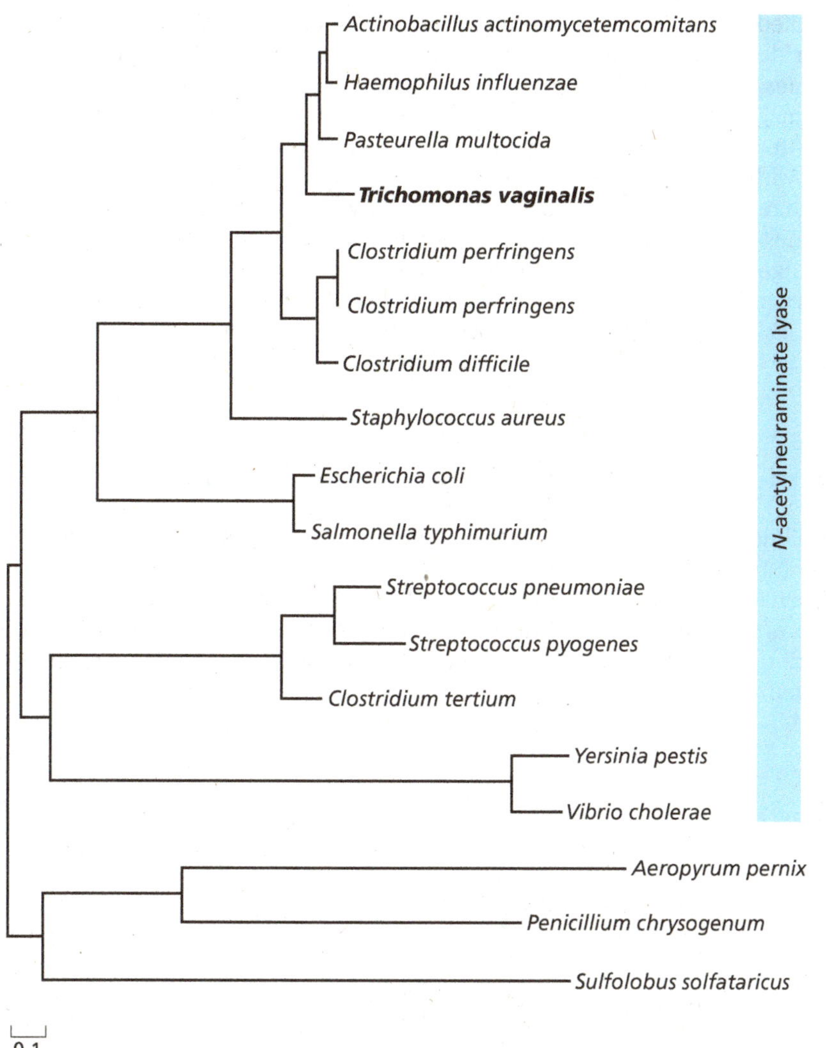


**Figure 7.16**
**Evidence of horizontal gene transfer of the gene for *N*-acetylneuraminate lyase from bacteria to the eukaryote flagellate *Trichomonas vaginalis*.** If an extra 24-residue N-terminal sequence is ignored, there is 80% identity at the amino acid level between the *Trichomonas* sequence and the sequences from the bacteria *Actinobacillus* and *Haemophilus*. The three sequences at the bottom of the tree are of different enzymes from the same protein superfamily and were used as an outgroup to root the tree. This tree gives strong evidence for the horizontal transfer of the gene from bacteria to *Trichomonas*. Whilst this is an unusual event, it is partly explained by the lifestyle of *Trichomonas*. (Adapted from A.P. de Koning et al. Lateral gene transfer and metabolic adaptation in the human parasite Trichomonas vaginalis. *Mol. Biol. Evol.* 17:1769–1773, 2000.)

prevalent in the other kingdoms of life: the Bacteria and the Archaea. When there is no HGT in the evolutionary history, the phylogenetic tree branches much in the way a real tree does, with branches remaining apart. If they could be correctly identified, HGT events would be represented in a phylogenetic tree as branches that rejoin. A tree of all life would then look something like Figure 7.15, although it should be noted that there is dispute between researchers about the degree to which HGT events have actually occurred.

Shortly after HGT, the sequence of the gene in the donor and recipient species will be very similar. Such pairs of genes are called **xenologous** genes. If such genes are included in a standard phylogenetic analysis, the resultant tree will have the gene from the recipient species appear in a much closer relationship to that of the donor species than should be the case (see Figure 7.16). Proving that HGT has occurred is usually very difficult, however. If the genome-wide base compositions of the two species involved are very different, it may be possible to deduce the direction of a recent transfer. A proper discussion of this topic is beyond the scope of this book, and the reader is referred to Further Reading.

## Major changes affecting large regions of the genome are surprisingly common

It might have been hoped that once whole genome sequences became available, it would be a relatively easy task to identify equivalent regions in different species (so-called **syntenic** regions containing the related genes in the same order) and hence identify the orthologous genes. However, many changes have been found in genomes at larger scales than that of individual genes. Duplication of entire chromosomes and even entire genomes has occurred. In addition, chromosomes have split into smaller fragments that have rejoined to make new chromosomes, sometimes with sections shuffled in order or inverted (see Figure 10.22). The frequency of such rearrangements is surprisingly high given the potential danger they pose to genome integrity and the organism's survival. One consequence of large-scale rearrangements is that orthologous genes usually do not occupy equivalent positions even in related species, although they often do retain a similar local environment. Hence complete genome sequences do not help to identify orthologous genes as much as might have been hoped.

## 7.3 Phylogenetic Tree Reconstruction

**Flow Diagram 7.3**
**In this section two example problems in phylogenetic tree reconstruction are presented to illustrate the process, how parameter choices can be justified, and the types of conclusions that can be drawn and tested.**

We will now explore the ways in which phylogenetic trees are reconstructed, using two examples. The first focuses on nucleotide data and the evolutionary relationships between species. The second uses a set of sequences from one protein superfamily, and examines how the resulting tree can be used to examine the history of protein evolution within the superfamily and to predict the function of uncharacterized superfamily members.

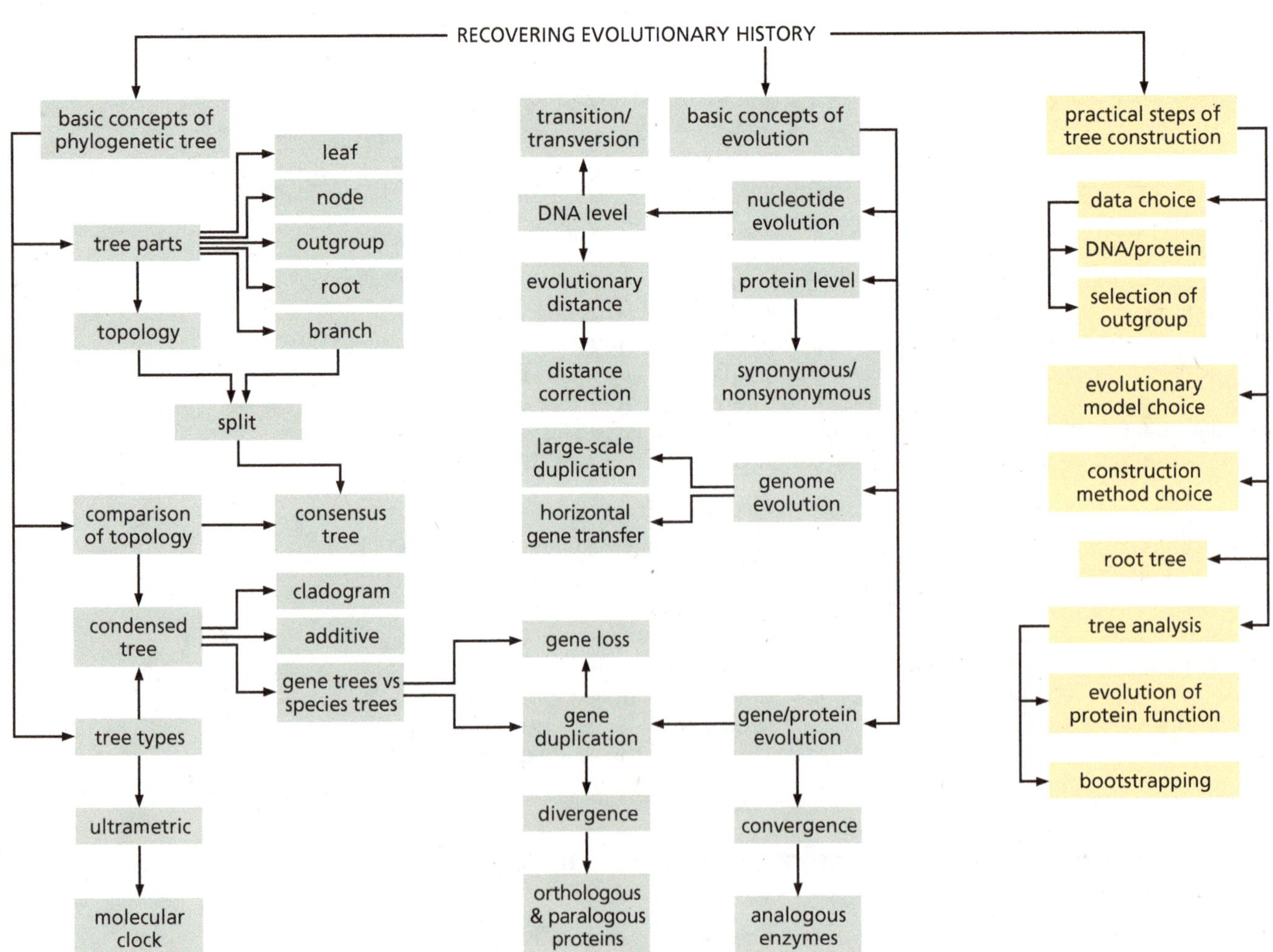

## Small ribosomal subunit rRNA sequences are well suited to reconstructing the evolution of species

If one is trying to reconstruct the evolutionary history of life to show where different species diverged, great care must be taken in the choice of the data. The ideal is a genomic region that occurs in every species but only occurs once in the genome. This avoids any potential misassignment of orthology. It is also crucial that there is little if any HGT within this region. The rate of change in this sequence segment must be fast enough to distinguish between closely related species, but not so fast that regions from very distantly related species cannot be confidently aligned. These contradictory requirements can be satisfied, in principle, by a single sequence that has some highly conserved regions and other regions that are more variable between species. When comparing sequences from two closely related species there will be no information in the highly conserved regions (they will be identical) but the more rapidly changing regions will have useful data. Conversely, when comparing data from two distantly related species, the rapidly changing regions will show almost uniform dissimilarity, but the more conserved regions will now have useful variation.

The DNA sequence specifying the small ribosomal subunit rRNA (called 16S RNA in prokaryotes) has been found to be one of the best genomic segments for this type of analysis, despite occurring in several copies in some genomes. This sequence formed the basis of the work by Carl Woese, which led to the original proposal that the prokaryotes comprised two quite distinct domains or superkingdoms and thus that all life should be split into the three domains: Bacteria, Archaea, and Eukarya (see Section 1.4). This work relied on the ability to correctly align sequences from very distantly related organisms. The 16S RNA sequence is now also the basis for a standard method of bacterial identification. A few protein-coding sequences have also been found suitable for determining the evolutionary relationships of species, and some are particularly useful for studying specific groups of species. Those that have been used in bacterial studies include enzymes such as the DNA gyrases GyrA and GyrB, and genes for chaperonins such as Hsp60. A major project is under way to identify for each known species a barcode of unique sequence. For the animal kingdom this project proposes using a 648-base pair segment of the gene for cytochrome *c* oxidase I, a component of the mitochondrial machinery involved in cellular aerobic respiration and found in all animals.

In the following example, we will reconstruct a phylogenetic tree for 38 species of bacteria from the phylum Proteobacteria, using the 16S RNA sequence. We will judge whether our reconstruction is a reasonable one by comparing it with Bergey's classification from 2001 (see Further Reading) in which the Proteobacteria are subdivided into five classes: Alphaproteobacteria, Betaproteobacteria, Gammaproteobacteria, Deltaproteobacteria, and Epsilonproteobacteria. Our 38 proteobacteria are spread across these five classes.

Four decisions must be made before starting the analysis, each of which is not necessarily independent of the others: which data to use, which method to use, which evolutionary model to use, and which (if any) tests to perform to assess the robustness of the prediction of particular tree features. We have already explained why we are using 16S RNA, and although it is advisable to consider possible tests of tree features at an early stage, the decision will be almost independent of the other choices, and so we will delay discussion of this until some trees have been generated.

## The choice of the method for tree reconstruction depends to some extent on the size and quality of the dataset

Numerous algorithms have been proposed for reconstructing phylogenetic trees from a multiple sequence alignment. Methods can be divided into two broad groups: those that first derive a distance measure from each aligned pair of sequences and then use these distances to obtain the tree; and those that use the multiple alignment directly, simultaneously evaluating all sequences at each alignment site, taking each site separately. A number of different methods have been developed in each of these groups, making this a potentially confusing subject for the beginner. Nearly all the methods are available as freeware that will run on almost any type of computer, and there are also some excellent commercial packages. Some applications run as a single program (for example MEGA3) with different methods corresponding to different menu choices, and others such as PHYLIP are suites of programs, each of which implements one method.

Commonly used distance-based methods that are mentioned in this chapter and will be discussed in Section 8.2 are **UPGMA** (unweighted pair-group method using arithmetic averages) and neighbor-joining (NJ). These both produce a single tree with defined branch lengths. Other methods of obtaining branch lengths from evolutionary distances are the **least-squares method** and the Fitch–Margoliash method, which are also discussed in Section 8.4. UPGMA makes the assumption that the sequences evolved at a constant equal rate (the molecular clock hypothesis), and produces rooted ultrametric trees with all sequences at the same distance from the last common ancestor. The NJ method belongs to the group of **minimum evolution methods**, and produces additive trees. It is derived from an assumption that the most suitable tree to represent the data will be that which proposes the least amount of evolution, measured as the total branch length in the tree. **Maximum parsimony** methods also involve a **minimum evolution** principle, but are based directly on the alignment and minimize the number of mutations required to reconstruct the phylogeny. The **maximum likelihood (ML)** method estimates the likelihood of a given tree topology to have produced the observed data assuming a given model of evolution, and reports the topology that produces the greatest likelihood as the most appropriate hypothesis of the evolutionary history. **Bayesian methods** take this statistical approach a step further, trying to include the likelihood of the tree topology. Table 7.1 lists some of the methods available, and gives a basic classification of them. Further details of methods and the underlying assumptions are given in Section 8.4. Some nondistance methods, such as the ML and Bayesian methods, can claim to produce results based on a stronger statistical foundation than others, such as maximum parsimony, and are certainly to be preferred to distance-based methods when adequate resources are available. However, all methods are known to give incorrect results under certain conditions, and each has its supporters and detractors.

Some methods are much more computationally demanding than others and may be impracticable for large datasets. The largest datasets can only be analyzed using distance methods, although improvements in algorithms for methods such as maximum likelihood are making their application more feasible. As an alternative, a subset of a large dataset could be chosen, either for a preliminary analysis to assess suitable models and parameters, or following the application of distance methods to examine some specific features with these more demanding techniques.

In general, the techniques that use distances produce a single tree, whereas most of those methods using the alignment directly can score and report a number of trees that best represent the data, and can also be used to evaluate tree topologies specified by the user. This latter feature allows these techniques to compare particular hypotheses about the data embodied in different trees and to get a quantitative measure of the differing degrees of fit to the sequence data. These considerations will also influence the choice of method.

As we shall see, although the data analyzed here give rather similar phylogenetic trees regardless of the method, the differences between trees can lead to significantly different conclusions. It is highly desirable to be able to justify the methods used and results obtained, and techniques such as maximum likelihood and Bayesian methods offer far more support than those such as neighbor joining. We

| Method name | Conversion of alignment to distances | Resulting tree type | Always results in a single tree | Tree scores | Testing of specified trees | Common implementations |
|---|---|---|---|---|---|---|
| UPGMA | Yes | Ultrametric | Yes | No | No | neighbor*, MEGA3 |
| Neighbor-joining (NJ) | Yes | Unrooted, additive | Yes | No | No | neighbor*, MEGA3 |
| Fitch–Margoliash | Yes | Unrooted, additive | Yes | No | No | fitch* |
| Minimum evolution | Yes | Unrooted, additive | No | Yes | Yes | MEGA3 |
| Quartet puzzling | No | Unrooted, additive | No | Yes | Yes | Tree-Puzzle |
| Maximum parsimony | No | Unrooted, additive/cladogram† | No | Yes | Yes | dnapars*, protpars*, MEGA3 |
| Maximum likelihood (ML) | No | Unrooted, additive | No | Yes | Yes | dnaml*, proml*, HyPhy, PHYML |
| Bayesian | No | Unrooted, additive | No | Yes | Yes | MrBayes |

*Part of the PHYLIP package.

†Maximum parsimony can determine the best tree topology without explicitly calculating branch lengths, which is done in a separate optional step.

All the listed programs can use the sequence alignment, generating distances where necessary. Some methods can report several trees if their scores are all very close to the optimum. These same methods can read a tree file and give a score for that particular topology, allowing hypothesis testing. All methods can be used to test the confidence in the reported topology by the bootstrap or other techniques. A comprehensive listing of relevant programs is available at http://evolution.genetics.washington.edu/phylip/software.html.

**Table 7.1**
**Some features of commonly used phylogenetic tree reconstruction methods and commonly available programs in which they are implemented.**

will use a maximum likelihood method to construct the 16S RNA tree. We chose the program PHYML because it is one of the fastest implementations available, and also because scripts are available, which make choosing the evolutionary model easier.

## A model of evolution must be chosen to use with the method

Many different evolutionary models have been proposed and Table 7.2 gives a summary of the most commonly used. The simplest ones assume little or no variation between sequence sites, and ignore base mutation preferences. They could be expected to perform badly if the actual evolutionary history of the data is more complex than they describe. At the other extreme are models that try to allow for many of the possible variations. These more complex models contain more parameters and, in principle, can fit data that have arisen under a greater range of evolutionary conditions. It should be noted that almost all tree-construction methods available assume that the same model and same model parameters apply at all times on all branches of the phylogenetic tree. The evolutionary model can be applied in two different ways. Those methods that distil the alignment into distances use the model to convert the percentage differences observed in the alignment into more accurate (corrected) evolutionary distances. No further use of the model is made during the reconstruction of the tree. In contrast, techniques that work directly with the sequence alignment use the evolutionary model at all stages of tree reconstruction.

The evolutionary distances used by some of the methods and the branch lengths produced by any method both measure the average number of substitutions (a change in the nucleotide or amino acid) per site. The uncorrected *p*-distance measured from the alignment is simply the fraction of sites that have a difference. When this distance is small, for example 0.1 substitution per site, it is a reasonably accurate measure of the true evolutionary distance. However, an uncorrected distance of

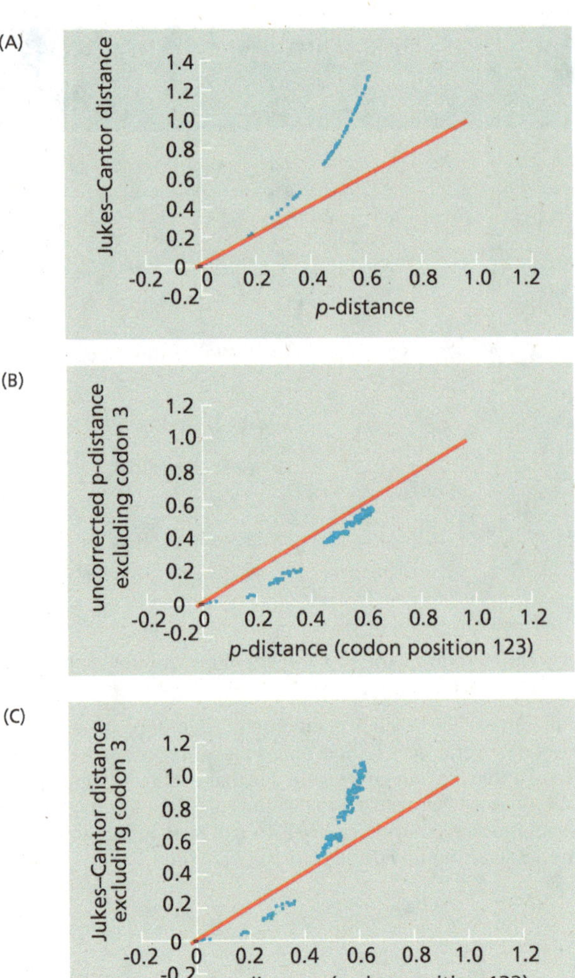


**Figure 7.17**
**Some examples of the corrections applied to *p*-distances and their effects.** In all three graphs the horizontal axis is the fraction of alignment positions at which base differences occur between a pair of sequences being compared (the uncorrected *p*-distance). Each blue dot represents an individual pairwise comparison taken from a large multiple alignment. The vertical axes represent corrected distances. In all the graphs, a plot of the uncorrected *p*-distances against each other would lie along the red line. (A) The standard Jukes–Cantor (JC) correction, with corrected distances (blue dots) always being larger than the *p*-distance. (B) A plot of the uncorrected *p*-distance calculated using all codon positions against the *p*-distance calculated using only the first and second codon positions. Because mutations at the third position in a codon are mostly synonymous, they accumulate at a faster rate than at the first and second positions. As a consequence, after a sufficiently long evolutionary period there might be no phylogenetic signal at these positions. In such circumstances it has been suggested that the third codon position is best ignored, giving altered *p*-distances as shown. For small uncorrected *p*-distances these altered distances are very small, as most of the accepted mutations have occurred in the third codon position. These altered distances are always smaller than the uncorrected value, as any changes at the third codon position are ignored. At larger *p*-distances the two values are highly correlated, so that this technique may result in very similar trees. An alternative solution to the problem of obtaining accurate large evolutionary distances would be to use protein sequences instead, still using a correction to the protein sequence *p*-distance. (C) The result of applying the JC correction to the *p*-distance measured without the third codon position. Such a correction results in large changes to the distances and may produce quite different phylogenetic trees. However, care must be taken to choose correction methods carefully, as they may not necessarily be appropriate for the particular dataset, as is discussed later in the chapter.

even 0.2 substitution per site will conceal some multiple substitutions at the same site, and hence underestimate the true evolutionary distance (see Figure 7.7). The models of evolution can be used to derive distance corrections, with equations such as those presented in Section 8.1, giving a corrected distance $d$ in terms of the uncorrected $p$-distance. Some examples of the effects of distance corrections are shown in Figure 7.17.

Regardless of the model chosen, it will have one or more parameters whose values will need to be specified. Among the parameters that might be required are the transition/transversion rate ratio $R$, the base composition, and parameters to specify the variation of mutation rates at different positions. Many of the parameter values can be determined from an analysis of the sequence alignment, and this is often done by default by the programs used to generate the trees. In this respect the maximum likelihood and Bayesian methods have an advantage, because they can both fit parameters along with the tree to obtain the best tree and model parameters simultaneously.

An apparently sensible approach to selecting the evolutionary model would be to try several different models, and select the tree that best fits the data. This has problems, however. For distance methods such as neighbor-joining there is no clear way of measuring goodness of fit, as such methods do not use the evolutionary model after calculating the distances. Thus, one could only judge the effectiveness of the model by looking for disagreement between the intertaxon distances in the data and in the trees generated by each model. The situation is a little easier for those

| Model name | Base composition | Different transition and transversion rates | All transition rates identical | All transversion rates identical | Reference |
|---|---|---|---|---|---|
| JC (JC69) | 1:1:1:1 | No | Yes | Yes | Jukes and Cantor (1969) |
| Felsenstein 81 (F81) | Variable | No | Yes | Yes | Felsenstein (1981) |
| K2P (K80) | 1:1:1:1 | Yes | Yes | Yes | Kimura (1980) |
| HKY85 | Variable | Yes | No | No | Hasegawa et al. (1985) |
| Tamura-Nei (TN) | Variable | Yes | No | Yes | Tamura and Nei (1993) |
| K3P (K81) | Variable | Yes | No | Yes | Kimura (1981) |
| SYM | 1:1:1:1 | Yes | No | No | Zharkikh (1994) |
| REV (GTR) | Variable | Yes | No | No | Rodriguez et al. (1990) |

In its basic form each of these models assumes that all sites in a sequence behave identically. A site-variation model can be imposed on any of these. The most common site-variation model used is the Gamma distribution, in which case often +G or +Γ is added to the model name, for example HKY85+Γ. It is also possible to allow some sites to be invariant, in which case '+I' is often added to the name. Commonly used alternative names are given in parentheses.

**Table 7.2**
**Models of molecular evolution that have been used in phylogenetic analysis.**

methods that work directly from the alignment, as there is a clearer connection between the model and the resulting tree. This is especially true in the case of maximum likelihood and Bayesian methods, where each tree can be given a score and thus can be compared to identify the best model. The statistical bases for the methods presented below that can be used to recognize when the tree produced using one model is significantly better than the tree of another model are beyond the scope of this book, and there is some debate as to their true merit. However, at present these statistical techniques are the only ones that can claim any rigor in choosing a model beyond personal preference and prejudice.

A way of comparing two evolutionary models has been proposed, called the **hierarchical likelihood ratio test** (**hLRT**), which can be used when one of the models is contained in, or nested in, the other. One model is nested within another if the latter model is more flexible than the former by having more parameters, and it must be possible to convert the more complex model into the simpler one by restricting specific parameter values. Examples of suitable pairs are JC and F81 (derived by Joseph Felsenstein in 1981), where the latter is the former plus variable base composition, and JC and Kimura-2-parameter (K2P), where the latter is the former plus transition/transversion rate ratios (see Table 7.2). In this method both models must be assessed with the same tree topology. Using the ML method for each evolutionary model in turn, the tree branch lengths are found that give the highest likelihood of having observed the sequences. These two likelihood values can then be compared using hLRT, which is a chi-squared test based on the difference in likelihoods and the difference in numbers of parameters in the two models. Figure 7.18 outlines the procedure for choosing a model.

There are problems with hLRT, in particular the fact that it is difficult to include some models in the scheme because of the requirement that models are nested. Also, to decide between a number of models requires several tests, which leads to problems in deciding the order of testing and the significance levels to use. Two other methods have been proposed, both based on a statistical analysis of the expected value of the maximum likelihood of trees, and taking account of the number of parameters in the evolutionary model used. The **Akaike information**

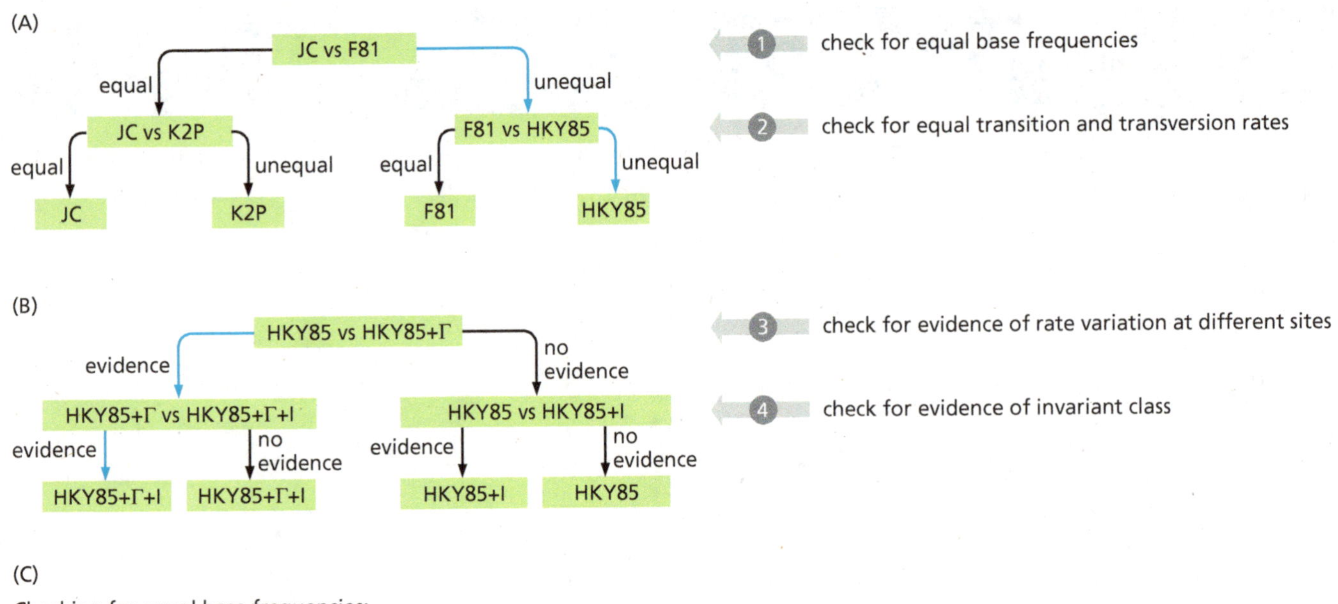


(C)

Checking for equal base frequencies:

| | | |
|---|---|---|
| $H_0$: JC | Log-likelihood = -19864.051 | 17 parameters |
| $H_1$ : F81 | Log-likelihood = -19859.027 | 20 parameters |

| | |
|---|---|
| Likelihood ratio statistic: | 10.048 |
| Degrees of freedom: | 3 |
| Probability: | 0.018156 (< **0.05**) |

Null hypothesis $H_0$ rejected: F81 chosen

**Figure 7.18**
**Examples of hierarchical likelihood ratio tests (hLRT) for determining the appropriate evolutionary model to use.** The values shown are those from the analysis of the 10-sequence subset of the 16S RNA dataset with the HyPhy program. (A) Initial tests are used to determine whether the 16S RNA data require an evolutionary model that includes unequal base composition and whether allowance must be made for differences in transition and transversion mutation rates. A series of hLRT model comparisons are run. The first models tested are Jukes–Cantor (JC) versus Felsenstein 81 (F81). Depending on the outcome of this test, the models used in the subsequent rates test will be different as shown. As the base composition rate is unequal according to the test (blue arrow), the JC model is discarded and F81 is tested against the HKY85 model by testing to see if transitions and transversions have different accepted mutation rates. HKY85 is an extension of the Kimura two-parameter model that takes into account the %GC ratio of the sequences being considered. The HKY85 model fits the data best. (B) Using hLRT, HKY85 is tested against HKY85+Γ, which incorporates the Gamma distribution to take into account mutation rate variation at different sites. Eventually, HKY85+Γ+I (which also includes the existence of a set of invariant sites) is chosen as the most appropriate model. Note that in practice many more models and tests were made than shown here. The full set of tests used resulted in the selection of the REV+Γ model. (C) The results of the first test, with the log-likelihoods of the trees, numbers of parameters involved in the models and trees, the statistical test values, and conclusion. In this case, the conclusion is that the data have significantly unequal base frequencies, leading to the rejection of the JC model.

**criterion** (**AIC**) is defined as $-2 \ln L_i + 2p_i$, where $L_i$ is the maximum likelihood value of the optimal tree obtained using the evolutionary model $i$ in a calculation which has $p_i$ parameters. These $p_i$ parameters are not just those of the model $i$, but also include the variables in the tree, normally the $N-3$ branch lengths of an unrooted fully resolved $N$-taxa tree. The **Bayesian information criterion** (**BIC**) depends on the size of the dataset $N$ as well as $p_i$ and is defined by $-2 \ln L_i + p_i \ln(N)$. However, the precise meaning of the dataset size $N$ in this context is not well defined, and has been associated with both the total number of alignment positions and also with just the number of variable alignment positions. Whichever is used, in most phylogenetic studies models with more parameters should be penalized more by the BIC than by the AIC. These methods do not require the evolutionary models to be nested, and also in contrast to the hLRT tests do not require the identical tree topology to be used in all calculations. Analysis using the AIC, which measures the support in the data for the given model, is straightforward: the model with the smallest AIC value is regarded as the most suitable. Any models having an AIC within two units of this are regarded as also being supported by the data, and those models up to seven units away as having a smaller level of support. Beyond this, the models are not supported. The analysis using the BIC proceeds in a similar manner. We will use the AIC method to choose an evolutionary model for the 16S RNA data. Note that there is evidence of a tendency of these methods to support more complex models when simpler models can produce better trees. Much more research is needed in this area to establish the usefulness of these methods.

## All phylogenetic analyses must start with an accurate multiple alignment

A multiple alignment of the 16S RNA data is required to construct the phylogenetic tree. Using the Web interface to a database that contains accurate multiple alignments of 16S RNA sequences, the Ribosomal Database Project (the RDP database), we can select a set of species and download the multiple alignment. This alignment is the starting point for the phylogenetic analysis and its accuracy is key to the whole process. Because the RDP database covers a very wide range of life, and because some of the set of 16S RNA sequences contain extensive insertions, there are numerous gaps in the alignment at many different positions. As a result, the downloaded alignment is extremely large compared with the actual lengths of the sequences. The multiple alignment for the entire database extends over 40,000 positions, but the small subset of chosen sequences has aligned bases at fewer than 2500 of these positions. It is useful therefore to start by removing all the columns that contain gaps only. This facility is present in almost every program that can view alignments.

## Phylogenetic analyses of a small dataset of 16S RNA sequence data

A protocol based on the AIC method for selecting an evolutionary model is freely available as a script called MrAIC. This script was used with the PHYML program to generate AIC values for a number of alternative models, using the set of 16S RNA sequences. The dataset used in this example is small enough to use in its entirety for the AIC analysis, but a subset might need to be chosen for some other datasets. Three evolutionary models were reported to have significant support as defined by AIC values within two units of the best-supported model, which in this case was REV+Γ, the **general time-reversible model** (known as GTR or REV) with an additional Gamma distribution model of site-rate variation (see Section 8.1). Other resources for such exploratory studies are available, including a program called ModelTest that can perform hLRT and AIC analyses based on likelihood values that can be obtained independently from any program of choice. The HyPhy program is slower than PHYML, but can perform a more complete analysis as it can calculate ML trees as well as apply AIC and nested model tests using the hLRT. The HyPhy results for a 10-sequence subset of this dataset are presented in Table 7.3 and give further support for the use of REV+Γ. The reason for using just the subset was purely due to the higher computational requirements of the HyPhy program.

Using the REV+Γ evolutionary model, an unrooted maximum likelihood phylogenetic tree is generated from our data using the program PHYML. This maximum likelihood tree is shown in three representations in Figure 7.19. If the current formal taxonomic classification reflects the true evolutionary history, then all five classes should be monophyletic. That is, the members of the class should form a group that includes only themselves and their last common ancestor. The unrooted tree reported by PHYML (see Figure 7.19A) shows three of the classes—the Alphaproteobacteria (green), Betaproteobacteria (blue), and Epsilonproteobacteria (yellow)—as monophyletic. The

**Table 7.3**
**Some of the data used to select the best model of evolution for use in further work on the 16S RNA dataset.**

| Model | Number of parameters | log-likelihood (ln L) | AIC |
|---|---|---|---|
| JC | 17 | −19864.051 | 39762.103 |
| F81 | 20 | −19859.027 | 39758.054 |
| HKY85 | 21 | −19779.596 | 39601.191 |
| HKY85+I | 22 | −19602.542 | 39249.084 |
| HKY85+Γ | 22 | −19462.357 | 38968.715 |
| HKY85+Γ+I | 23 | −19456.705 | 38959.411 |
| REV+Γ | 26 | −19426.581 | 38905.161 |

The values of the log-likelihood and AIC for the trees created by HyPhy using the 10-sequence subset of the 16S RNA dataset with the different evolution models listed. The number of parameters in the system is also shown, as it is used in hLRT tests as well as in calculating the AIC values.

Gammaproteobacteria (red) and the Deltaproteobacteria (magenta) are not monophyletic, however, according to this tree, but form **clades** with the Betaproteobacteria and the Epsilonproteobacteria, respectively. In fact, in phylogenetic studies, the Gammaproteobacteria (red) are often found to form a clade with the Betaproteobacteria, and it has been proposed that the two should be combined into a new taxonomic class called Chromatibacteria.

Most tree-viewing programs allow the user to choose a root for an unrooted tree. From other phylogenetic studies, the root of the proteobacterial tree is generally considered to lie either between the Deltaproteobacteria/Epsilonproteobacteria and the other classes, as shown in Figure 7.19B, or between the Epsilonproteobacteria and other classes. An alternative, but unreliable, way of choosing a root is to find the midpoint of the route between the most distantly related taxa, in this case *Buchnera aphidicola* and *Campylobacter jejuni*, which produces the tree shown in Figure 7.19C.

The rooted maximum likelihood trees show considerable variation in the distances from the leaves to the root, as shown by the branch lengths. One might suspect from this that using UPGMA, which assumes a molecular clock, to reconstruct the tree would lead to a significantly different tree. Figure 7.20 shows two rooted ultrametric trees reconstructed from the data using UPGMA and the JC distance correction model of evolution. The two trees differ in the way that alignment gaps are treated in the distance correction. The tree in Figure 7.20A ignores any alignment position that includes a gap, even if only one of the 38 sequences has a gap at this position. Since the distances are calculated between two sequences at a time, this means that any two sequences may have aligned sequence that is ignored because of gaps in other sequences. The tree in Figure 7.20B ignores only the positions with gaps in the particular pair of sequences for which the distance is being calculated, and thus uses more of the sequence data. Arguments can be made in favor of both methods of dealing with gaps, and it can be useful to try both. While the complete deletion method potentially ignores useful data, it can also avoid potentially misleading results from using regions that occur in only a subset of the taxa and whose evolution has been significantly different from the other parts of the sequences. Both of these UPGMA trees differ in several places from the presumably more accurate tree of Figure 7.19, and so it is not possible in this example to say which deletion method has in this case given the poorer result. For example, the Betaproteobacteria appear to evolve in a

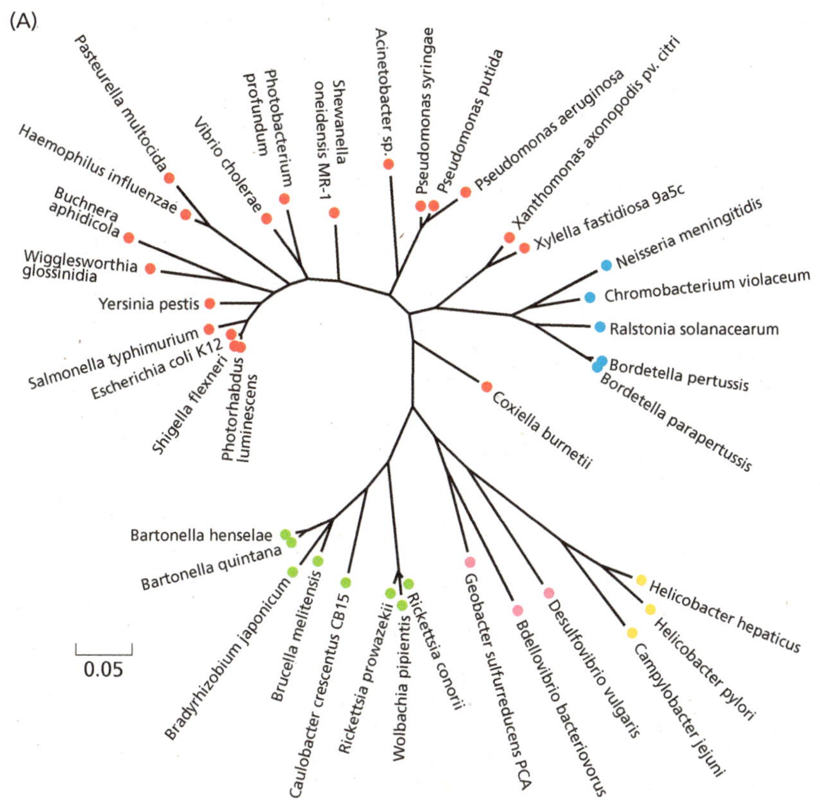


**Figure 7.19**
**The maximum likelihood tree for the 16S RNA dataset, as calculated using PHYML with the REV+Γ model.** (A) The unrooted tree initially produced by the program. The taxa are color-coded according to their class in the Bergey bacterial classification: green, Alphaproteobacteria; blue, Betaproteobacteria; red, Gammaproteobacteria; magenta, Deltaproteobacteria; and yellow, Epsilonproteobacteria. In this case the root has to be applied manually. (B) The same tree with the root chosen so that the Deltaproteobacteria and Epsilonproteobacteria diverge from the other classes first. (C) The same tree but now rooted at the midpoint between the two taxa furthest apart on the tree.

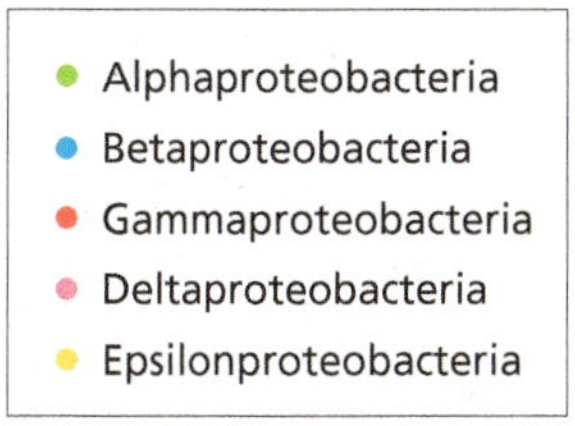

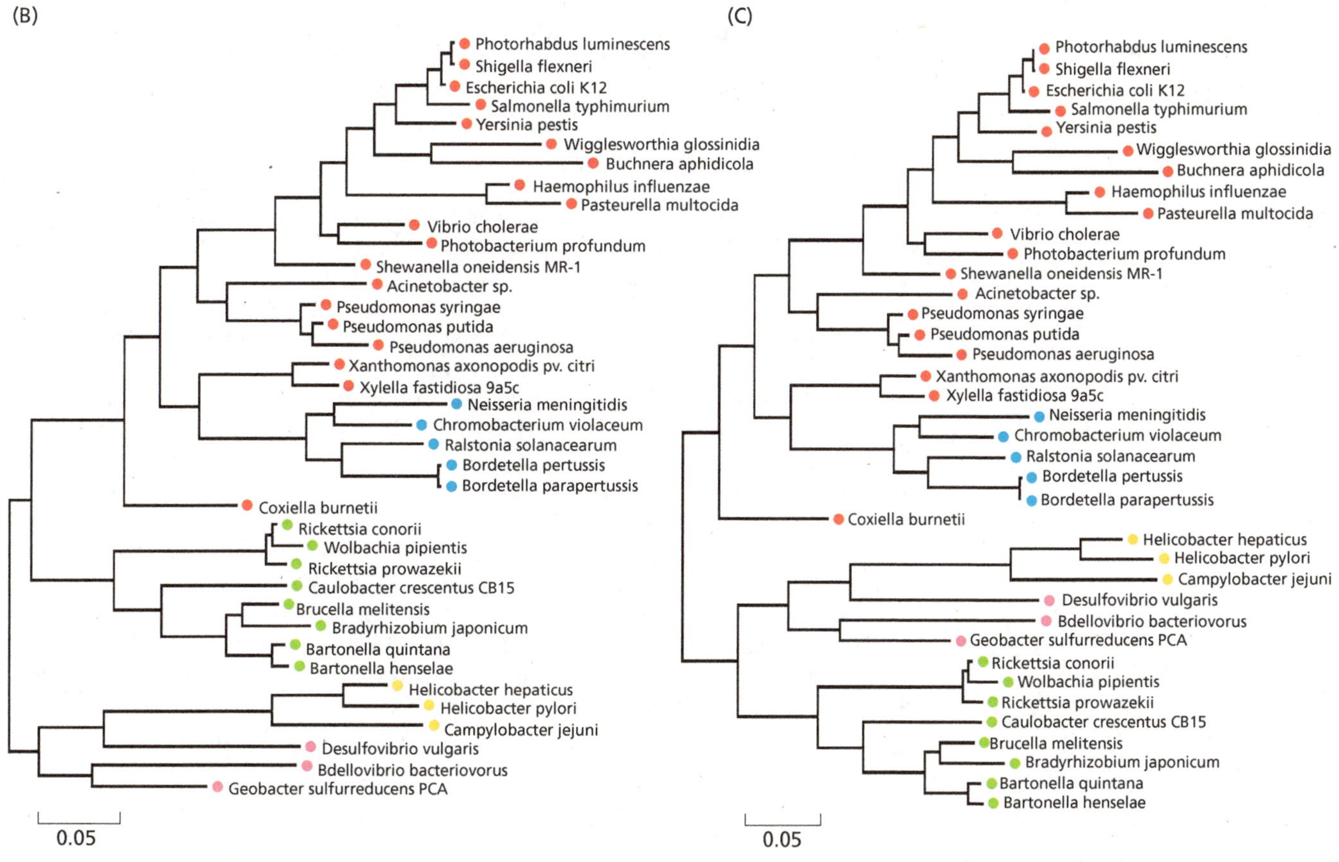

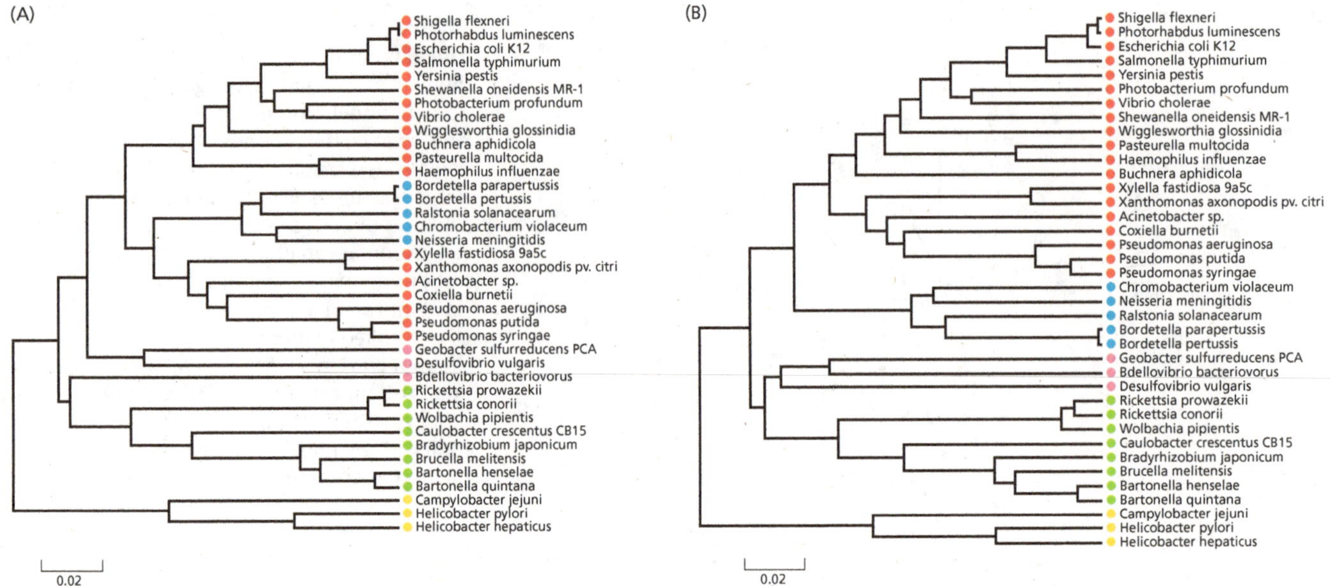


**Figure 7.20**
**UPGMA trees calculated for the 16S RNA dataset using the JC model distance correction in two different ways.** (A) Any alignment position at which any sequence that has a gap is ignored. (B) Alignment positions are only ignored when one of the two sequences being compared has a gap. The colors of the taxa labels are as in Figure 7.19.

different way from the Gammaproteobacteria in Figure 7.20A when compared with Figure 7.19. But both these trees differ from Figure 7.20B in which all five classes are monophyletic, which is how the tree would be expected to appear if both the tree and the preexisting taxonomic classification were an accurate representation of the evolutionary history of the species.

It would be useful to know how strongly certain features of the trees are supported by the data. The reliability of the topology can be tested by the bootstrap (see Figure 7.5). In Figure 7.21 a condensed tree resulting from Figure 7.20A is shown. In this case we have removed branches with a bootstrap value less than 75%, and this gives a tree which shows that within the limits of confidence given the five classes are monophyletic. The bootstrap test can thus give some indication of the reliability of conclusions drawn from the trees. Bootstrap values are in fact conservative, however, so that the 5% significance level will occur for branch values somewhat less than 95%. It should be noted that what the bootstrap measures is the degree of support within the data for the particular branch, given the evolutionary model and tree reconstruction method. The bootstrap values give no indication of the robustness of these features to changing the model or method.

In this example, we have rooted the tree using information from other studies, or by using the UPGMA method, which always produces rooted trees. There is no alternative in this case because our data do not contain any sequences from species that are more remote from the majority, known as the outgroup (see Figure 7.2D). When an accurately rooted tree is required, such groups should be included. The root can then be located between the outgroup and the remainder of the taxa.

This small study shows, albeit using evolutionary models and tree-reconstruction methods that differ greatly from each other, that the choice of model and reconstruction method can make a significant difference to the resulting tree and thus the conclusions that one draws from it. One should in any case be wary of using even the most carefully reconstructed single tree to deduce evolutionary relationships between species. And in all cases it is important to assess the reliability of the tree topology, using tests such as bootstrapping. In addition, in classification studies it is useful to compare trees generated for the same set of species with different genes. It is quite possible that some genes will have evolved in an anomalous way for some of the species, and this will become obvious when these trees are compared.

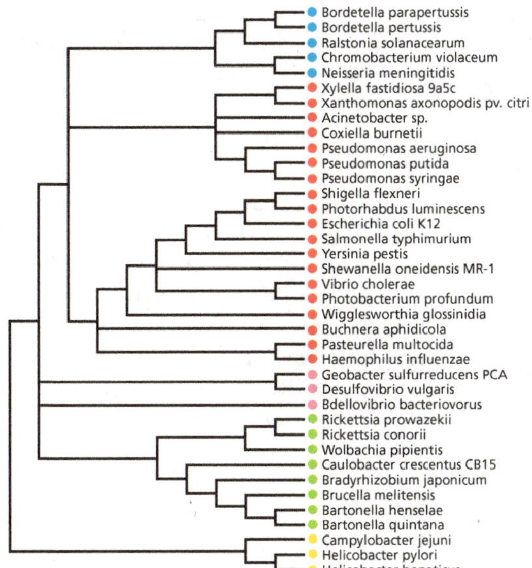


**Figure 7.21**
**The condensed tree obtained by a bootstrap analysis of the UPGMA tree of Figure 7.20.** Branches with bootstrap values below 75% have been removed. The colors of the taxa labels are as in Figure 7.19.

## Building a gene tree for a family of enzymes can help to identify how enzymatic functions evolved

As the number of known sequences has increased, many have been classified into families and **superfamilies** of homologs with related sequence and functions (see Box 7.4). In our second example, we are going to build a tree for a small selection of sequences taken from a large superfamily of enzymes. Note that the protein family databases generally classify the domains of proteins individually. The proteins used in this example all consist of three domains so will be identified as belonging to up to three entries in some databases.

In this example, we shall show the usefulness of phylogenetic analysis to address two common questions that are often asked about superfamilies. The first is predicting the function of a superfamily member if only its sequence is known. In principle, this might be done by multiple sequence alignment or a sequence pattern search (see Chapters 4 and 6), but is often helped by reconstructing a phylogenetic tree for superfamily members from a range of species and determining

### Box 7.4 Protein families and superfamilies

A gene or protein family is generally defined as a group of closely related homologs with essentially the same function. Note that function in this context usually refers only to biochemical function; in the case of enzymes, for example, proteins in the same family will carry out the same chemical reaction, have identical catalytic mechanisms, and act on rather similar substrates. For example, the acetolactate synthase (ALS) family contains enzymes that use the cofactor thiamine diphosphate (TDP). Homologous families are grouped into superfamilies. Because the protein families that make up a superfamily will have distinct (but usually related) functions, individual members of a superfamily will all be homologs but can have different functions from each other. For example, the ALS protein family belongs to a superfamily in which all the members bind and use the cofactor TDP, but their use of the additional cofactor flavin adenine dinucleotide (FAD) varies, and the reactions they catalyze are related but different. Many proteins consist of two or more distinct domains, and protein family databases usually classify domains individually into domain families. The nature of protein domains and their relation to the classification of protein structures will be discussed in more detail when we consider protein structure prediction in Section 14.1.

orthologs, as discussed earlier. The second question, relating particularly to superfamilies of enzymes, is how to elucidate the evolution of enzyme function within the superfamily. This can only be done using phylogenetic analysis.

The superfamily of enzymes we shall look at all use the cofactor thiamine diphosphate (abbreviated TDP or TPP after its alternative name of thiamine pyrophosphate). Some of the constituent families also use the cofactor flavin adenine dinucleotide (FAD), however. Unusually, in some cases, FAD does not participate in the reaction but it must still be bound for the enzyme to be active. We have selected enzymes from various families to build the tree, and should like to see what the phylogeny of these enzymes can tell us about how these properties evolved. The evolution of the superfamily must have involved either the development or loss of an FAD-binding site (or both), possibly on more than one occasion, and a correct reconstruction of the phylogenetic tree will reveal this history. A number of the bacterial enzymes included in this analysis have been studied experimentally, which enables a clear assignment of their catalytic function and cofactor use. In addition, structures are available for representatives of several families. All those determined have the same protein fold, but not all bind FAD. These experimental data will be extremely useful in analyzing the results of the phylogenetic reconstruction.

Out of the hundreds of sequences in this superfamily, we have selected a subset of sequences from enterobacteria (the gut bacteria such as *Escherichia coli* and *Salmonella* and their relatives), which are a subgroup of the Gammaproteobacteria. Some sequences that could have been included are only distantly related, causing problems with alignment, and are not well characterized experimentally; these were not included in the study. Eight experimental structures for the whole superfamily were available at the time of the analysis, and we have included the corresponding sequences, even though some are from yeast and not from bacteria. In total, 39 sequences were used. For reasons discussed below, no attempt was made to include an outgroup.

The first task is to obtain an accurate multiple sequence alignment. It is usually preferable to carry out phylogenetic analysis on nucleotide sequence data, in which synonymous as well as nonsynonymous mutations will be visible. An exception is when the sequences are so distantly related that the large number of mutations has resulted in a poor phylogenetic signal. In these cases amino acid sequences can be used. It is, however, more difficult to accurately align protein-coding nucleotide sequences than protein sequences because the methods do not recognize the existence of codons, a feature that, as has been discussed above, plays a very important role in the evolution of such sequences. In general, alignments of such nucleotide sequences should include only insertions or deletions of entire codons. As all the data in our example are protein-coding, the ideal alignment procedure is to translate the nucleotide sequence to protein, align this protein sequence, and then retranslate the alignment back to the nucleotide sequences before generating the tree. A number of programs conveniently perform this commonly required procedure, including the free programs BioEdit and MEGA3. The end result is an accurate multiple alignment of nucleotide sequences that can readily be retranslated to protein if required. Any additional information such as protein structure or key functional conserved amino acids should be included to try to make the alignment as accurate as possible, using manual adjustment to achieve this.

The complete evolutionary history of a large superfamily such as the one being used here is best studied in several steps, initially exploring the evolution of those sequences that are more closely related. Once different parts of the superfamily are better understood, attempts can be made to put the whole picture together. Because we have selected a group of closely related sequences, our dataset lacks clear outliers and a root cannot be justifiably identified. The lack of a root will, however, limit the conclusions that can be made.

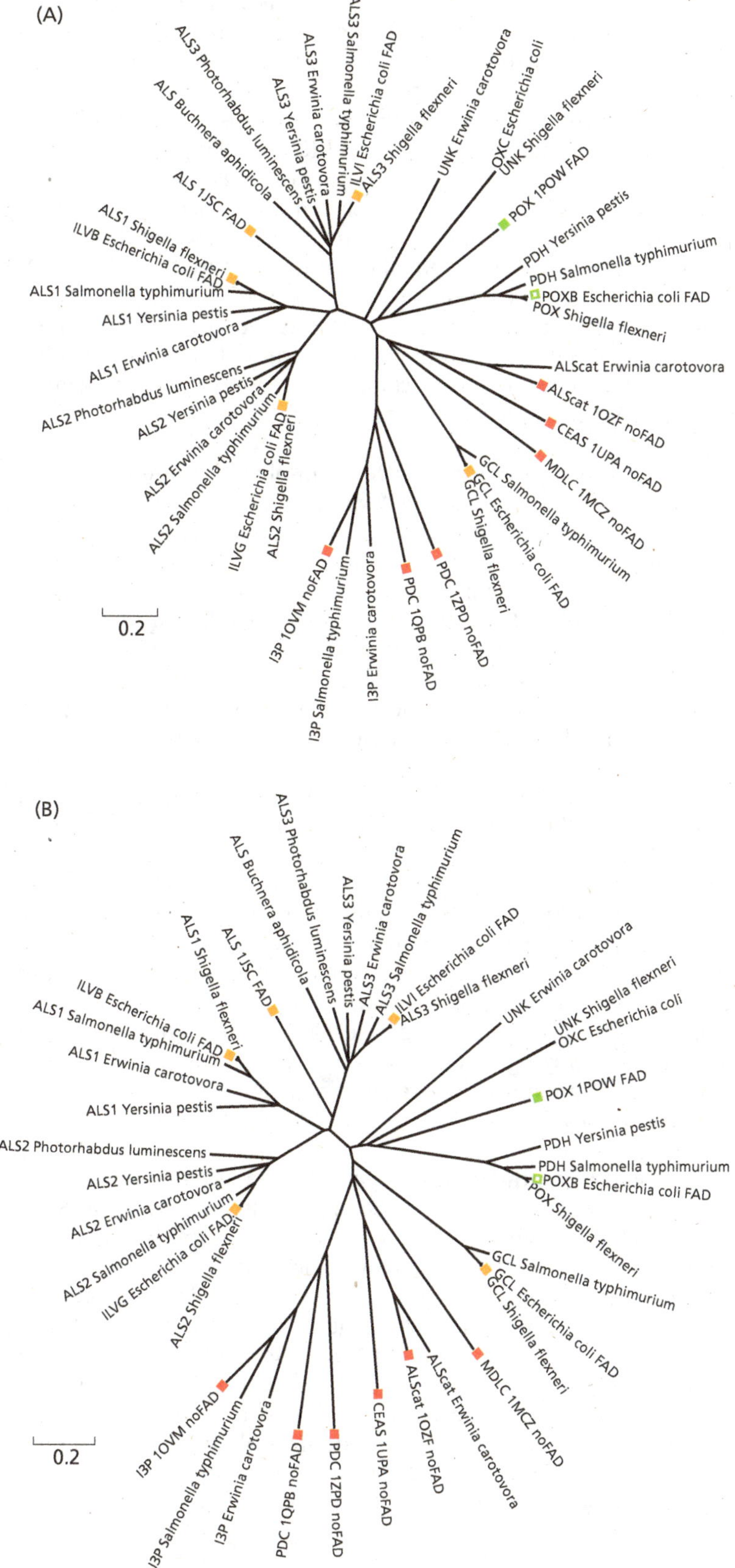


**Figure 7.22**

**The unrooted neighbor-joining (NJ) tree for the TDP-dependent enzyme superfamily dataset produced using two different distance corrections.** (A) Tree constructed using the JC model distance correction. (B) Tree constructed using the K2P model distance correction. Proteins for which an experimental structure is known that does not contain a bound FAD cofactor are indicated by red squares. Proteins known to bind FAD and to use it as an active cofactor in the reaction are indicated by green squares. Proteins known to bind FAD, but which do not use it in catalysis, are indicated by orange squares. Filled green or orange squares indicate proteins whose structures are known. The enzyme names have been abbreviated as follows: ALS, acetolactate synthase; CEAS, carboxymethylarginine synthase; ILV, acetolactate synthase; I3P, indolepyruvate decarboxylase; PDC, pyruvate decarboxylase; GCL, glyoxylate carboligase; MDLC, benzoylformate decarboxylase; OXC, oxalyl-CoA decarboxylase; POX, pyruvate oxidase; PDH, pyruvate dehydrogenase; UNK, proteins of unknown enzymatic function.

Once a multiple alignment has been generated, it is simple to use distance-based methods to generate trees, making arbitrary choices for the model and parameters. This approach can be used to take a very preliminary look at the data, but has the significant disadvantage that it can potentially be rather misleading. Figure 7.22 shows the unrooted trees that are produced using the neighbor-joining method with two simple evolutionary models: JC and K2P. Looking at individual clusters, many features are conserved between the two trees. For example, both contain three related clusters that separate the acetolactate synthase sequences ALS1, ALS2, and ALS3. However, the K2P tree (see Figure 7.22B) shows a split between those proteins known not to bind FAD (red labels) and those that do bind FAD (orange and green labels). This split does not occur for the JC tree (see Figure 7.22A). Because we cannot assign a root with any certainty, we cannot determine if the last common ancestor of these proteins did or did not bind FAD. Whichever is the case, the history represented by the JC tree would require more complex evolution of the potential/actual FAD-binding site than the K2P tree.

Using the same methods as for the 16S RNA example, the appropriate evolutionary model was again determined using PHYML and AIC to be REV with a Gamma-distribution mode of site mutation rate distribution, this time also including a parameter for the fully conserved (invariant) fraction of sites (REV+$\Gamma$+I). With this model and the maximum likelihood method of tree reconstruction, the tree in Figure 7.23A was obtained. From the perspective of evolution of the FAD-binding site, this tree shows simpler evolution than the NJ trees, as all three clusters (red, orange, and green) are monophyletic. While this might be an intuitive reason to prefer this tree, it is the method of model selection and use of maximum likelihood that give us the proper basis on which to prefer it to the trees in Figure 7.22.

This tree can now be used to try and predict protein functions for the two sequences from *Shigella flexneri* and *Erwinia carotovora* (highlighted in Figure 7.23A) for which no function has previously been assigned. The *Shigella* sequence is found very close to the OXC (oxalyl-CoA decarboxylase) of *Escherichia coli*, and will almost certainly have the same biochemical function. The *Erwinia* sequence is also in that region of the tree, but is not so closely associated with any protein whose function has been determined experimentally. From its position, it will probably function in a similar way to either OXC or pyruvate oxidase (POX), although it is also not far from the glyoxylate carboligase (GCL) cluster. This type of analysis has been given the name **phylogenomics**, and is of increasing importance because of the large number of genome sequences now available for which no experimental data on function are available.

If the tree is rooted along the branch separating the FAD-binding and non-FAD-binding proteins (see Figure 7.23B), the acetolactate synthase ALS1, ALS2, and ALS3 clusters can be clearly seen to be derived from two duplications that occurred before the species diverged. Note that (allowing for some gene loss) each cluster has almost identical species topology. From the order of the branching, one can see that the first duplication gave rise to the ALS3 gene lineage and another lineage that subsequently duplicated to give ALS1 and ALS2. The ALS 1JSC sequence in this part of the tree can be ignored in this context, as this protein is from yeast and is included as the closest structure known so far in this part of the superfamily. Examination of more sequences from the databases shows that the duplications of the ALS genes appear to have occurred in the enterobacteria only, and thus after the enterobacteria diverged from other Gammaproteobacteria.

We will not pursue this study further here, but there are some obvious ways to proceed. Firstly, alternative hypotheses as represented by trees such as those of Figures 7.22 and 7.23 can be compared, to see just how much confidence can be placed in the monophyly of the different groups of FAD-binding enzymes. This is easily done using maximum likelihood methods, as most ML programs allow the user to input their own tree topologies. Secondly, other regions of the superfamily could be examined to learn more about their evolutionary history. Finally, a complete tree for the superfamily could be attempted through careful selection and alignment of sequences. In this way the evolutionary history could be revealed, and the functional prediction of related sequences might be improved.

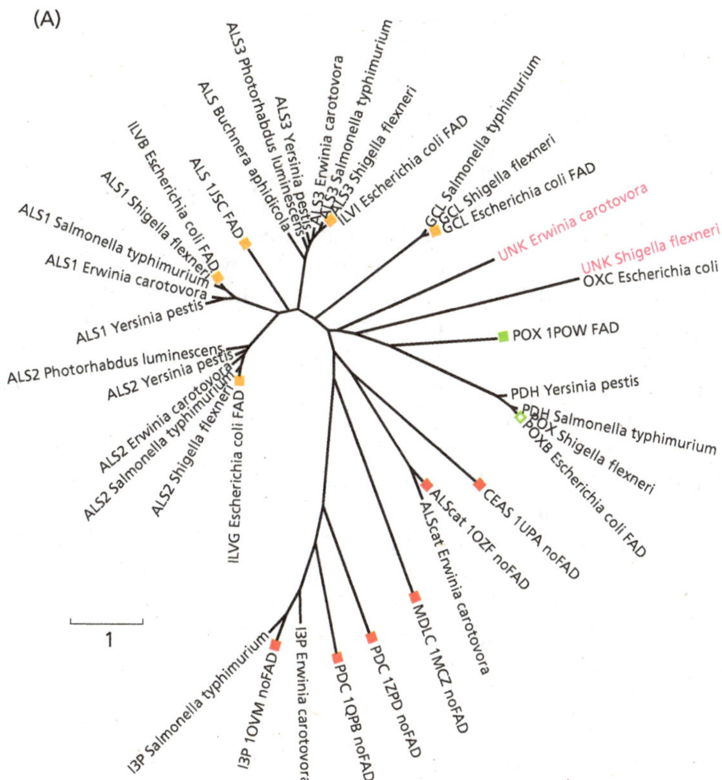


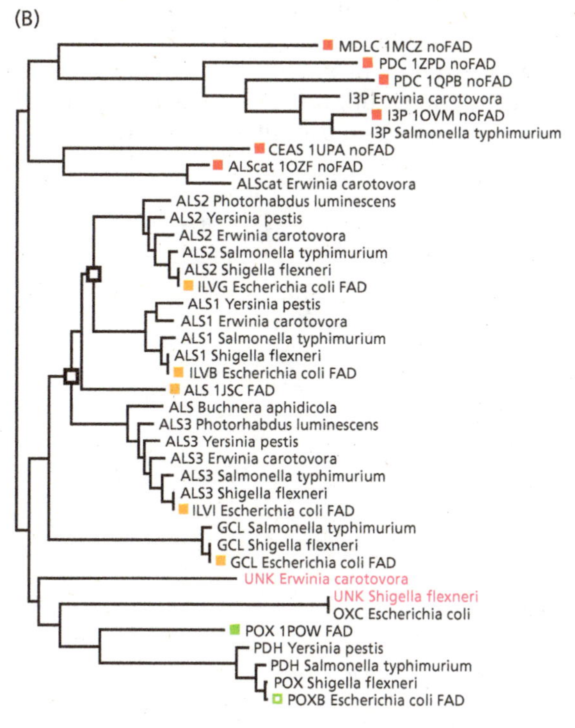


**Figure 7.23**
**The maximum likelihood (ML) tree for the TDP-dependent enzyme superfamily dataset using the REV+Γ+I model.** Two of the proteins are assigned in the sequence database as UNK, meaning that their function was not known or proposed on the basis of sequence analysis. Their positions within the tree have been highlighted, and possible function predictions are discussed in the text. (A) The unrooted tree, with enzyme abbreviations as in Figure 7.22. The colored squares relate to structural and cofactor use information as described in the legend to Figure 7.22. (B) The same tree rooted along the branch that separates the FAD-binding proteins (green and orange squares) from those that do not bind FAD (red squares).

## Summary

Phylogenetic trees reflect the evolutionary history of a group of species or a group of homologous genes and are constructed using the information contained in alignments of multiple homologous sequences. They work on the general assumption that the more similar two sequences are to each other, the more closely related they will be. The evolution of a set of homologous existing sequences can thus be represented in the form of a branching tree, with hypothetical common ancestral sequences at the base, the extant sequences at the tips of the branches, and intermediate hypothetical ancestors forming the internal nodes.

The choice of sequences is important and is determined by what type of tree is required. If an accurate tree reflecting the evolutionary history of a set of species is to be built, the sequences chosen to represent each species must be orthologs, and must also not include genes that might have been acquired by horizontal gene transfer. Gene family trees, in contrast, can include any homologous sequences that can be accurately aligned. Trees may also be unrooted or rooted. To obtain a root, one needs a known sequence or group of sequences that are remote from those of real interest, and yet are sufficiently related so that they can be aligned. In all cases, it is imperative to start with the best-possible multiple sequence alignment, which for protein-coding regions means not using nucleotide sequences for that step. However, as the nucleotide sequence contains the evolutionary history, it is quite usual to begin with a set of nucleotide sequences, convert them into protein-coding sequences for alignment, and then convert the alignment back to nucleotide sequences for actual tree construction.

There are numerous methods and algorithms for tree construction. Some work directly from the multiple sequence alignment, whereas others use evolutionary distances calculated between each pair of sequences in the alignment. Distance methods tend to produce a single tree, whereas other methods often produce a set of trees that then have to be compared and evaluated using additional techniques in order to choose the one most likely to be correct. In all cases, a method has to work in conjunction with a simplified model of molecular evolution, which aims to take account of some of the constraints on sequence evolution and the fact that not all the evolutionary history will be visible in the alignment. In general, different methods and models will produce different trees, so it is important to make the best choice for the data. Given that different methods give different trees, and not all features of the tree are equally well-supported by the data and methods used to generate them, care must be taken not to overinterpret phylogenetic trees.

# Further Reading


**General**

For further information on the theoretical background to the evolutionary models and the various techniques of tree construction, see Chapter 8 and the references therein.

A number of books have been written on this field, of which the following are particularly clear.

Felsenstein J (2004) Inferring Phylogenies. Sunderland, MA: Sinauer.

Nei M & Kumar S (2000) Molecular Evolution and Phylogenetics. New York: Oxford University Press.

## 7.1 The Structure and Interpretation of Phylogenetic Trees

**Phylogenetic trees reconstruct evolutionary relationships**

Gaucher EA, Thomson JM, Burgan MF & Benner SA (2003) Inferring the palaeoenvironment of ancient bacteria on the basis of resurrected proteins. *Nature* 425, 285–288.

**Tree topology can be described in several ways**

Maddison DR, Swofford DL & Maddison WP (1997) NEXUS: An extensible file format for systematic information. *Syst. Biol.* 46, 590–621.

NHX (New Hampshire Extended) http://www.phylogenomics.us/forester/NHX.html

**Consensus and condensed trees report the results of comparing tree topologies**

Iwabe N, Kuma K & Miyata T (1996) Evolution of gene families and relationship with organismal evolution: rapid divergence of tissue-specific genes in the early evolution of chordates. *Mol. Biol. Evol.* 13, 483–493.

Page RDM & Charleston MA (1997) From gene to organismal phylogeny: reconciled trees and the gene/species tree problem. *Mol. Phylogenet Evol.* 7, 231–240.

## 7.2 Molecular Evolution and its Consequences

**The rate of accepted mutation is usually not the same for all types of base substitution**

Janacek LL, Honeycutt RL, Adkins RM & Davis SK (1996) Mitochondrial gene sequences and the molecular systematics of the Arteriodactyl subfamily Bovinae. *Mol. Phylogenet. Evol.* 6, 107–119.

Nei M & Kumar S (2000) Molecular Evolution and Phylogenetics, chapter 4. New York: Oxford University Press.

**Different codon positions have different mutation rates**

Nei M & Gojobori T (1986) Simple methods for estimating the numbers of synonymous and nonsynonymous nucleotide substitutions. *Mol. Biol. Evol.* 3, 418–426.

Muto A & Osawa S (1987) The guanine and cytosine content of genomic DNA and bacterial evolution. *Proc. Natl Acad. Sci. USA* 84, 116–119.

Goldman N & Yang Z (1994) A codon-based model of nucleotide substitution for protein-coding DNA sequences. *Mol. Biol. Evol.* 11, 725–736.

Yang Z (2002) Inference of selection from multiple sequence alignments. *Curr. Opin. Genet. Dev.* 12, 688–694.

**Only orthologous genes should be used to construct species phylogenetic trees**

Brown JR (2003) Ancient horizontal gene transfer. *Nat. Rev. Genet.* 4, 121–132.

Cannon SB & Young ND (2003) OrthoParaMap: distinguishing orthologs from paralogs by integrating comparative genome data and gene phylogenies. *BMC Bioinformatics* 4, 35.

Doolittle WF (1999) Phylogenetic classification and the universal tree. *Science* 284, 2124–2128.

Fitch WM (2000) Homology: a personal view on some of the problems. *Trends Genet.* 16, 227–231.

Galperin MY, Walker DR & Koonin EV (1998) Analogous enzymes: independent inventions in enzyme evolution. *Genome Res.* 8, 779–790.

International Chicken Genome Sequencing Consortium (2004) Sequence and comparative analysis of the chicken genome provide unique perspectives on vertebrate evolution. *Nature* 432, 695–716.

de Koning AP, Brinkman FSL, Jones SJM & Keeling PJ (2000) Lateral gene transfer and metabolic adaptation in the human parasite *Trichomonas vaginalis*. *Mol. Biol. Evol.* 17, 1769–1773.

Loftus B, Anderson I, Davies R et al. (2005) The genome of the protist parasite *Entamoeba histolytica*. *Nature* 433, 865–868.

Ragan MA (2001) Detection of lateral gene transfer among microbial genomes. *Curr. Opin. Genet. Dev.* 11, 620–626.

Ridley M (2004) Evolution, 3rd ed,. pp 427–430. Oxford: Blackwell.

Tatusov RL, Koonin EV & Lipman DJ (1997) A genomic perspective on protein families. *Science* 278, 631–637.

Tatusov RL, Fedorova ND, Jackson JD, et al. (2003) The COG database: an updated version includes eukaryotes. *BMC Bioinformatics* 4, 41.

**Major changes affecting large regions of the genome are surprisingly common**

Blanc G, Hokamp K & Wolfe KH (2003) A recent polyploidy superimposed on older large-scale duplications in the Arabidopsis genome. *Genome Res.* 13, 137–144.

Blanc G & Wolfe KH (2004) Functional divergence of duplicated genes formed by polyploidy during Arabidopsis evolution. *Plant Cell* 16, 1679–1691.

Lynch M & Katju V (2004) The altered evolutionary trajectories of gene duplicates. *Trends Genet.* 20, 544–549.

## 7.3 Phylogenetic Tree Reconstruction

**Small ribosomal subunit rRNA sequences are well suited to reconstructing the evolution of species**

Cole JR, Chai B, Farris RJ et al. (2005) The Ribosomal Database Project (RDP-II): sequences and tools for high-throughput rRNA analysis. *Nucleic Acids Res.* 33, D294–D296.

Consortium for the Barcode of Life. http://www.barcoding.si.edu/

Woese C (1998) The universal ancestor. *Proc. Natl Acad. Sci. USA* 95, 6854–6859.

**The choice of the method for tree reconstruction depends to some extent on the size and quality of the dataset**

Jones M & Blaxter M (2005) Animal roots and shoots. *Nature* 434, 1076–1077.

**A model of evolution must be chosen to use with the method**

Felsenstein J (1981) Evolutionary trees from DNA sequences: a maximum likelihood approach. *J. Mol. Evol.* 17, 368–376.

Felsenstein J (2004) Inferring Phylogenies, chapter 19. Sunderland, MA: Sinauer.

Hasegawa M, Kishino H & Yano T (1985) Dating of the human-ape splitting by a molecular clock of mitchondrial DNA. *J. Mol. Evol.* 22, 160–174.

Jukes TH & Cantor C (1969) Evolution of protein molecules. In Mammalian Protein Metabolism (MN Munro, ed.) pp 21–132. New York: Academic Press.

Kimura M (1980) A simple model for estimating evolutionary rates of base substitutions through comparative studies of nucleotide sequences. *J. Mol. Evol.* 16, 111–120.

Kimura M (1981) Estimation of evolutionary distances between homologous nucleotide sequences. *Proc. Natl Acad. Sci. USA* 78, 454–458.

Nei M & Kumar S (2000) Molecular Evolution and Phylogenetics, pp. 154–158. New York: Oxford University Press. (*Expresses reservations about the use of AICs.*)

Posada D & Buckley TR (2004) Model selection and model averaging in phylogenetics: advantages of Aikaike Information Criterion and Bayesian approaches over likelihood ratio tests. *Syst. Biol.* 53, 793–808.

Rodriguez F, Oliver JF, Marin A & Medina JR (1990) The general stochastic model of nucleotide substitutions. *J. Theor. Biol.* 142, 485–501.

Tamura K & Nei M (1993) Estimation of the number of nucleotide substitutions in the control region of mitochondrial DNA in humans and chimpanzees. *Mol. Biol. Evol.* 10, 512–526.

Zharkikh A (1994) Estimation of evolutionary distances between nucleotide sequences. *J. Mol. Evol.* 39, 315–329.

**Phylogenetic analyses of a small dataset of 16S RNA sequence data**

Garrity GM (Editor-in-Chief) (2001) Bergey's Manual of Systematic Bacteriology, 2nd ed. New York: Springer-Verlag.

Taxonomy resources including numerous references at NCBI Taxonomy Website. http://www.ncbi.nlm.nih.gov/Taxonomy/taxonomyhome.html/

**Building a gene tree for a family of enzymes can help to identify how enzymatic functions evolved**

Eisen JA (1998) Phylogenomics: improving functional predictions for uncharacterised genes by evolutionary analysis. *Genome Res.* 8, 163–167.

Naumoff DG, Xu Y, Glansdorff N & Labedan B (2004) Retrieving sequences of enzymes experimentally characterized but erroneously annotated: the case of the putrescine carbamoyltransferase. *BMC Bioinformatics* 5, 52.

Sjölander K (2004) Phylogenomic inference of protein molecular function: advances and challenges. *Bioinformatics* 20, 170–179.

# BUILDING PHYLOGENETIC TREES

8

THEORY CHAPTER

**When you have read Chapter 8, you should be able to:**

Describe the models for estimating evolutionary distances from aligned sequences.

Describe the models for estimating time-dependent probabilities of specific mutations.

Reconstruct trees using evolutionary distances between sequences.

Contrast the methods for generating tree topologies.

Reconstruct trees using alignments directly.

Evaluate the methods for calculating branch lengths.

Contrast the methods for the comparative evaluation of alternative tree topologies.

Show the ways of quantifying the reliability of tree features.

This chapter deals with the theoretical basis of the techniques and algorithms that are used to reconstruct phylogenetic trees from sequence-alignment data. These techniques are mostly based on particular models of molecular evolution, which were briefly introduced in Chapter 7. In the so-called distance methods of tree construction, the models are used to obtain estimates of the evolutionary distances between sequences, and these distances become the raw data for tree generation, producing a single tree as the reconstruction of the evolutionary history. Other methods of tree construction use the sequence alignment directly, together with a particular evolutionary model. As we shall see, these latter methods explore many different tree topologies and potentially produce several possible trees. All nondistance methods have in common the ability to give a score for any tree topology and we shall look at the variety of functions used to give these scores. Finally, we shall discuss ways in which the resulting trees can be tested to determine the likelihood that certain features are correct. Section 7.1 described the basic types of phylogenetic trees and their nomenclature, and Section 7.2 presented some key concepts in molecular evolution. This knowledge will be assumed here. All discussions in this chapter also assume that the set of aligned sequences submitted for analysis are homologous.

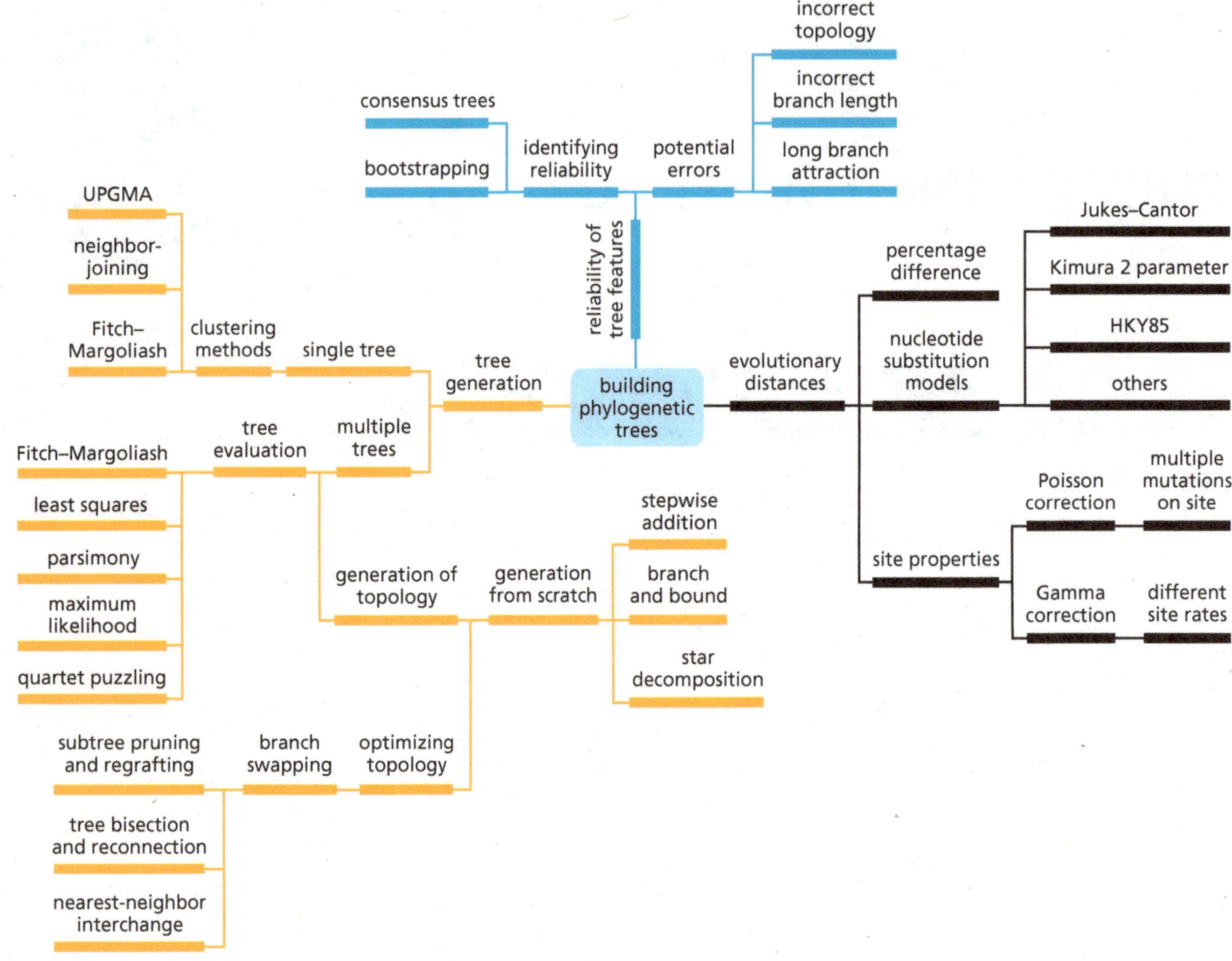


**Mind Map 8.1**
**A mind map representation of the techniques involved in building phylogenetic trees, including tree generation, calculation of evolutionary distance, and reliability of trees.**

## 8.1 Evolutionary Models and the Calculation of Evolutionary Distance

Analysis of reliable sequence alignments has revealed some general rules governing the ways in which mutation leads to changes in DNA sequences over time (see Section 7.2). These observations have been incorporated into models of molecular evolution, which are used in all methods of tree-building (see Flow Diagram 8.1). Our focus in this chapter will be on DNA rather than protein sequences, and most of the models of mutation and evolution discussed here will refer to nucleotide sequences. Very similar schemes are, however, used to model amino acid sequence evolution, similar to that described in Section 5.1 in the context of PAM scoring matrices. Some features of nucleotide sequence evolution can easily be included in very simple models. Others have proved much more difficult to incorporate in a practical way, and are frequently ignored. After describing some simple models in common current use, we will briefly discuss a few features that are at present usually not modeled properly.

### A simple but inaccurate measure of evolutionary distance is the *p*-distance

As explained in Chapter 7, the evolutionary distance between two sequences is an estimate of the number of mutations that has occurred since those sequences

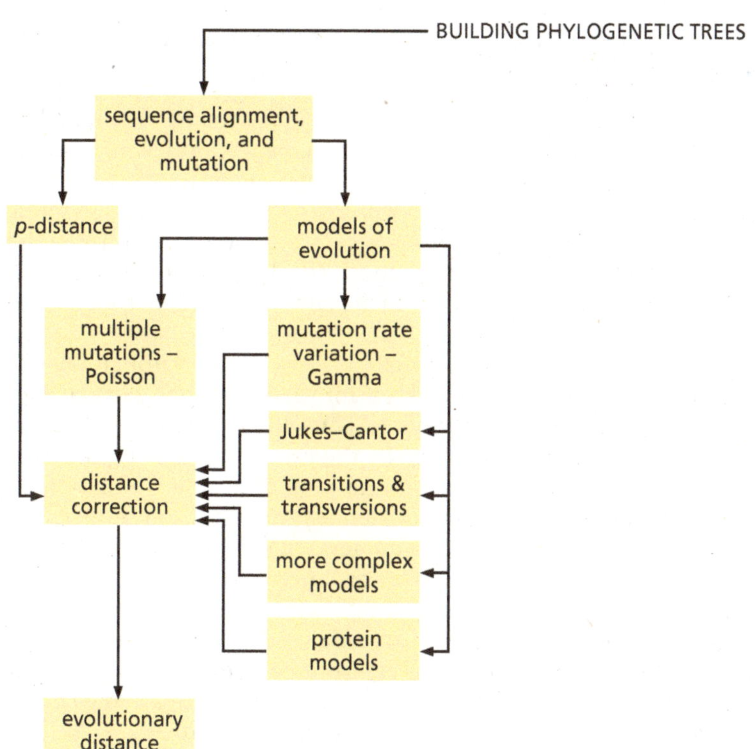


**Flow Diagram 8.1**
**Two key concepts are introduced in this section.** Firstly, the observed sequence differences in an alignment usually need to be corrected to make them linearly dependent on evolutionary distance. Secondly, a series of alternative but often related models of sequence evolution have been proposed.

diverged from their common ancestor. The simplest, if unrealistic, measure of evolutionary distance is the fractional alignment difference; that is, the observed fraction of sites in two aligned sequences that have different bases. If an alignment of two sequences has $L$ positions (without counting positions at which one sequence has a gap), of which $D$ differ, then the **fractional alignment difference**, usually called the $p$-distance, is defined as

$$p = \frac{D}{L} \tag{EQ8.1}$$

This is an evolutionary distance, but can also be considered as an estimate of the probability of a site being mutated and is used in this way in some of the later discussion.

This measure is usually too inaccurate for serious work, as was explained in Section 7.2. For low rates of mutation and/or short evolutionary times, very few sequence differences are observed. Thus, for the relatively short sequence lengths that are often used to reconstruct phylogeny there is considerable statistical variation between sequences, leading to potentially large errors in $p$-distance estimates. Much more frequently encountered is the case that at longer evolutionary times many sites have undergone mutation more than once. Observed differences are thus an underestimate of the total number of mutations that have occurred over that time, and it is the total number we are interested in for phylogenetic tree construction. Finally, not all species evolve at the same rate, and neither do all genes in the same species, because they are under different evolutionary pressures at different times. There may be fortuitous averaging out for sufficiently long times but this is by no means always the case. To take account of these sources of error in the $p$-distance, the evolutionary distance measure needs to be modified by the application of a more sophisticated evolutionary model.

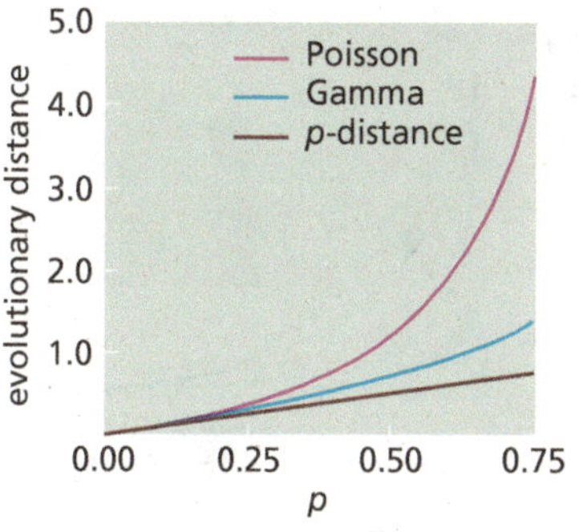


**Figure 8.1**
**Comparison of three methods of estimating the evolutionary distance between sequences.** The $p$-distance, $p$, of Equation EQ8.1 (fraction of different aligned residues) is plotted against itself for comparison with the values obtained by applying the Poisson $d_P$ (Equation EQ8.3) and Gamma $d_\Gamma$ (Equation EQ8.4) corrections. The value of $a = 0.70$ was used for the Gamma distance (see Figure 8.2). The vertical axis gives the corrected distance which, especially in the case of the Gamma distance, can be considerably different from $p$. The corrections only have a small effect for $p$ values below 0.2.

## The Poisson distance correction takes account of multiple mutations at the same site

A simple correction to the *p*-distance can be derived by assuming that the probability of mutation at a site follows a Poisson distribution, with a uniform mutation rate *r* per site per time unit. After a time *t*, the average number of mutations at each site will be *rt*. The formula $e^{-rt}(rt)^n/n!$ gives the probability of *n* mutations having occurred at a given site during time *t*. We want to derive a formula that relates the observed fraction of sites that have mutated (the *p*-distance) to the actual number of mutations that have occurred, which is not measurable from the data.

Consider two sequences that diverged time *t* ago. The probability of no mutation having occurred at a site is given by $e^{-rt}$ for each sequence, given the assumption of a Poisson distribution of mutations. Thus the probability of neither sequence having mutated at that site is given by the expression $e^{-2rt}$. We also assume that no situation has occurred in which several mutations at a site have resulted in both sequences being identical, such as identical mutations at both sequence sites. In this case this probability can be equated with the observed fraction of identical sites, given by $(1 - p)$ where *p* is the *p*-distance.

Because each sequence has evolved independently from the common ancestor, they are an evolutionary distance 2*rt* from each other, which we will write as *d*. This evolutionary distance *d* is measured in terms of the average number of mutations that have occurred per site, not the time since divergence. This leads to the equation

$$1 - p = e^{-d} \quad \text{(EQ8.2)}$$

from which we can derive the **Poisson corrected distance**

$$d_P = -\ln(1 - p) \quad \text{(EQ8.3)}$$

This corrected evolutionary distance, $d_P$, starts to deviate noticeably from the *p*-distance for $p > 0.25$ (see Figure 8.1).

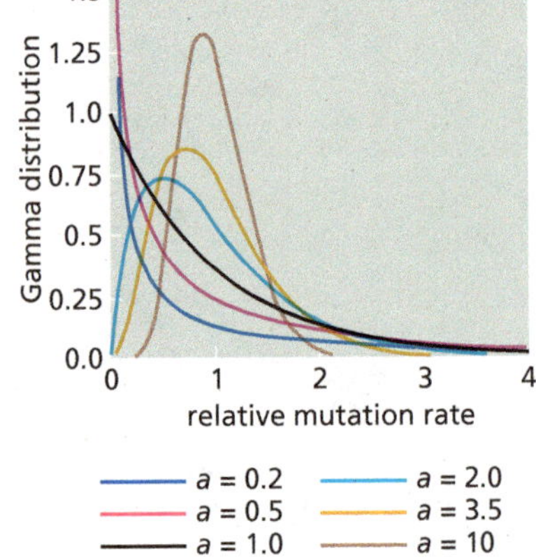


**Figure 8.2**
**The Gamma distribution model of the variation in relative mutation rate at different sequence sites.** The Gamma distributions shown have a parameter *a*, and have been adjusted to have a mean value of 1. At high values of *a* the distribution approaches the situation of all sites having an identical mutation rate. Low values give a considerable variation in mutation rate. For protein-sequence data the estimated values of *a* often lie in the range 0.2–3.5.

## The Gamma distance correction takes account of mutation rate variation at different sequence positions

Another assumption shown to be highly questionable is that of an equal rate of mutation at different positions in the sequence; that is, the assumption that there is only one value of *r*, which applies to the whole sequence. This is far from true for protein-coding and other functional sequences. In 1971, Thomas Uzzell and Kendall Corbin reported that a **Gamma distribution (Γ)** can effectively model realistic variation in mutation rates. Such a distribution can be written with one parameter, *a*, which determines the site variation (see Figure 8.2). Using this, it is possible to derive a corrected distance, referred to as the **Gamma distance** $d_\Gamma$:

$$d_\Gamma = a\left[(1 - p)^{-1/a} - 1\right] \quad \text{(EQ8.4)}$$

where *p* is the *p*-distance. Values of *a* have been estimated from real protein-sequence data to vary between 0.2 and 3.5. The Gamma distance is not significantly different from the *p*-distance for $p < 0.2$ (see Figure 8.1) but can diverge markedly for greater values of *p*.

## The Jukes–Cantor model reproduces some basic features of the evolution of nucleotide sequences

The mutation models described so far do not include any information relating to the chemical nature of the sequences, which means that they can be applied directly to both nucleotide and protein sequences. Some evolutionary models have been constructed specifically for nucleotide sequences. One of the simplest models of this type is that due to Thomas Jukes and Charles Cantor. It makes a number of assumptions that are known to be incorrect in most cases, but despite this it has proved useful. Its simplicity allows us to derive the formula for the evolutionary distance between two sequences. Some more complex models will be presented later for which the equivalent results will be given but not proved.

In the Jukes–Cantor (JC) model all sites are assumed to be independent and to have identical mutation rates. Furthermore, all possible nucleotide substitutions are assumed to occur at the same rate $\alpha$ per unit time. A matrix can represent the substitution rates as follows:

$$\begin{array}{c} \\ \text{A} \\ \text{C} \\ \text{G} \\ \text{T} \end{array} \begin{array}{c} \begin{array}{cccc} \text{A} & \text{C} & \text{G} & \text{T} \end{array} \\ \begin{pmatrix} -3\alpha & \alpha & \alpha & \alpha \\ \alpha & -3\alpha & \alpha & \alpha \\ \alpha & \alpha & -3\alpha & \alpha \\ \alpha & \alpha & \alpha & -3\alpha \end{pmatrix} \end{array} \qquad \text{(EQ8.5)}$$

where the rows and columns of the matrix represent the four different bases as shown. The mutations occur from the $i$th row to the $j$th column. Each row sums to zero, meaning that the number of sites (that is, sequence length) remains constant. In addition, each column sums to zero, which means that the numbers of each base (that is, the base composition) remains constant.

Suppose that an ancestral sequence diverged time $t$ ago into two related sequences. After this time $t$, the fraction of identical sites between the two sequences is $q(t)$, and the fraction of different sites is $p(t)$, so that $p(t) + q(t) = 1$. We can calculate the fraction of identical sites after time $t + 1$, written $q(t+1)$. If we restrict ourselves to terms in $\alpha$ ($\alpha$ is usually so small that terms in $\alpha^2$ are negligible by comparison) then there are only two ways of getting an identical site at time $t + 1$. Since a fraction $\alpha$ of sites will mutate to each of the three other bases between $t$ and $t + 1$, the fraction of sites having no mutations occurring in this time period is $(1 - 3\alpha)$. Thus the probability of an aligned pair of identical residues not mutating in this time is $(1 - 3\alpha)^2$, which is $(1 - 6\alpha)$ ignoring terms in $\alpha^2$. Since a fraction $q(t)$ of sites were identical at time $t$, we can expect that a fraction $(1 - 6\alpha)q(t)$ remain identical at time $t + 1$. The other route to identical sites at time $t + 1$ is if the aligned residues are different at time $t$, which is the case for a fraction $p(t)$, but one mutates to become identical to the other. A fraction $\alpha$ of any base will mutate to each of the other three bases in that time period, but only one of these is the same as the aligned base of the other sequence. This site on the other sequence must not have mutated, which, as already discussed, is the case for a fraction $(1 - 3\alpha)$. Furthermore, this event can happen in two equivalent ways, depending on which of the two aligned bases mutates. Thus, the probability of such an event occurring between times $t$ and $t + 1$ is $2\alpha(1 - 3\alpha)p(t)$. Ignoring terms in $\alpha^2$, this is $2\alpha p(t)$. Therefore the fraction of identical sites at time $t + 1$, $q(t+1)$, is

$$q(t+1) = (1 - 6\alpha)q(t) + 2\alpha p(t) \qquad \text{(EQ8.6)}$$

This can be manipulated [recall that $p(t) = 1 - q(t)$] to estimate the derivative of $q(t)$ with time as

$$\frac{dq(t)}{dt} = q(t+1) - q(t) = 2\alpha - 8\alpha q(t) \qquad \text{(EQ8.7)}$$

It is easily confirmed by substitution into Equation EQ8.7 that

$$q(t) = \frac{1}{4}\left(1 + 3e^{-8\alpha t}\right) \qquad \text{(EQ8.8)}$$

which includes the condition that at time $t = 0$ all equivalent sites on the two sequences were identical [$q(0) = 1$]. Note that at very long times $q_\infty = 1/4$, so this model predicts a minimum of 25% identity even on aligning unrelated nucleotide sequences. Finally, $3\alpha t$ mutations would be expected during a time $t$ for each sequence site on each sequence. At any time each site will be a particular base, which will mutate to one of the other three bases at the rate $\alpha$. Hence the evolutionary distance between the two sequences under this model is $6\alpha t$. By rearrangement this corrected distance ($d_{JC}$) can be obtained from Equation EQ8.8 in terms of $p$:

$$d_{JC} = -\frac{3}{4}\ln\left(1 - \frac{4}{3}p\right) \qquad \text{(EQ8.9)}$$

To obtain a value for the corrected distance, one simply substitutes $p$ with the observed proportion of site differences in the alignment. This attempt to account for multiple mutations at the same site is more sophisticated than the Poisson correction, but there is a clear similarity between Equations EQ8.3 and EQ8.9.

## More complex models distinguish between the relative frequencies of different types of mutation

Progressive improvements have been made to the JC model, of which only a few will be discussed here. While it is easy to identify models that are formally more realistic, these are not necessarily more effective in representing real data. Their success will depend on the actual evolutionary history of the dataset under study, and how to choose the best model in practice is discussed in Section 7.3.

One improvement over the JC model involves distinguishing between rates of transitions and transversions, as was described in Section 7.2 and illustrated in Figure 7.8. To distinguish the different rates of transitions and transversions, they are assigned the values $\alpha$ and $\beta$, respectively. If this is the only modification made to the JC model, the resulting model is called the Kimura two-parameter (K2P) model, and has the rate matrix

$$\begin{array}{c} \\ \text{A} \\ \text{C} \\ \text{G} \\ \text{T} \end{array} \begin{array}{c} \begin{array}{cccc} \text{A} & \text{C} & \text{G} & \text{T} \end{array} \\ \begin{pmatrix} -2\beta-\alpha & \beta & \alpha & \beta \\ \beta & -2\beta-\alpha & \beta & \alpha \\ \alpha & \beta & -2\beta-\alpha & \beta \\ \beta & \alpha & \beta & -2\beta-\alpha \end{pmatrix} \end{array} \qquad \text{(EQ8.10)}$$

This model results in a corrected distance, $d_{K2P}$, given by

$$d_{K2P} = -\frac{1}{2}\ln(1 - 2P - Q) - \frac{1}{4}\ln(1 - 2Q) \qquad \text{(EQ8.11)}$$

where $P$ and $Q$ are the observed fractions of aligned sites whose two bases are related by a transition or transversion mutation, respectively. These values are obtained from the alignments of the real data. Note that the $p$-distance, $p$, equals $P + Q$. The transition–transversion ratio, commonly written $R$, is defined as $\alpha/2\beta$ for the K2P model. As explained in Box 8.1, $R$ is directly obtainable from the data, but is not required in order to apply the distance correction defined by the above equation.

Both the JC and the K2P models considered so far predict a 1:1:1:1 equilibrium base composition. This is almost never the case in reality; multicellular eukaryotes have an average genomic GC content of 40%, and bacteria show a wide range of variation, with measured GC contents ranging from approximately 25% to 74% (see Figure 7.9). The rate matrices can easily be modified to account for any particular base composition $\pi_A:\pi_C:\pi_G:\pi_T$. When applied to the K2P model this modification gives a model called HKY85 (after M. Hasegawa, H. Kishino and T. Yano, who published it in 1985) and has the matrix

$$\begin{array}{c} \\ A \\ C \\ G \\ T \end{array} \begin{array}{c} \begin{array}{cccc} A & C & G & T \end{array} \\ \begin{pmatrix} (-2\beta-\alpha)\pi_A & \beta\pi_C & \alpha\pi_G & \beta\pi_T \\ \beta\pi_A & (-2\beta-\alpha)\pi_C & \beta\pi_G & \alpha\pi_T \\ \alpha\pi_A & \beta\pi_C & (-2\beta-\alpha)\pi_G & \beta\pi_T \\ \beta\pi_A & \alpha\pi_C & \beta\pi_G & (-2\beta-\alpha)\pi_T \end{pmatrix} \end{array} \qquad \text{(EQ8.12)}$$

More complex models have been proposed in which more, or even all, types of base substitutions have their own rate constants. A selection of the more commonly used models is listed in Table 7.2 (page 253), but many more have been described. It should be noted, however, that these models still suffer from the incorrect assumption that all sites have the same mutation rate. To help overcome this, the Gamma distribution can be applied. If it is applied to the JC model (the JC+Γ model) with Γ parameter $a$, the corrected distance equation becomes

$$d_{JC+\Gamma} = \frac{3}{4}a\left[\left(1 - \frac{4}{3}p\right)^{-1/a} - 1\right] \qquad \text{(EQ8.13)}$$

Evolutionary models can be used in two ways in methods for phylogenetic tree reconstruction. For some methods, distance formulae such as Equations EQ8.9, EQ8.11, and EQ8.13 are derived and are applied to the sequence-alignment data to obtain corrected evolutionary distances. Other methods use a formula such as Equation EQ8.8 to calculate the probability of particular mutations having occurred along a tree branch of a particular length.

For these latter methods, values for the parameters in the models, such as $\alpha$, $\beta$, and the base composition, must be specified. Although many tree-building programs allow parameters to be specified by the user, it is also common practice for programs to calculate parameter values whenever possible from the data under study. The estimation of one of these parameters—the transition/transversion rate ratio $R$—is discussed in Box 8.1 to illustrate that in many cases the calculation of these quantities is not difficult, but still not trivial.

## Box 8.1 Calculating the transition/transversion rate ratio *R*

There are good reasons for wanting to estimate the transition/transversion rate ratio *R*. Not all evolutionary models include this mutational preference, and it would be useful if the sequence data could be analyzed to detect any signs that such a preference had occurred. This would help in selecting an appropriate model and also in suggesting parameter values where required. The simplest evolutionary model to include a transition/transversion rate ratio is the Kimura two-parameter (K2P) model, and we will focus exclusively on this model in the following discussion. Note that if the model is only to be applied as a distance correction the value of *R* is not required, as the correction formula does not involve this parameter directly (Equation EQ8.11).

Given an alignment of two nucleotide sequences, it is easy to distinguish those sites at which a transition or a transversion mutation would convert between the two aligned bases. The fractions of these two types of sites in the alignment are usually called *P* and *Q*, respectively, as in Equation EQ8.11. The ratio *P*/*Q* could be used as the value of the transition/transversion rate ratio *R* in models that require it, such as K2P, but in fact, as will now be explained, that is generally a poor estimate of this parameter.

Recall that in the K2P model the actual transition rate ($\alpha$) and transversion rate ($\beta$) are used (Equation EQ8.10), in terms of which *R* is given as $\alpha/2\beta$. However, following the K2P analysis the distance correction is expressed in terms of *P* and *Q* alone, as they are directly observable from the sequence alignment. To see why there is a problem with using *P*/*Q* as an estimate of *R* we will use the K2P model equations that are the equivalents of the JC model (Equation EQ8.8), giving the variation with time *t* of the fraction of sites *P* and *Q* as defined above, namely

$$P(t) = \frac{1}{4}\left(1 - 2e^{-4(\alpha+\beta)t} + e^{-8\beta t}\right) \qquad \text{(BEQ8.1)}$$

and

$$Q(t) = \frac{1}{2}\left(1 - e^{-8\beta t}\right) \qquad \text{(BEQ8.2)}$$

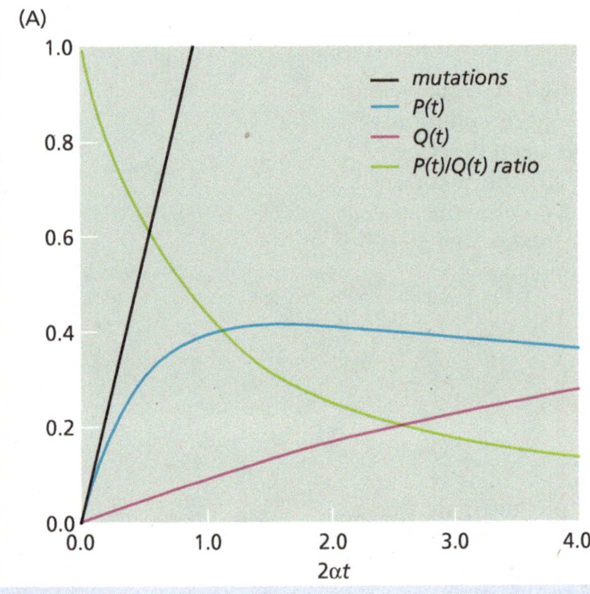


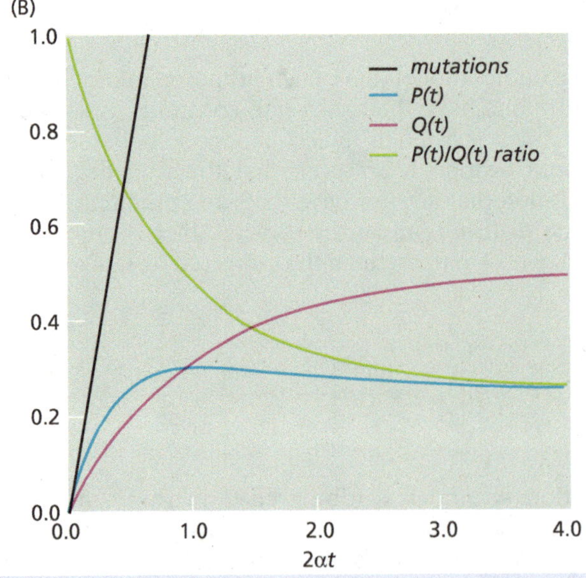


**Figure B8.1**
**A demonstration of the unsuitability of using the measurement *P*/*Q* as an estimator of the transition/transversion rate ratio *R*.** The Kimura two-parameter model is assumed throughout. The expected fraction of sites in the sequence alignment where a transition (blue line) or transversion (pink line) mutation would convert between the two aligned bases according to the K2P model is plotted against $2\alpha t$, the actual number of transitions that has occurred per site in the two sequences since divergence time *t* ago. Two different values of *R* were used to calculate these curves: (A) $R = 10$ and (B) $R = 2$. The black line represents the total number of all mutations that have actually occurred per site. The green line shows the $P(t)/Q(t)$ ratio, a possible estimator of the value of *R* as a fraction of the correct value of *R*. A value of 1 for the green curve means that at this point $P(t)/Q(t)$ gives the correct value; a value of 0.8 indicates that it is only 80% of the correct value, and so on. The ratio *P*/*Q* is only close to the correct value of *R* when there have been very few mutations.

## Box 8.1 Calculating the transition/transversion rate ratio *R* (continued)

Using these formulae we can calculate the expected values of $P$, $Q$, and $R$ as they vary over time (see Figure B8.1). Notice that for extremely long divergence times, regardless of the absolute or relative values of $\alpha$ and $\beta$, $P$ will tend to $1/4$, and $Q$ will tend to $1/2$, with a remaining $1/4$ of sites identical between the two sequences. Hence in this model the value of $P/Q$ will always be close to ½ for highly divergent sequences whatever the actual transition/transversion preference. Shorter divergence times will give estimates of $P$ and $Q$ from these equations that reveal the underlying mutational preferences, but as seen in Figure B8.1, only measurements obtained from the most recently diverged sequences will give estimates of $R$ that are close to the true value. However, the small number of mutations in closely related sequences is likely to make such an estimate highly unreliable, except when very long sequences are available, because of the statistics of the mutations observed.

By rearranging equations BEQ8.1 and BEQ8.2, the following formula can be derived to estimate $R$:

$$R_{\text{K2P}} = \frac{-\ln(1-2P-Q)}{-\ln(1-2Q)} - \frac{1}{2} \qquad \text{(BEQ8.3)}$$

This formula is reliable at all times in estimating $R$, assuming the sequence data are sufficient to obtain good sampling statistics.

However, even this formula has been criticized. The calculated value of $R$ relates only to the two sequences selected. When analyzing a real dataset, this value could be different for each pair of sequences in the multiple alignment. A possible solution is to use maximum likelihood methods to estimate the best single value of $R$ to use. This can be done for all the data simultaneously, and in parallel with the optimization of other model parameters and the reconstruction of the tree (see Further Reading).

One of the very common assumptions present in models of evolution is that sequence positions behave identically and independently in respect to mutation. Almost no sequence segments are truly random, however, as they almost invariably exhibit some mutation correlation between positions. This is particularly the case for protein-coding genes, for DNA that specifies functional RNAs such as tRNAs and rRNAs, and for sequences that recognize and bind gene-regulatory and other proteins. All these are subject to evolutionary pressure to conserve functionally important parts of the sequence, and thus display different mutation rates at these sites compared with the rest of the sequence.

By their very nature these correlations are sequence-specific, and for this reason cannot be properly included in a model of evolution without making that model sequence-specific too. One partial solution is to include in the model a method that assigns sequence positions to blocks that have similar evolutionary properties, such as mutation rates, with a procedure to assign the block boundaries. An example of such a method is the hidden Markov model used in combination with a variant of the Gamma correction by Joseph Felsenstein and Gary Churchill (see Further Reading).

## There is a nucleotide bias in DNA sequences

If sequences were truly random, one would expect the four bases to occur equally frequently. Taking account of the difference in thermal stability between GC and AT base pairs, one might expect base preferences, but these would have a simple dependence on temperature. In fact, the true situation is far more complex, as shown by the range of GC content of bacterial genomes (see Figure 7.9). Not all the bacteria shown in that figure live at exactly the same temperature, but none lives in extreme conditions of heat or cold. In addition, there are regions of different GC content within genomes, a feature that is clearly important in humans, where regions of distinct GC composition, which are known as **isochores**, have distinct properties, as discussed in Section 10.4.

Whatever the reasons for this variation in base composition, it is extremely hard to model realistically. Detailed studies have shown that base composition varies in a nontrivial way throughout evolution, so that each branch of a phylogenetic tree may show a different trend in composition variation. All the models discussed above use a fixed and universal base composition. Some methods have been proposed to deal with this problem but have not yet been widely applied (see Further Reading).

### Models of protein-sequence evolution are closely related to the substitution matrices used for sequence alignment

The discussion so far has focused on nucleotide sequences, although some concepts, such as the Poisson correction, can be easily applied to protein sequences. Any models of protein evolution will share many similarities with the models presented above, but have a 20 × 20 rate matrix to account for the 20 different amino acids. A protein model related to the JC nucleotide model in having just a single rate for all possible mutations can be readily analyzed to yield a distance correction equation

$$d_{JCprot} = -\frac{19}{20}\ln\left(1 - \frac{20}{19}p\right) \qquad \text{(EQ8.14)}$$

where $p$ is now the fraction of identical sites in an alignment of two protein sequences.

It is now common practice to use empirical matrices based on the work that derived some of the substitution scoring schemes described in Chapters 4 and 5. In particular, the PAM matrices apply an evolutionary model. One of the most popular protein matrices for phylogenetic studies—the JTT matrix derived by David Jones, Willie Taylor, and Janet Thornton—is the result of attempts to update the PAM matrix work. These models have been developed to take account of the specific amino acid composition of the datasets (see Further Reading).

## 8.2 Generating Single Phylogenetic Trees

There are two main classes of techniques for reconstructing phylogenetic trees. Clustering methods gradually build up the tree, starting from a small number of sequences and adding one sequence at each step. The output from these methods is a single tree that attempts to recover the evolutionary relationships between the sequences (see Flow Diagram 8.2). In the second group of methods, many different tree topologies are generated and each is tested against the data in a search for those that are optimal or close to optimal according to particular criteria. We will discuss clustering methods first, followed by the other techniques.

### Clustering methods produce a phylogenetic tree based on evolutionary distances

The strength of clustering methods is their speed and robustness, and their ability to reconstruct trees for very large numbers, even thousands, of sequences. In contrast, many of the techniques that search tree topology are unable to process more than about 100 sequences. Along with this strength come weaknesses, however. There is no associated measure of how well the resulting trees fit the data, so that alternatives (for example, trees reconstructed using the same technique but different distance corrections) cannot be easily compared. As a consequence, it is

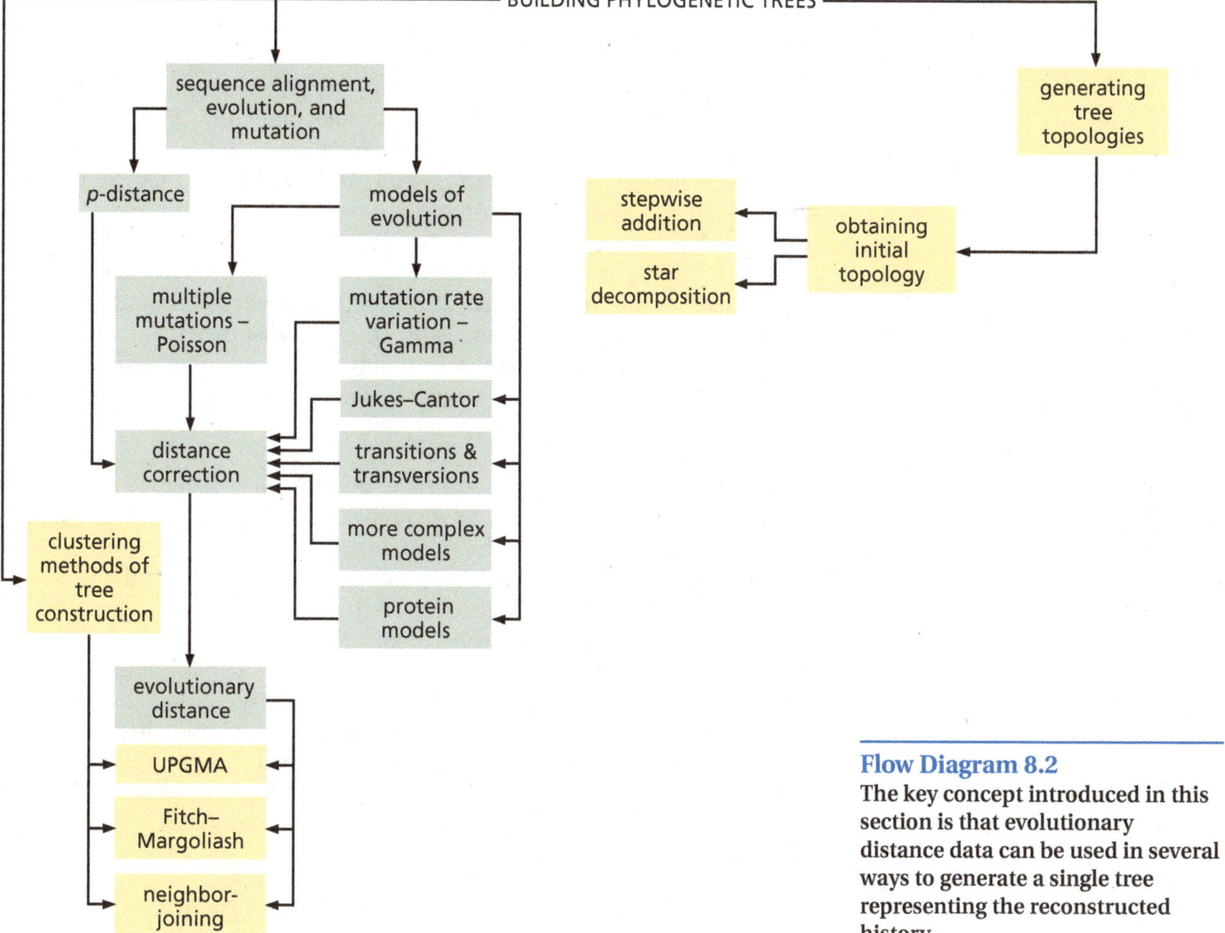


**Flow Diagram 8.2**
The key concept introduced in this section is that evolutionary distance data can be used in several ways to generate a single tree representing the reconstructed history.

not possible to test evolutionary hypotheses by comparing a particular tree topology with that identified by the method as the best. This is not the case with the topology search methods, and is their strength.

Most clustering methods reconstruct a phylogenetic tree for a set of sequences on the basis of their pairwise evolutionary distances. Derivation of these distances will have involved the application of an equation such as Equations EQ8.9, EQ8.11, or EQ8.13, which use some basic measures (for example, the *p*-distance) taken from the alignment together with some parameters that may have been previously specified or, alternatively, derived from the data (for example, the transition/transversion rate ratio *R* as described in Box 8.1).

Obtaining evolutionary distances is quite straightforward, but there are potential problems. First, the alignments may contain errors, leading to incorrect distances. Second, some assumptions (such as identical rates of change at all sites) made in the evolutionary models may not hold for the particular dataset being used. As a consequence, the corrected evolutionary distance equations used may not be appropriate and may lead to errors in the distances. One important point to be aware of with evolutionary distance methods of tree construction is that the complete set of distances will almost certainly not be perfectly self-consistent, as this method of generating distances has limitations even if the model exactly reflects the evolution of the sequence data. This is because the formulae are exact only in the limit of infinitely long sequences. These limitations mean that evolutionary distances cannot always be recovered exactly, even for test sequences generated by the same model. As a consequence, any reconstructed tree will almost certainly not reproduce all the distances correctly.

## The UPGMA method assumes a constant molecular clock and produces an ultrametric tree

The first distance method we will look at—UPGMA—is simple to apply, but has the disadvantage that it assumes a constant molecular clock. This method is one of the oldest, having been devised by Robert Sokal and Charles Michener in 1958. The name UPGMA is an acronym of unweighted pair-group method using arithmetic averages, a description of the technique used. Assumption of a constant rate of evolution has important consequences for a dataset of sequences that are all associated with the same evolutionary time point, namely the present day, as it dictates that the same number of substitutions will have occurred in each sequence since the time of the last common ancestor. Thus, the distance from any node to any leaf that is its descendant will be the same for all descendants. The trees produced by this method are rooted and ultrametric (see Figure 7.2C), and all the leaves are at the same distance from the root.

The two sequences with the shortest evolutionary distance between them are assumed to have been the last to diverge, and must therefore have arisen from the most recent internal node in the tree. Furthermore, their branches must be of equal length, and so must be half their distance. This is how the construction of the tree is started. The method must recover all the internal nodes in the tree and at each step another internal node is recovered.

The sequences are grouped into clusters as the tree is constructed, each cluster being defined as the set of all descendants of the new node just added. Initially, all sequences are regarded as defining their own cluster. At each stage, the two clusters with the shortest evolutionary distance are combined into a new cluster. The tree is complete when all the sequences belong to the same cluster, whose node is the root of the tree.

The distance between two clusters is defined as follows. Consider the construction of a tree for $N$ sequences and suppose that at some stage you have clusters X containing $N_X$ sequences and Y containing $N_Y$ sequences. Initially, each cluster will contain just one sequence. The evolutionary distance ($d_{XY}$) between the two clusters X and Y is defined as the arithmetic average of the distances between their constituent sequences, that is

$$d_{XY} = \frac{1}{N_X N_Y} \sum_{i \in X, j \in Y} d_{ij} \qquad \text{(EQ8.15)}$$

where $i$ labels all sequences in cluster X, $j$ labels all sequences in cluster Y, and $d_{ij}$ is the distance between sequences $i$ and $j$. When two clusters X and Y are combined to make a new cluster Z there is an efficient way of calculating the distances of other clusters such as W to the new cluster. The new distances can all be defined using existing cluster-to-cluster distances without the need to use the constituent sequence-to-sequence distances, using the equation

$$d_{ZW} = \frac{N_X d_{XW} + N_Y d_{YW}}{N_X + N_Y} \qquad \text{(EQ8.16)}$$

This method is very straightforward to apply, and can be used to construct trees for large sets of sequences. The method is illustrated in Figure 8.3 for six sequences. In the first step, sequences A and D are found to be the closest ($d_{AD}$ = 1) and are combined, creating cluster (and node) V at a height of ½ (= $d_{AD}/2$) (see Figure 8.3A). Following calculation of the distances of V from the other sequences, the closest pair is E and V, which are combined into cluster W at a height of 1 (= $d_{EW}/2$) (see Figure 8.3B). Continuing in this way, the final tree is obtained (see Figure 8.3E).

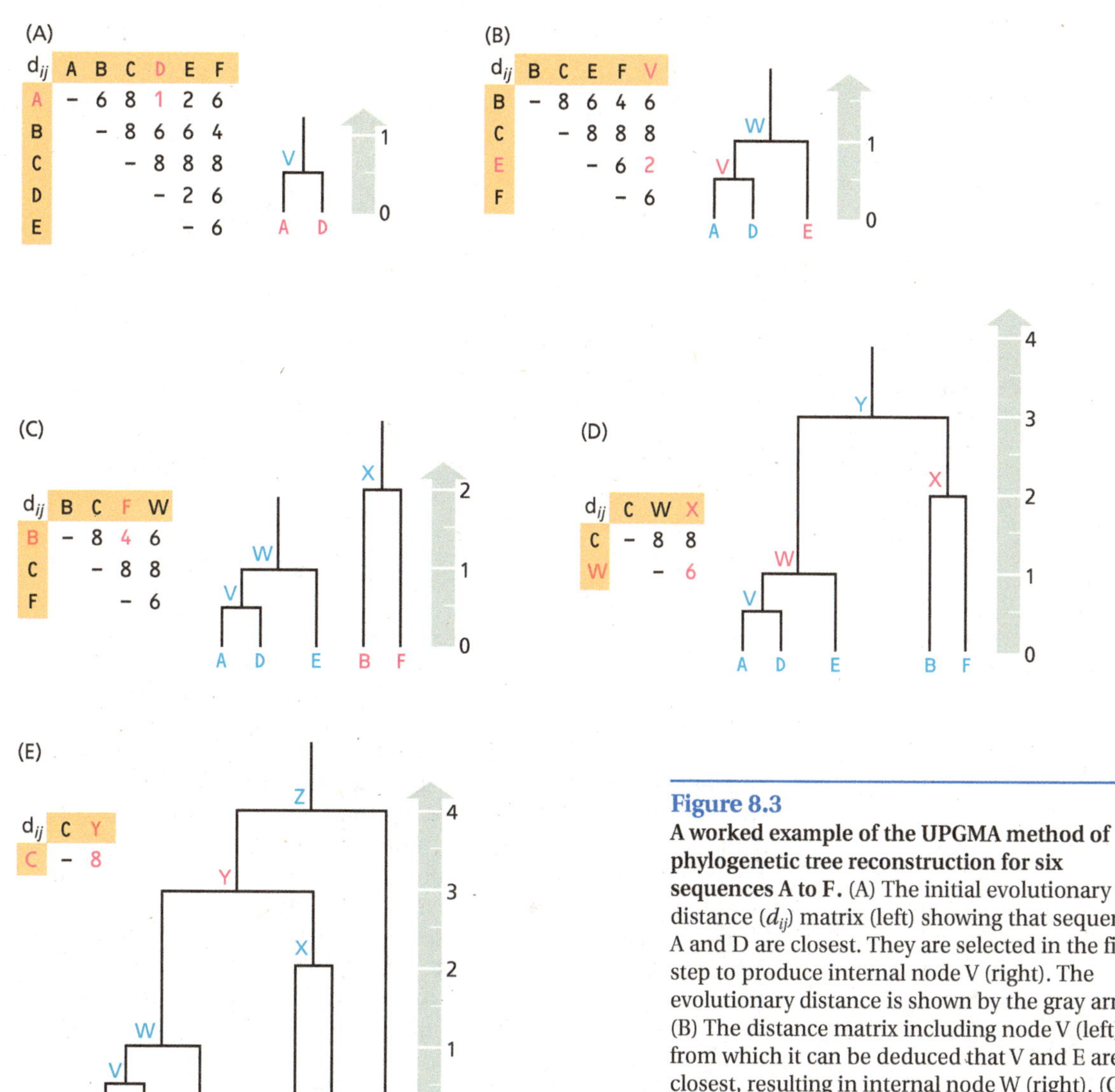


**Figure 8.3**
**A worked example of the UPGMA method of phylogenetic tree reconstruction for six sequences A to F.** (A) The initial evolutionary distance ($d_{ij}$) matrix (left) showing that sequences A and D are closest. They are selected in the first step to produce internal node V (right). The evolutionary distance is shown by the gray arrow. (B) The distance matrix including node V (left), from which it can be deduced that V and E are closest, resulting in internal node W (right). (C,D) Subsequent steps defining nodes X, Y, and Z and resulting in the final tree (E).

If the data being analyzed did not evolve under conditions of a molecular clock, however, the tree produced by UPGMA could be seriously in error. A dataset can, however, be tested beforehand for likely compatibility with the method. For an ultrametric tree to be appropriate for the dataset, for all sets of three sequences A, B, and C, the three distances $d_{AB}$, $d_{AC}$, and $d_{BC}$ should either all be equal or two should be equal and the third distance be the shortest. This is the case for the dataset of Figure 8.3.

## The Fitch–Margoliash method produces an unrooted additive tree

We shall now look at some methods that do not make the assumption of constant mutation rate, but do assume that the distances are additive. The Fitch–Margoliash method is based on analysis of a three-leaf tree as shown in Figure 8.4. The distances $d_{ij}$ between leaves A, B, and C are trivially given in terms of the branch lengths by the formulae

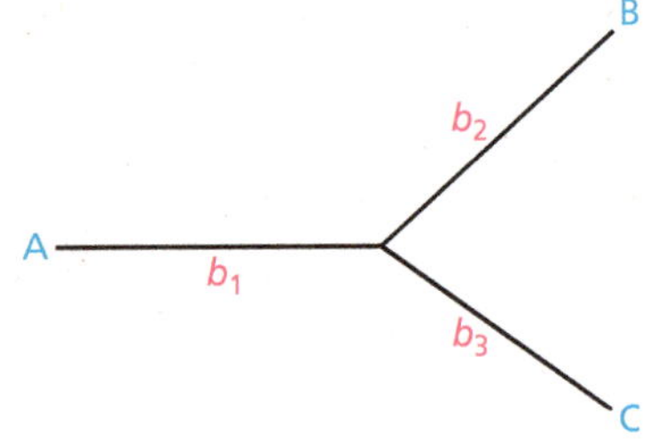


**Figure 8.4**
**The small tree from which the Fitch–Margoliash method equations (Equation EQ8.18) are derived.** The leaves A, B, and C are all connected to the same internal node by the branches $b_1$, $b_2$, and $b_3$. The same formulae apply when there are other internal nodes of the tree on the paths between the three leaves and the common internal node, so long as the $b_i$ are interpreted as the sum of all path branch lengths.

$$d_{AB} = b_1 + b_2$$
$$d_{AC} = b_1 + b_3$$
$$d_{BC} = b_2 + b_3 \qquad \text{(EQ8.17)}$$

This means the tree is being treated as additive. We can readily derive formulae for the branch lengths in terms of the distances

$$b_1 = \frac{1}{2}(d_{AB} + d_{AC} - d_{BC})$$
$$b_2 = \frac{1}{2}(d_{AB} + d_{BC} - d_{AC})$$
$$b_3 = \frac{1}{2}(d_{AC} + d_{BC} - d_{AB}) \qquad \text{(EQ8.18)}$$

which are estimates of the branch lengths based solely on the evolutionary distances.

Trees with more than three leaves can be generated in a stepwise fashion similar to that used by the UPGMA method. At every stage three clusters are defined, with all sequences belonging to one of the clusters. The distance between clusters is defined (as in UPGMA) by a simple arithmetic average of the distances between sequences in the different clusters (see Equation EQ8.15). At the start of each step we have a list of sequences not yet part of the growing tree and of clusters representing each part (if there is more than one) of the growing tree. The distances between all these sequences and clusters are calculated, and the two most closely related (those with the shortest distance) are selected as the first two clusters of a three-leaf tree. A third cluster is defined that contains the remainder of the sequences, and the distances to the other two are calculated. Using the above equations one can then determine the branch lengths from this third cluster to the other two clusters and the location of the internal node that connects them (see Figure 8.4). These two clusters are then combined into a single cluster with distances to other sequences again defined by simple averages. There is now one less sequence (cluster) to incorporate into the growing tree. By repetition of such steps this technique is able to generate a single tree in a similar manner to UPGMA. Because the method of selecting closest cluster pairs is identical to that used in UPGMA, the tree topology produced from a set of data is the same for both methods. The differences in the trees produced are the branch lengths and that the UPGMA tree is ultrametric whereas the Fitch–Margoliash tree is additive.

A worked example of the Fitch–Margoliash method is shown in Figure 8.5, which also illustrates the minor extra complication involved in finding branch lengths between internal nodes. This example also illustrates a weakness of the method in using the evolutionary distances directly to select nearest neighbors. If there are very different evolutionary rates along different tree branches the two closest sequences as measured by evolutionary distance may not really be neighbors. This situation occurs in the data of Figure 8.5 for the sequences A and C, as will be proved when the same data are analyzed using the neighbor-joining method presented below. In this case it leads to a negative branch length. The true evolutionary history cannot give rise to a tree with negative branch lengths, so this is clearly an error of the method. In addition, the tree produced does not exactly reproduce the distances between sequences. There are many occasions when the rate variation is not so great as to produce these effects, in which case the method is capable of reconstructing the correct tree. Another application of the Fitch–Margoliash method is to calculate branch lengths for any given tree topology, which is useful when comparing different trees, as we shall see in Section 8.4.

(A) STEP 1 ($N = 5$)

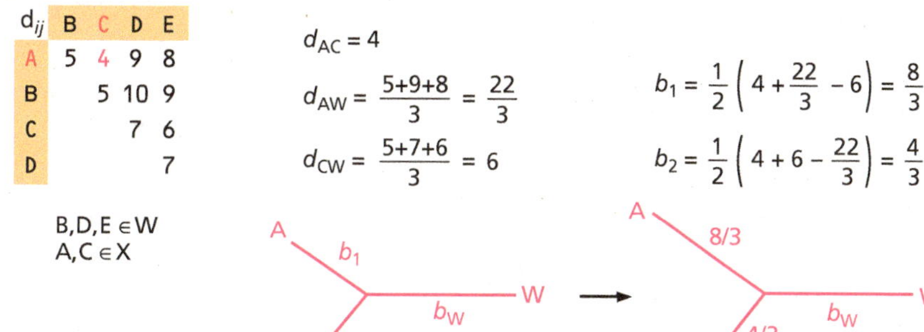

| $d_{ij}$ | B | C | D | E |
|---|---|---|---|---|
| A | 5 | 4 | 9 | 8 |
| B | | 5 | 10 | 9 |
| C | | | 7 | 6 |
| D | | | | 7 |

B,D,E ∈ W
A,C ∈ X

$$d_{AC} = 4$$

$$d_{AW} = \frac{5+9+8}{3} = \frac{22}{3}$$

$$d_{CW} = \frac{5+7+6}{3} = 6$$

$$b_1 = \frac{1}{2}\left(4 + \frac{22}{3} - 6\right) = \frac{8}{3}$$

$$b_2 = \frac{1}{2}\left(4 + 6 - \frac{22}{3}\right) = \frac{4}{3}$$



(B) STEP 2 ($N = 4$)

| $d_{ij}$ | D | E | X |
|---|---|---|---|
| B | 10 | 9 | 5 |
| D | | 7 | 8 |
| E | | | 7 |

A,C ∈ X
D,E ∈ Y
B,X ∈ Z

$$d_{XB} = 5$$

$$d_{XY} = \frac{8+7}{2} = \frac{15}{2}$$

$$d_{BY} = \frac{10+9}{2} = \frac{19}{2}$$

$$b_3 = \frac{1}{2}\left(\frac{8}{3} + \frac{4}{3}\right) = 2$$

$$(b_3 + b_4) = \frac{1}{2}\left(5 + \frac{15}{2} - \frac{19}{2}\right) = \frac{3}{2}$$

$$b_4 = \frac{3}{2} - b_3 = \frac{3}{2} - 2 = -\frac{1}{2}$$

$$b_5 = \frac{1}{2}\left(5 + \frac{19}{2} - \frac{15}{2}\right) = \frac{7}{2}$$

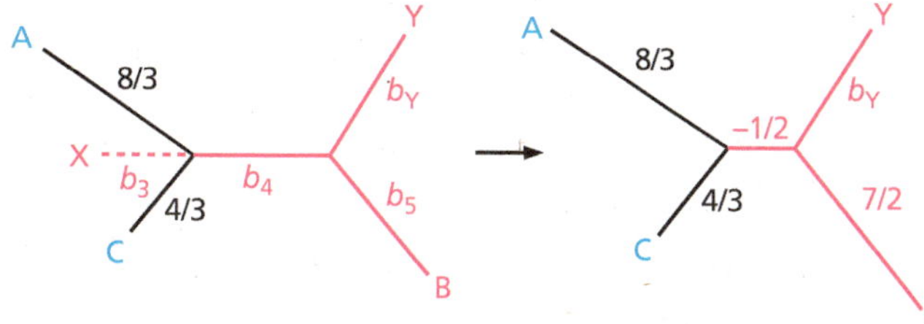



(C) STEP 3 ($N = 3$)

| $d_{ij}$ | E | Z |
|---|---|---|
| D | 7 | 26/3 |
| E | | 23/3 |

A,B,C ∈ Z

$$d_{DE} = 7$$

$$d_{DZ} = \frac{26}{3}$$

$$d_{EZ} = \frac{23}{3}$$

$$(b_6 + b_7) = \frac{1}{2}\left(\frac{26}{3} + \frac{23}{3} - 7\right) = \frac{14}{3}$$

$$b_6 = \frac{1}{3}\left(\left[\frac{8}{3} - \frac{1}{2}\right] + \frac{7}{2} + \left[\frac{4}{3} - \frac{1}{2}\right]\right) = \frac{13}{6}$$

$$b_7 = \frac{14}{3} - b_6 = \frac{14}{3} - \frac{13}{6} = \frac{5}{2}$$

$$b_8 = \frac{1}{2}\left(7 + \frac{26}{3} - \frac{23}{3}\right) = 4$$

$$b_9 = \frac{1}{2}\left(7 + \frac{23}{3} - \frac{26}{3}\right) = 3$$

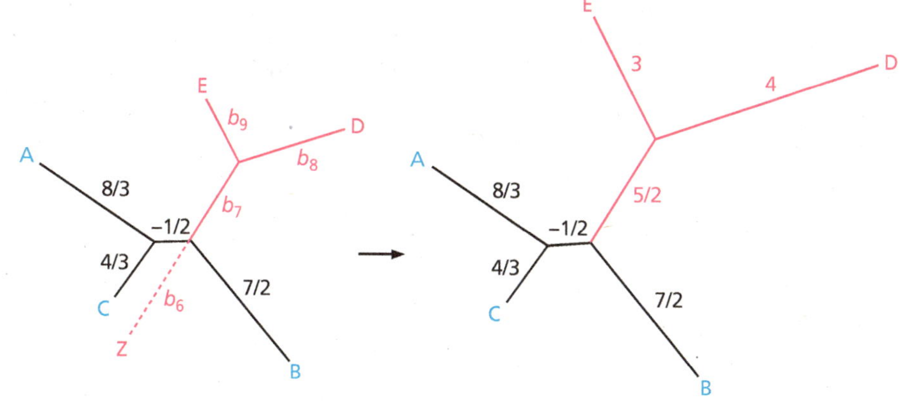



**Figure 8.5**
**A worked example of the Fitch–Margoliash method reconstruction of a phylogenetic tree for five sequences ($N = 5$).** At each step the three-leaf tree that is the equivalent of Figure 8.4 is shown in red on the left-hand tree. (A) In the first step the shortest distance is used to identify the two clusters (A, C) which are combined to create the next internal node. A temporary cluster W is defined as all clusters except these two, and the distances calculated from W to both A and C. The method then uses Equations EQ8.18 to calculate the branch lengths from A and C to the internal node that connects them. (B) In the second step, A and C are combined into the cluster X and the distances calculated from the other clusters. After identifying B and X as the next clusters to be combined to create cluster Z, the temporary cluster Y contains all other sequences. X is a distance $b_3$ from the new internal node, and the distance between the internal nodes is $b_4$. Branch length $b_4$ is calculated to be negative, which is clearly not realistic. However, in the further calculations this branch is treated like all the others. (C) Combining sequences A, B, and C into cluster Z, the sequences D and E are added to the tree in the final step. (D) The final tree has a negative branch length, and also does not agree with all the original distance data. The tables list the patristic distances $\Delta_{ij}$—the distances measured on the tree itself—and the errors $e_{ij}$. This tree has the wrong topology, as becomes clear later using neighbor-joining (see Figure 8.7), which in this case can successfully recover the correct tree.

(D) patristic distance matrix $\Delta_{ij}$ from the tree and errors $e_{ij}$

| $\Delta_{ij}$ | B | C | D | E |
|---|---|---|---|---|
| A | 5.7 | 4.0 | 8.7 | 7.7 |
| B | | 5.3 | 10.0 | 9.0 |
| C | | | 7.3 | 6.3 |
| D | | | | 7.0 |

| $e_{ij}$ | B | C | D | E |
|---|---|---|---|---|
| A | 2/3 | 0 | −1/3 | −1/3 |
| B | | 1/3 | 0 | 0 |
| C | | | 1/3 | 1/3 |
| D | | | | 0 |

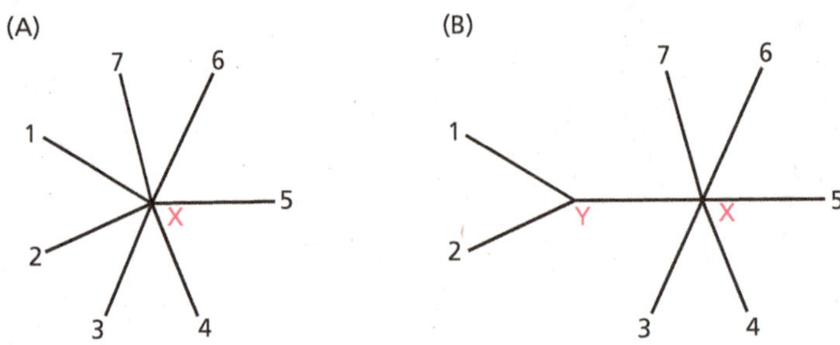


**Figure 8.6**
**The first step of the neighbor-joining method.** (A) The situation at the start of neighbor-joining, with a star tree in which all sequences are joined directly to a single internal node X with no internal branches. (B) After sequences 1 and 2 have been identified as the first pair of nearest-neighbors, they are separated from node X by an internal branch to internal node Y. The method calculates the branch lengths from sequences 1 and 2 to node Y to complete the step.

## The neighbor-joining method is related to the concept of minimum evolution

The neighbor-joining (NJ) method proposed by Naruya Saitou and Masatoshi Nei in 1987 does not assume all sequences have the same constant rate of evolution over time, and so like the Fitch–Margoliash method is more generally applicable than UPGMA. The basis of the method lies in the concept of minimum evolution, namely that the true tree will be that for which the total branch length, $S$, is shortest. This has its origin in the idea that any model used to explain data should be as simple as possible, which in the case of evolution equates to a model with as few mutation events as are consistent with the data. As we will see in the discussion of parsimony methods later in the chapter, some other applications of the principle of minimum evolution require the calculation of many different tree topologies, from which the one with the smallest $S$ is chosen. The neighbor-joining technique is a simple approximation, but one that is very effective at constructing trees for large datasets. The resulting tree is not rooted and is additive, a property that is assumed in deriving the formulae for its construction.

Neighbors in a phylogenetic tree are defined as a pair of nodes that are separated by just one other node. As with the methods described above, pairs of tree nodes are identified at each step of the method and used to gradually build up a tree. The way in which pairs of clusters are identified as neighbors in each step is different from that used by Fitch–Margoliash and UPGMA, resulting in the potential to generate a different topology. As will be shown by example, the method used to identify neighbors is more effective than that used by Fitch–Margoliash, so that this method is more robust with data having a large range of rates of evolution. The method is best explained by starting with all sequences arising from the same node and gradually distinguishing pairs of nodes that are neighbors.

To derive the neighbor-joining equations, consider $N$ sequences, labeled 1 to $N$. The two trees of Figure 8.6 illustrate one step of the neighbor-joining technique. The tree in Figure 8.6A is a star tree, in which all sequences are related directly to a single ancestral sequence at node X. Figure 8.6B shows a closely related tree in which neighbor sequences 1 and 2 have been separated from node X by another node, Y. Note that because these trees are not rooted, the direction of evolution along the branch connecting X and Y is not clear. The distance between two sequences is written $d_{ij}$ for sequences $i$ and $j$. The length of a branch of a tree between leaf (or node) $e$ and node $f$ is called $b_{ef}$, and the total branch length $S$ of a star tree such as that in Figure 8.6A is given by

$$S = \sum_{i=1}^{N} b_{i\mathrm{X}} = \frac{1}{N-1}\sum_{i<j}^{N} d_{ij} \qquad \text{(EQ8.19)}$$

where the second equality follows from assuming that the tree correctly reproduces the $d_{ij}$.

The total branch length of the tree in Figure 8.6B, where sequences 1 and 2 have been removed from the central node X by internal node Y, is given by

$$S_{12} = b_{1\mathrm{Y}} + b_{2\mathrm{Y}} + b_{\mathrm{XY}} + \sum_{i=3}^{N} b_{i\mathrm{X}} \qquad \text{(EQ8.20)}$$

which needs to be converted into a form that uses the sequence distances $d$. To perform this conversion, three equivalences should first be noted. Firstly, the last term of Equation EQ8.20 is equivalent to Equation EQ8.19 with $N = N - 2$; secondly, $b_{1\mathrm{Y}} + b_{2\mathrm{Y}} = d_{12}$, and thirdly

$$b_{\mathrm{XY}} = \frac{1}{2(N-2)}\sum_{i=3}^{N}\left(d_{1i} + d_{2i}\right) - \frac{1}{2}\left(b_{1\mathrm{Y}} + b_{2\mathrm{Y}}\right) - 2\sum_{i=3}^{N} b_{i\mathrm{X}} \qquad \text{(EQ8.21)}$$

Making appropriate substitutions we obtain

$$S_{12} = \frac{1}{2(N-2)}\sum_{i=3}^{N}\left(d_{1i} + d_{2i}\right) + \frac{d_{12}}{2} + \frac{1}{N-3}\sum_{3\le i<j}^{N} d_{ij} - \frac{1}{N-2}\sum_{i=3}^{N} b_{i\mathrm{X}} \qquad \text{(EQ8.22)}$$

It remains to convert the last term to a sum over $d_{ij}$ using Equation EQ8.19, after which it can be combined with the penultimate term to give

$$S_{12} = \frac{1}{2(N-2)}\sum_{i=3}^{N}\left(d_{1i} + d_{2i}\right) + \frac{1}{N-2}\sum_{3\le i<j}^{N} d_{ij} + \frac{d_{12}}{2} \qquad \text{(EQ8.23)}$$

This can be simplified if we define the following terms:

$$U_1 = \sum_{i=1}^{N} d_{1i} \qquad \text{(EQ8.24)}$$

$$U_2 = \sum_{i=1}^{N} d_{2i} \qquad \text{(EQ8.25)}$$

and

$$d_{sum} = \sum_{i<j}^{N} d_{ij} \qquad \text{(EQ8.26)}$$

Substitution and careful manipulation give

$$S_{12} = \frac{2d_{sum} - U_1 - U_2}{2(N-2)} + \frac{d_{12}}{2} \qquad \text{(EQ8.27)}$$

Every pair of sequences $i$ and $j$, if separated from the star node as shown in Figure 8.6B, for the case of sequences 1 and 2 will produce a tree of total branch length $S_{ij}$. According to the minimum evolution idea, the tree that should be chosen is that with the smallest $S_{ij}$. Noting that the sum of the evolutionary distances for all sequence pairs in the dataset, $d_{sum}$, is constant for the data being studied and independent of the particular sequences $i$ and $j$, in this case the minimum evolution principle can be shown to be equivalent to finding the pair of sequences with the smallest value of the quantity $d_{ij}$ defined by

**Figure 8.7**
**A worked example of the neighbor-joining method reconstruction of a phylogenetic tree for five sequences ($N = 5$).** This example uses the same five sequences as in Figure 8.5. At each step the evolutionary distances $d_{ij}$ are converted to $\delta_{ij}$ using Equations EQ8.24 and EQ8.28. The values $(N-2)\delta_{ij}$ have been listed to make the working clearer. At each step the pair of clusters with smallest (i.e., least positive) $\delta_{ij}$ are identified and combined into the next cluster. The final tree is in exact agreement with the original data for the distances between taxa. The topology of the tree differs from that produced by the Fitch–Margoliash method in Figure 8.5.

(A) STEP 1 ($N = 5$)

$d_{ij}$

| | B | C | D | E |
|---|---|---|---|---|
| A | 5 | 4 | 9 | 8 |
| B | | 5 | 10 | 9 |
| C | | | 7 | 6 |
| D | | | | 7 |
| E | | | | |

$U_i$

| | $U_i$ |
|---|---|
| A | 26 |
| B | 29 |
| C | 22 |
| D | 33 |
| E | 30 |

$3\delta_{ij}$

| B | C | D | E | |
|---|---|---|---|---|
| −40 | −36 | −32 | −32 | A |
| | −36 | −32 | −32 | B |
| | | −34 | −34 | C |
| | | | −42 | D |
| | | | | E |

D and E are neighbors through internal node W with $d_{DW} = \frac{1}{2}\left(7 + \frac{33-30}{3}\right) = 4$ and $d_{EW} = 7 - 4 = 3$.

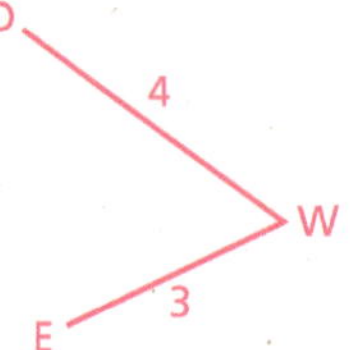


(B) STEP 2 ($N = 4$)

$d_{ij}$

| | B | C | W |
|---|---|---|---|
| A | 5 | 4 | 5 |
| B | | 5 | 6 |
| C | | | 3 |
| W | | | |

| | $U_i$ |
|---|---|
| A | 14 |
| B | 16 |
| C | 12 |
| W | 14 |

$2\delta_{ij}$

| B | C | W | |
|---|---|---|---|
| −20 | −18 | −18 | A |
| | −18 | −18 | B |
| | | −20 | C |
| | | | W |

C and W are neighbors through internal node X with $d_{CX} = \frac{1}{2}\left(3 + \frac{12-14}{2}\right) = 1$ and $d_{WX} = 3 - 1 = 2$.

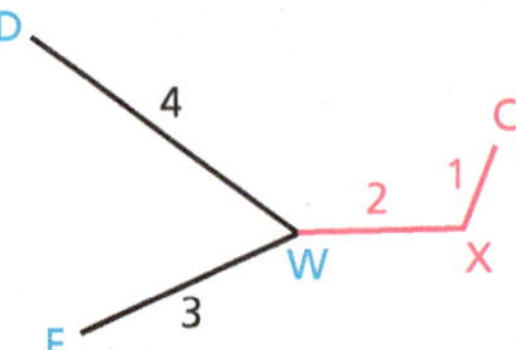


(C) STEP 3 ($N = 3$)

$d_{ij}$

| | B | X |
|---|---|---|
| A | 5 | 3 |
| B | | 4 |
| X | | |

| | $U_i$ |
|---|---|
| A | 8 |
| B | 9 |
| X | 7 |

$\delta_{ij}$

| B | X | |
|---|---|---|
| −12 | −12 | A |
| | −12 | B |
| | | X |

Three alternatives (of which here we choose one of the two with an internal node):
A and X are neighbors through internal node Y with $d_{AY} = 2$ and $d_{XY} = 1$ or
B and X are neighbors through internal node Y with $d_{BY} = 3$ and $d_{XY} = 1$.
Whichever is chosen, the remaining distance $d_{AY}$ or $d_{BY}$ will be found in the next $d_{ij}$ matrix.

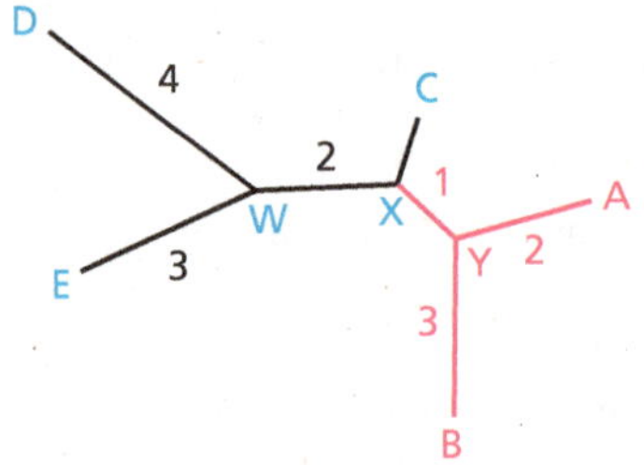

$$\delta_{ij} = d_{ij} - \frac{U_i + U_j}{N-2} \qquad \text{(EQ8.28)}$$

Once this pair has been found, the distances to the new node Y (see Figure 8.6B) must be calculated. From the tree in Figure 8.6B, the following equality is seen to hold:

$$b_{1Y} = \frac{1}{2}\left( d_{12} + \frac{\sum_{i=3}^{N} d_{1i} - \sum_{i=3}^{N} d_{2i}}{N-2} \right) = \frac{1}{2}\left( d_{12} + \frac{U_1 - U_2}{N-2} \right) \qquad \text{(EQ8.29)}$$

Thus, once the neighbors $i$ and $j$ have been identified from calculating $\delta_{ij}$, the new node Y can be added to the tree such that

$$b_{iY} = \frac{1}{2}\left( d_{ij} + \frac{U_i - U_j}{N-2} \right) \qquad \text{(EQ8.30)}$$

and

$$b_{jY} = d_{ij} - b_{iY} \qquad \text{(EQ8.31)}$$

We now require the distances from Y to the other sequences, $k$, which can be seen from inspection of Figure 8.6B to be easily calculable using

$$d_{Yk} = \frac{1}{2}\left( d_{ik} + d_{jk} - d_{ij} \right) \qquad \text{(EQ8.32)}$$

This formula is exactly equivalent to Equation EQ8.18 of the Fitch–Margoliash method. To add more nodes, we now repeat the process, starting with the star tree formed by removing sequences $i$ and $j$, to leave a star tree with node Y as a new leaf (that is, 1 and 2 are omitted from Figure 8.6B). Note that at each step, the value of $N$ in the formulae decreases by 1. A worked example of neighbor-joining is shown in Figure 8.7. The data for this example are identical to those used for the Fitch–Margoliash example in Figure 8.5, and neighbor-joining is much more successful at recovering the correct tree topology and distances. In particular the neighbors A and B (as opposed to A and C) are correctly identified, leading to a tree with no negative branch lengths that exactly reproduces the given distance data.

Some variants of the neighbor-joining method use weighted distances derived from estimates of variance in the corrected distance. For many evolutionary models, it is possible to derive formulae that estimate this variance, which can be used as a measure of the confidence in the distance and converted into a weighting. Two examples of weighted neighbor-joining related methods are the weighbor and BIONJ methods (see Further Reading).

## Stepwise addition and star-decomposition methods are usually used to generate starting trees for further exploration, not the final tree

Two other more general methods are also frequently used to generate a starting tree for further exploration. They differ from the methods discussed so far in that they do not necessarily use the evolutionary distances to construct the tree. Because of the large number of possible tree topologies for even quite small datasets, methods that explore alternative topologies almost always start from an initial tree and explore some similar tree topologies, rather than try every single possible tree. Neighbor-joining is often used to provide the initial tree, but alternatives exist, of which we will discuss two: **stepwise addition** and **star decomposition**. As with neighbor-joining, these methods could be used to simply generate a single tree with no further optimization. However, this is rarely if ever done.

Stepwise addition proceeds from an initial three-sequence tree, adding one sequence at a time. At each addition the tree is evaluated using an optimizing function of the type that will be described in Section 8.4, which returns an optimum value as a score for that topology. In both this method and star decomposition a number of different functions can be used to obtain an optimal value, such as total branch length, total number of mutations, or the likelihood of a specific tree given a particular evolutionary model. We will use the symbol $S$ for the value of the function used.

Each time a sequence is added, it is added at each possible branch of the tree, and the new tree with the optimal value of $S$ is retained for subsequent steps. Several variants of the method exist, which differ in how one chooses the next sequence to add. Sequences can simply be added in any order, but this is not likely to produce a satisfactory tree. Alternatively, a number of randomized orderings of the sequences can be examined, and the best tree chosen. Probably the best method, although more time-consuming, is to try all remaining sequences on all branches and choose the best of these trees.

The star-decomposition method starts with all sequences linked to the same single internal node X, as in the neighbor-joining method discussed above (see Figure 8.6). As before, two sequences are selected and joined together via their own internal node. For the neighbor-joining method, the function being optimized is $S_{12}$, but in this case other functions are used for evaluating which sequences to detach from the X node next. All possible pairs are examined, and the tree with the optimal value of $S$ is retained. At each subsequent step, another two sequences or other internal nodes joined to node X are combined to create another new internal node, until node X has only three branches. Only the tree with optimal $S$ is retained each time.

## 8.3 Generating Multiple Tree Topologies

Unlike the techniques discussed previously, the other tree-construction algorithms we describe in this chapter involve two processes. One generates multiple possible tree topologies, which are then evaluated in a subsequent procedure to obtain branch lengths and measures of optimality. The generation of the topology is, in principle, separate from the evaluation, but in many cases the measure of optimality is calculated as tree-building proceeds, and is used as feedback during the construction process to exclude some topologies from consideration. The algorithms for constructing tree topologies will be presented in this section (see Flow Diagram 8.3), and the evaluation phase will be discussed in more detail in Section 8.4. We do, however, need to introduce here two important concepts used in tree evaluation: maximum parsimony and maximum likelihood. Maximum parsimony evaluates trees by looking for the one that uses the least possible number of mutations to reconstruct the phylogeny. Maximum likelihood evaluates a number of comparable trees on the criterion of which is most likely to be a correct representation of the data. All the tree-generation methods discussed here require a starting topology, which is obtained using one of the techniques described in Section 8.2, such as neighbor-joining, stepwise addition, or star decomposition. We will mainly discuss the generation of unrooted trees, as this is the situation most often encountered in practice. Usually, if a rooted tree is required, the root is determined separately from

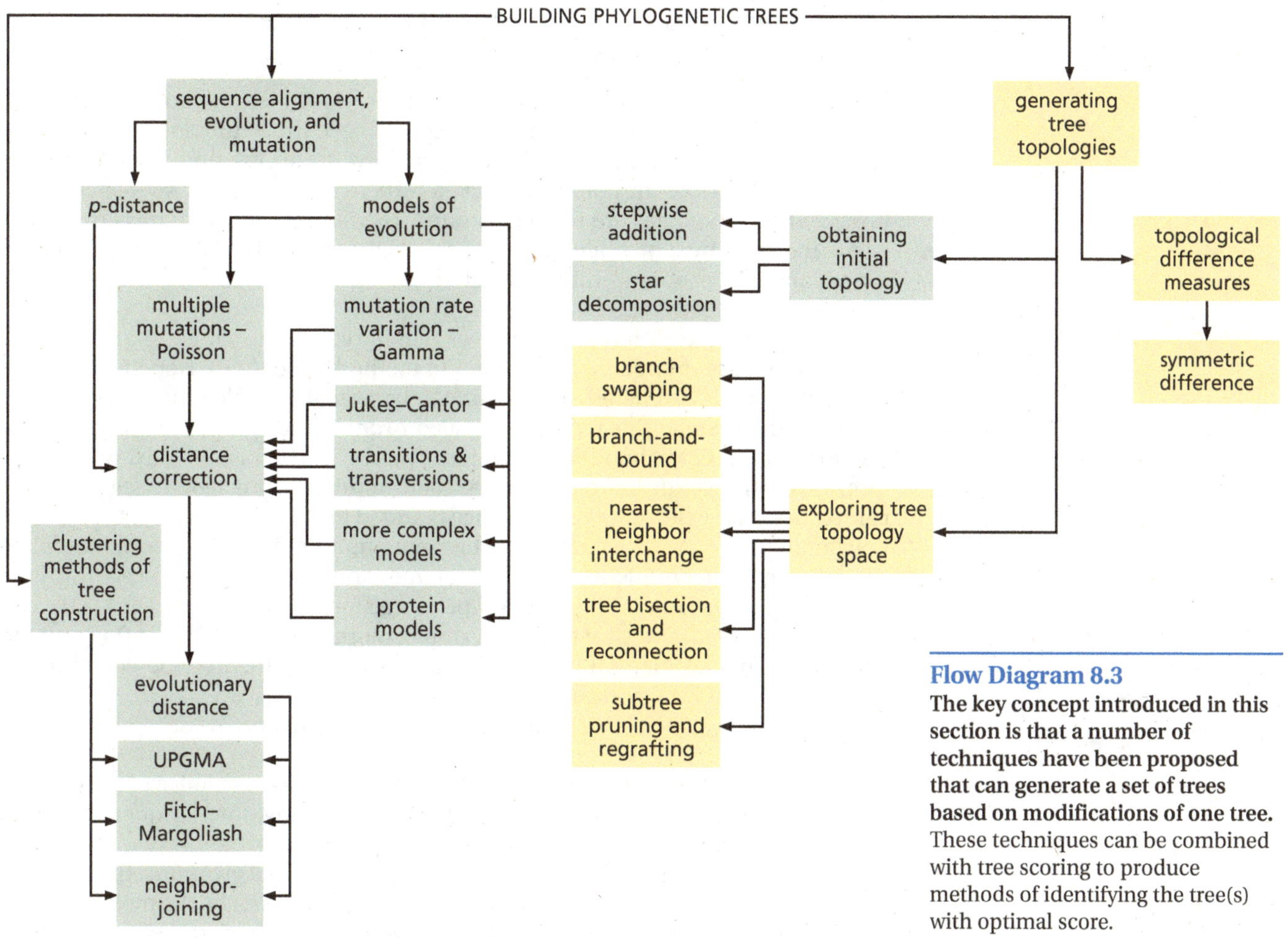

**Flow Diagram 8.3**
**The key concept introduced in this section is that a number of techniques have been proposed that can generate a set of trees based on modifications of one tree.** These techniques can be combined with tree scoring to produce methods of identifying the tree(s) with optimal score.

the overall topology and branch lengths. The techniques used to determine the root will be discussed at the end of this section.

An unrooted tree of $N$ sequences that only has bifurcating nodes has $(2N - 3)$ branches. For three or more sequences, the number of distinct unrooted trees, $T_N$, is given by

$$T_N = 1.3.5....(2N - 5) = \frac{(2N - 5)!}{2^{N-3}(N - 3)!} \qquad \text{(EQ8.33)}$$

which increases quickly with the number of sequences. There are only 15 different trees with five sequences, but 2,027,025 with 10 sequences. There are $T_{N+1}$ different rooted trees with $N$ sequences. An exhaustive search of all possible tree topologies is obviously impractical for even modest numbers of sequences, and methods have been proposed that limit the search while, hopefully, including all relevant tree topologies.

The measure of optimality used by these methods is given the symbol $S$, which, as noted in Section 8.2, will be qualitatively different for different kinds of measures used. For example, in parsimony methods it is related to the total branch length or total number of mutations; in maximum-likelihood methods it is the likelihood of that particular tree given the model. In the former case $S$ will be minimized, and in the latter it will be maximized. For the present discussion, this is all we need to know about $S$.

## The branch-and-bound method greatly improves the efficiency of exploring tree topology

As there is only one topology for a tree with three sequences, by choosing any three sequences we can generate a partial tree of the correct topology. Adding one sequence at a time in all possible different ways to an existing partial tree can generate all possible unrooted trees. Even if there are only a few sequences, if all possible tree topologies are to be generated, the number quickly grows to a point where the trees must be evaluated using the measure $S$ as they are created, because they cannot all be stored simultaneously.

In general, we are only interested in trees that have extreme values of $S$ (maximum or minimum, as required by the evaluation method to be used). Depending on the way in which $S$ is estimated, a shortcut can be available, which improves the efficiency of the search. This technique, called **branch-and-bound**, identifies sets of tree topologies that cannot have suitable $S$ values. In the case of maximum parsimony, we can exploit the fact that the total number of nucleotide substitutions cannot decrease when another sequence is added to a partial tree. So, keeping with maximum parsimony, if a partial tree has more mutations than a known complete tree, adding further branches to the partial tree cannot possibly produce an optimal tree. If a record is kept of the best complete tree found so far, of length $S_m$, we can identify such dead-end situations and stop exploring related trees. In this way the branch-and-bound method performs an exact search of the tree topologies in the sense that no relevant trees are omitted from the study. However, all trees that have a value $S$ sufficiently close to $S_m$ may be stored for further analysis.

The branch-and-bound technique can benefit from judicious choice of the order in which sequences are added. For example, the starting three-sequence tree is selected by calculating the branch lengths of all such trees and picking the one (in the case of parsimony) with the most substitutions. This will give you a tree with the three most divergent sequences, even though the tree will still be the most parsimonious one for those sequences. A fourth sequence is then added to this tree, exploring all remaining sequences in all possible topologies, and again choosing the topology that gives the maximum parsimony tree with the most substitutions. This is repeated until all sequences have been added and a complete tree produced. This should be contrasted with many of the methods discussed above, in which the most closely related sequences are selected first.

The rationale is that this order will rapidly produce trees with many substitutions. Once the partial tree's $S$-value approaches that of the current optimum tree, the suboptimal partial trees will often have values exceeding $S_m$ and further branch addition can be ruled out. If a new complete tree is generated with a smaller $S$ than $S_m$, it becomes the new current optimal tree, and $S_m$ is updated accordingly.

## Optimization of tree topology can be achieved by making a series of small changes to an existing tree

When there are many sequences, it becomes impractical to perform a full search of all possible tree topologies, even taking advantage of the branch-and-bound method. In such cases a suboptimal complete tree can be generated and used as a starting point for subsequent exploration of similar tree topologies. The methods we will discuss involve making small modifications to an existing tree and seeing if it is better according to criterion $S$. There may, however, be local extrema for $S$, which can prevent us finding the tree with the globally optimal $S$ (see Appendix C for discussion on the question of finding global minima in other contexts).

As well as comparing trees on the basis of their $S$-values, it is useful to have a quantitative measure of the difference between the topology of two different trees that is independent of the branch lengths. One such measure, called the **symmetric difference**, is described in Box 8.2.

One method of making small differences to a tree is to switch the order of the branches in a procedure known as **branch swapping**. There are three main branch-swapping techniques. The **nearest-neighbor interchange (NNI) method** creates modified trees with a symmetric difference of 2 from the starting tree. Any internal branch of a bifurcating tree will have four neighboring nodes (see Figure 8.8), each of which may represent an individual sequence or a subtree. Without changing the four nodes, there are three possible different topologies for this branch, of which one is present in the starting tree. In the NNI method, the other two topologies of this internal branch are produced and examined to see if they have more optimal $S$ values. Usually all $(N-3)$ internal branches are examined, generating $2(N-3)$ alternative trees. When the best of the new trees has been found it becomes the next

## Box 8.2 Measuring the difference between two trees

If two different trees have been proposed to represent the evolutionary history of the same data, it is useful to have a quantitative measure of the difference between them. If the two trees have the same topology, the branch lengths can be compared to highlight differences. If they have different topologies, however, the branch lengths will not give a full measure of the discrepancies. What is required is a quantitative topological measure.

The difference between tree topologies can be quantified through the use of splits (Figure B8.2; see also Section 7.1). A topological distance between two trees can be defined using the number of splits that differ. This is known as the **symmetric** or **Robinson–Foulds difference**. For unrooted trees of $N$ sequences, there are $N-3$ internal branches, and the number of different splits between two trees will therefore range from 0 to $2(N-3)$.

In the example shown in the Figure B8.2, eight sequences were used to build the tree and there will therefore be a maximum of 10 possible different splits. The number of different splits produced is sometimes divided by the total number of splits (in this case 10) to give a clearer measure of the relative extent of the difference.

It can be shown that the number of different splits is equal to the minimum number of elementary operations—operations where nodes are merged or split—that are needed to convert one topology to the other. Alternative definitions of distances have been proposed that measure the minimum number of operations required when using the nearest-neighbor interchange (NNI), the **subtree pruning and regrafting (SPR)**, or the **tree bisection and reconnection (TBR)** methods of branch swapping to modify trees (see Further Reading).

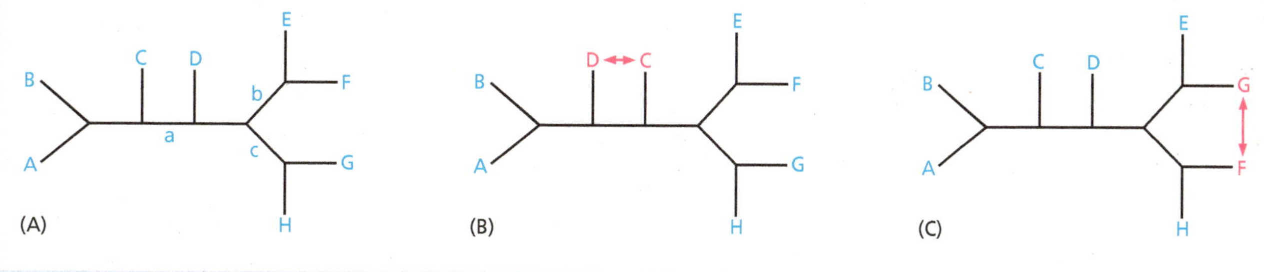

**Figure B8.2**
**Three trees constructed from the same set of data to illustrate how the topological difference between trees can be calculated in the form of the symmetric difference.** The symmetric distance is calculated as the number of splits that differ between the two trees. (A) A tree for the eight sequences A–H, with three internal branches labeled a, b, and c. (B) Tree as in (A) but with the positions of nodes C and D switched, as indicated by the red arrows. (C) Tree as in (A) but with the positions of nodes F and G switched. The symmetric distance between trees (A) and (B) is 2 because they have different splits for branch a — (A,B,C)(D,E,F,G,H) compared with (A,B,D)(C,E,F,G,H), respectively. Tree (C) is a distance of 4 from (A), because of alternative splits for branches b and c, and it is 6 from (B) because of alternative splits for branches a, b, and c.

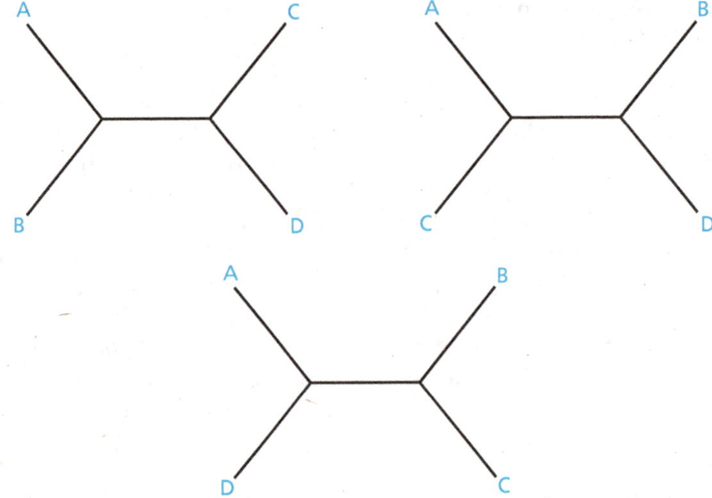


**Figure 8.8**
**The basic structure around an internal branch of a phylogenetic tree.** Each such branch will have four neighboring nodes, which may be individual sequences or subtrees. There are three distinct topologies for this branch.

starting tree. This technique is easy to apply but only makes small changes in the tree, and so is susceptible to local extrema. It is possible to extend the technique by altering several internal branches simultaneously to create larger modifications that may overcome these problems. The search is continued until no further improvement in the $S$ value has been obtained after some prespecified number of attempts.

Subtree pruning and regrafting (SPR) can produce more drastic changes in topology. One branch of the current tree is cut to produce two subtrees (see Figure 8.9). Each of these is then reattached to the other subtree using the cut branch. Attachment is explored at all possible branches of the other subtree. This method can produce substantially more new trees than NNI at each step; in fact, a number proportional to the square of the number of branches.

The third technique, tree bisection and reconnection (TBR), is related to SPR, but now the cut branch is removed from the subtrees, and they are rejoined by connecting any two branches, one from each subtree. This produces even more new trees. As before, the optimal new tree is used as the starting tree in the

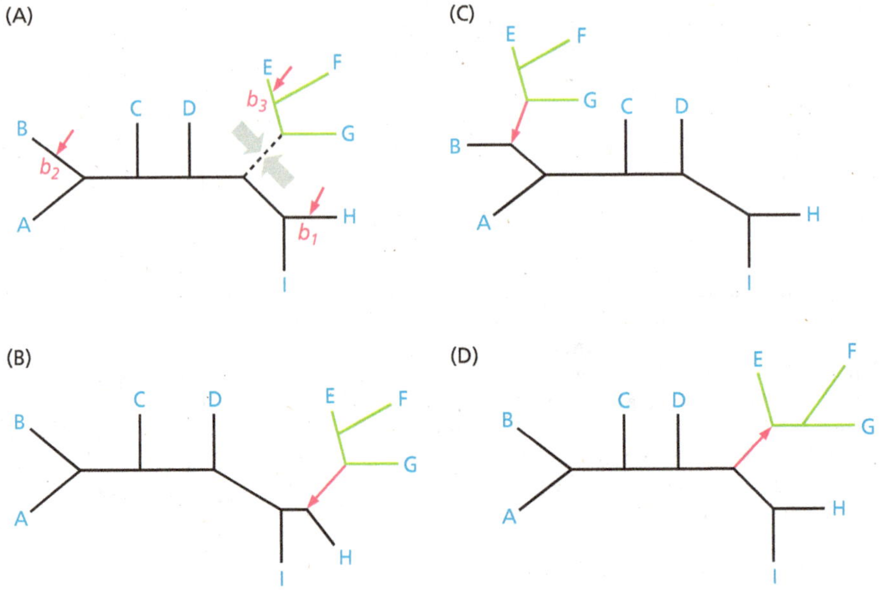

**Figure 8.9**
**An example of subtree pruning and regrafting (SPR).** (A) The initial tree has nine sequences labeled A–I. A branch of the tree is cut at the point indicated by the gray arrows, creating a large subtree and the smaller subtree with taxa E, F, and G. (B) In this tree the subtree has been placed on the external branch $b_1$ to taxon H, thus creating a new node. The symmetric difference (see Box 8.2) between this tree and tree A is 2. (C) In this tree the subtree has been placed on the external branch $b_2$ to taxon B. The symmetric difference between this tree and A is 6. (D) This tree has been produced by reattaching the larger subtree to the smaller subtree E, F, G at its original location but this time creating a new node on branch $b_3$ between E and F. This tree has a symmetric difference of 2 from A. In general, the symmetric difference between the initial and new trees depends on many factors relating to the size of the tree and subtrees and where the regrafting occurs relative to the pruning.

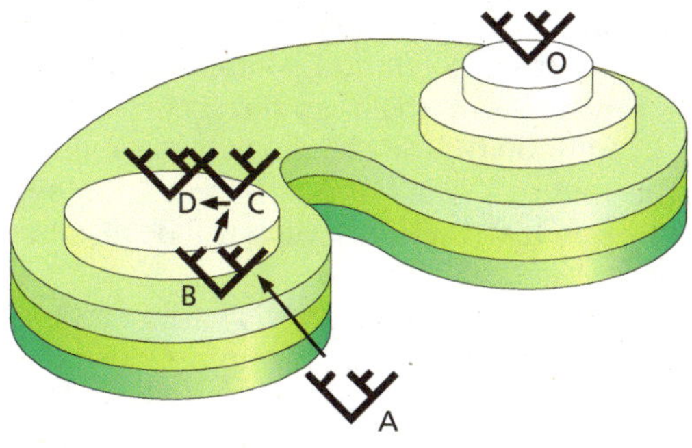


**Figure 8.10**
**An imaginary landscape surface of criterion *S*.** In this case the tree which best describes the data will be that with the highest value of *S*. There are several maxima in the landscape. The methods described in this chapter to optimize S move over the landscape in small steps, following a path such as represented by trees A to D. If they find tree D, most methods will terminate and not locate tree O, which is an even better fit.

following step. These two methods, SPR and TBR, produce new trees that have greater variation from the starting tree than the nearest-neighbor interchange, and so may be better at finding the globally optimal tree.

To understand the practical problems that may occur during optimization we must imagine a surface for the criterion *S*. Given a topological distance measure such as the symmetric distance (see Box 8.2) we can now conceive of a spatial arrangement of the trees according to this measure (see Figure 8.10). This space will have a region in which one tree has a more optimal *S*-value than the surrounding trees, and will appear as a peak in Figure 8.10. Unfortunately, in most cases there will be more than one such region, indicating that the optimal tree in this region may be only locally optimal. We are searching for the globally optimal tree, however. The branch-swapping methods allow us to explore this space, but at each step they can only move a small distance. If this step size is smaller than required to move from a locally optimal tree out of the group of nearby less optimal trees, the search will not be able to escape from this region, and the globally optimal tree will not be found. Two simple techniques can help us to avoid this situation. We can take bigger steps, although this can make it harder to home in on the optimal tree. Alternatively, we can perform many searches using different starting trees, in the hope that at least one of these will lead to the best tree. This multiple minima problem is presented in more detail in Appendix C for other optimization problems. In regard to trees, however, it is distinct from the examples discussed there, in that tree topology can only be changed in finite steps.

The tree-construction methods that search multiple different tree topologies often give a similar optimal *S*-value for several trees. It may be that there is benefit in reporting them all individually, but often it is more useful to obtain a consensus tree that illustrates those features of the trees that are well conserved and thus may be regarded as more reliable (see Figure 7.6). Consensus trees are not the only way of assessing and representing alternative or uncertain topologies, and methods using networks instead of trees have also been developed. Ways of assessing how strongly certain tree features are supported by the data are discussed in Section 8.5.

## Finding the root gives a phylogenetic tree a direction in time

Many of the techniques discussed in this chapter produce unrooted trees, yet it is often desirable to know where the root lies, so that all the branches can be assigned a direction in time. There are several ways of producing a root, but they all involve inexact assumptions and it is usually advisable to resist defining the root when possible. One way of defining a root on any tree is to arbitrarily choose the midpoint between the two most distantly related sequences. This is very speculative, however, and not to be recommended.

The most usual method of rooting a tree is to include a set of sequences that is known to be an outgroup; that is, sequences relatively distant from all the other sequences (see Figure 7.2D). It is preferable to use several outgroup sequences, and see that they cluster together, and also to try to use an outgroup that is still relatively close to the other sequences. This method is often used, but requires care. If the outgroup is very distantly related to the main set, accurate sequence alignment can be problematic, and in addition the long length of the branches connecting them to the other parts of the tree can cause errors in tree topology, as a result of the phenomenon of **long-branch attraction**, which will be discussed in Section 8.5.

Another proposed technique for rooting trees is based on detecting gene duplication events. This method has its origins in attempts to create reconciled trees, which combine species and gene phylogeny (see Figure 7.12). As discussed in Section 7.2, gene duplications and other evolutionary events can cause the phylogeny of genes to differ from the phylogeny of the species they have come from. Methods have been proposed to reconcile these differences; they are based

## Box 8.3 Going back to the origins of the tree of life

There is no difficulty in producing a phylogenetic tree that connects the main domains of life if the sequences chosen are similar enough for a meaningful alignment to be created. It is quite common, however, to find that the details of tree topology change significantly when different genes are used. Assuming the alignments are correct, there are two key reasons for such disagreements.

Firstly, there have been numerous horizontal gene transfer (HGT) events in the history of life, which result in some species appearing in unexpected locations on trees if those genes are used (see Figure 7.16). There have even been suggestions that in the very early stages of evolution there was so much gene transfer that it may be almost impossible to resolve the deepest evolutionary connections. It has been observed that genes involved in functions such as the processing of genetic information, as in transcription and translation, give more consistent trees, possibly because they are less likely to have undergone gene transfer.

The second reason for different tree topologies is due to the combination of two different genomes, a phenomenon known as genome fusion. Familiar examples are the presumed symbioses between a proto-eukaryotic cell and prokaryotes that gave rise to eukaryotic cells containing mitochondria and chloroplasts. There is also a suggestion that the eukaryotic cell itself arose from the fusion of two distinct prokaryotes. If this is the case, different eukaryotic genes will have originated in one or other of these prokaryotes, and thus the eukaryotes as a group will be found to have a different origin depending on the genes used to reconstruct their phylogeny.

The standard tree-reconstruction techniques presented in this chapter cannot model these evolutionary processes and will produce incorrect trees. Careful analysis of the tree can reveal HGT events (see Figure 7.16) while a genome fusion would appear in a tree as two ancestors joining to form the descendant. James Lake and Maria Rivera have proposed a method called conditioned reconstruction that is able to produce such trees, and have proposed a ring of life instead of a root (see Figure B8.3). While the idea of a genome fusion origin for eukaryotes is quite well accepted, it is too soon to know how well this proposed ring of life will be received.

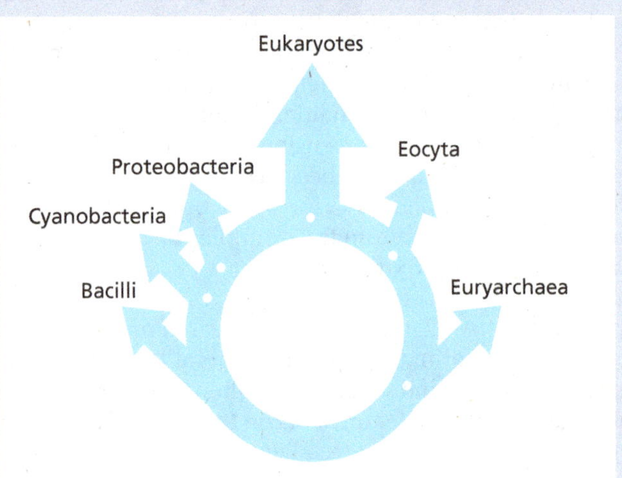


**Figure B8.3**
**A proposed schematic diagram of the ring of life.** According to this scheme, the eukaryotes have been derived by genome fusion of an ancestor of the present-day bacterial lineage Proteobacteria (or possibly a common ancestor of the Proteobacteria and Cyanobacteria) with an ancestor of the archaeal lineage Eocyta (Crenarchaeota). The bacterial ancestor is assumed to have provided the operational genetic content of the eukaryotic cell (genes and structures involved in amino acid biosynthesis, energy generation, and so on), while the eocyte provided the informational content (genes involved in transcription, translation, and related processes). (Adapted from M.C. Rivera and J.A. Lake, The ring of life provides evidence for a genome fusion origin of eukaryotes, *Nature* 431:152-155, 2004.)

on identifying a unifying phylogeny that involves the fewest gene duplications and gene losses necessary to explain the data. The same concept can be used to root trees. If gene duplications are present in the phylogenetic history, a root can be sought that minimizes the number of duplications. The **speciation duplication inference** (**SDI**) method of Christian Zmasek and Sean Eddy is one such method, and requires knowledge of the evolutionary tree for the species as well as the presence of several gene duplications within the dataset.

One important issue in evolutionary studies is the root of all known life—the last universal common ancestor—and the evolutionary events leading to the separation of major domains of life, such as the eukaryotes. This area is briefly discussed in Box 8.3 as it involves problems and proposed solutions not usually encountered in normal phylogenetic analysis.

## 8.4 Evaluating Tree Topologies

Each tree topology generated must be evaluated to see how consistent it is with the data. There are a number of methods for doing this, each of which quantifies the agreement between the data and the tree through application of a mathematical function such as the estimated number of mutations involved in the reconstructed phylogeny (see Flow Diagram 8.4). As in the previous section, the symbol $S$ will be used to represent the value of any of these functions. In some methods, such as maximum parsimony, the optimal value is the minimum value for $S$, while in others it is the maximum. The optimization of $S$ has two separate steps. First, the optimal value of $S$ is found for each given tree topology. This exercise will involve calculating the branch lengths, except for the parsimony approach, which calculates the total length of all branches. In the second step, trees are ranked according to their value of $S$. If several trees are identified as relevant, because they all have near-optimal values of $S$, it can be helpful to construct a consensus tree (see Figure 7.6).

One group of methods uses the evolutionary distances between sequences as the basis for their optimizing functions. These distances are the same as those used by the techniques described previously, obtained by application of a formula derived using a model of evolution. In contrast to techniques such as UPGMA and neighbor-joining, however, the methods applied here can give a score for any submitted tree topology, and are also able to identify several equally or almost equally good alternative topologies. After these distance-based methods, we will discuss techniques that use the sequence alignment directly in evaluating the tree.

### Functions based on evolutionary distances can be used to evaluate trees

When functions involving evolutionary distances are used to optimize a tree topology, the function optimized during the calculation of branch lengths for a particular tree may not necessarily be the same function that will be used to compare different trees. For example, branch lengths may be calculated by a least-squares fit of the evolutionary distances, but trees might be compared according to their total branch length. We will start by explaining some of the ways in which branch lengths are derived for given tree topologies, and then look at some of the functions used in comparing topologies.

The Fitch–Margoliash method as presented in Figure 8.5 can be used to calculate branch lengths for a tree of specified topology, as long as all internal nodes of the tree are resolved so that they all have three branches. By selecting one node at a time from the specified topology, the formulae of Equation EQ8.18 can be used to calculate branch lengths and so reconstruct the tree. In principle, the neighbor-joining method can be applied in an equivalent way.

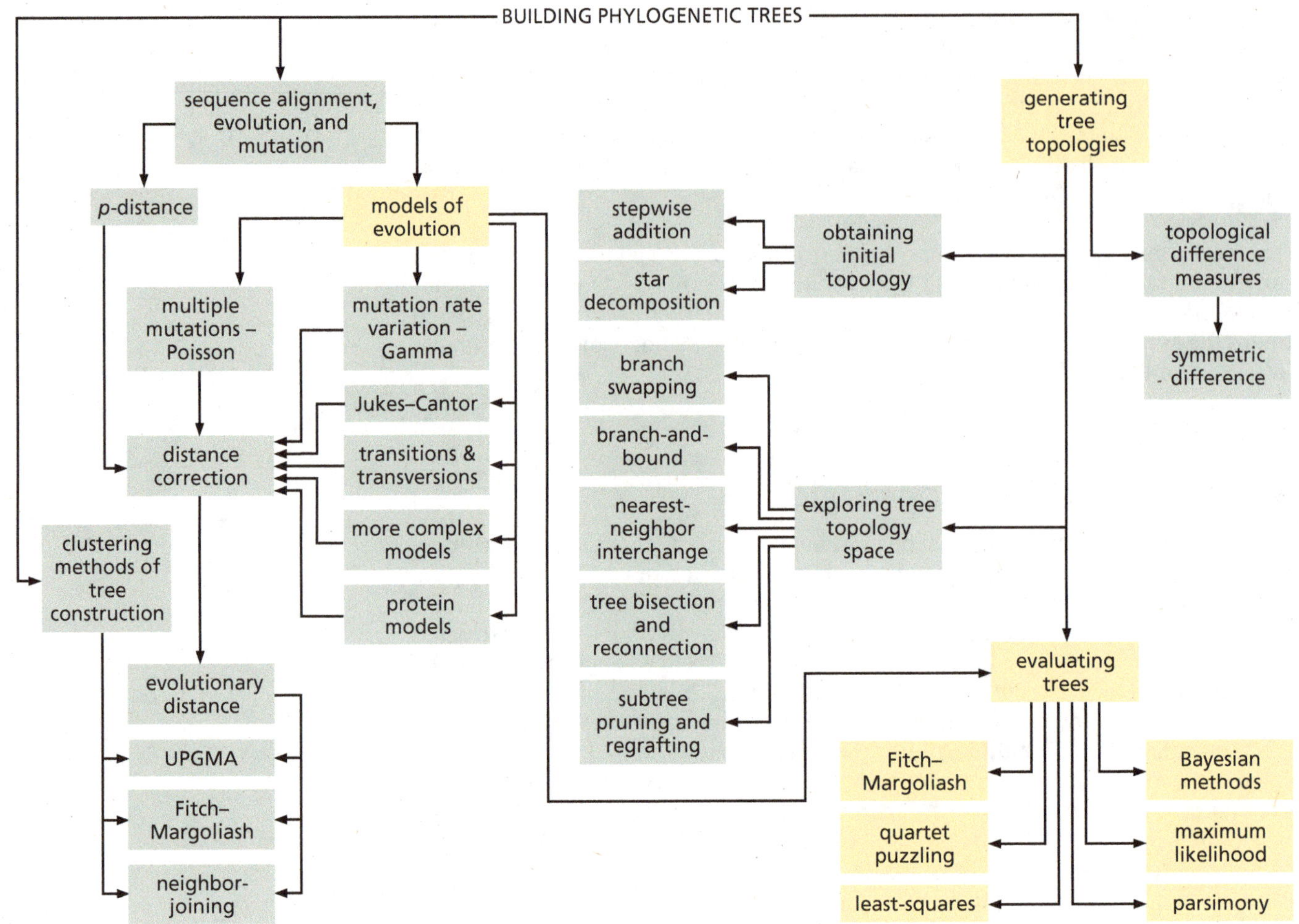


**Flow Diagram 8.4**
**In this section a number of different techniques are presented that can be used to assign a score to a given tree topology based on the evolutionary information present in a sequence dataset.**

Another method, which can be employed when calculating branch lengths, obtains a best fit to the distances derived from analysis of the sequence alignment by applying a least-squares fitting approach. The aim is to minimize a function of the errors in the tree. A criterion of the form

$$S = \sum_{i<j} w_{ij}\left(d_{ij} - \Delta_{ij}\right)^m \qquad \text{(EQ8.34)}$$

is used, where the $d_{ij}$ are the distances between sequences $i$ and $j$ calculated from the alignment, the $\Delta_{ij}$ are the sequence distances calculated from the tree (sometimes referred to as **patristic distances**) and $m$ is an integer, usually 2. The weights $w_{ij}$ although sometimes not used (all assigned the value 1) are often a function of the distance $d_{ij}$. Sometimes the inverse of $d_{ij}$ is used, implying that the errors and uncertainty in the data are proportional to the distances.

The least-squares method can determine all branch lengths simultaneously through the use of matrix algebra, although this technique is computationally expensive. Suppose we have the tree given in Figure 8.11 and suppose that the distances $\Delta_{ij}$ in the tree between nodes $i$ and $j$ may not be the exact evolutionary distances $d_{ij}$ obtained from the sequence alignment. Then any pair of sequences $i$ and $j$ have a patristic distance which in the case of Figure 8.11 can be written as

$$
\begin{aligned}
\Delta_{AB} &= b_1 + b_2 \\
\Delta_{AC} &= b_1 + \phantom{b_2 +} b_3 + \phantom{b_4 +} b_5 \\
\Delta_{AD} &= b_1 + \phantom{b_2 + b_3 +} b_4 + b_5 \\
\Delta_{BC} &= \phantom{b_1 +} b_2 + b_3 + \phantom{b_4 +} b_5 \\
\Delta_{BD} &= \phantom{b_1 +} b_2 + \phantom{b_3 +} b_4 + b_5 \\
\Delta_{CD} &= \phantom{b_1 + b_2 +} b_3 + b_4
\end{aligned}
\qquad \text{(EQ8.35)}
$$

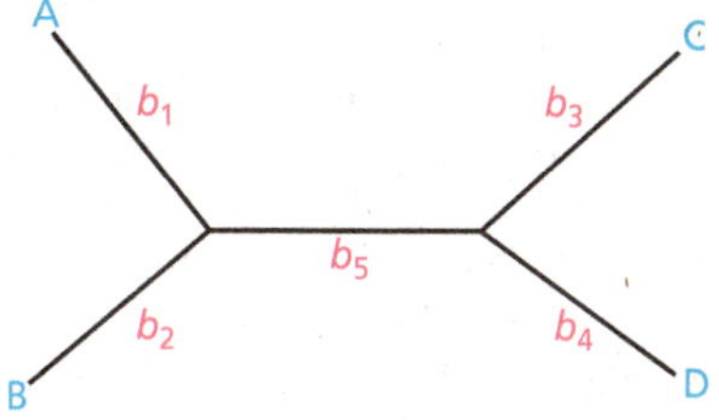


**Figure 8.11**
**Phylogenetic tree used in the worked example of the least-squares method.** The patristic distances are specified in Equation EQ8.35, followed by the theory, which can be used to determine the values of the branch lengths $b_1$ to $b_5$.

This can be written in matrix notation as

$$\boldsymbol{\Delta} = \boldsymbol{T}\boldsymbol{b} \qquad \text{(EQ8.36)}$$

where $\boldsymbol{\Delta}$ and $\boldsymbol{b}$ are column vectors, $\boldsymbol{\Delta}$ with six elements in the order shown above and $\boldsymbol{b}$ with five elements in the order $b_1$ to $b_5$. $\boldsymbol{T}$ is a matrix which holds the tree topology, defined as

$$\boldsymbol{T} = \begin{bmatrix} 1 & 1 & 0 & 0 & 0 \\ 1 & 0 & 1 & 0 & 1 \\ 1 & 0 & 0 & 1 & 1 \\ 0 & 1 & 1 & 0 & 1 \\ 0 & 1 & 0 & 1 & 1 \\ 0 & 0 & 1 & 1 & 0 \end{bmatrix} \qquad \text{(EQ8.37)}$$

We want there to be no errors in the $\Delta_{ij}$ so that when the weights are also included we have

$$w_{ij}\left(d_{ij} - \Delta_{ij}\right) = 0 \qquad \text{(EQ8.38)}$$

for all sequences $i$ and $j$, which can also be represented using vectors and matrices as

$$\boldsymbol{w}\left(\boldsymbol{d} - \boldsymbol{\Delta}\right) = \boldsymbol{w}\left(\boldsymbol{d} - \boldsymbol{T}\boldsymbol{b}\right) = 0 \qquad \text{(EQ8.39)}$$

$\boldsymbol{d}$ is a column vector equivalent to $\boldsymbol{\Delta}$ but containing the distances from the alignment, $\boldsymbol{w}$ is a square matrix with only the diagonal elements non-zero, in this case

$$\boldsymbol{w} = \begin{bmatrix} w_{AB} & 0 & 0 & 0 & 0 & 0 \\ 0 & w_{AC} & 0 & 0 & 0 & 0 \\ 0 & 0 & w_{AD} & 0 & 0 & 0 \\ 0 & 0 & 0 & w_{BC} & 0 & 0 \\ 0 & 0 & 0 & 0 & w_{BD} & 0 \\ 0 & 0 & 0 & 0 & 0 & w_{CD} \end{bmatrix} \qquad \text{(EQ8.40)}$$

To obtain the least-squares estimate of the branch lengths the equations must be solved for $\boldsymbol{b}$. Using standard matrix algebra techniques, we first multiply by the transpose matrix $\boldsymbol{T}^{\mathrm{T}}$ (element $t_{ij}^{\mathrm{T}} = t_{ji}$).

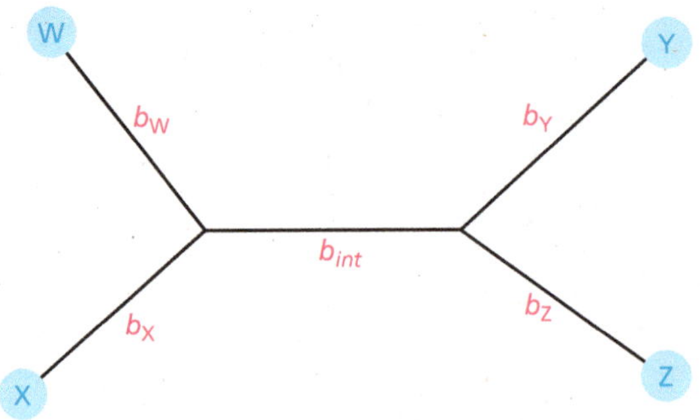


**Figure 8.12**
**The tree structure for four clusters W, X, Y, and Z used in describing the least-squares method of Bryant and Waddell.** The clusters, drawn as circles, represent one or more sequences, with all sequences belonging to one of the four clusters.

$$\boldsymbol{T}^{\mathrm{T}}\boldsymbol{wd} - \boldsymbol{T}^{\mathrm{T}}\boldsymbol{wTb} = 0 \tag{EQ8.41}$$

Because $\boldsymbol{T}^{\mathrm{T}}\boldsymbol{wT}$ is square, we can now obtain its inverse, and hence the solution is

$$\boldsymbol{b} = \left(\boldsymbol{TwT}^{\mathrm{T}}\right)^{-1}\left(\boldsymbol{T}^{\mathrm{T}}\boldsymbol{wd}\right) \tag{EQ8.42}$$

Using the matrix method of least-squares for a large number of topologies is demanding of computer resources, and there are simpler and faster methods. One of these will be described here, but others are described in or referenced by David Bryant and Peter Waddell (see Further Reading). It was realized that for fully resolved trees the determination of any given branch length involved very similar formulae of two kinds, one for internal and one for external branches. These formulae can be written down without the need to create and invert the matrix $\boldsymbol{T}$, considerably speeding up the calculation. For internal branches, the tree is simplified to one with four clusters of sequences, as shown in Figure 8.12. Labeling the clusters W, X, Y and Z, a distance $d_{YZ}$ is defined as the arithmetic mean of all inter-cluster distances. If the clusters have $n_W$, $n_X$, $n_Y$, and $n_Z$ sequences respectively, we have

$$d_{\mathrm{YZ}} = \frac{\sum\limits_{i \in \mathrm{Y}, j \in \mathrm{Z}} d_{ij}}{n_{\mathrm{Y}} n_{\mathrm{Z}}} \tag{EQ8.43}$$

Note that this is identical to the cluster distance defined for UPGMA in Equation EQ8.15. The internal branch length $b_{int}$ can then be expressed as

$$b_{\mathrm{int}} = \frac{1}{2}\left[\gamma(d_{\mathrm{WY}} + d_{\mathrm{XZ}}) + (1-\gamma)(d_{\mathrm{XY}} + d_{\mathrm{WZ}}) - d_{\mathrm{WX}} - d_{\mathrm{YZ}}\right] \tag{EQ8.44}$$

where

$$\gamma = \frac{(n_{\mathrm{X}} n_{\mathrm{Y}} + n_{\mathrm{W}} n_{\mathrm{Z}})}{(n_{\mathrm{W}} + n_{\mathrm{X}})(n_{\mathrm{Y}} + n_{\mathrm{Z}})} \tag{EQ8.45}$$

For exterior branches, the length of the branch to cluster W is given by

$$b_{\mathrm{W}} = \frac{1}{2}\left(d_{\mathrm{WX}} + d_{\mathrm{WY}} - d_{\mathrm{XY}}\right) \tag{EQ8.46}$$

exactly analogous to the formula of the Fitch–Margoliash method (Equation EQ8.18).

The methods described so far in this section determine the branch lengths for a tree of defined topology. This is the first step in assessing a given tree topology. However, when comparing alternative topologies the trees are not necessarily ranked on the basis of these calculations. Two different concepts have been applied to the problem of assessing trees based on evolutionary distance data. The minimum-evolution method attempts to identify the tree with the smallest total branch length, as this will be the tree with fewest mutations required to explain the data. It uses as its criterion

$$S = \sum_i b_i \qquad \text{(EQ8.47)}$$

where the $b_i$ are the lengths of the branches of the tree and the summation is over all the tree branches; that is, $S$ is the total length of all the tree branches. In some variants the magnitude of each branch length is used in the formula. This is because the correct tree (assuming the data consist of evolved sequences correctly aligned) should have no negative branch lengths, yet the formula as written in Equation EQ8.47 actually rewards such instances. The optimum tree is that with the smallest value of $S$, although several trees may have $S$-values close to this optimum. Note that the neighbor-joining method is closely related to minimum evolution (compare Equations EQ8.47 and EQ8.19).

The second way of assessing the trees with their branch lengths involves applying the idea of least-squares, attempting to minimize the discrepancy between the evolutionary distances measured on the tree and those of the data. The criterion used is of the form previously shown in Equation EQ8.34, the value of $S$ being used to rank the different tree topologies to identify the tree that most closely agrees with the data evolutionary distances. The original Fitch–Margoliash method of tree reconstruction used an inverse-square weighting scheme at this stage to compare alternative topologies whose branch lengths had been determined as described in Section 8.2.

## Unweighted parsimony methods look for the trees with the smallest number of mutations

The parsimony method of tree evaluation differs from the methods discussed previously in that it uses the actual sequence alignment rather than distance measures. Although most parsimony methods use the nucleotide sequence, modifications have been proposed that enable the technique to be applied to protein sequences. The idea behind the parsimony method is to give a tree based on as simple an evolutionary history as possible, which in this case is equated with the minimum number of mutations required to produce the observed sequences. The philosophy of the approach is similar to that of the methods of minimum evolution and neighbor-joining. Parsimony does not use the evolutionary models described in Section 8.1, and so there is a potential risk of underestimating the number of mutations between distantly related sequences, as a single mutation event will be preferred at a given site unless the data give clear evidence for multiple mutations. It is possible to weight mutations so that they are not all regarded as equal. We will first discuss **unweighted parsimony**, followed by the different techniques that are required for weighted parsimony methods.

In parsimony methods each position (site) in the multiple alignment is considered separately. For a given tree topology we must determine the minimum number of mutations required to reproduce the observed data for that site. The total for all sites for that topology is the measure $S$, which is used to compare alternative topologies; the tree with the smallest value of $S$ is taken as the predicted tree. One unusual feature of the method is that it is not necessary to calculate the branch lengths in order to calculate the number of mutations. The calculation of branch lengths can be applied just to the optimal trees.

The first stage is to distinguish between **informative sites** that can help to distinguish between alternative topologies and other sites. Sites that are not informative require the same minimum number of mutations regardless of the tree topology, for example because they are fully conserved (**invariable sites**), or because only one base (or amino acid residue) is present more than once in the alignment (**singleton sites**). Singleton sites are not informative because, regardless of the tree topology proposed, each unique base or residue will require just a single substitution. Once informative sites have been identified, we need to determine the minimum number of mutations required to reproduce the data. This is easier to do for rooted than for unrooted trees, and so a root is arbitrarily assigned for the purposes of the calculation. This can be done because, fortunately, the number of minimal mutations is the same regardless of root position. This arbitrary root is only used for these calculations.

In unweighted parsimony all mutations are given the same weight. The minimal number of changes at a given position in the sequence at an ancestral internal node can be counted using a simple method called the Fitch algorithm, which gradually moves back to the root from the leaf sequences. This is often referred to as **post-order traversal**. At each node a set of bases is assigned; just a single base is assigned to each leaf unless there is uncertainty at that sequence position. If the two immediate descendants of an internal node have some bases of their sets in common, these bases are also assigned to the ancestral node. If the descendants have no bases in common, then all bases of both descendants are assigned to the internal node. Each time there is no base in common in the descendants a mutation must have occurred and is added to the count for the tree.

The process is illustrated in Figure 8.13, which shows a tree for six sequences, with the base present at a particular alignment position shown at the leaves. Thus, the immediate descendant sequences of node V in this figure have C and T at the given position. Following the post-order traversal method, both C and T are assigned to node V, and one mutation is counted. Similarly, internal node X is assigned A and T with addition of another mutation. Working further back, the internal node of which V and X are descendants, node Y, can then be assigned a T, the base in common between V and X. This procedure is carried out at each alignment site for each tree topology under consideration. The mutations can be summed in a straightforward manner for all alignment positions to give the final total $S$ for that tree. The trees with different topologies can be ranked according to the value of $S$.

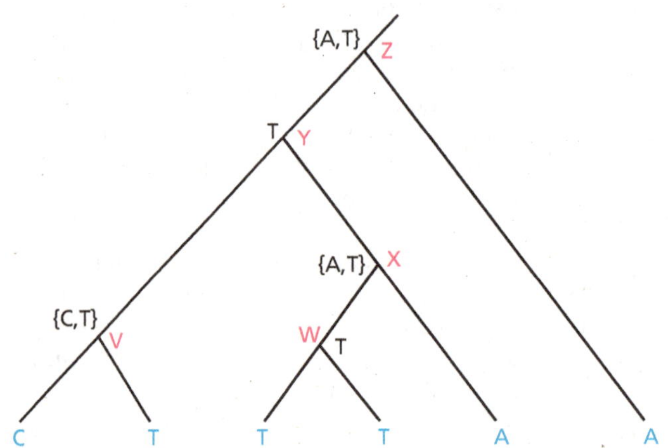


**Figure 8.13**
**A worked example of post-order traversal to determine the states at the ancestral nodes of a tree for six sequences.** The base is shown that occurs in each sequence at a particular alignment position where three T bases are aligned with two A and one C. Node V is assigned the possible set of bases {C, T} because its daughter nodes do not have any bases in common, so that the assignment must contain all bases assigned to the daughter nodes. Internal node Y is assigned base T only, as T is common to its daughter nodes V and X. The total mutation count for the tree is 3.

While this method correctly counts the minimal number of mutations, it does not give the information required to calculate branch lengths. It turns out that the sets of bases assigned at each internal node are not a complete list of those that might have occurred for the minimal number of mutations. To obtain the complete list of possible base assignments to the internal nodes, which is a prerequisite for determining the branch lengths, two calculation steps are needed. The first is a modified version of the post-order traversal described above, counting mutations and gathering information about possible node assignments. In the second step, which proceeds from the tree root to the leaves, all possible sets of consistent internal node assignments are obtained.

In the modified first step, each internal node $i$ is assigned two sets of bases F($i$) and H($i$) according to the assignment F($j$) for its immediate descendant, or daughter nodes $j$. We sum the numbers of each base $k$ in the daughter nodes F($j$) to get $n_i(k)$. For example, if there were two daughter nodes with sets of base assignments {A,C} and {C}, then $n_i(\mathrm{A}) = 1$, $n_i(\mathrm{C}) = 2$, $n_i(\mathrm{G}) = n_i(\mathrm{T}) = 0$. F($i$) is assigned the set of all bases that have the largest value of $n_i$, in this case {C}. The second set H($i$) is assigned the set of all bases whose $n_i$ is one less than this maximum, in this case H($i$) = {A}. In addition, during this step, a cost $S(i)$ is assigned to each node, which is the cost of the tree of all descendants from node $i$. The cost $S(i)$ is calculated using the formula

$$S(i) = \left[ \sum_{\text{descendants } j} \left( S(j) + 1 \right) \right] - \max_{k=\mathrm{A,C,G,T}} \left[ n_i(k) \right] \qquad \text{(EQ8.48)}$$

For initialization of the calculation, the leaves (that is, the present-day sequences) are assigned $S(j) = 0$ and their set F($j$) is simply the base at that sequence site. A worked example is given in Figure 8.14, using the same tree as in Figure 8.13. As shown in Figure 8.14B, the sets F created are identical to the assignments from the simpler technique discussed above and shown in Figure 8.13. The value of $S$ at the tree root is the cost for the whole tree, 3 for the example shown, in agreement with the post-order traversal calculation.

**Figure 8.14**
**A worked example of the technique for identifying all possible states of the internal nodes of a tree with unweighted parsimony.** (A) The tree of Figure 8.13. (B) The first step in which each internal node is assigned its base sets F($i$) and H($i$) as described in the text, proceeding from the leaves to the root. (C) The second step, in which the internal node assignments are refined working from the root to the leaves. (D) A summary of the possible sets of base assignments for the internal nodes.

(A)

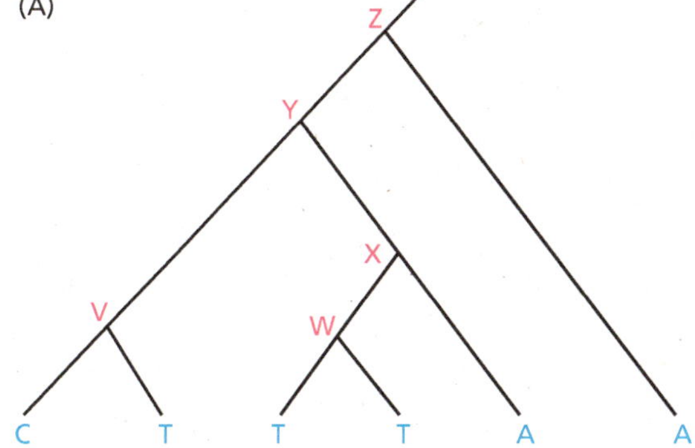


(B) FIRST STEP (calculated from leaves to root in order of table):

| internal node i | $n_i$(A) | $n_i$(C) | $n_i$(G) | $n_i$(T) | F($i$) | H($i$) | S($i$) |
|---|---|---|---|---|---|---|---|
| V | 0 | 1 | 0 | 1 | {C,T} | {A,G} | 1 |
| W | 0 | 0 | 0 | 2 | {T} | { } | 0 |
| X | 1 | 0 | 0 | 1 | {A,T} | {C,G} | 1 |
| Y | 1 | 1 | 0 | 2 | {T} | {A,C} | 2 |
| Z | 1 | 0 | 0 | 1 | {A,T} | {C,G} | 3 |

(C) SECOND STEP:

| parent node | parent assignment | daughter assignment set | comment | daughter assignment set | comment |
|---|---|---|---|---|---|
| Z | T | Y = {T} | T∈F(Y) | | |
| Z | A | Y = {A,T} | A∈H(Y) | | |
| Y | T | X = {T} | T∈F(X) | V = {T} | T∈F(V) |
| Y | A | X = {A} | A∈F(X) | V = {A,C,T} | A∈H(V) |
| X | T | W = {T} | T∈F(W) | | |
| X | A | W = {T} | A∉F(W)∪H(W) | | |

(D) Summary of alternative assignment sets:

| | V | W | X | Y | Z |
|---|---|---|---|---|---|
| 1 | T | T | T | T | T |
| 2 | T | T | T | T | A |
| 3 | A | T | A | A | A |
| 4 | C | T | A | A | A |
| 5 | T | T | A | A | A |

The second step involves starting at the tree root and working toward the leaves. For each possible assignment [taken from F($i$)] of a base $k$ to a parent node $i$, one or more possible assignments can be given to the daughter nodes $j$. If base $k$ is in the set F($j$), the node $j$ is assigned this base too. (For example the third line of Figure 8.14C.) If it is not in F($j$) but is in H($j$), then $j$ has a set of possible assignments that includes all the bases in set F($j$) and base $k$ in addition. (For example the second line of Figure 8.14C.) If $k$ is not in either of sets F($j$) and H($j$) then node $j$ is assigned all the bases in set F($j$). (For example the sixth line of Figure 8.14C.) This leads to a set of assignments for all the internal nodes, each of which will have the cost S, as illustrated in the worked example of Figure 8.14. To see that this technique recovers more possible assignments than the method of Figure 8.13, note that originally node Y was assigned only base T. Now, in the case of node Z being assigned base A, Y is assigned a new set of bases {A,T}, as a result of which the possible assignment of A at node Y arises.

All possible sets of internal node assignments are given in Figure 8.14D. From Figure 8.13 we might easily have found sets 1 and 2, but the others, especially set 3, would not have been found. It is particularly important to have all these sets when calculating branch lengths, when all possible patterns of mutation should be considered. If the second step of working upward from the root to the leaves is omitted, we can only determine the most parsimonious tree topology, but not determine actual branch lengths. When calculating branch lengths, the location of mutations in each of the alternative assignment sets must be considered. If alternatives exist (for example, see Figure 8.14D) the number of mutations assigned to a branch is usually taken as the average of all these. The uninformative singleton sites must also be included in the branch length calculation, as they are only uninformative in comparing alternative topologies.

## Mutations can be weighted in different ways in the parsimony method

Unweighted parsimony uses the total number of nucleotide substitutions required for the tree as the criterion $S$. However, not all mutations are equally likely and a weighting scheme may produce better results. In unweighted parsimony all nucleotide substitutions are weighted equally (see Figure 8.15A), while in weighted parsimony the substitutions are given different values that reflect, for example, the transition/transversion bias (see Figure 8.15B). Because transitions usually occur at a greater rate than transversions, over long times it is possible that sufficient transitions will have occurred to lose any clear phylogenetic signal. It has therefore been proposed that such data are best analyzed with a weight of 0 for transitions, so that only transversions are counted (transversion parsimony).

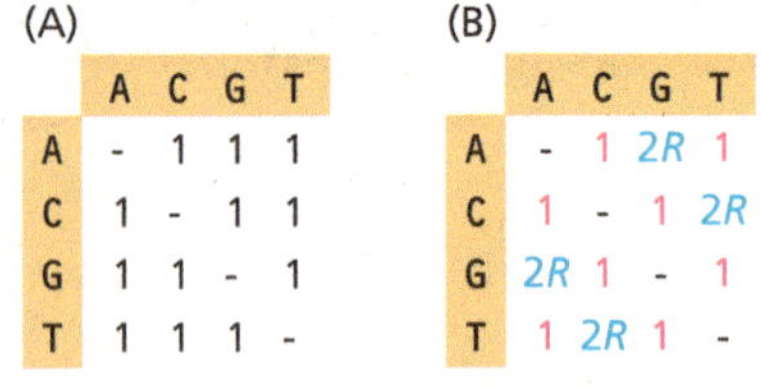

(A)

| | A | C | G | T |
|---|---|---|---|---|
| A | - | 1 | 1 | 1 |
| C | 1 | - | 1 | 1 |
| G | 1 | 1 | - | 1 |
| T | 1 | 1 | 1 | - |

(B)

| | A | C | G | T |
|---|---|---|---|---|
| A | - | 1 | $2R$ | 1 |
| C | 1 | - | 1 | $2R$ |
| G | $2R$ | 1 | - | 1 |
| T | 1 | $2R$ | 1 | - |

**Figure 8.15**
**Weight matrices for use in maximum parsimony to distinguish between different types of mutation.** (A) This matrix has all nucleotide substitutions weighted equally, as in unweighted parsimony. (B) This matrix is weighted to favor transitions over transversions by a factor of $R$, which represents the excess of transitions over transversions (see Box 8.1).

A different form of weighting is to weight different sites in the alignment. In protein-coding sequences, for example, different weights are given to mutations at different codon positions. If protein sequences themselves are being studied, the mutations observed are at the amino acid level, and can correspond to one, two, or three nucleotide mutations. Weighting schemes have been proposed that try to account for this.

When weights are applied, the **Sankoff algorithm** should be used to evaluate maximum parsimony. This algorithm is a form of dynamic programming, a technique that was introduced in Section 5.2 in the context of pairwise sequence alignment. As in the dynamic programming methods of Chapter 5, there are two stages in the calculation. The first is to calculate the lowest total mutation cost of the tree (equivalent to the sequence alignment score) and the second is traceback to determine the sets of possible node assignments and thus the branch lengths (equivalent to determining the actual alignment as described in Section 5.2). The initial stage of this algorithm is a post-order traversal. At each node each of the possible bases is considered in turn. For each of these bases, the lowest possible cost is calculated of obtaining the observed descendant leaves. This cost is the weighted count of mutations, *S*. The calculation relies on having the cost (weight) of each mutation specified, denoted here by $w_{ij}$ for a mutation from base *i* to base *j*. The $w_{ij}$ values will not necessarily be independent of alignment site, but this will not be indicated in the notation for simplicity. Each internal node will be the ancestor of two daughter nodes, with calculated costs at the three nodes being written as $S_a$, $S_{d1}$, and $S_{d2}$, respectively. The minimal cost at the ancestral node when the base *i* is assigned to it will be identified as $S_a(i)$. This can be calculated from values at the daughter nodes as follows:

$$S_a(i) = \min_{j=A,C,G,T}\left[w_{ij} + S_{d1}(j)\right] + \min_{k=A,C,G,T}\left[w_{ik} + S_{d2}(k)\right] \quad \text{(EQ8.49)}$$

All possible base assignments must be considered: that is, having any possible base at either daughter node give rise to any possible base at the ancestral node. The calculations must be started at the tree leaves where the base assignments are unambiguous, and the connecting internal nodes can then be calculated, followed by other nodes back to the root. At the leaves the observed base should be assigned a cost of 0, and all other possibilities infinite costs, forcing all subsequent calculations to use only the observed bases in their cost estimates. All those assignments of the two daughter nodes that produce a minimal cost for a given ancestral node assignment should be stored for later use. In Figure 8.16 the calculation previously outlined for the other parsimony methods has been redone using the Sankoff method. In this case, all the weights have been assigned a value of 1, so that the results should be identical to those previously achieved, so that all the methods can be readily compared. The costs associated with assigning each of the four bases to the internal nodes are given, together with a list of base pairs that contributed to this score. For a base pair listed as ab, the left-hand daughter node, as the tree is drawn, had the a assignment, and the right-hand daughter node, the b.

Once all costs have been calculated back to the root, the score associated with this alignment site and this tree topology is given by the smallest score, in this case the

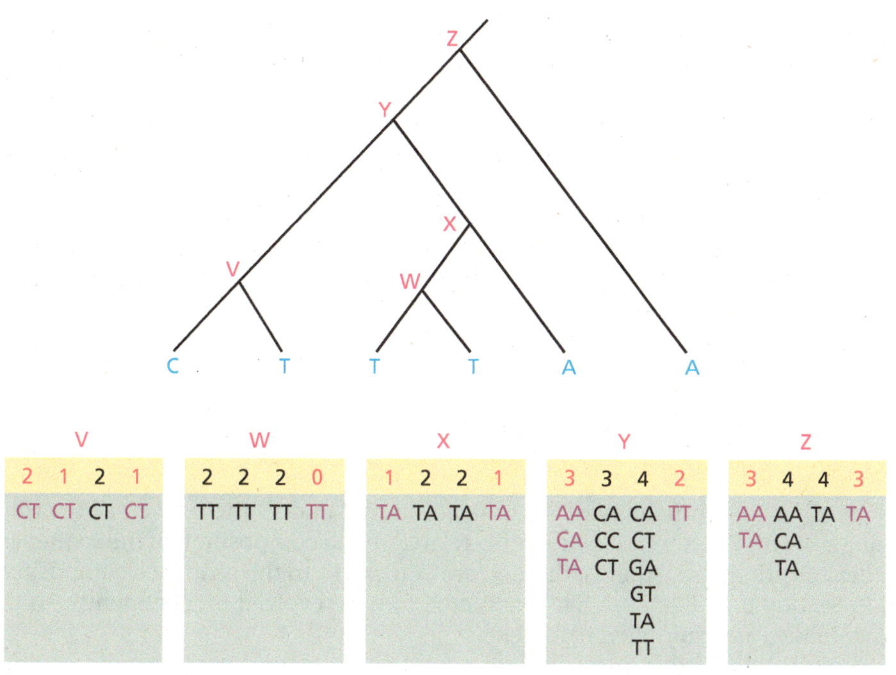


**Figure 8.16**
**Application of the Sankoff algorithm for weighted parsimony.** The algorithm is applied to the same tree as used in Figures 8.13 and 8.14. Note that although differing weights could have been used, in this case all the weights are 1 so that the result is exactly identical to that of Figures 8.13 and 8.14. At each of the five internal nodes V to Z (calculated in that order) the score is given for each of the four possible bases in order (A,C,G,T). Underneath each base is a list of the combinations of bases in the daughter nodes that lead to the given score. The base combinations show the bases on each of the two daughter nodes, with the left base corresponding to the left daughter node as the tree is drawn. Thus for node Y, if it is assigned as A with a score of 3, this could arise from node V being assigned as A, C, or T and node X always being assigned as A. These base combinations are the equivalent of the traceback arrows in the dynamic programming matrices of Section 5.2. All the scores and base combinations involved in the traceback are colored red. If all possible traceback options are explored, the table of Figure 8.14D is reproduced exactly.

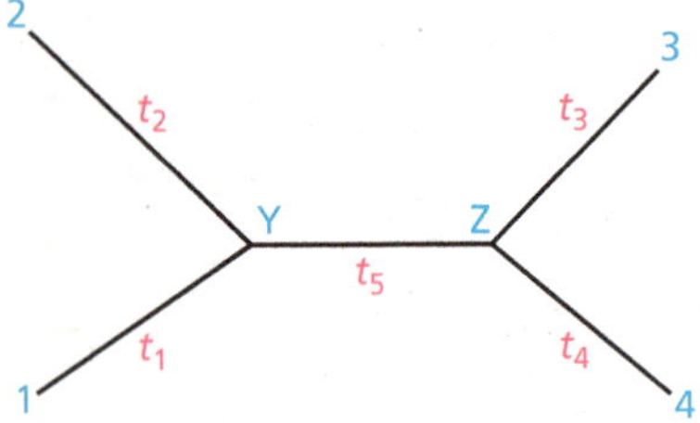


**Figure 8.17**
**Example phylogenetic tree for which the maximum likelihood equations in the text are derived.** The branches are labelled $t_i$ here as opposed to $b_i$ in other figures because the lengths are modeled as evolutionary time in the equations.

value 3, given for assignments at the root of bases A or T. The sets of possible base assignments at the internal nodes can be obtained by a traceback procedure, following the different contributing assignments. These are all colored red in Figure 8.16. If this procedure is followed, all the sets listed in Figure 8.14D will be recovered, so that the method is in full agreement with the previous results.

## Trees can be evaluated using the maximum likelihood method

The probability of the occurrence of a particular evolutionary history as represented by a particular tree can also be calculated with the assumption of a specific model of evolution. This is called the likelihood of the model (see Appendix A). The tree that has the maximum likelihood—that is, the highest probability assuming the model—will be predicted to be the correct tree. If the model is a reasonable representation of the evolutionary processes that have occurred then this should be the case. The likelihood is defined in terms of a particular tree topology and branch lengths together with a specific model of evolution. In this discussion we will assume that the model used is time-reversible, a property of all the models discussed in this chapter and Chapter 7. Time-reversible models give the same likelihood regardless of which time direction is assumed for a branch, so that consequently the location of the root is unimportant in the calculation.

We will first see how the likelihood of a tree can be calculated, using the example in Figure 8.17. This is a tree with four leaves at nodes labeled 1 to 4, with observed sequences at these leaves, but unknown ancestral sequences at the internal nodes Y and Z. The probabilities calculated depend, among other things, on the tree topology ($T$). Each branch will be considered separately, and the probability of base $i$ at any position mutating into base $j$ in a time $t$ will be written $P(j|i,t)$. The precise mathematical form of this will depend on the evolutionary model used. In the JC case with mutation rate parameter $\alpha$, we have

$$P\left(j\middle|i,t\right) = \begin{cases} \frac{1}{4}\left(1+3e^{-4\alpha t}\right) & \text{for } j=i \\ \frac{1}{4}\left(1-e^{-4\alpha t}\right) & \text{for } j \neq i \end{cases} \qquad \text{(EQ8.50)}$$

Note that this differs from Equation EQ8.8 in having terms in $e^{-4\alpha t}$ rather than $e^{-8\alpha t}$ because that equation referred to the simultaneous evolution of two sequences.

As with parsimony, we are considering the sequence data themselves, and look at each alignment position in turn, ignoring sites with gaps. Another feature in common with parsimony is the arbitrary and temporary use of a tree node as the root during the calculation. The likelihood of specific bases $x_Y$ and $x_Z$ occurring at a particular sequence position at internal nodes Y and Z of the tree with topology $T$ and branch lengths $t_i$ (see Figure 8.17) is given by

$$\begin{aligned} &P(x_1,x_2,x_3,x_4,x_Y,x_Z|T,t_1,t_2,t_3,t_4,t_5) = \\ &q_{x_Y} P(x_1|x_Y,t_1)P(x_2|x_Y,t_2)P(x_Z|x_Y,t_5)P(x_3|x_Z,t_3)P(x_4|x_Z,t_4) \end{aligned} \qquad \text{(EQ8.51)}$$

where $q_{x_Y}$ is the probability of the base at internal node Y (here taken as the root) being $x_Y$. (This is assumed to be simply related to the composition of the sequences under consideration—that is, these $q_a$ are equivalent to the sequence composition $q_a$ in Section 5.1.) Each branch is assumed to have evolved independently, so the probabilities are multiplied together.

The internal nodes Y and Z could have any possible base for the $i$th position and this must be taken into account. The likelihood $L_i$, of the observed sequences $x_1$ to $x_4$ is obtained by considering all possible bases (residues) at $x_Y$ and $x_Z$ to obtain

$$L_i = \sum_{x_Y = A,C,G,T} q_{x_Y} \mathrm{P}(x_1|x_Y,t_1)\mathrm{P}(x_2|x_Y,t_2)\left[\sum_{x_Z = A,C,G,T} \mathrm{P}(x_Z|x_Y,t_5)\mathrm{P}(x_3|x_Z,t_3)\mathrm{P}(x_4|x_Z,t_4)\right] \tag{EQ8.52}$$

Note that at each node these alternative bases are mutually exclusive possibilities, and so are added together. Considering all the positions, we can obtain the total likelihood $L$ by multiplying these individual terms together; or alternatively by taking logarithms, we obtain

$$\ln L = \sum_i \ln L_i \tag{EQ8.53}$$

where the sum is over all sequence positions $i$. Note that node Z could equally well have been used as the root, and the same result would have been obtained.

For the given tree topology we still need to find the tree with maximum likelihood by varying the available parameters. The parameters to be optimized include the branch lengths and possibly also the parameters specific to the evolutionary model used, for example substitution rate $\alpha$ in the case of Jukes–Cantor. The partial derivatives with respect to the parameters of the individual terms of the equations for likelihood are readily derived, and these can be used in standard function minimization routines to obtain the optimal parameter values and the maximum likelihood.

A search of tree topologies must be made as described previously, calculating the maximum likelihood $L$ for each. As with parsimony, there may be several trees with similar values of $L$ that may all be regarded as relevant, and some form of consensus tree may be produced.

As an example of this method, consider the determination of the maximum likelihood tree for four sequences of one base each (two being A and the other two C) under the (unrealistic but here usefully simplifying) constraint of all five branches having the same length $t$, using the Jukes–Cantor evolutionary model. The base composition will be assumed to be equal for all four bases, that is 25% each. The two possible tree topologies are shown in Figure 8.18. This problem has only one parameter to determine, namely the branch length.

As the sequences are only one base long there is only one site, and the likelihood of a tree is given by Equation EQ8.52 with each P term replaced by one from Equation EQ8.50. All possible bases must be considered for the internal nodes Y and Z, giving 16 different terms in the equation. All bases are assumed to be equally present; that is, $q_A = q_C = q_G = q_T = 1/4$. Because all branches are the same length, only two different terms will occur in the equations: $E$ is the probability of a branch of length

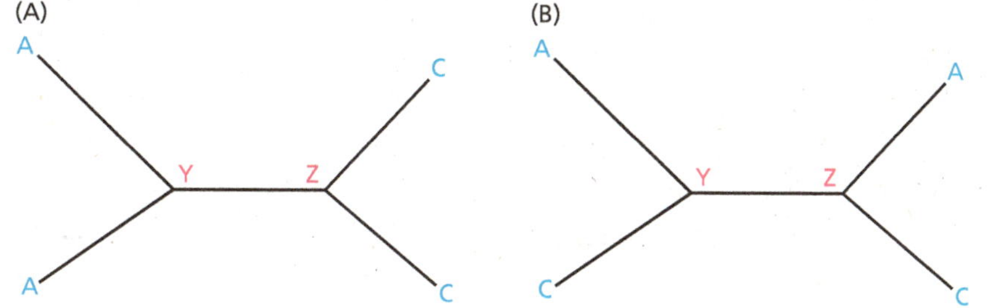


**Figure 8.18**
**Two possible trees that could be generated by the maximum likelihood method for four sequences composed of one base each (A, A, C, and C).** Note that all five branch lengths are (unrealistically) constrained to be equal.

$3\alpha t$ (where $\alpha$ is the rate parameter of the JC model, and $t$ is time) between nodes with identical bases

$$E = \frac{1}{4}\left(1 + 3e^{-4\alpha t}\right) \tag{EQ8.54}$$

and $D$ is the probability of a branch of length $3\alpha t$ between nodes with different bases

$$D = \frac{1}{4}\left(1 - e^{-4\alpha t}\right) \tag{EQ8.55}$$

For the tree in Figure 8.18A, if the nodes Y and Z have bases A and C, respectively, four of the branches have the same base at each end, giving an $E$ term, and that joining Y and Z has a $D$ term. The likelihood for this particular choice of bases and topology is given by $1/4E^4D$ where the initial 1/4 arises from the $q_A$ term. In a similar way, the likelihood for the tree summing over all base choices for nodes Y and Z is given by

$$L_I = \frac{1}{4}(E^4D + 2E^3D^2 + 4E^2D^3 + 2ED^4 + 7D^5) \tag{EQ8.56}$$

where the first term is from base choice (A,C) and the second from base choices {(A,A) and (C,C)} where the first base is on node Y and the second on node Z. The tree in Figure 8.18B results in the equation

$$L_{II} = \frac{1}{4}(2E^3D^2 + 2E^2D^3 + 10ED^4 + 2D^5) \tag{EQ8.57}$$

where now the first term arises from base choices {(A,A) and (C,C)} and the second from {(A,C) and (C,A)}.

To predict the branch length and tree topology, we need to know the maximum likelihood, which in this simple case we can determine by plotting these functions with varying $\alpha t$ (see Figure 8.19). It is clear from comparing the curves in Figures 8.19A and 8.19B that the optimal tree (that with greatest $L$) will have the topology shown in Figure 8.18A and $\alpha t$ of approximately 0.103, giving branch lengths of approximately $3\alpha t = 0.31$. Note that although the likelihood of this tree is dominated by the contribution from the first term of Equation EQ8.56, other trees with different bases at Y and Z contribute about one-sixth of the total. From Figure 8.19B, the topology

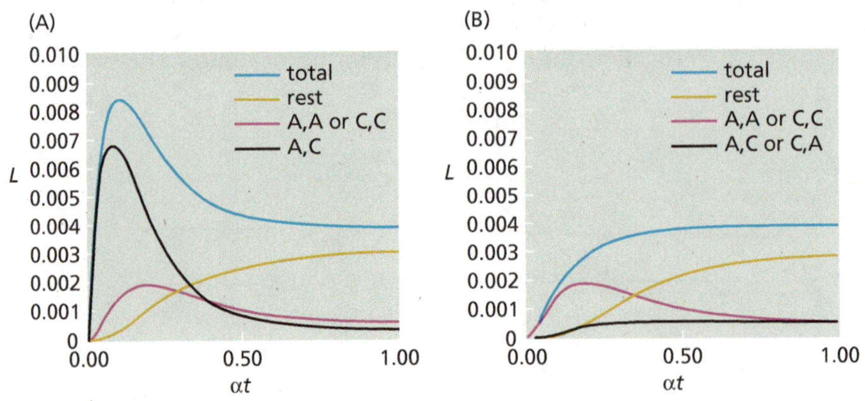


**Figure 8.19**
**The variation of the maximum likelihood $L$ and some of its components for the two trees of Figure 8.18.** (A,B) Plots of the total likelihood $L$ (light blue) and selected components for the trees of Figures 8.18A and 8.18B, respectively, plotted against $\alpha t$, where $\alpha$ is the rate parameter of the JC model, and $t$ is time. These correspond to Equations EQ8.56 and EQ8.57, respectively. Two components have been separated from the total. That labeled "A,A or C,C" (red) is the likelihood term when both internal nodes Y and Z are base A or both are base C. The component "A,C" in (A) (black) is the likelihood term when node Y has the base A and node Z has the base C. In (B) the black line includes both this tree and also the tree in which node Y has the base C and node Z has the base A. The other components of the likelihood are shown by the yellow line.

shown in Figure 8.18B never reaches half the likelihood value for the optimal tree and so that topology can be discarded.

In realistic cases, only the top-scoring topology, its branch lengths, and likelihood are reported. This simple case allows us to investigate individual contributions and understand the model a little better. The dominant base choice for the base assignment to the node Y and Z of the tree of Figure 8.18A is (A,C) as might have been guessed for this simple problem. At sufficiently long times for many mutations to have occurred ($\alpha t$ approaching 1), ultimately all base choices for nodes Y and Z become equally likely to have produced any base at an external node. At these long times, the tree is essentially indistinguishable from random. The plots in Figure 8.19 confirm this and show that the total likelihood at large $\alpha t$ converges to the same value for both topologies.

This presentation is intended to give an insight into the nature of the science behind a maximum likelihood calculation. The details of practical algorithms have been omitted as beyond the scope of this book. However, without an efficient algorithm the method would be impractical because of the significant computational effort required. For further insights into the algorithmic developments that have made the maximum likelihood method viable, see Further Reading. As well as optimizing branch-length calculation algorithms, the improvements involve careful attention to the efficiency of topology sampling.

## The quartet-puzzling method also involves maximum likelihood in the standard implementation

Quartet puzzling was proposed as a response to the practical limitations imposed by the computational demands of maximum likelihood methods. A quartet is any set of four taxa chosen from a tree. Each quartet of a tree will have one of the three topologies of Figure 8.8. If the topologies of all quartet subtrees are known, it is possible to reconstruct the correct topology of the complete tree. This property of quartets is exploited to determine the topology of the complete tree.

In the first of three steps all possible quartet subtrees of the data are generated, considering all three possible topologies for each quartet. The maximum likelihood method is used to determine the best tree from each set of three. In fact, the method tries to allow for the lack of a clear preference for one of the topologies, although in the next step only one of any possible alternatives is used. The next step only requires the topology of the quartets, and involves generation of a complete tree. Taxa are added one at a time in such a way as to minimize the disagreement between the resultant tree and the splits present in the quartet trees. This is the process known as quartet puzzling, and results in a tree topology that is often dependent on the order of addition of taxa to the tree. To overcome this bias, many trees are determined adding the taxa in different orders. In the final step these trees are summarized into a majority rule consensus tree.

An example of the method is given in Figure 8.20 using six taxa. The correct tree that is to be reconstructed from the data is shown in Figure 8.20A. It has 15 possible quartet trees with the splits obtained from maximum likelihood analysis listed in Figure 8.20B. One of these is chosen at random as the starting point. We will use quartet 1 (AD)(BC) and follow the construction of one complete tree. Two further taxa must be added to this tree, and in general will be selected in random order. Here we will add E first and then F.

Four quartets contain information about the tree with taxa A–E that is not present in the starting tree. These quartets are 2, 4, 7, and 11 of Figure 8.20B. For each quartet we consider the addition of E at every possible branch of the starting tree. Whenever adding E at a branch would result in a different split from that of the quartet, the

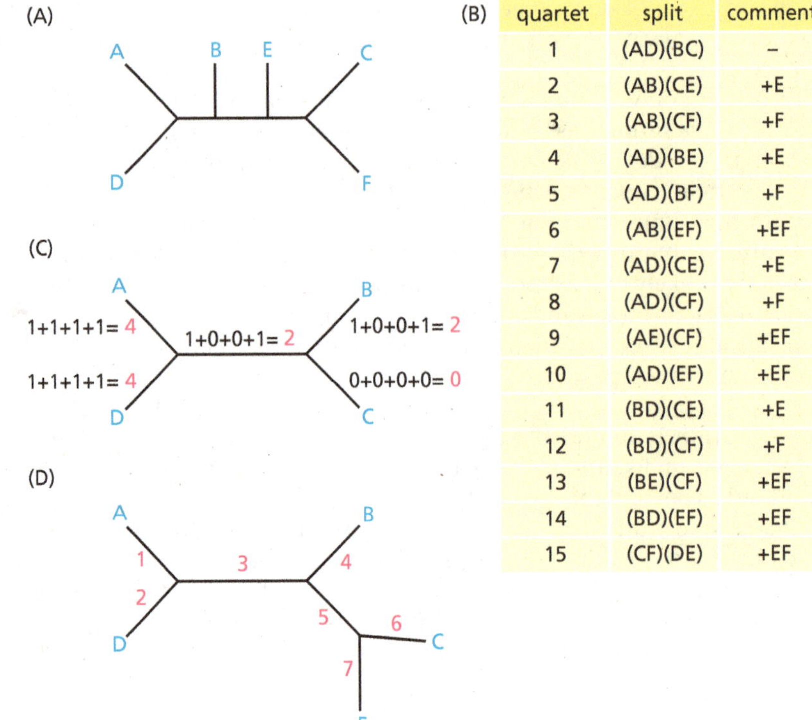


| quartet | split | comment |
|---|---|---|
| 1 | (AD)(BC) | – |
| 2 | (AB)(CE) | +E |
| 3 | (AB)(CF) | +F |
| 4 | (AD)(BE) | +E |
| 5 | (AD)(BF) | +F |
| 6 | (AB)(EF) | +EF |
| 7 | (AD)(CE) | +E |
| 8 | (AD)(CF) | +F |
| 9 | (AE)(CF) | +EF |
| 10 | (AD)(EF) | +EF |
| 11 | (BD)(CE) | +E |
| 12 | (BD)(CF) | +F |
| 13 | (BE)(CF) | +EF |
| 14 | (BD)(EF) | +EF |
| 15 | (CF)(DE) | +EF |

**Figure 8.20**
**A worked example of the quartet-puzzling method of tree reconstruction.** (A) The correct tree whose topology is to be reconstructed from the data. (B) A list of the quartet splits of this tree, with a comment to indicate the taxa not present in the ABCD quartet that is used as the starting point for building the tree. In a real case the splits will have been determined from the data using maximum likelihood, and would include errors as well as possibly some splits not fully resolved. (C) The starting tree with taxa A–D, showing the branch weights used to determine the position of taxon E. (D) The tree with E added, and the branches numbered as in the text discussion.

branch is penalized with a value of 1. If the split of the quartet is correctly given, that branch receives no penalty. The penalties are added up for the four quartets, and the branch with the lowest penalty is chosen as the site to add E to the tree. In the case of a tie, the branch is selected at random from the favored branches. Figure 8.20C shows the process in action, resulting in the tree of Figure 8.20D.

It now remains to add taxon F. This is done using the same technique, now involving all quartets that include F and three of the taxa A–E. There are 10 of these, as listed in Figure 8.20B. The branch penalties clearly disfavor branches 1–4 as numbered in Figure 8.20D. Branch 6 is (correctly) favored over 5 and 7, as a result of the splits of quartets 9, 13, and 15. These each penalize branches 5 and 7, but not branch 6, resulting in total penalties for branches 5, 6 and 7 of 3, 0, and 3, respectively. The quartet splits in this example were all consistent, leading to a clear result that is independent of the order of addition of taxa. However, data errors will lead to some inconsistencies and to different final topologies that are dependent on the addition order. To cope with this, many trees are constructed adding the sequences in different orders, and the consensus tree is reported.

## Bayesian methods can also be used to reconstruct phylogenetic trees

The maximum likelihood method calculates the optimal tree to represent the data by calculating the probability that the assumed evolutionary model (the hypothesis H) produced the data D, written P(D|H). This is the likelihood of the model. In contrast, Bayesian methods involve statistical analysis that includes estimates of the probability that a given model has occurred, based on prior assumptions. (See Appendix A, where the key concepts of Bayesian methods are presented.) The assumed model in this case is in fact not simply an evolutionary model as defined and discussed in Section 8.1, but also includes the tree topology. The reason for this becomes clear on realizing that the speciation and gene duplication events that give rise to the bifurcations at the nodes are not included in the likelihood calculations discussed above, and so must be included in some other way. Ideally, we would like to know the posterior probability of the model given the data, written P(H|D), and would then choose the model with the highest such probability. As shown in Appendix A, to calculate such quantities we need an estimate of the prior probability of the model, which includes the tree topology.

There are currently no generally agreed prior probabilities for tree topology, and a number of different aspects need specifying in order to apply Bayesian methods, for example the distribution of speciation events in time. The issues involved are very different from the others described in this book, and the interested reader is directed to Further Reading for details. Some skepticism about Bayesian methods has been expressed because of the lack of clear evidence that the prior probabilities used are realistic. An additional problem is that the method ultimately needs to calculate the posterior probability of the model, but this calculation is very difficult as a result of the nature of the formula. It requires a weighted average of values calculated (in principle) on all possible tree topologies. A technique called **Markov chain Monte Carlo (MCMC)** is used to achieve this. In essence, the method runs for many thousands of cycles, each cycle consisting of two steps. In the first step an attempt is made to change the model, typically the tree topology. This change is either accepted or rejected using a test that involves the prior probability distribution of the model as well as the calculated likelihood P(D|H). The current tree—the unchanged tree if the change was rejected—is then included in the set of trees used in the second step to calculate the average mentioned above. Typically, because only relatively small topology changes are made in each cycle, consecutive trees are highly correlated. The initial tree will probably be very different from the optimal topology, which it may take several hundred or even tens of thousands of cycles to approach. These early trees are not truly representative of the data, and should be omitted from the averaging process. Hundreds of thousands of cycles are typically run in the averaging phase, but the number depends on many factors including the size of the dataset and the properties of the topology change step.

The Bayesian method just described appears, together with maximum likelihood, to be among the most accurate available for constructing trees, and is proving increasingly popular. The result is usually reported as a summary of the different trees produced, and thus automatically has indications of feature reliability commonly only obtained by running a bootstrap analysis, which is described in the next section.

## 8.5 Assessing the Reliability of Tree Features and Comparing Trees

If sufficient computer resources are available in most cases any of the techniques described in this chapter can be applied to any dataset to produce one or more phylogenetic trees. Failures are most likely with methods using corrected evolutionary distances, for example if using Equation EQ8.9 for the JC model and encountering more than 75% base differences between two aligned nucleotide sequences. This particular problem can often be overcome by using a series expansion of the logarithm in this equation. However, even when it is the best tree found in a thorough topology search, a reported tree is not necessarily correct in all details, and further tests must be carried out to explore the reliability of the tree features. Most trees produced from real data by any of the methods discussed in this chapter will contain some features that are strongly present in the data and very reliable, and others that are not well supported and possibly spurious. It is important to be able to estimate the strength of evidence for different tree features, and some methods for doing this are discussed next (see Flow Diagram 8.5).

It is useful to briefly consider the reasons why an incorrect tree may have been produced. The method used may have involved the assumption of an evolutionary

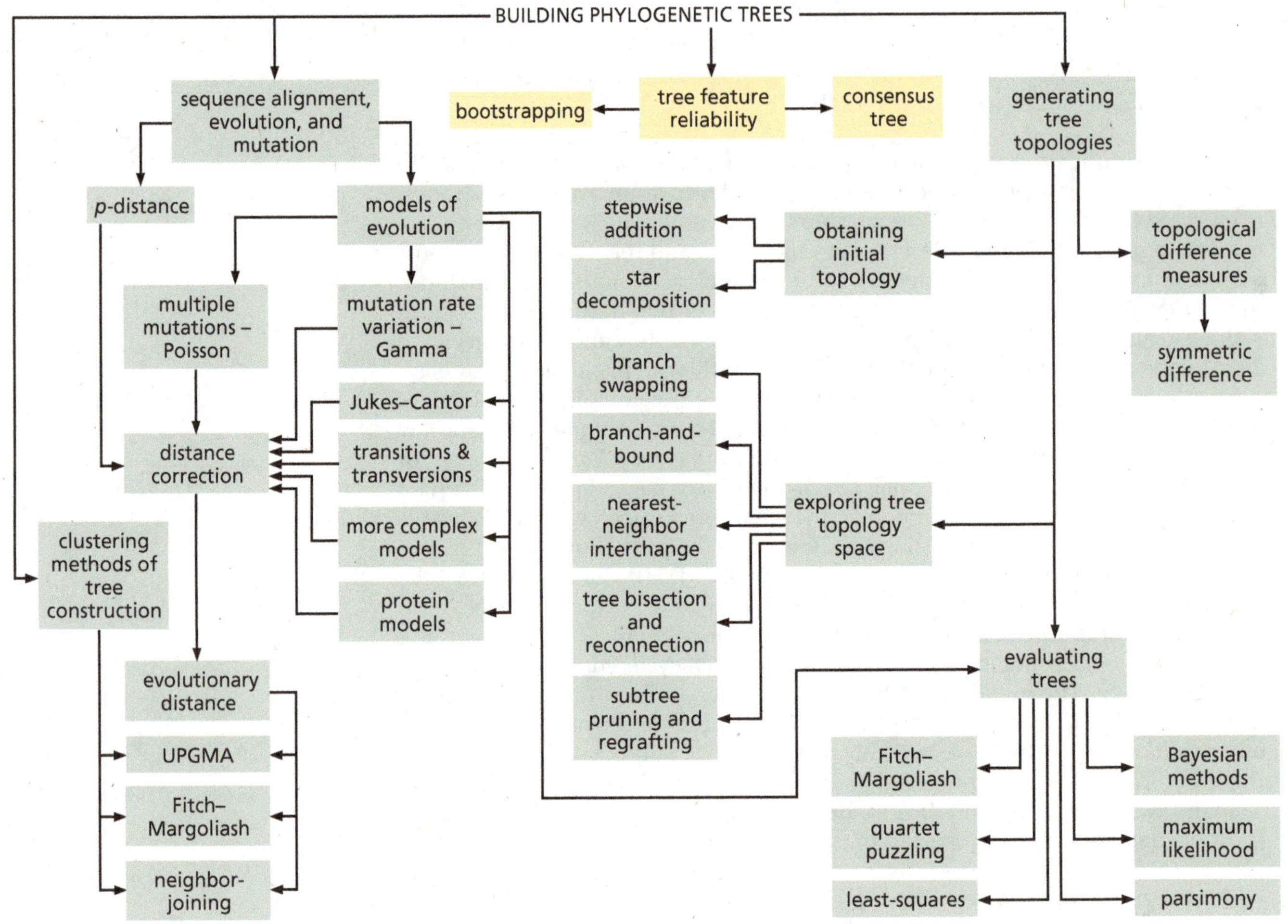


**Flow Diagram 8.5**
**The key concept introduced in this section is that features of phylogenetic trees can be assessed to see how strongly they are supported by the data.**

model that was inconsistent with the data, either because of incorrect parameterization or because the model was too simple. Alternatively, the data themselves can cause problems if there is insufficient sequence or if inconsistencies or errors are present. Highly variable mutation rates can also cause problems. Many methods are vulnerable to certain evolutionary features that can lead to particular types of errors occurring in the tree. One of these is long-branch attraction.

Another situation where further analysis is required occurs when some of the suboptimal trees are reported. In this case, even if each tree has a score, such as the number of mutations from maximum parsimony analysis, this score does not easily translate into a measure of the relative support in the data for the alternative trees. Methods for trying to assess tree support are therefore required. The same techniques can be applied to examine competing hypotheses presented as two or more alternative tree topologies.

## The long-branch attraction problem can arise even with perfect data and methodology

As the amount of data used increases, one would hope that all methods would become more likely to determine the correct tree, and in general this is the case. However, there is at least one exception, often referred to as long-branch attraction (or the **Felsenstein zone**), which results in a tree that is inconsistent with the data. This potential problem involves trees in which some branches are considerably longer than others, as illustrated by the trees in Figure 8.21.

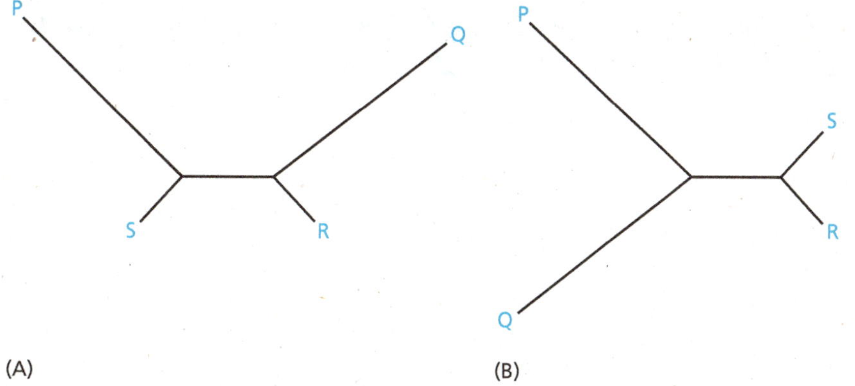


**Figure 8.21**
**Two trees that illustrate the potential problem of long-branch attraction.** In trees where some branches are considerably longer than others, maximum parsimony may incorrectly favor a tree with the wrong topology. In a situation where the tree in A is the true tree, the method of maximum parsimony will always prefer the tree in B instead. To see why this will happen, consider a single site that has base X in the sequences R and S. Unless this site is not X and the same base in both sequences P and Q it will be an uninformative site. However, if it is not X, parsimony will propose that the tree in B is to be preferred.

This example involves four sequences, with the phylogenetically correct tree being that in Figure 8.21A. Considering the parsimony method, one can readily see how that method will produce the incorrect tree, Figure 8.21B, in which the two long branches are neighbors. The two short branches are assumed to be sufficiently short that at most sites in the sequences S and R no mutations have occurred and the sequences have the same base. Using parsimony, a site will only be informative, and thus be included in the analysis, if the bases at that site in the two long branch sequences (P and Q) are different from that at the short branches and yet the same as each other. Otherwise the site will be a singleton. In this case, however, maximum parsimony will propose that the long branches are neighbors, and the tree in Figure 8.21B will be proposed. The same problem can also arise with other phylogenetic tree methods. It is hard to determine the extent to which this problem arises in real data, although there are a number of cases where it has been suspected.

## Tree topology can be tested by examining the interior branches

When investigating the reliability of the features of a phylogenetic tree it is particularly important to assess the accuracy of splits that identify key evolutionary events such as when clades diverge from each other. The methods for doing this fall into two categories. Some explore the range of branch lengths for a given confidence interval, usually to see if the range includes zero, in which case the associated split may be poorly supported. An alternative, the bootstrap technique (see Box 8.4), involves deriving a series of trees and looking at the fraction that contain the same split.

The normal bootstrap test method generates a large number of trees with potentially different topologies. These are usually analyzed on the basis of the topology, ignoring the branch lengths. Each split in the original tree is searched for in the new trees, and the percentage occurrence in the set of trees is noted. This percentage is taken to indicate the reliability of the split. It has been noticed, however, that there is a difference between 95% confidence intervals and a split present in 95% of the bootstrap trees. In general, the bootstrap values underestimate strong support for splits. Methods that are more complex have been proposed as improvements, of which currently the best is possibly the AU (approximately unbiased) method (see Further Reading for a detailed discussion of this issue).

Analytical methods have been derived for calculating the standard deviations of the branch lengths obtained by some of the distance-based methods, usually involving considerably more calculation than is needed to produce the tree. If the branch-length estimates in these methods are assumed to follow a normal distribution, a simple $z$-test (see Section 16.4) can be used to determine the possible range of lengths for a given confidence interval. If the length range includes zero length, the related split will be assigned correspondingly low confidence.

## Box 8.4 The bootstrap method

As well as obtaining a result from applying a technique to a set of data, we frequently want to have an estimate of the possible variability or uncertainty in that result. On some occasions it is possible to obtain more data, for example by taking more measurements, but often, as in the case of sequence data, this is not an option. More new sequences can be added, but there is no real equivalent of a repeat measurement for a given sequence. In 1979, Bradley Efron proposed a method he called the **bootstrap analysis**, which uses the original dataset to generate estimates of the variability of the results.

The method is remarkably simple and yet usually very effective. If we knew the true distribution from which the data came we could calculate any properties of interest from it. The key to the technique is to assume that the original dataset is sampled in an unbiased manner from the real but unknown distribution. A new dataset can be produced by sampling from the original one. If it is sampled correctly, it will also be an unbiased sample of the true distribution, and can be used to estimate property variances. Typically the identical analysis is run on the new datasets as was run on the original one. Normally several hundred new datasets—bootstrap replicates—are created and analyzed.

To generate unbiased replicate datasets, one simply randomly selects data points from the original data, making each selection with equal probability of choosing any of the data points. If there are $N$ data points in the original dataset, normal bootstrapping involves selecting $N$ replicate data points. A crucial feature of the selection process is that every selection is from the complete set of the original data. Some points will not be selected for the replicate, and others will be selected more than once. This is referred to as sampling with replacement.

Variations on this theme exist. The technique just described is nonparametric bootstrapping. It suffers from the disadvantage that biases present in the original data will simply reappear in the replicates. In favorable circumstances an alternative called parametric bootstrapping can be used. In this case, the model fitted to the original data (here the phylogenetic tree and evolution model) is used to generate replicate datasets. These datasets are then reanalyzed as before. However, if the fitted model has serious errors compared with reality, this will propagate through to the replicates.

A procedure has been proposed that can be used to obtain a numerical estimate of the branch length standard deviations. Called the **bootstrap interior branch test**, it involves generating new datasets by sampling the original one, and then obtaining the branch lengths on the same tree topology with the new data. Note that the replicate datasets are not analyzed exactly as the original data were, since the only tree topology analyzed is that produced by the original data. In this way, sample distributions are obtained for each branch length. These can be used to estimate the confidence of the interior branches being non-zero. A further set of tests has been proposed for examining whether there has been a constant molecular clock (rate of evolution) for the period of evolution covered by the data (see Further Reading).

### Tests have been proposed for comparing two or more alternative trees

There are many occasions when there are competing hypotheses concerning the evolutionary origins of particular groups of species. If two trees are generated based on the two competing hypotheses, then some of the methods described earlier can be used to score both trees. However, it is not immediately apparent how to assess the relative weight to be given to each tree on the basis of, for example, their calculated likelihoods. The likelihood ratio tests described in Section 7.3 when choosing evolution models cannot be applied here, because both trees will have the same number of degrees of freedom, so there are no degrees of freedom for the chi-squared test comparing them.

One approach to this problem is the set of methods known as paired-site tests. The basis of these tests is to examine which tree is given support by each sequence site. If both trees are equally well supported, the relative measure of support at each site should average to zero for all sites. A variety of statistical tests have been proposed for the distribution of this site support. For example, in the RELL method, rather than make assumptions about the properties of the distribution, bootstrapping is used to estimate the significance of the results.

A similar problem can arise when more than two trees need to be compared. Each pair of trees can be analyzed separately with methods such as RELL, but there are serious problems with such multiple testing. One proposed solution, in much the same spirit as the RELL test, is the SH test in which the actual distribution is sampled with bootstrapping. This is still a very difficult problem, however, and these methods are not straightforward to apply.

## Summary

In this chapter we have presented many of the theoretical and technical aspects relevant to deducing phylogenetic trees from sequence data. All interpretations require a model of evolution, and a number of these are presented. The simplest models are clearly incorrect, for example in assuming that substitutions occur at the same rates at all sequence positions. Nevertheless, these models seem to contain sufficient realism to permit some useful analysis. In tree-building methods, the sequence data are used in two key forms—the individual positions in the alignment itself or in the form of estimates of evolutionary distance. In the latter case, application of an evolutionary model only requires the equation for distance correction.

The quickest and most simplistic tree-generating techniques result in a single tree. These can be useful, and even quite accurate, and can handle very large numbers of sequences. More sophisticated methods involve optimization of certain criteria for a large number of different tree topologies. There are many ways to generate tree topologies, some being more efficient at exploring the possible range of topologies. However, some rare events in evolution such as genome fusions require topologies that are outside the range normally considered. The parsimony method attempts to identify all the mutations that have occurred, but does not use evolution models. In contrast, the maximum likelihood technique makes explicit use of the time-dependent mutation probabilities provided by these models to estimate the most likely representation of the evolutionary history of the data, as do the quartet-puzzling and Bayesian methods.

The large number of different techniques available indicates that none has yet demonstrated clear superiority, and there is still much debate in the research community about their relative performance. Some techniques can only analyze a few dozen sequences at most, and cannot rigorously search all tree topologies. Furthermore, there are difficulties in determining the statistical confidence that can be placed on tree features. Despite the problems, methods exist that attempt to indicate the reliability of certain tree features. In addition, tests have been proposed that allow two or more trees to be compared for their support by the data, the most commonly used of which is the bootstrap.

# Further Reading

The following books cover a much broader range of topics than could be covered here and are highly recommended. In particular, several go into details of the evolutionary models, and discuss some other methods of tree construction. They also present a statistical analysis of the parameterization of the models.

Felsenstein J (2004) Inferring Phylogenies. Sunderland, MA: Sinauer.

Hillis DM, Moritz C & Mable BK (eds) (1996) Molecular Systematics. Sunderland, MA: Sinauer.

Li WH (1997) Molecular Evolution. Sunderland, MA: Sinauer.

*Methods in Enzymology* vol 183 (1990) and vol 266 (1996). (*Contain several important papers.*)

Nei M & Kumar S (2000) Molecular Evolution and Phylogenetics. Oxford: Oxford University Press.

## 8.1 Evolutionary Models and the Calculation of Evolutionary Distance

For models of evolution see Further Reading for Section 7.3.

**The Gamma distance correction takes account of mutation rate variation at different sequence positions**

Felsenstein J & Churchill GA (1996) A hidden Markov model approach to variation among sites in rate of evolution. *Mol. Biol. Evol.* 13, 93–104.

Uzzell T & Corbin KW (1971) Fitting discrete probability distributions to evolutionary events. *Science* 172, 1089–1096.

**More complex models distinguish between the relative frequencies of different types of mutation**

Wakeley J (1996) The excess of transitions among nucleotide substitutions: new methods of estimating transition bias underscore its significance. *Trends Ecol. Evol.* 11, 158–163.

**There is a nucleotide bias in DNA sequences**

Foster PF (2004) Modeling compositional heterogeneity. *Syst. Biol.* 53, 485–495.

**Models of protein-sequence evolution are closely related to the substitution matrices used for sequence alignment**

Felsenstein J (2004) Inferring Phylogenies, chapter 14. Sunderland, MA: Sinauer.

Goldman N & Whelan S (2002) A novel use of equilibrium frequencies in models of sequence evolution. *Mol. Biol. Evol.* 19, 1821–1831.

## 8.2 Generating Single Phylogenetic Trees

**The UPGMA method assumes a constant molecular clock and produces an ultrametric tree**

Sneath PHA & Sokal RR (1973) Numerical Taxonomy. San Francisco: Freeman.

**The Fitch–Margoliash method produces an unrooted additive tree**

Fitch WM & Margoliash E (1967) Construction of phylogenetic trees. *Science* 155, 279–284.

**The neighbor-joining method is related to the concept of minimum evolution**

Bruno WJ, Socci ND & Halpern AL (2000) Weighted neighbor-joining: a likelihood-based approach to distance-based phylogeny reconstruction. *Mol. Biol. Evol.* 17, 189–197.

Gascuel O (1997) BIONJ: an improved version of the NJ algorithm based on a simple model of sequence data. *Mol. Biol. Evol.* 14, 685–695.

Saitou N & Nei M (1987) The neighbor-joining method: a new method for reconstructing phylogenetic trees. *Mol. Biol. Evol.* 4, 406–425.

## 8.3 Generating Multiple Tree Topologies

**Optimization of tree topology can be achieved by making a series of small changes to an existing tree**

Allen BL & Steel M (2001) Subtree transfer operations and their induced metrics on evolutionary trees. *Ann. Combinatorics* 5, 1–15.

Felsenstein J (2004) Inferring Phylogenies, chapter 20. Sunderland, MA: Sinauer.

**Finding the root gives a phylogenetic tree a direction in time**

Lake JA & Rivera MC (2004) Deriving the genomic tree of life in the presence of horizontal gene transfer: conditioned reconstruction. *Mol. Biol. Evol.* 21, 681–690.

Rivera MC & Lake JA (2004) The ring of life provides evidence for a genome fusion origin of eukaryotes. *Nature* 431, 152–155.

Stechmann A & Cavalier-Smith T (2003) The root of the eukaryotic tree pinpointed. *Curr. Biol.* 13, R665–R666.

Wolf YI, Rogozin IB, Grishin NV & Koonin EV (2002) Genome trees and the tree of life. *Trends Genet.* 18, 472–479.

Zmasek CM & Eddy SR (2001) A simple algorithm to infer gene duplication and speciation events on a gene tree. *Bioinformatics* 17, 821–828.

## 8.4 Evaluating Tree Topologies

**Functions based on evolutionary distances can be used to evaluate trees**

Bryant D & Waddell P (1998) Rapid evaluation of least squares and minimum evolution criteria on phylogenetic trees. *Mol. Biol. Evol.* 15, 1346–1359.

**Trees can be evaluated using the maximum likelihood method**

Felsenstein J (2004) Inferring Phylogenies, chapter 16. Sunderland, MA: Sinauer.

Guindon S & Gascuel O (2003) A simple, fast, and accurate algorithm to estimate large phylogenies by maximum likelihood. *Syst. Biol.* 52, 696–704.

Stamatakis A, Ludwig T & Meier H (2005) RAxML-III: a fast program for maximum likelihood-based inference of large phylogenetic trees. *Bioinformatics* 21, 456–463.

**The quartet-puzzling method also involves maximum likelihood in the standard implementation**

Strimmer K & von Haeseler A (1996) Quartet puzzling: a quartet maximum likelihood method for reconstructing tree topologies. *Mol. Biol. Evol.* 13, 964–969.

Strimmer K, Goldman N & von Haeseler A (1997) Bayesian probabilities and quartet puzzling. *Mol. Biol. Evol.* 14, 210–211.

**Bayesian methods can also be used to reconstruct phylogenetic trees**

Felsenstein J (2004) Inferring Phylogenies, chapter 18. Sinauer, Sunderland, MA.

Holder M & Lewis PO (2003) Phylogeny estimation: traditional and Bayesian approaches. *Nat. Rev. Genet.* 4, 275–284.

## 8.5 Assessing the Reliability of Tree Features and Comparing Trees

**The long-branch attraction problem can arise even with perfect data and methodology**

Hendy MD & Penny D (1989) A framework for the quantitative study of evolutionary trees. *Syst. Zool.* 38, 297–309.

**Tree topology can be tested by examining the interior branches**

Efron B (1979) Bootstrap methods: another look at the jackknife. *Ann. Stat.* 7, 1–26.

Efron B (1985) Bootstrap confidence intervals for a class of parametric problems. *Biometrika* 72, 45–58.

**Tests have been proposed for comparing two or more alternative trees**

Felsenstein J (2004) Inferring Phylogenies, chapters 19–21. Sunderland, MA: Sinauer.

Kishino H, Mayata T & Hasegawa M (1990) Maximum likelihood inference of protein phylogeny and the origin of chloroplasts. *J. Mol. Evol.* 31, 151-160. (RELL).

Nei M & Kumar S (2000) Molecular Evolution and Phylogenetics, chapters 9 and 10. Oxford: Oxford University Press.

Shimodaira H (2002) An approximately unbiased test of phylogenetic tree selection. *Syst. Biol.* 51, 492–508.

Sitnikova T (1996) Bootstrap method of interior-branch test for phylogenetic trees. *Mol. Biol. Evol.* 13, 605–611.

# PART 4

# GENOME CHARACTERISTICS

The genome encodes all the machinery and information for any living organism, from the most simple to the most complex. Knowing the whole genome of a species and what each part of the genome does would increase our understanding of how the organism works. Recently, the genomes of whole organisms have been sequenced and mapped. Computational techniques play a vital role both in the assembly of genomic sequences and in the analysis. Comparing genomes provides important information both for studying evolution and medical science.

This part of the book consists of two chapters divided into an introduction plus Applications Chapter and a more detailed Theory Chapter. The first chapter describes the most easily obtainable techniques to locate, predict, and annotate genes within a genomic sequence, and also shows how complex some of the output from these predictions can be. It deals both with prokaryotic and eukaryotic genomes. The next chapter surveys the major methods used for gene detection and genome annotation and also describes how models of whole genes can be built from individual components such as ribosome-binding sites.

# REVEALING GENOME FEATURES

9

APPLICATIONS CHAPTER

**When you have read Chapter 9, you should be able to:**

Identify regions with nonprotein gene features.

Discuss how homology aids gene prediction.

Explain why prokaryotic gene prediction is generally easier.

Search for separate gene components such as exons.

Identify splice sites.

Describe the varying success of promoter prediction programs.

Compare and contrast methods for measuring the accuracy of predictions.

Use experimental results as well as prediction in genome annotation.

This chapter will describe the practical use of some common gene-prediction and associated programs that are accessible online. There has been a need for such analysis since the earliest nucleotide sequencing experiments. However, the technological advances that enabled the human genome sequencing project to be completed so quickly have resulted in many more multi-megabase sequences being produced. It is important to have automated accurate analysis methods, which can identify the functional elements of these sequences as fast as they are being produced. In this chapter we will explore some of the practical aspects of such analysis, whereas in Chapter 10 the basic concepts and techniques underlying gene-prediction methods will be discussed.

We will limit our presentation in this chapter to identifying genes and gene control regions, and to obtaining the sequences of any related proteins. This is only the beginning of an analysis of the genes found in a nucleotide sequence, which should ultimately involve many of the techniques described in other chapters of this book. For example, the protein sequences should be subjected to the methods described in Chapter 4 to identify homologous sequences and propose protein function on the basis of conserved sequence patterns. Depending on the results of such work, further analysis may be performed, either to investigate evolutionary relationships, as described in Chapter 7, or to model the protein structure and function using the techniques of Chapters 11, 13, and 14. Finally, expression studies of the kind described in Chapter 15 can be useful to help confirm gene predictions, for example by confirming that predicted mRNA molecules are indeed expressed.

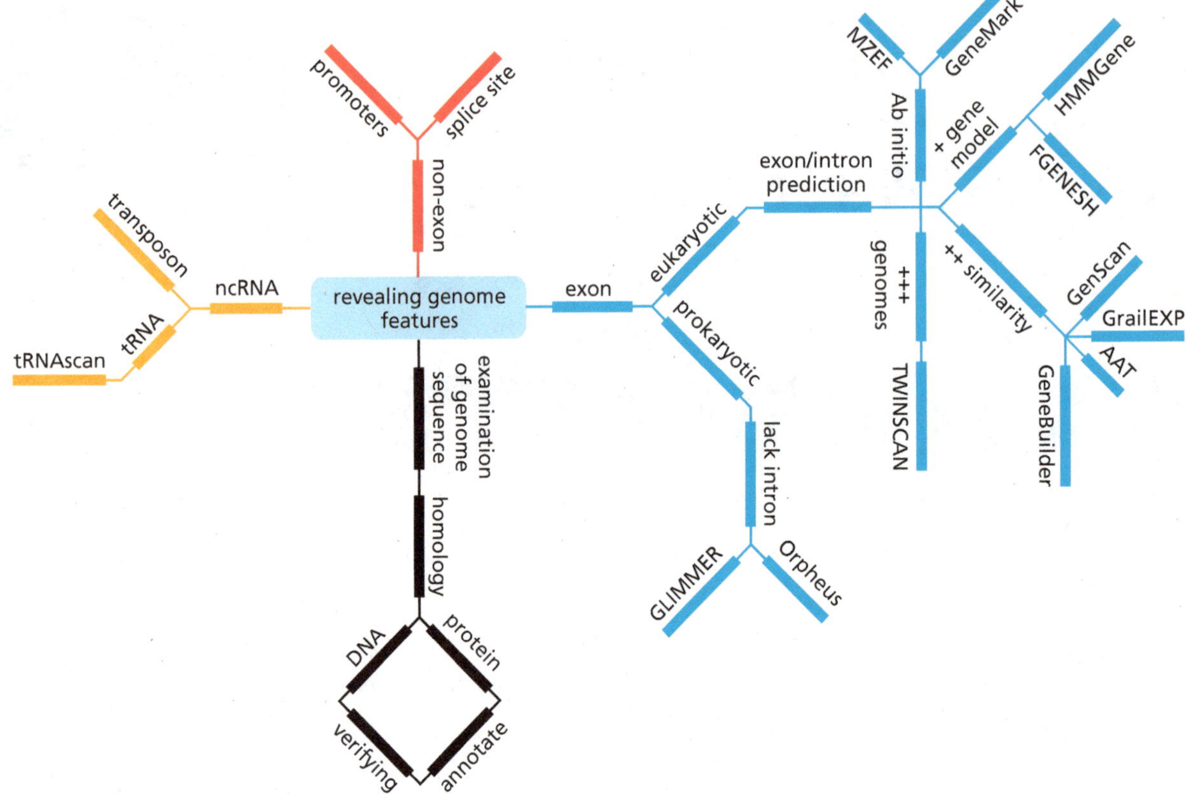


**Mind Map 9.1**
**A mind map to help the reader remember the steps that can be taken in predicting genes from DNA sequences.**

This chapter starts with a description of the techniques that can be applied to identify tRNA molecules and genes in prokaryotes. In both areas there are well-established methods, which have a high degree of success. Following that we will explore the methods used to identify eukaryotic genes and control regions. Because the gene structure is more complex in eukaryotes many different methods have been proposed, and they tend to result in more modest success. The methods described in this chapter are only a subset of those available in this active area of research. Before reading this chapter, the reader is advised to review the basic structural features of genes and their control, as presented in Chapter 1.

## 9.1 Preliminary Examination of Genome Sequence

As described in Chapter 1, regions of DNA that code for proteins are first transcribed into messenger RNA and then translated into protein. During translation, codons consisting of three nucleotides determine which amino acid will be added to the growing protein chain and at what point translation starts and stops. The region that is translated into a protein is called an open reading frame (**ORF**). Every stretch of DNA has six possible reading frames, three in each direction. A particular reading frame will establish the sequence of amino acids encoded by that sequence, and in the vast majority of cases a gene is translated in only one reading frame. An ORF ends with a stop codon and usually starts with a methionine codon (see Section 1.2 and Table 1.1). Therefore, the most common method of finding protein-coding regions in DNA is to look for ORFs in a genomic sequence. In a prokaryotic genome, ORFs for individual proteins are relatively easy to detect, as the protein-coding region between the methionine codon and the stop codon is not interrupted by noncoding introns. ORFs in prokaryotes can thus be identified by searching for a region of DNA that has an uninterrupted sequence of codons that code for amino acids and a start and a stop codon in the same reading frame. The longer the ORF, the more likely it is to be a true protein-coding sequence.

In eukaryotes the presence of introns, which break up the coding sequence into separate exons, makes the situation more complicated (see Section 1.3). Regions of DNA that do not code for protein, such as introns, usually have very short runs of amino acid encoding codons and many stop codons. Introns can be numerous and very long, much longer than the exon sequences. It is therefore necessary to predict not only where the whole gene starts and finishes, but where each exon starts and finishes. It is possible to distinguish the exons in the form of relatively long runs of amino acid encoding codons. In addition, there are splice sites at the beginning and end of each intron, which have characteristic sequence motifs that can aid prediction. Many eukaryotic genes can, however, be spliced in different ways to give different proteins (see Section 1.3), which is a further complication.

In addition to the coding sequence, a gene also has a transcription start site and a basal (or core) promoter region at which transcription of the gene is initiated (see Figures 1.12 and 1.13). Many genes also have a more extensive promoter region upstream of the core promoter, sometimes extending as much as 200 base pairs (bp) farther upstream. Basal promoters, located within 40 bp of a transcription start site, are usually recognizable and can be used in identifying an ORF. Eukaryotic genes are generally each preceded by their own promoter. In prokaryotes, however, several genes are often arranged in a single transcription unit—an operon—which is preceded by a single promoter region and a single transcription start site (see Figure 1.15).

A region of genomic sequence rarely has more than one function. That function might be to encode a protein, a tRNA or other functional RNA, a transposon, or one of several other genetic elements. The major exception to the one sequence–one function rule is some viral genomes, where the same sequence segment uses two different reading frames to encode two or more proteins. In all other genomes only small overlaps, if any, occur.

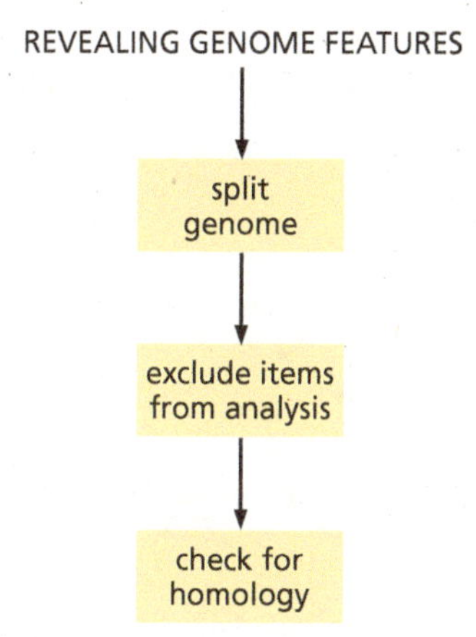


**Flow Diagram 9.1**
**In this section the steps are presented that should be explored before applying the more sophisticated gene detection methods that form the main subject of this chapter.**

## Whole genome sequences can be split up to simplify gene searches

Most gene analysis programs cannot deal with more than a few megabases of sequence at one time. Therefore, especially in the case of eukaryotes, the genome must be split into many smaller fragments for analysis (see Flow Diagram 9.1). The splitting is further justified by the existence of regions of significantly different GC content, as many gene prediction programs use different parameters according to the GC content. In practice, the splitting is usually automated using scripts so that the sequence is divided into manageable fragments, each of which is then analyzed by gene prediction. This genome annotation pipeline approach can give the impression of full genome analysis at the press of a button. The pipeline can consist of a number of different analysis programs whose results are combined into the final prediction. These pipeline programs are not usually available for Web-based prediction, however, and the user has to find their own method of dividing the sequence. Because prokaryotic genomes are that much smaller than eukaryotic genomes, they often do not require a splitting step.

## Structural RNA genes and repeat sequences can be excluded from further analysis

Genes for which RNA, rather than protein, is the functional product are often known as **noncoding RNA (ncRNA) genes**. Identifying noncoding RNA genes and repeat sequences has been found to be easier than identifying protein-coding genes. It is therefore advantageous when analyzing a genome sequence to identify those regions that are not coding for proteins first.

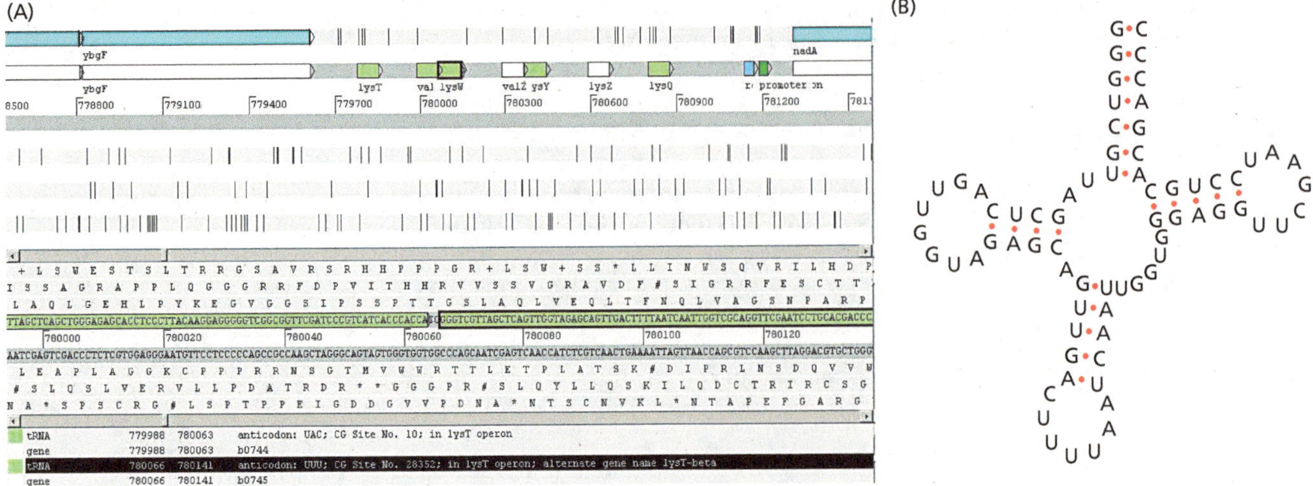


**Figure 9.1**
**A detailed view of a tRNA coding region and the secondary structure of the tRNA molecule.** (A) A tRNA gene (lysW) identified within a genomic DNA sequence. The tRNA gene sequence is shown in the lower right part of the figure, with a green background surrounded by thick black lines, starting GGGTCGTT... A detailed description of this sequence representation can be found in the legend of Figure 9.3. (B) A schematic illustration of the secondary structure of lysW tRNA. the sequence starts on the left of the upper helix, now shown as RNA, i.e., GGGUCGUU...

We shall illustrate the detection of noncoding RNAs with the case of the tRNA genes in *E. coli.* As discussed in Section 10.1, all tRNAs have a highly conserved structure, which is based on base-pairing (see Figure 1.5); knowledge of this structure can be used to help interpret sequence. Programs are available on the Web that attempt to specifically locate tRNA genes. The annotated genome of *E. coli* illustrated in Figure 9.1A shows the tRNA genes for various amino acids as green-shaded boxes. The gene for tRNA$^{Val}$ and the *lysW* gene for tRNA$^{Lys}$ are highlighted by sequence. To

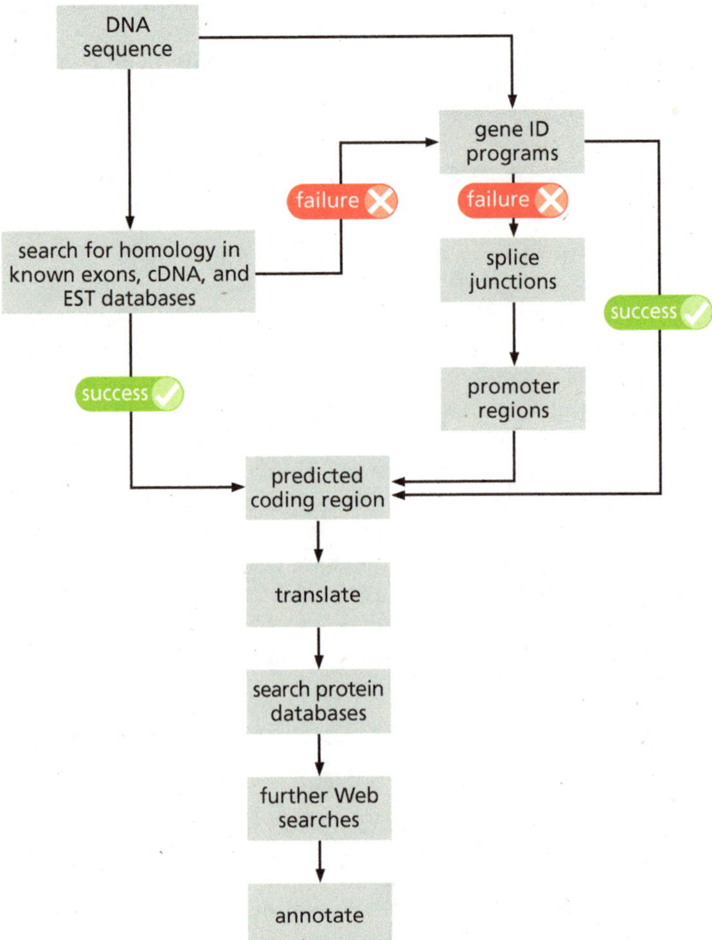


**Figure 9.2**
**A flow chart of steps involved in the identification and annotation of gene sequences.** The first step is to search for homology, if no homologs are found, gene prediction programs must be used. Both methods can be explored to verify the predictions.

predict the tRNA genes, a region of *E. coli* sequence containing both protein-coding ORFs and tRNA genes was submitted to the specialist program tRNAscan, which correctly picks out the genes coding for the tRNAs, gives the type of amino acid the tRNA carries, and gives a diagram of the secondary structure of the molecule (see Figure 9.1B). Repeat sequences can be identified using programs like SEG, as discussed in Section 4.7 and Box 5.2.

## Box 9.1 Expressed sequence tags

Expressed sequence tags (ESTs) are short sequences in the 3′ untranslated regions (UTR) of complementary DNA (cDNA). 5′ ESTs can also be generated, but 3′ ESTs, because they tend to form part of the UTR, will have less cross-species conservation. Thus an EST is a small—usually 200 to 500 nucleotides long—portion of an intact gene that can be used to help identify unknown genes and map their positions within a genome. This is done by using the ESTs as hybridization probes to fish for the gene in genomic DNA. A number of EST databases exist, such as dbEST, which is the EST part of NCBI's GenBank. The ESTs in these databases are annotated as soon as new information is available.

When a genomic sequence is aligned to an EST, alignment of the rest of the cDNA will reveal the intron–exon pattern. Figure B9.1 shows a DNA segment that has three known exons (160–263, 524–605, and 1164–1236) which have been submitted to a BLAST search through the human EST database. It is immediately evident that three EST segments have been found that match well with the unknown sequence. These three segments correspond well with the actual exons. The color coding of the homology search signifies the strength of the hits, ranging from red (a strong hit, strong possibility of it being correct) to blue (a weak hit, probably not a correct match).

If a search through the various EST databases does not give a significant match, then the genomic sequence must be run through gene identification programs. If these predictions give relatively consistent results then the sequence can be translated to protein sequence and checked. If very inconsistent prediction or no predictions are obtained, however, then further analysis of the genome sequence, such as prediction of splice junctions and promoter regions, can be helpful to delineate exon regions. Once a predicted exon has been obtained it must be translated to a protein sequence. This will check the correctness of the prediction and allow for annotation and further online analysis.

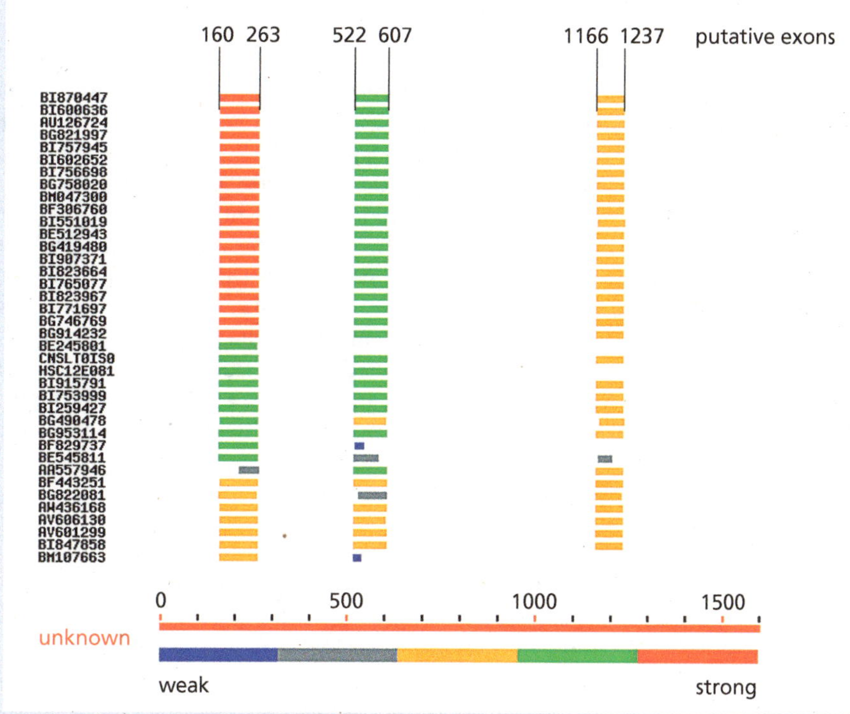


**Figure B9.1**
**Homology search with the PISSRLE DNA sequence that contains three exons (5 to 7) against the EST database.** The search identifies three consecutive hits that illustrate the three exons. The color coding of the homology search signifies the strength of the hits ranging from red to blue.

## Homology can be used to identify genes in both prokaryotic and eukaryotic genomes

Figure 9.2 shows a flow chart of the steps that can be followed in the process to identify and annotate a new genomic sequence. A new DNA sequence can be sent for a homology search against a number of databases. Now that many prokaryotic genomes have been completely sequenced, searching for homologous sequences is a powerful tool for gene prediction. A newly completed bacterial genome can, for example, be sent for a homology search against other prokaryotic genome databases. One can expect to identify more than half the genes in a prokaryotic genome by homology. Database searching and the alignment of homologous sequences are described in Chapters 4 to 6. If a significant hit is found, especially with a sequence for which the protein and its function are known, then it is possible to infer that the unknown sequence is a homolog and very probably has a similar function.

For eukaryotes, because the protein sequence being used for the search may correspond to separate exons in the DNA sequence, a homology search is more complicated. A useful aid is the databases of expressed sequences by which protein-coding genes can be identified by means of ESTs (see Box 9.1).

# 9.2 Gene Prediction in Prokaryotic Genomes

Although introns do exist in prokaryotes they are extremely rare and therefore are ignored by the programs discussed below. We will not discuss their identification in this chapter, but one should be aware of their possible presence.

In prokaryotes the promoter regions are usually well defined. Thus, because of the general lack of introns, the beginnings and ends of genes are easier to detect. The relative simplicity of bacterial gene structure has led to some very successful gene-prediction techniques for prokaryotes that use functional signals such as the ribosome-binding site, the stop codon that signals the end of translation, and other well-defined features.

We will illustrate gene prediction in prokaryotes (see Flow Diagram 9.2) with a region of approximately 4000 bases from the *E. coli* genome. Figure 9.3 shows the annotation of this region, where the blue-colored bars denote protein-coding genes, which are identified by a gene name (for example *tolQ*). Although not shown in the figure, information about protein homology and other functional predictions is available for each of these genes. The light green bars in the "annot for all" line show the location of tRNA genes (see Section 9.1). In the lower (more detailed) part of the figure the bright green-colored base sequence is the promoter of the *pal* gene starting at position –35 relative to the start of its transcription. (We shall deal with the details of promoter prediction later in the chapter.) Above this, is the end of the *tolB* gene, which ends with a TGA stop triplet that signals the end of translation. One feature that is immediately apparent is that the promoter for the *pal* gene overlaps the end of the *tolB* gene (which may cause problems for prediction programs). There is an ATG triplet (coding for a methionine, M) signaling the start of translation at the beginning of the *pal* gene.

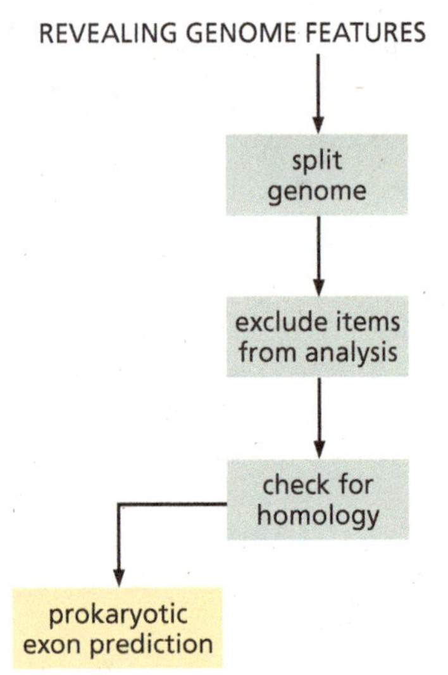


**Flow Diagram 9.2**
**In this section some of the practical aspects of prokaryotic gene prediction are described.**

Due to their relatively small size it is possible to analyze a complete prokaryotic genome as a single sequence. Table 9.1 lists a selection of some of the prokaryotic genomes that are now fully sequenced.

A number of gene-prediction programs are available and are routinely used within major research groups. Even among prokaryotes, different species can have species-specific features such as a bias in the use of particular codons. When available, it is preferable to choose a prediction program that has been trained on the

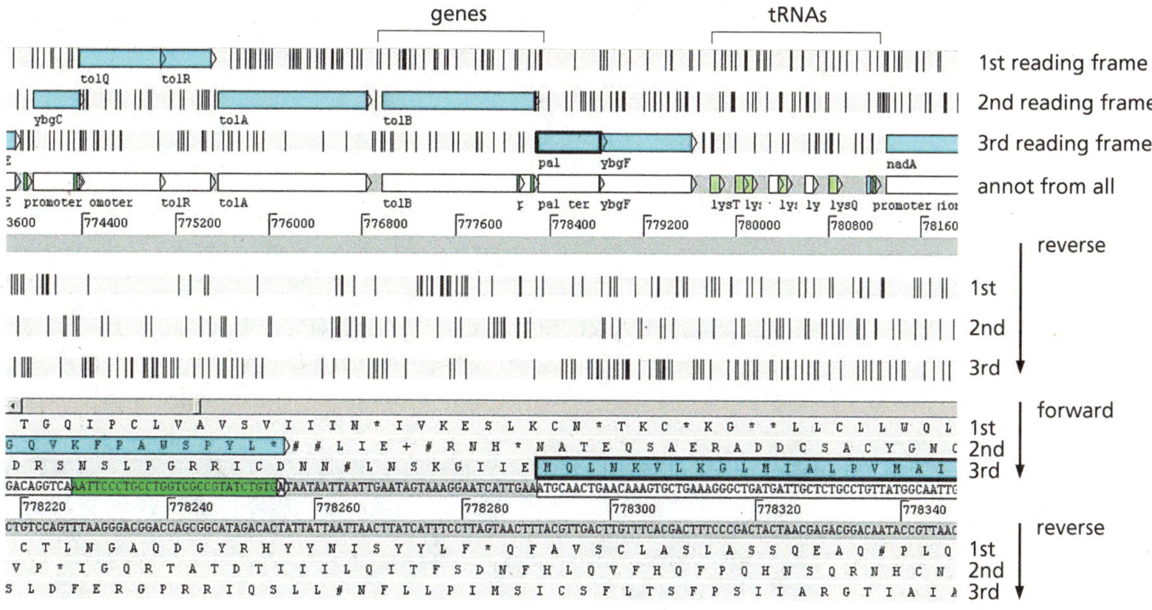


**Figure 9.3**
**Annotations.** A segment of the *E. coli* genome that has been fully annotated, illustrated using the Artemis program. In the top half, there is a ruler showing the base numbers (approximately from base 773,800 to 782,800). Above and below the ruler are summary lines of the annotation for the 5′3′ and 3′5′ strands. In this region of the genome all the annotations are on the 5′3′ strand. The three lines above and below these summary lines show the different reading frames, with every stop codon shown by a vertical black line. The protein-coding genes (shown in blue boxes) are free of stop codons. tRNA genes are shown as green boxes. The lower part of the figure gives the same information but in protein translation and also at a higher scale, in this case from base 778,218 to 778,362. All reading frames are translated, and colored boxes are shown matching those in the upper part of the figure. The region of the end of the *tolB* gene and start of the *pal* gene is shown. The dark green box shows an identified promoter upstream of the *pal* gene.

particular species under study. The most popular programs of recent years have been ORPHEUS and GLIMMER, which are discussed in Section 10.3. These are not available on the Web but can be downloaded and run on a UNIX machine. Some programs can be used for both prokaryotic and eukaryotic gene prediction. Grail, for example, has specific *E. coli* gene-prediction features as one of its datasets. However, these programs are very rarely used by those groups that specialize in prokaryotic genome analysis. Small sequence sections can be submitted to Web-based programs, as described later, to determine all the ORFs, which can subsequently be used in database searches for homologous sequences. This will usually not be as effective for gene prediction as the genome-scale programs mentioned above but can still be of use for small-scale work.

## 9.3 Gene Prediction in Eukaryotic Genomes

Compared to prokaryotes, gene prediction in eukaryotes is a much more daunting task (see Flow Diagram 9.3). To identify protein-coding genes in eukaryotic genomic DNA, a number of additional factors have to be taken into account. These include low gene density in many regions of the genome and the more complicated gene structure. These are explained in detail in Section 10.4. In eukaryotic genes, both exons and introns have to be located by a combination of exon prediction and splice-site prediction. The predicted exons then have to be translated to see if together they can generate a protein sequence that is homologous to a known functional protein. Figure 9.4 shows an example of eukaryotic gene structure and the resulting expressed protein. There are a few eukaryotic genomes in which very few genes have any

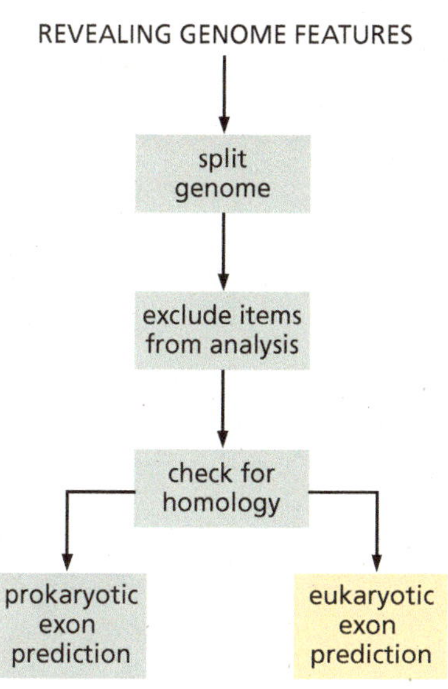


**Flow Diagram FD9.3**
**In this section the various different methods of eukaryotic gene prediction are surveyed.**

introns, for example the budding yeast *Saccharomyces cerevisiae*. In such cases the prokaryotic gene-prediction programs may be sufficient for initial analysis of the sequence. Homology searches can be used to confirm the gene predictions.

## Programs for predicting exons and introns use a variety of approaches

A number of methods are available for the prediction of exon DNA regions. Most programs available on the Web use a combination of two or more approaches, for example database similarity searches and statistical pattern recognition. The use of the programs summarized below will be illustrated by two examples of eukaryotic gene sequences: exons 1 and 2 of the human *ALDH10* gene, encoding a fatty aldehyde dehydrogenase, and exons 5 to 7 of the human protein kinase gene *CDK10* (see Box 9.2).

**Table 9.1**
**A few examples of known prokaryotic genomes.** At the time of writing, 18 archaeal, 141 bacterial, 124 bacteriophage, and about 850 viral genomes have been sequenced. Compared to the size of the human genome (3,140,575,725 bp in all 24 chromosomes), archaeal and bacterial genomes are relatively small, while phage and virus genomes are even smaller.

| Archaea | Length (bp) |
|---|---|
| *Aeropyrum pernix* | 1,669,695 |
| *Halobacterium* sp. NRC-1 | 2,014,239 |
| *Methanococcus maripaludis* | 1,661,137 |
| *Pyrococcus abyssi* | 1,765,118 |
| *Pyrococcus furiosus* DSM 3638 | 1,908,256 |
| *Thermoplasma acidophilum* | 1,564,906 |
| *Thermoplasma volcanium* | 1,584,804 |
| **Bacteria** | |
| *Chlamydophila pneumoniae* | 1,229,853 |
| *Enterococcus faecalis* V583 | 3,218,031 |
| *Escherichia coli* CFT073 | 5,231,428 |
| *Escherichia coli* K12 | 4,639,221 |
| *Helicobacter hepaticus* ATCC 51449 | 1,799,146 |
| *Helicobacter pylori* 26695 | 1,667,867 |
| *Listeria innocua* Clip11262 | 3,011,208 |
| *Pseudomonas aeruginosa* PAO1 | 6,264,403 |
| *Salmonella enterica* subsp. enterica serovar Typhi | 4,809,037 |
| **Phages** | |
| Alteromonas phage PM2 | 10,079 |
| Bacillus phage GA-1 | 21,129 |
| Bacteriophage 77 | 41,708 |
| Bacteriophage 933W | 61,670 |
| **Viruses** | |
| Abelson murine leukemia virus | 5,894 |
| Acute bee paralysis virus | 9,491 |
| African swine fever virus | 170,101 |
| Avian sarcoma virus | 3,718 |
| Bovine leukemia virus | 8,419 |

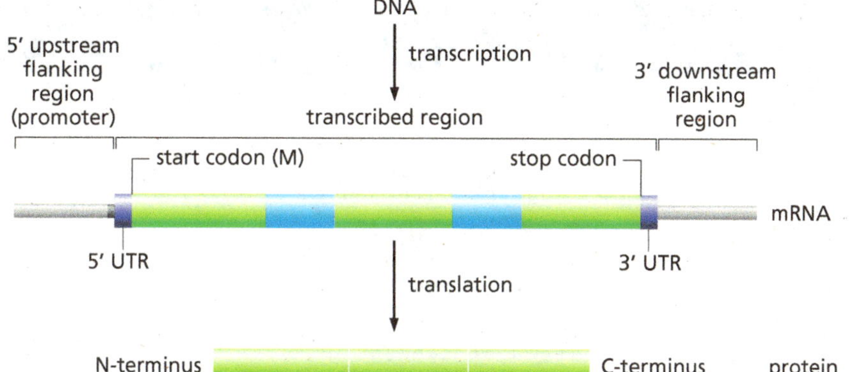


**Figure 9.4**
**From eukaryotic DNA to protein.** The transcribed mRNA gene structure is shown with exons in green and noncoding intron regions in blue. Translation starts from a promoter region. The promoter region, which can be sequentially upstream or downstream from the gene, contains sequences to enhance or inhibit the transcription of a gene. The introns are noncoding and are excised after transcription. Each end of the transcribed gene is flanked by untranslated regions (UTRs), which have specific roles in the formation of the mature mRNA transcript. The mature mRNA is translated by the ribosome into the protein product.

Where a choice of parameters was available, the default settings for that particular prediction program were used for both examples. We have chosen the particular exons in each gene to give examples of the different types of exons that will be encountered in a eukaryotic gene. A gene that is a complex of multiple exons not only contains internal exons separated from their adjacent exons by introns (such as *CDK10* exons 5, 6, and 7), but also an initial exon (such as exon 1 of *ALDH10*) and a terminal exon. These first and last exons are less easy to detect accurately, as they are flanked on one side by regulatory regions signaling the start or stop of transcription and translation, which are not easy to predict.

## Gene predictions must preserve the correct reading frame

When predicting exons, one must remember that they will have to be spliced together to produce a complete protein-coding sequence all in the same reading frame. Because the genetic code involves three-base codons it is important that the correct frame is retained at the splice sites, which may even be in the middle of a codon. If this does not occur, then the following exon will be translated in the wrong frame. While this idea is simple to understand, interpreting it in practice is more complicated than might be expected. The outcome depends on two properties of each predicted exon: its starting point and its length. The effects of errors in these two

**Table 9.2**
**The possible consequences for the translated protein of mistakes in the prediction of an exon.** The term "incorrect, correct frame"in the "length of exon" column means that, although the exon has the wrong length, the number of bases is such that the resulting relative frameshift in the following exon is correct.

| Start of exon | Length of exon | Effect on translation of this exon | Effect on translation of correctly starting next exon |
|---|---|---|---|
| Correct | Correct | Correct | Correct |
| | Incorrect, correct frame | Correct but extra or missing residues | Correct except possibly the first residue |
| | Incorrect, wrong frame | Correct but extra or missing residues | Incorrect |
| Incorrect, correct frame | Correct | Correct but extra or missing residues | Correct except possibly the first residue |
| | Incorrect, correct frame | Correct but extra or missing residues | Correct except possibly the first residue |
| | Incorrect, wrong frame | Correct but extra or missing residues | Incorrect |
| Incorrect, wrong frame | Correct | Incorrect | Incorrect |
| | Incorrect, correct frame | Incorrect | Incorrect |
| | Incorrect, wrong frame | Incorrect | Possibly correct if the two first exon frameshifts cancel |

## Box 9.2 Two different gene sequences for prediction

```
.. .. .. .. .. .. .. .. .. .. .. .. .. .. .. .. .. .. .. .. .. .. .. ..
cgatccaagg agcccagcgc ctagggcgga cccgcgggag cgtctattga gtaaccgttg
ttataggaga cgaagcccgg gaaggagctt tcgcgcctgc gccgcggggc cgtccgcgtc
tgcgcctgcg cgcaagagag gcggggccag cgctcggcat ggcggagcca gatctggagt  EXON 1
gcgagcagat ccgtctgaag tgtattcgta aggagggctt cttcacggtg cctccggaac
acagggtgcg cggggtgcca cccgggcagc tctgcccgcc tcgctagcgg cactgcccgg
ctgggtctgg ggagcctcgt gtcgcgctgc cgcgccgagg cttccggcac gggcgggaac
gacagtccca gagttccccg cggcgggggc ggaagccggg gcggggcggg ctcaggaccc
ccgacagccg gtccctggag atctgagggg ccggggcggg ctcagggatg cttcgcgccc
gcggagagac gggcccggga cttgggaaga gcaggctccg ggatccagct cgggcgctgc
tgggttcagc gcccgagctg ggctttgcag gctgagccgc gagccacttg tttgtgggga
gaatttacac ccgcgacgag ttcgagcttg aaggctcccg ctgggcttgg gctgcatgga
gcggggtcca gcgctgctgg cggcttcacg cgaaccctgg gcgcgtccct tgcggaggtg
ccggtgcctc cctctgccgg cttcacgcga accctgggcg tcccttgcgg aggtgccggt
gcctccctct gcagagacca ggagagagcc tctgggcgtg tgggcggtgt tgcccaggtg
cacggctggt gggggtggag agccctgaac tttggccgct ggtttgtgtt tttaactgct
ggctggtgct ctgagaggcc aagccacgtg ttcagtaaga atcattaaca gatcgtggct
cgggcaggct gctgcagctg ggaaccctga gcttttagta actccacagg ggcagcagag
tcaggcctca gccgaatttg tgcttccaac tctgcctagc tgtcacccac ctgtggataa
tgaagttcct gtggcagata ctggaggccc cggagcacct gcagcacgat gcacttgacc
tgtgcagagg gtcagaggta ccagtcaggg tgagcctgca gggggatgac aacaggagcc
acactttact catccagacg tgcagcaggc gctcaccgcg ttgagagccg tgtgctggac
tctggggtga agtagtgagt tggcctttcc tttttttgttg tttttttgag acagggtcta
gcgctctcgc ccaggctgga atgcagtggc gtgatcacgg ctcagtgcat cgtcagcctc
ctgggctcaa gtgatcttcc tgcctcggcc tcccaagtgc tggaattaca ggcgtgagcc  EXON 2
accgccccca gcctggcctg gcatttcttt gagttcagga agtgtgacaa ggatttggac
acccagaaat aagcgtgtcg agaagagcac aagcagagga tgtgagaagg . .. .. ..
.. .. .. .. .. .. .. .. .. .. .. .. .. .. .. .. .. .. .. .. .. .. .. ..
.. .. .. .. .. .. .. .. .. .. .. .. .. .gtttca gctgggacga tgccggagtg  EXON 3
tgaaggagtt tgagaagctg aaccgcattg gagagggtac ctacggcatt gtgtgtgatg
.. .. .. .. .. .. .. .. .. .. .. .. .. .. .. .. .. .. .. .. .. .. .. ..
.atcgggccc gggacaccca gacagatgag attgtcgcac tgaagaaggt gcggatggac  EXON 4
aaggagaagg atg .. .. .. .. .. .. .. .. .. .. .. .. .. .. .. .. .. .. ..
.. .. .. .. .. .. .. .. .. .. .. .. .. .. .. .. .. .. .. .. .. .. ..cag
gcatccccat cagcagcttg cgggagatca cgctgctgct ccgcctgcgt catccgaaca  EXON 5
tcgtggagct gaaggaggtg gttgtgggga accacctgga gag .. .. .. .. .. .. ..
.. .. .. .. .. .. .. .. .. .. .. .. .. .. .. .. .. .. .. .. .. .. .. ..
catcttcctg gtgatgggtt actgtgagca ggacctggcc agcctcctgg agaatatgcc  EXON 6
aacacccttc tcggaggctc ag.. .. .. .. .. .. .. .. .. .. .. .. .. .. ..
.. .. .. .. .. .. .. .. .. .. .. .. .. .. .. .. .. .. .. .. .. .. .. ..
gtcaagtgca tcgtgctgca ggtgctccgg ggcctccagt atctgcacag gaacttcatt  EXON 7
atccacag .. .. .. .. .. .. .. .. .. .. .. .. .. .. .. .. .. .. .. ..
.. .. .. .. .. .. .. .. .. .. .. .. .. .. .. .. .. .. .. .. .. .. .. ..
cagggacctgaag gtttccaact tgctcatgac cgacaagggt tgtgtgaaga cag. ..  EXON 8
.. .. .. .. .. .. .. .. .. .. .. .. .. .. .. .. .. .. .. .. .. .. ..cag
cggatttcgg cctggcccgg gcctatggtg tcccagtaaa gccaatgacc cccaaggtgg  EXON 9
tcactctctg. .. .. .. .. .. .. .. .. .. .. .. .. .. .. .. .. .. .. ..
.. .. .. .. .. .. .. .. .. .. .. .. .. .. .. .. .. .. .. .. .. .. .. ..
gtaccgagcc cctgaactgc tgttgggaac caccacgcag accaccagca tcgacatgtg  EXON 10
.. .. .. .. .. .. .. .. .. .. .. .. .. .. .. .. .. .. .. .. .. .. .. ..
ggctgtgggc tgcatactgg ccgagctgct ggcgcacagg cctcttctcc ccggcacttc
cgagatccac cagatcgact tgatcgtgca gctgctgggc acgcccagtg agaacatctg  EXON 11
gccg. .. .. .. .. .. .. .. .. .. .. .. .. .. .. .. .. .. .. .. .. ..
.. .. .. .. .. .. .. .. .. .. .. .. .. .. .. .. .. .. .. .. .. .. .. ..
ggcttttcca agctgccact ggtcggccag tacagcctcc ggaagcagcc ctacaacaac  EXON 12
ctgaagcaca agttcccatg gctgtcggag gccgggctgc gcctgctgca cttcctgttc
atgtacgacc ctaagaaaag.. .. .. .. .. .. .. .. .. .. .. .. .. .. .. ..
.. .. .. .. .. .. .. .. .. .. .. .. .. .. .. .. .. .. .. .. .. .. .. ..
ggcgacggcc ggggactgcc tggagagctc ctatttcaag gagaagcccc tac. .. ..  EXON 13
.. .. .. .. .. .. .. .. .. .. .. .. .. .. .. .. .. .. .. .. .. .. .. ..
cctgtgagcc ggagctcatg ccgacctttc cccaccaccg caacaagcgg gccgccccag  EXON 14
ccacctccga gggccagagc aagcgctgta aaccctg. .. .. .. .. .. .. .. .. ..
```

**Figure B9.2A**

**The PISSRLE DNA sequence (*CDK10* gene) with the exons highlighted in red illustrating the large intron regions that intersperse the exon segments.** The start codon ATG is found 12 nucleotides into the delineated exon (highlighted by a red box). The 5′ UTR is shown in yellow letters. Only the first intron sequence is shown in full.

## Box 9.2 Two different gene sequences for prediction (continued)

The example sequences used in this chapter are exons 1 and 2 of the human *ALDH10* gene, encoding a fatty aldehyde dehydrogenase, and exons 5 to 7 of the human cyclin-dependent kinase (cdk) gene *CDK10* (initially know as PISSLRE, Swiss-Prot ID: CDK10_HUMAN). This encoded protein is related to other cyclin-dependent kinases and is important in certain cancers. This gene has many homologs in the databases and it is highly probable that it or its homologs were used in the training of the prediction programs; thus it is used as an easy eukaryotic example throughout this chapter. The complete kinase gene consists of 14 exons separated by large introns (Figure B9.2A). The exons are highlighted in red illustrating the large intron regions that intersperse the exon segments. The start codon ATG is found 12 nucleotides into the delineated exon (highlighted by a red box). The 5′ UTR is shown in yellow.

*ALDH10* codes for a fatty aldehyde dehydrogenase. The protein-coding region consists of 10 exons (Figure B9.2B). Exons 1 and 2 were chosen to illustrate the difficulties of predicting the first exon, especially when it contains more then just a protein-coding segment. The *ALDH10* gene is a relatively new entry in the databases and most programs will not have included it in their training datasets.

**Figure B9.2B**
**A schematic representation of the *ALDH10* gene with the exons colored blue.**

properties on that particular exon and the following one are given in Table 9.2. This is further illustrated in Figure 9.5, which compares the region including the first two exons of *ALDH10* as determined by experiment and by a gene-prediction program.

These types of problems can be circumvented by translating the individual predicted exons and checking the translations against a database search. Coding regions can be translated in three different reading frames and also both in the 5′ to 3′ and 3′ to 5′ direction. The most suitable protein candidate for database searches should be the segment that contains no or the least number of stop codons in its sequence. Figure 9.6 shows the predicted exon regions of exons 5, 6, and 7 of the kinase gene. When the predicted segment of DNA for exon 5 is translated there are two translations that contain only one stop codon (the other translations have more). One is a 5′ to 3′ translation and the other 3′ to 5′. When both of these are submitted to a database search of the Swiss-Prot database, only the 5′ to 3′ translation gives hits that match the same region in a set of homologous proteins. The 3′ to 5′ translation has hits that span various sequence segments and the set of proteins it finds are not obviously homologous. Therefore, we would choose the 5′ to 3′ frame 3 translation as the most likely to be correct. Exons that are part of the same gene can be translated in different reading frames.

One further issue illustrated in Figure 9.7 is the effect of the correct location of the start codon (see FGENESH prediction below). In eukaryotic genes the first transcribed ATG codon is not always the translation start site (for more detail see Section 10.6). In this example it has been experimentally determined that the second ATG is the start site and the first exon has a 5′ UTR of about 250 bases. The UTR is poorly predicted by most current prediction programs.

## Some programs search for exons using only the query sequence and a model for exons

We will now describe two methods that identify exons using the properties of the query sequence. In both cases no attempt is made to generate a multi-exon gene

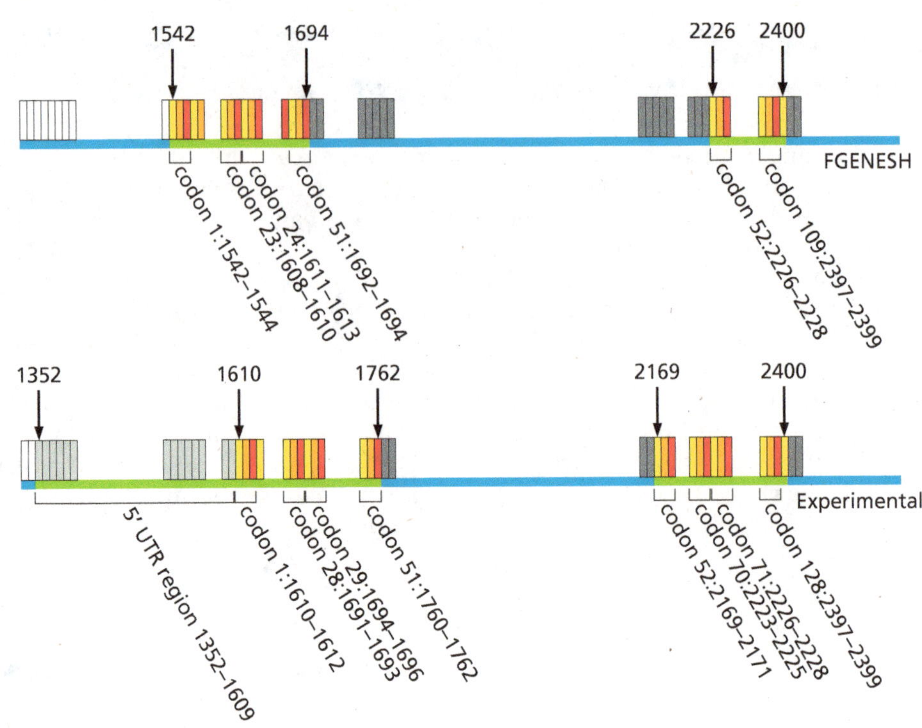


**Figure 9.5**
**Comparison of the prediction of FGENESH with the experimental data for the first two exons of the *ALDH10* gene.** Exons are shown as green bars while the introns and intragene regions are shown as blue bars. Individual bases are shown as narrow vertical boxes. The 5′ UTR region is shown as light gray bases, and the introns are shown as dark gray bases. The three reading frames are illustrated by coloring successive protein-coding bases yellow, orange, and red. The base positions are as given in Table 9.4 and are indicated above the bars. The splice sites as predicted by FGENESH and their experimental positions are shown by numbered arrows. FGENESH does not predict a 5′ UTR region, and has an incorrect start position for exon 1, which has an incorrect reading frame compared with experiment. As a consequence, predicted codon 24 is one base downstream of actual codon 1. However, the length of the predicted first exon is identical to the experimental protein-coding region for exon 1, so both first introns start immediately after codon 51, and start the second exon in reading frame 1. FGENESH does not predict the correct 5′ end of exon 2, but as it is 57 bases (i.e., a multiple of 3) downstream of the correct start, the predicted codons match the experimental data, and predicted codons 52–109 are identical to experimental codons 71–128.

structure, let alone identify the 5′ or 3′ UTR regions. They are true *ab initio* methods in that neither of them uses any prior knowledge of existing genes and gene products.

The program MZEF is designed specifically to predict internal protein-coding exons. It starts with a potential internal exon, which is defined to include the AG of the 3′ splice site of the preceding intron followed by an ORF and then by the GT of the 5′ splice site of the next intron. It employs statistical analysis of sequence patterns that enables exons to be discriminated from the rest of the DNA sequence. Sequence analysis of large numbers of known exons has highlighted certain recurring sequence characteristics. For example, there is a preference in exons compared to introns for particular six-nucleotide sequences, which are known as hexamers, **codon-pairs**, or **dicodons** (see Sections 10.2 and 10.3). A dicodon preference score for exons can be calculated on the basis of these observations. In addition, the statistical preferences observed at, for example, the 5′ and 3′ splice sites can be used to derive a preference score for these and other regions. The program calculates scores for nine discriminant variables (Table 9.3). When the scores are plotted for known exons and introns they appear in distinct regions of the graph. When an unknown sequence is submitted to the program, it is similarly scored and plotted. If the score falls within the exon region the sequence is predicted to be an exon, and vice versa for an intron. The program is based on quadratic discriminant analysis and is explained in more detail in Section 10.5 and Figure 10.16.

By default, the current minimum ORF size is 18 bp and the maximum is 999 bp. Because the genome of the model plant *Arabidopsis thaliana* (see Box 9.3) has been found to contain some very long ORFs, the default maximum is increased to 2000 bp for this species.

The GeneMark gene-prediction algorithm uses a modified version of HMM, the inhomogeneous Markov Chain (IMC) models, which are described in Chapter 10. Like MZEF it is based on statistical analysis of short nucleotide sequences but in this case up to 8 bp in length. Essentially, GeneMark generates a maximum likelihood parse of the DNA sequence into coding and noncoding regions. It can be used for both prokaryotic and eukaryotic genes, and its application to prokaryotic genes is discussed in Section 10.3. It is one of the few programs designed for prokaryotic gene prediction that has been found useful for eukaryotic sequence analysis.

**Figure 9.6**
**Genie predictions.** Genie-predicted exons 5 to 7 of the *CDK10* DNA (green), their translated product, and the various searches through the Swiss-Prot database. (A) The actual exons as given in Figure B9.2A are shown in capital letters. Most prediction programs will search for an ATG start codon in the first exon (red). The correct hit of exon 5 is illustrated by the matching of many homologous proteins in the same region of the unknown sequence (C). The blue lines indicate low confidence in the match. The correct translation of the actual exons as given in Figure 9.15 is highlighted in yellow in part (B). The incorrect translation of exon 5 (3′5′) is also shown in part (B). When this incorrect translatopn is submitted to a Swiss-Prot database search, it gives matches that are not associated with the same region, and are from different types of proteins (illustrated in box D). Exon 7 has not been correctly identified. Translation of that region gives two segments with no stop codon. When these translated segments are submitted to a BLAST search low homology hits are found with a citrus virus protein and a hypothetical protein, respectively.

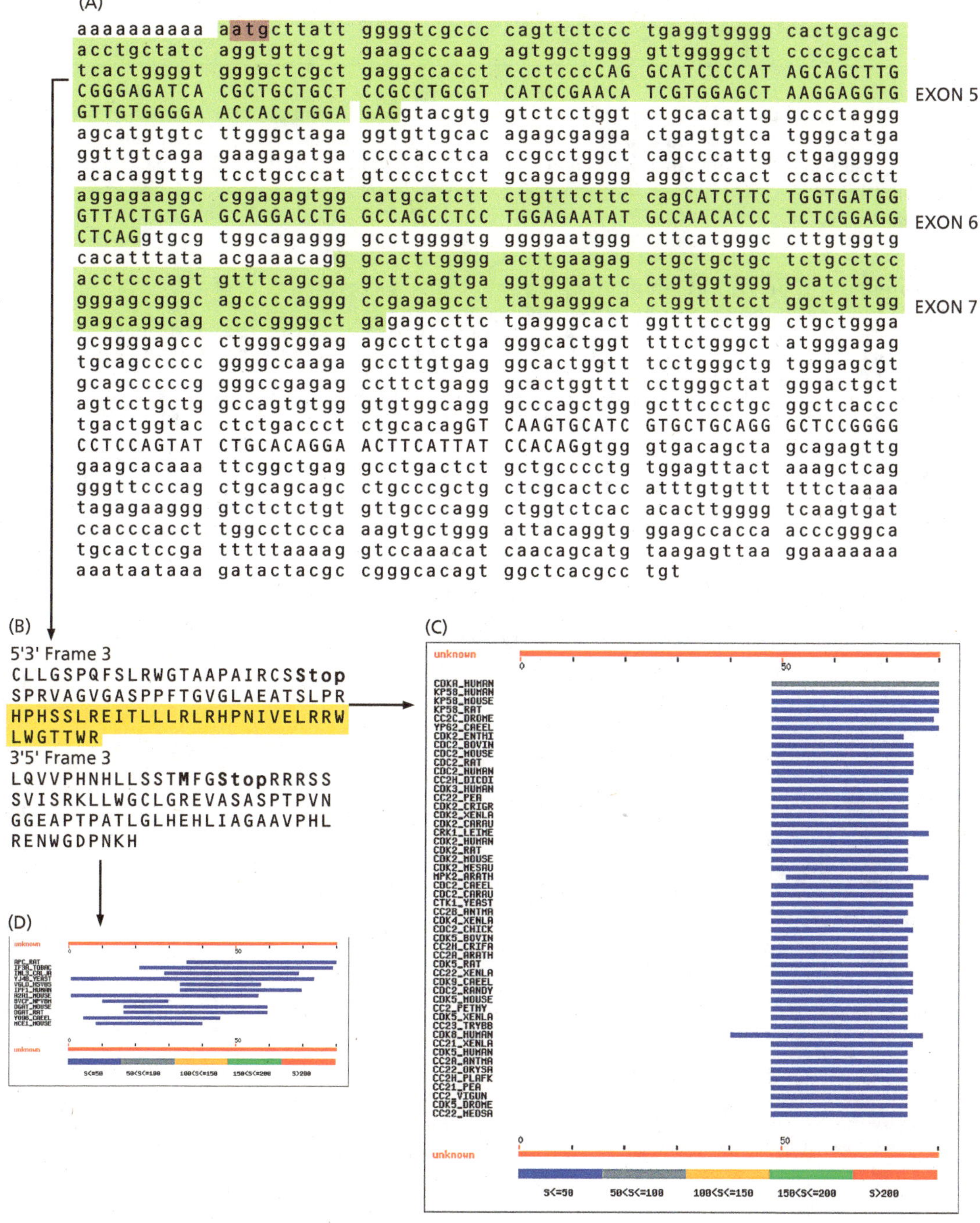

## Box 9.3 ***Arabidopsis thaliana*: a model plant**

*Arabidopsis thaliana* is a small, flowering plant that is widely used as a model organism in plant biology. *Arabidopsis* is a member of the mustard family, which also includes cultivated species such as cabbage and radish. *Arabidopsis* offers important advantages for basic research in genetics and molecular biology because it is small and has a rapid life cycle, is easy to grow, and there are a large number of mutant lines. Its genome (114.5 Mb out of 125 Mb total) was sequenced in 2000 and consists of five chromosomes.

**Figure B9.3**
***Arabidopsis thaliana.*** (Courtesy of A. Davis, Bioimaging JIC.)

**Figure 9.7**
**In contrast to the *CDK10* gene, in the *ALDH10* gene, translation does not begin at the first ATG codon (boxed in red) in the first exon.** It has been experimentally determined that the second ATG codon (bold capitals) is used. This is 62 bases downstream of the first ATG codon, so that if the first ATG codon was predicted as the start of translation, the wrong reading frame would be used. The experimentally determined exons are highlighted in red, and translated bases are written in capital letters. Note that this sequence corresponds to the lower line of Figure 9.5.

```
tttaaaaggg aggagcgggc tggaggggaa agagggagaa catggtcatt actgaatcca     60
cacattgcac aaatagaaaa aggaacaggc aggggaatag tcaattatgt atttgcctcc    120
tgctgtgtaa atcagcactt cagtaagata aggtgaggac agagcagcta cctgtgggga    180
catttaacct tttatctgta gctatctgct tagggacata gagaaaggca gnttcttgca    240
tgactcagct ttttgcttaa tttttccttt ttggcatatg aattgagctc ccacgngntt    300
ttggttggtt ttgggcataa gtggagagtt caattggggc cagggccccg agttattttc    360
ttttcacaaa tatacccttt agagttcaga ggaaaggctg ggattagagc cttctttgag    420
cactattcat ggattacatg agggagtngc tgtgaaaaga gggccccgga ctaagccctg    480
ggcacctgaa caagtgagag gtcaggaagg ggagaaaaat ccagcaaagg aaccccagag    540
gacccggcca gtagctaaaa agaaaagtgc ggaagagtgc agggaaatgg taaagaaaag    600
aaagctttcc acaagacagg accgatcagc tgagccaact gctgctgagt acagtgatga    660
gagctgaagt ctctcgcccg aaaggcaata aggcgggagg gggaaacctg agggcagcgg    720
gttctctctt gggagcggga acacggaacc atgggcagcg aggcaaatgc agcctgggga    780
gaagctctga gtaacggacg tgcgaggatg atgcttgtgc aagaaggaaa tgaccgctgt    840
cccggctccg cgggagacga accccggctt cccgccctca gggactagct ctccagggaa    900
ctgggacggt cagtgccggt cgaggcagct cctcgctgaa ggaaggagca ggggacaggg    960
aacggaatgg ggagcgctag cccccaacta cgtccatctg gccatgtttg aagagccana   1020
aaatggagga gggaacccct agagcgtgcc agacggagac tgctctccgt ngcagtcggg   1080
gcgcttccgg cagggcgccn actcccagcc gagcgccctc cgcctgctcc tccaggattc   1140
ctcttcgccc tttctggggc cgccccgggg cgctctcagn aggatggcca acaccttccc   1200
tccatcccta caccccgccg ccccctgccc gtggccgcgc tcggctcccg cactgctcac   1260
tccacccccct acatcccagc ccgctgccag agccggggag agggcggggg ccgcgtgggc   1320
gagaccgtga acagcggctg tcacgtgggc cgcccaggcc aatagggtg aggctttggg   1380
tccagctcag tcctcccccg gcgcctccga ctggcagtgg gactcagcgg gcgtggaggt   1440
cgcggctgag cgagcgagcc ctgggcgagt gaattgtggc tgtgggttga cggtggagac   1500
acccccccgga gggaggcgga gggaagggag gcgaggcctg cacctgcatg cttcccgcct   1560
cccactcccc agcgcccccg gaccgtgcag ttctctgcag gaccaggccA TGGAGCTCGA   1620
AGTCCGGCGG GTCCGACAGG CGTTCCTGTC CGGCCGGTCG CGACCTCTGC GGTTTCGGCT   1680
GCAGCAGCTG GAGGCCCTGC GGAGGATGGT GCAGGAGCGC GAGAAGGATA TCCTGACGGC   1740
CATCGCCGCC GACCTGTGCA AGgtancacg cgtgcggcgg ggtgtgggga aactggcccc   1800
cgccgngcac ttgtggactg gagtcttcgg ctgggttttg tttttgcttt tacatttngg   1860
attactccac cactgggagt atgatctcca gcgatacaga taaagccaaa gttcccgcag   1920
actttccagg tcctctagca ctcagaaggg catatgttac ctagcttctg tggttccttt   1980
tctgtatatt agagaattag caagccctta ccagggcgtg aagggtgcaa aaggagtctg   2040
aatggcaaac agctagtctg ataatgccag ttgttgtcac tacaggtgta cctggtnnnt   2100
gttctgacat tnagggccaa gtgtatcata cttacnctgn aagnttaact gtgattctct   2160
tataacagAG TGAATTCAAT GTGTACAGTC AGGAAGTCAT TACTGTCCTT GGGGAAATTG   2220
ATTTTATGCT TGAGAATCTT CCTGAATGGG TTACTGCTAA ACCAGTTAAG AAGAACGTGC   2280
TCACCATGCT GGATGAGGCC TATATTCAGC CACAGCCTCT GGGAGTGGTG CTGATAATCG   2340
GAGCTTGGAA TTACCCCTTC GTTCTCACCA TTCAGCCACT GATAGGAGCC ATCGCTGCAG   2400
gtctggttgc caccttatgt ctatatacct ttttagggag gcttattttc tcatattaat   2460
tggnattaag gatagtggct aattaaatac atttacttgg tgatttgcct ttgtttacac   2520
caccagtgta ctggaattca tacatccata cata
```

**Table 9.3**
**The scores used in the MZEF exon-prediction program**

| Scores | Method of calculation |
|---|---|
| Exon length | $Log_{10}$ of the actual length in base pairs (bp) |
| Intron–exon transition | (Intron hexamer frequency preference in 54-bp window to the left of the 3′ splice site) – (exon hexamer frequency preference in 54-bp window to the right of 3′ splice site) |
| Branch-site score | Maximum branch score in –54 to –3 window with respect to 3′ splice site |
| 3′ splice site score | Based on 3′ flanking splice-site characteristics |
| Exon score | Position-dependent triplet frequency preference for true 3′ splice sites versus pseudo-3′ splice sites in –24 to 3 window |
| Strand score | Hexamer frequency preference for the forward strand versus the reverse strand |
| Frame score | Frame-specific hexamer frequency preference for exon versus intron in frame *i* |
| 5′ splice site score | Based on 5′ flanking splice site characteristics |
| Exon–intron transition | (Exon hexamer frequency preference in 54-bp window to the left of the 5′ splice site) – (intron hexamer frequency preference in 54-bp window to the right of 5′ splice site) |

Both of our example genes were submitted to both programs. Tables 9.4 and 9.5 show the results for *ALDH10* and *CDK10*, respectively. From Table 9.5, we can see that exons 5, 6, and 7 of the kinase gene were located exactly with MZEF. In the case of the *ALDH10* gene, MZEF predicted one exon position that located the protein-coding part of the first exon (see Table 9.4). This is interesting, as in theory MZEF would be expected to be more successful in predicting the second (internal) exon than the first (initial) exon. It did not, however, find the second exon at all. GeneMark correctly predicts exon 5 of the kinase gene and overpredicts the length slightly of exons 6 and 7 (see Table 9.5). Because exon 6 is predicted 22 bases too long (not a multiple of three) the sequence of the seventh exon will be in the wrong reading frame if these predicted exons are used to construct the whole protein-coding

| Program | Exon 1 | Exon 2 |
|---|---|---|
| **Experimental** | **1352–1762** | **2169–2400** |
| MZEF | 1601–1762 | – |
| GeneMark | 1610–1762 | 2169–2400 |
| HMMGene | 1610–1762 | 2169–2400 |
| FGENESH | 1542–1694 | 2226–2400 |
| GenScan | 1610–1762 | 2169–2400 |
| GrailEXP | 1601–1762 | 2169–2459 |
| AAT | 1607–1762 | 2169–2459 |
| GeneBuilder | 1601–1783 | 2169–2400 |
| GeneWalker (Gene 1) | – | 2226–2400 |
| GeneWalker (Gene 2) | 1601–1762 | 2086–2148 |
| TWINSCAN | 1610–1762 | 2169–2400 |
| **Average predicted** | **1596–1759** | **2169–2400** |

**Table 9.4**
**Comparison of predictions of exons 1 and 2 in the human *ALDH10* gene.** The numbers indicate nucleotide position. The Experimental row shows the real exon boundaries. Color coding is according to the main type of techniques used by the prediction programs. The bottom row shows the average predicted region from all the programs in the table.

**Table 9.5**
**Comparison of predictions of exons 5, 6, and 7 in the gene for the human cyclin-dependent protein kinase *CDK10*.**

| Program | Exon 5 | Exon 6 | Exon 7 | Comments |
|---|---|---|---|---|
| **Experimental** | **161–263** | **524–605** | **1169–1236** | |
| MZEF | 161–263 | 524–605 | 1169–1236 | Choice of organism needed |
| GeneMark | 161–263 | 524–627 | 1164–1266 | Need to choose species carefully |
| HMMGene | 161–263 | 524–605 | 1169–1279 | |
| FGENESH | 167–269 | 530–611 | 1175–1242 | |
| GenScan | 161–263 | 524–605 | 1169–1236 | Translates predicted gene |
| GrailEXP | 161–263 | 524–605 | 1169–1236 | Predicts internal, all as excellent |
| AAT | 161–263 | 524–605 | 1169–1279 | With database search |
| GeneBuilder | 161–263 | 524–605 | 1169–1236 | Translates predicted gene |
| GeneWalker | 161–263 | 524–605 | 1133–1156 | Need to register. Only deals with human genes |
| TWINSCAN | 161–263 | 524–605 | 1169–1236 | |

These predictions may look very good but that is due to the fact that the example sequence has many homologs in the databases and these may have been used in training the programs. The example used here is to illustrate how to use the methods rather than to test their accuracy, although some do perform better than others.

sequence (see Figure 9.8). If the overprediction was a length divisible by three, then the subsequent exons would be correctly translated.

With *ALDH10*, GeneMark correctly identifies the translated region of the first exon and the whole of the second exon. In addition, it predicts the first exon to be an initial exon and the second an internal exon (see Table 9.4). The length of the first exon is such that the reading frame of the second exon will be correct (that is, the difference from the correct length is a whole number of exons).

A method called FirstEF has been derived that is similar to MZEF but specifically used for predicting the first exons. This recognizes the fact that most gene-prediction programs only identify the protein-coding region and miss the UTR. When run on the *ALDH10* sequence, FirstEF predicts the first exon to be bases 1437 to 1762, which is still missing some of the UTR but is the best of the predictions reported here for this exon. It does not report which ATG codon is the translation start site.

## Some programs search for genes using only the query sequence and a gene model

Two programs both using the HMM method will be used to illustrate the prediction of complete genes with a gene model and a query sequence. These programs, HMMGene and FGENESH, differ in the way the different components of the genes are identified and the relative weights given to each.

The HMMGene program predicts whole genes and can predict several whole or partial genes in one sequence using the HMM method. HMMGene can also be used to predict splice sites and start or stop codons. If some aspects of the sequence to be predicted are known, such as hits to ESTs, proteins, or repeat elements, these regions can be locked as coding or noncoding, and the program will then find the best gene structure using these as constraints. Note that any such constraints based on sequence similarity must be obtained using other programs.

(A)

MAEPDLECEQIRLKCIRKEGFFTVPPEHRLGRCRSVKEFEKLNRIGEGTYGIVYRARDTQTDEIVALKKVR
MDKEKDGIPISSLREITLLLRLRHPNIVELKEVVVGNHLESIFLVMGYCEQDLASLLENMPTPFSEAQVKC
IVLQVLRGLQYLHRNFIIHRDLKVSNLLMTDKGCVKTADFGLARAYGVPVKPMTPKVVTLWYRAPELLLGT
TTQTTSIDMWAVGCILAELLAHRPLLPGTSEIHQIDLIVQLLGTPSENIWPGFSKLPLVGQYSLRKQPYNN
LKHKFPWLSEAGLRLLHFLFMYDPKKRATAGDCLESSYFKEKPLPCEPELMPTFPHHRNKRAAPATSEGQS
KRCKP

(B) GeneMark Prediction

5'3' Frame 3
I P I S S L R E I T L L L R L R H P N I V E L K E V V V G N H L E || S I
F L V M G Y C E Q D L A S L L E N M P T P F S E A Q V R G R G A W || D R
S S A S C C R C S G A S S I C T G T S L S T G G **Stop** Q L G R V G S ||

C) FGENESH Prediction

5'3' Frame 3
I S S L R E I T L L L R L R H P N I V E L K E V V V G N H L E R Y || V L
V M G Y C E Q D L A S L L E N M P T P F S E A Q V R || C I V L Q V L R G
L Q Y L H R N F I I H R W

D) GeneWalker Prediction

5'3' Frame 3
I P I S S L R E I T L L L R L R H P N I V E L K E V V V G N H L E || S I
F L V M G Y C E Q D L A S L L E N M P T P F S E A Q || A H P D W Y L
**Stop**

**Figure 9.8**
**Figure illustrating another way in which prediction programs can result in the wrong reading frame being used for translation.** Part (A) shows the correct gene translation of the *CDK10* gene, composed of a series of exons. Exons 5, 6, and 7 are colored red, blue, and green respectively. The resulting translations of three prediction programs for exons 5 to 7 are shown in parts (B) to (D). The predicted exons are separated in the translated protein by a || symbol. Correctly translated parts of exons are colored according to part (A). Black underlined residues are shown when the exon is predicted to extend further at the 3′ end than is correct. FGENESH predicts the correct reading frame for all three exons despite incorrectly predicting the exon start and stop positions. GeneMark, because it predicts an extension to exon 6 that is a number of bases not divisible by three, gets the reading frame wrong for exon 7. A similar result is seen for GeneWalker, but this time on predicting premature termination of exon 6.

FGENESH is a gene-prediction algorithm that is one of a set of similar methods; among these are FGENES, FGENESH, and FGENESH+. FGENES is basically a pattern-based gene predictor using dynamic programming, FGENESH is similar but uses the HMM method, while FGENESH+ also includes sequence similarity. FGENESH is based on recognition of sequence patterns of different types of exons, promoter sequences, and the polyadenylation signals that occur at the 3′ end of many protein-coding genes in eukaryotes. Using dynamic programming, the algorithm finds the optimal combination of these patterns in the given sequence and constructs a set of gene models as is illustrated for the rice genome (see Box 9.4).

HMMGene correctly predicts the location of exons 5 and 6 in the kinase gene; however, exon 7 is slightly overpredicted in length (see Table 9.5). An overprediction of 43 bases (not divisible by three) occurs in the seventh exon, which will have an effect on subsequent exons (see Table 9.2). HMMGene predicts both exons for the *ALDH10* sequence (see Table 9.4) identically to GeneMark.

FGENESH predicted the three exons of the kinase gene slightly downstream of their true locations. All are shifted by six bases, but are the correct length and therefore the correct reading frame is conserved. However, the first two amino acids of each exon are lost, but two extra residues at the 3′ end of the exon are added (see Figure 9.8C). The gene prediction for *ALDH10* by FGENESH gives the wrong length for both exons but in both cases a correct multiple of base content (MBC) is kept (see Figure 9.5). The first exon is predicted to start just upstream of an ATG codon, which happens not to be the experimentally determined translation start site. However, when the two exons are joined together and translated in all reading frames, this first ATG codon is the start of an ORF that includes all of the second exon. Scanning this predicted protein against the protein database produces high-scoring hits to *ALDH10* homologs, but only for the second part of the sequence (see Figure 9.9). This part of the sequence corresponds to the second exon, which has been correctly translated. However, by starting translation at the wrong ATG codon, the first exon is incorrectly translated and has no high-scoring homologs. This error in interpreting the first exon could be overcome in one of two ways. Alternative ATG codons could be explored, in this case revealing only one alternative—the correct translation start site—or the equivalent region of a high-scoring homolog could be used to identify similar protein-coding regions in the *ALDH10* DNA sequence.

## Genes can be predicted using a gene model and sequence similarity

As has just been illustrated with the predictions of FGENESH, known sequences can prove very useful in helping to predict new genes. Various sequence databases can be used to augment the *ab initio* predictions. For some organisms, databases of experimental ESTs and cDNAs can be used to identify protein-coding regions, introns, and even alternative splicing. While one might hope to find exact sequence matches, experimental techniques (especially for ESTs) give significant sequencing errors, leading to potential ambiguities. Where sequences are not available for the same organism as the query sequence, use of homologous sequences can still be of benefit.

The program GenScan (see Section 10.5) is based on a probabilistic model of human genomic sequences. The model used by GenScan approximates widespread structural and compositional features of human genes such as common promoter elements, exons, introns, and splice sites. Specialized features specific to very few genes are omitted, but general features, such as the presence of the promoter TATA-box sequence, are included. For further details on the gene model used by GenScan see Figure 10.19. This means that although the model was designed for human genes, it should model any eukaryotic DNA equally well with suitable parameterization. GenScan uses this model to predict the locations of exons in genomic DNA sequence. GenScan can predict the presence of more than one gene in a sequence, and can deal with incomplete genes. It also analyzes both of the DNA strands, and thus can predict the presence of genes on either strand.

With GrailEXP, a list of the most likely protein-coding regions in the submitted genomic DNA sequence is predicted by integrating pattern-recognition information of the type discussed earlier with information on those regions with similarity to EST sequences. The outcome of this analysis is then evaluated using a **neural network method** to calculate the likelihood that these coding regions are true exons. GrailEXP is a versatile program suite that can also be used to locate

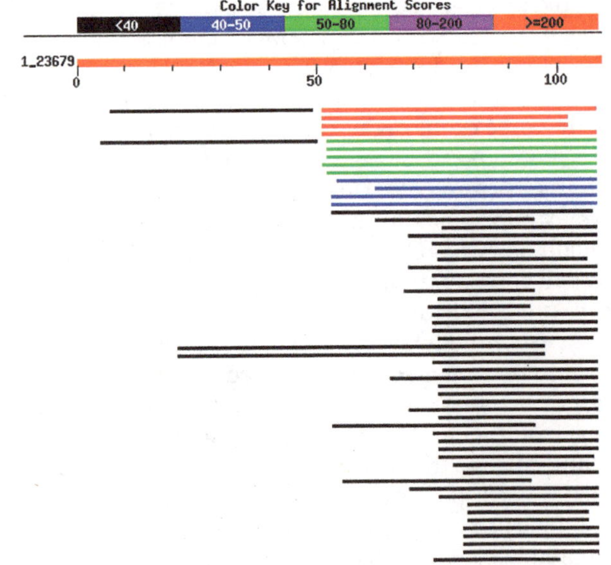


**Figure 9.9**
**A combination of prediction errors described and illustrated in Figures 9.5 and 9.8.** For the *ALDH10* gene, FGENESH predicts the wrong length for both exons but in both cases a correct multiple of base content (MBC) is kept (see Figure 9.5). The first exon is incorrectly predicted to start just upstream of an ATG codon (see Figure 9.5), resulting in the wrong reading frame for the first exon. However, if the predicted coding region is translated in all reading frames and submitted in a database search, the correct product can sometimes be identified. In the search results shown here the reading frame used was such that the query sequence has the correct translation of the second exon. As a consequence, some database hits are reported that have a significant score, and only align with that region.

## Box 9.4 FGENESH and the rice genome

Rice is one of the most important crops in the world, feeding a large proportion of the global population. The program FGENESH was one of the programs used to predict and annotate the rice genome. It was found to be very successful, generating 53,398 gene predictions with initial and terminal exons.

The quality of the predictions was measured by using base-level specificities and sensitivities (see Section 9.6) which showed that false-negative predictions were more likely to be a problem than false-positive predictions. In other words FGENESH was more likely to miss an exon fragment than to label something as part of an exon incorrectly. Exon-level sensitivities and specificities were much worse, meaning that the exon–intron boundaries were not precisely delineated, even though the existence of a gene was correctly detected.

The predicted genes were then compared to the *Arabidopsis* genome to obtain some of the annotations: 80.6% of predicted *Arabidopsis* genes had a homolog with rice, but only about 50% of the predicted rice genes had a homolog in *Arabidopsis*.

**Figure B9.4**
**Flowering rice.** (Courtesy of A. Davis, Bioimaging JIC.)

EST/mRNA alignments, certain types of promoters, polyadenylation sites, CpG islands, and repetitive elements.

The AAT (Analysis and Annotation Tool) Web site has a program that combines database searches with gene predictions. It will work best on human sequences, because the gene-sequence statistics were collected from human DNA. Basically it identifies genes in a DNA sequence by comparing the query sequence against cDNA (GAP2 program) and protein sequence (NAP program) databases. Then the prediction program combines the matches with gene-prediction results. The gene-location algorithm, called GSA2, computes exon sequence statistics using the method of the MZEF program.

GeneBuilder predicts functional DNA sites and coding regions using several approaches, which it then combines with similarity searches in protein and EST databases. The models of the gene structure are obtained using the dynamic programming method (see Section 5.2). A number of parameters can be used for the prediction. Different exon homology levels with a protein sequence can improve the prediction.

GeneBuilder has two options for predictions: the predict exon option and the gene model option. The gene option is used for predicting the full gene model while the exon option is used to select the exons with the best scores. We have run the predictions on the kinase and *ALDH10* genes using both modes. Using the gene option for *ALDH10*, only one exon is predicted, which spans the protein-coding segment of the first exon and is in the correct frame. The exon option predicts seven potential exons. Four predicted exons are before the first exon, the fifth and sixth predicted exons span the protein-coding region of exon 1, and the seventh predicted exon correctly identifies exon 2 (as shown in Table 9.4 and discussed below). GeneBuilder correctly identified all three exons of the kinase gene in the gene mode (as shown in Table 9.5) but the exon mode predicted five exons, of which none was correct.

GeneWalker is a Web-based program developed by Japan Science and Technology Corporation. It locates general exonic regions in human DNA by analyzing the DNA sequence for tell-tale signals such as TATA boxes. In addition, it uses similarity matching to look for sequence fragments that may delineate exon-specific sites such as start and stop codons, and 5′ and 3′ splice sites. The program then calculates a coding potential (CP) at each nucleotide position, on the basis of a statistical analysis of the local segment. The CP then gives the likelihood that the nucleotide forms part of a coding region.

Subsequently, GeneWalker predicts the likely promoter, exon, or terminator regions, and compiles this information into a global prediction of the DNA sequence. The GeneWalker Web viewer gives the final predictions and other information predicted by the program, such as likely signal sequences, start and stop codons, and splice sites.

Identical correct predictions of 5, 6, and 7 exons were obtained for the kinase gene by all the above programs, with the exception of AAT and GeneWalker, which have errors in the seventh exon prediction. AAT predicts the seventh exon to extend 43 bases downstream, but is otherwise correct. GeneWalker predicts a very short (24 base) exon that is just upstream of the experimental seventh exon (see Figure 9.8D).

These programs do not all predict the same start for the first exon of *ALDH10*, but they all start at or just before the correct translation start site. All, apart from GeneBuilder, correctly predict the 3′ end of the first exon, but GeneBuilder predicts 21 extra bases, giving rise to an extra seven amino acids in the predicted protein. All the programs predict the first exon to end in the same frame and, apart from GeneWalker, all correctly predict the start of the second exon. GenScan and GeneBuilder correctly predict the second exon, but AAT and GrailEXP predict an extra 59 bases at the 3′ end. As well as predicting another 19 amino acid residues in both cases, this is likely to lead to problems with the third exon frame. GeneWalker predicts two possible genes from the *ALDH10* DNA sequence (see Table 9.4). The first is a gene consisting of only one exon, which contains the region of the second true exon (2226–2400) and is identical to the FGENESH prediction. The second GeneWalker gene prediction consists of three exons, of which the first lies upstream of the first real exon (1136–1344), the second spans the region of the protein-coding region of the *ALDH10* exon (see Table 9.4), and the third (from nucleotides 2086–2148) is predicted just before the location of the second real exon (2169–2400).

## Genomes of related organisms can be used to improve gene prediction

Sequences with a functional role, which currently are thought to constitute only about 5–10% of the human genome, will be conserved in evolution, whereas other sequences will usually not be. There are exceptions; for example, some segments of noncoding DNA (also referred to as junk DNA) exceeding 200 bases in length are perfectly conserved across several eukaryotic species. Examples of these are some of the genetic elements known as **transposons**. Initially it was thought that transposon function was purely selfish. However, recent research has shown that transposons play an important role by their ability to change the way DNA code is edited before translation by introducing new splice sites into a gene (see Box 9.5). As a consequence of these findings there is active research into the functions of these elements.

When comparing equivalent regions of the genomes of two related organisms, sequence segments showing similarity can be taken to indicate that these regions have a function. Exons and promoters, for example, would be expected to share similarity to a greater extent than introns. This observation has been exploited by a number of recent prediction programs to improve gene-prediction accuracy. Note that as well as requiring two genomic sequences, this approach assumes that the gene is present in both. TWINSCAN is an example of this type of program and has been used here to predict the *CDK10* and *ALDH10* genes. Both predictions were carried out using the human and mouse genomes. The kinase exons are all correctly predicted, as well as identified as internal genes. TWINSCAN successfully

## Box 9.5 Transposons and repeated elements

Transposons (sometimes known as jumping genes or selfish DNA) are segments of DNA that can move around to different positions in the genome of a cell. In the process, they may cause mutations or increase or decrease the amount of DNA in the genome.

The first transposons were discovered in the 1940s by Barbara McClintock. She worked with maize (*Zea mays*) and found that transposons were responsible for a variety of gene mutations with unusual behavior. She was awarded the Nobel Prize in 1983 for her discovery.

One of the most highly repeated elements in the human genome are the Alu elements, which are relics of a type of transposon known as a retrotransposon, which is derived from RNA. HIV-1—the cause of AIDS—and other human retroviruses act similarly to retrotransposons, inserting a DNA copy of their RNA genome into the host cell DNA. Alu elements are common in introns and some are recognized by the spliceosome and can be spliced into mRNA, thus creating a new exon that will be transcribed into a new protein product. As alternative splicing can provide not only the new mRNA but also the old, nature can test new proteins without getting rid of the old ones (see Figure B9.5).

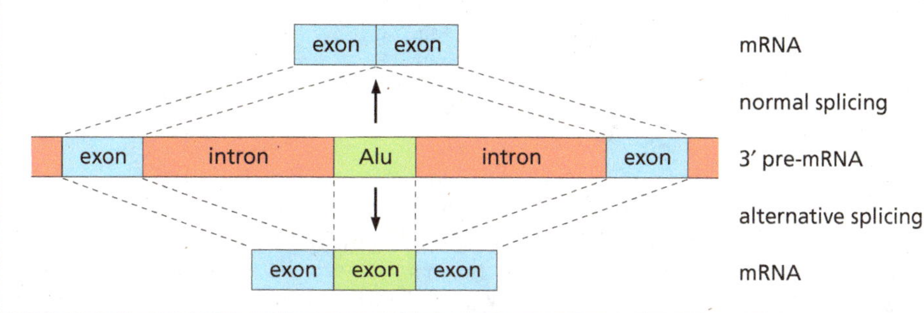


**Figure B9.5**
**Intronic Alu elements, which can be spliced into mRNA creating a new exon.** (Adapted from J.W. Kimball, http://users.rcn.com/jkimball.ma.ultranet/BiologyPages/T/Transposons.html)

predicts the protein-coding region of the first exon and second exon in the *ALDH10* gene, as well as identifying the first exon as an initial exon.

The prediction programs described in this chapter were chosen because of their easy availability and to demonstrate the use of various gene-finding schemes. These programs change in algorithms and parameterization, and databases are updated daily, so that the results given here may not be reproducible when repeated after publication of this book. These results should be taken as examples that illustrate general principles, and not of the success or failure of specific methods. More and better programs, often with more user-friendly interfaces, are continually becoming available.

## 9.4 Splice Site Detection

A full characterization of most eukaryotic genes requires a prediction of their splicing structure; that is, the way their RNA will be spliced to remove introns and produce a continuous coding sequence (see Flow Diagram 9.4). As splice sites, by definition, demarcate intron–exon boundaries, some of the exon-prediction programs discussed earlier incorporate splice-site information to help locate exons. There are some programs, however, that are specifically designed to recognize splice sites. In addition, one will want to predict at which end of the exon the splice junction is located. In other words, is it an exon–intron (donor) or an intron–exon (acceptor) boundary (see Figure 9.10).

Examples of the sequence signals for these splice sites are shown in Figure 10.11. The signal is dominated by dinucleotides in both cases, with weak signals extending

**Flow Diagram 9.4**
**In this and the following section some of the practical aspects of splice site and promoter prediction are presented.**

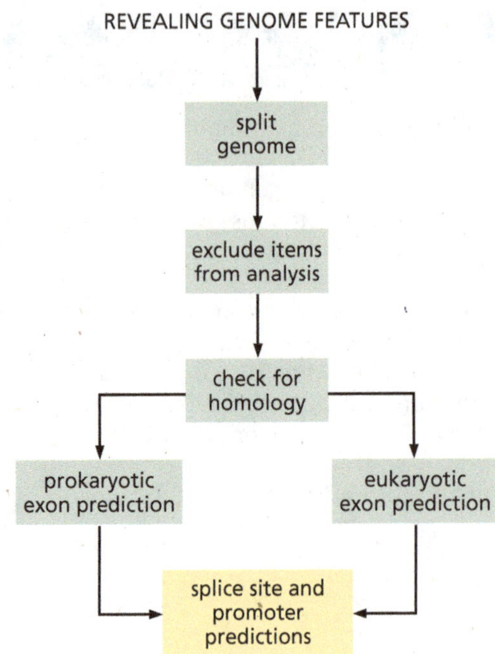


mostly on the intron side. If one tries to identify splice sites solely on the basis of the GT and AG dinucleotides this will lead to so many false-positive predictions that it will be impractical. For this reason, more sophisticated methods to identify splice sites have been devised. Multilayer neural networks (discussed in Section 12.4), decision trees, and position-specific oligonucleotide counts have all been used.

### Splice sites can be detected independently by specialized programs

Four splice-site recognition programs were used to predict splice sites for our *ALDH10* sequence. The results are illustrated in Figure 9.10. The program SpliceView has an exceptionally user-friendly interactive splice-site analysis tool (see Figure 9.11). It illustrates the predicted regions as a bar chart. The user can then choose the threshold above which to view the splice sites (in Figure 9.11 only sites that are excellent have been chosen). The exact position for each site can then be found interactively.

## 9.5 Prediction of Promoter Regions

Many of the programs used above to predict exons also incorporate information about the location of the core promoter regions in the vicinity of the transcription start site (TSS). Some programs focus exclusively on these and other promoter

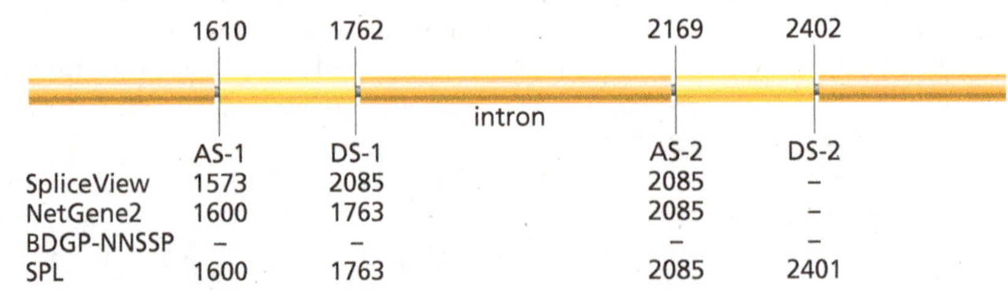


**Figure 9.10**
**A schematic of acceptor and donor sites (AS and DS, respectively) and intron–exon, exon–intron boundaries on a DNA sequence.** The top numbers give the true exon 1 and 2 nucelotide positions of the *ALDH10* gene, while the numbers at the bottom give the splice prediction sites. The predictions are shown for those splice sites that had a reasonable score and fitted with the exon prediction as given in Table 9.3. A "–" indicates no satisfactory prediction was found.

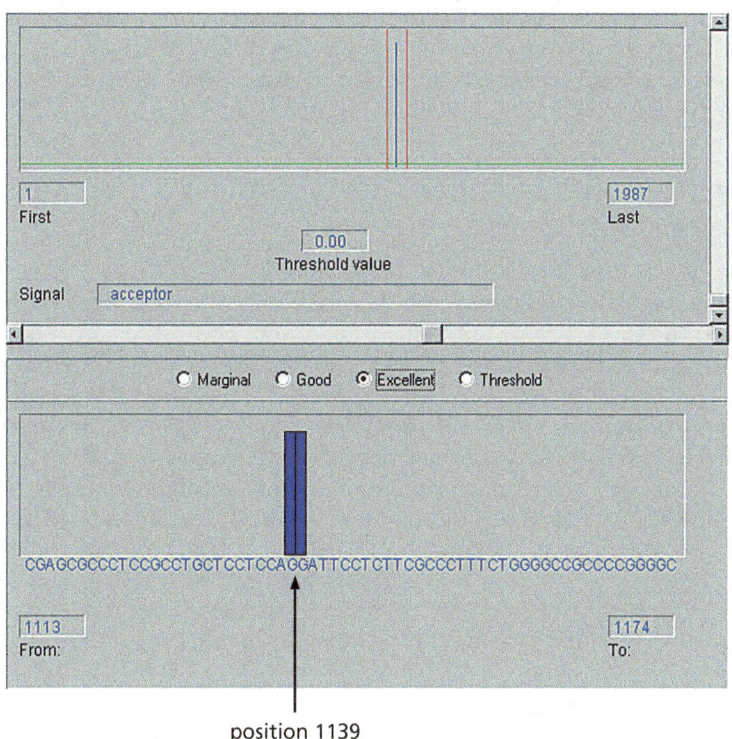


**Figure 9.11**
A snapshot illustration of the interactive interface provided by the SpliceView program.

signals. In addition to predicting the TSS, some of these programs can suggest regions involved in controlling expression of these particular genes. The independent prediction of these promoter regions can help analyze information about gene or protein expression gained experimentally.

Promoters are the regions of DNA where RNA polymerase and associated general transcription factors bind to initiate the transcription of a nearby coding sequence. For prokaryotic genes, and for genes transcribed by RNA polymerase II in eukaryotes, a core promoter site is typically a short distance upstream from the transcription start site. The sequence patterns are better defined for the core promoter regions and a database of eukaryotic promoters for RNA polymerase II has been created—the Eukaryotic Promoter Database (EPD). In the case of eukaryotic genes there are additional, more extensive promoter regions that may cover thousands of base pairs and contain binding sites for numerous other gene-regulatory proteins that determine when and where the gene will be switched on in the organism.

## Prokaryotic promoter regions contain relatively well-defined motifs

In prokaryotes such as *E. coli*, the RNA polymerase is composed of two functional components: the core enzyme and a so-called sigma factor ($\sigma$). The sigma factor plays an important part in promoter sequence recognition. There are several different sigma factors, each recognizing specific subsets of promoters that have different nucleotide sequences.

Comparison of *E. coli* promoters identified four conserved patterns that can be used to find promoter regions in most prokaryotes: the Pribnow box, which is a TATAAT segment at position –10 from the transcription start site; a TTGACA box at position –35; a somewhat conserved transcription start site; and an AT-rich region before the TTGACA box. These conserved patterns can be used to predict promoter regions. These are similar to the core promoters identified in eukaryotes but have slightly different consensus sequences and positions. In addition, there are some prokaryotic promoters found within a few hundred bases upstream of the gene that are involved in the control of specific genes or classes of genes. These would form part of a system that could, for example, cause expression of a set of genes in response to external conditions such as temperature.

## Eukaryotic promoter regions are typically more complex than prokaryotic promoters

Promoter prediction in eukaryotes is more difficult, especially as some of the regions that regulate eukaryotic gene expression can be thousands of nucleotides away from the transcription start site. In addition, control regions may be found either upstream or downstream of genes. Therefore, locating all the control sites for a eukaryotic gene is no simple matter. There has, however, been some success in predicting the well-characterized promoter regions associated with genes transcribed by RNA polymerase II, which comprise the majority of the protein-coding genes. In addition, eukaryotic promoters contain multiple elements or control regions, binding various transcription factors, and these elements are often found in more than one promoter region.

Various methods have been explored for the prediction of promoter regions. Some are concerned with recognizing the regions of the core promoter of genes transcribed by RNA polymerase II, such as TATA boxes, CAAT boxes, and the transcription initiation region (Inr). Other methods extend this by also looking for combinations of other common gene-regulatory motifs.

## A variety of promoter-prediction methods are available online

This section will briefly note some promoter-prediction programs that are accessible online. Some methods, which can be used for both prokaryotes and eukaryotes, use a score or weight matrix to search and score a sequence for segments that are specific for promoter regions, such as the Pribnow box in prokaryotes. It should be noted, however, that different matrices must be used for different groups of organisms. The principles underlying the application of weight matrices in eukaryotic promoter prediction is dealt with in more detail in Section 10.5. For example, the FunSiteP algorithm uses a weight matrix to predict promoters by locating putative binding sites for transcription factors. In addition to giving the predicted site of the promoter, by detecting different sets of binding sites, the program predicts which class the promoter will belong to as classified in the EPD. This is given in the last column of the output file from FunSiteP. For example, the number 6.1.5 corresponds to promoters of regulatory proteins in the EPD.

Neural network methods have also been used to predict promoter regions. The network is trained to distinguish between promoter and nonpromoter sequences. The algorithm NNPP (Neural Network Promoter Prediction), which is discussed in Section 10.5, was developed for the Berkeley *Drosophila* Genome Project (BDGP), and uses time-delay neural network architecture and recognition of specific sequence patterns to predict promoters (see Figure 10.14 for details). The Promoter 2.0 algorithm uses a combination of perceptron-like neural networks and genetic algorithms to optimize the weights used in the neural networks.

TSSG and TSSW algorithms predict potential transcription start sites (TSSs) using the linear discriminant method. Predictions are based on characteristics describing functional motifs and oligonucleotide composition of the promoter sites. The difference between TSSG and TSSW is that the model composition of the promoter sites is based on different promoter databases. CorePromoter is also based on discriminant analysis (in this case, quadratic discriminant anlysis) of human core-promoter sequences (see Figure 10.16). It is similar in concept to the MZEF program described in Section 9.3.

The ProScan promoter-prediction program is based on the analysis of promoter-specific binding sites that are also found by chance in nonpromoter regions. The ratio of the densities of these sites in promoter and nonpromoter sequences has been calculated from experimental data, and these ratios were then used to build a scoring profile called the Promoter Recognition Profile. ProScan uses this profile, in combination with a Phillipe Bucher weighted scoring matrix for a TATA box (see Figure 10.12), to differentiate between promoter and nonpromoter regions. This program has been parameterized using only primate data, and therefore may not perform as well on data from other species.

As is shown in Section 10.5, there is a biased distribution of short oligonucleotides in regions containing promoters relative to other regions of the genome. These are used in the program PromoterInspector to identify the location of promoter regions, achieved by sliding a window along the DNA sequence and classifying its content. The distribution of a number of these oligonucleotides is used to predict the site of promoters but does not identify the strand involved.

## Promoter prediction results are not very clear-cut

The promoter-prediction methods discussed above are available on the Web and have been run on our example sequence from *ALDH10* (the kinase example does not contain the promoter sequence). The results are given in Table 9.6. A promoter prediction was considered correct if it was within 200 bp 5′ or 100 bp 3′ of the experimentally determined transcription start site. The predictions shown in Table 9.6 are those with the highest scores. Only TSSW, ProScan, and PromoterInspector fall within the criteria for a correctly predicted promoter. FunSiteP predicts 14 possible promoter sites, all of which fall within the criteria, but all were defined as low scoring.

Even though some of the programs failed to find the correct promoter region in this example, they should not be dismissed. All the methods described have strengths as well as weaknesses, and will perform better on certain types of DNA than on others. For a raw genomic DNA sequence, a range of programs should be used, and the consensus used as the best prediction.

Although the prediction of promoter regions may not necessarily help in gene identification, putative promoter sites may be used in the analysis of gene or protein expression experiments. Genes or proteins that show similar expression may be controlled by the same regulatory proteins. To test this hypothesis, the promoter regions of the genes in question can be predicted and inspected for similarities. Similar promoter regions may indicate coregulation. For example, operons (sets of adjacent genes in bacteria that form a single transcriptional unit) such as those coding for flagella, have been implicated in pathogenicity. Virulence factors of pathogenic bacteria such as toxins or invasins can be encoded by specific regions

| Promoter predictions | TSS 1416 |
|---|---|
| BDGP-NNPP | 444–494 |
| CorePromoter | 1034 |
| TSSG | 0 |
| TSSW | 1372 |
| ProScan | 1078–1328 |
| FunSiteP | 14 sites |
| Promoter 2.0 | 800 & 1200 |
| PromoterInspector | 1205–1400 |

**Table 9.6**
**Comparison of the promoter prediction in the *ALDH10* sequence.** The numbers indicate the nucleotide position of the predicted start site. The experimentally determined transcription start site (TSS) is given in white at the top.

of the prokaryotic genome termed "pathogenicity islands." For example, more than 10 pathogenicity islands have been recognized in *Salmonella*.

Similarly, if a set of proteins is coexpressed, then the proteins can be identified, the sequences reverse translated into DNA, the genes located, and the upstream regions analyzed for similar promoter sequences. This is an especially effective approach in prokaryotes, where the promoter regions are more easily identified. In eukaryotes, where promoter regions are not easily identified, one can analyze the upstream sequence for common patterns.

## 9.6 Confirming Predictions

It is important to be able to confirm gene predictions or have an idea of their accuracy. In this section we first describe how the prediction accuracy of a specific program can be measured; this is particularly useful information for potential programmers but also for users, to have a general idea of the quoted accuracy of the program of their choice. Subsequently, we look at how translation of the gene product can confirm a correct gene prediction (see Flow Diagram 9.5).

### There are various methods for calculating the accuracy of gene-prediction programs

Authors often give a numerical value for the accuracy of their prediction program and it is important to have at least some idea of what these numbers mean. Being able to calculate an accuracy measure that can be compared to other programs is important for program developers, especially when automated annotation of genomic sequences is being considered. This is because the steps during automatic annotations cannot be checked by a user and running automatic annotation implies annotation on a grand scale making user-checks more difficult.

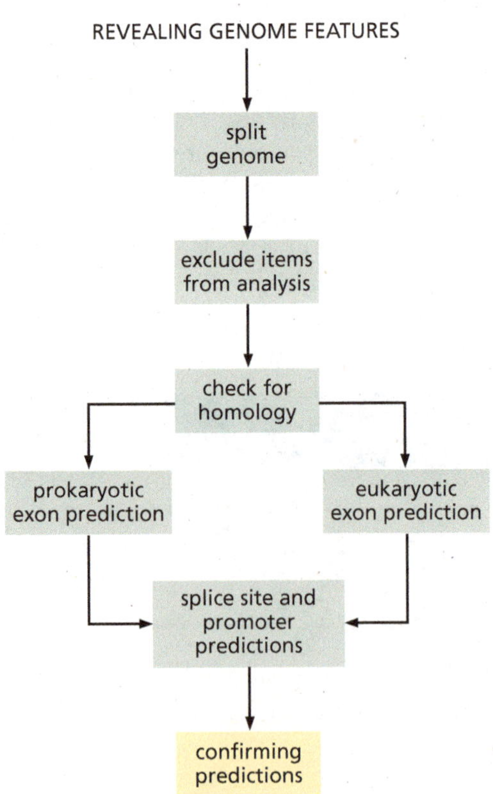


**Flow Diagram 9.5**
**In this section some of the methods that can be used to check the predictions of genes and control signals are described.**

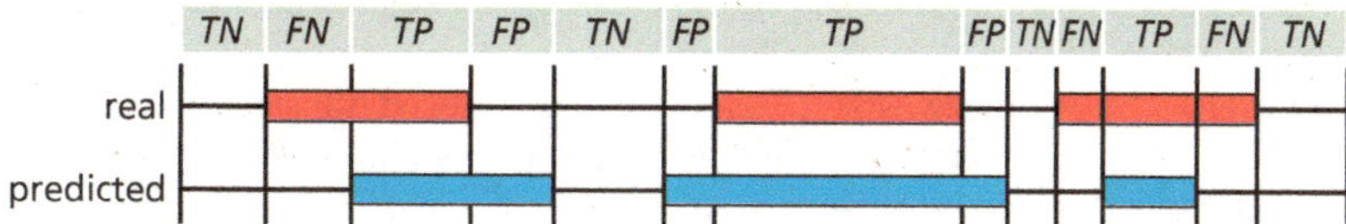


**Figure 9.12**
**The various ways a prediction can be defined.** The figure illustrates the meaning of true negative (*TN*), true positive (*TP*), false negative (*FN*), and false positive (*FP*) predictions based on the overlap of predicted versus true genes. Exons are said to be correctly predicted when the overlap between the actual exon (red) and the predicted exon (blue) is greater than or equal to a specific threshold. It also shows how even when an exon is correctly located, part of it can be incorrectly predicted; for example, it can be underpredicted or overpredicted on either end, or shorter than the actual exon.

The accuracy of exon predictions can be measured at three different levels: coding nucleotide (base level); exonic structure (exon level); and protein product (protein level). At the nucleotide level, prediction accuracy for each nucleotide can be defined by discriminating between true positive (*TP*), true negative (*TN*), false positive (*FP*), and false negative (*FN*) predictions (see Box 10.1). Figure 9.12 shows some examples of the nucleotide-defined prediction terms. These measure the accuracy by comparing the predicted coding value, whether coding or noncoding, with the actual coding value for each nucleotide. These are expressed by **sensitivity** (*Sn*) and **specificity** (*Sp*).

At the exon level, determining prediction accuracy depends on the exact prediction of exon start and end points. Here, *Sn* is the proportion of the real exons that is correctly predicted while *Sp* is the proportion of predicted exons that is correctly predicted. At the level of protein sequence, prediction accuracy can be measured as percent similarity between the translated predicted gene product and the actual gene product. This, however, depends on obtaining a correct translation frame (see Section 9.3).

*Sn* and *Sp* values are often quoted on papers and Web sites describing a specific exon-prediction program. In this chapter we have been using qualitative descriptions of the prediction and the translated products rather than these values, as this chapter deals mainly with the user's view of gene prediction, and the user, working with unknown sequences, cannot calculate the *Sn*, *Sp*, or any of the above values. However, the user should be aware of the meaning of these values, and the future program developer should be able to calculate them by reference to how well a program works on known genes.

## Translating predicted exons can confirm the correctness of the prediction

The most reliable method of checking the accuracy of exon prediction is by detection of homology of the resulting protein sequence to sequences already present in the protein-sequence databases. Any predicted exon must be translated to a protein sequence for confirmation of the correctness of the prediction. Translating predicted exons into protein sequence, using one of the programs available on the Web, will give segments of protein sequence that can be pooled together and run using a sequence search program against a protein database. For example BLASTX translates the nucleotide sequence in all six reading frames and then automatically compares the translated products against a protein sequence database. Figure 9.13 illustrates the result of a search against a correct exon-to-protein translation. All the significant matches are spanning the same region of the sequence and the proteins found by the test sequence are homologs. Any DNA sequence can, in principle, be translated in three different reading frames, and also in both the 5′ to 3′ and 3′ to 5′ direction; to allow for the fact that the coding sequence may be on the complementary DNA strand, you should test each option.

## Constructing the protein and identifying homologs

Once the promoters, possible introns, exons, and splice sites have been predicted, it is necessary to combine these results and predict the protein sequence that would be expressed. A generally useful exercise is to take the most commonly occurring exon predictions as correct, or an average of the predictions when there

**Figure 9.13**

***CDK10* gene translation searches.** (A) The translated *CDK10* exons are given in a table with the direction and frame of translation. (B) This illustrates the BLAST search results through the Swiss-Prot database of the spliced translated exons. All the searches are significant (red colored lines) and all are to CDK-like kinases.

(A)

| EXON | Dir | Frame | Translated amino acid sequence |
|---|---|---|---|
| 1 | 5'3' | 3 | MAEPDLECEQIRLKCIRKEGFFTVPPEHR |
| 2 | 5'3' | ? | |
| 3 | 5'3' | 3 | LGRCRSVKEFEKLNRIGEGTYGIV |
| 4 | 5'3' | 3 | RARDTQTDEIVALKKVRMDKEKD |
| 5 | 5'3' | 3 | RHPHSSLREITLLLRLRHPNIVELRRWLWGTTWR |
| 6 | 5'3' | 2 | IFLVMGYCEQDLASLLENMPTPFSEAQ |
| 7 | 5'3' | 1 | VKCIVLQVLRGLQYLHRNFIIH |
| 8 | 5'3' | 2 | RDLKVSNLLMTDKGCVKT |
| 9 | 5'3' | 3 | ADFGLARAYGVPVKPMTPKVVTL |
| 10 | 5'3' | 2 | YRAPELLLGTTTQTTSIDM |
| 11 | 5'3' | 2 | WAVGCILAELLAHRPLLPGTSEIHQIDLIVQLLGTPSENIWP |
| 12 | 5'3' | 1 | GFSKLPLVGQYSLRKQPYNNLKHKFPWLSEAGLRLLHFLFMYDPKK |
| 13 | 5'3' | 2 | RATAGDCLESSYFKEKPLR |
| 14 | 5'3' | 3 | PCEPELMPTFPHHRNKRAAPATSEGQSKRCKP |

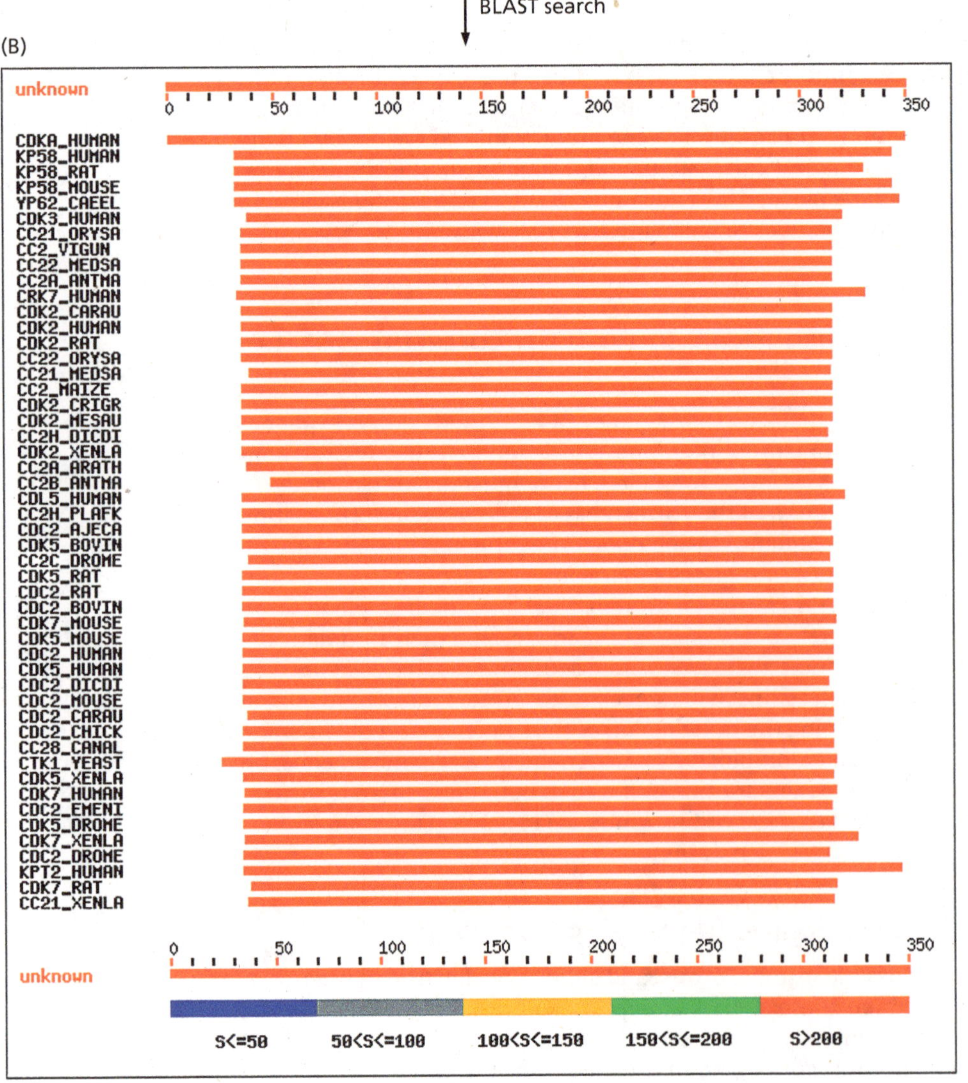

is no clear preference (for example, in *ALDH10* exon 1 this would be 1610 or 1601 to 1762), and to calculate an average or a consensus prediction of all the hits for the promoter region. From our results for *ALDH10* this consensus gives us an exon 1 from position 1610 to 1759, and an exon 2 from 2169 to 2400. Comparing this with the experimentally determined exons it is quite a good prediction, with a putative promoter around position 1289. That is all very well, but we still, in principle, do not know what the protein is, what it does, and if the exons are correct.

The next step is to take the predicted exons and run them through the various translation programs available. The translation should be performed for all three reading frames and in both directions (5′ to 3′ and 3′ to 5′). Easy-to-use translation programs that perform all six translations are found on the ExPASy site, and TRANSLATOR is on the JustBio site. Figure 9.14 shows the result of the Frame 3 translation of the first predicted *ALDH10* exon in the 5′ to 3′ direction, using the ExPASy translator. Once the translations have been obtained, the ones that contain the least number of stop codons (if any) are submitted to a database search program. Here we have used the default BLAST. The Frame 3 translation BLAST search through the Swiss-Prot database found the correct protein (ALDH) and its close homologs as the highest-scoring hits (see Figure 9.14).

Thus from having an unknown piece of DNA, we now know that the region predicted as exon 1 is at least partially correct, and that the gene is most likely to encode a member of the fatty aldehyde dehydrogenase family of proteins. It may now be possible to delineate the exact exon position by taking the protein sequence of the homolog and using the program GeneWise to align this with our *ALDH10* DNA sequence. GeneWise allows the comparison of a DNA sequence to a protein sequence. This comparison enables the simultaneous prediction of gene structure (exons and introns) with homology-based alignment. GeneWise uses a gene-prediction model and an HMM of a protein domain to do this.

**Figure 9.14**
***ALDH10* translation and searches.** The DNA sequence of the exon prediction for the first exon of *ALDH10* is taken and submitted to translations, of which the correct result is shown. This translated segment is then sent off for a homology search through a protein database (in this case BLAST through the Swiss-Prot database). The best-scoring hits are shown, colored according to their alignment scores S, as given by the key along the base of the figure.

atggagctcgaagtccggcgggtccgacaggcgttcctgtccggccggtcgcgacctctgcggtttcggc
tgcagcagctggaggccctgcggaggatggtgcaggagcgcgagaaggatatcctgacggc

translate (5′3′Frame3)

A MetELEVRRVRQAFLSGRSRPLRFRLQQLEALRRMetVQEREKDILT

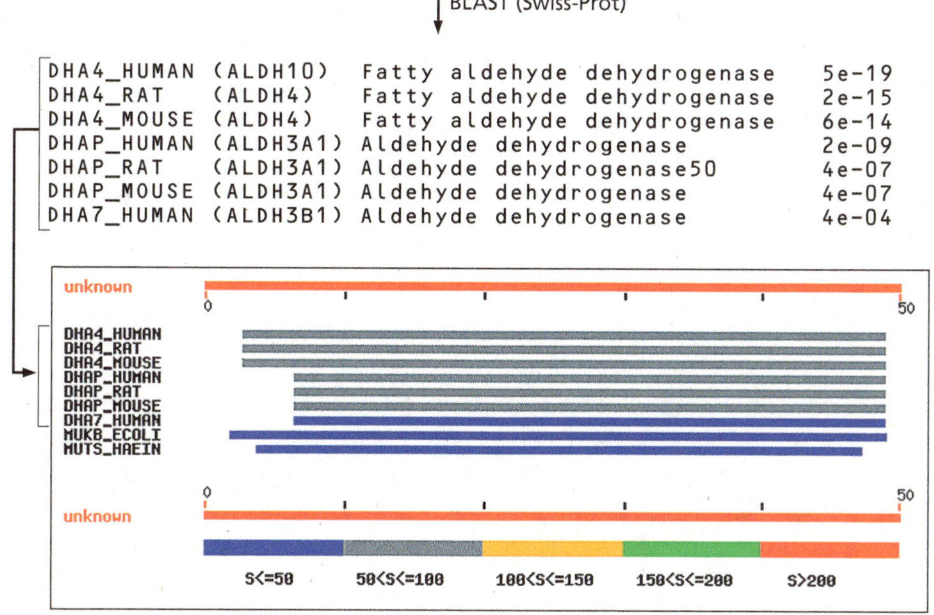

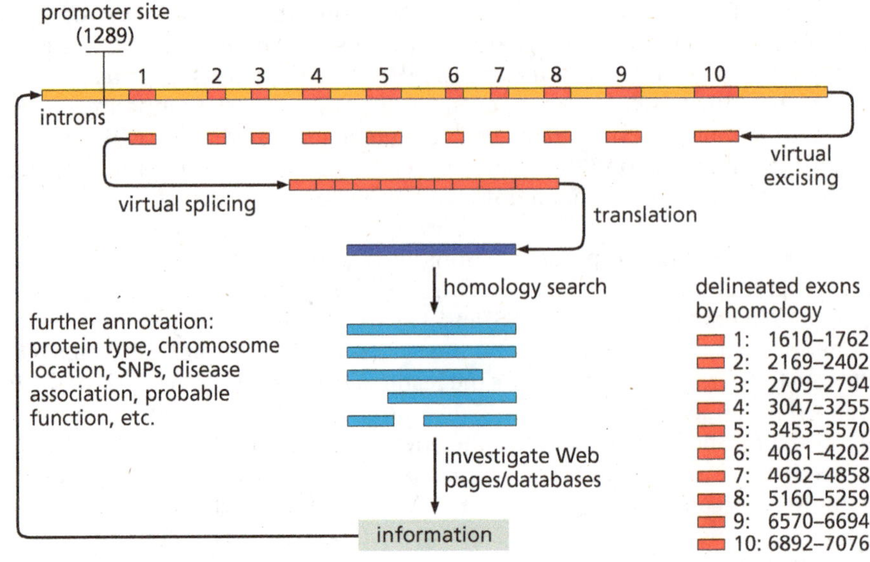


**Figure 9.15**
**Annotating the gene.** A schematic of the annotated gene-structure DNA sequence of *ALDH10* and illustration of further steps to be taken to obtain more information for a more detailed annotation(such as single nucleotide polymorphism (SNP) and chromosome location) once the gene structure has been predicted.

We chose the rat protein homolog to compare with our target *ALDH10* DNA sequence. The first exon is now delineated precisely—from 1610 to 1762—and the second exon from 2169 to 2400. Submitting the whole *ALDH10* DNA sequence consisting of 10 exons (7709 bp) for alignment with the rat protein homolog provides a complete exon–intron map (see Figure 9.15). The accuracy of such an analysis will depend on the degree of similarity between the protein homolog found by BLAST and the gene under investigation. Thus, if the only homolog found was a betaine aldehyde dehydrogenase from *E. coli*, then the GeneWise analysis would not delineate the correct exon. Comparison of the *E. coli* protein with exons 1 and 2 of human *ALDH10* finds a putative exon in the region 2301 to 2402, which does span the 3′ end of the true second exon, but it does not give a clear-cut answer such as one obtains with rat fatty aldehyde dehydrogenase.

A typical series of steps in eukaryotic gene prediction and confirmation is shown in Table 9.7.

## 9.7 Genome Annotation

Once all the genes have been predicted, it remains to determine what function the encoded proteins might play. The obvious way to start to determine gene function is by sequence analysis of the kind discussed in Chapter 4, although often that step will be part of the exon (gene) identification method. If they have significantly

**Table 9.7**
**A typical series of steps in eukaryotic gene prediction.**

| | |
|---|---|
| 1 | Submit DNA sequence to exon prediction programs |
| 2 | Take average or consensus exon prediction |
| 3 | Translate predicted exon into protein sequences in all frames and directions |
| 4 | Take translation with least or no stop codons |
| 5 | Search protein database with the translated segment |
| 6 | If hit found use protein homolog in GeneWise to delineate exon(s) |
| 7 | Repeat for all other exons |
| 8 | Annotate and splice exon to obtain putative protein sequence |

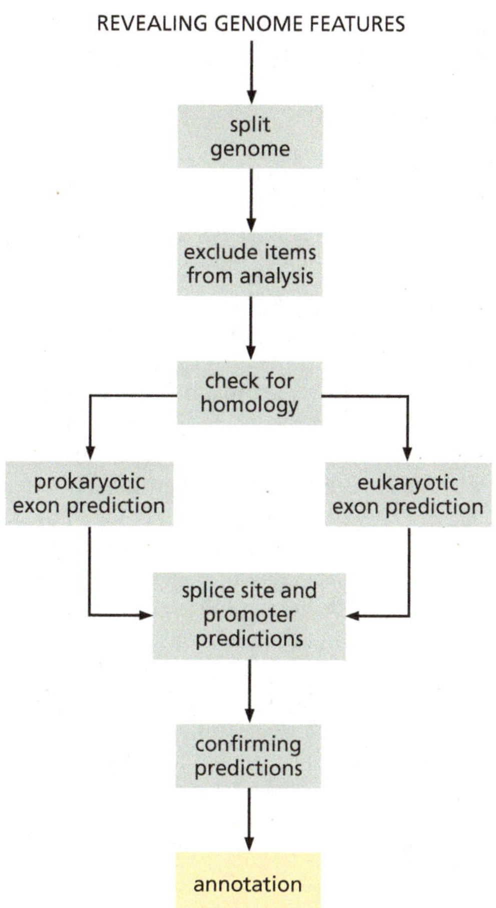


**Flow Diagram 9.6**
**In this section the technique of genome annotation is described to explain how the prediction methods presented in this chapter are complemented by other work.**

scoring hits in searches against sequence and pattern databases, the sequences can be predicted with considerable confidence to have similar function and other properties. In all genomes to date, however, many gene products do not yield to such analysis, and other steps are required. (In any case, the similarity may be to only a part of the protein, such as a single domain.) In most cases, the only realistic way of assigning functions is by further experimental work. Thus, there is often a large effort following sequencing to discover the phenotype on disrupting an individual gene. The phenotype may be described by gross features, such as overall cell size, but often will be in terms of the different gene or protein expression as compared with the wild type. Such experiments can also demonstrate linked control with other genes whose function is better understood. Such methods can yield gene functions at several levels of description.

## Genome annotation is the final step in genome analysis

There is more to genomes than just the genes, and a full genome annotation must cover all the aspects (see Flow Diagram 9.6). The location of tRNA genes has already been discussed, and the rRNA genes also need to be located. In addition, there are repeat sequences that can be of interest and can have functional roles. A significant fraction of almost any genome is repetitive. Repetitive sequences fall primarily into three classes: local repeats (tandem repeats and simple sequence repeats), families of dispersed repeats (mostly transposable elements and retrotransposed cellular genes), and segmental duplications (duplicated genomic fragments). Repetitive sequences are so numerous that simply annotating them well is an important problem in itself. There are a few Web-based packages available to identify repeat sequences within a genome; RECON and RepeatMasker are good examples, and both programs are largely based on alignment procedures.

Beyond this, there are features that encompass several genes, such as operons and pathogenicity islands. In addition, other genes may show signs of recent lateral transfer. It is beyond the scope of this book to cover the methods by which these and other features can be identified.

In addition, the use of pathway information can aid gene and genome annotation, especially in later stages, when after a first round of annotation many proteins have been identified as hypothetical. These are predicted gene products for which no function or identification is yet possible. They can be true ORFs or mispredictions. Some are conserved through many species (conserved hypothetical) indicating that they are true ORFs. Quite large sections of a genome that has undergone annotation are classed as hypothetical; for example, around 50% of the *Neisseria meningitidis* ORFs are of unknown function. Comparing a new genome to a well-annotated and functionally defined one (such as that of *E. coli*) can aid the analysis of specific pathways [such as those obtained from the Kyoto Encyclopedia of Genes and Genomes (KEGG)] and may identify missing components or blanks in pathways. These blanks can mean either that no equivalent protein exists in the genome under investigation or that one of the hypothetical proteins is the blank (see Figure 9.16). Figure 9.16A shows the general gluconeogenesis pathway, while Figure 9.16B illustrates that of human (*Homo sapiens*); the green boxes illustrate those enzymes identified in the human genome that are associated with this pathway, with *ALDH10* highlighted. Figure 9.16C and 9.16D show the same pathway for *Bifidobacterium longum* (BL), which also has the *ALDH10* enzyme, and for *Neisseria meningitidis* (NM), which does not. The pathways could be used to try and find the gene products that are in the same pathway as *ALDH10* in *B. longum* by careful examination of the sequence of the other enzyme members, or could be used to explore the possibility that *N. meningitidis* has an *ALDH10*-like enzyme. More careful study of the hypothetical ORFs can identify missing pathway members.

## Gene ontology provides a standard vocabulary for gene annotation

One of the important aspects of genome annotation has been the recognition of the importance of gene ontology (see Section 3.2). In this context, gene ontology is a set of standardized and accepted terms that encompass the range of possible functions and can be found on the Gene Ontology Consortium's Web site. Wherever possible these terms should be used, as they make it easier to do meaningful further analysis such as a search for genes of similar proposed function. To enable exchange of annotation information, a common description language should also be used, for example the GFF (General Feature Format).

It is also important to be able to compare (or exchange) results between groups that are involved in annotating a genome. Currently, much of the valuable information is spread out over a multitude of Web sites and databases that are not integrated. Some recent attempts have been made by the coordinated effort of a number of scientific communities to create annotation system consortia (for example, the Human Genome Project consortium). A new method of sharing annotation has recently been set up; the Distributed Annotation System (DAS) allows for the integration of data from different and distributed databases. It works with the concept of layers, where each layer contains particular annotation data. These third-party

**Figure 9.16**
**Gluconeogenesis pathways.** (A) A general overview of all possible molecules involved in this pathway. All are not colored as this picture is not associated with any organism. (B) The gluconeogenesis pathway in humans. Green-filled boxes show proteins that have been identified in humans and are associated with this pathway. The numbers in the boxes give the EC (enzyme family) numbers.

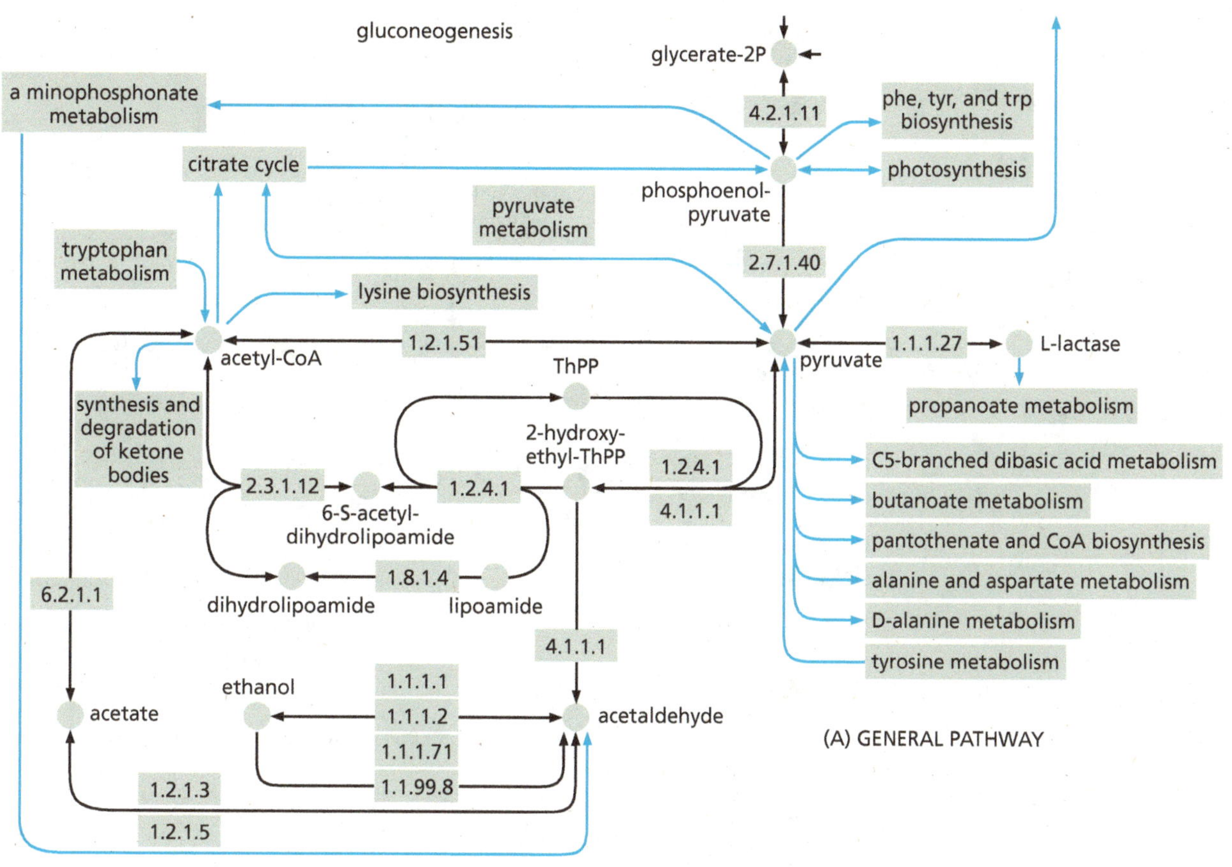
gluconeogenesis
glycerate-2P
a minophosphonate metabolism
4.2.1.11
phe, tyr, and trp biosynthesis
citrate cycle
photosynthesis
phosphoenol-pyruvate
pyruvate metabolism
tryptophan metabolism
2.7.1.40
lysine biosynthesis
1.2.1.51
1.1.1.27
L-lactase
acetyl-CoA
pyruvate
ThPP
propanoate metabolism
synthesis and degradation of ketone bodies
2-hydroxy-ethyl-ThPP
1.2.4.1
C5-branched dibasic acid metabolism
2.3.1.12
1.2.4.1
6-S-acetyl-dihydrolipoamide
4.1.1.1
butanoate metabolism
pantothenate and CoA biosynthesis
1.8.1.4
alanine and aspartate metabolism
6.2.1.1
dihydrolipoamide
lipoamide
D-alanine metabolism
4.1.1.1
tyrosine metabolism
1.1.1.1
ethanol
1.1.1.2
acetate
acetaldehyde
1.1.1.71
1.2.1.3
1.1.99.8
1.2.1.5
(A) GENERAL PATHWAY

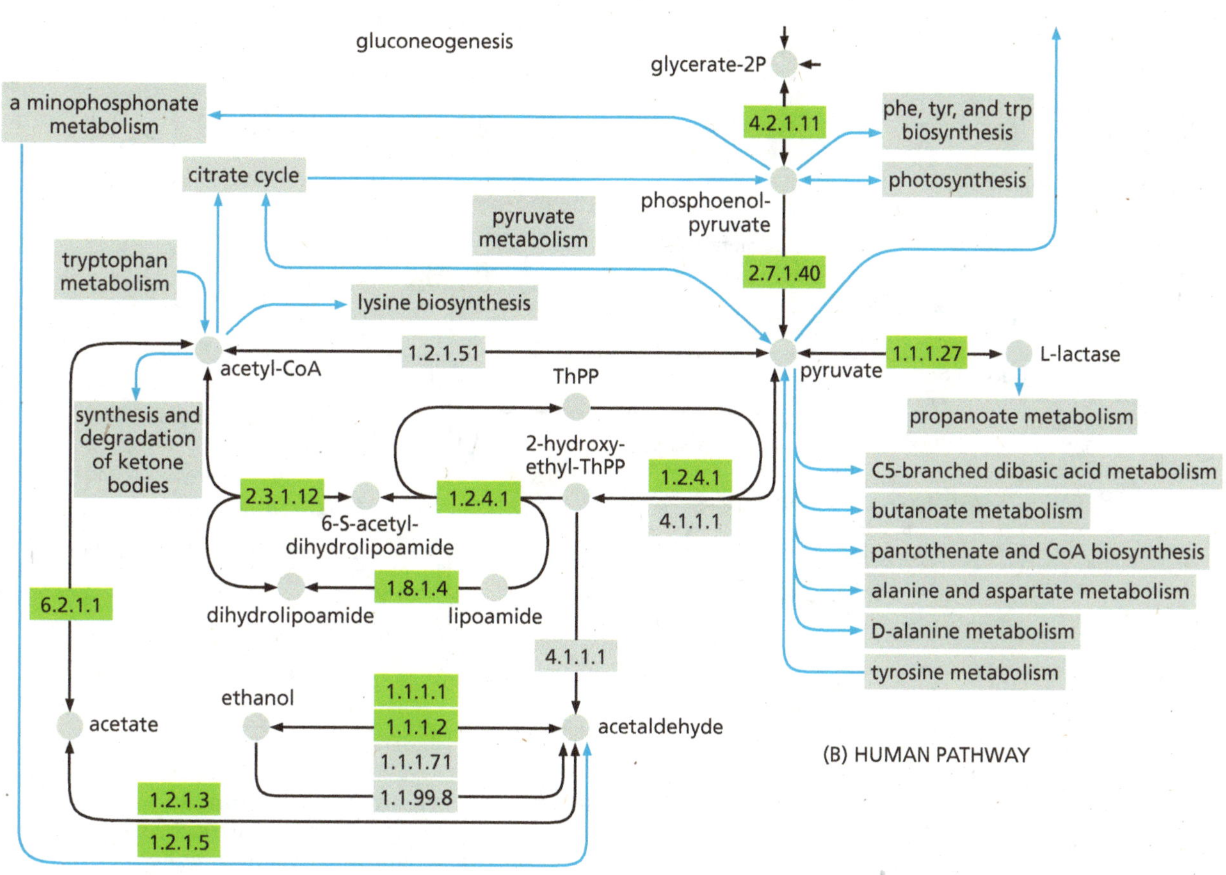
gluconeogenesis
glycerate-2P
a minophosphonate metabolism
4.2.1.11
phe, tyr, and trp biosynthesis
citrate cycle
photosynthesis
phosphoenol-pyruvate
pyruvate metabolism
tryptophan metabolism
2.7.1.40
lysine biosynthesis
1.2.1.51
1.1.1.27
L-lactase
acetyl-CoA
pyruvate
ThPP
propanoate metabolism
synthesis and degradation of ketone bodies
2-hydroxy-ethyl-ThPP
1.2.4.1
C5-branched dibasic acid metabolism
2.3.1.12
1.2.4.1
6-S-acetyl-dihydrolipoamide
4.1.1.1
butanoate metabolism
pantothenate and CoA biosynthesis
1.8.1.4
alanine and aspartate metabolism
6.2.1.1
dihydrolipoamide
lipoamide
D-alanine metabolism
4.1.1.1
tyrosine metabolism
1.1.1.1
ethanol
1.1.1.2
acetate
acetaldehyde
1.1.1.71
1.2.1.3
1.1.99.8
1.2.1.5
(B) HUMAN PATHWAY

gluconeogenesis
glycerate-2P
a minophosphonate metabolism
4.2.1.11
phe, tyr, and trp biosynthesis
citrate cycle
photosynthesis
pyruvate metabolism
phosphoenol-pyruvate
tryptophan metabolism
2.7.1.40
lysine biosynthesis
1.2.1.51
1.1.1.27
L-lactase
acetyl-CoA
pyruvate
ThPP
propanoate metabolism
synthesis and degradation of ketone bodies
2-hydroxy-ethyl-ThPP
1.2.4.1
C5-branched dibasic acid metabolism
2.3.1.12
1.2.4.1
butanoate metabolism
6-S-acetyl-dihydrolipoamide
4.1.1.1
pantothenate and CoA biosynthesis
1.8.1.4
alanine and aspartate metabolism
6.2.1.1
dihydrolipoamide
lipoamide
D-alanine metabolism
4.1.1.1
tyrosine metabolism
ethanol
1.1.1.1
acetate
1.1.1.2
acetaldehyde
1.1.1.71
1.2.1.3
1.1.99.8
1.2.1.5

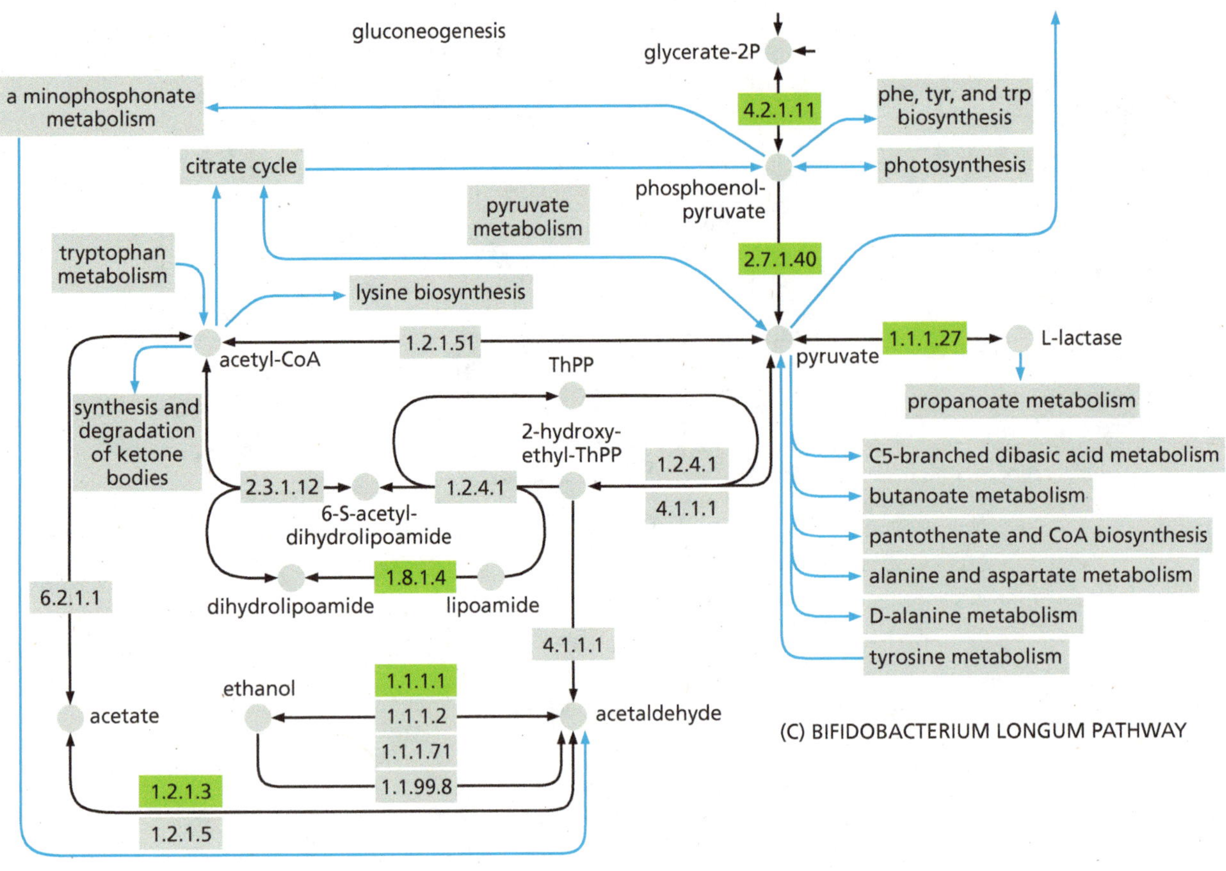

(C) BIFIDOBACTERIUM LONGUM PATHWAY

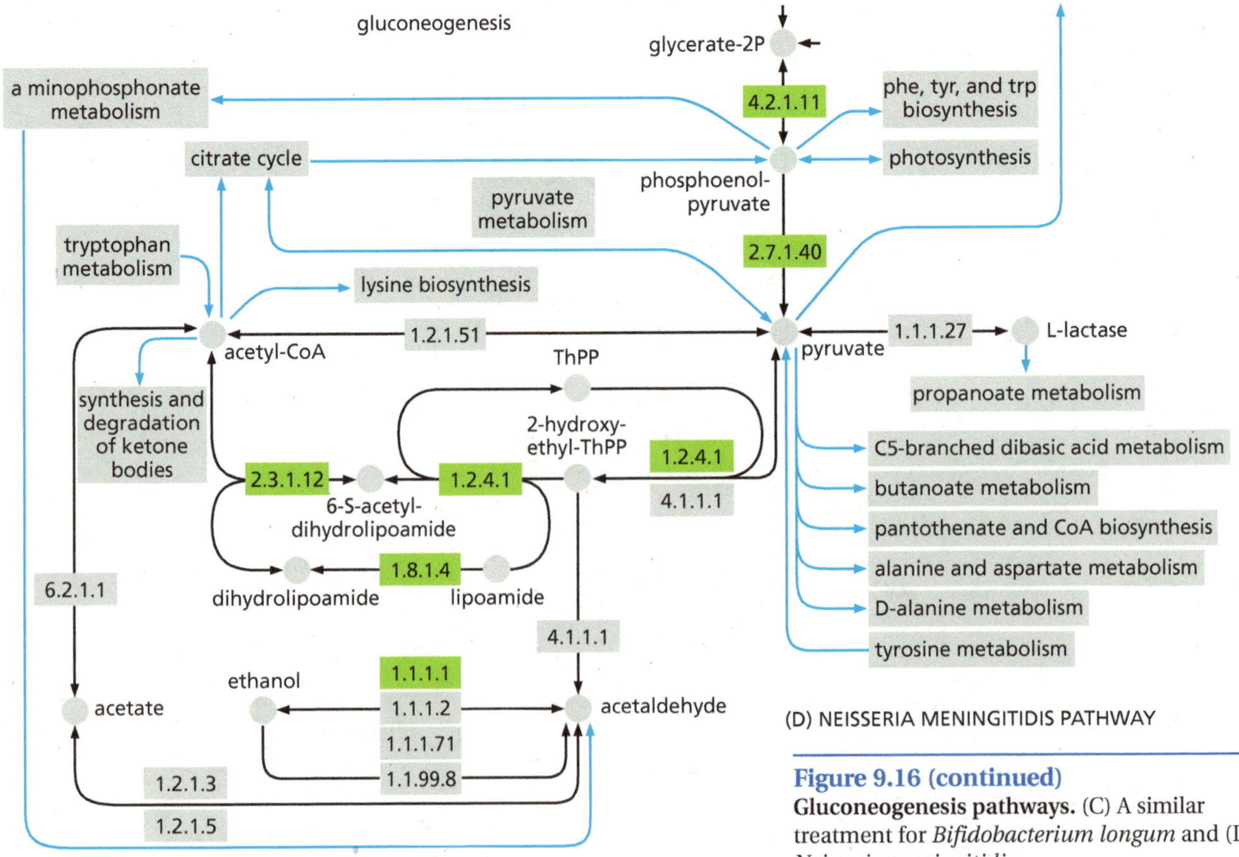

gluconeogenesis
glycerate-2P
a minophosphonate metabolism
4.2.1.11
phe, tyr, and trp biosynthesis
citrate cycle
photosynthesis
pyruvate metabolism
phosphoenol-pyruvate
tryptophan metabolism
2.7.1.40
lysine biosynthesis
1.2.1.51
1.1.1.27
L-lactase
acetyl-CoA
pyruvate
ThPP
propanoate metabolism
synthesis and degradation of ketone bodies
2-hydroxy-ethyl-ThPP
1.2.4.1
C5-branched dibasic acid metabolism
2.3.1.12
1.2.4.1
butanoate metabolism
6-S-acetyl-dihydrolipoamide
4.1.1.1
pantothenate and CoA biosynthesis
1.8.1.4
alanine and aspartate metabolism
6.2.1.1
dihydrolipoamide
lipoamide
D-alanine metabolism
4.1.1.1
tyrosine metabolism
ethanol
1.1.1.1
acetate
1.1.1.2
acetaldehyde
1.1.1.71
1.2.1.3
1.1.99.8
1.2.1.5

(D) NEISSERIA MENINGITIDIS PATHWAY

**Figure 9.16 (continued)**
**Gluconeogenesis pathways.** (C) A similar treatment for *Bifidobacterium longum* and (D) *Neisseria meningitidis*.

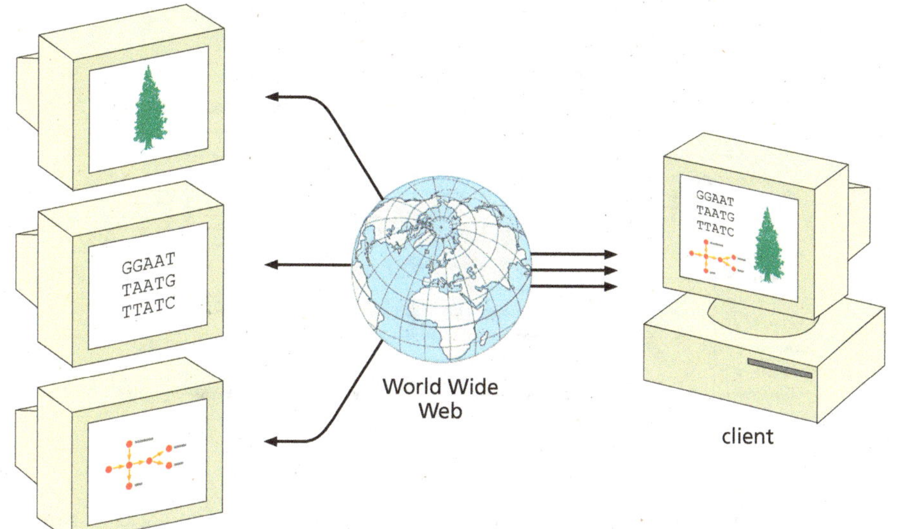


**Figure 9.17**
**Basic Distributed Annotation System (DAS) architecture.** Separate servers provide annotations relative to the reference sequence (sequence server such as University of Washington Genome Center) and the client can fetch the data from multiple servers (such as Ensemble and Pathways) and automatically generate an integrated view on the client.

servers are controlled by individual annotators (or groups), but this system works on the basis of a common data-exchange standard—the DAS-XML. This allows for the provision of layers from various servers, which are overlaid to produce a single integrated view by a DAS client program (see Figure 9.17).

The *ALDH10* DNA sequence will be used as an example, as we know with some degree of confidence from the analysis described in the previous sections where the exons, introns, and putative promoters are. The exons can be spliced together to obtain the coding sequence and translated to the protein product. From the previous searches the protein was shown to be homologous to fatty aldehyde dehydrogenase. Therefore it should be possible to extract the putative function of the DNA product using the DAS server and other Web pages.

For example, from a Web BLAST search at Swiss-Prot we find that this gene is the cause of a serious disease, the Sjögren–Larsson syndrome (see Box 9.6). The occurrence and effects of allelic variations can be investigated and looked for in the query

## Box 9.6 The Sjögren-Larsson syndrome

Also known as the ichthyosis, spasticity, oligophrenia syndrome, this is a genetic (inherited) disease characterized by three criteria: ichthyosis (thickened fish-like skin), spastic paraplegia (spasticity of the legs), and oligophrenia (mental retardation). The gene defect linked to Sjögren–Larsson syndrome has been located on chromosome number 17 (17p11.2). The mutation is recessive: the presence in a person of one copy of the mutated gene is not detrimental, but if two of these carriers have children, the risk for each child of receiving both defective genes and having the Sjögren–Larsson syndrome is 0.25. Mutations in the gene lead to a deficiency of the enzyme fatty aldehyde dehydrogenase 10 (FALDH10). The syndrome, therefore, is due to a deficit of FALDH10.

The Sjögren–Larsson syndrome was described by Torsten Sjögren (1896–1974), a professor of psychiatry at the Karolinska Hospital in Stockholm and a pioneer in modern psychiatry and medical genetics. Sjögren–Larsson syndrome is also known as SLS; other terms are fatty alcohol: $NAD^+$ oxidoreductase deficiency (FAO deficiency); fatty aldehyde dehydrogenase deficiency (FALDH deficiency); and fatty aldehyde dehydrogenase 10 deficiency (FALDH10 deficiency).

To find out more, search the ExPASy site for *ALDH10* and then link to the MIM database.

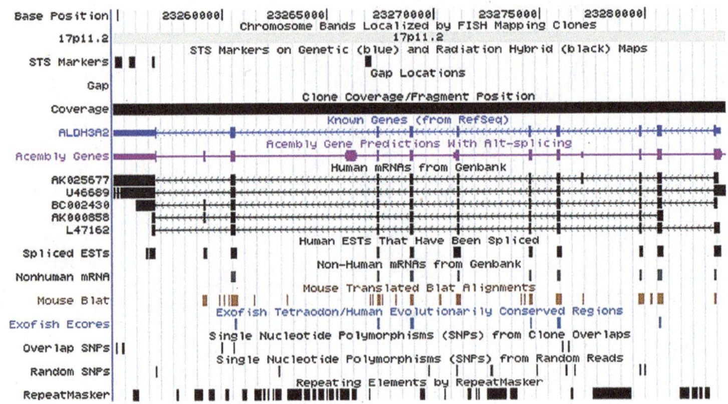


**Figure 9.18**
**A snapshot of the Genome Browser interactive window.** Information from many other databases and Web pages is summarized in one view. Clicking on any of the information will open a more detailed description.

sequence using the MIM Web site and associated links. Other good sites for further investigation are the Genome Browser at the University of California, San Francisco, Gene Ontology consortium, and Ensemble Web sites.

Once the gene product has been identified (*ALDH10* in the working example) one can go to the UCSC Genome Browser and search the human genome with the name. Searching with *ALDH10*, we find that it is located on chromosome 17 and an interactive graphical representation is provided with further information (see Figure 9.18) where much of the information is summarized. Clicking on one of the links will lead to another Web page with more detail. Links from the Genome Browser lead to the Online Mendelian Inheritance in Man (OMIM) Web site, which provides a catalog of human genes and genetic disorders, and to GeneCards, which gives further links and summaries of information gathered from each linked Web site. These links should be visited to explore the gene product and gather more information on its function and medical importance.

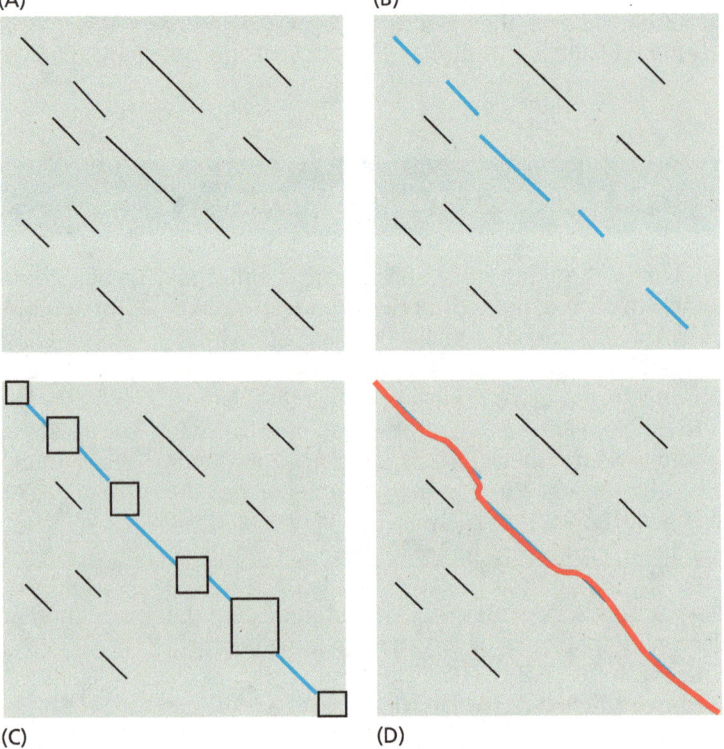


**Figure 9.19**
**Graphical illustration of the LAGAN method.** Initially, LAGAN generates local alignments between the two sequences (A) and assigns a weight to each aligned fragment. It then joins the local alignments if the end of one precedes the start of the next local alignment (B). Finally, it calculates optimal alignments in the limited area (boxed) around the local-alignment anchors using the Needleman–Wunsch algorithm (C). Figure D shows the final product.

## 9.8 Large Genome Comparisons

Although it is not within the scope of this chapter to go into detail of whole-genome comparisons, many of the methods used are similar to those for part-genome comparisons. New methods are becoming available, however, that allow whole-genome alignments (see Section 5.5) and also whole- or part-genome annotation. Two such tools, available by Web access, are LAGAN and Multi-LAGAN. These methods allow the global alignment of relatively large genomic sequences, correctly aligning protein-coding exons between distant homologs such as the human and chicken sequences. LAGAN first generates local alignments between the sequences, subsequently constructs a rough global map of the aligned local sets, and finally calculates a final global alignment using the Needleman–Wunsch algorithm (see Sections 5.2 and 5.5) by choosing the best alignment near the area around the rough global map (see Figure 9.19). After alignment, the aligned sequences can be annotated and graphically represented using the program VISTA. To use the program with annotation, the user needs to have the sequence files and at least annotation for one of the sequences. Figure 9.20 illustrates the alignment of the whole human *ALDH10* gene with a mouse sequence. The homologous parts of the alignment are shown as peaks in the graph and contain the annotation for the exons of the human gene. Therefore, once we have identified the *ALDH10* gene, the identification can either be verified using alignments, or the same genes can be identified or predicted in other organisms.

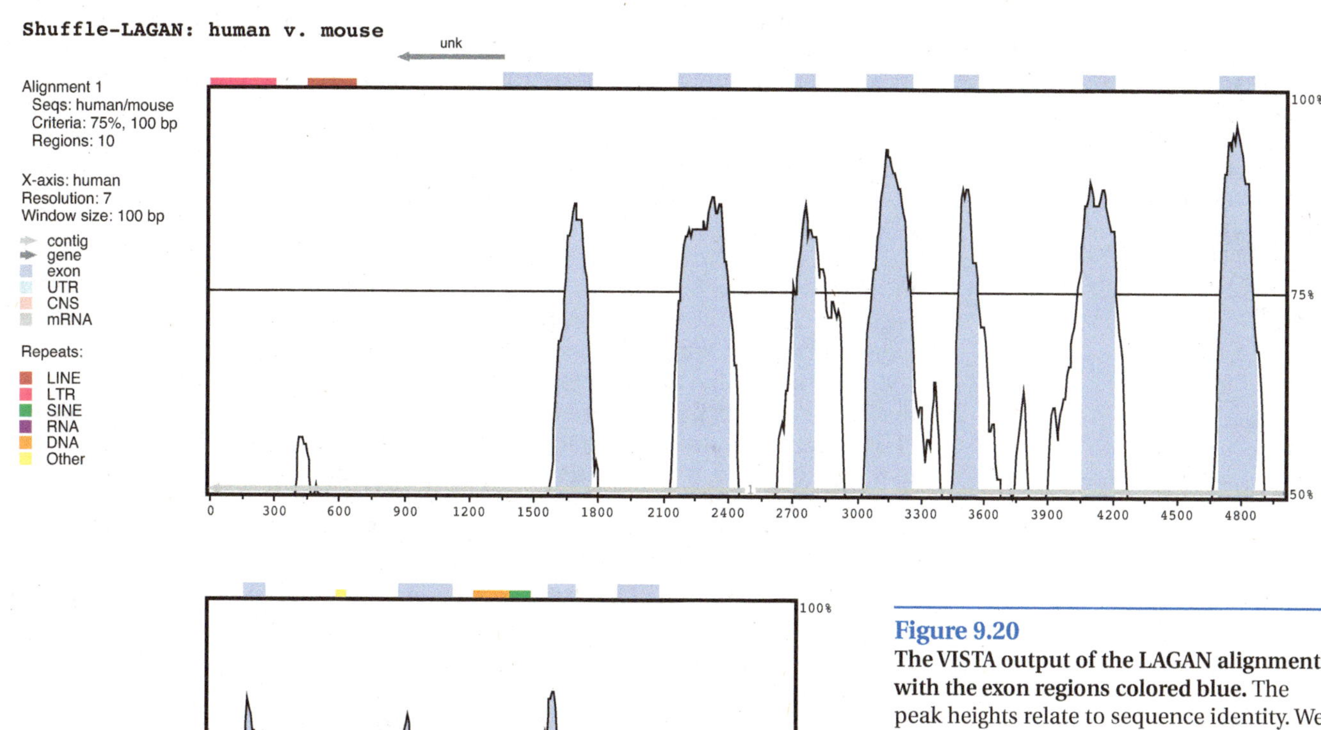

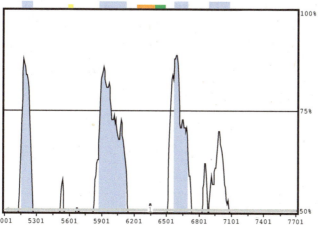

**Figure 9.20**
**The VISTA output of the LAGAN alignment with the exon regions colored blue.** The peak heights relate to sequence identity. We can see that the exonic areas and areas around the human exons are highly conserved in the mouse sequence. Other colored annotations can be visualized as well; these include UTRs, conserved noncoding sequences (CNS), mRNA, contigs, and genes. For example, a gene labeled unk (given by thin dark-gray arrows above the sequence bar) is seen top left, and a SINE repeat is shown as a green bar just upstream of the penultimate exon.

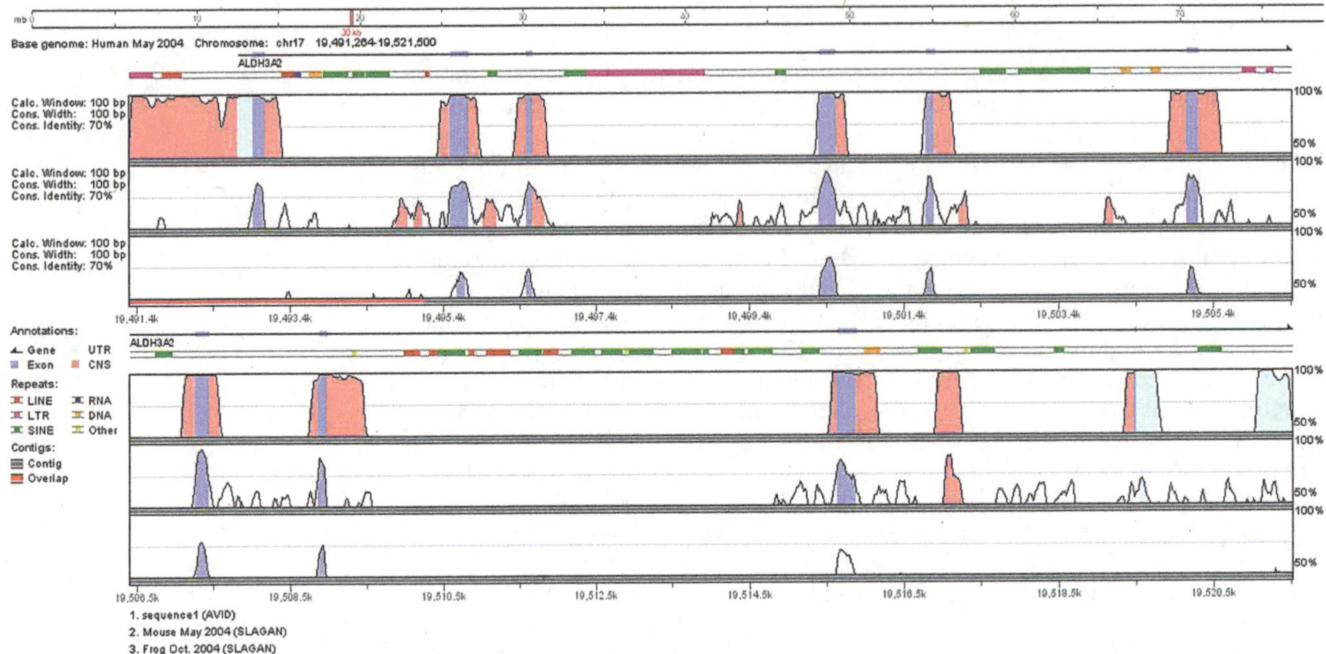


**Figure 9.21**
**The human sequence of the *ALDH10* gene matched to other *ALDH10* or *ALDH10*-like genes in mouse and frog genomes, as shown by the VISTA server.** We can see that all the exons are present in the more closely related mouse genome, but the first and last exons in the figure are missing from frog. Exons are shown in dark blue, UTRs in light blue and conserved non-coding sequences (CNS) in pink.

The VISTA server consists of a number of programs and databases that allow comparative analysis of genomic sequences. The set of programs allows the user to align sequences from different species to compare a user-specified sequence with whole genome assemblies (see Figure 9.21). This makes it possible to consider the phylogenetic relationships and graphically visualize the results. However, the VISTA server accepts only 20 kilobase sequence lengths.

## Summary

In this chapter we have discussed the process of analyzing an unannotated DNA sequence and discovering the protein-coding regions (if any) within the DNA, followed by identifying the encoded protein product. Initially the DNA sequence should be submitted for homology searches through a DNA or EST database as this may provide a quick and comprehensive answer if the DNA or a close homolog has already been sequenced and analyzed and is in the database. It may provide a clue to the gene structure or function if a less close homolog is available. However, if no match is found then the next step is to try and predict the exon–intron structure of the DNA. The predicted exons can be translated and sent for a homology search through a protein database. Finding a significant hit at this level does not only enable a more accurate delineation and therefore annotation of the DNA sequence but also will often provide clues to the function and cellular location of the gene product. Through a series of translation and search steps it should be possible to obtain a relatively accurate map of intron–exon sites. The predicted and translated exons can be spliced together to form a putative protein sequence, which can be further analyzed as described in the preceding and following chapters.

# Further Reading

## 9.1 Preliminary Examination of Genome Sequence

For a general description of the processes involved in DNA to protein read:

Alberts B, Johnson A, Lewis J, Raff M, Roberts K & Walter P (2008) Molecular Biology of the Cell, 5th ed. New York: Garland Science.

## 9.2 Gene Prediction in Prokaryotic Genomes

Frishman D, Mironov A & Gelfand M (1999) Starts of bacterial genes: estimating the reliability of computer predictions. *Gene* 234, 257–265.

Wolf YI, Rogozin IB, Kondrashov AS & Koonin EV (2001) Genome alignment, evolution of prokaryotic genome organization, and prediction of gene function using genomic context. *Genome Res.* 11, 356–372.

## 9.3 Gene Prediction in Eukaryotic Genomes

Borodovsky M & McIninch J (1993) GenMark: Parallel gene recognition for both DNA strands. *Comput. Chem.* 17, 123–134.

Burge CB & Karlin S (1998) Finding the genes in genomic DNA. *Curr. Opin. Struct. Biol.* 8, 346–354.

Claverie J-M (1997) Computational methods for the identification of genes in vertebrate genomic sequences. *Hum. Mol. Genet.* 6, 1735–1744.

Dong S & Searls DB (1994) Gene structure prediction by linguistic methods. *Genomics* 23, 540–551.

Fickett J (1996) Finding genes by computer: The state of the art. *Trends Genet.* 12, 316–320.

Fickett JW & Hatzigeorgiou AB (1997) Eukaryotic promoter recognition. *Genome Res.* 7, 861–878.

Larizza A, Makalowski W, Pesole G & Saccone C (2002) Evolutionary dynamics of mammalian mRNA untranslated regions by comparative analysis of orthologous human, artiodactyl and rodent gene pairs. *Comput. Chem.* 24, 479–490.

Makalowski W (2000) Genomic scrap yard: How genomes utilize all that junk. *Gene* 259, 61–67.

Makarov V (2002) Computer programs for eukaryotic gene prediction. *Brief. Bioinform.* 3, 195–199.

Marcotte EM (2000) Computational genetics: Finding protein function by nonhomology methods. *Curr. Opin. Struct. Biol.* 10, 359–365.

Milanesi L, D'Angelo D & Rogozin IB (1999) Genebuilder: interactive *in silico* prediction of gene structure. *Bioinformatics* 15, 612–621.

Solovyev VV, Salamov AA & Lawrence CB (1994) Predicting internal exons by oligonucleotide composition and discriminant analysis of spliceable open reading frames. *Nucleic Acids Res.* 22, 5156–5163.

Zhang MQ (1997) Identification of protein coding regions in the human genome based on quadratic discriminant analysis. *Proc. Natl Acad. Sci. USA* 94, 565–568.

Zhang MQ (2002) Computational prediction of eukaryotic protein-coding genes. *Nat. Rev. Genet.* 3, 698–709.

## 9.4 Splice Site Detection

Nakata K, Kanehisa M & DeLisi C (1985) Prediction of splice junctions in mRNA sequences. *Nucleic Acids Res.* 13, 5327–5340.

## 9.5 Prediction of Promoter Regions

Bucher P (1990) Weight matrix descriptions of four eukaryotic RNA polymerase II promoter elements derived from 502 unrelated promoter sequences. *J. Mol. Biol.* 212, 563–578.

Hacker J & Kaper JB (2000) Pathogenicity islands and the evolution of microbes. *Annu. Rev. Microbiol.* 54, 641–679.

Scherf M, Klingenhoff A & Werner T (2000) Highly specific localization of promoter regions in large genomic sequences by PromoterInspector: A novel context analysis approach. *J. Mol. Biol.* 297, 599–606.

## 9.6 Confirming Predictions

Bajic V (2000) Comparing the success of different prediction software in sequence analysis: a review. *Brief. Bioinform.* 1, 214–228.

Baldi P, Brunak S, Chauvin Y et al. (2000). Assessing the accuracy of predicition algorithms for classification: an overview. *Bioinformatics* 16, 412–424.

Burset M & Guigó R (1996) Evaluation of gene structure prediction programs. *Genomics* 34, 353–357.

Rogic S, Mackworth AK & Ouellette FB (2001) Evaluation of gene-finding programs on mammalian sequences. *Genome Res.* 11, 817–832.

## 9.7 Genome Annotation

Burge C & Karlin S (1997) Prediction of complete gene structures in human genomic DNA. *J. Mol. Biol.* 268, 78–94.

Domingues FS, Koppensteiner WA & Sippl MJ (2000) The role of protein structure in genomics. *FEBS Lett.* 476, 98–102.

Teichmann SA, Murzin AG & Chothia C (2001) Determination of protein function, evolution and interactions by structural genomics. *Curr. Opin. Struct. Biol.* 11, 354–363.

## 9.8 Large Genome Comparisons

Brudno M, Do CB, Cooper GM et al. (2003) LAGAN and Multi-LAGAN: Efficient tools for large-scale multiple alignment of genomic DNA. *Genome Res.* 13, 721–731.

Frazer KA, Pachter L, Poliakov A et al. (2004) VISTA: computational tools for comparative genomics. *Nucleic Acids Res.* 32, W273–W279.

Mayor C, Brudno M, Schwartz JR et al. (2000) VISTA: Visualizing global DNA sequence alignments of arbitrary length. *Bioinformatics* 16, 1046.

## Box 9.5

McClintock B (1950) The origin and behavior of mutable loci in maize. *Proc. Natl Acad. Sci. USA* 36, 344–355.

McClintock B (1956) Controlling elements and the gene. *Cold Spring Harb. Symp. Quant. Biol.* 21, 197–216.

# GENE DETECTION AND GENOME ANNOTATION

THEORY CHAPTER

**When you have read Chapter 10, you should be able to:**

- Recount the methods of detecting tRNA genes.
- Show how protein coding regions can be identified from subtle signals.
- Explain why prokaryotic gene detection is much easier than eukaryotic.
- Describe how the binding of components of transcription and translation often leave detectable sequence signals.
- Explain why gene detection can be assisted by identifying sequence homologs in databases.
- Describe how separate techniques are often employed to identify particular gene components.
- Show that gene models are composed of combinations of gene components.
- Describe how splice-site predictions focus on eliminating false positives.
- Discuss how genome comparison can help resolve ambiguities in gene prediction.

As more whole-genome sequences become available, these very long DNA sequences must be analyzed to delineate and identify the protein-coding genes and other genetic elements in these largely uncharacterized genomes. There are often some experimental data available to assist in this task, such as sequences of previously characterized genes and gene products and EST sequences. The EST data, which should be from the same organism as the genome sequence, can be aligned with the sequence to identify translated regions of genes. Reported sequences of genes and gene products can also be aligned, and even homologous sequences from other organisms can be useful in identifying potential homologs in the genome. However, the set of genes revealed by such analysis will almost certainly be incomplete and, in any case, other features of genomes such as repeat and control sequences will be missing. In order to obtain a full description of genomes, accurate automated methods of interpreting the nucleotide sequence have been developed to define the genes, functional RNA molecules, repeats, and control regions, and to characterize the gene products. This whole procedure is known generally as **annotation**, and frequently runs in parallel with the completion of sequencing and sequence assembly (see Figure 10.1).

**Mind Map 10.1**
**A mind map showing the applications and the heory behind the applications of gene prediction and annotation.**

To date, most genome annotation projects have been restricted to the major sequencing laboratories. For bacterial genomes, these large teams can take just 2 or 3 months from finishing the sequence to generating an annotation and analysis for publication. The genomes of higher organisms are not so straightforward to

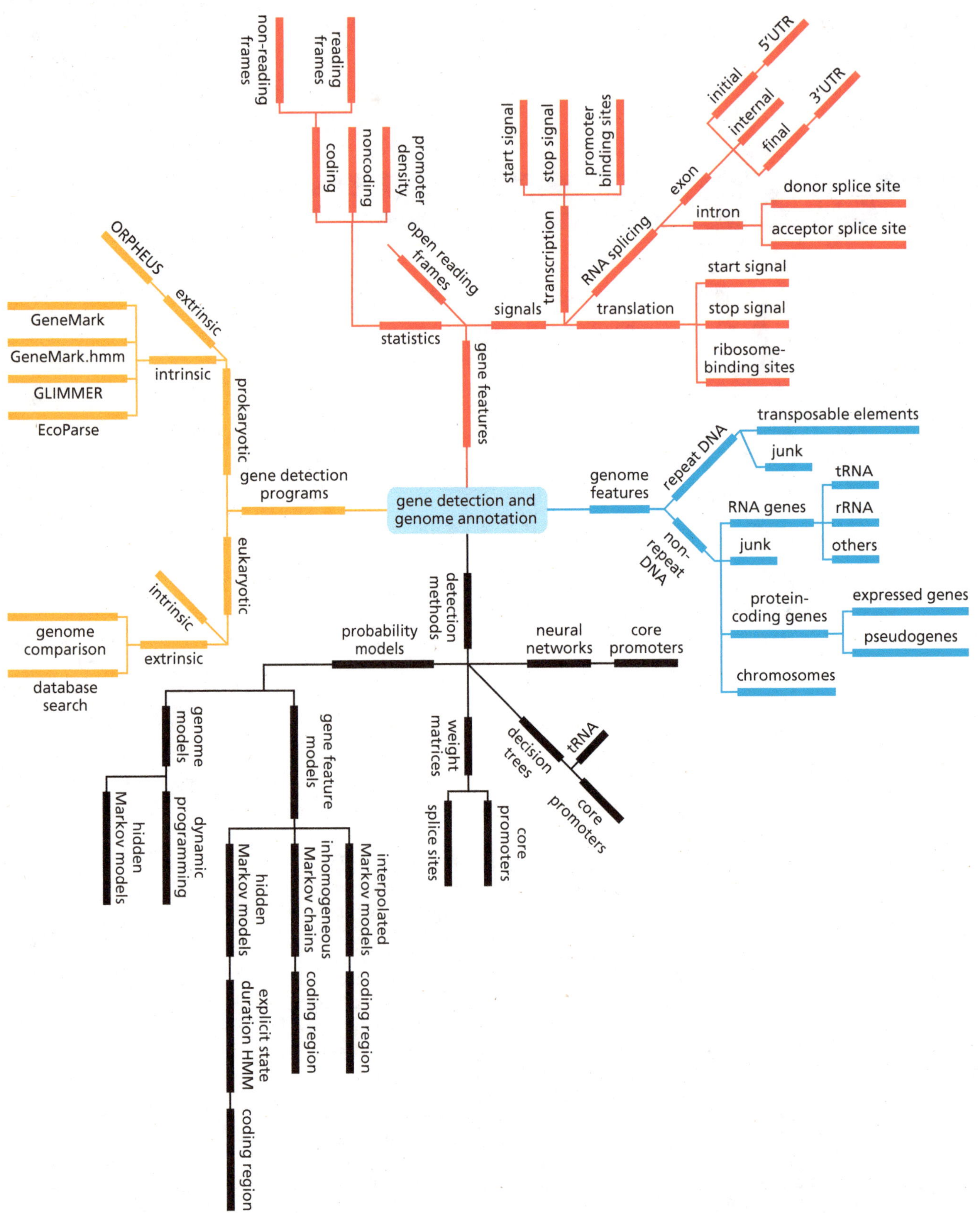

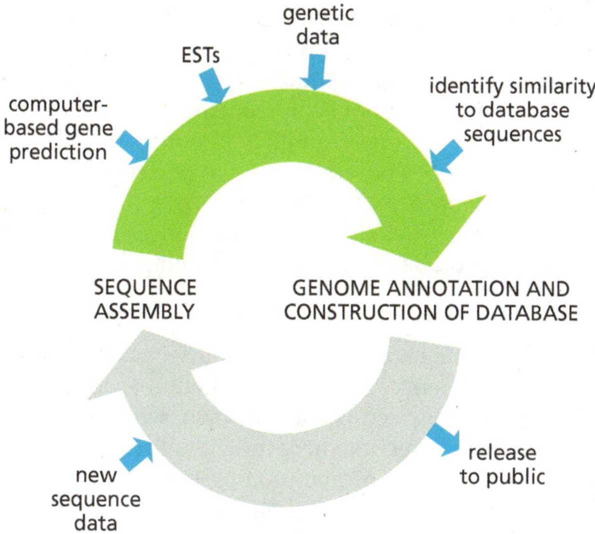


**Figure 10.1**
**A general scheme for a genome sequencing project focusing on the major computational analysis steps.** The process of constructing and annotating a genome is often cyclic, starting before the entire sequence has been assembled. The figure is very simplistic, omitting many details, but shows how the computer-based gene-prediction methods discussed here fit into the overall scheme. Note that experimental data are added at several points, namely the new sequence data, ESTs, and genetic data. The latter will be information relating to the effects of gene disruption and/or mutation. The database construction referred to after annotation could be an organism-specific database, or simply the relevant entries in the standard databases such as EMBL and UniProt.

analyze, and it will be some time before the human genome is annotated to the same accuracy as is routine for prokaryotes. In addition, the availability of many annotated bacterial genomes and the speed and ease with which such genomes can be sequenced has encouraged many new studies, in particular comparing genomes to investigate strain differences and the evolution of closely related species. These developments are making genome annotation projects much more common, and taking them outside the domain of the major laboratories.

This chapter takes a detailed look at some of the typical algorithms used in gene detection and genome annotation. The practical aspects of using gene-prediction programs, with these or similar algorithms, are covered in Chapter 9. Many of the algorithms used in genome annotation are very complex, often involving several separate components such as neural nets, hidden Markov models (HMMs), and decision trees that are then combined. In this chapter we shall describe only some of these in detail. Neural nets are dealt with in detail in Section 12.4 on secondary structure prediction and only the general structures will be presented here. The HMMs used for genome annotation are more complex than those introduced in Section 6.2 for sequence alignment, but the same principles hold, and only their general structures will be given here. The other types of algorithms will be presented in detail.

Because of significant differences both in biological features and practical methodology, eukaryotic and prokaryotic genome annotation will be considered separately, starting with the prokaryotes. In both cases, however, the task can be considered as comprising two steps. Firstly, genes and other functional components must be located in the genome sequence. The second step involves identifying the likely properties, especially the biochemical and cellular function, of the gene products. Some annotation tools perform both of these steps. This second step will be given relatively little attention in this chapter, because it involves the application of techniques described elsewhere in this book such as sequence similarity in Chapter 4. Before reading this chapter you are advised to look at Chapter 1, especially Section 1.3, to familiarize yourself with the details of gene structure (including the control elements) and functional RNA molecules. Many of the techniques described below take advantage of known consensus sequences and RNA secondary structures to detect these features.

The identification of genes in prokaryotic genomes would appear to be relatively simple, as one can easily locate all ORFs. If we examine all ORFs as long as or longer than the shortest gene then the set will necessarily include all the genes. However, when looking for short genes such a procedure will generate many false positives; that is, predicted genes that are not genuine genes and are not expressed. There can be several reasons why an ORF is not expressed; most commonly, it is not associated with regulatory sequences that are required to initiate transcription. Some evidence for the incorrect prediction of a putative gene can be deduced from gene expression experiments, suggested by a lack of expression at any life stage under many different conditions. However, such work is time-consuming and largely unproductive, so significant attention has been paid to reducing the prediction error rates as much as possible.

Because the control of gene expression occurs in several stages, each of which has a signal present in the sequence, improved prediction accuracy can be obtained by paying attention to the presence of each known signal, building up a complex model of the gene. Ways of detecting these individual components will be discussed first, followed by the models of protein-coding genes. The genes specifying RNA molecules are significantly different in their control, but due to their high level of sequence conservation, other techniques can be used to locate them.

Eukaryotic genomes present a far more difficult case and their analysis can result in extremely high false-positive rates of ORF prediction unless great care is taken. This is because in eukaryotes a single protein is often encoded by several exons, and examples are known of very short exons coding for only a dozen residues. In addition in some regions of eukaryotic genomes only a few percent of bases are protein-coding. Thus short ORFs must be considered as potentially in exons, but many such ORFs will occur in those regions that do not contain genes.

Eukaryotic genes contain much more structure than prokaryotic genes (for example, introns and exons and their splice sites, longer and more complex promoter regions, and 3′ signals for poly(A) addition). However, these extra signals are often less distinct and harder to identify than the signals in prokaryotic genomes. For example, upstream signals for promoters and enhancers are much harder to identify, in part because in eukaryotes some act much further away from the start of transcription (often several kilobases upstream) and also because of the large number of different potential molecules involved. Furthermore, as already mentioned, the lengths of coding ORFs in eukaryotes tend to be far less distinct from those of random ORFs than in prokaryotes, as the eukaryotic genomes are much larger and the coding sequences are often only a very small proportion of the whole genome. The derivation of methods that can accurately identify the individual components of eukaryotic gene structure, or combine them into an accurate gene-prediction program is a very active area of research. In this chapter we can only summarize the general approaches taken and show some of the types of model investigated to date.

It has been found that even in highly efficient prokaryotic genomes, in which almost the entire length of one or other strand is part of a gene, there is no extensive overlapping between genes. The major exception to this occurs in viral genomes, which need to be extremely efficient since they must use the least possible amount of nucleic acid, which has to be contained in a very small capsid. In prokaryotes, however, small overlaps are quite common and prokaryotic gene-prediction programs have been designed to recognize this feature and deal with it.

Repeat sequences are very rarely found in protein-coding genes. Therefore, the first step in the analysis of a new genome sequence is usually to identify the repeat sequences, so that these regions can be excluded from the subsequent analysis for protein-coding genes. Usually the second step is to identify and exclude those regions that code for functional RNA molecules. This is because they can be readily identified due to their sequence conservation.

A distinction must be made between those methods that attempt to identify genes without reference to known sequences (sometimes called **intrinsic** methods) and those that take advantage of similarity to database sequences (sometimes called **extrinsic** methods). The former can, in principle, discover any new gene (that is, one that is not related to another known sequence), and as it is common for 20% of the genes in an organism not to be homologous to any other known sequence this can be an important advantage. Perhaps in time, when many genomes over the full range of the tree of life are known, such methods will not be so important.

Another recent innovation is to use two or more equivalent regions from the genomes of related organisms, e.g., rat and mouse. When the two sequences are sufficiently similar this can make it easier to identify those regions that are more conserved and thus probably have a function. In addition, such methods give an immediate appreciation of the evolutionary relationship between the two sequences.

## 10.1 Detection of Functional RNA Molecules Using Decision Trees

Whether the genome is prokaryotic or eukaryotic, an important class of genes that should be located before using the gene-detection methods discussed below are those that are not translated into protein, such as rRNA and tRNA (see Flow Diagram 10.1). rRNA sequences are sufficiently well characterized and well conserved that they can be identified by sequence similarity with little difficulty, and so will not be discussed further. There will be many tRNA genes (more than 50 in a bacterial genome, and even more in larger genomes), and it would be useful to know the set of tRNA molecules available to an organism, as this might suggest that some codons are not used, which would be valuable information when it comes to detecting protein-coding sequences by the methods presented below. The detection of tRNAs is more complicated but techniques exist that effectively solve the problem. The method described here in detail is implemented in a program called tRNAscan, which is one of the most commonly used programs for analysis of prokaryotic genomes. Because of the false-positive rate, this method cannot be usefully used for eukaryotes, but another called tRNAscan-SE (described briefly below) is available that has the required accuracy.

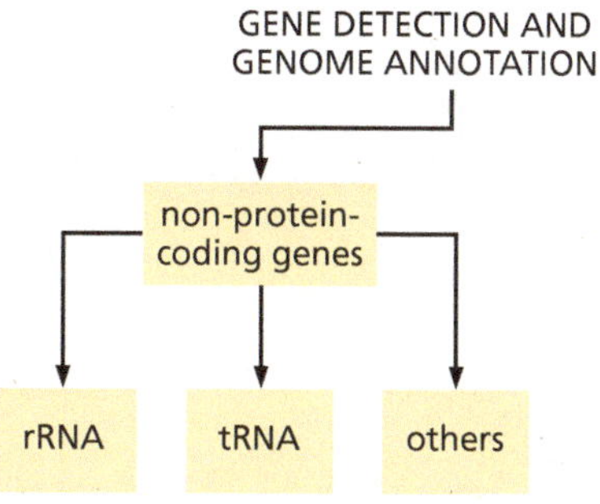


**Flow Diagram 10.1**
**The key concept introduced in this section is that specific methods have been proposed to identify functional RNA molecules encoded in genomes.**

### Detection of tRNA genes using the tRNAscan algorithm

tRNA molecules have a highly conserved structure with a number of short double-helical stems and short loops, forming a cloverleaf secondary structure (see Figure 10.2). Alignment of many known tRNA sequences has identified regions of particularly high sequence and structure conservation that have been used to construct a very specific tRNA-detection algorithm. The algorithm consists of a sequence of steps, each testing for the presence of a specific element of the tRNA sequence and/or structure, such as conserved bases or conserved base-pairing (see Figure 10.3). This algorithm is a simple form of a **decision tree**, where at each step the sequence has to pass a test.

Often, decision trees involve several further steps regardless of the outcome of the current step, whereas in this case all but one step terminates the search if the sequence fails to pass the test. In addition to the basic test at each step of the algorithm, a note is made of whether the sequence performs particularly well, such as having three of four expected invariant bases, instead of just the minimum two. A record is kept of how often the sequence passes a test at a higher level of stringency, counting the number of steps at which this is achieved, the value of which is referred to as SG. The details of each step are given in Figure 10.3. The final test before accepting that sequence as a tRNA gene depends on the value of SG.

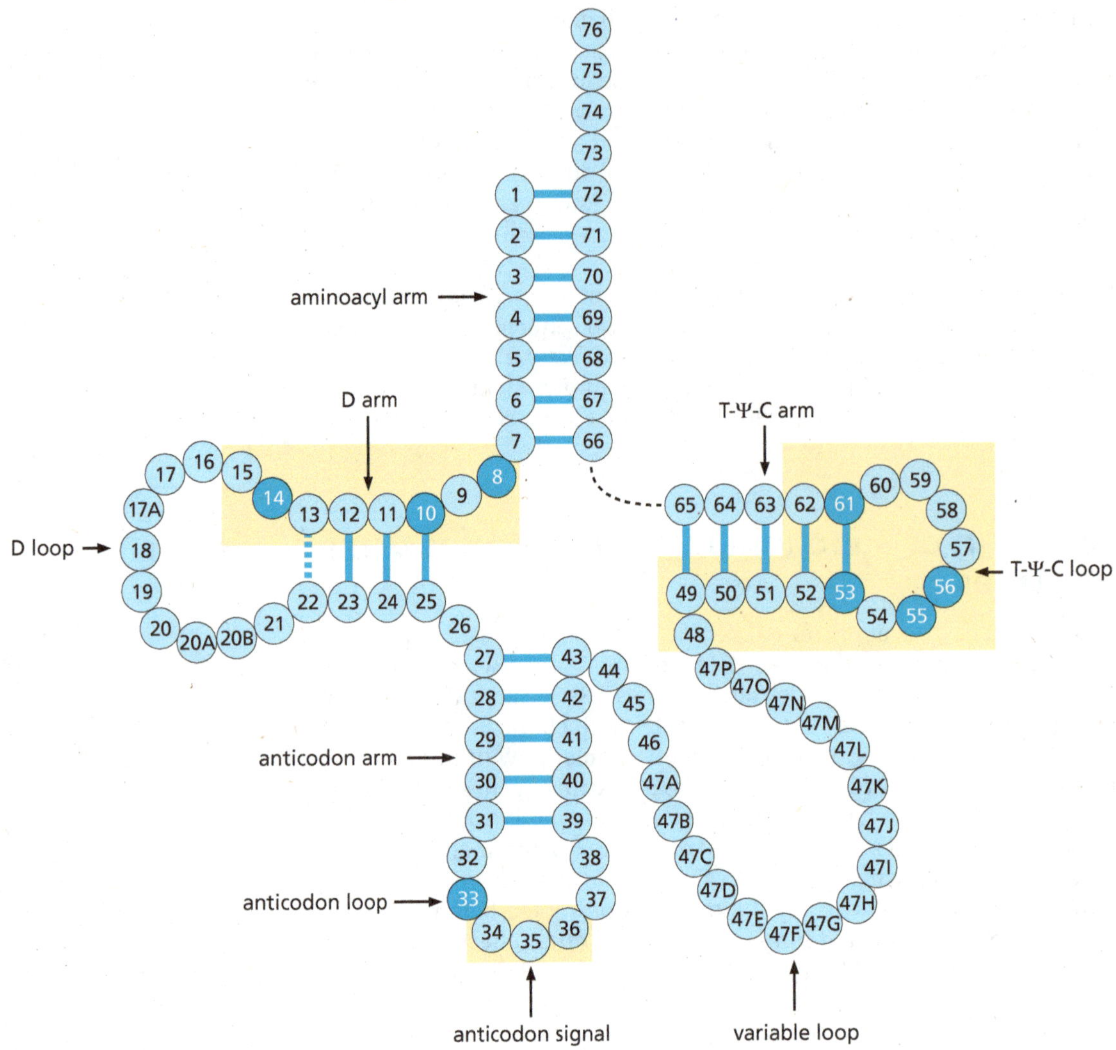


**Figure 10.2**
**The cloverleaf structure of a tRNA molecule, showing those features that are used in tRNAscan for detection.** The bases are numbered, allowing for insertions such as occur in the variable loop. The different parts of the structure that are involved in the tRNAscan algorithm (see Figure 10.3) are identified, including in dark blue the key invariant bases and the base-pairings, which are expected to be very highly conserved. (Adapted from G.A. Fichant and C. Burks, Identifying potential tRNA genes in genomic DNA sequences, *J. Mol. Biol.* 220:659–671, 1991.)

This method has proved to be extremely accurate in predicting tRNA genes, correctly assigning about 97.5% of them. It is estimated to predict only one false positive per 3 million bases, which means just one false positive (if that) in a bacterial genome.

## Detection of tRNA genes in eukaryotic genomes

The error rate for tRNAscan is estimated at one per three megabases, which means very few if any false positives for a bacterial genome. However, this is too high a false-positive rate to use in detecting tRNA genes in eukaryotes, producing about 1000 false positive tRNA genes from the human genome sequence. (This is almost as many false positives as expected real tRNA genes.) To circumvent this, Todd Lowe and Sean Eddy proposed a more sophisticated algorithm called tRNAscan-SE.

The tRNAscan-SE method combines several tRNA-prediction programs into a single highly accurate prediction program. Each of the programs has a quite different method of predicting tRNA genes. The programs used include tRNAscan, which as shown above looks for specific conserved bases at specified relative distances. Another used was developed by Angelo Pavesi and colleagues and

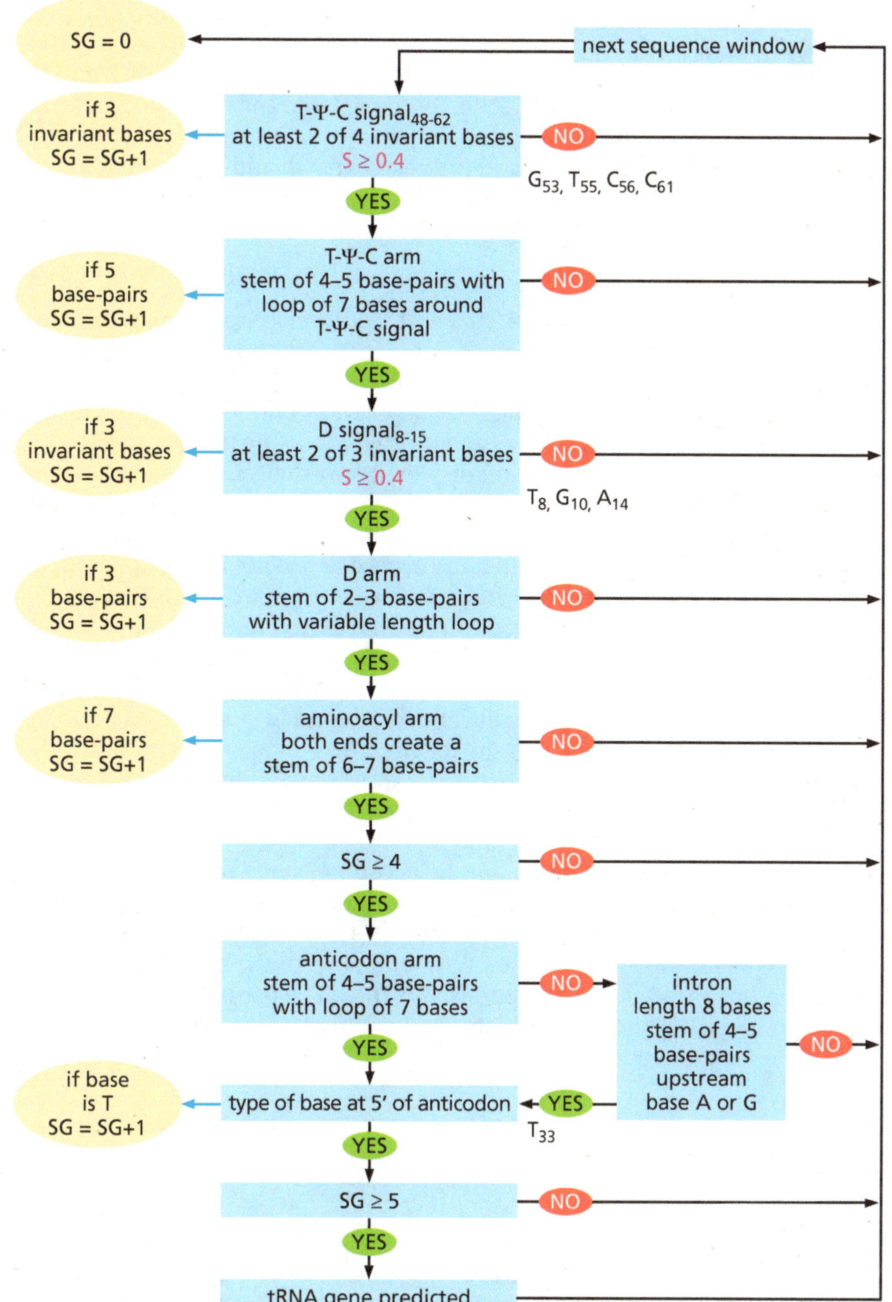


**Figure 10.3**
**The steps involved in the tRNAscan algorithm.** This is a decision tree with the addition of a general score SG that allows for some restricted leniency in matching the idealized tRNA sequence. The base numbers and molecular structures referred to are identified in Figure 10.2. Note that as the DNA sequence is searched, the algorithm refers to thymine bases; in the actual tRNA molecule these will be uracil. Two steps involve identifying the T-Ψ-C and D signals by scoring with frequency matrices. (Frequency matrices record the frequency of occurrence of each base at each position in the signal.) The score S used in these two steps is the ratio of the sums of the observed and highest frequencies at each position ignoring any invariant bases. If the score S equals or exceeds 0.4 the signal is regarded as present. The required invariant bases are listed next to the step which involves them. (Adapted from G.A. Fichant and C. Burks, Identifying potential tRNA genes in genomic DNA sequences, *J. Mol. Biol.* 220:659–671, 1991.)

searches for sequences related to the control of RNA polymerase III, the enzyme responsible for transcribing tRNA genes. A third method by Sean Eddy and Richard Durbin uses covariance models (a more advanced form of HMM) that take account of the sequence and secondary structure of tRNAs. This method is the most accurate of the three, but far too slow for full-scale application to the genomes of higher organisms. Each method has distinct strengths and weaknesses. It was noticed that the first two methods identified effectively all real tRNA genes, in addition to many false positives. The predictions of these first two methods are combined together and then further analyzed in a decision tree that includes the covariance models in a way that successfully identifies almost all the false-positive predictions. In this way a false-positive rate of less than one per 15,000 megabases is achieved.

## 10.2 Features Useful for Gene Detection in Prokaryotes

Much of the early work on gene detection used bacterial genes as the model, largely because of the considerable amount of reliable data available. The complete sequences of many bacterial genes were already known and genes are packed into a bacterial genome at relatively high density, with little noncoding intergenic DNA. The relative simplicity of bacterial gene structure led to some very successful gene-prediction techniques for prokaryotes.

Before presenting the background concepts and their implementation in gene-detection programs, it is worth pointing out that some problems are trivial. For example, one can easily enumerate all the potential ORFs present in the genome, and thus all possible protein products. (By a potential ORF we mean any stretch of sequence beginning at a start codon and proceeding to the first stop codon.) If the intended application is identification of proteins by their molecular masses (as in proteomic studies that utilize mass spectroscopy as an identification tool; see Section 15.2) then such a list may be sufficient, despite containing many nonexistent proteins. It should be noted that the true gene will not necessarily start at the beginning of such potential ORFs, and may begin at a start codon downstream of this, a situation which as will be seen can be difficult to predict, but is nevertheless trivial to overcome for such proteomic studies by identifying all possible start codons and listing all possible proteins. The longer the potential ORF, the more likely it is to really be a gene, as the likelihood of internal stop codons occurring in a random sequence increases with its length. Thus a key problem in prokaryote gene detection is to distinguish the true and false genes in the set of short potential ORFs of, say, 150 bases or fewer.

A consequence of this situation is that by accepting some false positives, a gene-detection method can achieve very high rates of detection. Put another way, these methods should really be detecting the false ORFs. One must be wary of some of the high success rates quoted (even over 98%), and false-positive rates would be more informative, but are often not quoted (see Box 10.1). The reason for this is simple: we are only slowly getting to the point where we know with some certainty the full range of true genes in a limited sample of species, and thus cannot as yet distinguish false

**Flow Diagram 10.2**
**The key concept introduced in this section is that prokaryotic genes have certain features and signals that can be used to identify genes in uninterpreted sequence.**

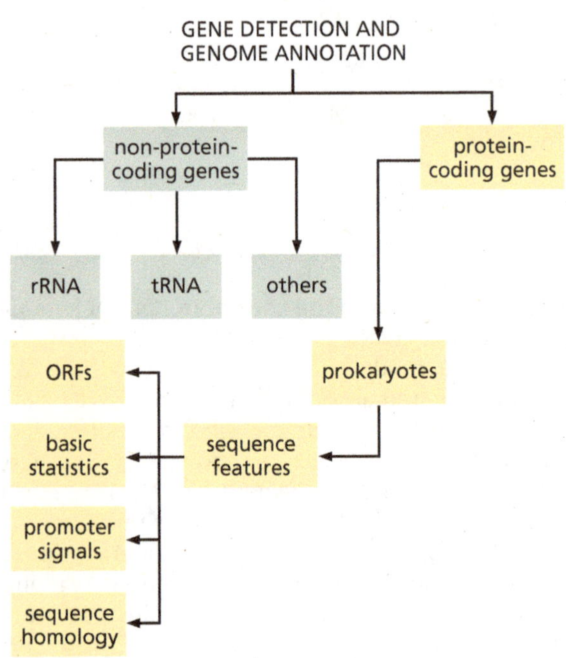

## Box 10.1 Measures of gene prediction accuracy at the nucleotide level

Authors often report a numerical value for the accuracy of their prediction program and it is useful to have an understanding of what these numbers mean. The ability to calculate an accuracy measure that enables the comparison of rival programs is important, especially when they are often the major method of predicting genes for new genomic sequences.

In the field of gene prediction, accuracy can be measured at three different levels: (1) coding nucleotides (base level); (2) exon structure (exon level); and (3) protein product (protein level). The second of these is only relevant to eukaryotes and will be discussed in Box 10.3. Here we will examine the first of these, accuracy at the nucleotide level. This subject was discussed extensively by Moises Burset and Roderic Guigo, and we will present their analysis and definitions, which are largely accepted within the field.

Figure B10.1 shows the four possible comparisons of real and predicted genes. Every base is assigned as gene or non-gene by the prediction program, and comparison with the real genes allows us to label each base. Considering first those bases correctly predicted, those that are within genes are labeled as a true positive (*TP*), while those predicted correctly not to be in a gene are labeled as a true negative (*TN*). Looking now at the bases whose prediction was incorrect, those wrongly assigned as within a gene are labeled as a false positive (*FP*), while those incorrectly omitted from a gene are labeled as a false negative (*FN*). Usually the base assignment is to be in a coding or noncoding segment, but this analysis can be extended to include noncoding parts of genes, or any other functional parts of the sequence.

The fraction of the bases in real genes that are correctly predicted to be in genes is the sensitivity (*Sn*) defined as

$$Sn = \frac{TP}{TP + FN} \qquad \text{(BEQ10.1)}$$

This can also be interpreted as the probability of correctly predicting a nucleotide to be in a gene given that it actually is. The fraction of those bases which are predicted to be in genes that actually are is called the specificity (*Sp*) defined as

$$Sp = \frac{TP}{TP + FP} \qquad \text{(BEQ10.2)}$$

This can also be interpreted as the probability of a nucleotide actually being in a gene given that it has been predicted to be.

Note that this last definition is different from that usually used in other fields for specificity:

$$Sp_{\text{standard}} = \frac{TN}{TN + FP} \qquad \text{(BEQ10.3)}$$

This is because in eukaryotes the proportion of nucleotides in genes is rather small, so that *TN* is very large, and the normal formula results in uninformative values. A program that simply predicts the human genome to contain no genes will give a high value for the normally defined specificity despite not being of any use in gene prediction.

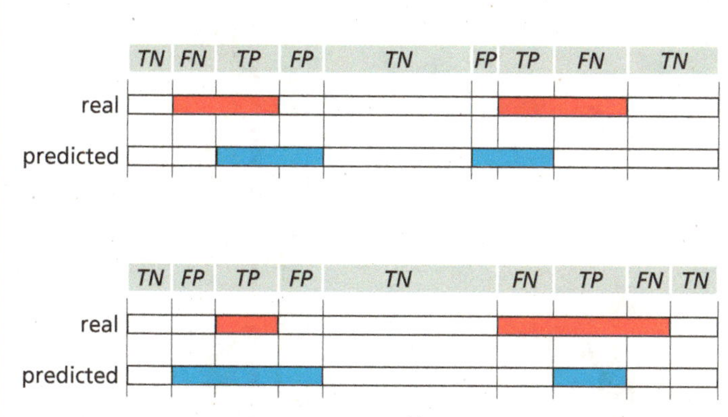


**Figure B10.1**
**Comparison at the base level of actual and predicted genes.** Red bars are entire actual genes in the case of prokaryotes or exons in the case of eukaryotes. Blue bars are entire predicted genes in the case of prokaryotes or exons in the case of eukaryotes. Regions of the sequence are assigned as true positive (*TP*), true negative (*TN*), false positive (*FP*), or false negative (*FN*). The accuracy of the predictions can be described using terms and formulae of Equations BEQ10.1 and BEQ10.2.

## Box 10.1 Measures of gene prediction accuracy at the nucleotide level (continued)

Care has to be taken in using these two values to assess a gene-prediction program because, as with the normal definition of specificity, extreme results can make them misleading. For example if all the sequence is predicted as a gene, we have $Sn = 1.0$ although $Sp$ may be small. If only a very few nucleotides are predicted as in genes, but all of them are in real genes, $Sp$ can be 1.0 while $Sn$ is small. Burset and Guigo proposed the approximate correlation coefficient ($AC$) as a single measure to circumvent these difficulties. This is defined as

$$AC = 2 \times ACP - 1 \tag{BEQ10.4}$$

where ACP (the average conditional probability) is defined as

$$ACP = \frac{1}{4}\left[\frac{TP}{TP+FN} + \frac{TP}{TP+FP} + \frac{TN}{TN+FP} + \frac{TN}{TN+FN}\right] \tag{BEQ10.5}$$

Note that the first two terms of $ACP$ are the sensitivity and specificity, and the third is the usual definition of specificity. $AC$ varies between −1 and +1, and is like a normal correlation coefficient. The standard Pearson correlation coefficient could have been used to obtain almost identical results, but has problems if both the real and predicted sequences do not contain genes and non-gene regions. The $AC$ does not have problems in such circumstances.

positives from potential new (previously unknown) genes. This point is discussed further later in this chapter.

The interactions of components of the transcription and translation machinery with the nucleotide sequence, coupled with the constraints imposed on protein-coding nucleotide sequence, have resulted in a number of distinct features that can be used to identify genes on the basis of their sequence (see Flow Diagram 10.2). Some of these aspects have been discussed in Section 1.3. Almost all gene-detection methods utilize one or more of these sequence features. For this reason, and because the methods require a more technical analysis than was presented in Chapter 1, they are briefly reviewed here.

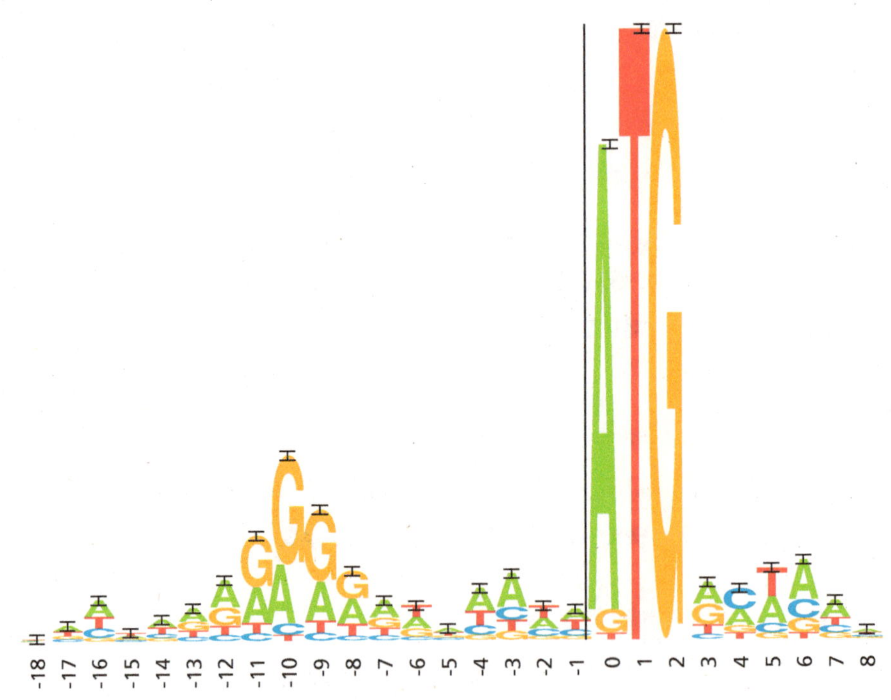


**Figure 10.4**
**A logo of the ribosome-binding sites and start codon in *E. coli* genes.** (See Section 6.1 for a definition of logos.) One hundred and forty-nine sequences were aligned on their start codons. Note that the number of bases between the ribosome-binding site (RBS) around position −10 and the start codon is not fixed, so the RBS signal is possibly stronger than shown here, as those sites are not properly superposed. The bases at positions 1 and 2 are fully conserved as T and G, respectively. However, at position 0, although the vast majority of observed bases are A, in addition G and to a very small degree T are also observed. Thus a small number of start codons in *E.coli* are TTG. (From T.D. Schneider and R.M. Stephens, Sequence logos: a new way to display consensus sequences, *N.A.R.* 18 (20):6097–6100, 1990, by permission of Oxford University Press.)

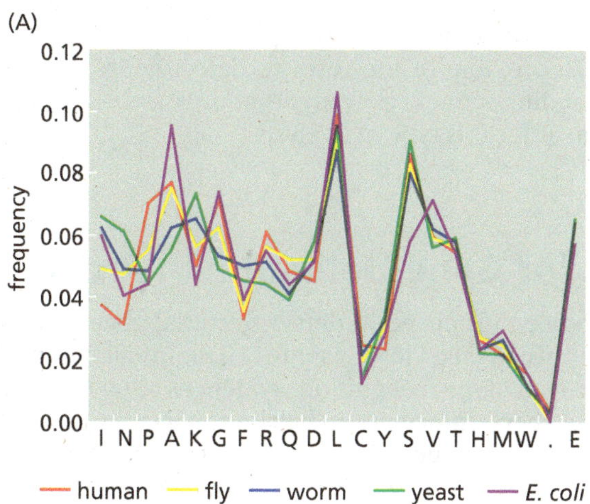


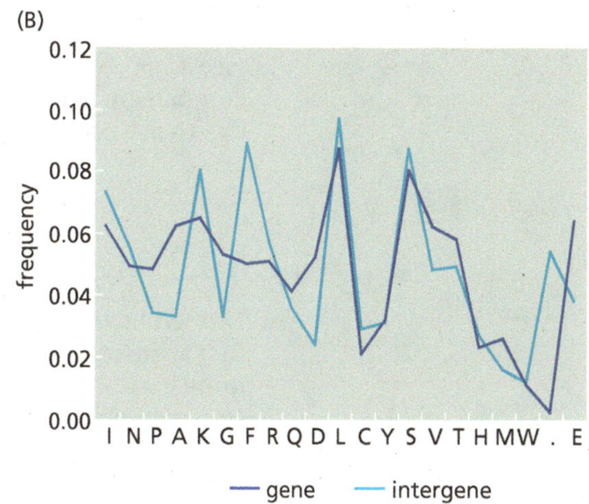


**Figure 10.5**

**Plot showing the frequency of occurrence of different amino acid codons in genes and intergenic DNA.** The amino acids are represented by their one-letter code, and the three stop codons are combined and represented by the dot. (A) shows a comparison of the amino acid codon frequencies in genes for humans, *D. melanogaster* (fly), *C. elegans* (worm), *S. cerevisiae* (yeast), and *E. coli* (*E. coli*). The amino acids have been ordered to show most difference between species on the left side of the graph. (B) shows the amino acid codon frequencies for *C. elegans* in genes and intergenic DNA. Clearly some of the amino acid codons show a considerable difference in occurrence in coding and noncoding segments. (Adapted from N. Echols et al., Comprehensive analysis of amino acid and nucleotide composition in eukaryotic genomes, comparing genes and pseudogenes, *N.A.R.* 30:2515–2523, 2002.)

The most basic characteristic of a gene is that it must contain an open reading frame (ORF) that begins with a start codon (ATG, coding for methionine) and ends with a stop codon (one of TAA, TAG, or TGA). In accordance with convention, we give here the codons found on the so-called coding or sense DNA strand, which has the same sequence as the RNA, which is physically transcribed from the complementary template strand: the mRNA will have the start codon AUG. If other recognition sequences are being looked for, care is needed to prevent confusion between the two DNA strands. It is important to appreciate that there are exceptions to these standard codons. For example, *E. coli* uses GTG for 9% and TTG for 0.5% of start codons (see Figure 10.4), and many mitochondria (with the exception of plant mitochondria) translate TGA as "tryptophan" rather than "stop." Once the genetic code relevant to the data has been ascertained, it is a simple matter to identify ORFs longer than some minimum length.

An ORF can have only a single stop codon, and that must be right at the 3′ end. As a result the statistics of relative frequency of codon occurrence are distinct for coding as opposed to noncoding sequences and can be used as a gene-prediction feature. Codon frequency will also be influenced by the amino acid composition of the encoded protein and by any codon bias imposed by tRNA availability. Examination of known genes has confirmed the validity of this premise, and distinct codon frequencies have been observed in the coding reading frame, other reading frames, and in noncoding sequence. In this context, noncoding sequence refers to sequence that does not have a gene in any reading frame. Figure 10.5 shows the observed frequencies for genes and noncoding sequence in the case of four eukaryotes. Practical applications of this gene feature tend to focus on the frequency distributions of hexanucleotides—stretches of six nucleotides, also called a dicodon—as this has been found to give better predictions than looking at individual codons.

As illustrated in Figure 1.12 there are three distinct recognition sequences in prokaryotic genes: two promoter-specific sequences just 5′ of the coding sequence and a translation-termination sequence 3′ of the stop codon. Characteristic consensus sequences were given for these in Section 1.3, but they can vary between organisms, and care has to be taken to use the best available information for the species whose sequence is being analyzed. Note also that prokaryotic genes are often arranged in operons (see Figure 1.15), with one promoter serving a cluster of individual protein-coding sequences.

One other property used in prokaryotic gene detection is the persistence of significant sequence similarity even after very long evolutionary periods. It is common to be able to detect protein-sequence homology between mammalian and bacterial species despite the considerable time since divergence from a common ancestor (see Chapters 4 and 5). Hence, one of the ways to detect genes is by sequence comparison with known coding sequences from other organisms. It is more effective to use protein sequences for this type of analysis.

## 10.3 Algorithms for Gene Detection in Prokaryotes

Numerous methods have been proposed to detect bacterial genes ever since the first sequences were determined (see Flow Diagram 10.3). Initial techniques were quite basic, as little was known of the recognition sequences or codon-distribution statistics noted above. As more information became available, gene-detection methods developed to take advantage of them. We will examine three methods that are among the most successful currently available: GeneMark and its successor GeneMark.hmm, GLIMMER, and ORPHEUS. While these do not cover the full range of techniques that have been applied, they are representative of most published methods. In addition, we will briefly present the model used in the EcoParse method, as it is an HMM model with basic similarities to those described in Section 6.2.

### GeneMark uses inhomogeneous Markov chains and dicodon statistics

GeneMark uses a **Markov chain** model to represent the statistics of coding and noncoding reading frames. The method uses the dicodon statistics to identify coding regions. Consider the analysis of a sequence $\boldsymbol{x}$ whose base at the $i$th position is called $x_i$. The Markov chains used are fifth-order, and consist of terms such as

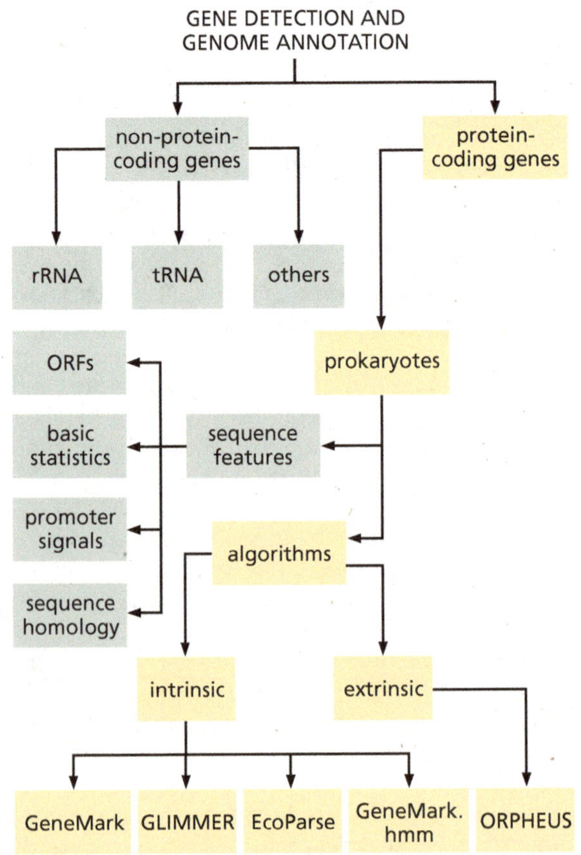


**Flow Diagram FD10.3**
**In this section the methods used to detect prokaryotic gene features are described.**

$P(a|x_1x_2x_3x_4x_5)$, which represent the probability of the sixth base of sequence $\boldsymbol{x}$ being $a$ given that the previous five bases in the sequence $\boldsymbol{x}$ were $x_1x_2x_3x_4x_5$, resulting in the first dicodon of the sequence being $x_1x_2x_3x_4x_5a$. These terms must be defined for all possible pentamers with the general sequence $b_1b_2b_3b_4b_5$. The values of these terms can be obtained by analysis of training data, consisting of nucleotide sequences in which the coding regions have been accurately identified. When there are sufficient data for good statistics without the need to resort to pseudocounts (see Section 6.1) they are given by

$$P(a \mid b_1b_2b_3b_4b_5) = \frac{n_{b_1b_2b_3b_4b_5a}}{\sum_{a=\mathrm{A,C,G,T}} n_{b_1b_2b_3b_4b_5a}} \qquad \text{(EQ10.1)}$$

where $n_{b_1b_2b_3b_4b_5a}$ is the number of times the sequence $b_1b_2b_3b_4b_5a$ occurs in the training data. This is the maximum likelihood estimator of the probability from the training data.

GeneMark assumes each reading frame has unique dicodon statistics, and thus has its own model probabilities. These are labeled $P_1(a|b_1b_2b_3b_4b_5)$, $P_2(a|b_1b_2b_3b_4b_5)$, and so on, where the coding reading frame is labeled 1, the other two reading frames in the same direction are labeled 2 and 3, and the three reading frames on the complementary strand are labeled 4–6. Note that the reading frames 1–3 are identified here by the codon position of the last base of the pentamer, so that $b_5$ is the $i$th base in a codon for the $P_i$ term, and in a similar fashion for reading frames 4–6. The regions of sequence that do not code for a protein are assumed to have the same statistics on all six reading frames, so they only have one set of parameters, labeled as $P_{nc}(a|b_1b_2b_3b_4b_5)$. These sets of parameters form seven distinct models, which can be used to estimate the likelihood that any given sequence is coding or noncoding, and, if coding, which reading frame is involved.

The probability of obtaining a sequence $\boldsymbol{x} = x_1x_2x_3x_4x_5x_6x_7x_8x_9$ in a coding region in the translated reading frame 3 (that is, $x_1x_2x_3$ is a translated codon) is given by

$$\begin{aligned} P(\boldsymbol{x} \mid 3) = {} & P_2(x_1x_2x_3x_4x_5)\, P_2(x_6 \mid x_1x_2x_3x_4x_5)\, P_3(x_7 \mid x_2x_3x_4x_5x_6) \\ & \times P_1(x_8 \mid x_3x_4x_5x_6x_7)\, P_2(x_9 \mid x_4x_5x_6x_7x_8) \end{aligned} \qquad \text{(EQ10.2)}$$

where the first term is the probability of finding the pentamer $x_1x_2x_3x_4x_5$ in the coding frame 2, which will result in $x_9$, the last base of $\boldsymbol{x}$, being in reading frame 3. Note the cycling of successive terms through $P_1(a|...)$, $P_2(a|...)$, and $P_3(a|...)$. Such models are called periodic, phased or inhomogeneous Markov models (see Figure 10.6). (The noncoding model is a homogeneous Markov model.) The probability of obtaining the sequence $\boldsymbol{x}$ on the complementary strand to the coding sequence can be obtained by a similar formula using cycling terms $P_4(a|...)$, $P_5(a|...)$, and $P_6(a|...)$.

What we really want is the likelihood that the segment of sequence $\boldsymbol{x}$ is in coding frame 3, which we will write $P(3 \mid \boldsymbol{x})$, the probability that model 3 applies given that we have sequence $\boldsymbol{x}$. Using Bayes formula (see Appendix A) we can derive

$$P(3 \mid \boldsymbol{x}) = \frac{P(\boldsymbol{x} \mid 3)P(3)}{P(\boldsymbol{x} \mid \mathrm{nc})P(\mathrm{nc}) + \sum_{m=1}^{6} P(\boldsymbol{x} \mid m)P(m)} \qquad \text{(EQ10.3)}$$

where P(3) and P(nc) are the *a priori* probabilities of the coding frame 3 and noncoding models, respectively and $m$ is the index of the coding frame. Equivalent formulae give the likelihood for each of the other models.

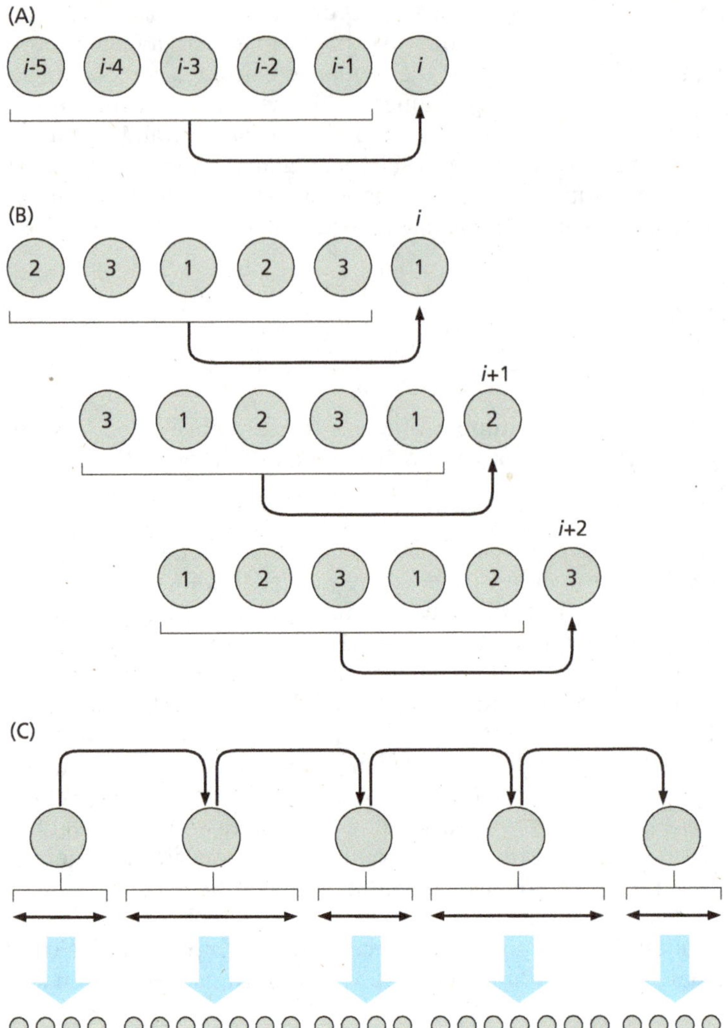


**Figure 10.6**
**Illustration of the structures of three different types of Markov model that output a nucleotide sequence.** (A) shows a homogeneous fifth-order Markov model, with the five states *i*–5 to *i*–1 generating state *i*. Each state corresponds to a nucleotide. (B) shows three periodic fifth-order Markov models, each modeling a different DNA reading frame. Each model is similar to that shown in (A), except that the probabilities are dependent on the position of the base within the codon. Each state is labeled with the codon position of the represented base. For example, the top model (generating state *i* at codon position 1) has the first base at codon position 2. (C) shows a hidden semi-Markov model. The circles in the top row now represent sequence features such as an exon or an intergenic region. These have specific length distributions and the first step in the output of sequence is generating their length, in this instance from a probability distribution. The lengths are represented by the double arrows in the middle row. Then, for each sequence feature in turn, sequences of the required length are generated according to models, which in the case of exons might be similar to the model shown in (B). These sequences are represented by the bottom row of circles. After generating a sequence feature the model moves to the next sequence feature, the move being represented by the black arrows at the top of the diagram. (Adapted from C.B. Burge and S. Karlin, Finding the genes in genomic DNA, *Curr. Opin. Struct. Biol.* 8:346–354, 1998.)

In GeneMark, P(nc)—the *a priori* probability of the sequence being noncoding—was assumed to be 1/2, and P(1)–P(6) assumed to all be 1/12. Sliding windows of 96 nucleotides were scored in steps of 12 nucleotides. If $P(i \mid \boldsymbol{x})$, where $i = 1$–6 exceeds a chosen threshold, the window is predicted to be in coding reading frame *i*. Note that GeneMark will only predict one reading frame to be a gene, so gene overlaps are not allowed. This could be potentially problematic, as bacterial genomes contain many small overlaps between genes. The final predicted gene boundaries should be defined by start and stop codons in that reading frame. The start codons are not always clearly predicted in GeneMark, which just reports the window containing the 5′ end and this could have more than one ATG codon.

*E. coli* genes can be divided into three groups according to their codon usage. Most genes are assigned to class I. Those in class II are expressed at an exceptionally high level, and those in class III have compositional characteristics that suggest they have been acquired by horizontal gene transfer (see Section 7.2). Using class I genes to provide the parameters for the models resulted in a poor prediction of the class III genes. Parameterization based on some class III genes improved their prediction, but lowered performance on the other gene classes. This shows the importance of selecting a good dataset for parameterization, as well as the problem of assuming that all genes can be described with the same model, even in the same organism.

## GLIMMER uses interpolated Markov models of coding potential

The GLIMMER program also uses Markov models to predict genes, but these models are slightly different in structure from those used in GeneMark and are called interpolated Markov models. The fifth-order Markov models described above require a large database of training sequences, because every dicodon must occur several times in order to obtain accurate estimates of the parameters. Each fifth-order Markov model in GeneMark has 4096 dicodon probabilities, plus the probabilities of occurrence of the pentamers. In practice, some hexamers will occur infrequently, while some longer oligomers may be more common, and even occur sufficiently frequently to permit their use in higher-order models. The GLIMMER program tries to tailor the Markov model according to the strengths and weaknesses present in the training data.

To make the method as automatic and objective as possible, GLIMMER selects its own training data from the sequence to be annotated. It searches for long ORFs that do not overlap any other long ORF. For example in the initial reported analysis of the bacterium *Haemophilus influenzae* genome the minimum ORF length used was 500 bases long. The intention is to select those ORFs that have a very high likelihood of being real genes. If the sequence is GC-rich, even longer ORFs may be needed, as there will be a lower probability of stop codons occurring by chance. It is also possible to intervene in selecting ORFs for the training set, for example to take account of genes already identified with confidence.

Using this training set, GLIMMER generates parameters for Markov models of increasing order from zeroth- to eighth-order. At every stage a check is made of the number of observations of each sub-sequence (pentamers for the fifth-order model, and so on). If there are at least 400 observations in the training data the parameterization occurs as given in Equation EQ10.1. If fewer occurrences are found, the resulting values are weighted according to how they compare with related values of a lower-order model. Suppose in parameterization of the *i*th-order model the sub-sequence $x_{j-i}x_{j-i+1}\ldots x_{j-1}$ does not occur sufficiently often. The four probabilities for the next base $x_j$ are compared with the values of the *i*–1th order model for the sub-sequence $x_{j-i+1}x_{j-i+2}\ldots x_{j-1}$ using a $\chi^2$ test. If this test, which accounts for the number of observations, shows the probabilities to be consistent with the lower-order model, the parameters of the *i*th-order model are weighted by zero. When the $\chi^2$ test supports a significant difference between the parameters of the two models, weighting is applied to the *i*th-order parameters reflecting both the confidence according to the $\chi^2$ test and also the number of observations.

The resultant interpolated Markov model can involve a weighted sum of terms from all model orders. However, if a higher-order term was derived from over 400 observations, the weighting is such that all lower-order terms are ignored. In that way preference is rightly given to the higher-order models when they are based on real data. Rather than scoring the complete sequence with a fixed window, all ORFs longer than some minimum are scored in all six reading frames as for GeneMark. If the model for the correct reading frame (in other words, the frame that includes the start and the stop codon) scores greater than a threshold value, the ORF is predicted to be a gene, assuming it passes a final test concerning gene overlap.

In GLIMMER the sub-sequences for which probabilities are given of a subsequent base are referred to as context strings, in that they give the context in which that base exists. All the Markov models mentioned so far use context strings that immediately precede the following base, but this need not be the case. Later versions of GLIMMER use what the authors describe as an interpolated context model. In this case the context string (still from 0 to 8 bases long) may not immediately precede the following base. The model uses those bases up to eight bases before the following base that have correlated distributions with it.

If two ORFs score well but overlap, an attempt is made to see if one ORF can be discounted on the basis of comparison of scores in the overlap region and ORF size. Because the overlap may arise from choosing a start codon upstream of the true start site, making the predicted gene longer than the true gene, GLIMMER tries to find alternative start codons for the shorter overlapping ORF that remove the overlap. If no suitable alternative start codon is found, the shorter ORF is rejected and predicted to be a noncoding sequence. GLIMMER looks carefully at overlaps between predicted genes, and selects a set of predicted ORFs that can overlap up to a specified maximum extent.

GLIMMER has proved to be very effective at automatically predicting the genes in bacterial genomes. It often correctly predicts 95% or more of the genes, with relatively few false positives and false negatives, even though it does not use any extra signal information such as ribosome-binding sites or homology with known genes. Many of the errors are due to short genes intentionally disregarded by the program. (The dicodon statistics of short ORFs cannot be determined with confidence.) The success of GLIMMER indicates the strong signal given by the base statistics.

## ORPHEUS uses homology, codon statistics, and ribosome-binding sites

The ORPHEUS program attempts to improve on the methods presented so far by using information that those programs ignored. One of the key differences is that ORPHEUS uses database searches to help determine putative genes, and is thus an extrinsic method. This initial set of genes is used to define the coding statistics for the organism, in this case working at the level of codons, not dicodons. These statistics are then used to define a larger set of candidate ORFs. From this set, those ORFs with an unambiguous start codon end are used to define a scoring matrix for the ribosome-binding site (RBS), which is then used to determine the 5′ end of those ORFs where alternative starts are present.

The starting point for ORPHEUS is to determine an initial set of high-confidence genes by sequence homology to known proteins. Ideally, care should be taken that the protein database used only contains entries that are very reliable. Many hypothetical proteins from previous genome annotation projects have been entered into protein databases, and in time some will almost certainly be found to be erroneous. The aligned regions are extended to the nearest start and stop codons, and this set of ORFs is used to determine codon statistics.

The frequency $f(x_i x_{i+1} x_{i+2})$ of occurrence of each codon $x_i x_{i+1} x_{i+2}$ in the ORF set is measured, as well as the base composition of the whole genome (frequencies $q_A$, $q_C$, $q_G$, and $q_T$). The codon frequencies are converted to $\log[f(x_i x_{i+1} x_{i+2})]$. The average value of $\log[f(x_i x_{i+1} x_{i+2})]$ will depend on the composition, and is given by

$$\mu = \sum_{\text{codon}} q_{x_i} q_{x_{i+1}} q_{x_{i+2}} \log\left[f(x_i x_{i+1} x_{i+2})\right] \quad \text{(EQ10.4)}$$

where the summation is over all codons in this set of high-confidence ORFs. The variance $\sigma^2$ of $\log[f(x_i x_{i+1} x_{i+2})]$ is defined by

$$\sigma^2 = \sum_{\text{codon}} q_{x_i} q_{x_{i+1}} q_{x_{i+2}} \left\{\log\left[f(x_i x_{i+1} x_{i+2})\right] - \mu\right\}^2 \quad \text{(EQ10.5)}$$

with the summation as before.

With these quantities, measures can be defined and used to predict ORFs in those regions not already covered by ORFs defined by sequence similarity. The normalized coding potential of the sequence segment $x_m...x_{m+3n-1}$, $n$ codon starting at base $x_m$, is then defined as

$$R(x_m...x_{m+3n-1}) = \frac{Q(x_m...x_{m+3n-1}) - \mu n}{\sigma\sqrt{n}} \qquad \text{(EQ10.6)}$$

where $Q(x_m...x_{m+3n-1})$ is called the coding potential, and is the sum of the $n$ terms $\log[f(x_i x_{i+1} x_{i+2})]$, one for each codon in the sequence. If the values $\log[f(x_i x_{i+1} x_{i+2})]$ had a Gaussian distribution this formula would represent the number of standard deviations of the codon frequencies of the region from the mean. This formula also takes account of sequence length, allowing values for different sequence lengths to be meaningfully compared.

The value of $R$ is obtained for the ORF reading frame, and also for the sequence in the alternative reading frames, $R(x_{m-1}...x_{m+3n-2})$ and $R(x_{m+1}...x_{m+3n})$. The amount by which $R(x_m...x_{m+3n-1})$ in the ORF reading frame exceeds the higher of the two alternative reading frame values is called the coding quality $\Omega(x_m...x_{m+3n-1})$. If $\Omega(x_m...x_{m+3n-1})$ exceeds a given threshold (and the sequence is longer than some minimum, typically 100 codons) that sequence defines a candidate ORF. Note that if two ORFs have significant overlap, the one with the higher coding potential $Q$ is chosen for further analysis.

At this stage an attempt is made to extend the candidate ORF in the 5′ direction. ATG, GTG, and TTG are all considered to be potential start codons (see Figure 10.4). Possible extensions are only considered if they do not lead to an overlap of more than six bases with another ORF. For an extra section to be added so that the ORF would begin at $x_{start}$, the coding quality $\Omega$ $(x_{start}...x_{start+98})$ of a 99-base section would have to exceed a second threshold value.

Some of the candidate ORFs only have one possible start codon and have upstream ORFs within 30 bases. This subset of ORFs has relatively reliable 5′ ends and is now used to define the RBS weight matrix. This involves alignment of the sequences up to 20 bases upstream of the start codon. The Shine–Dalgarno sequence (see Section 1.3) will not necessarily occur at the same distance from the start codon in all cases, and the method of defining the scoring scheme includes both trying to locate the signal on each sequence and calculating the base preferences at each position. An iterative method is used to locate the pattern in each sequence and to obtain the weight matrix.

Once the RBS scoring scheme has been refined, it is used to determine the start points for those ORFs that have alternative 5′ ends. If the RBS score does not exceed a threshold, the next potential start codon is considered. In this way ORFs with no known homology are defined including a translation start signal when possible.

It was often found that shorter ORFs would have higher coding potential than overlapping longer ORFs, leading to the latter being rejected early in the search. To overcome this problem, initial ORF searches ignore ORFs shorter than, say, 2000 bases. Once the longer ORFs have been predicted, further runs are made, gradually lowering the thresholds on the minimum ORF length. At each stage, the regions in which ORFs have been assigned are excluded from further analysis. In this way large overlaps between ORFs are avoided.

## GeneMark.hmm uses explicit state duration hidden Markov models

The original GeneMark method used nucleotide statistics to locate potential ORFs, but did so in a sequence window, and there was no further analysis to define the

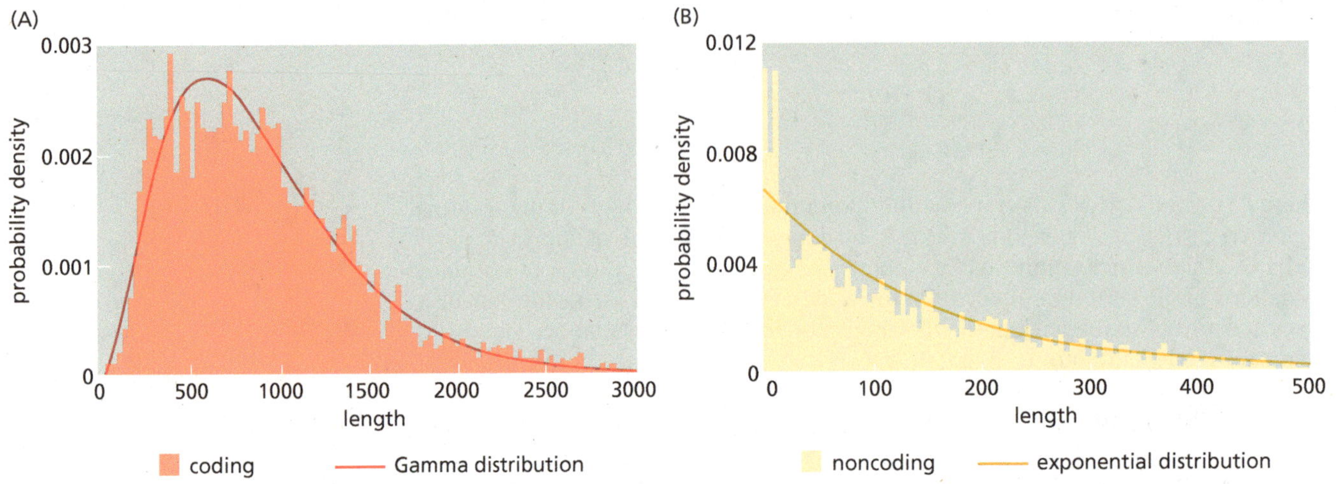


**Figure 10.7**
**Length distribution probability densities of protein-coding (A) and noncoding (B) sequences of *E. coli*.** The actual data are shown by the histograms. These have been fitted by simple algebraic distributions (a Gamma distribution for the protein- coding regions, and an exponential distribution for the noncoding regions) and illustrate how easily an experimentally observed complex length distribution can be represented. The lengths are given in nucleotides. (Adapted from A.V. Lukashin and M. Borodovsky, GeneMark.hmm: new solutions for gene finding, *N.A.R.* 26:1107–1115, 1998.)

gene boundaries. In developing an improved method, called GeneMark.hmm, the Markov models were transformed into a hidden Markov model of gene structure. Further improvements included the specific use of RBS signals to refine the 5′ end of the gene.

The HMM used in this work is different from that presented in Section 6.2, and is called a **semi-Markov model**, **HMM with duration**, or **explicit state duration HMM**. As discussed in Chapter 6, the kind of HMM presented there has a particular distribution of output sequence lengths associated with the number of states in the model and the transition probabilities between states. Examples of this are shown in Figure 6.10. In GeneMark.hmm, one state in the model will be used to generate many different genes. These genes will have a variety of lengths, with a particular length distribution that may not be well represented by the distributions shown in Figure 6.10. Compare this with the situation in profile HMMs, where most (all) sequences will be expected to have roughly the same profile length; for example, globin domains do not show much length variation. This length distribution can be imposed on an HMM with some relatively small modifications to the algorithms given in Section 6.2.

The details will not be given here, but the basic principle is that the length distribution must be specified such that it can be sampled. On entering such a state in an HMM, the first task is to determine from this length distribution the duration for that particular visit, for example, the length of this particular gene. The model then uses state emission probabilities to emit that number of bases, before a transition is made to another state of the model (see Figure 10.6C). In GeneMark.hmm, the emission probabilities are taken from the inhomogeneous Markov models that

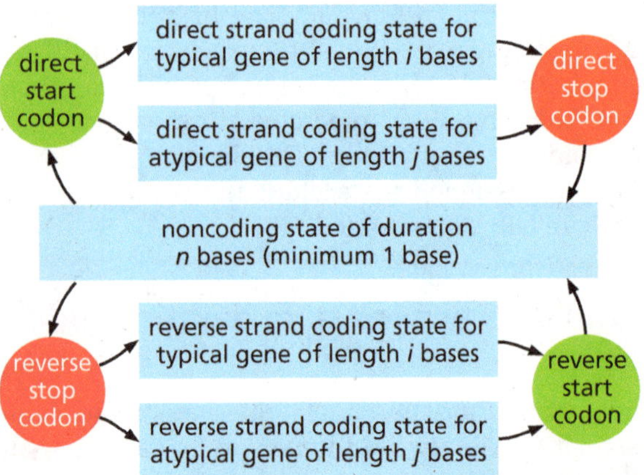


**Figure 10.8**
**The general structure of the HMM used in GeneMark.hmm, showing the individual states.** Each of the five main states (shown in blue) has a specific length distribution, specified by an algebraic distribution obtained from data such as those shown in Figure 10.7. The average durations are given, showing that the forward and reverse strand states have identical distributions. Note that the order of the states for reverse strand genes is opposite that for forward strand genes. This model does not just predict genes, but also assigns them as typical or atypical.

were previously described. The state durations were taken from a study of the length distributions of coding and noncoding regions of *E. coli* and were fitted to simple theoretical distributions to facilitate their incorporation into the model (see Figure 10.7).

The structure of the hidden Markov model used is given in Figure 10.8. Both strands are represented by their own states, and a distinction is made between typical and atypical genes. These correspond to the different classes of genes discussed above from *E. coli*, so that the atypical genes are those with different coding statistics, possibly because they are the result of a recent horizontal gene transfer. The model includes explicit states to represent the start and stop codons. It has a structure that prevents overlap of genes on the two strands, and in fact the noncoding state has a minimal duration of one base. The sequence emission path is assigned by using the Viterbi method, slightly modified from that given previously in Section 6.2 to account for the explicit duration states.

As small gene overlaps are quite common in bacterial genomes, a further step is added after this that uses information on the RBS to refine the position of the 5′ end. In GeneMark.hmm the RBS is defined as a five-position nucleotide-probability matrix based on an alignment of sites in 325 *E. coli* genes whose RBS signals have been annotated in GenBank. This matrix, which has a consensus sequence of AGGAG, was used to search upstream of any alternative start codons for genes predicted by the HMM. Those start codons with a sufficiently high-scoring RBS were accepted as the 5′ end, which leads to some small gene overlaps.

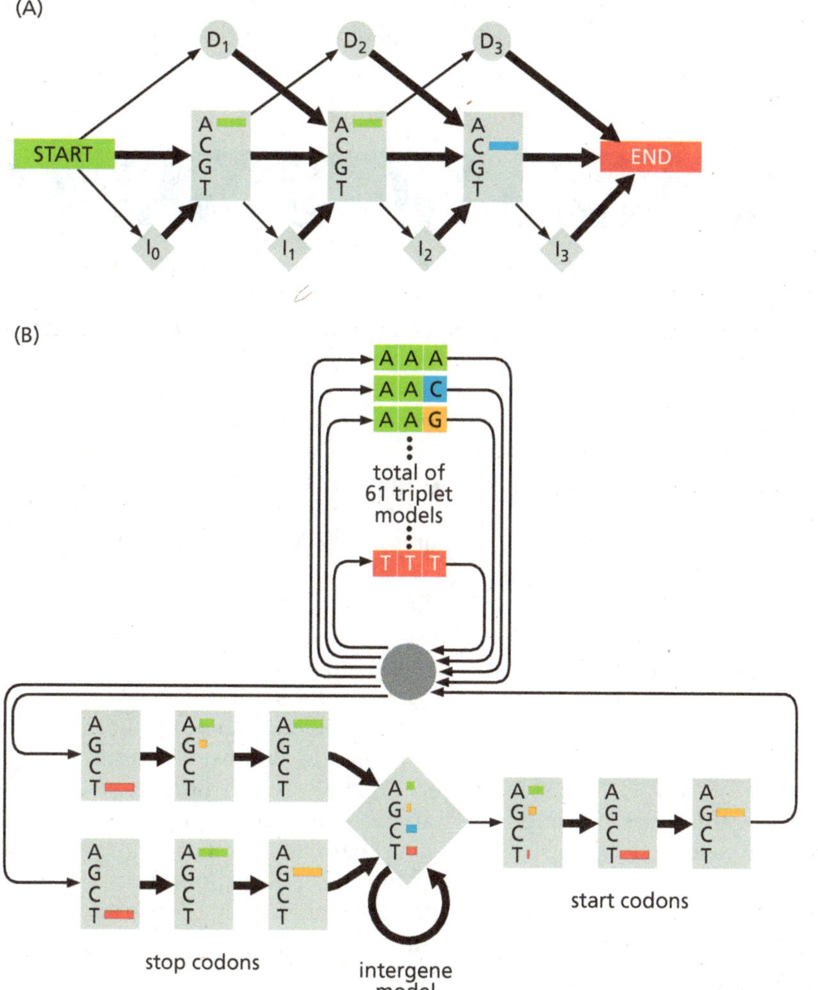


**Figure 10.9**
**The HMM structure for the EcoParse gene model of *E.coli*.**
(A) The HMM structure for an individual codon (in this case "AAC") of EcoParse. The state transitions are shown by arrows whose thickness is related to their probability. Note that there are fewer state transitions than in the HMM model shown in Figure 6.7. The match states show the probabilities for emission of the four bases. Each match state has only one non-zero emission probability, shown by the colored bar. (B) The complete HMM model for EcoParse, with the codon models and intergene model shown in abbreviated form. The intergene model has the same basic structure to the codon model, but may be over ten times the length. Note that the probabilities of the bases in the first match state of the start codon correspond to the data shown in Figure 10.4.

This method still has problems determining the 5′ ends of genes, and still has problems finding genes where there is extensive overlap. When tested on 10 bacterial genomes other than *E. coli*, more than 75% of genes were exactly predicted, and almost 95% of genes had at least the 3′ end correctly predicted. (These percentages are of genes annotated in the public databases; that is, they are the best available annotations including the experimental data where known.) In addition, approximately 10% of additional new genes were predicted. The status of these new genes could not easily be resolved, and they might ultimately be found to be false positives, although some did show homology to other database sequences.

## EcoParse is an HMM gene model

The methods described above are those that have been found to be most accurate, but they do not cover the full range of possible techniques. To introduce these techniques, we will look at one more method, which uses a hidden Markov model of a gene to detect genes in a nucleotide sequence. Unlike the methods already discussed, it allows for potential errors in the sequence. Due to the relative ease of use and success of newer methods, especially the ability of some to be easily reparameterized, it is not in common use at present, but it should be possible in principle to derive a related method that could compete with the best available. The model presented was designed for use with *E. coli* sequences, but could be reparameterized for other species. The potential advantage of allowing for sequence errors has been of little relevance with the accuracy standards achieved in many genome projects. However, it may be more useful in the future as some groups start investigating more complex systems such as mixed bacterial populations (see Box 10.2).

The HMM model used is very similar to those discussed in Section 6.2 for sequence profile alignment, and has match, insert, and delete states. Each of the 61 possible non-stop codons has its own model, the basic form of which is shown in Figure

### Box 10.2 Sequencing many genomes at once

Almost all genome projects to date have involved careful sequencing of the genome of a specific strain of a specific species, taking strenuous efforts to avoid contamination by foreign DNA. This is particularly important because of the need to break the genome into many small segments, as current sequencing technology can only work on a limited length of nucleic acid. In many projects care has also been taken to know quite accurately, in advance of performing the actual sequencing step, from which part of the genome each small segment has been obtained. To obtain this information requires a lot of extra effort in splitting up the genome. The rationale behind this method was that (especially in higher organisms) repeat sequences could make it impossible to reconstruct the whole genome correctly from the small sequence segments. Contamination from other genomes would only exacerbate the problem.

The success of the shotgun sequencing procedure has shown that, in practice, a less careful approach is possible. The main concept behind shotgun sequencing is that such concerns are unfounded, and that with suitable algorithms a set of short sequences can be assembled into a complete genome without the need for any extra positional information. This is the technique that was used by Celera in their work on the human genome.

Further work has shown (at least in the case of the small genomes of bacteria) that the assembly of shotgun sequences is even more robust, and with care can cope well with mixed genomes. A powerful demonstration of this is the investigation of the microbial populations in the surface waters of the Sargasso Sea by J. Craig Venter and colleagues. The sample contained a mixture of about 1800 different genomes at different levels of abundance. Although some of these genomes could be reconstructed with high accuracy and coverage (i.e., repeat sequencing) of three times or more, this did not apply to most of the sequence obtained. In this case the reported genes from the less-well-covered sequence were predicted on the basis of homology to existing bacterial database sequences. However, approaches such as EcoParse that can specifically allow for a low level of sequencing error (by allowing insertions and deletions in the codon models) may have advantages in these circumstances.

10.9A. The thickness of the transition arrows is proportional to the transition probabilities, so that insertions and deletions are rare events, but not impossible. In principle, this allows the model to account for sequencing errors, something not considered in the methods previously presented. The emission probabilities are very simple, with just one base having non-zero probability (that is, 1) at each match state. All the 61 residue-coding codon models begin and end at the same central silent state (see Figure 10.9B), and their transition probabilities from this state are related to codon usage in *E. coli*. The central state is reached from the noncoding intergene state via the start codon model, and leads eventually to the stop codon models. The start and stop codon models do not allow for insertions and deletions, and have emission probabilities related to the frequency of occurrence of these alternatives as seen in *E. coli*. For example, the start codon model produced the proportions of A, G, and T at first base seen in Figure 10.4. Several different intergene models were examined. To allow for variation in the intergene length, there are significant insertion-state probabilities in this part of the model. In principle, the TATA box and other sequence signals may be represented, as might more subtle signals not yet identified. The parameters of the model were fitted with a set of *E. coli* sequences.

## 10.4 Features Used in Eukaryotic Gene Detection

The eukaryotic gene-detection methods will be presented in a different way from those for prokaryotes discussed above. Instead of treating each program separately, we will focus on particular predictive features in eukaryotic genes (see Flow Diagram 10.4), and discuss a number of different programs under each heading. The reason for this is that there are a relatively small number of successful prokaryotic gene-detection programs that dominate the field and share many characteristics. By contrast, the eukaryotic gene-detection field is much more diverse, perhaps due in part to the relative lack of success of any one program to give a complete and accurate analysis (see Chapter 9). We will first discuss the techniques used to detect individual gene components before looking at the ways these have been combined to predict complete genes. Finally, we will look at the ways in which homology to the sequence databases has been used to aid gene detection and annotation. The emphasis will be on showing the range of methods that have been used, rather than giving full details for any one method.

### Differences between prokaryotic and eukaryotic genes

Many of the principles that apply to the detection of genes in prokaryotes also apply to gene finding in eukaryotes. For example, the coding regions of eukaryotic genomes have distinct base statistics similar to those found in prokaryotes. In addition, although the signals differ, there are equivalent transcription and translation start and stop signals. The detection of tRNA molecules is as easy here as for bacterial systems. However, as mentioned above, the considerably larger genome sizes in eukaryotes require detection systems with far lower rates of false positives.

A crucial difference in gene structure causes eukaryotic gene detection to be far harder: there are numerous introns present in many genes. As a result the length of the protein-coding segments (exons) is on average smaller in eukaryotes than in prokaryotes, resulting in poorer base statistics, and making their detection more difficult. Furthermore, in contrast to prokaryotes, whose ORFs will always end in a stop codon, this is not the case for exons, which may end in a splice site at the exon–intron junction. As will be seen, these splice sites do not have very strong signals, making their detection problematic.

An additional difference that can also cause difficulties is that the density of genes in most segments of eukaryotic genomes is significantly less than in prokaryotes.

**Flow Diagram 10.4**
**The key concept introduced in this section is that eukaryotic genes have certain features and signals that can be used to identify genes in uninterpreted sequence.**

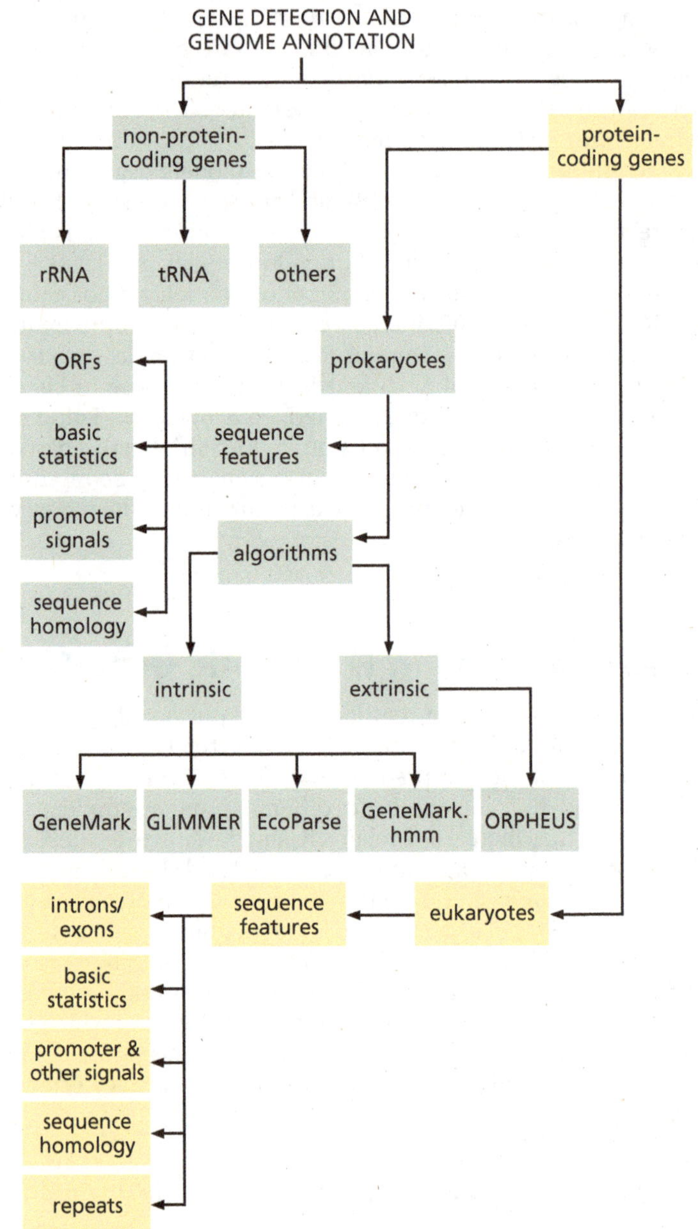


*E. coli* has on average one gene with a single ORF approximately every 1100 bases. The average vertebrate gene has six exons spanning about 30 kb, and intergenic sequences can be very large. For example in human chromosome 6 there are approximately 1600 genes in over 165 megabases. However, this is an oversimplification, as eukaryotic genomes often have regions of different GC content called isochores, which correspond to regions of higher and lower gene density. The regions of highest GC content approach the gene density found in bacteria, whereas those of lowest GC content can have less than a tenth the number of genes per megabase. Using this knowledge, we can have an idea of where to look for genes and roughly how many to expect. In addition, some programs use different parameterizations for different GC levels.

The large regions of the genomes of higher organisms that do not encode any genes have been referred to as junk DNA. Much of this junk DNA has repetitive sequences, and hence the first step in detecting eukaryotic genes is usually to detect these repeats using a program such as RepeatMasker. This can be coupled with the initial identification of all RNA genes to eliminate large sections of the genome from consideration as potential protein-coding genes, as functional regions rarely overlap.

## Introns, exons, and splice sites

In Figure 10.10 the length distributions are given for vertebrate introns and different classes of exons. By comparison with the *E. coli* data of Figure 10.7, exons are on average considerably shorter than prokaryotic genes. As a consequence, eukaryotic coding regions tend to be harder to recognize. Furthermore, whereas a start and a stop codon always delimit prokaryotic coding sequences, most exons are delimited by splice signals. In fact, exons are not the exact eukaryotic equivalent of a prokaryotic ORF because they will also include any 5′ or 3′ untranslated regions (UTRs), and it is not uncommon for the initial exon not to contain any coding sequence. To make matters worse, the splice signals at intron–exon boundaries are quite variable (see Figure 10.11), making them hard to locate accurately. Because eukaryotic gene structure is much more complex and variable than that of prokaryotes, the models used in eukaryotic gene detection must be more complex. As well as detecting the individual components they must allow for a variable number of exons in each gene.

To avoid potential confusion, it is worth noting that many people working in the field of gene detection use the term exon as a general term for a region of coding sequence. This is strictly speaking incorrect, as the initial and final exons will always contain some untranslated regions. Moreover, it is quite possible for an initial or

**Figure 10.10**
**The distribution of lengths of introns and exons in the human genome.** The number observed is plotted against the length in nucleotides for (A) introns, (B) initial exons, (C) internal exons, and (D) terminal exons. The intron distribution has been fitted with an exponential function, whereas empirical smoothed distributions were used to fit the exon data. Note that these data were analyzed before the complete human genome sequence was available, and the distributions may be different when applied to the complete genome. This is especially the case for the initial and terminal exons, where there were least data. (Adapted from C. Burge and S. Karlin, Prediction of complete gene structures in human genomic DNA, *J. Mol. Biol.* 268:78–94, 1997.

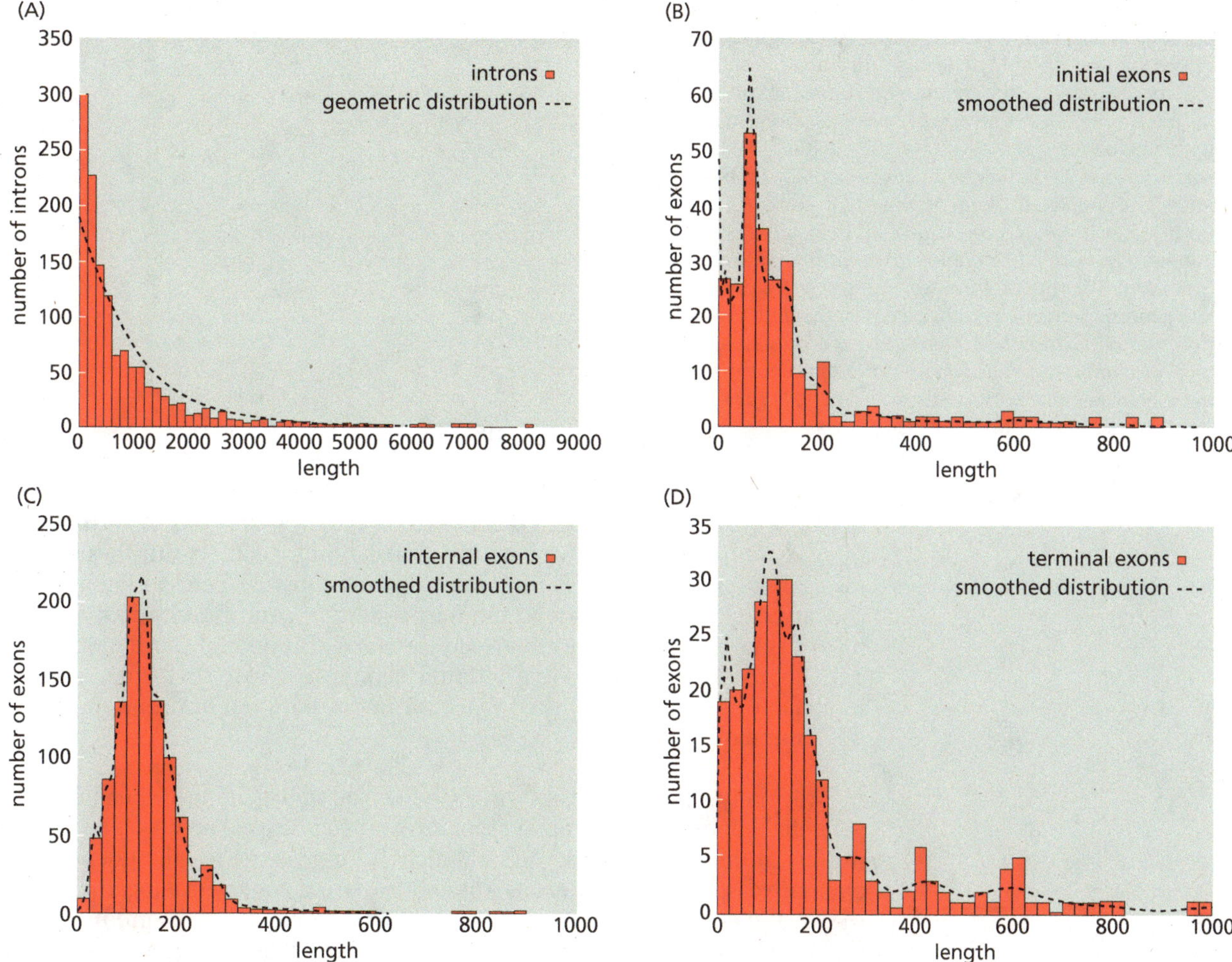

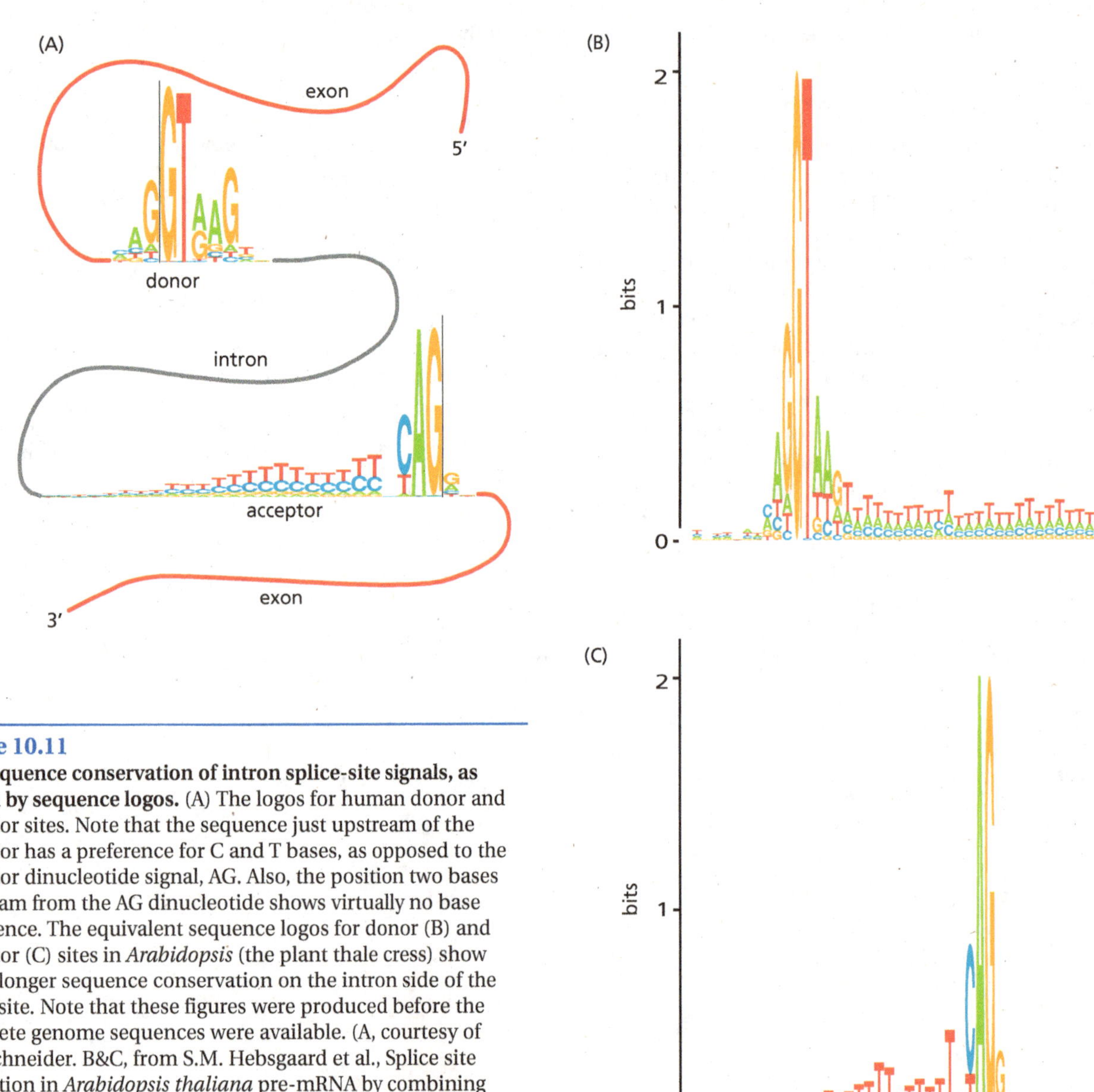


**Figure 10.11**
**The sequence conservation of intron splice-site signals, as shown by sequence logos.** (A) The logos for human donor and acceptor sites. Note that the sequence just upstream of the acceptor has a preference for C and T bases, as opposed to the acceptor dinucleotide signal, AG. Also, the position two bases upstream from the AG dinucleotide shows virtually no base preference. The equivalent sequence logos for donor (B) and acceptor (C) sites in *Arabidopsis* (the plant thale cress) show much longer sequence conservation on the intron side of the splice site. Note that these figures were produced before the complete genome sequences were available. (A, courtesy of T.D. Schneider. B&C, from S.M. Hebsgaard et al., Splice site prediction in *Arabidopsis thaliana* pre-mRNA by combining local and global sequence information, *N.A.R.* 24 (17):3439-3452, 1996, by permission of Oxford University Press.)

final exon to contain no coding sequence whatsoever. The reason for this lies with the substantially different way that the translation machinery attaches to the RNA transcripts in eukaryotes as opposed to prokaryotes. As seen in Figure 10.4, the prokaryotic RBS is very close to the start codon. In contrast, eukaryotic genes can have a considerable distance between the transcription and translation start sites and no discernable ribosome-binding sequences. Just to further complicate matters, it has been established that translation does not always start at the first ATG codon of the mRNA.

A particularly difficult problem can arise in eukaryotic genomes when moving from gene detection to protein prediction, a trivial step in prokaryotes. The splicing of introns in the RNA is not always identical for a given gene—the phenomenon of alternative splicing—with exons sometimes being removed altogether from the mRNA and not translated. Alternative splicing can give rise to the production of two or more different proteins from the same gene, and these are often known as **splice variants**. A well-known example of this is the alternative splicing that leads to the production of membrane-bound immunoglobulins in B lymphocytes, compared with the production of the same immunoglobulins in secreted form (antibodies) after the B lymphocytes have differentiated into plasma cells. The location of splice sites in a eukaryotic gene can therefore be hard to interpret in terms of a basic intron–exon structure, without a detailed knowledge of the different protein variants that can be produced from the gene. At present the only practical method of determining alternative splicing variants is to compare the whole sequence with the EST databases (see Box 9.1), relying on experimental data rather than attempting to predict this.

These factors have led to a greater emphasis on the use of homology and database searching to help identify eukaryotic genes. As well as locating or confirming genes through the use of protein-sequence homology, cDNA and EST databases are frequently used, the latter being especially useful for confirming the positions of introns and the existence of splice variants.

### Promoter sequences and binding sites for transcription factors

A further difference between prokaryotic and eukaryotic gene structures is that the sequence signals in the upstream regions are much more variable in eukaryotes, both in composition and position. The control of gene expression is more complex in eukaryotes than prokaryotes, and can be affected by many molecules binding the DNA in the region of the gene. This leads to many more potential promoter binding signals spread over a much larger region (possibly several thousand bases) in the vicinity of the transcription start site.

One feature worth noting is that there is strict control of gene expression, and some genes are known to be poorly expressed because high levels of expression would be damaging (for example, genes for growth factors). Such genes sometimes lack the TATA box that is characteristic of the promoters for many eukaryotic protein-coding genes (see Figure 1.13). Thus if certain promoter signals such as the TATA box are missing, this does not necessarily mean there is no gene or the gene is inactive. This can lead to extra problems in prediction and interpretation.

To date, as will be discussed below, these features have not proved particularly useful in eukaryotic gene detection, although it is not yet clear if this is due to weak signals or a current lack of knowledge of the actual signals. Despite differences in the fine details, the same kinds of signal detection methods are found useful for eukaryotic gene detection as were used for prokaryotes.

## 10.5 Predicting Eukaryotic Gene Signals

As mentioned above, there are many different sequence signals present in eukaryotic genomes that can be used to help in gene prediction (see Flow Diagram 10.5). These signals are more variable than in prokaryotes in that they are often absent (or as yet undetected) and the promoters and other regulatory systems bind to regions that can be far removed from the site of transcription. As a result, detection of these signals is usually more difficult in eukaryotes. Even those signals that are present in both prokaryotes and eukaryotes have different sequence preferences, and are often located using different techniques. In this section some of these methods will be presented to show the great variety of approaches that can be used.

### Detection of core promoter binding signals is a key element of some eukaryotic gene-prediction methods

The absence of ribosome-binding signals in eukaryotic genes means that ideally we would like to detect the transcription start site (TSS) so as to be sure of the 5′ end of the mRNA. This requires knowledge of the core promoter sites, the sequences that bind the general transcription-initiation factors and the RNA polymerase, and methods of finding these sites will be discussed here. In addition, there are numerous other conserved sites in the extensive promoter regions of eukaryotic genes that bind specific gene-regulatory proteins and are found in many different genes. These sites are usually found by similar methods to those presented here.

The control of transcription initiation is highly complex in eukaryotes, and there is great variation in the sequence signals present. As described in Section 1.3, there are

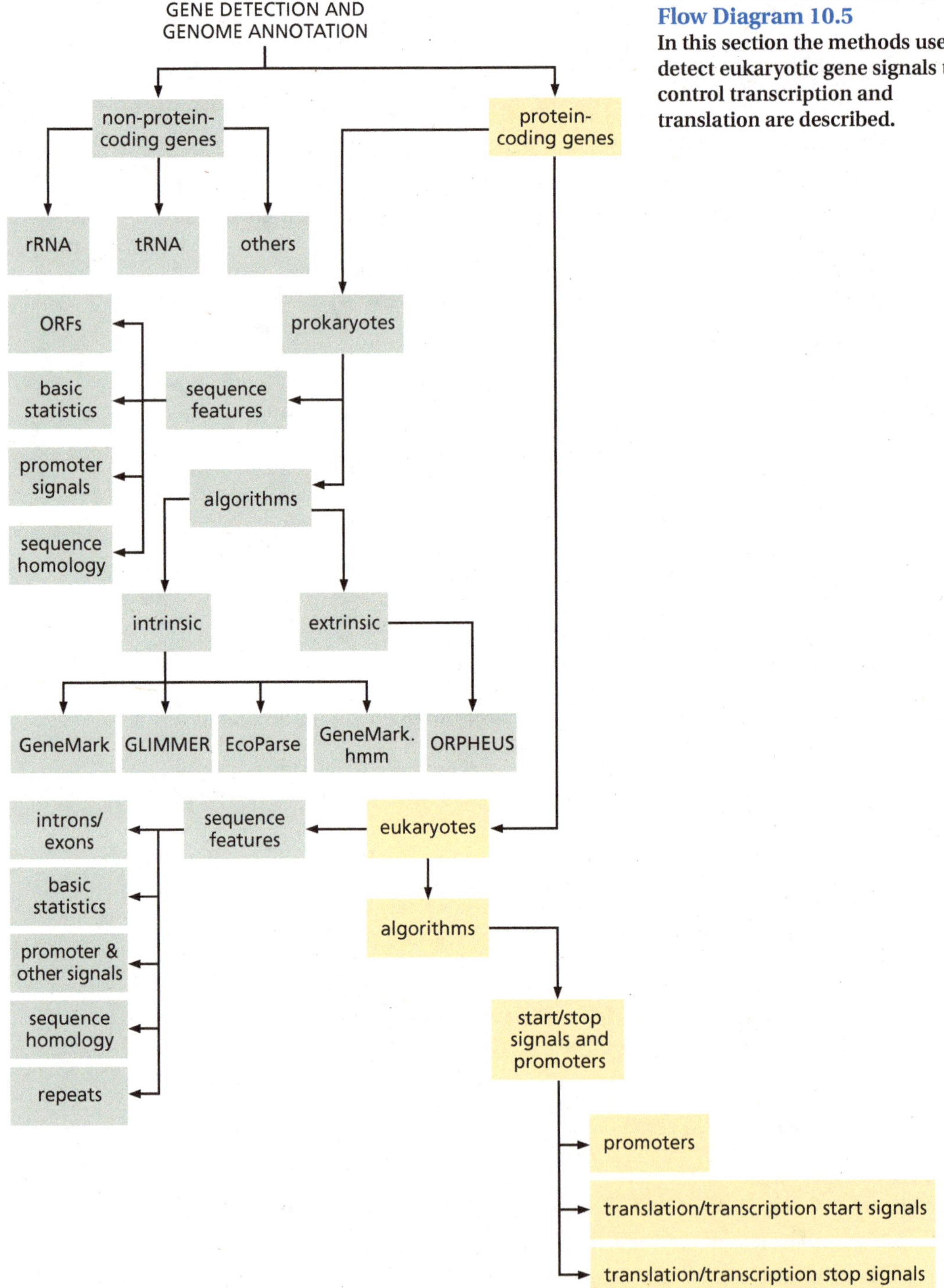


**Flow Diagram 10.5**
**In this section the methods used to detect eukaryotic gene signals that control transcription and translation are described.**

three RNA polymerases involved in transcription in eukaryotes, each transcribing a different class of genes. Most protein-coding genes whose transcription is regulated are transcribed by RNA polymerase II and have a similar type of core promoter sequence that includes the TATA element, which binds the general transcription factor TFIID, and on which the transcriptional complex including RNA polymerase II is assembled. However, 30% of protein-coding genes do not have the TATA element. In addition, the location of the TATA element relative to the TSS is quite variable. The same applies to all the other core promoter sequences, so that it has not proved possible to generate a model of eukaryotic core promoter signals that does not miss a significant percentage of known genes. For this reason, many gene-detection programs do not incorporate promoter predictions, and the study of these upstream sequence signals is often carried out separately (see Section 9.5).

The first step in generating models of the promoter signals is to collect together sequences experimentally verified as involved in binding of the transcription-initiation complex to the promoter, and to detect the degree of sequence similarity. In Chapter 6, methods of sequence pattern detection were presented, and those discussed below are very similar, many being variants of position-specific scoring matrices (PSSMs) and logos (see Sections 6.1 and 6.3). Two key sets of techniques will be presented here: those that aim to identify the precise location of the promoter signals, and those that only attempt to locate the general region containing the TSS.

## A set of models has been designed to locate the site of core promoter sequence signals

In 1990 Philippe Bucher derived weight matrices to identify four separate RNA polymerase II promoter elements, namely the TATA box, the cap signal or initiator (Inr), the CCAAT box and the GC box. Using more than 500 aligned eukaryotic sequences, the weights of different bases $a$ at position $u$ in a signal sequence were obtained from the general equation

$$w_u(a) = \ln\left(\frac{n_u(a)}{e_u(a)} + \frac{c}{100}\right) + c_u \qquad \text{(EQ10.7)}$$

where $n_u(a)$ is the number of occurrences of base $a$ at position $u$ in the alignment, $e_u(a)$ is the expected number of bases $a$ at position $u$, $c$ is a small number (often 2) to prevent numerical problems when base $a$ is not observed at this position, and $c_u$ is adjusted to make the greatest $w_u(a)$ zero. The value of $e_u(a)$ depends on whether mononucleotides or dinucleotides are used as the reference. In the former case the values are independent of position $u$, and reflect the base composition of the sequences, whereas the latter case pays attention to the observed preferences of base $a$ to be preceded or followed by particular bases.

The sequences used for this study had to be very carefully curated to ensure the information was reliable and unbiased. Only sequences whose transcription start site was experimentally determined and which were thought to use the RNA polymerase II were included. These were further pruned to remove those with significant sequence similarity in the region upstream of the TSS.

It should be noted that the alignment of these sequences prior to deriving the weight matrices could not be done using the methods described in Chapters 5 and 6. As the sequences of interest are short and are DNA they will occur by chance with relatively high probability. Furthermore, as they represent binding sequences, gap insertion within the patterns is not appropriate. For these reasons, a special technique was developed that jointly optimized the alignment, the definition of the pattern, and the weight matrices. A key feature of this method is to produce the best possible signal for the promoter sequence relative to the background.

The weights for the 15-base TATA-box signal are given in Figure 10.12A together with the frequencies of occurrence of the bases and the cut-off score that must be exceeded for a sequence to be assigned as a TATA box. In Figure 10.12B the position distribution of this signal from the TSS is plotted, which indicates it is found between 20 and 36 bases upstream of the TSS (measured from the second T; position 0 in Figure 10.12B). Although the signal sequence (as defined by the weights) occurs at other positions in some sequences, the background occurrence is low, so that this TATA sequence is distinctive and a good promoter signal. In Figures 10.12C and 10.12D the equivalent data are given for the 8-base cap signal. This has a very narrow distribution about the experimental TSS (1–5 bases downstream), but has a higher background frequency, no doubt partly on account of being shorter and thus more likely to occur by chance. Unfortunately, even the low background of the TATA-box matrix does not prevent false negatives occurring, often at a relatively high rate of one per kilobase of sequence.

Both the TATA box and cap signal have been found in vertebrate and non-vertebrate sequences, whereas the CCAAT box and GC box are only found in vertebrates and, moreover, have a rather high level of background noise, despite being 12 and 14 bases long, respectively, and both having some very highly conserved positions. Another feature of the CCAAT box and GC box is the wide variation in their distance

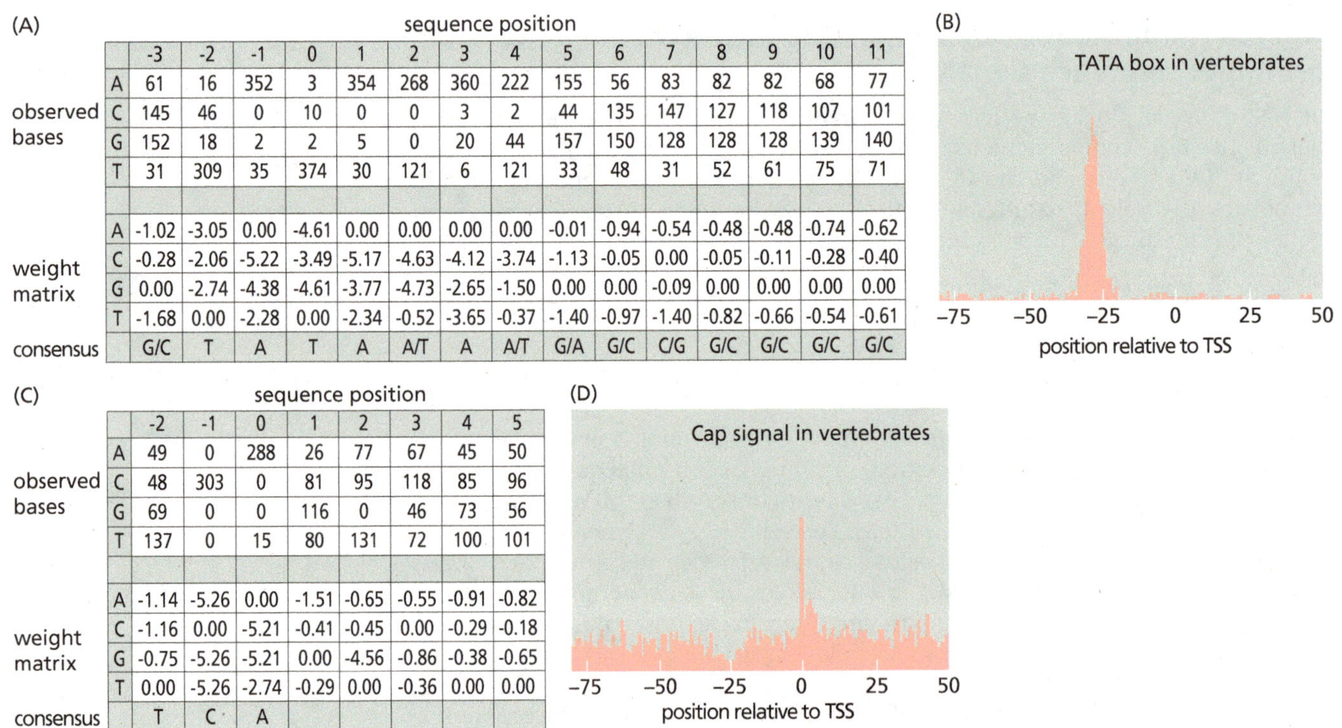


(A)

| | | -3 | -2 | -1 | 0 | 1 | 2 | 3 | 4 | 5 | 6 | 7 | 8 | 9 | 10 | 11 |
|---|---|---|---|---|---|---|---|---|---|---|---|---|---|---|---|---|
| observed bases | A | 61 | 16 | 352 | 3 | 354 | 268 | 360 | 222 | 155 | 56 | 83 | 82 | 82 | 68 | 77 |
| | C | 145 | 46 | 0 | 10 | 0 | 0 | 3 | 2 | 44 | 135 | 147 | 127 | 118 | 107 | 101 |
| | G | 152 | 18 | 2 | 2 | 5 | 0 | 20 | 44 | 157 | 150 | 128 | 128 | 128 | 139 | 140 |
| | T | 31 | 309 | 35 | 374 | 30 | 121 | 6 | 121 | 33 | 48 | 31 | 52 | 61 | 75 | 71 |
| weight matrix | A | -1.02 | -3.05 | 0.00 | -4.61 | 0.00 | 0.00 | 0.00 | 0.00 | -0.01 | -0.94 | -0.54 | -0.48 | -0.48 | -0.74 | -0.62 |
| | C | -0.28 | -2.06 | -5.22 | -3.49 | -5.17 | -4.63 | -4.12 | -3.74 | -1.13 | -0.05 | 0.00 | -0.05 | -0.11 | -0.28 | -0.40 |
| | G | 0.00 | -2.74 | -4.38 | -4.61 | -3.77 | -4.73 | -2.65 | -1.50 | 0.00 | 0.00 | -0.09 | 0.00 | 0.00 | 0.00 | 0.00 |
| | T | -1.68 | 0.00 | -2.28 | 0.00 | -2.34 | -0.52 | -3.65 | -0.37 | -1.40 | -0.97 | -1.40 | -0.82 | -0.66 | -0.54 | -0.61 |
| consensus | | G/C | T | A | T | A | A/T | A | A/T | G/A | G/C | C/G | G/C | G/C | G/C | G/C |

(C)

| | | -2 | -1 | 0 | 1 | 2 | 3 | 4 | 5 |
|---|---|---|---|---|---|---|---|---|---|
| observed bases | A | 49 | 0 | 288 | 26 | 77 | 67 | 45 | 50 |
| | C | 48 | 303 | 0 | 81 | 95 | 118 | 85 | 96 |
| | G | 69 | 0 | 0 | 116 | 0 | 46 | 73 | 56 |
| | T | 137 | 0 | 15 | 80 | 131 | 72 | 100 | 101 |
| weight matrix | A | -1.14 | -5.26 | 0.00 | -1.51 | -0.65 | -0.55 | -0.91 | -0.82 |
| | C | -1.16 | 0.00 | -5.21 | -0.41 | -0.45 | 0.00 | -0.29 | -0.18 |
| | G | -0.75 | -5.26 | -5.21 | 0.00 | -4.56 | -0.86 | -0.38 | -0.65 |
| | T | 0.00 | -5.26 | -2.74 | -0.29 | 0.00 | -0.36 | 0.00 | 0.00 |
| consensus | | T | C | A | | | | | |

**Figure 10.12**
**The position-specific nucleotide preferences of the vertebrate TATA box and cap signal.** (A) The observed base frequencies for the region –3 to +11 of the TATA box, and the weight matrix obtained from them. (The "0" position is the second T in the sequence.) A simple representation of the base preferences is shown at the bottom of each column. The weight matrices are used to score for the presence of a TATA box. The score, summed for all 15 positions, must exceed –8.16 for the region to be said to have a TATA box. (B) The distribution of TATA boxes in a test dataset identified using the weight matrix shown in (A). The horizontal axis is the base number relative to the transcription start site. (C) The observed base frequencies for the region –2 to +5 of the cap signal, and the weight matrix obtained from them. The weight matrices are used to score for the presence of a cap signal. The score, summed for all eight positions, must exceed –3.75 for the region to be said to have a cap signal. (D) The distribution of cap signals in a test dataset identified using the weight matrix shown in (C). The horizontal axis is the base number relative to the transcription start site, and shows a clear peak at 0 as would be expected. (Adapted from P. Bucher, Weight matrix descriptions of four eukaryotic RNA polymerase II promoter elements derived from 502 unrelated promoter sequences, *J. Mol. Biol.* 212:563-578, 1990.)

from the TSS. This difference can be as much as 150 bases; this means that even after the signals have been detected, the location of the TSS is still much in doubt.

In GenScan (a gene-detection program based on HMMs, which is discussed more fully below) the promoter detection component uses Bucher's TATA-box and cap-signal models. In an attempt to avoid missing those genes that lack the TATA box, the model allows for both possibilities. The allowed alternatives are a TATA box and cap signal with a 14–20-base (random) sequence between them or a 40-base random sequence (see Figure 10.13). It has been acknowledged that this model is rather basic, in that if a TATA box is absent, no signals are searched for. Given that it is just one part of a larger algorithm, only limited resources could be devoted to it, and a lack of greater understanding of the biological mechanism prevented the construction of more specific models.

The NNPP (Neural Network for Promoter Prediction) program uses a very similar model for promoters to that in GenScan, namely a TATA box and Inr (initiator) signal separated by a variable number of bases. In this case, however, a neural network is used to identify potential promoter sites (see Figure 10.14).

The key features of neural nets that are needed to understand their use here will now be briefly described. The detailed description of neural network architectures and training techniques will be left until Section 12.4, where they are presented in their application to secondary structure prediction. The models consist of layers of units, which usually communicate only with units in other layers, communication involving passing a value to one or more other units. The value emitted by a unit depends in some way on all the inputs it receives, for example only emitting a signal if the sum of all input signals exceeds a threshold. In addition, the connections have weights associated with them. One of the unit layers (the input layer) emits values that (in applications based on predicting properties of a sequence) depend on the small window of sequence. (For examples see Figures 10.14 and 10.15.) Another key layer is the output layer, which usually has only a few (often just one) units. The emission from the output layer, usually in the range 0 to 1, is interpreted as the prediction for the sequence window or central residue/base. There are usually one or more other layers between these two, called hidden layers, which modulate the emission from the input layer and participate in transforming it into the output prediction. Networks are trained with a set of data to determine suitable values for any weights and thresholds in the model.

This particular neural network of NNPP has one set of hidden units to detect the TATA box and another set to detect the Inr signal, and uses a time-delay neural network (TDNN). Each sequence position is represented by four input units, each representing one base, with a total of 51 bases in the input layer. Each TATA-box hidden unit has inputs from 15 consecutive bases, and the Inr hidden units also receive signals from 15 consecutive bases. However, the region of the sequence scanned by the TATA-box hidden units (at the upstream end of the input layer) is different from the equivalent region for the Inr hidden units (at the downstream end of the input layer). The same weighting scheme is applied to all the TATA-box hidden units and similarly (with another set of weights) for the Inr hidden units. All the hidden units feed into a single output unit whose signal is used to determine the prediction. The weights were obtained by training with a set of reliable

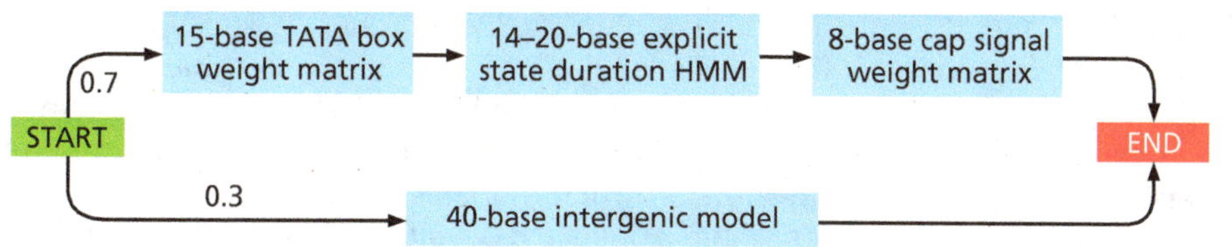


**Figure 10.13**
**The promoter prediction method in GenScan, showing two alternative and mutually exclusive paths.** The TATA box and cap signals are identified using the weight matrices given in Figure 10.12. The sequence between them is allowed to be between 14 and 20 bases long. The alternative to this is a 40-base region without any specific sequence signals identified. This latter model is chosen with a probability of 0.3, as opposed to 0.7 for the TATA box and cap signal pathway.

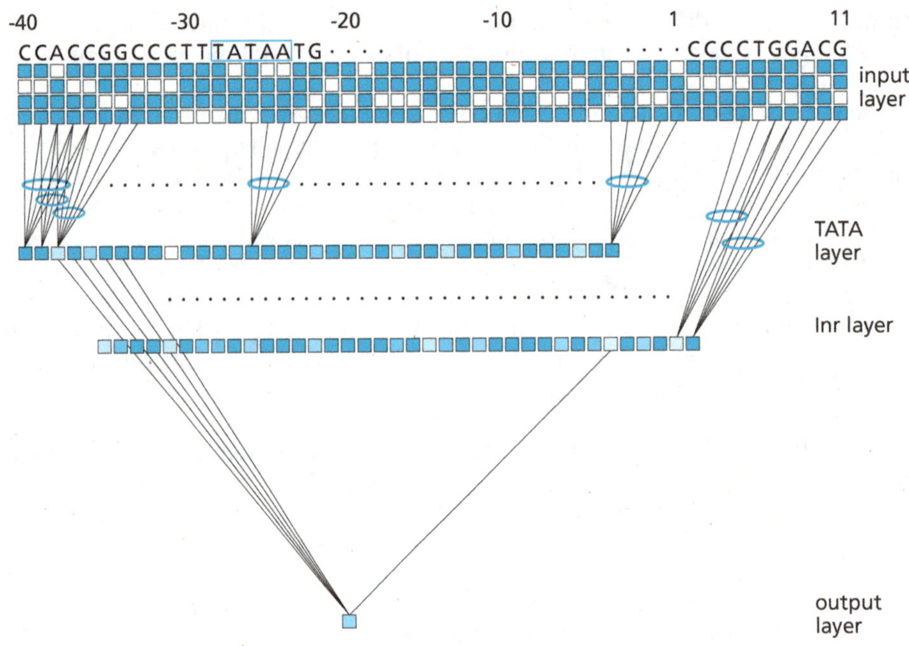


**Figure 10.14**
**The neural network used by NNPP to locate TATA-box and Inr signals in a time-delay neural network.** The four layers of squares underneath the sequence represent the input layer units and encode the sequence in a very simple way. Each unit will have a value of 0 or 1 related to the base at that position in the sequence, and the units are colored according to their values; blue corresponds to 0 and white to 1. Each of the units in the two hidden layers (the TATA and Inr layers) receives input from a consecutive set of 15 bases. The sets for each hidden layer unit are represented by an oval, and for clarity only a few representative ones are shown. The connections between the input layer units and hidden layer units are indicated by a line, and again only some representative lines are shown. There is a set of weights associated with these connections, so each connection can make a distinct contribution to the value of the hidden units. All the hidden units make a contribution to the value of the single unit of the output layer. If this output unit value exceeds a threshold, the 51-base window is identified as containing a TATA box and Inr signal, although the precise locations are not specified. If the output unit value is less than the threshold, no sites are identified in this window. The sequence shown has a TATA-box signal, as indicated by the boxed segment, but no Inr signal. (From M.G. Reese, Computational prediction of gene structure and regulation in the genome of *Drosophila melanogaster*, PhD Thesis, University of Hohenheim, 2000.)

sequences that had been carefully annotated by a human curator. One difference between this technique and the HMM approach of GenScan is that, in general, the precise location of the TATA box and Inr signal are not readily available, and NNPP just reports that the two were found within the 51-base window. This method has been incorporated into the complete gene-prediction package Genie.

Another program that uses a neural network to predict promoters is Grail, which like GenScan is a complete gene-prediction package. In this case, instead of using a neural net that scans the sequence, the sequence is scanned with matrices to identify the four elements identified by Bucher. Within a given sequence window the number of such elements found, their score, and the distances between certain of them are all used as inputs to the neural network (see Figure 10.15) together with the GC content of the window. The neural network then emits the promoter score, which is used together with predictions of the 5′ and 3′ ends of coding regions to predict the location of promoters.

As well as the core promoters found in many genes, a large number of transcription factors bind in the upstream region of genes and regulate gene activity. In eukaryotes these binding sites frequently occur within 500 bases of the TSS, although even more distant binding sites are not uncommon, as are regulatory sites downstream of the TSS. A number of databases have been compiled listing the known binding sites for transcription factors.

The ProScan (PROMOTER SCAN) algorithm tries to use these signals to assist in locating the promoter site. The principle of the method is to identify regions containing a number of such signals known to occur upstream of the TSS, in addition to locating the TATA box using Bucher's matrix. The locations of putative transcription factor binding sites are identified in a training sequence database that distinguishes between non-promoter sequence and the two strands of promoter sequence. This is done for a sequence window (250 bases in the standard implementation) and is converted into a density (the number of sites per base). A promoter-recognition profile is calculated as the ratio of the densities of forward and reverse promoter strands to the non-promoter density, each transcription factor signal giving a separate component of the profile. (To avoid numerical problems, the profile score assigned is 50 where no sites occur in the non-promoter sequences, and where no promoter sequence has a site the density weight of 0.001

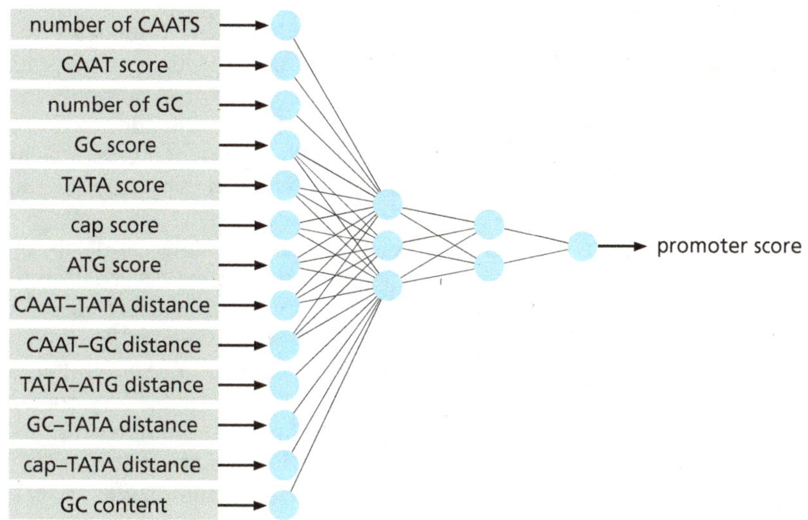


**Figure 10.15**
**The neural network used by the Grail program to predict RNA polymerase II promoters.** A variety of measures, such as the weight matrix score for TATA-box signal and simple measurements such as the GC content and the distance between different signals, are used to assign values to the units in an input layer of a neural network. These feed via two hidden layers to an output layer unit, which predicts the presence of RNA polymerase II promoters in the sequence window. (Adapted from E.C. Uberbacher, Y. Xu and R.J. Mural, Discovering and understanding genes in human DNA Sequence using GRAIL, *Methods in Enzymology* 266:259–281, 1996.)

is assigned.) In this way preferences to be upstream or downstream of the TSS are quantified. When predicting the promoter site for a query sequence, the same sequence window size is used, and any transcription factor signals identified are scored with the profile density. In addition, the 50 bases at the 3′ end of the window are tested for a TATA box with the Bucher matrix, scoring 30 if one is found. A threshold score is used to determine whether the window is predicted to contain a promoter. For high thresholds, at which only 50–60% of promoters were identified, the false-positive rate could be as low as one per 15–25 kilobases. A lower threshold found 80% of the promoters but gave a false positive every 2590 bases; given the length of eukaryotic genes this means that the majority of the predictions were false.

## Predicting promoter regions from general sequence properties can reduce the numbers of false-positive results

The large number of false-positives that arise when searching eukaryotic genomes for short sequence signals is simply a reflection of the fact that a short sequence will occur quite often by chance. Modifying the parameters to reduce the number of false positives does not necessarily solve this problem, as often this results in fewer true positives. To get round this difficulty, some techniques attempt to identify just the general region of the sequence containing the promoter on the basis of its general sequence properties. In fact, the accurate location of even the general region containing promoters is useful, as it could be used to separate a genome into individual genes, and thus assist in the general problem of gene detection.

The PromFind program devised by Gordon Hutchinson is one such program. It is based on the observed frequencies of hexamers in promoter, coding, and noncoding sequence regions. A differential measure $D_1(b_1b_2b_3b_4b_5b_6)$ comparing promoter and noncoding regions was calculated using

$$D_1(b_1b_2b_3b_4b_5b_6) = \frac{f_{\text{promoter}}(b_1b_2b_3b_4b_5b_6)}{f_{\text{promoter}}(b_1b_2b_3b_4b_5b_6) + f_{\text{noncoding}}(b_1b_2b_3b_4b_5b_6)} \tag{EQ10.8}$$

where $f_{\text{promoter}}(b_1b_2b_3b_4b_5b_6)$ is the observed frequency of hexamer $b_1b_2b_3b_4b_5b_6$ in the promoter regions. An equivalent measure $D_2(b_1b_2b_3b_4b_5b_6)$ was obtained for promoter and coding regions. These parameters vary from 0 to 1, with higher values occurring for hexamers preferentially found in promoters. Some of the hexamers found with greatest $D_1(b_1b_2b_3b_4b_5b_6)$ in the promoter regions corresponded to known patterns, such as TATAAA, and many contained CpG dinucleotides. To make a prediction, the $D_1(b_1b_2b_3b_4b_5b_6)$ and $D_2(b_1b_2b_3b_4b_5b_6)$ values are summed over a 300-base window that is moved 10 bases each time along the sequence. Subject to the two sums exceeding empirically determined thresholds, the window with the highest $D_1(b_1b_2b_3b_4b_5b_6)$ score is predicted to be a promoter region. This rather simple algorithm was reasonably successful in identifying promoter regions.

The PromoterInspector program developed by Matthias Scherf and colleagues uses a related method to PromFind in that specific short sequences are used to classify a region. Rather than search for hexamers, this method looks for nucleotide sequences of the general form A-$N_1$-B or A-$N_1$-B-$N_2$-C, where A, B, and C are short oligonucleotides of specific sequence (length 3–5 bases), and $N_1$ and $N_2$ are short stretches of any sequence (up to 6 bases). An example of such a sequence would be ACTNNCCG (where N is any base). As in PromFind, a sequence window must be used, in this case of 100 bases, and successive windows are four bases apart. If at least 24 consecutive windows are identified as having promoter characteristics, that region, which will be at least 192 bases long, is predicted to contain promoter sequences. In the initial tests the predicted promoter regions averaged 270 bases in length. This method does not predict the strand, so that the TSS could occur at either end of the prediction. From the few comparisons available, this method appears to correctly predict fewer promoter regions, but also makes significantly fewer false-positive predictions. Results from PromoterInspector are shown in Section 9.5.

McPromoter, devised by Uwe Ohler and colleagues, predicts a 300-base region to contain the promoter, with the TSS 50 bases from the 3′ end. The prediction is based on an interpolated Markov model similar to that used in GLIMMER; that is, it relies on distinct coding statistics in the promoter region. In this case the test sequences were classified into promoter, intron, and coding regions, and interpolated Markov models were generated for each group. The method was effective at distinguishing promoter regions from coding sequence, but fared worse in distinguishing promoters from introns. Despite the attention paid to the problem, the prediction of many false positives remains.

The final promoter-prediction method we will examine is CorePromoter, which is based on discriminant functions. In this context, a discriminant function is a numerical combination of measured properties of a sequence region whose value can be used to distinguish, for example, promoters from introns. In this particular case, the function involves linear and quadratic terms of the properties. The basic principle of a quadratic discriminant function is given in Figure 10.16A. The measurements were based on pentamer frequencies in a 30- or 45-base window relative to those in surrounding windows. Pentamers were chosen instead of hexamers because they required less data to achieve accurate statistics. Eleven windows labeled $w_i$ surrounding the TSS were used to define the variables (see Figure 10.16B), with the score of pentamer $b_1b_2b_3b_4b_5$ in window $w_i$ defined by

$$S_i(b_1b_2b_3b_4b_5) = \frac{f_i(b_1b_2b_3b_4b_5)}{f_i(b_1b_2b_3b_4b_5) + f_{av}(b_1b_2b_3b_4b_5)} \quad \text{(EQ10.9)}$$

where $f_i(b_1b_2b_3b_4b_5)$ is the frequency of occurrence of $b_1b_2b_3b_4b_5$ in the window, and $f_{av}(b_1b_2b_3b_4b_5)$ is the average of $f_{i-1}(b_1b_2b_3b_4b_5)$ and $f_{i+1}(b_1b_2b_3b_4b_5)$. An initial linear discriminant model based on windows $w_4$ to $w_7$ gave good signals that peaked at the TSS, but there were also many spurious signals. The larger model shown was found to have far fewer false positives and be less susceptible to changes in window position or size.

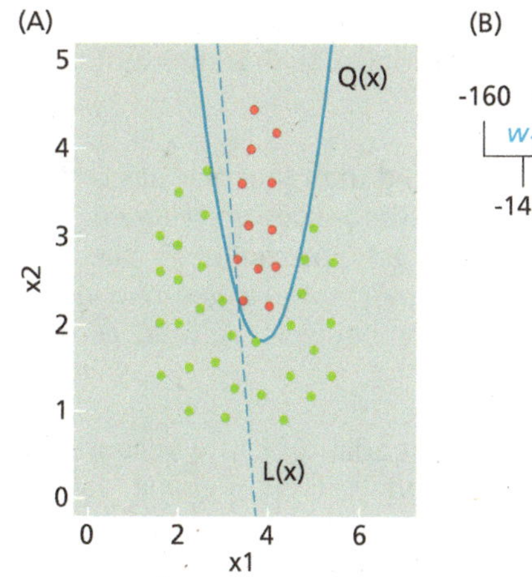


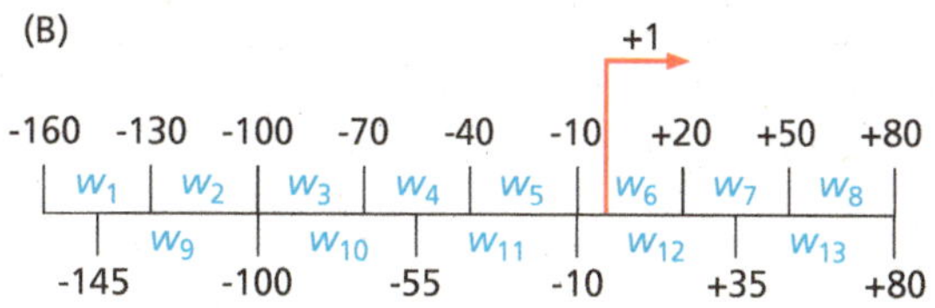


**Figure 10.16**
**An illustration of the principles of linear and quadratic discriminant analysis (LDA and QDA, respectively) and their application to locating transcription start sites.** (A) Two properties x1 and x2 are measured for every member of a set of samples (sequences in the present context, but in principle any object). Each of the samples is in this case known to belong to one of two groups. In the present context the properties may be base statistics, and the groups may be those that code for proteins and those that do not. When the values of these properties are plotted, distinguishing the two groups here by green and red dots, the groups are found to lie in distinct regions of the graph. (This will only be the case for certain properties that are in some way related to the grouping.) In fortunate cases a suitable line [L(x)] or quadratic curve [Q(x)] can be found that divides the graph into two regions corresponding to these sample groups. Here only the quadratic curve can do this. Once the curve is known, the group membership of an unknown sample can be predicted. The group is assigned according to which side of the quadratic curve the x1 and x2 values would be plotted. (B) Part of the CorePromoter method for predicting the location of the transcription start site (TSS) involves a QDA in the region surrounding the TSS (located at position +1). A base statistic measure is calculated for 13 windows (sequence segments) labeled $w_1$ to $w_{13}$, the regions being shown in the figure. The values calculated are used with a formula which is a quadratic discriminant to predict if there is a TSS within the 240-base segment. (A, adapted from M.Q. Zhang, Identification of protein coding regions in the human genome by quadratic discriminant analysis, *Proc. Natl. Acad. Sci. U.S.A.* 94:565_568, 1997.)

### Predicting eukaryotic transcription and translation start sites

Although exons are usually defined by their coding statistics, using techniques not too dissimilar to those used in prokaryotes to detect coding regions, there is still a difficulty in defining the 5′ end of translation. There is no defined ribosome-binding site in eukaryotic genes, as the ribosome simply attaches to the 5′ end of the mRNA and then slides along until it finds a suitable transcription start site. The region around the ATG start codon is found to affect the efficiency of translation, presumably according to the strength with which it binds the ribosome, and a number of examples exist where translation starts at the second or third ATG codon from the 5′ end of the mRNA. GenScan uses a 12-base weight matrix to detect suitable translation start sites with the ATG in positions 7–9. There are problems in locating the start codons in some genes because the sequence seems designed to prevent easy translation. This seems to be the case in genes for growth factors and transcription factors, for example, whose products must be under tight control.

Predicting the position of the transcription start site requires the identification of some core promoters, especially the cap (Inr) and TATA-box signals. The cap signal occurs at the point of transcription initiation. If the TATA box can be located transcription will be expected to start 20 to 36 bases downstream. However, as mentioned above, not all genes contain these signals and they are also difficult to detect. As prediction of the gene product does not require the location of the transcription start site, most programs do not attempt a prediction.

### Translation and transcription stop signals complete the gene definition

As in prokaryotes, the eukaryotic translation stop signal is simply one of the three stop codons, and thus is easily recognized in the coding region. The transcription stop signal is the polyadenylation signal, with the consensus AATAAA. In GenScan a standard weight matrix is used to model this. In Grail the matrix extends 6 bases upstream and 59 bases downstream. Many methods do not attempt to locate the 3′ end of the mRNA, and focus exclusively on the protein-coding region of the gene.

## 10.6 Predicting Exons and Introns

As the protein-coding regions of eukaryotic genes are interspersed with noncoding introns, the relatively simple ORF identification techniques that were employed for prokaryotic genomes must be radically modified. When the exons are long enough, some of the base statistic techniques can still be used, but these need to be combined with methods to identify splice sites in order to define their limits (see Flow Diagram 10.6).

Because the signals for splice sites are not well defined their location has been an area of considerable research, with many different methods tried and refined. There is still a need for significant improvements to assist gene prediction. It should be noted that this apparent lack of clear definition may have a genuine role in processes such as alternative splicing, which is now being recognized as much more common than initially realized.

The initial and terminal exons present a further potential difficulty because they may also contain noncoding segments (the 5′ and 3′ UTR regions). Specific methods have been developed to try to identify these specific types of exon.

## Exons can be identified using general sequence properties

All internal introns and exons in a eukaryotic gene are delimited by the splice sites at which introns are cut out of the RNA transcript and the exon sequences joined together. These splice sites have distinct sequence signals. There are programs that predict splice sites without information about introns and exons, and other programs that predict introns and exons without reference to splice sites (see Sections 9.3 and 9.4 for some examples and their results). The former methods, focusing on the splice sites, will be covered in the subsequent section. In this section we will examine the latter methods, some of which take advantage of the base statistic properties of the protein-coding regions. It should be noted that the structure of these genes has led to some alternative definitions of the prediction accuracy based on the introns and exons instead of looking at the level of individual nucleotides (see Box 10.3).

In some cases the prediction of coding and noncoding regions per se closely follows the techniques already discussed for prokaryotes, namely the use of statistics based on dicodons, and distinguishing the reading frames in coding regions. Note that the initial and terminal exons also include the 5′ and 3′ UTRs, which will have noncoding statistics, thus complicating the analysis. A further complication is that exons tend to be shorter than the average prokaryotic ORF, and can even be less than 10 bases long. The base statistics methods cannot identify these very short exons. There seems to be a minimal length for introns that varies according to species but is around 50 bases.

Of the programs we have touched on already, GenScan identifies eukaryotic coding regions by dicodon statistics, as in the prokaryotic examples given earlier, but in GenScan's case it uses an explicit state duration HMM based on the observed length distribution of real exons (see Figure 10.10). The length of the potential exon is generated from this distribution, and its sequence generated with probabilities based on the dicodon statistics.

Neural networks can also be used to predict coding regions, as in programs such as Grail and NetPlantGene. In NetPlantGene, which as its name suggests was developed specifically for use in plants, six separate networks are used, taking the average of their output as the final predictive result. Four of these networks have 804 input units representing 201 bases (each sequence position being represented by four units, one for each possible base), connected to 15 units in a single hidden layer that feed their signals to a single output layer unit. These four networks differ in their initial weights and hence in their final trained weights. The other two networks have input layers corresponding to 101 and 251 bases. Each of these six networks produces an output value between 0 and 1, which is a measure of the likelihood that the central base (in the input window sequence) is in a protein-coding

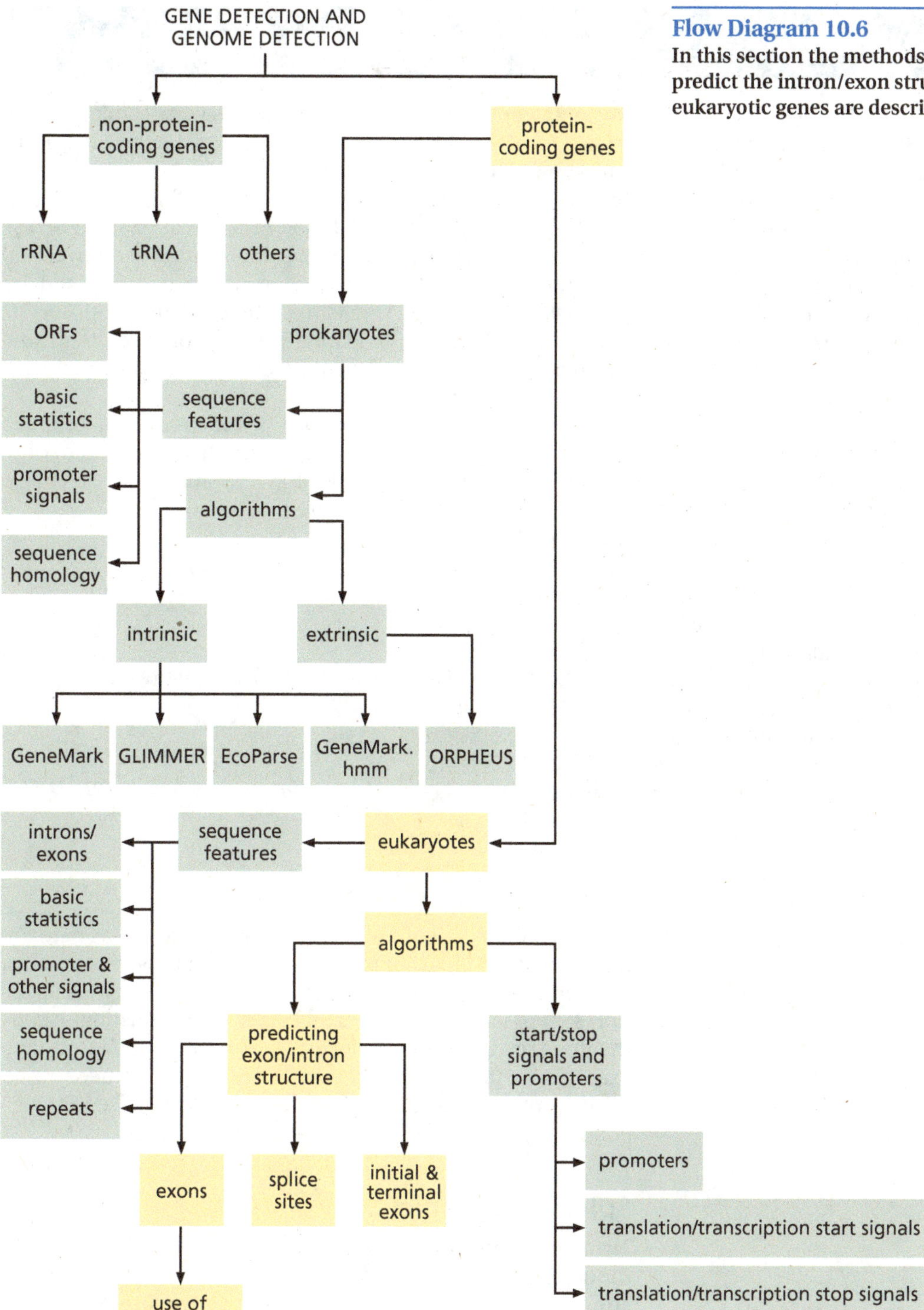


**Flow Diagram 10.6**
**In this section the methods used to predict the intron/exon structure of eukaryotic genes are described.**

region. The query sequence is scanned to produce the prediction of which regions are protein coding. The final prediction is made on the basis of the average of the six separate values. Prediction values >0.5 are regarded as predicting a coding base, whilst lower values predict the base to be non coding. An example output for the NetPlantGene coding prediction is shown in the top section of Figure 10.17.

One completely different aspect of coding regions in eukaryotes compared to prokaryotes is that the exons in the transcribed RNA will be spliced together to produce the final protein-coding sequence. This means that the reading frames of

## Box 10.3 Measures of gene prediction accuracy at the exon level

As discussed in Box 10.1, it is important to be able to quantify the accuracy of predictions so that rival programs can be meaningfully compared. The exon structure present in most eukaryotic genes complicates this task. For example, it would be useful to know if a program was partially successful in predicting a gene because it failed to predict one of the exons but correctly predicted all the others, as opposed to identifying all the exons but not predicting any of them exactly. Such information is also of great value when deriving new methods, as it can highlight specific deficiencies that would merit further attention. In Box 10.1 the accuracy measurements based on individual nucleotides were presented. Here we will look at the problem at the exon level. As before, we will present the analysis of Moises Burset and Roderic Guigo.

At the exon level, determining prediction accuracy depends on the exact prediction of exon start and end points (see Figure B10.2). The number of actual (real) exons in the data is represented by the symbol *AE*. The number of predicted exons is represented by the symbol *PE*. In the measures presented here, four kinds of exons are distinguished. Correct exons (total number represented by *CE*) are those that are predicted exactly, i.e., with the exactly correct splice sites. Missing exons (*ME*) are those that occur in reality but have no overlap with any predicted exons. Wrong exons (*WE*) are the converse, namely the predicted exons that have no overlap with any real exons. Finally, pairs of predicted and real exons that have an overlap but are not exactly identical are ignored. (It should be noted that other measures have been proposed that, for example, consider the case of predicted exons with only one of the splice sites correctly located.)

There are two measures each of sensitivity and specificity used in the field, each of which measures a different but useful property. The sensitivity measures used are

$$Sn_1 = \frac{CE}{AE} \quad \text{and} \quad Sn_2 = \frac{ME}{AE} \qquad \text{(BEQ10.6)}$$

The specificity measures used are

$$Sp_1 = \frac{CE}{PE} \quad \text{and} \quad Sp_2 = \frac{WE}{PE} \qquad \text{(BEQ10.7)}$$

the exons must be consistent, which is referred to as being in-frame. This requirement must be incorporated into gene-prediction programs. For example, if one exon ends at codon position 2, the next must start with its first base at codon position 3. In practice this means that exons can be distinguished according to which of the three reading frames they use. This is probably best understood with a practical example, as in Figures 9.5 and 9.8. This does not need to be accounted for when initially predicting exons, but must be taken account of when predicting the complete gene, as is covered later in Section 10.7.

### Splice-site prediction

Turning to splice-site prediction, most introns start (i.e., have a 5′ end) with a GU dinucleotide in the RNA (GT in the DNA) at what is referred to as the **donor splice site**. The 3′ end of introns (**acceptor splice site**) is mostly an AG dinucleotide. Locating occurrences of AG and GT would identify all possible splice sites with these sequence properties, but in addition there would be about 30 to 100 false predicted sites for every true one. As with promoters, this problem must be resolved by trying to use the properties of the surrounding sequence to reduce the false-negative prediction rate to a manageable level. Figure 10.11A shows sequence logos for the regions of the donor and acceptor splice sites for human introns, and Figures 10.11B and 10.11C show equivalent logos for *Arabidopsis*. These data indicate that there are extensive sequence signals available to help detect splice sites, and that these can be species specific. It should be noted that the informative sites are mainly on the intron side of the splice site. The methods presented below ignore the existence of the rare U12-type introns (see Section 1.3); this is an acceptable approximation, as far less than 1% of all introns are of this type.

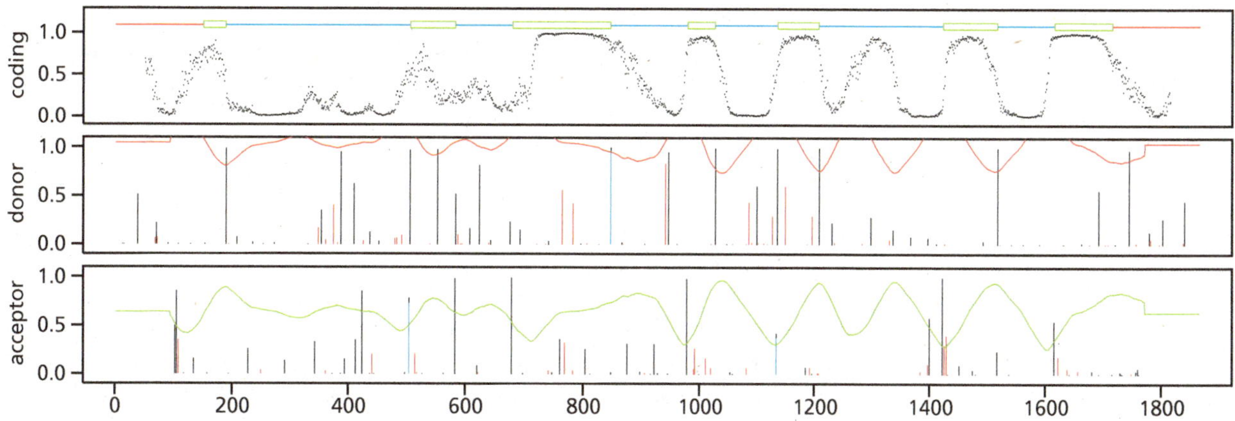


**Figure 10.17**
**The NetPlantGene prediction for the *A. thaliana* Rha1 gene sequence.** The top plot shows the score for the coding prediction networks, and plotted against this the actual intron and exon locations drawn as a horizontal line showing the correct exons as green bars, separated by blue introns, with the intergene regions colored magenta. (Base numbers are given at the bottom of the figure.) Six of the seven exons are well predicted. However, the second exon is poorly predicted, and there is an additional incorrect prediction around bases 1250–1350. The two lower plots show the predictions for donor and acceptor splice sites. The changing threshold for splice-site prediction is shown as the red and green lines, respectively. The magenta predictions have been rejected during further filtering as mentioned in the text. The cyan predictions have received strengthened scores according to the score modifications mentioned in the text. The final prediction has 11 of 12 sites correct with five false positives. Note how the variable thresholds help correctly predict the lower-scoring acceptor site at base 1134 rather than the higher-scoring donor sites at base 1135 and nearby. There are no suitable splice sites predicted in the region 1250–1350, so the overall prediction is (correctly) for no exon in that region. (From S.M. Hebsgaard et al., Splice site prediction in Arabidopsis thaliana pre-mRNA by combining local and global sequence information, *N.A.R.* 24 (17):3439–3452, 1996, by permission of Oxford University Press.)

## Splice sites can be predicted by sequence patterns combined with base statistics

The SplicePredictor method makes predictions on the basis of the sequence around the splice site and base-composition differences on either side of the splice site. The concept behind the model is that a variety of sequence factors influence the splicing efficiency, such that splicing occurs with probability $P_{\text{splice}}$, and that this can be represented using a formula of the form

$$\ln \frac{P_{\text{splice}}}{1 - P_{\text{splice}}} = \alpha + \beta\Delta_{\text{U}} + \gamma\Delta_{\text{GC}} + W_{\text{signal}} \qquad \text{(EQ10.10)}$$

where $\alpha$, $\beta$, and $\gamma$ are coefficients fitted using a set of test data, $\Delta_{\text{U}}/\Delta_{\text{GC}}$ is the difference in fractional U/GC content between the 50 bases upstream and the 50 bases downstream of the conserved dinucleotide in the primary RNA transcript, and $W_{\text{signal}}$ is a measure of the three upstream and four downstream bases surrounding the dinucleotide. $W_{\text{signal}}$ can be represented by a standard weight matrix based on observed base frequencies at the positions, or the weights of base $a$ at position $i$ can be fitted with the other parameters. Any site scoring $P_{\text{splice}}$ above a threshold is predicted to be a splice site. Note that all GU and AG dinucleotides in the sequence are considered, but only them, so the rare (<1%) splice sites that do not have the conserved dinucleotide will be missed.

In deriving the final model, some of these terms were initially omitted, allowing an assessment of their importance, especially in reducing the number of false positives. If the threshold is set to predict all known splice sites (i.e., give no false negative results), the $W_{\text{signal}}$ measure on its own only eliminates half the false positives, leaving about 15 per true site; adding the base-composition terms ($\Delta_{\text{U}}$ and $\Delta_{\text{GC}}$) only halves that number again. If the threshold is reduced so as to miss 5% of the true sites (a sensitivity of 95%) then false-positive predictions are reduced to, at best, just below two per true site. To improve on this, the splice sites had to be classified into subgroups. For donor sites, those with GGU (about 80%) were distinguished from the others, which were called HGU (H signifying not-G). For acceptor sites, the two subgroups were CAG (about 70%) and DAG (D signifying not-C). This resulted in some improvement, largely in predicting acceptor sites, but false positives still outnumbered true positives for 95% sensitivity. This method was initially developed for plants, in which introns tend to be U(T)-rich and exons GC-rich.

The method was further extended to account for the fact that splice sites must occur as a donor and acceptor pair, and that there may be other nearby sites predicted with a high score. This requires including in the model some idea of the size of introns, which in the case of plants was set at 600 bases, estimated at the time to include over 95% of known plants introns. Subsequently, this method has been modified to replace the $P_{splice}$ score given above with a score based on observed dinucleotide frequencies and a Bayesian statistical model.

## GenScan uses a combination of weight matrices and decision trees to locate splice sites

In GenScan, a more sophisticated variant of weighted matrices is applied to the prediction of donor splice sites. This scheme, called maximal dependence decomposition (MDD), attempts to improve on the assumption of independence of all positions in the signal. As the signals act as a unit, in this case to bind the U1 snRNA component of the spliceosome, the individual positions are not truly independent, and compensatory changes can be expected, leading to strong correlation between positions. Using a test set of donor sites, this correlation was determined using $\chi^2$ tests, and the position with greatest correlation to other positions was identified. The test set was then split into two subsets: one consists of those sites with the consensus base at that position and the other of all other sites. This process was repeated on the new subsets until either there were insufficient examples of a subset or no further correlation was found. The end result was a group of donor-site subsets, each of which is then used to generate a weight matrix, and a decision tree that can be used to decide which matrix to use for a given sequence (see Figure 10.18). This model is not used in GenScan for the acceptor sites because there was a lack of correlation in the data.

## GeneSplicer predicts splice sites using first-order Markov chains

The program GeneSplicer takes the MDD method and uses it to classify the splice sites as in GenScan. However, instead of determining a weight matrix for each subset, a first-order Markov chain is used. This can account for correlation between each base and the base immediately upstream. The parameters are chosen in a similar manner to that discussed for the GLIMMER coding-region prediction method. The donor site is represented by a 16-base window, and the acceptor site by a 29-base window. The sequences are scored with the difference between the log-odds score for a true and a false splice-site model, the latter obtained from sequences known not to have splice sites. In addition to this, simple coding and noncoding models were constructed as second-order Markov models. Such models assign probabilities for the following base according to the last two bases, thus involving a codon (three bases) at each step. These second-order models are a simpler variant of the fifth-order models described in Section 10.3, but can also distinguish coding and noncoding regions.

Putting these models together, the score of a donor splice site at sequence position $j$ is given by

$$S_{donor}(j) = S_{signal}(j,16) + \left(S_{code}(j-80) - S_{noncode}(j-80)\right) + \left(S_{noncode}(j+1) - S_{code}(j+1)\right) \quad \text{(EQ10.11)}$$

and that of an acceptor splice site by

$$S_{acceptor}(j) = S_{signal}(j,29) + \left(S_{noncode}(j-80) - S_{code}(j-80)\right) + \left(S_{code}(j+1) - S_{noncode}(j+1)\right) \quad \text{(EQ10.12)}$$

$S_{signal}(j,n)$ is the score obtained with the MDD and first-order Markov models, while $S_{code}(j)$ and $S_{noncode}(j)$ are the scores of the second-order Markov models for coding and noncoding regions, respectively, starting at base $j$ and of length 80 bases. A

(A)

sequence position

| observed bases | | -3 | -2 | -1 | +1 | +2 | +3 | +4 | +5 | +6 |
|---|---|---|---|---|---|---|---|---|---|---|
| | A% | 33 | 60 | 8 | 0 | 0 | 49 | 71 | 6 | 15 |
| | C% | 37 | 13 | 4 | 0 | 0 | 3 | 7 | 5 | 19 |
| | G% | 18 | 14 | 81 | 100 | 0 | 45 | 12 | 84 | 20 |
| | U% | 12 | 13 | 7 | 0 | 100 | 3 | 9 | 5 | 46 |
| | consensus | A/C | A | G | G | U | A/G | A | G | U |

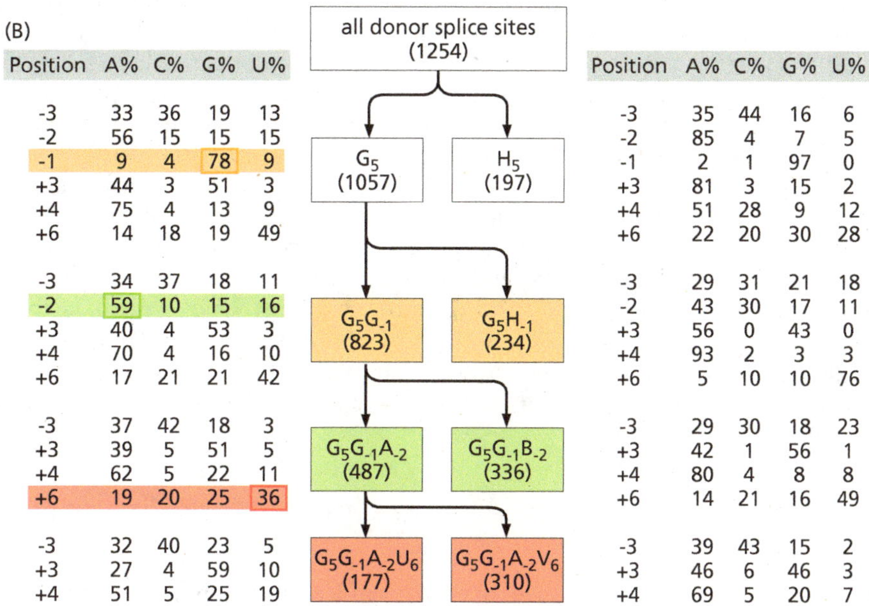


**Figure 10.18**
**Analysis of the composition of 1254 donor splice sites by the maximal dependence decomposition method as used in the GenScan prediction technique.** (A) Summary of the observed bases at different positions relative to the splice site. The GU dinucleotide is at position +1 and +2, with splicing between –1 and +1. Note that positions +1 and +2 are invariable in this dataset. (B) The classification of the sequences into subgroups is shown, according to the maximal dependence decomposition method. The subgroups are named according to their fixed bases; thus the $G_5G_{-1}A_{-2}$ group has G at positions –1 and 5 and A at position –2. H means not-G, B means not-A, and V means not-U. The number of sequences that occur in each group is given in parentheses after the group name. The base frequencies at the variable positions within each subgroup are shown alongside. Note that some positions have quite different compositions in some subgroups, e.g., positions –2 and +3 in $H_5$ and $G_5$. The base selected to define the next two subgroups is always the general consensus at that position, shown highlighted. (Reprinted from *J. Mol. Biol.*, 268, C. Burge and S. Karlin, Prediction of complete gene structures in human genomic DNA, 78–94, 1997, with permission from Elsevier.)

threshold value is used to determine the score required to predict a splice site. This model often gave false-positive predictions close to true positives but with lower scores. To remove these spurious predictions, lower-scoring sites within 60 bases of a predicted splice site were rejected.

## NetPlantGene combines neural networks with intron and exon predictions to predict splice sites

NetPlantGene, which was discussed earlier in this section in the context of coding-region prediction, also predicts splice sites. It uses neural networks to obtain scores, but then uses extensive filtering of the resultant sites to reduce the false-positive predictions. The donor sites are modeled with a set of 10 networks, each with 23 bases in the input layer (represented by four units each, as in Figure 10.14), a hidden layer of 10 units, and a single output layer unit. The 10 networks were separately trained from different starting points, and their outputs (from 0 to 1) averaged to get the final score. A similar set of 10 networks was used to score acceptor sites, except that these have a 61-base input window and 15 hidden layer units. In contrast to the neural nets used for coding-region prediction, in the case of splice-site prediction the threshold output value used to distinguish prediction from nonprediction of a splice site is varied during training so that 95% of true sites are predicted.

To improve on these predictions, it was argued that splice-site predictions should correlate with coding region predictions, in that the splice sites should be at the edges of coding regions. The set of neural networks used to predict coding regions in NetPlantGene has as output a score for each base of its likelihood of being in a coding region. The gradient of this coding-region score was calculated using a crude numerical approximation, and this was used to modify the threshold for splice-site prediction (see Figure 10.17). The resulting thresholds are lowered at the boundaries of coding regions as desired.

Further filtering is based on experimental observations. False splice sites have been found to occur in regions where the coding or noncoding prediction is strong (output layer prediction value close to 1 or 0, respectively) over a segment of sequence. Identifying such regions of coding or noncoding prediction, and removing any predicted splice sites contained in them, successfully reduced the number of false positives by about 10% in the test set without removing a single true site. In addition, there are often sites close downstream of the true acceptor site that are not usually active, but can become active if the true site is disrupted. Therefore, the model was altered to reject all predicted acceptor sites up to 20 bases downstream of another acceptor prediction, removing another 10% of false positives. Similarly, predicted donor sites are often clustered, and so all other donor site predictions within 15 bases of a more strongly predicted donor site were rejected. In all, this reduced the false positives by an impressive 65%, with almost no true sites being rejected. Further modification of the scores involved recognizing that in the absence of alternative splicing donor and acceptor sites must alternate along the gene, and that there is a known distribution of intron and exon lengths. The score modifications are used to remove predictions that result in exons or introns that are outside the known length limits.

## Other splicing features may yet be exploited for splice-site prediction

So far little account has been taken of the known length distributions of exons and introns. Another aspect of the molecular biology of splicing not mentioned so far is the location of the branch point in the intron, which is a functionally significant sequence (see Figure 1.14). Despite these omissions, thresholds could be set such that 57% of all donor sites and 21% of all acceptor sites were predicted with no false positives at all. At 95% sensitivity levels the number of false positives using NetPlantGene was still low enough to make this method very useful.

A few programs, GenScan being one example, try to use the sequence signal for the branch point as an aid to prediction (see Section 1.3 for a summary of the intron structure and the role of the branch point). However, in general the signal is so weak that there is insufficient information, which is why many methods ignore it completely. To illustrate its weakness, a study in *Arabidopsis* found that the signal was only seven bases long, with only two positions having more than 1 bit of information, one being the conserved adenine base.

## Specific methods exist to identify initial and terminal exons

The methods above that combine an exon prediction with splice-site predictions assume that the entire exon codes for protein. This is only the case for internal exons, and those at the 5′ and 3′ ends of genes will also have an untranslated region. (There are several cases where the untranslated region extends to some internal exons as well.) It is clearly important to detect these noncoding exons, but because of their different characteristics they require alternative methods. The Zhang group has paid particular attention to this problem, using quadratic discriminant analysis (see Figure 10.16A) as in their internal exon program MZEF. The program to detect first exons (FirstEF) uses a combination of decision trees and discriminant functions that examine potential splice donor sites, CpG regions, and promoter regions largely on the basis of oligonucleotide statistics. The discriminant functions used in the terminal exon program JTEF attempt to locate the splice sites, terminal codon, and the polyadenylation signal.

### Exons can be defined by searching databases for homologous regions

The eukaryotic methods discussed so far have not attempted to use knowledge from existing sequence databases to assist in gene prediction. When the databases are reliable, the detection of homology can greatly help to identify genes, and when mRNA-related sequences are also known these can aid identification of the exon/intron structure.

Several different approaches can be taken to incorporate homology information into gene detection. Predicted exons can be searched against the database to identify related proteins, which can then be used to assist in finding other exons by homology to those regions of the database entry not in the original exon. In a similar way, homology to ESTs and cDNAs can be used to confirm exon predictions. Although, in principle, standard alignment programs can be used, it has been found advantageous to develop specialized methods that anticipate the large gaps corresponding to introns, and the dinucleotide splice site signals.

When the ESTs or cDNAs are from the same organism as the genome sequence they can help identify a gene and its exon structure even when the function of the gene or protein is not known. EST databases can also be useful in providing evidence of alternative splicing, which otherwise can be difficult to identify. It may not be immediately apparent that alternative protein sequences derive from the same gene, as opposed to two genes related by a duplication event.

One can also search for similar database sequences, although this is usually only used when trying to identify genes of particular families, rather than attempting to locate all genes in a new sequence. Thus for example using a representative or representative sequences from a particular protein family can identify many (possibly all) family members present in the genome. Often this homology information is incorporated into the general gene-detection scheme with an appropriate scoring method.

Many *ab initio* gene-prediction programs have been modified to take account of homology to experimental gene sequences. GenScan was modified to allow BLAST hits to the genome sequence to be used as an extra guide. Because of the probabilistic basis of GenScan (it is an HMM) the BLAST hits have to be converted into a probability. This is made relatively easy, as the *E*-values (see Section 4.7 and Section 5.4) give guidance, although it should be noted that the probabilities under discussion here are very different from those mentioned in relation to database searching. Account is taken of the fact that some of the BLAST hits may be artifacts. The GenScan probability for a particular gene feature prediction is modified to account for a BLAST hit or the lack of one and in this way account is taken of known apparent homologies. The resultant method is called GenomeScan.

## 10.7 Complete Eukaryotic Gene Models

Once all the separate components of a gene have been predicted it is possible to put them all together to predict a complete gene structure. A number of programs exist that can display a variety of predictions of individual components together, so that a manual interpretation can be made of the gene structure (see Flow Diagram 10.7). This may be useful in fine-tuning a gene structure when some extra experimental information becomes available, but as a general technique is not practicable, not least because it leads to subjective results that may not be reproducible.

Many programs have been written that predict individual gene components and then combine these automatically into a gene prediction. Many of the early programs would only predict a single gene for a query sequence, but newer programs do not have this limitation. Most programs use one of two general techniques to combine the elements into a predicted gene: hidden Markov models or dynamic programming (see Sections 6.2 and 5.2, respectively). The details of the application of these methods vary between programs, but we will discuss the general features of this approach with reference to two programs: GenScan and

**Flow Diagram 10.7**
**In this section some complete eukaryotic gene models are described.**

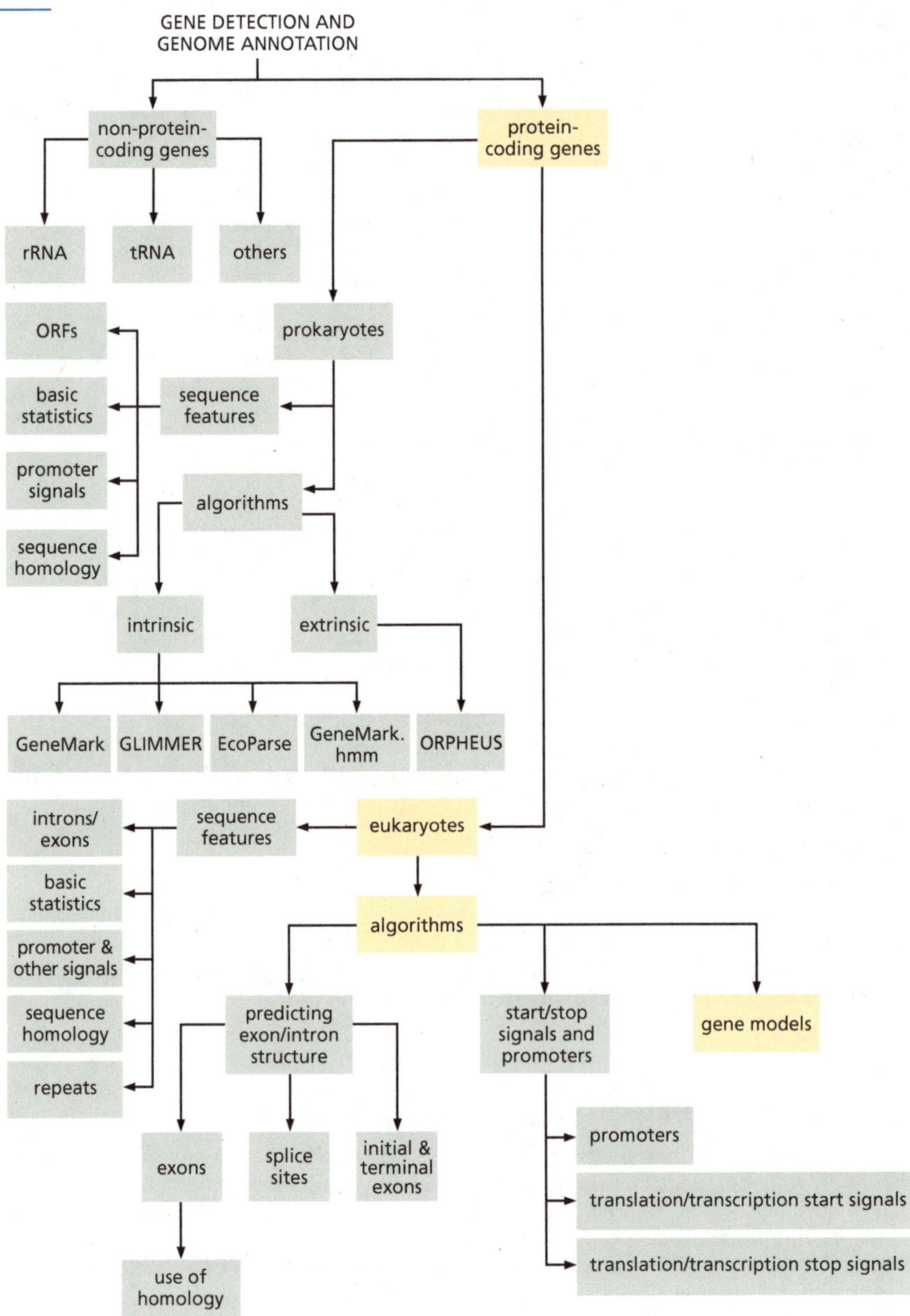

GAZE. Note that even programs such as Grail, which base component prediction largely on neural networks, cannot use neural networks for this stage. Many of these programs use a model that prevents any gene overlap. Although this is often thought to be of little importance in eukaryotic genomes, 2% of the *Caenorhabditis elegans* genes in WormBase occur in introns of other genes, and are referred to as **nested genes**.

GenScan uses an HMM (see Figure 10.19) that considers both the forward and reverse strands of DNA simultaneously. Note that the gene components on the reverse strand are located in reverse order when compared to the forward strand. This complex model is composed of several different submodels that identify each component. Many of the submodels have been discussed above, such as the MDD method for donor sites and the explicit state duration HMM for coding regions. A key feature of the submodels is that they provide probability scores to the HMM states representing them. This complete-gene model accounts for the requirement for successive exons to maintain the reading frame, and can also detect several genes in one input sequence. However, it cannot detect overlapping genes.

The GAZE program uses dynamic programming to combine the different component predictions into whole genes. In this view of the problem a gene is seen as a sequence of features with a defined order, each of which can be assigned a score. The predicted features in the unknown sequence are aligned to the gene model and the features that give the highest-scoring gene are reported. The program is designed to require simply the probabilistic scores of the individual component predictions, so that it can readily accept output from any prediction methods that can produce such scores. Prediction scores for these individual components can be read by GAZE from a file in the standard General Feature Format (GFF) and must include a score for each prediction. For dynamic programming to work, scores must be associated with all potential assignments of sequence segments, including intergene regions. The scores for intergene regions, introns, and exons can be taken from observed length distributions, and they may be defined within the GAZE program itself or read in with other predictions from a GFF file.

The GAZE system is, in principle, very flexible in that it can use many different gene models. These models are defined in a separate file using the markup language XML, so the gene model can be changed simply by editing this file. GAZE reads this model file and then uses dynamic programming to determine the best path through the model, and hence the optimal gene structure for the given sequence. A model for multiple genes on both strands is given in Figure 10.20. By comparison with the GenScan model in Figure 10.19, this model shows the splice sites explicitly, giving it a different appearance from the GenScan model, but they are exactly equivalent.

## 10.8 Beyond the Prediction of Individual Genes

There is much more to annotating a genome than identifying the location of all protein-coding genes and other sequence features (see Flow Diagram 10.8). At the simplest level, the annotation should include an analysis of any predicted protein products, based on a sequence analysis of the kind described in Chapter 4. In addition, it should include experimental data that can be assigned to particular sequence features. Further analysis may involve comparison with the genomes of other organisms. This is frequently done by alignment of the two genomes or regions of them, and comparing the features present in each. Both of these aspects of genome annotation are discussed briefly below. Finally, this section looks at how such annotations require reevaluation as further experimental data are obtained (possibly on the basis of the predicted genes and gene products).

**Flow Diagram 10.8**
**In this section several more advanced topics of gene prediction and genome annotation are described.**

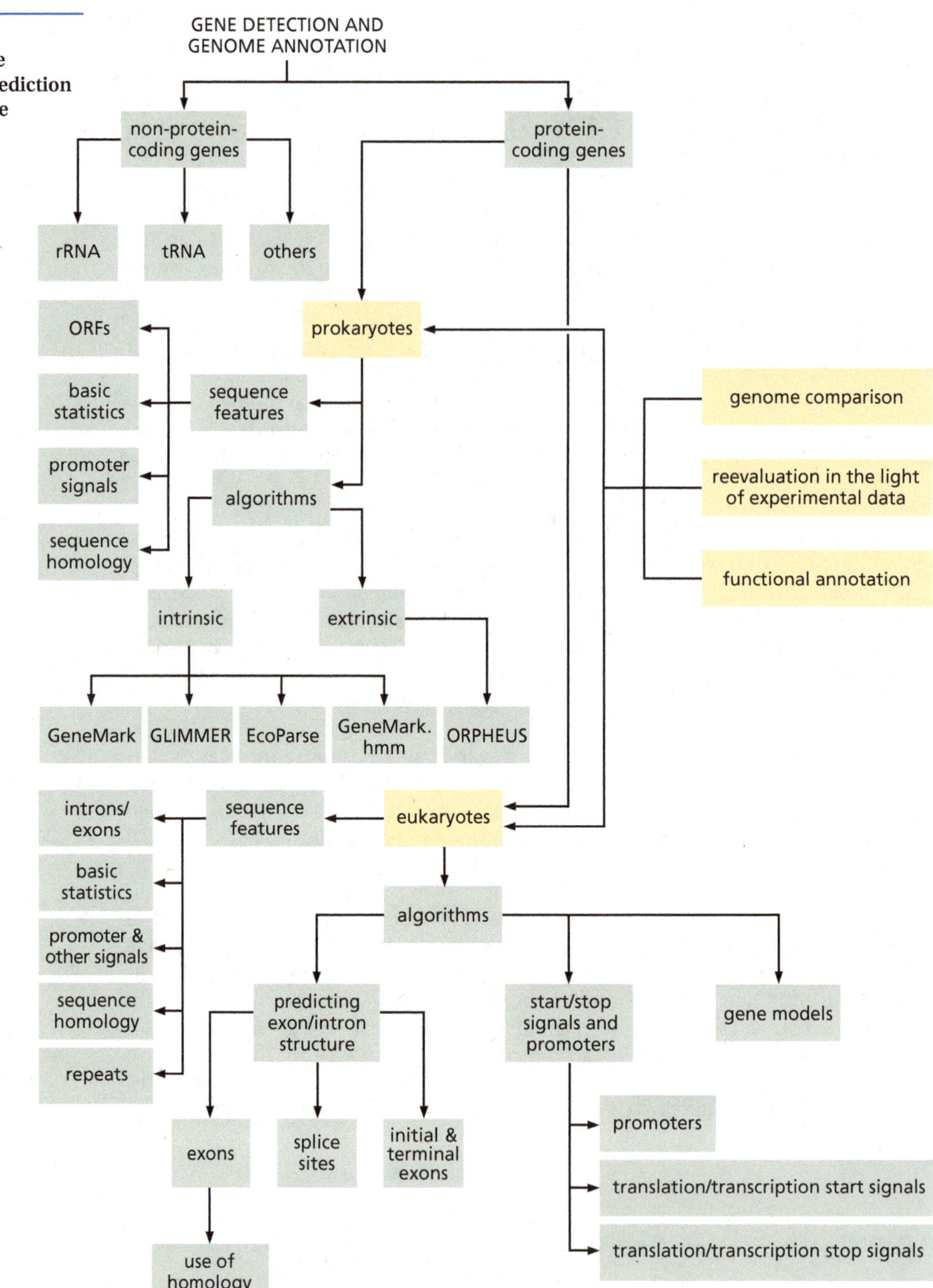


## Functional annotation

Once all the genes have been predicted, there is still a need to determine what functions the encoded proteins might play. The obvious way to start to predict gene function is by sequence analysis. If the encoded protein has one or more significant matches against sequence and pattern databases the function and other properties can be predicted with considerable confidence to be similar to those of the matches. The majority of gene products can be assigned a function on this basis, although usually the function is not as specific as is needed for real analysis. For example, identifying a gene product as a kinase is useful, but unless the substrates

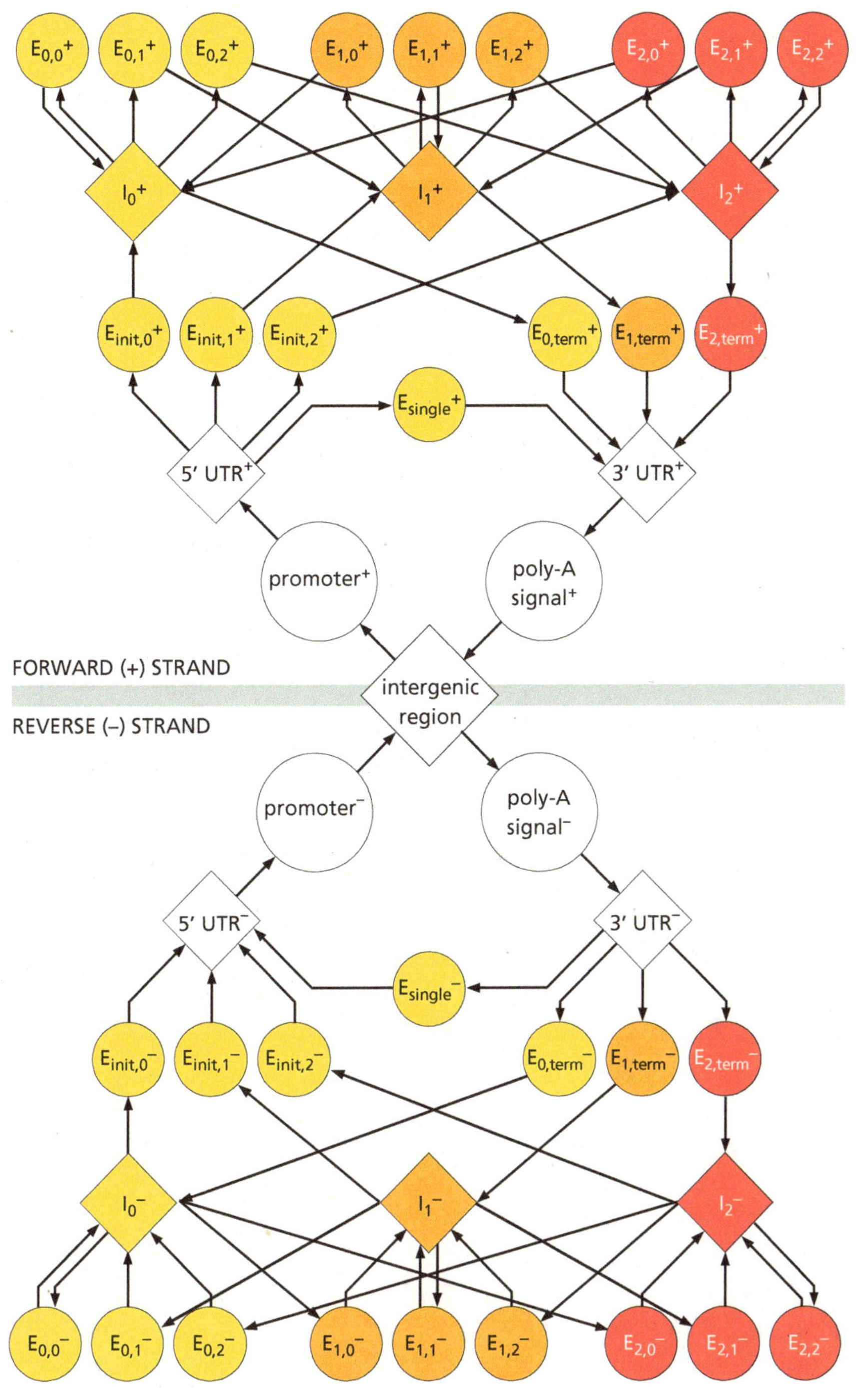


**Figure 10.19**
**Simplified representation of the HMM model structure used in GenScan.** The sequence is scanned in one direction (the forward or + strand). From the central intergenic state a gene on the + strand is identified starting with the promoter and moving forward to the poly(A) signal, whereas a gene on the reverse (–) strand starts with the poly(A) signal at the end of the gene and continues back to the promoter. No gene overlap is allowed, whether on the same or both strands. The phase of the exons must be maintained, and is indicated by the subscripts 0, 1, and 2. Exons are colored according to their phase, introns being colored according to the phase of the following exon (following being defined by the 5′ to 3′ direction). $E_{0,1}$+ is an exon of a gene on the forward strand that is in frame 0 (i.e., the previous exon ended with a complete codon) and ends with the first base of a codon. Such an exon must be followed by one that starts in frame 1, e.g., $E_{1,2}^+$ or $E_{1,term}^+$. The situation is similar for genes on the reverse strand, except that because the exons are encountered in reverse order, $E_{0,1}^-$ must be followed (when scanning the forward strand) by an exon that ends with a complete codon, e.g., $E_{2,0}^-$ or $E_{init,0}^-$. Each object in this model is itself a model, some with considerable complexity. The promoter signal has been described in Figure 10.13. The exon model includes splice sites such as the donor model in Figure 10.18, as well as the protein-coding regions. The protein-coding regions use a model similar to Figure 10.6B. The exon and intron models use explicit state duration HMMs to reproduce the known length variations (see Figure 10.10). The possibility of a single exon gene is allowed for by the $E_{single}$ model. The introns and intergenic region are also modeled by homogeneous fifth-order Markov models, as in Figure 10.6A. This model can identify many genes on both strands. (Adapted from C. Burge and S. Karlin, Prediction of complete gene structures in human genomic DNA, *J. Mol. Biol.* 268:78–94, 1997.)

and expression control are known, the true role of this gene in the organism will remain relatively undefined. Additionally, all sequenced genomes to date have a significant proportion (10% and upward) of gene products that do not yield to such analysis, and in numerous other cases the similarity is to only a part of the protein, such as a single domain.

The only realistic way of assigning functions is by further experimental work. Thus, there is often a large effort following sequencing to discover the phenotype on disrupting an individual gene. The phenotype may be described by gross features such as overall cell size, but often will be in terms of the different gene or protein expression as compared with the normal phenotype. Such experiments can also

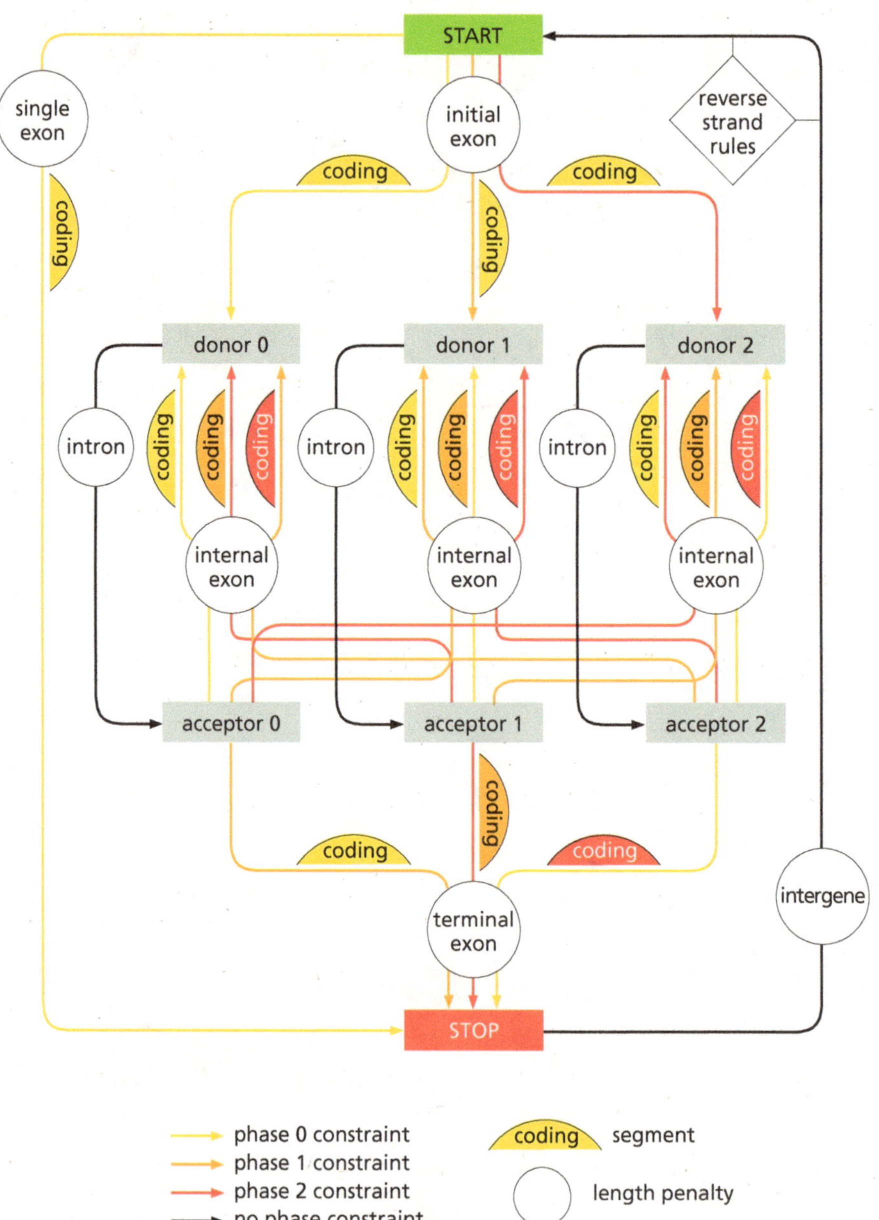


**Figure 10.20**
**Simplified representation of the HMM model structure used in GAZE.** This is the equivalent model layout to that shown in Figure 10.19 for GenScan. In this case, the underlying technique is dynamic programming, so that at each stage alternative interpretations of the next sequence segment must be scored. The features in rectangles are scored using weight matrices in the published version of GAZE, although other methods could also be incorporated. The scores of protein-coding and noncoding segments (in the white circles) are based on their lengths with reference to known length distributions, and are referred to as length penalties. The protein-coding regions have in addition a score associated with the coding potential. The exon phase is maintained in the same way as for GenScan, and the coding segments have been colored with the same scheme. There is an equivalent structure for the genes on the reverse strand, as in GenScan, but these are here only illustrated by the box in the top right corner. This model can identify many genes on both strands. (Adapted from K.L. Howe, T. Chothia and R. Durbin, GAZE: a generic framework for the integration of gene-prediction data by dynamic programming, *Genome Res.* 12:1418–1427, 2002.)

demonstrate linked control with other genes whose function is better understood. Such methods can yield gene functions at several levels of description. Many of these experiments involve gene expression arrays and/or proteomic studies, whose analysis is discussed in Chapters 15 and 16.

One of the important aspects of genome annotation has been the recognition of the importance of gene ontology (see Section 3.2). In this context, gene ontology is a set of standardized and accepted terms that encompass the range of possible functions. Wherever possible these terms should be used, as they make it easier to do further analysis such as search for genes of similar proposed function.

There is more to genomes than just the genes, and a full genome annotation must cover all the aspects. The location of tRNA molecules has already been discussed, and the rRNA molecules also need to be located. In addition there are repeat sequences that can be of interest, and can have functional roles. Beyond this, there are features that encompass several genes, such as operons and pathogenicity

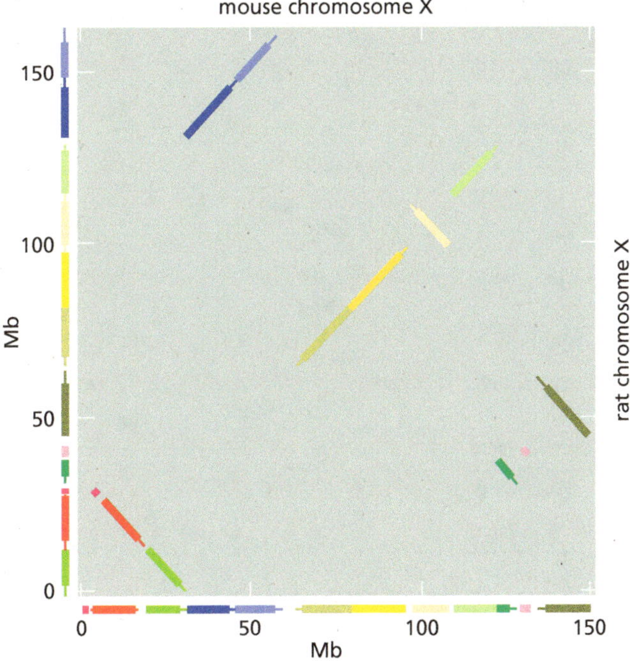


**Figure 10.21**
**Large-scale sequence alignment of the mouse and rat X chromosomes.** The time of the last common ancestor of the mouse and rat is estimated at 12–24 million years; not a long time in evolutionary terms, and yet there have been many large-scale rearrangements, including sequence inversions. At least five events have occurred in each of the two species since they diverged. In addition to these events, many more smaller-scale changes have also occurred. Hence, alignment of the two sequences is not as trivial as might at one time have been hoped. However, alignment is also quite straightforward, although the lengths of the sequences and specific nature of the problem mean that specific algorithms have been developed, as discussed in Section 5.5. (Adapted from R.A. Gibbs et al., Rat Genome Sequencing Project Consortium: Genome sequence of the Brown Norway rat yields insights into mammalian evolution, *Nature* 428:493–521, 2004.)

islands. The latter are large-scale insertions from other species, often with a set of genes coding for transport machinery and toxins. In addition other genes may show signs of recent lateral transfer. Information relating genetic changes to inherited conditions exists for a number of organisms, especially humans and domesticated animals. Inclusion of such information is extremely important, especially to assist research in medicine and veterinary science. It is beyond the scope of this book to cover the methods whereby these and other features can be identified and included in the genome annotations.

There are two schools of opinion concerning the quality of the information that should be included in genome annotations. Some groups have taken the view that only the most reliable information should be included, so as to minimize the number of incorrect assignments, in an attempt to prevent misleading future workers. An example of the application of such a careful approach is the Ensembl human genome annotation project. The alternative approach is to include as much information as possible, even when there is little if any supporting evidence. Clearly such data need to be presented very carefully to ensure that the degree of certainty is given wherever possible. The advantage of such an approach is that for a gene that has no strong evidence in favor of a specific function, some indication, however uncertain, might suggest experimental work that could ultimately clarify the role of the gene or its product.

## Comparison of related genomes can help resolve uncertain predictions

When the genome sequence of a reasonably closely related organism is available, comparison of the two sequences can be a very powerful tool in determining the status of uncertain gene predictions. This comparison can help identify potential incorrect sequence that can be further examined by resequencing the region. It should be noted that aligning the two genomes is not necessarily trivial, as large-scale rearrangements are common, but it should be possible to find regions of **synteny** where the gene structure is sufficiently similar as to make their common evolutionary ancestry apparent. An example of this large-scale rearrangement is shown by the alignment of the X chromosomes of mouse and rat (see Figure 10.21),

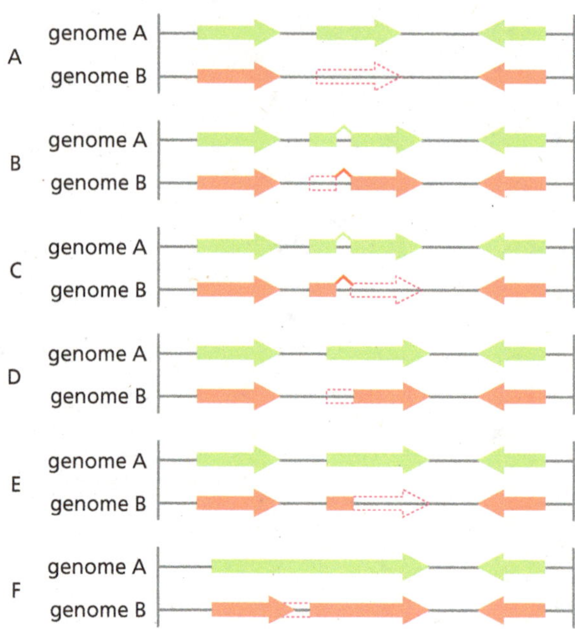


**Figure 10.22**
**Some of the possible differences in annotation between the syntenic regions of two genomes.** Genome A is taken as the reference, and used as a guide to the identification of genes in genome B. However, unless genome A was exceptionally well characterized experimentally, it would be unwise not to also question the annotation of genome A. The observed differences in A to F are discussed in detail in the text. (Adapted from Figure 1 of Brachat et al., 2003.)

which shows that almost all of the sequences are represented in both chromosomes (the rat chromosome is slightly larger than that of the mouse) but the order and often direction are changed.

Comparing syntenic regions can reveal errors in one or other genome annotation (see Figure 10.22). Note that at present a significant fraction of most genome annotations is not fully verified by experiment, so that care has to be taken with applying such knowledge to other genomes. Therefore experimental evidence should be sought whenever possible to confirm any modified predictions. The discussion below is based on Figure 10.22 and is from a study of the *Saccharomyces cerevisiae* genome. Because this genome contains mostly single ORF genes and some genes with only a few exons, it does not describe the full range of possible situations for higher eukaryotes with more complex gene structures. However, extension to such genomes is straightforward.

If a particular ORF is only identified as an expressed gene in the annotation of one species whilst the surrounding regions are in good agreement (see Figure 10.22A), the assignment of that ORF as a gene needs to be reassessed. It could be that the ORF has been overlooked in one species, possibly because a sequencing error led to the ORF being split into two sequences that were each too short to have been considered as potential genes. In such a case resequencing the region should clarify if indeed an error was made, as it is also possible that a mutation has disrupted the gene.

Figures 10.22B and 10.22C show cases of failure to identify an intron at the 5′ or 3′ end of an exon, leading to omitting the previous or succeeding exon. This situation could arise simply due to the failure of splice-site predictions in the case of genome B. Errors in sequencing can lead to incorrect identification of ORFs with missing 5′ or 3′ ends (see Figures 10.22D and 10.22E). As before, resequencing can resolve the uncertainty. In some circumstances, such errors can lead to the misidentification of a single long ORF as two shorter ORFs (see Figure 10.22F). For a full explanation of how both experimental and computer-based methods can be applied to these problems see Further Reading.

The above discussion is based on work that was largely reliant on human intervention in the examination of identified differences between two annotated genomes. Even for a 13-megabase genome this is a substantial amount of work. Many of the ideas have been used to construct algorithms to automate much of the work. Many of these use techniques that are related to ones described above, but now operating on two sequences simultaneously. The details will not be presented here, but some references are provided under Further Reading.

### Evaluation and reevaluation of gene-detection methods

As yet no organisms (except perhaps viruses) have had experimental confirmation of every gene and satisfactory confirmation of the lack of expression of all other ORFs. But more evidence is accumulating all the time, and can be used to reevaluate prediction methods and current genome annotations. Much of these data are from nonsequencing experiments, such as gene-expression analysis. However, additional information can also be obtained from comparisons with the new genome sequences of related organisms. We will briefly explore some of the approaches that have been used to improve on the initial annotations. Attention usually focuses on the shorter ORFs as these are most likely to be in error; longer ORFs are very unlikely to occur by chance.

One of the best-studied genomes from this perspective is that of *Saccharomyces cerevisiae*. This contains 5000–6000 genes, but the exact number is possibly only known to within 100 genes. A 2001 reevaluation by Valerie Wood and co-workers estimated there are 5570 genes in total, compared with the 1996 estimate of 5885 made immediately after the sequencing was completed. However, even this is not a final answer, and the genome annotation is continually under review, especially as genome sequences are published for relatively close species.

The initial annotation of *Saccharomyces cerevisiae* attempted to avoid problems of assessing short ORFs by restricting attention to ORFs at least 100 codons in length, of which there are about 7500 in the yeast genome. If the length limit is reduced to 20 codons the number of ORFs to consider rises to more than 41,000. A number of ORFs, known as orphans (or ORFans), were found to have no known homologs in the databases, and no function could be assigned to them by searching, for example, for PROSITE patterns (see Section 4.9). These might be true genes, either ones whose homologs have yet to be identified, or that have evolved to completely new functions, or they may be false gene predictions.

In the 2001 reannotation Wood and colleagues paid attention to these genes, to see if further data could be found to support or contradict the prediction. In some cases homologs were found in newly sequenced genomes, or experimental evidence was obtained that the gene was expressed. Care was taken to examine possible overlaps with other genes, which was taken as generally indicative of the absence of a gene. Another property used to flag up dubious assignments was the GC content of the ORF, as if this deviated considerably from the genome average the prediction was likely to be false. In general, the orphan ORFs were annotated as very hypothetical (193 genes) or spurious (370 genes). Another aspect of the reevaluation was to look carefully at alignments of homologs to the sequence, to see if they might suggest possible frameshifts or sequencing errors that had escaped attention. In total 46 such modifications were discovered, including three new genes.

## Summary

In this chapter we have looked at the problem of identifying functional regions in nucleotide sequences. The RNA molecules can be dealt with in almost identical ways regardless of the organism under consideration, with the exception that the larger eukaryotic genomes require a method with a lower false-positive rate to be useful. However, when we consider locating protein-encoding genes the situation is very different between eukaryotes and other species.

In both cases, genes have to be identified on the basis of their component parts, many of which are detectable only by short sequence patterns that have barely any signal. Thus each component is difficult to detect without making errors that can carry over to the model of the whole gene. In prokaryotes, whose genes have a simpler structure, this is less problematic. In eukaryotes, current methods still make a large number of prediction errors, so that genome annotation is an art, often requiring considerable specialist knowledge both of the idiosyncrasies of the programs and of the particular species under study. The availability of related single gene and whole genome sequences has resulted in a further refinement of the process.

Further experimental work and reanalysis is gradually improving the accuracy of genome annotations and helping to identify problems with current automatic methods. It is only once the annotation has been made that attention can be paid to the ORFans (sequences with no known homolog) to try to establish if they are indeed expressed. Meanwhile the issue of alternative splicing still requires further study to understand the mechanisms of control, and how they might be identified in the sequence so that they can be predicted. This field is still evolving rapidly, and will continue to do so for some time to come, as there is still much potential improvement.

# Further Reading

## 10.1 Detection of Functional RNA Molecules Using Decision Trees

### tRNA detection


Fichant GA & Burks C (1991) Identifying potential tRNA genes in genomic DNA sequences. *J. Mol. Biol.* 220, 659–671.

Lowe TM & Eddy SR (1997) tRNAscan-SE: a program for improved detection of transfer RNA genes in genomic sequences. *Nucleic Acids Res.* 25, 955–964.

Pavesi A, Conterio F, Bolchi A et al. (1994) Identification of new eukaryotic tRNA genes in genomic DNA databases by a multistep weight matrix analysis of transcriptional control regions. *Nucleic Acids Res.* 22, 1247–1256.


### Detection of other RNA genes


Eddy SR (2002) Computational genomics of noncoding RNA genes. *Cell* 109, 137–140.


## 10.2 Features Useful for Gene Detection in Prokaryotes


Echols N, Harrison P, Balasubramanian S et al. (2002) Comprehensive analysis of amino acid and nucleotide composition in eukaryotic genomes, comparing genes and pseudogenes. *Nucleic Acids Res.* 30, 2515–2523.

Schneider TD & Stephens RM (1990) Sequence logos: A new way to display consensus sequences. *Nucleic Acids Res.* 18, 6097–6100.


### Identifying protein-coding regions using base statistics


Fickett JW & Tung C (1991) Assessment of protein coding measures. *Nucleic Acids Res.* 20, 6441–6450.


In addition, most references in Section 10.3 have useful material.

## 10.3 Algorithms for Gene Detection in Prokaryotes

### GeneMark, GeneMark.hmm, and further developments


Besemer J, Lomsadze A & Borodovsky M (2001) GeneMarkS; a self-training method for prediction of gene starts in microbial genomes. Implications for finding sequence motifs in regulatory regions. *Nucleic Acids Res.* 29, 2607–2618.

Borodovsky M, Rudd KE & Koonin EV (1994) Intrinsic and extrinsic approaches for detecting genes in a bacterial genome. *Nucleic Acids Res.* 22, 4756–4767.

Lukashin AV & Borodovsky M (1998) GeneMark.hmm: new solutions for gene finding. *Nucleic Acids Res.* 26, 1107–1115.


### GLIMMER


Delcher AL, Harmon D, Kasif S et al. (1999) Improved microbial gene identification with GLIMMER. *Nucleic Acids Res.* 27, 4636–4641.

Salzberg SL, Delcher AL, Kassif S & White O (1998) Microbial gene identification using interpolated Markov models. *Nucleic Acids Res.* 26, 544–548.

Salzberg SL, Pertea M, Delcher AL et al. (1999) Interpolated Markov models for Eukaryotic gene finding. *Genomics* 59, 24–31.

**ORPHEUS**

Frishman D, Mironov A, Mewes H-W & Gelfand M (1998) Combining diverse evidence for gene recognition in completely sequenced bacterial genomes. *Nucleic Acids Res.* 26, 2941–2947.

**EcoParse**

Krogh A, Mian IS & Haussler D (1994) A hidden Markov model that finds genes in *E.coli* DNA. *Nucleic Acids Res.* 22, 4768–4778.

**Prokaryotic genomes**

Ermolaeva MD, White O & Salzberg SL (2001) Prediction of operons in microbial genomes. *Nucleic Acids Res.* 29, 1216–1221.

Lio P & Vannucci M (2000) Finding pathogenicity islands and gene transfer events in genome data. *Bioinformatics* 16, 932–940.

**Markov models**

Burge CB & Karlin S (1998) Finding the genes in genomic DNA. *Curr. Opin. Struct. Biol.* 8, 346–354.

## 10.4 Features Used in Eukaryotic Gene Detection

Statistics of intron and exon length distributions and other such generalized gene structure statistics are reported in several genome sequence papers.

**Preliminary analysis for human genes**

Burge C & Karlin S (1997) Prediction of complete gene structures in human genomic DNA. *J. Mol. Biol.* 268, 78–94. (*Genscan*)

Hebsgaard SM, Korning PG, Tolstrup N et al. (1996) Splice site prediction in *Arabidopsis thaliana* pre-mRNA by combining local and global sequence information. *Nucleic Acids Res.* 24, 3439–3452.

Stephens RM & Schneider TD (1992) Features of spliceosome evolution and function inferred from an analysis of the information at human splice sites. *J. Mol. Biol.* 228, 1124–1136.

## 10.5 Predicting Eukaryotic Gene Signals

**Initial analysis of core promoter sequences**

Bucher P (1990) Weight matrix descriptions of four eukaryotic RNA polymerase II promoter elements derived from 502 unrelated promoter sequences. *J. Mol. Biol.* 212, 563–578.

FitzGerald PC, Sturgill D, Shyakhtenko A et al. (2006) Comparative genomics of *Drosophila* and human core promoters. *Genome Biol.* 7, R53.

Penotti FE (1990) Human DNA TATA boxes and transcription initiation sites. A statistical study. *J. Mol. Biol.* 213, 37–52.

**Algorithms of core promoter detection**

Burge C & Karlin S (1997) Prediction of complete gene structures in human genomic DNA. *J. Mol. Biol.* 268, 78–94. (*Genscan*)

Matis S, Xu Y, Shah M et al. (1996) Detection of RNA polymerase II promoters and polyadenylation sites in human DNA sequence. *Comput. Chem.* 20, 135–140. (*Grail*)

Reese MG (2000) Genome annotation in *Drosophila melanogaster*. PhD thesis. University of Hohenheim, Germany. (*NNPP, Genie*)

**Promoter recognition**

Hutchinson G (1996) The prediction of vertebrate promoter regions using differential hexamer frequency analysis. *Comput. Appl. Biosci.* 12, 391–398. (*ProFind*)

Ohler U, Harbeck S, Niemann H et al. (1999) Interpolated Markov chains for eukaryotic promoter recognition. *Bioinformatics* 15, 362–369. (*McPromoter*)

Prestridge DS (1995) Predicting Pol II promoter sequences using transcription factor binding sites. *J. Mol. Biol.* 249, 923–932. (*PromoterScan, ProScan*)

Scherf M, Klingenhoff A, Werner T (2000) Highly specific localization of promoter regions in large genomic sequences by PromoterInspector: A novel context analysis approach. *J. Mol. Biol.* 297, 599–606. (*PromoterInspector*)

Uberbacher EC, Xu Y & Mural RJ (1996) Discovering and understanding genes in human DNA sequence using GRAIL. *Methods Enzymol.* 266, 259–281. (*Grail*)

Zhang MQ (1998) Identification of human gene core promoters in silico. *Genome Res.* 8, 319–326. (*CorePromoter*)

## 10.6 Predicting Exons and Introns

Zhang MQ (1997) Identification of protein coding regions in the human genome by quadratic discriminant analysis. *Proc. Natl Acad. Sci. USA* 94, 565–568.

A number of the papers listed above, especially for Section 10.5, also describe exon and intron prediction methods.

## 10.7 Complete Eukaryotic Gene Models

Howe KL, Chothia T & Durbin R (2002) GAZE: A generic framework for the integration of gene-prediction data by dynamic programming. *Genome Res.* 12, 1418–1427.

A number of the papers listed in the other sections also describe complete gene models.

## 10.8 Beyond the Prediction of Individual Genes

**Detailed reexamination of the annotation of a complete genome**

Brachat S, Dietrich FS, Voegeli S et al. (2003) Reinvestigation of the *Saccharomyces cerevisiae* genome annotation by comparison to the genome of a related fungus: *Ashbya gossypii*. *Genome Biol.* 4, R45.

Wood V, Rutherford KM, Ivens A et al. (2001) A re-annotation of the *Saccharomyces cerevisiae* genome. *Comp. Funct. Genomics* 2, 143–154.

**Large-scale changes in chromosomes**

Gibbs RA, Weinstock GM, Metzker ML et al. (2004) Genome sequence of the Brown Norway rat yields insights into mammalian evolution. *Nature* 428, 493–521.

In 2005 an international effort examined the accuracy of the available methods in a program called the ENCODE Genome Annotation Assessment Project. Many different methods were used to annotate approximately 1% of the human genome. Amongst these methods are several that use information obtained from the alignment of two or more large genomic segments from related species. The results from this exercise were published in Volume 7 Suppl 1 of *Genome Biology* in 2006.

## Box 10.1 Measures of gene prediction accuracy at the nucleotide level

Burset M & Guigo R (1996) Evaluation of gene structure prediction programs. *Genomics* 34, 353–367. *(Further details about ways to evaluate the accuracy of gene-detection programs.)*

## Box 10.2 Sequencing many genomes at once

Venter JC, Remington K, Heidelberg JF et al. (2004) Environmental genome shotgun sequencing of the Sargasso Sea. *Science* 304, 66–74.

## Box 10.3 Measures of gene prediction accuracy at the exon level

Burset M & Guigo R (1996) Evaluation of gene structure prediction programs. *Genomics* 34, 353–367. *(Further details about ways to evaluate the accuracy of gene-detection programs.)*

# PART 5

# SECONDARY STRUCTURES

DNA, RNA, and protein molecules almost always require a specific three-dimensional structure in order to perform their function. In particular, proteins fold into three-dimensional structures that are formed from small, regular, repeating structures called secondary structures. The information from a structure aids in further understanding the function of a molecule. In addition, such structural information can be used to improve sequence alignments.

This part of the book describes how to predict the secondary structures of biological molecules. The first chapter gives an introduction to available methods and discusses how to implement them and interpret the results. It also deals with specialized predictions, such as prediction of membrane-spanning regions and the secondary structure of RNA.

The second chapter describes in depth the methods used for secondary structure predictions and the underlying principles of these methods, such as neural network and hidden Markov model techniques.

# 11

# OBTAINING SECONDARY STRUCTURE FROM SEQUENCE

APPLICATIONS CHAPTER

**When you have read Chapter 11, you should be able to:**

Predict the location of secondary structure elements from a protein's sequence alone.

Assess the accuracy of prediction programs.

Compare how prediction algorithms perform on proteins of different secondary-structure classes.

Compare commonly used prediction approaches and algorithms.

Predict the location of transmembrane regions in membrane proteins.

Predict the existence of helices that can form coiled coils.

A protein's activity or biochemical function is determined by its three-dimensional shape, or fold, as described in Chapter 2. The fold of a polypeptide chain brings together amino acids from different parts of the chain such that chemical groups are positioned in a configuration that can confer catalytic activity, as in an enzyme's active site, or form a binding site for another protein or small molecule. Such configurations can also increase the protein's structural integrity.

An important aspect of bioinformatics deals with the question: given a protein sequence, what is its structure? The only way at present to elucidate the three-dimensional conformation of a protein is by experimental methods such as X-ray crystallography or nuclear magnetic resonance (NMR). However, it is not practical to determine the three-dimensional structure of every protein experimentally. Genome sequencing projects have, on the other hand, given rise to an explosion in the number of known protein sequences. Searching the databases for sequence homologs of known structure and function can provide helpful insights into an unknown protein's structure and its biochemical activity. There are, however, many protein sequences for which no homologs of known structure can be found. In such circumstances, being able to predict at least some structural features from the protein sequence is useful in providing clues as to the protein's overall structure and function.

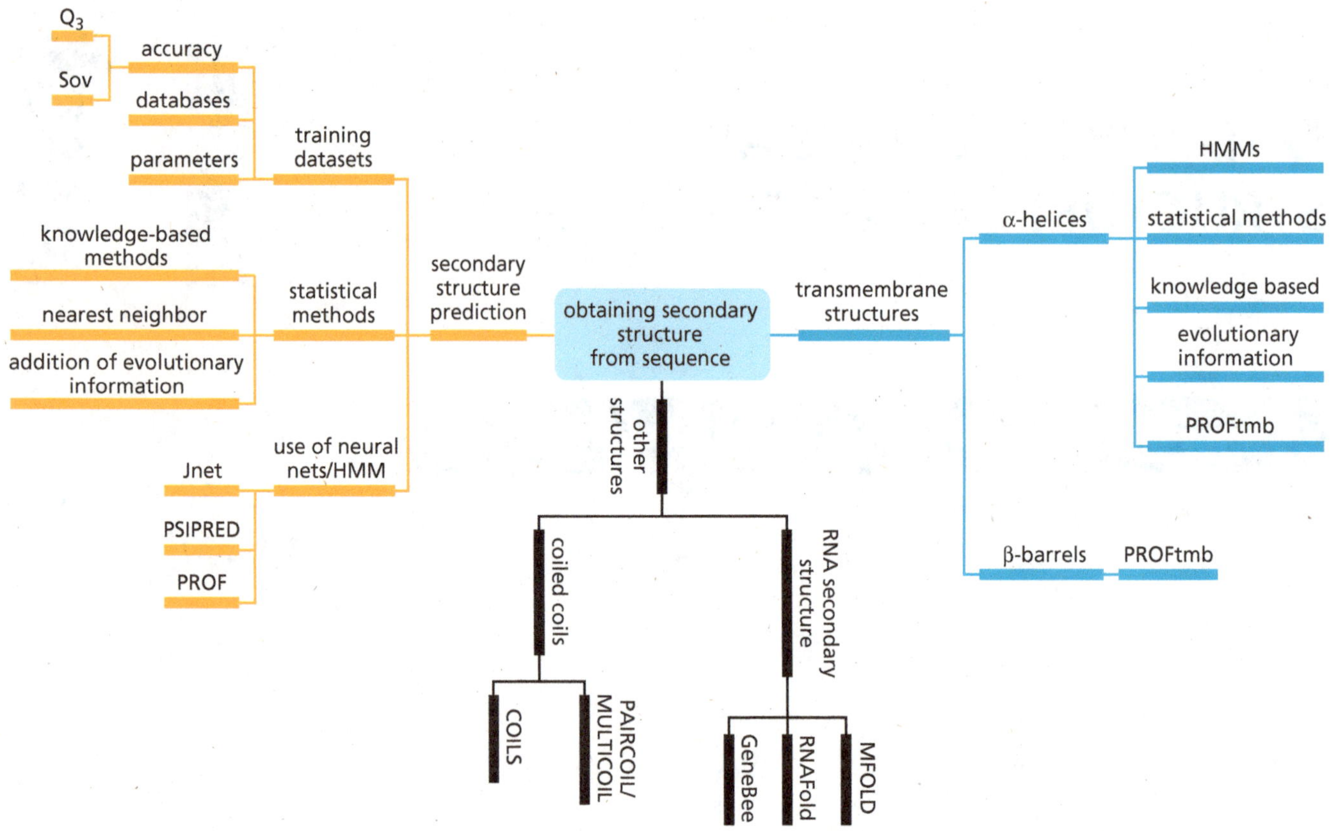

obtaining secondary structure from sequence
secondary structure prediction
training datasets
accuracy
$Q_3$
Sov
databases
parameters
statistical methods
knowledge-based methods
nearest neighbor
addition of evolutionary information
use of neural nets/HMM
Jnet
PSIPRED
PROF
transmembrane structures
α-helices
HMMs
statistical methods
knowledge based
evolutionary information
PROFtmb
β-barrels
PROFtmb
other structures
coiled coils
COILS
PAIRCOIL/MULTICOIL
RNA secondary structure
GeneBee
RNAFold
MFOLD

**Mind Map 11.1**
**A mind map showing the main topics covered in this chapter, which deals with how to predict structural features from sequence.**

It is generally accepted that the information necessary for a protein to fold into its native form is contained in its amino acid sequence. In the 1950s Christian Anfinsen (see Figure 11.1), studying bovine pancreatic ribonuclease (RNase), showed that the amino acid sequence of the protein determines its three-dimensional structure. This very important observation was published in the *Journal of Biological Chemistry* in 1954. Anfinsen unfolded the RNase enzyme and then observed how the enzyme refolded spontaneously back to its original form under natural conditions. He and his colleagues received the Nobel Prize for this work on protein folding in 1972. But the question of why a particular sequence adopts a specific fold is still far from being solved. A computational method of protein folding that can be applied to a sequence to predict its tertiary structure accurately *a priori* does not yet exist. Nevertheless, it has proved possible to predict secondary structure elements—α-helices, β-strands, β-turns, and random coil—with some success using statistical and neural network methods that rely on parameters derived from analyzing large numbers of sequences with known structures.

In this chapter we will deal firstly with the main secondary structure prediction methods in use today, and in the later parts of the chapter describe methods to predict special structural aspects such as the transmembrane regions of a protein. In the accompanying Chapter 12 the theory underlying the prediction programs will be described in depth. Approaches to predicting the overall fold of a protein are discussed in Chapter 13.

Methods have also been developed to identify other biologically important sites on proteins, such as sites for posttranslational modification, signal sequences, and sites of interactions with other proteins. Most of these methods can be obtained through the ExPASy Web server and will not be discussed further here, as they essentially involve the recognition of protein sequence motifs, as described in Sections 4.8–4.10 and Section 6.6, rather than the prediction of structure from sequence.

**Figure 11.1**
**Photograph of Christian Anfinsen.**

In this chapter we will also deal with the prediction of coiled coils, in which α-helices from two protein chains wind around each other to give a rod-like structure. Coiled coils are characteristic of a class of transcription factors—the **leucine zipper** transcription factors—and thus prediction of the ability to form a coiled coil will give clues to the function of an unknown protein.

There have been some attempts to predict from sequence alone the active sites of proteins and, more recently, protein-interaction sites. These types of prediction are concerned with the structure–function relationship of a protein and will be dealt with in Chapter 14. Finally, we will take a brief look at secondary structure prediction on nucleotide sequences, as illustrated by tRNA structure.

OBTAINING SECONDARY STRUCTURE FROM SEQUENCE

different ways to predict

nearest neighbor

statistical methods

machine learning

neural nets

HMMs

mixture of the above

**Flow Diagram 11.1**

**The key concept introduced in this section is that many different approaches have been taken in deriving methods for predicting protein secondary structure.**

## 11.1 Types of Prediction Methods

Many prediction programs are designed to recognize just three different regular structural states: α-helices, β-strands, and β-turns (see Figure 11.2). Hence, when predicting secondary structure, it is common to produce a four-state prediction, where each residue is predicted as an α-helical, β-strand, β-turn, or random coil conformation. In this context a coil conformation is one that is not an α-helix, β-strand, or β-turn and will include unstructured loops and the irregular regions of

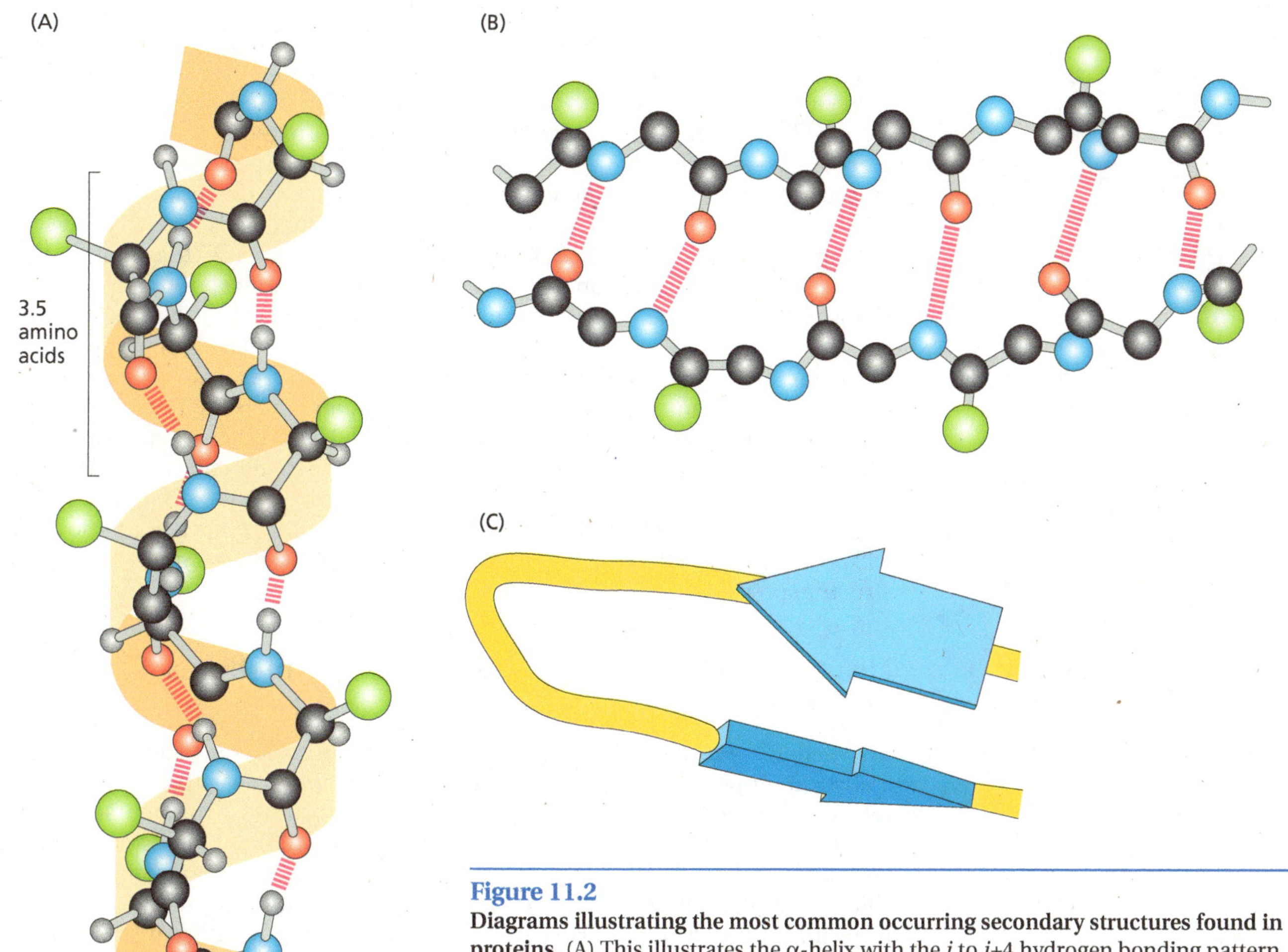


**Figure 11.2**

**Diagrams illustrating the most common occurring secondary structures found in proteins.** (A) This illustrates the α-helix with the *i* to *i*+4 hydrogen bonding pattern. (B) This shows a β-sheet consisting of two antiparallel β-strands. The β-strands form specific hydrogen bonds between them to form the sheet. (C) This shows a β-turn. In parts A and B the residue side chains are represented as green spheres.

protein chain that link elements of secondary structure. Other programs use a three-state prediction where the β-turn is not predicted. Turns can occur between consecutive β-strands in a β-sheet, for example. There are some algorithms that even try to predict different kinds of helical or turn conformations, such as π-helices, $3_{10}$-helices, and polyproline helices, and type I, II, and other types of turns.

Methods for predicting protein secondary structure can be broadly divided into the following categories: statistical analyses, also referred to as probabilistic analyses; knowledge-based analyses; **machine-learning methods**; and those mainly based on neural networks. In addition there are consensus methods, which take an average of a set of different predictions. Most automated methods in use today use a mixture of these techniques, and all of them incorporate some form of statistical analysis.

## Statistical methods are based on rules that give the probability that a residue will form part of a particular secondary structure

The probabilities are derived from analyzing structure and sequence data from large sets of proteins of known structure, and correlating structural and sequence features to form statistical rules for secondary structure assignment. The early statistical methods suffered from lack of data because of the small number of experimentally solved three-dimensional structures. This is not such a problem today for globular proteins using simple statistical models, but remains so in other cases.

For many years, the most widely used method of this type was that of Chou and Fasman, which was developed in the 1970s. It was based initially on the analysis of a small set of proteins—only 15 structures in the first set—and simply assigned individual amino acids as α- or β-formers, indifferent, and α- or β-breakers (see Figure 11.3). Short segments composed of formers together with an absence of many breakers were assigned as the core of an α-helix or β-strand. The boundaries of secondary structure elements were delineated by the presence of strong breakers. In 1989 the initial dataset was extended to include 64 proteins and subsequently further extended to 144 nonhomologous proteins with a total of 33,118 residues from which to obtain new parameters. This last reparameterization changed the assignment of some of the residues. Statistical analysis of the largest dataset showed that there is still an uncertainty of more than 10% in the values of the parameters (see Section 12.2). In the past few years, this method has been superseded by more accurate statistical methods, such as the GOR method described later, which make assignments on the basis of stretches of residues, thus enabling local interactions between residues that will influence secondary structure to be taken into account. The latest version of GOR incorporates statistical information from a very much larger dataset than the original GOR method and it will be shown that a larger dataset can and does improve the accuracy of secondary structure prediction (see Section 11.4).

## Nearest-neighbor methods are statistical methods that incorporate additional information about protein structure

Knowledge-based methods are optimized statistical methods: in addition to using statistical propensities of amino acids to form particular structures, they incorporate knowledge of the physics and chemistry of protein structure, such as the shapes, sizes, and physicochemical properties of the different amino acid residues. One such method described later is PREDATOR, which also uses an approach, often called nearest neighbor, that involves finding short sequences of known structure from the databases that closely match stretches of the query sequence. Some of these methods use multiple sequence alignments to take account of the degree of conservation of residues. Prediction from a multiple alignment of protein sequences is a good way to improve prediction accuracy as during evolution

| Amino Acid | helix | | strand | |
|---|---|---|---|---|
| | Designation | *P* | Designation | *P* |
| Ala | F | 1.42 | b | 0.83 |
| Cys | I | 0.70 | f | 1.19 |
| Asp | I | 1.01 | B | 0.54 |
| Glu | F | 1.51 | B | 0.37 |
| Phe | f | 1.13 | f | 1.38 |
| Gly | B | 0.61 | b | 0.75 |
| His | f | 1.00 | f | 0.87 |
| Ile | f | 1.08 | F | 1.60 |
| Lys | f | 1.16 | b | 0.74 |
| Leu | F | 1.21 | f | 1.30 |
| Met | F | 1.45 | f | 1.05 |
| Asn | b | 0.67 | b | 0.89 |
| Pro | **B** | **0.57** | **B** | **0.55** |
| Gln | f | 1.11 | h | 1.10 |
| Arg | I | 0.98 | I | 0.93 |
| Ser | I | 0.77 | b | 0.75 |
| Thr | I | 0.83 | f | 1.19 |
| Val | f | 1.06 | F | 1.70 |
| Trp | f | 1.08 | f | 1.37 |
| Tyr | b | 0.69 | F | 1.4 |

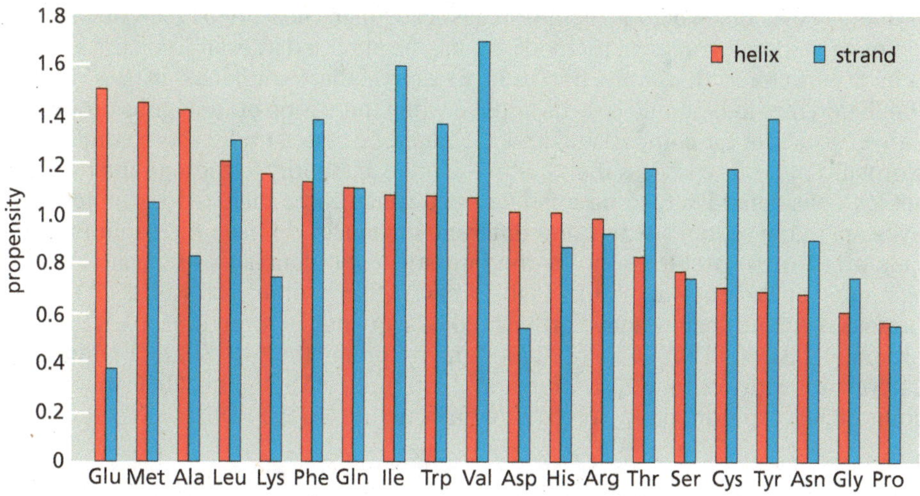


**Figure 11.3**
**Chou and Fasman Propensities (*P*).** F stands for strong former, f weak former, while B and b stand for strong and weak breaker, respectively. I (indifferent) indicates residues that are neither forming nor breaking helices or strands. We can see that Pro has the lowest propensity for forming a helix and a low one for strands as well. However, many other residues that are either weak or indifferent have been reclassified since the propensities shown here have been reparameterized as more data have become available. Data from P.Y. Chou and G.D. Fasman, Prediction of the secondary structure of proteins from their amino acid sequence, *Adv. Enzymol.* 47:45–148, 1978.

residues with similar physicochemical properties are conserved if they are important to the fold or function of the protein.

## Machine-learning approaches to secondary structure prediction mainly make use of neural networks and HMM methods

Machine-learning methods train a neural net or other learning algorithm to acquire structure–sequence relationships, which can then be applied to predict structure from a protein sequence. The essential difference from statistical and knowledge-based approaches is that instead of using the sequence dataset to derive rules or parameters, a neural network is fed the sequence dataset as input, and during a learning period the connections between input and output (the parameters) are adjusted until the output—the structure assignment—is as accurate as possible. The method is then ready to be used to predict secondary structure assignments from an amino acid sequence. There is no need for the neural network to correspond to any clear physical model of the sequence–structure relationship. It is possible to obtain successful prediction networks that do not lend themselves to further analysis that might reveal new underlying concepts. Nevertheless, these methods provide some of the most accurate predictions currently available.

## 11.2 Training and Test Databases

The parameters that any method uses to make predictions are derived from a database of proteins for which the structure is known, the **training dataset**. The performance and accuracy of the prediction method is then tested on another independent dataset of structures, the **test dataset**. The appropriate selection of training datasets is crucial to achieving the greatest prediction accuracy possible, while a good test dataset will illustrate the ability of the prediction program.

The general aim is to include a range of different and unrelated sequences and structures in the dataset. The inclusion of many similar structures in the training set may unduly bias the prediction algorithm parameters to predict that type of structure relatively more accurately. In practice, α-helices and random coils are more prevalent in proteins than are β-strands and β-sheets, and this imbalance needs to be avoided when choosing the training dataset. A simple solution is available when devising methods using neural networks. In these cases training procedures are used that oversample structures with β-strands to compensate for their relative scarcity.

In addition, no protein in the testing set should be homologous to a protein in a training set, as this would give an unrealistically high measure of accuracy. A true estimate of the accuracy of a method can only be obtained by examining the prediction of structures that were not used to derive the parameters of the model. Prediction methods should only be applied when there is no homologous protein of known structure for comparison, and so should be tested under these conditions. Normally, proteins with less than 25–30% sequence identity to those in the training set are considered likely to be unrelated and thus can be included in the test set. Even so, many proteins with quite different sequences do have the same fold. To help avoid these problems, nonredundant structural databases are available (for example, the Protein Data Bank's PDB_SELECT) that contain a subset of all known protein structures below a given threshold of sequence or structural similarity (for example, the October 2004 set of nonredundant protein in PDB_SELECT contained 2485 chains with less than 25% sequence identity). As of September 2006, there were some 38,882 experimentally solved protein structures in the PDB as a whole. There

**Flow Diagram 11.2**
**The key concepts introduced in this section are that there is no universally approved definition of protein secondary structure and that prediction methods are parameterized and tested by using datasets of proteins of known structure.**

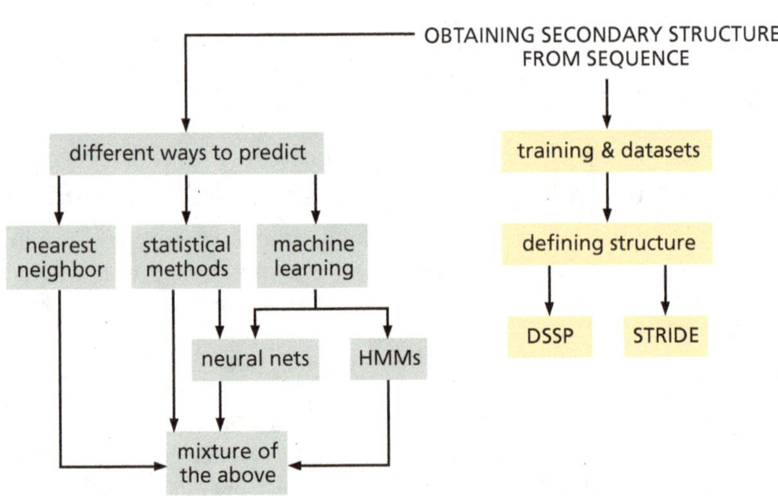

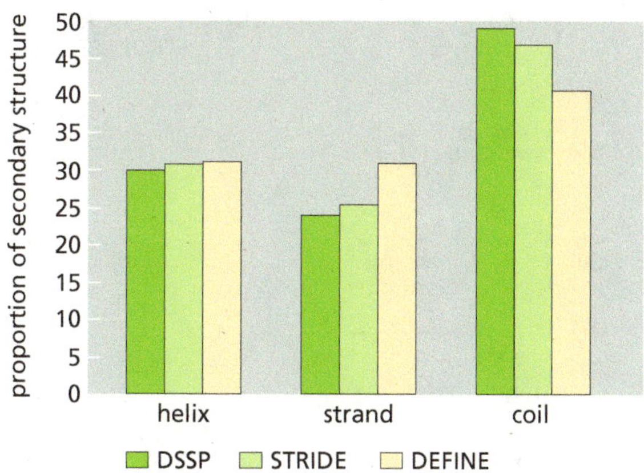


**Figure 11.4**
**Different methods for assignment of secondary structure.** The proportions of secondary structure assigned by three different automatic methods from the same dataset of protein structures.

are around 2500 protein structures that have less than 25% identity. Thus only about 8% of the solved structures are truly independent data.

### There are several ways to define protein secondary structures

There is no single best method of exactly defining protein secondary structural elements from the atomic coordinates. Secondary structure is assigned in experimentally determined protein structures according to parameters such as the torsional angles (dihedral angles) of the residues ($\phi$, $\psi$), their hydrogen-bonding patterns, and the positions of $C_\alpha$ atoms (see Section 2.1). Various automated methods have been developed that assign secondary structure from the atomic coordinates according to specific rules, using one or more of these parameters.

The most commonly used of these programs is DSSP, which assigns secondary structure according to hydrogen-bond patterns. Another, STRIDE, uses both hydrogen-bond energy and backbone dihedral angles. DEFINE matches the inter-atomic distances in the protein with those from idealized secondary structures. However, these programs do not always produce identical results from the same data, giving slightly different secondary structure assignments (see Figure 11.4). The differences are almost exclusively at the ends of structural elements. These differences in defining secondary structure elements can affect the apparent accuracy of secondary structure prediction methods. A prediction method should be trained and subsequently tested using training and testing datasets whose structural features were defined using the same assignment method.

## 11.3 Assessing the Accuracy of Prediction Programs

Predicting the secondary structure of a protein is very useful but we have to have an idea of the accuracy of the prediction from a given program. The measured accuracy of a prediction algorithm is used to help us estimate its likely performance when presented with a sequence of unknown structure. Accuracy can be measured either in respect of individual residue assignments or in relation to the numbers of helices and strands that are correctly predicted. The $\mathbf{Q_3}$ and **Sov** measures will be presented here, and discussed in more detail in Section 12.1.

### $Q_3$ measures the accuracy of individual residue assignments

One commonly used measure of accuracy is called $Q_3$. It is applied at the residue level and is given by calculating the percentage of correctly predicted residues within a given sequence of known secondary structure. Note that the $Q_3$ measure of a method's accuracy is given by the average $Q_3$ over a test dataset not used in

**Flow Diagram 11.3**
**The key concept introduced in this section is that there are several ways of measuring the accuracy of a protein secondary structure prediction, each of which has strengths and weaknesses.**

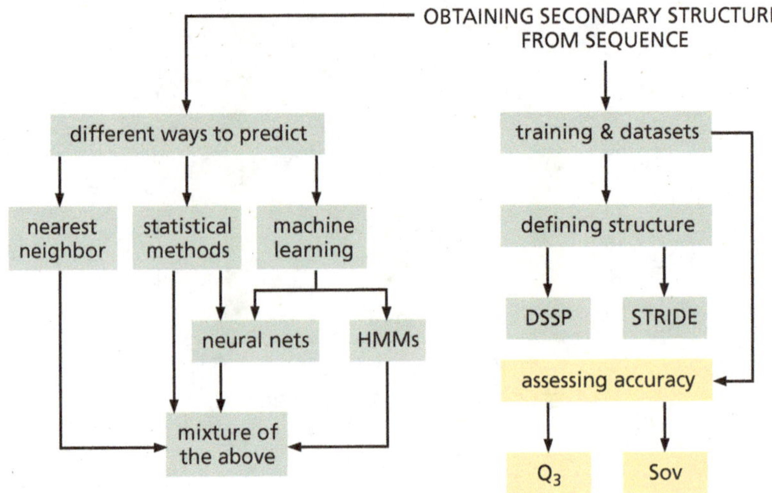


training or parameterization. If the training dataset is used it will suggest the method is far more accurate than it really is.

The values of $Q_3$ can range between 0 and 1, with 1 indicating a perfect prediction. The value for a random prediction will depend on the percentages of the different states present in known protein structures (modified to account for the presence of homology). If there were equal amounts of helix, strand, and coil, a random prediction would give a value of 0.33 for $Q_3$. As an example in a study by Barry Robson and colleagues, the value of $Q_3$ for random prediction was calculated to be 0.38 (38%). Care must be taken in blindly using $Q_3$ values to assess the performance of a prediction method. As shown in Figure 11.5, it is possible for two predictions to give the same $Q_3$ value despite one being much less useful.

## Secondary structure predictions should not be expected to reach 100% residue accuracy

Although it might at first seem that 100% correct prediction of the secondary structure of each residue in a sequence is a desirable and attainable goal, this is not the case. There are two reasons for this, one relating to the difficulty in defining secondary structures, the second to the variability observed in the structures of homologous proteins.

The uncertainties in secondary structure definition, discussed earlier in Section 11.2, mean that less than 100% of all residues can be defined with absolute certainty. Therefore, the maximum $Q_3$ value that can be expected for a prediction program to attain (when run against a complete database) will be less than 100%.

Homologous proteins do show some variability despite the fact that almost all secondary structures are preserved in homologous proteins. There are often small differences in the positions of the ends of regular elements (see Figure 11.6). Additionally, if a loop is not of functional or structural importance, small secondary structure elements can be found within it in some homologous proteins.

$$Q_3 = \frac{\text{number of residues correctly predicted}}{\text{total number of residues}}$$

VLHQASGNSVILFGSDVTVPGATNAEQAR amino acid sequence 29 residues long

HHHHHCCCCEEEECCCEEECCCCCHHHHH actual secondary structure

CHHHCCCCEEEECCCCCEEECCCHHHHHH prediction 1: **$Q_3$ = 22/29 = 76%:** useful

HHHHHCCCCHHHHCCCHHHCCCCCHHHHH prediction 2: **$Q_3$ = 22/29 = 76%:** terrible

**Figure 11.5**
**Using the $Q_3$ accuracy measurement has its problems.** A hypothetical sequence of 29 residues with an actual secondary structure is used as an example. Prediction 1 gives a useful prediction, predicting the correct number of secondary structures in the correct regions, only the ends of the predictions are incorrect. The $Q_3$ for this prediction is 76%. Prediction 2 only predicts the terminal helices correctly; the strands are predicted as helices. The $Q_3$ for this prediction is also 76%. [Note the correct prediction of coil (loop) regions is also counted.] Therefore both the terrible and useful predictions give identical $Q_3$ levels giving no indication of which prediction is correct. E indicates strand residues, H denotes helix residues and C denotes coil.

```
HQKVILVGD GAVGSSYAFAMVLQGI        AQEIGIVDI
GARVVVIGA GFVGASYVFALMNQGI        ADEIVLIDA
RCKITVVGV GDVGMACAISILLKGL        ADELALVDA   multiple alignment
YNKITVVGV GAVGMACAISILMKDL        ADEVALVDV
DNKITVVGV GQVGMACAISILGKSL        TDELALVDV
PIRVLVTGAAGGQIAYSLLYSIGNGSVFGKDQPIILVLLDI

CCCBBBCCC CHHHHHHHHHHHHCC         CCCBBBCCC
CCBBBBBCC CHHHHHHHHHHCCCC         CCBBBBBCC
CCBBBBBCC CHHHHHHHHHHCCCC         CCBBBBBCC
CCBBBBBCC CHHHHHHHHHHHCCC         CCBBBBBCC   DSSP assignment
CCBBBBBCC CHHHHHHHHHHCCCC         CCBBBBBCC
CCCBBBCCC CHHHHHHHHHHHHCC         CCCBBBCCC
CCBBBBBCCCCHHHHHHHHHHCCCCCCCCCBBBBBBCC

CCCBBBCCCCCHHHHHHHHHHCCCCCCCCCCBBBBCCC   minimum consensus

CBBBBBBCCCCHHHHHHHHHHHHCCCCCCCCBBBBBBCC   maximum consensus
```

**Figure 11.6**
**Not all homologous proteins are assigned identical secondary structure assignments.**
The strand assignments are green and the helix red. This discrepancy will affect the training of secondary structure programs. (Adapted from R.B. Russell and G.J. Barton, The limits of protein secondary structure prediction accuracy from multiple sequence alignment, *J. Mol. Biol.* 234:951–957, 1993.)

Calculations on large datasets of proteins show that, because of this variability, even an excellent secondary-structure prediction may give a $Q_3$ of only 80%.

## The Sov value measures the prediction accuracy for whole elements

It is more useful to predict the correct number, type, and order of secondary structure elements than to predict some elements well and miss others completely. For some distinctive protein folds, the order of secondary structure elements can be a strong clue to the overall fold and thus to the class and possible function of the protein (see Box 11.1). The accuracy of this aspect of structure prediction is assessed by a measure known as the fractional overlap of segments (Sov). This measures the percentage of correctly predicted secondary structure segments rather than individual residue positions, and it pays less attention to small errors in the ends of structural elements, which is a common problem in prediction algorithms. The $Q_3$ provides an overall impression of the likely accuracy of a prediction, while the Sov value indicates how well a method performs in predicting the correct type of secondary structure (see Figure 11.7). For a detailed description of both $Q_3$ and Sov calculations see Section 12.1.

## CAFASP/CASP: Unbiased and readily available protein prediction assessments

In most cases until recently, when a new secondary structure prediction method was published its accuracy was reported measured against a reference dataset that often differed from others previously used for the purpose. As a consequence it has proved difficult to compare the accuracy of new methods against those previously published. This problem has now been solved with the advent of international assessment methods that are readily available to the whole community and are designed to be as bias-free as possible. These are known as CAFASP (Critical Assessment of Fully Automated Structure Prediction) and CASP (Critical Assessment of Structure Prediction).

CASP and CAFASP use structural information that has just been determined and is not yet published. The different sequences are carefully analyzed to determine the degree of homology to published protein structures, which is used to assess the likely degree of difficulty. The sequences are provided for predictions of secondary and tertiary structure, and following publication of the structures the predictions can then be compared to the experimentally determined structures.

```
x-ray:  CCCEEEECCHHHHHHHHHCCCCEEEEECCCCCCCCCCHHHHHHCCCC
pred:   CCCEEECCCCHHHHHHHHHCCCEECCECCCCCCCCCCHCHCHCHHCC
```

**Figure 11.7**
**Sov looks for correct segments.**
A $Q_3$ accuracy value will take all matched residue predictions to the X-ray structure as correct. Sov looks for predicted segments of the correct type. Therefore, the first strand and helix will have similar accuracy values; the second strand will have a reasonable $Q_3$, although Sov will have a low value because the prediction is fragmented. The same applies to the last helix. Strands are generally denoted by the letter B (as in Figure 11.6), or E.

## Box 11.1 Are secondary structure predictions useful?

The short answer is yes: they can be useful in classifying proteins based on secondary structure predictions in the context of genome analysis, aspects of protein function predicted based on expert analysis of secondary structure, and one can use secondary structure prediction to obtain or improve tertiary structure. An example of successful secondary structure prediction applications is described in this box.

Graham P. Davies and colleagues have looked at type I DNA restriction enzymes for which there was little structural knowledge. They aligned the R domain of the EcoKI enzyme and used the PHD methods of secondary structure prediction (see main text) to produce an alignment of the R subunits from various proteins with EcoKI.

The prediction suggested an alternating α-helix–β-strand structure throughout the region that contains a conserved and functional DEAD-box motif. (The DEAD-box motif is probably involved in ATP hydrolysis.) From the prediction and alignment the authors predicted that these specific R domains contain helicase folds 1A and 2A. Although no structural verification exists yet for this prediction, a sequence search through the structural database picks other proteins that contain this type of fold.

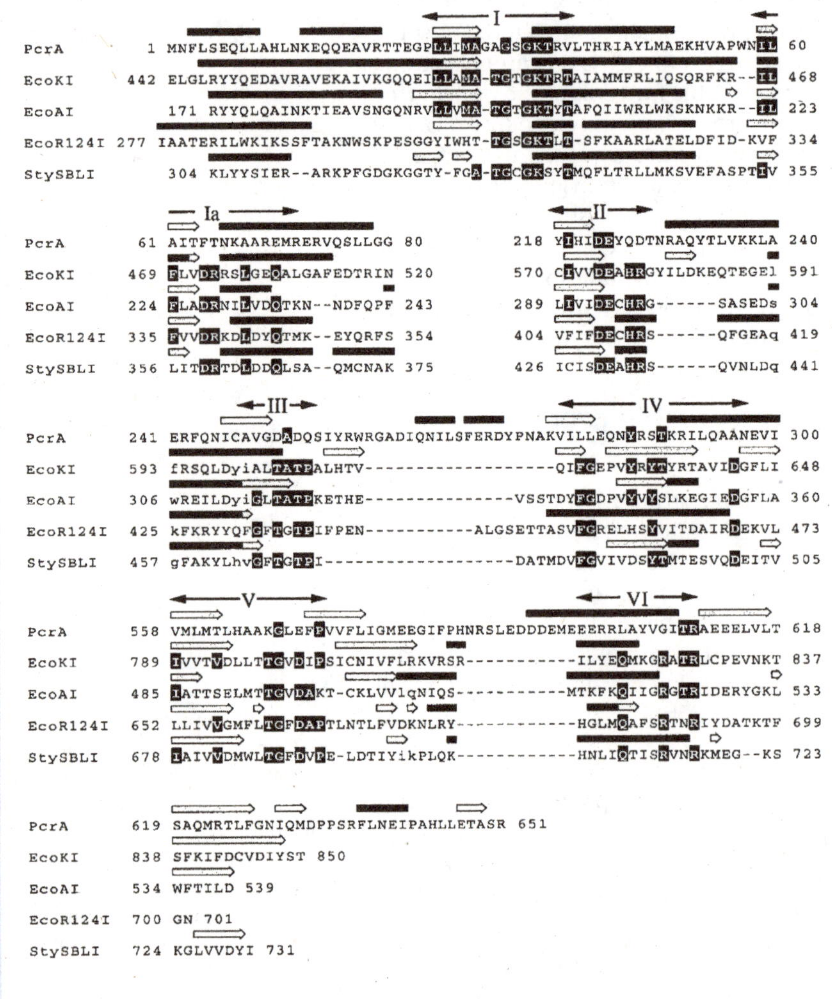

**Figure B11.1**
**Prediction of type I R subunits from EcoKI, EcoAI, EcoR124I, and StySBLI.** An alignment of amino acid sequence and predicted secondary structure of the region of type I R subunits from EcoKI, EcoAI, EcoR124I, and StySBLI containing the DEAD-box motifs, with domains 1A and 2A of the PcrA helicase. The motifs are indicated above the sequences by the numbered bold black arrows, β-strands are indicated by light arrows, and α-helices by black rectangles. (Reprinted from *J. Mol. Biol.*, 290, G.P. Davies et al., On the structure and operation of type I DNA restriction enzymes, 565–579, 1999, with permission from Elsevier.)

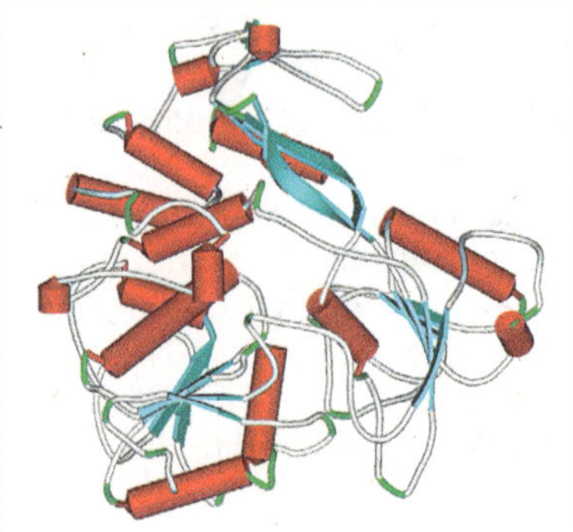

**Figure B11.2**
**A crystal structure that contains the helicase folds 1A and 2A.**

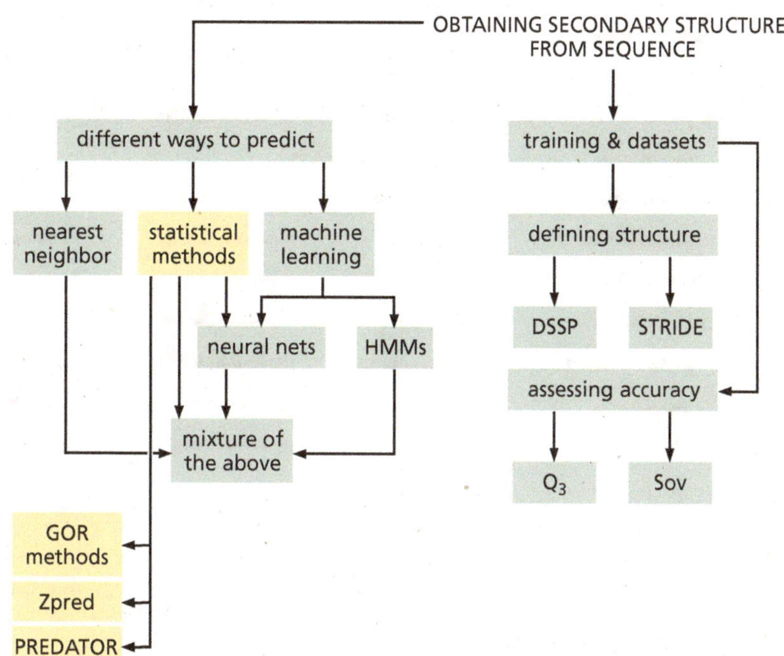


**Flow Diagram 11.4**
**In this section some of the statistical and knowledge-based prediction methods are discussed.**

## 11.4 Statistical and Knowledge-Based Methods

To illustrate the methods, we will use an example set of four small proteins whose sequence and structure have been determined and which represent the four main classes of protein tertiary structure (all α, all β, α+β, and α/β) (see Figure 11.8).

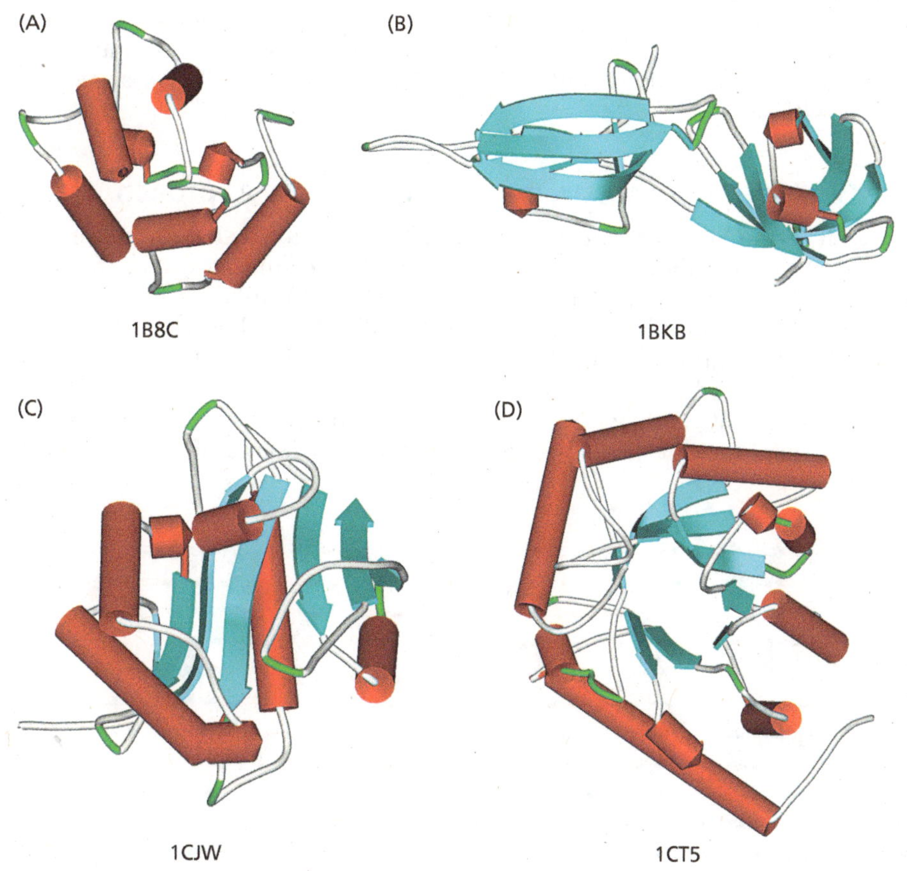


**Figure 11.8**
**Three-dimensional representations of the proteins used in the example predictions.** (A) Parvalbumin (1B8C) is a calcium-binding protein with an all α-helical conformation (helices are represented by red-colored cylinders). (B) Translation Initiation Factor 5A (1BKB) is classified as all β-sheet (strands are represented by blue arrows). (C) A serotonin *N*-acetyltransferase (1CJW) is an enzyme that falls into the α+β class containing both α-helices and β-strands. (D) A hypothetical protein YBL036C (1CT5), from baker's yeast and classified as an α-helix/β-strand alternating fold with a TIM α/β-barrel. (The four-character identifiers after each protein name are the Protein Databank PDB entry names.)

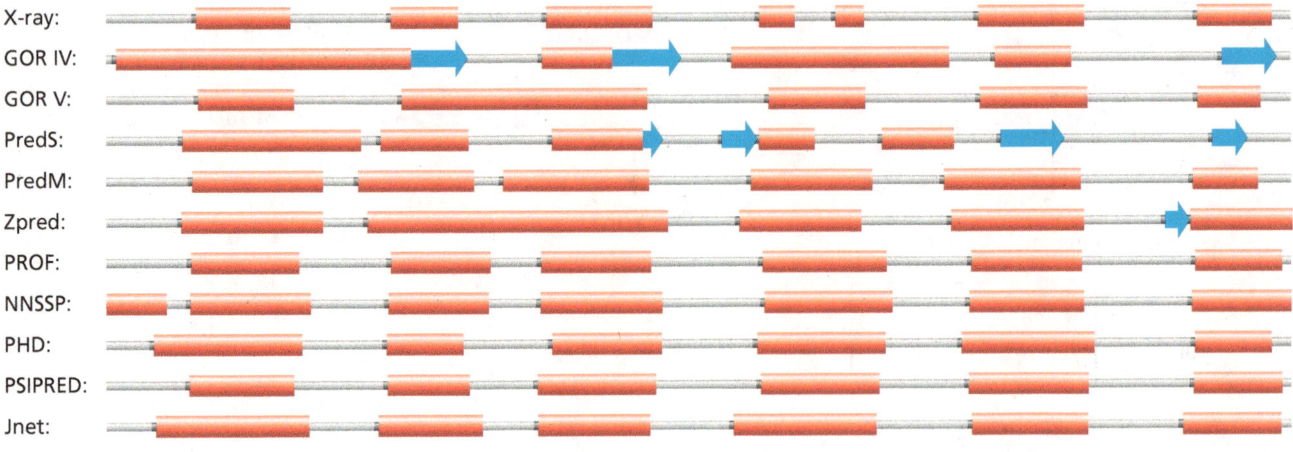


**Figure 11.9**
**A schematic diagram of the secondary structure prediction by various programs compared to the observed X-ray structure of the all α-fold protein (1B8C).** The helices are represented by a red cylinder, the blue arrow indicates β-strands and a gray stick illustrates a coil. The prediction algorithms shown are the GOR algorithm (GOR IV) and the most recent version of the GOR algorithm with evolutionary information (GOR V), PREDATOR using a single sequence (PredS), PREDATOR with multiple sequences (PredM), the Zpred multiple alignment prediction (Zpred), and the neural network programs PROF (son of DSC), NNSSP, PHD, PSIPRED, and Jnet.

Figures 11.9 to 11.12 compare the secondary structure predictions for the four example proteins for all the methods described in this chapter. The secondary-structure assignments determined by X-ray crystallography are given at the top of each figure.

Table 11.1 gives the $Q_3$ and Sov values of the predicted secondary structures. As we shall see, no method achieves its published accuracy value for every protein, emphasizing the fact that the published values are averages, and that any individual prediction, even on a method with a high published accuracy, will always have to be treated with caution.

## The GOR method uses an information theory approach

The GOR method is a widely used statistical method named after its authors Jean Garnier, David Osguthorpe, and Barry Robson. However, the latest version, GOR V, does not use statistical parameters alone. The statistical-based GOR methods predicted three conformations: helix (H), extended (β-strand; E), and coil (C), and when published was reported to have a $Q_3$ value of 64.4% (for GOR IV).

The basis of the GOR method lies in incorporating the effects of local interactions between amino acid residues by taking successive **windows** of 17 residues and

**Figure 11.10**
**Secondary structure prediction for the β-fold (1BKB) protein.** Some methods can be seen to suffer from overprediction of α-helices. However, only Zpred predicts a small N-terminal α-helix, which all the other methods miss, but due to the rest of the prediction inaccuracy, Zpred barely reaches 45%. Nearly all methods suffer from overprediction of α-helices at the C-terminal end of the protein. The best method here is Jnet, which predicts both the α-helical region and the strands at the end correctly.

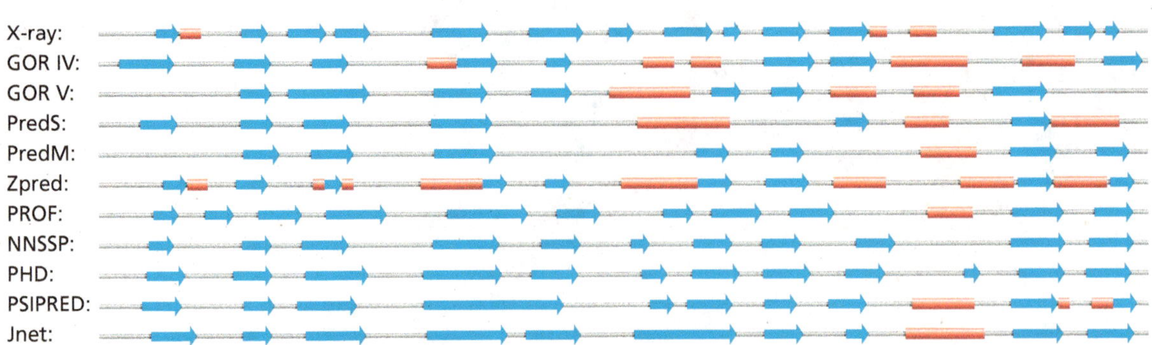

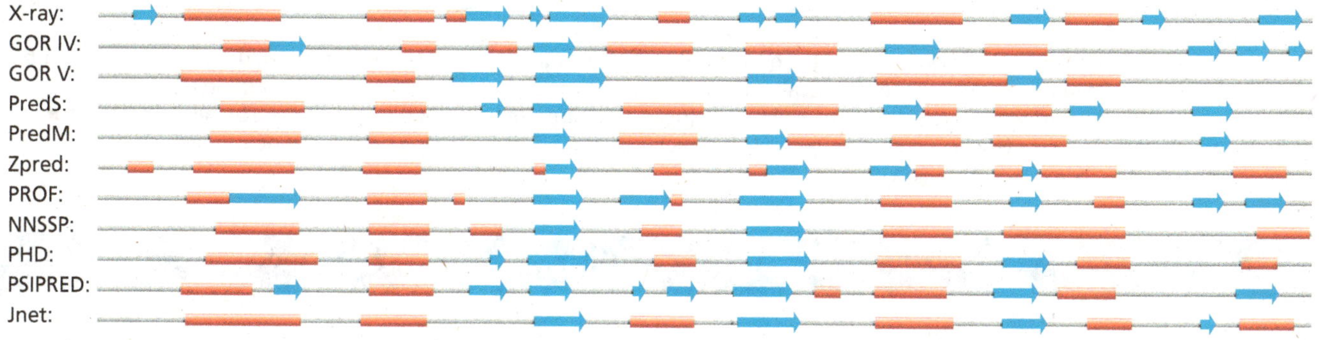


**Figure 11.11**
**A schematic of the secondary structure prediction for the α+β class of protein (1CJW).**

considering the effect of residues from positions *j*–8 to *j*+8 on the conformation of the residue at position *j*. Thus, for each residue there is a 17-residue profile that quantifies the contribution each residue makes to the probability of a given structural state for *j* (see Figures 11.13 and 11.14).

To determine this profile, GOR uses three types of information: **self-information**, **directional information**, and **pair information**. Self-information is the information a residue carries about its own conformation, and is related to the Chou–Fasman propensities. Directional information is the information about the conformation at position *j* carried by a residue at position $i \neq j$ and is independent of the type of residue found at *j*. Pair information takes account of the type of residue at *j*. Figure 11.13 illustrates the effect that directional information for a proline residue five residues carboxy-terminal to *j* has on the probability of *j* adopting α-helical conformation. Proline is a helix breaker, and a comparison with Figure 11.14 shows that its presence will lower the probability of *j* adopting α-helical conformation, even if the residue at position *j* is a helix-forming residue.

The original GOR method has been improved over the years, mainly by using larger training datasets and more detailed statistics, which account not only for amino acid composition, but also for amino acid pairs and triplets. These improvements gave rise to four versions of the algorithm, the latest of these being GOR IV. GOR IV was based on statistics from a database that contained 267 nonhomologous structures. GOR V in addition includes evolutionary information obtained by using PSI-BLAST (see Section 6.1) to align the sequences first and use the information from the alignment to improve the prediction accuracy. (This type of improvement has been used with the GOR method before and is described below.) The Sov measure reported for GOR V was 70.7% and the prediction accuracy $Q_3$ was 73.5% compared to the original value of $Q_3$ of 64.4%.

To illustrate how a single program can develop and improve, Figure 11.15 shows predictions with the original GOR method, GOR IV, and the latest GOR V. In this

**Figure 11.12**
**A schematic representation of the secondary structure prediction of the α/β-fold protein (1CT5).**

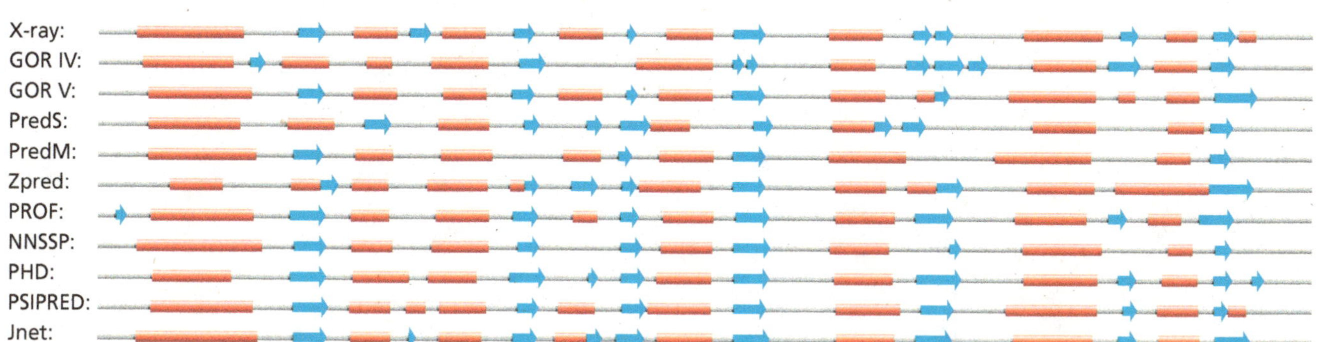

**Table 11.1**
The percentage accuracy ($Q_3$) and percentage segment overlap (Sov) are given for each prediction algorithm discussed in the chapter. The prediction methods in the white area are statistical or medium/long range parameter algorithms without the use of multiple sequences. The gray area indicates incorporation of multiple sequence data into the prediction algorithms. The blue area shows methods that in addition to statistical, medium/long-range interactions, and other parameters, are using neural networking techniques to improve the prediction accuracy.

| Method | | all α (1B8C) | all β (1BKB) | α + β (1CJW) | α/β (1CT5) |
|---|---|---|---|---|---|
| GOR V | $Q_3$ | 59.0% | 55.1% | 39.2% | 67.2% |
| | Sov | 46.4% | 53.3% | 34.3% | 62.3% |
| Chou–Fasman | $Q_3$ | 50.9% | 42.6% | 38.0% | 52.4% |
| | Sov | 50.0% | 2.2% | 34.7% | 41.6% |
| PREDATOR | $Q_3$ | 55.5% | 55.9% | 49.0% | 66.4% |
| | Sov | 52.7% | 51.0% | 45.1% | 67.8% |
| GOR V | $Q_3$ | 84.3% | 60.3% | 69.9% | 73.8% |
| | Sov | 79.5% | 59.9% | 62.8% | 77.1% |
| Predator Multiple Seq. | $Q_3$ | 80.5% | 64.0% | 54.2% | 80.1% |
| | Sov | 80.4% | 53.5% | 49.7% | 81.4% |
| Zpred | $Q_3$ | 76.9% | 44.9% | 60.8% | 66.0% |
| | Sov | 73.2% | 48.1% | 63.6% | 71.3% |
| PROF | $Q_3$ | 90.7% | 58.1% | 65.1% | 77.9% |
| | Sov | 91.7% | 57.3% | 63.2% | 83.7% |
| NNSSP | $Q_3$ | 81.5% | 73.5% | 66.9% | 82.0% |
| | Sov | 85.0% | 72.9% | 65.6% | 81.4% |
| PHD | $Q_3$ | 87.9% | 73.5% | 75.9% | 78.1% |
| | Sov | 92.6% | 72.9% | 76.0% | 85.0% |
| PSIPRED | $Q_3$ | 94.4% | 65.4% | 67.8% | 82.8% |
| | Sov | 92.8% | 61.9% | 66.8% | 87.4% |
| Jnet | $Q_3$ | 84.2% | 74.3% | 70.5% | 84.4% |
| | Sov | 91.5% | 72.2% | 65.8% | 91.5% |

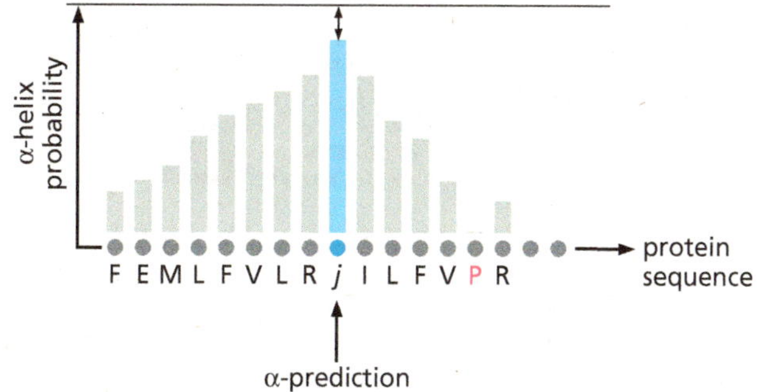


**Figure 11.13**
**The effect of an α-helix breaker (Pro) at position *j*+5.** The proline, which substitutes methionine, diminished the overall additive propensity of residue j to form a helix (indicated by the bottom black arrow).

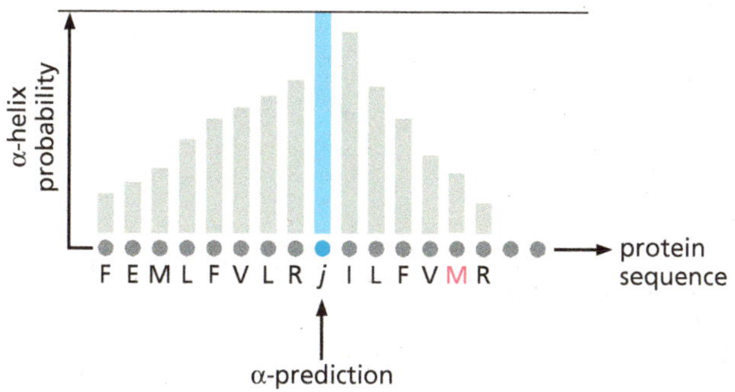


**Figure 11.14**
**The effect of a non-α-helix breaker (Met) at position *j*+5.** The methionine improves the additive propensity of residue j to form a helix (indicated by the bottom black arrow).

chapter GOR IV is used to illustrate the use of information theory without the additional information obtained from aligned sequences. GOR IV is substantially better than the original GOR method. This shows the importance of having a large training dataset. The evolutionary information from aligned sequences used in GOR V improves the accuracy of prediction greatly and indicates the importance of conservation of protein structure and its usefulness in prediction algorithms.

All GOR versions are easy to use and provide results that are easy to interpret (see Figure 11.16). However, the accuracy is still relatively low (especially with the earlier versions of GOR methods), and this has driven the search for additional information that can be coded into statistical prediction algorithms. With the advent of more powerful computers, secondary structure prediction evolved. However, most secondary structure prediction methods use a statistical analysis of known protein structures for some or all of the parameters they use.

## The program Zpred includes multiple alignment of homologous sequences and residue conservation information

It has been recognized for quite some time that information obtained from aligned homologous sequences enhances prediction results. This is because generally during evolution residues are conserved if they are important to the function or fold of the protein. The effect of this is that secondary structure elements such as α-helices and β-strands are more conserved than loops unless the loops are involved in functional tasks. For the same reason insertions or deletions are less likely to occur in α-helices and β-strands than in loops. The information present in a multiple sequence alignment can be utilized in two ways to help improve the prediction accuracy. A secondary structure prediction can be obtained for each sequence in the alignment. The predictions for each residue in an alignment column can then be averaged to obtain the prediction at that position. Alternatively, at each alignment position some type of measure such as a conservation value can be calculated and used to modify the prediction (see Figure 11.17).

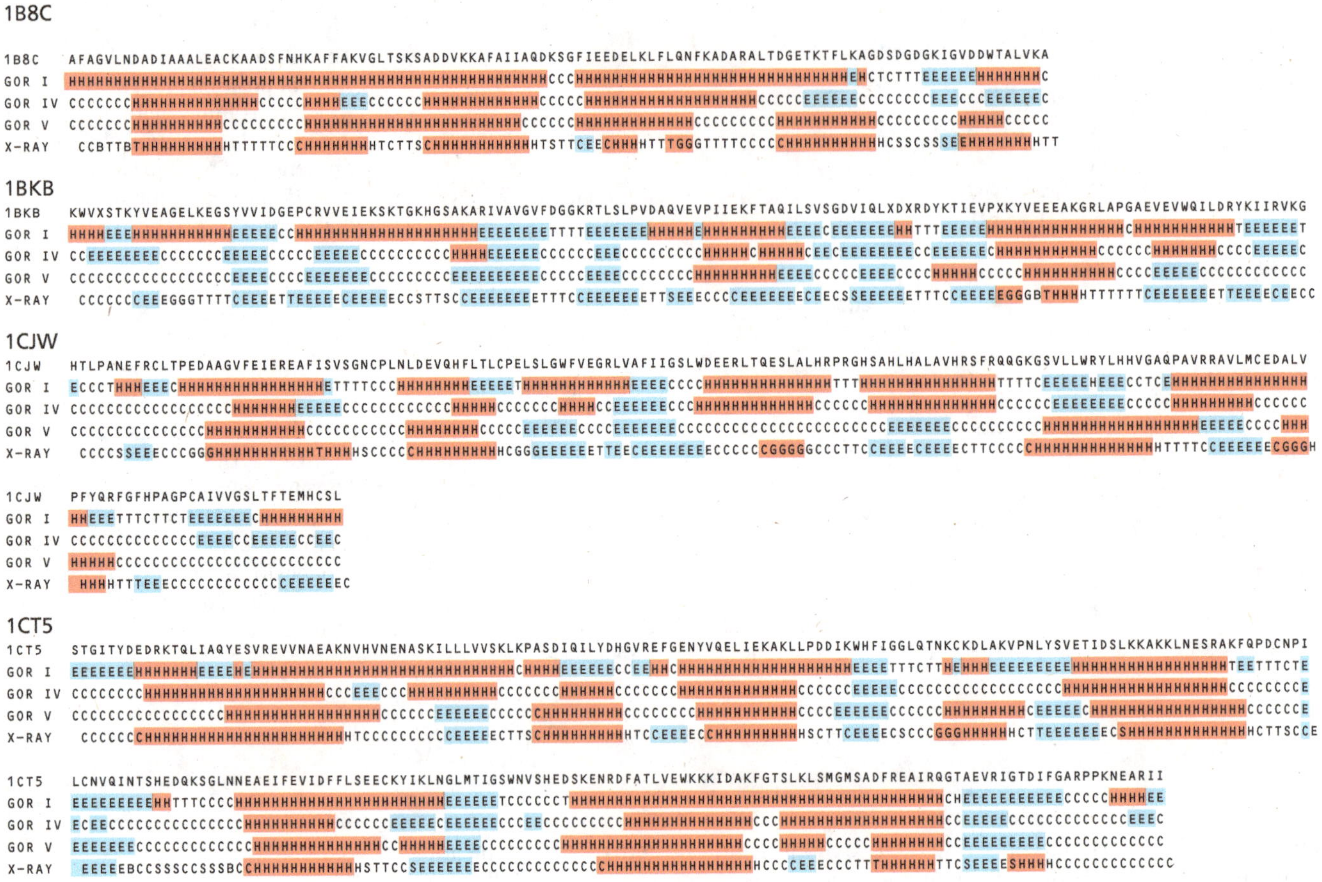

1B8C

```
1B8C    AFAGVLNDADIAAALEACKAADSFNHKAFFAKVGLTSKSADDVKKAFAIIAQDKSGFIEEDELKLFLQNFKADARALTDGETKTFLKAGDSDGDGKIGVDDWTALVKA
GOR I   HHHHHHHHHHHHHHHHHHHHHHHHHHHHHHHHHHHHHHHHHHHHHHHHHHHHHHCCCHHHHHHHHHHHHHHHHHHHHHHHHHHHHHHEHCTCTTTEEEEEEHHHHHHHC
GOR IV  CCCCCCCHHHHHHHHHHHHHHCCCCCHHHHEEECCCCCCHHHHHHHHHHHHHCCCCCHHHHHHHHHHHHHHHHHHHCCCCCEEEEEECCCCCCCCEEECCCEEEEEEC
GOR V   CCCCCCCHHHHHHHHHHCCCCCCCCCHHHHHHHHHHHHHHHHHHHHHHHHCCCCCHHHHHHHHHHHHHCCCCCCCCCHHHHHHHHHHHCCCCCCCCCHHHHHCCCCC
X-RAY    CCBTTBTHHHHHHHHHHTTTTTCCCHHHHHHHHTCTTSCHHHHHHHHHHHHTSTTCEECHHHHTTTGGGTTTTCCCCCHHHHHHHHHHHCSSCSSSEEHHHHHHHHTT
```

1BKB

```
1BKB    KWVXSTKYVEAGELKEGSYVVIDGEPCRVVEIEKSKTGKHGSAKARIVAVGVFDGGKRTLSLPVDAQVEVPIIEKFTAQILSVSGDVIQLXDXRDYKTIEVPXKYVEEEAKGRLAPGAEVEVWQILDRYKIIRVKG
GOR I   HHHHEEEHHHHHHHHHHHEEEEECCHHHHHHHHHHHHHHHHHHHHEEEEEEEETTTTEEEEEEEHHHHHEHHHHHHHHHEEEECEEEEEEEHHTTTEEEEEHHHHHHHHHHHHHHHCHHHHHHHHHHHTEEEEEET
GOR IV  CCEEEEEEEECCCCCCCEEEEECCCCEEEEECCCCCCCCCCHHHHEEEEEECCCCCEEECCCCCCCCCCHHHHHCHHHHHCEECEEEEEEEEECCEEEEEECHHHHHHHHHHHCCCCCCHHHHHHHCCCCEEEEEC
GOR V   CCCCCCCCCCCCCCCCCCCEEEECCCCEEEEEEECCCCCCCCCEEEEEEEEEECCCCEEEECCCCCCCCCHHHHHHHHHEEEECCCCCEEEECCCHHHHHCCCCCHHHHHHHHHHCCCCEEEEECCCCCCCCCCCC
X-RAY    CCCCCCCEEEGGGTTTTCEEEETTEEEEECEEEEECCSTTSCEEEEEEEEETTTCEEEEEEETTSEEECCCCEEEEEEEECEECSSEEEEEETTTCEEEEEGGGBTHHHHTTTTTTCEEEEEEETTEEEECEECC
```

1CJW

```
1CJW    HTLPANEFRCLTPEDAAGVFEIEREAFISVSGNCPLNLDEVQHFLTLCPELSLGWFVEGRLVAFIIGSLWDEERLTQESLALHRPRGHSAHLHALAVHRSFRQQGKGSVLLWRYLHHVGAQPAVRRAVLMCEDALV
GOR I   ECCCTHHHEEECHHHHHHHHHHHHHHHHETTTTCCCHHHHHHHHEEEEETHHHHHHHHHHHHEEEECCCCHHHHHHHHHHHHHHTTTHHHHHHHHHHHHHHHTTTTCEEEEEHEEECCTCEHHHHHHHHHHHHHHH
GOR IV  CCCCCCCCCCCCCCCCCCCHHHHHHHEEEEECCCCCCCCCCCCHHHHHCCCCCCHHHHCCEEEEEEECCCHHHHHHHHHHHHHCCCCCCHHHHHHHHHHHHHHCCCCCEEEEEEEECCCCCHHHHHHHHHCCCCCC
GOR V   CCCCCCCCCCCCCCCHHHHHHHHHHHCCCCCCCCCCCHHHHHHHHCCCCCEEEEEECCCCEEEEEEECCCCCCCCCCCCCCCCCCCCCCCEEEEEEECCCCCCCCCCHHHHHHHHHHHHHHHHHEEEEECCCCHHH
X-RAY    CCCCSSEEECCCGGGHHHHHHHHHHTHHHHSCCCCCHHHHHHHHHHCGGGEEEEEETTEECEEEEEEEEECCCCCCGGGGGCCCTTCCEEEECEEEECTTCCCCCHHHHHHHHHHHHHHTTTTCCEEEEEEECGGGH

1CJW    PFYQRFGFHPAGPCAIVVGSLTFTEMHCSL
GOR I   HHEEETTTCTTCTEEEEEEECHHHHHHHHH
GOR IV  CCCCCCCCCCCCCCEEEECCEEEEECCEEC
GOR V   HHHHHCCCCCCCCCCCCCCCCCCCCCCCCC
X-RAY    HHHHTTTEEECCCCCCCCCCCCCEEEEEEC
```

1CT5

```
1CT5    STGITYDEDRKTQLIAQYESVREVVNAEAKNVHVNENASKILLLVVSKLKPASDIQILYDHGVREFGENYVQELIEKAKLLPDDIKWHFIGGLQTNKCKDLAKVPNLYSVETIDSLKKAKKLNESRAKFQPDCNPI
GOR I   EEEEEEEHHHHHHHEEEEHEHHHHHHHHHHHHHHHHHHHHHHHHHHHHHCHHHHEEEEEECCEEHHCHHHHHHHHHHHHHHHHHHHEEEETTTCTTHEHHHEEEEEEEEEHHHHHHHHHHHHHHHHHTEETTTCTE
GOR IV  CCCCCCCHHHHHHHHHHHHHHHHHHHHCCCEEECCCHHHHHHHHHHCCCCCCCHHHHHHCCCCCCCHHHHHHHHHHHHHCCCCCCEEEEECCCCCCCCCCCCCCCCCCHHHHHHHHHHHHHHHHHHHCCCCCCCCE
GOR V   CCCCCCCCCCCCCCCCCHHHHHHHHHHHHHHHHHCCCCCCEEEEEECCCCCCHHHHHHHHHCCCCCCCCHHHHHHHHHHHCCCCEEEEEECCCCCHHHHHHHHHCEEEEECHHHHHHHHHHHHHHHHHHCCCCCCE
X-RAY    CCCCCCCHHHHHHHHHHHHHHHHHHHHHHHTCCCCCCCCCCEEEEECTTSCHHHHHHHHHHTCCEEEECCHHHHHHHHHHSCTTCEEEECSCCCGGGHHHHHHCTTEEEEEEECSHHHHHHHHHHHHHHCTTSCCE

1CT5    LCNVQINTSHEDQKSGLNNEAEIFEVIDFFLSEECKYIKLNGLMTIGSWNVSHEDSKENRDFATLVEWKKKIDAKFGTSLKLSMGMSADFREAIRQGTAEVRIGTDIFGARPPKNEARII
GOR I   EEEEEEEEEEHHTTTCCCCHHHHHHHHHHHHHHHHHHHHHHHEEEEEETCCCCCTHHHHHHHHHHHHHHHHHHHHHHHHHHHHHHHHHHHHHHHHHCHEEEEEEEEEEECCCCCHHHHEE
GOR IV  ECEECCCCCCCCCCCCCCCHHHHHHHHHHCCCCCCEEEEECEEEEEECCCEECCCCCCCCCHHHHHHHHHHHHHHCCCHHHHHHHHHHHHHHHHHHCCEEEEECCCCCCCCCCCCCEEEC
GOR V   EEEEEEECCCCCCCCCCCCCHHHHHHHHHHHHHHCCHHHHHEEEECCCCCCCCCHHHHHHHHHHHHHHHHHHHHCCCCHHHHHCCCCCHHHHHHHHCCEEEEEEEEECCCCCCCCCCCCC
X-RAY    EEEEEBCCSSSCCSSSBCCHHHHHHHHHHHHSTTCCSEEEEEEECCCCCCCCCCCCCCHHHHHHHHHHHHHHHHHCCCCEEECCCTTTHHHHHHTTCSEEEEESHHHHCCCCCCCCCCCC
```

**Figure 11.15**
**The progression from GOR to GOR V method.** The X-ray structure assignments are compared here to: the original GOR method, the later version (GOR IV), and the GOR method that uses multiple alignment to incorporate information from evolution (GOR V). The predictions improve with the later versions of GOR. Red indicates helices, blue indicates strands.

Zpred is an automated procedure based on the GOR algorithm that gives a 9% improvement in prediction accuracy over GOR alone when information from multiple sequences is included. It uses both the average prediction and the conservation value approach (see Figure 11.17). The first step in a Zpred prediction incorporates the information from the alignment of the homologous sequences by averaging the α, β, and coil parameters at each position in the alignment. Insertions and deletions in the alignment (gaps) are assigned the value 0. As well as taking an averaged value of the GOR prediction at each aligned residue position, Zpred uses an encoded Venn diagram representation of amino acid conservation (see Figure 11.18 and Table 11.2) to calculate a conservation value to be assigned to the averaged prediction. This **Zvelebil conservation number** (**$C_n$**) is 1.0 for all aligned residues being identical; if all the residues aligned are within the same set (for example positive = KR) but not identical then the conservation value is 0.9. The conservation value then decreases with decreasing similarity (increasing number of different properties) of the amino acids. For example, from Table 11.2 if we align Arg and Leu then there are five differences between these two types of residues. The conservation value then would be $C_n = 0.9 - (0.1 \times 5) = 0.4$. Because gaps are normally associated with the variable parts of a protein structure and are thus less likely to occur where there is a conserved structural segment, they are given all the properties; this has the effect of decreasing the conservation value greatly.

## There is an overall increase in prediction accuracy using multiple sequence information

With the exception of the example of the all-β class protein, the incorporation of information obtained from multiple alignment and from the physicochemical properties of the aligned residues improves the secondary structure prediction in comparison to

```
AFAGVLNDADIAAALEACKAADSFNHKAFFAKVGLTSKSADDVKKAFAII
CCCCCCCHHHHHHHHHHHHHHCCCCCHHHHEEECCCCCCHHHHHHHHHHH
AQDKSGFIEEDELKLFLQNFKADARALTDGETKTFLKAGDSDGDGKIGVD
HHCCCCCHHHHHHHHHHHHHHHHHHHCCCCCEEEEEECCCCCCCCEEECC
DVTALVKA
CEEEEEEC
```

sequence length: 108

GOR IV:

alpha helix (Hh) : 50 is 46.30%

beta sheet (Ee) : 18 is 16.67%

random coil (Cc) : 40 is 37.04%

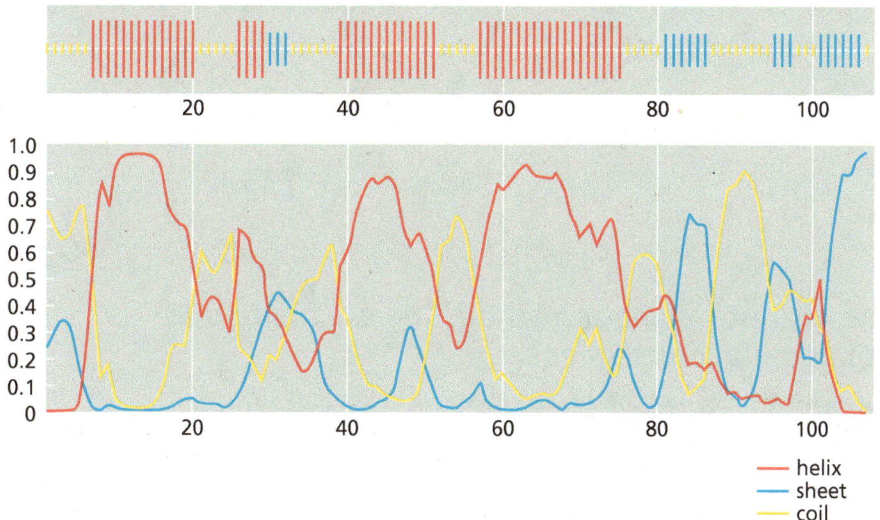


**Figure 11.16**
**A snapshot of the Web-based results from GOR IV prediction of 1B8C.** First the sequence with a three-state prediction is shown. This is followed by the predicted structural content summary, a schematic and a probability profile for each prediction state. The profile illustrates the strength of the prediction. In this example there are few discrepancies; most of the helical prediction is way above the other states, and only around residues 95–105 could the prediction be open to interpretation. It is this region that differs between GOR and GOR IV. In GOR the segment is predicted to be helical, while in GOR IV it is predicted to adopt the β-structure.

methods that use single sequence information. This improvement is especially evident in relation to the α+β protein fold (see Figure 11.11) The method still suffers from overprediction of α-helical residues and underprediction of β-strands, as is evident when the prediction is applied to the all-β fold (see Figure 11.10). In general, Zpred predicts the location of secondary structure (α-helix or β-strand) correctly, but has problems differentiating between α- or β-secondary structures.

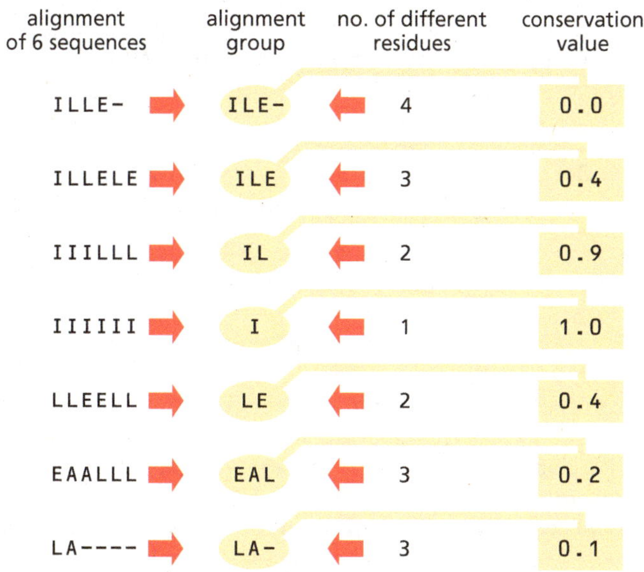


**Figure 11.17**
**Calculation of the conservation value in Zpred.** The final conservation value is not only dependent on the number of different residues in the alignment but also on the type of residues. The conservation value is calculated based on the Venn diagram of the physicochemical properties of the residues (Figure 11.18). The conservation value can then be used to modify the prediction of the residues at an aligned position. The aligned residues are grouped into the number of different residues (a gap is counted as a residue, illustrated as a dash). Then using the tabulated Venn diagram as shown in Table 11.2 a conservation value is calculated for the group. The number of times a residue appears in the group is also taken into account in later versions of the program.

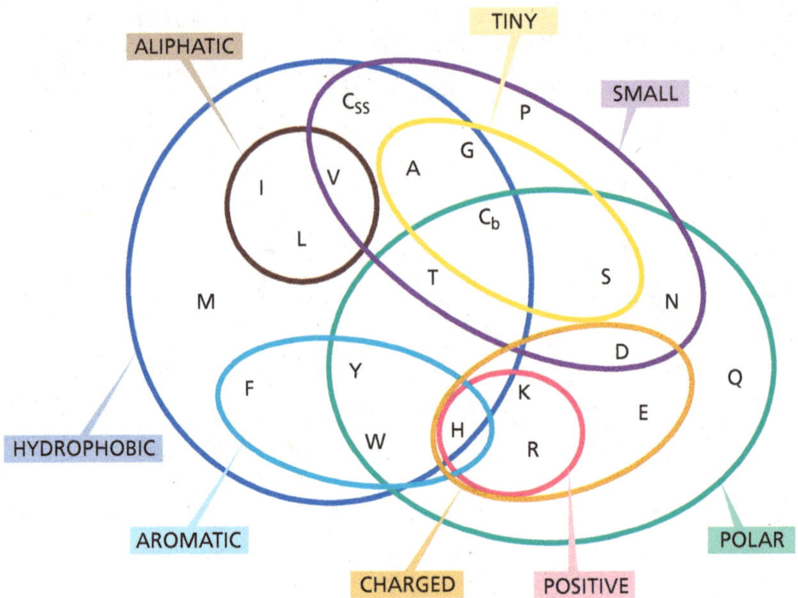


**Figure 11.18**
**A Venn diagram representing the relationship of the 20 naturally occurring amino acids based on a selection of physicochemical properties.** The amino acids are divided into two major sets, the POLAR and HYDROPHOBIC groups. The set that includes fully charged amino acids is divided into the subset positive, the negative subset being defined by implication. Because of its unique backbone properties proline is separated from the rest of the amino acids. $C_{ss}$ is a disulphide forming cysteine, $C_b$ is a free one.

## The nearest-neighbor method: The use of multiple nonhomologous sequences

The **nearest-neighbor method** (sometimes incorrectly referred to as the homolog method) is a commonly used approach to secondary structure prediction, and is incorporated into several different programs. It takes the principle that proteins with similar sequences share the same fold and applies it at the level of short stretches of sequence. It makes the assumption that even in a set of non-homologous proteins, short stretches of similar sequence may have the same secondary structure.

A list of short sequence segments is produced by sliding a window of a given length (usually 16 residues) along a set of database sequences of known structure. The secondary structure of the central amino acid in each segment is recorded, producing a secondary structure dataset. A sliding window of the same size is then applied to the query sequence, each segment is compared to the dataset segments, and a given number of best-matching segments are identified. The frequencies of the known secondary structure of the middle residue in each of these matching segments are then used to predict the probability of each type of secondary structure for that residue in the query segment. A final prediction for each residue is made using either a rule-based approach or a neural network.

## PREDATOR is a combined statistical and knowledge-based program that includes the nearest-neighbor approach

The GOR methods use structure-forming propensities derived from the analysis of short segments of protein sequence; that is, local interactions. However, the formation of secondary structure in proteins does not only depend on local interactions. In particular, β-sheets can be made up of β-strands that are separated by some distance in the polypeptide chain (see Figure 11.19). The program PREDATOR is a statistical and knowledge-based program that makes use of specific long-distance interactions, or sequence information, related to the formation of secondary structures. It predicts secondary structure by identifying potentially hydrogen-bonded residues in the amino acid sequence. The β-strands are predicted by delineating different classes of β-bridges on the basis of the analysis of hydrogen bonding between particular residues in known structures. The α-helices are recognized on

| Amino acid | Property | | | | | | | | | |
|---|---|---|---|---|---|---|---|---|---|---|
| | Hydrophobic | Positive | Negative | Polar | Charged | Small | Tiny | Aliphatic | Aromatic | Proline |
| Ile | Yes | | | | | | | Yes | | |
| Leu | Yes | | | | | | | Yes | | |
| Val | Yes | | | | | Yes | | Yes | | |
| Cys | Yes | | | | | Yes | | | | |
| Ala | Yes | | | | | Yes | Yes | | | |
| Gly | Yes | | | | | Yes | Yes | | | |
| Met | Yes | | | | | | | | | |
| Phe | Yes | | | | | | | | Yes | |
| Tyr | Yes | | | Yes | | | | | Yes | |
| Trp | Yes | | | Yes | | | | | Yes | |
| His | Yes | Yes | | Yes | Yes | | | | Yes | |
| Lys | | Yes | | Yes | Yes | | | | | |
| Arg | | Yes | | Yes | Yes | | | | | |
| Glu | | | Yes | Yes | Yes | | | | | |
| Gln | | | | Yes | | | | | | |
| Asp | | | Yes | Yes | Yes | | | | | |
| Asn | | | | Yes | | Yes | | | | |
| Ser | | | | Yes | | Yes | Yes | | | |
| Thr | Yes | | | Yes | | Yes | | | | |
| Pro | | | | | | Yes | | | | Yes |
| Gap | Yes | Yes | Yes | Yes | Yes | Yes | Yes | Yes | Yes | Yes |
| Unk | Yes | Yes | Yes | Yes | Yes | Yes | Yes | Yes | Yes | Yes |

**Table 11.2**
**The physicochemical properties of the amino acids as used by the Zpred prediction program are given in this table.** The properties of Leu and Arg are given in blue and are discussed in the text. Unk indicates an unidentified residue type in the sequence of amino acids, and is treated the same as a gap.

the basis of the occurrence of residues in hydrogen-bonded pairs in *i* to *i*+4 interactions (see Section 2.1 and Figure 11.2).

Secondary structure propensities are also inferred from sequence similarity using the nearest-neighbor approach. PREDATOR calculates seven independent secondary structure probabilities for each residue: parallel and antiparallel sheets and helices as predicted by long-range interactions; sheet, helix, and coil propensities predicted by the nearest-neighbor method; and turn propensities. These different propensities are then converted into a final secondary structure prediction.

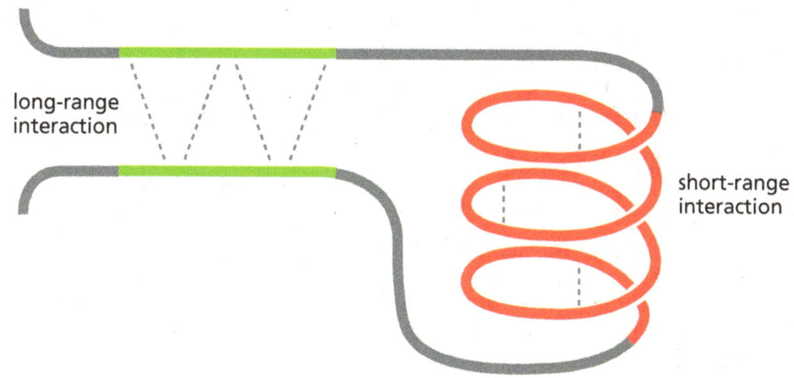


**Figure 11.19**
**A schematic illustration of long- and short-range interactions.** Hydrogen bonding between β-strands (green lines) can include interactions that are separated by large sections of sequence, while those within an α-helix (red coil) are not.

In addition PREDATOR can use a set of homologous sequences. However, it does not create a multiple alignment, but uses pairwise alignments to improve the structural prediction. The results of both single- and pairwise PREDATOR are illustrated in Figures 11.9 to 11.12. For the single-sequence prediction, the incorporation of long-range interactions, nearest-neighbor propensities, and turn propensities into the prediction program gives a marked improvement over GOR IV in the $\alpha$+$\beta$ protein (see Table 11.1). Predictions for the other structural classes are not much better than GOR IV. When a set of sequences is used, the prediction is much improved for all classes with respect to GOR IV. With respect to GOR V the results are similar in the all-$\alpha$ and all-$\beta$ classes of proteins, but seem to be better in the $\alpha/\beta$ class and less accurate for the $\alpha$+$\beta$ class. In general, from GOR V, Zpred, and pairwise PREDATOR we can see that incorporating information from local or global sequence homology improves the accuracy of secondary structure predictions.

## 11.5 Neural Network Methods of Secondary Structure Prediction

The remaining prediction programs we shall discuss use neural network methods. Artificial neural network techniques were initially developed to simulate information processing and learning in the brain. They are an example of machine-learning techniques, which are good at discarding redundant information; for example, a neural network can store common features that apply to many data items, and not allocate individual parameters to single sequence patterns. The aim of machine-learning algorithms is to automatically fit a model value to a known value as closely as possible. In other words, when applied to secondary structure prediction, the algorithm will learn by iterative changes to its parameters until the predicted structure is as similar to the observed secondary structure as possible.

Neural networks operate by processing information through so-called layers (see Figure 11.20). The simplest neural network is a two-layered network called a **perceptron**. The first layer is the input layer and the second is the output layer. Each layer can have many **nodes** or **units**. The firing of a node in a neural network is

**Flow Diagram 11.5**
**In this section some of the prediction methods that use neural networks are discussed.**

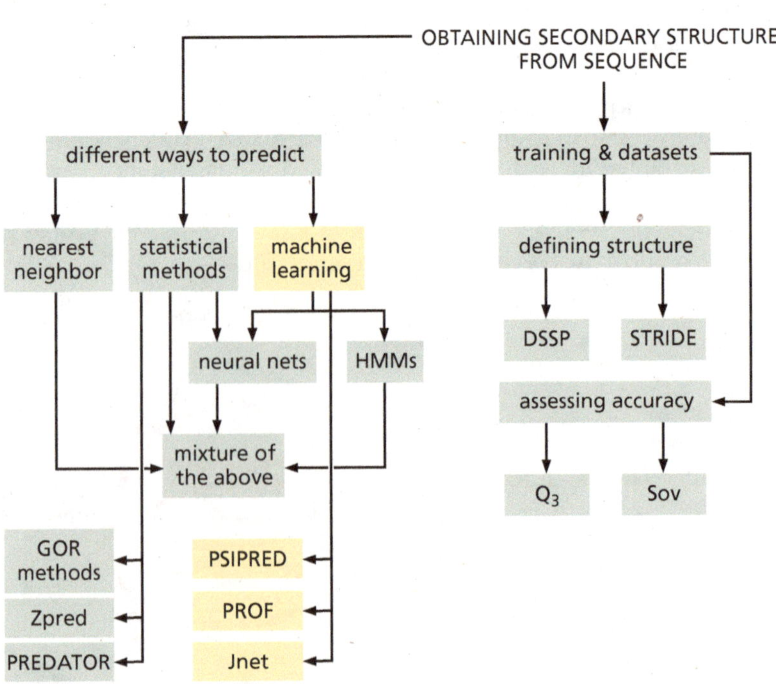

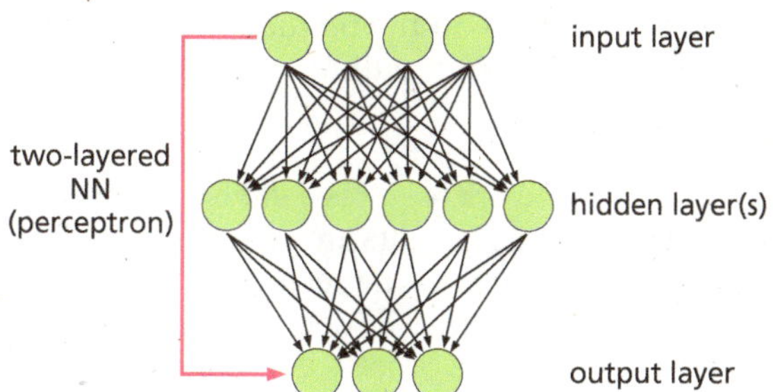


**Figure 11.20**
**A schematic diagram of a two-layered (red arrow) and multiple layered neural network (NN) representation.** The first layer is always an input layer, followed in a multiple layered network by one or more hidden layers and lastly the output layer.

simulated by assigning the binary values of 1 or 0 to its output. The value of 1 is produced when the weighted sum of inputs exceeds a predetermined threshold value. A more complex and more commonly used neural network is one that has one or more layers between the input and output layers, the so-called **hidden layers** (see Figure 11.20). The hidden layer enables joint and conditional actions to be performed on the information passed to it from the input layer because there is more than one path to an output node (see Figure 11.20). The number of units in any layer is at the discretion of the designer and independent of the number in any other layer in the network. In most of the applications considered here, the input layer takes as its input a representation of the protein sequence, and the final output is a representation of the secondary structure prediction.

The input signal for an amino acid is often a group of 20 units in the input layer. All of these will represent an individual residue type. In most prediction programs the sequence is sampled by a sliding window, with the central residue being the one whose secondary structure is predicted. A network that takes as input successive 13-residue windows will have an input layer of 260 (13×20) units. When only a single sequence is used as the input, the signals of the input layer units will be 0 except that representing the particular residue, which will have a value of 1. When using multiple aligned sequences the input layer signals will be related to profiles based on these alignments (see Section 6.1 for details of these profiles). Alternative encoding is possible, such as using residue properties or additional units to represent gaps.

The output layer usually consists of as many units as there are alternative conformations to predict (usually three). Each of these will produce a signal that will be in the range 0 to 1, and usually the highest of these is taken as the prediction (see Figure 11.21). If three output units represent helix, strand, and coil, respectively, the output (1,0,0) would be a perfect helix prediction, but (0.76, 0.55, 0.37) would

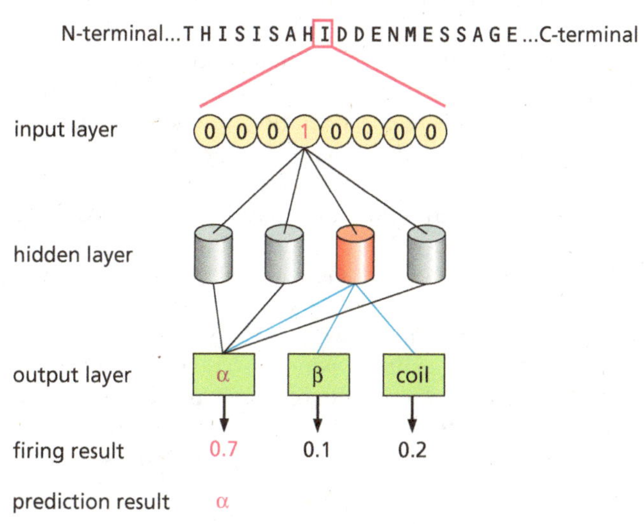


**Figure 11.21**
**A simplified representation of a feed-forward multilayer neural network configuration as could be applied to an actual secondary structural prediction.** A number of nodes are present in the input layer, which can be fired by certain types of residues, e.g., the D (Asp). There are often 20 nodes per residue, with just one having the value 1 (which means it is activated). The nodes that are activated then pass a signal to the hidden layers, where conditional and additive calculations are performed on the information. Nodes receiving signals above a certain value, such as the red one, will fire to the output layer. Similar calculations are made in the output layer, which in this case results in the α node producing the highest signal; α-helix is the secondary structure predicted for the central residue of the window.

also predict a helix. The confidence of the prediction can be related to how close the highest value is to 1 and also to the difference between the highest and the other values.

Parameterization of a neural network requires a dataset whose secondary structure is known. To train a network, initial weights are assigned, usually by taking random numbers. The network is used to predict the secondary structures of all the training sequences. The initial weights are adjusted such that the prediction becomes closer to the actual secondary structure. The process of prediction and weight adjusting is repeated until there is good agreement or no further improvement in the prediction. For further details on the parameterization and operation of these neural networks see Section 12.4.

## Assessing the reliability of neural net predictions

The reliability of methods that use neural networks is judged by computing a secondary structure **prediction confidence level (PCL)**. This is not the same as the accuracy, $Q_3$, of a prediction method, although, like $Q_3$, it is based on statistical analyses of prediction accuracy. A neural network predicts the three secondary structure types (α, β, and coil) using real numbers from the output units that make up the output layer. A prediction for the given residue is assigned by choosing the unit with the highest number. However, all the numbers can be used to extract additional information. The PCL is calculated using these numbers. It is defined as a function of the difference between the output unit with the highest value (the winner unit) and the output unit with the next highest value. This difference is used to define a **reliability** or **confidence index** for each residue prediction. Usually the index is scaled to have a value between 0 (lowest reliability) and 9 (highest). In practice, the confidence index is a useful indicator of key regions that are expected to be predicted with a high level of accuracy. A PLC of 5 or above obtained with the Jnet method described below, for example, will, on average, signify a residue prediction accuracy of 84%.

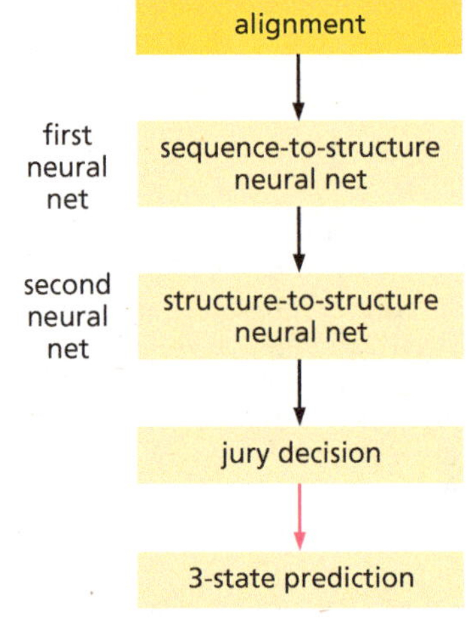


**Figure 11.22**
**A diagram illustrating the steps involved in the PHD secondary structure prediction program.** There are two neural nets, which take as part of the input information from multiple alignments; the output from the first neural net is then fed into the second neural net. Finally a numerical average (jury decision) is taken over a number of different level 2 networks to give the final prediction.

## Several examples of Web-based neural network secondary structure prediction programs

In this section we look at programs that use a combination of statistical parameters, nearest neighbors, and neural network techniques. These programs are illustrated below using the four protein sequences already described.

The output from one neural net can easily be used as input to another. Prediction methods such as PHD have a second neural net that takes as its input the secondary structure signals output by the first neural net (see Figure 11.22). The justification for this structure is that it permits correlation between the conformations of neighboring residues to be included, which occurs in the second-level network. The PHD (jury decision neural networks) program uses information from aligned homologous sequences, local statistical information, and long-range structural parameters encoded in two neural networks. The first network is a **sequence-to-structure network** whose input represents the local alignment, residue conservation, and additionally long-range sequence information (see Figure 11.22). The long-range information consists of the residue composition of the whole protein, the protein length, and the distance of the sliding window to each end of the protein.

The output of this first neural network is the structural state (α, β, or coil) of the residue at the center of the sliding window. The output of the first network forms the input to the second neural network, a **structure-to-structure network**. The reason for this is that it permits correlation between the conformations of neighboring residues to be incorporated. Finally PHD uses a simple average of the predictions as the final result.

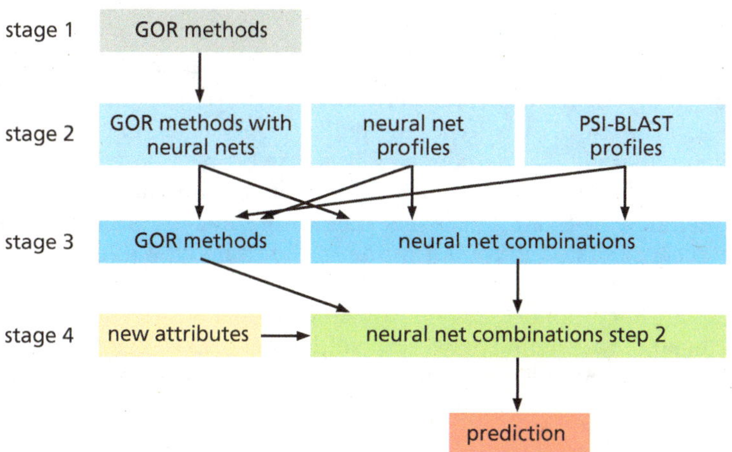


**Figure 11.23**
**A schematic diagram showing the steps involved in the prediction cascade used in the PROF secondary structure prediction program.**

NNSSP (Neural Net Nearest Neighbor) is a neural network development of a statistical and knowledge-based program, SSP, which, like PREDATOR, used the nearest-neighbor approach. NNSSP has an advanced scoring system that combines sequence similarity, local sequence information, and knowledge-based information on β-turns and the amino- and carboxy-terminal properties of α-helices and β-strands. In addition, the program uses multiple sequence alignment to further improve accuracy.

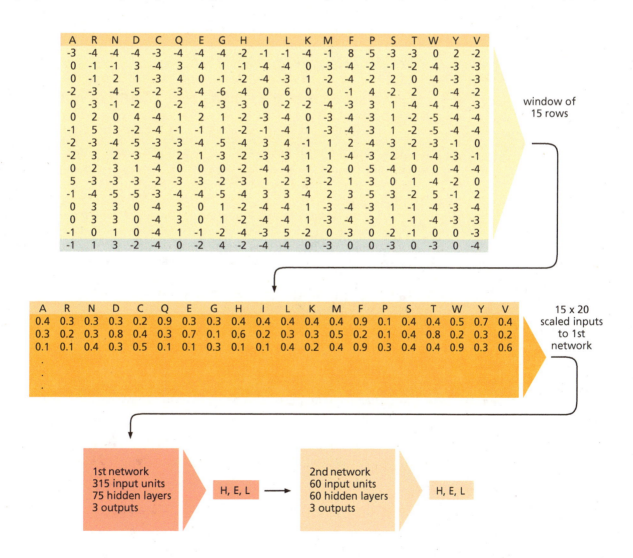

| A | R | N | D | C | Q | E | G | H | I | L | K | M | F | P | S | T | W | Y | V |
|---|---|---|---|---|---|---|---|---|---|---|---|---|---|---|---|---|---|---|---|
| -3 | -4 | -4 | -4 | -3 | -4 | -4 | -4 | -2 | -1 | -1 | -4 | -1 | 8 | -5 | -3 | -3 | 0 | 2 | -2 |
| 0 | -1 | -1 | 3 | -4 | 3 | 4 | 1 | -1 | -4 | -4 | 0 | -3 | -4 | -2 | -1 | -2 | -4 | -3 | -3 |
| 0 | -1 | 2 | 1 | -3 | 4 | 0 | -1 | -2 | -4 | -3 | 1 | -2 | -4 | -2 | 2 | 0 | -4 | -3 | -3 |
| -2 | -3 | -4 | -5 | -2 | -3 | -4 | -6 | -4 | 0 | 6 | 0 | 0 | -1 | 4 | -2 | 2 | 0 | -4 | -2 |
| 0 | -3 | -1 | -2 | 0 | -2 | 4 | -3 | -3 | 0 | -2 | -2 | -4 | -3 | 3 | 1 | -4 | -4 | -4 | -3 |
| 0 | 2 | 0 | 4 | -4 | 1 | 2 | 1 | -2 | -3 | -4 | 0 | -3 | -4 | -3 | 1 | -2 | -5 | -4 | -4 |
| -1 | 5 | 3 | -2 | -4 | -1 | -1 | 1 | -2 | -1 | -4 | 1 | -3 | -4 | -3 | 1 | -2 | -5 | -4 | -4 |
| -2 | -3 | -4 | -5 | -3 | -3 | -4 | -5 | -4 | 3 | 4 | -1 | 1 | 2 | -4 | -3 | -2 | -3 | -1 | 0 |
| -2 | 3 | 2 | -3 | -4 | 2 | 1 | -3 | -2 | -3 | -3 | 1 | 1 | -4 | -3 | 2 | 1 | -4 | -3 | -1 |
| 0 | 2 | 3 | 1 | -4 | 0 | 0 | 0 | -2 | -4 | -4 | 1 | -2 | 0 | -5 | -4 | 0 | 0 | -4 | -4 |
| 5 | -3 | -3 | -3 | -2 | -3 | -3 | -2 | -3 | 1 | -2 | -3 | -2 | 1 | -3 | 0 | 1 | -4 | -2 | 0 |
| -1 | -4 | -5 | -5 | -3 | -4 | -4 | -5 | -4 | 3 | 3 | -4 | 2 | 3 | -5 | -3 | -2 | 5 | -1 | 2 |
| 0 | 3 | 3 | 0 | -4 | 3 | 0 | 1 | -2 | -4 | -4 | 1 | -3 | -4 | -3 | 1 | -1 | -4 | -3 | -4 |
| 0 | 3 | 3 | 0 | -4 | 3 | 0 | 1 | -2 | -4 | -4 | 1 | -3 | -4 | -3 | 1 | -1 | -4 | -3 | -3 |
| -1 | 0 | 1 | 0 | -4 | 1 | -1 | -2 | -4 | -3 | 5 | -2 | 0 | -3 | 0 | -2 | -1 | 0 | 0 | -3 |
| -1 | 1 | 3 | -2 | -4 | 0 | -2 | 4 | -2 | -4 | -4 | 0 | -3 | 0 | 0 | -3 | 0 | -3 | 0 | -4 |

window of 15 rows

| A | R | N | D | C | Q | E | G | H | I | L | K | M | F | P | S | T | W | Y | V |
|---|---|---|---|---|---|---|---|---|---|---|---|---|---|---|---|---|---|---|---|
| 0.4 | 0.3 | 0.3 | 0.3 | 0.2 | 0.9 | 0.3 | 0.3 | 0.4 | 0.4 | 0.4 | 0.4 | 0.4 | 0.9 | 0.1 | 0.4 | 0.4 | 0.5 | 0.7 | 0.4 |
| 0.3 | 0.2 | 0.3 | 0.8 | 0.4 | 0.3 | 0.7 | 0.1 | 0.6 | 0.2 | 0.3 | 0.3 | 0.5 | 0.2 | 0.1 | 0.4 | 0.8 | 0.2 | 0.3 | 0.2 |
| 0.1 | 0.1 | 0.4 | 0.3 | 0.5 | 0.1 | 0.1 | 0.3 | 0.1 | 0.1 | 0.4 | 0.2 | 0.4 | 0.9 | 0.3 | 0.4 | 0.4 | 0.9 | 0.3 | 0.6 |
| . | | | | | | | | | | | | | | | | | | | |
| . | | | | | | | | | | | | | | | | | | | |
| . | | | | | | | | | | | | | | | | | | | |

15 x 20 scaled inputs to 1st network

1st network
315 input units
75 hidden layers
3 outputs

H, E, L

2nd network
60 input units
60 hidden layers
3 outputs

H, E, L

**Figure 11.24**
**A graphical representation of PSIPRED dual network prediction.** First a raw profile generated by PSI-BLAST is taken and scaled to a 0–1 range. A window of 15 elements is fed to the 1st network, which performs the initial secondary structure prediction using various residue parameters. This initial prediction is fed into a 2nd neural network where it is filtered to produce the final three-state secondary structure prediction. (Adapted from D.T. Jones, Protein secondary structure prediction based on position-specific scoring matrices, *J. Mol. Biol.* 292:195–202, 1999.)

## PROF: Protein forecasting

PROF is based on cascading different types of prediction or alignment algorithms within one program using neural networks to choose the final prediction (see Figure 11.23). The use of many types of prediction rationales, or classifiers, is based on the theory that all of the evidence relevant to a prediction should be used in making that prediction. Therefore, a combination of different types of algorithms, training methods, or training datasets should improve the overall prediction, as long as the classifiers used are not prone to the same type of errors; that is, the prediction algorithms used do not all overpredict a certain type of structure. The PROF program incorporates all the principles embodied in the GOR prediction methods together with neural networks, multiple sequence alignment information, the use of PSI-BLAST alignment profile (see Section 6.1), and three additional attributes of secondary structure.

## PSIPRED

This method is similar to the PHD prediction algorithm described above in that it uses a multiple alignment profile to obtain values for the input layer units of the first of two successive neural nets. PSIPRED is a three-stage method beginning with the generation of a multiple alignment and PSI-BLAST profile for the query sequence, generation of the initial secondary structure, and finally filtering of the initial prediction. The program uses a two-stage neural net with input taken in a window of 15 residues and outputs a three-state prediction for the central residue (helix, strand, and coil) which is used as the input for the second neural net. The second neural net also uses a 15-residue window and gives the final three-state prediction of the central residue (see Figure 11.24). The output includes the confidence level of the prediction (see Figure 11.25).

## Jnet: Using several alternative representations of the sequence alignment

Jnet is made up of four sets of sequence-to-sequence and structure-to-structure neural networks. Each of the four sequence-to-structure networks receives input from one of four alternative representations of a set of aligned sequences, using a 17-residue window and a conservation number for each residue. The output from these are input into the structure-to-structure networks (see Figure 11.26), using a window of 19 residues and a conservation number. The output for a given residue from the set of neural nets, a three-state prediction, is taken as the final prediction if they are all in agreement. Where there is disagreement between the predictions from the nets the final prediction is obtained by applying a third neural network (see Figure 11.26).

In general, the programs based on neural networks predicted the secondary structure more accurately than the other methods, especially in the all-$\alpha$ and $\alpha/\beta$ classes (see Figures 11.9 to 11.12 and Table 11.1). However, the examples discussed here and given in Table 11.1 only deal with one representative of each class of protein and therefore this is not an assessment of the prediction methods as such, only a guide for the student. Figure 12.6 shows the range of $Q_3$ and Sov values for the PSIPRED method applied to a large set of proteins. The variation of values shows that one example cannot be taken as indicative of the general performance of any method.

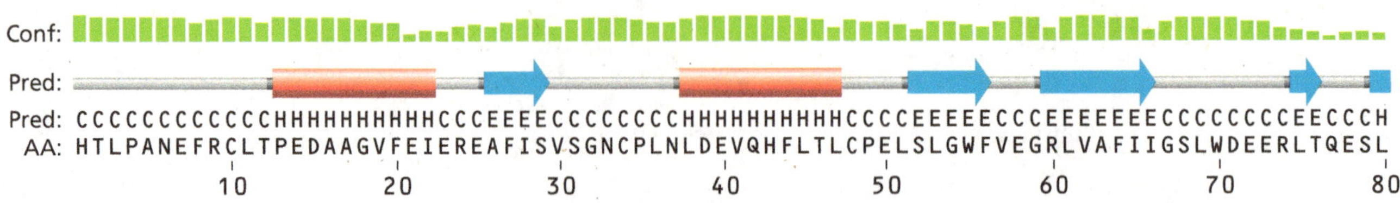


**Figure 11.25**
**An example of the output for PSIPRED prediction.** In addition to the three-state prediction given underneath each residue (AA), a cartoon representation is provided as well, where arrows indicate β-strands and cylinders the α-helices. A bar-chart representation (Conf) of the confidence level of each prediction is also given.

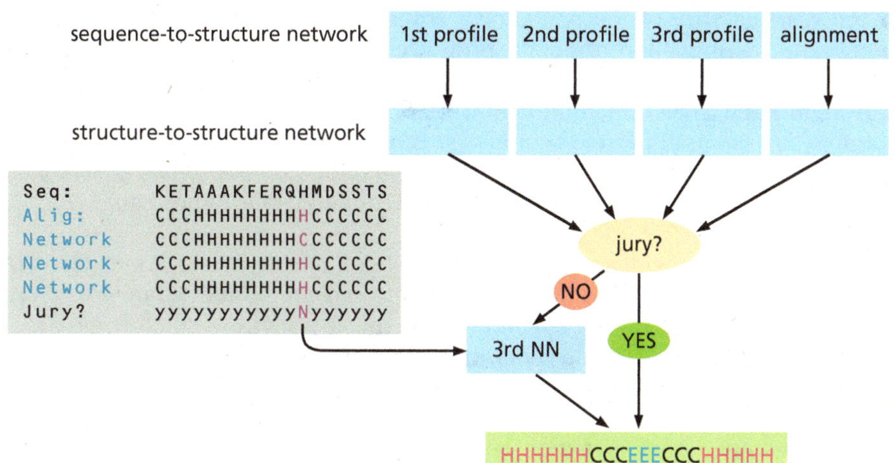


**Figure 11.26**
**A schematic and simplified representation of the Jnet neural network prediction program.** Blue boxes represent the four sets of paired neural networks, each receiving input from a different representation of a sequence alignment. Predictions are compared at each residue position. At those positions where unanimous agreement does not occur (an extreme form of jury decision) a third neural network is used to determine the final prediction.

## 11.6 Some Secondary Structures Require Specialized Prediction Methods

The methods described so far in this chapter are designed to predict the common types of secondary structure of globular protein domains that are not associated with membranes. However, these are not the only forms of secondary structure that occur in proteins. For example, alternative helical structures called $3_{10}$- and $\pi$-helices have been identified in short stretches, and are usually predicted as $\alpha$-helix by the prediction programs. Another form, more commonly associated with fibrous proteins, is the multi-stranded coiled coil formed by several closely interacting $\alpha$-helices. Proteins which span membranes do so using $\alpha$-helices or $\beta$-strands, but the properties of these are very distinct from those studied so far in this chapter. Finally, single-stranded RNA molecules can also have secondary structure, as described in Section 1.1.

In the remaining sections of this chapter we will first survey the secondary structure prediction methods employed for transmembrane proteins, and then look at methods designed to study coiled coils and RNA.

**Flow Diagram 11.6**
**In the following sections some of the methods developed for more specialized secondary structure prediction are discussed.**

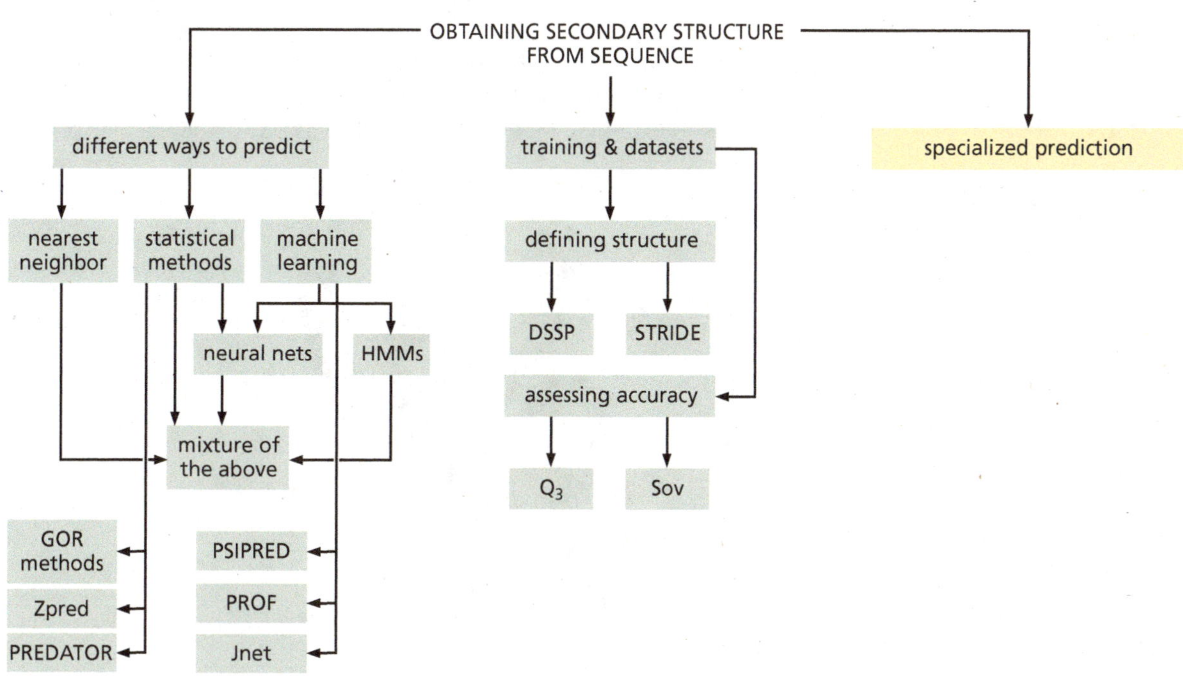

In the vast majority of the few known **transmembrane protein** structures, the membrane-spanning portions of the protein chain are in the form of α-helices, and most prediction methods are designed to find helices. This is not the only type of transmembrane structure found, however. The water-transporting porins, for example, are channel proteins composed of a transmembrane β-sheet structure, curved round to form a cylindrical β-barrel. To date, no other types of secondary structure have been found in membrane-embedded regions of proteins.

Transmembrane proteins range in topological complexity from those such as the epidermal growth factor receptor (EGFR), in which the protein chain crosses the membrane only once, with a single helix, to the G-protein-coupled receptors, in which the protein chain threads back and forth across the membrane to give seven transmembrane helices (see Box 11.2).

In the remainder of this section some specific features relevant to predicting secondary structure and topological arrangement of transmembrane regions will be explored.

## Transmembrane proteins

Membrane proteins exist simultaneously in different environments; parts of the protein reside in the aqueous environment on either side of the membrane,

### Box 11.2 Membrane proteins are functionally important

Epidermal growth factor receptor (EGFR) is one of the receptor tyrosine kinases (RTKs) that exist as cell surface transmembrane glycoproteins and constitute the launch sites for some of the complex signal transduction pathways that take place within the cell. Structurally, RTKs consist of a very variable extracellular ligand-binding domain, which gives the receptor its specificity, a transmembrane spanning component consisting of a single α-helix, and a cytoplasmic domain, which possesses tyrosine kinase activity (see Figure B11.3A).

Nearly all RTKs are found in the plasma membrane, monomeric and inactive until they bind their respective ligand. Then a succession of molecular events follows, which leads to full activation. These include receptor dimerization (the association of two distinct receptor molecules) and receptor tyrosine trans-autophosphorylation to give the enzymatically active form of the protein.

G-protein-coupled receptor is one of a class of integral membrane proteins belonging to the 7TM superfamily of transmembrane receptors that possess seven membrane-spanning elements (see Figure B11.3B). Other examples are receptors of the olfactory epithelium that bind odorants and receptors of the neurotransmitter serotonin in the mammalian brain. Upon ligand binding, these receptors activate G proteins.

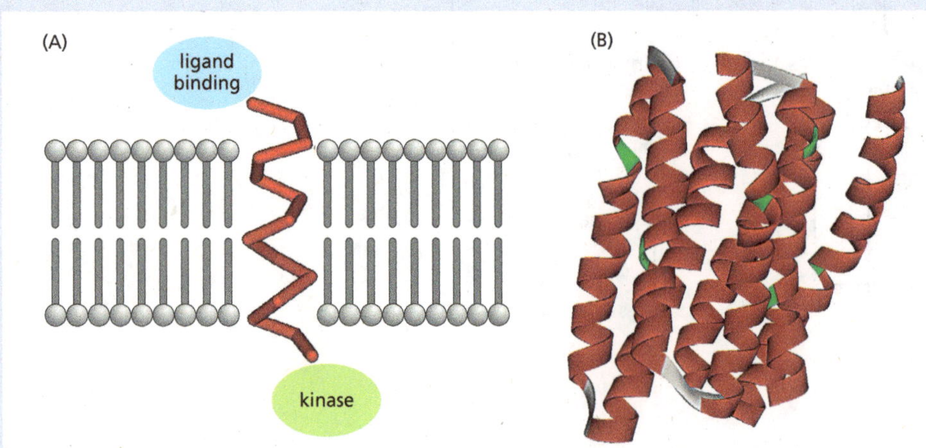


**Figure B11.3**
**(A) Schematic structure of a receptor tyrosine kinase including a single α-helix spanning the membrane, and (B) an example of a 7-transmembrane spanning molecule.**

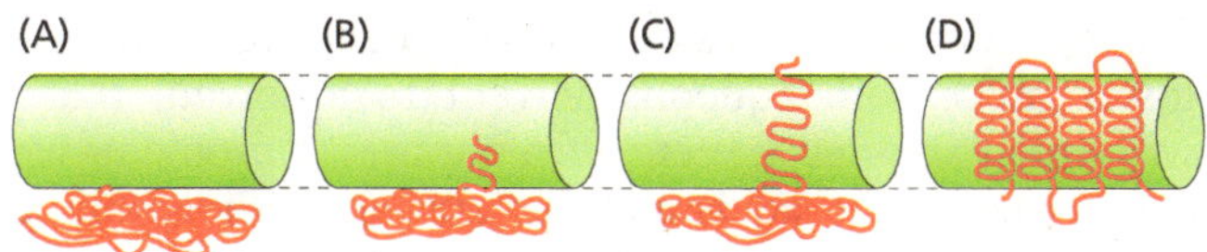


**Figure 11.27**
**The four main ways in which proteins may be attached to a membrane.** (A) Attachment by ionic interactions between the protein and the cytosolic face of the lipid bilayer. (B) Attachment via a lipid or prenyl anchor, which is added to the protein posttranslationally and is inserted into the cytosolic leaflet of the lipid bilayer. Proteins that attach to membranes in this way have no specialized structural or sequence features that can be recognized. (C,D) Transmembrane proteins have part of the protein chain embedded in the lipid bilayer. (C) In bitopic membrane proteins the protein chain crosses the membrane once only. (D) In polytopic membrane proteins, the protein chain threads back and forth across the membrane multiple times.

whereas other parts are located in the hydrophobic environment of the membrane interior. In addition, the surface of the membrane is highly ionic. These different environments impose structural constraints on membrane proteins that are reflected in their sequence and can be used to distinguish them from cytosolic proteins. These properties can also be used to predict secondary and other structural aspects of membrane proteins.

Membrane proteins can be categorized by their degree of interaction with the membrane, as shown in Figure 11.27. Some are only anchored to one side of the membrane, either by ionic interactions (see Figure 11.27A) or via a hydrocarbon chain that has been added posttranslationally (see Figure11.27B). Proteins with this type of membrane attachment follow, for the most part, the structural rules for cytosolic proteins and will not be discussed further in this chapter.

The remaining membrane proteins comprise the integral or transmembrane proteins, in which part of the protein is embedded entirely within the lipid bilayer. Transmembrane proteins can be either bitopic (see Figure 11.27C) or polytopic (Figure 11.27D). Bitopic proteins cross the membrane once, having a part of the protein on each side of the membrane. Polytopic proteins cross the membrane many times. The general structural elements of transmembrane proteins are summarized in Table 11.3.

## Quantifying the preference for a membrane environment

As the membrane interior is hydrophobic, it can be expected that the residues that are located within the membrane are nonpolar and hydrophobic in nature. Therefore, identifying regions of hydrophobicity in the amino acid sequence should help in predicting the structure of a transmembrane protein and give some clues to its general function. Amino acids vary in the hydrophobicity of their side chains. **Hydrophobic scales** have been constructed that assign values to the hydrophobicity of each amino acid. Many hydrophobic scales have been generated from solution studies, crystallographic data, or a combination of both. No particular method is in general better than the others, and all have been used in transmembrane helix prediction programs.

**Table 11.3**
**General structural elements of transmembrane proteins.**

- Helices are about 15–30 residues long
- Transmembrane structures are predominantly apolar
- Residues that interact within the core of a multi-helix transmembrane protein do not have to be hydrophobic as they can form salt-bridges with each other. Therefore, charged residues can occur in a coordinated fashion where a positively charged residue in one position is complemented by a negative residue
- Helices are tightly packed and can form a coiled-coil structure
- β-strands within a transmembrane sheet contain a hydrophobic residue every second position, which faces the lipid, while the inward (into the sheet) facing residues can be either polar or apolar
- β-strands in transmembrane sheets are often flanked by aromatic residues

Solution-based hydrophobicity scales are usually calculated by measuring the free energy of transfer ($\Delta G_t$) from aqueous solvent to a solvent that mimics the membrane environment. However, there are a number of problems with such estimations, because some amino acids are insoluble in the membrane-mimicking solvents. In addition, it is not easy to decide which solvent mimics the membrane environment the best.

Hydrophobicity scales from crystallographic data are obtained using a computer program which calculates how hydrophobic a residue is by rolling a water-mimicking sphere (usually 1.4 Å radius) over a protein surface and identifying which residues can make van der Waals contacts with the solvent (see Figure 14.19A). To improve the accuracy of the hydrophobicity scales some scientists have combined both the crystallographic and solution-based methods.

Other hydrophobicity scales can be obtained by, for example, the GES (Goldman, Engelman, and Steitz) method, which uses the solvent exposure of a particular residue in a polyalanine helix of 20 residues. Another scale is based on the frequency that a residue is found in a membrane-spanning segment. Yet another method uses the free energy of transfer of tri- or pentapeptides where the middle amino acid is the one that is being examined.

## 11.7 Prediction of Transmembrane Protein Structure

Most transmembrane proteins span the membrane in the form of one or more α-helices, and we shall discuss these first. If more than one helix is present, the helices usually form a compact bundle. The properties of the lipid bilayer force a number of structural constraints on the α-helical transmembrane segments. The first is length. An average membrane thickness is 30 Å, which corresponds to an α-helix of between 15 and 30 residues.

Transmembrane regions are therefore generally predicted by looking for hydrophobic regions of specific length in the sequence. A hydrophobicity scale is

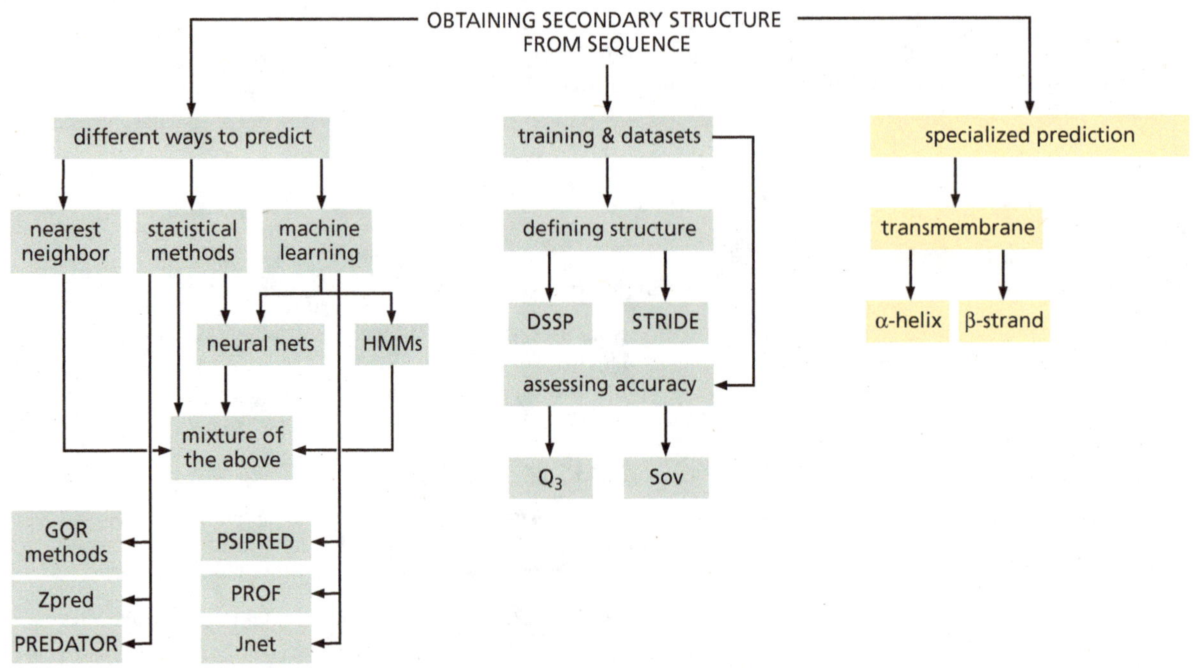


**Flow Diagram 11.7**
**In this section some of the methods developed for the prediction of transmembrane proteins are discussed.**

used to assign values to individual residues and the values are converted into a **hydropathic profile** by using a sliding window to average the values over a number of residues. This identifies hydrophobic stretches effectively. The window is set to a length that corresponds to the average length of a transmembrane segment, from 15 up to 30 residues in length. A smaller window is used when searching for the ends of the transmembrane helix. As no one hydrophobicity scale is best, you should ideally submit the sequence to a number of transmembrane-prediction programs that use different hydrophobicity profiles. Those segments that are consistently predicted to be transmembrane by all the programs are most likely to be correct. Hydropathic profiles are usually sufficient to predict the correct location of the transmembrane segment in single-pass membrane proteins.

## Multi-helix membrane proteins

If a membrane protein has more than one transmembrane segment, it is not only important to predict which segments will form the transmembrane helix but also to predict the relative orientation of the helices in the membrane and the side-chain interactions. Residues that interact within the core of a multi-helix transmembrane protein do not have to be hydrophobic, as they can form salt bridges with each other. Therefore, charged residues can occur in a coordinated fashion in the helices, where a positively charged residue in one helix is complemented by a negative residue in the interacting helix.

The helices in such structures contain both hydrophobic and charged residues, forming a structural element that has a different character on each side—an **amphipathic helix**. The use of hydropathic profiles only will not usually predict these

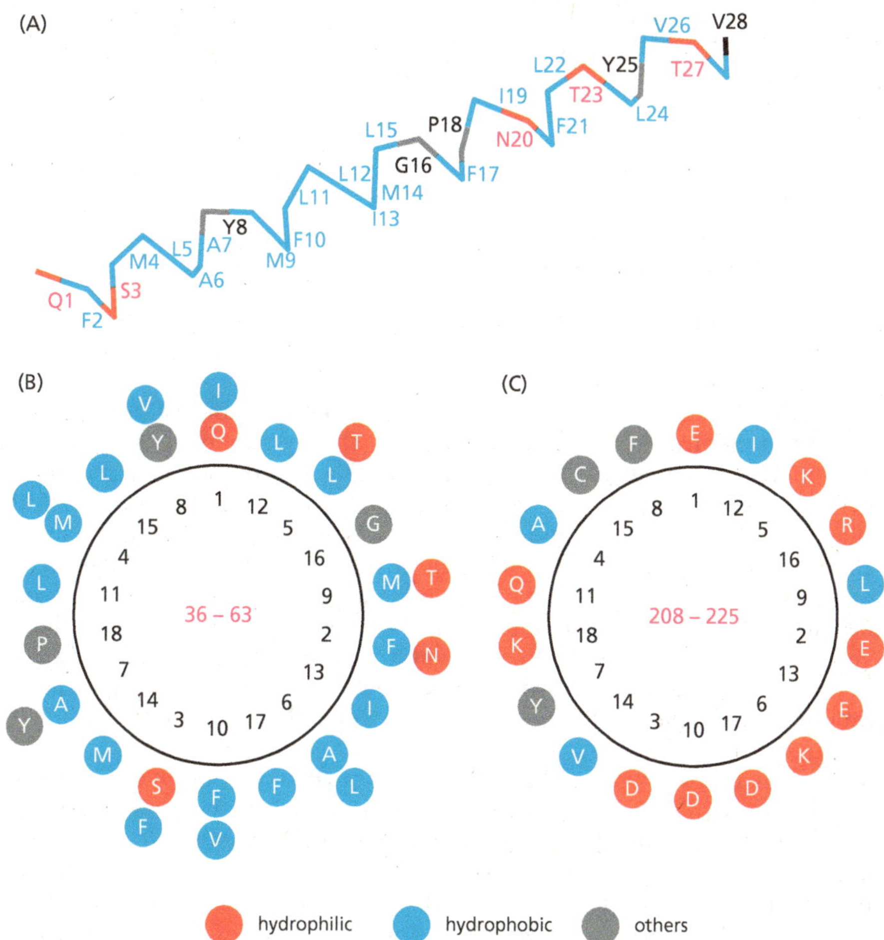


**Figure 11.28**
**Transmembrane helices can be represented by helical wheel diagrams.** (A) The first transmembrane helix of bovine rhodopsin, with hydrophobic residues in blue, hydrophilic in red and others in gray. The 27 amino acids in the helix are numbered from the first one within the membrane. There are no strongly hydrophilic charged residues in this helix at all. The five residues designated as hydrophilic here have uncharged polar side chains and are only weakly hydrophilic. (B) A helical wheel representation of the same helix, with the amino acids color-coded as in (A). With the exception of a serine near the beginning of the helix, the hydrophilic and neutral residues are all on one face of the helix, thus making the helix amphipathic. Most of the hydrophilic residues are clustered in one part of the chain and are arranged so that they fall on one side of the helix. This hydrophilic area is part of an interface between this helix and other helices within the membrane. (C) A helical wheel representation of a non-transmembrane helix from a phosphoinositol kinase. Note the high proportion of charged, and thus strongly hydrophilic, residues [for example, aspartic acid (D), glutamic acid (E), lysine (K), and arginine (R)].

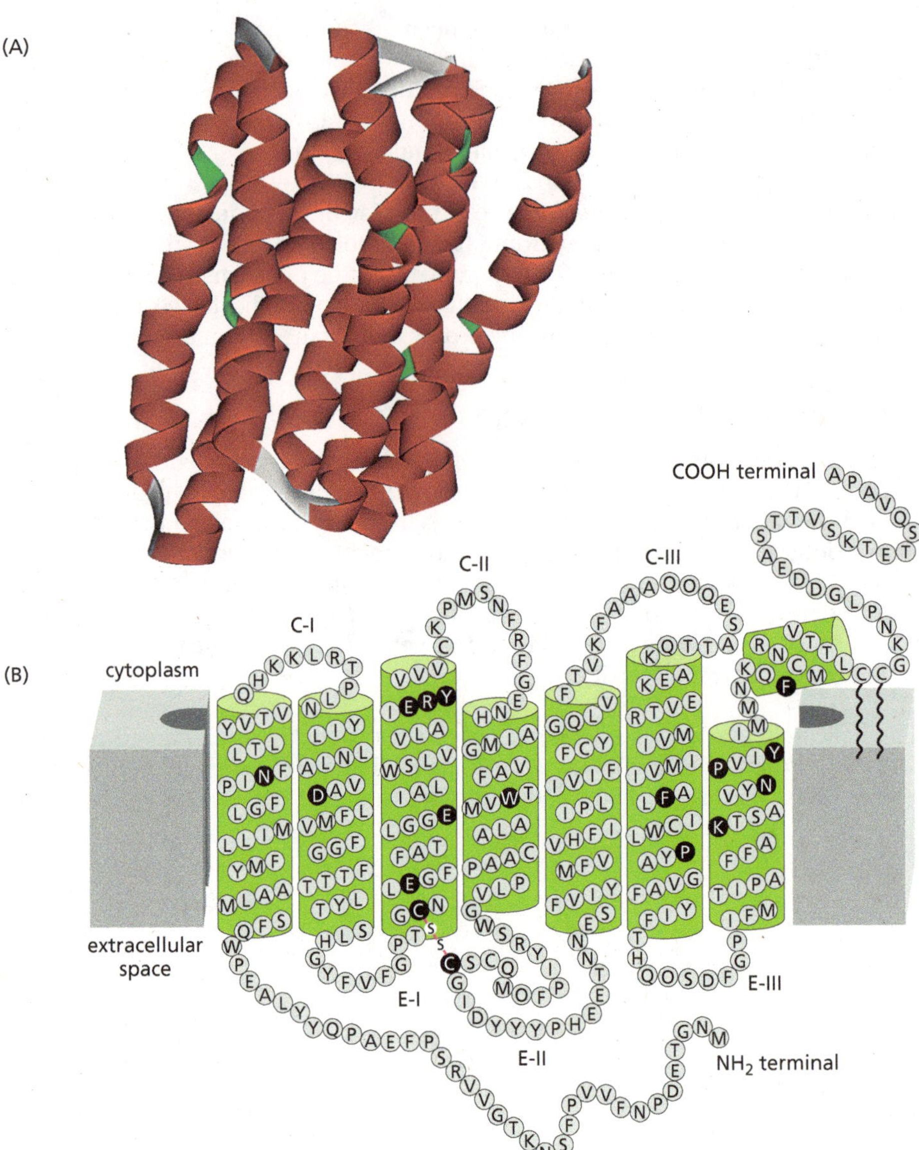


**Figure 11.29**
**The three-dimensional structure of bovine rhodopsin.** (A) A ribbon diagram of the experimentally derived structure. The seven transmembrane helices are colored red. PDB code 1F88. Green is used to indicate the locations of proline residues, which cause distortions in these helices. (B) A schematic representation of the topological arrangement of rhodopsin in the membrane. Note the cytoplasmic and extracellular loops connecting the helices. Rhodopsin is located in the distal membranes of rod photoreceptor cells, a specialized form of endoplasmic reticulum, so in this case extracellular corresponds to the lumen of the discs. (B, from K. Palczewski et al., Crystal structure of rhodopsin: a G protein-coupled receptor, *Science* 289:739–745, 2000. Reprinted with permission from AAAS.)

structures well. In such cases, a measure of the amphipathic nature of a helix, known as the **hydrophobic moment**, is used. The hydrophobic moment is the hydrophobicity of a peptide measured for different angles of rotation per residue. It is calculated for all angles of rotation from 0 to 180 degrees. The hydrophobic moment is an aid in the recognition of an amphipathic helix by identifying when the residues on one side of the structure are more hydrophobic than on the other. The hydrophobic moment is sensitive enough to distinguish between transmembrane helices and amphipathic helices found in globular protein domains. Surface helices have, in general, high hydrophobic moment while membrane helices (both single-pass and multi-pass) have low hydrophobic moment and high hydrophobicity, and hydrophilic helices have low hydrophobic moment and low hydrophobicity.

The hydrophobic moment can be represented visually by a helical wheel. Figures 11.28A and 11.28B illustrate the first transmembrane helix of the protein rhodopsin and its helical-wheel representation. Rhodopsin is a member of the G-protein-coupled receptor family, which all have seven transmembrane helices (see Box 11.2 and Figure 11.29). From Figure 11.28B we can see that there are no charged residues in the helix, and only five polar (weakly hydrophilic) residues. The amphipathicity

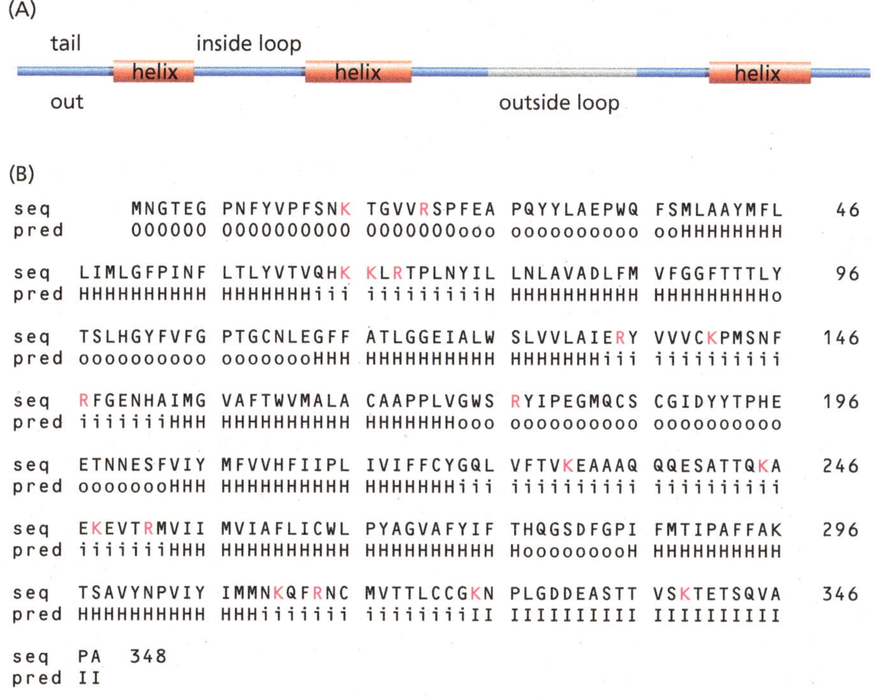

(A)
tail
inside loop
helix
helix
helix
out
outside loop
(B)
seq MNGTEG PNFYVPFSNK TGVVRSPFEA PQYYLAEPWQ FSMLAAYMFL 46
pred 000000 0000000000 0000000ooo oooooooooo ooHHHHHHHH
seq LIMLGFPINF LTLYVTVQHK KLRTPLNYIL LNLAVADLFM VFGGFTTTLY 96
pred HHHHHHHHHH HHHHHHHiii iiiiiiiiiH HHHHHHHHHH HHHHHHHHHo
seq TSLHGYFVFG PTGCNLEGFF ATLGGEIALW SLVVLAIERY VVVCKPMSNF 146
pred oooooooooo oooooooHHH HHHHHHHHHH HHHHHHHiii iiiiiiiiii
seq RFGENHAIMG VAFTWVMALA CAAPPLVGWS RYIPEGMQCS CGIDYYTPHE 196
pred iiiiiiiHHH HHHHHHHHHH HHHHHHHooo oooooooooo oooooooooo
seq ETNNESFVIY MFVVHFIIPL IVIFFCYGQL VFTVKEAAAQ QQESATTQKA 246
pred oooooooHHH HHHHHHHHHH HHHHHHHiii iiiiiiiiii iiiiiiiiii
seq EKEVTRMVII MVIAFLICWL PYAGVAFYIF THQGSDFGPI FMTIPAFFAK 296
pred iiiiiiiHHH HHHHHHHHHH HHHHHHHHHH HooooooooH HHHHHHHHHH
seq TSAVYNPVIY IMMNKQFRNC MVTTLCCGKN PLGDDEASTT VSKTETSQVA 346
pred HHHHHHHHHH HHHiiiiiii iiiiiiiiII IIIIIIIIII IIIIIIIIII
seq PA 348
pred II


**Figure 11.30**
**The transmembrane helices and the locations of tails and loops relative to the membrane as predicted for bovine rhodopsin by the HMMTOP program.** (A) A schematic showing the structures and locations distinguished by the HMMTOP prediction program. The helices are located within the membrane; loops can be either inside or outside the cell. The loops are divided into segments immediately flanking each helix and that are close to the membrane (called tails here), and those that are truly outside the membrane. In addition the amino- and carboxy-terminal ends of the protein need to be distinguished. (B) The predicted assignments for bovine rhodopsin as given by the prediction program HMMTOP. Predictions are o, outside tail; O, outside loop; H, membrane helix; i, inside tail; I, inside loop. This figure also illustrates the positive-inside rule established by von Heijne and colleagues showing that the intracellular loops between transmembrane helices have a higher content of arginines (R) and lysines (K), colored red, than do the extracellular loops.

is immediately apparent, as these hydrophilic residues are mainly on one side of the helix. The side chains of the hydrophilic residues project outward from the helix and allow these residues to form electrostatic interactions with polar groups in amino acids from adjacent helices. The single serine at the hydrophobic side of the wheel is near the end of the transmembrane helix. Thus, a helical wheel can be used to examine the amphipathicity of a transmembrane helix and its accuracy of prediction. Figure 11.28C shows a helical wheel for a non-transmembrane helix from a cytoplasmic phosphoinositol kinase. The difference in amphipathicity between a transmembrane and a cytoplasmic helix is quite marked.

Predictions of the orientation of α-helices in the membrane use the **positive-inside rule** established by von Heijne and colleagues. This rule states that intracellular loops between transmembrane helices have a higher content of arginines (R) and lysines (K) than do extracellular loops (see Figure 11.30). The positive-inside rule reflects the observation that nonmembrane regions inside have more positively charged residues than the regions outside the cell. This rule also applies to proteins in internal membranes such as those of the endoplasmic reticulum, in which intracellular corresponds to the cytosolic side of the membrane and extracellular to the luminal side. This rule was initially described and used in the TopPred prediction program and has since been incorporated in many more recent programs.

Additional observations incorporated into various prediction methods include the finding that the residues that flank each transmembrane region (often this segment is 15 residues long), sometimes known as helix tails, are related to the orientation of the protein. If the carboxy-terminal portion of the protein flanking the transmembrane segment is more positively charged than the amino-terminal portion, then the carboxy-terminal portion will be on the cytosolic side.

## A selection of prediction programs to predict transmembrane helices

A number of algorithms designed to predict transmembrane helices from the amino acid sequence have been developed, and there are various user-friendly

Web-based tools that will do this and also predict the orientation of the helices in the membrane. These methods can identify around 90–95% of all transmembrane segments. The use of hydrophobicity profiles to predict potential transmembrane regions has been in use for many years, but it was not until about 1992 that von Heijne designed the first method (TopPred) to predict the complete topology of transmembrane proteins. TopPred and some other algorithms set a threshold of hydrophobicity above which a segment is considered a transmembrane helix, but have the drawback that some helices may be missed because they fall just short of the threshold.

```
X-RAY   MNGTEGPNFY VPFSNKTGVV RSPFEAPQYY LAEPWQFSML AAYMFLLIML   50
HMMTOP  MNGTEGPNFY VPFSNKTGVV RSPFEAPQYY LAEPWQFSML AAYMFLLIML
SOSUI   MNGTEGPNFY VPFSNKTGVV RSPFEAPQYY LAEPWQFSML AAYMFLLIML
DAS     mngtegpnfy vpfsnktgvv rspfeapqyy laepwqfsML AAYMFLLIML
TMHMM   MNGTEGPNFY VPFSNKTGVV RSPFEAPQYY LAEPWQFSML AAYMFLLIML
TMpred  MNGTEGPNFY VPFSNKTGVV RSPFEAPQYY LAEPWQFSML AAYMFLLIML
PHDhtm  MNGTEGPNFY VPFSNKTGVV RSPFEAPQYY LAEPWQFSML AAYMFLLIML
TMAP    MNGTEGPNFY VPFSNKTGVV RSPFEAPQYY LAEPWQFSML AAYMFLLIML

X-RAY   GFPINFLTLY VTVQHKKLRT PLNYILLNLA VADLFMVFGG FTTTLYTSLH   100
HMMTOP  GFPINFLTLY VTVQHKKLRT PLNYILLNLA VADLFMVFGG FTTTLYTSLH
SOSUI   GFPINFLTLY VTVQHKKLRT PLNYILLNLA VADLFMVFGG FTTTLYTSLH
DAS     GFPINFLTLY Vtvqhkklrt plnyILLNLA VADLFMVFGG FTTTLytslh
TMHMM   GFPINFLTLY VTVQHKKLRT PLNYILLNLA VADLFMVFGG FTTTLYTSLH
TMpred  GFPINFLTLY VTVQHKKLRT PLNYILLNLA VADLFMVFGG FTTTLYTSLH
PHDhtm  GFPINFLTLY VTVQHKKLRT PLNYILLNLA VADLFMVFGG FTTTLYTSLH
TMAP    GFPINFLTLY VTVQHKKLRT PLNYILLNLA VADLFMVFGG FTTTLYTSLH

X-RAY   GYFVFGPTGC NLEGFFATLG GEIALWSLVV LAIERYVVVC KPMSNFRFGE   150
HMMTOP  GYFVFGPTGC NLEGFFATLG GEIALWSLVV LAIERYVVVC KPMSNFRFGE
SOSUI   GYFVFGPTGC NLEGFFATLG GEIALWSLVV LAIERYVVVC KPMSNFRFGE
DAS     gyfvfgptgc nlegffatlg geIALWSLVV LAIERYVvvc kpmsnfrfge
TMHMM   GYFVFGPTGC NLEGFFATLG GEIALWSLVV LAIERYVVVC KPMSNFRFGE
TMpred  GYFVFGPTGC NLEGFFATLG GEIALWSLVV LAIERYVVVC KPMSNFRFGE
PHDhtm  GYFVFGPTGC NLEGFFATLG GEIALWSLVV LAIERYVVVC KPMSNFRFGE
TMAP    GYFVFGPTGC NLEGFFATLG GEIALWSLVV LAIERYVVVC KPMSNFRFGE

X-RAY   NHAIMGVAFT WVMALACAAP PLVGWSRYIP EGMQCSCGID YYTPHEETNN   200
HMMTOP  NHAIMGVAFT WVMALACAAP PLVGWSRYIP EGMQCSCGID YYTPHEETNN
SOSUI   NHAIMGVAFT WVMALACAAP PLVGWSRYIP EGMQCSCGID YYTPHEETNN
DAS     nhaimGVAFT WVMALACAap PLVGWSRYIP EGMQCSCGID YYTPHEETNN
TMHMM   NHAIMGVAFT WVMALACAAP PLVGWSRYIP EGMQCSCGID YYTPHEETNN
TMpred  NHAIMGVAFT WVMALACAAP PLVGWSRYIP EGMQCSCGID YYTPHEETNN
PHDhtm  NHAIMGVAFT WVMALACAAP PLVGWSRYIP EGMQCSCGID YYTPHEETNN
TMAP    NHAIMGVAFT WVMALACAAP PLVGWSRYIP EGMQCSCGID YYTPHEETNN

X-RAY   ESFVIYMFVV HFIIPLIVIF FCYGQLVFTV KEAAAQQQES ATTQKAEKEV   250
HMMTOP  ESFVIYMFVV HFIIPLIVIF FCYGQLVFTV KEAAAQQQES ATTQKAEKEV
SOSUI   ESFVIYMFVV HFIIPLIVIF FCYGQLVFTV KEAAAQQQES ATTQKAEKEV
DAS     esfVIYMFVV HFIIPLIVIF FCYGQLVftv keaaaqqqes attqkaekev
TMHMM   ESFVIYMFVV HFIIPLIVIF FCYGQLVFTV KEAAAQQQES ATTQKAEKEV
TMpred  ESFVIYMFVV HFIIPLIVIF FCYGQLVFTV KEAAAQQQES ATTQKAEKEV
PHDhtm  ESFVIYMFVV HFIIPLIVIF FCYGQLVFTV KEAAAQQQES ATTQKAEKEV
TMAP    ESFVIYMFVV HFIIPLIVIF FCYGQLVFTV KEAAAQQQES ATTQKAEKEV

X-RAY   TRMVIIMVIA FLICWLPYAG VAFYIFTHQG SDFGPIFMTI PAFFAKTSAV   300
HMMTOP  TRMVIIMVIA FLICWLPYAG VAFYIFTHQG SDFGPIFMTI PAFFAKTSAV
SOSUI   TRMVIIMVIA FLICWLPYAG VAFYIFTHQG SDFGPIFMTI PAFFAKTSAV
DAS     tRMVIIMVIA FLICWLPYAG VAFYIfthqg sdfgpIFMTI PAFfaktsav
TMHMM   TRMVIIMVIA FLICWLPYAG VAFYIFTHQG SDFGPIFMTI PAFFAKTSAV
TMpred  TRMVIIMVIA FLICWLPYAG VAFYIFTHQG SDFGPIFMTI PAFFAKTSAV
PHDhtm  TRMVIIMVIA FLICWLPYAG VAFYIFTHQG SDFGPIFMTI PAFFAKTSAV
TMAP    TRMVIIMVIA FLICWLPYAG VAFYIFTHQG SDFGPIFMTI PAFFAKTSAV

X-RAY   YNPVIYIMMN KQFRNCMVTT LCCGKNPLGD DEASTTVSKT ETSQVAPA   348
HMMTOP  YNPVIYIMMN KQFRNCMVTT LCCGKNPLGD DEASTTVSKT ETSQVAPA
SOSUI   YNPVIYIMMN KQFRNCMVTT LCCGKNPLGD DEASTTVSKT ETSQVAPA
DAS     ynpviyimmn kqfrncmvtt lccgknplgd deasttvskt etsqvapa
TMHMM   YNPVIYIMMN KQFRNCMVTT LCCGKNPLGD DEASTTVSKT ETSQVAPA
TMpred  YNPVIYIMMN KQFRNCMVTT LCCGKNPLGD DEASTTVSKT ETSQVAPA
PHDhtm  YNPVIYIMMN KQFRNCMVTT LCCGKNPLGD DEASTTVSKT ETSQVAPA
TMAP    YNPVIYIMMN KQFRNCMVTT LCCGKNPLGD DEASTTVSKT ETSQVAPA
```

**Figure 11.31**
**Comparison of the prediction of transmembrane helices in bovine rhodopsin by different prediction programs: HMMTOP, SOSUI, DAS, TMHMM, TMpred, PHDhtm, and TMAP.** The location of transmembrane helices in the protein structure as determined by X-ray crystallography is given in the top line (X-RAY). The transmembrane helices are highlighted in yellow. Loops that are extracellular are colored in black, while cytoplasmic loops are given in blue. Boxed sequences are those that were finally predicted as transmembrane based on the consensus results of the seven prediction methods.

A number of approaches have been designed to improve this shortcoming and general prediction accuracy, including programs such as MEMSAT, which incorporates hydrophobicity, separate propensity scales for amino acids at the tail and head region of the membrane, and model parameters that include the number of membrane-spanning segments, the topology, the length, and the location to optimize prediction of the orientation of the transmembrane helices.

The methods available on the Web for predicting transmembrane segments and any additional information on their topology will be illustrated using the polytopic membrane protein bovine rhodopsin (see Figure 11.29). The predictions of a number of commonly used programs are given in Figure 11.31.

## Statistical methods

Like the statistical methods of secondary structure prediction discussed in the previous sections, the program TMpred predicts membrane-spanning segments and their orientation using an algorithm that is based on a statistical analysis of a protein database, in this case a database of known transmembrane protein sequences (TMbase). The prediction is made by scoring the sequence using a combination of several scoring matrices derived from analysis of TMbase. The user can set minimum and maximum lengths for the transmembrane region to be predicted; the default is a minimum of 17 and a maximum of 33 residues.

The output gives a number of possible transmembrane helices in the sequence and their predicted topology; in other words whether the helix is going across the membrane from outside to the inside, or vice versa. It also highlights the topology with the highest level of significance. This is the prediction that should, generally, be used.

## Knowledge-based prediction

Physicochemical properties of amino acid sequences, such as hydrophobicity and charge, are used in prediction methods such as SOSUI. This method is based on certain assumptions; the main one states that a membrane protein must have at least one very hydrophobic primary transmembrane helix. Further transmembrane helices can exist in a multispanning membrane protein and their hydrophobicity can be similar to hydrophobic segments found in soluble proteins. The primary transmembrane helices are stabilized by a combination of amphiphilic side chains at the helix ends as well as having a hydrophobic central region. Four physicochemical parameters are used in this method: a hydropathy index, an amphiphilicity index, an index of amino acid charges, and each sequence length.

The output from SOSUI gives the segments that are predicted to be transmembrane helices, the hydropathy profile, helical-wheel diagrams, and a picture of the membrane topology orientation (see Figure 11.32). Compare the transmembrane topology diagram output by the program with the one derived from the X-ray structure (see Figure 11.29). It can be seen that although SOSUI locates the segment representing the first transmembrane helix in the correct region of sequence, the fact that the start of the helix is not accurately predicted (it is three residues too late) affects the folding of the helix and therefore the arrangement of hydrophobic and nonhydrophobic residues on the helix (compare the helical wheel for this prediction to that in Figure 11.28 to see that the amphipathic nature in the prediction is lost). Therefore the prediction of the detailed topology is much more difficult and less accurate than prediction of the general location of the transmembrane segments and the overall topology that tells us which loops are on the cytoplasmic side and which are on the extracellular side.

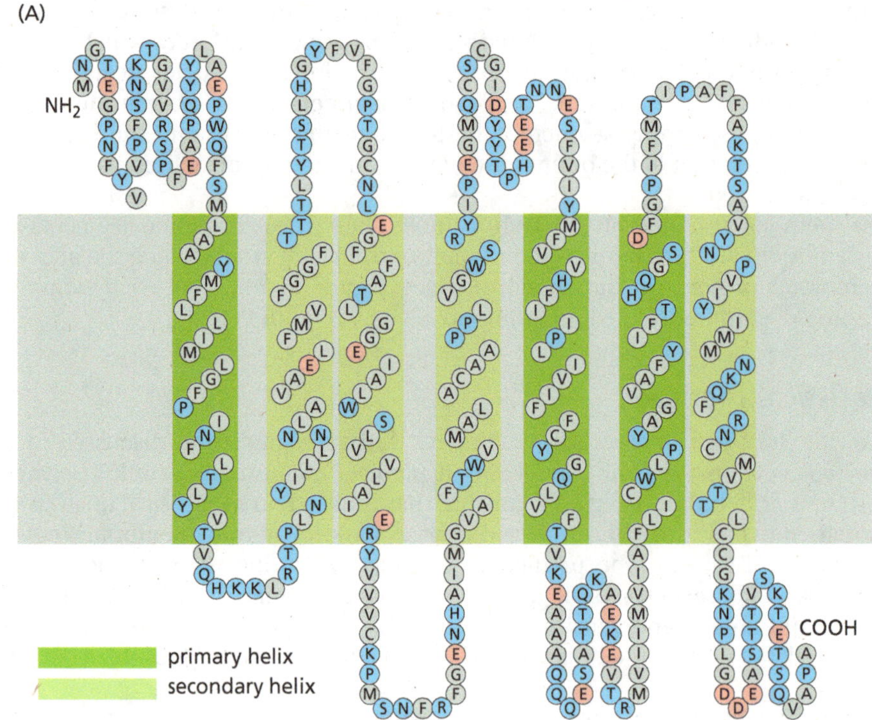


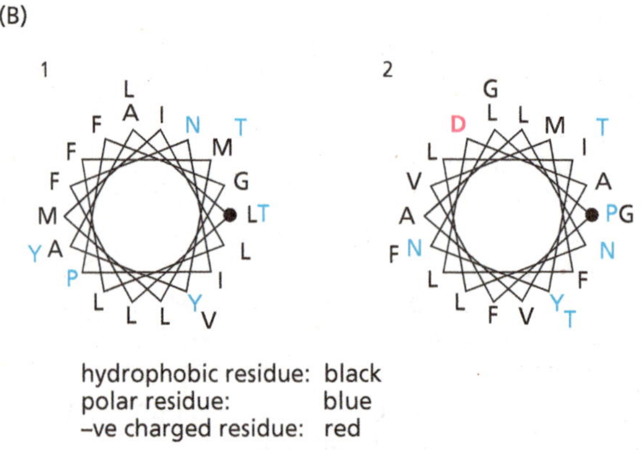


**Figure 11.32**
**Prediction results for rhodopsin obtained by SOSUI.** (A) Schematic of the locations of transmembrane helices output by the program. (B) Helical wheels for the predicted first and second transmembrane helices. Note the difference in amphipathicity that results from the underprediction of the beginning of the first helix.

## Evolutionary information from protein families improves the prediction

As with prediction of secondary structure the use of evolutionary information generally improves the prediction. Several membrane helix prediction methods use information from the alignment of protein sequences. One such method is TMAP, which can use as input either a single sequence or, preferably, a multiple alignment of the query sequence and its homologs. The method uses two sets of propensity values, one set for the middle hydrophobic part of the putative transmembrane helix and another for the terminal regions. These propensity values are assigned for each position along the alignment and then an average calculated. The average value is weighted based on the similarity of the sequences to each other. The propensity values are based on the statistical analysis of residue types found in transmembrane helices from the Swiss-Prot database that were designated as transmembrane segments. Multiple alignments improve prediction accuracy. However, for 20–30% of all proteins, there are still no homologs in current databases. In response to this situation, the so-called dense alignment surface (DAS) method was developed.

The dense alignment surface method (DAS) is used to improve comparisons between a collection of nonhomologous membrane proteins and the query protein using a specially derived scoring matrix based on neighborhood selectivity (NS) of amino acid pairs. The matrix only gives high scores when aligning two transmembrane helix segments. Every 10-residue window has a final score that is the average of the gapless alignment scores with all segments of a set of about 40 membrane protein sequences. The segment is predicted as transmembrane if the score is above a certain cut-off.

The results consist of a list of predicted transmembrane helices and a DAS profile scoring curve where transmembrane helices are those segments that peak above a DAS profile cut-off (see Figure 11.33). The segments above the lower cut-off were taken to represent the transmembrane helices.

## Neural nets in transmembrane prediction

PHDhtm is a neural network method that uses evolutionary information to improve the prediction of transmembrane segments. Patterns of amino acid substitutions between sequences of proteins in the same structural family are highly correlated with the tertiary structure for that family. PHDhtm uses this evolutionary information as input to a neural network to predict the transmembrane helices. The first step in a PHDhtm prediction is to find homologs for the query sequence and then generate a multiple sequence alignment. The alignment is put into a neural network in which transmembrane locations are encoded by two codes: one that designates residues as being in a transmembrane helix, and the other code designates residues as nonmembrane bound. These preferences are used as input into a dynamic programming algorithm that searches for the best model to represent the best transmembrane localization. Finally the positive-inside rule is applied to predict the topology.

The multiple alignment generated by PHDhtm can also be submitted to the TMAP server, for example, to make a confirmatory prediction.

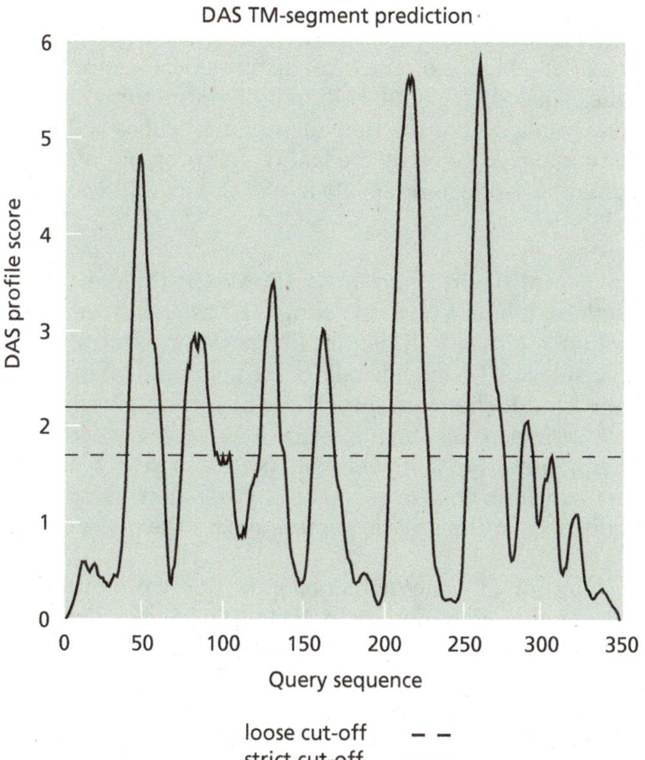


**Figure 11.33**
**A profile of the DAS scoring values.** This shows the regions of the bovine rhodopsin sequence that are above and below the DAS cut-off. Those regions above the cut-off are predicted to be transmembrane helices.

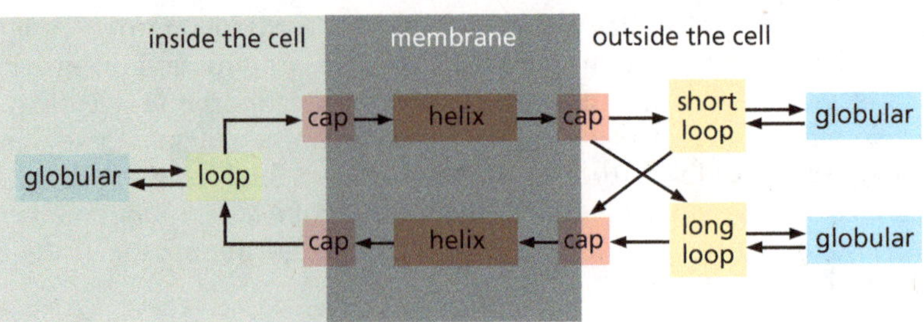


**Figure 11.34**
**The HMM states used by TMHMM.** The core transmembrane helices; helical tails (caps) on either side of the membrane; loops on the inside; short and long loops outside the cell; and globular-domain-like structures in the middle of each loop.

## Predicting transmembrane helices with hidden Markov models

The hidden Markov model (HMM) is a good technique for prediction of transmembrane helices because it can incorporate hydrophobicity, charge bias, helix lengths, as well as grammatical constraints into one model for which algorithms for parameter estimation and prediction already exist. The basic principle of the HMM approach in transmembrane prediction is to define a set of states; each residue is predicted to be in one of the states. In the simplest case, three states can be designated: one for the inside loops, one for the outside loops, and one for the transmembrane segments. Each of these states has an associated probability distribution over the 20 amino acids that describes the variability of each of these amino acids in the region modeled. The states are connected to each other in a biologically reasonable manner; for example, the state for the inside loop is connected to itself because there can be more than one amino acid involved in such a loop, and also to the transmembrane helix state because after an inside loop another helix will, usually, begin. Then a state-transition probability is associated with each transition. The amino acid and transition probabilities are learned by the HMM method. By defining these states and connecting them in a cycle, an architectural model can be designed that closely resembles the biological system to be modeled. The path of a protein sequence through the states with the highest computed probability should predict the true topology.

The HMM-based program TMHMM designates a number of possible different HMM states (see Figure 11.34): (1) the core transmembrane helix; (2,3) helical tails (caps) on either side; (4) loops on the cytoplasmic side; (5) short loops and (6) long loops outside the cell; and (7) globular-domain-like structures in the middle of each loop. The actual prediction of the transmembrane helices is done by finding the most probable topology as given by the HMM. The program returns a list of transmembrane segments, inside and outside loops, and a graphical probability profile (see Figure 11.35).

Another similar HMM-based program is HMMTOP (hidden Markov model for topology prediction). It builds on a very similar HMM architecture, but the method used for prediction is different. It defines five structural states: (1) the transmembrane helix; (2) inside and (3) outside tails of the helix; and (4) inside and (5) outside loops (see Figure 11.36). There are several prediction options. Firstly, it is possible to submit either a single query sequence or a query sequence as part of a set of homologous sequences. It is, in general, better to use a set of homologous sequences. One can also choose to make a prediction using a standard set of parameters or one that optimizes the parameters for the query sequence.

If the actual localization of some segments of the query protein is known, then the localization-of-sequence-part option in HMMTOP allows the user to delineate these segments. The prediction is then performed using these segments as additional restrictions. The syntax for defining the topology of a given segment is: begin_position-end_position-type, where begin_position and end_position are the sequence numbers, while type is the one-letter code of the location of the structural

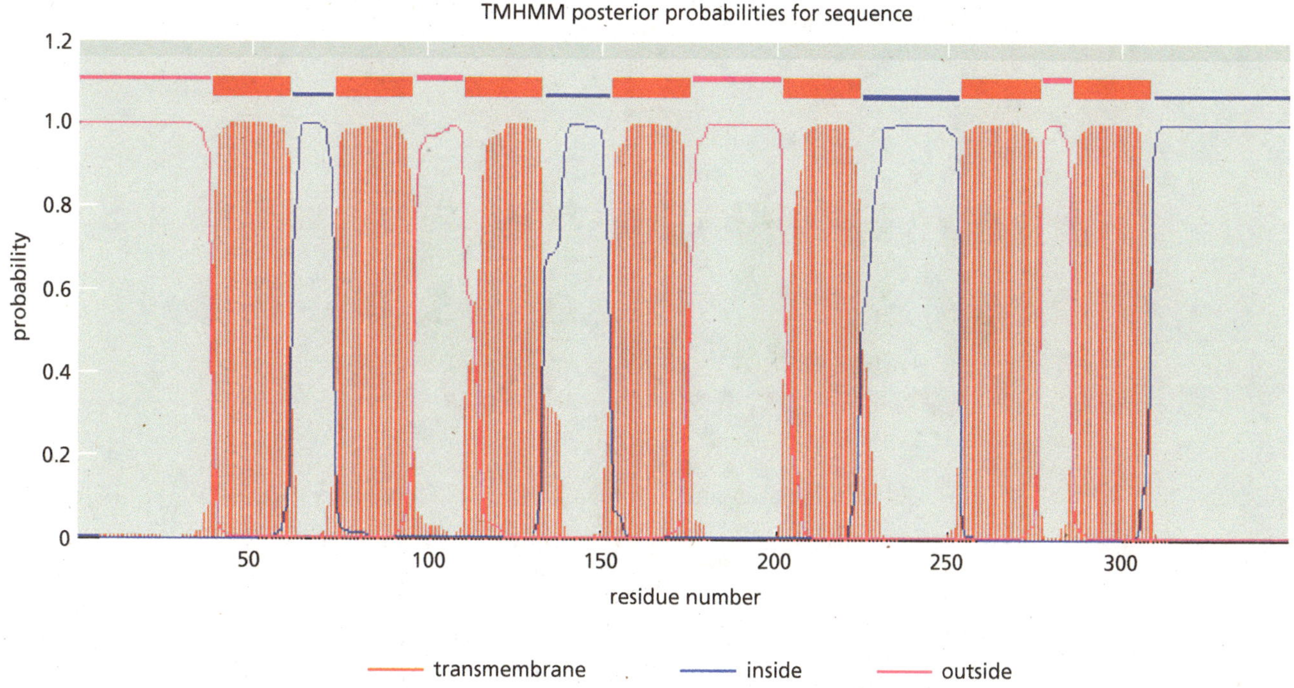


| | | | |
|---|---|---|---|
| Sequence | outside | 1 | 38 |
| Sequence | TMhelix | 39 | 61 |
| Sequence | inside | 62 | 73 |
| Sequence | TMhelix | 74 | 96 |
| Sequence | outside | 97 | 110 |
| Sequence | TMhelix | 111 | 133 |
| Sequence | inside | 134 | 152 |
| Sequence | TMhelix | 153 | 175 |
| Sequence | outside | 176 | 201 |
| Sequence | TMhelix | 202 | 224 |
| Sequence | inside | 225 | 253 |
| Sequence | TMhelix | 254 | 276 |
| Sequence | outside | 277 | 285 |
| Sequence | TMhelix | 286 | 308 |
| Sequence | inside | 309 | 348 |

**Figure 11.35**
**The output of the TMHMM program for the bovine rhodopsin sequence.** The table lists the residue number ranges associated with each predicted segment.

parts described above. For example 56-76-I generates a prediction where the sequence segment between positions 56 and 76 will be in the cytosol (inside). The results returned consist simply of the amino acid sequence with the prediction given underneath each amino acid (see Figure 11.30B). In general O or o denotes outside, while I or i gives the inside loop or tail and H codes for the transmembrane. The TMHMM and HMMTOP predictions are given in Figure 11.31.

## Comparing the results: What to choose

In general the methods all give similar predictions on the one protein we have used as an example (see Box 11.3). Some methods tend to predict longer helices, such as HMMTOP, TMAP, and, occasionally, SOSUI. TMAP predicts shorter helices at the amino and carboxy termini of the protein sequence, which leads to a less accurate prediction of the start of the first and end of the last helix. The DAS method seems to predict shorter transmembrane segments overall. Most of the methods predicted the overall topology correctly. Only a selection of programs available is illustrated here to demonstrate the various methods used in the prediction programs.

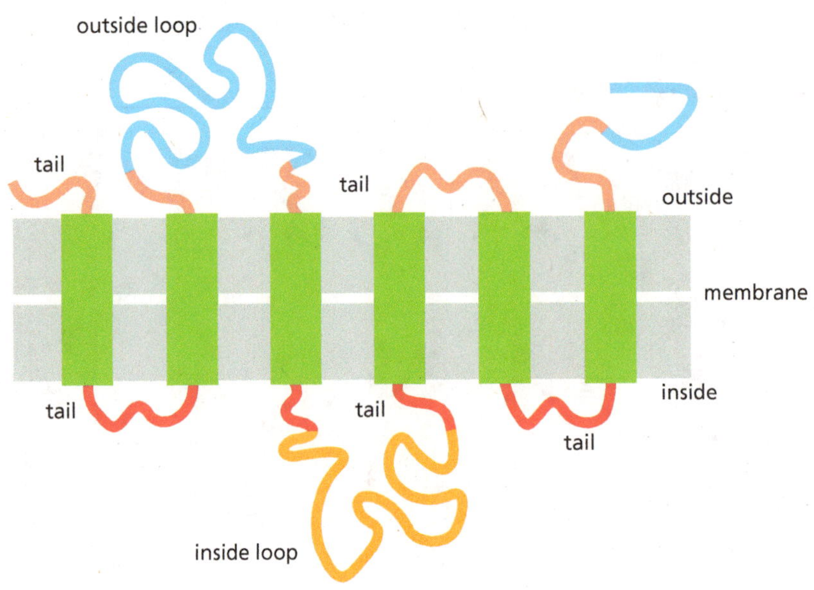


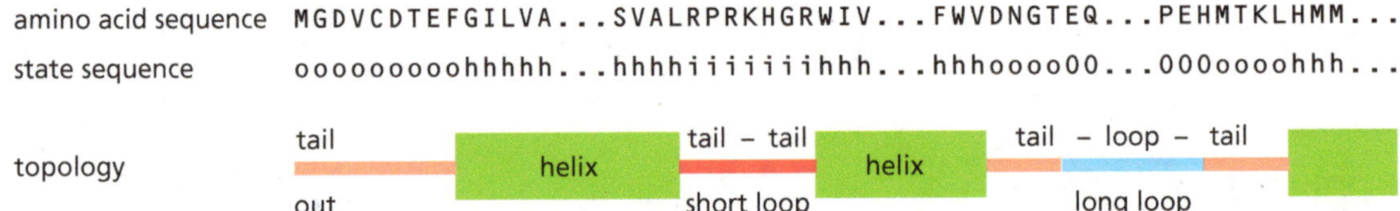


**Figure 11.36**
**Five states of membrane-spanning structure in the HMMTOP program.** The five states are inside loop (I), inside tail (i), membrane helix (h), outside tail (o), and outside loop (O). Tails (red and pink) are thought to interact with the inside or outside parts of the membrane, while loops (brown and cyan) do not. Two tails between helices can form short loops, but longer loops are formed by tail–loop–tail sequences. (Adapted from G.E. Tusnady and I. Simon, Principles governing amino acid composition of integral membrane proteins: application to topology prediction, *J. Mol. Biol.* 283:489: 506, 1998.)

To obtain the most accurate prediction, one should run a number of prediction algorithms, as they all use different hydrophobicity profiles and/or algorithms to predict the amphipathicity, which loops are in and out of the cytosol, the ends of the helices, and the overall topology. The boxed segments, illustrated on Figure 11.31, show a user's choice of transmembrane segments based on the predictions from all seven methods. This is obtained by aligning the results and extrapolating from the most frequently occurring prediction at each residue to obtain a consensus. Additionally, it is good to test the helical possibilities using the helical-wheel representation, looking mainly for patterns of polar residues on one side rather than interspersed throughout.

## What happens if a non-transmembrane protein is submitted to transmembrane prediction programs

If a sequence does not code for a transmembrane protein, such as the kinase protein p110α, and it is submitted to the transmembrane prediction programs then the results one usually obtains are:

- No transmembrane helices are predicted (HMMTOP, TMHMM, and SOSUI, which predicted the protein correctly to be soluble)
- Only one or very few helices are predicted to be transmembrane
- Predicted helices are usually very short – not long enough to span a membrane
- Helical wheels are quite different; they may be more charged and the distribution of the polar/charged residues is not as one-sided as for transmembrane helices (see Figure 11.28)

## Prediction of transmembrane structure containing β-strands

β-barrel structures in a membrane consist of a large antiparallel sheet rolled into a cylindrical structure. The sequences of the β-strands have a hydrophobic residue at

## Box 11.3 Predicting transmembranes–an example

Vitamin K epoxide reductase (VKOR) catalyzes the conversion of vitamin K 2,3-epoxide into vitamin K. Recently, the gene encoding the catalytic subunit of VKOR was identified as an integral membrane protein. These vitamin K-dependent proteins are important as coagulation factors, and are involved in bone metabolism and signal transduction. In order to understand the structure–function relationship of these proteins, it is important to understand the membrane topology.

Jien-Ke Tie and colleagues first used a number of prediction programs to predict the transmembrane structure of VKOR and subsequently determined the topology experimentally. They used seven programs to predict the topology; PHD, TMHMM, TopPred, TMpred, DAS, SOSUI, and MEMSAT. Five of the seven programs predicted three transmembrane helices. PHD predicted only two and MEMSAT predicted four (see Figure B11.4A). Five also predict that the C terminus of VKOR is in the cytoplasm. From these results the authors predict that VKOR has three transmembrane helices with the C terminus located in the cytoplasm and the N terminus inside. The transmembrane predictions were checked using an *in vitro* glycosylation assay. These experimental results strongly suggest that VKOR is a type III membrane protein with the C terminus outside and the N terminus inside. Other experiments confirmed that the first helix spans residues 10 to 29, which corresponds with the predictions. More experiments were performed to test the various predictions. In conclusion, the experiments suggest that there are three transmembrane helices, which agree with most of the prediction programs. The predicted topology both using prediction programs and experimental data is illustrated in Figure B11.4B.

(A)

| programs | TM no. | C terminus | TM1 | TM2 | TM3 | TM4 |
|---|---|---|---|---|---|---|
| PHD | 2 | In | | 85–109 (1.0) | | 119–143 (0.87) |
| TMHMM 2.0 | 3 | In | 10–29 (1.0) | | 101–123 (0.90) | 127–149 (0.93) |
| TopPred 2 | 3 | In | 9–29 (0.65) | 78–98 (0.67) | 109–129 (1.0) | |
| TMpred | 3 | In | 9–29 (0.81) | 75–97 (0.57) | 101–129 (1.0) | |
| DAS | 3 | | 12–27 | 83–96 | 102–146 | |
| SOSUI | 3 | | 11–31 (primary) | 75–97 (secondary) | | 116–138 (primary) |
| MEMSAT | 4 | In | 13–29 (0.76) | 81–97 (0.60) | 104–124 (1.0) | 131–148 (0.68) |

(B)

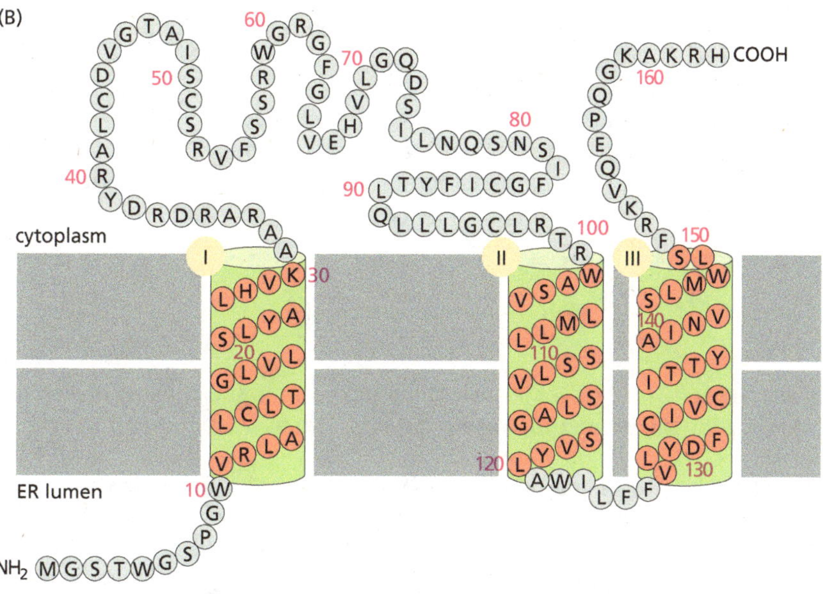


**Figure B11.4**
**(A) A table showing the prediction results from various programs.**
**(B) A schematic of the transmembrane prediction of VKOR.**
(A, from J.K. Tie et al., Membrane topology mapping of vitamin K epoxide reductase by *in vitro* translation/cotranslocation. *J. Biol. Chem.* 280: 16410–16416, 2005.)

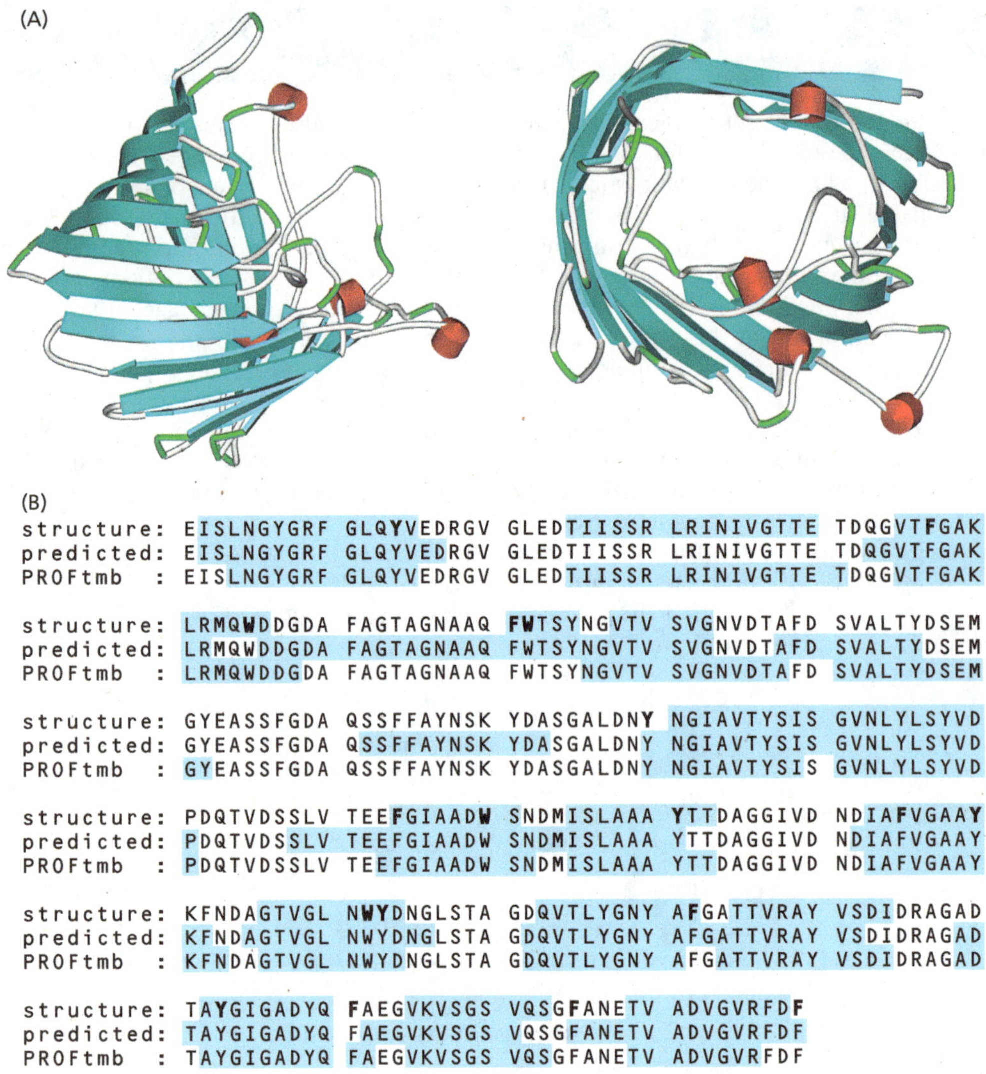


**Figure 11.37**
**Prediction of the transmembrane β-barrel.** (A) The crystallographic structure of a transmembrane β-barrel, porin. (B) The amino acid sequence of porin. Note the aromatic residues (bold) flanking the membrane β-strands. The top line shows the sequence color-coded with transmembrane strands (shaded blue) according to the X-ray structure determination. The lower lines show the predicted structure using the Eisenberg hydrophobicity scale and the Web-based program PROFtmb.

every second position in the sequence so that the outside of the cylinder will be nonpolar and the inside polar. In addition, it has been observed that the transmembrane β-strands are generally flanked in the sequence by aromatic residues. It is also unlikely that a single extended strand will exist in the hydrophobic environment of the bilayer, as the hydrogen-bonding backbone NH and CO groups will be isolated. Usually, to identify transmembrane β-strands a sided hydrophobicity $H_S$ is calculated using the Eisenberg hydrophobicity scale $h$ and the following equation:

$$H_s(i) = \frac{1}{4}[h(i-2) + h(i) + h(i+2) + h(i+4)] \qquad \text{EQ11.1}$$

If the residue at $i$–2 or $i$+4 is aromatic, the hydrophobicity value is increased to 1.6, thus biasing the prediction toward strands with aromatic residues at the ends.

An example of a β-stranded transmembrane protein for which the structure has been solved is the membrane channel protein porin (see Figure 11.37A) from the bacterium *Rhodopseudomonas blastica* (PDB code 1PRN). Equation EQ11.1 was used to calculate a prediction profile for the transmembrane β-segments (see Figure 11.37B). The results show that generally the β-strands are identified but not very

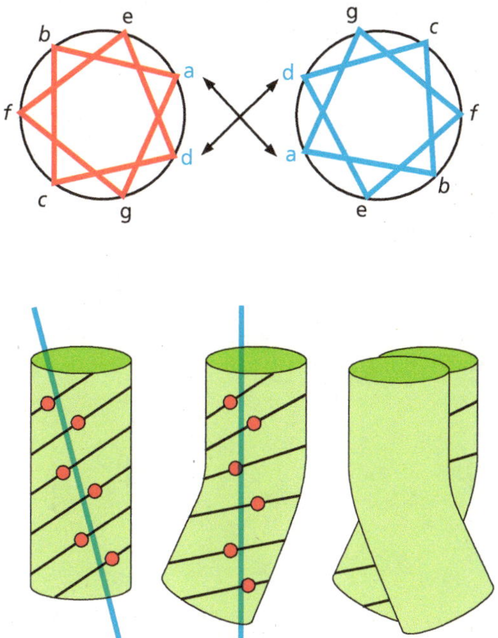


**Figure 11.38**
**Coiled-coil geometry.** As there are 3.6 residues to each turn of the α-helix, the a and d residues form a hydrophobic seam (blue letters and red dots), which, as each heptad is slightly under two turns, slowly twists around the helix. The coiled coil is formed by component helices coming together to bury their hydrophobic seams. As the hydrophobic seams twist around each helix, so the helices also twist to coil around each other, burying the hydrophobic seams and forming a supercoil.

accurately. Recently a new method called PROFtmb was developed that predicts transmembrane β-barrels and is available over the Web. This prediction method is based on an HMM model using a set of 56 transmembrane β-barrel structures to obtain the parameters used in the HMM (see Chapter 12 for more details). From Figure 11.37B we can see that the prediction is much better than the one using hydrophobicity profiles only.

## 11.8 Coiled-Coil Structures

Coiled-coil conformation is obtained when two (or more, often three) α-helices intertwine, forming a superhelical twist (see Figure 11.38). Such coils are found in both transmembrane helix bundles and intracellular proteins. They form between helices with particular distributions of hydrophobic residues. Generally, the sequence of an α-helix that forms a coiled coil will display a periodicity of a repeated unit of seven amino acids, which is called a heptad pattern. If those positions are designated by the letters abcdef and g then positions a and d are hydrophobic and e and g are often charged (see Figure 11.38A). This arrangement results in a hydrophobic seam of residues a and d, which runs in a shallow helix along the surface of the α-helix. The repeating seven-residue pattern results in a twist within the helix, as positions a and d come together to bury their hydrophobic nature from the surrounding solvent (Figure 11.38B). When the helical components are juxtaposed, they twist around each other to form a tight supercoil, as a result of the hydrophobic interaction of a and d residues. Algorithms that predict coiled-coil regions are based on these preferences. When positions a and d are repeating leucines, the structure is called a leucine zipper. In the leucine zipper families of transcription factors, leucine zipper dimerization domains occur alongside DNA-binding domains.

Coiled coils are usually strong structures and occur in the keratins (which form intermediate filaments in cells), in the motor protein myosin II (which is a dimer held together by a coiled-coil tail), in fibrin (which is important in blood clotting)

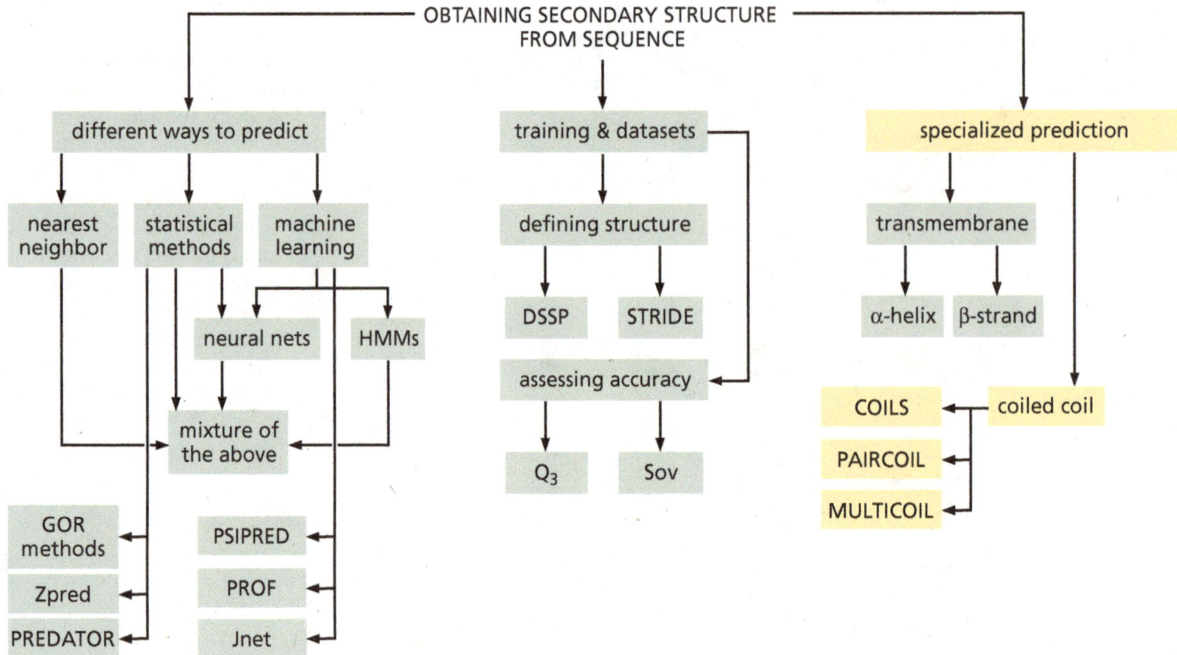


**Flow Diagram 11.8**
**In this section some of the methods developed for the prediction of coiled coils are discussed.**

and as triple helices in the extracellular matrix protein collagen. Programs that predict coiled-coil structure in proteins and are available on the Web will be illustrated with the sequence from the structure of the effector domain of the protein kinase PKN/PRK1. This coiled-coil domain binds to the G protein Rho (see Figures 11.39 and 11.40).

## The COILS prediction program

COILS compares the query sequence to a database of known parallel two-stranded coiled coils and derives a similarity score. By comparing this score to the distribution of scores in globular and coiled-coil proteins, the program calculates the probability that the query sequence will be in a coiled-coil conformation.

The program allows the user to choose whether to weight the a and d positions of the coiled-coil. This allows the user to assign the same weight to the two hydrophobic positions as to the five hydrophilic positions within the heptad repeat. Assignment of the same weights permits the identification of false positives. In general it is recommended to run a weighted and an unweighted prediction and compare the outputs. A drop of more than 20–30% in the probability when an unweighted prediction is run indicates a false-positive prediction due to a highly charged protein sequence.

The COILS output is a trace of coiled-coil prediction probability (see Figure 11.41A). The trace for the effector domain sequence using the unweighted prediction is

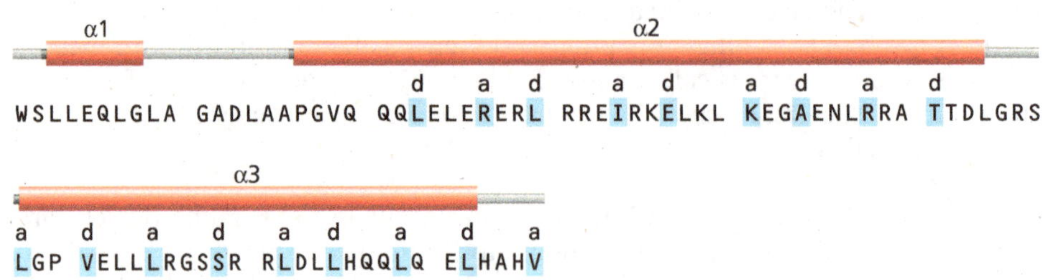


**Figure 11.39**
**Sequence of the PKN1 coiled-coil domain.** The secondary structural elements (α1–α3) of the PKN domain are indicated at the top. The a–d heptad repeats of helices α2 and α3 are displayed above the sequence in the coiled-coil domain.

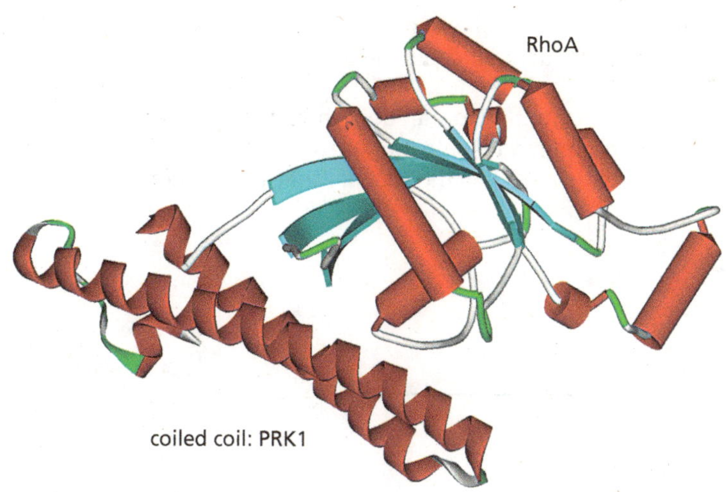


**Figure 11.40**
**The ribbon diagram of the X-ray structure of the coiled-coil domain PRK1 bound to RhoA protein.**

shown for window lengths of 14, 21, and 28 residues. The program clearly predicts a two-helix coiled coil when a 14- or a 21-residue window is used (the weighted prediction is not much different in this case). The 28-residue window, however, fails to predict a second region. In Figure 11.41B we can see why. The crystal structure has identified two helices: helix A is 38 residues long and will therefore be identified using all the window lengths; helix B, however, is only 25 residues long and will therefore fail to be identified when a window of 28 residues is used. The trace in Figure 11.39A gives only an approximate location of the helices; however, one can investigate the probabilities in detail by looking at the numerical data (see Figure 11.41C) that are also provided by the program. The higher the score the more probable it is that the residue is predicted as part of a coiled-coil helix. Exploring the numerical format predicts that helix A runs from residue 19 to 49 and helix B spans residues 62 to 82.

## PAIRCOIL and MULTICOIL are an extension of the COILS algorithm

A newer program than COILS is the PAIRCOIL program, which predicts the location of coiled-coil regions using pairwise residue correlations obtained from a two-stranded coiled-coil database. As described previously, in COILS coiled coils have been mainly identified by the occurrence of hydrophobic residues spaced every four residues apart. In PAIRCOIL this is extended to include pairwise residue correlations in known coiled coils. Each sequence in the coiled-coil database was used to tabulate the frequency of occurrence of each pair of amino acids at each pair of positions in a heptad repeat. These frequency values in the coiled-coil database are used to estimate the probability that a given residue pair exists in a given pair of heptad-repeat positions in a coiled coil. The probabilities are then used to compute a score for each residue, which corresponds to the likelihood that this residue is in a coiled coil. The MULTICOIL program is based on the same algorithm but has been modified to locate dimeric and trimeric coiled coils.

## Zipping the leucine zipper: A specialized coiled coil

The leucine zipper is a special type of coiled coil in which leucines are the repeating hydrophobic residues (see Figure 11.42A). This domain occurs mostly in regulatory proteins and thus in many oncogenic proteins. The 2ZIP method is a Web server prediction program designed especially to look for leucine zipper repeats. The 2ZIP program combines standard coiled-coil prediction techniques as described above with a search for the characteristic leucine repeat.

The output is a text formatted prediction shown in Figure 11.42B. If the query sequence does not contain a leucine zipper the program will nevertheless predict other coiled-coil regions and leucine repeats that do not correspond to a leucine zipper.

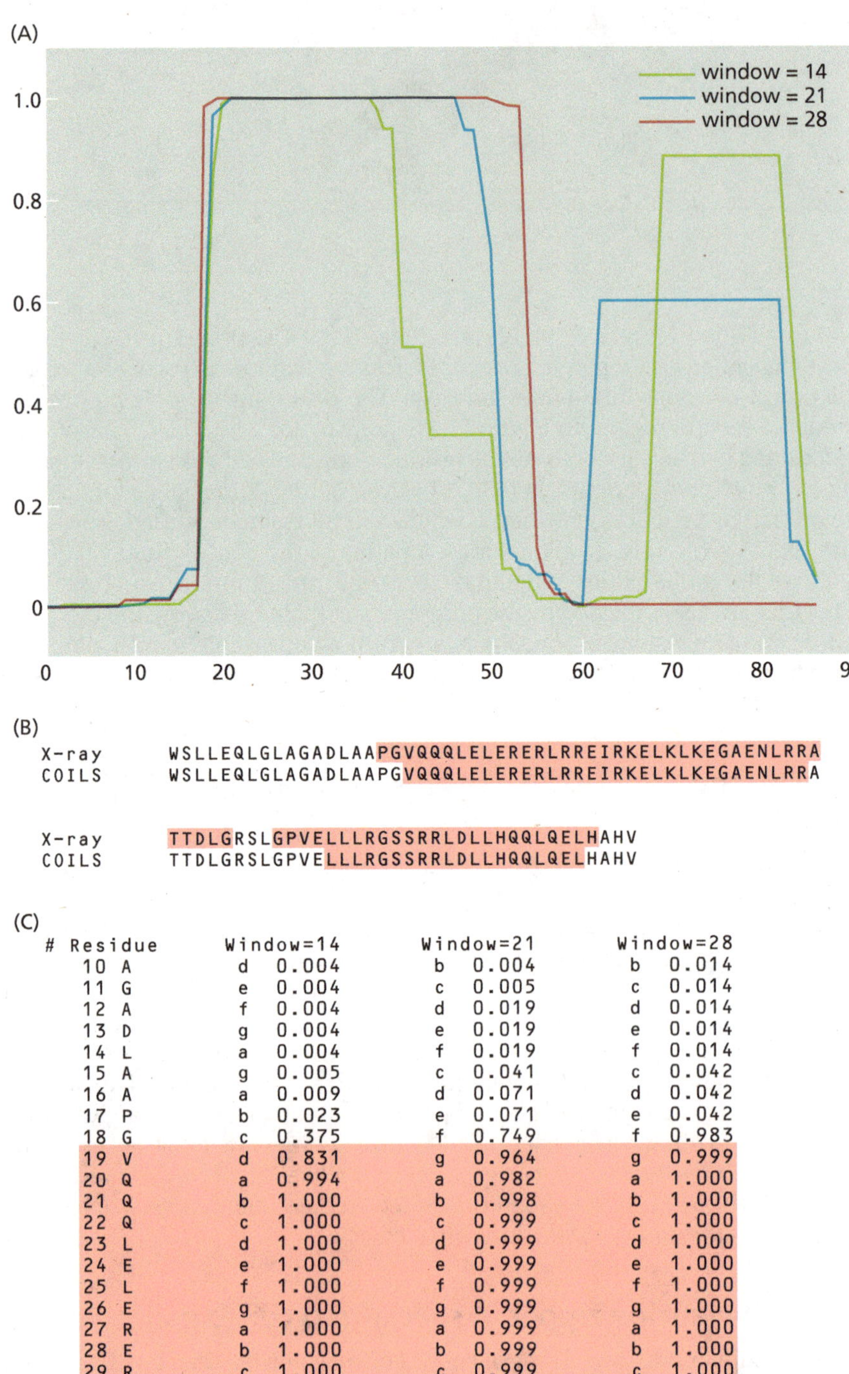


| # | Residue | Window=14 | | Window=21 | | Window=28 | |
|---|---|---|---|---|---|---|---|
| 10 | A | d | 0.004 | b | 0.004 | b | 0.014 |
| 11 | G | e | 0.004 | c | 0.005 | c | 0.014 |
| 12 | A | f | 0.004 | d | 0.019 | d | 0.014 |
| 13 | D | g | 0.004 | e | 0.019 | e | 0.014 |
| 14 | L | a | 0.004 | f | 0.019 | f | 0.014 |
| 15 | A | g | 0.005 | c | 0.041 | c | 0.042 |
| 16 | A | a | 0.009 | d | 0.071 | d | 0.042 |
| 17 | P | b | 0.023 | e | 0.071 | e | 0.042 |
| 18 | G | c | 0.375 | f | 0.749 | f | 0.983 |
| 19 | V | d | 0.831 | g | 0.964 | g | 0.999 |
| 20 | Q | a | 0.994 | a | 0.982 | a | 1.000 |
| 21 | Q | b | 1.000 | b | 0.998 | b | 1.000 |
| 22 | Q | c | 1.000 | c | 0.999 | c | 1.000 |
| 23 | L | d | 1.000 | d | 0.999 | d | 1.000 |
| 24 | E | e | 1.000 | e | 0.999 | e | 1.000 |
| 25 | L | f | 1.000 | f | 0.999 | f | 1.000 |
| 26 | E | g | 1.000 | g | 0.999 | g | 1.000 |
| 27 | R | a | 1.000 | a | 0.999 | a | 1.000 |
| 28 | E | b | 1.000 | b | 0.999 | b | 1.000 |
| 29 | R | c | 1.000 | c | 0.999 | c | 1.000 |
| 30 | L | d | 1.000 | d | 0.999 | d | 1.000 |

**Figure 11.41**
**Output from the COILS program for predicting coiled-coil regions as applied to the effector domain of the protein kinase PKN/PRK1.** (A) The prediction trace for the probability of coiled-coil regions using sliding windows of 14 (green), 21 (blue), and 28 (red) residues. The 14- and 21-residue windows give two regions predicted to be able to form coiled-coil structure; the 28-residue window, however, gives only one. (B) The sequence of the protein and the secondary structure as obtained from the X-ray structure (top line) and the prediction (bottom line). The reason for the failure of the 28-residue window can now be seen; the second helix is only 25 residues long. (C) The first part of the tabulated result given by COILS. In (B) and (C) helical segments are color-coded in pink.

(A)

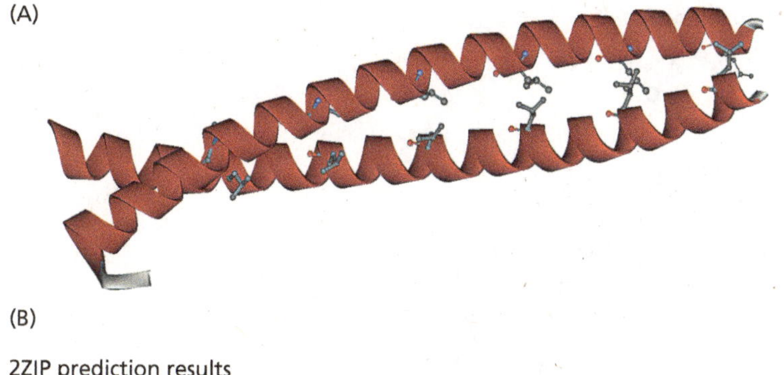

(B)

2ZIP prediction results
1) number of potential LEUCINE ZIPPERS: 1

```
1---------11--------21--------31--------41--------51-------
RMKQLEDKVEELLSKNYHLENEVARLKKLVGDLLNVKMALDIEIATYRKLLEGEESRIS
 CCCCCCCCCCCCCCCCCCCCCCCCCCCCCCCCCCC
    L------L------L------L------L------L
    LZLZLZLZLZLZLZLZLZLZLZLZLZLZLZLZ
```

**Figure 11.42**
**Leucine zipper prediction.** (A) The crystal structure of GCN4 leucine zipper, with the heptad repeat leucines shown in a ball-and-stick representation. (B) The results of the prediction from the program 2ZIP.

## 11.9 RNA Secondary Structure Prediction

Proteins are not the only biological macromolecules that form highly complex three-dimensional structures. RNAs (tRNA, mRNA, and rRNA) are long chains of nucleotides that fold into various secondary structures as a result of intrastrand base pairing and base stacking (see Section 1.1). As is the case for proteins, the three-dimensional structure of an RNA molecule affects its biological function. Messenger RNAs (mRNA) carry information as well as being involved in translational control. Other RNAs seem to play an important role in the regulatory function of the cell. RNA is also able to catalyze cellular processes. Secondary structure is important for some of these functions and can shed light on the structure–function relationships

**Flow Diagram 11.9**
**In this section some of the methods developed for the prediction of RNA secondary structure are discussed.**

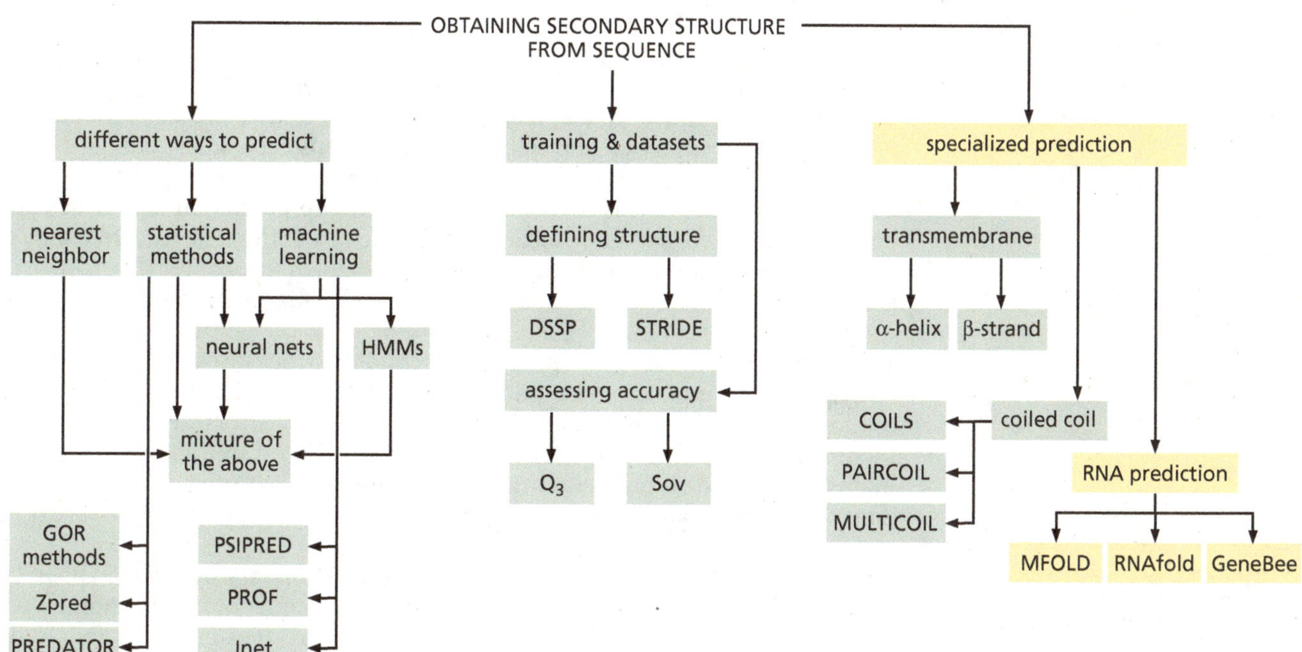

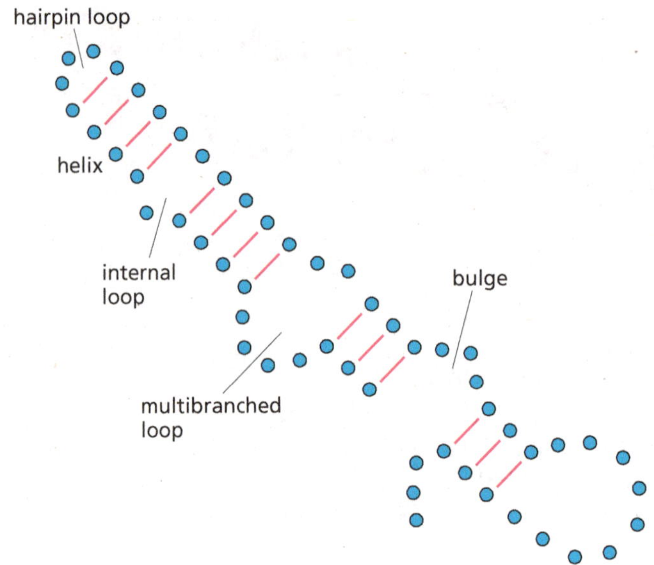


**Figure 11.43**
**Types of secondary structure present in RNA molecules.** Nucleotides are represented by blue circles, with the hydrogen bonding shown as a single red line joining two circles. The structures formed include double helix, internal loops, hairpin loops, bulges, and multibranched loops.

of RNA, and also aid in the automated identification of genes for particular RNAs in genomic DNA (see Section 10.1).

In RNA, most base pairing occurs by classical Watson–Crick pairing between cytosine and guanine by means of three hydrogen bonds and between uracil and adenine via two hydrogen bonds. In addition, a single hydrogen bond can be formed between guanine and uracil. Intrastrand base pairing gives rise to a variety of secondary structures. The most frequent are double helix (which will form when two RNA strands are perfectly base-paired for more than a few nucleotides), stem loop, internal loop, hairpin loop, bulge, multibranched loop (see Figure 11.43), and pseudoknots. Single-stranded DNA can also form these types of secondary structure.

Generally, the likelihood of forming a particular RNA secondary structure can be expressed as the amount of free energy released or used up by forming the base pairs. The greater the free energy the more work is needed to form a specific configuration and, therefore, the less likely it is that such a structure will be formed. As described in Appendices B and C the calculation of molecular free energies is difficult, and in practice usually requires significant assumptions. One of the key assumptions made that is relevant to RNA structure prediction is that base-pair energies can be considered in isolation from the rest of the structure. In reality the energy of a base pair is modified by other base pairs in the vicinity. In practice this interaction is ignored, and the free energies of base pairs are simply added together.

Thus, the total free energy of a particular secondary structure can be determined by adding up the component base-pair energies. The more negative (that is, the lower) the total free energy of a structure, the more likely this structure is to be formed. Identifying a total base-pair configuration with the lowest free energy is the aim of many RNA structure prediction methods. These types of prediction method use empirical energy parameters to compute a free energy (Appendix B).

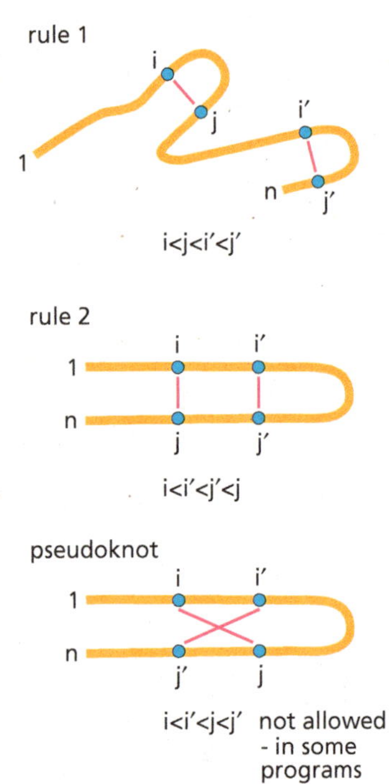


**Figure 11.44**
**The most common rules used in prediction algorithms for RNA structure.** Blue filled circles represent nucleotides at positions i, i′, j, and j′, respectively. The lines joining them represent hydrogen bonding.

Several approaches are often combined to predict secondary structure in RNA. One is to investigate a set of homologous nucleotide sequences. If no homologous sequences are available then a set of constraints and rules has to be followed to help calculate the energy of folding. Some or all of the following basic rules have been used in various programs for the prediction of RNA secondary structure. First, the secondary structure is considered as a collection of base pairs that construct the structure. If i and j, respectively, represent a base (see Figure 11.44) at a specific position in the RNA sequence, then a secondary structure is a set of ordered pairs i.j where the rule that $1 \le i < j \le n$ holds true and the following rules apply (see Figure 11.44).

- *i* and *j* are more than four nucleotides apart in the sequence

If there are two base pairs (i.j and i′.j′) then:

- i.j and i′.j′, base pair i.j comes before i′.j′

or

- the structural element formed by i.j has a base pair in it formed by i′.j′

The last requirement prevents pseudoknots from forming. These can occur with two base pairs and possible crossovers. The rule that disallows pseudoknots has been incorporated because some programs that calculate minimum-energy structures cannot cope with this type of fold. However, pseudoknots do occur and are quite important structural elements in RNA and there are a number of current programs that do allow for the prediction of pseudoknots, such as GeneBee described below.

Most prediction algorithms calculate some form of folding energy. The most basic method would be to construct numbers of different possible structures, assign an energy value for each base pair, sum them up for each structure, and then choose

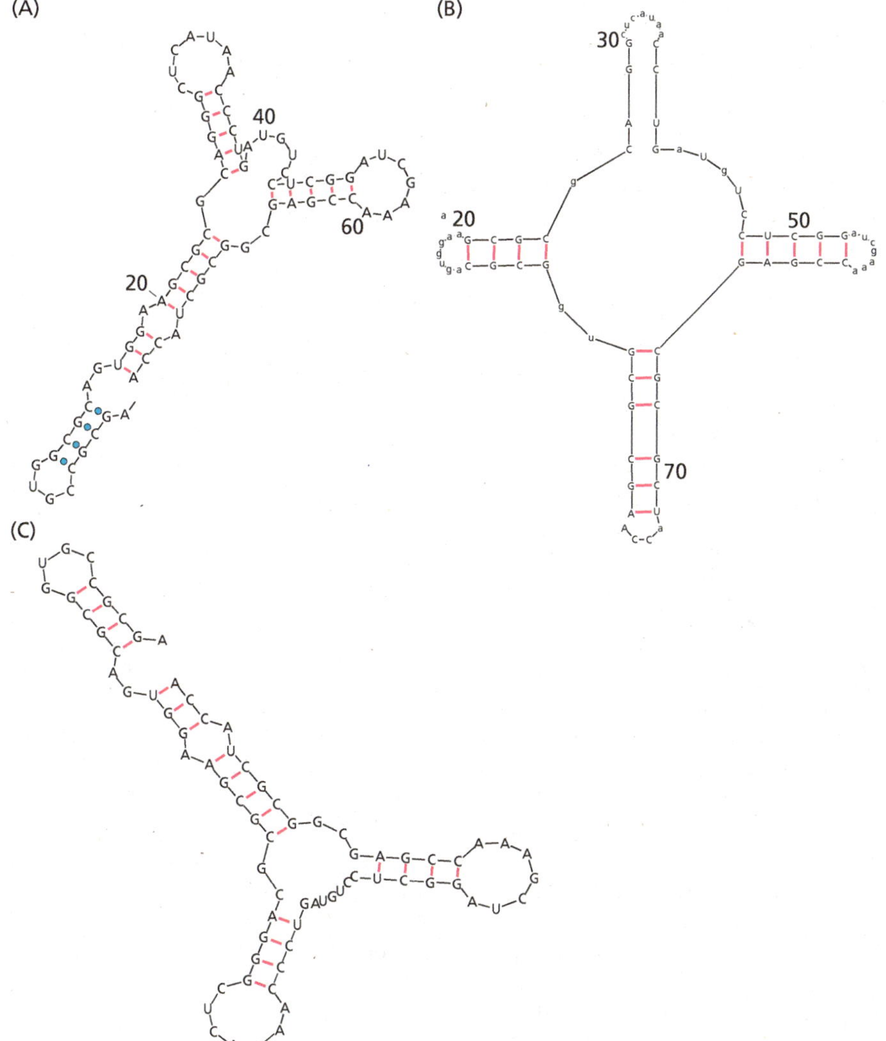


**Figure 11.45**
**Predicted structures for a yeast initiator tRNA.** (A) MFOLD; (B) GeneBee; and (C) RNAfold. Hydrogen bonding in base pairs is shown as a red line. GeneBee also gives the calculated free energy for each separate segment.

the structure with the lowest energy. However, this is not satisfactory and more complicated energy calculations, which incorporate rules based on observation of known RNA structures, are usually used.

The current Web-based RNA prediction servers mostly use energy parameters to calculate a possible structure. All provide a picture of the secondary structure with possible base-pairings and various free-energy estimates. The sequence of yeast initiator tRNA has been submitted to three different Web servers: FOLD, RNAfold, and GeneBee. The predicted results for MFOLD (see Figure 11.45A), GeneBee (see Figure 11.45B) and RNAfold (see Figure 11.45C) were compared to the generic tRNA structure (Figure 10.2). All three correctly predicted the anticodon and T-Ψ-C loops. However, GeneBee also predicted the aminoacyl arm and the D loop in the correct positions. A further set of RNA prediction methods has been proposed based on context-free grammars, which are too complex to discuss here. See Further Reading.

## Summary

Protein secondary structure prediction is a fast and relatively easy method of obtaining preliminary structural information about a protein, and may be the only source of information if no homolog is available with an experimentally determined structure. Prediction methods can be roughly subdivided into statistical, knowledge-based, neural network, and consensus methods. This subdivision illustrates how the methods have developed over time and how new structural information has been incorporated in combination with improvements in computer technology. In this chapter we have discussed methods that include residue statistics with the addition of information from surrounding residues (GOR methods), incorporation of long-range interaction parameters and nearest-neighbor information (PREDATOR), use of information from aligned sequences (Zpred, GOR V), and neural network/HMM technology applied to a variety of secondary structure parameters and information (NNSSP, PHD, PROF, PSIPRED, and Jnet). A large improvement in $Q_3$ accuracy came with the use of multiple alignments in prediction methods. The more advanced methods perform considerably better than methods based on statistical information only. However, all prediction methods fare better with proteins containing all or high α-helical content than those with β-strands. The discrepancy between the all α- and all β- structures has been explained by the fact that the formation of an α-helical fold depends on short-range (local) amino acid interactions while a β-containing fold depends on long-range interactions, which are more difficult to predict.

There is no single best prediction method, and it is therefore best to use a number of different methods, especially those that use multiple sequence information. PSIPRED and the methods found on the NNSSP prediction server form good starting points for secondary structure prediction. Common sense should be applied to prediction results, along with any experimental knowledge about the protein. If a certain fold is suspected or inferred from the function of the protein, that knowledge can be used to choose the prediction closest to the suspected structure. Secondary structure methods can only provide a putative structural arrangement, often to about 75% accuracy, but predictions should be considered as a stepping stone to further experimental or theoretical work.

Also in this chapter we have dealt with the prediction of specialized protein secondary structures found in transmembrane regions of integral membrane proteins and in coiled-coil folds. For such structures, the general secondary structure prediction algorithms are not sufficient. Because of the constraints imposed by the hydrophobic interior of the lipid bilayer, transmembrane regions of proteins are either in the form of hydrophobic α-helices or form an intramembrane β-barrel. No other type of transmembrane secondary structure has been found. At present, only transmembrane helical regions can be predicted with any reliability. Submitting the protein rhodopsin, which has seven transmembrane helices, to various commonly used prediction programs available over the Web shows that the best result is obtained by running a number of programs and choosing a consensus prediction. The differences in predictions between different programs are due to the fact that they use different ways of calculating the hydrophobicity of a sequence—its hydropathic profile—and also use different rules and methods to calculate helical state and conformation. There is at present no best way of calculating the hydropathic profile.

Coiled coils and leucine zippers are formed when two α-helices with particular properties twist around each other. Prediction of sequences that will be able to form coiled coils depends on the recognition of repeating patterns of hydrophobic residues. RNA molecules form secondary structure through base-pairing and the prediction of RNA structure from sequence was also briefly discussed in this chapter.

## Further Reading

### 11.1 Types of Prediction Methods

Rost B & Sander C (2000) Third generation prediction of secondary structures. *Methods Mol. Biol.* 143, 71–95.

### 11.2 Training and Test Databases

**PDB**

Berman HM, Westbrook J, Feng Z et al. (2000) The Protein Data Bank. *Nucleic Acids Res.* 28, 235–242. *Also look at http://www.wwpdb.org*

**STRIDE**

Frishman D & Argos P (1995) Knowledge-based protein secondary structure assignment. *Proteins* 23, 566–579.

**DSSP**

Kabsch W & Sander C (1983) Dictionary of protein secondary structure: pattern recognition of hydrogen-bonded and geometrical features. *Biopolymers* 22, 2577–2637.

**DEFINE**

Richards FM & Kundrot CE (1998) Identification of structural motifs from protein coordinate data: secondary structure and first-level supersecondary structure. *Proteins* 3, 71–84.

### 11.3 Assessing the Accuracy of Prediction Programs

Brylinski M, Konieczny L & Roterman I (2005) SPI – structure predictability index for protein sequences. *In Silico Biol.* 5, 227–237.

### 11.4 Statistical and Knowledge-Based Methods

Chou PY & Fasman GD (1974) Prediction of protein conformation. *Biochemistry* 13, 222–245.

Frishman D & Argos, P (1995) Knowledge-based secondary structure assignment. *Proteins* 23, 566–579.

Frishman D & Argos P (1996) Incorporation of long-distance interactions into a secondary structure prediction algorithm. *Protein Eng.* 9, 133–142.

Frishman D & Argos P (1997) Seventy-five percent accuracy in protein secondary structure prediction. *Proteins* 27, 329–335.

Garnier J, Osguthorpe DJ & Robson B (1978) Analysis of the accuracy and implications of simple methods for predicting the secondary structure of globular proteins. *J. Mol. Biol.* 120, 97–120.

Garnier J, Gibrat JF & Robson B (1996) GOR method for predicting protein secondary structure from amino acid sequence. *Methods Enzymol.* 266, 540–553.

Gibrat JF, Garnier J & Robson B (1987) Further developments of protein secondary structure prediction using information theory. New parameters and consideration of residue pairs. *J. Mol. Biol.* 198, 425–443.

Kloczkowski A, Ting KL, Jernigan RL & Garnier J (2002) Combining the GOR V algorithm with evolutionary information for protein secondary structure prediction from amino acid sequence. *Proteins* 49, 154–166.

Sen TZ, Jernigan RL, Garnier J & Kloczkowski A (2005) GOR V server for protein secondary structure prediction. *Bioinformatics* 21, 2787–2788.

Zvelebil MJ, Barton GJ, Taylor WR & Sternberg MJE (1987) Prediction of protein secondary structure and active site using the alignment of homologous sequences. *J. Mol. Biol.* 195, 957–967.

### 11.5 Neural Network Methods of Secondary Structure Prediction

Cuff JA & Barton GJ (2000) Application of enhanced multiple sequence alignment profiles to improve protein secondary structure prediction. *Proteins* 40, 502–511.

Rost B (1996) PHD: predicting one-dimensional protein structure by profile based neural networks. *Methods Enzymol.* 266, 525–539.

Salamov AA & Solovyev VV (1995) Prediction of protein secondary structure by combining nearest-neighbor algorithms and multiple sequence alignment. *J. Mol. Biol.* 247, 11–15.

## 11.7 Prediction of Transmembrane Protein Structure

Bagos PG, Liakopoulos TD & Hamodrakas SJ (2005) Evaluation of methods for predicting the topology of beta-barrel outer membrane proteins and a consensus prediction method. *BMC Bioinformatics,* 7.

Buchanan SK (1999) β-Barrel proteins from bacterial outer membranes: structure, function and refolding. *Curr. Opin. Struct. Biol.* 9, 455–461.

Cserzö M, Wallin E, Simon I et al. (1997) Prediction of transmembrane a-helices in prokaryotic membrane proteins: the dense alignment surface method. *Protein Eng.* 10, 673–676.

von Heijne G (1992) Membrane protein structure prediction. *J. Mol. Biol.* 225, 487–494.

Hirokawa T, Boon-Chieng S & Mitaku S (1998) SOSUI: classification and secondary structure prediction system for membrane proteins. *Bioinformatics* 14, 378–379.

Hofmann K & Stoffel W (1993) TMBASE – a database of membrane spanning protein segments. *Biol. Chem.* 374, 166.

Jones DT, Taylor WR & Thornton JM (1994) A model recognition approach to the prediction of all-helical membrane protein structure and topology. *Biochemistry* 33, 3038–3049.

Krogh A, Larsson B, von Heijne G & Sonnhammer EL (2001) Predicting transmembrane protein topology with a hidden Markov model: application to complete genomes. *J. Mol. Biol.* 305, 567–580.

Kyte J & Doolittle RF (1982) A simple method for displaying the hydropathic character of a protein. *J. Mol. Biol.* 157, 105–132.

Ng P, Henikoff J & Henikoff S (2000) PHAT: a transmembrane-specific substitution matrix. *Bioinformatics* 16, 760–766.

Rost B, Casadio R, Fariselli P & Sander C (1995) Prediction of helical transmembrane segments at 95% accuracy. *Protein Sci.* 4, 521–533.

Sonnhammer EL, von Heijne G & Krogh A (1998) A hidden Markov model for predicting transmembrane helices in protein sequences. *Proc. Int. Conf. Intell. Syst. Mol. Biol.* 6, 175–182.

Tusnady GE & Simon I (1998) Principles governing amino acid composition of integral membrane proteins: application to topology prediction. *J. Mol. Biol.* 283, 489–506.

Tusnady GE & Simon I (2001) Topology of membrane proteins. *J. Chem. Inf. Comput. Sci.* 41, 364–368.

Tusnady GE & Simon I (2001) The HMMTOP transmembrane topology prediction server. *Bioinformatics* 17, 849–850.

## 11.8 Coiled-Coil Structures

Berger B, Wilson DB, Wolf E et al. (1995) Predicting coiled coils by use of pairwise residue correlations. *Proc. Natl Acad. Sci. USA,* 92, 8259–8263.

Bornberg-Bauer E & Rivals E (1998) Vingron Computational approaches to identify leucine zippers. *Nucleic Acids Res.* 26, 2740–2746.

Lupas A, Van Dyke M & Stock J (1991) Predicting coiled coils from protein sequences. *Science* 252, 1162–1164.

Maesaki R, Ihara K, Shimizu T et al. (1999) The structural basis of Rho effector recognition revealed by the crystal structure of human RhoA complexed with the effector domain of PKN/PRK1. *Mol. Cell* 4, 793–803.

Wolf E, Kim PS & Berger B (1997) MultiCoil: a program for predicting two- and three-stranded coiled coils. *Protein Sci.* 6, 1179–1189.

## 11.9 RNA Secondary Structure Prediction

Brodskii LI, Ivanov VV, Kalaidzidis IaL et al. (1995) GeneBee-NET: Internet-based server for analyzing biopolymers structure. *Biokhimiia* 60, 1221–1230.

Mathews DH, Sabina J, Zucker M & Turner H (1999) Expanded sequence dependence of thermodynamic parameters provides robust prediction of RNA secondary structure. *J. Mol. Biol.* 288, 911–940.

Zuker M (2003) Mfold web server for nucleic acid folding and hybridization prediction. *Nucleic Acids Res.* 31, 3406–3415.

## Box 11.1

Davies G P, Martin I, Sturrock SS et al. (1999) On the structure and operation of type I DNA restriction enzymes. *J. Mol. Biol.* 290, 565–579.

## Box 11.3

Tie JK, Nicchitta C, von Heijne G & Stafford DW (2005) Membrane topology mapping of vitamin K epoxide reductase by in vitro translation/cotranslocation. *J. Biol. Chem.* 280, 16410–16416.

# PREDICTING SECONDARY STRUCTURES

# 12

THEORY CHAPTER

**When you have read Chapter 12, you should be able to:**

- Describe the assignment of protein secondary structure.
- Outline the special measures of prediction accuracy.
- Describe the structural preferences of residues and their modification by local sequence.
- Show how to improve prediction by using information from homologous sequences.
- Explain nearest-neighbor prediction based on local sequence similarity.
- State the basic concepts of neural network architecture and training.
- Show how neural networks are applied to secondary structure prediction.
- Show how HMM methods are applied to transmembrane protein prediction.
- Show how support vector machines predict secondary structure.
- Discuss the prediction of other features such as nonfolded regions and functional sites.

A full understanding of the function of a protein, including the basis for any catalysis, substrate specificity, and any regulation of its action requires knowledge of the protein's three-dimensional structure. This is why so much effort has been invested in the experimental determination of protein structures. Despite these efforts the structure is known for only a very small proportion of protein sequences. Using sequence analysis as described in Chapters 4, 5, and 6, many sequences for which no experimental structure is available can be aligned with ones of known structure. In this way, structures can be proposed based on homology, a process described in detail in Chapter 13. However, there are still many sequences that lack a homolog of known structure. These are the sequences for which secondary structure prediction can be especially useful. Chapter 11 is a practical guide to applications in this area. In this chapter we will present the principles behind the many different methods of protein secondary structure prediction.

The protein fold of globular proteins is described in terms of the packing together of the secondary structural elements described in Section 2.1. Analysis of protein structures reveals regularities, for example α-helices often pack against each other at a limited set of distinct orientations, and the packing of successive secondary structures in protein folds shows a preferred handedness. Evidence has emerged for the existence of a limited set of protein folds, which suggests that if the secondary structural elements of a sequence are known, prediction of the three-dimensional structure might be feasible. (The use of energy calculations to determine protein folds *ab initio*—discussed in Box 13.1—while making progress, is as yet not a possibility except possibly for the smallest of domains.) This use of secondary structures as a step toward protein fold prediction remains a key motivation for the development of secondary structure prediction.

As more structures are determined and other features identified, further prediction methods are being developed. Some of these features are functional, such as sites of phosphorylation. The importance of membrane proteins has been recognized, both in terms of their roles in living systems and in medicine, and predictive methods for transmembrane helices and complete chain topologies have been developed. Another structural feature, initially not anticipated but subsequently recognized as important, is that certain regions of the peptide chain can take up alternative conformations under certain conditions, which can be related to the protein function or regulation, but can also be associated with disease states as in the case of prions. In this chapter we will survey the techniques that have been employed to predict these structural features.

**Mind Map 12.1**
**There are many different types of programs used in predicting the structural features from primary sequence.** This mind map shows the main programs and necessary data used in prediction applications.

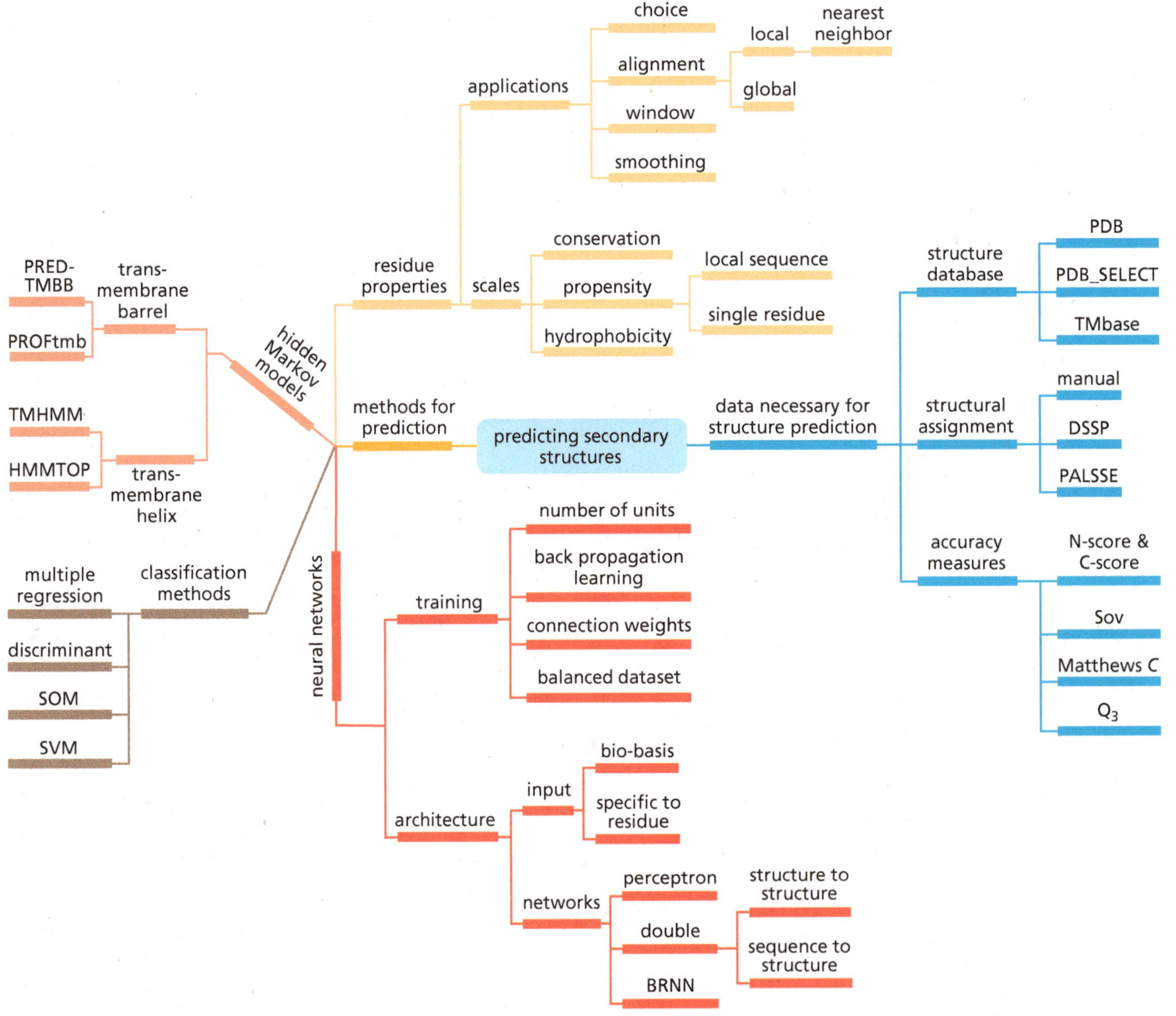

Before presenting any prediction methods it is necessary to define the secondary structural elements. In Section 2.1 they were presented with perfect geometrical regularity, but in globular proteins they are often distorted, and the boundaries between these regular structures and other regions are not easily identified. Several schemes will be described for secondary structure assignment, as will be the consequences of the considerable level of disagreement between them. Following this, the calculation of prediction accuracy will be presented, and will be shown to be definable in several ways.

The next section of this chapter will describe early prediction methods for secondary structures and transmembrane helices. These are based on identifying the structural preferences (propensities) of individual residue types and their modulation by other residues nearby on the chain. It has been found that significant improvements in prediction accuracy are made when account is also taken of homologous sequences. An alternative to this approach is to identify similar short sequence segments in proteins of known structure and to predict the same structure in the query sequence, known as the nearest-neighbor approach.

One of the most successful techniques applied to the secondary structure prediction problem is the use of neural networks. The key aspects of this technique are presented, followed by details of some of the networks in common use. The technique has also been applied to other problems such as β-turns and unfolded regions. Following this, the application of hidden Markov models (HMMs) is presented, which has proved particularly successful at predicting the secondary structures and topologies of membrane proteins. The final part of the chapter looks at other more general techniques of data classification, which can be used in this area to distinguish between sequence segments with and without certain structural properties. Such methods have been used for secondary structure prediction, but also for predicting unfolded regions, phosphorylation sites, and other features.

## 12.1 Defining Secondary Structure and Prediction Accuracy

Most biochemistry textbooks give simple definitions of α-helices and β-strands based on the presence of hydrogen-bonding patterns such as between residue pairs ($i$, $i$+4). In these presentations the structures are usually shown as perfectly regular, with linear strands and helices following a straight axis. However, these secondary structures are usually found to be distorted in globular proteins, both in terms of their axis and the hydrogen-bonding patterns. β-strands in particular are almost always curved, and inserted residues that do not participate in β-sheet interactions (β-bulges) are relatively common. In addition, the divergence from regularity often increases toward the ends of these structures. As a result the identification of all residues that are part of secondary structures in globular protein folds is not trivial. This secondary structure assignment is often regarded as best performed by visual inspection, but this introduces an element of subjectivity and nonreproducibility. To circumvent this requires automated secondary structure identification.

In principle, secondary structures can be defined on the basis of the residue torsion angles, hydrogen-bonding patterns, overall curvature of the polypeptide backbone, energetically favorable interactions, or any combination of these. Based on the original idealized structure definitions, a definition utilizing hydrogen bonding might appear at first to be the most productive approach. However, as already mentioned, the observed distortions can lead to nonidealized bonding patterns. The situation is further aggravated by the lack of a clear-cut definition of hydrogen-bonding interactions. Hydrogen bonds are usually defined in terms of a set of

**Flow Diagram 12.1**
**The key concepts introduced in this section are that irregularities in most protein secondary structures make their exact definition difficult in globular proteins, and that there are several different ways of measuring the accuracy of a secondary structure prediction.**

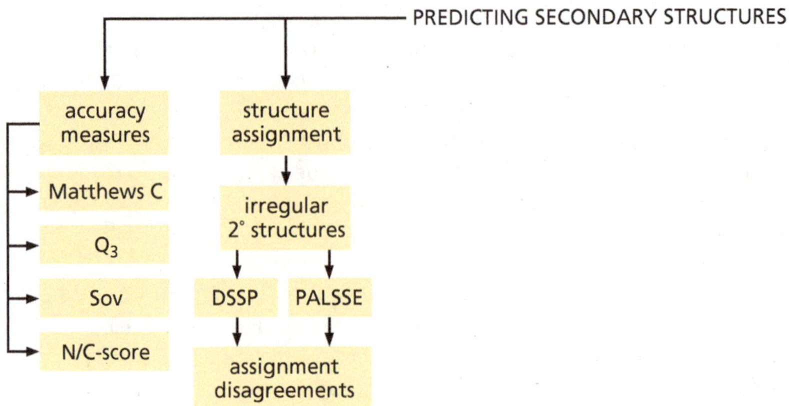


distance and angle constraints, which should include all interactions that have a moderately attractive energy, but in distorted structures they can often occur with geometries outside these constraints, as a consequence of which some potential weak hydrogen bonds may be missed. In this section we will describe some of the approaches used to automate secondary structure definition, and also compare some of the resulting assignments.

The difficulty in defining secondary structures for globular protein folds has consequences when measuring the success of prediction methods. Furthermore, it is often more important to predict the correct number and approximate location of helices and strands than to identify exactly which residues are involved. This is especially the case when trying to identify folds. Some special measures of prediction accuracy have been developed that place more emphasis on the helices and strands than the residues and these are presented at the end of this section.

## The definitions used for automatic protein secondary structure assignment do not give identical results

The original definitions of regular structures in globular protein folds strongly emphasized the hydrogen-bonding patterns involving the main chain groups. These patterns can be identified by defining a hydrogen bond in terms of the geometry of the atoms involved, possibly through calculating the energy of interaction. Typical definitions assign a hydrogen bond only if the atomic coordinates have the O and N atoms within a specified distance range (for example ≤3.5 Å) and if, in addition, the angle between the C=O and N-H bonds falls within a specified range (for example ≤60°). When the definition involves calculating energies, any interaction more favorable than a given cut-off value is designated as a hydrogen bond.

The residues of ideal geometry α-helices have a main chain group hydrogen bonded to another main chain group four residues distant; this is often referred to as an ($i$, $i$+4) interaction. The C=O group of residue $i$ bonds to the N-H group of residue $i$+4. Residues at the start (N-terminal end) of helices only have these bonds from their C=O group while those residues toward the C-terminal end only have these bonds from their N-H group. Residues in the middle of long α-helices have both interactions. The less common $3_{10}$ and π-helices have ($i$, $i$+3) and ($i$, $i$+5) interactions, respectively. Any helix will have a run of consecutive residues with these interactions. Structure assignment methods that use hydrogen-bonding patterns have a minimum number of such consecutive residues (often two), which defines the shortest possible helix they can assign. One of the most commonly used secondary structure assignment methods, DSSP (Define Secondary Structure of Proteins), proposed by Wolfgang Kabsch and Christian Sander in 1983, takes the approach just described to assign helices. However, if hydrogen bonds are identified

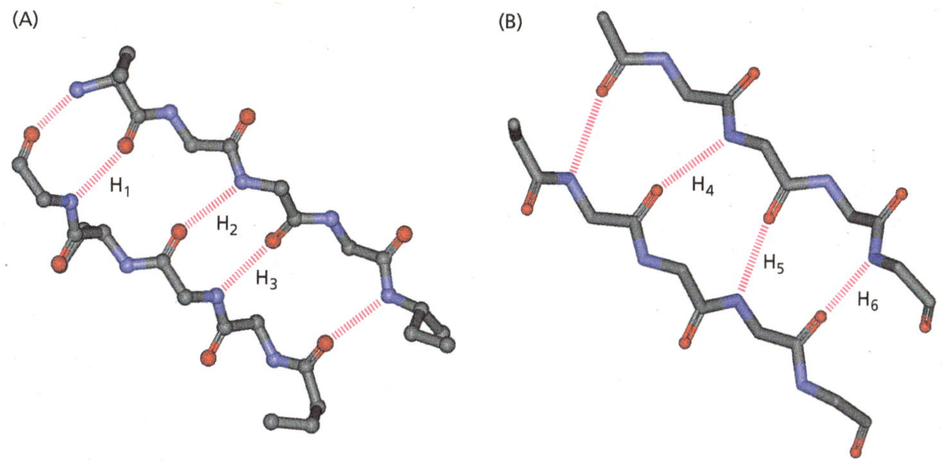


**Figure 12.1**
**The types of hydrogen-bonding patterns used by some methods such as DSSP to identify β-sheets.** (A) Part of an antiparallel β-sheet, showing only nonhydrogen atoms, with hydrogen bonds drawn as dashed magenta lines and labeled $H_1$, $H_2$, and $H_3$. DSSP looks for tripeptides linked together either by $H_1$ and $H_2$, or by $H_2$ and $H_3$, as indicative of antiparallel β-strands. (B) Part of a parallel β-sheet, with hydrogen bonds labeled $H_4$, $H_5$, and $H_6$. DSSP looks for tripeptides linked together either by $H_4$ and $H_5$, or by $H_5$ and $H_6$, as indicative of parallel β-strands.

for residue pairs ($i$–1, $i$+3) and ($i$, $i$+4) these are taken to define a minimal (that is shortest possible) 4-residue α-helix composed of the residues $i$ to $i$+3. The minimal $3_{10}$ and π-helices (three and five residues, respectively) are defined by DSSP in a similar manner. When an isolated pair of residues is found they define a turn, with the classic β-turn involving one ($i$, $i$+2) hydrogen bond.

The identification of β-strands from hydrogen-bonding information proceeds in a similar way to that described for helices, except that now the pair of individual residues is replaced by a pair of sequence segments. These two segments can occur at any separation within the sequence, as β-sheets are often formed by interaction between sequentially remote β-strands. A specific pair of hydrogen bonds is sought between the two segments, there being four different pairings that define parallel and antiparallel strands (see Figure 12.1). In DSSP, pairs of suitably hydrogen-bonded segments are called bridges, and each bridge forms an entity called a ladder together with any consecutive bridges of the same type. The ladders are pairs of β-strands that interact. In a final phase, DSSP joins ladders that share residues (i.e., β-strands) into β-sheets.

The details described above for DSSP do not allow for significant deviation from the idealized structures. To allow for irregularities such as β-bulges the rules were modified, but they still require more regular structure at the edges of the helix or strand. Despite this potential shortcoming DSSP has remained a popular choice for defining secondary structures, partly perhaps because of the wide availability of the program, but also because for a long time there has not been a program that seemed to offer significantly better assignments. It reports seven different structural states with an eighth state (coil) assigned to any residue that is not assigned one of these (see Table 12.1). Typically these states are combined to reduce them to

| Structural state | Description |
|---|---|
| B | β-bridge |
| C | not B, E, G, H, I, S or T (i.e., coil) |
| E | β-strand |
| G | $3_{10}$-helix |
| H | α-helix |
| I | π-helix |
| S | bend |
| T | β-turn |

**Table 12.1**
**The eight structural states reported by the DSSP method, with their descriptions.**

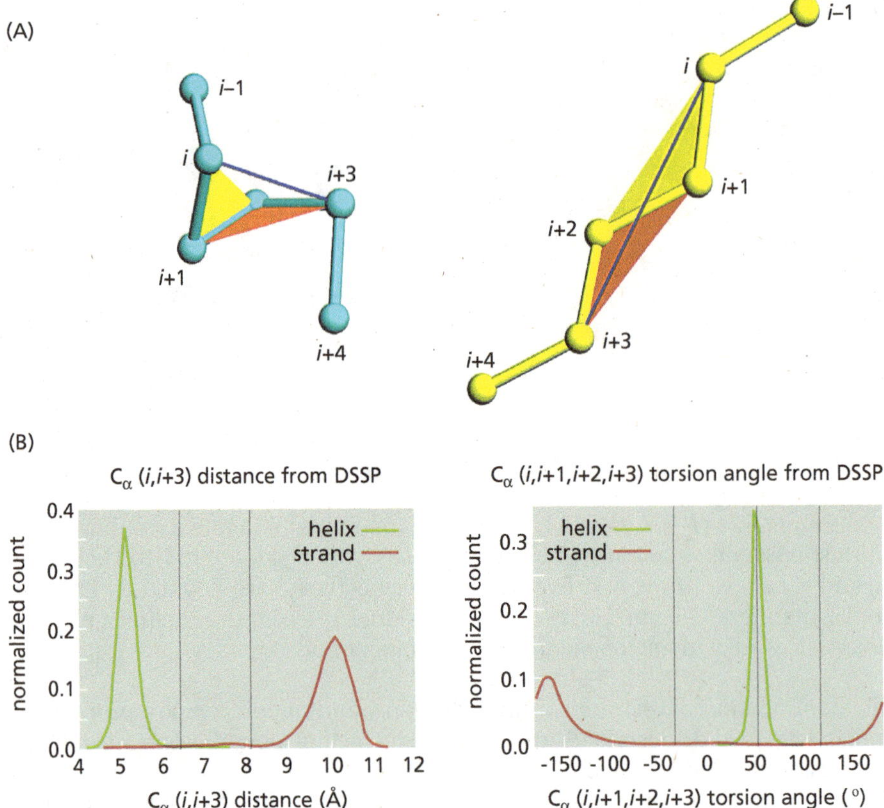


**Figure 12.2**
**Two geometrical parameters used in the PALSSE method to distinguish α-helices and β-strands.** (A) The distance between $C_\alpha$ atoms of residues *i* and *i*+3 shows distinct ranges for the two structural states. The distributions are shown for pairs of residues (*i*, *i*+3) that have been assigned as in an α-helix (green) or β-strand (red). (B) The torsion angle for the $C_\alpha$ atoms of the four residues *i* to *i*+3 also shows distinct ranges. Note that in these graphs the structural states of the residues were defined using the DSSP method. The DSSP method is relatively cautious in assignment and these measures may not be as discriminating if other assignments are used. (From I. Majumdar, S. Sri Krishna and N.V. Grishia, PALSSE: a program to delineate linear secondary structural elements from protein structures, *BMC Bioinformatics* 6:202, 2005.)

the three states normally predicted: helix (H), strand (E), and coil (C). There are several ways of making this reduction, the most common being to assign any DSSP G or H state residue to state H; any DSSP B or E state residue to state E; and all other DSSP state residues to state C. An alternative is to assign any DSSP G, H, or I state to state H, thereby including all identified $3_{10}$, α-, and π-helix residues. It is also possible to make more complex reassignments, for example removing very short helices and strands by defining the residues of any two-residue strand or four-residue α-helix as coil state C.

As was mentioned in the introduction to this section, there are many ways in which secondary structures can be identified without using hydrogen bonding, mostly involving analysis of the geometry defined by the $C_\alpha$ atoms. Two commonly used measurements based on the coordinates of the $C_\alpha$ atoms are the distances between residue pairs such as (*i*, *i*+3) and the torsion angles defined by four successive $C_\alpha$ atoms. Many such measures have the potential to discriminate between α-helices and β-strands (see Figure 12.2). Individual helices and strands are built up from consecutive residues in a similar manner to that described above for hydrogen bond methods. The hydrogen bonding between β-strands is often replaced by requirements of close approach of the $C_\alpha$ atoms of the two strands. A number of detailed protocols have been proposed, most involving several steps to try to overcome the difficulties of the distortions from ideal geometry. The most recent of these are PALSSE (Predictive Assignment of Linear Secondary Structure Elements) by Nick Grishin and co-workers, and β-Spider by Marc Parisien and Francois Major.

Numerous secondary structure assignment methods assign approximately 50–60% of all residues in globular protein structures as belonging to an α-helix and β-strand (see Figure 11.4). (This relates to a subset of known structures designed to remove the known bias in the current structure databases due to the presence of many homologs for some protein families.) The different methods are in better agreement

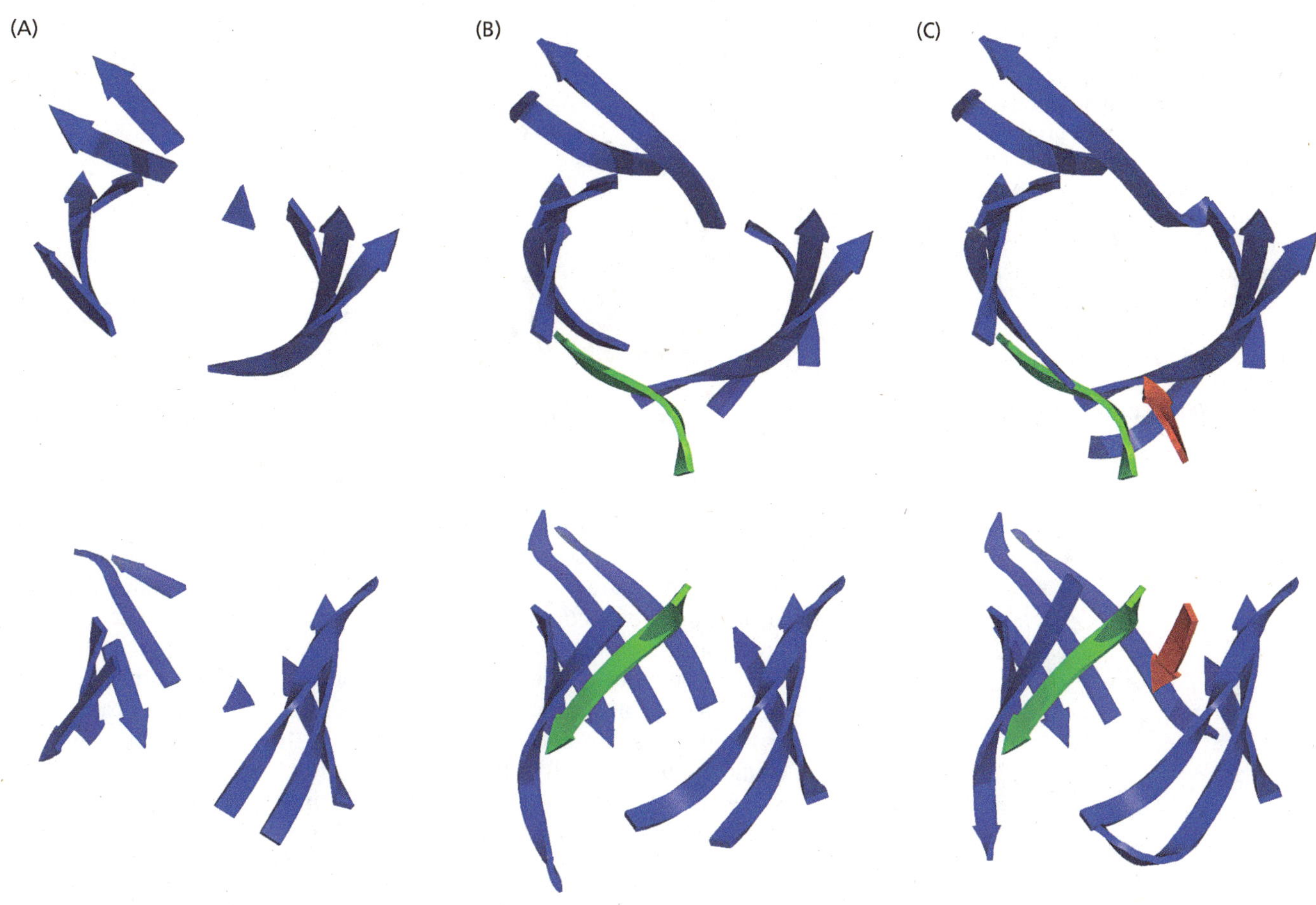


on the fraction of α-helical residues, often close to 30%, and less consistent in the β-strand content. It is noticeable in a number of cases that some methods such as DSSP fail to identify all the β-strands (see Figure 12.3). This may indicate that these methods have been designed to err on the side of caution rather than overassign secondary structural elements. Recently methods have appeared that assign significantly more residues as α-helix and β-strand; for example PALSSE assigns almost 80% of all residues as belonging to a secondary structure element. This can lead to identifying the full set of secondary structural elements (see Figure 12.3) but at the risk of including some spurious elements. From the perspective of fold identification this might be preferable to missing elements, but the methods have not yet been properly assessed.

Figure 12.4 shows the distribution of lengths of α-helices and β-strands in nonmembrane globular proteins as given by the DSSP assignment modified to have a minimum of five and three residues, respectively. The data were taken from a bias-free subset of structures, and show that most β-strands are rather short, few exceeding 10 residues in length. In contrast α-helices are on average about 10 residues long, and occasionally exceed 20 residues. Although the different assignment methods mostly identify the same secondary structures, they often do not agree on the initial and final residues for each helix and strand. (For this comparative discussion we will not include the recent methods that assign about 80% of the sequence to secondary structures, since as shown in Figure 12.3 they clearly identify additional elements.) For α-helices this translates into only 70–90% of the helical residues being consistently so assigned. In the case of β-strands, their shorter length results in only 50–60% of strand residues being consistently assigned. There has been a general perception that β-strands are hard to identify, as is illustrated by the example shown in Figure 12.3.

**Figure 12.3**
**Alternative β-strand assignment for the β-barrel in *E. coli* pyruvate formate-lyase (PDB entry 1H16).** Assignments have been made using (A) DSSP, (B) PALSSE, and (C) β-Spider. The protein is shown in two orthogonal orientations. DSSP misses two β-strands of the barrel, one of which (colored green) is identified by PALSSE, and both of which are assigned by β-Spider. Note that the lengths of equivalent β-strands vary considerably according to the method used.

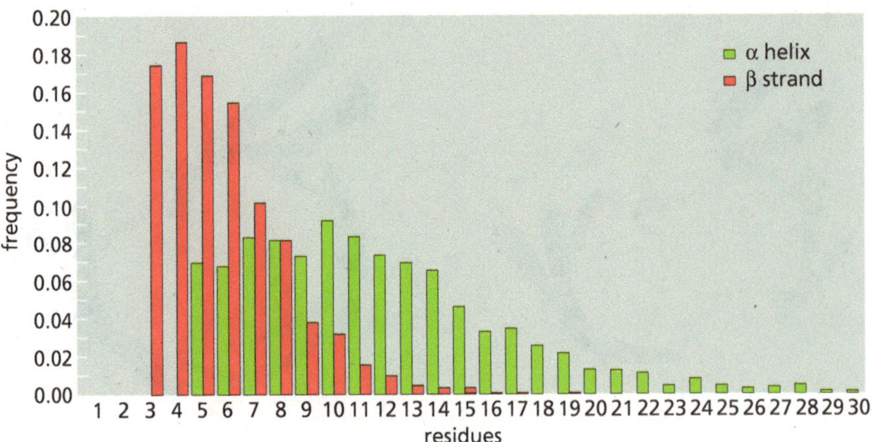


**Figure 12.4**
**The distribution of lengths of α-helices (green) and β-strands (red) in soluble globular proteins, as determined from experimental structures.** The secondary structure definitions used were those of DSSP, but taking the minimum length of an α-helix to be five residues, and of a β-strand to be three residues. (Data taken from S.C. Schmidler et al., Bayesian segmentation of protein secondary structure, *J. Comput. Biol.* 7:233–248, 2000.)

The lack of consistent assignments is one of the reasons why prediction methods should not be expected to achieve 100% accuracy, and indeed would suggest that any such method would be unduly biased to a particular assignment method. Another reason for lowering our expectations of prediction accuracy is the observation that while folds are strongly conserved over long periods of evolution the precise locations of the secondary structural elements have been found to shift slightly (see Figure 11.6). However, it is possible that in many cases these shifts are more indicative of the difficulty of structural assignment than true changes in the fold. Revisiting this topic with a less cautious method such as PALSSE might result in different conclusions. In particular, since the average PALSSE β-strand length is probably greater in this method than for DSSP, whose length distribution is shown in Figure 12.4, the proportion of consistently assigned residues will probably be increased when comparing similarly high percentage assigning methods. However, this will require retraining of the prediction programs based on the assignments from these newer methods.

Secondary structures in membrane proteins are also distorted and therefore subject to the same difficulties in assignment as have already been discussed. The same assignment methods have been applied to them, and Figure 12.5 shows the length distribution for transmembrane helices as assigned using DSSP. Note that they are usually much longer than the α-helices found in soluble globular proteins, which makes their prediction easier.

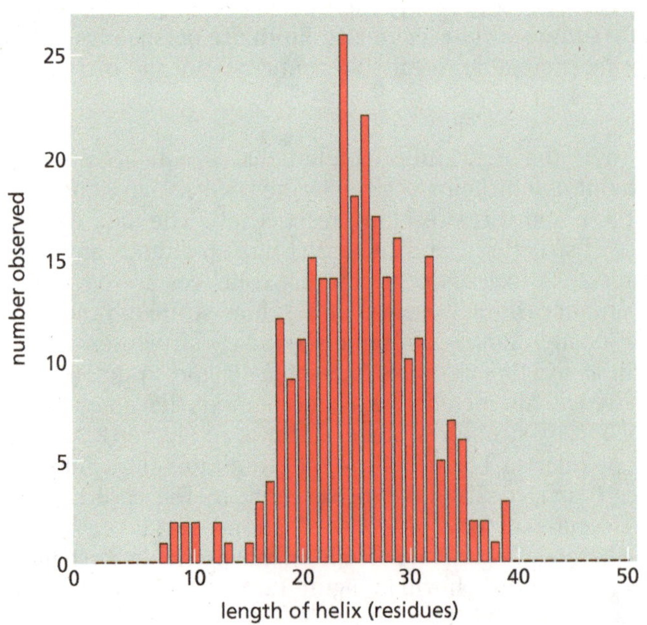


**Figure 12.5**
**The distribution of transmembrane helix lengths as found using the DSSP definition.** These data are for a nonredundant structure dataset containing 268 helices. The average length is 25.3 residues with a standard deviation of 5.9 residues. Note the existence of a small number of shorter helices, which cannot span the membrane. These half-TM often occur in pairs, when they are often found to be full-length helices with a localized distortion to nonhelical structure. (Adapted from J.M. Cuthbertson, D.A. Doyle and M.S.P. Sansom, Transmembrane helix prediction: a comparative evaluation and analysis, *Protein Eng. Des. Sel.* 18:295–308, 2005.)

## There are several different measures of the accuracy of secondary structure prediction

It is necessary to assess the accuracy of a set of predictions to determine the most accurate method available and also to identify successful improvements when developing and fitting parameters for new methods. Alternative measures have been proposed, some of which look at the level of individual residues while others focus on complete helices and strands. (This situation is analogous to that for exon/intron prediction discussed in Box 10.3.) In general the measures are only meaningful if they refer to the prediction of a large number of nonhomologous structures because the methods are intended to be widely applicable.

The most commonly used residue measure is $Q_3$, which is described in Section 11.3, and is defined as the fraction of all the residues that are correctly predicted, considering all possible states including nonprediction states such as coil, and is often quoted as a percentage varying from 0% to 100%. A completely random prediction of a dataset of structures with equal amounts of three possible states would give a $Q_3$ value of 33%. Note that the $Q_3$ measure can be used with any number of possible states; in the case of transmembrane helix predictions there are often only two predicted states but $Q_3$ is still an appropriate measure. As a consequence of the lack of consistent secondary structure assignment mentioned previously, assuming the prediction method is not biased toward a particular assignment method $Q_3$ values much above about 80% cannot be expected when calculated for a large set of structures. Individual structures will have a greater range of values. The variation in $Q_3$ obtained for a range of proteins in a test database using the PSIPRED method is given in Figure 12.6A. This clearly shows the range of accuracy achieved, so that the reported value of 76% for the average $Q_3$ of the method does not guarantee a prediction of this quality for all sequences. In this case the standard deviation of the $Q_3$ values is 7.8%. Often only the average $Q_3$ for the whole test dataset is quoted as the accuracy of the method. When the reference secondary structures have been assigned using a cautious method, such as DSSP, almost half the residues can be in the coil state. In this case, overprediction of coil will give a noticeably larger $Q_3$ than overprediction of helix or strand. Hence attention must be paid to the properties of the set of proteins used to determine the average $Q_3$. The measure is not without problems, as illustrated in Figure 11.5.

Another measure frequently used, often referred to as the **Matthews correlation coefficient** *C* (also written MCC), is defined for each predicted state, for example helix, as

$$C_{helix} = \frac{TP.TN - FN.FP}{\sqrt{(TP+FP)(TP+FN)(TN+FP)(TN+FN)}} \quad \text{(EQ12.1)}$$

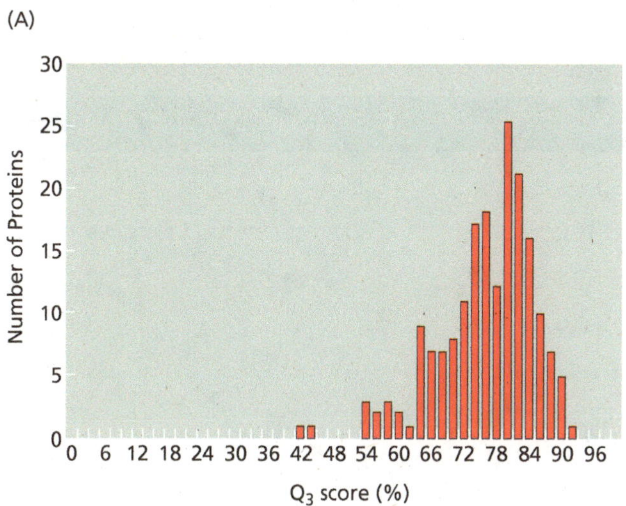


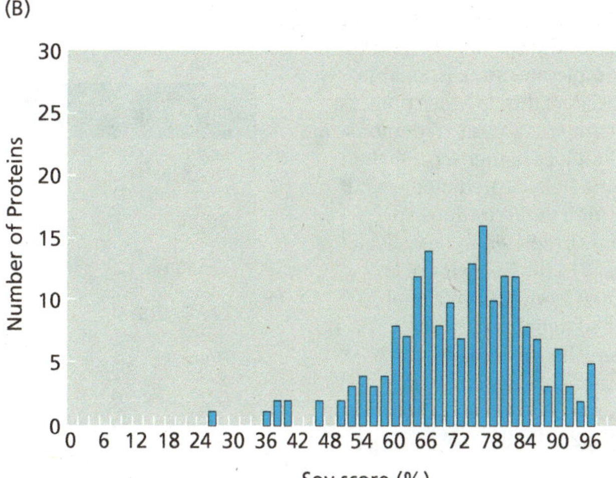


**Figure 12.6**
**The accuracy of the PSIPRED secondary structure prediction on a dataset of 187 proteins, as given by the $Q_3$ and Sov measures.** (A) The distribution of the $Q_3$ values for individual proteins, which has a mean of 76.0% and a standard deviation of 7.8%. (B) The distribution of the Sov values for individual proteins, which has a mean of 73.5% and a standard deviation of 12.7%. (Adapted from D.T. Jones, Protein secondary structure prediction based on position-specific scoring matrices, *J. Mol. Biol.* 292:195–202, 1999.)

where *TP* is the number of true positive residues, i.e., the number of residues correctly predicted as helical, as opposed to the number of residues incorrectly predicted to be helical, which is *FP*, the number of false positives. *TN* (true negative) and *FN* (false negative) are the equivalents for nonhelical residues. The four terms in the square root are thus the total number of residues predicted as helical, the number of actual helical residues, the number of actual nonhelical residues, and the total number of residues predicted as nonhelical, respectively. This gives a more balanced weighting of the prediction outcomes, and has values in the range –1 to 1, with the latter being a perfect prediction. A completely random prediction would give a $C_{helix}$ value of 0, with a value of –1 corresponding to the exact prediction of all helical residues as nonhelical and vice versa.

It would be much more beneficial to predict the correct number, type, and order of secondary structure elements than to predict some elements well and miss others completely. This is because it would give us a better opportunity to determine the protein fold and hence overall structure. The $Q_3$ and *C* measures do not address this. Burkhard Rost and colleagues proposed a measure based on the fractional overlap of segments (Sov). This is sensitive to the amount of overlap between predicted and observed structures, but is designed to try to allow for the observed variation in element boundaries which was discussed above. Like $Q_3$, Sov is defined on a percentage scale and treats both the observed and predicted structures equally.

If all observed segments of helices, strands, and coil are labeled $s_{obs}$ and all predicted segments of these states $s_{pred}$, then $S_o$ is the set of all overlapping pairs of $s_{obs}$ and $s_{pred}$ where the segments are the same state. (Note that if, for example, several predicted helices overlap with one observed one, each overlap will result in a pair in $S_o$ and each of these will contain the observed helix.) The set $S_n$ is defined to include any segments $s_{obs}$ that are not overlapped by a predicted segment of the same state, and therefore do not appear in the set $S_o$. The length in residues of any segment $s_{obs}$ is given by len($s_{obs}$). For any segment pair in $S_o$ of a given state, the length of the actual overlap is called minov($s_{obs}$,$s_{pred}$) and the total extent for which at least one residue is that state is called maxov($s_{obs}$,$s_{pred}$).

Given these definitions the segment overlap measure is defined as

$$\text{Sov} = \frac{100}{N_{\text{Sov}}} \sum_{S_o} \left[ \frac{\text{minov}(s_{obs}, s_{pred}) + \delta(s_{obs}, s_{pred})}{\text{maxov}(s_{obs}, s_{pred})} \text{len}(s_{obs}) \right] \tag{EQ12.2}$$

The terms in the summation represent the fraction of the full extent of the segment pair that the observed and predicted states agree, augmented by a factor $\delta(s_{obs},s_{pred})$ to allow for some variation in segment boundaries as observed in protein structures. $\delta(s_{obs},s_{pred})$ is defined as

**Table 12.2**
**A comparison of two prediction accuracy measures, $Q_3$ and Sov.** Five possible predictions of a single observed helix are shown. Predictions 1 and 2 have the same number of helical residues, and yet 1 is completely unrealistic. Although Sov registers this, $Q_3$ does not. Predictions 2, 3, 4, and 5 demonstrate that Sov gives higher scores when only one helix is predicted even if fewer individual residues are correctly predicted to be helical, contrary to the behavior of $Q_3$. (Adapted from A. Zemla et al., A modified definition of Sov, a segment-based measure for protein secondary structure prediction assessment, *Proteins* 34:220–223, 1999.)

| | | Sov | $Q_3$ |
|---|---|---|---|
| Observed | CHHHHHHHHHHC | | |
| Prediction 1 | CHCHCHCHCHCC | 12.5 | 58.3 |
| Prediction 2 | CCCHHHHHCCCC | 63.2 | 58.3 |
| Prediction 3 | CHHHCHHHCHHC | 40.6 | 83.3 |
| Prediction 4 | CHHCCHHHHHCC | 52.3 | 75.0 |
| Prediction 5 | CCCHHHHHHCCC | 80.6 | 66.7 |

$$\delta(s_{obs}, s_{pred}) = \min \begin{cases} (\text{maxov}(s_{obs}, s_{pred}) - \text{minov}(s_{obs}, s_{pred})) \\ \text{minov}(s_{obs}, s_{pred}) \\ \text{int}[\text{len}(s_{obs})/2] \\ \text{int}[\text{len}(s_{pred})/2] \end{cases} \qquad \text{(EQ12.3)}$$

where int[$x$] means the integer truncation of $x$, i.e., int[3.9] = 3. The summation terms are also weighted by the observed segment length. $N_{Sov}$ is the sum of the number of observed residues (i.e., in $s_{obs}$) in all overlap pairs in $S_o$ (including duplications if present) plus the number of residues in the segments of $S_n$.

Table 12.2 shows some example pairs of observed and predicted α-helices together with their Sov and $Q_3$ scores, showing that only predictions which have

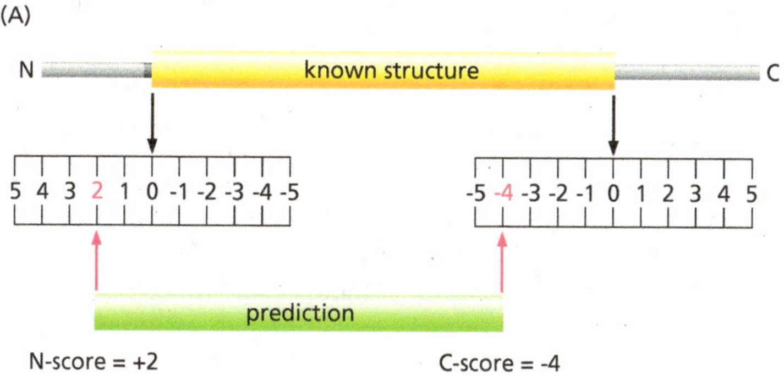


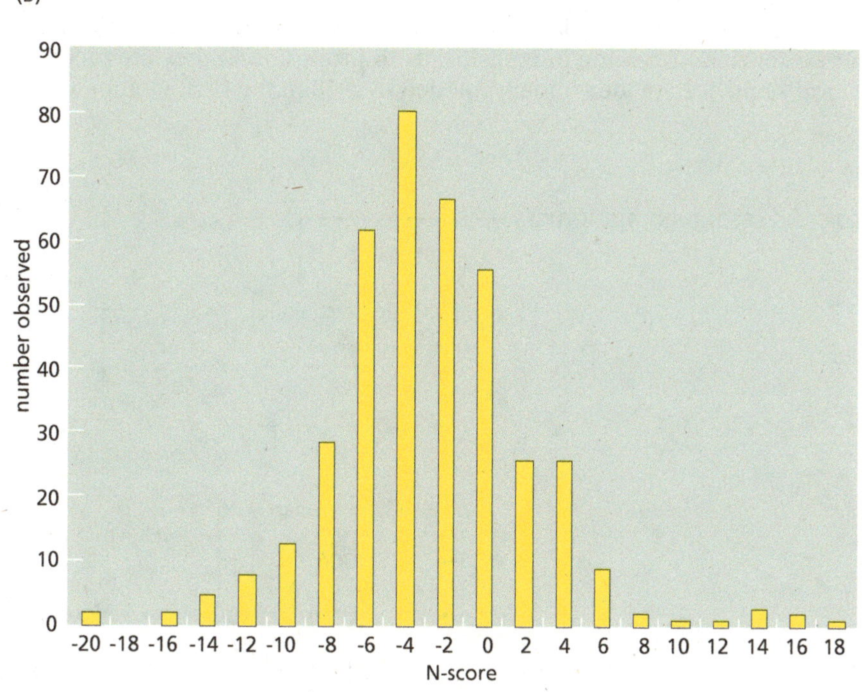


**Figure 12.7**
**A proposed measure of the accuracy of transmembrane helix prediction based on the location of the helix ends.** (A) Definition of the accuracy measure in terms of the incorrect location of the helix start and end. The score is the sum of the absolute values of the N-score and C-score. (B) Application of the measure to TMHMM2 predictions. The N-score is plotted, indicating that there is a tendency to mispredict the start of the helix as several residues toward the C terminal relative to its correct position. (Adapted from J.M. Cuthbertson, D.A. Doyle and M.S.P. Sansom, Transmembrane helix prediction: a comparative evaluation and analysis, *Protein Eng. Des. Sel.* 18:295–308, 2005.)

the right number of helices in approximately the right locations have a high Sov score, whereas the $Q_3$ measure is much more forgiving of such errors. Figure 12.6B shows the range of Sov values obtained with the PSIPRED method. Comparing these with the $Q_3$ equivalents shows that there are more Sov scores below 60%, although in this case the average value of 73.5% is only slightly lower than the $Q_3$ average.

In the field of transmembrane helix prediction the $Q_3$ measure is often used, but there are also some specific measures at the level of the helices rather than the residues, in some ways analogous to Sov. The number of correctly predicted helices is sometimes defined, usually as any predicted helix that has an overlap with an observed helix of a minimum number of residues. The number of residues required is often quite small relative to the size of transmembrane helices (see Figure 12.5) with values of 3 and 5 common. A recent study used the errors in residue numbers of predictions of the helix termini (see Figure 12.7) to examine the accuracy of methods. Figure 12.7B reveals that the prediction method used (TMHMM2) has a tendency to miss the first turn or so of the helix.

## 12.2 Secondary Structure Prediction Based on Residue Propensities

As soon as a few protein structures had been determined to a resolution that permitted the location of individual residues and identification of the secondary structural elements, it became clear that there were some correlations between these two aspects. For example, glycine residues were often found in the tight turns now referred to as β-turns; proline residues were clearly disfavored in α-helices but frequently found in β-turns. These compositional preferences were subsequently quantified as residue propensities, and can be assigned for any identifiable structural features as will be shown below.

**Flow Diagram 12.2**
**The key concept introduced in this section is that amino acids have structural preferences, which can be combined in various ways to design a secondary structure prediction method.**

The residue propensities immediately suggest a possible way to predict structural elements based on the local compositional biases, a concept first formulated into a practical method by Peter Chou and Gerald Fasman in 1974 to predict α-helices and β-strands. The principle has since been applied to many other structural states, including those associated with membrane protein secondary structure. A logical extension of residue propensities is to propose that they are modified by the local sequence, an idea initially applied in 1978 in the GOR method that gave

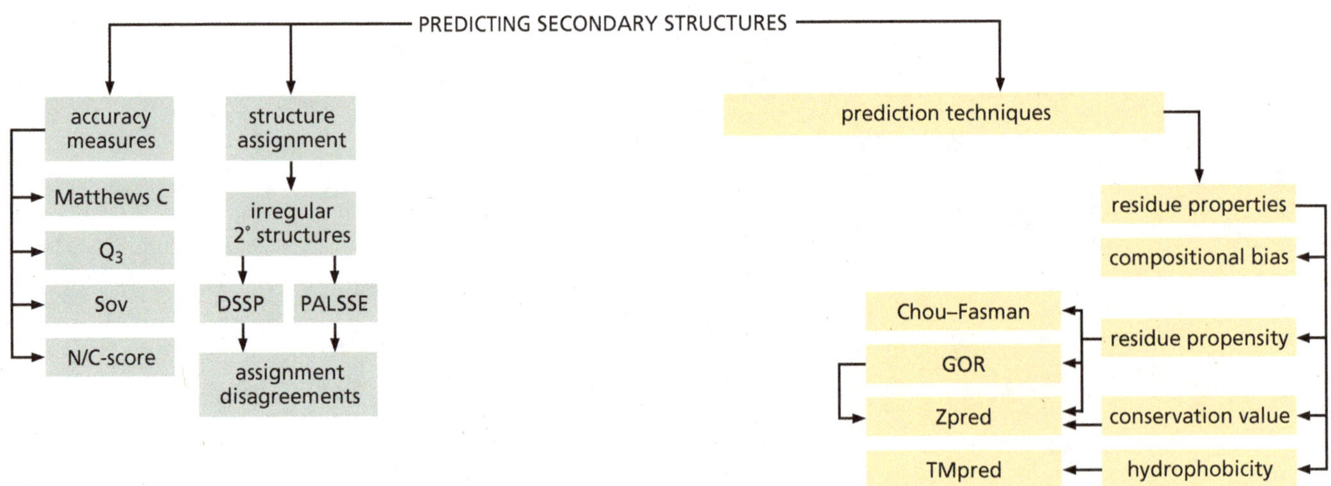

significantly more accurate predictions than the Chou–Fasman method and will be described in detail. This was related to the observation that sufficiently long identical sequence segments had only been observed in a single conformation. However, exceptions to this have now been found and some effort has been spent to quantify the degree to which local sequence alone determines the secondary structure. As these methods were being developed and refined there was a great increase in the number of known sequences and structures, and in particular it was found that use could be made of the sequences of known and putative homologs to make further significant improvements in prediction accuracy.

Many of the methods discussed in this section are rarely used at present, as more complex techniques described later in this chapter are thought to be more accurate. However, many of these other methods have features that are closely related to those presented here, so that it is instructive to compare them. In addition it is by no means certain that further work on propensities will not result in new methods of higher accuracy, for example the update in 2002 of the GOR method to use homologous sequence data as described later in this section.

## Each structural state has an amino acid preference which can be assigned as a residue propensity

The calculation of amino acid preferences for a given structural state is dependent on being able to unambiguously define the structural state for each residue in a protein structure. The problem of structural assignment has already been discussed at length in Section 12.1, so we will restrict mention of it here to the observation that any set of propensities is reliant to some degree on the choice of assignment method. However, of more critical concern is the set of protein structures used to derive the propensities, because it is important that they have minimal compositional and structural biases that might contaminate the propensities we are trying to measure. Ideally the set of structures will span the range of known protein folds so as to reproduce the fold composition of the protein universe while also avoiding any structures that are close homologs. There are still protein folds for which no experimental structure has yet been determined, leading to an unavoidable structural bias, but action can be taken to reduce the bias present in the set of known structures. One attempt to derive such datasets, called PDB_SELECT, uses sequence similarity as the basis for selecting the subset of structures. In this case all structures to be included in the unbiased subset are required to have a sequence identity less than a threshold value (typically 25%) for any aligned sub-sequences longer than 80 residues. The strong bias in known structures is shown by the observation that the fraction of them included in the subset is 1–4% depending on the chosen threshold. However, the resulting database subset is not bias-free, with for example many immunoglobulin fold domains. Another approach to this problem is to use the fold libraries as guides to the redundancy problem. For a brief presentation on these libraries see Section 14.1. For further insight into this problem see Further Reading.

Given a dataset of protein structures with assigned structural states, the derivation of residue propensities is relatively straightforward because there are now sufficient experimental structures to obtain values within the accuracy that is likely to be required. Possible exceptions to this might occur if many alternative structural states are assigned, or for rare states such as $\pi$-helices, and especially when studying membrane proteins. If there is a problem of insufficient data the techniques described in Section 6.1 can be applied in an attempt to minimize the effects. The following derivation will assume there are sufficient data available.

Firstly the fraction of all the dataset residues that is of type $a$ is calculated, written $p_a$ as in Chapters 5 and 6. Then, the same residue type $a$ fraction is calculated but for the subset of all residues that are of structural type $s$, written $p_{s,a}$. The propensity of residues of type $a$ to occur in proteins in structural type $s$ is then given by

$$P_{s,a} = \frac{p_{s,a}}{p_a} \qquad \text{(EQ12.4)}$$

These parameters were first calculated for α-helices and β-strands by Peter Chou and Gerald Fasman in 1974, at which time there were insufficient data (less than 2500 residues) to accurately determine the values. As well as obtaining numerical values for the propensities, each residue type was classified as a strong or weak helix former or breaker or indifferent and the equivalents for strands (see Figure 11.3). The propensities were recalculated several times as more data became available, in 1998 using a dataset of over 33,000 residues (see Figure 12.8). There are some notable differences between the 1974 and 1998 values, such as $P_{\alpha,Arg}$ and $P_{\beta,Met}$, which might be due to the lack of data or that the 1974 data showed strong biases; two of the 15 proteins analyzed were hemoglobin and myoglobin, which have a common fold.

By 2004 the protein structure datasets were large enough to derive residue propensities at different positions within α-helices. The motivation for this was to try to improve the prediction of the ends of α-helices. Even after removing any proteins with >25% sequence residue identity there were over 8000 α-helices for analysis. The propensities were calculated using Equation EQ12.4, for example dividing the fraction of helices that start with a glycine residue by the fraction of glycine residues in the complete database. Many residue types show a clear periodic variation in propensity. Figure 12.9 gives the position-dependent helix propensities at the N-terminal end of helices for the aromatic residues and an averaged ln(propensity)

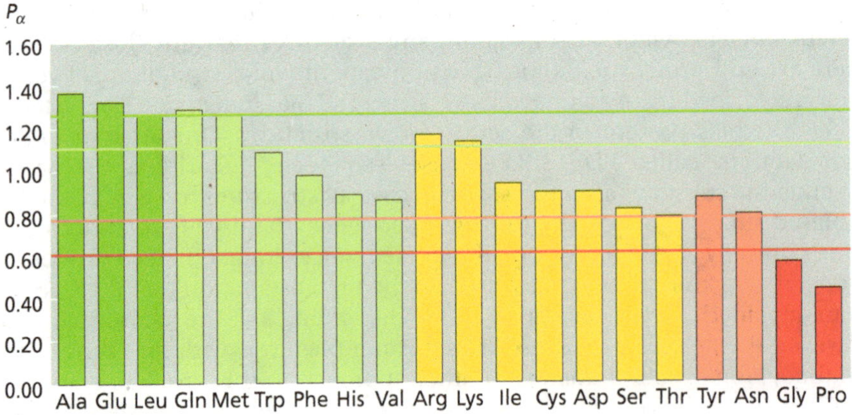


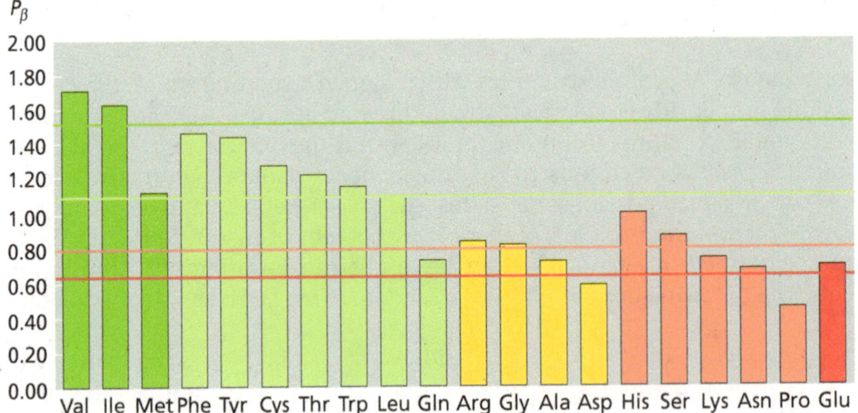


**Figure 12.8**
**The Chou–Fasman α-helix and β-strand propensities, as calculated in 1998 using data on over 33,000 residues.** The residues had previously been classified by Chou and Fasman in 1974 on the basis of their propensities as strong formers, formers, indifferent, breakers, and strong breakers of the secondary structure, colored dark green, light green, yellow, light red, and dark red, respectively. The colored lines indicate the 1974 values of the division between these classes. Although the two sets of values are strongly correlated there are many variations in the detail, such as $P_{\beta,Asp}$.

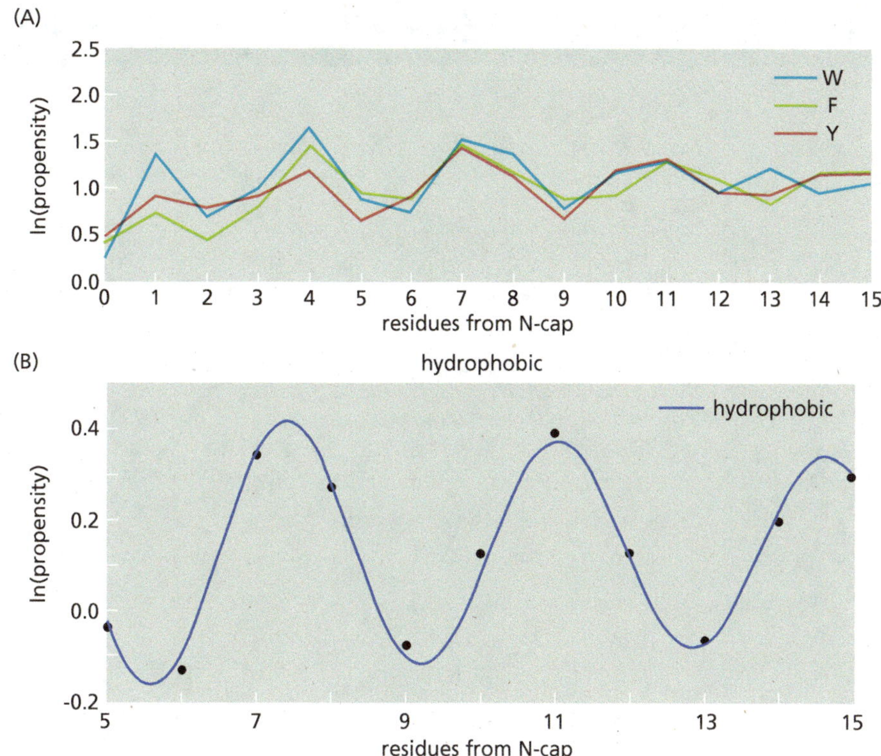


**Figure 12.9**
**The variation of Chou–Fasman-type α-helix propensity with distance from the N-terminal end of the α-helix.** The N-cap is the residue immediately before the first residue of the α-helix, and is assigned the residue number 0. (A) The position-specific propensities for aromatic residues (identified by their one-letter code) to be located in the first 15 residues of an α-helix. A clear periodicity can be seen. (B) The periodicities averaged over all hydrophobic residues show a preference for hydrophobic residues to occur on the same side of the helix as the first residue. In this case ln(propensity) has been plotted, which shows the periodicity more clearly. (Reprinted from *J. Mol. Biol.*, 337, D.E. Engel and W.F. DeGrado, Amino acid propensities are position-dependent throughout the length of α-helices, 1195–1205, 2004, with permission from Elsevier.)

for all hydrophobic residues, showing a clear period of 3.6 residues. This variation is thought to have its origins in the many α-helices that have one surface buried in the protein core and the other exposed to the solvent.

Residue propensities were derived for β-turns, and the Chou–Fasman predictions usually included them as well as the helices and strands. The four different positions were distinguished, usually labeled $i$ to $i$+3, with proline the most common residue found at position $i$+1 and glycine the most common at $i$+2 and $i$+3. One of the most accurate determinations of these residue propensities was by Gail Hutchinson and Janet Thornton in 1994 based on almost 4000 β-turns (see Table 12.3). Further work by Patrick Fuchs and Alain Alix has defined propensities for different types of β-turns, and includes values for residues at positions flanking the turn.

Early methods for predicting transmembrane helices did not use transmembrane helix residue propensities, but preferred residue hydrophobicity scales. Since the interior of membranes is very hydrophobic it was expected that transmembrane helices would have stretches of hydrophobic residues, and furthermore that hydrophobic residues would prefer to be in the membrane than in the aqueous environment. The hydrophobicity scales that developed were usually based on physical chemistry experiments such as the partition between water and octanol. Many such scales have been proposed, from which two relatively popular ones (KD and GES) are given in Table 12.4. An additional scale is given that was obtained by optimizing the scale for prediction purposes. This last scale does not therefore have a direct experimental basis, but has proved significantly more accurate for transmembrane helix prediction.

The lack of experimental membrane protein structures hindered the derivation of transmembrane helix propensities for many years. Despite this there was considerable experimental evidence for the location of some transmembrane helices, data that were collected in annotated sequence databases. This information was used in 1994 to derive propensities for the prediction program MEMSAT. Because

**Table 12.3**
**β-turn propensities.** Data from Hutchinson & Thornton (1994).

| Residue | *i* | *i*+1 | *i*+2 | *i*+3 |
|---|---|---|---|---|
| Ala | 0.81 | 0.96 | 0.66 | 0.89 |
| Arg | 0.69 | 0.93 | 0.75 | 0.93 |
| Asn | 1.54 | 1.02 | 2.14 | 1.06 |
| Asp | 1.56 | 1.24 | 1.86 | 0.99 |
| Cys | 1.42 | 0.73 | 0.98 | 1.20 |
| Gln | 0.89 | 0.94 | 0.93 | 1.01 |
| Glu | 0.87 | 1.35 | 0.92 | 0.89 |
| Gly | 1.09 | 1.04 | 2.14 | 1.64 |
| His | 1.25 | 0.95 | 1.16 | 0.93 |
| Ile | 0.66 | 0.61 | 0.42 | 0.68 |
| Leu | 0.73 | 0.67 | 0.47 | 0.78 |
| Lys | 0.80 | 1.22 | 0.94 | 1.10 |
| Met | 0.70 | 0.48 | 0.41 | 0.68 |
| Phe | 0.98 | 0.66 | 0.96 | 0.95 |
| Pro | 1.48 | 2.45 | 0.63 | 0.96 |
| Ser | 1.29 | 1.23 | 1.06 | 1.03 |
| Thr | 1.08 | 0.79 | 0.94 | 1.20 |
| Trp | 0.62 | 0.65 | 0.76 | 0.79 |
| Tyr | 1.04 | 0.75 | 0.83 | 1.07 |
| Val | 0.72 | 0.70 | 0.54 | 0.84 |

of concern about the accuracy of the data, a distinction was made between proteins with a single transmembrane helix and those with more, and the data shown here will relate to the former only. Using the sequences of the 285 single transmembrane helix proteins propensities were derived, but rather than using the definition of Equation EQ12.4 above, the logarithm of the ratio was taken to obtain the log-likelihood ratio:

$$P'_{s,a} = \ln\left(\frac{p_{s,a}}{p_a}\right) \qquad \text{(EQ12.5)}$$

(Note the similarity to the substitution matrix formula of Equation EQ5.2.) Five different structural states were distinguished: three transmembrane helical (inside, middle, and outside) and two non-transmembrane helical (inside and outside). The values for these are shown in Figure 12.10. As might be expected the "helix middle" values correlate quite well with the amino acit hydrophobicity scales of Table 12.4.

All the measures shown here can be regarded as properties of the different amino acids. Many of these quantitative amino acid properties have been collected into a database called AAindex. This has several hundred indices, including over 30 hydrophobicity scales.

## The simplest prediction methods are based on the average residue propensity over a sequence window

If a strong compositional preference exists for a structural state, and that state extends over several consecutive residues, then most regions which fold to that

| Residue | KD | GES | Zviling |
|---|---|---|---|
| Ile | 4.5 | 3.1 | 9.9 |
| Val | 4.2 | 2.6 | 13.4 |
| Leu | 3.8 | 2.8 | 14.5 |
| Phe | 2.8 | 3.7 | 4.6 |
| Cys | 2.5 | 2.0 | 3.8 |
| Met | 1.9 | 3.4 | 7.0 |
| Ala | 1.8 | 1.6 | 4.3 |
| Gly | −0.4 | 1.0 | 0.1 |
| Thr | −0.7 | 1.2 | −11.4 |
| Ser | −0.8 | 0.6 | −0.3 |
| Trp | −0.9 | 1.2 | −4.2 |
| Tyr | −1.3 | −0.7 | −3.4 |
| Pro | −1.6 | −0.2 | −35.3 |
| His | −3.2 | −3.0 | −19.2 |
| Glu | −3.5 | −8.2 | −49.4 |
| Gln | −3.5 | −4.1 | −29.6 |
| Asp | −3.5 | −9.2 | −40.7 |
| Asn | −3.5 | −4.8 | −52.5 |
| Lys | −3.9 | −8.8 | −44.8 |
| Arg | −4.5 | −12.3 | −64.5 |

**Table 12.4**
**Three representative amino acid hydrophobicity scales.** KD is the Kyte–Doolittle scale; GES is reported by Engelman, Steitz, and Goldman, and the Zviling scale is one of three alternatives reported by Zviling, Leonov, and Arkin. Although there is a significant correlation between the scales, there are some differences such as each scale having a different residue as the most hydrophobic.

state will be expected to have residues that have preferential propensities. Therefore if a stretch is found containing residues that have suitably high propensities, that state could be predicted to occur in that region. Any stretch of a specified length that had an average propensity exceeding a chosen threshold could be predicted as being in that state. Even residues with very unfavorable propensities are likely to occur occasionally, so careful choice of the precise number of residues over which to average and the threshold required for prediction are required to obtain accurate predictions.

The number of residues over which to average is called the **window**, and the length depends on the structure being predicted. The window must be short enough to enable detection of the shortest structures, but if it is too short it could lead to many spurious assignments. β-strands are usually shorter than α-helices (see Figure 12.4) and thus require shorter windows. For example, in the Chou–Fasman method β-strands are initially identified using windows of four or five residues, whereas α-helices are initially identified from six-residue windows. Transmembrane helix predictions generally use even larger windows, often of 15–20 residues.

The simplest prediction methods of this kind are those using hydrophobicity scales for transmembrane helix prediction. All windows with average hydrophobicity exceeding the threshold are assigned as transmembrane helices. When windows overlap, they are combined into a single helix prediction. The example shown in Figure 12.11 illustrates a moderately successful prediction of one transmembrane helix, with the peak in average hydrophobicity not exactly centered on the correct location.

The Chou–Fasman method for predicting α-helices, β-strands, and β-turns is much more complicated. In part this is because they used different window sizes

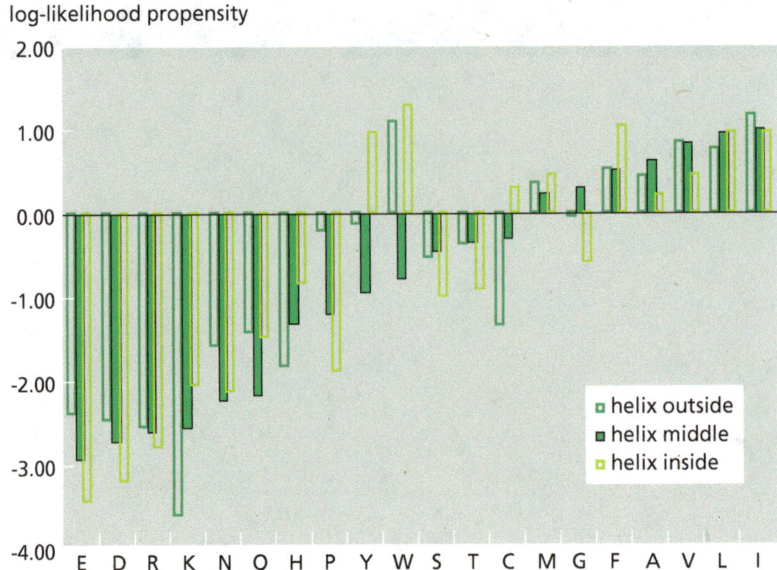


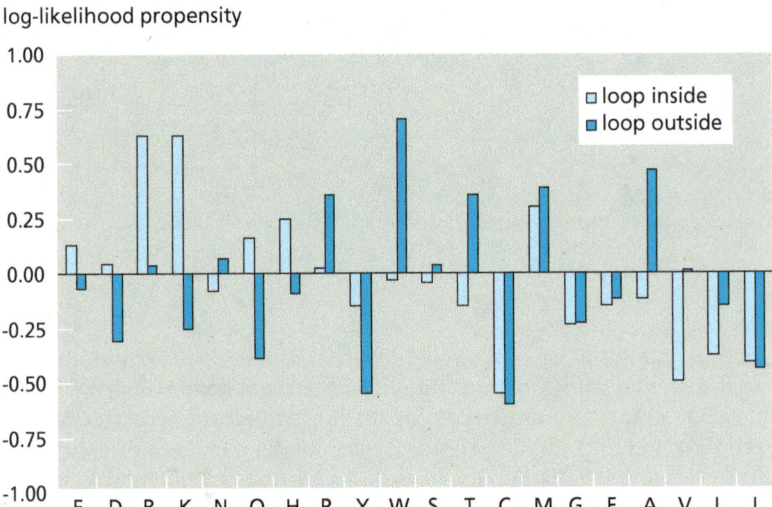


**Figure 12.10**
**Transmembrane propensities for the different amino acids, calculated as log-likelihoods.** Five different structural states (environments) are distinguished: the loops (peptide segments) not associated with the transmembrane helix are divided into the inside (cytoplasmic) and outside; the transmembrane helix is differentiated into the inside, middle, and outside. Note the different scales of the two plots. (Data from D.T Jones, W.R. Taylor and J.M. Thornton, A model recognition approach to the prediction of all-helical membrane protein structure and topology, *Biochemistry* 33:3038–3049, 1994.)

and thresholds to initially identify and then to extend α-helices and β-strands. However, the method also has to choose between alternatives. In the transmembrane helix example the only alternatives were the presence or absence of the helix, whereas the Chou–Fasman method has to adjudicate between possible predictions of α-helix and β-strand for the same window. When looking to initially identify a helix or strand, the choice made depends on which state has the larger average propensity. However, the rules in regions where structures are extended are quite complex. The Chou–Fasman method uses a different technique for β-turn prediction. The structure extends over four residues, each position having a different propensity, as was shown for a different set of propensities in Table 12.3. To predict a β-turn at a given position the propensities for the residues at the four different positions are multiplied together, and a β-turn is predicted if the product exceeds a threshold.

The COILS method of predicting coiled coils uses a similar principle, taking the geometric mean of the propensity over the window. Because coiled coils have a repeating heptad the propensities are separately defined for each heptad position, and the window size used is 14, 21, or 28 residues. All possible heptad positions are

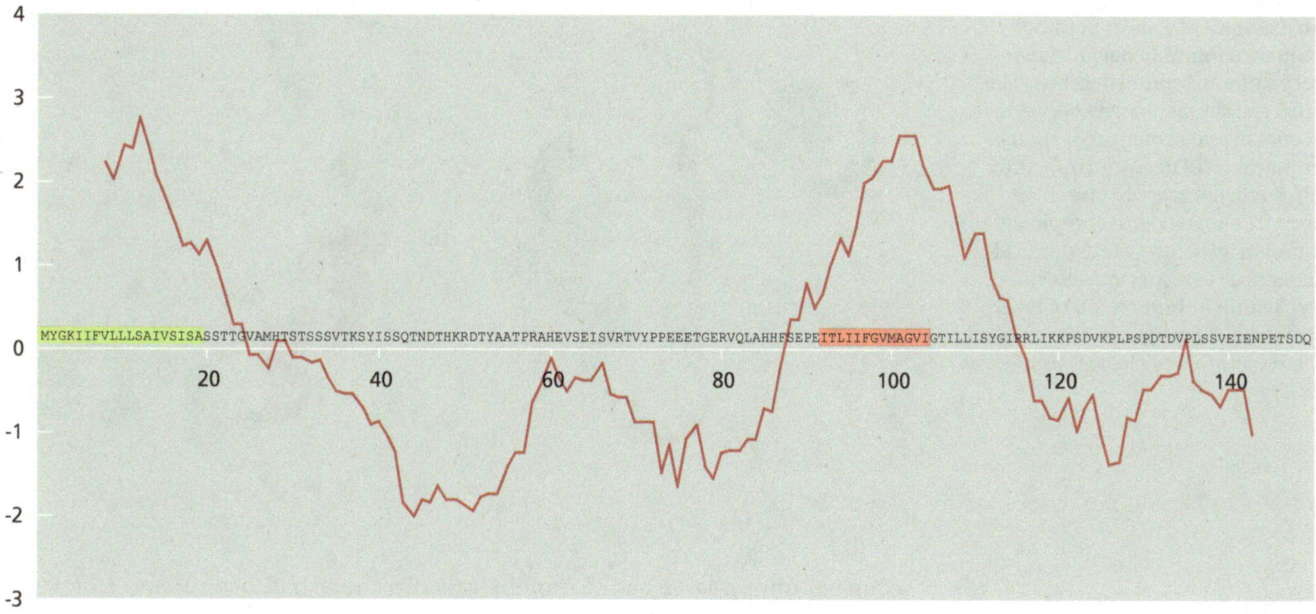


**Figure 12.11**
**An example of a transmembrane prediction using the average of a hydrophobicity scale over a window.** The KD (Kyte–Doolittle) hydrophobicity scale has been used with a window length of 15 residues. The protein in this example is human glycophorin A, which has a transmembrane helix located at residues 92–114, shown in light red. This is approximately the location of the only peak in hydrophobicity for this sequence, except for a peak at the N terminal, which corresponds to a nonpolar signal peptide shown in green.

explored for each residue, choosing the highest scoring as the prediction, and then converting this into a probability. Regions that form a coiled coil usually have probabilities close to 1.

The MEMSAT program for transmembrane helix prediction uses the five propensities shown in Figure 12.10. As was the case for the Chou–Fasman method, the prediction is given by the structural assignment that maximizes the sum of the residue propensities. In order to determine this assignment a dynamic programming method (as described in Section 5.2) was used to identify the optimal prediction.

Although these methods were usually the first to be applied and were sufficiently successful to encourage further efforts, in general they were not sufficiently accurate and were soon superseded by more complex methods, such as will be described later in this chapter.

## Residue propensities are modulated by nearby sequence

Although secondary structure prediction has been moderately successful when using residue propensities as described above, it is thought likely that the folded structural state of a residue is strongly influenced by the nearby sequence. The reasoning for this is that if an entire peptide chain always folds up to the same structure, presumably shorter segments will also always adopt the same structure. For very short segments this clearly does not hold; certainly at the limit of individual residues they occur in many possible structural states. By quantifying this local modulation of the residue propensities new prediction methods have been proposed that are significantly more accurate than those based on individual residue preferences.

Before describing the prediction methods themselves, we will look at what is known of these local sequence effects. It is difficult to clearly define when a sequence has folded into two different structures, but a potential solution is to specify that no residue should have the same state assigned in both structures. Although this is perhaps an unduly severe definition, Rita Casadio and colleagues used it with DSSP assignments in 2000 to identify just one eight-residue and 16 seven-residue

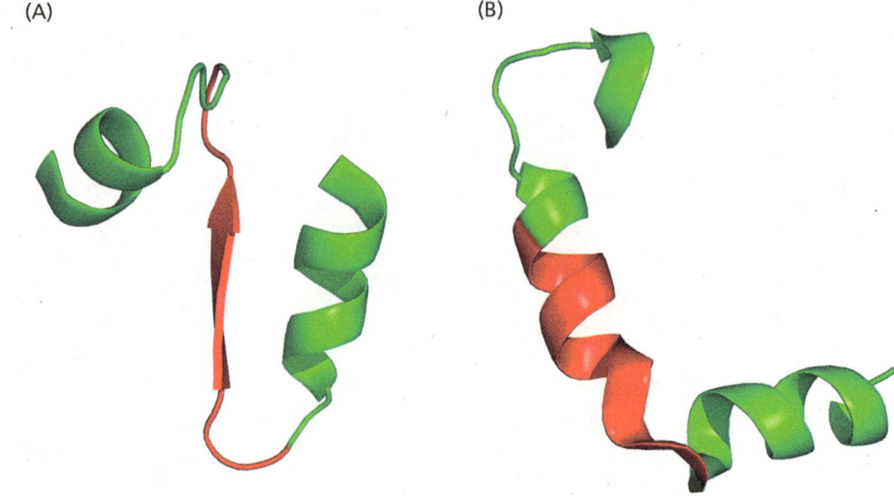


**Figure 12.12**
**An example of a short peptide sequence found to occur in two very different conformations.** The nine-residue sequence KGVVPQLVK occurs in two proteins, (A) mouse importin α (PDB entry 1IAL) and (B) *E. coli* pyruvate kinase (PDB entry 1PKY), but with completely different structures. The identical sequence segments are shown in red, within a short region of local structure. After the first two residues the structures are clearly completely different.

segments from a large set of protein structures that had been found to take up significantly different structures. Subsequently a nine-residue sequence has been found that has different structures (α-helix and β-strand) in two proteins (see Figure 12.12). These results suggest that the influence of the nearby sequence dominates local structure, and that a method which accounts for the effects of five or more residues to either side of a central residue has the potential to give a highly accurate prediction.

Other studies have shown that there certainly is information about the structural state in the local sequence, although they have also given a clear indication that nonlocal or long-range effects are also important. Gavin Crooks and Steven Brenner analyzed the information present in neighboring residues and concluded that residues up to four positions distant may have an influence. However, they also found that the use of this local information did not achieve prediction accuracies close to 100%, suggesting that some nonlocal effects must also be involved. Daisuke Kihara analyzed the correlation between the accuracy of secondary structure predictions and the degree of spatial contact with sequentially more remote residues. The data suggest that residues with fewer nonlocal spatial contacts are on average predicted more accurately.

Taking all of these studies together, there is significant information about secondary structure present in the local sequence, but there is additional nonlocal information too. The definition of residue propensity presented above can be readily extended to quantify the local effects. An example of such residue propensities is shown in Figure 12.9. The GOR series of methods, which contain such local information, are based on an information theory derived by Barry Robson in 1974, which is presented below. The only way such methods might include nonlocal effects would be by increasing the size of the window by a considerable factor.

The central concept of the GOR methods is that the protein sequence contains the information that is translated into the protein conformation. The $j$th residue of the sequence $\boldsymbol{x}$ is written $x_j$. The structural state of the $j$th residue is written as $S_j$. Thus the $x_j$ are one of 20 types, and the $S_j$ one of three or four, depending on the particular GOR version, as some versions predict β-turn residues as well as β-strand, α-helix, and coil. The probability of the $j$th residue being in structural state $s$ is written $P(S_j = s)$. $P(S_j = s|\hat{\boldsymbol{x}})$ is the probability of the $j$th residue being in structural state $s$ given that the sequence contains residue(s) $\hat{\boldsymbol{x}}$. Here $\hat{\boldsymbol{x}}$ refers to a specific part of the protein sequence, e.g., $x_j$; $x_jx_{j+1}$; $x_{j-8}\ldots x_{j+8}$. These two probabilities will only be identical if the residues in $\hat{\boldsymbol{x}}$ have no influence at all on the structural state $S_j$.

The Fano definition of the mutual information that sequence $\hat{\boldsymbol{x}}$ contains about the structural state of the $j$th residue being $s$ is given by

$$\mathrm{I}(S_j = s\,;\hat{\boldsymbol{x}}) = \log\left(\frac{\mathrm{P}(S_j = s|\hat{\boldsymbol{x}})}{\mathrm{P}(S_j = s)}\right) \tag{EQ12.6}$$

(see Appendix A). If $\hat{\boldsymbol{x}}$ has no influence on $S_j$, then the two terms in the logarithm are identical and $\mathrm{I}(S_j = s\,;\hat{\boldsymbol{x}})$ is 0; if $\hat{\boldsymbol{x}}$ favors/disfavors $S_j$ being $s$, then $\mathrm{I}(S_j = s\,;\hat{\boldsymbol{x}})$ is positive/negative.

If sequence $\hat{\boldsymbol{x}}$ can be split into two parts $\hat{\boldsymbol{x}}_1$ and $\hat{\boldsymbol{x}}_2$, e.g., $x_i$ and $x_{i-1}x_{i+1}$, then we can define two quantities. The information that $\hat{\boldsymbol{x}}_1$ and $\hat{\boldsymbol{x}}_2$ contain together about the conformation $S_j$ being $s$ is written

$$\mathrm{I}(S_j = s\,;\hat{\boldsymbol{x}}_1,\hat{\boldsymbol{x}}_2) = \log\left(\frac{\mathrm{P}(S_j = s|\hat{\boldsymbol{x}}_1,\hat{\boldsymbol{x}}_2)}{\mathrm{P}(S_j = s)}\right) \tag{EQ12.7}$$

The information that $\hat{\boldsymbol{x}}_2$ alone carries about $S_j$ given that $\hat{\boldsymbol{x}}_1$ also occurs is written

$$\mathrm{I}(S_j = s;\hat{\boldsymbol{x}}_2|\hat{\boldsymbol{x}}_1) = \log\left(\frac{\mathrm{P}(S_j = s|\hat{\boldsymbol{x}}_1,\hat{\boldsymbol{x}}_2)}{\mathrm{P}(S_j = s|\hat{\boldsymbol{x}}_1)}\right) \tag{EQ12.8}$$

These three terms of Equations EQ12.6 to EQ12.8 are related by the following equation, as is easily confirmed by substitution:

$$\mathrm{I}(S_j = s\,;\hat{\boldsymbol{x}}_1,\hat{\boldsymbol{x}}_2) = \mathrm{I}(S_j = s\,;\hat{\boldsymbol{x}}_1) + \mathrm{I}(S_j = s\,;\hat{\boldsymbol{x}}_2|\hat{\boldsymbol{x}}_1) \tag{EQ12.9}$$

Consider the mutually exclusive events of the $j$th residue either being in conformational state $s$ or not in this state, which will be written $\bar{s}$. Expressions can be written for $\mathrm{I}(S_j = (s:\bar{s});\,\hat{\boldsymbol{x}})$, the preference (in information content) of $\hat{\boldsymbol{x}}$ for $s$ over $\bar{s}$.

$$\mathrm{I}(S_j = (s{:}\bar{s});\hat{\boldsymbol{x}}) = \mathrm{I}(S_j = s\,;\hat{\boldsymbol{x}}) - \mathrm{I}(S_j = \bar{s};\hat{\boldsymbol{x}}) = \log\left(\frac{\mathrm{P}(S_j = s|\hat{\boldsymbol{x}})}{\mathrm{P}(S_j = s)}\right) - \log\left(\frac{\mathrm{P}(S_j = \bar{s}|\hat{\boldsymbol{x}})}{\mathrm{P}(S_j = \bar{s})}\right) \tag{EQ12.10}$$

Note that because $s$ and $\bar{s}$ are mutually exclusive

$$\mathrm{P}(S_j = \bar{s}) = 1 - \mathrm{P}(S_j = s) \tag{EQ12.11}$$

and

$$\mathrm{P}(S_j = \bar{s}|\hat{\boldsymbol{x}}) = 1 - \mathrm{P}(S_j = s|\hat{\boldsymbol{x}}) \tag{EQ12.12}$$

so we can derive

$$\mathrm{I}(S_j = (s{:}\bar{s});\hat{\boldsymbol{x}}) = \log\left(\frac{\mathrm{P}(S_j = s|\hat{\boldsymbol{x}})}{1 - \mathrm{P}(S_j = s|\hat{\boldsymbol{x}})}\right) - \log\left(\frac{\mathrm{P}(S_j = s)}{1 - \mathrm{P}(S_j = s)}\right) \tag{EQ12.13}$$

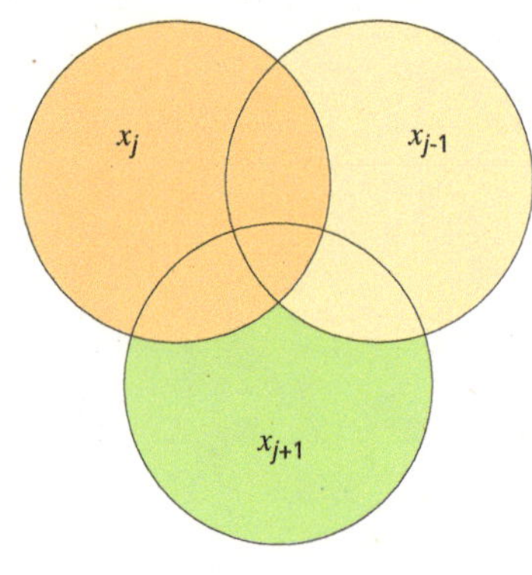


**Figure 12.13**
**Diagram illustrating the information three residues $x_{j-1}$, $x_j$, and $x_{j+1}$ hold about the structural state of one of them, $x_j$.** The information about the structural state present in each of three consecutive residues $x_{j-1}$ to $x_{j+1}$ is shown as overlapping circles. The orange-shaded area represents the term $I(S_j = (s:\bar{s}); x_j)$; the yellow-shaded area represents the term $I(S_j = (s:\bar{s}); x_{j-1}|x_j)$; the green-shaded area represents the term $I(S_j = (s:\bar{s}); x_{j+1}|x_{j-1}x_j)$. This is an illustration of Equation EQ12.15.

A similar expression to Equation EQ12.9 holds for the terms for preference of $s$ over $\bar{s}$ when $\hat{\boldsymbol{x}}$ is split into two components $\hat{\boldsymbol{x}}_1$ and $\hat{\boldsymbol{x}}_2$:

$$I(S_j = (s{:}\bar{s}); \hat{\boldsymbol{x}}_1, \hat{\boldsymbol{x}}_2) = I(S_j = (s{:}\bar{s}); \hat{\boldsymbol{x}}_1) + I(S_j = (s{:}\bar{s}); \hat{\boldsymbol{x}}_2 | \hat{\boldsymbol{x}}_1) \qquad \text{(EQ12.14)}$$

Figure 12.13 shows a representation of the information that a tripeptide sequence $x_{j-1}x_jx_{j+1}$ carries about residue $x_j$ being in state $S_j$. Terms such as $I(S_j = (s:\bar{s}); x_j)$ and $I(S_j = (s:\bar{s}); x_{j+1})$ are represented by the circles labeled $x_j$ or $x_{j+1}$, respectively. The yellow area of the circle labeled $x_{j-1}$ that is outside the circle labeled $x_j$ represents the term $I(S_j = (s:\bar{s}); x_{j-1}|x_j)$. This is because, as explained above, that term is the information about $S_j$ contained in $x_{i-1}$ given that $x_i$ has already been accounted for. Similarly, the green area of the circle labeled $x_{j+1}$ outside the circles labeled $x_j$ and $x_{j-1}$ represents the term $I(S_j = (s:\bar{s}); x_{j+1}|x_{j-1}x_j)$. The identification of these terms in the figure allows us to readily see that

$$I(S_j = (s{:}\bar{s}); x_{j-1}x_jx_{j+1}) = I(S_j = (s{:}\bar{s}); x_j) + I(S_j = (s{:}\bar{s}); x_{j-1} | x_j) + I(S_j = (s{:}\bar{s}); x_{j+1} | x_{j-1}x_j) \qquad \text{(EQ12.15)}$$

which could also have been derived by repeated application of Equation EQ12.14. There are as many alternative forms of this equation as there are orders in which the residue terms can be chosen. This concept can readily be extended to include the effects of more residues, although a graphical representation is not easy.

The first term on the right-hand side of Equation EQ12.15 refers to a single residue in the sequence. As written, the term refers to the mutual information of the residue about its own structural state, but in the alternative forms of the equation it may relate to the conformation of a residue some distance away. The second and third terms involve two and three residues, respectively. Since there are 20 different residues, there will be 400 of the second and 8000 of the third terms. The more residues whose effects are considered, the more complex the terms involved. It is not feasible to obtain so many parameters from the limited data available, and therefore approximations are required. Referring to Figure 12.13, if we assume that there are no overlaps we obtain the expression:

$$I(S_j = (s{:}\bar{s}); x_{j-1}x_jx_{j+1}) = I(S_j = (s{:}\bar{s}); x_{j-1}) + I(S_j = (s{:}\bar{s}); x_j) + I(S_j = (s{:}\bar{s}); x_{j+1}) \qquad \text{(EQ12.16)}$$

Alternatively we could assume that all the circles only overlap with that of the residue whose structural state is being considered, $x_j$, to obtain:

$$I(S_j = (s{:}\bar{s}); x_{j-1}x_jx_{j+1}) = I(S_j = (s{:}\bar{s}); x_j) + I(S_j = (s{:}\bar{s}); x_{j-1} | x_j) + I(S_j = (s{:}\bar{s}); x_{j+1} | x_j) \qquad \text{(EQ12.17)}$$

Three types of terms can be distinguished: self-information, directional information, and pair information. The information a residue carries about its own structural state—$I(S_j = (s:\bar{s}); x_j)$—is called self-information. The information about the structural state at the $j$th position carried by residue $x_i$ (where $i \neq j$)—$I(S_j = (s:\bar{s}); x_i)$—is called directional information, and is independent of the type of residue found at $j$. In contrast to this, the term $I(S_j = (s:\bar{s}); x_i | x_j)$ takes account of the type of residue at the $j$th position, and is referred to as pair information. This is the information carried by $x_i$ about the structural state at the $j$th position given that the residue at that position is of type $x_j$.

In the GOR methods eight residues are considered to either side of the central residue $a_j$. The two approximations above now become:

$$I(S_j = (s{:}\bar{s})\,;x_{j-8}...x_{j+8}) = \sum_{m=-8}^{8} I(S_j = (s{:}\bar{s})\,;x_{j+m}) \tag{EQ12.18}$$

and

$$I(S_j = (s{:}\bar{s})\,;x_{j-8}...x_{j+8}) = I(S_j = (s{:}\bar{s})\,;x_j) + \sum_{m=-8}^{8} I(S_j = (s{:}\bar{s})\,;x_{j+m}\,|\,x_j) \tag{EQ12.19}$$

In Equation EQ12.19 the summation limits of $m$ have been simplified to include the term for $m = j$ because it is zero anyway. The GOR I and GOR II methods use Equation EQ12.18, whereas GOR III uses Equation EQ12.19. Note that only the first term in Equation EQ12.19 is common to both expressions. GOR IV and GOR V use another approximation where each term involves two residues, but unlike in Equation EQ12.19 the terms involve all possible pairs of residues in the region $j\pm8$ residues.

To calculate these quantities for given residue types and structural states we return to the probability formulae given earlier, and use a suitable structure database to derive the values. $P(S_j = s)$ is simply the fraction of residues of the database in structural state $s$, written $p_s$. To obtain values for terms such as $P(S_j = s\,|\,x_j)$ we use

$$P(S_j = s\,|\,x_j) = \frac{P(S_j = s\,,\,x_j)}{P(x_j)} = \frac{p_{s,j}}{p_j} \tag{EQ12.20}$$

where the first equality comes from Equation EQA.4 of Appendix A. Note that this particular term is the Chou–Fasman propensity (see Equation EQ12.4). For example, the self-information terms can be calculated from Equations EQ12.13 and EQ12.20 as

$$I(S_j = (s{:}\bar{s});x_j) = \log\left(\frac{p_{s,j}}{1-p_{s,j}}\right) - \log\left(\frac{p_s}{1-p_s}\right) \tag{EQ12.21}$$

Terms such as $P(S_j = s\,|\,x_{j+m})$ involve the fraction of residue pairs $m$ residues apart (with residue types specified) in which the structural state of $x_j$ is $s$. To parameterize GOR IV a database containing over 63,000 residues was employed.

Depending on the version of GOR, using an approximation—Equation EQ12.18, EQ12.19, or another—at the $j$th position, the term $I(S_j = (s:\bar{s}); x_{j-8}...x_{j+8})$ or $P(S_j = (s\,;x_{j-8}...x_{j+8})\,/\,P(S_j = (\bar{s}\,;x_{j-8}...x_{j+8})$ is calculated for all the possible structural states $s$ for all sequence positions. At a given sequence position the structural state with the largest value is taken as the predicted state (see Figure 12.14). In early versions these values could be modified by addition of constants, in an attempt to introduce experimental estimation of the overall structural state composition for the protein of interest, an option now discontinued. Unlike the other versions, GOR V also includes information from homologous sequences, an aspect that will be discussed in the following section.

The GOR method only considers the structural state of one residue at a time. However, it is clear that structural states are correlated as, because of the very nature of the structures involved, helical and strand residues occur in stretches. The theory presented above has been extended to consider a consecutive pair of structural states simultaneously. Each possible pair of structural states must be separately parameterized, resulting for three states in a nine-fold increase in parameters, which places very heavy demands on the dataset size for parameterization. A further complication is that it is no longer possible to simply calculate a value for

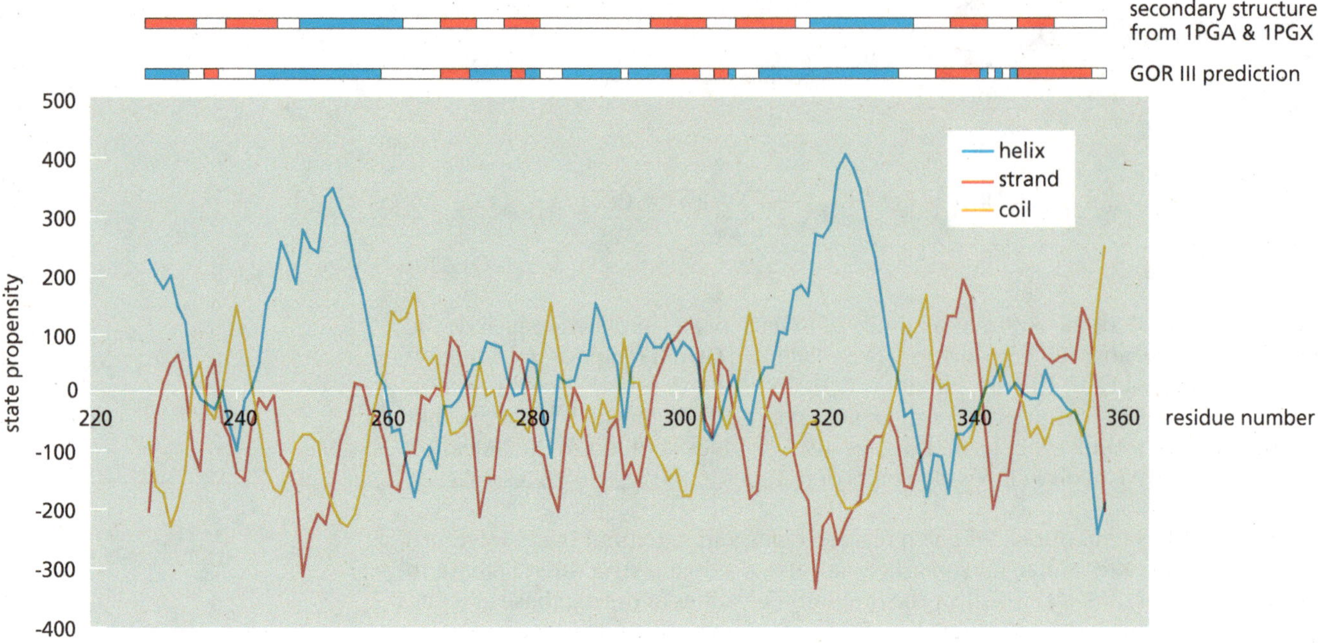


**Figure 12.14**
**The GOR III prediction for residues 228–358 of IgG binding protein G from *Streptococcus*.** (This region of sequence has been chosen because experimental structures are available.) The three state propensities are plotted in the figure, and the prediction (the largest propensity at each residue position) is shown in the lower bar, colored blue for α-helix and red for β-strand. Above is the bar showing the structures found experimentally. Note the two correctly predicted α-helices are located at much stronger prediction signals than the incorrect ones. At several places secondary structures are predicted that are only one residue long, especially a stretch "HCHCH" about position 345. These would normally be removed from the final prediction by a filter. The experimental structure shown is as defined in the two PDB entries for Streptococcus IgG binding protein G, with database ids 1PGA and 1PGX.

each state at each residue position. To overcome this problem a dynamic programming technique is used to optimize the prediction. When implemented with nine-residue windows the method seems to be an improvement on GOR IV according to the average $Q_3$ measure. For more detail of this method see Further Reading.

Choosing the largest structural state value at each sequence position can lead to secondary structures just a single residue long. Usually the prediction is modified by a filter (**prediction filtering**), which imposes a minimal length for the helices and strands, typically four and two, respectively. This is a common last stage in many prediction methods that lack any other constraint that prevents short secondary structures.

## Predictions can be significantly improved by including information from homologous sequences

The accuracy of the GOR method described above increased with each successive version, reaching an average $Q_3$ of 64% for GOR IV. The lack of structural data is a severe limit on further development based on using approximations to calculate terms such as $I(S_j = (s : \bar{s}) ; x_{j-8} \ldots x_{j+8})$. An alternative approach that leads to significant improvement in prediction accuracy is to include information from related sequences. Instead of just using the sequence of the protein whose secondary structure prediction is wanted, a multiple alignment is constructed following a search for similar sequences, and the multiple alignment information is used to predict the secondary structure. The first such method, called Zpred, was published in 1987 and was a modification of GOR II. Details of the way that Zpred uses multiple sequences are presented in Section 11.4.

The key observation that justifies this approach is that the protein fold is more highly conserved than the sequence, so that segments of a multiple alignment are expected to correspond to equivalent secondary structural elements. As has been noted (see Figure 11.6), the correspondence is not exact, but it is clearly good enough to be useful. By comparison to the same prediction method based on a single sequence the increase in the average $Q_3$ can be as much as 9%.

The GOR V and Zpred methods use the same method to incorporate the information from homologous sequences in the prediction. A secondary structure prediction is obtained for each separate homologous sequence. These are then combined using the sequence alignment to identify equivalent residues. The structural state prediction score is averaged over all the sequences at each position of the query sequence. The final prediction at each position is the structural state with the highest average score. This could be subsequently modified by a filter to remove short strands and helices.

An alternative method is possible, which takes the prediction for each sequence to its conclusion of specifying the state of each residue, even applying the filter to each sequence prediction. At this point, using the alignment to identify equivalent positions in the sequences, the most frequently predicted structural state at each position can be taken as the final prediction. This method uses the concept of the so-called jury or majority voting technique. A further application of the filter could be made to this averaged prediction.

Other ways of introducing information from homologous sequences have been used in nearest-neighbor methods (see Section 12.3) and neural networks (see Section 12.4).

## 12.3 The Nearest-Neighbor Methods are Based on Sequence Segment Similarity

The secondary structure prediction methods described so far involve analyzing known structures to determine individual residue or residue pair structural preferences and prescribing a technique for combining these into a prediction. In contrast nearest-neighbor methods directly apply the raw data of the observed structure for a given segment of sequence. If the structural state of a given residue was completely determined by a surrounding sequence segment that was sufficiently short for all possible segment sequences to be present in the structure database, prediction would be a trivial process. It would suffice to locate the relevant sequence in the database, which would then determine the structural state of the central residue of the segment.

The main practical problem with this approach is the lack of structural information on all possible segment sequences so that similar sequences need to be examined.

**Flow Diagram 12.3**

**The key concept introduced in this section is that similarity of short sequence segments can be used as a predictive tool if at least one of the segments is of known structure.**

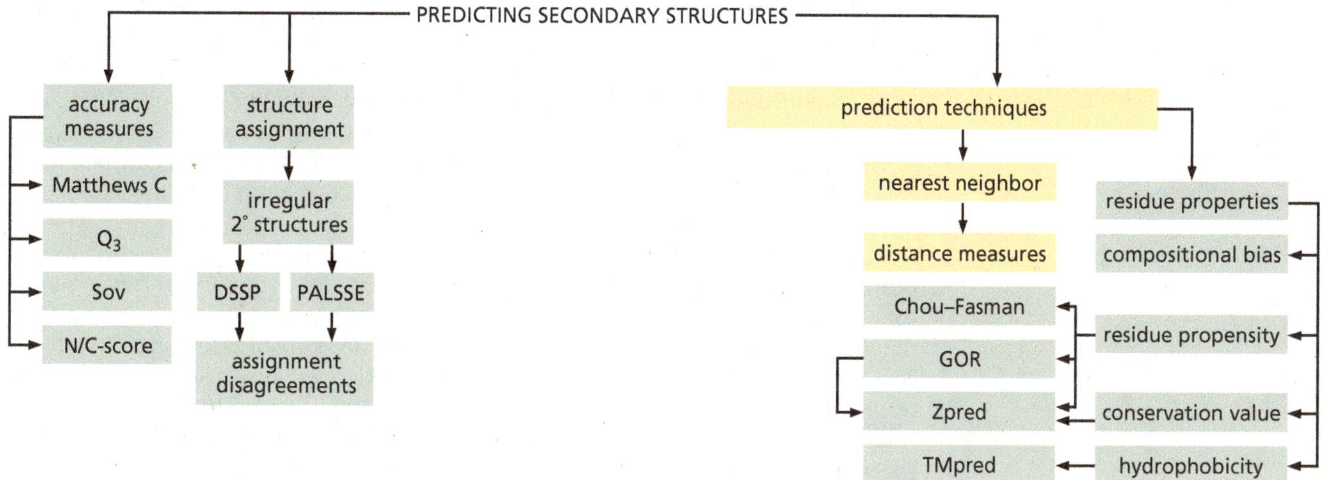

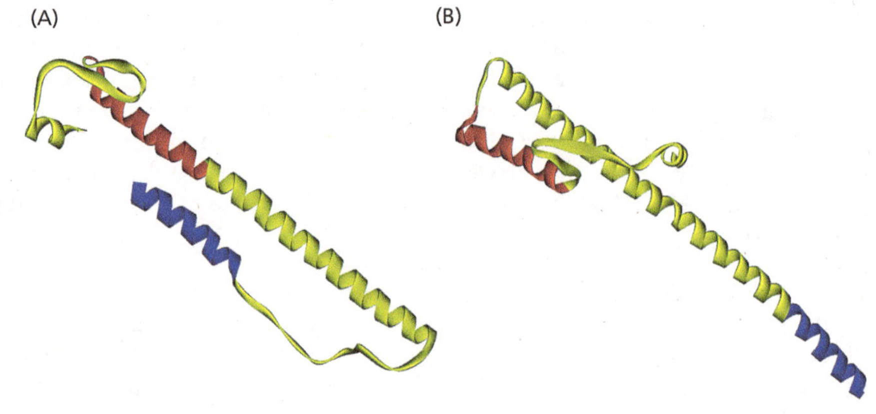


**Figure 12.15**
**Two alternative conformations for the same sequence segment from influenza virus hemagglutinin.** The environment of the protein is the same except for a change of pH, which leads to a switch in conformation that is part of the mechanism for fusing the viral membrane to that of its host, allowing the virus to invade the host cell. The two structures are taken from PDB entry 3HMG (A) and PDB entry 1HTM (B). They show the identical stretch of sequence, and the regions are colored similarly to help identify equivalent parts in the two conformations.

Hence the description of these methods will dwell at length on measures of sequence similarity. Furthermore, usually the prediction is based on an average of structures from a number of these similar segments, so the methods of averaging are also discussed.

An additional difficulty, which affects all secondary structure prediction methods based on local sequence, is that nonlocal effects do influence protein structure, so that the size of segment that it is necessary to employ is difficult to define. In the previous section it was noted that examples exist of seven-, eight-, and nine-residue segments that have alternative structures (see Figure 12.12). However, the problem is more serious than these observations might suggest, as a number of proteins are known to have alternative folds with some regions having significantly different conformations under different conditions. This situation is believed to apply to only a small minority of proteins. The amyloidogenic proteins are a medically important class of such proteins, and an approach to the prediction of their behavior will be presented below. Other examples may have normal physiological functional importance, such as the hemagglutinin example shown in Figure 12.15, in which a pH-dependent structural change is necessary for the influenza virus to invade the host cell. All the current secondary structure prediction methods only propose a single structure and clearly cannot reproduce these features. The difficulty is most apparent in a discussion of nearest-neighbor methods because they use the segment structures directly, and so will have either the alternative conformations in their database or just one. Note that all the databases described in this chapter, which are careful selections of known structures, intentionally remove the structures of proteins whose sequences are closely similar, and so are unlikely to include these alternative folds. It is perhaps not coincidental that the nearest-neighbor technique has been used to try to predict amyloidogenic sequence segments, which might have alternative folds, details of which will be presented later in this section.

Note that the sequence similarity used by the nearest-neighbor methods is over only short regions of perhaps 19 residues, but often fewer, so that there is no expectation or requirement that the two sequences are true homologs. This is in contrast to the use of sequence alignments in other secondary structure prediction methods.

Before describing the more technical details of nearest-neighbor methods we will outline the basic procedure (see Figure 12.16). For each segment of the sequence (often referred to as the window) the best gapless alignments are sought from a database of protein sequences whose structures are known. This often involves a scoring scheme developed specifically for the task, as will be presented below. A set of the highest-scoring alignments (often 50 or more) are used to predict the structural state of the central residue of the window. The prediction may involve taking the most commonly found states of the aligned residue, but may also involve weighting by the score.

**Figure 12.16**
**Secondary structure prediction of Phe261 of chymosin B using the NNSSP nearest-neighbor method with a 17-residue window and 50 nearest neighbors.** The three-state secondary structure is shown beneath each sequence segment, with h, e, and c meaning α-helix, β-strand, and coil, respectively. The N-terminal residue number is given for each neighbor. The scoring scheme takes account of the residue environment as well as a standard alignment scoring scheme. Note that all three structural states occur at the central window residue in this example. The final prediction for Phe261 will depend on a weighted average of the 50 different central residue states, and in this case is β-strand.

## Short segments of similar sequence are found to have similar structure

The existence of similar protein folds for homologous proteins is well established in the structural databases. Exceptions to this such as presented by amyloidogenic proteins are rare enough to be remarked on even when without clear physiological or medical importance. However, the identification of substructures with general sequence characteristics has only occurred relatively recently.

A number of attempts have been made to create a database of structural segments with correlated sequences, of which just one will be described here to give a feel for the results. The I-sites library of David Baker and co-workers contains approximately 250 identified short segments (3 to 19 residues in length) that show clear correlation of structure and sequence for all occurrences within a dataset of nonredundant protein structures. An example entry is shown in Figure 12.17 for an α-helix that ends with a proline residue. As well as displaying the structural similarity of the segments in this entry, the figure also shows the sequence similarity and its relation to the structure. The I-sites entries are not completely independent, and there are many relationships between entries. For example an α-helix in a given structure might have a segment of the helix matching one entry, while the segments covering the N and C termini of the helix may match other entries, such that the entries overlap. In addition, the identified entries do not fully cover the range of

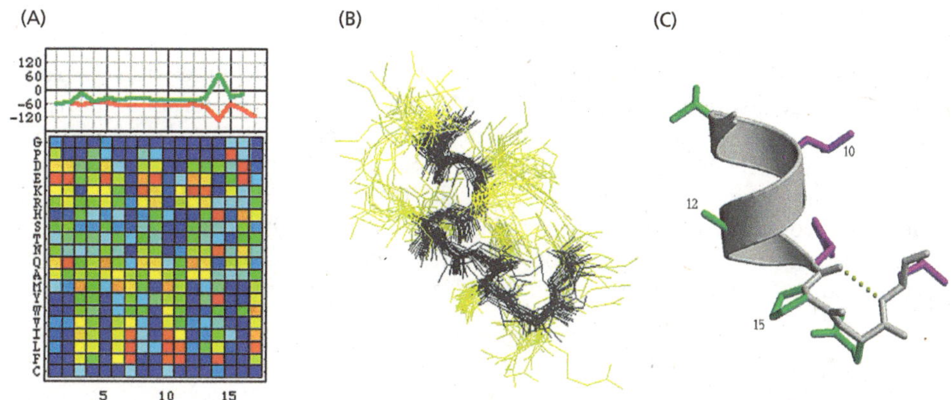


**Figure 12.17**
**An entry from the I-sites sequence-structure motif database—the proline-containing α-helix C-cap.** (A) The average ϕ (red) and ψ (green) dihedral angles of each of the 17 residues of the entry and color-coded matrix of the log-odds scores for each residue type at each position, showing for example the strong preference for proline at position 15. Dark blue indicates least favored amino acids, and red most favored at that particular position in the motif. (B) Thirty members of the set are shown superposed, with the backbone atoms drawn black and the side chains in green. (C) A representation of the key structural features of the entry, shown using residues 24–31 from PDB entry 2CTC (corresponding to positions 9–16 in part A), the α-helix shown as a broad gray ribbon ending at the proline. Highly conserved polar residues are shown in green and nonpolar residues in purple. The dotted line shows a conserved hydrogen bond. (Reprinted from *J. Mol. Biol.*, 281, C. Bystroff and D. Baker, Prediction of local structure in proteins using a library of sequence-structure motifs, 565–577, 1998, with permission from Elsevier.)

structure and sequence encountered. Only about one third of all the database residues were classified into one of the approximately 80 entries with the strongest sequence–structure correlations.

The I-sites database and the chameleon sequence results mentioned earlier show that although there is a definite relationship between sequence and structure it is difficult to define this relationship for short segments. The addition of more protein structures has, if anything, made the situation more complex, as there are now many cases of nonhomologous sequences folding to the same protein domain. As a consequence there are many cases of similar structure resulting from apparently unrelated sequences, possibly obscuring the sequence–structure correlation. Despite this, as will now be discussed, an approach to secondary structure prediction based on such correlations has been quite successful.

## Several sequence similarity measures have been used to identify nearest-neighbor segments

The nearest-neighbor methods mostly use gapless alignments, so that the generation of all possible alignments is simple and feasible. The key difficulty is to define the nearest-neighbor sequences by a quantitative measure. Despite the availability of many substitution-scoring matrices used in sequence alignment (see Sections 4.3 and 5.1), most of these secondary structure prediction methods frequently derived their own scoring schemes. In this section we will briefly survey some of the measures used, the alternative ways in which the structure database search was restricted, and the choice of nearest neighbors to use in the prediction step. Note that this is not an exhaustive list, but is illustrative of the range of methods used.

The SIMPA96 scoring method is one of the few to use standard scoring matrices, in this case BLOSUM-62. All segments whose alignment scores exceeded a cut-off value are included in the prediction. A window of 15 residues is used, with a score cut-off value of 13. In this method the entire set of known protein structures is searched, and the original work in 1996 reported that on average around 3000 segments contributed to each residue prediction.

Tau-Mu Yi and Eric Lander restricted their method to use only the 50 highest-scoring 19-residue segments to produce the prediction for the central residue. The scoring system they used to identify the nearest neighbors was an equally weighted sum of the Gonnet substitution matrix score (derived for normal sequence alignment problems) and a score made up of propensities in the log-likelihood form of Equation EQ12.5. The propensities referred to 15 different states—five environmental states for each of the three structural states (helix, strand, and coil), defined using the known structure in terms of the solvent accessible fraction of the residue side chain surface and the fraction of the side chain surface that is contacted by polar atoms or solvent molecules. The rationale behind this scheme was that methods based on such environmental classifications had proved useful in predicting protein folds by alignment of nonhomologous proteins.

The NNSSP method uses a modified version of this scoring scheme with 12 structural states by distinguishing between the N- and C-caps and internal sections of helices and strands, as well as the regions just to the N- and C-terminal sides of these structures, β-turns, and coil. In addition NNSSP uses six different environmental states, making 72 different residue states in all. However, the major difference is that the set of protein structures to be searched for nearest neighbors is strictly limited. Two methods were proposed to choose which proteins to include in the search. One is a measure of the difference in amino acid composition, defined as

$$D_{comp} = \sum_{a=1}^{20} \left( p_a^{query} - p_a^{database} \right)^2 \qquad \text{(EQ12.22)}$$

where the terms in the summation refer to the frequencies of occurrence of residue type $a$ in the query and database sequences, respectively. The other measure is based on the Chou–Fasman propensities, and is given by

$$D_{CF} = \sum_{s} \left( \frac{\sum_{j=1}^{N_{query}} P_{s,x_j}}{N_{query}} - \frac{\sum_{j=1}^{N_{database}} P_{s,y_j}}{N_{database}} \right)^2 \quad \text{(EQ12.23)}$$

where the summation is over the structural states $s$ (just helix, strand, and coil), and the query and database sequences have $N_{query}$ and $N_{database}$ residues, and their $j$th residues are written $x_j$ and $y_j$, respectively. This measure reflects the average difference in propensity over the sequence for each structural state. Only those database structures that were among the closest 90 to the query sequence according to both measures (i.e., the smallest 90 values of each measure) were included in the search for nearest-neighbor segments. There is evidence for a correlation between amino acid composition and protein structural class, which might explain why this selection procedure was found to marginally improve the accuracy of the secondary structure prediction.

Steven Salzberg and Scott Cost proposed an alternative distance measure also based on the idea of Chou–Fasman propensities. For a pair of residue types $a$ and $b$ they suggested a distance measure

$$\delta_{ab} = \sum_{s} \left| P_{s,a} - P_{s,b} \right| \quad \text{(EQ12.24)}$$

where the structural states are labeled $s$ as previously, and the sum is over the signless propensity differences. Note that smaller values of $\delta_{ab}$ indicate more similar residues. The distance between two sequence segments is given by adding the $\delta_{ab}$ terms for all aligned positions, but it is also multiplied by a weight for the database segment. This weight is based on previous effectiveness of the segment in correct predictions, and is defined as the ratio of the number of successful predictions with the segment to the total number of predictions made with the segment. The segment scores are therefore smallest for the nearest neighbors.

All the methods mentioned above use a fixed-length sequence window although, as is mentioned below, in some cases several different window lengths are used in separate predictions that are then averaged to produce the final prediction. In the SSPAL method the nearest neighbors are of variable length, and are high-scoring gapped local alignments. Only 90 protein structures were searched for nearest neighbors, selected using Equation EQ12.23. The 50 highest-scoring nonintersecting local alignments (as can be obtained using the techniques described in Section 5.2 and illustrated in Figure 5.17) are identified for each structure. The scoring scheme used is very similar to that of NNSSP. All query sequence positions will hopefully be aligned in a number of these local alignments, and to predict any residue only the 50 highest-scoring alignments are used (see Figure 12.18). The potential advantage of this method is that long alignments are likely to be highly significant indicators of a similar protein fold, and might easily be missed by restricting the alignments to short windows.

It is possible to include information from homologous sequences in nearest-neighbor methods, and as for other methods this resulted in improved prediction accuracy. However, the way this information has been applied here is significantly different from the use in other secondary structure methods. All other such methods

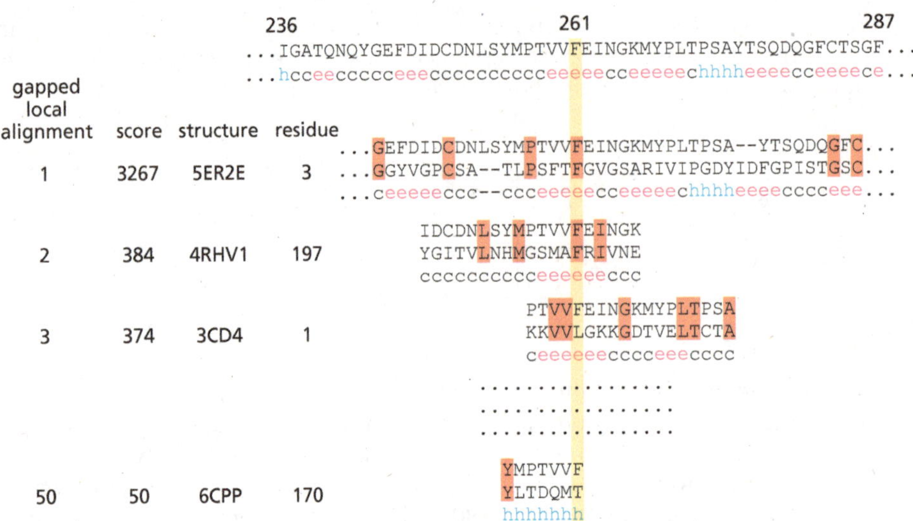


**Figure 12.18**

**Secondary structure prediction of Phe261 of chymosin B using the SSPAL nearest-neighbor method with 50 nearest neighbors.** The 50 nearest neighbors for each position will be those highest-scoring nonintersecting local alignments that include that position, taken from the large set of such alignments (up to 50 for each of 90 database structures). The N-terminal residue number is given for each neighbor. The three-state secondary structure is shown beneath each sequence segment, with h, e, and c meaning α-helix, β-strand, and coil, respectively. The scoring scheme takes account of the residue environment as well as a standard alignment scoring scheme. The final prediction for Phe261 will depend on a weighted average of the 50 different central residue states, and in this case is β-strand. This prediction is dominated by the highest-scoring local alignment score, which is so much higher than the others that it dominates the prediction. (Adapted from A.A. Salamov and V.V. Solovyev, Protein secondary structure prediction using local alignments, *J. Mol. Biol.* 268:31–36, 1997.)

use homologous sequences of the query sequence whose secondary structure is to be predicted. NNSSP and SSPAL use the HSSP database of homologous sequences to known protein structures to augment the database of potential nearest neighbors. Each of the aligned sequences in the HSSP entry for the database structure is scored for the segment, and the average score is used instead of the score for the structure itself. A similar procedure is used in SIMPA96 but the average over the aligned sequences involves a sequence weight, using the scheme described in Figure 6.1.

## A weighted average of the nearest-neighbor segment structures is used to make the prediction

The simplest way to obtain a prediction based on nearest neighbors is to use only the single nearest neighbor, assigning its structure to the equivalent segment of the query protein. This is done by Salzberg and Cost, who use this closest segment to predict the structural state of just the central residue of the query sequence window. This results in no conflicting residue state predictions.

Most published methods only predict the structural state of this central residue of each window. As a result the prediction for each sequence window is independent of other windows. This is not the case for SSPAL (see Figure 12.18). Therefore even when a set of nearest-neighbor segments is considered for a given window it is only necessary to define the method for deciding between the predictions of the different segments. One method is to simply count up the number of occurrences of each structural state for the central residue of each nearest-neighbor segment, and to predict the most commonly found state as that of the query sequence residue. This weights all neighbors equally, despite the availability of a measure to distinguish between them—the measure already used to determine the nearest neighbors. NNSSP and SSPAL, in common with some other methods, use the similarity measure to weight the predictions.

There is a potential problem in simply selecting the most frequent structural state for the prediction, in that the states are not present in the database in equal frequencies, for example usually having more α-helical than β-strand states. This bias can cause errors in the prediction. In SIMPA96 the fractions of α-helix, β-strand, and coil are multiplied by 0.84, 1.11, and 0.79, respectively, before choosing the highest value to make the prediction. In developing NNSSP it was found that this balanced prediction improved the β-strand prediction, but at the cost of overall accuracy as measured by $Q_3$. The difference between the highest and next-highest values can be used to construct a measure of the confidence of the prediction.

Some methods make several predictions using different window sizes and similarity measures, and then combine the different predictions into a final one. NNSSP combines the predictions by simple majority, and also applies a filter to remove helices and strands shorter than three and two residues, respectively. Yi and Lander use a neural network to achieve the final prediction. In both cases this technique succeeds in improving the average $Q_3$ by about 1%.

Leszek Rychlewski and Adam Godzik developed a method that uses all the nearest-neighbor sequence to predict all the query sequence segment states. This uses the 30 nearest 16-residue neighbors for prediction, but because all segment residues are predicted, each residue of the query sequence is predicted from 480 neighbor segments (16 sets of 30). Each neighbor prediction is weighted by its alignment score, which was used initially for neighbor detection. The accuracy of the prediction was found to be only slightly lower at the edges of the segments than at the center, suggesting that most other methods may be ignoring useful information by restricting the prediction to the central residue. However, the average $Q_3$ prediction accuracy of approximately 72% is similar to that of SIMPA96 and NNSSP. The highest reported accuracy of the methods discussed is 73.5% for SSPAL when including homologous sequence information.

## A nearest-neighbor method has been developed to predict regions with a high potential to misfold

An interesting application of the nearest-neighbor method has been reported, which shows promise in detecting sequence segments that can misfold. Such proteins are of medical importance, as a number of diseases such as Alzheimer's seem to be a consequence of such misfolded proteins. It is thought that the misfolding region always misfolds to a β-strand-rich region. The HβP method is of interest not just because of its novel application but also because of the modifications applied to the nearest-neighbor technique.

Long-range contacts with the sequence segment clearly have an influence on the folded conformation, as is evident from the failure of any sequence window-based prediction method yet proposed to be capable of perfect predictions for a database of proteins. A simple measure of the long-range contacts is given by identifying all nonhydrogen atom interresidue contacts of ≤4 Å where the two residues involved are separated by at least four residues in the sequence. All such interactions can be summed for each residue to obtain the TC (tertiary contact) measure.

The HβP method looks for seven-residue segments that have different conformational preferences for different degrees of long-range interaction as measured by the TC values. The segment size is intentionally short to give a significant opportunity to identify nearest neighbors that have different structural states for the central residue. The method depends on finding such structural variation in the neighbors to identify potential misfolding regions. The nearest neighbors were restricted to be seven-residue segments in the structure database that have the same central residue as the query segment. These are scored using the PAM30 substitution scoring matrix, and the top 30 selected, provided their score exceeds 15.

The large variation in the size of amino acid side chains and differences in hydrophobicity cause different residue types to have distinct ranges of TC, with the average over a large protein structure database varying from about 6 to 25. The TC values of the central five residues of each neighbor were analyzed, and the neighbors classified into those with low, intermediate, or high TC values (see Figure 12.19). The fractions were determined of the low-TC neighbors with a central residue in α-helical state, $P(\alpha|low)_{NN}$, and of the high-TC neighbors with a central residue in β-strand state, $P(\beta|high)_{NN}$. Instead of assuming these values to apply to the query sequence, extensive analysis of the test data identified clear correlations between these values and the number of nearest neighbors assigned as low TC and

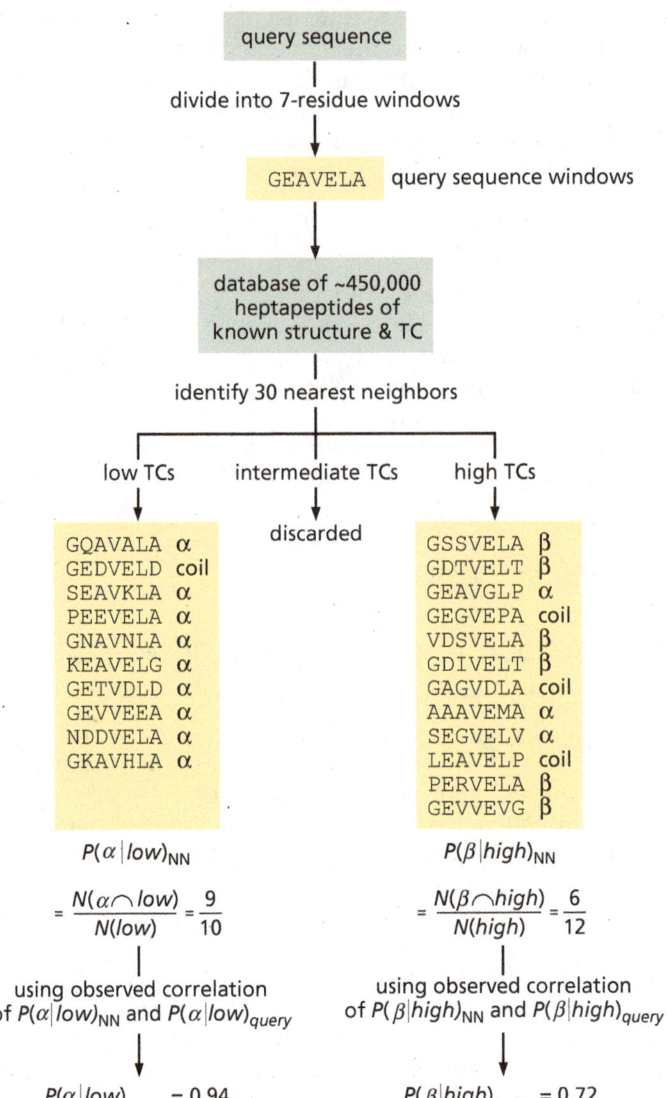


**Figure 12.19**
**The HβP method for predicting amyloidogenic regions.** This is based on the nearest-neighbor concept, in which neighbors are classified according to the amount of tertiary contact (TC). The α-helical preference of the low-TC neighbors and β-strand propensity of the high-TC neighbors are used to make the prediction. The final prediction is based on correlations observed between the number of neighbors and proportion in the α-helical or β-strand state and the known structure in test cases. (Adapted from S. Yoon and W.J. Welsh, Detecting hidden sequence propensity for amyloid fibril formation, *Protein Sci.* 13:2149–2160, 2004.)

high TC, respectively. The correlation curves were used to obtain more accurate values for $P(\alpha|low)_{query}$ and $P(\beta|high)_{query}$. Sequence segments were predicted to have a significant potential to misfold into β-strand-rich structures if they did not fold as β-strands in the native state and also had high values of these two calculated values. An example prediction is shown for the $hIAPP_{4-34}$ sequence associated with amyloid fibril formation in type II diabetes (see Figure 12.20).

## 12.4 Neural Networks Have Been Employed Successfully for Secondary Structure Prediction

We will now look at the application of neural networks in this area, most of which are of the feed-forward type. As will be discussed in detail below, such networks consist of layers of units with communication between units in consecutive layers. All the methods discussed so far in this chapter rely on some form of carefully designed model and statistical analysis of the existing protein structure data to obtain the prediction. The neural network technique appears to lack this element of design and to use a brute force method to achieve the results. Some care is needed to design the network structure, although in most cases this only seems to correlate closely with

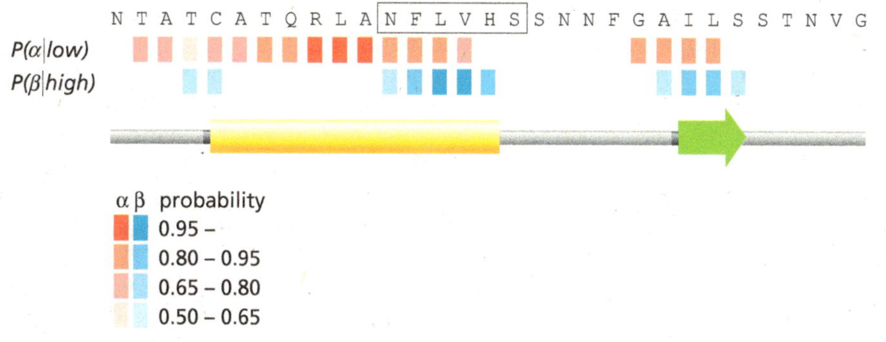


**Figure 12.20**
**The HβP prediction method applied to the hIAPP$_{4-34}$ sequence associated with amyloid fibril formation in type II diabetes.** The predictions of $P(\alpha|low)$ and $P(\beta|high)$ are shown below the sequence, with more intense color indicating a stronger prediction. The line below gives the PHD secondary structure prediction, with a yellow α-helix and a green β-strand. The method predicts the central small segment to be amyloidogenic, as both $P(\alpha|low)$ and $P(\beta|high)$ are high. There is evidence that the six boxed residues are actively involved in the fibril formation, verifying the prediction. (Adapted from S. Yoon and W.J. Welsh, Detecting hidden sequence propensity for amyloid fibril formation, *Protein Sci.* 13:2149–2160, 2004.)

the form of the input and output layer units. However, the form of communication between units seems to lack any element of intentional design. For this reason the application of neural networks can appear to not require careful thought. The studies presented in this section show this to be a misconception, although a certain degree of success can be achieved for minimal effort.

The neural network approach is potentially very powerful because a three-layer feed-forward network can in principle represent a vast range of possible functional forms. The term functional form is meant here in a very general sense, as the relationship

PREDICTING SECONDARY STRUCTURES

accuracy measures
Matthews C
$Q_3$
Sov
N/C-score

structure assignment
irregular 2° structures
DSSP
PALSSE
assignment disagreements

prediction techniques
nearest neighbor
distance measures
Chou–Fasman
GOR
Zpred
TMpred
residue properties
compositional bias
residue propensity
conservation value
hydrophobicity

neural networks
PHD
BRNN
training
architecture
residue representation in input layer
network prediction averaging
sequence to structure
structure to structure
hidden layer

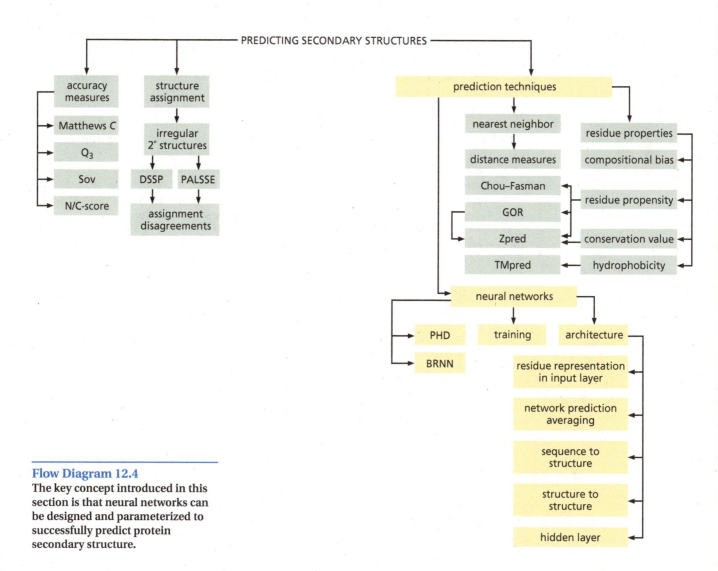

**Flow Diagram 12.4**
**The key concept introduced in this section is that neural networks can be designed and parameterized to successfully predict protein secondary structure.**

between some input data and the output. We cannot write down an exact equation relating protein sequence to secondary structure, but the neural networks considered here are certainly capable of representing forms much more complex than for example those of Equations EQ12.18 and EQ12.19. Furthermore, the networks can be parameterized by giving test sequences and their correct secondary structure without any other information. The ability of the network to represent the link between sequence and structure is not limited by our ability to express the link in theoretical terms. Unfortunately it is difficult if not impossible to extract theoretical insights from the successful neural networks we will discuss below. Therefore, while their use has resulted in progress being made in accurately predicting secondary structure it has not led to a better understanding of the underlying relationships.

In this section, we will begin by describing the details of layered feed-forward neural networks in general terms, both of their structure and how they can be parameterized. Following this the network structures frequently used for protein secondary structure prediction will be described, including a survey of the ways of representing the sequence in the input layer. The final part of this section will look at some more complicated examples of the application of neural networks in this area, especially the use of other network structures.

## Layered feed-forward neural networks can transform a sequence into a structural prediction

The basic structural element of a neural network is the **unit** or **node**. Units communicate with other units, receiving signals and sending out signals based on those received. We will start by examining layered feed-forward networks, as they are the most commonly used and the easiest to understand. In this case all the units are arranged in layers, and there are restricted communications between units, in that there is a clear direction to the units' signals. Starting from the units in the input layer, communication is from these units to those in the next layer, until the output layer is reached (see Figure 11.20). In most of the networks we will consider in detail all units in one layer communicate with all units in the next; these are often referred to as fully connected networks. The consequences of some of these connections being absent are trivial for the discussion that follows. The input data (in this case the protein sequence) are encoded into the **input layer** units such that the information is transmitted on by their signals to the next layer. The output of the **output layer** units is the result (in this case the secondary structure prediction).

The simplest layered feed-forward neural networks have just two layers—the input and output layers—and are usually called **perceptrons**. Their simplicity makes their potential application more limited, but it also makes it easier to interpret their features, as will be seen later. Any additional layers are called **hidden layers** and although several could be used, in most cases there is just one, resulting in a total of three layers. Each layer has a specific number of units, usually not related to the number in any other layer. The number in the input layer is determined by the nature of the input data. For example, if protein sequence is to be represented using one unit for each residue type, for a 13-residue window this will require 260 (13 × 20) units. The output layer typically contains one unit for each possible secondary structure state in the prediction. The number of units in any hidden layers is usually determined by trial and error to produce the best results.

The signals emitted by a unit have values in a range, here taken as 0 to 1, and are a response to the signals they receive. In the case of the input layer, the input data can be regarded as the signal. The details of how this is done for protein sequences will be presented later. Figure 12.21 shows the communications of a single unit in a hidden layer, the $v$th unit of layer $\lambda$. This layer has $N_\lambda$ units, each of which receives signals from all the $N_{\lambda-1}$ units of the preceding layer (labeled $\lambda$–1). The $u$th unit of this previous layer sends a signal $\sigma_u^{(\lambda-1)}$ to every unit of layer $\lambda$, the signal to the $v$th

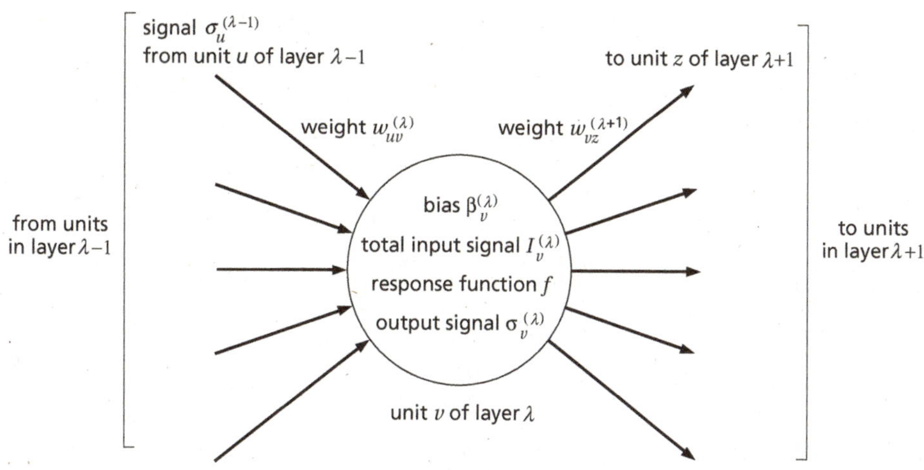


**Figure 12.21**
**A unit in a middle layer λ of a layered feed-forward neural network, showing the connections to other units and the signals and parameters involved.**

unit being modified by a **weight** $w_{uv}^{(\lambda)}$. Thus, the total input signal of the $v$th unit of layer $\lambda$ is given by

$$I_v^{(\lambda)} = \sum_{u=1}^{N_{\lambda-1}} w_{uv}^{(\lambda)} \sigma_u^{(\lambda-1)} \qquad \text{(EQ12.25)}$$

Each non-input-layer unit also has a bias (sometimes called a threshold) which shifts the total input; the bias of the $v^{\text{th}}$ unit of layer $\lambda$ is $\beta_v^{(\lambda)}$. In addition each non-input-layer unit has a **response function** (also called a **transfer function**) which converts the bias-shifted input signal into the output signal. There is no requirement for all units in a network to use the same function, but it is unusual for units in the same layer to use different functions. In the following discussion we assume all units have the same response function, which is therefore simply written $f$. The general equation relating the input and output signals is therefore

$$\sigma_v^{(\lambda)} = f\left[I_v^{(\lambda)} + \beta_v^{(\lambda)}\right] \qquad \text{(EQ12.26)}$$

For the type of application considered here (classifying a residue as having one of a set of structural states) $f$ is usually a step function (see Figure 12.22). This signal is then sent to every unit in the layer $\lambda+1$, but the signal is modified for each unit of this layer by multiplying by the weight $w_{vz}^{(\lambda+1)}$ for the $z$th unit.

The layers are processed in order from input to output. The output unit signals, here usually in the range 0 to 1, are interpreted to obtain the result. In the case of secondary structure prediction, usually each different possible state is associated with a single unit in the output layer, and the signal strength of that unit is taken as the measure of support for that state. A three-state prediction network of this type will have three output layer units, which might produce signals 0.74, 0.34, 0.47 for the α-helix, β-strand, and coil units, respectively. The prediction of this type of network is the state whose unit emits the highest signal, sometimes referred to as the **winner takes all** strategy, in this case giving a prediction of α-helix. Note that if the output layer units have response functions of the kind shown in Figure 12.22, the signals are not constrained to sum to any particular value and cannot be interpreted as probabilities. It is possible to ensure the output layer signals sum to 1 by using the normalized exponential response function (also called **softmax**)

$$f\left[I_z^{(output)} + \beta_z^{(output)}\right] = \frac{e^{-\left(I_z^{(output)} + \beta_z^{(output)}\right)}}{\sum_z e^{-\left(I_z^{(output)} + \beta_z^{(output)}\right)}} \qquad \text{(EQ12.27)}$$

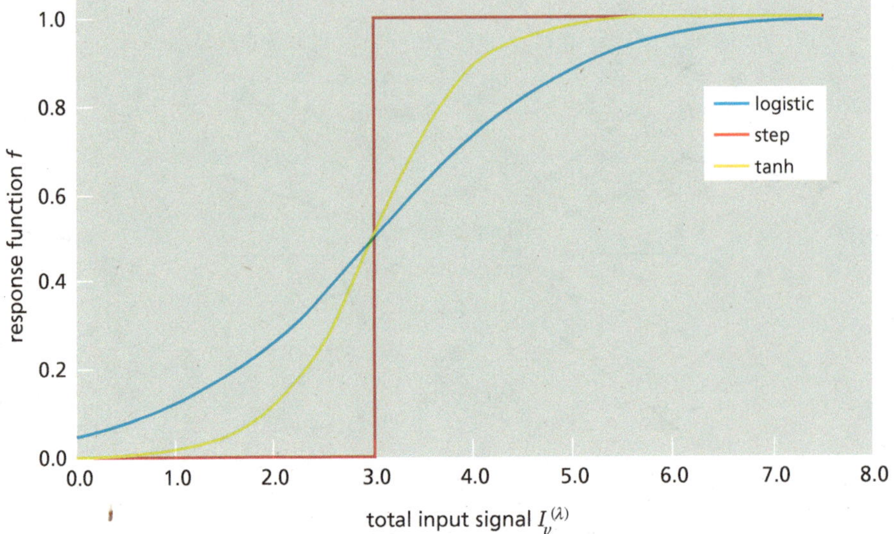


**Figure 12.22**
**Three typical response functions of a neural network unit.** All are shown for the vth unit of network layer $\lambda$ with a bias $\beta_v^{(\lambda)}$ of –3. The horizontal axis is the total input signal $I_v^{(\lambda)}$ before the bias is applied. The blue line is the logistic function given by

$$f(x)=\frac{1}{1+e^{-sx}}$$

where the gain $s$ is a constant often, as here, taken as 1; the red line is the step function; the yellow line is the tanh function

$$f(x)=\frac{\tanh(x)+1}{2}$$

where $x$ is the bias-adjusted input signal (i.e., $x=I_v^{(\lambda)}+\beta_v^{(\lambda)}$).

where the sum is over all output layer units $z$. In this case the output values can be interpreted as probabilities. The networks described here use a sequence window as input and interpret the output as the predicted state of the central residue.

The parameters of neural networks, namely the weights and biases of the units, need to be specified before use in prediction. This is in addition to deciding on the architecture, how many units to have in each layer, and the response function(s). The values of parameters are usually determined by a process of training the network using a set of data for which the correct output signals are known. In the current context, this requires a dataset of protein sequences and the associated secondary structure states of each residue. The training process is described in detail in Box 12.1.

One of the most important aspects of neural networks is the way the input data are encoded within the network. In this section we will look at representing a single sequence, leaving the discussion of multiple sequences for the following section. There is a simple way to represent a single protein sequence, although as will be discussed below it is not necessarily the best. The input layer units represent a window of sequence. At each sequence position any of the residue types could occur, so by having 20 input layer units for each residue position, the residue type can be encoded in a straightforward way. Usually the unit corresponding to the residue present at that position will have an output signal of 1, and the other 19 units will have signals of 0. These signals will be modified by the unit weights before being transmitted to the next layer. This design is often referred to as **orthogonal encoding**. Most networks also have another unit, often called the **spacer unit**, at each position to account for the absence of a residue, which will occur at the N or C terminus, although this can also be encoded by all 20 units having an output of 0. While this representation contains all the sequence information, it requires a large number of input layer units and often results in a network with many parameters to fit. Several of the secondary structure prediction networks discussed below have over 10,000 parameters, most associated with the input layer. An example of a neural network for nucleotide sequence analysis that uses orthogonal encoding is illustrated in Figure 10.14.

One of the first applications of neural networks to protein secondary structure prediction was due to Ning Qian and Terrence Sejnowski. They made a detailed study of the use of perceptrons and fully connected networks with one hidden layer, all using the orthogonal encoding for the residues. (They tried other encoding

## Box 12.1 Neural networks must be parameterized by training before use for prediction

Neural networks contain many parameters, such as the biases $\beta_v$ and weights $w_{uv}$, in some cases over 10,000 parameters in total. These are fitted using a training dataset composed of sequences with known secondary structures in a procedure called **supervised learning**. This involves using the network to predict secondary structures for the training dataset and learning a good parameter set by step-by-step improvements based on the prediction errors. Two aspects of the training of neural networks will be explored. Firstly, the details of the scheme to update parameters will be described. Secondly, the importance of having a balanced training dataset will be discussed, which involves careful consideration of the secondary structure content and sequence redundancy of the training data.

The method commonly used to iteratively modify the parameters to improve the prediction accuracy is called the **back-propagation method**. The essence of this method is to use the network to predict sequences of known secondary structure, from which the errors $E$ in the output can be calculated. More precisely, a sequence window $W$ is taken from the training database, and the secondary structure of the central residue predicted and compared with the known state. If we can calculate the influence of the weights on this error, i.e., all the terms $\partial E / \partial w_{uv}^{(\lambda)}$, these can be used to derive new weights that should result in a reduced error. This is a standard function optimization problem (minimizing $E$) using first derivatives, dealt with in Appendix C, and discussion here will be restricted to aspects particular to training neural networks. The steepest descent method would use the form

$$\Delta w_{uv}^{(\lambda)} = -\eta \frac{\partial E}{\partial w_{uv}^{(\lambda)}} \tag{BEQ12.1}$$

where $\eta$ is a constant, here called the **learning rate**, and $\Delta w_{uv}^{(\lambda)}$ is the correction to be applied to $w_{uv}^{(\lambda)}$. However, it is more common (because faster convergence is usually achieved) to use a form related to conjugate gradients

$$\Delta w_{uv}^{(\lambda)}(t) = -\eta \frac{\partial E}{\partial w_{uv}^{(\lambda)}}(t) + \varepsilon \Delta w_{uv}^{(\lambda)}(t-1) \tag{BEQ12.2}$$

where $\varepsilon$ is a constant called the momentum term, and the labels ($t$–1) and ($t$) refer to two consecutive estimations of the weights. Typical values of $\eta$ and $\varepsilon$ used in training secondary structure prediction networks are 0.0001–0.05 and 0.2–0.9, respectively. Note that similar schemes can also be derived for other parameters such as the unit biases, but these will not be given here.

To derive specific formulae for the derivatives $\partial E / \partial w_{uv}^{(\lambda)}$ we first note that the error $E$ must be a function of the output signals $\sigma_v^{(\lambda)}$, which in turn are functions of the input signals $I_v^{(\lambda)}$, which are functions of the weights $w_{uv}^{(\lambda)}$. By application of the chain rule

$$\frac{\partial E}{\partial w_{uv}^{(\lambda)}} = \frac{\partial E}{\partial \sigma_v^{(\lambda)}} \frac{\partial \sigma_v^{(\lambda)}}{\partial I_v^{(\lambda)}} \frac{\partial I_v^{(\lambda)}}{\partial w_{uv}^{(\lambda)}} \tag{BEQ12.3}$$

Equation EQ12.25 yields

$$\frac{\partial I_v^{(\lambda)}}{\partial w_{uv}^{(\lambda)}} = \sigma_u^{(\lambda-1)} \tag{BEQ12.4}$$

Note that this involves the signals from the previous layer in the network. From Equation EQ12.26 the middle term on the right-hand side of Equation BEQ12.3 can be written $f'_v[I_v^{(\lambda)}]$, where the prime symbol indicates the first derivative of the function $f$. This is usually easily calculated, in the specific case of the logistic response function (see Figure 12.22) being

$$\frac{\partial \sigma_v^{(\lambda)}}{\partial I_v^{(\lambda)}} = \frac{\partial}{\partial I_v^{(\lambda)}}\left(\frac{1}{1+e^{-\left(I_v^{(\lambda)}+\beta_v^{(\lambda)}\right)}}\right) = \frac{e^{-\left(I_v^{(\lambda)}+\beta_v^{(\lambda)}\right)}}{\left(1+e^{-\left(I_v^{(\lambda)}+\beta_v^{(\lambda)}\right)}\right)^2} = \sigma_v^{(\lambda)}\left(1-\sigma_v^{(\lambda)}\right) \tag{BEQ12.5}$$

Note that when calculating the bias error gradients $\partial E / \partial \beta_v^{(\lambda)}$, the partial derivatives $\partial \sigma_v^{(\lambda)} / \partial \beta_v^{(\lambda)}$ are needed and are identical to $\partial \sigma_v^{(\lambda)} / \partial I_v^{(\lambda)}$. Leaving the response function unspecified, Equation BEQ12.3 becomes

$$\frac{\partial E}{\partial w_{uv}^{(\lambda)}} = \frac{\partial E}{\partial \sigma_v^{(\lambda)}} f'_v\left[I_v^{(\lambda)}\right]\sigma_u^{(\lambda-1)} \tag{BEQ12.6}$$

The terms $\partial E / \partial \sigma_v^{(\lambda)}$ have not yet been defined. Because the signals $\sigma_v^{(\lambda)}$ are sent to all units of the next network layer (except in the case of the signals of the output layer units), the error $E$ in the final network output will have components from all these units. All these components must be added together, so that the $\partial E / \partial \sigma_v^{(\lambda)}$ can be expressed in terms of the equivalent values for the units of the layer next-nearest to the output layer using the chain rule as for Equation BEQ12.3:

$$\frac{\partial E}{\partial \sigma_v^{(\lambda)}} = \sum_{\substack{\text{units } z \text{ of} \\ \text{layer } \lambda+1}} \frac{\partial E}{\partial \sigma_z^{(\lambda+1)}} \frac{\partial \sigma_z^{(\lambda+1)}}{\partial I_z^{(\lambda+1)}} \frac{\partial I_z^{(\lambda+1)}}{\partial \sigma_v^{(\lambda)}} \tag{BEQ12.7}$$

Using Equation EQ12.25, the last right-hand side term is, $w_{vz}^{(\lambda+1)}$ which on substitution gives

continued ...

## Box 12.1 Neural networks must be parameterized by training before use for prediction (continued)

$$\frac{\partial E}{\partial \sigma_v^{(\lambda)}} = \sum_{\substack{\text{units } z \text{ of} \\ \text{layer } \lambda+1}} \frac{\partial E}{\partial \sigma_z^{(\lambda+1)}} f_z'\left[I_z^{(\lambda+1)}\right] w_{vz}^{(\lambda+1)} \qquad \text{(BEQ12.8)}$$

The values $\partial E / \partial \sigma_v^{(\lambda)}$ are easily defined for the output layer, which then allows the terms for the other layer units to be calculated by back propagation of this formula, hence the name of the technique.

There is a choice of definition for the network prediction error $E$, which leads to different forms of the terms $\partial E / \partial \sigma_v^{(\lambda)}$ of the output layer units. A simple and commonly used definition of $E$ is the sum of squared deviations from the correct values, which for the prediction of the structural state of the central residue of a particular sequence window $W$ is given by

$$E_W = \sum_{\substack{\text{units } v \text{ of} \\ \text{output layer}}} \left(\sigma_{v,W}^{(output)} - c_{v,W}\right)^2 \qquad \text{(BEQ12.9)}$$

where $c_{v,W}$ is the desired output of the output layer unit $v$—0 or 1 according to whether that unit corresponds to the actual structural state of the central residue of window $W$. There are occasions when it can be argued that the relative entropy (see Equation EQA.9) between the signals and desired output is a more appropriate measure. Using each window $W$ to generate updated parameters does not lead to stable optimization, and often a batch of several hundred windows is used, summing the errors.

To train a network, initial weights are assigned, usually by taking random numbers in a range such as ±0.2. The training data are predicted using the network, updating the parameters for each batch. Because the predictions for consecutive residues in a protein sequence are highly correlated it is usual practice to randomly select the order of windows for training. This training cycle is repeated, predicting the training data many times, until the error is sufficiently small or cannot be further reduced.

Considerable care is required in selecting the training data, beyond the usual issue of ensuring there is minimal homology between any of the proteins involved. The dataset will contain uneven proportions of the structural states, for example 47% coil, 32% helix, and only 21% strand. If these data are used in these proportions it will bias the network, as the training would be twice as sensitive to coil as compared to strand errors. To circumvent this, **balanced training** involves selection of equal numbers of examples of the different structural states. A sample of the data also needs to be kept for testing the trained network, in order to obtain a realistic assessment of the prediction accuracy.

The final training issue we will examine involves overtraining the network. Particularly when the network contains a large number of parameters, there is a tendency for the network to be trained beyond reproducing general features, and to start reproducing the particular features of the training data (see Figure B12.1). The characteristic sign of this is that the prediction accuracy as measured using the test data decreases despite the accuracy of the training dataset prediction having increased. There are two ways to respond to this. The system can be frequently monitored, and the optimally performing network used as the working version. Alternatively, training can be stopped as soon as the test data performance decreases. Neither of these is entirely satisfactory, as they suggest the true accuracy will be less than that measured on the test data. The optimal solution is to remodel the network to use fewer parameters, which should reduce the ability to overfit the data.

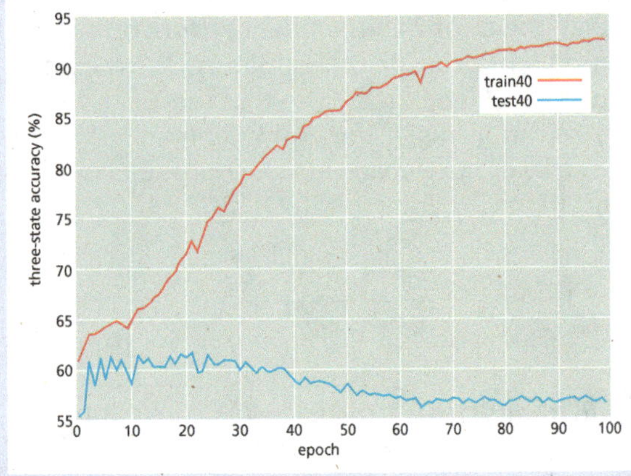


**Figure B12.1**
**An illustration of overtraining a network.** As the network is trained it is tested to assess the current accuracy. The training time is measured in epochs, an epoch being the presentation of the complete training dataset. Thus by the end of the graph, the data have all been presented 100 times to the network. The accuracy measured on the training data continues to increase, meaning that the parameters are still not fully optimized to fit the training data. However, the accuracy measured on the test data (a set that is completely independent of the training data) reached a maximum after approximately 20 epochs, and subsequently the performance worsened. This is due to the network learning specific features of the training dataset as opposed to general features. (Adapted from S.K. Riis and A. Krogh, Improving prediction of protein secondary structure using structured neural networks and multiple sequence alignments, *J. Comput. Biol.* 3:163–183, 1996.)

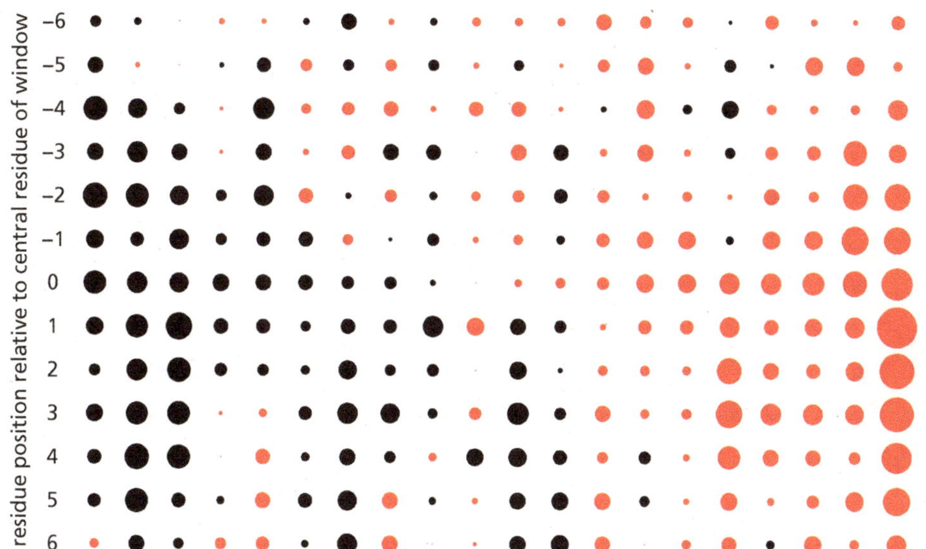


**Figure 12.23**
**The weights applied to signals arriving at the α-helix output layer units from the input layer units in the neural network of Qian and Sejnowski.** The areas of the circles are proportional to the values of the weights, with red circles for negative weights and black circles for positive weights. There is a general agreement in terms of residue preferences with the residue propensities described in Section 12.2, although there are differences with data such as used by the GOR method. This form of weight representation is called a Hinton diagram.

methods but could not obtain any accuracy gains with them.) One network they trained used a 13-residue window, represented as a 273-unit input layer of 13×21 units), all of which were connected directly to three output layer units. In principle this network could reproduce the early GOR method as described by Equation EQ12.18. The resulting network achieved an average $Q_3$ of 62.5%, and the input layer units had weights to the α-helix output layer unit as shown in Figure 12.23. The simplicity of the network allows a straightforward interpretation of these weights in terms of the residue α-helix preferences. The residues are ordered in Figure 12.23 according to their weight in the central position of the window (marked 0). A comparison with Figure 12.8 shows that this order correlates very strongly with the order given by the Chou–Fasman α-helix propensities, yet this has arisen purely through training the network.

When the network included a hidden layer there was only a very marginal improvement in the average $Q_3$, but the training of the network was more efficient. The most accurate network had 40 hidden layer units. Recognizing that the network had no way of correlating the secondary structure of neighboring residues, an attempt was made to correct for this omission by using the output from the network as input to a further network. The general structure of this network architecture is shown in Figure 12.24. It has been used in several studies with modifications, including PSIPRED and several versions of PHD. The first network is often called a **sequence-to-structure network**, and the second a **structure-to-structure network**, the names deriving from the meaning of the input and output. Using this cascaded network architecture, Qian and Sejnowski obtained a system that gave an average $Q_3$ of 64.3%.

Many of the modifications to the basic scheme of Figure 12.24 involve extra input data and extra output layer units for the sequence-to-structure network. In the 1996 version of PHDsec, apart from the units for a 13-residue sequence window, the input layer of both networks has a further 32 units, 20 of which describe the residue composition of the sequence, 4 of which give approximate information about the sequence length, and 8 of which give an indication of how close the particular window is to the N and C termini. The last 12 of these are simple 0/1 signals; for example only one of the four units reporting sequence length will have a non-zero signal (with a value 1) indicating whether the length is ≤60, ≤120, ≤240, or >240 residues. Thomas Petersen and colleagues used a similar scheme for the sequence-to-structure network input layer, and also modified the output layer of

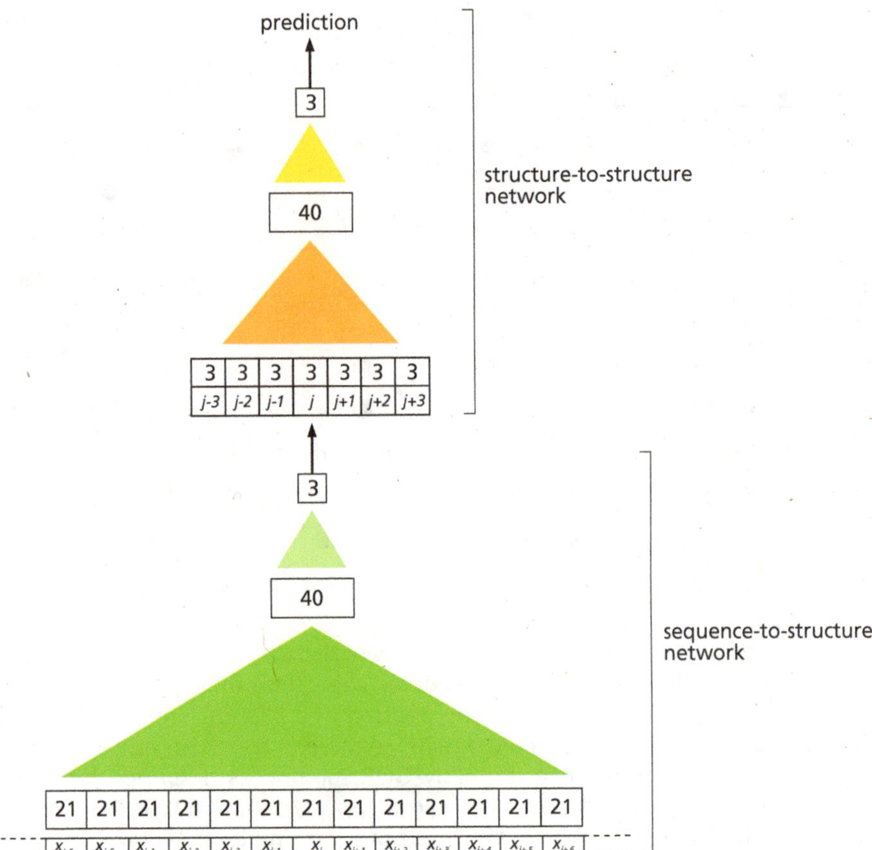


**Figure 12.24**
**Schematic representation of the double neural network architecture frequently used in protein secondary structure prediction.** The network architecture details are taken from the work of Qian and Sejnowski. The large colored triangles indicate schematically all the connections (these are fully connected networks), and the black boxes immediately beneath them represent units in a network layer, with the numbers of units given in each box. A single 13-residue window (residues $x_{j-6}$ to $x_{j+6}$) is shown for the sequence-to-structure network, resulting in the initial α, β, and coil structure prediction for the central residue ($x_j$) represented by the black arrow. All the communications between units associated with this network are drawn in green. The intermediate predictions for seven positions $j$–3 to $j$+3 are the input to the structure-to-structure network, which produces the final prediction for residue $x_j$.

the same network. As well as trying architectures with three output layer units as in Figure 12.24, they also tried networks with nine output layer units corresponding to the prediction of the central residue of the window and the two residues immediately adjacent. They called this **output expansion**, and, when used, all nine signals were passed as input to the structure-to-structure network. In another common modification of the model shown in Figure 12.24, a spacer unit is added to the input to the structure-to-structure network, which usually results in four units per window position.

Ideally the representation of the sequence in the input layer will only contain information on those features which strongly correlate with the secondary structure. For example, in a network trying to predict just α-helices, having a single unit at each window position with as output signal the value of the Chou–Fasman helical propensity, might be an effective and efficient representation. The drastic reduction in input layer units and thus parameters should make it easier to train the network. However, such an approach risks imposing our misconceptions and simplifications on a system that in theory can improve on them. It is possible to achieve a similar effect while still allowing the network to identify the encoding. Søren Riis and Anders Krogh proposed a network in which each set of 20 input units representing one window position is represented in a first hidden layer by three units (see Figure 12.25). All 20 input units send signals to all three hidden units, involving a total of 63 parameters—60 weights and 3 biases. Since the intention is for the three hidden units to provide an encoding of the amino acids, this structure is repeated at all window positions, with the same values used in all cases for the 63 parameters. This technique of using the same weights for equivalent parts of a network is called **weight sharing**. The first hidden layer will contain 39 units for a 13-residue window, and these send signals to units in a second hidden layer, which then signals the output layer (see Figure 12.26). At the expense of this extra layer requiring just 63 parameters, the 260 units of an orthogonal encoded

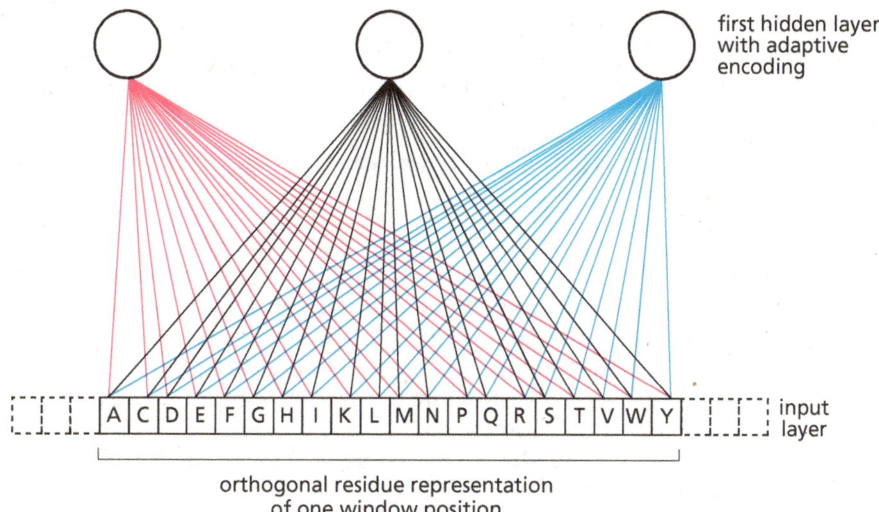


**Figure 12.25**
**The neural network architecture employed by Riis and Krogh to implement adaptive encoding.** The encoding converts the 20 signals at one window position into three in the first hidden layer. The three hidden layer units receive the same signals from the input layer, but apply different weights to them. The structure and weights are reproduced at each window position.

network have been replaced by just 39 units. As a consequence the total number of parameters has been reduced by a factor of 20 or more.

Riis and Krogh applied these networks to the protein secondary structure problem in a novel manner. Separate networks were used to predict the three states—α-helix, β-strand, and coil—and, for each, several alternative networks were used with slight differences such as different numbers of units in the hidden layer or different training. In the α-helix networks a 3-residue pattern of connections was used rather than full connectivity (see Figure 12.26), attempting to mimic something of the structural regularity of helices. These predictions were combined in a similar manner to the structure-to-structure network in Figure 12.24 to obtain the final prediction with an average $Q_3$ accuracy in excess of 66%.

There are alternative ways to average over the predictions of several networks. For example in PHDsec where the output layer of the structure-to-structure network has three units for helix, strand, and coil states, the arithmetic mean of the particular structure state signals of all networks is calculated. This is referred to as **jury decision**, and the final prediction of the residue structural state is taken as the state with the highest average. An alternative method suggested by Thomas Petersen and colleagues, and called **balloting probabilities**, uses the difference between the two highest signals from the units of the output layer of the structure-to-structure network as a measure of the prediction confidence. Only a subset of the separate network predictions is used to make the final prediction, chosen as those networks

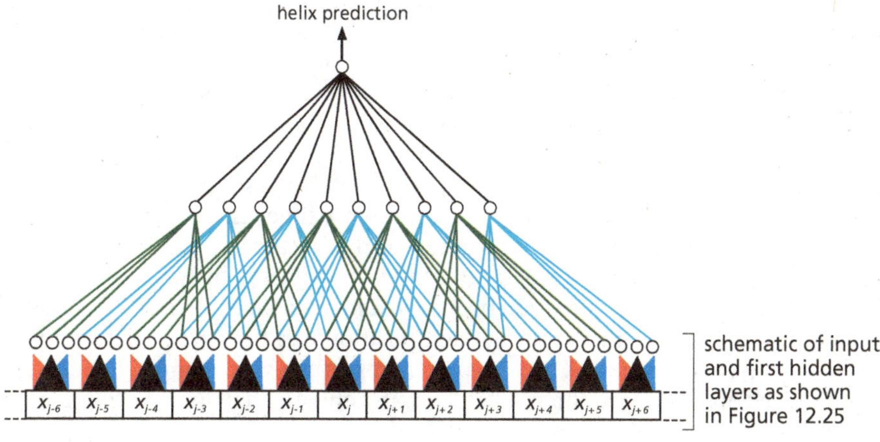


**Figure 12.26**
**Example of a Riis and Krogh adaptive encoding neural network architecture for α-helix prediction.** The repeated red, blue, and black triangles represent the adaptive encoding as shown in detail in Figure 12.25. The communication from the first hidden layer to the second has a repeat pattern intended to mimic the helical structure, and has been colored for clarity. However, the weights are all independent, and do not relate in any way to the colors.

with the highest average prediction confidence over the query sequence. The final prediction is a weighted average of the output signals for this subset of networks, weighted by the average confidence for the network over the sequence.

## Inclusion of information on homologous sequences improves neural network accuracy

All the neural network methods mentioned above have been described as applied to the problem of prediction from a single protein sequence. As has been explained previously in relation to other prediction techniques, including information from homologous sequences can produce significant improvements in prediction accuracy. The same has proved to be the case for neural networks, and many of the networks presented above have been modified to incorporate knowledge of homologous sequences. As was the case for GOR V and Zpred, the information used involves sequences related to the query sequence. These sequences must be provided to the neural network as a multiple alignment including the query sequence.

Before exploring how this information is used in the prediction it is necessary to briefly consider how it is obtained. Secondary structure prediction can only be improved by homologous sequences if they are correctly identified and also correctly aligned, or else they are likely to reduce the accuracy instead. If the prediction program is to be truly automated, it needs to begin by identifying the sequences homologous to a query sequence. This involves the methods described in detail in Chapters 4, 5, and 6. Care must be taken not to erroneously include unrelated sequences while still trying to include all homologous sequences, as there will not be any manual intervention before the prediction when such problems could be identified. Most methods currently use PSI-BLAST (see Section 6.1) and try to avoid the problems just mentioned by restricting the search to three iterations, having strict E-value thresholds, and by masking low complexity sequence segments (see Box 5.2). Some researchers have proposed masking all predicted coiled coils and transmembrane helix segments or even removing from the sequence search database any entries with such regions.

There are two fundamentally different ways of including sequence homology information in a prediction. Each homologous sequence can be used separately to obtain a secondary structure prediction, followed by averaging these single-sequence predictions on the basis of the alignment to produce the final prediction. This is the method used by GOR V and Zpred, and presented in Section 12.2, but has also been applied to neural networks by Riis and Krogh as discussed below. However, most neural network prediction methods use the alternative technique, in which the information present in all sequences is used to modify the input to the neural network, resulting in a single prediction for the entire alignment.

Riis and Krogh used each homologous sequence independently to predict a secondary structure, arguing that any other way of using the aligned sequences as input to the neural network would lose information on the correlations between nearby residues in the sequence. These correlations are precisely the information used as the basis of the GOR methods (see Section 12.2). Once the predictions have been made for all the sequences, they are averaged using the sequence alignment. Each sequence is assigned a weight, based in part on a measure of the entropy of the alignment. Two forms of prediction average are possible. The individual structure state signals can be weighted and then averaged, taking the largest resulting value at each position as the predicted state. Alternatively, and used by Riis and Krogh, the individual sequence predictions can be resolved to states at each position, and then the weighted average of the states at each position can be used to determine the alignment-based prediction. A final step in their method was to apply a filter, in this case a 15-residue window three-layer neural network that took account of the degree of conservation at each position and the presence of indels. This method produced a $Q_3$ improvement of 5% over the single sequence prediction, and an average $Q_3$ accuracy of 71.3%.

Almost all other neural network secondary structure prediction methods include homologous sequence information by using the multiple alignment to modify the signals in the input layer. When the input layer uses orthogonal encoding for the residues this has the advantage that all the information can be input using the same network architecture as for single sequence work. In the 1996 version of the PHDsec prediction method each sequence position is represented in the neural network by 24 units. The residues at each alignment position are represented by their frequencies, which directly replace the 0/1 representation of the single sequence. There are an additional four other units—three directly related to the multiple sequence alignment and one spacer unit as described previously. One unit has a signal that corresponds to the number of gaps (deletions) in the alignment at that position. Another unit's signal corresponds to the number of insertions at that position, since only the alignment positions in which the query sequence is present are explicitly modeled in the network. A further unit signal is related to the residue conservation observed in the alignment at that position. The PHDsec method achieved an average $Q_3$ accuracy above 72%.

The PSIPRED method of David Jones uses a PSSM created using PSI-BLAST as the source of information on homologous sequences, as shown in Figure 11.24. Since the PSSM contains residue alignment scores rather than frequencies (see Section 6.1) there are both positive and negative values, and these need to be converted to be suitable signals for the input layer units. The PSIPRED method uses a scaled logistic function for this (see Figure 12.22), where terms $e^{-X}$ are replaced by terms $e^{mX}$ with $m$ a constant. The average $Q_3$ accuracy for this method exceeded 75%. The PSI-BLAST PSSM is currently acknowledged as the optimal way to introduce homologous sequence data into the prediction, although there are variations in the fine details of implementation.

## More complex neural nets have been applied to predict secondary and other structural features

A number of studies have used neural networks to predict β-turns, with some even trying to predict which class of β-turn. The BTPRED method for predicting an unclassified β-turn uses a feed-forward network with a single hidden layer, very similar in architecture to the sequence-to-structure networks mentioned above. However, the input layer (representing a nine-residue window) uses 23 units per residue—20 for the sequence using orthogonal encoding (homologous sequences are not used in this method) and three that are from the output of PHDsec for the central residue. PHDsec produces reliability indices, which are a scaled measure of the difference between the highest and next highest output signals, in the range 0–9 in PHDsec, but converted to the 0–1 range here. The problem of balanced training sets was explored in some detail during the training of this network.

The Betaturns method has separate networks to predict the different classes of β-turn. This method can use PSI-BLAST PSSM data as input, and uses a two network system similar in general structure to Figure 12.24. This method also includes the secondary structure prediction, this time from PSIPRED, and this time that information is included in the second network together with a single output unit signal from the first network. Thus each residue position is represented in the input to the second network by four units.

An iterative architecture known as a cascade-correlation neural network is used in the DESTRUCT method for helix, strand, and coil prediction. This method also predicts the ψ torsion angle of each residue, which is known to discriminate quite

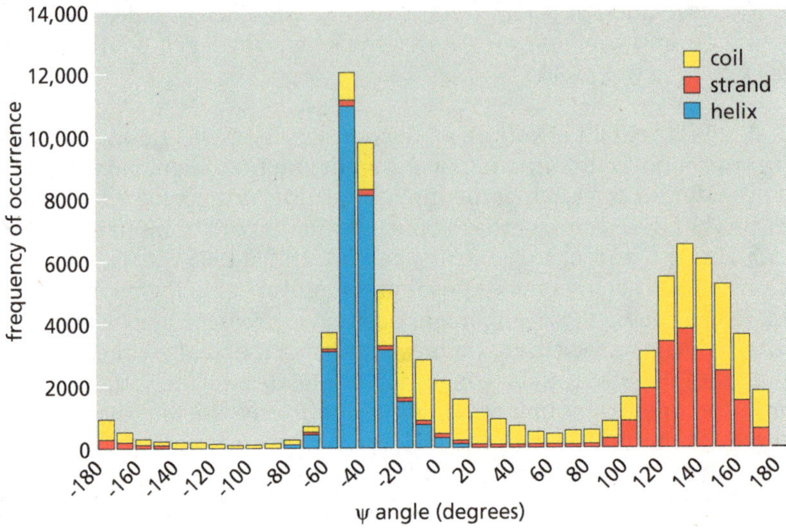


**Figure 12.27**
**The correlation of secondary structure with ψ angle.** Almost all α-helical residues have a ψ angle in the region –60° to 0°, and almost all β-strand residues have a ψ angle of 100–180°. (Compare with Figure 12.2.)

well between residues in helices and strands (see Figure 12.27). Using a 15-residue window and a PSI-BLAST PSSM, two separate networks with one hidden layer were used to obtain preliminary predictions of the three structural states and the ψ angle of the central residue. These initial values were then used in an iterative network (see Figure 12.28) to make the final predictions. Three iterations were found to give the most accurate results, improving the average $Q_3$ by 5% from the initial prediction. The network was trained using a technique involving second derivatives (see Appendix C) and incorporated a method for adding hidden layer units in a controlled way to improve performance. This method is reported to have an average $Q_3$ prediction accuracy in the region of 80%.

Another network architecture that has been found useful for secondary structure prediction is the bidirectional recurrent neural network (BRNN), as implemented in SSpro. This network design (see Figure 12.29) has been used for normal three-state predictions, and also to predict all eight DSSP structural states (see Table 12.1). Only a brief overview can be presented here, and the original papers must be read for further details. The prediction for the central residue of a window does not only depend on the sequence of that window, but also on two hidden layers, one (*F*) derived using the sequence N terminal and the other (*B*) derived using the sequence C terminal of the window. These two hidden layers (called the forward and backward contexts) are derived by iteration along the sequence, one position at a time from the ends, and have as index the sequence position of the residue whose structure prediction they directly influence. The structural prediction of residue $x_i$ is determined by signals from the output layer units that have direct connection with the units of contexts $F_i$ and $B_i$ and also from the hidden layer, which receives the window input $I_i$. $F_i$ is obtained iteratively, starting at $F_0$, with $F_i$ being the direct result of context $F_{i-1}$ and $I_i$. $B_i$ is also iteratively determined, in this case starting at the C terminal, with $B_i$ being the direct result of context $B_{i+1}$ and $I_i$. This method has an average $Q_3$ accuracy of approximately 78%.

## 12.5 Hidden Markov Models Have Been Applied to Structure Prediction

The prediction of protein secondary structure can also be achieved using hidden Markov models (HMMs). The basic concepts behind HMMs were presented in Section 6.2 in the context of sequence profiles, but examples of the use of such models were also given for gene annotation work. (See Figures 10.9 and 10.19.) We

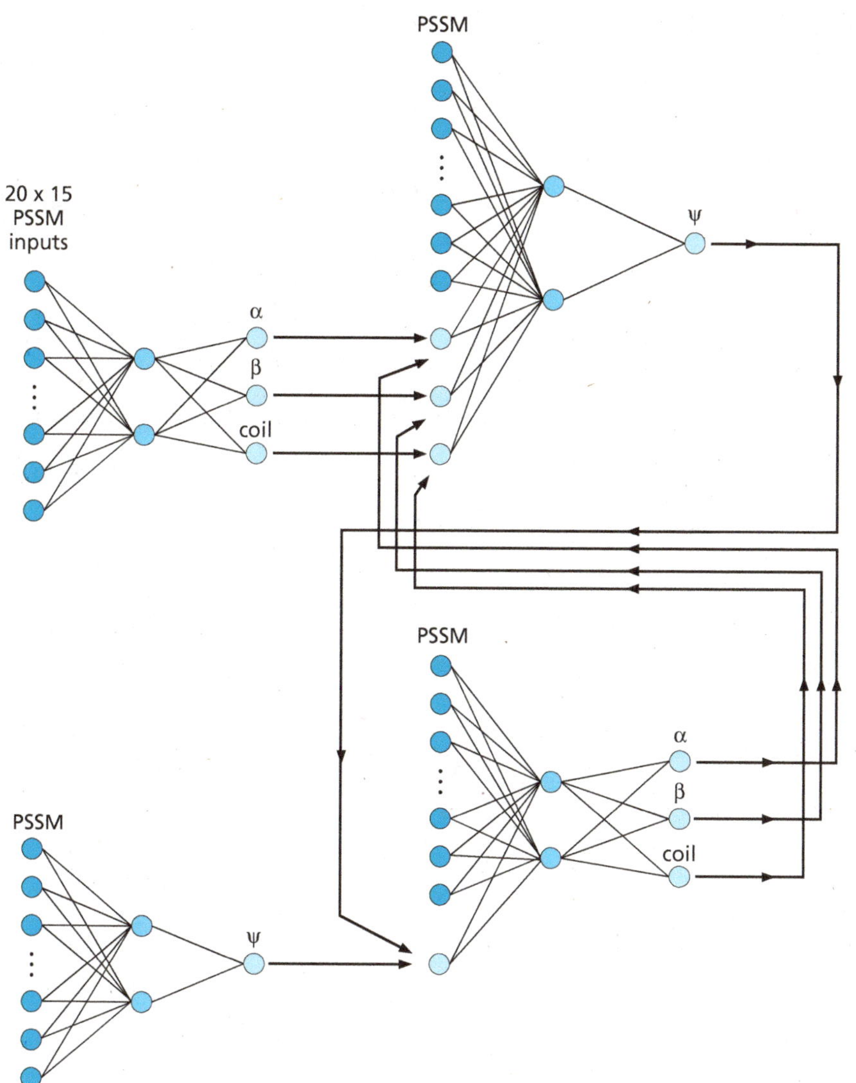


**Figure 12.28**
**The architecture of the neural network used in the DESTRUCT method.** On the left side are the two networks that produce (upper network) the initial secondary structure and (lower network) the ψ predictions. These both have as input only the sequence information from the PSSM, in a 15-residue window. The right side of the figure shows the iterative network architecture, which consists of two independent networks. The upper one predicts ψ from the PSSM and three-state secondary structure prediction. The lower one predicts the three-state secondary structure from the PSSM and ψ prediction. The networks are run iteratively to convergence. (Adapted from Fig. 1 of Wood & Hirst, 2005.)

shall present some HMMs that have been designed to predict a number of different specific structures. In this subject area, they are most commonly used in the prediction of secondary structural elements of membrane proteins, although their use in globular proteins will also be described. The models used for membrane protein work are notable for the explicit representation of the protein architecture in the structure of the model itself.

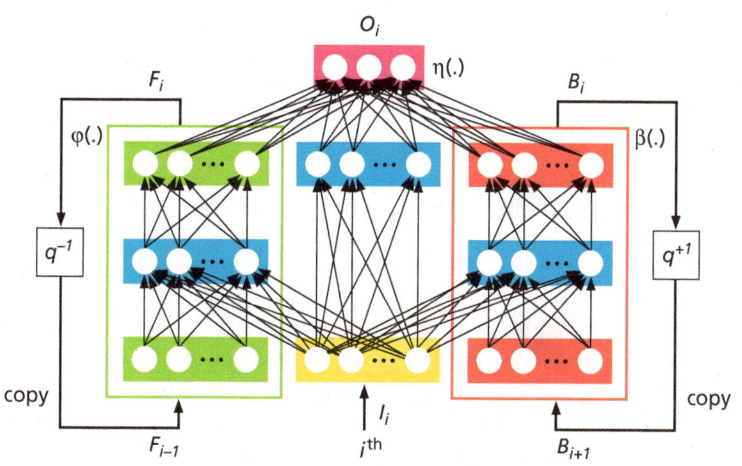


**Figure 12.29**
**The bidirectional recurrent neural network (BRNN) used in the SSpro prediction method of Pollastri et al., 2002.** The working of this complex neural network is briefly described in the main text. The left part of the network is called the forward context ($F$), and the right part the backward context ($B$). The output layer ($O_i$) has three units, corresponding to predicting α-helix, β-strand, or coil for residue $i$. The hidden layers are all colored blue. (From P. Baldi et al., Exploiting the past and the future in protein secondary structure prediction, *Bioinformatics* 15:937–946, 1999, by permission of Oxford University Press.)

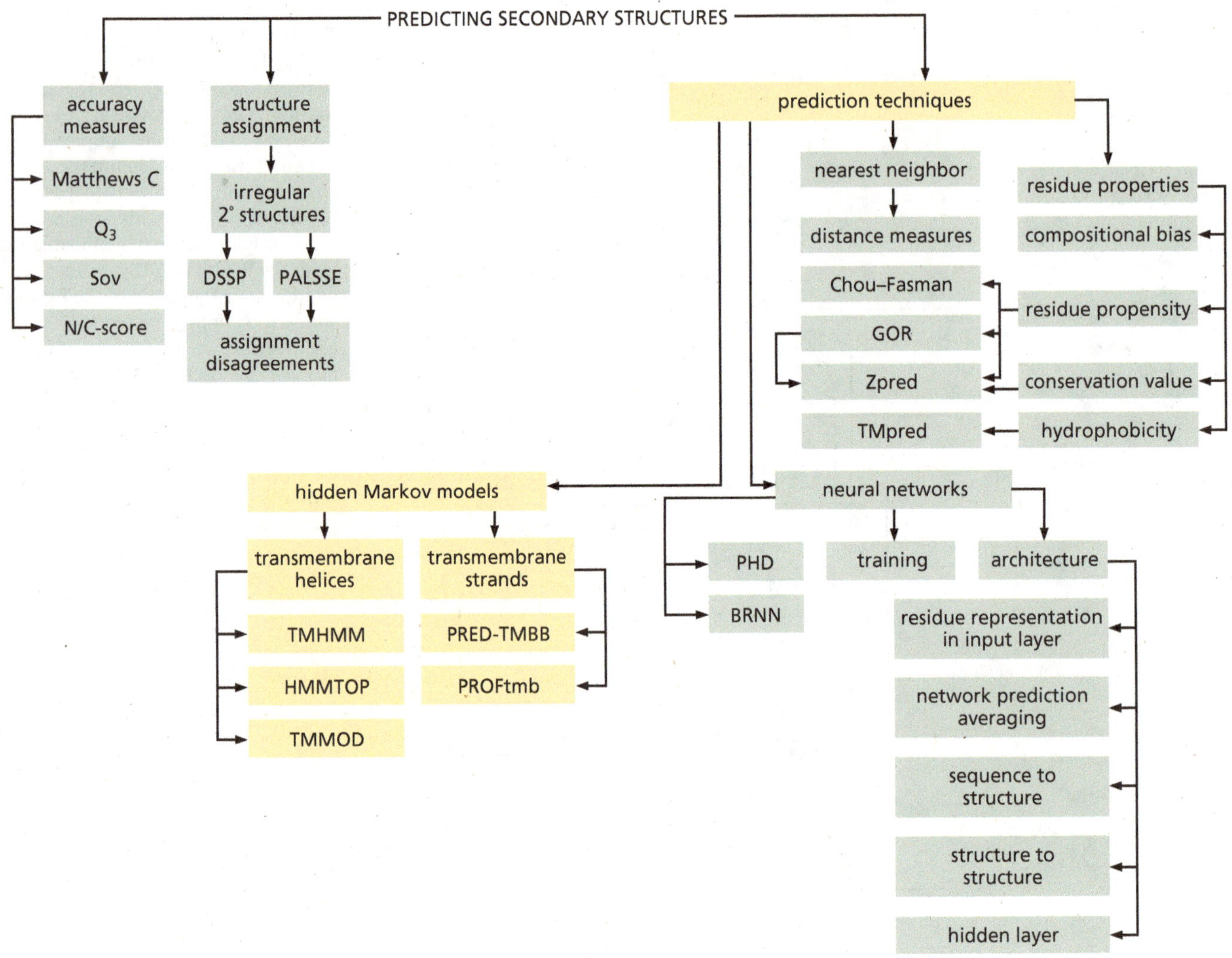


**Flow Diagram 12.5**
**The key concept introduced in this section is that membrane protein secondary structure and topology can be predicted by the application of hidden Markov models.**

## HMM methods have been found especially effective for transmembrane proteins

Membrane proteins typically have several membrane-spanning segments (either helices or strands) connected by segments that lie outside the membrane in the extracellular (outside) or cytoplasmic (inside) space. The segments outside the membrane are commonly referred to by the prediction methods as loops, which may be short loops like the segments called loops in globular proteins, but may also be larger segments, even multidomain globular folds. The HMM prediction methods for membrane proteins use models that explicitly include these different segments, and in addition distinguish between the central part of transmembrane helices and the end regions, usually called caps.

Gábor Tusnády and István Simon constructed the HMMTOP model (see Figure 12.30) of membrane proteins with transmembrane helices. This model has three key components that model the helix, tail segments at either end of the helix, and loop regions (see Figure 11.36). In addition a distinction is made between extracellular and cytoplasmic segments. All possible transitions between states are shown in Figure 12.30. For the helix and tail segments this structure enforces the observed length distribution. The length distribution for the loop segments will be similar to that shown in Figure 6.10 for the HMM model of Figure 6.9. The HMM parameters were estimated using known structures in an equivalent way to that discussed in Section 6.2 for fitting profile HMMs with aligned sequences. The probabilities for emitting residues were obtained from the observed residue frequencies in the different states, modified by pseudocounts. Homologous sequences can be included in the method by multiplying together the prediction probabilities for each sequence.

The TMHMM prediction model has a very similar architecture to HMMTOP, differing mainly in the treatment of loops (see Figure 11.34). The tail states of HMMTOP are

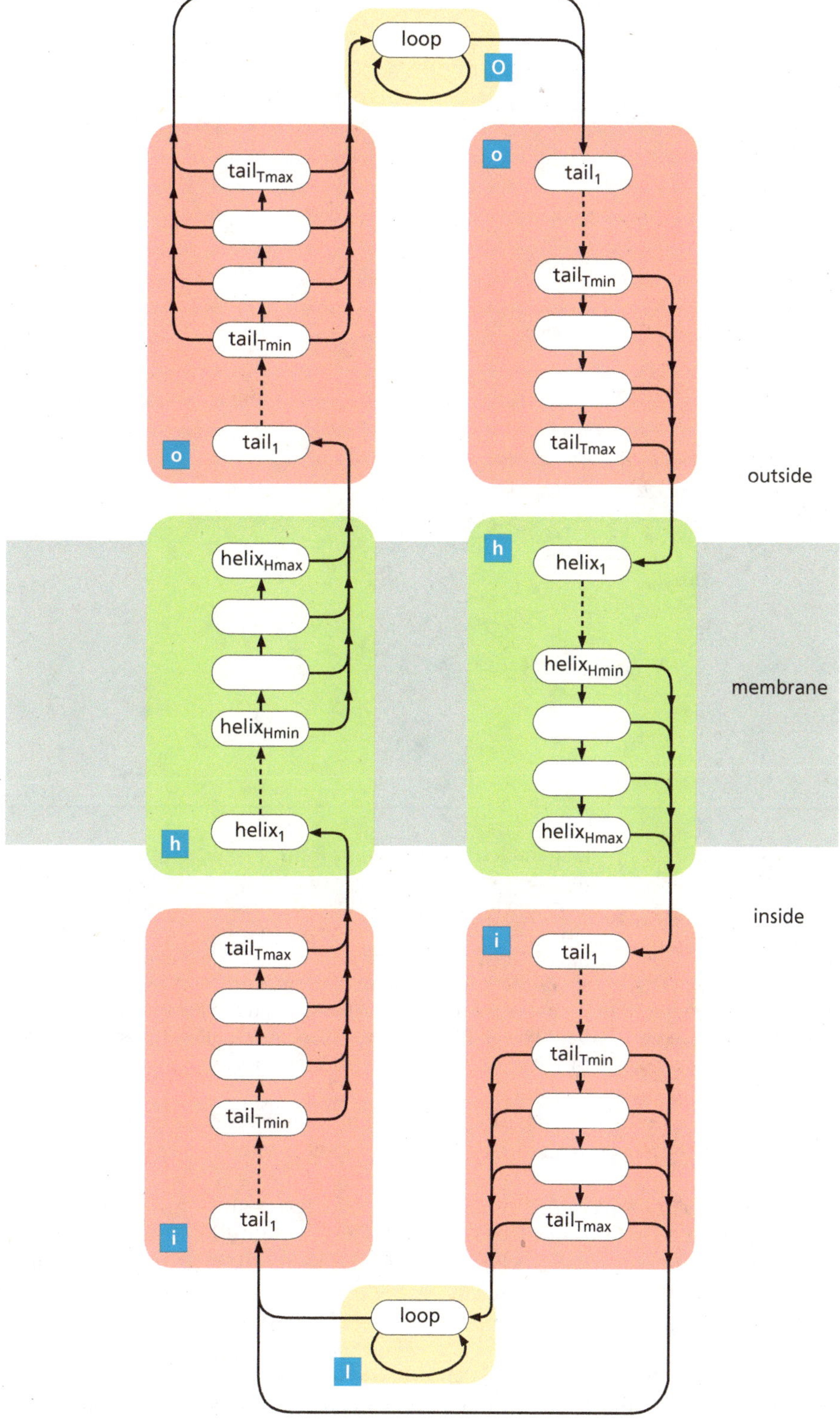


**Figure 12.30**
**The architecture of the HMMTOP model for transmembrane helix prediction.** The transmembrane helices have a length of 17–25 residues corresponding to Hmin and Hmax, respectively. The inside and outside tails have a length of 1–15 residues corresponding to Tmin and Tmax, respectively. The blue letters indicate the reported structural states, so that a segment from an inside loop to an outside loop might be written "IIIiihhhhhhhhhhhhhhhhhhhhooooooOOOO."

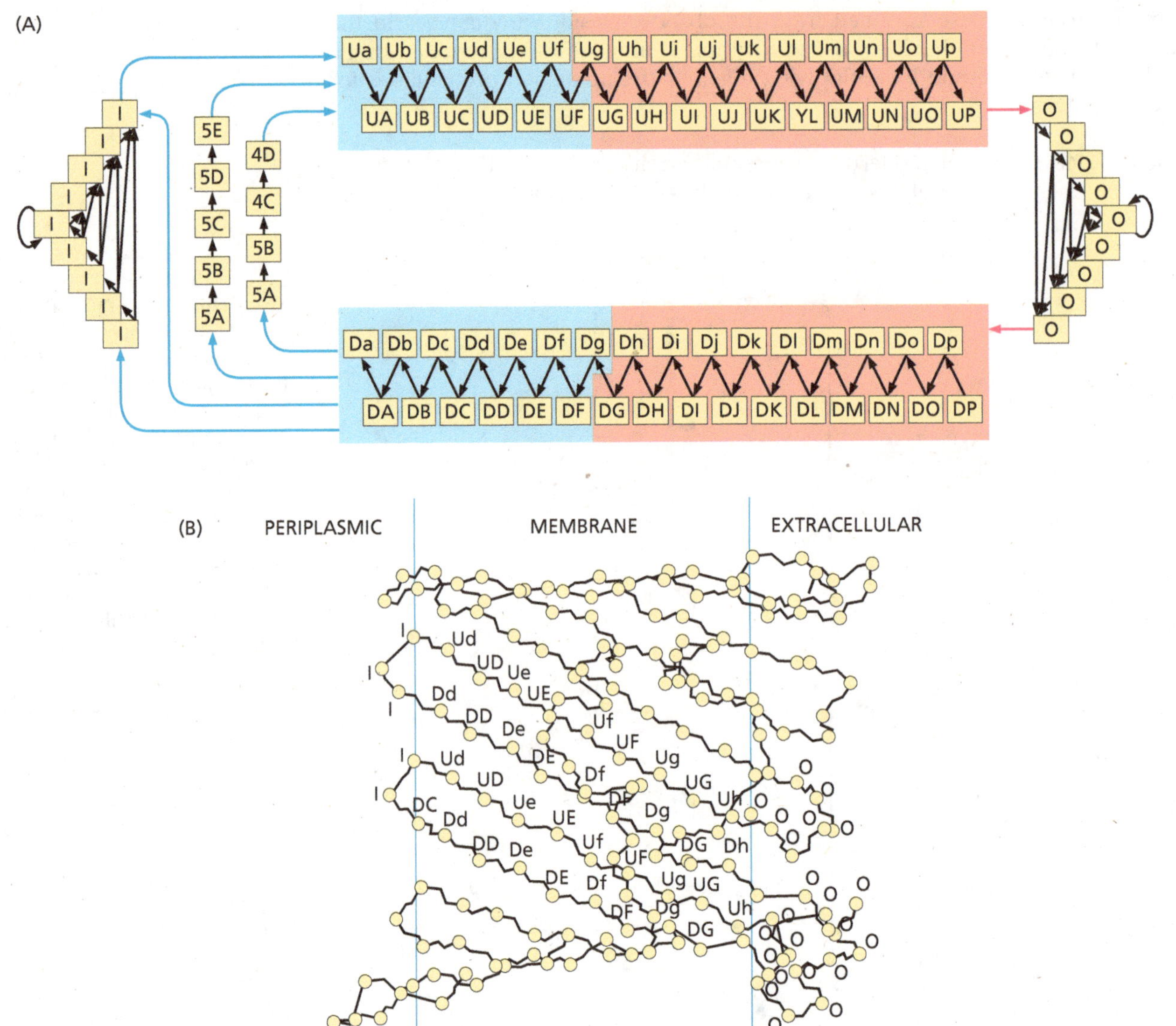


**Figure 12.31**
**The PROFtmb HMM model for transmembrane β-barrel prediction.** (A) The architecture of the PROFtmb HMM model. Two β-strands are explicitly represented, an "up" and a "down" strand, each of which has alternating distinct hydrophobic and hydrophilic environments. Three alternative cytoplasmic loops can be modeled—four- and five-residue β-turns and larger loops, all shown on the left side of the model. The extracellular model does not include the β-turn models. To allow for different β-strand lengths, the blue transitions to and from the β-turn and loop models can have come from or go directly to any state in the blue box. Similarly, the outside loop connects directly with any state in the red box. (B) A schematic of the *R. blastica* porin structure showing, for part of the molecule, how the residues have been assigned structural states. Note that the β-strands are significantly shorter than the maximum length of the model. (A and B, from H.R. Bigelow *et al.*, Predicting transmembrane beta-barrels in proteomes, *Nucleic Acids Res.* 32:2566–2577, 2004, by permission of Oxford University Press.)

incorporated in the loop model of TMHMM, but the five residues at the ends of the transmembrane helices (called caps) are explicitly modeled. The helix model of TMHMM has a length of 5–25 residues excluding these caps. The cytoplasmic and short extracellular loops in the model have specific states and transitions that allow for a length of 0–20 residues, with a single globular state like the loop state on HMMTOP. In addition there is a long extracellular loop model up to 100 residues in length designed to represent entire globular domains. This has 10 loop states leading to a three-state globular domain model, followed by another 10 loop states.

It was noted that the signal peptide, whose function is to initiate the insertion of the membrane protein into the membrane, has similarities to the transmembrane helix

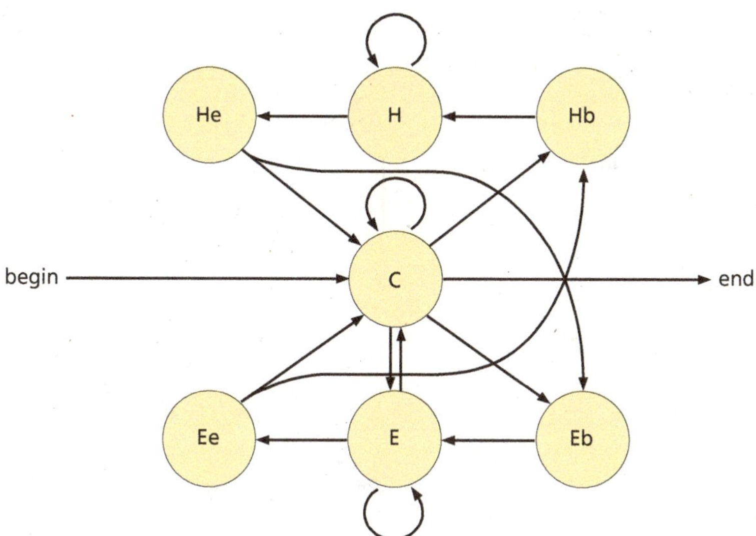


**Figure 12.32**
**The architecture of the hidden neural network of YASPIN.** The initial neural network has a seven-state output, namely α-helix (H), β-strand (E), coil (C), and the single residue at the beginning and end of a helix or strand (Hb, He, Eb, and Ee, respectively). The architecture allows consecutive stretches of residues to be in the H, E, or C states only. The output from the model contains only three states—H, E, and C—with the Hb and He elements emitting state H, and the Eb and Ee elements emitting state E. (Adapted from from Lin *et al.*, 2005.)

segments. The signal peptide often includes a hydrophobic stretch of residues (See Figure 12.11), although usually only 7–15 residues in length, shorter than most transmembrane helices. Despite the difference in length, this can often lead to incorrect prediction of these segments as transmembrane helices. The Phobius method tries to circumvent this problem by including a model for the signal peptide, and is an augmented version of TMHMM. The model includes the membrane segment as well as the N-terminal region and the cleavage site. Phobius has been found to be more accurate than both HMMTOP and TMHMM even when they are combined with a separate signal peptide prediction method.

A number of models have appeared in recent years for single-chain β-barrel membrane proteins. These barrels have an even number of β-strands of 6–22 residues with an average length of 12 residues. Figure 12.31 shows the architecture of the PROFtmb HMM model, which in common with other proposed models includes two β-strands, one in each direction across the membrane, as well as loop and β-turn models for the nonmembrane segments. All the different regions of the model can have variable lengths, in an attempt to model the observed length distributions. The PRED-TMBB architecture is similar to this, but does not have the β-turn models, instead having explicit N- and C-terminal models on the cytoplasmic side.

## Nonmembrane protein secondary structures can also be successfully predicted with HMMs

Hidden Markov models have also been applied to the prediction of globular protein secondary structures. YASPIN is one of the most accurate three-state prediction methods, and uses a combination of neural networks and an HMM called a **hidden neural network** (HNN). A 15-residue sequence window is used with 20 input layer units per residue representing the PSSM and an additional spacer unit. These feed signals to a hidden layer of 15 units that in turn send signals to the output layer of seven units. The seven output states from the neural network are α-helix (H), β-strand (E), coil (C), the beginning and end of α-helix (labeled Hb and He, respectively), and the beginning and end of β-strand (labeled Eb and Ee, respectively). The four beginning and end states are each a single residue in length. The HMM is used as a filter of the neural network output, converting this seven-state signal into a three-state prediction. The HMM architecture is shown in Figure 12.32. The shortest possible helices and strands in this model are three residues long, which could be problematic as some strands are shorter than this.

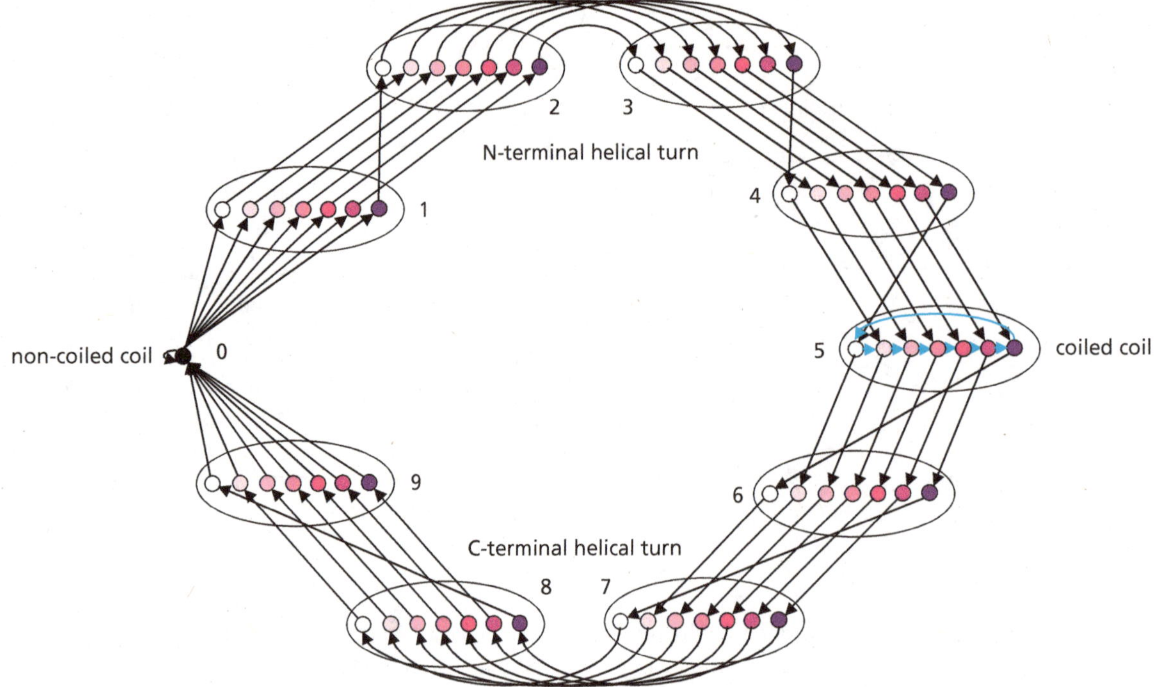


**Figure 12.33**
**The HMM architecture of MARCOIL coiled-coil predictor.** States in groups 1–4 represent the positions of the four residues of the first helical turn of the coiled coil, and those in groups 6–9 the positions of the four residues of the last helical turn of the coiled coil. In each of these groups there are seven color-coded heptad states. Note that only the transitions corresponding to the standard progression of heptad states are shown, but in addition all heptad states have transitions of smaller probability to the other six heptad states of the next group. The states of the group labeled "5" represent the central segment of the coiled coil, whose standard progression is represented by the blue transitions. The state labeled "0" corresponds to all the residues not in the coiled coil, and has a transition back to itself.

An HMM method has also been used for coiled-coil prediction in the program MARCOIL. The architecture of this HMM is shown in Figure 12.33, although for clarity not all transitions are shown. Coiled coils have heptad repeats, with each of the seven positions having characteristic residue preferences. However, the structure can begin and end at any of the seven positions. The HMM explicitly models the first and last four residues of the coiled coil in the eight groups of heptad states labeled 1–4 and 6–9. The central part of the coiled coil is modeled with the heptad group 5, which is the only group that has transitions between heptad states in the same group. All possible transitions occur between the heptad states of consecutive groups, but they have low probabilities except for those that cycle through the heptad states as in standard coiled coil. Only these standard transitions have been drawn in the figure. One potential benefit of this model is that it does not use a sequence window. Other coiled-coil prediction methods have often used a 28-residue window to reduce false-positive predictions, which limits the sensitivity to detect short coiled coils. This model can in principle predict coiled coils of nine residues or more. The probabilities for the transitions from group 5 states to group 6 states will determine the length distribution of predicted coiled coils. The transition probabilities of the model were predefined, and only the emission probabilities fitted to data from known structures. Despite not being fully optimized, MARCOIL has comparative accuracy with rival coiled-coil prediction methods.

## 12.6 General Data Classification Techniques can Predict Structural Features

The protein secondary structure prediction problem is essentially one of classification of each residue as belonging to one of a limited set of states. Therefore it is not surprising that some of the standard classification methods have been found useful, of which the example of neural networks has already been encountered. In this section we will survey some of these applications to give an impression of the variety of methods available. Some of the methods are described in more detail in Chapter 16 where the problem of classification is covered more thoroughly as applied to the problem of gene expression analysis.

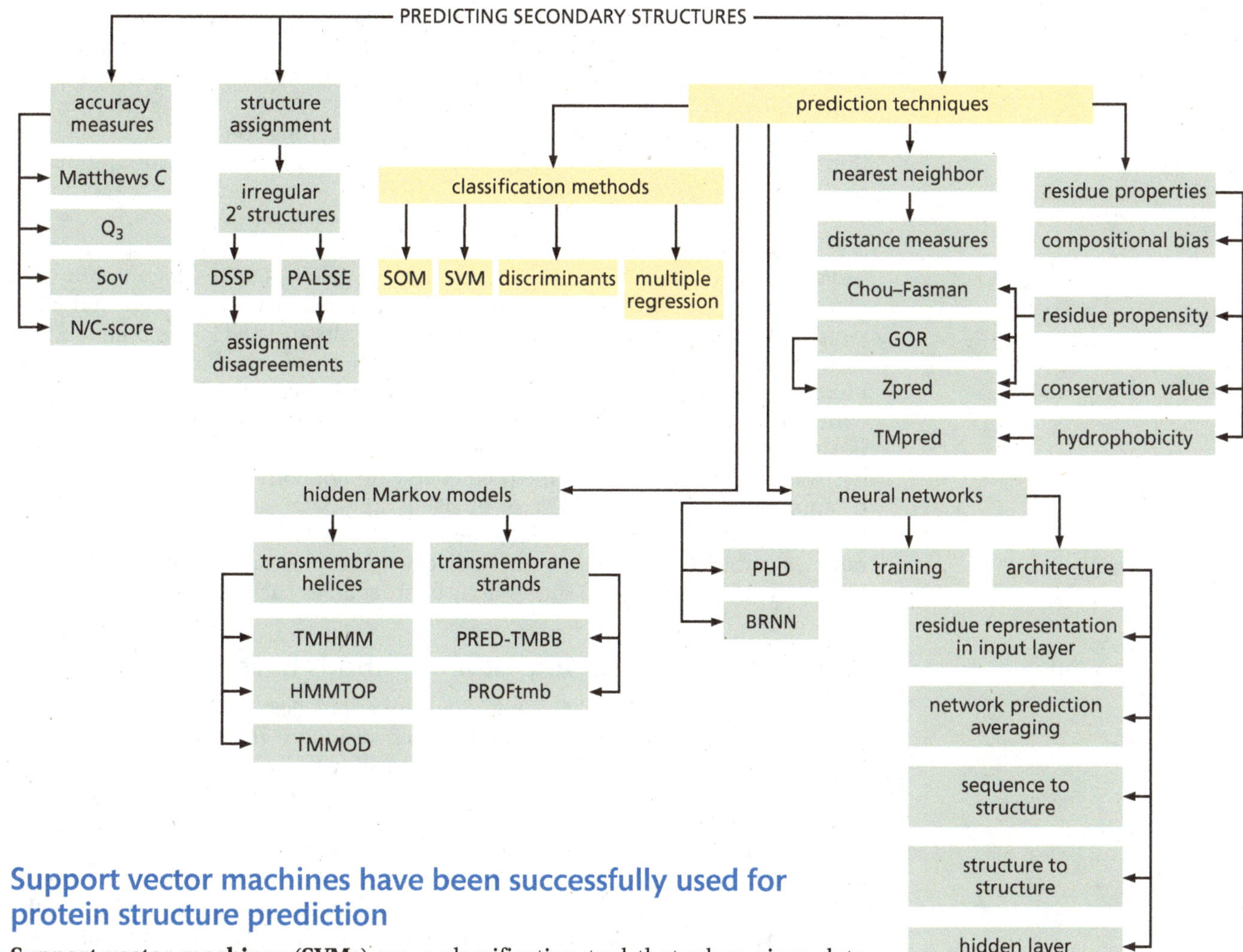


**Flow Diagram 12.6**
**The key concept introduced in this section is that some general data classification techniques can be usefully applied to predict protein secondary structure.**

## Support vector machines have been successfully used for protein structure prediction

**Support vector machines** (**SVMs**) are a classification tool that when given data input in the form of a vector produces an output that predicts to which of a limited set of classes the data belongs. We will focus here on the specific aspects of this method as applied to protein secondary structure prediction. Further details of the technique are provided in Section 16.5. The most common application of SVMs is as binary classifiers, in which role they distinguish between two alternative classes. When dealing with three-state prediction (helix, strand, coil) these can be of the form one-versus-one, for example distinguishing between helix and coil states, or alternatively of the one-versus-rest form, which might distinguish helix from not-helix.

Before describing how these binary SVMs can be combined to give a three-state prediction, the method of representing the input will be discussed. A sequence window is used as input, often a 15-residue segment. This is similar to the window sizes used in previously discussed techniques, and clearly these methods are analyzing equivalent information. Each residue position is represented by 20 values, often with an additional value equivalent to the spacer unit of neural networks. Thus a 15-residue window is represented by a 315-value vector. When using a single sequence as input, the 20 values representing a residue are filled with the same 0/1 system of neural network orthogonal encoding. Aligned sequences are usually represented using a scaled version of the PSSM from PSI-BLAST, although early work used the residue frequencies, which may have been a factor in the lower accuracy achieved.

In marked contrast to neural networks, these methods have few explicit parameters to fit, usually leading to fewer problems when determining their optimal values.

One of the key parameterization decisions relates to the kernel function (see Further Reading). In almost all cases of secondary structure prediction a radial kernel function has been used. The issue of using balanced training datasets is as important as ever. During training the support vectors are determined, which in this area are reported to be numerous, possibly leading to significant computational requirements. SVMs use supervised training like neural networks, in which the correct answers are provided.

A set of binary SVMs can be combined to give a three-state output in a number of ways using a technique called a **directed acyclic graph (DAG)**, which is a form of decision tree, as shown in Figure 12.34. Each SVM shown is separately and independently used to classify the window, and the results are used in combination with the decisions to arrive at the final state prediction. This is not the only way of using binary SVMs for this problem, as the final prediction can also be based on direct comparison of the outputs of a set of three SVMs as shown in Figure 12.35. Usually the output from these architectures needs to have a filter applied, similar to neural networks, as consecutive residue predictions are not correlated. This can be done by application of a further SVM, or by using other filter methods as was done for neural networks. The accuracy of the best of these models was found to be comparable with the best alternative methods described previously. The SVM methodology has also been applied successfully in other areas such as β-turn prediction.

## Discriminants, SOMs, and other methods have also been used

The DSC method uses linear discriminant functions (see Figure 10.16) to predict protein secondary structure. Firstly for each residue a three-state prediction is obtained (reparameterized GOR-type prediction with Equation EQ12.18) using a 17-residue window. Further residue properties are obtained, including the distance from the N or C terminus, hydrophobicity measures, and measures of residue conservation, insertions, and deletions in an alignment of homologous sequences. A total of 10 such measures is obtained for each residue, converted into distances from the property mean in units of standard deviations (an equivalent of the z-score in the statistics of normal distributions). A further 10 measures are obtained by smoothing these values using nearby residues, giving 20 measures for each residue. Three linear discriminants were defined, one for each of the helix, strand, and coil states, each being a linear combination of the 20 measures. The initial residue prediction is made by taking the state whose discriminant function returns

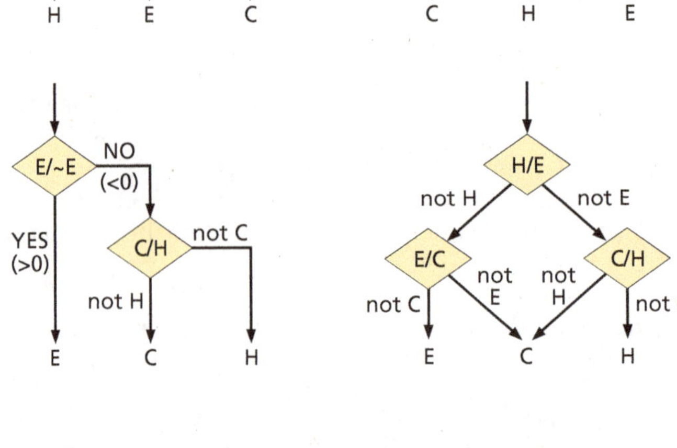

**Figure 12.34**
**Some of the possible architectures of SVM methods for three-state secondary structure prediction.** The diamonds represent the SVM, with labels to indicate the classifier. For example, "E/C" means an SVM that distinguishes strand and coil. The arrows are annotated with the conditions under which the path will be taken.

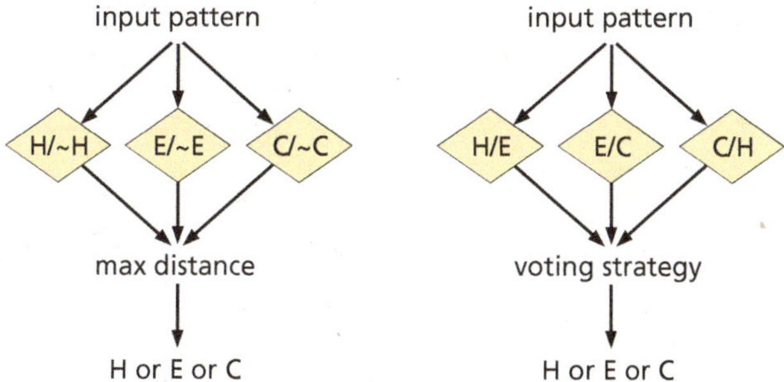


**Figure 12.35**
**Alternative architectures for SVM prediction of protein secondary structure, which combine the output from three SVMs to determine the prediction.** In the first case, the prediction is that state whose SVM calculates it to be furthest from the decision border. In the second case, the prediction is based on the strongest vote.

the largest value. The prediction is refined by adding seven further measures that relate to the whole protein, the fractional composition for certain polar residues, and the initially predicted helical and strand content. These 27 measures are then used with a second set of three linear discriminants to improve the prediction. In the last step the final prediction is obtained by applying a rule-based filter. The average $Q_3$ accuracy of the method is 72%.

Linear discriminants have also been found useful in predicting unfolded segments in proteins. Initial work had indicated a simple function based on the mean hydrophobicity (using the Kyte–Doolittle scale) and the mean net charge would distinguish folded and unfolded proteins. The FoldIndex method uses a 51-residue window to make such predictions, with apparently comparable accuracy to alternative methods.

Another technique of data classification that is becoming popular is the Kohonen **self-organizing map (SOM)**, details of which will be given in Section 16.3. This is a form of neural network, but with architecture very distinct from those discussed previously in this chapter. The output layer consists of a set of units, usually in a grid, only one of which will generate a signal for a given input. During training, the weights of the SOM adjust such that (if successful) the different classes of data give output unit signals in distinct regions. The GPI-SOM method uses this technique to identify glycosylphosphatidylinositol (GPI) anchor attachment signals in a sequence. The output layer is a grid of 40×40 units (see Figure 12.36) that accurately identifies these signals. The method required more computationally efficient

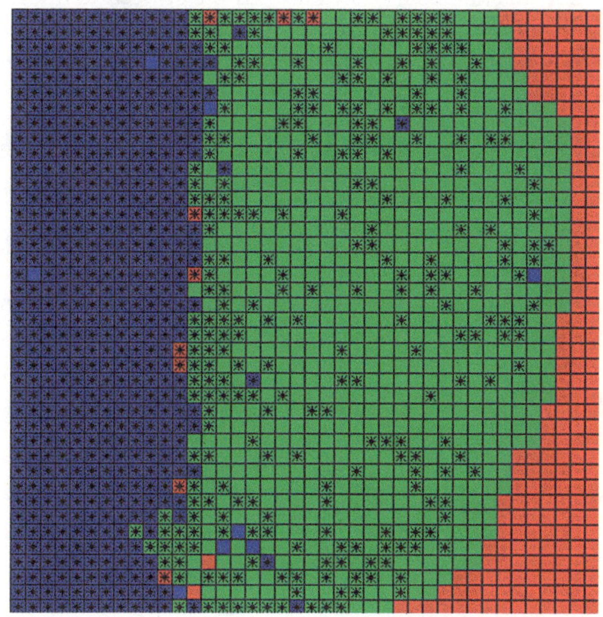

**Figure 12.36**
**A map of the output layer of the GPI-SOM for predicting GPI anchor attachment signals in a sequence.** It consists of 40×40 units. Those colored green indicate the prediction of presence of a GPI anchor attachment signal. The blue squares indicate the absence of such a signal and the red squares are no prediction. The stars represent squares that produce the prediction for *Saccharomyces cerevisiae* proteins. (From N. Fankhauser and P. Mäser, Identification of GPI anchor attachment signals by a Kohonen self-organizing map, *Bioinformatics* 21:1846–1852, 2005, by permission of Oxford University Press.)

sequence representations than those previously discussed in this chapter, which probably explains the relatively recent application of this method to these problems.

The final classification method whose use in structure prediction will be described is multiple linear regression. The PSIMLR method uses PSI-BLAST PSSMs scaled using a logistic function and a 17-residue window. The prediction is based on a two-stage application of multiple linear regression. Each of the three structural states has a linear equation associated with it that includes the PSSM of the window region and terms representing the interaction of all pairs of residues in the window. The first regression application predicts the structural state and solvent-accessibility for each residue. These values are then added to the original terms and a second regression applied to obtain the final prediction. This method is reported to have an average accuracy comparable to that of the best neural network methods.

## Summary

The prediction of protein secondary structure has been accomplished with varying degrees of success using a wide range of techniques. Despite the availability of an increasing set of experimental structures the predictions are still of great use. Hence methods are still being improved, and the latest techniques are exploited to try to improve the prediction accuracy. The last 10 years have seen improvements, but the limits to the possible accuracy have yet to be achieved. These limits are set by difficulties in unambiguously assigning secondary structures due to distortions from the ideal geometries.

The methods whose basis can be most clearly understood involve the statistical analysis of residue preferences and their modification by their surroundings. The local sequence up to 8–10 residues distant clearly holds the key information, but longer-range effects that are much harder to model also have an influence. When direct analysis as employed in the GOR and nearest-neighbor methods failed to achieve perfect results, attention moved to techniques commonly employed in data classification. Neural networks have the potential to attain greater accuracy, although their success was also due to improved ways of including information from homologous sequences.

The prediction of secondary structures in membrane proteins started more recently because it was necessary to wait until the structural information became available that is required to parameterize the models. The thickness of the membrane leads to larger average α-helix and β-strand lengths than in nonmembrane-associated globular protein folds, which makes the identification of the structures a little easier. However, the methods became more ambitious in trying to predict topology as well as locate structures. The HMM methods prove especially effective for this problem, and considerable success has been achieved.

There are many other structural and functional features that can be predicted, such as amyloidogenic regions, coiled coils, and various functional signals. These have also been tackled with a variety of methods and varying degrees of success. Any protein property that can be localized to a sequence segment and has some form of signal within the sequence can, in principle, be predicted using the techniques discussed in this chapter.

# Further Reading

## 12.1 Defining Secondary Structure and Prediction Accuracy


**DSSP**

Kabsch W & Sander C (1983) Dictionary of protein secondary structure: pattern recognition of hydrogen-bonded and geometrical features. *Biopolymers* 22, 2577–2637.

**PALSSE**

Majumdar I, Sri Krishna S & Grishin NV (2005) PALSSE: A program to delineate linear secondary structural elements from protein structures. *BMC Bioinformatics* 6, 202.

**β-Spider**

Parisien M & Major F (2005) A new catalog of protein β-sheets. *Proteins* 61, 545–558.

**Limits of prediction accuracy**

Russell RB & Barton GJ (1993) The limits of protein secondary structure prediction accuracy from multiple sequence alignment. *J. Mol. Biol.* 234, 951–957.

**TM properties and accuracy measures**

Cuthbertson JM, Doyle DA & Sansom MSP (2005) Transmembrane helix prediction: a comparative evaluation and analysis. *Protein Eng. Des. Sel.* 18, 295–308.

**PSIPRED ($Q_3$ and Sov variation)**

Jones DT (1999) Protein secondary structure prediction based on position specific scoring matrices. *J. Mol. Biol.* 292, 195–202.

**Matthews correlation coefficient**

Matthews BW (1975) Comparison of the predicted and observed secondary structure of T4 phage lysozyme. *Biochim. Biophys. Acta* 405, 442–451.

**Sov**

Zemla A, Venclovas C, Fidelis K & Rost B (1999) A modified definition of Sov, a segment-based measure for protein secondary structure prediction assessment. *Proteins* 34, 220–223.

**SS assignment comparison**

Colloc'h N, Etchebest C, Thoreau E et al. (1993) Comparison of three algorithms for the assignment of secondary structure in proteins: the advantages of a consensus assignment. *Protein Eng.* 6, 377–382.

**Length distributions (non-TM)**

Schmidler SC, Liu JS & Brutlag DL (2000) Bayesian segmentation of protein secondary structure. *J. Comput. Biol.* 7, 233–248.


## 12.2 Secondary Structure Prediction Based on Residue Propensities


**PDB_SELECT**

Hobohm U, Scharf M, Schneider R & Sander C (1992) Selection of representative protein data sets. *Protein Sci.* 1, 409–417.

Hobohm U & Sander C (1994) Enlarged representative set of protein structures. *Protein Sci.* 3, 522–524.

**In-depth analysis of unbiased structural datasets**

Chu CK, Feng LL & Wouters MA (2005) Comparison of sequence and structure-based datasets for nonredundant structural data mining. *Proteins* 60, 577–583.

**Hydrophobicity scales**

Engelman DM, Steitz TA & Goldman A (1986) Identifying nonpolar transbilayer helices in amino acid sequences of membrane proteins. *Annu. Rev. Biophys. Biophys. Chem.* 15, 321–353.

Kyte J & Doolittle RF (1982) A simple method for displaying the hydropathic character of a protein. *J. Mol. Biol.* 157, 105–132.

Zviling M, Leonov H & Arkin IT (2005) Genetic algorithm-based optimization of hydrophobicity tables. *Bioinformatics* 21, 2651–2656.

**Chou–Fasman**

Chou PY & Fasman GD (1974) Conformational parameters for amino acids in helical, β-sheet and random coil regions calculated from proteins. *Biochemistry* 13, 211–222.

Chou PY & Fasman GD (1974) Prediction of protein conformation. *Biochemistry* 13, 222–245.

Chou PY & Fasman GD (1978) Prediction of the secondary structure of proteins from their amino acid sequence. *Adv. Enzymol. Relat. Areas Mol. Biol.* 47, 45–147.

Engel DE & DeGrado WF (2004) Amino acid propensities are position-dependent throughout the length of α-helices. *J. Mol. Biol.* 337, 1195–1205.

Kyngä J & Valjakka J (1998) Unreliability of the Chou–Fasman parameters in predicting protein secondary structure. *Protein Eng.* 11, 345–348.

**COILS**

Lupas A (1996) Prediction and analysis of coiled-coil structures. *Methods Enzymol.* 266, 513–525.

**MEMSAT**

Jones DT, Taylor WR & Thornton JM (1994) A model recognition approach to the prediction of all-helical membrane protein structure and topology. *Biochemistry* 33, 3038–3049.

**Local and nonlocal effects**

Crooks GE & Brenner SE (2004) Protein secondary structure: entropy, correlations and prediction. *Bioinformatics* 20, 1603–1611.

Kihara D (2005) The effect of long-range interactions on the secondary structure formation of proteins. *Protein Sci.* 14, 1955–1963.

**β-turn propensities**

Hutchinson EG & Thornton JM (1994) A revised set of potentials for β-turn formation in proteins. *Protein Sci.* 3, 2207–2216.

**β-turn propensities and use of PSSMs**

Fuchs PFJ & Alix AJP (2005) High accuracy prediction of β-turns and their types using propensities and multiple alignments. *Proteins* 59, 828–839.

**AAindex**

Kawashima S & Kanehisa M (2000) AAindex: amino acid index database. *Nucleic Acids Res.* 28, 374.

**GOR theory**

Robson B (1974) Analysis of the code relating sequence to conformation in globular proteins. *Biochem. J.* 141, 853–867.

**GOR I**

Garnier J, Osguthorpe D & Robson B (1978) Analysis of the accuracy and implications of simple methods for predicting the secondary structure of globular proteins. *J. Mol. Biol.* 120, 97–120.

**GOR II**

Garnier J & Robson B (1989) The GOR method for predicting secondary structures in proteins. In: Prediction of Protein Structure and the Principles of Protein Conformation, chap. 10 (GD Fasman, ed.), pp. 417–465. New York: Kluwer Academic/Plenum Press.

**GOR III**

Gibrat JF, Garnier J & Robson B (1987) Further developments of protein secondary structure prediction using information theory. New parameters and consideration of residue pairs. *J. Mol. Biol.* 198, 425–443.

**GOR IV**

Garnier J, Gibrat JF & Robson B (1996) GOR method for predicting protein secondary structure from amino acid sequence. *Methods Enzymol.* 266, 540–553.

**GOR V (includes other sequences)**

Kloczkowski A, Ting KL, Jernigan RL & Garnier J (2002) Protein secondary structure prediction based on the GOR algorithm incorporating multiple sequence alignment information. *Polymer* 43, 441–449.

**Using consecutive pair of structural states**

Sadeghi M, Parto S, Arab S & Ranjbar B (2005) Prediction of protein secondary structure based on residue pair types and conformational states using dynamic programming algorithm *FEBS Lett* 579, 3397–3400.

**Zpred**

Zvelebil MJ, Barton GJ, Taylor WR & Sternberg MJE (1987) Prediction of protein secondary structure and active sites using the alignment of homologous sequences. *J. Mol. Biol.* 195, 957–961.

**Treatment of gaps in multiple alignments during prediction**

Simossis VA & Heringa J (2004) The influence of gapped positions in multiple sequence alignments on secondary structure prediction methods. *Comput. Biol. Chem.* 28, 351–366.

## 12.3 The Nearest-neighbor Methods are Based on Sequence Segment Similarity

Rychlewski L & Godzik A (1997) Secondary structure prediction using segment similarity *Protein Eng.* 10, 1143–1153.

**NNSSP**

Salamov AA & Solovyev VV (1995) Prediction of protein secondary structure by combining nearest-neighbor algorithms and multiple sequence alignments. *J. Mol. Biol.* 247, 11–15.

**SSPAL**

Salamov AA & Solovyev VV (1997) Protein secondary structure prediction using local alignments. *J. Mol. Biol.* 268, 31–36.

Saltzberg S & Cost S (1992) Predicting protein secondary structure with a nearest-neighbor algorithm. *J. Mol. Biol.* 227, 371–374.

Yi T-M & Lander ES (1993) Protein secondary structure prediction using nearest-neighbor methods. *J. Mol. Biol.* 232, 1117–1129.

**I-sites**

Bystroff C & Baker D (1998) Prediction of local structure in proteins using a library of sequence–structure motifs. *J. Mol. Biol.* 281, 565–577.

**SIMPA96**

Levin JM (1997) Exploring the limits of nearest neighbor secondary structure prediction *Protein Eng.* 10, 771–776.

**Correlation between amino acid composition and protein structural class**

Chou KC & Zhang CT (1994) Predicting protein folding types by distance functions that make allowances for amino acid interactions. *J. Biol. Chem.* 269, 22014–22020.

**HβP**

Yoon S & Welsh WJ (2004) Detecting hidden sequence propensity for amyloid fibril formation. *Protein Sci.* 13, 2149–2160.

## 12.4 Neural Networks Have Been Employed Successfully for Secondary Structure Prediction

**PHDsec**

Rost B & Sander C (1993) Prediction of protein secondary structure at better than 70% accuracy. *J. Mol. Biol.* 232, 584–599.

Rost B (1996) PHD: Predicting one-dimensional protein structure by profile-based neural networks. *Methods Enzymol.* 266, 525–539.

**PHDpsi**

Przybylski D & Rost B (2002) Alignments grow, secondary structure prediction improves. *Proteins* 46, 197–205.

**PSIPRED**

Jones DT (1999) Protein secondary structure prediction based on position-specific scoring matrices. *J. Mol. Biol.* 292, 195–202.

Petersen TN, Lundegaard C, Nielsen M, Bohr H et al. (2000) Prediction of protein secondary structure at 80% accuracy. *Proteins* 41, 17–20.

Qian N & Sejnowski TJ (1988) Predicting the secondary structure of globular proteins using neural network models. *J. Mol. Biol.* 202, 865–884.

Riis SK & Krogh A (1996) Improving prediction of protein secondary structure using structured neural networks and multiple sequence alignments. *J. Comput. Biol.* 3, 163–183.

**Back-propagation learning**

Rumelhart DE, Hinton GE & Williams RJ (1986) Learning representations by back-propagating errors. *Nature* 323, 533–536.

**BTPRED**

Shepherd AJ, Gorse D & Thornton JM (1999) Prediction of the location and type of β-turns in proteins using neural networks. *Protein Sci.* 8, 1045–1055.

**Betaturns**

Kaur H & Raghava GPS (2004) A neural network method for prediction of β-turn types in proteins using evolutionary information. *Bioinformatics* 20, 2751–2758.

**DESTRUCT**

Wood MJ & Hirst JD (2005) Protein secondary structure prediction with dihedral angles. *Proteins* 59, 476–481.

**SSpro**

Baldi P, Brunak S, Frasconi P et al. (1999) Exploiting the past and the future in protein secondary structure prediction. *Bioinformatics* 15, 937–946.

Pollastri G, Przybylski D, Rost B & Baldi P (2002) Improving the prediction of protein secondary structure in three and eight classes using recurrent neural networks and profiles. *Proteins* 47, 228–235.

## 12.5 Hidden Markov Models Have Been Applied to Structure Prediction

**HMMTOP**

Tusnády GE & Simon I (1998) Principles governing amino acid composition of integral membrane proteins: Application to topology prediction. *J. Mol. Biol.* 283, 489–506.

**TMHMM**

Krigh A, Larsson B, von Heijne G & Sonnhammer ELL (2001) Predicting transmembrane protein topology with a hidden Markov Model: Application to complete genomes. *J. Mol. Biol.* 305, 567–580.

**Phobius**

Käll L, Krogh A & Sonnhammer ELL (2004) A combined transmembrane topology and signal peptide prediction method. *J. Mol. Biol.* 338, 1027–1036.

**PROFtmb**

Bigelow HR, Petrey DS, Liu J et al. (2004) Predicting transmembrane beta-barrels in proteomes. *Nucleic Acids Res.* 32, 2566–2577.

**PRED-TMBB**

Bagos PG, Liakopoulos TD, Spyropoulos IC & Hamodrakas SJ (2004) A Hidden Markov Model method, capable of predicting and discriminating β-barrel outer membrane proteins. *BMC Bioinformatics* 5, 29.

**YASPIN**

Lin K, Simossis VA, Taylor WR & Heringa J (2005) A simple and fast secondary structure prediction method using hidden neural networks. *Bioinformatics* 21, 152–159.

**MARCOIL**

Delorenzi M & Speed T (2002) An HMM model for coiled-coil domains and a comparison with PSSM-based predictions. *Bioinformatics* 18, 617–625.

## 12.6 General Data Classification Techniques can Predict Structural Features

Guo J, Chen H, Sun Z & Lin Y (2004) A novel method for protein secondary structure prediction using dual-layer SVM and profiles. *Proteins* 54, 738–743.

Hua S & Sun Z (2001) A novel method of protein secondary structure prediction with high segment overlap measure: Support vector machine approach. *J. Mol. Biol.* 308, 397–407.

Kim H & Park H (2003) Protein secondary structure prediction based on an improved support vector machines approach. *Protein Eng.* 16, 553–560.

Nguyen MN & Rajapakse JC (2003) Multi-class support vector machines for protein secondary structure prediction. *Genome Inform.* 14, 218–227.

Ward JJ, McGuffin LJ, Buxton BF & Jones DT (2003) Secondary structure prediction with support vector machines. *Bioinformatics* 19, 1650–1655.

Zhang Q, Yoon S & Welsh WJ (2005) Improved method for predicting β-turn using support vector machine. *Bioinformatics* 21, 2370–2374.

**DSC**

King RD & Sternberg MJE (1996) Identification and application of the concepts important for accurate and reliable protein secondary structure prediction. *Protein Sci.* 5, 2298–2310.

**FoldIndex**

Prilusky J, Felder CE, Zeev-Ben-Mordechai T et al. (2005) FoldIndex: a simple tool to predict whether a given protein sequence is intrinsically folded. *Bioinformatics* 21, 3435–3438.

**GPI-SOM**

Fankhauser N & Mäser P (2005) Identification of GPI anchor attachment signals by a Kohonen self-organizing map. *Bioinformatics* 21, 1846–1852.

**PSIMLR**

Qin S, He Y & Pan X-M (2005) Predicting protein secondary structure and solvent accessibility with an improved multiple linear regression method. *Proteins* 61, 473–480.

# PART 6

# TERTIARY STRUCTURES

The folded form of the protein is the active and functioning biological unit, and often provides information about the function and binding of a specific protein. There are experimental techniques, such as X-ray crystallography and nuclear magnetic resonance methods, for obtaining three-dimensional structures of molecules. However, it is not currently possible to obtain the three-dimensional structure of all biological molecules using these techniques. Therefore computational methods have been developed to predict the three-dimensional structure of biological molecules, especially proteins.

The first chapter introduces some of the concepts necessary to understand the methods available for predicting the structure of proteins. The rest of this chapter is then devoted to a detailed description of how to model a three-dimensional protein structure based on a homologous protein with an experimental structure available. The second chapter deals with the computational methods for obtaining and analyzing further information from the tertiary structure of proteins: how to use it to verify or predict a protein's function, how to find binding sites, and how to dock small molecules to a protein.

# MODELING PROTEIN STRUCTURE

13

APPLICATIONS CHAPTER

**When you have read Chapter 13, you should be able to:**

Summarize how to determine a protein's three-dimensional structure.

Show how potential energy functions are used to describe protein conformational energies.

Show how statistical potential functions are used in protein structure prediction.

Describe the difficulty of modeling protein 3-D conformation *ab initio*.

Predict protein fold using threading.

Discuss the sequence and structure requirements for successful homology modeling.

State the principles of homology modeling and the stages in making a model.

Explain automated homology modeling.

Model the structure of the phosphatidylinositol kinase subunit p110α.

The three-dimensional structure—or tertiary structure—of a protein can be used much more effectively than sequence alone to understand the protein's function, mechanism of action, and its structure–function relations. For example, residues that are far away in the linear amino acid sequence can be very close together in the actual folded protein. Knowledge of the three-dimensional structure is also essential for the rational design of site-directed mutations and drugs, and can explain experimental observations such as binding specificities or antigenic properties.

There are two experimental methods for obtaining three-dimensional structures: X-ray crystallography and nuclear magnetic resonance (NMR) spectroscopy. At their best, these methods provide detailed and accurate structural information. In general, X-ray crystallography is more accurate and can deal with larger structures, but can only provide a static snapshot of the molecule. NMR, on the other hand, although it can be less accurate, gives coordinates for structures over time, thus providing the additional information of the protein's internal motion. Both these techniques are laborious, however, and require expensive equipment and elaborate technical procedures. In addition, many proteins fail to crystallize, a prerequisite for X-ray crystallography, or are insufficiently soluble for NMR studies.

Because of the cost and time involved in determining protein structures experimentally, reliable methods of predicting a protein's three-dimensional structure from its protein sequence would be a great help now that vast amounts of sequence data are available for many different organisms. A protein's three-dimensional structure is determined by its amino acid sequence, but how a particular sequence folds into a given structure is not yet completely understood.

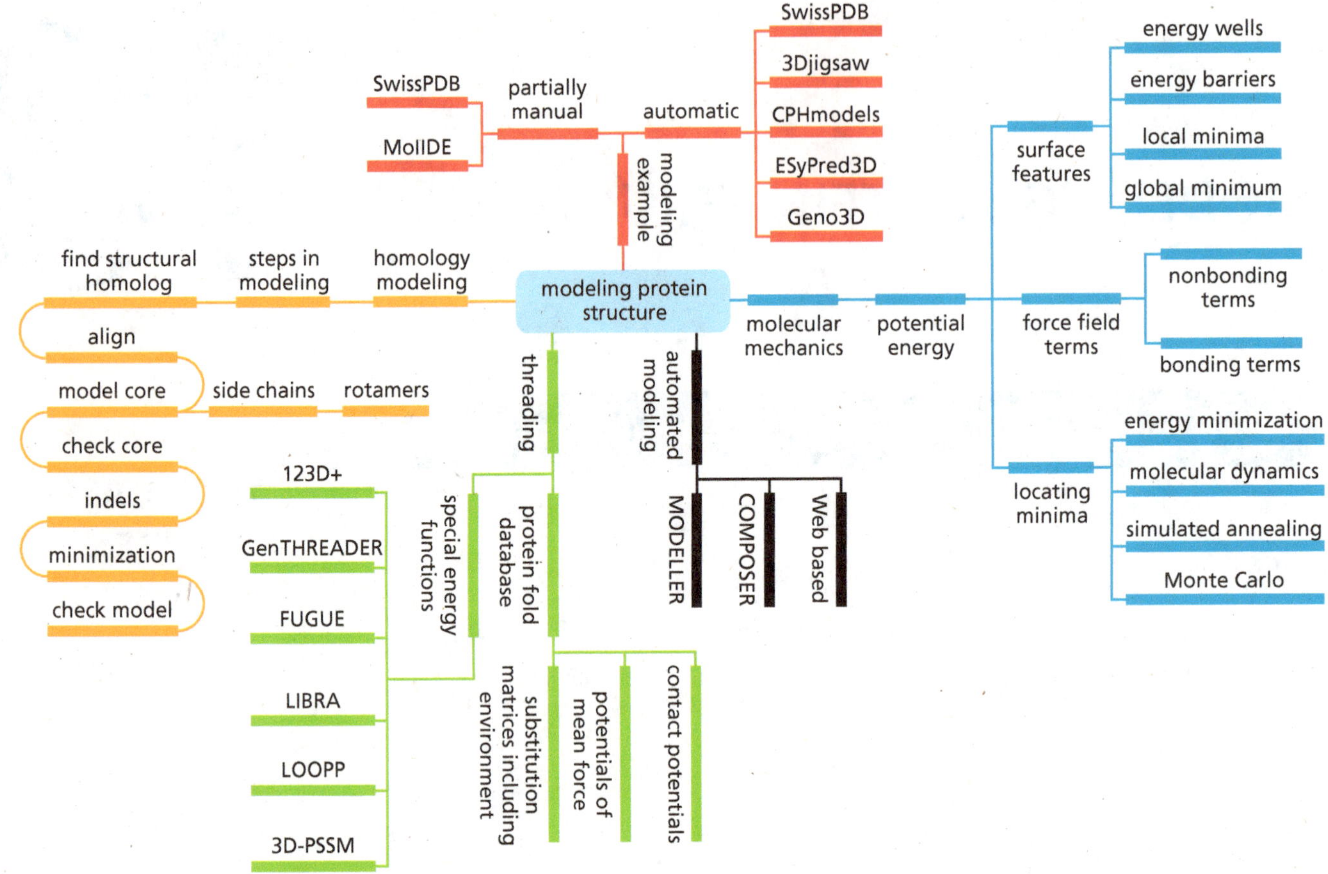


**Mind Map 13.1**
**A mind map illustrating how to obtain a tertiary (three-dimensional) structure from a protein sequence.**

Both homology modeling and threading make use of principles derived from what is known about the natural folding of proteins to a stable structure. Out of all possible structures, a protein adopts as its stable **native structure** or **native state** the one that involves the lowest free energy. The entropic part of the free energy is difficult to calculate, but it has been found that useful results can be obtained even when the entropy is ignored. In this case, the native state is predicted to be that of lowest enthalpy. This is referred to as the minimum potential energy; **potential energy** is the enthalpic energy stored in the system in the form of the energy in covalent bonds and noncovalent interactions, for example. Therefore, in theory, it should be possible to compute the folding of a protein sequence into its native structure using a **potential energy function** or **force field** that is derived from the interactions between all atoms in the protein and between the protein and the solvent.

The ***ab initio* approach** involves predicting the structure from first principles using thermodynamic and physicochemical theory. In this approach all possible conformations of a protein sequence should be evaluated to identify the minimum energy structures. In practice only a subset of conformations is sampled, but the sampling is designed to efficiently identify the low-energy conformations so that all important conformations are examined. This approach can currently predict only small single-domain proteins with sufficient accuracy to be useful for further work. Moreover, this is often not an option due to a lack of available computing power. However, these limitations may recede in the near future. *Ab initio* prediction is briefly described in Box 13.1

The most commonly used and best-developed methods for tertiary structure prediction are of two main types. The first is called **homology modeling** (also known as **comparative** or **knowledge-based modeling**), and relies on modeling the structure of the unknown protein with respect to the known structure of a homologous protein. We shall cover this method in detail in this chapter. The second approach, known generally as **threading** or **fold recognition**, does not need a homologous protein structure; instead, it attempts to model the sequence onto all known protein folds to see which one it fits best. Once a possible fold has been

## Box 13.1 *Ab initio* methods of tertiary structure prediction are too computationally intensive to be used routinely

This chapter describes the prediction of protein structures based on knowledge of the structures of their homologs or known folds. In principle if an accurate force field is available to calculate the energy of any conformation, the structure can be predicted completely independently. This is the *ab initio* approach, and requires identification of the conformation of the global energy minimum without having any prior fold information to bias the search.

The complete set of conformations available to a protein is vast, and the potential energy surface will have numerous local minima. This problem is therefore a challenge in terms of the computer resources required, and the most efficient conformational search methods must be employed. Molecular dynamics and simulated annealing are effective for conformational sampling, but still very computationally demanding. Alternative methods based on **Monte Carlo methods** (see Appendix C) have been found to provide the best combination of search efficiency and computational speed. Often the geometry of the protein is simplified and the solvent is not explicitly represented, as these approximations can significantly speed up the calculations as compared to all-atom protein models including solvent molecules. Some dramatic simplifications have been tried, including the representation of a residue or side chain by just one atom. Even then, success has been very limited, and currently only single-domain proteins have yielded useful results.

Some research groups have successfully predicted the structures of a few single-domain proteins *ab initio*. One such technique involves two key stages. At first only backbone torsion angles can vary, with all bond lengths and angles fixed at idealized geometries and side chains represented by single atoms. Thousands of conformational searches are performed using homologous sequences and Monte Carlo methods, which are designed to explore all the folds that are likely to be favored by the query sequence. The query sequence is then threaded (see Section 13.2) onto approximately 500 models representing the range of folds created. At this point all atoms are represented including hydrogen atoms, and a force field is used that models solvent effects. Following further Monte Carlo calculations, the prediction is made as the model with the lowest energy. The results of two such calculations, each taking many days, are shown in Figure B13.1.

Intermediate methods between homology modeling and *ab initio* methods have also been proposed, such as the ROSETTA/HMMSTR method. In this technique the query sequence is searched against the I-sites database (see Figure 12.17) to propose small folding motifs. These are then refined using molecular mechanics calculations to obtain the final model.

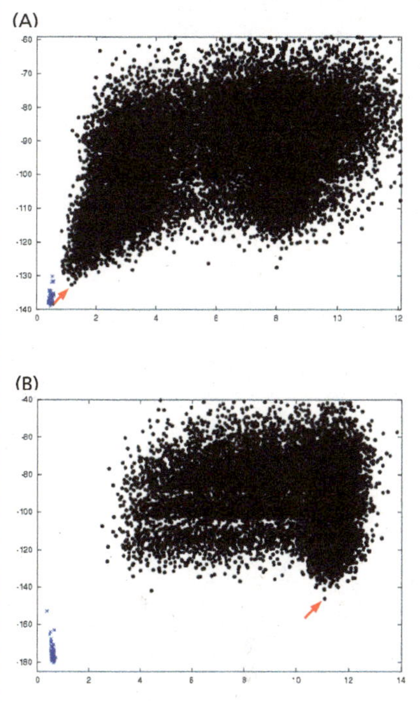


**Figure B13.1**
**Typical plots of the energy of predicted protein structures and RMSD (Root Mean Square Deviation) in Ångstroms from the correct structure of two single-domain proteins.** The black dots represent many independent *ab initio* prediction runs. The blue dots show the results of structural energy refinement starting from the correct structure. The lowest-energy *ab initio* run is indicated by the red arrow. In (A) this is within 1 Å RMSD of the correct structure; a successful prediction. In (B) the lowest-energy prediction is unsuccessful at approximately 11 Å RMSD. The proteins studied were E.coli RecA (A) and human Fyn tyrosine kinase SH3 domain (B). (From P. Bradley et al., Toward high-resolution de novo structure prediction for small proteins, *Science* 309:1868–1871, 2005. Reprinted with permission from AAAS.)

identified, a more realistic structural model for the query protein can be obtained by the techniques of homology modeling.

Following a brief look at how the energy of a protein conformation is calculated (see Section 13.1), we will discuss how threading methods can be used to predict the fold in the absence of structural homologs (see Section 13.2). In the remainder of the chapter the homology modeling process will be described in detail, starting with a description of the principles underlying homology modeling (see Section 13.3), the stages in a homology modeling process (see Section 13.4), and some of the individual programs available, including the automated packages (see Section 13.5). The last section of the chapter (see Section 13.6) is a worked example, modeling the phosphatidylinositol kinase p110α subunit on the structure of the related protein p110γ.

## 13.1 Potential Energy Functions and Force Fields

When modeling a protein structure, whether *ab initio* or in relation to another protein fold, the aim is to obtain a structure of lowest possible energy that satisfies the known stereochemical constraints on protein structures, such as allowable values for backbone torsion angles $\phi$ and $\psi$ (see Section 2.1), and appropriate packing of side chains. To explore the possibilities and assess whether a given conformation is energetically favorable, the geometry of a protein conformation, in terms of its atomic coordinates, is related to its potential energy by means of a collection of equations known as potential energy functions. These represent all the components that contribute to the overall potential energy of the protein. The combination of all these energy functions for a given conformation is called the force field.

Two types of force field can be distinguished: those that calculate the potential energy of a given conformational state that might include other molecules such as solvent, and those that calculate an energy which includes statistically averaged environmental effects. In this section we will briefly discuss both of these.

Once a force field has been specified the energy of any conformation can be defined, and any energetically unfavorable regions identified, as can any strongly favorable interactions. This can be useful in assessing proposed structures and complexes. There are a number of techniques that can be used to predict native conformations on the basis of identifying energy minima. They are described below, emphasizing their relative strengths and weaknesses, as the details of the algorithms are given in Appendix C. These methods are often collectively referred to as **molecular mechanics**.

**Flow Diagram 13.1**
**The key concepts introduced in this section are that the energies of protein systems can be described using a number of relatively simple terms, and that techniques have been developed to identify optimal conformations based on these terms.**

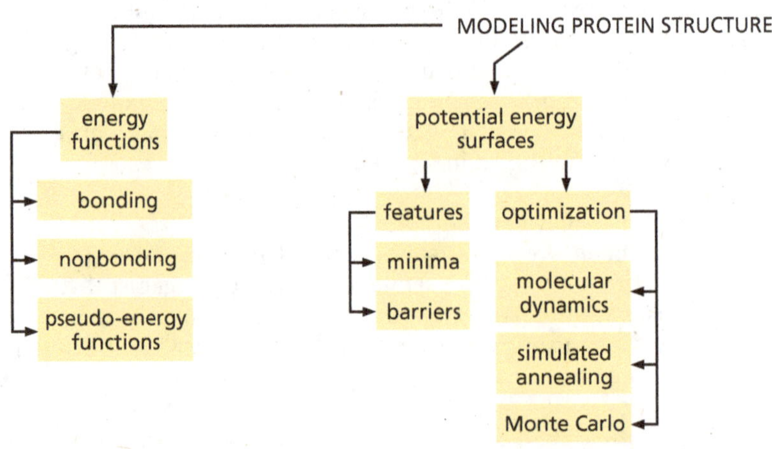

## The conformation of a protein can be visualized in terms of a potential energy surface

The potential energy (enthalpy) of a molecule depends on the atomic coordinates and the particular atom types, and can be defined for any conformation. In the case of small molecules the conformation can be defined by just a few variables, often the three coordinates of each atom. In the case of larger molecules such as a protein, there are often hundreds or thousands of variables needed to define the conformation. It is possible to draw a surface that represents the variation of the potential energy as the conformation varies: the **potential energy surface**. An example of this is the Ramachandran plot (see Figure 2.8). Although this is usually shown with allowed and disallowed areas, these correspond to relatively lower and higher potential energy, respectively.

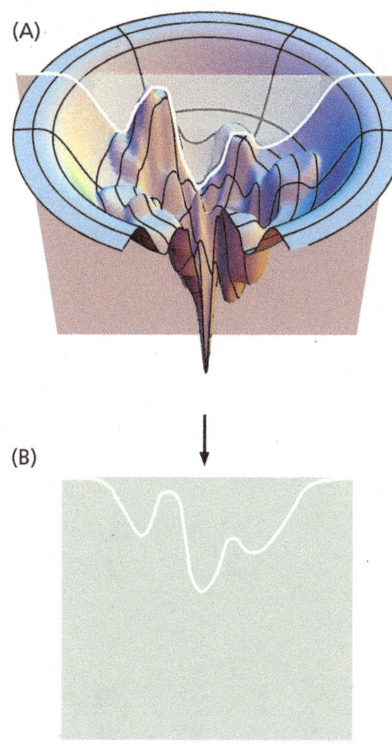


**Figure 13.1**
**A rugged energy landscape with energy barriers, showing multiple energy minima.** The coordinate system is centered on the global minimum. (A) The horizontal plane represents arbitrary conformational coordinates and the vertical axis is the energy of the system. (B) A cross-section of this energy landscape. (Courtesy of K. Dill and L. Schweitzer.)

According to thermodynamic theory, molecular systems will mostly be found in those conformations that have the lowest free energy. This free energy includes the contribution of entropy as well as the potential energy. The free energy surface of such systems can be shown graphically in a figurative way (see Figure 13.1). The surface is very complicated, reminiscent of a landscape in which the valleys and peaks represent the location of energy minima and maxima, respectively. The frequently occurring conformations will be those that lie at the bottom of the deepest valleys—the point or points of the global free energy minimum. Diagrams such as this are often referred to as a folding funnel in the case of protein systems, when the conformations corresponding to points on the surface far from the global energy minimum correspond to the protein in an unfolded state. Even when looking at a simple cross-section of the surface, such as Figure 13.1B, it is clear that locating the global minimum is not trivial. This point will be further discussed later in this section, and also in Appendix C, where some techniques will be described for locating minima.

It is very difficult to calculate the entropic component of the free energy, especially for solvated systems, because it requires averaging the energy over very many states of the system. However, in many cases it has been found that this component can be assumed constant, and thus ignored while still obtaining useful results. There are problems such as modeling protein folding where this assumption cannot be made, but when modeling protein structure, the focus of this chapter, the approximation appears to provide adequate results. Under this assumption only the potential energy is calculated, which is done using simple terms as will now be described.

## Conformational energies can be described by simple mathematical functions

A molecular geometry can be defined in two distinct ways: external coordinates, that is the values of $x,y,z$ for each atom, and internal coordinates that are based on simple chemical concepts such as bond length, bond angle, and torsion angles (see Section 2.1). The conformation is the geometry as well as the atom types and often a definition of the covalent structure, by which we mean which atoms are bonded to each other, and the bond order (single, double, partially double, etc.). The potential energy of the molecule depends exclusively on the conformation. Thus, all the components that define the conformation can be viewed as parameters that are used to define the potential energy.

For the calculations of interest here, which do not involve modeling chemical reactivity, the potential energy can be usefully separated into two components: that due to the covalent bonding structure, and that—often referred to as nonbonding—which is due to other interactions between atoms that are separated by at least two covalent bonds. Hydrogen bonding is often represented by explicit terms, in which case they are grouped with the other bonding terms. The nonbonding component is usually restricted in protein modeling to the terms representing the electrostatic interaction between atomic charges and the terms for the dispersion and repulsion interactions (see Appendix B for further details). These are of necessity between every pair of atoms in the system and thus use the external coordinate system.

In contrast, the covalent interactions are modeled in potential energy calculations of proteins by using the internal coordinate system. This is very efficient, there being approximately as many bond length terms as numbers of atoms, and often fewer bond angle and torsion angle terms. (If external coordinates were being used there would be a term to account for every atom interacting with every other atom, resulting in many more terms.) Theories have been developed, for example using a **harmonic approximation**, which lead to simple mathematical forms that describe the variation of energy of chemical entities such as bond lengths (see Appendix B for further details). Experimental evidence from the spectroscopy of small molecules suggests that to a fair approximation these terms relating to individual internal coordinates contribute independently to the molecular energy. As a consequence the bonding component of the potential energy of a molecule is often expressed as a sum of many simple terms, each representing a particular bond length, bond angle, or torsion angle. The spectroscopic results also suggest how greater accuracy could be achieved, by adding cross-terms to the sum, these being simple products of the individual terms.

The collection of algebraic terms and parameters in both the bonding and nonbonding components is usually referred to as a force field. Usually all atoms are defined as belonging to a limited set of atomic types—often 20–30—which makes parameterization easier. These will typically distinguish between atoms that are the same element but in different valence states. All bond length terms between two atoms of given types will have identical parameters. A variety of different force fields are available, including some specifically designed for use with proteins, such as the programs CHARMM and AMBER. Typically the set of parameters is refined over time, usually maintaining the same set of algebraic terms and atom types.

Because the goal of the calculations discussed here is to determine the conformation with the lowest potential energy, it is not necessary to obtain the absolute value of the energy. All we need to know is the difference in energy between alternative conformations. The force field terms do just this, so that for example bond length terms give the energy relative to that of an assumed ideal bond length, which is another parameter of the force field.

Because these force fields are not calculating absolute energy values, may use different terms, and differ in parameterization, the same force field must be used for all calculations in a study, as the values produced by different force fields will usually not be directly comparable. In general, force fields can only be compared by comparing their predictions.

## Similar force fields can be used to represent conformational energies in the presence of averaged environments

The potential functions described above represent the enthalpic energy of a molecular conformation with all of the atoms of the system explicitly represented. Thus, if the system contains only a protein molecule and no solvent molecules, the energy calculated will be that of the protein *in vacuo*. To calculate solvated energy would require explicit representation of the solvent molecules. It is possible to derive potential functions—**pseudo-energy functions**—that do not have this requirement, and represent the missing atoms by energy terms that reproduce a statistical average of their effect.

In protein structure prediction there are two different cases where pseudo-energy functions are useful, representing either the solvent or local protein environment.

An explicit representation of the solvent will usually require several thousand water molecules as well as some ions, and additionally suffers from the need to average over the vast range of possible solvent configurations. Often in structure prediction the solvent is either ignored or the force field terms are modified to give better agreement with solvated structures without using an explicit representation. This usually requires modifying the nonbonding terms, which otherwise can lead to the protein structure being too compact. Note that these potentials are not suitable for an in-depth study of solvation or of any conformations that involve specific solvent interactions, unless these particular solvent molecules are explicitly represented.

The pseudo-energy functions that represent statistically averaged local protein structure are used in threading methods, whose detailed presentation will be found in Section 13.2. They embody such information as the propensity of particular amino acids, or a short sequence of amino acids, to be found in particular secondary structures, the likelihood that particular residues will form hydrophobic interactions with each other, and so on. They are all derived from statistical analysis of databases of protein structure. The application of these potentials is almost completely limited to threading, as their energies are not accurate enough for detailed analysis of interactions.

## Potential energy functions can be used to assess a modeled structure

When using potential functions that do not include statistically averaged effects, the individual energy terms can be used to analyze any given structure. For example, if a bond length strongly deviates from ideality (as defined in the force field parameters), it will have a strongly unfavorable energy. Looking for such force field terms will identify any regions of the structure that are unlikely to be realistic. (There are cases in enzyme catalysis of unfavorable ligand conformations stabilized by the enzyme binding site, but in general the local geometry as represented by bond lengths, bond angles, and torsion angles is very close to ideal in protein structures.) There are programs, for example PROCHECK or MolProbity, which search structures to identify energetically unfavorable regions based on stereochemical geometry validation. Their application is discussed in more detail in Section 13.4.

A detailed analysis of the energy terms for a given conformation may also reveal particular interactions to be highly stabilizing or destabilizing, giving insight into the molecular function. While this is especially the case for a protein–ligand complex, the key forces stabilizing a protein fold can also be revealed. The application of this sort of analysis will be discussed more in Chapter 14.

## Energy minimization can be used to refine a modeled structure and identify local energy minima

As will be discussed in Section 13.4, even when a structure is modeled using a template from a homologous protein, parts of the protein backbone, especially the loops, will be different. There may also be stereochemical clashes in the positions of some side chains, where different residues occupy the same position in **target protein** and **template protein**. To help solve these problems molecular mechanics techniques are used that will now be described. These methods, using a force field to calculate energies, are able to remove unfavorable interactions and thus improve the molecular geometry. In addition, they can explore available conformations to locate those at energy minima.

The simplest molecular mechanics method for exploring the potential energy surface is **energy minimization**. This procedure is made up of many steps, during each of which the conformation is modified to give a new conformation of lower energy. The series of conformations produced will gradually descend the nearest

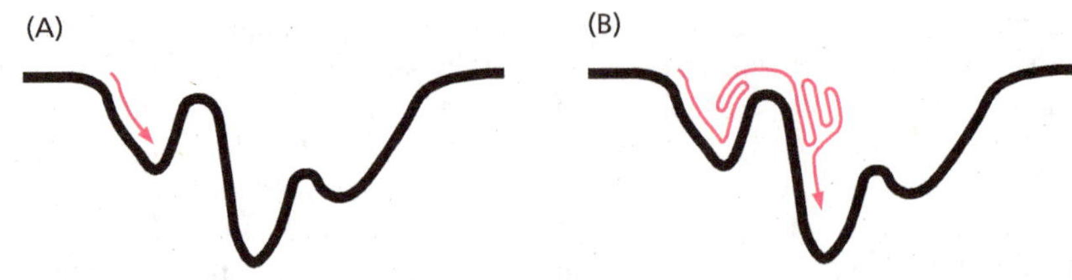


**Figure 13.2**
**An illustration of the movement of the system over the energy surface during the application of optimization methods.** (A) The path on the potential energy surface taken during energy minimization, leading to the nearest energy minimum, in this case a local minimum. The path is drawn here on a single slice of the energy surface, but in practice will not be limited in this way. (B) The path on the potential energy surface taken during molecular dynamics or a simulated annealing run. This is more likely to lead to the global energy minimum.

energy well until it reaches the bottom (see Figure 13.2A). Once at the local energy minimum the process is stopped, reporting the conformation at the minimum as the predicted optimal conformation. It is of particular use in homology modeling as it allows side chains in the protein core to be relaxed so they can pack together without overlapping. Often, the initial modeling produces a protein core in which some of the side chains overlap. Energy minimization can relieve this bad geometry, usually without making substantial changes in the other regions of the structure.

Energy minimization is the fastest of the molecular mechanics minimization procedures, but at the cost of only locating the local energy minimum, namely the bottom of the well that contains the initial conformation. As can be seen from Figure 13.1, protein potential surfaces have many separate wells, of which most (if not all but one) will have an energy at the well bottom that is not the lowest possible. It is the global energy minimum that is required to produce the correct prediction. One way of trying to increase the chances of locating the global energy minimum with energy minimization is to make many runs—perhaps several hundred or thousands—each starting from a slightly different conformation, and then to keep the lowest energy conformation(s) from all these runs.

There are several alternative methods of achieving energy minimization. All of them need an estimate of the **energy gradient** at the current conformation, being the change in energy for a small change in each of the coordinates defining the conformation. This can be understood in Figure 13.1 as the slope of the surface. The changes applied to produce the new conformation are related to the gradient (see Appendix C for further details). The simplest energy minimization scheme in common use is the **steepest descent method**. This involves estimating the energy gradient at the current conformation, and changing the coordinates to move directly down this gradient. The amount of change is varied proportional to the magnitude of the gradient, on the basis that near the minimum the gradients will be small. A more sophisticated scheme, the **conjugate gradient method**, uses the directions of two successive gradients to try to make a more intelligent guess at the location of the minimum. Other techniques such as the **Newton–Raphson method** require the gradient of the gradient—the second derivatives—and are therefore much more costly in computer time, but do perform much better than the other two methods in the vicinity of a minimum.

At any minimum the energy gradient is zero by definition, and the current value of the gradient can be used as a measure of how close the conformation is likely to be to that at the minimum. Usually the process is continued until the gradient of the current conformation is less than a specified threshold, typically around 0.01 kcal mol $Å^{-1}$. If the gradient in any cycle is less than this value, the calculation is said to have achieved convergence and is terminated. Sometimes convergence is not reached, and often a limit is set for the number of steps (also called cycles) allowed.

## Molecular dynamics and simulated annealing are used to find global energy minima

In the real world, proteins are not rigid structures as might be wrongly deduced from the common representations of protein folds. They have considerable flexibility, and due to thermal energy the atoms are vibrating about the mean positions that are listed as the atomic coordinates. The technique of **molecular dynamics** involves solving the equations that predict the motion of the atoms in the molecule over time (see Appendix C for further details). Thus at any point in time, the atoms in the system have defined positions and velocities, and an estimate of the acceleration of each atom is needed to determine its position a short instant of time later. Usually the time step used is 1 femtosecond ($10^{-15}$ second) and the calculation is frequently run for several tens of thousands of steps. Such a calculation requires a considerable computational effort, requiring large computer systems and many hours for each calculation.

Molecular dynamics calculations are very useful in homology modeling because the technique is able to cross small energy barriers that separate adjacent wells in the potential surface. In this way a molecular dynamics run can escape local minima and have a greater chance than energy minimization of finding the global energy minimum well (see Figure 13.2B).

The parameter in molecular dynamics that determines the vibrational energy of the system is the temperature, as this is related to the kinetic energy that the molecules have. In standard molecular dynamics runs, this temperature (often 300 K) is specified at the start and kept constant. In the technique known as **simulated annealing** the temperature is varied during the run. At first a very high temperature such as 1000 K is used, which gives the system sufficient vibrational energy to easily cross high energy barriers. In this way many more potential surface wells can be sampled, increasing the chance of entering the global energy minimum well. As the run progresses the temperature is lowered, so that gradually the system is trapped in a well, hopefully leading to the identification of the global minimum energy. In the case of simulated annealing only the last part of the run is sampled, as that is when the global minimum conformation might be found. If the computational resources are available several runs from different starting points are used to increase the chance of finding the global minimum.

## 13.2 Obtaining a Structure by Threading

Often no homologous protein with an experimentally solved structure can be found to match the target sequence. Therefore structure prediction methods that do not depend on homology have to be used. *Ab initio* methods are very computer intensive (see Box 13.1) and rarely applied. From analysis of many experimentally solved structures it has been proposed that there are only a limited number of different ways a protein can fold, possibly less than 2000. If this is true, some sequences with no detectable homology will have the same protein shape. This hypothesis has been confirmed, as several such examples have already been observed experimentally. Therefore many currently used structure prediction procedures depend on fitting the target sequence to known protein folds and selecting the one that seems the most energetically and stereochemically favorable.

As we saw in Chapter 11, the same secondary structure elements can be formed by many different sequences, and this is also true for tertiary structure. Numerous examples are known of proteins with no detectable sequence similarity, but which have strikingly similar tertiary structures. For example there are similar elongated folds in the very distantly related tumor and nerve growth factors or the SH3-like folds in sequentially unrelated proteins (see Figure 13.3). There are also many supersecondary structures—distinctive combinations of secondary structure elements—such as β-barrels, jelly-roll motifs, and Greek key motifs (see Chapter 2) that are found in many different proteins.

Because fold recognition techniques do not depend primarily on sequence comparison, a structural relationship between proteins may be recognized even if

**Flow Diagram 13.2**
**The key concept introduced in this section is that the technique of threading can be used to predict protein structures because it identifies the most appropriate representative protein fold from a library of folds.**

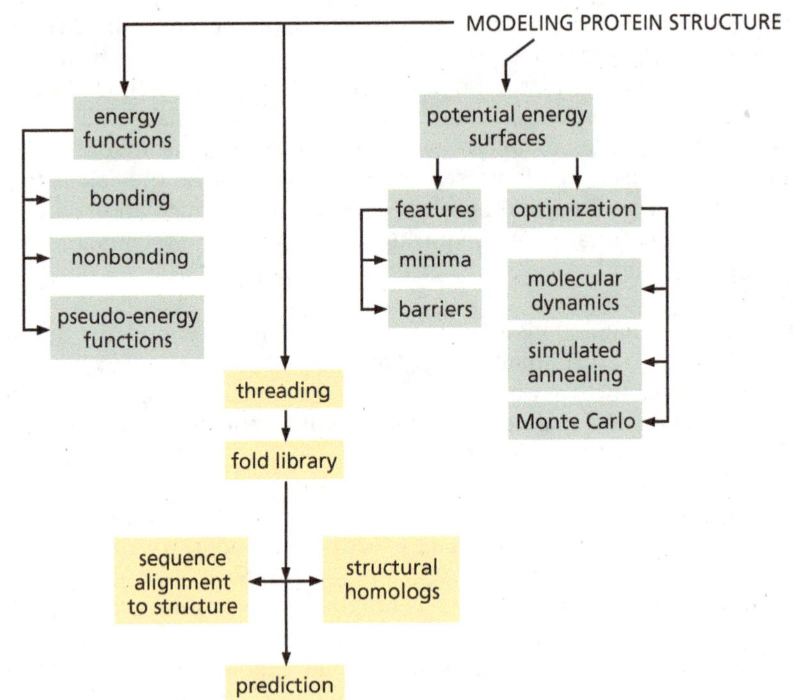


the sequence similarity is very low or nonexistent. This conservation of structure can be due both to common ancestry and to the fact that physical constraints limit the number of folds that proteins can adopt. Therefore, the same fold can occur in a wide variety of different proteins.

The method of protein fold recognition or **threading** attempts to find folds that are compatible with the sequence of the target protein (see Figure 13.4). It can be visualized as pulling a string of amino acids through the fold and for each possible set of aligned positions examining the compatibility of each amino acid with that specific fold. The target sequence is aligned to each protein structure in a library of folds and the compatibility of the sequence for that structure is calculated. This procedure is repeated for all the folds in a library. If a template structure is found to have a significantly high score, it is assumed that the target sequence folds in much the same way as that structure. This can be contrasted with a related technique commonly known as inverse protein folding, which involves searching a database

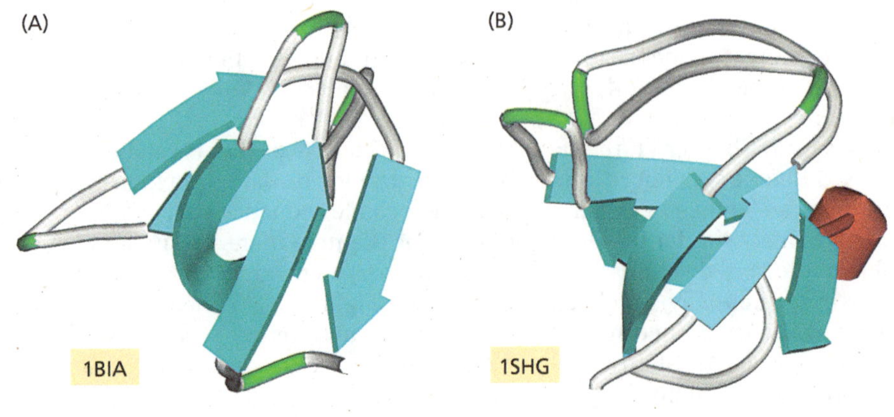


(C)

```
1BIA: ..FINRPVKLIIGDKEIFG-ISRGID-KQGALLLEQDGIIKPWMGGEISLRSAEK---------
1SHG:   MDETGKELVLALYDYQEKSPREVTMKKGDILTLLNSTNKDWWKVEVNDRQGFVPAAYVKKLD
```

**Figure 13.3**
**The ribbon representation of the structures of an SH3 domain.** (A) dihydrofolate reductase (1BIA) and (B) a kinase (1SHG). The sequence identity of these two domains is only 14.5%. Normal sequence alignment programs would not identify these structures as having a similar fold. (C) A sequence alignment based on the structural superposition.

of protein sequences with a known protein structure to predict which sequences are most likely to adopt that fold (see Figure 13.4).

## The prediction of protein folds in the absence of known structural homologs

There are many proteins in the PDB that have been shown to have a similar fold, including many homologs. It is unnecessary to use all these structures to predict the fold of a target sequence. Therefore, libraries of unique folds have been developed, some of which will be described in this section. Producing these libraries requires a method for structure comparison. There are many such methods, but we will not describe their algorithms here as they are not directly relevant to the threading problem (see Further Reading). The evaluation of the compatibility of the target sequence and a protein fold requires a scoring scheme. The schemes used are related to some described in previous chapters, but are sufficiently distinct that we will discuss some of their general aspects. The target sequence can be aligned with a protein fold in many ways, and the threading method must determine the best for each of the library folds. The algorithms used to obtain the optimal threading alignment are related to those for sequence alignment, but some differences will be presented. There are several ways in which different methods make their final prediction of a fold for the target sequence, not just based on the optimal alignment scores. At the end of this section some of the ways of selecting the final prediction will be discussed.

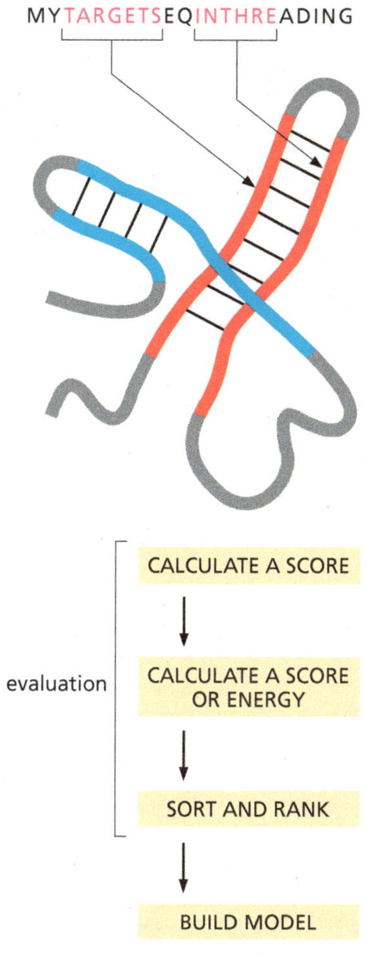


**Figure 13.4**
**Diagrammatic representation of the threading procedure.** First, segments of sequence are structurally aligned (threaded) on to a fold and a score/energy is obtained for each alignment. A dynamic programming technique is used to find the alignment that has the best score/energy. This is done for each fold in the fold library, and the results are ranked. The folds giving the best-scoring results are then selected for use in modeling the query sequence.

## Libraries or databases of nonredundant protein folds are used in threading

Currently there are around 35,000 protein structures in the PDB/RCSB/MSD database. If we had to look at all these structures to predict the target this would be very time consuming. However, many of these structures are of the same protein with different bound ligands or point mutations, or of proteins with similar folds. Therefore libraries that contain only unique folds have been developed. Using these libraries radically reduces the number of structures that have to be explored. Currently, only 700–1500 structures are needed to represent those experimentally determined.

There are several nonredundant fold libraries; CATH and SCOP are described here as representative examples. These differ in how they classify protein folds, and also in how individual structure folds are identified. CATH classifies protein folds at four levels: Class (C), Architecture (A), Topology (T), and Homologous superfamily (H) (see Figure 14.6). The level of interest in this section is the topology, which is effectively a description of a fold. In the topology level the structures are classified into fold families according to their overall shape and the connectivity of the secondary structures using a structure comparison algorithm. In mid-2005 there were approximately 900 different topology entries in CATH. The SCOP (Structural Classification Of Protein) library classifies protein structures into three main levels: folds, superfamilies, and families. The fold level is equivalent to the CATH topology level, and by 2005 there were more than 1000 different fold entries in the SCOP library. Figure 13.5 shows that the number of newly identified folds is decreasing or at least leveling off. This suggests that the libraries might be approaching completeness, at least for describing nonmembrane protein globular folds.

## Two distinct types of scoring schemes have been used in threading methods

Evaluation of the compatibility of the target sequence aligned with a protein fold requires a quantitative measure of the structural environment. Many different measures have been proposed, but they mostly fall into two classes. One class is a

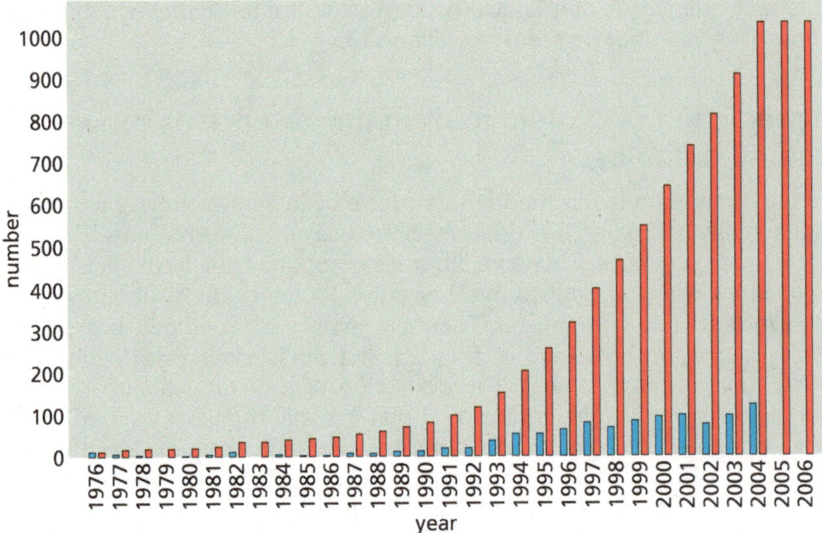


**Figure 13.5**
**A histogram of the number of new protein folds identified by SCOP each year.** The data were obtained from the RCSB Web site. The red bars signify the cumulative number of folds, the blue bars signify the number of new folds identified in that year.

modification of the amino acid substitution score matrices discussed in Section 5.1. These take account of the likelihood of a substitution given the nature of the environment. The second class explicitly includes details of the structure in the vicinity of each residue, involving inter-atomic distances or numbers of residues within a specified distance.

The substitution matrices used for threading differ in one major respect from those used for sequence alignments. In this case the two aligned residues are not equivalent, in that one is at a particular location in the structural fold, while the other is in the target sequence and in an unknown structural location. Because of this the substitution matrices used are asymmetric, as substitution of an alanine in the fold with a lysine from the target sequence is not equivalent to substituting a lysine in the fold with an alanine. In the first case, the fold may not have sufficient space or polar environment to accommodate the lysine. In the second case, the substitution may cause a cavity and a hydrophobic residue in a polar environment.

The FUGUE method described by Tom Blundell and colleagues uses 64 distinct scoring matrices, each defining a specific environment for the residue in the fold. The environment is defined by three properties: four classes of main chain conformation, two classes of solvent accessibility, and eight classes of hydrogen bonding distinguishing between bonding involving side chain, main chain NH, and main chain CO groups. These three properties are independent of each other and result in the 64 defined environments. The substitution matrices were defined from a structural database in an analogous way to that used to define the BLOSUM matrices (see Section 5.1). The LOOPP method of Ron Elber and colleagues also uses environment-dependent substitution matrices but these are derived differently from those of FUGUE.

Knowledge of the location of secondary structure elements in the template sequence can be used to define variable gap penalties. A distinction can be made between insertion and deletion with respect to the known fold. Insertions and deletions in the middle of secondary structure elements are given high penalties whereas those outside secondary structure elements are given very low penalties.

The second class of scoring scheme involves functions that all depend, to some degree, explicitly on inter-residue distances in the protein fold. GenTHREADER and LOOPP include terms that define the potential of mean force. These terms assign an energy that depends on the distance between specified atoms. The backbone N and O atoms and the side chain $C_\beta$ of each residue are used in these terms, distinguishing in addition interactions involving residues of short-(<11 residues), medium-, and

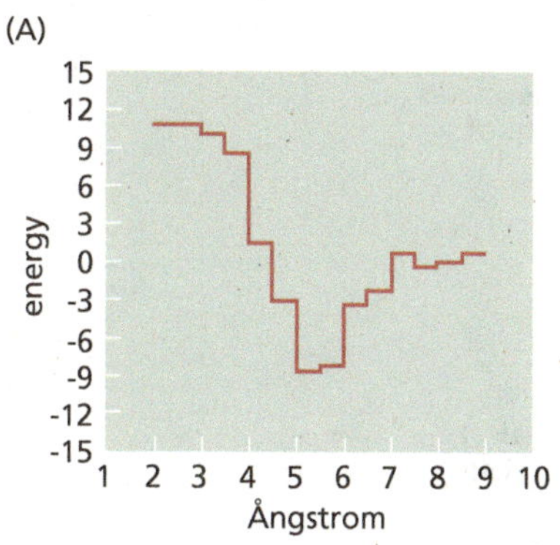


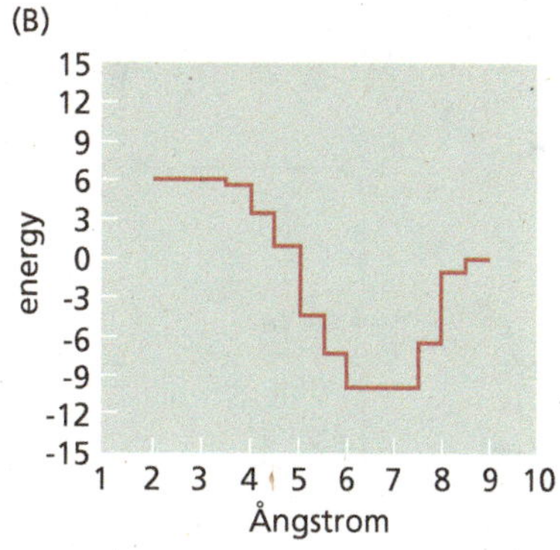


**Figure 13.6**
**An example of energy terms derived from observed protein structures, as used in threading programs such as LOOPP.** The plots show the interaction energy for a specific pair of amino acids as a function of distance (in Ångstroms). (A) shows the interaction energy for interacting Val-Leu residue pairs, and (B) the energy for interacting Phe-Trp residue pairs.

long-range (>22 residues) sequence separation. The number of observed specific atom pairs ($C_\beta \rightarrow C_\beta$ for example) at a given distance is converted to an energy (see Figure 13.6).

GenTHREADER in addition uses a solvation potential for each residue which depends on the number of $C_\beta$ atoms located within 10 Å of its $C_\beta$ atom. The number of $C_\beta$ atoms has been found to correlate well with the residue solvent accessibility. Many other scoring functions have been derived involving the number of atoms within a specified distance. One of the most common is the pairwise contact potential (PCP), which is similar to the potential of mean force. Another is the contact capacity potential (CCP), which is intended to represent the hydrophobic contribution to the free energy of folding. These last two terms involve the fraction of pairs of residue types in contact relative to the total of such pairs. The potentials are defined as the logarithm of a ratio, in a similar way to the propensities described by Equation EQ12.5. Many variations of these types of potentials have been used in programs such as LIBRA and 123D+.

3D-PSSM uses a combination of the two classes of scoring schemes described above. Instead of using substitution matrices it uses PSSMs (see Section 6.1) to account for environmental effects along the sequence. PSI-BLAST is used to generate a profile for each fold in the library. Where the library contains identical folds for nonhomologous sequences, separate profiles are generated for each of these folds. The final profile is generated by combining positions at equivalent locations in the different folds using structure superposition to determine the equivalences (see Figure 13.7). In addition to this profile a solvation potential similar to that of GenTHREADER is used.

## Dynamic programming methods can identify optimal alignments of target sequences and structural folds

The algorithms used by threading methods to obtain the optimal alignment of the target sequence with each of a library of folds are closely related to the dynamic programming techniques described in Section 5.2. However, particularly when potentials of mean force are used in the scoring scheme, significant modifications of the basic sequence alignment methods are required.

Some programs, like 3D-PSSM, only involve scoring terms that relate to one sequence position. These techniques can use the standard dynamic programming methods described in Section 5.2 to determine the optimal alignment of the target sequence to the template structure. Because the two sequences are often of very different lengths the end gaps are not penalized at all. This is often called global–local dynamic programming. In the case of 3D-PSSM, profiles are constructed for both the fold and the target sequence. The target sequence is matched to the fold profile and the template sequence is aligned to the target profile. The highest-scoring alignments from these are retained.

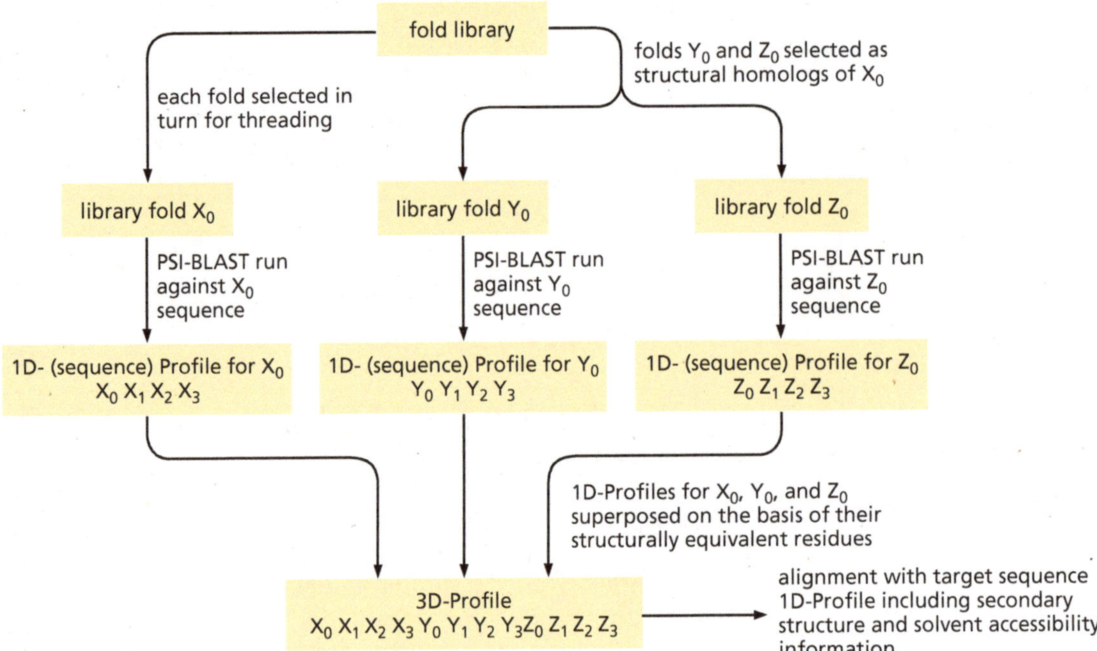


**Figure 13.7**
**A flow diagram illustrating the steps involved in generating the fold 3D-Profiles used in the 3D-PSSM algorithm.** Each sequence $X_0$ in the fold library is aligned using PSI-BLAST with its homologs $X_1$, $X_2$, etc. The SCOP database is used to identify other structural homologs $Y_0$ and $Z_0$, which are also aligned with their homologs using PSI-BLAST. In this way PSSMs are created for the different sequence families that are associated with the same protein fold as $X_0$. These are combined with data on secondary structures and solvent accessibility to produce a 3D-Profile of the fold. An equivalent profile is constructed for the query sequence, also based on homologous sequences and (in this case predicted) secondary structures. Dynamic programming methods are used to align the query profile with the 3D-Profile of each fold in the fold library.

Methods that have scoring terms dependent on several sequence positions cannot use the standard dynamic programming methods. GenTHREADER for example uses double dynamic programming, a technique proposed by Willie Taylor and colleagues. Other methods use dynamic programming schemes involving iteration to obtain the optimum alignment.

## Several methods are available to assess the confidence to be put on the fold prediction

The highest-scoring alignment can be used to immediately propose the fold of the target sequence. As was the case for sequence alignment it is necessary to evaluate the significance of the score obtained and to use this to estimate the confidence of the prediction. In addition some schemes use more complex methods to determine the final fold prediction, in some cases involving neural networks.

The significance of the score obtained is usually determined by comparison with the scores obtained using a set of randomized sequences. This is often known as the **shuffle test**. The analysis is less rigorous than that used for sequence database searching (see Sections 4.7 and 5.4), being usually based on assuming a normal distribution of scores.

An alternative test is the **Sippl test**, named after Manfred Sippl, one of the originators of the threading technique. This method evaluates the accuracy of the threading prediction for a target sequence of length *N* by generating a set of wrong structures with the same sequence. To generate these incorrect structures the sequence is fitted without gaps onto a database of folds that are of length *N*. If the folds are larger, they are cut into *N*-length structural segments and threaded with the target sequence. A score is calculated for all these *N*-length segments, and these values are used to obtain a mean score and standard deviation assuming a normal distribution. The score of the predicted fold is calculated relative to this distribution. If the prediction is correct it should have the best score.

In GenTHREADER the alignment score is just one of six properties or parameters, which are fed into a feed-forward neural network that evaluates whether the target

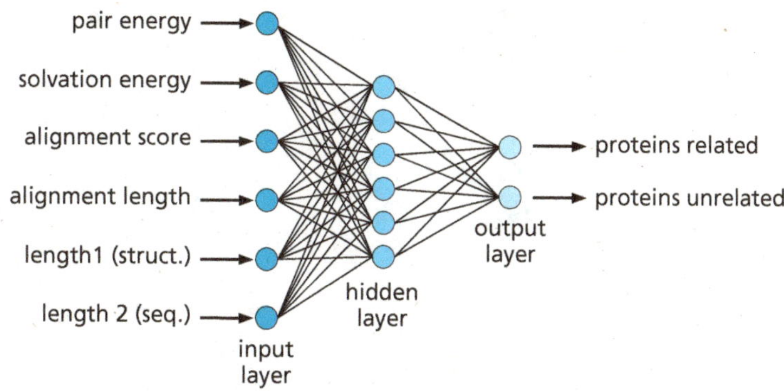


**Figure 13.8**
**The architecture of the neural network used in the GenThreader algorithm.** Instead of using only the energy scores to predict the fold, they are combined with the initial sequence alignment score and the lengths of query and target sequences, as well as the number of aligned residues to improve the fold prediction. The neural network used is a standard feed-forward model as described in Section 12.4, with six hidden layer units and two output units. The prediction is based on the higher-scoring output unit. (Adapted from D.T. Jones, GenTHREADER: an efficient and reliable protein fold recognition method for genomic sequences, *J. Mol. Biol.* 287:797–815, 1999.)

sequence has that fold (see Figure 13.8). The output layer of this network has two units, with signals between 0 and 1, one signaling the fold is a possible candidate and the other that it is unsuitable. If the signal from the former is greater than 0.34 the confidence of the prediction is described as MEDIUM, if greater than 0.79 it is described as HIGH, and if 1.0 is described as CERTAIN.

## The C2-like domain from the Dictyostelia: A practical example of threading

The C2-like domain will be used as an example to illustrate the strengths and weaknesses of the most commonly used threading algorithms. The C2 domain is an all-β phospholipid-binding domain distinct from the catalytic domain of PI3 kinases. It occurs in many different proteins and in some cases is known to mediate interactions with cell membranes. A C2 domain-like protein from the Dictyostelia was chosen as an example. No significant homology is found by using BLAST on the PDB sequences.

Several Web-based programs are used below with the C2 domain-like protein to illustrate general aspects of these types of predictions. To run the prediction programs, in the default modes, all that is needed is a target sequence. The results are returned in various formats: some as text e-mails, while others also give a link to HTML formatted results, along with figures and various scores. To simplify the correct identification many programs, in addition to a numerical score, give a verbal descriptor of the confidence of the prediction, such as HIGH in GenThreader or CERTAIN in FUGUE. Two programs, PHYRE and LOOPP, provide a finished model, making model building unnecessary if one is happy with the alignment. LOOPP uses MODELLER (described in later sections in this chapter) to model its threading results (see Figure 13.9)

The threading programs PHYRE, FUGUE, and 123D+ all identified the C2 domain of cytosolic phospholipase A. Figure 13.9B shows the alignments from each method between this template and the target sequence. Although there are small variations in the alignments, the predictions of the β-strands are the same. The prediction of the one helix present in the structure is different between PHYRE and the other two programs. This difference in prediction leads to the helix being lost in the PHYRE calculation (see Figure 13.9A). All the strands have been predicted, although in the diagrams in Figure 13.9A some strands have not been drawn due to the cartoon illustration using different secondary structure assignments (see Figure 12.3). GenTHREADER predicts the top fold to be a C2 domain from phosholipase E. The β-strands are similarly predicted, although there are more small insertions throughout the alignment. The same residues are predicted as β-strands as in FUGUE, PHYRE, and 123D+. LOOPP identifies as the top-scoring hit a C2 domain from Rabphilin-a. A large insertion and a deletion are present within the LOOPP prediction. This causes quite a different structural folding, although the β-core is

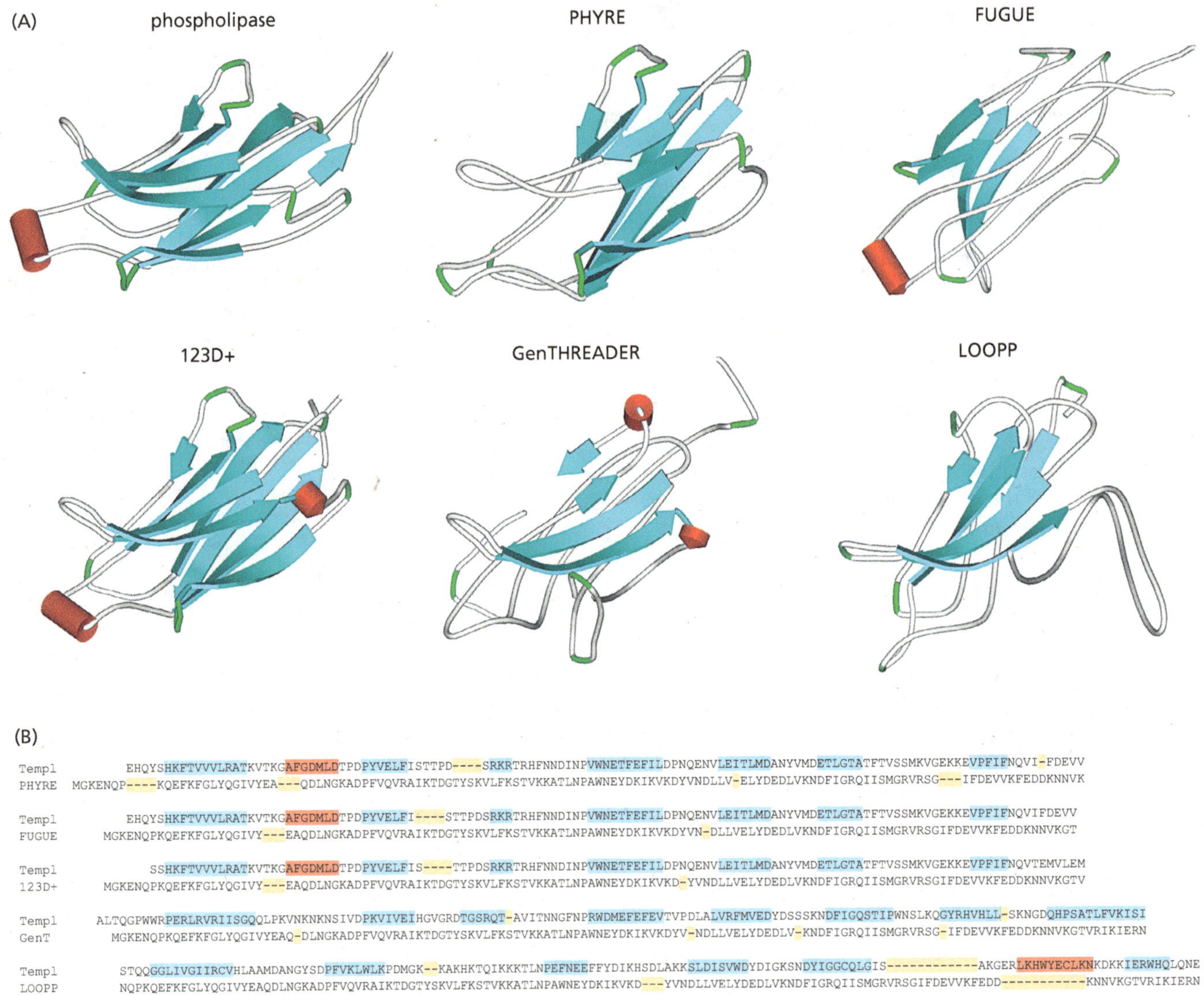


**Figure 13.9**
**Prediction results from different threading programs.** (A) A ribbon representation of the modeled predicted fold by different threading programs compared to the fold that the programs have used as a template, which is the C2 domain from phospholipase. (B) The structural alignment given by each program with the template. Strands are colored in blue, helices in red, and gaps in the alignment are highlighted in yellow.

conserved and obviously C2-like. In addition, a couple of different regions are predicted as β-strands with respect to the other methods. As we can see from Figure 13.9A all the core structures are similarly predicted, but there are large variations in the loop regions.

One method failed to identify the target sequence as a C2 domain fold. LIBRA predicts the sequence to be caspase-8, an α/β Rossmann fold. Figure 13.10 illustrates the fold, the prediction, and alignment. It is very different from the C2-like domain.

Even though there was no significant sequence homology—the percentage identity ranges from 10% to 16%—between the target protein and the sequences in the structural database, in general the fold prediction was successful. The C2 fold was identified more than once. Therefore this fold would be used with a high degree of confidence even had we not known that our target was a C2 domain. Once the fold is identified, it is useful to use all the alignments provided by the threading programs to correctly align the target protein. If more than one threading program aligns regions similarly, these should be used as the core (or conserved) regions. The rest of the sequences should be aligned on the basis of these core regions. Once a relatively accurate alignment is obtained, homology

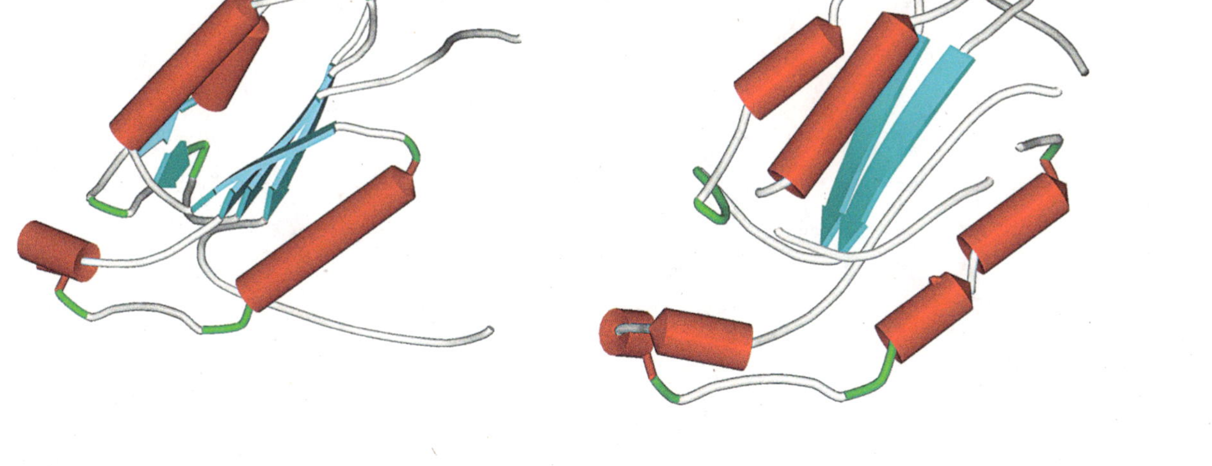


```
Templ  -DKVYQMKSK-PRGYCLIINNHNFAKAREKVPKLHSIRDRNGTHLDAG-ALTTTFEELHFE--IKPHDDCT-VEQIYEILKIYQLMDHSNMDCFICCILSHGDKGIIYGTDGQEAPIYELTSQFTGLKCPSLAGKPKVFFIQACQGDNYQ
Target MGKENQPKQEFKFGLYQGIVYE-AQDLNGKADPFVQVRAIKTDGTYSKVLFKSTVKKATLNPAWNEYDKIKVKDYVNDLLVELYDEDLVKNDFIGRQIISM---GRVR----SGI-FDEVVKFED--DKNNVKGTVRIKIERN-------
```

**Figure 13.10**
**Similar ribbon diagram as in Figure 13.9 for the prediction from LIBRA, which did not identify a C2 domain as the fold.** The predicted fold (B) is an α/β Rossman fold from caspase as shown in (A).

modeling as described below can be carried out to generate a three-dimensional representation of the threaded sequence.

To use threading methods to their full potential the following should be taken into account. If the target protein is suspected to consist of multiple domains—for example, if the amino acid sequence is very long—then the protein should be divided into putative single domains. In some cases domains can be distinguished by sequence features or experimental knowledge.

The same sequence will often give different results with different threading programs, so run the sequence through as many as possible. If a fold appears in the top 10 or 20 hits in different programs then it is more likely to be the correct answer.

If sequences homologous to your target are available, perform a threading study in them as well. Different homologous sequences can also give different results from each other. So, if the same family of folds is given for a homologous sequence, and is also found by different methods, then it is very likely to be the correct fold. Even if different folds are returned by different programs or homologous sequences, it is often possible to choose a most likely fold by consensus.

Use whatever knowledge is available about the protein, such as its function, and evaluate the protein folds that have been found by threading in the light of that knowledge. If there is a functional similarity it again increases the likelihood that the protein has been correctly identified.

Before modeling the predicted fold, check the alignment given by the program. If possible edit the alignment to improve the areas of insertions and deletions with respect to the core structure. Human insight and knowledge are powerful tools.

## 13.3 Principles of Homology Modeling

In the absence of an experimental structure for a protein, model-building on the basis of a known three-dimensional structure of a homologous protein is, at present, the most reliable method of obtaining structural information. This

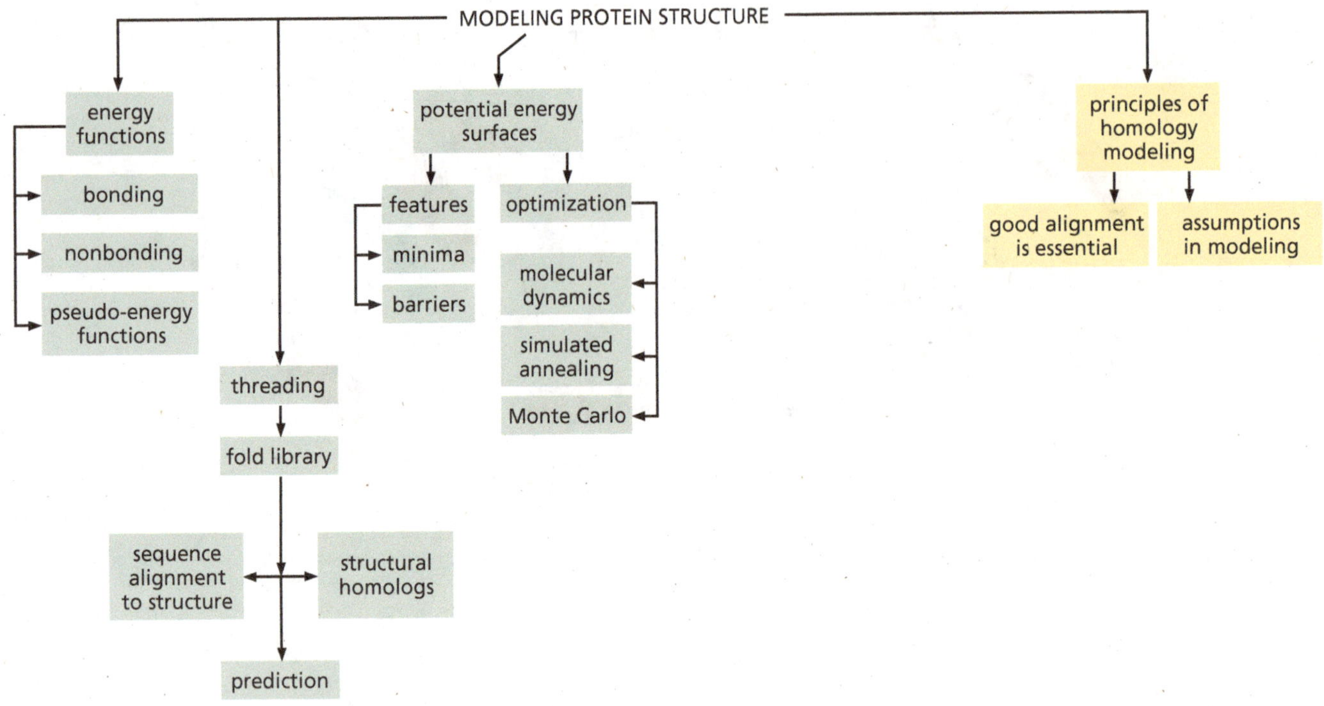


**Flow Diagram 13.3**
**This section describes some of the basic principles of homology modeling.**

**homology modeling** approach is also referred to as **comparative modeling** or **knowledge-based modeling**. It depends on the fact that the structures of homologous proteins are better conserved during evolution than the amino acid sequences. Not only will most proteins with very similar sequences have almost identical backbone structures, but even proteins with quite dissimilar sequences can have similar conformations.

The first homology models were built as early as the 1960s using wire and plastic models of bonds and atoms (see Figure 13.11). The first published homology model structure, in 1969, was of the small globular protein α-lactalbumin, which was modeled on the basis of the structure of hen egg white lysozyme as the template

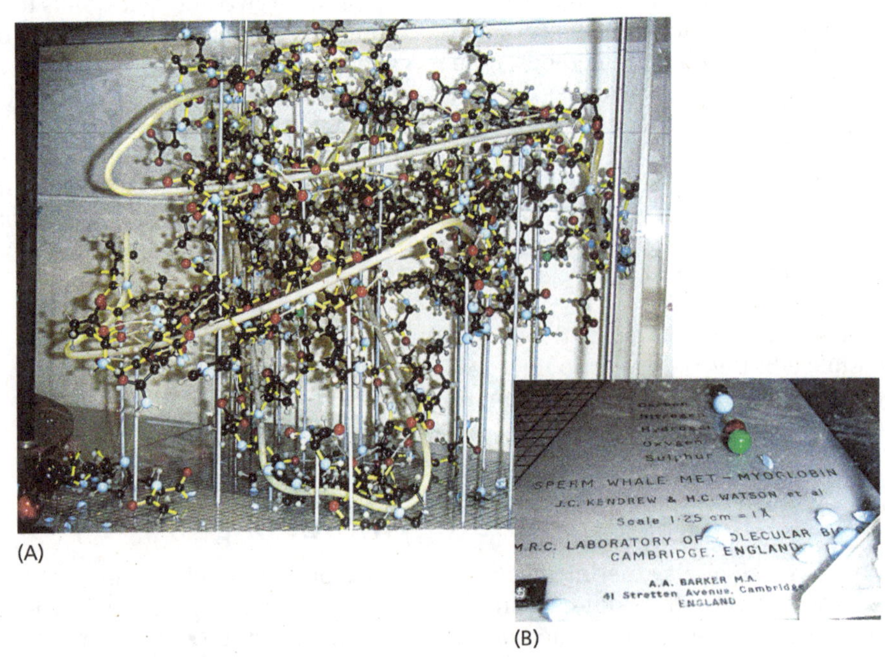


**Figure 13.11**
**A photograph of a wire model of sperm whale myoglobin built by John Cowdery Kendrew and colleagues.** This structure is based on low-resolution X-ray data. It shows the complexity of a protein structure and illustrates how time consuming it was to build such a precise representation. (With thanks to Birkbeck College, London, UK, for permission to take this picture.)

(A)

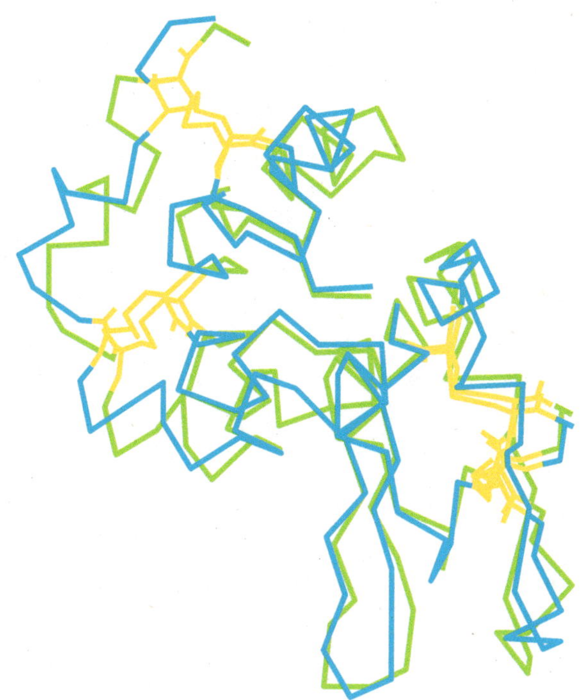

(B)

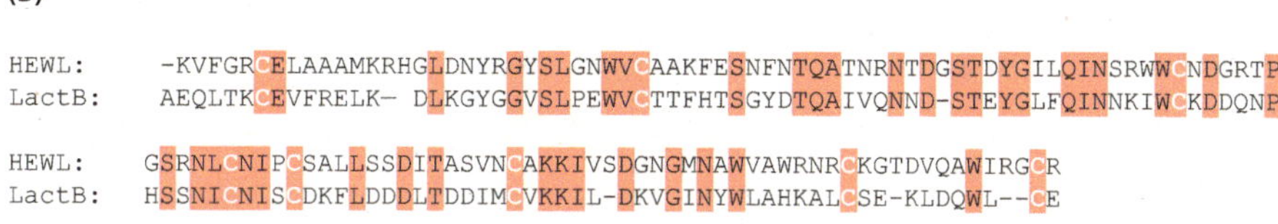


**Figure 13.12**
**Superposition of the $C_\alpha$ trace of lysozyme and α-lactalbumin.** (A) The X-ray structure of lysozyme (green) superposed on that of α-lactalbumin (blue). The disulfide bonds are highlighted in yellow and do indeed have the same binding pattern as predicted by their sequence separation and therefore as modeled. (B) The alignment of lysozyme and α-lactalbumin; residues that are identical in both sequences are highlighted in red and the cysteines are in white.

protein; the sequence identity between these two proteins is 39%. In modeling terminology, α-lactalbumin is known as the target protein. The structure of lysozyme was modified by hand to accommodate those amino acids of α-lactalbumin that did not match those in lysozyme. In addition the two proteins contained an identical pattern of cysteines, suggesting a similar arrangement of disulfide bonds. When the structure of α-lactalbumin was later solved by X-ray crystallography, the model turned out to be essentially correct apart from the carboxy-terminal end, which is different in structure to that of lysozyme. Figure 13.12 shows the X-ray structure of α-lactalbumin superposed on that of lysozyme. The disulfide bonds are highlighted in yellow in the structures and do indeed have the same binding pattern.

## Closely related target and template sequences give better models

Model building by homology is a multistep process (see Figure 13.13). At almost all stages, whether using automated or manual modeling, rational choices are made on the basis of prior experimental knowledge. The number of choices available and the accuracy of the model depend strongly on the sequence similarity between the protein to be modeled—the target structure—and the protein or proteins that are used as the template.

Not surprisingly, the more closely related the target and template sequences are, the more accurate the model and the easier the model-building procedure. If the template and target have greater than 90% sequence identity, then the backbone of the model may be as good as a crystallographic structure. When the sequence

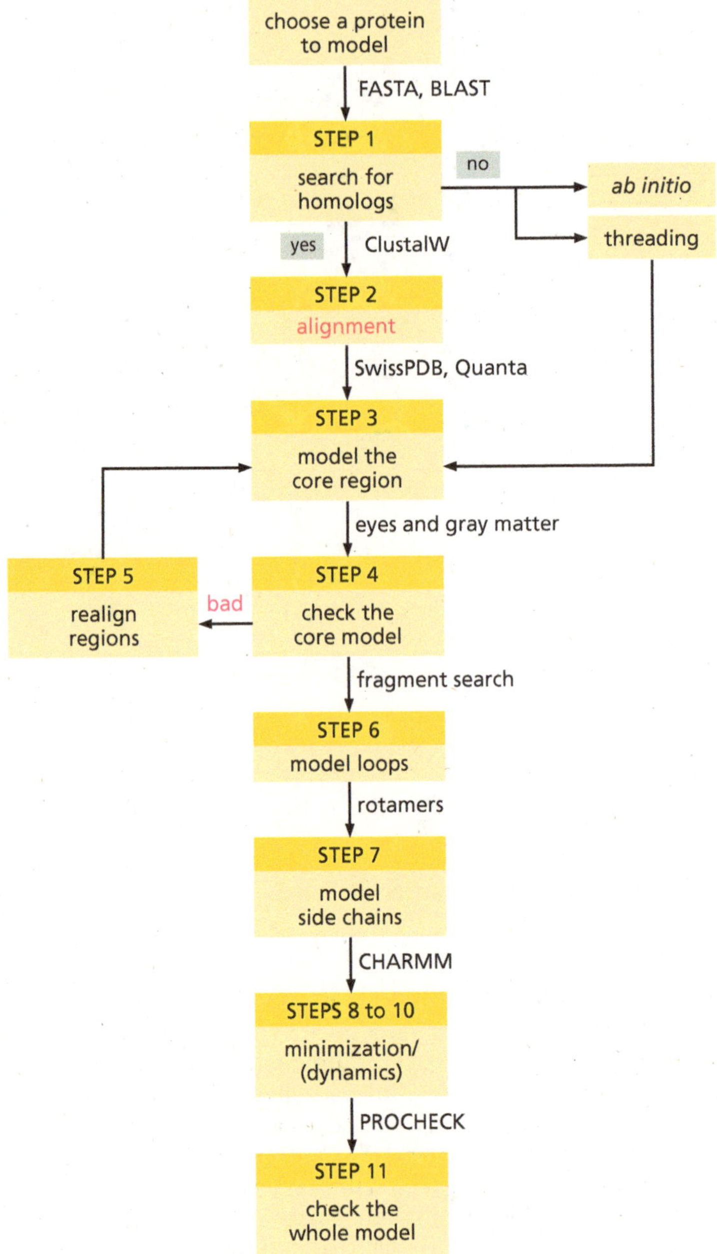


**Figure 13.13**
**A flow diagram showing the steps involved in modeling a target protein based on a template structure, as well as the possible steps to take when no template structure is identified.**

identity falls below 25%, large errors are often made and, in general, modeling is not advisable unless there are additional experimental data that can help in determining the three-dimensional structure.

## Significant sequence identity depends on the length of the sequence

Studies of the relation between the sequence similarity and three-dimensional structure for the cores of globular proteins have indicated that the cut-off point for successful modeling is around 25% sequence identity (see Figure 13.14). However, percentage identity is dependent on the length of the alignment, and so homology thresholds for structurally reliable alignments are calculated as a function of alignment length. From Figure 13.15 it can be seen that only proteins more than 100 amino acids long can be modeled with any degree of accuracy at the 25% identity level. For smaller proteins (60–100 amino acids) the percentage identity has to be over 30%. Peptides smaller than 30 amino acids should be modeled with caution,

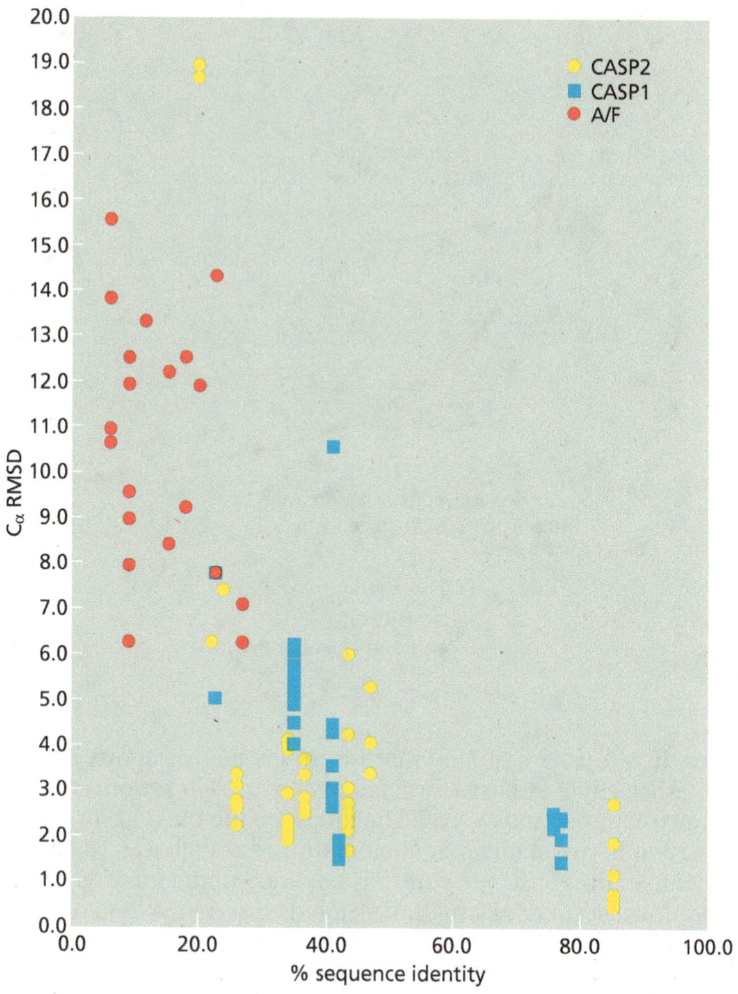


**Figure 13.14**
**A plot showing the results for comparative modeling from the CASP1 (Critical Assessment of techniques for protein Structure Prediction) (blue squares) and CASP2 (yellow circles) modeling experiments.** The results of *ab initio* and fold-recognition modeling for low-sequence-identity targets are also shown by red circles. As the sequence identity increases, the RMSD between the experimental structure and predicted structure decreases. In the *ab initio* predictions, no model is below 6 Å RMSD. (Adapted from A.C.R. Martin, M.W. MacArthur and J.M. Thornton, Assessment of comparative modelling in CASP2, *Protein Struct. Funct. Genet.* Suppl. 1:14–28, 1997.)

even at 100% identity, even though at 45% identity some similarity to the structure of the aligned partner can be assumed. The one exception is when the peptide to be modeled is a ligand in a protein, as the binding site and environment will add additional constraints to the putative structure.

## Homology modeling has been automated to deal with the numbers of sequences that can now be modeled

Homology modeling is becoming an important stage in interpreting genomic data and assigning possible functions to genes. A structural model is such a great aid to understanding biological function that it has led to efforts to ensure that all gene products should have either an experimentally solved structure or a modeled structure. To model so many proteins the tasks of producing accurate sequence alignments and building three-dimensional models from the alignments have been fully automated. In yeast, all the open reading frames (ORFs) with a relatively high degree of identity to template proteins of known structure have been modeled by homology using the program MODELLER. Automated homology modeling for proteins of lower sequence similarity is still not advisable without close analysis of the final model.

## Model building is based on a number of assumptions

Two main assumptions are made in homology modeling. First, it is assumed that the polypeptide backbone of regions conserved between template and target have identical spatial coordinates. Although this assumption is adequate for modeling, it is not strictly true; conserved regions will have similar, but usually not identical,

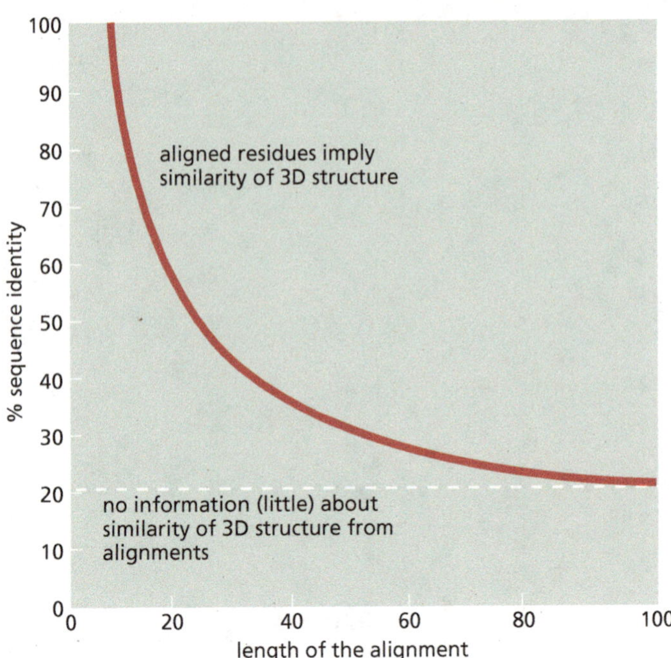


**Figure 13.15**
**A graph showing the dependence of a structurally reliable alignment on both sequence identity and alignment length.** An alignment falling to the right of the red line will imply similarity in three-dimensional structure, whereas one falling to the left and below the red line will provide little or no information. The white line at around 20% sequence identity is the threshold below which alignments, whatever the length, should not be used for homology modeling. This easily visualized length–homology correlation gives the modeler a convenient slide rule to assess whether a given alignment will provide a reliable model. (Adapted from C. Sander and R. Schneider, Database of homology-derived protein structures and the structural meaning of sequence alignment, *Proteins* 9:56–68, 1991.)

coordinates. It is further assumed that insertions and deletions in the sequence alignment, especially if extensive, will fall mainly in loop regions, which consist of relatively unstructured random coil. The loops are the most likely regions to differ in conformation between target and template and are allowed to be more flexible in the model than the conserved core. It is not uncommon for even close homologs to have additional secondary structure elements inserted in loop regions.

## 13.4 Steps in Homology Modeling

The process of modeling a protein three-dimensional structure is composed of a number of separate steps, which are shown in Figure 13.13 and described below. For simplicity, we will work through the case where a single homolog is used as the template. For modeling a protein semi-manually according to the steps outlined below one can use programs such as Swiss-PdbViewer, or MolIDE, which are freely available but have to be downloaded to a local machine.

Before embarking on the modeling process, a few general decisions have to be made. When there is more than one homolog with a solved structure, the main decision that a modeler has to take is whether to base the model on just the template that is most similar to the target in sequence, to use a template that is an average structure based on all the templates, or to use different fragments from each structure to make up a template, as described later.

Each of these options has advantages and disadvantages. Models built on a template based on a single structure may give the most accurate results if the target and template are closely related in sequence. The advantage of using only one structure rather than the average of several is that distortion due to different structural conformations will not be incorporated into the model. Some structures that are similar in sequence (around 45–50% identity) show as much as 1.5–1.8 Å root mean square deviation (**RMSD**) in the relative positions of the atoms.

By its very nature an average template is not true to any one of the structures. On the other hand, this can make it a more appropriate choice, as the model is that of

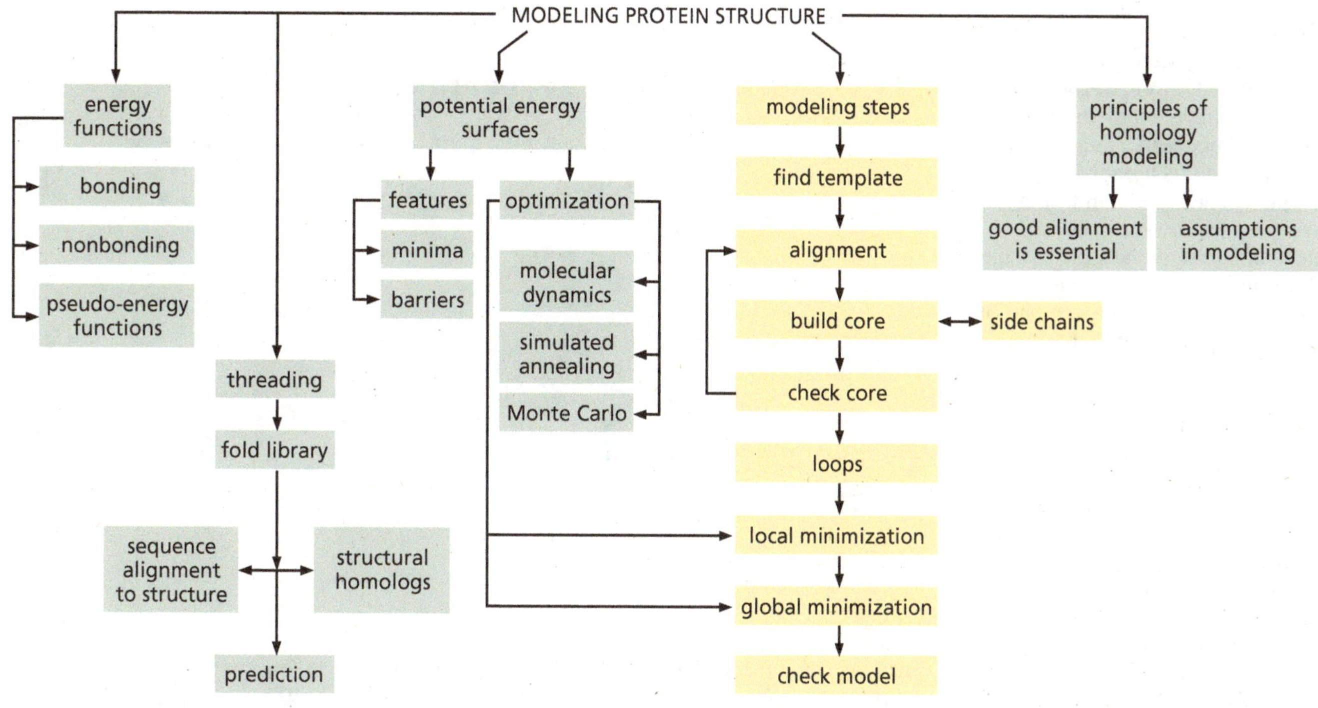


**Flow Diagram 13.4**
**In this section the key steps in the process of homology modeling are described.**

an unknown protein and an average may give the closest representation. The structural model based on separate fragments from a set of homologous structures has the disadvantage of needing to join the fragments together, and the joins are the parts of a model where most of the errors arise. When more than one homologous structure is available, probably the most logical approach is to base the target-conserved regions on one homolog and the loops (insertions and deletions) on fragments from the other homologs in the set, as far as possible.

## Structural homologs to the target protein are found in the PDB

The first step is to find structural homologs to the target protein. Most experimentally solved structures have been placed in the Protein Data Bank (PDB). A database with the sequences of the PDB structures is searched with the target sequence for homologs using one of the sequence-search programs such as BLAST or FASTA, as described in Sections 4.6 and 4.7. The protein with the highest similarity score and the highest sequence identity over the largest stretch of amino acids is chosen as the template. Where the functional family of the target protein is known, the ideal template protein would be one belonging to the same family.

If no structural homolog is found there are four possibilities open to the modeler: give up; perform an *ab initio* prediction (see Box 13.1); try a threading method to find a possible template; or search for short regions of homology, especially those that contain functional residues (see Section 4.10). If the last option is successful, you can at least make an incomplete model to obtain at least some, if only partial, structural information. But if neither this nor the threading option is successful, the modeling exercise must be put on hold to await publication of the target structure or that of a homolog.

## Accurate alignment of target and template sequences is essential for successful modeling

Once a suitable template protein has been identified, the target and template sequences must be aligned. This alignment is the most crucial step in the modeling

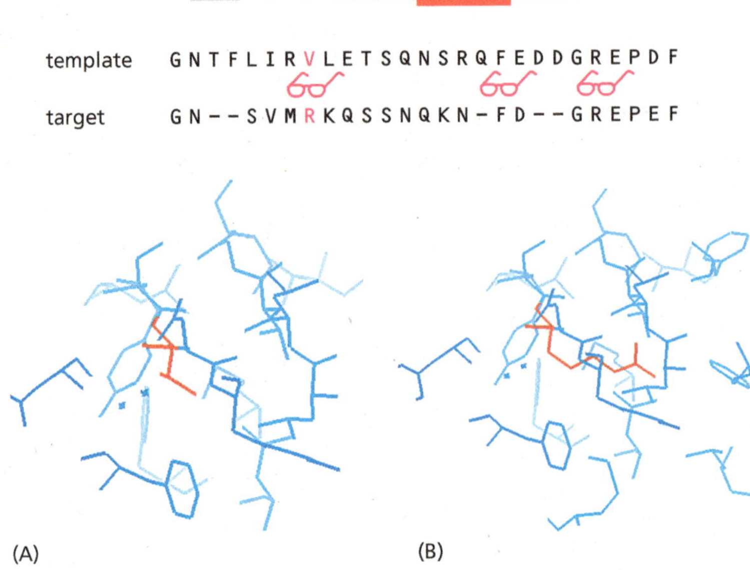


**Figure 13.16**
**The effect alignment can have on a modeled structure.** The top part of the figure shows a hypothetical alignment between a target sequence and its homologous template. The top line indicates structural features associated with the template sequence; the zigzag line indicates a β-strand and the cylinder an α-helix. The first pair of spectacles in the alignment marks a misalignment at a single position. The positively charged arginine highlighted in red in the target sequence is matched to a hydrophobic valine (also in red) instead of to the arginine one residue to its left. This misalignment has occurred as a result of two gaps that have been inserted amino-terminal to the arginine. The panels show (A) the template structure and (B) the target model as modeled using the above alignment. As can be seen in (A), the original valine (in red) points into the hydrophobic core of the protein, and when an arginine is modeled at the same coordinates it will also point into the core (B). But now there is a buried charged residue within the core, which is energetically unfavorable, and the larger side chain of the arginine also clashes with many other side chains. The second pair of spectacles indicates the effect of a deletion in the middle of a secondary structure element—an α helix. The last pair of spectacles shows the correct alignment of functionally important residues.

process. Consider the model of a protein as the three-dimensional representation of the sequence alignment. If the alignment is wrong, then the three-dimensional structure will be wrong as well. Even a one-residue misalignment can lead to incorrect analysis. Take the case illustrated in Figure 13.16, where a mismatched residue has caused a one-residue misalignment of an arginine downstream of where the mismatch occurred. In the template structure (see Figure 13.16A) the valine is oriented towards the hydrophobic core, whereas the arginine of the target sequence model should point toward the outside of the protein. Because of the structure of the homolog in this region, the incorrect alignment results in the arginine side chain being modeled as entering the mainly hydrophobic core of the structure where it cannot be accommodated.

Regions that are difficult to align should be visually reinspected when the model of the protein core has been created as described in the next step. The modeled core should conform to general structural rules, such as that most charged residues should be on the surface of the protein unless they can be accommodated by other charges in the structure or there is a functional reason for a charged residue to be buried. If the core does not conform, then it will be necessary to realign the sequences and remodel the core.

In general an alignment can be improved in the following ways. Insertions and deletions should not fall in the middle of secondary structure elements as far as possible. If available, multiple homologous sequences should be used to make the alignment, even if only one structure will be used as the template. And any other information known about the protein should be used, such as what type of residues constitutes the active site and should be matched to the template.

## The structurally conserved regions of a protein are modeled first

The structurally conserved regions—the core—are modeled first. Modeling the core is simply achieved by transferring the x, y, and z coordinates of every matched atom within an aligned residue from the template to the target molecule. The backbone atoms are then joined together to form peptide bonds at the correct angles. It is usually possible to copy only some of the side-chain coordinates, as many side chains in the target will not be identical to those in the template. Regions with insertions and deletions are left for later. In most computer programs and graphical

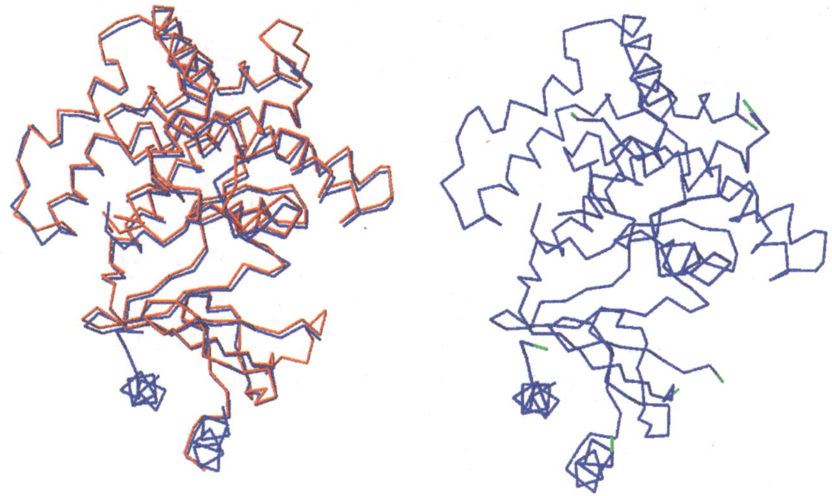

**Figure 13.17**
**Inspection of the modeled SCR (Structurally Conserved Region) shows where the insertions and deletions that were incorporated into the alignment are located in the target conformation with respect to the template.**
(A) This shows the target structure superposed on the template structure. (B) This shows only the target model with the end of the core regions where an insertion has to be modeled. The modeled core region should be carefully analyzed to see if any insertions (or deletions) will cause structural disruption.

packages the modeling of backbone and side chains occurs simultaneously, but for clarity we will deal with the backbone first and will describe the procedure and theory behind the modeling of side chains later. Figure 13.17 shows a $C_\alpha$ wire representation of the modeled structure of a kinase catalytic domain in blue and the template kinase structure in red. Note that at this point the insertions and deletions have not been modeled, so that the core structure is a set of discontinuous chains.

## The modeled core is checked for misfits before proceeding to the next stage

It is important to check the core for misfits now, as this is the last chance to test the correctness of the model before the time-consuming processes of loop building and energy minimization. Once the core of the structure has been built, the backbone conformation should be inspected for regions of insertions or deletions, and regions that were difficult to align. Regions where the alignment requires insertions should not be facing into the structure, nor be part of or disrupt secondary structure elements (see Figure 13.17). In regions that were difficult to align, one should examine whether moving the insertion by one or two residues would position it in a more favorable conformation.

## Sequence realignment and remodeling may improve the structure

If some insertions will disrupt the core or secondary structures, or if a particular region would seem to benefit from a slightly different alignment, then it is necessary to step back and realign the target to the template, remodel the core, and check it again. At this point it is advisable to save all the core structures that have been built. These can then be superposed on each other and the best core model chosen, using the criteria described above. On the other hand, if a particular region does not improve after a couple of tries and it does not seem to interfere with any of the known functional areas, use any of the core structures and continue with the modeling.

## Insertions and deletions are usually modeled as loops

Once a good alignment has been obtained, the other major procedure in model building is modeling the loops. These are the regions that usually contain insertions or deletions and are the most variable in sequence. Because of their variability in both sequence and length, loops are generally the most difficult regions to model.

However, loops are often involved in ligand recognition, ligand binding, and even in the active sites of proteins, for example, the active-site loop in protein kinases. Therefore, it is important to model loops as accurately as possible. To add to the problem, loops are often the most mobile part of a protein structure and may not even have coordinates in the crystallographic structure.

If the target protein contains an insertion in a loop sequence relative to the template structure, there will be no template coordinates from which to model the insertion. The easiest way round this problem is if there are other structural homologs with the same insertion. It is then possible to model the missing part using their coordinates. However, an insertion is often unique to the target protein. In this case, the most widely used method for modeling loops, in both manual and automatic procedures, is to search for fragments of the same length in a database of high-resolution structures.

The fragments are then incorporated, or annealed, to the core structure, as shown in Figure 13.18. Both ends of the core structure to which the loop has to be modeled are taken as the **anchor points**. The anchor points can range from one to many residues, although the standard length is two residues at either end. A search is made for a fragment the length of the loop plus the anchor points. The additional residues at the end of the fragment are then fitted onto the anchor points and the RMSD of their $C_\alpha$ atoms calculated. Loops that have the lowest RMSD are selected for further evaluation (see Figure 13.19). If, for example, an insertion of five residues has to be modeled, the fragment database is searched for a fragment of five residues describing the loop and two-residue anchors at each end, making a nine-residue fragment. The fragment that gives the lowest RMSD, interferes least with the core structure, and has the closest sequence similarity with the target is chosen and annealed to the structure. This process is repeated for all insertions.

Short deletions can be dealt with by local energy minimization, which brings the loop boundaries together. Large deletions are extremely difficult to deal with. Basically a large chunk of the structure has been deleted, and unless the structural boundaries flanking the deleted region are close together they cannot be joined. This may imply that the overall structure is incorrect.

Some programs, such as COMPOSER (described later in the chapter), try to model loops using only homologous structures. This approach is based on an analysis that found that loops from homologous structures provide a more accurate model than those from non-homologs. To model loops, COMPOSER searches all homologous structures for loops that are similar in sequence, and uses them even if they differ by one or two amino acids. The one or two residues left to model are then obtained by the fragment search method described above.

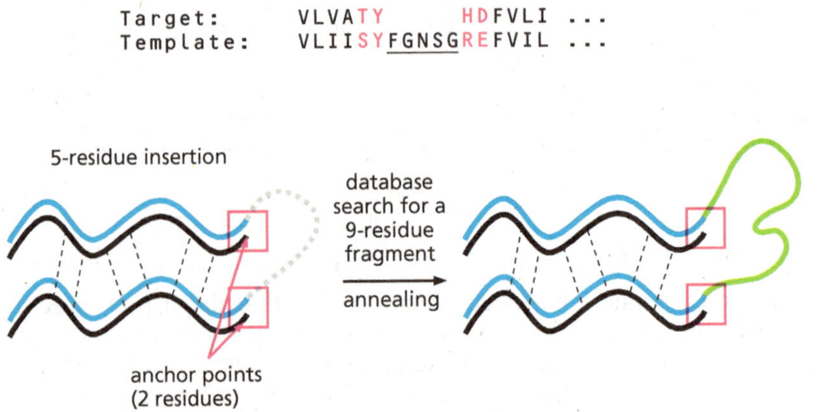


**Figure 13.18**
**A schematic illustration showing the database search method for building a loop.** A hypothetical alignment between a target sequence and its homologous template with a five-residue insertion in the target is shown. First, two two-residue anchor regions are identified on each end of the regular core structure (boxed in red). Then a database search of high-resolution fragments is performed for a fragment nine residues long: five for the insertion and four for the anchor points. Once a fragment is found that does not interfere with the core structure it is selected to be pasted into the structure giving a modeled loop region.

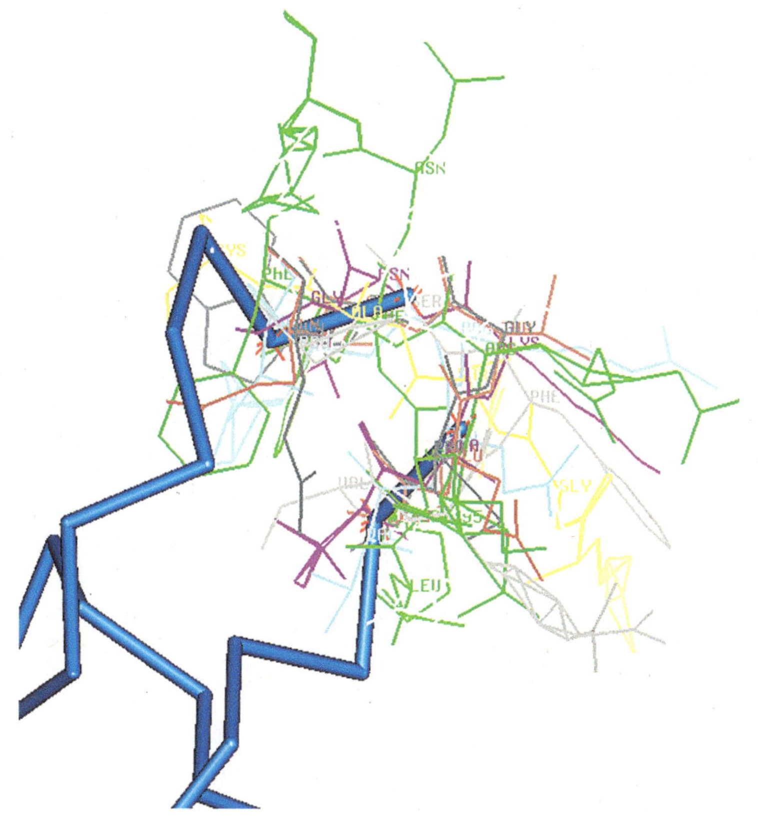


**Figure 13.19**
**A real example of building loops from a database search.** Ten loops have been selected on the basis of lowest RMSD for further evaluation depending on their conformation (core-disruptive potential) and sequence homology. The user then chooses one of the ten to paste into the modeled structure.

## Nonidentical amino acid side chains are modeled mainly by using rotamer libraries

The amino acid side chains confer the distinct characteristics of a protein, in both structure and function. To be able to build the side chains in a model, however, it is necessary to have some understanding of the conformations they can adopt.

A study of the atomic coordinates of side-chain atoms of conserved residues in proteins with similar three-dimensional structure has found that in more than 90% of cases the side chains have the same conformations in the two proteins. Therefore, in the simplest case where the aligned residues between the template and target structure are identical, all the atomic coordinates of the template structure can be transferred to the target structure directly, as shown in Figure 13.20A.

To predict the side-chain conformation when the aligned residues are not identical, **rotamer** libraries are used. Exhaustive analysis of conformations of side chains within proteins of known structure, and calculation of the energies of side-chain packing, have shown that most side chains are limited to a relatively small number of the many possible **dihedral angle** ($\chi_1$,$\chi_2$) energy minima (see Figure 13.21). For example, although leucine has nine allowed $\chi_1$,$\chi_2$ **conformers** (side chain conformations), just two account for 88% of leucine side chains found in proteins. Some side chains exhibit rotamer preference (one of a set of conformers arising from restricted rotation about one single bond) that depends on the main-chain secondary structure. Tryptophan, for example, has 75% of its $\chi_1$ values near 180° in α-helices, while 62% of the $\chi_1$ values are near –60° in β-sheets. Martin Karplus and colleagues have shown that the side-chain conformation also depends on the main-chain conformation; in other words, whether the main chain is in a β-strand, α-helix, turn, or coil conformation. Subsequently, backbone-dependent rotamer libraries have been compiled that, in conjunction with energy calculations, enable modeling packages to model nonidentical side chains

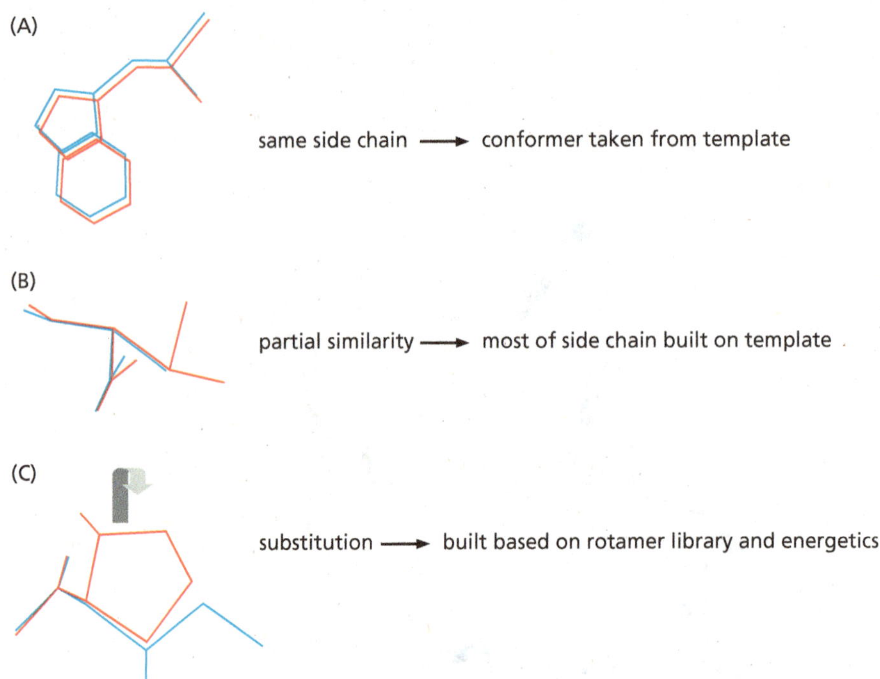


**Figure 13.20**
**Various methods employed by most programs to build side chains.** (A) When the side chains at a given position are identical, the conformation of the target side chain can be taken directly from the template. (B) When the side chains are similar but not identical, most of the target side chain can be built from the template. (C) When the side chains are quite different, the conformation of the target side chain has to be deduced from a library of rotamer structures and an assessment of the energetics.

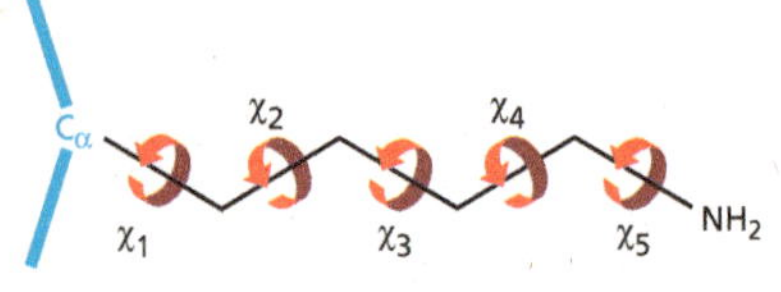


**Figure 13.21**
**The side-chain torsion angles are named chi1 ($\chi_1$), chi2 ($\chi_2$), etc.** The above shows part of the backbone, in blue, and from the $C_\alpha$ the side chain in black lines, with all the angles labeled. This is the side chain of lysine.

(see Figure 13.20C). Various side-chain modeling algorithms have been developed, which use different versions of rotamer libraries and different energy functions. Which method is used in practice will often depend on what is available in the program used for the modeling.

## Energy minimization is used to relieve structural errors

Once a model consisting of the core and loop region has been constructed, energy minimization is applied to the modeled loops. These areas often have bad geometries as a result of splicing together structures from two different proteins. Local energy minimization of these regions, allowing only the atoms involved in the loops, the joins, and short extensions either side of the joins to move, will improve the conformation of the loop–core hinge regions. Some programs also allow for the movement of atoms in close proximity to the selected loop. Fifty to 100 steps of the steepest descent method described earlier are usually sufficient to produce reasonable geometries at this stage.

Global energy minimization is performed to alleviate bad bond angles and to improve local geometries and atomic interactions. Usually it is good practice to initially start with 50 to 100 steps of steepest descent minimization, after which one can switch to the conjugate gradient (or similar) method until convergence is achieved.

## Molecular dynamics can be used to explore possible conformations for mobile loops

Simple molecular dynamics can be performed to investigate possible conformations of large insertions or mobile loops. Molecular dynamics calculation cannot be done with the widely used free graphical package Swiss-Pdb Viewer, which is described later, but commercial packages such as Quanta and Insight II do provide this option. Many different calculations can be performed, and there are a variety of physical models to choose from, but details of these procedures are beyond the scope of this book.

## Models need to be checked for accuracy

It is important to verify the accuracy of the model, and to estimate the likelihood and magnitude of potential errors. To estimate errors in a model, the structural parameters of the model are compared with those derived from crystallographically solved structures.

Many researchers have looked for energetic criteria that could distinguish incorrectly folded structures from correct ones. The researchers deliberately misfolded a protein sequence into a totally different conformation from its native one. For example, the sequence of a protein that has an all-α-helical structure could be made to form an all-β structure, and vice versa. Then the energy of the incorrect hypothetical fold was calculated to see if the energy terms could distinguish between the correctly and incorrectly folded proteins. Unfortunately, these energy criteria were not sensitive enough, and other ways of distinguishing correct from incorrect folds have had to be found. In addition a segment of a protein could be locally stressed even though the global energy of the whole native protein is at its minimum. This often occurs in the loops. The correct conformation of a loop may have high local interaction energy.

A variety of statistical criteria derived from empirical observations of known structures provide a standard against which a model can be measured. Parameters investigated and used as standards include torsion angles, bond angles and bond lengths, the distributions of polar and hydrophobic residues, and inter-residue contacts. There are several programs that can calculate such structural factors, which then can be compared against the norm; that is, the statistical average.

The program PROCHECK assesses the stereochemical quality of a given structure. It looks at how normal the geometry of the residues in a given protein structure is, as compared with stereochemical parameters derived from well-refined, high-resolution structures. Unusual regions highlighted by PROCHECK are not necessarily errors, but may be unusual features for which there is a reasonable explanation, such as distortions due to ligand binding in the protein's active site. Nevertheless, they indicate areas that should be checked carefully. It is good practice to run programs like PROCHECK on both the model and the template(s) upon which the model is based. Errors highlighted by PROCHECK may have been transferred from the template and therefore do not necessarily indicate a bad model. Programs like PROCHECK require a coordinate file and often the estimated resolution of the structure. When the coordinate file comes from a theoretical model, the resolution used should be that of the template structure plus a value that allows for the fact that it is a model (0.5–1.0 Å depending on the homology between the template and target). The output of PROCHECK is a set of PostScript files with graphs that describe the $\phi$, $\psi$ torsion angles in a Ramachandran plot (see Chapter 2), individual residue Ramachandran plots, chi-squared plots, main-chain parameters (see Figure 13.22), side-chain parameters, residue properties, the distribution of main-chain and side-chain bond length, root mean square distances from residue-type planarity, and plots illustrating distorted geometry. A similar Web-based program that checks the stereochemical validity of a structure is MolProbity. This program analyzes all atom contacts as well as steric variables.

The WHAT_CHECK program (part of the WHAT IF suite) compares the local contact patterns with the average contact patterns for similar residue–residue contacts found in the database. It provides an output that highlights amino acids that have unusual bond angles, bond lengths, hydrogen-bond donors that are not solvent accessible and do not form hydrogen bonds within the protein, residues that have nonbonded overlap of van der Waals radii with other residues in the structure (bumps), improper chirality, nonplanar groups in amino acids where planar groups are expected, proline puckering, residue-packing quality, and an analysis of torsion angles. Although the outputs from all these checks are long lists rather than pretty plots as obtained with PROCHECK, it is good practice to run the model coordinates through more than one program and then compare the outputs. Table 13.1 gives programs for assessing tertiary structure prediction.

ANOLEA uses a knowledge-based approach to verifying how lifelike models are. Basically it contains information about the preferred spatial relationships between residue types in real protein structures. These are used statistically to assess the similarity of the spatial distribution of residues in the model to that observed in real

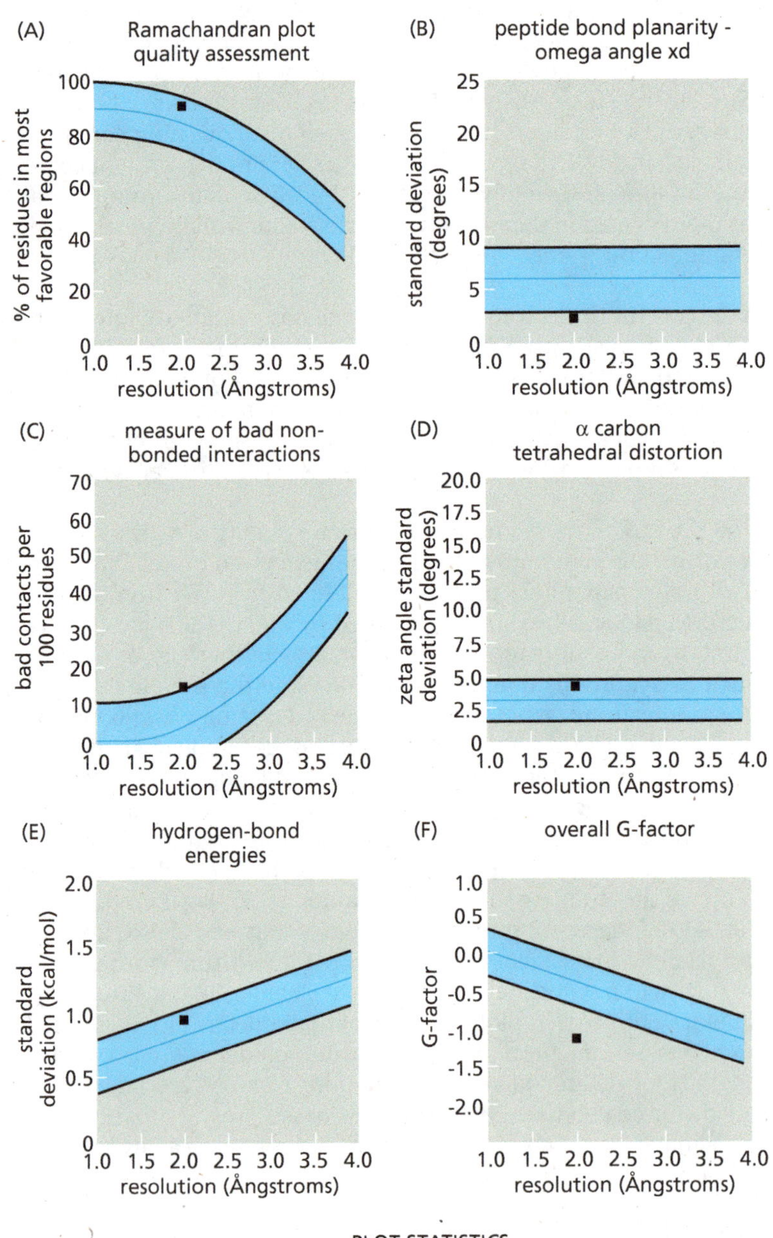


PLOT STATISTICS

| stereochemical parameter | no. of data pts | parameter value | comparison values typical value | band width | no. of band widths from mean | |
|---|---|---|---|---|---|---|
| (A) % residues in A, B, L | 256 | 71.9 | 83.8 | 10.0 | -1.2 | inside |
| (B) omega angle at dev | 281 | 0.6 | 6.0 | 3.0 | -1.8 | BETTER |
| (C) bad contacts/100 residues | 3 | 1.1 | 4.2 | 10.0 | -0.3 | inside |
| (D) zeta angle at dev | 262 | 0.9 | 3.1 | 1.6 | -1.4 | BETTER |
| (E) H-bond energy at dev | 173 | 0.7 | 0.8 | 0.2 | -0.5 | inside |
| (F) overall G-factor | 282 | 0.2 | -0.4 | 0.3 | 2.0 | BETTER |

**Figure 13.22**
**An example output of the main-chain parameters for an SH2 domain model from PROCHECK.** Parameter values that fall within the blue band are within expected measures for a structure at that particular resolution; values below or above the blue band signify better or worse than expected, depending on the parameter value. The black square indicates where the predicted structure's parameters fall. The nearer to the blue central line the better.

**Table 13.1**
**A list of programs for checking the correctness of modeled structures.**

| Program | Description |
|---|---|
| ANOLEA | Atomic Non-Local Environment Assessment |
| BVSPS | Biotech Validation Suite for Protein Structures |
| EVA | EValuation of Automatic protein structure prediction |
| PROCHECK | Verification of the stereochemical quality of a predicted structure |
| MolProbity | Verification of the stereochemical quality |
| WHAT IF | Protein structure analysis program |

proteins. An atomic mean force potential (AMFP) score using nonlocal interactions (see Figure 13.23) is calculated. Sequence separation as well as distance is used to differentiate between local and nonlocal interactions. In general, all the short-range interaction energies between atoms are calculated and expressed in the form of total energies or energy scores. High energies denote incorrectly folded protein or protein segments. The output ANOLEA gives is a list of residues with high energies and a color-coded sequence.

There is one more check to be performed: seeing is believing. Look at the structure. If it does not look right then there may, indeed, be something wrong.

## How far can homology models be trusted?

Incorrect sequence alignment is the main source of serious errors in homology modeling. Therefore if you trust your alignment, your model will be quite accurate, at least within the conserved regions. Even partial models are valuable for suggesting functional mechanisms that can be tested experimentally. For ligand or drug design, more accurate models are often necessary, and such models have already been used successfully in ligand design experiments. For example, analyses of the sequence of the HIV virus led to the discovery that the virus contains an aspartic protease (HIV-PR). Inactivation of HIV-PR produced immature, noninfectious viral particles.

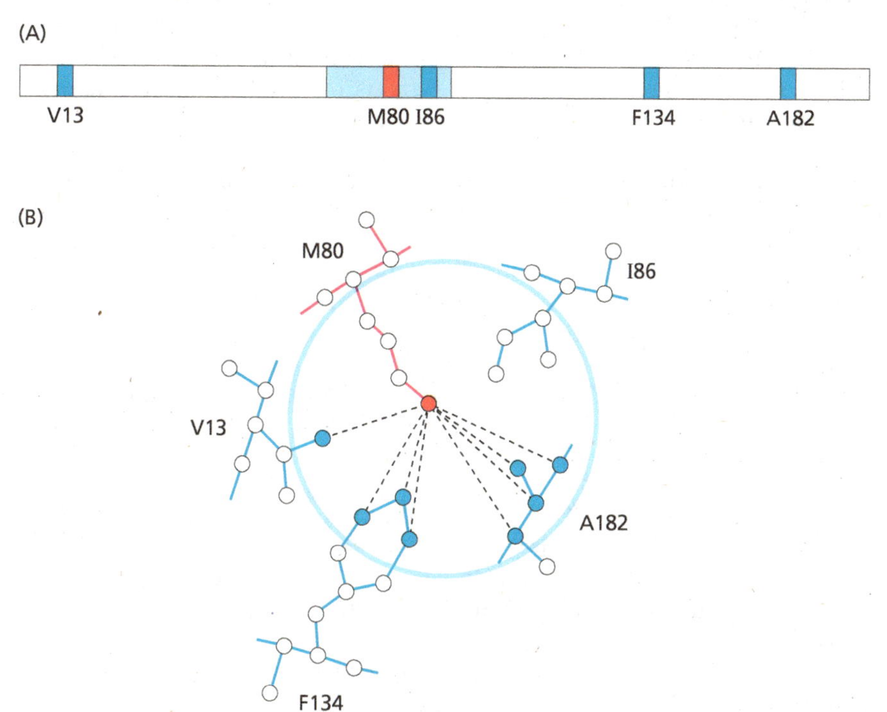


**Figure 13.23**
**Nonlocal environment of an atom.** (A) Representation of a protein sequence. Met80 $C_\varepsilon$ (red ball in B) is the atom for which the nonlocal interaction energies are evaluated. The light blue area in the sequence represents the amino acid residues that are considered as local for Met80 (11 residues around M80 in the sequence) and therefore are not considered in the calculation. V13, F134, and A182 are examples of nonlocal interacting amino acid residues for Met80. (B) A schematic representation of part of the protein chain illustrated in (A). The circle represents a sphere of 7 Å radius centered on $C_\varepsilon$ of Met80. All the atoms inside this sphere that belong to Met80 and the atoms of the local residue Ile86 are shown as white circles and are not considered in the calculation. All atoms of nonlocal amino acids that are located within the sphere—Val13, Phe134, and Ala182—are shown as blue circles. These atoms are included in the nonlocal interaction energy calculation. (Adapted from F. Melo and E. Feytmans, Assessing protein structures with a non-local atomic interaction energy, *J. Mol. Biol.* 277:1141–1152, 1998.)

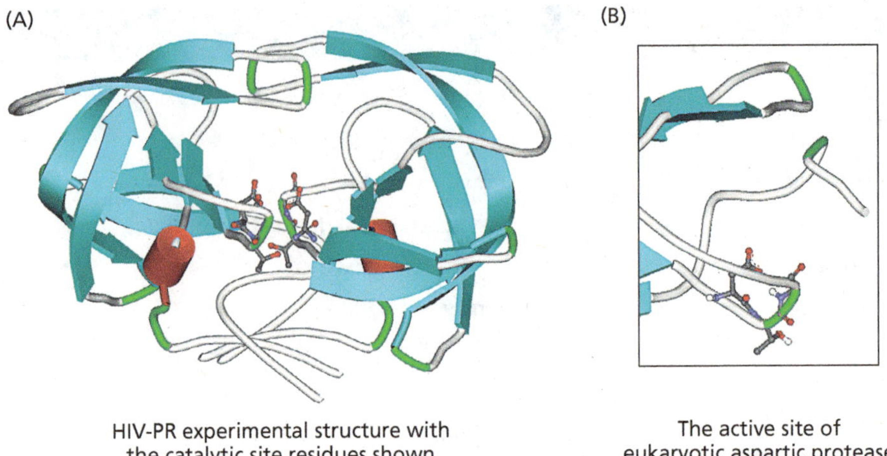


**Figure 13.24**
**(A) The experimental structure of the protease from HIV with the catalytic site residues shown as ball-and-stick, and (B) the catalytic residues of a eukaryotic aspartic protease.** Both structures have the same loop and residue arrangement within the catalytic region.

Therefore HIV-PR has been identified as one of the prime targets for structure-assisted drug design (see Chapter 14). But initially there was no experimental structure of the HIV-PR. The structure of HIV-1 PR was modeled using the known structures of eukaryotic aspartic proteases and later Rous sarcoma virus protease as templates. These models could be used as starting points for drug discovery. When the experimental crystal structures of HIV-PR became available they confirmed that the models were essentially correct and that the active site of HIV-PR closely resembles the active site of the aspartic proteases (see Figure 13.24). The correctness of a model can only really be judged when compared with an experimental structure.

## 13.5 Automated Homology Modeling

All the main steps outlined in Section 13.4 hold true to some extent for both manual and automatic modeling. Automated model building by homology has revolutionized modeling, turning it into an almost routine technique for obtaining at least some insight into the three-dimensional structure of a new protein. Even manual modeling is now done within a graphics software package where most of the time-consuming and difficult parts of model building, such as the transfer of coordinates from homologous structures, building side chains, or searching for the fragments to build loops, are performed by subroutines with little manual input.

Fully automatic model-building programs, such as Swiss-Model and MODELLER, have the advantages of objectivity, rapidity, and, in the case of Swiss-Model, simplicity. However, user intervention is often required and although most modeling packages allow changes in the default alignment and cut-off values, the process becomes immediately more complicated for the user than the totally automatic process. In addition, modeling with a black-box technique does not allow the user to gain insight and understanding of either the target structure or the modeling steps involved. When problems do arise one is often unable to identify the region or step where modeling has gone wrong and it is therefore more difficult to rectify the situation than when modeling manually.

Some commonly used modeling packages and programs will be briefly described below. Some of these programs are free but have to be installed on a local machine, for example Swiss-Pdb Viewer and MolIDE. Some can be run remotely, such as Swiss-Model, but do not allow much user interaction. Others cost a small amount, such as the complete WHAT IF modeling package, and others are expensive and versatile commercial packages, such as Quanta or Insight II from Accelrys, MOE from the Chemical Design Group, or SYBYL from Tripos.

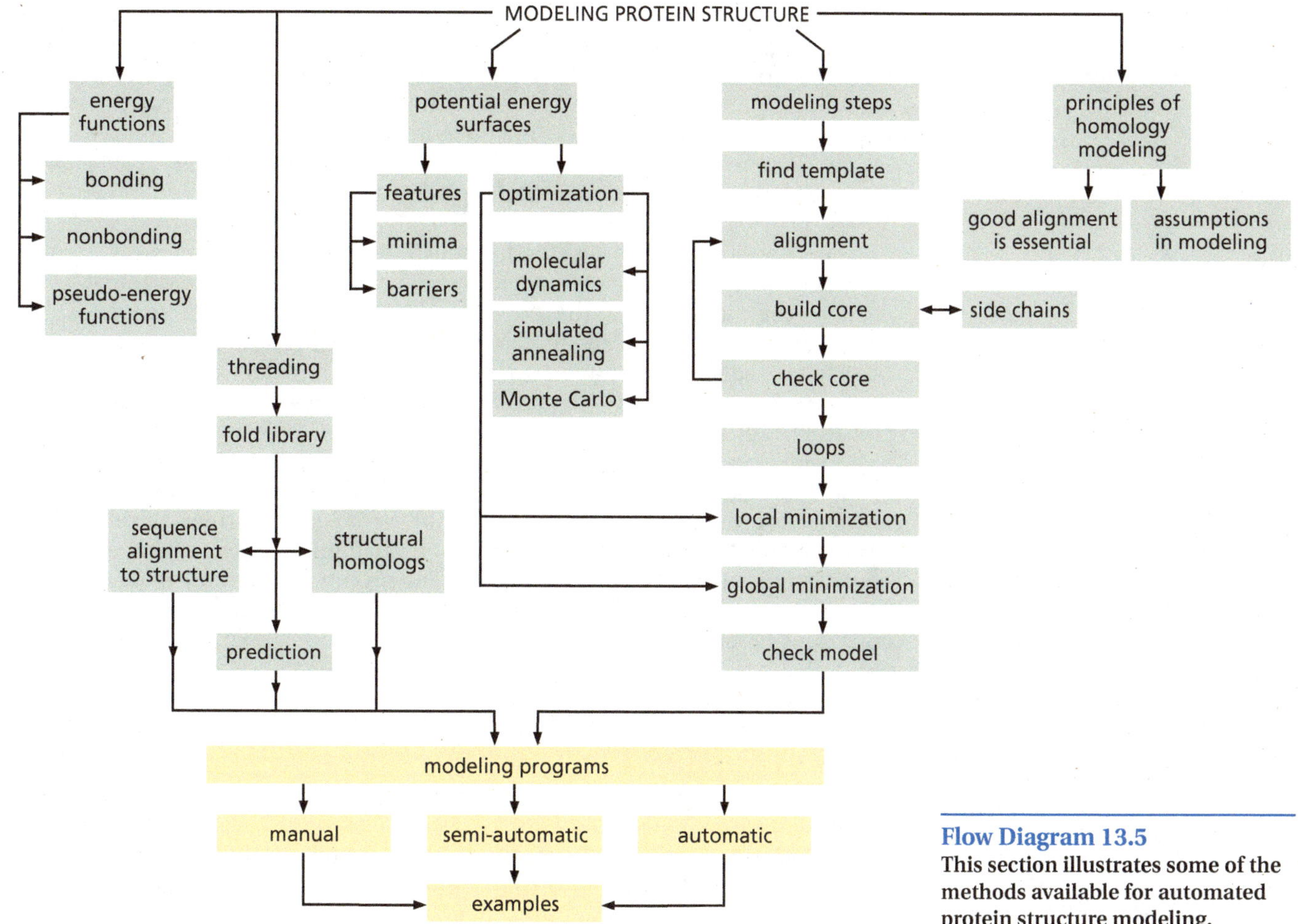


**Flow Diagram 13.5**
**This section illustrates some of the methods available for automated protein structure modeling.**

## The program MODELLER models by satisfying protein structure constraints

MODELLER creates a theoretical target structure on the basis of a known template structure using the idea of satisfying constraints. First, these spatial constraints in the form of atom–atom distances and dihedral angles are extracted from the template structure(s) and used in the target. These are combined with general rules of protein structure such as bond length and angle preferences. The alignment is used to determine equivalent residues between the target and the template. Finally, the model for the target is optimized until a model that best satisfies the spatial constraints is obtained (see Figure 13.25). Although MODELLER is usually used for homology modeling of protein three-dimensional structure, because the program deduces structure by satisfaction of spatial constraints the program can also be used in experimental structure determination by NMR. Generally, the output of MODELLER is a tertiary structure of a protein that satisfies a set of constraints as well as possible.

## COMPOSER uses fragment-based modeling to automatically generate a model

Another way of obtaining a model structure is by the assembly of rigid fragments of structure from a set of proteins with similar sequences to that of the target protein. This is the method used by the program COMPOSER. As an initial step COMPOSER searches a database of tertiary structure to find sequences homologous to the target

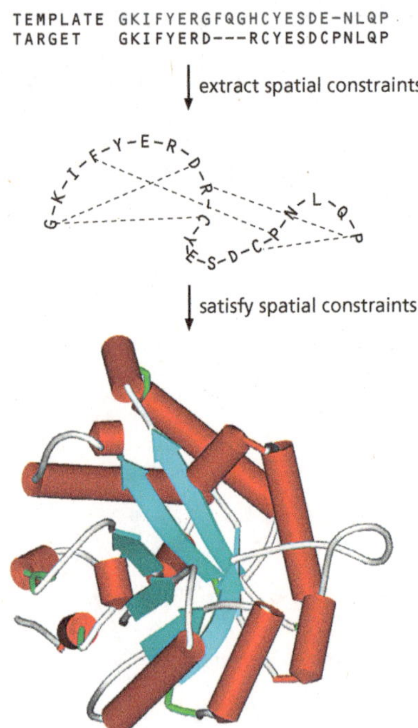


**Figure 13.25**
**A simplified schematic figure of the steps involved in the automated modeling program, MODELLER.** First the target is aligned to template(s), then constraints are extracted from the template(s) and used in the target. After that equivalent atoms between target and template are identified using the alignment, and the target is optimized until a model that best satisfies all the constraints is found.

protein and performs an alignment. A structural alignment is performed by initially specifying at least three topologically equivalent residues found in all the homologs. The structures are aligned or superposed using these equivalent residues and used to find other topologically equivalent residues to realign the structure and improve the fit. This is repeated until no more topologically equivalent residues are found. These residues then define the structurally conserved regions (SCRs) and an average framework is defined from these SCRs.

The contribution of each structural template to this average framework is calculated and a weighted value assigned to each. The highest-weighted fragment is then used as the structure for the corresponding conserved target sequence. The weighting of each template fragment will usually also take into account the percentage sequence identity between target and template, with the most similar sequence being given the highest weighting. The SCRs of the model are then constructed by fitting fragments of homologs to the templates and mutating them to the sequence of the target protein. Finally, the loops are modeled as described in Section 13.4. Side chains in COMPOSER are built in a similar manner as described above using energy calculation in addition to specific rules developed for COMPOSER (1200 rules), which define the probabilities of their orientation in the equivalent position in a homolog and depend on whether they are in an α-helix or a β-sheet. The final model can be improved by energy minimization, dynamics, or constraint-based modeling.

## Automated methods available on the Web for comparative modeling

A number of homology modeling servers are now available via the Web: Swiss-Model, 3Djigsaw, CPHmodels, ESyPred3D, Geno3D, and SDSC1. Swiss-Model, one of the first, is an automated knowledge-based protein modeling package. Swiss-Model first uses BLASTP2 to search the ExNRL-3D sequence database (derived from the PDB) to identify similar sequences to the target that also have structural coordinates. The search is refined by the program SIM, which selects templates with sequence identities greater than 25% and that are longer than 20 residues. Swiss-Model then generates models using the ProModII package. This includes procedures that superpose the related three-dimensional structures. It also generates a multiple alignment with the target sequence, creates a framework for the new sequence, builds loops, completes and corrects the main chain, corrects and builds side chains, verifies the modeled structure, checks the packing, and refines the structure by energy minimization and molecular dynamics.

The program Geno3D uses spatial constraints, such as distances and dihedral angles, to build the model once a structure with similar sequence has been found. When structural homologs have been identified the user chooses which ones are to be used as templates or template. Then for each template, Geno3D first performs a secondary structure prediction, and calculates the percentage agreements in secondary structure between the templates and the target sequence. Thus even if the sequence homology is low, but the secondary structure agreement is good, a model will be derived.

CPHmodels and ESyPred3D are neural network-based modeling servers. ESyPred3D uses neural networks to obtain an alignment and then uses MODELLER to actually build the three-dimensional structure.

## Assessment of structure prediction

The Critical Assessment of Structure Prediction (CASP) and the Critical Assessment of Fully Automated Structure Prediction (CAFASP) (see also Section 11.3) are two

## Box 13.2 Antibody modeling

Antibodies are the soluble proteins of the immune system that bind to pathogens and their toxic products, preventing their harmful action and labeling them as material to be destroyed by the body. The molecules bound by antibodies are called **antigens**, and any given antibody molecule is specific for a unique molecular structure or **epitope**, which the antibody binds via its so-called **antigen-binding site**. Each antibody recognizes just one epitope, but collectively antibodies recognize a vast range of chemical structures, including protein surface structures, carbohydrates, lipids, and even small artificial molecules. The repertoire of possible antigen-binding sites that can be produced in the body runs into billions.

To understand how antibodies bind their antigens with such specificity, and how different amino acid sequences in the **variable region** of the antibody molecule (see Figure B13.2) produce binding sites of such a vast range of affinities and specificities, one would like to compare many different structures of antigen-binding sites and their epitopes. The structures of only a few hundred antibodies have been solved by X-ray crystallography and there are even fewer structures of antibody–antigen complexes. There are, however, many more antibody sequences in the databases, and homology modeling can be a useful tool in extending the number of structures of antigen-binding sites. Being able to model the structures of antigen-binding sites can help in guiding the synthesis of novel antibody variable regions for potential therapeutics and laboratory reagents.

Antibody, or immunoglobulin, molecules are roughly Y-shaped and consist of four protein chains linked by disulfide bonds (see Figure B13.3). Each antibody molecule has two identical antigen-binding sites, which are located at the ends of the arms of the Y and which are the parts that vary from antibody to antibody. The rest of the molecule is almost identical, in both sequence and structure, in all antibodies of a given class (for example, IgG).

An antibody molecule consists of four chains of two different types: two identical heavy chains, and two identical, smaller, light chains. Each chain is composed of a series of discrete domains of around 110 amino acids with a characteristic structure known as the immunoglobulin fold. The amino-terminal domains of both heavy and light chains vary in length and in amino acid sequence between antibodies and are therefore referred to as the variable domains ($V_H$ and $V_L$, respectively). They form the ends of the Y and contain the antigen-binding sites. The remaining domains are the constant (C) domains.

The antigen-recognition site of an antibody consists of loops, also referred to as CDRs (complementarity-determining regions). There are six loops in the $V_L$ and $V_H$ regions that make up the CDRs: three are from the heavy chain (H1 to H3) and three from the light chain (L1 to L3). Although these loop sequences vary considerably between antibodies there is some structural similarity, which has been exploited to classify the CDRs into families. The members of a family have similar

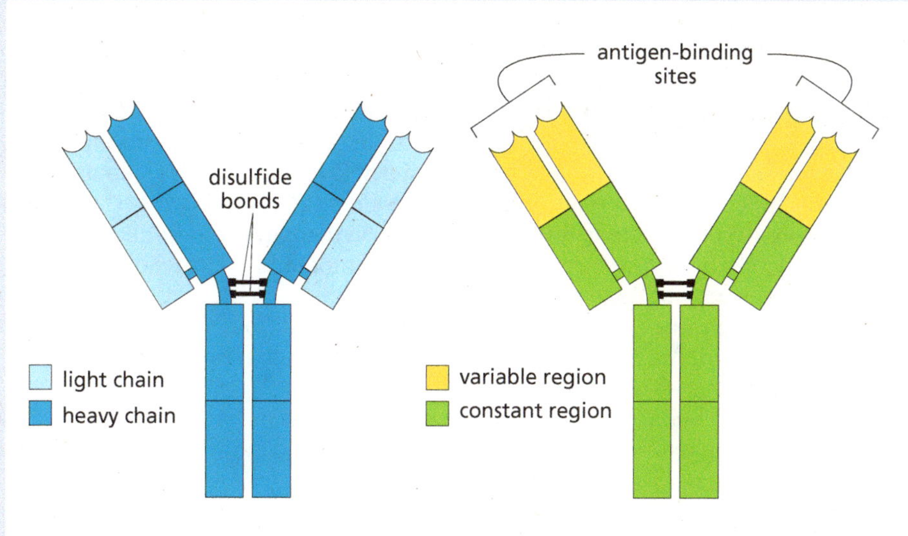


**Figure B13.2**
**The structure of an immunoglobulin, illustrating the roughly Y-shaped conformation of the molecules.**

## Box 13.2 Antibody modeling (continued)

packing constraints, and rules based on these constraints have been formulated for model-building packages that are specific for antibodies. One CDR (H3) does not conform to these rules, however, and yet is critical for binding antigen. Seven structural classes of the CDR-H3 loop have been defined, but still some H3 CDR are 13 amino acids or more in length and do not fit into this classification. The CDR classifications are part of the commercial antibody-modeling program AbM, but a Web-based antibody modeling server, called WAM, based on the AbM package is available.

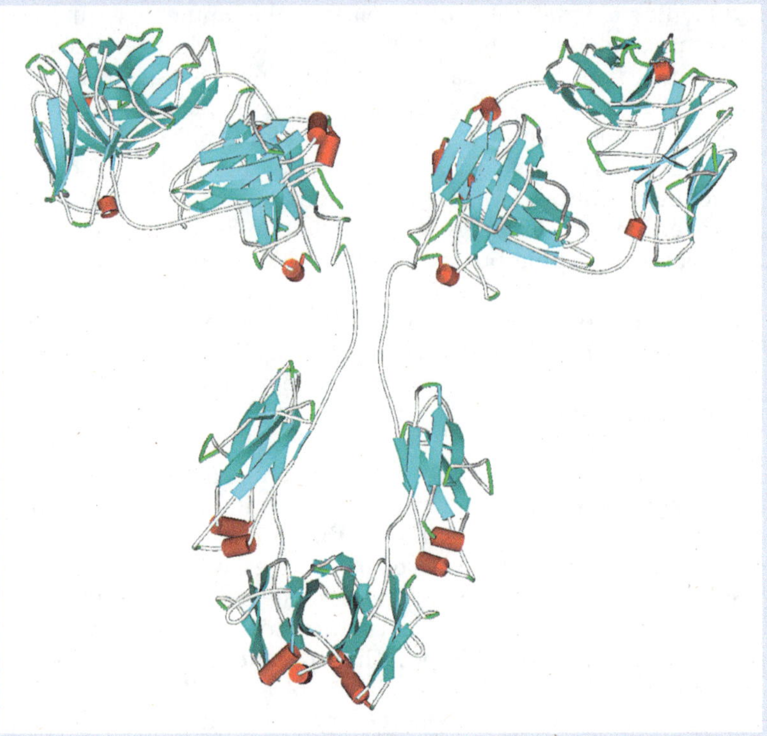

**Figure B13.3**
**Representation of an immunoglobulin showing secondary structures.**

open competitions, which attempt to provide a measure of the progress of structure prediction. This is done by trying to measure, in a quantitative way, the prediction success of a set of predefined and solved but not yet released structures. An independent group or individual obtains protein targets for use in CASP and/or CAFASP some time in advance of their public availability. The targets are either NMR or X-ray protein structures comprising one or more domains either determined and not published or expected to be determined in time for review. The sequence of these targets is given to the groups competing in CASP or CAFASP. The individual groups then make a set of predictions of the three-dimensional structure based on the protein sequence and submit the results for independent and comparative review by human assessors for CASP results, or to automated evaluation programs for the CAFASP results. Once a year there is a meeting held in Asilomar, USA, to discuss and compare the results. This type of annual assessment helps to identify the areas that need more development as well as the progress made in three-dimensional structure prediction by *ab initio* methods, homology modeling, or threading. Such periodic assessments not only identify the better prediction and modeling techniques but also highlight the problems that still need to be addressed. In addition it highlights the best criteria to use for assessing structure predictions generally.

In general it has been shown that, when all types of targets are taken into account, currently no single prediction approach is absolutely accurate. However, homology modeling with a high-resolution structural template gives by far the best results, of sufficient quality that they can be used in further theoretical and experimental analysis. When no structural homolog is available, the *ab initio* and to a lesser extent the threading method are the only options.

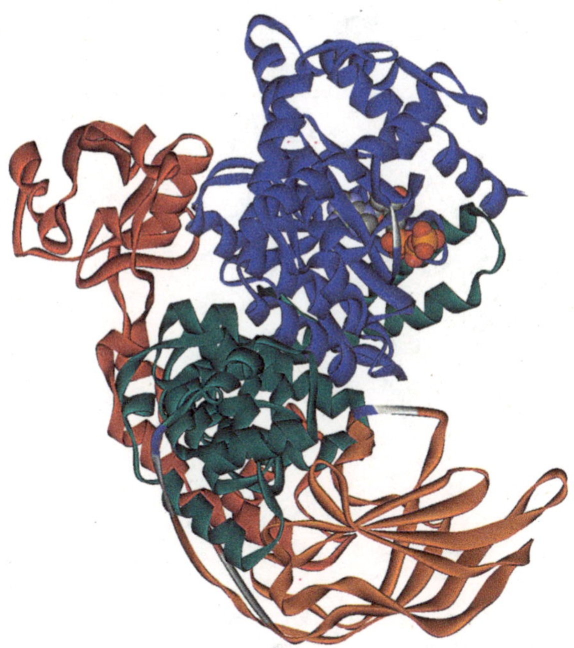

**Figure 13.26**
**A ribbon representation of the p110γ structure.** The structure is color-coded according to the domains of the protein. In blue is the catalytic domain with a bound molecule of ATP (represented by a space-filling model). The catalytic domain is the one used in the modeling examples.

## 13.6 Homology Modeling of PI3 Kinase p110α

Phosphoinositide 3-kinases (PI3Ks) are a group of signal transduction enzymes that add phosphate groups to lipids in cell membranes. These lipids then initiate further protein-mediated signal transduction cascades that lead to alterations in cell behavior such as the initiation or inhibition of cell division. The misregulation of such pathways is one of the causes of some types of cancer. PI3 kinases share short conserved amino acid motifs in the catalytic domain with other kinases, such as protein kinases, but in other regions the PI3 kinases vary greatly in sequence from the protein kinases. The human PI3 kinase p110α subunit is of particular interest as it is involved in signaling pathways that are associated with many diseases, including cancer. Some early models of the human PI3 kinase p110α catalytic domain were published but the sequence identity with protein kinases of known structure was so low that the models could not be analyzed with confidence. The structure of the related pig PI3 kinase p110γ (PDB code 1E8X) has now been solved (see Figure 13.26). All PI3 kinases are multidomain proteins (see Figure 13.26), but here we will describe just the modeling of the catalytic domain of p110α using the p110γ catalytic domain as template.

### Swiss-Pdb Viewer can be used for manual or semi-manual modeling

First search the PDB sequence database to find out if the sequence of the catalytic domain of p110α has a structural homolog. This can be done from one of the database search Web pages, such as BLAST at the European Bioinformatics Institute (EBI) or NCBI. The search finds 10 structures but three are human p110γ homologs and seven are pig p110γ homologs. The sequences are saved and aligned with the target sequence (p110α) using an alignment program such as ClustalW. This alignment is then saved.

To model the structure we shall first use the package Swiss-Pdb Viewer, and subsequently MolIDE. Both packages can be obtained free of charge for most computer platforms. The Swiss-Pdb Viewer program is suitable for small-scale semi-manual

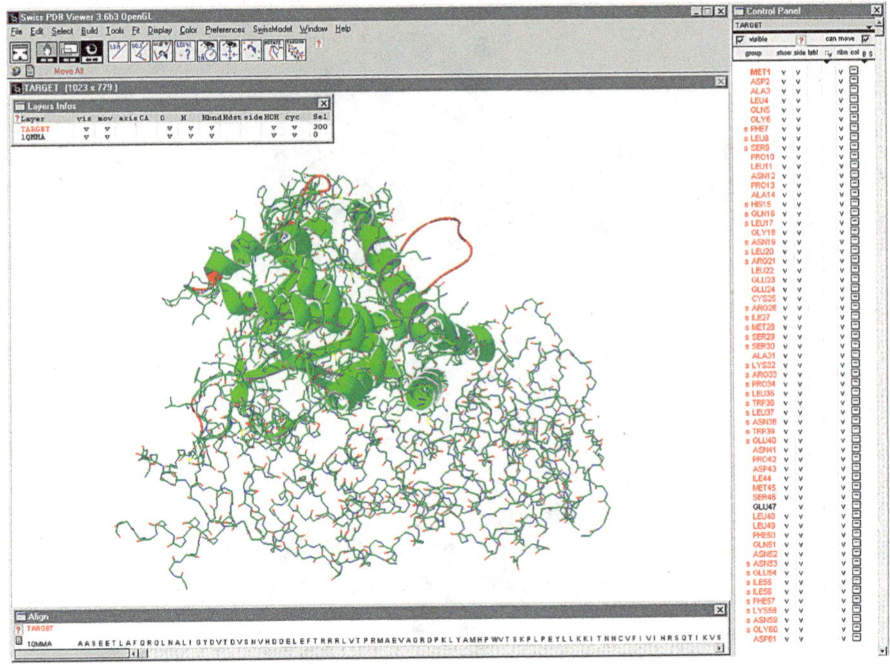


**Figure 13.27**
**A snapshot of the Swiss-Pdb Viewer modeling interface.** The main screen shows the modeled and template structure. To the right of the main screen is a control panel that allows easy selection of a number of viewing options.

modeling although it does not have the power and scope of commercial packages. Figure 13.27 shows the use of Swiss-Pdb Viewer in modeling p110α on the p110γ structure.

## Alignment, core modeling, and side-chain modeling are carried out all in one

The spatial coordinates file of the template structure (p110γ) is opened in Swiss-Pdb Viewer. As this is a multidomain protein, it is advisable to edit the coordinate file to contain only the coordinates for the catalytic domain before modeling. It is always easier to work with just the domain that one is interested in. Loading the p110α target sequence is done using the command "Load Raw Sequence to Model." The sequence alignment with p110γ is done using the commands "Fit Raw Sequences," and "Magic Fit." The alignment should be checked after the first "Fit

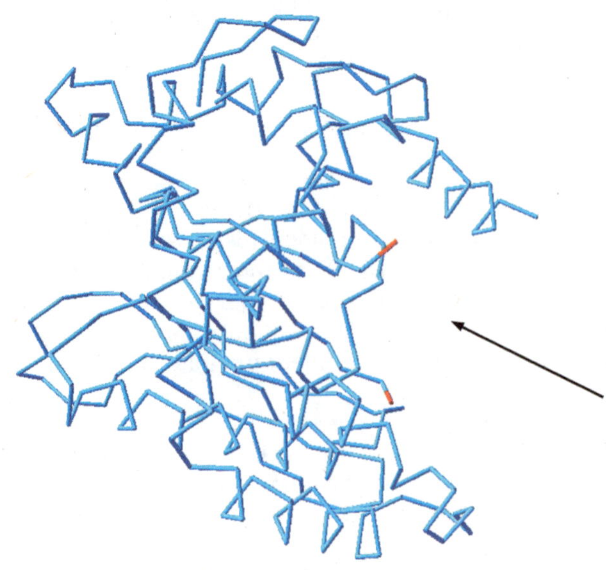

**Figure 13.28**
**The $C_\alpha$ representation of the core model of the p110α catalytic domain (blue).** The loops have not been modeled in yet. The ends of a large insertion in p110α with respect to p110γ are indicated by ends colored in red and also indicated by an arrow.

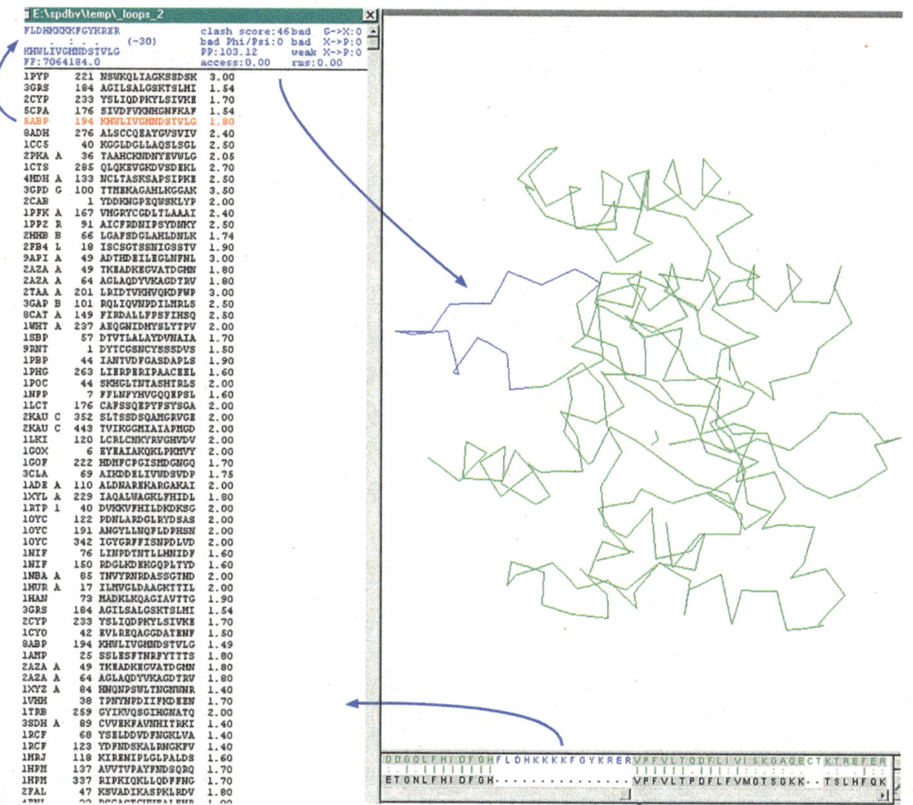

**Figure 13.29**
**Building a loop in Swiss-Pdb Viewer.** Once insertions are identified, the "Scan Loop Database" option provides a list of possible loops. The chosen loop for p110α is highlighted in red in the list. In the top window the loop fragment is aligned with the sequence of p110α and a number of parameters are also provided, such as energy, number of clashes, and number of bad dihedral angles. The chosen loop is fitted into the model automatically. One can scroll down all the loops to find the best one. Once chosen, another loop can be built. After all the loops have been built it is necessary to subject the model to energy minimization. The blue arrows indicate the loop that is being used as an example.

Raw Sequences" step. As you change the alignment you will see the structure change as well. This also generates coordinates for matched backbone and side-chain atoms. Save the modeled p110α coordinates. At this stage you have the core structure of the p110α with gaps where insertions are to be included, as illustrated in Figure 13.28 using the $C_\alpha$ backbone only.

## The loops are modeled from a database of possible structures

The loops have still to be built—in this case amounting to four stretches (three insertions and one deletion). At this point the model and alignment should be checked carefully to ensure that the insertions and deletions have been placed correctly. There is one large insertion (15 residues long). This insertion, which is in the target with respect to the template, is due to the fact that that particular loop in the p110γ structure is too disordered to be fitted into the crystallographic electron density. Therefore the loop does exist in p110γ but there are no structural data for it. Selecting "Build Loop" from the build menu, anchor residues can be designated at either end of the loop searches through a list of possible loop candidates. The file of candidates must be downloaded separately from the Swiss-Model Web page. A dialog box appears with possible loops. Choose the most suitable one based on the count of the clashes (displayed at the top of the window), energy information (shown after the "FF" text), and a mean force potential value (PP) computed from a mean-force potential (see Figure 13.29). A second way of modeling a loop is possible by choosing "Scan loop database"; a list of loops is provided with slightly different parameters. This procedure is repeated until all loops have been modeled.

## Energy minimization and quality inspection can be carried out within Swiss-Pdb Viewer

Once all the insertions have been modeled and the file saved, proceed to "Energy Minimize." In Swiss-Pdb Viewer a global energy minimization is performed once all

(A)

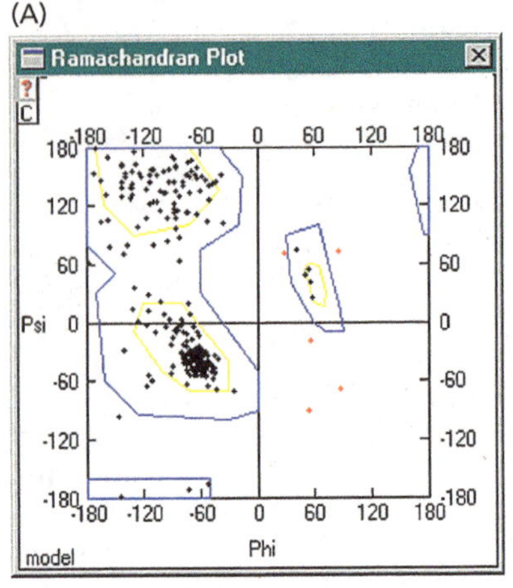


(B)

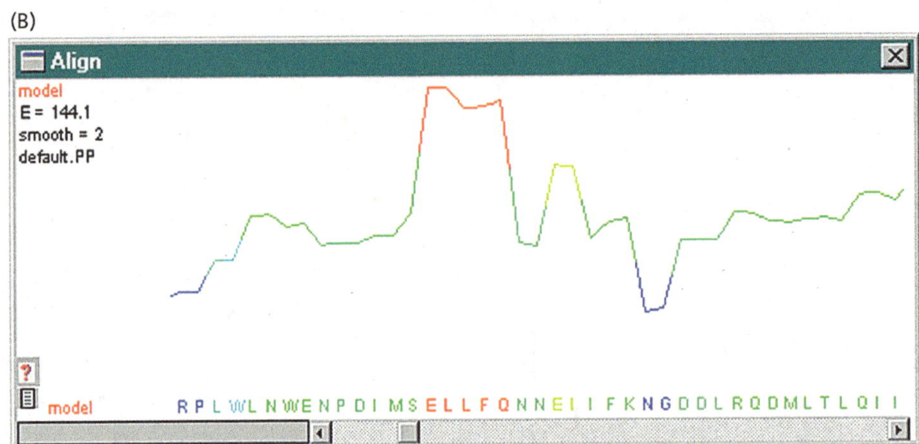


**Figure 13.30**
**(A) The Ramachandran plot of p110α.** Six residues are in the disallowed region of the plot; all of these are in loop segments. (B) Graph of structure potential energy against the sequence. One region well above 0 is colored red; this region is probably not correct. Again this set of residues are to be found within a loop.

the loops have been built. After minimization, some side chains may be listed as energetically unfavorable. Use the tool "Fix Selected Side Chains" to improve these automatically, and then minimize again.

Swiss-Pdb Viewer can be used to inspect the quality of the model. The first step is to view the Ramachandran plot (see Section 2.1) with all selected residues and see how many, and which, residues fall outside the allowed regions. In the p110α model, six amino acids fall in the disallowed areas (see Figure 13.30A). All of these are in loop regions, and while this situation is not ideal, there is no simple way of improving the loop models and therefore they are often ignored. Check for amino acids that cause clashes with other side-chain and backbone atoms. It is also possible to check the potential energy of each residue, although this tool should be used with caution. Expand the alignment window, and a curve depicting how each residue likes its surroundings will be displayed (see Figure 13.30B). If a residue is probably correct, its energy will be below zero, whereas an probably incorrect residue will have an energy above the zero axis. In any case, the modeled coordinates should be saved and sent to the various packages, like PROCHECK, to obtain as much information about its correctness as possible.

## MolIDE is a downloadable semi-automatic modeling package

MolIDE is a new graphical front-end that makes the generation of models using a degree of manual control much easier than with programs like Swiss-Model. The graphical front-end provides a reasonably user-friendly way of interacting with the various packages it uses to generate a model and integrates several useful bits of functionality. The process of modeling with MolIDE is based on the steps outlined above: find a template, align the sequence to the template, build an initial model, build the loops, then refine the model. MolIDE simply uses particular programs, such as PSI-BLAST, to achieve each step, or relies on user input.

In order to identify a suitable template, MolIDE uses PSI-BLAST in a two-stage process. First it iteratively scans a sequence database and generates a profile for the query sequence. Then it uses this profile to scan a sequence database derived from the PDB. This generates a list of structures that may be similar enough to the target to use for building a model. The PSIPRED secondary structure prediction method (see Chapters 11 and 12) is then used to predict the secondary structure of the target. This information is subsequently used in generating an alignment between the template and the target, along with the PSI-BLAST alignment.

All these steps are performed with the minimum input from the user. The starting point is to open the target sequence in the MolIDE interface, and initially run

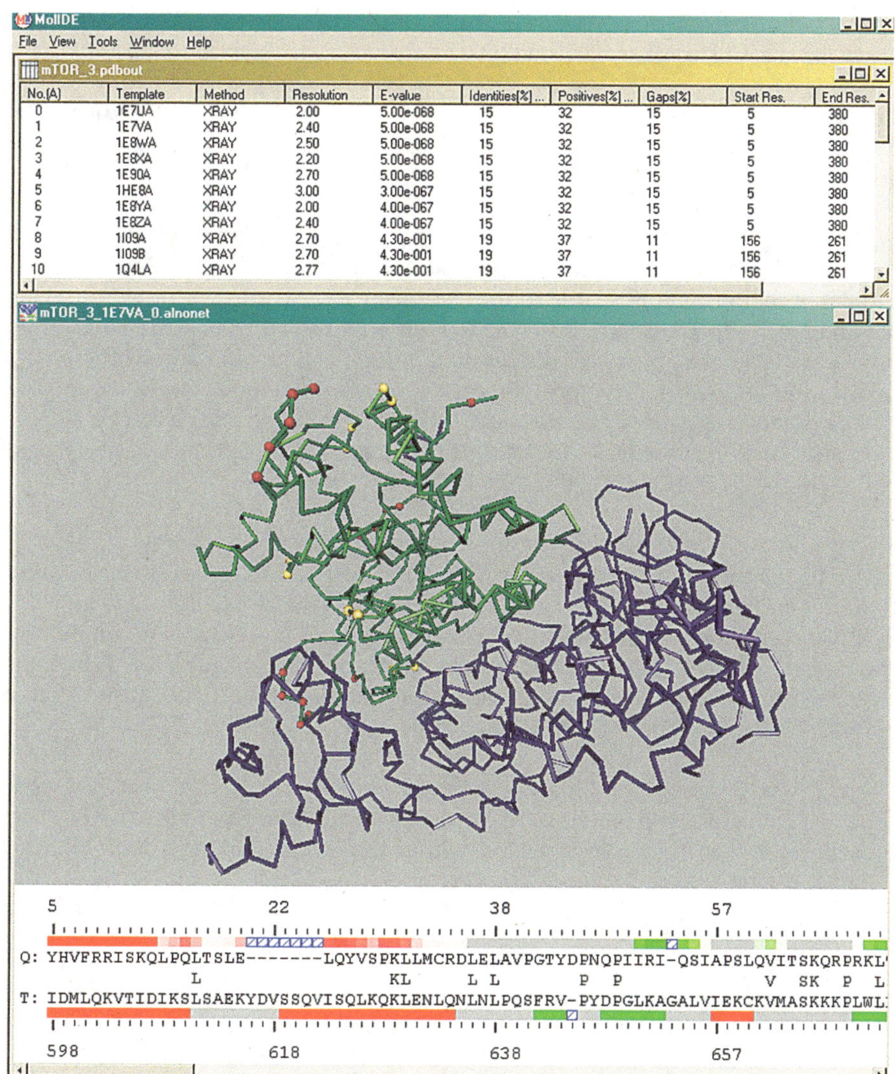


**Figure 13.31**
The MolIDE interface between the programs used in modeling and the user.

PSI-BLAST followed by PSIPRED. All output files can be read into MolIDE and investigated. Once these two programs have been run, the user opens the generated "PDB Hits Alignment File" and selects a target from the list of possible targets. The program then displays the model, along with the alignment between the template and the target (see Figure 13.31).

The side chains are then generated using the program SCWRL3—a program for predicting side-chain conformations given a backbone model and sequence. At this point the SCR has been modeled but the insertions and deletions still need to be dealt with. Via the MolIDE interface the user can edit the alignment and set positions for which loops are to be generated. Each loop is generated one at a time for the model using a program called Loopy. Modeled parts of the structure within the region thus selected are permitted to move, improving the realism of the loop. If too much of the structure is permitted to move, however, it will almost certainly generate errors.

## Automated modeling on the Web illustrated with p110α kinase

Although Swiss-Model was the first program that could be used as a fully automated method for generating a model via a Web page, MolIDE can generate a model with hardly any user input as well. Other automatic modeling approaches are 3Djigsaw, CPHmodels, ESyPred3D, and Geno3D.

The easiest and least interactive way of modeling p110α is to choose "First Approach Mode" in the Swiss-Model Web site. Paste in the target sequence, enter your e-mail address, and wait.

While we are waiting let us look at some of the options that can be defined. First, the BLAST limit, which restricts the likelihood of matching the p110α sequence to an unrelated sequence. It is related to the significance measures discussed in Sections 4.7 and 5.4. If the first search with the default limit setting of 0.00001 does not find anything, try increasing the BLAST limit, although you risk modeling on the wrong protein fold. However, if the protein structures related to your target sequence are already known, as is p110γ in this case, the code for this structure can be specified using the "Provide One Or More Of Your Own Templates" option and the program will build the structure using that particular template.

The results from Swiss-Model are returned in a number of separate e-mails. The first informs the modeler that a hit with the p110γ structure has been found and that the sequence identity is 47.8%. Another e-mail has the results file as an attachment, and contains the alignment and the coordinates for both the template(s) and the target. To view the results from Swiss-Model you need Swiss-Pdb Viewer installed on your own computer. Additional result files include the output from the structural analysis program WHAT IF.

Open Swiss-Pdb Viewer and load in the results file. Figure 13.32 shows the $C_\alpha$ coordinates of the model (in green) superposed on the template structure (in red). It can be seen that even the large insertion within the model has been built. However,

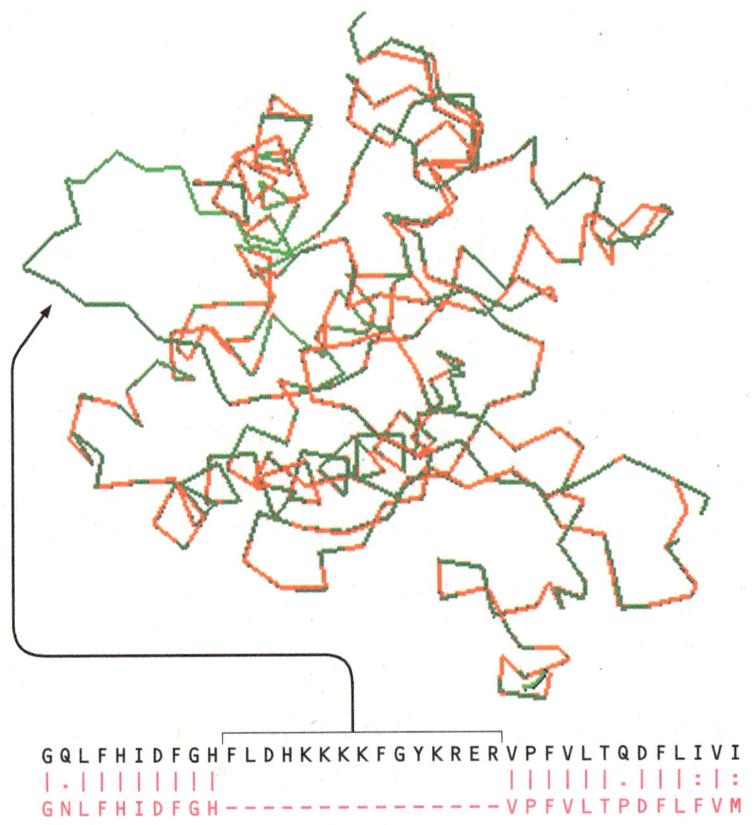


**Figure 13.32**
**The $C_\alpha$ coordinates of the model (in green) superposed on the template structure (in red).** The large loop, which is part of the substrate-binding loop, has been modeled automatically by Swiss-Model. However, even in the X-ray coordinates, this loop is not defined because of its mobile nature. Therefore, models of large loops should be considered with great caution.

such large insertions should be regarded with caution, as in this case the loop in p110γ is so flexible that it is not visible in the electron-density map. Within the Swiss-Pdb Viewer it is possible to realign the sequences and inspect the quality of the target. If changes are made, it is possible to resubmit the alignment for further modeling through the Web pages.

All the other prediction programs (3Djigsaw, CPHmodels, ESyPred3D, and Geno3D) generate a model for p110α. Some allow for some interactive setting of parameters, others even for changes in alignment. In the most automatic mode all the user needs to do is paste in the sequence and press the "Submit" button. Geno3D not only returns the coordinates for the model but also all the PROCHECK files, enabling the user to check the model.

## Modeling a functionally related but sequentially dissimilar protein: mTOR

All the automatic prediction programs modeled the structure of p110α based on the template(s) p110γ. However, when we try and model the low-level homologous protein mTOR (~15% identity) using all the above described automated programs, Geno3D and Swiss-Model do not find a satisfying template for the sequence using the default settings. Table 13.2 shows which programs returned a model for mTOR. The RMSD values indicate how similar the models are to p110γ and to each other. This shows that models generated by different methods can differ significantly and the models should be investigated further with the checking programs.

Using MolIDE it is possible to generate a model accurate enough that it is being used in drug-related ligand docking. Even though many of the steps in MolIDE are automatic, importantly it identifies p110γ proteins as homologous. It is then possible using this interface, and the capability to modify the alignment and choose where the insertions/deletions will be put, to create a reliable model. Therefore, unless one has a template that has high sequence identity with the target, the model must be built using at least partially manual steps.

(A)

| p110α | Swiss-Model | ESyPred3D | Geno3D | CPHmodels |
|---|---|---|---|---|
| **p110γ** | 2.96 | 2.94 | 3.69 | 1.64 |
| Swiss-Model | | 1.49 | 2.23 | 1.66 |
| ESyPred3D | | | 2.71 | 2.81 |
| Geno3D | | | | 3.18 |

(B)

| mTOR | Swiss-Model | ESyPred | Geno3D | CPHmodels |
|---|---|---|---|---|
| **p110γ** | X | 4.54 | X | 4.76 |
| Swiss-Model | | X | X | X |
| ESyPred3D | | | X | 7.33 |
| Geno3D | | | | X |

**Table 13.2**
**(A) The RMSD backbone values between the p110γ (template) structure and the various automatic models of p110α.** Also RMSD values are reported between all the models. There are differences between the structures (shown by the RMSD value; the higher the value the more difference between the structures). These differences occur mainly in the loop regions of the model. The non-loop region shows ~1.4 Å RMSD values. This means that there are slight differences in the models nonetheless. It is important to check the automatically generated models carefully before using them for further studies. (B) As above but using mTOR as the target. Because mTOR is a low-homology model, the differences between the template structure and the models are larger. X denotes that no model could be generated by this method using default settings.

### Generating a multidomain three-dimensional structure from sequence

Multidomain proteins can be predicted with some success by docking individual domains onto a known structure of the whole complex. The complete p110$\gamma$ protein has a modular organization, where a phospholipid-binding C2 domain, the kinase catalytic domain, and a binding domain for the intracellular signaling protein Ras are arranged round a helical spine. Earlier predictions on the p110$\alpha$ sequence predicted the catalytic domain and the C2 domain. With the availability of the p110$\gamma$ structure, all the domains can be modeled and assembled to form the complete multidomain complex.

## Summary

To properly understand the structure–function relationship of the increasing number of protein sequences in the databases requires determining the structures of the proteins they encode. Experimental structure determination by X-ray crystallography and NMR is still too slow and expensive to keep up with the rate of sequence generation, and various methods of predicting a protein's structure from its sequence alone are needed. *Ab initio* methods that aim to derive a protein's structure from its sequence by first principles are still, in most cases, both unreliable (as the principles underlying protein folding are not yet fully understood), and impractical (as they require vast computer resources). However, methods that predict the tertiary fold of an amino acid sequence on the basis of fold recognition—commonly known as threading methods—are available and relatively easy to use and have been described in this chapter in some detail.

Homology modeling avoids the problems of *ab initio* protein structure prediction and threading by using the structure of a known homologous protein as the template. It is at present the most reliable way of determining a protein's likely tertiary structure when only its amino acid sequence is known. The technique requires at least one other protein (the template) that is homologous with the query protein (the target) and which has an experimentally solved structure, either by X-ray crystallography or NMR. The first and most important step in modeling is to make an accurate alignment between the target and template sequences. Accurate alignment is aided if multiple homologous sequences are available, and the success of homology modeling increases the more closely related the target and template sequences are.

Conserved regions of the alignment are modeled first, by simply transferring the atomic coordinates of the template backbone. After the conserved regions of the backbone have been built, side chains in the core are modeled. After the conserved parts of the protein have been built, the variable regions are modeled, paying particular attention to the fact that any insertions or deletions in the target sequence relative to the template are most likely to fall in loop regions. They are often built by searching specialized databases of short structural fragments, and fitting the best fragments into the modeled structure.

Models can be improved by energy minimization techniques that can relieve unfavorable inter-atomic interactions. Energy minimization procedures use well-established thermodynamic and physicochemical principles to find the best conformation with the least energy.

Homology modeling can be carried out manually or semi-manually using interactive graphics programs such as Swiss-Pdb Viewer or MolIDE and proceeding step-by-step, or by using fully automated Web-based packages in which all that is required is to enter the target sequence. The use of various packages has been illustrated by modeling PI3 kinase p110$\alpha$ on the template structure of PI3 kinase p110$\gamma$.

Homology modeling is going to become more important as the numerous genome sequencing projects are completed and more and more representative structures are experimentally solved to use as templates. Large-scale automated or semi-automated modeling is becoming a routine way of obtaining initial information on the structures of most of the proteins now being discovered.

Once a good model has been obtained, it can be investigated just like a real structure; mutations can be incorporated to explore structure–function relationships and the structure can be used to suggest further experimental projects, which in turn can be analyzed using the modeled structure. If the model is really accurate, it can be used to investigate ligand docking, ligand–protein interactions, and the design of potential drugs.

## Further Reading


**Modeling: General**

Forster MJ (2002) Molecular modelling in structural biology. *Micron* 33, 365–384.

### 13.1 Potential Energy Functions and Force Fields

Bowie JU, Luthy R & Eisenberg D (1991) A method to identify protein sequences that fold into a known three-dimensional structure. *Science* 253, 164–170.

Dill KA & Chan HS (1997) From Levinthal to pathways to funnels. *Nat. Struct. Biol.* 4, 10–19.

***Ab initio* modeling**

Bradley P, Misura KMS & Baker D (2005) Toward high-resolution de novo structure prediction for small proteins. *Science* 309, 1868–1871.

Klepeis JL, Wei Y, Hecht MH & Floudas CA (2005) *Ab initio* prediction of the three-dimensional structure of a de novo designed protein: a double-blind case study. *Proteins* 58, 560–570.

### 13.2 Obtaining a Structure by Threading

Jones DT & Hadley C (2000) Threading methods for protein structure prediction. In Bioinformatics: Sequence, structure and databanks. (D Higgins, WR Taylor eds), pp 1–13. Heidelberg: Springer-Verlag.

**123D+**

Alexandrov NN, Nussinov R & Zimmer RM (1996) Fast protein fold recognition via sequence to structure alignment and contact capacity potentials. *Pac. Symp. Biocomput.*, 53–72.

Thiele R, Zimmer R & Lengauer T (1999) Protein threading by recursive dynamic programming. *J. Mol. Biol.* 290, 757–779.

**GenTHREADER**

Jones DT, Taylor WT & Thornton JM (1992) A new approach to protein fold recognition. *Nature* 358, 86–89.

Jones DT (1999) GenTHREADER: an efficient and reliable protein fold recognition method for genomic sequences. *J. Mol. Biol.* 287, 797–815.

**3D-PSSM**

Kelley LA, MacCallum RM & Sternberg MJE (2000) Enhanced genome annotation using structural profiles in the program 3D-PSSM. *J. Mol. Biol.* 299, 499–520.

**FUGUE**

Shi J, Blundell TL & Mizuguchi K (2001) FUGUE: Sequence-structure homology recognition using environment-specific substitution tables and structure-dependent gap penalties. *J. Mol. Biol.* 310, 243–257.

**LIBRA**

Ota M & Nishikawa K (1997) Assessment of pseudo-energy potentials by the best-five test: a new use of the three-dimensional profiles of proteins. *Protein Eng.* 10, 339–351.

Ota M & Nishikawa K (1999) Feasibility in the inverse protein folding protocol. *Protein Sci.* 8, 1001–1009.

**LOOPP**

Meller J & Elber R (2001) Linear programming optimization and a double statistical filter for protein threading potentials. *Proteins* 45, 241–261.

Teodorescu O, Galor T, Pillardy J & Elber R (2004) Enriching the sequence substitution matrix by structural information. *Proteins* 54, 41–48.

Tobi D & Elber R (2000) Distance-dependent, pair potential for protein folding: results from linear optimization. *Proteins* 41, 40–46.

**LIBELLULA**

Juan D, Graña O, Pazos F, Fariselli P, Casadio R & Valencia A (2003) A neural network approach to evaluate fold recognition results. *Proteins* 50, 600–608.

**SCOP**

Andreeva A, Howorth D, Brenner SE, Hubbard TJP, Chothia C & Murzin AG (2004) SCOP database in 2004: refinements integrate structure and sequence family data. *Nucleic Acids Res.* 32, D226–D229.

Brenner SE, Chothia C, Hubbard TJP & Murzin A (1996) Understanding protein structure: using SCOP for fold interpretation. *Methods Enzymol.* 266, 635–643.

Murzin A, Brenner SE, Hubbard TJP & Chothia C (1995) SCOP: Structural Classification of Proteins database for the investigation of sequences and structures. *J. Mol. Biol.* 247, 536–540.

## 13.5 Automated Homology Modeling

Blundell TL, Carney D, Gardner S et al. (1988) Knowledge-based protein modeling and design. *Eur. J. Biochem.* 172, 513–520.

Srinivasan N & Blundell TL (1993) An evaluation of the performance of an automated procedure for comparative modelling of protein tertiary structure. *Protein Eng.* 6, 501–512.

Sutcliffe MJ, Haneef I, Carney D & Blundell TL (1987) Knowledge-based modeling of homologous proteins, Part I: Three-dimensional frameworks derived from the simultaneous superposition of multiple structures. *Protein Eng.* 1, 377–384.

Sutcliffe MJ, Hayes FR & Blundell TL (1987) Knowledge-based modeling of homologous proteins, Part II: Rules for the conformations of substituted sidechains. *Protein Eng.* 1, 385–392.

**MolIDE**

Canutescu AA & Dunbrack RL Jr (2005) MolIDE (Molecular Integrated Development Environment): a homology modeling framework you can click with. *Bioinformatics* 21, 2914–2916.

**SCWRL**

Canutescu AA, Shelenkov AA & Dunbrack RL Jr (2003) A graph-theory algorithm for rapid protein side-chain prediction. *Protein Sci.* 12, 2001–2014.

**PSIPRED**

Jones DT (1999) Protein secondary structure prediction based on position-specific scoring matrices. *J. Mol. Biol.* 292, 195–202.

**LOOPY**

Xiang Z, Soto CS & Honig B (2002) Evaluating conformational free energies: the colony energy and its application to the problem of protein loop prediction. *Proc. Natl Acad. Sci. USA* 99, 7432–7437.

**MODELLER**

Fiser A & Sali A (2003) Comparative protein structure modeling with MODELLER: a practical approach. *Methods Enzymol.* 374, 463–493.

Sali A & Blundell TL (1993) Comparative protein modelling by satisfaction of spatial restraints. *J. Mol. Biol.* 234, 779–815.

Sanchez R & Sali A (2000) Comparative protein structure modeling. Introduction and practical examples with MODELLER. *Methods Mol. Biol.* 143, 97–129.

**Assessing models**

Lund O, Nielsen M, Lundegaard C & Worning P (2002) CPHmodels 2.0: X3M a computer program to extract 3D models. Abstract presented at the Second Biennial UCSC-QB3 Symposium on Bioinformatics: Predicting the structure and function of proteins (Informal post-CASP Workshop) Earth & Marine B206 University of California, Santa Cruz. http://www.cbs.dtu.dk/services/CPHmodels/

Melo F, Devos D, Depiereux E & Feytmans E (1997) ANOLEA: a www server to assess protein structures. *Proc. Int. Conf. Intell. Syst. Mol. Biol.* 5, 187–190.

Melo F & Feytmans E (1997) Novel knowledge-based mean force potential at atomic level. *J. Mol. Biol.* 267, 207–222.

Melo F & Feytmans E (1998) Assessing protein structures with a non-local atomic interaction energy. *J. Mol. Biol.* 277, 1141–1152.

Sander C & Schneider R (1991) Database of homology-derived protein structures and the structural meaning of sequence alignment. *Proteins* 9, 56–68.

**MolProbity**

Lovell SC, Davis IW, Arendall WB 3rd et al. (2003) Structure validation by C-alpha geometry: phi, psi, and C-beta deviation. *Proteins* 50, 437–450.

**PROCHECK**

Laskowski RA, MacArthur MW, Moss DS & Thornton JM (1993) PROCHECK: a program to check the stereochemical quality of protein structures. *J. Appl. Cryst.* 26, 283–291.

Morris AL, MacArthur MW, Hutchinson EG & Thornton JM (1992) Stereochemical quality of protein structure coordinates. *Proteins* 12, 345–364.

**CASP**

Fischer D, Barret C, Bryson K et al. (2003) CAFASP-1: critical assessment of fully automated structure prediction methods. *Protein Eng.* 16, 157–160.

Martin ACR, MacArthur MW & Thornton JM (1997) Assessment of comparative modelling in CASP2. *Proteins* (Suppl 1), 14–28.

**Antibody and HIV examples**

Blundell T & Pearl L (1989) Retroviral proteinases. A second front against AIDS. *Nature* 337, 596–597.

Lesk AM & Tramontano A (1993) Antibody structure. In Structure of Antigens (MHV Van Regenmortel ed.), Ann Arbor, MI: CRC Press.

Pearl LH & Taylor WR (1987) A structural model for the retroviral proteases. *Nature* 329, 351–354.

Whitelegg NRJ & Rees AR (2000) WAM: an improved algorithm for modelling antibodies on the WEB. *Protein Eng.* 13, 819–824.

# 14

# ANALYZING STRUCTURE–FUNCTION RELATIONSHIPS

APPLICATIONS CHAPTER

**When you have read Chapter 14, you should be able to:**

Compare the conservation of structure in proteins of divergent sequence.

Use fold libraries to classify structure.

Determine the domain structure of a protein.

Interpret functional information returned by structural comparisons.

Discuss how structural comparisons can give clues to function.

Describe the different types of structural comparison methods.

Search the databases for structural and functional information, using the protein Cbl as an example.

Locate and analyze binding sites.

In the previous chapters we have learned how to predict and model the structure of a protein. The structure itself can reveal a great amount of information about the function of the molecule. However, in many cases the understanding of the protein's function is significantly enhanced by a detailed comparison with other proteins whose structure and structure–function relationship are well characterized. In this chapter we will see how knowledge of protein structure and of evolutionary relationships can provide answers to biological questions concerning the structure–function relationship of a protein.

A protein's activity or function is determined by its three-dimensional fold in the sense that the residues needed for a specific activity are brought together in the right geometry. Once the three-dimensional structure is known, one can start to explore whether the sequence and structure together give any information about the protein's mechanism of action or its biochemical and biological functions. Knowing the fold structure can lead to a better understanding of the function, for example by highlighting which residues are actually involved in substrate binding or other interactions. Once a structure has been determined, structurally similar homologs can often be found that cannot be picked up by sequence comparison. Active sites can be defined, and in some cases the biochemical function of the protein, even clues about its function in the cell, can be deduced.

When a function is identified by sequence searches alone, for example if a cDNA search with BLAST picks up hits that are homologous to kinase-related sequences, it can be postulated that the target sequence will also have a kinase-like function

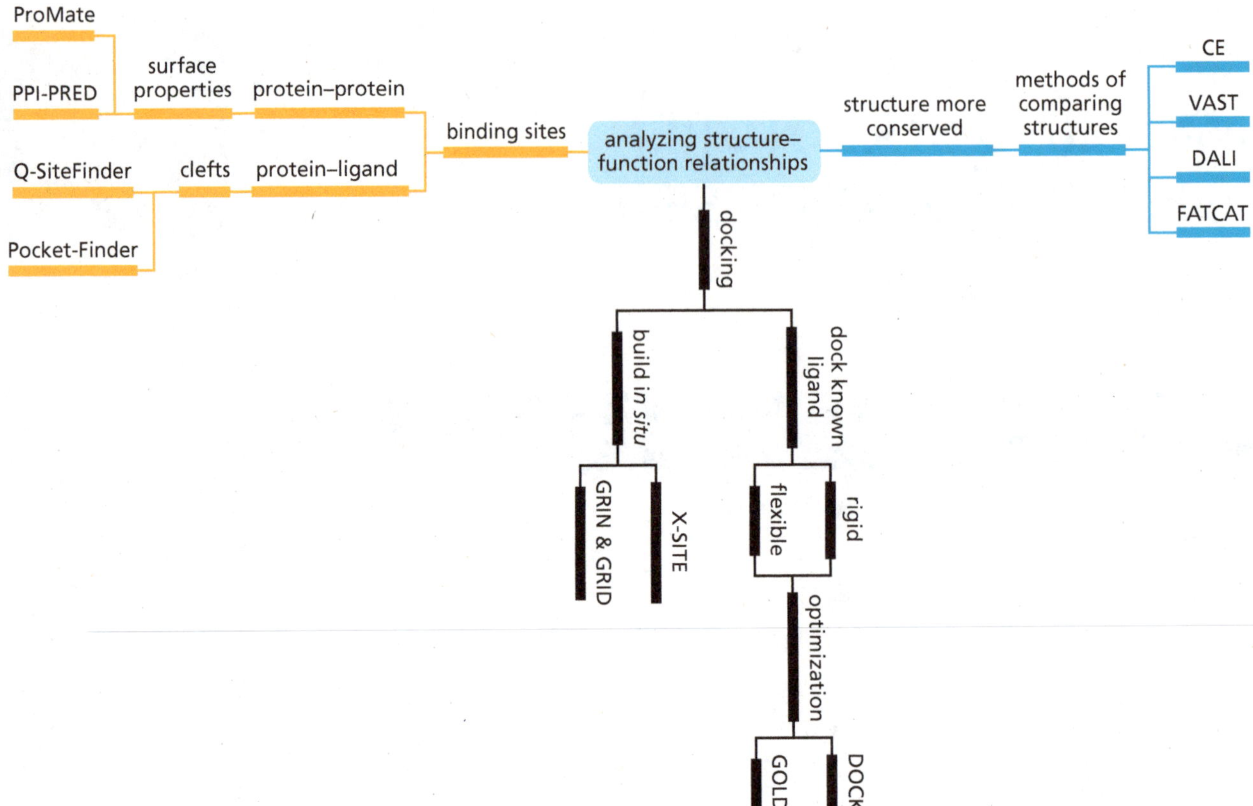


**Mind Map 14.1**
**Once we have a tertiary structure, various analyses can be carried out. This mind map shows the general analytical topics covered in this chapter.**

and this can be verified experimentally. However, structural information can provide more detailed information about the function of the protein and can aid in identification of binding motifs or catalytic centers. This is an important reason for modeling the structure using one of the methods described in Chapter 13 when no experimental structure is available. In this chapter we will look at how we can analyze the structure of a protein to obtain more information about its function and biological role.

## 14.1 Functional Conservation

The function of a protein depends primarily on its structure. As more structures become available due to the development of X-ray and NMR techniques, the relationship between function and structure can be better understood. In addition, there is an increase in sequence data due to the completion of large-scale genomic sequencing programs. Many of these proteins will not have their structure solved experimentally or undergo detailed experimental analysis in the near future. However, initiatives have been set up specifically to determine the protein structures

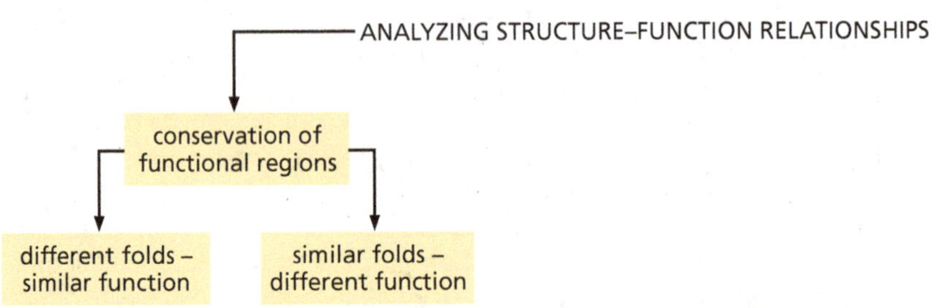


**Flow Diagram 14.1**
**The key concepts introduced in this section are that although similar protein structure is suggestive of similar function, it is not always the case, and vice versa.**

of particular genomes, often referred to as Structural Genomics. Despite this, there still will be many proteins for which structures are currently unavailable; therefore their functional elucidation will depend on the analysis of their sequence and structural information.

As will be described, in many cases proteins that share a structural fold have a similar function. Libraries have been constructed based on known protein structures and sequences that can greatly assist in predicting protein function in the absence of experimental data. However, a cautious approach is advisable, as there are numerous examples of proteins with different functions yet the same fold, and vice versa.

## Functional regions are usually structurally conserved

In the previous chapters we discussed sequence and structural homologs. However there is another class of homologs—the functional homolog. Functional homologs catalyze very similar reactions, have the same fold, the same catalytic mechanism, and have ligand-binding sites at the same locations. There are many cases where a protein shares no or little sequence homology and yet is a functional homolog. This means that comparisons of structures will often highlight functionally important regions and motifs that cannot be deduced from sequence alone.

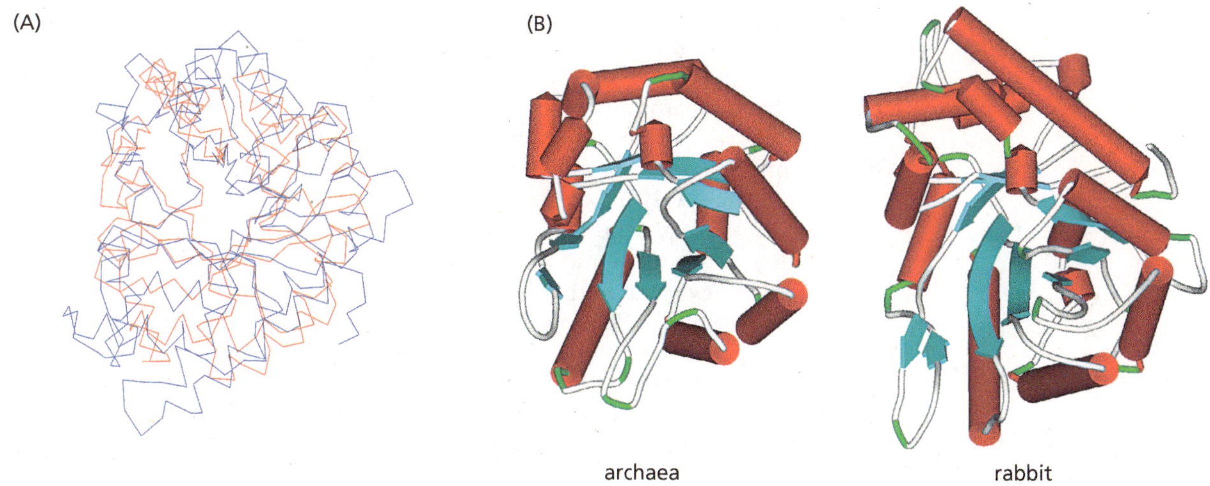


(C)

```
Rabbit  :   1 KELSDIAHRIVA-PGKGILAADESTGSIAKRLQSIGTENTEENRRFYRQLLLTADDRVNPCIGGVILFHE
              111111111111 1111111111111111111111          111111111 1111  111111111
Archaea:    1 NLTEKFLRIFARRGKSIILAYDHGIEHGPADFMDNPDS-------ADPEYILRLARDAG--FDGVVFQRG

Rabbit  :  70 TLYQKADDGRPFPQVIKSKGGVVGIKVDKGVVPLAGTNGETTTQGLDGLSERCAQYKKDGADFAKWRCVL
              11111      11111     1111111111       1111111      1111111111111111111
Archaea:   62 IAEKY------YDGSV-----PLILKLNGKTTLYN---GEPVSVANC----SVEEAVSLGASAVGYTIYP

Rabbit  : 140 KIGEHTPSALAIMENANVLARYASICQQNGIVPIVEPEILPDGDHDLKRCQYVTEKVLAAVYKALSDHHI
                     1 1111111111111111111111111111111    11  11111111111111111
Archaea:  114 -------GSGFEWKMFEELARIKRDAVKFDLPLVVWSYPRGGKVVNE-TAPEIVAYAARIALELGA----

Rabbit  : 210 YLEGTLLKPNMVTPGHACTQKYSHEEIAMATVTALRRTVPPAVTGVTFLSGGQS--EEEASINLNAINKC
                  111111111             1111111111 11111  1111111111  11111111111111
Archaea:  172 ----DAMKIKYTG-------------DPKTFSWAVK-VAGKV--PVLMSGGPKTKTEEDFLKQVEGVLEA

Rabbit  : 278 PLLKPWALTFSYGRALQASALKAWGGKKENLKAAQEEYVKRALANSLACQ
                   111111111111111                11111111111111
Archaea:  222 -----GALGIAVGRNVWQRR----------------DALKFARALAELVY

Sequence id = 19.7%
```

**Figure 14.1**
**Comparison of the archaeal and rabbit FBPA domains.** (A) The $C_\alpha$ representation and superposition of the archaeal and rabbit FBPA IA fold. (B) A secondary structure schematic of the two FBPA IA structures with helices as red cylinders and strand as blue arrows. The left is the archaeal structure while the fold on the right is from a rabbit. (C) The structural alignment of the two FBPA IA folds, with helices colored in red and strands in blue, showing the matching of secondary structure elements, even when there is hardly any sequence homology. Regions containing the number 1 in between the aligned sequence denote those residues that are structurally aligned.

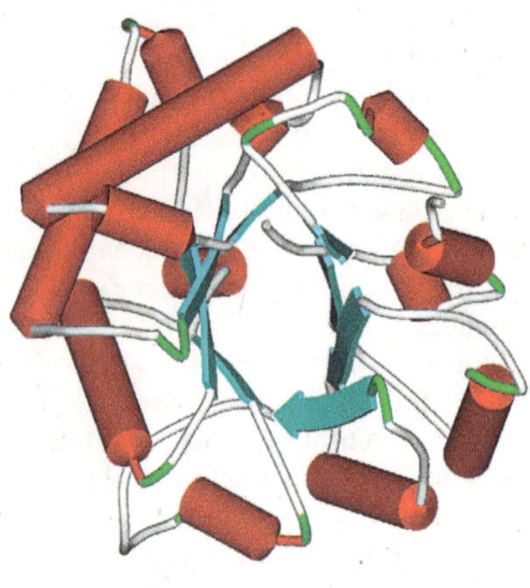

**Figure 14.2**
**The FBPAs have a classical TIM barrel fold.** The TIM barrel from triosephosphate isomerase (TIM) fold is illustrated here with a secondary structure schematic. The fold contains a parallel β-sheet barrel (β–α–β units). It is classed as an α/β protein.

An example of this can be found in the fructose-1,6-bisphosphate aldolases (FBPAs). The cleavage and formation of FBP (fructose 1,6-bisphosphate) from GAP (glyceraldehyde 3-phosphate) and DHAP (dihydroxyacetone phosphate) by FBPA is an important step in glycolysis. Two different classes of FBPA have been identified, which share no significant sequence identity: Class I FBPA and Class II FBPA. Class I is found mainly in eukaryotes and Class II in bacteria. The activities of both types of FBPA have been detected in archaea but no genes with sequence homology to either class have been detected in fully sequenced archaeal genomes. However, using biochemical, cloning, and recombinant expression methods a gene was found to code for an archaeal FBPA protein similar in catalysis to the Class I type. This protein was called FBPA IA, and is also a member of a divergent family of archaeal proteins with overall sequence identity between archaeal members as low as 20%.

Recently the crystal structure was determined of a FBPA IA from the hyperthermophilic crenarchaeote *Thermoproteus tenax* (Tt-FBPA, PDB code 1OJX). The FBPA IA monomer adopts a parallel *αβ*-barrel fold (see Figure 14.1) similar to those found in the Class I and II FBPA proteins. With the crystal structure of the FBPA IA from *T. tenax* solved, it can be seen that all three FBPA sequence families adopt the TIM barrel fold (see Figure 14.2).

However, it must be borne in mind that although proteins with similar structural folds are often functionally related, this is not necessarily the case. There are quite a number of proteins that share a similar topology but have no functional relationship. Proteins within the same fold type can also have secondary structures and turn regions that are different, occurring as insertions within the fold or as entire extra domains (see Figure 14.3). Having the same fold does not necessarily mean that the proteins have a common evolutionary origin. The structural similarities between proteins that share a common fold may arise from the fact that proteins pack into a finite number of favored arrangements. The TIM barrel also provides an example of this. In a recent CATH survey there were about 900 structures that were classified as TIM barrel. Although, most of these function as metabolic enzymes they catalyze quite different reactions and are involved in different pathways (see Figure 14.4). Even though they catalyze diverse reactions, their active site is always found at the C-terminal end of the barrel, which suggests that these proteins did diverge from a common ancestral TIM barrel. In addition, these enzymes often have extra domains that precede, interrupt, or follow the barrel, indicating domain shuffling during protein evolution. Therefore if a protein is identified as having a TIM-barrel fold, one cannot predict its function unless there is also considerable sequence homology.

Similarity of biochemical function then, is most likely if proteins have both sequence and structural similarity, strongly indicating descent from a common ancestor. This is the case for proteins from different species, where sequence has diverged somewhat as a result of speciation but structure and function have been retained. Similarly, proteins that have arisen relatively recently from gene duplication within a genome, such as the mammalian α- and β-globins, are also likely to retain the same function.

## Similar biochemical function can be found in proteins with different folds

In contrast to proteins such as the FBPAs, there are instances where proteins with the same biochemical function have quite different overall structures. Examples of this are found among the serine proteases and the pyridoxal phosphate-dependent aminotransferases. In these cases, however, the enzymatic mechanism is similar and the active sites have converged on similar arrangements.

Another example of a function that can be carried out by two different protein domains is recognition and binding of phosphorylated tyrosine residues. Binding of one protein to another via a phosphotyrosine residue on its surface is an important general mechanism by which activated receptors and other signal transduction proteins can be regulated or recruit the next protein partner in a signaling pathway. Two quite different protein domains can bind phosphotyrosines: these are the Src-homology 2 domain family (SH2) and the phosphotyrosine-binding domain family (PTB) (see Figures 14.5A and 14.5B, respectively). There is, however, a subtle difference between the activities in these two domains. SH2 domains are selective for phosphotyrosines flanked by a particular amino acid sequence on the carboxy-terminal side, whereas the specificity of PTB domains is based on the sequence immediately on the amino-terminal side of the phosphotyrosine. There is no discernible sequence or structural homology between these two domains, although their function is similar. On the other hand, the PTB domain has been found to have structural homology to the pleckstrin homology (PH) domains (see Figure 14.5C), which are known to bind phospholipids. Biochemical experiments have confirmed the ability of the PTB domain to bind phospholipids as well.

## Fold libraries identify structurally similar proteins regardless of function

In Chapter 13 fold libraries were introduced as an aid in predicting the tertiary structure of a protein. Fold libraries also contain a large amount of information

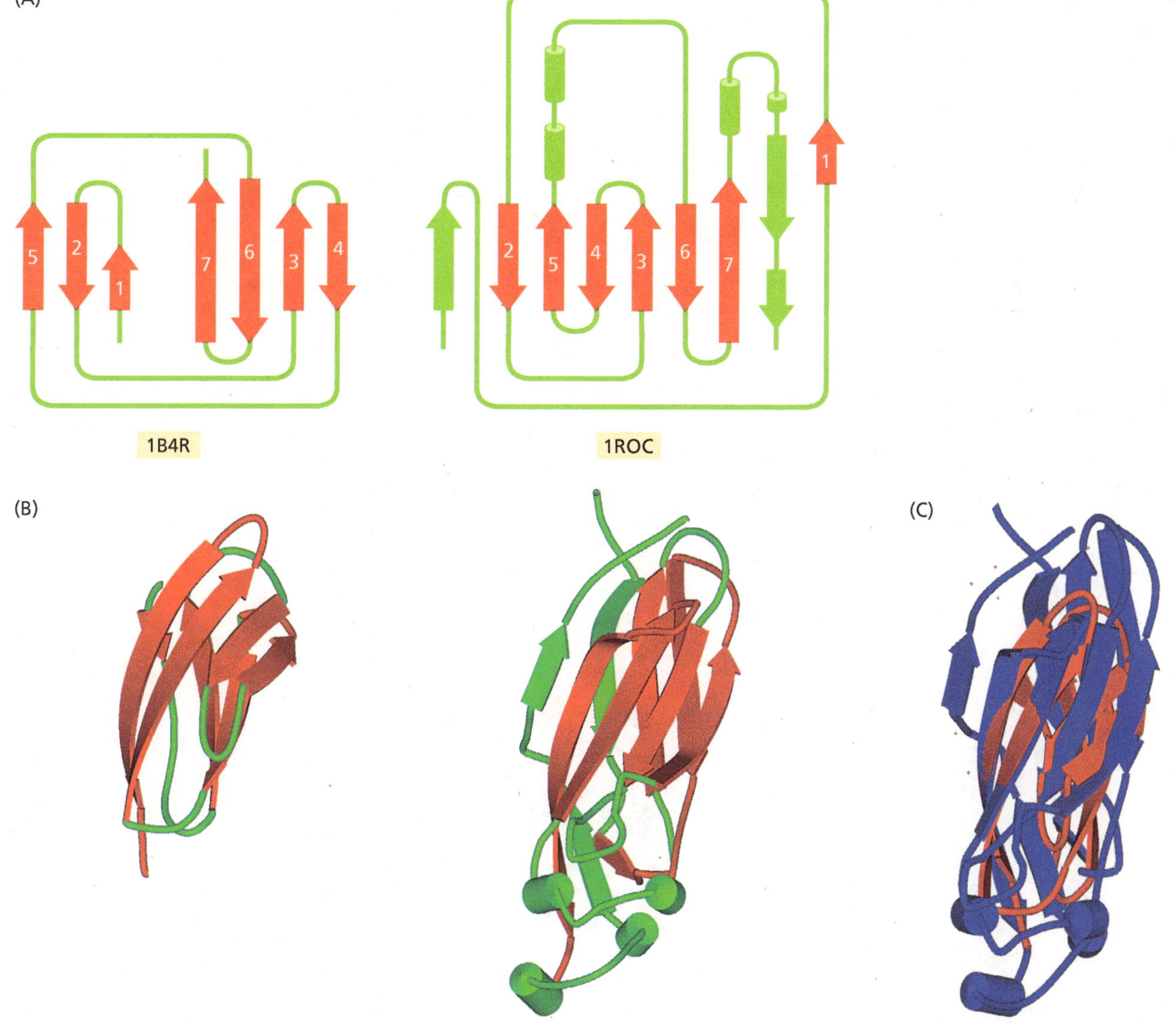


**Figure 14.3**
**Two immunoglobulin-like beta-sandwich folds.** The immunoglobulin fold is found in many proteins. In this figure the fold from the PKD domain of the human polycystein-1 protein (1B4R), which is involved in polycystic kidney disease, and the histone deposition protein (1ROC) are shown. (A) This illustrates a schematic diagram of the fold where arrows indicate β-strands and cylinders α-helices. Those β-strands that form part of the immunoglobulin fold in both proteins are colored red. Both proteins have the same fold and fall into the same fold family. However, 1ROC has two large insertions between the fold-forming structures (in green). (B) A $C_\alpha$-cartoon representation of the above, and (C) the two structures superposed.

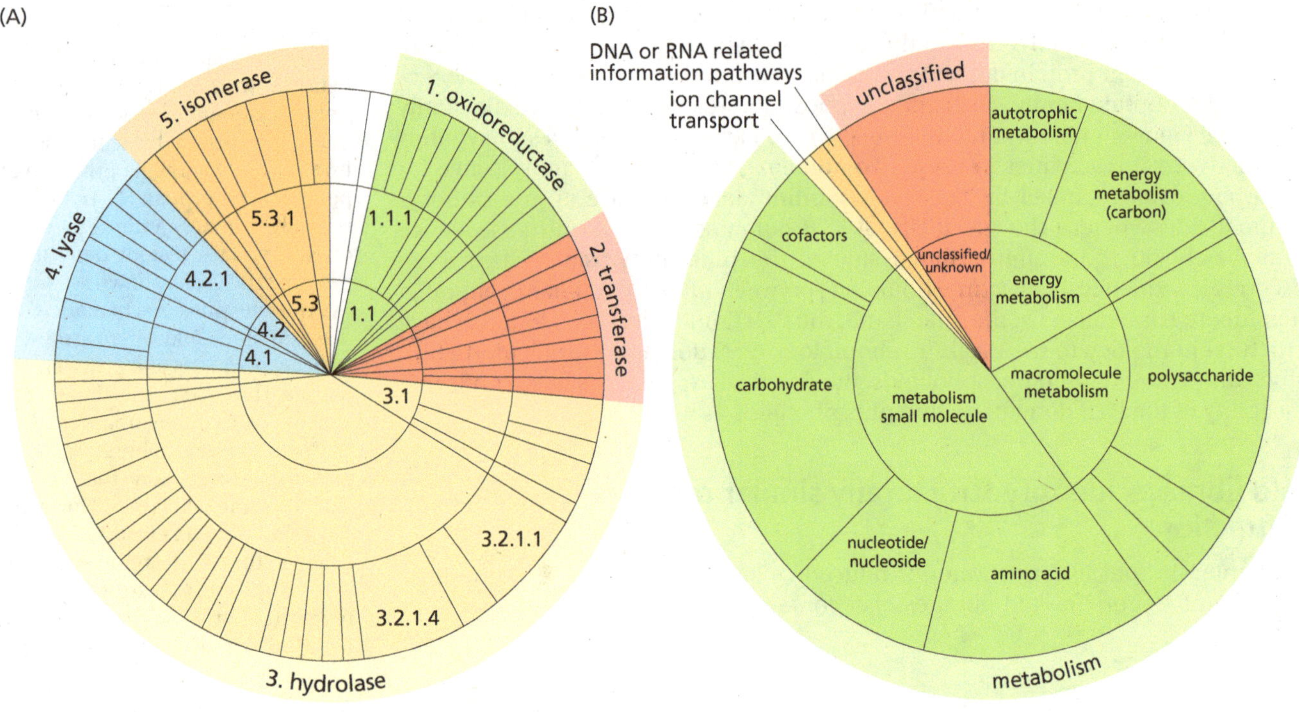


**Figure 14.4**
**The different types of function of TIM illustrated in a functional wheel diagram.** (A) The wheel shows the distribution of TIM functions, as given by the type of reaction, as a set of pie charts. The colored slices represent different enzymatic functions while the white slice shows that a fraction of TIMs have nonenzyme functions as well. (B) This shows the pie chart representation of the biological/pathway function of the various TIM proteins. (Based on N. Nagano, C.A. Orengo and J.M. Thornton, One fold with many functions: the evolutionary relationships between TIM barrel families based on their sequences, structures, and functions, *J. Mol. Biol.* 321:741–765, 2002, with permission from Elsevier.)

about protein function and the structure–function relationship, and therefore can be very useful for identifying the potential function of the structure. The most common fold libraries are CATH, SCOP, and FSSP.

Generally, all the fold libraries classify the proteins according to criteria that are based on the structural arrangement of the protein and on their sequence homology. The SCOP database was created by expert manual curation of protein structures and is supported by automated methods. It aims to provide a detailed and comprehensive description of the structural and evolutionary relationships

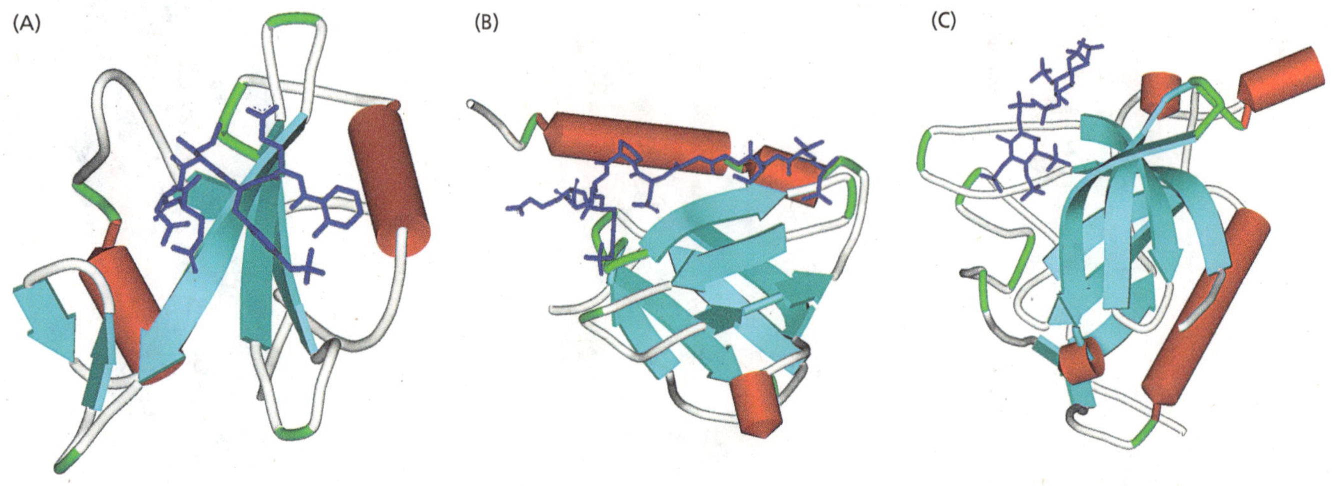


**Figure 14.5**
**Different types of phosphopeptide-binding proteins.** (A) The Src homology domain (SH2), (B) the phosphotyrosine-binding (PTB) domain, and (C) the pleckstrin homology (PH) domain. The different folds all bind a phosphotyrosine peptide, drawn in dark blue.

between all proteins whose structure is known. The proteins are classified by family, superfamily, and fold. Proteins classified as families are undoubtedly evolutionarily related. In SCOP this usually means that the pairwise residue identities between proteins are 30% or higher, but there are exceptions. In some cases, where similar functions and structures provide definitive evidence of common evolutionary descent, the proteins are assigned to the same family despite the absence of high sequence identity. An example of such a family provided by the SCOP Web site, are the globins; these proteins form a family although some members have sequence identities as low as 15%. Proteins that fall into the superfamily class are described as having a probable common evolutionary origin. These have low sequence identities, but similar structural and functional features. Classification into the fold class

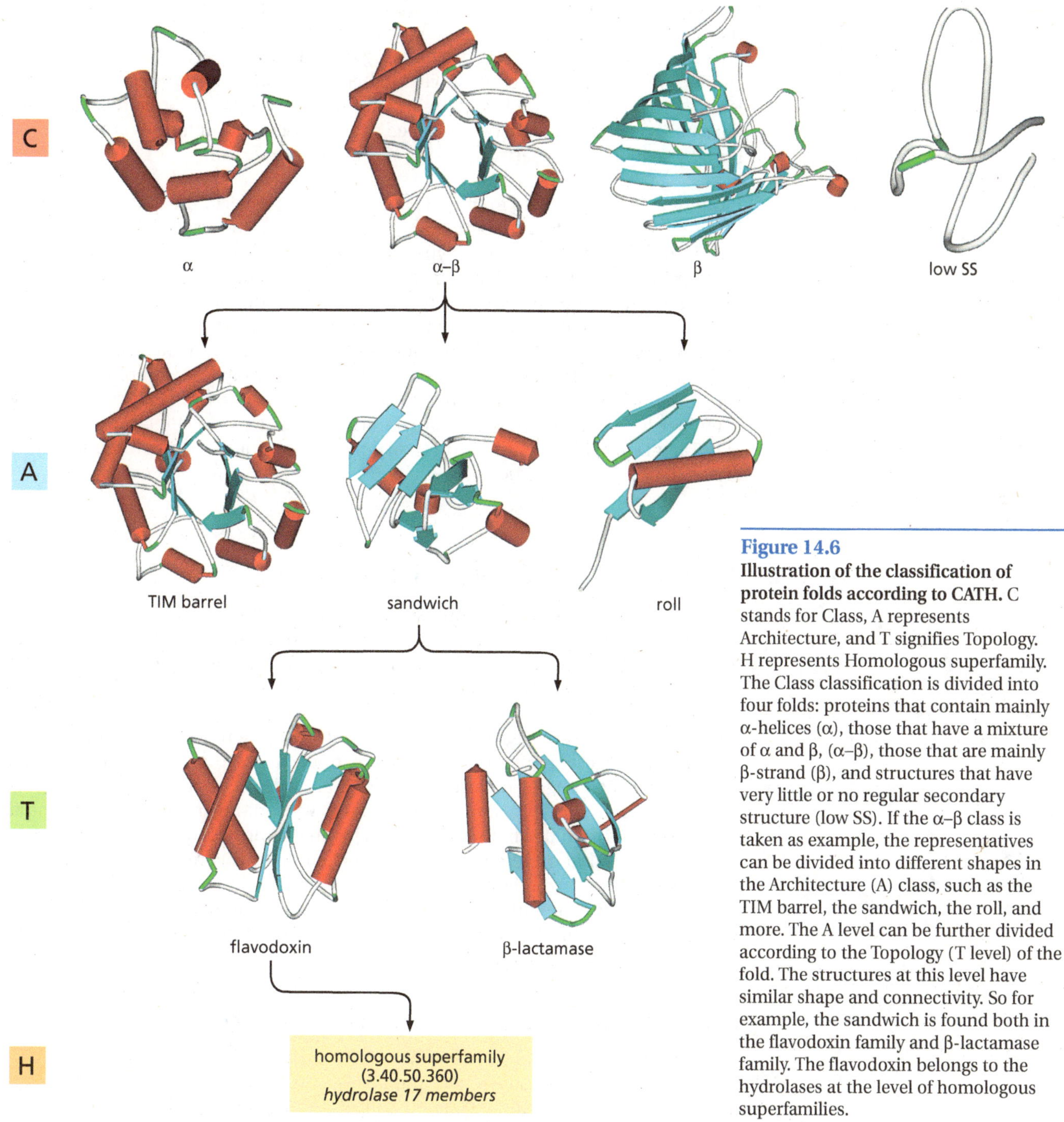


**Figure 14.6**
**Illustration of the classification of protein folds according to CATH.** C stands for Class, A represents Architecture, and T signifies Topology. H represents Homologous superfamily. The Class classification is divided into four folds: proteins that contain mainly α-helices (α), those that have a mixture of α and β, (α–β), those that are mainly β-strand (β), and structures that have very little or no regular secondary structure (low SS). If the α–β class is taken as example, the representatives can be divided into different shapes in the Architecture (A) class, such as the TIM barrel, the sandwich, the roll, and more. The A level can be further divided according to the Topology (T level) of the fold. The structures at this level have similar shape and connectivity. So for example, the sandwich is found both in the flavodoxin family and β-lactamase family. The flavodoxin belongs to the hydrolases at the level of homologous superfamilies.

means that there is major structural similarity between the proteins. In other words they have the same major secondary structures in the same arrangement and with the same topological connections.

The CATH database is a hierarchical classification of NMR and crystal structures solved to resolution better than 3.0 Å. CATH divides the proteins into four major levels: Class, Architecture, Topology (fold family), and Homologous superfamily (hence the name CATH). Proteins are classified into the same class according to the secondary structure composition and packing (see Figure 14.6–C). Class is usually assigned automatically. The proteins are classified into three main classes: mainly-α, mainly-β, and α–β. The α–β class describes both alternating α/β structures and α + β structures. A fourth class has now been recognized—a class containing protein domains that have low secondary structure content. Proteins in the same architecture (see Figure 14.6–A) have a similar overall shape of the domain structure that is not dependent on the connectivity between the secondary structures. Structures that are grouped into **topological families** (see Figure 14.6–T) have similar overall shape and connectivity of their secondary structures. This is equivalent to the SCOP fold families. Proteins within this level probably share a common ancestor and can therefore be described as homologous. Similarities are identified both by sequence comparisons and by structure comparison.

Proteins are classified into four levels in FSSP. The top level of the fold classification is based on secondary structure composition and supersecondary structural motifs, recognizing five distinct types. The next level of classification is fold type; the third level of the classification deduces evolutionary relationships based on structural similarities in addition to functional or sequence similarities. The last level of the classification is a representative subset of the Protein Data Bank extracted using a 25% sequence identity threshold. All-against-all structure comparison was carried out within the set of representatives. Homologs are only shown aligned to their representative. Alignments available from FSSP may be very useful for further analysis. Other fold libraries constructed in similar ways include 3DEE and Dali, which contain structural domain definitions for all proteins with an experimentally solved structure.

## 14.2 Structure Comparison Methods

There are several alternative approaches to comparing two protein structures, each of which will be discussed in this section. Before comparing protein structures it is often useful to identify the domains and to analyze each independently. This can be done either by visual inspection or using automated methods.

The simplest structure alignment method treats the structures as rigid, and tries to identify the largest group of equivalent pairs of atoms that can be superposed within a specified degree of accuracy (see CE and VAST). An alternative method is based on identifying equivalent rigid segments without regard for their sequential order along the peptide chain (see DALI). The final approach to structural comparison recognizes that protein folds are not rigid, often having regions that can act as hinges separating two relatively rigid regions. Typically, but not exclusively, these are interdomain segments, making the comparison of multidomain proteins more accurate. Such methods can superpose two structures by identifying several equivalent rigid regions (see FATCAT).

### Finding domains in proteins aids structure comparison

The identification of protein domains can be done by visual inspection. However this is time consuming, subjective, and needs expert knowledge. It would be much more efficient to have an automated computer program that can divide protein

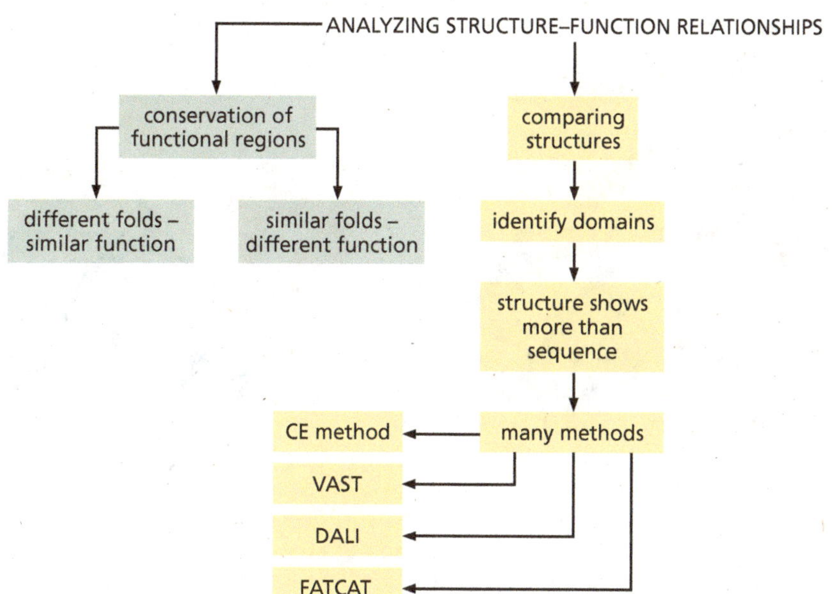


**Flow Diagram 14.2**
**In this section a number of methods for comparing protein structures are described.**

structures into domains, and can minimize the errors that can arise by visual analysis. There are several automatic methods to identify domains from the atomic coordinates; for example, PDP, the Protein Domain Parser server, or the program DIAL, which identifies domains based on clustering of distances between secondary structure elements. Both programs need the protein coordinates to be uploaded to the server. The PDP returns a list of domains while DIAL returns a list and a color-coded coordinate file.

The human protein Cbl (PDB code 2CBL) will be used as an example to illustrate the resources available for domain identification, searching the databases for similar structures, and how such structure comparisons can be used to find information on the possible function of a protein domain. Cbl and its homologs in other species are intracellular signaling proteins that have, among a number of other cellular functions, a role in inhibiting the antigen-receptor signaling pathway in T lymphocytes. The Cbl gene is also a known proto-oncogene: a truncated and oncogenic version of Cbl (v-Cbl) is carried by a mouse tumor virus. Thus its mechanism of action in the cell is of considerable interest. The first three-dimensional structure of the conserved amino-terminal region of Cbl was published in 1999, and since then a combination of structure and sequence comparisons and experimental biochemistry and cell biology has revealed much about its activity.

When the structure was solved it revealed the presence of an unusual SH2-like domain in the amino-terminal region of Cbl, which had not been detected by sequence analysis. When the Cbl structure was used to guide the alignment of the Cbl sequence with other SH2-domain sequences, important functional motifs typical of SH2 domains could be identified in the Cbl sequence. This explained the protein's known ability to bind to phosphorylated tyrosines in intracellular signaling proteins such as the ZAP-70 protein tyrosine kinase from T cells. Another domain of the Cbl structure—the RING-finger domain—identified Cbl as a probable E3 ubiquitin ligase, following experimental work that associated RING-finger domains of this type with ubiquitin ligase activity. This activity was subsequently confirmed experimentally, and Cbl is now generally considered primarily as an E3 ubiquitin ligase that targets receptor tyrosine kinases and other proteins for degradation.

The structure of the Cbl amino-terminal region shows that it consists of three domains: a mainly α-up–down bundle, a mainly α-orthogonal bundle, and an α–β

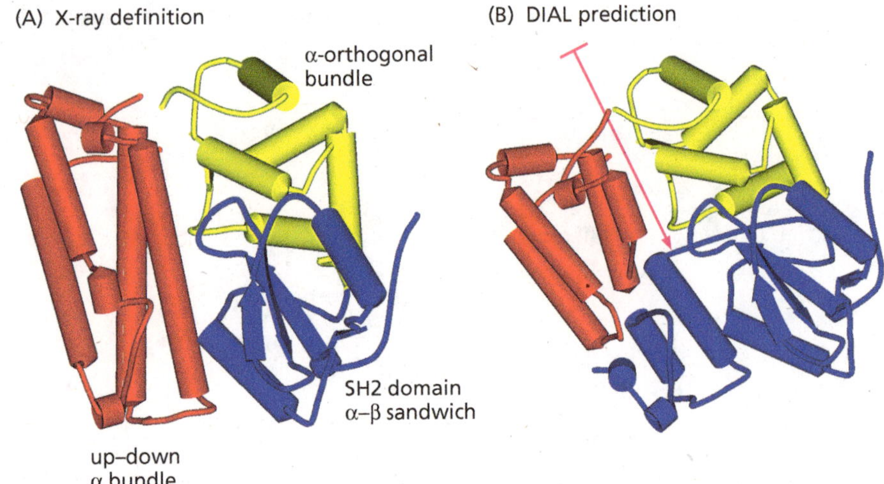


**Figure 14.7**
**Alternative definitions of the Cbl domains.** (A) The three domains of Cbl as defined in the PDB X-ray structure file. The first domain (red) is mainly an α-up–down bundle. The second domain (green) is mainly α-orthogonal bundle, and the third domain (blue) is the α–β sandwich with structural similarity to an SH2 domain. (B) This shows the predicted domain architecture by DIAL. Part of the first domain is predicted as the last. The red arrow in (B) shows where the third domain could be split into two domains based on further visual inspection.

sandwich (see Figure 14.7A). It is the third domain that we will use as an example to illustrate the identification of the SH2 domain. So that each domain can be identified and analyzed separately, the coordinate file of 2CBL was submitted to the DIAL domain identification program. DIAL identified three domains, but it divided the α-up–down bundle and assigned a part of it to domain 3 (see Figure 14.7B). In this instance, it may make it harder for the fold-recognition programs to identify domain 3 as a SH2-like domain. Visual inspection of the predicted domain organization suggests that the third domain, identified as one large domain, should be split as shown by the arrow in Figure 14.7B into two domains. Even though this does not give the correct number of domains, it would enable the correct analysis of the SH2-like domain. The PDP server also identified three domains: domain 1 correctly from residues 47 to 175; domain 2 terminated a few residues early as did the prediction for domain 3.

Once the number of domains in a protein has been identified and their start and stop positions delineated, each domain can be submitted separately to the programs described below to identify its class or fold.

## Structural comparisons can reveal conserved functional elements not discernible from a sequence comparison

In order to determine the presence of the SH2 domain in Cbl the following steps can be taken. Firstly, the Cbl sequence is analyzed with the sequence databases to identify similar sequences (see Sections 4.6 and 4.7). Previously we have used DIAL to identify the domains of Cbl. To simplify the sequence search we will take only the region we have identified to include the SH2-like domain: residues 250 to 350. No sequence homologs other than Cbl proteins from other species are found (see Figure 14.8). Thus the sequence alone cannot give us any clues to the function of this region. The next step is to turn to the solved structure of the Cbl conserved amino-terminal region, and use the structural coordinates of the region of interest to look for similar structures.

## The CE method builds up a structural alignment from pairs of aligned protein segments

The CE (Combinatorial Extension) algorithm compares two structures by splitting them up into smaller segments usually eight residues in length. The method aligns pairs of segments, one from each protein, with respect to their structural similarity based on local geometry. Pairs of aligned segments that represent possible continuous alignment paths are extended or discarded until a single optimal alignment has been reached.

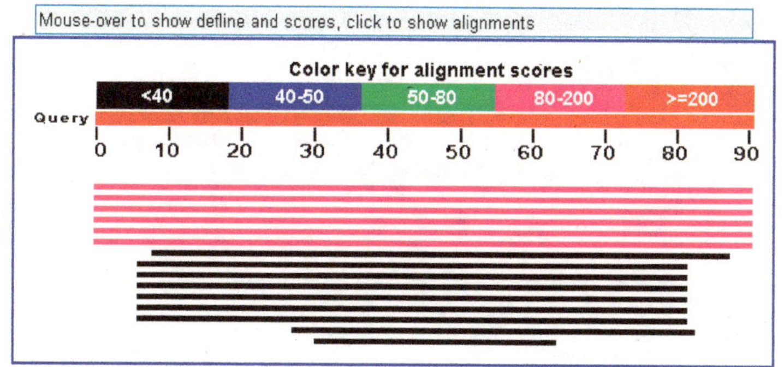


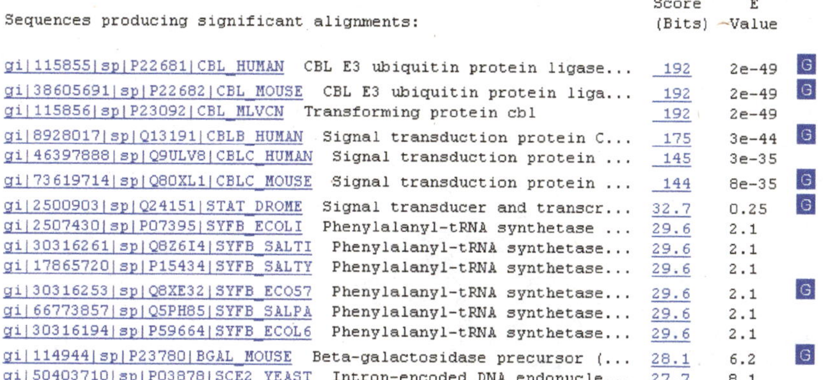

Sequences producing significant alignments:

| Sequence | Description | Score (Bits) | E Value | |
|---|---|---|---|---|
| gi\|115855\|sp\|P22681\|CBL_HUMAN | CBL E3 ubiquitin protein ligase... | 192 | 2e-49 | G |
| gi\|38605691\|sp\|P22682\|CBL_MOUSE | CBL E3 ubiquitin protein liga... | 192 | 2e-49 | G |
| gi\|115856\|sp\|P23092\|CBL_MLVCN | Transforming protein cbl | 192 | 2e-49 | |
| gi\|8928017\|sp\|Q13191\|CBLB_HUMAN | Signal transduction protein C... | 175 | 3e-44 | G |
| gi\|46397888\|sp\|Q9ULV8\|CBLC_HUMAN | Signal transduction protein ... | 145 | 3e-35 | |
| gi\|73619714\|sp\|Q80XL1\|CBLC_MOUSE | Signal transduction protein ... | 144 | 8e-35 | G |
| gi\|2500903\|sp\|Q24151\|STAT_DROME | Signal transducer and transcr... | 32.7 | 0.25 | G |
| gi\|2507430\|sp\|P07395\|SYFB_ECOLI | Phenylalanyl-tRNA synthetase ... | 29.6 | 2.1 | |
| gi\|30316261\|sp\|Q8Z6I4\|SYFB_SALTI | Phenylalanyl-tRNA synthetase... | 29.6 | 2.1 | |
| gi\|17865720\|sp\|P15434\|SYFB_SALTY | Phenylalanyl-tRNA synthetase... | 29.6 | 2.1 | |
| gi\|30316253\|sp\|Q8XE32\|SYFB_ECO57 | Phenylalanyl-tRNA synthetase... | 29.6 | 2.1 | G |
| gi\|66773857\|sp\|Q5PH85\|SYFB_SALPA | Phenylalanyl-tRNA synthetase... | 29.6 | 2.1 | |
| gi\|30316194\|sp\|P59664\|SYFB_ECOL6 | Phenylalanyl-tRNA synthetase... | 29.6 | 2.1 | |
| gi\|114944\|sp\|P23780\|BGAL_MOUSE | Beta-galactosidase precursor (... | 28.1 | 6.2 | G |
| gi\|50403710\|sp\|P03878\|SCE2_YEAST | Intron-encoded DNA endonucle... | 27.7 | 8.1 | |

**Figure 14.8**
**A BLAST search through the Swiss-Prot database with the Cbl-SH2 domain sequence.** No SH2-like proteins are found. Only hits with Cbl homologs give significant *E*-values. This particular result will change as the database is updated.

In the example discussed here, the PDB coordinates of the SH2-like domain of Cbl have been submitted for a structural comparison against the whole database. A list of 153 protein chains that have been structurally aligned was returned by the program. Of these, 135 are SH2-like domains. The results returned give the RMSD (root mean square deviation) between the chain-pairs as well as a ***z*-statistic**. Chains with the lowest RMSD and highest *z*-scores should be chosen; in practice proteins with a *z*-score of 3.5 or better are usually found to have a similar fold. In addition, one should look at the length of the structure that has been matched; the more residues matched the better. Figure 14.9A shows the top results for the Cbl search on the CE Web site. The top hit is to the SH2 domain of SHP2 (also called Syp). This protein consists of three domains, and only the first domain has an SH2-like fold, which was the fold picked by the CE method. Figure 14.9B shows the fit of the Cbl domain 3 and the SH2 of SHP2 superposed on each other. If our example consisted only of domain 3, a hit on a multidomain protein would provide us with additional information. The regions of the SH2-like domain that are in contact with the other domains cannot be involved in the binding of its ligand peptide.

## The Vector Alignment Search Tool (VAST) aligns secondary structural elements

The VAST (Vector Alignment Search Tool) algorithm computes units of structural similarity between pairs of secondary structures that are of a similar type and have similar orientation and connectivity. The method uses an algorithm based on graph theory to identify the equivalent pairs of secondary structural elements in the two structures, and an optimization technique to refine the alignment. Precomputed structural alignments made by VAST are stored in the Entrez 3D database.

Given a user-specified set of 3D coordinates VAST can be used as a tool to search for folds that are structural homologs. The VAST structural alignment picked the STAT-1 SH2 domain as the most significant match after Cbl itself (see Figure 14.10).

(A)

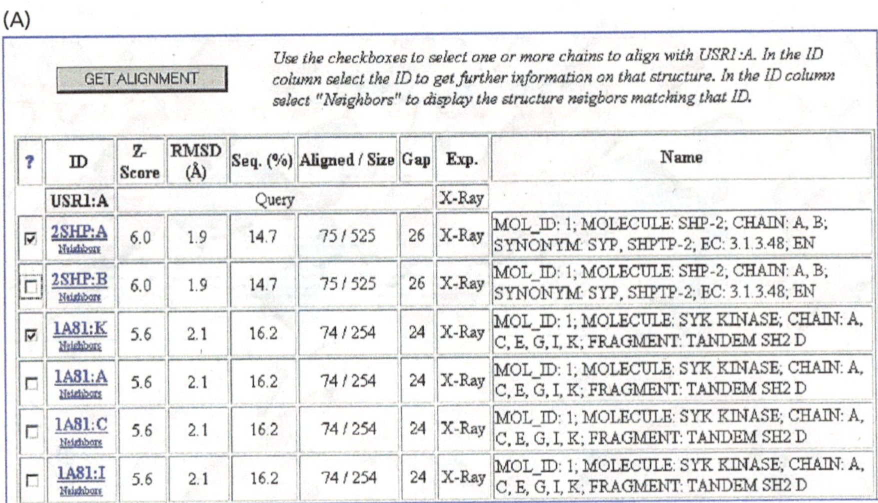

GET ALIGNMENT

*Use the checkboxes to select one or more chains to align with USR1:A. In the ID column select the ID to get further information on that structure. In the ID column select "Neighbors" to display the structure neigbors matching that ID.*

| ? | ID | Z-Score | RMSD (Å) | Seq. (%) | Aligned / Size | Gap | Exp. | Name |
|---|---|---|---|---|---|---|---|---|
| | USR1:A | Query | | | | | X-Ray | |
| ☑ | 2SHP:A Neighbors | 6.0 | 1.9 | 14.7 | 75 / 525 | 26 | X-Ray | MOL_ID: 1; MOLECULE: SHP-2; CHAIN: A, B; SYNONYM: SYP, SHPTP-2; EC: 3.1.3.48; EN |
| ☐ | 2SHP:B Neighbors | 6.0 | 1.9 | 14.7 | 75 / 525 | 26 | X-Ray | MOL_ID: 1; MOLECULE: SHP-2; CHAIN: A, B; SYNONYM: SYP, SHPTP-2; EC: 3.1.3.48; EN |
| ☑ | 1A81:K Neighbors | 5.6 | 2.1 | 16.2 | 74 / 254 | 24 | X-Ray | MOL_ID: 1; MOLECULE: SYK KINASE; CHAIN: A, C, E, G, I, K; FRAGMENT: TANDEM SH2 D |
| ☐ | 1A81:A Neighbors | 5.6 | 2.1 | 16.2 | 74 / 254 | 24 | X-Ray | MOL_ID: 1; MOLECULE: SYK KINASE; CHAIN: A, C, E, G, I, K; FRAGMENT: TANDEM SH2 D |
| ☐ | 1A81:C Neighbors | 5.6 | 2.1 | 16.2 | 74 / 254 | 24 | X-Ray | MOL_ID: 1; MOLECULE: SYK KINASE; CHAIN: A, C, E, G, I, K; FRAGMENT: TANDEM SH2 D |
| ☐ | 1A81:I Neighbors | 5.6 | 2.1 | 16.2 | 74 / 254 | 24 | X-Ray | MOL_ID: 1; MOLECULE: SYK KINASE; CHAIN: A, C, E, G, I, K; FRAGMENT: TANDEM SH2 D |

(B)

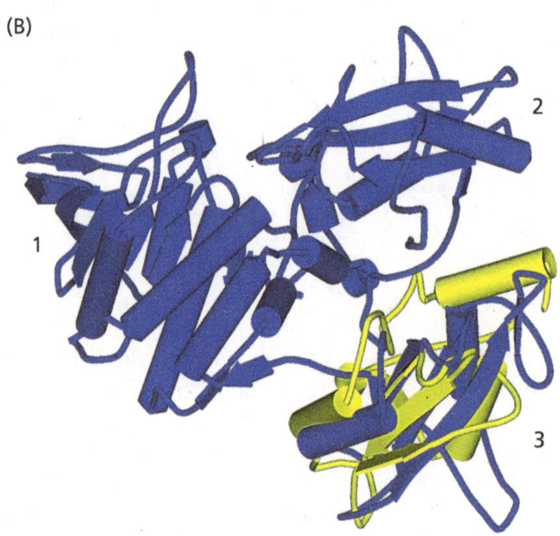


**Figure 14.9**
**Results from the structural search by the program CE with the Cbl SH2- like domain.** (A) The output from the CE program, listed according to z-score. All results shown identify an SH2-containing protein. The top hit is with a tyrosine phosphatase (SHP2). (B) SHP2 consists of three domains, shown here. The Cbl SH2-like domain (green) has been superposed on the SHP2 SH2 domain (number 3).

Analysis of the structural alignment shows that there is a conserved pattern of residues around the essential arginine residue (R294 in Cbl) that makes the main interactions with the target phosphotyrosine (see Figure 14.10A). However, submitting the Cbl sequence for pattern searches (see Section 4.9) does not pick up the SH2 similarity (except to itself). In this case it is only the availability of structural coordinates and structural similarity that enables the identification of an SH2-like domain in Cbl.

## DALI identifies structure superposition without maintaining segment order

The DALI algorithm uses the atomic coordinates of two proteins to calculate and compare residue–residue ($C_\alpha$–$C_\alpha$) distance matrices for each protein. Many possible alignments are obtained at the same time, leading to the definition of optimal and sub-optimal alignments. The method allows insertion of gaps of any length between rigid segments. An unusual feature of this method is that it permits the reversal of chain direction and the aligned segments to be combined in orders other than those given by the sequence.

The DALI precomputed structural neighbors of any protein already in the PDB are available in the FSSP database, which is accessible through the DALI Web page. DALI results of a structural search between the user's coordinates and the structural database are returned as a list of top-scoring hits and structural alignments. The

(A)

```
Cbl_A   YMAFLTYDEVKARLQKFihKPGSYIFRLSCTRl-GQWAIGYVTAD-----GNILQTIphn
SH2_A   IMGFISKERERALLKDQ--QPGTFLLRFSESSreGAITFTWVERSqnggePDFHAVEpyt

Cbl_A   ------kpLFQALIDGfr--------eGFYLFPDGRNQNPD
SH2_A   kkelsavtFPDIIRNYkvmaaenipeNPLKYLYPNIDKDH
```

(B)

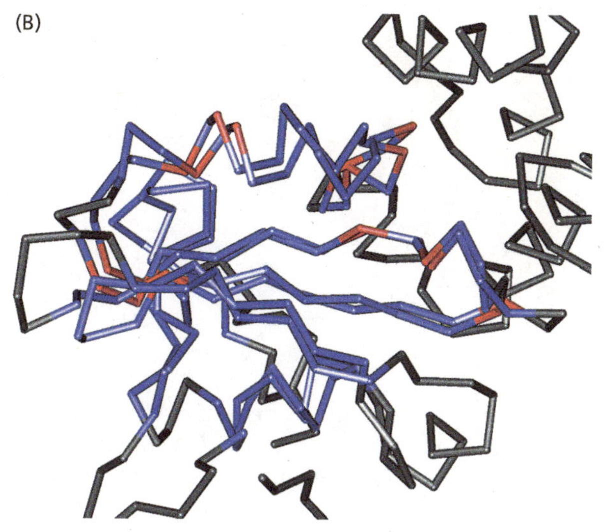

**Figure 14.10**
**The results from the structural alignment algorithm VAST.**
(A) A sequence alignment between Cbl and the STAT-1 SH2 domain. The structurally aligned regions are given in blue with sequence conserved residues illustrated in red. The residues in black lower case are not included in the structural alignment. (B) The $C_{\alpha}$ backbone superposition of the two SH2 domains color coded as in (A) with gray regions representing those that are not structurally aligned.

top-scoring hit for the Cbl structure (apart from itself) was the structure of the SH2 domain of Grb10. The RMSD between the 81 aligned residues was 2.0 Å with only 9% identity between the two sequences. Figure 14.11 shows the structure of Cbl and Grb10 next to each other in the same orientation.

## FATCAT introduces rotations between rigid segments

The structural alignment program FATCAT first identifies equivalent rigid segments, at least eight residues in length, in both structures. It then uses a dynamic programming

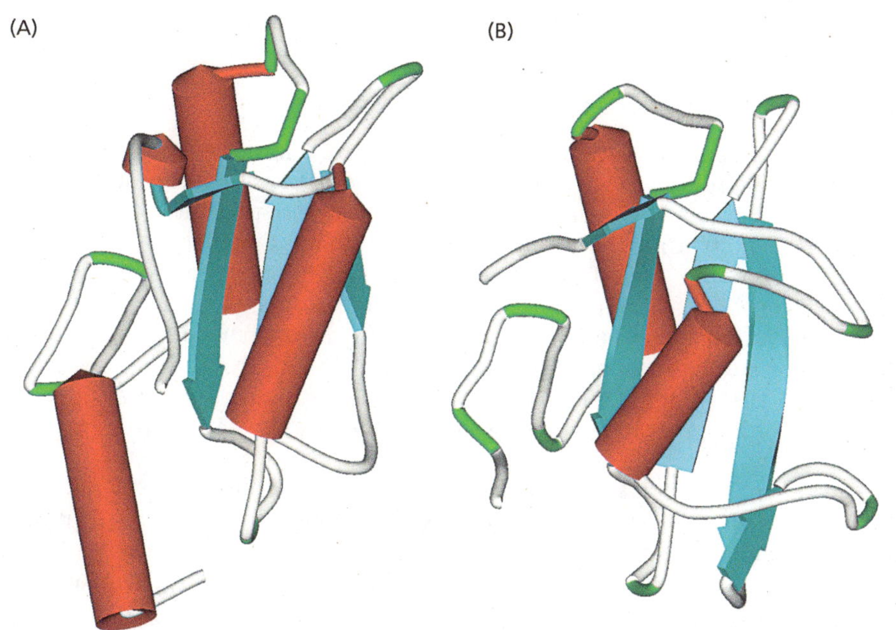


**Figure 14.11**
**Results from a DALI search.** The top-scoring fold with (A) Cbl was the Grb10 SH2 domain (B). Shown here are the secondary structure ribbon representations of both in the same orientation. The structural similarity is noticeable.

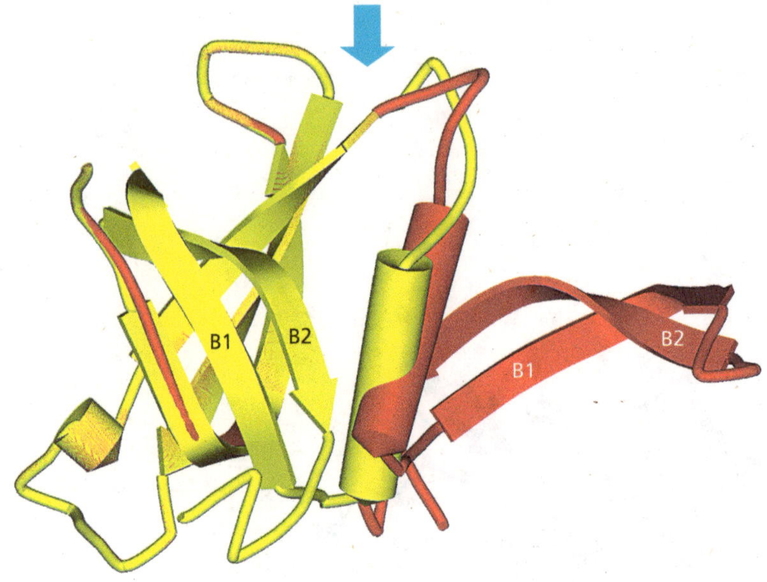


**Figure 14.12**
**The program FATCAT allows the insertion of twists around structural elements as is illustrated here.** In the 21st hit—which is an OB fold—a twist was introduced around a helix, which flips strands B1 and B2 such that they now will align with the Cbl fold. The figure here shows the original OB fold (red) and the OB fold with the twist (yellow/green).

algorithm (see Chapter 5) to combine the segments, if necessary using rotations, called twists, about hinges. The algorithm optimizes a function of the RMSD between equivalent atoms, insertions, and deletions in the alignment, and any twists that were introduced.

When the Cbl structure is submitted to FATCAT, Grb10 SH2 domain is once more the top hit (apart from Cbl itself). The first 19 hits are all SH2-like folds and none include any twists, meaning that the structures were treated as rigid. The 20th hit has an HPr-like fold and only 1% sequence identity with Cbl but no twists were introduced. The 21st hit is the first hit where a twist is introduced to structurally align Cbl with an OB-like fold. To align the OB-like fold with Cbl, a twist was introduced near a helix, which flips a β-sheet around (see Figure 14.12).

In the case of Cbl, the programs all returned similar results, with an SH2-like domain as the top hit or hits. It is always desirable to confirm the results by using a number of different algorithms. One should submit the target structure to more than one program and, if there is any disagreement, take the majority result.

## 14.3 Finding Binding Sites

Proteins do not act in isolation. To carry out their functions they must form transitory or stable complexes with other molecules; these may be other proteins, small-molecule substrates or regulators, DNA, RNA, or membrane lipids, depending on the protein. Often proteins contain specialized domains that bind to specific molecules important to their function. For example, many intracellular signaling proteins consist of multiple protein-interaction domains through which the protein is located at the right intracellular site and through which it interacts with the appropriate target. Gene regulatory proteins bind to DNA through a variety of specialized DNA-binding domains with distinctive structures, while also binding regulatory proteins or small molecules via other domains.

The protein p53 is a transcription factor present at low levels in the cell nucleus in an inactive state. Its levels increase rapidly in response to DNA damage, hypoxia, and nucleotide deprivation, triggering a cascade of molecular events through extended regulatory associations with certain genes. These interactions delay the cell cycle, giving time for the cellular DNA repair mechanisms to operate, or when

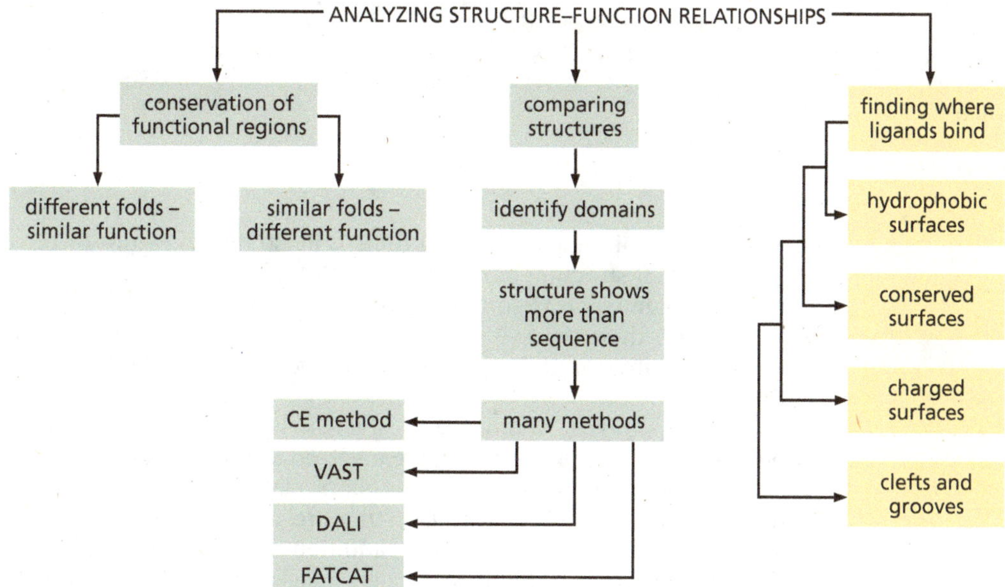


**Flow Diagram 14.3**
**In this section some of the methods available for identifying binding sites on proteins are described.**

the DNA is severely impaired to initiate apoptosis, the programmed cell death. p53 is mutated in more than 50% of invasive tumors and is therefore an important target in cancer research. There are an estimated 88 direct protein interactions and 87 transcriptional associations (indirect links) to p53 (see Figure 14.13). All these interactions cannot occur at once and they cannot all bind to the same binding site. p53 is a multidomain protein with different binding partners to different domains (see

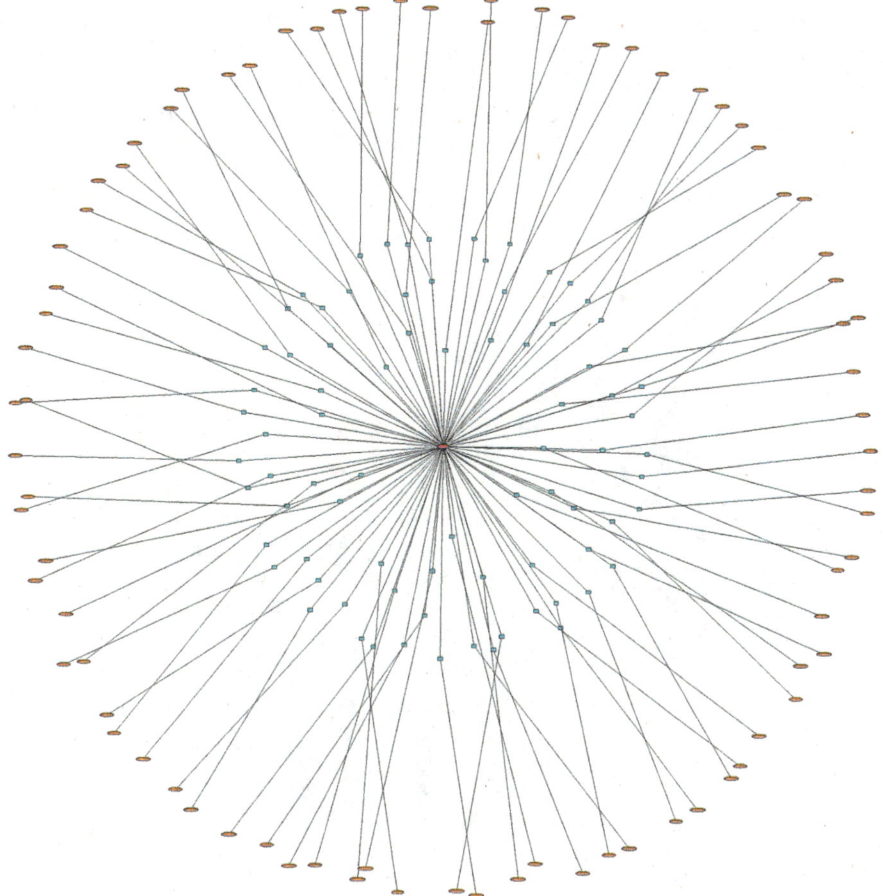

**Figure 14.13**
**An interaction diagram of p53 with its many direct-binding partners.** Each line connects to different proteins for which there is experimental evidence that the two interact. The central red area denotes p53; yellow ovals are the proteins that associate with p53.

| transactivation | SH3 | DNA binding | TET | REG |
|---|---|---|---|---|
| TFIID, p62, TAF, MDM2, RPA, Pin1, CREB, FAK, and more | Sin3, WOX1, IκBα | ASPP-family, Brca1, Bcl2, BAK, and more | TFIIH, WRN, TBP, Grp75, PTEN, STK15, and many more | |

**Figure 14.14**
**A schematic representation of the various regions of the p53 protein and some of the proteins that bind to the specific regions.** It is unlikely that many of these proteins could bind simultaneously to the same region.

Figure 14.14). The binding of different units is illustrated in Figure 14.15 where an RNA polymerase II structure (gray) binds to DNA as well as other RNA polymerase subunits at the same time. Determining the types of interactions a protein can make at any one time is therefore a vital part of analyzing the structure–function relationships and increasing our understanding of the overall function of the protein.

There are two main types of binding sites. The first type occurs at protein–protein or protein–DNA interfaces, usually large areas on the surface of the protein structure. The second type forms ligand pockets or clefts. These binding pockets can penetrate quite deep within the protein structure, although they are usually accessible from the surface via a channel. In addition many proteins bind metal ions. Metal ions are important in biological processes as they can mediate an interaction between protein and ligand and they can also act as a nucleophilic catalyst or in an electron transfer role.

## Highly conserved, strongly charged, or hydrophobic surface areas may indicate interaction sites

Fold library searches can indicate on the basis of similarity to homologous proteins where putative interaction sites in a protein are located. However, there are many other methods that have the potential to identify interaction sites. If we have a protein for which the type of substrate or binding partner is known, exploration of the protein surface can identify putative interaction sites. A multiple alignment of related proteins can be constructed, based either on sequence or structure. When this is superposed onto the structural coordinates of the query protein it can be used to identify shared features of the surface residues. Any small clusters of residues that are conserved throughout the alignment are highlighted on the three-dimensional

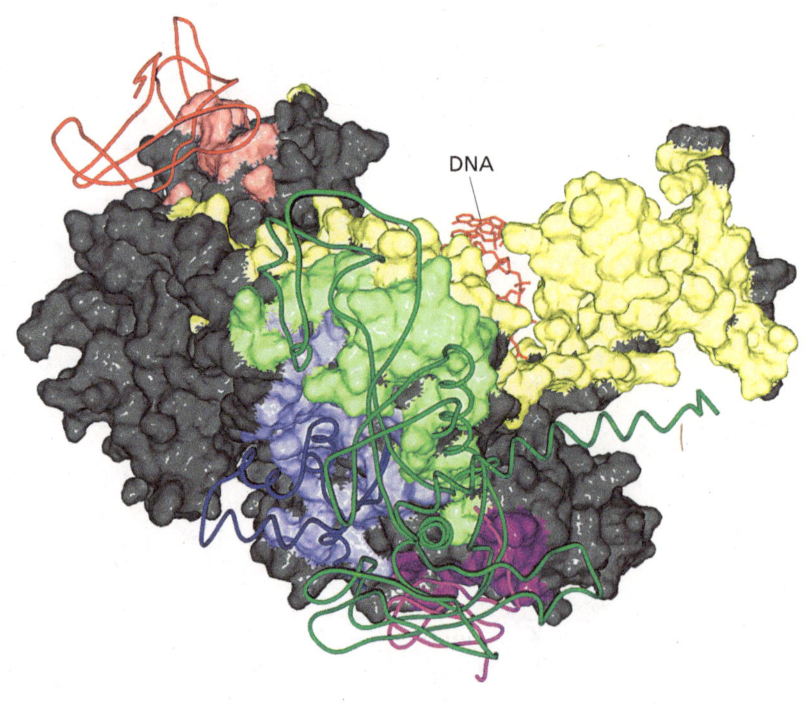


**Figure 14.15**
**The binding architecture of the RNA polymerase II subunit.** Other RNA polymerases that bind to this protein are each shown as a $C_\alpha$-trace in different colors. Each part of the surface on RNA polymerase II that interacts with the other polymerases is color-coded accordingly. For example, the red-colored surface of RNA polymerase II interacts with the red $C_\alpha$-trace. A strand of DNA is just visible in red. This illustrates the complexity of binding in some proteins.

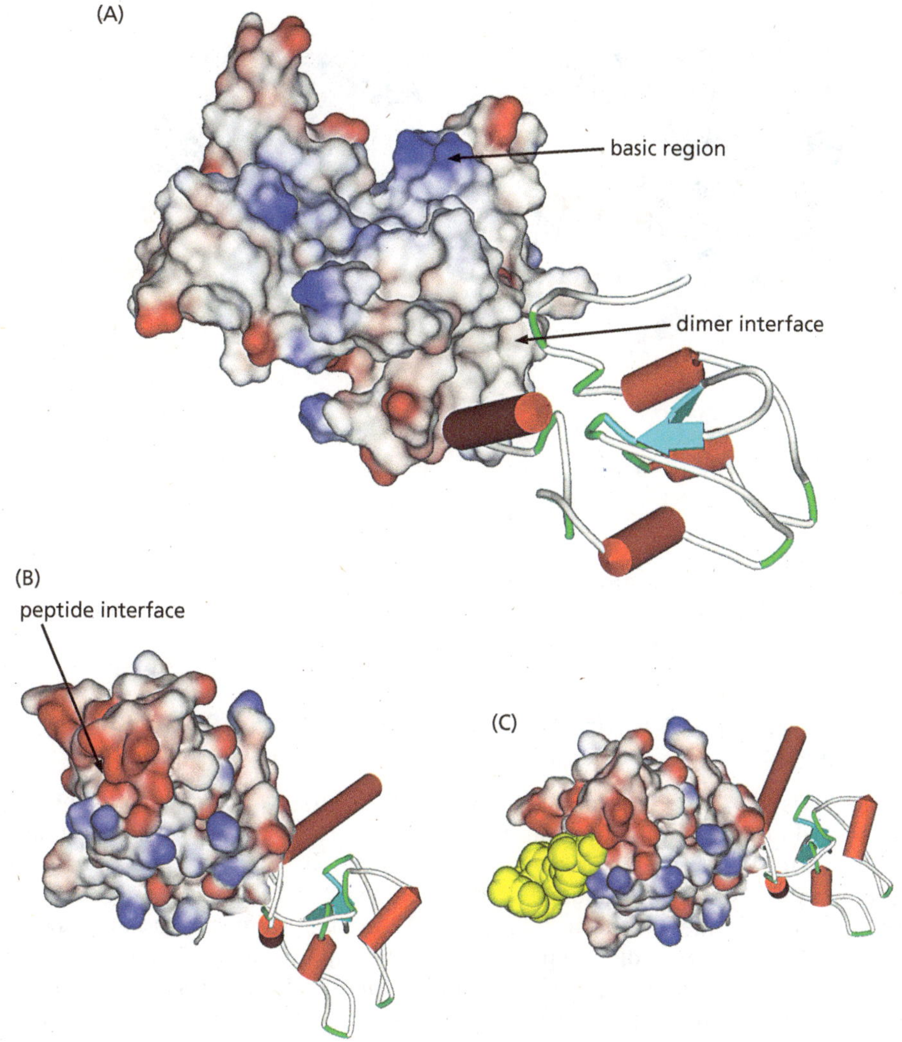


**Figure 14.16**
**The interaction regions of survivin.** (A) A monomer of survivin represented as a surface colored according to charge, where red regions are negative and blue regions are positive (basic). Survivin forms a dimer—the hydrophobic (white) dimer interface is indicated by an arrow. (B) The peptide interface situated on the opposite side of the molecule from the interface shown in (A), is negatively charged. (C) The survivin–peptide complex with the peptide colored yellow/green.

representation of the protein; this often shows that conserved residue clusters that are spread out in the sequence come together in the folded structure. If such conserved patches occur on the surface of a protein they may represent one or more interaction sites. In some cases, the type of interaction can be deduced from the character of the surface patch. For example, if it contains many charged residues, especially of the same charge (positive or negative), it can be deduced that the interaction will be mainly through electrostatic attraction. On the other hand, a conserved hydrophobic patch, which will be unusual on the surface of a typical globular protein, points to an interaction based on hydrophobic forces. For example, the crystal structure of survivin (1XOX), a mammalian cell-cycle regulatory protein that inhibits apoptosis, has revealed an extensive hydrophobic interface for dimerization along one of the surfaces of the survivin monomer, a basic patch that acts as a sulfate- or phosphate-binding module, and a solvent-accessible acidic/hydrophobic patch in the carboxy-terminal region that is involved in protein–protein interaction with the Smac/Diablo peptide (see Figure 14.16).

To analyze surface properties an interactive structure display program is needed (see Section 2.2). Given a suitable program, the structure can be colored according to the physicochemical properties of the residues or their degree of conservation. Patches of conserved or hydrophobic or polar residues should be further investigated as possible binding sites. As binding sites for substrates are part of the active site, similar techniques can be applied to find enzyme active sites. A number of

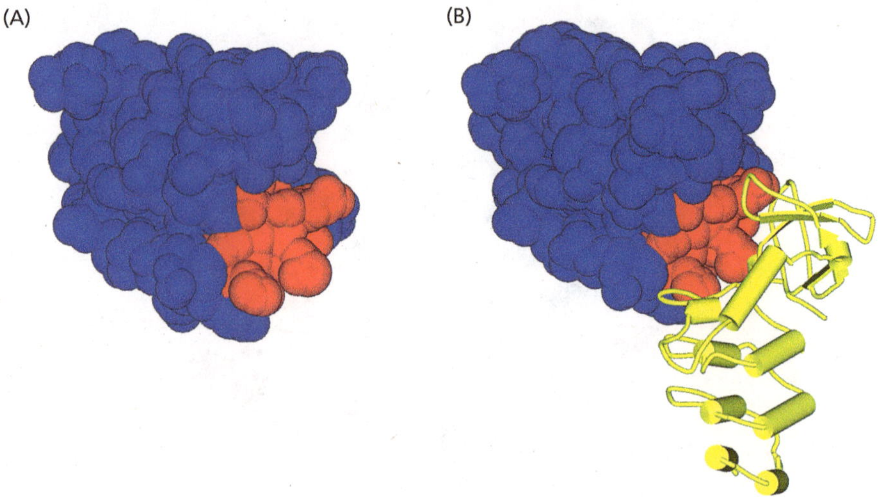


**Figure 14.17**
**Protein–protein interaction sites of the DNA-binding domain of p53 as predicted with the program ProMate.** (A) One region is predicted as a binding site (in red). The crystal structure of the p53–ASPP2 complex is available. (B) We can see that ASPP2 (in green) binds to the predicted patch (which is also known to bind the DNA).

programs are available on the Web to analyze surface properties and predict regions of proteins that may bind other proteins or contain ligand-binding sites. A few are illustrated below.

## Searching for protein–protein interactions using surface properties

An analysis of 57 unique protein structure surfaces that were known to be involved in heteromeric temporary protein–protein interactions showed that mainly β-sheets or long loops are involved in the interaction interface but no α-helices. Additionally, aromatic residues were preferred, with clusters of either hydrophobic or aromatic residues at the interface. These findings were incorporated into an automatic interface-prediction program called ProMate. p53 (PDB code 1YCS) was submitted to ProMate for identification of interaction sites. One site was identified as a potential protein–protein interaction region (see Figure 14.17A). A crystal structure exists of p53 in complex with ASPP2, one of the proteins known to interact with p53. Figure 14.17B shows that ASPP2 binds to the site identified by ProMate. This site in addition is known to interact with DNA.

Another similar program is called PPI-PRED, and is based on analysis of the properties of protein interaction interfaces identified in a set of 180 protein and 56 protein–DNA complexes. When p53 was analyzed using PPI-PRED two major sites were identified (see Figure 14.18). One is the ASPP2/DNA-binding site. The other

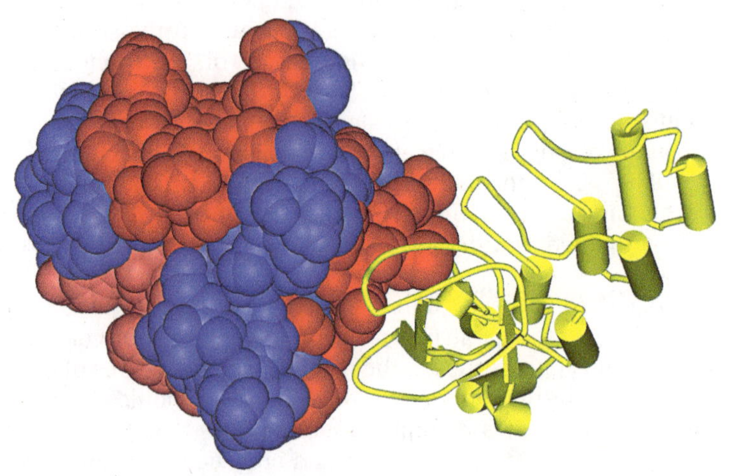

**Figure 14.18**
**Protein–protein interaction sites of the DNA-binding domain of p53 as predicted with the program PPI-PRED.** This predicts more binding sites than ProMate (three sites); these are color-coded according to probability of being a binding site, with red the most probable. One of the sites is the ASPP2/DNA-binding site. The actual binding partner (of any) of this second site is presently unknown. As this is only the DNA-binding domain of the p53, the second region can be occupied by one of the other p53 domains. However, there are no structures available for the whole protein.

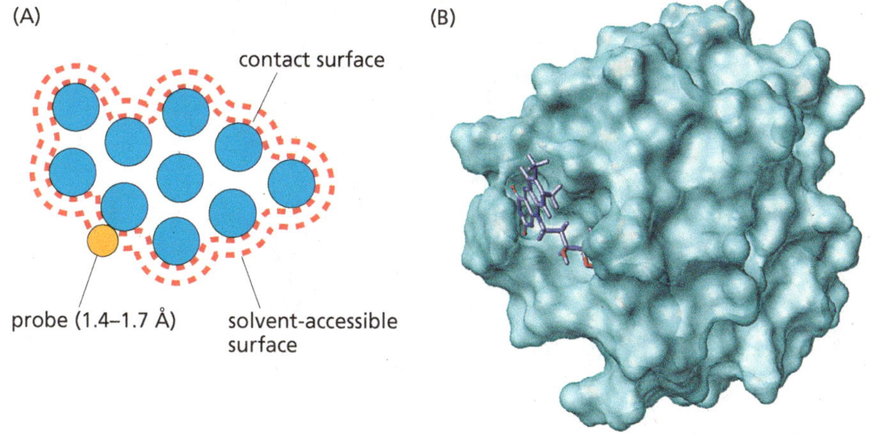


**Figure 14.19**
**Calculation of solvent-accessible surface.** The surface of a molecule can be defined to be the part of the molecule that is accessible to solvent. (A) One method to calculate the surface is to have the solvent molecule (water) represented by a sphere that has a radius of 1.4 Å to 1.7 Å. This is called the probe sphere (the orange ball). The solvent-accessible surface is defined as the trace of the probe sphere center as it rolls over the molecule (outer red dashed line). The contact surface is that part of the molecular surface that can be touched by the edge of the probe sphere (inner red dashed line). There are variations of how a surface can be calculated. (B) Illustration of the solvent-accessible a surface of *Clostridium beijerinckii* flavodoxin protein using a sphere probe of 1.4 Å. The ligand flavin mononucleotide (FMN) bound to the protein is also shown.

may well be involved in binding another protein or is involved in the probable trimerization structure.

## Surface calculations highlight clefts or holes in a protein that may serve as binding sites

Most computational methods search for binding sites by analyzing the surface of a protein to locate pockets. First, a surface, usually the solvent-accessible surface areas of the protein, is calculated (see Figure 14.19) and then the surface is explored for clefts and cavities that may accommodate a ligand. Usually, the largest pocket is taken as the binding site. A number of programs are available for free download, such as SURFNET. However, two programs to locate putative binding sites—Pocket-Finder and Q-SiteFinder—are available via a Web interface.

Pocket-Finder is based on the LigSite algorithm. The Pocket-Finder program defines potential binding sites based solely on the geometry of the receptor. The method is based on the observation that if a line passes through a cavity or cleft a section of the line will lie within the protein, followed by a section in the cavity or cleft, followed by another section within the protein. Such a pattern will indicate the possible presence of a cavity in the region of the central line segment. For each of a grid of points within the protein, seven directions are examined to see if they have this pattern. A count is made of the number of these directions whose patterns identify it as in a pocket. The grid point is predicted to be in a cavity if it achieves a count of seven, with smaller non-zero numbers suggesting a location within a pocket at correspondingly shallower depth. The pockets are defined by all those connected grid points that have counts exceeding a threshold value.

The Q-SiteFinder method also uses a grid of points, but in contrast to Pocket-Finder identifies putative binding pockets by locating regions that would have favorable interaction energy with a probe that represents part of a general ligand. The interaction of the probe with the protein is defined by the nonbonding terms (see Appendix B) of a methyl group. At each grid point the interaction energy is calculated of the probe at that position with the protein. Only those grid points whose interaction energy is more attractive than a defined threshold are kept. Finally, potential binding pockets are defined by combining the grid points that remain if they are within a specified distance.

Submitting the enzyme dihydrofolate reductase (DHFR) without its inhibitor or cofactor NADPH to Pocket-Finder identifies as the top site most of the site where the NADPH and inhibitor bind, as does Q-SiteFinder (see Figure 14.20). Q-SiteFinder predicts a lower volume (smaller) pocket as compared to other programs including Pocket-Finder (see Figures 14.20D and 14.20E). Often, when applied to large proteins the algorithms that depend only on analysis of the surface calculate binding sites

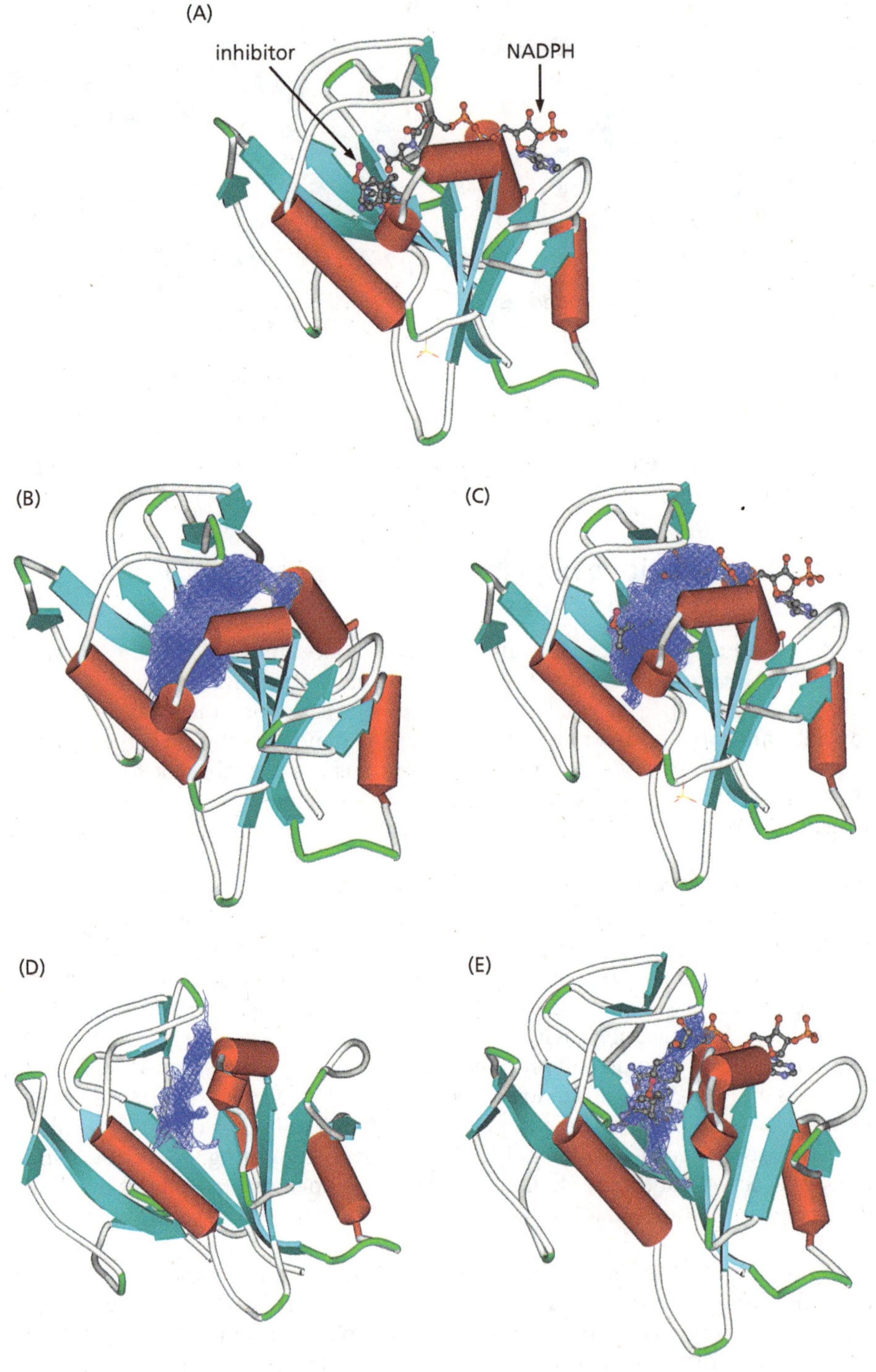


**Figure 14.20**
**Ligand binding-site programs find probable cavities where ligands can bind.** The results from Q-SiteFinder and Pocket-Finder are shown. (A) The crystal structure of dihydrofolate reductase with its inhibitor and cofactor NADPH. (B) The best site found by Q-SiteFinder and (C) with the NADPH and cofactor shown. (D) The best site found by Pocket-Finder and (E) with the NADPH and cofactor shown. The sites are quite well defined.

that are too large. The smaller volume binding site predictions using Q-SiteFinder were found to be more similar to the ligand they contain and independent of the size of the protein. Both programs also identify the residues that are part of the binding site. These can be used in the programs described below to dock the ligand.

## Looking at residue conservation can identify binding sites

Amino acids that are important functionally and also for binding tend to be conserved across species during evolution. This observation has been used to predict binding and functional residues and is referred to as the evolutionary trace method. A number of methods have been proposed that calculate conservation scores for

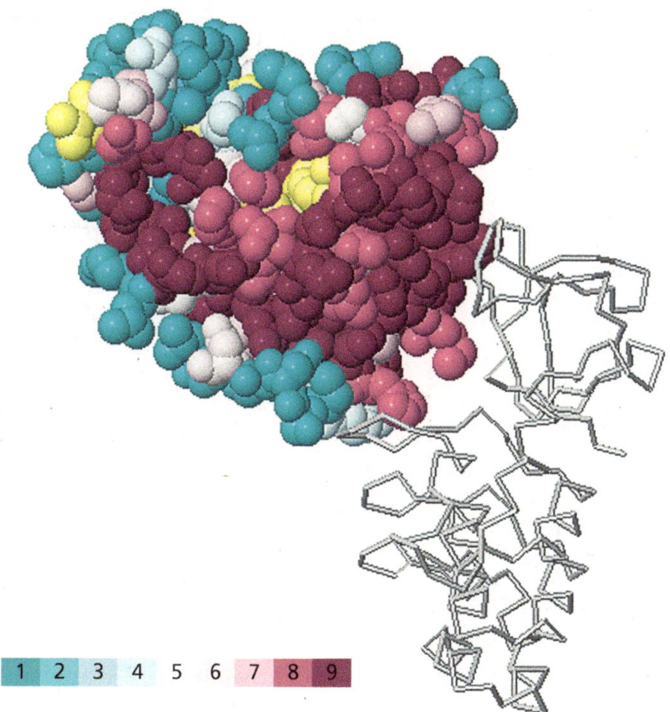


**Figure 14.21**
**Identification of binding sites using ConSurf.** The p53 was submitted to the ConSurf server to visualize areas conserved during evolution which are therefore possible interaction sites. The dark purple colors indicate conserved regions. The ASPP2/DNA-binding site is within the conserved region.

each residue based on the sequence alignment of a family of proteins. These types of scoring often depend on the evolutionary relationship between the protein under study and its homologous proteins using substitution matrices (see Section 5.1). For example the ConSurf program takes the sequence of the submitted protein and performs a PSI-BLAST search (see Section 6.1) through the Swiss-Prot database. After it has extracted the sequences found from the search it does a multiple alignment using ClustalW (see Section 6.4). Based on this alignment a phylogenetic tree (see Chapters 7 and 8) is constructed. Subsequently it calculates position-specific conservation scores and divides these into a nine-color gradient for visualization. Current research is exploring the promise of combining the two methods of surface searching (for example SURFNET) and conservation analysis (such as ConSurf) to give more specific and accurate binding sites.

The p53 protein was submitted to the ConSurf program, which calculated evolutionary scores that were then used to color the residues of the structure (see Figure 14.21). The highly conserved residues within the p53 family fall mainly into the ASPP2/DNA-binding region as described above. An evolutionary trace method from Accelrys was also applied to analyze the DHFR residues (see Further Reading). In this study the residues that are involved in cofactor and folate binding were successfully identified using this approach.

## 14.4 Docking Methods and Programs

Once an active site or a probable binding site has been identified, the binding of small-molecule ligands such as enzyme substrates, inhibitors, and cofactors can be modeled. If the type of ligand is known or suspected, then the next step is to either model in the ligand itself or to find a potential ligand structure from a database of small molecules and fit it into the binding site. In addition there are programs that enable the user to design ligands specific for a particular binding site. Analyzing the interactions between ligand and protein leads to a better understanding of how ligand binding might affect the protein's structure and function. Strong binding is generally determined by the shape and chemical properties of both the binding site

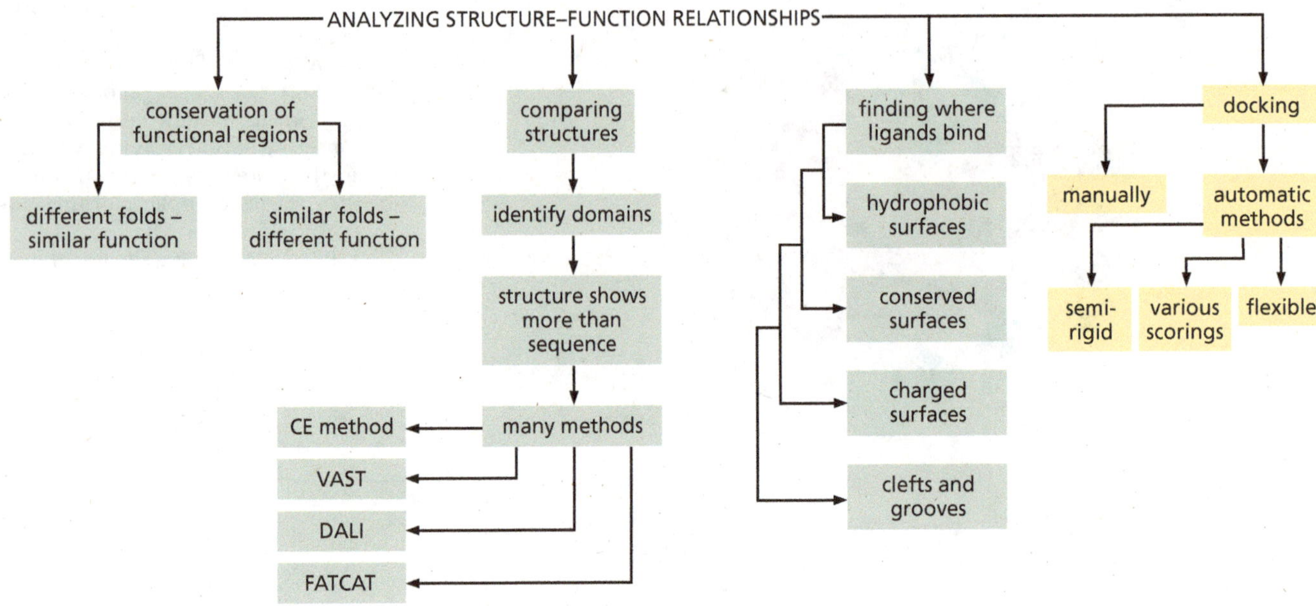


**Flow Diagram 14.4**
**In this section some of the processes available for predicting the binding mode of other molecules to a protein structure are described.**

and substrate and by the orientation of the ligand. In addition, analysis of the docked ligand and the interacting residues can suggest ways in which a ligand might be modified to make it into a useful therapeutic drug (see Box 14.1). Molecular modeling of this sort is now used as a high-throughput way of rapidly investigating the binding of potential small-molecule drugs to the target protein and identifying those that are worth following up further. There is insufficient space in this book to give many details of the techniques of molecular modeling that are applied to the problem of ligand binding. Some details of the energy terms and methods that can be used are given in Appendices B and C. For further details, see the references in those appendices.

## Simple docking procedures can be used when the structure of a homologous protein bound to a ligand analog is known

The procedure known as **docking** attempts to model the structure of the protein–ligand complex. The simplest strategy is manual docking, which can be performed with many computer graphics programs. For useful results, this almost always requires that ligand-binding information is available for a homologous structure. As homologous proteins generally preserve the same fold to bind the same molecules, comparison between homologous proteins may reveal structural conservation within the binding pocket.

Simple superposition, or structural alignment, of the target structure and the homolog bound to its ligand enables the same or an analogous ligand to be docked to the target structure. Once the two protein structures are superposed, a potential ligand can be fitted by reference to the template ligand bound to the homologous protein. The structure can then be submitted to some form of energy minimization procedure or molecular dynamics analysis (as described in Appendix C) to test and improve the fit. However, this type of docking is subjective and frequently the ligand binds in a different orientation than that found in the homolog.

## Specialized docking programs will automatically dock a ligand to a structure

In traditional drug discovery, thousands of compounds have to be screened *in vitro* to find a tiny handful of potential drug leads that can be taken on to the next stage. This is both time consuming and costly. It was therefore of considerable commercial

## Box 14.1 Rational Drug Design

Drugs are ligands that bind to a specific protein and either increase its activity (an agonist) or decrease/inhibit the protein's activity (an antagonist). Those ligands that decrease or prevent a protein's activity are called inhibitors. The ligands often bind in a reversible manner, competing with the natural ligand (competitive binding), but they can also bind irreversibly. There are many aspects in addition to binding to their target that determine whether a ligand might make a suitable drug, such as how soluble the molecule is, how stable, and especially how selective. Most compounds bind to molecules other than just the protein that is the desired target; in other words, they are not very selective and give rise to unforeseen and sometimes potentially dangerous side effects.

Most drugs have been discovered by chance or by experimental as well as computer-based (virtual) large-scale screening methods. In virtual screening methods, potential target compounds (also known as leads) from a database of small molecules are automatically docked into the target protein and the resultant complex is assigned a score. The best-scoring compounds are then selected for rational drug design or *in vitro* and *in vivo* testing. The small-molecule databases used in such work often contain several thousand to millions of different compounds, permitting a very broad range of ligand chemistry to be explored.

In rational drug design the structure of the target protein and its natural ligand is used to identify or design other putative ligands. The potential drugs can be designed *de novo* or by modification of lead compounds found through large-scale screening methods.

The first drug to be designed in such a fashion was Relenza®, which is used to treat flu. This drug was developed by selecting molecules that were likely to bind to the conserved regions of the enzyme neuraminidase (see Figure B14.1A). Neuraminidase is an enzyme produced by the flu virus to release newly formed virus from infected cells. Hence an inhibitor of this enzyme would stop new viruses from being released into the body. A similar structure-based design approach was used for the design of inhibitors of the parainfluenza virus hemagglutinin-neuraminidase.

Other examples of the practical application of these techniques include drugs developed to treat HIV, such as ritonavir (see Figure B14.1B) and indinavir, where rational drug-design methods were used to find ligands that inhibit viral proteases involved in the correct assembly of viral proteins. Viagra was initially developed using these techniques to try to treat hypertension by inhibiting phosphodiesterase, which was expected to lead to increasing vasodilation. Of course all potential drugs, however found or designed, have to undergo stringent initial and clinical trials before being released on the market. Testing Viagra as a treatment for hypertension and related effects in patients with severe angina proved to be disappointing. However, further trials carried out at the same time with high doses of the drug revealed side effects that included frequent erections in the male participants. This led to the research team testing the drug as a possible treatment for impotence, which proved to be much more successful.

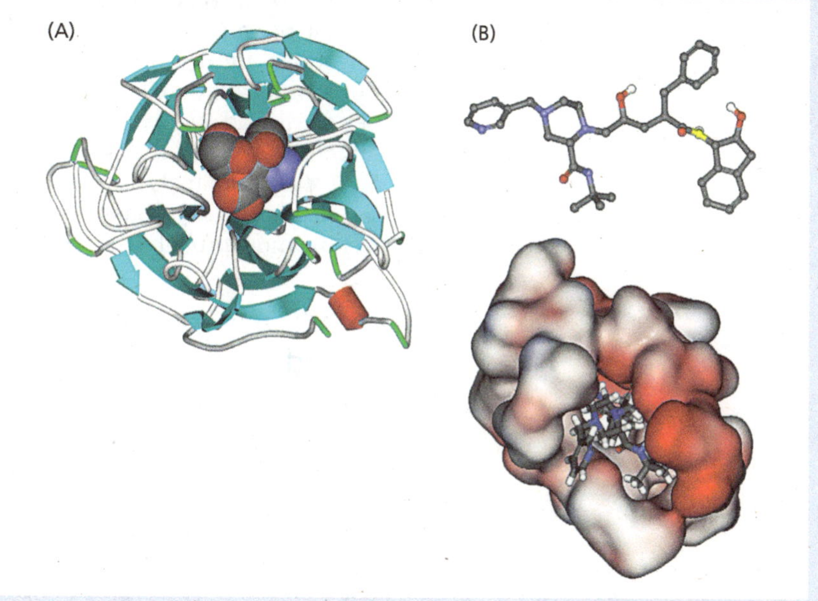


**Figure B14.1**
**Two modeled protein–ligand complexes.** (A) The ribbon representation of neuraminidase with Relenza® shown as a space-filling model docked into the pocket. (B) A space-filling representation of the HIV protease pocket with potential drug ligands (in stick representation).

importance to develop automated programs for identifying ligands for a given target protein *in silico*.

Potential *in silico* ligands are either extracted from chemical and structural databases of small molecules or generated by *de novo* design to fit the binding site. The task of identifying candidate ligands is not simple, as either the enzyme or substrate, or both, can change conformation during binding. Mathematical models that describe such molecular interactions with any accuracy are highly computer intensive and so the problem is usually simplified (see Appendix B). In the early docking programs both the ligand and protein were static rigid bodies. In reality this is far from true as both the ligand and protein are flexible. Most techniques now employ at least partial ligand flexibility.

In more sophisticated docking methods, known generally as **conformational flexible docking**, the ligand, at least, is not constrained to be rigid and conformational degrees of freedom have to be taken into account. Monte Carlo methods in conjunction with simulated annealing (see Appendix C) can be used to search all possible conformations of the ligand. At each step of the Monte Carlo procedure the conformation of the ligand is changed by rotations around a bond or by translating or rotating the entire molecule. The energy of the ligand in the binding pocket is calculated and the movement is either accepted or discarded. Some procedures remember the conformational space already sampled, thereby preventing the same conformation from being calculated again. The most advanced programs do not constrain the protein binding pocket and allow partial or free movement of specific residue atoms, side chains, or bonds.

## Scoring functions are used to identify the most likely docked ligand

The automated docking methods generate a large number of possible ligands and protein–ligand structures, and give each structure a score. The score is used to analyze how well the ligands bind to the protein and therefore to select the most suitable binding partners. Therefore docking programs contain a search algorithm and a scoring function.

The scoring function is usually made up of a number of descriptors, including some form of interaction energy between the protein and ligand. The scoring of the docked ligands is very important in identifying which docked ligand to select. However, a rigorous scoring function is very computer intensive and therefore simplifications are made to nearly all scoring methods. There are basically two types of scoring functions. The first type is generally based on binding energies between the ligand and the protein, using a collection of terms called the force field (see Appendix B). The other scoring type uses approximations of binding energies based on statistical analysis of structurally determined protein–ligand complexes and is called **knowledge-based scoring**. Programs can use a combination of these to select the most probable docked ligand.

## The DOCK program is a semirigid-body method that analyzes shape and chemical complementarity of ligand and binding site

One of the most basic characteristics of binding complexes is that the ligand and the binding site are complementary in shape. One docking program that uses this criterion is DOCK, which also takes into account potential chemical interactions. DOCK explores the ways in which the ligand and the protein can fit together. It generates many possible orientations and conformations of a putative ligand within a user-selected region of the protein structure. The orientations are scored using several schemes designed to measure steric and/or chemical complementarity of the protein–ligand complex. The scores are then used to choose the most likely candidate from a search through a set of ligands or to select the best binding orientation of a single ligand.

DOCK first uses atomic coordinates to generate a solvent-accessible surface area of the target protein. Only the surface of the designated binding site needs to be generated. Spheres defined by the shape of cavities in the protein surface are then generated and placed into the binding site. An energy grid for the target protein is calculated. Each grid point has a score and charge associated with it. The positions of the center of the spheres act as the potential location sites for the ligand atoms. The ligand atoms are then matched with the sphere centres to determine possible ligand orientations, and many orientations are generated for each ligand. Finally, the orientations for each ligand are scored on the basis of shape, electrostatic potential, and force-field potential (see Appendix B). The top-scoring orientation for each ligand is saved and used to compare all ligands investigated.

## Fragment docking identifies potential substrates by predicting types of atoms and functional groups in the binding area

Some docking approaches first identify favorable sites for different types of interactions within the binding site and then search for ligands with functional groups that match these interaction sites.

The GRIN and GRID programs identify those regions within a binding site that have a high affinity for certain types of physicochemical probes; these include single atoms and functional chemical groups, such as methyl or carboxyl groups, with different charge or hydrophobicity. A three-dimensional contour map of the energy surface of the site is obtained, from which the regions most favorable for binding the different probes can be determined. GRID has been used successfully to design inhibitors of the influenza virus based on its crystal structure which eventually led to GlaxoSmith Kline's anti-flu drug Relenza® (see Box 14.1).

The LUDI method uses a library of small fragments to analyze how well these fragments fit the active site of a specific protein. It fits the fragments in such a way that hydrogen bonds can be formed with the protein and that hydrophobic pockets are matched to hydrophobic groups. Subsequently, the fragments are joined to form a complete single molecule, which can be the putative inhibitor or give clues to the chemist how to design a good inhibitor.

The program X-SITE is also based on the small-fragment principle but the X-SITE fragment data are derived from high-resolution protein structures in the PDB. The data used by X-SITE are based on an analysis of the contact preferences of different atomic types with three-atom fragments of each of the 20 residue side chains found in proteins. A three-dimensional distribution of preferred location of specific atomic types around the three-atom fragments is obtained (see Figure 14.22). This information is then used to predict the most favorable location of particular atoms in a specified region. These atom-type locations can then be used to either predict the orientation of a known ligand or to design a novel ligand for a desired pocket.

## GOLD is a flexible docking program, which utilizes a genetic algorithm

More accurate and faster docking techniques are being developed to speed up the search for putative lead compounds for drug discovery. Some methods make use of genetic algorithms (GA algorithms). A genetic algorithm, in general, uses the principles of evolution to find an optimal solution for a computational problem and is described in more detail in Section 6.5. The GOLD (Genetic Optimization for Ligand Docking) docking method uses a genetic algorithm to find the best docked ligands as well as full ligand flexibility and partial protein flexibility.

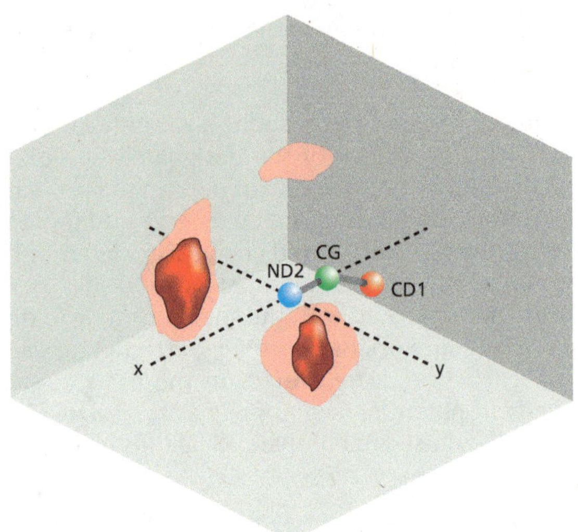


**Figure 14.22**
**An example of a three-dimensional distribution around an ND2 atom of the side chain of Asn used in X-SITE.** The red clouds show the distribution of oxygen atoms that are in the vicinity of the ND2 atom. (Data obtained from Laskowski RA, Thornton JM, Humblet C & Singh J. X-SITE: use of empirically derived atomic packing preferences to identify favourable interaction regions in the binding sites of proteins. *J.Mol. Biol.* 259:175–201.)

As an example, the inhibitor of dihydrofolate reductase was docked using the GOLD program into the pocket identified by Q-SiteFinder (see Section 14.3). The results are shown in Figure 14.23. Seven out of the ten saved orientations of the docked inhibitor fit quite well on the inhibitor from the structural studies. Three are fitted into the pocket where NADPH would bind.

There are many other methods for docking a ligand and optimizing its docked structure. Many new methods are still being developed; however, many of these share the same basic techniques with added novel extensions. In general, the methods either allow for the *de novo* design of a ligand within a specified pocket, as in X-SITE, or a structurally known/predicted ligand is docked as an entity into the pocket, as in GOLD. The docking methods first perform a search of possible binding modes followed by optimization and then need a scoring system to rank each conformation. Many of the methods available are described in a review by Richard Taylor and colleagues (see Further Reading).

## The water molecules in binding sites should also be considered

Proteins *in vivo* do not exist in a vacuum; most are located in an aqueous environment. Water molecules surround the protein surface forming hydrogen bonds with polar atoms of side chains and the main chain. Water molecules are also found in

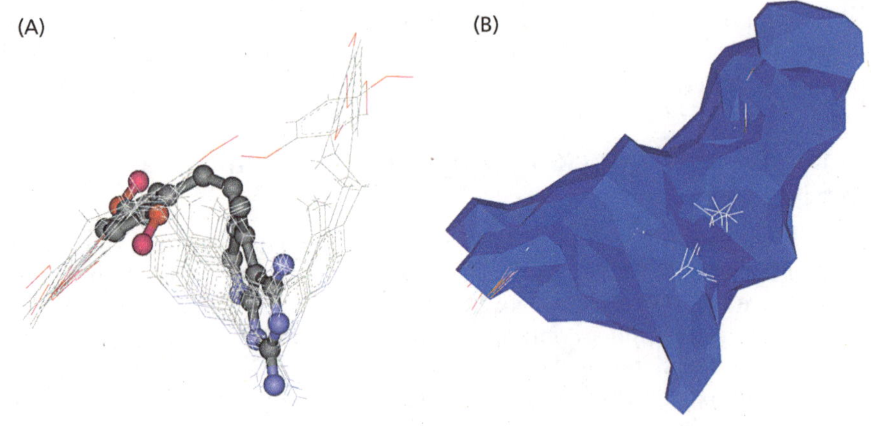


**Figure 14.23**
**Results from the program GOLD when fitting NADPH into dihydrofolate reductase.** (A) The docked NADPH (thin lines) fits very well on the structurally determined ligand (ball-and-stick representation). (B) The fitted ligand in the pocket as determined by Q-SiteFinder.

binding sites and at many biomolecular interfaces. Sometimes the water is replaced by chemical groups of the ligand upon binding, but this is not always the case. Protein-bound water molecules have been shown to be important in substrate recognition in a number of protein structures, contributing to ligand binding and catalysis. They can, for example, be a bridge between a ligand atom and an atom in the binding site. In the well-studied complex of DNA with the bacterial repressor protein for the tryptophan operon, the base-specific binding of the Trp repressor to DNA occurs through a number of water-mediated hydrogen bonds between the bases and the protein.

Bound water molecules often contribute to the structural stability of a ligand–protein complex by forming hydrogen-bond networks and by lining grooves on solvent-exposed protein surfaces. Water can also allow promiscuity of binding partners. This is the case for the major histocompatibility complex (MHC) proteins of the immune system, which are central to immune responses and also determine tissue type. There are hundreds of different variants of MHC proteins in the human population and each can bind a number of different peptide ligands; to accommodate this variety, water molecules bridge the gaps between the atoms of the binding site and the peptide.

When docking ligands into binding sites, water molecules should not be forgotten, therefore, but water has tended to be ignored in the past. One useful tool is the program CONSOLV, which predicts the conservation of water molecules upon ligand binding. The algorithm is used to identify those water molecules in the binding site of the ligand-free protein structure that are likely to be conserved when a ligand binds.

## Summary

In this chapter we have seen that the protein fold can give a very good indication of the possible function of a protein. This close connection between three-dimensional structure and function is often referred to as the structure–function relationship. However, the relationship is often rather complex, so that cases are known of similar structures with different functions, and vice versa. Despite this, with care, much useful information can be obtained.

To be able to analyze the structure–function relationships it is necessary to be able to compare protein structures. Therefore Section 14.2 described a variety of structure comparison methods that can be applied. The results of such comparisons can be used to gain insight into aspects of protein function such as which family they might belong to (for example proteinases), whether the catalytic residues are structurally conserved, and how the binding pockets have been modified to accommodate their specific ligands. For example, if our target protein belongs to a class of proteinases that has a known catalytic triad we can examine the protein to see if a similar triad occurs in a suitable orientation. This is more powerful than a purely sequence-based analysis, as ultimately it is the spatial orientation that is vital for function. Structural comparisons of this type can also identify particular protein-interaction domains that carry binding sites that cannot be distinguished from the sequence alone.

Once the function has been deduced, it is highly desirable to locate the ligand-binding sites that affect the protein's function and, through that, can have a cascade of consequences. For example, the inhibition of *abl* tyrosine kinase by a drug called Gleevec® (Novartis) is used to combat cancers such as leukemia and has recently been found to have an effect on rheumatoid arthritis. Once binding sites have been identified, modeling how ligands dock into them can lead to an analysis of how small changes to the ligands will affect the binding, which can be used to propose new potential drugs and cures.

## Further Reading


**General**

Andersen JN, Del Vecchio RL, Kannan N et al. (2005) Computational analysis of protein tyrosine phosphatases: practical guide to bioinformatics and data resources. *Methods* 35, 90–114.

Bajorath J (2001) Rational drug discovery revisited: interfacing experimental programs with bio- and chemo-informatics. *Drug Discov. Today* 6: 989–995.

Leach A (2000) Molecular Modelling: Principles and Applications. Harlow: Prentice Hall.

Taylor RD, Jewsbury PJ & Essex JJW (2002) A review of protein-small molecule docking methods. *J. Comput. Aided Mol. Des.* 16, 151–166.


### 14.1 Functional Conservation


**Structure–function relationships**

Choi KH, Shi J, Hopkins CE et al. (2001) Snapshots of catalysis: the structure of fructose-1,6-(bis)phosphate aldolase covalently bound to the substrate dihydroxyacetone phosphate. *Biochemistry* 40, 13868–13875.

Liu Z, Sun C, Olejniczak ET et al. (2000) Structural basis for binding of Smac/DIABLO to the XIAP BIR3 domain. *Nature* 408, 1004–1008.

Lorentzen E, Pohl E, Zwart P et al. (2003) Crystal structure of an archaeal class I aldolase and the evolution of (βα)8 barrel proteins. *J. Biol. Chem.* 278, 47253–47260.

Lorentzen E, Siebers B, Hensel R & Pohl E (2004) Structure, function and evolution of the Archaeal class I fructose-1,6-bisphosphate aldolase. *Biochem. Soc. Trans.* 32, 259–263.

Nagano N, Orengo CA & Thornton JM (2002) One fold with many functions: the evolutionary relationships between TIM barrel families based on their sequences, structures and functions. *J. Mol. Biol.* 321, 741–765.

**Fold recognition programs**

Gibrat JF, Madej T & Bryant SH (1996) Surprising similarities in structure comparison. *Curr. Opin. Struct. Biol.* 6, 377–385.

Madej T, Gibrat JF & Bryant SH (1995) Threading a database of protein cores. *Proteins* 23, 356–369.

Shindyalov IN & Bourne PE (1998) Protein structure alignment by incremental combinatorial extension (CE) of the optimal path. *Protein Eng.* 11, 739–747.

Ye Y & Godzik A (2004) FATCAT: a web server for flexible structure comparison and structure similarity searching. *Nucleic Acids Res.* 32, W582–W585.

**Domain identification**

Sowdhamini R & Blundell TL (1995) An automatic method involving cluster analysis of secondary structures for the identification of domains in proteins. *Protein Sci.* 4, 506–520.

Sowdhamini R, Rufino SD & Blundell TL (1996) A database of globular protein structural domains: clustering of representative family members into similar folds. *Fold.Des.* 1, 209–220.

Vinayagam A, Shi J, Pugalenthi G et al. (2003) DDBASE2.0: Updated domain database with improved methods for the identification of structural domains. *Bioinformatics* 19, 1760–1764.

**Viewing programs**

Krieger E, Koraimann G & Vriend G (2002) Increasing the precision of comparative models with YASARA NOVA – a self-parameterizing force field. *Proteins* 47, 393–402.

Pettersen EF, Goddard TD, Huang CC et al. (2004) UCSF Chimera–a visualization system for exploratory research and analysis. *J. Comput. Chem.* 25, 1605–1612.


### 14.3 Finding Binding Sites


Bradford JR & Westhead DR (2005) Improved prediction of protein–protein binding sites using a support vector machines approach. *Bioinformatics* 21, 1487–1494.

Klon AE, Heroux A, Ross LJ et al. (2002) Atomic structures of human dihydrofolate reductase complexed with NADPH and two lipophilic antifolates at 1.09 a and 1.05 a resolution. *J. Mol. Biol.* 320, 677–693.

Laurie AT & Jackson RM (2005) Q-SiteFinder: an energy-based method for the prediction of protein-ligand binding sites. *Bioinformatics* 21, 1908–1916.

Neuvirth H, Raz R & Schreiber G (2004) ProMate: a structure based prediction program to identify the location of protein–protein binding sites. *J. Mol. Biol.* 338, 181–199.

**Evolutionary trace methods**

Glaser F, Morris RJ, Najmanovich RJ et al. (2006) A method for localizing ligand binding pockets in protein structures. *Proteins* 62, 479–488.

Landau M, Mayrose I, Rosenberg Y et al. (2005) ConSurf 2005: the projection of evolutionary conservation scores of residues on protein structures. *Nucleic Acids Res.* 33, W299–W302.

Lichtarge O, Bourne HR & Cohen FE (1996) Evolutionarily conserved Galphabetagamma binding surfaces support a model of the G protein-receptor complex. *Proc. Natl Acad. Sci. USA* 93, 7507–7511.

Lyons T Evolutionary trace analysis of dihydrofolate reductase (DHFR). A pharmaceutical case study. Accelrys (http://www.accelrys.com/reference/cases/studies/dhfr.pdf). San Diego.

Mihalek I, Res I & Lichtarge O (2006) Evolutionary trace report maker: a new type of service for comparative analysis of proteins. *Bioinformatics* 22, 1656–1657.

## 14.4 Docking Methods and Programs

Jones G, Willett P & Glen RC (1995) Molecular recognition of receptor sites using a genetic algorithm with a description of desolvation. *J. Mol. Biol.* 245, 43–53.

Kuntz ID, Meng EC & Shoichet BK. (1994) Structure-based molecular design. *Acc. Chem. Res.* 27, 117–123.

Laskowski RA, Thornton JM, Humblet C & Singh J (1996) X-SITE: use of empirically derived atomic packing preferences to identify favourable interaction regions in the binding sites of proteins. *J. Mol. Biol.* 259, 175–201.

Meng EC, Shoichet BK & Kuntz ID (1992) Automated docking with grid-based energy evaluation. *J. Comput. Chem.* 13, 505–524.

Shoichet BK, Bodian DL & Kuntz ID (1992) Molecular docking using shape descriptors. *J. Comput. Chem.* 13, 380–397.

Sousa SF, Fernandes PA & Ramos MJ (2006) Protein–ligand docking: current status and future challenges. *Proteins* 65, 15–26.

# PART 7

# CELLS AND ORGANISMS

Looking at how a whole cell or organism works and responds to outside stimuli is important in biomedical science in order to understand how cells, organs, and organisms function and what happens when things go wrong and disease sets in. Elegant experimental techniques have been developed to measure the expression of each gene (or protein) in a specific sample. However, these experiments generate large amounts of data and new computational methods have been developed to help with the analysis. The first two chapters in this part deal with the methods and the underlying principles of the most commonly used techniques for analyzing expression data. The first chapter has introductory and application material, while the second chapter contains detailed and mathematical explanations of the techniques and statistical analysis.

Knowing the component parts of the system and the expression levels is useful, but does not give us the whole picture. Ideally, it would be helpful to know what effect each of the parts has on the others. We need a dynamic representation of a system that can be manipulated. The methods and principles behind this type of study are introduced in the last chapter: the study of systems biology.

# PROTEOME AND GENE EXPRESSION ANALYSIS

15

**When you have read Chapter 15, you should be able to:**

- Explain gene expression analysis.
- Describe the clustering of large-scale data.
- Identify groups of co-expressed genes.
- Explain protein expression analysis.
- Summarize the identification of proteins in expression experiments.
- Explain how experiments can identify differences in expression.

Many techniques have become available recently that produce vast amounts of quantitative biological data. These techniques include, amongst others, RNA interference (RNAi), various gene expression techniques, and protein expression analysis. We will only use the common gene expression (microchips, microarrays) and protein expression (two-dimensional gels) methods as examples to look at the various ways one can analyze high-throughput data. In this chapter we will deal with the analysis of the large amount of data generated by these types of experiments. We will focus on the types of conclusion that can be reached and how they appear in the different analytical methods. The next chapter will describe in more detail the bases of these techniques.

The vast amount of data coming from the genome sequencing programs will be of limited use unless it can be linked to ways of measuring when, and at what level, all these genes are expressed. Measuring the expression of selected genes for particular purposes has been a key methodology in cell biology and developmental biology for years, but techniques have been developed more recently that enable the simultaneous measurement of the expression of thousands of genes, providing a snapshot of the total gene expression in a cell or tissue at a given moment. Comparison of gene expression patterns under different conditions—for example healthy and diseased tissue—should shed light on the function of different genes and how their expression varies in particular diseases. There are hopes that identification of specific gene expression patterns can be linked to disease diagnosis, prognosis, and treatment.

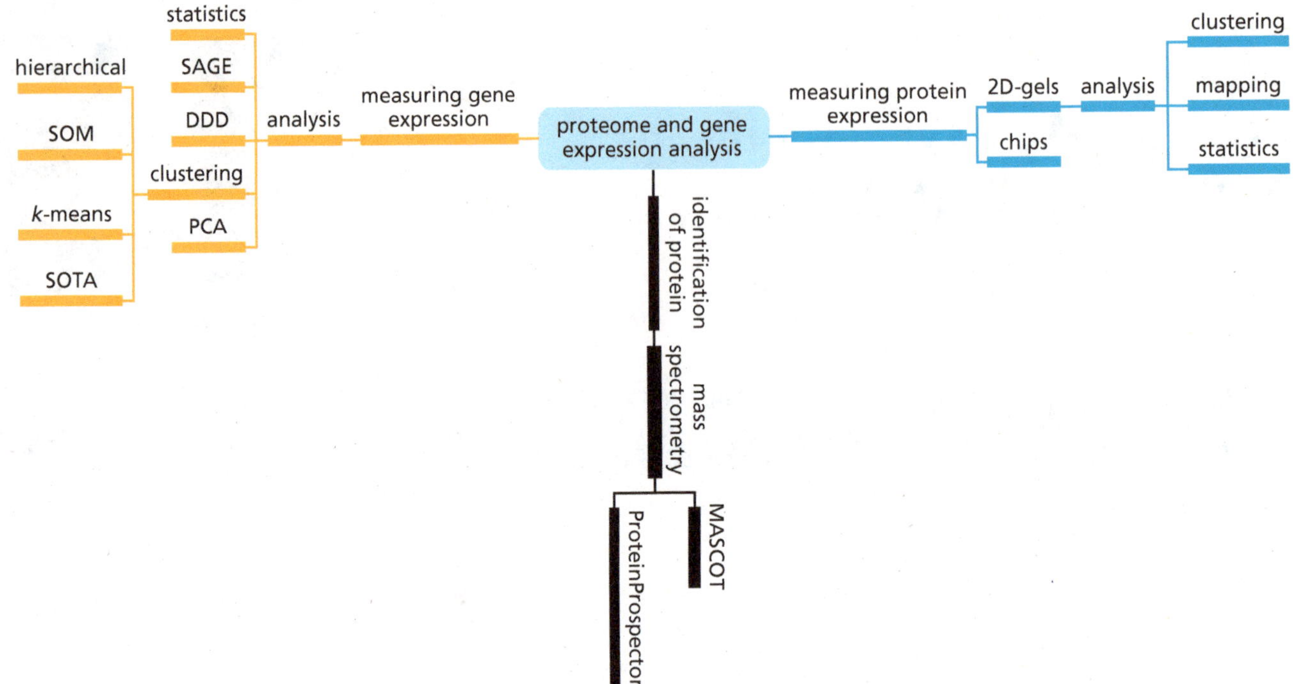


**Mind Map 15.1**
**A mind map representation of how to analyze expression data either from microarrays or protein expression experiments.**

Gene expression begins when genes are transcribed into messenger RNAs (mRNAs), which are then translated to produce proteins (see Section 1.2). Total gene expression in cultured cells or a tissue sample can be detected in two main ways. One is the detection and quantification of total messenger RNA—the **transcriptome**—by **DNA microarray technology**. The other detects the total protein composition of the sample—the **proteome**—by separating the protein products of the genes by **two-dimensional gel electrophoresis** or **chromatography**, followed by identification of their constituent peptides by **mass spectrometry**. In both these cases a single experiment produces enormous amounts of raw data, and new techniques had to be devised for data collection, storage, and analysis. The transcriptome and proteome, unlike the genome, are changeable in response to conditions, and depend on the state of development, the environment, and the type of tissue.

The success of large-scale functional genomics also depends on robust and efficient systems for both tracking and managing material and information flow. The data must be stored in a database and in addition it is necessary to have some type of Laboratory Information Management System (LIMS) that can track all samples and what happens to them from the moment they are submitted for analysis.

Monitoring the simultaneous expression of multiple genes provides information that cannot be obtained by monitoring the expression of one or a few genes at a time. By revealing which genes are expressed together, or **co-expressed**, for example, these techniques can identify genes that may be functionally related, such as the various members of a multiprotein complex or a metabolic or signaling pathway. This information can be used to help assign possible functions to unidentified genes with the same expression patterns. Co-expression can also help indicate which genes are under the control of the same regulatory system. An additional power of microarrays is that the mRNAs in two different samples can be compared in a single experiment.

Measuring mRNA, however, does not tell us the whole story of gene expression. To obtain a functional protein, mRNAs have to be translated, and the protein products often undergo a variety of permanent or temporary posttranslational modifications that influence their function. The simultaneous measurement and analysis of large numbers of different proteins is the province of proteomics, and some of the applications are listed in Table 15.1. There are many more proteins than there are genes

**Table 15.1**
**Applications of proteomics.**

| |
|---|
| • The measurement of protein composition and protein levels and their comparison in different types of tissue, such as normal and cancerous |
| • The analysis of differential protein expression in different cell types |
| • Analysis of changes in protein expression over time or after stimulation |
| • The analysis of changes in protein expression induced by drugs and other ligands |
| • The identification of the absence or presence of proteins in different cells |
| • Detection of posttranslational modifications of proteins such as glycosylation and phosphorylation |
| • The mapping of protein expression in tissues |
| • The identification of unknown proteins and their annotation |

in a genome: many transcripts can be spliced in various ways to give different mRNAs, and thus different protein products, from the same gene, and proteins can be modified after translation to give yet more variety (see Chapter 1).

Many hitherto unknown proteins and their isoforms are being identified by genomic and proteomic techniques and thousands more may yet be discovered. Large numbers of enormous datasets are being generated and the databases containing proteome information are growing in size and in the sophistication and complexity of the information they contain.

## 15.1 Analysis of Large-scale Gene Expression

High-throughput, whole-genome DNA microarrays have become a very useful tool in biological research. However, the interpretation of the large amount of data produced by a microarray experiment or a series of experiments can be time consuming. It is also difficult because different methods can yield alternative conclusions. The aim of these experiments is usually to extract some biological or functional meaning from the lists of genes, either by identifying critical genes that might be responsible for a biological effect, or by finding patterns within the genes that point to an underlying biological process and annotating each one of the genes.

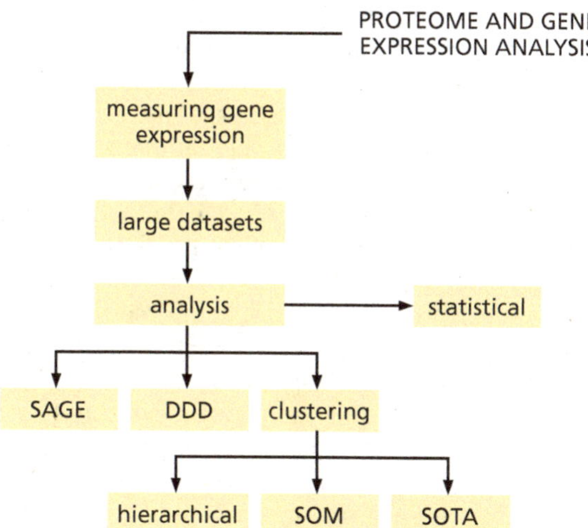


**Flow Diagram 15.1**
**In this section common experimental aspects of gene expression and of the analysis of the resulting data are described.**

## The expression of large numbers of different genes can be measured simultaneously by DNA microarrays

DNA microarrays and chips are composed of short fragments of DNA attached to a surface or synthesized directly on the surface, such as a glass microscope slide, in a predetermined arrangement, so that the sequence of the DNA fragment at any position is known. In the most basic form of a microarray experiment, the mRNAs in the sample to be tested are labeled with fluorescent tags and mixed with the array. RNAs in the sample that are complementary to fragments on the array will base-pair or **hybridize** with the fragments. Unbound sample is washed away, and the microarray is scanned with a fluorescence imager. RNAs that have bound their complementary array fragment are detected as fluorescent spots at specific positions, which give their identities, while the intensity of the fluorescence measures the level of the RNAs in the original sample. In practice, the sample mRNA is first converted to cDNA by reverse transcription, and is also labeled in this reaction, or the RNA is amplified by *in vitro* transcription and then labeled, and this labeled RNA is hybridized to the array. For small-scale DNA arrays where high sensitivity is required, sample RNA can also be directly labeled with a radioactive tag without amplification.

A DNA array can contain from tens or hundreds to hundreds of thousands of different sequences, depending on the purpose for which it is to be used. The DNA fragments that make up the array will either be cDNAs reverse transcribed from cellular mRNA, separated, and spotted onto the surface by contact printing or ink-jet technology, or will be oligonucleotides that have been synthesized *in situ* on the surface, as in the well-known Affymetrix GeneChip® arrays. The surface to which the DNA molecules are attached may be a glass slide (used in cDNA microarray), often with a chemical coating to which the fragments can be covalently attached, or a silica wafer (used in production of the Affymetrix GeneChip® arrays). The arrayed DNA fragments are usually referred to as the probes or the probeset and the sample RNAs (or cDNAs) as the targets. Microarray technology is not only used to study gene expression: with an appropriate probeset derived from genomic DNA, and genomic DNA as the sample, they are being used, for example, to characterize the single-nucleotide polymorphisms in human and animal populations. In this chapter we will limit our discussion to gene expression microarrays.

## Gene expression microarrays are mainly used to detect differences in gene expression in different conditions

Most gene expression microarray experiments are intended not simply to detect the genes being expressed at a given time, but to detect differences in gene expression under different conditions; for example, in healthy versus diseased tissue, in one tissue compared to another, at several different developmental stages, or after treatment with a drug or other agent. In one common method of using spotted cDNA microarrays to compare gene expression in two different conditions, two samples (for example, treated and untreated or reference) are labeled with two different fluorescent dyes (see Figure 15.1). The cDNA from one sample is labeled with Cy5 (a dye that is observed as a red color) and the cDNA from the other sample with Cy3 (a dye that is seen as green). The two fluorescently labeled samples are then mixed together and allowed to hybridize competitively to the probes on the array.

If we assume that the transcription level of a specific gene is estimated accurately in the amount of its mRNA in the sample, then the transcription level of a gene is proportional to the intensity of the fluorophore signal left on the complementary probe. The intensity of the signal from each spot is captured by a laser scanner using different wavelengths for each fluorophore and converted to an electronic image. In commercial software the images are overlaid and pseudo-colored, with red for Cy5 and green for Cy3, for visual comparison. The microarray image can thus be analyzed to extract the ratio of one labeled target to the same target labeled

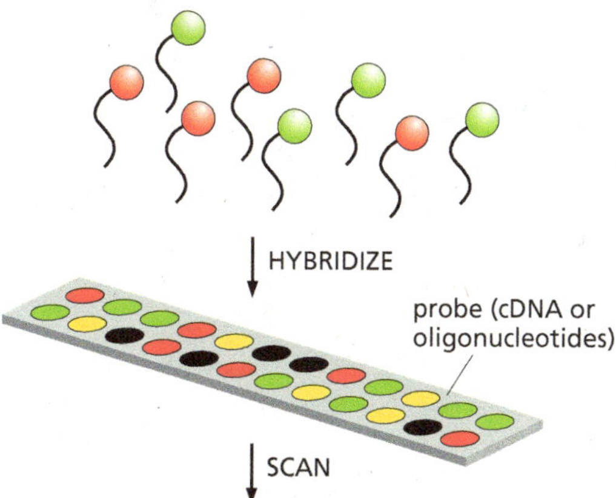
cDNA from sample A labeled with Cy5, + cDNA from sample B labeled with Cy3, gives rise to different colors on chip
HYBRIDIZE
probe (cDNA or oligonucleotides)
SCAN
relative proportion of each cDNA determined from level of fluorescent signal from each dye

**Figure 15.1**
**The principle of a two-color DNA microarray experiment.** The microarray itself consists of a glass slide or other solid substrate carrying a regular array of spots of cDNAs or oligonucleotides of known sequence (the probes) in predetermined positions. In a typical experiment to detect differences in gene expression in two different conditions, the total cellular mRNA from the two samples (A and B) is converted to cDNA and one sample is labeled with the fluorescent dye Cy3 and the other with Cy5, which fluoresces at a different wavelength. For example, Cy3 can be represented by the green spot while Cy5 is represented by a red spot. Cy3- and Cy5-labeled cDNA is mixed and hybridized to the microarray. After washing away unbound cDNA, the microarray is scanned with a laser scanner to detect the probes that have become fluorescently labeled. The resulting image of fluorescent spots is analyzed by specialized software that calculates the relative proportions of A and B cDNA in the bound cDNA at each spot. These data are then further analyzed to determine, for example, which genes are expressed more strongly in sample A and which in sample B.

with a different fluorophore. If a target spot is red then the Cy5-labeled sample is expressing more of that gene than the Cy3-labeled sample and the strength of color will reflect the extent of the difference (see Figure 15.2). The reverse is true for a green spot. If the spot is yellow then the amount of expression is equivalent in both samples. The pixel intensity of each spot is taken and subtracted from the background of the image before saving it for further data analysis. Because the spots are set out in a specific microarray grid the identities of the genes in the spots are known and can therefore easily be identified. Clustering the data into clusters of gene sets or sample sets with distinct and similar gene-expression patterns is the next step in analyzing these types of data.

There are two basic approaches in microarray technology: a one-color technique, where a single sample is hybridized to each microarray after it has been labeled with a single fluorophore; and the two-color procedure, where two samples are labeled with different fluorophores and hybridized together on a single microarray, as described above. Both approaches have advantages and disadvantages. When using the two-color approach, the hybridization of both samples to the same microarray makes a direct comparison possible. This leads to a reduction of the data variability, therefore improving the accuracy when analyzing differential expression between sample pairs. The main advantage of the one-color approach is simplicity in experimental design. Hybridization of a single sample on a single microarray enables comparisons across many microarrays. Data variability arising from using this technique can be reduced by performing replicate runs.

However, biological variation and technical difficulties in probe binding and the measurement of fluorescence intensities make the process of distinguishing signals from noise in the microarray data problematic. It is therefore important to run replicate experiments and to perform statistical analysis (see Chapter 16) to estimate significance of differential expression in the microarray data prior to detailed cluster analysis. Programs that enable the user to remove sources of error

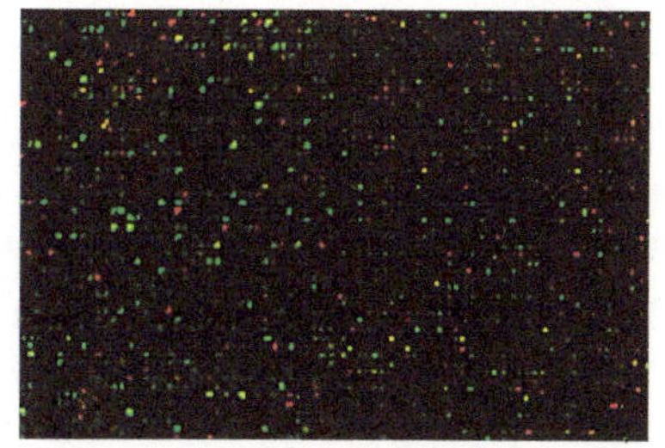

**Figure 15.2**
**A typical raw image of a scanned microarry.** This is the type of image obtained in an experiment like that described in Figure 15.1. The red spots represent Cy5-labeled cDNAs (sample A) and the green spots Cy3-labeled cDNAs (sample B). A red spot means that the gene corresponding to the probe at that position is overexpressed in sample A while a green spot indicates overexpression in sample B. A yellow spot indicates that the gene is expressed at the same level in both samples.

include BioConductor and various software packages listed on the Stanford Microarray Database pages.

## Serial analysis of gene expression (SAGE) is also used to study global patterns of gene expression

One alternative to microarrays for investigating patterns of gene expression is a technique known as **serial analysis of gene expression (SAGE)**. It has both an experimental and a bioinformatics component. It is based on the following observations:

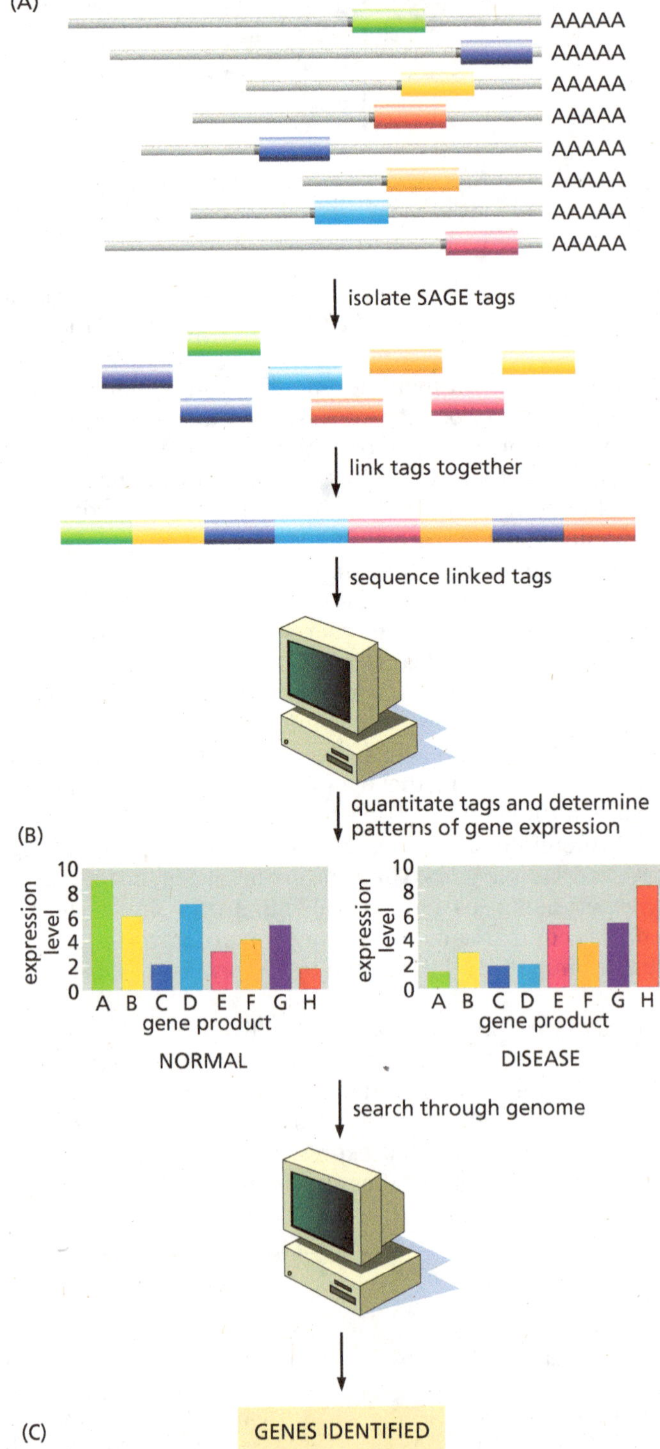


**Figure 15.3**
**An outline of the SAGE method for comparing levels of gene expression.** (A) Short sequence tags (10–14 bp) are obtained from a unique position within each transcript. The sequence tags are isolated and are linked together to produce long DNA molecules that can be cloned and sequenced. (B) Once sequenced, the abundance of each tag can be calculated, and this value is converted to a value that gives the expression level of the corresponding transcript. For example, there is less of transcript A (green bar) in the diseased state than in the normal state, while transcript H is more abundant (red bar) in the diseased than in the normal state. (C) The tags are then used to search the appropriate genome to identify the corresponding genes.

first, that a short sequence (a tag) contains enough information to uniquely identify a gene (provided that the tag is obtained from a unique position within each gene); second, that the sequence tags from the total cellular RNA (converted into cDNA) can be linked together to form long DNA molecules, called concatemers. This DNA sequence is read and counted. The total number of times a particular tag is observed in the concatemers approximates the expression level of the corresponding gene (see Figure 15.3). The data produced by SAGE include a list of the tags with their corresponding counts, providing a digital output of cellular gene expression that can easily be analyzed further. The SAGE analysis programs SAGEmap and xProfiler are available on the NCBI website and allow the user to specify which organ is to be investigated. Libraries consisting of gene lists organized by the various types of tissues or cell lines are provided for further choice. The expression associated with these gene lists can be divided into two groups and compared with each other. The output from SAGE provides the SAGE tag, the UniGene ID (Identification number), the gene description, and color- and letter-coded differences in expression levels.

## Digital differential display uses bioinformatics and statistics to detect differential gene expression in different tissues

Another alternative to microarrays for looking at differential gene expression in some circumstances is purely computational. Digital differential display (DDD) is a method for comparing EST-based expression profiles in different tissues or conditions from various libraries or between pools of EST libraries. An EST library contains short sequences cloned from the total cellular mRNA (converted to cDNA) of a particular tissue or particular condition. The theory is that genes expressed at a high level will be represented by more ESTs than those expressed at a lower level. Genes whose expression levels differ significantly from one set of EST libraries to the next are identified using a statistical test.

The NCBI's UniGene database forms the core of the DDD method. In UniGene, all the human EST sequences in the databases have been put into distinct clusters, where each cluster represents a single gene. The DDD methods then compare the number of sequences from each EST library assigned to a particular UniGene cluster, and identifies those differences between the clusters that are likely to be biologically significant. The user can select EST libraries from a list on the DDD Web page and may combine selected libraries into specific pools. Figure 15.4 shows a DDD analysis of two selected pools. Each of the three columns on the left represents a particular pool, and the rows represent UniGene clusters. On the right is the gene description, which gives the name of the cluster, and the UniGene ID number for that cluster. Clicking on this ID provides a summary report for that cluster.

| normal | | cancer | | |
|---|---|---|---|---|
| 0.00000 | 0.05528 | 0.00000 | Hs.85844 | neurotrophic tyrosine kinase receptor, type 1 (NTRK1) |
| A<B | B>A B>C | C<B | | |
| 0.04628 | 0.00000 | 0.00000 | Hs.9408 | paired immunoglobulin-like receptor beta [PILR(BETA)] |
| A>B A>C | B<A | C<A | | |
| 0.04488 | 0.05528 | 0.00000 | Hs.74561 | alpha-2-macroglobulin (A2M) |
| A>B A>C | B<A | C<A | | |
| 0.00000 | 0.00000 | 0.04353 | Hs.75290 | ADP-ribosylation factor 4 (ARF4) |
| A<C | B<C | C>A C>B | | |

**Figure 15.4**
**The results returned by the DDD program.** The results are from a DDD calculation on two pools of normal prostate ESTs (A, represented by the column on the left and B, represented by the column on the right) and a pool of tumor prostate ESTs (C). Each row gives you the number of sequences in each pool that mapped to the UniGene cluster represented by the gene named on the far right together with its UniGene ID (in blue). The value above each circle is the fraction of sequences within that pool that mapped to the cluster shown, and the size and shade of the circle is a visual reflection of that value. Beneath each circle, the relationship between the expression levels for that gene in different pools is summarized. For example, for the spot in the second row, pool A contains a greater number of sequences than pools B or C (A > B, A > C) for the paired immunoglobulin-like receptor beta.

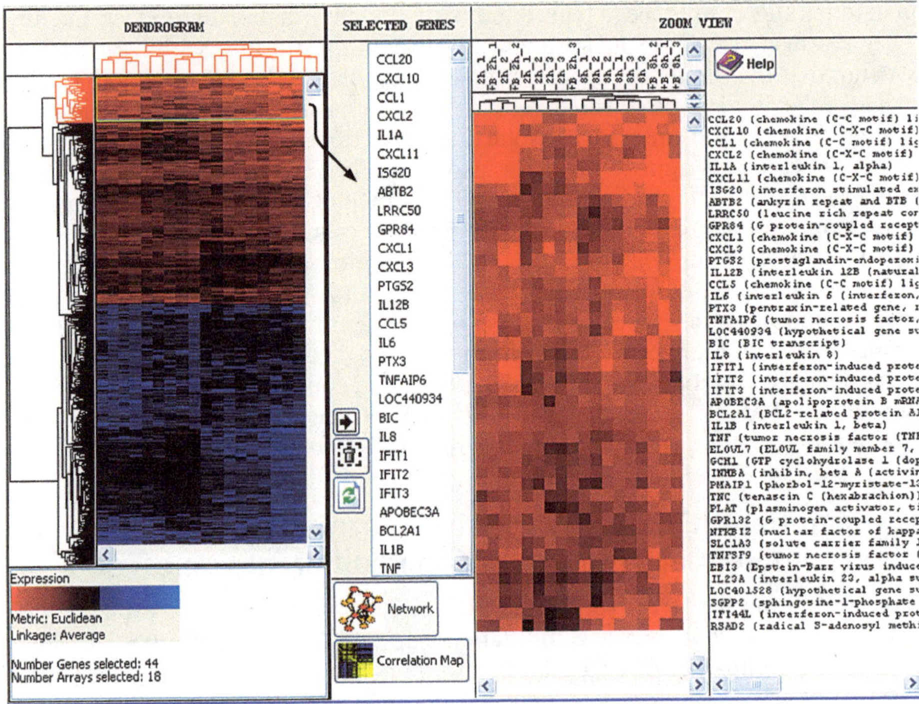


**Figure 15.5**
**Hierarchical clustering.** A dendrogram with a color-coded map of hierarchical clustering performed both on the samples and the genes. Each column of the data corresponds to a different sample, and each row to a different gene. The red indicates upregulation of genes while blue indicates downregulation of genes. A subset of genes selected for closer investigation is shown on the right.

Figure 15.4 shows an example output from DDD run on two pools of ESTs from normal prostate tissue and a pool of ESTs from prostate tumors. Of the four clusters displayed here, only the gene for ADP-ribosylation factor 4 shows a significantly greater representation in the cancer pool compared to both normal tissues. The other clusters show that there are, in fact, significant differences in gene expression between the two normal pools. This selection of a number of similar pools to compare is good practice in order to avoid false-positive results, both in the digital world and the real world of gene (or protein) expression.

## Facilitating the integration of data from different places and experiments

In general it is difficult to compare and integrate data from different laboratories and experiments. Therefore a group of scientists have set up a consortium (the MGED society) to standardize the output and annotation of microarray data, which will facilitate sharing the data and creating a consolidated database. The set of standardization rules is called MIAME, which stands for Minimum Information About a Microarray Experiment. This includes information that is essential for someone else to interpret the results of the experiment and even to reproduce the experiment. The journal *Nature* and other *Nature* research journals will in general only accept articles dealing with microarray data that comply with MIAME. The public repositories ArrayExpress at the EBI and GEO at NCBI have been set up to store and distribute MIAME-compliant microarray data. In addition a MicroArray Quality Control (MAQC) project is under way to assess the quality of DNA microarray data. This project has recently concluded that with careful experimental design, data transformation, and analysis, microarray data can be compared between different formats and laboratories.

## The simplest method of analyzing gene expression microarray data is hierarchical cluster analysis

The main aim of gene expression analysis is the identification of common patterns of gene expression; for example, which genes are being co-expressed, and which

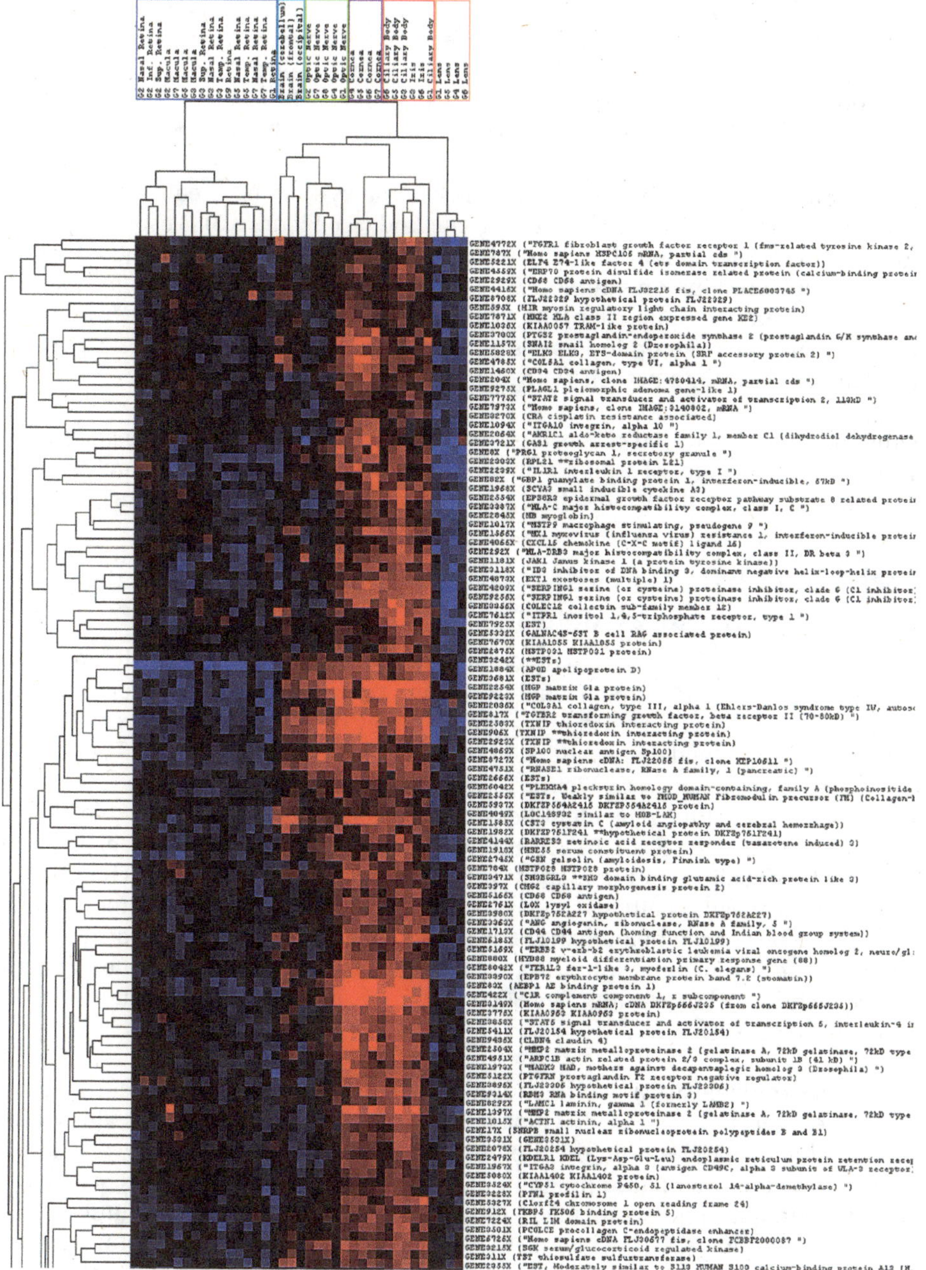

**Figure 15.6**
**Clustering of the eye parts.** Hierarchical clustering of the experiment described by Diehn and colleagues where the various parts of the eye were investigated to identify specific gene signatures. All the parts of the eye cluster together and the differences between, for example, genes in the cornea with respect to the lens are obvious. (Data from J.J. Diehn et al., Differential gene expression in anatomical compartments of the human eye. *Genome Biol.* 6:R74, 2005.)

genes have been downregulated or upregulated in one sample compared to the other. **Hierarchical clustering** is the most widely used method for analyzing patterns of gene expression in microarray data.

The results of a hierarchical cluster analysis of gene expression microarray data are typically displayed as a dendrogram with a color-coded grid. The grid of colored squares contains rows that represent a gene for which expression was detected in one or other of the samples, and the columns represent the different samples or conditions (see Figure 15.5). The precise shade of each square represents the fold increase or decrease of the expression level of the gene in that sample in relation to some reference level, often the median level of expression of the gene over all the samples. Typically, higher expression is shown by red, lower expression by green or blue, and expression similar to the reference as black. In the display, it is easy to see how the rows of genes are arranged into distinct blocks composed of genes with similar expression patterns in the same tissue. In the experiment depicted in

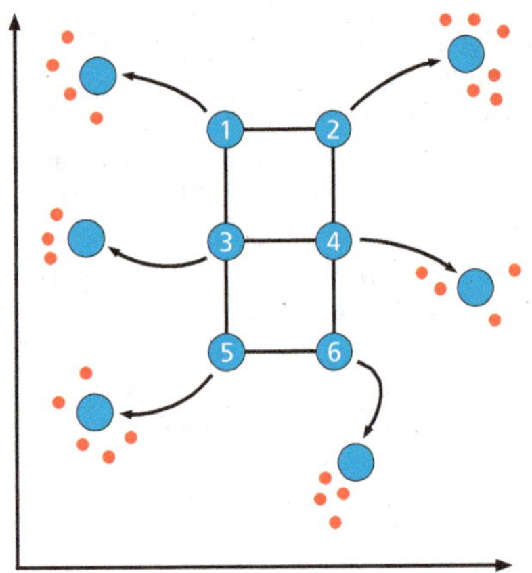


**Figure 15.7**
**The principles of a self-organizing map (SOM).** A random initial arrangement in two-dimensional space of a predetermined number of nodes (potential clusters) is made, in this case a rectangular 3 × 2 grid, with the nodes represented by the blue circles. Data points (red dots) are successively placed on the map and each time the nodes are allowed to move in the direction of the data points based on similarity. This procedure is repeated many times until each node describes a cluster of data points.

Figures 15.5 and 15.6, the aim was to detect sets of genes whose expression provides a unique signature for each eye compartment—lens, retina, cornea, and so on—with the eventual aim of using these signatures to identify candidate genes for genetic diseases that affect the eye. In this example, it is clear that the pattern of gene expression in the cornea, say, is very different from that in the lens.

We have already seen how hierarchical cluster analysis can be used to build phylogenetic trees in Section 8.2. The UPGMA method (see Figure 8.3) is especially close to those described below. In microarray or any other type of expression studies, either genes/proteins or samples, or both can similarly be arranged into a tree structure where the branch lengths represent the degree of similarity between the clusters. The dendrogram in Figures 15.5 and 15.6 shows the results of clustering the samples according to their gene expression patterns.

Other more sophisticated clustering methods have been applied to the analysis of gene-expression data, of which ***k*-means clustering** and techniques such as self-organizing maps (SOMs) and **self-organizing tree algorithms (SOTA)** are just a few.

## Techniques based on self-organizing maps can be used for analyzing microarray data

SOMs are a form of unsupervised neural network that consist of a fixed number of nodes, often in a two-dimensional grid, each node representing a gene cluster. Unlike the networks described in Section 12.4 they do not consist of layers. During the training step the data are added to the map one by one, and used to adjust the positions of the nodes (see Figure 15.7). After all the expression data have been processed the resulting map identifies which genes are associated with which node, defining the gene clusters. The clustering is according to the pattern of expression that can be visualized in the map. A disadvantage of the basic SOM technique (as well as the *k*-means method) is that the number of clusters to accommodate the data has to be chosen beforehand. The advantage of SOM techniques over hierarchical clustering techniques is that SOMs clearly define distinct gene clusters, unlike hierarchical clustering. The SOM and *k*-means techniques are described in more detail in Section 16.3.

One freely available SOM package for analyzing gene expression microarray data, called GeneCluster2, can be downloaded from the developers' Web site along with data to illustrate and try out the method. Other free SOM packages are also available such as the JAVA-based program Cladist. Figure 15.8 illustrates the SOM clustering technique applied to microarray data from macrophages stimulated by lipopolysaccharide (LPS), which acts in the cell like a Toll-like receptor agonist. Macrophages are white blood cells that are involved in mounting a defense in animals against foreign bodies. The Toll-like receptor is a membrane-bound protein that recognizes pathogens and starts the cell's response to activate the immune system. In this study the expression levels of various genes were measured at different time points after stimulation with LPS. This type of clustering illustrates the patterns of gene expression over either time (as in this case) or different types of stimulations. A 3 × 3 grid is used in Figure 15.8A (giving a maximum of nine clusters) and a 2 × 2 grid in Figure 15.8B (giving a maximum of four clusters). The number of genes comprising the cluster is given for each. For example, Node 0 in Figure 15.8A has 13 genes. Clicking on a particular cluster gives the list of the genes and their descriptions. These can be sent for further analysis to a protein–protein interaction map program as described below.

The structure of patterns of gene expression can be explored by varying the geometry of the SOM; that is, organizing it around grids with different numbers of nodes, such as 3 × 3 versus 6 × 4. The appropriate number of preset clusters to use will vary

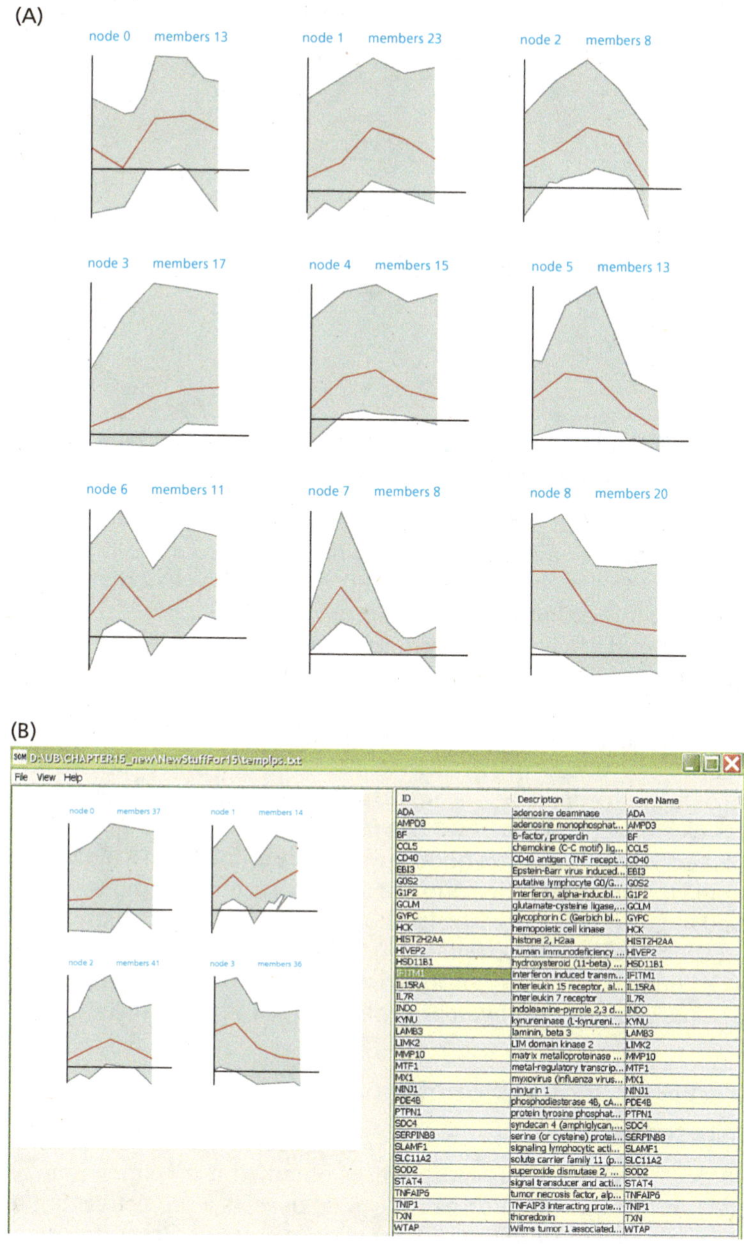


**Figure 15.8**
**An example output from a SOM module of the program Cladist.** This uses the SOM method to cluster the expression patterns of different genes. (A) In this experiment there were five samples corresponding to five different time points after stimulation of macrophages with lipopolysaccharide. The clusters are derived from the data using a 3 × 3 grid (see text), which sets the number of clusters to nine. Each box defines a cluster and contains the mean expression profile (across the samples) of a set of genes in red with the range of actual expression profiles of each gene in the cluster in gray. The clusters are numbered node 0, node 1, and so on. Each cluster also shows the number of genes it includes: node 0 has 13 genes, for example. (B) When a smaller grid (2 × 2) is used, the genes are forced into fewer clusters, which may either lead to indistinct clusters or combine similar clusters into one. The text window gives the names and accession numbers of all the genes in a selected cluster (node 3).

with different experiments. In this case, the nine-cluster-nodes SOM has too many nodes as some of the patterns are repeated; for example, nodes 1, 2, 4, and 5 all show a very similar expression pattern. The 2 × 2 geometry gives four distinct clusters of gene expression patterns. As nodes are added, distinctive and tight clusters emerge. No new nodes should be added when an increase in the grid size does not produce any significant new patterns. To analyze the clusters further, it is necessary to look at the individual gene elements assigned to a particular cluster.

Problems are often encountered, however, when using simple hierarchical clustering methods to analyze data. One of these problems is that it is difficult to decide when a group of genes forms a cluster significantly distinct from all other genes in the tree. The technique does not provide any guidance on the length a branch has to be in order to separate significantly different clusters. This is in contrast to the SOTA method.

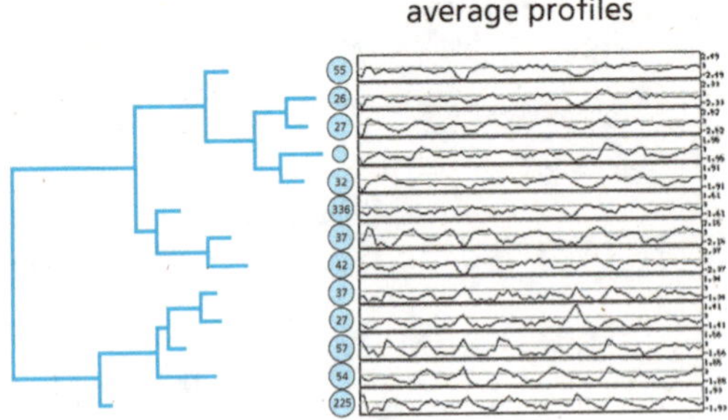


**Figure 15.9**
**An example of a SOTA output.** In this example, the expression profiles of 800 genes have been reduced to 13 clusters. The average expression profile for each cluster is plotted alongside. Unlike SOMs, the relationship between the clusters can be visualized in a hierarchical tree representation.

## Self-organizing tree algorithms (SOTAs) cluster from the top down by successive subdivision of clusters

A clustering method that uses both hierarchical and neural network techniques is the self-organizing tree algorithm (SOTA). It clusters from the top down (that is, starting with all the data in a single cluster) by successive subdivision of clusters, in contrast to the bottom-up clustering of simple hierarchical clustering. The highest hierarchical levels are resolved first and then the details of the lower level are examined.

In the SOM method the number of clusters is predetermined, which makes SOMs a somewhat subjective exercise. In addition, the lack of a tree structure prevents the detection of higher-order relationships; that is, the relationship between SOM clusters is less clear. The SOTA method combines the advantages of both hierarchical clustering and SOMs. A series of nodes are initially arranged in a binary tree and are then adapted to the characteristics of the dataset. The output nodes—the clusters—are allowed to grow until either the level of variability in all the branches is below a given threshold, or until a specified number of clusters is reached. Figure 15.9 shows an example of a SOTA output where 800 genes have been divided into 13 clusters. For each cluster the average gene expression profile is shown.

## Clustered gene expression data can be used as a tool for further research

Clustered data on gene expression patterns obtained from either gene expression microarrays or genome bioinformatics can be used as a predictive tool to identify new transcription factors or other cell-regulatory proteins. Regulatory elements are identified using both gene expression patterns and the clustering of genes according to function, based on functional annotations obtained experimentally or from sequence homology. The clustered genes (or proteins) can be analyzed with respect to protein–protein interaction data to see if the genes can form a functionally related pathway. For example, Figure 15.10 illustrates how a cluster identified by SOM clustering (node 0, see list in Figure 15.8B) has been subjected to the pSTIING database to obtain an interaction map. We can see that many of the genes submitted (in red) are connected either directly or through another interaction partner. Therefore it is likely these genes are part of a specific functional pathway. Some of the genes form individual clusters. They may be connected via a gene product (protein) that has not been identified as significantly changed, or may not be on the actual chip. The map can be extended to see if other connections can be found (see Figure 15.10B). In such ways a more complete pathway or interaction map can be generated. These interaction maps also form the starting point for system biology modeling, as will be described in Chapter 17.

A vast collection of data from many gene expression and protein expression experiments is now freely available on the Web and can be mined for biological reanalysis or used as reference data for the development of new bioinformatics tools. For example, the L2L tool is a repository of microarray data that users can search with their own up- or downregulated gene list to see if anyone else has got a similar gene expression pattern.

There are now a number of databases both for microarray experiments, such as the whole genome of the yeast *S. cerevisiae*, or general experiments. The data repositories can be searched with different criteria, such as using ORF name or gene name. Figure 15.11 shows an example result of a search using the ORF name YDL037c. It returns the results of all experiments that include this gene on a chip. Other prominent repositories are the Stanford Genomic Resources, or ArrayExpress at EBI.

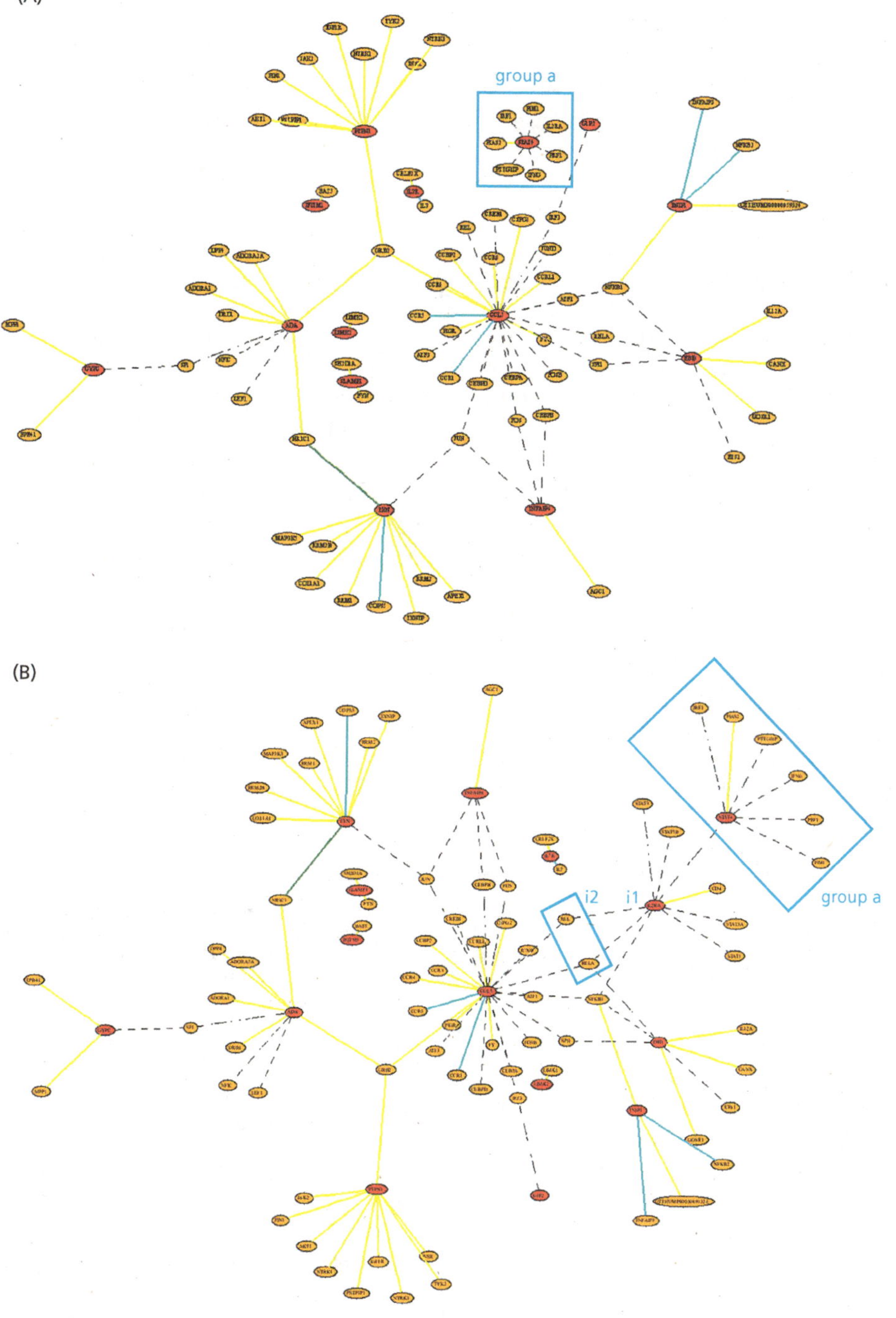


**Figure 15.10**
**An example of protein networks from clustering.** Once a cluster of interesting genes (or proteins) has been obtained the members of that cluster can be submitted to a protein–protein (gene–gene) interaction map finder, such as pSTIING. (A) This shows the genes submitted from the node 0 cluster from Figure 15.8B (red circles). Those genes that interact either directly (solid line) or by transcriptional activation (dotted line) are considered to be functionally associated and may be part of a specific pathway. Sometimes, the genes will interact but through further intermediary units, and the map has to be extended. (B) This illustrates how two unconnected groups can be connected by including further interactions. Group a now joins the rest of the network via an intermediary i1 which connects group a and the rest of the network through other proteins (i2).

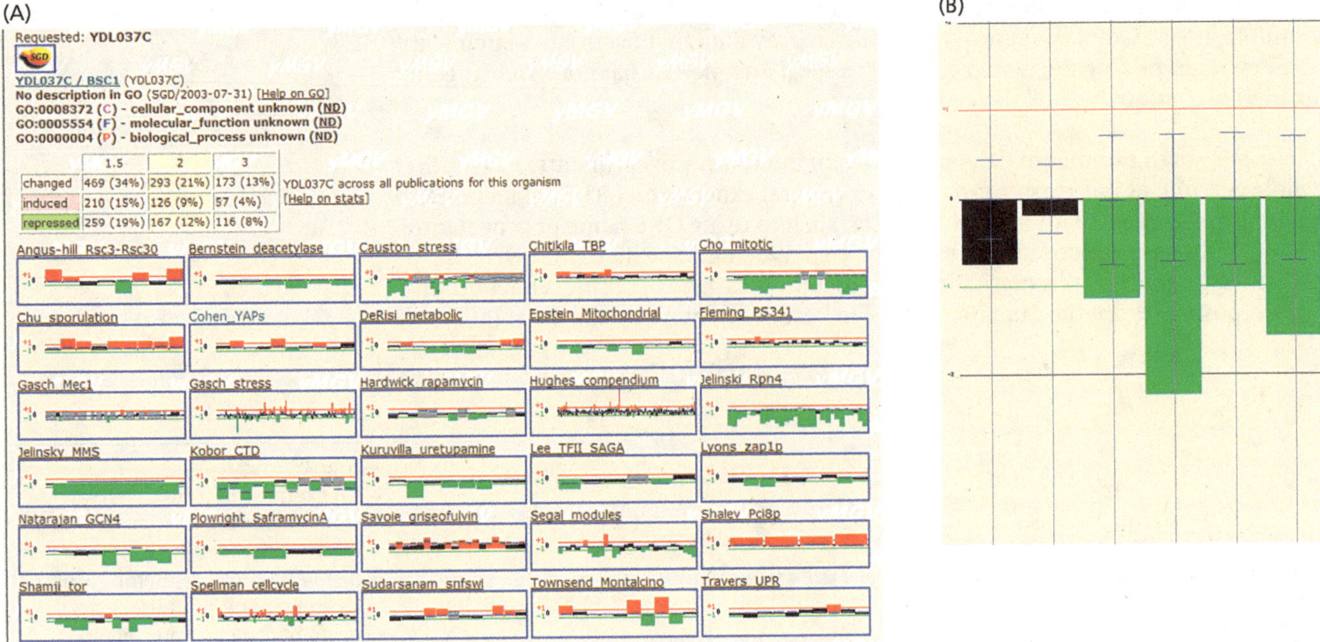


**Figure 15.11**
**Output from the yeast yMGV database of gene-expression microarray data.** (A) These data can be searched using an ORF name or gene name and the search will return a graphical representation of their expression within each experiment. (B) Clicking on any of the boxes will give a more detailed view of the expression profile of that gene in a particular experiment. Clicking on the title of each experiment will provide further details about the study itself.

## 15.2 Analysis of Large-scale Protein Expression

The proteome refers to all the proteins that make up an organism or, on a smaller scale, the total number of proteins found in a particular cell type at a specific point in time and under specific conditions. An organism will have different protein expression in various parts of its body. The protein expression will also differ between

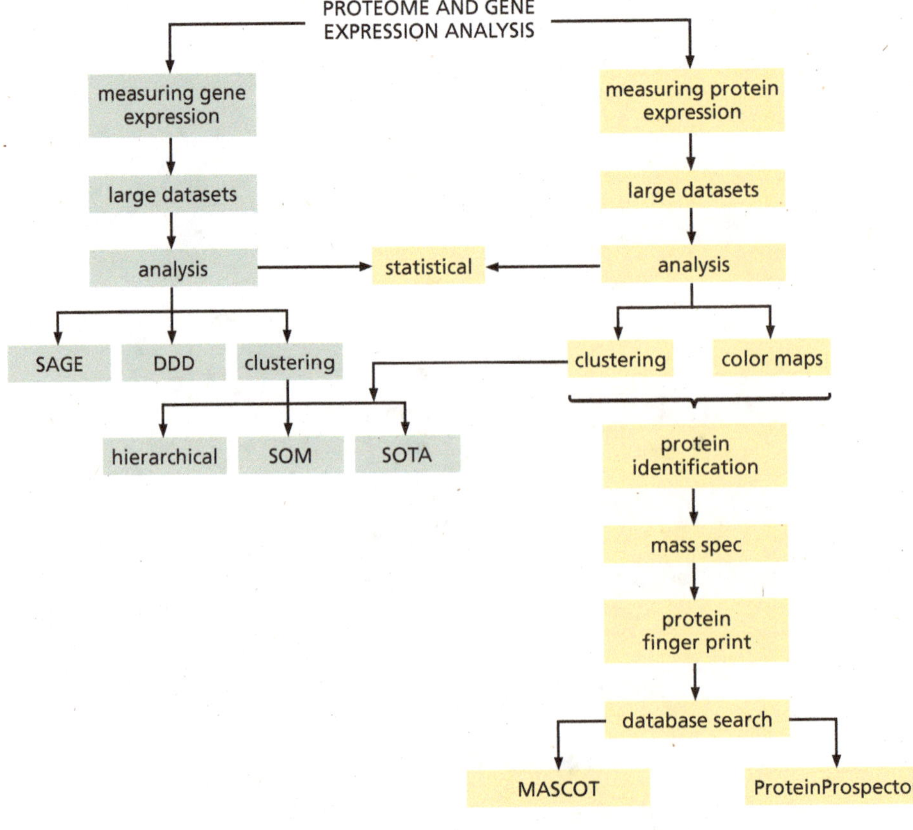


**Flow Diagram 15.2**
**This section describes some experimental aspects of protein expression and of the analysis of the resulting data, showing the overlap between gene and protein expression.**

the separate stages of an organism's life cycle and under different environmental conditions. To understand how an organism or a cell functions both under normal and abnormal (such as disease) conditions it is important to know how protein expression is affected.

Using 2D gels and other techniques many proteins and their isoforms have been identified and thousands more remain to be discovered. Numerous sets of data have been created and more and more databases containing proteome information are being populated. Such datasets need to be integrated for proper data mining and analysis, and for comparison to data such as metabolic and signaling pathways and protein interactions.

## Two-dimensional gel electrophoresis is a method for separating the individual proteins in a cell

Proteomics experiments that aim to characterize the proteome of a cell type or tissue at a particular point in time commonly use the technique of 2D gel electrophoresis, either on its own or followed by mass spectrometry (MS) of the separated proteins in order to identify them. Two-dimensional gel electrophoresis separates proteins in two dimensions according to two independent properties. In the first step, isoelectric focusing (IEF) of the protein mixture separates proteins on the basis of their isoelectric points (or pI/pH). Electrophoresis [in the presence of the detergent sodium dodecyl sulfate (SDS)] of the separated proteins in a direction along the gel at right angles to the IEF then resolves each spot into its constituent proteins on the basis of their molecular weights (see Figure 15.12). A 2D gel thus provides a kind of map (with pH on one axis and molecular weight on the other) on which, in theory, any given protein will always occur as a spot at the same location,

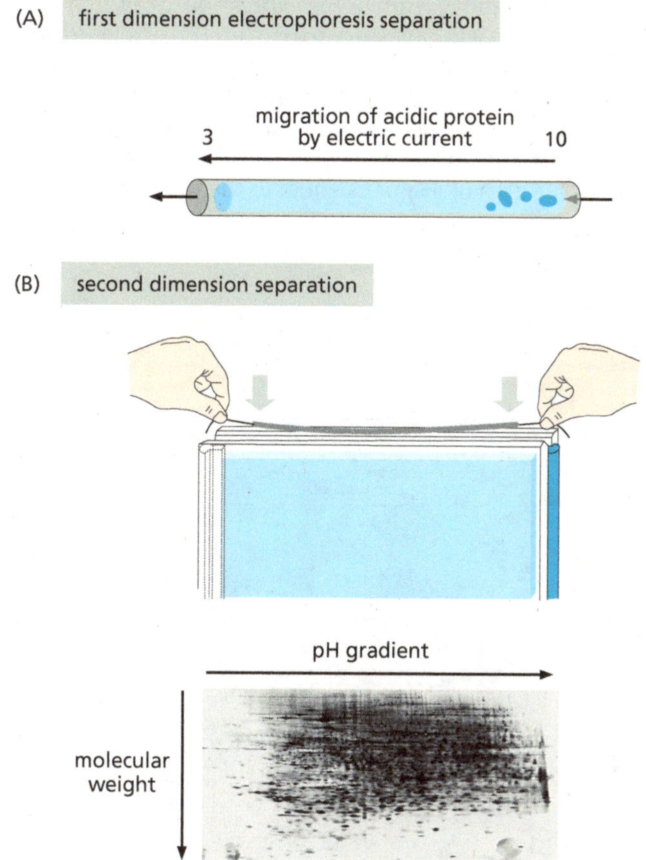


**Figure 15.12**
**Schematic of a 2D gel run.** The proteins are (A) first separated according to their pI, where the proteins migrate due to an electric current being applied to the sample in a tubular gel. The acidic proteins will migrate as shown. (B) Then the tubular gel is loaded onto another gel and the proteins are separated according to their molecular weight in the second dimension.

whatever the sample being analyzed. Once the gel has been run and the proteins separated, it is stained and scanned into an image-analysis program to detect and identify the spots.

The problem with this technique is that on even the best gel only about 3000 spots are visible. There are many more different proteins than that in a cell: some estimates are as high as 100,000. The limitations of 2D gels are due to a number of factors, one being that the gel is not large enough to separate proteins that have very similar molecular weights or pH range. In addition, membrane proteins often become insoluble during the first-dimension run and low-abundance proteins are often not detectable.

## Measuring the expression levels shown in 2D gels

Once 2D gels have been run they have to be scanned to obtain images that can be analyzed on a computer. A number of processes have to be performed on the images before the results can be studied. This is usually done by a combined 2D gel spot-detection and image-analysis program. There are quite a number of commercial programs available; in the examples in this chapter the program Melanie was used as well as the freely available viewing program Flicker.

The first step is detection of spots. This is especially important if dyes invisible to the eye have been used to stain the separated proteins. There are many algorithms for detecting protein spots, and all have variable success rates. Such algorithms work by detecting the edges of spots, or by detecting the center of the spot and working outward until a lighter edge is reached. The problems with most spot-detection algorithms are false detection of stained areas that are not protein spots (such as dust particles, and so on), and an inability to resolve spots that lie close together on a set of similar gels (see Figure 15.13).

To enable a quantitative comparison of the expression levels of the same protein on a number of different gels, it is necessary to convert the pixel intensity of the spot to some meaningful value, such as the protein volume. This is done by quantification algorithms. Quantification is based on the measurement of spot intensity (pixel intensity inside the detected shape). This intensity is then used to estimate the amount of protein, as judged, for example, by the area of the spot. For example, Melanie calculates the area of a spot as equal to the number of pixels × pixel area.

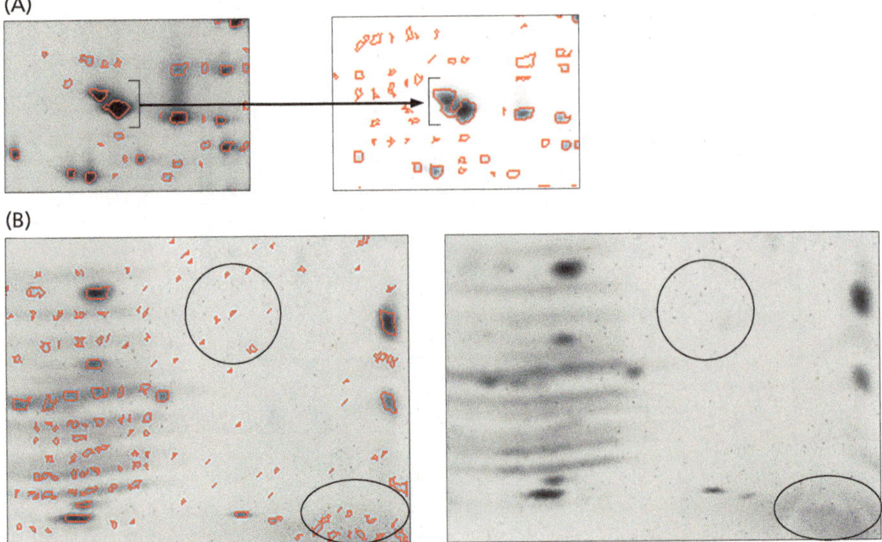


**Figure 15.13**
**Problems encountered in spot detection on 2D gels.** (A) A mismatch between two gels because a pair of spots on one gel (left) have been correctly detected as two spots (indicated by the black bracket) while on a second gel (right) the same spots have been detected as a single continuous feature. (B) Dust particles and a smudge in the gel (right panel, ringed areas) can be mistakenly identified as protein spots (left panel, ringed areas).

Protein spots can be detected using different methods, such as silver staining, fluorescence, or radioactivity, and the different methods will give different intensity readings. Thus two gels using different staining methods are difficult to compare. To some extent, such problems can be circumvented by measuring relative values, such as volume ratios (spot volume divided by total spot volumes) or relative optical density.

If protein spots on two or more gels are to be compared, the gels have to be matched so that the corresponding spots can be identified. One of the main applications of 2D gels is to compare protein expression in different samples by comparing the relative intensities of the corresponding spots on the gels run from each sample. This requires correct matching, or registration, of the corresponding spots. Matching is one of the most important tasks of a gel image-analysis program, and one of the most difficult. Different programs match the spots in different ways: some align the gels first, and use a set of easily identifiable spots as landmarks to match the rest. Melanie first aligns the gel images, and as the images are often not exactly identical in size or shape, as a result of distortion of the gel, the program has to warp the images to fit. Spots in all four corners of one gel are first selected as landmarks and the corresponding spots are identified on the gels to be aligned. Melanie then uses these fixed points to align the gels by least-squares minimization and image transformation.

Once the gels are aligned the spots can be matched. Melanie uses the approach of associating each spot on the gel—called a center spot—with a surrounding cluster of spots within a fixed radius of the center spot. The radius chosen depends on the size of the image, the number of spots in the image, and a minimum number of spots that have to be in the cluster. Matching then compares these clusters. Mismatching of spots occurs to varying degrees, depending on the complexity of the sample being analyzed.

## Differences in protein expression levels between different samples can be detected by 2D gels

The aim of the 2D gel experiments is to calculate different expression levels of proteins in different samples. One way of avoiding the problem of matching different gels as described above uses a strategy similar to that of the two-color microarrays described earlier. Two different dyes are used to tag proteins in two different samples and the two samples are run on the same gel (see Figure 15.14). The gels are then analyzed using spot-detection and image-analysis instruments and software with the capacity to deconvolute the signals from the differently tagged proteins within each spot. This technique circumvents the problem of mismatching, but only two or three samples can be run on one gel.

Probably the simplest way to initially analyze the differences between spot volumes in two gels or two groups of gels is a **scatterplot** (see Figure 15.15). This plots the paired values and places a regression line through the plotted points. The linear dependence of one set of data to the other set of matched data estimates the relationship between the values and the regression line—the best-fitting straight line through the paired values—gives the linear dependence. A correlation coefficient gives the goodness-of-fit of the line. So when looking for pairs of spots that differ, one looks for spots that lie away from the regression line and whose correlation coefficient tends to zero.

## Clustering methods are used to identify protein spots with similar expression patterns

As with gene expression data, clustering is a useful method of grouping similar gels or spots together and extracting protein expression patterns that can indicate biological differences or similarities between samples. Many of the clustering methods used for microarray data can be applied to 2D gel data. Figure 15.16 shows an unrooted hierarchical clustering tree constructed from 2D gel data from Swiss 3T3 cells that were stimulated with different growth factors at different times. Protein synthesis after the various stimulation regimes, or with no stimulation, was measured and compared using 2D gels and then analyzed by hierarchical clustering. In this instance, the gels clearly fall into two groups (see Figure 15.16). One group clusters on the same main branch as the unstimulated sample, suggesting that these treatments had little effect on protein expression. The other branch clusters together those samples that were stimulated for a longer time and those that were stimulated by more than one growth factor and platelet-derived

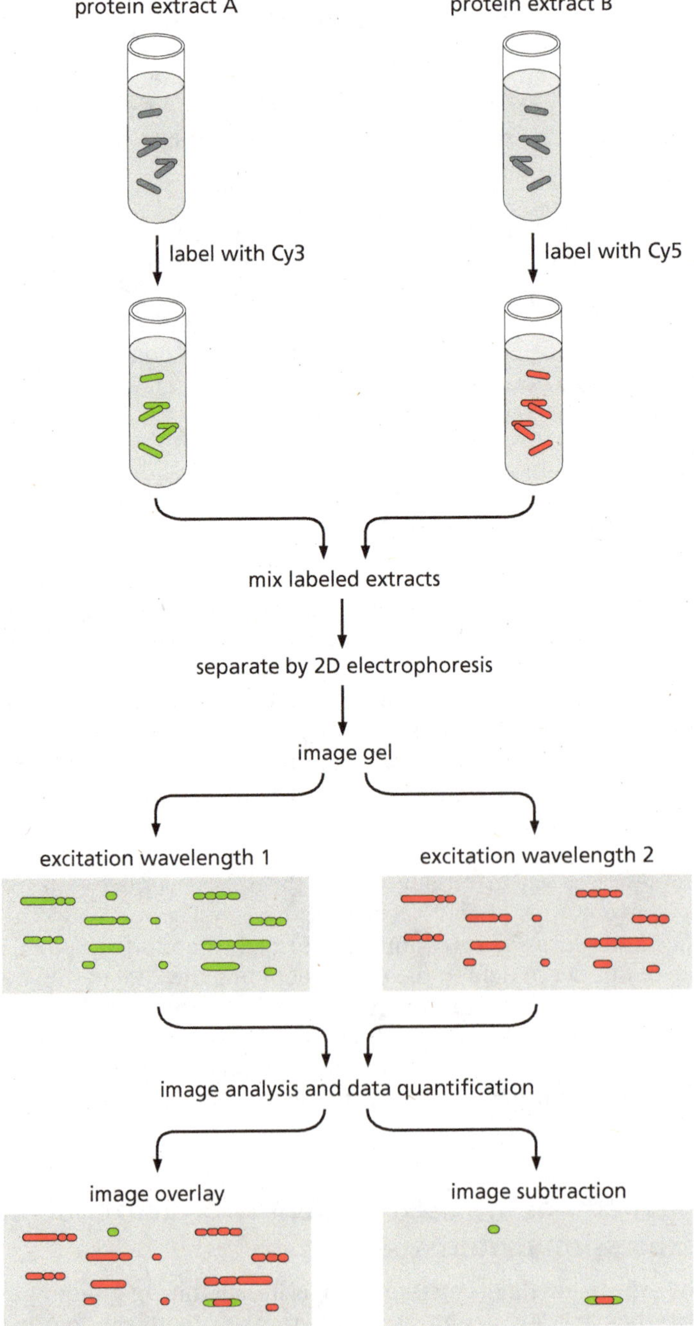


**Figure 15.14**
**The use of two different fluorescent labels to study differential protein expression by 2D gel analysis.** The proteins in one sample are tagged with Cy5 and in a second sample with Cy3. The samples are then mixed and run on a 2D gel system, which separates proteins according to their molecular weight and pI. The image is then scanned using different wavelengths. One wavelength detects Cy3-labeled proteins while the other detects Cy5-labeled proteins. The images are then analyzed in a computer analysis program; when the images are matched and placed on top of each other, those spots that changed are highlighted.

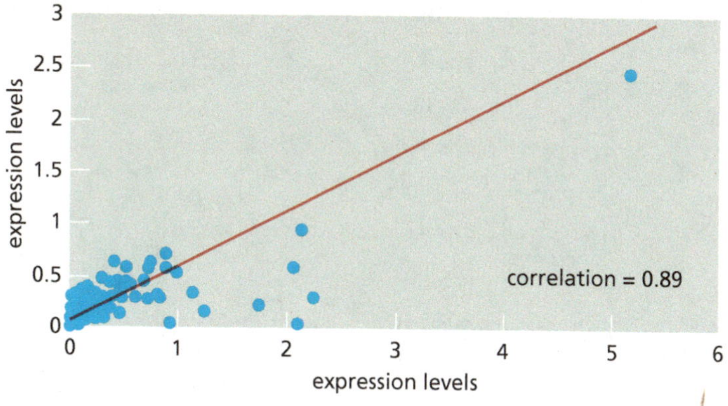


**Figure 15.15**
**Scatterplot of protein expression data from cells treated with different combinations of growth factors.** Cells were treated with epidermal growth factor (EGF), insulin growth factor (IGF), platelet-derived growth factor (PDGF), and combinations of these. Protein content and levels in each cell sample were estimated after protein separation on 2D gels. Outlying points represent those spots for which expression levels are very different between the two gels. This scatterplot tells us that, overall, the expression of most proteins is similar between the two samples, as the correlation coefficient is near to 1.

growth factor (PDGF). Further investigation showed that the treatments represented by the second group of gels had, indeed, a more pronounced effect on protein synthesis than the treatments in the other group.

The program ChiClust first defines a value between spot volumes above which spots are designated as similar. Spots are then selected in one biologically relevant group of gels, for example gels run from samples of a particular cancer, which are similar. This defines the similarity pattern for that group of gels. These protein features found to be similarly expressed in the subset of gels are then used to group, or cluster, the remaining spots in additional gels. The results can be displayed as a simple relational tree. In this tree, similar gels are connected by short branches (edges), while the longer edges join less-similar gels (see Figure 15.17A). The results of such a clustering can indicate families of gels that would not be picked out if using all spots, as any relevant information would be lost due to the variability of the data. Such a subset of spots can be submitted to other types of clustering such as a SOM for expression-pattern analysis (see Figure 15.17C), and to programs that visualize patterns of spot expression.

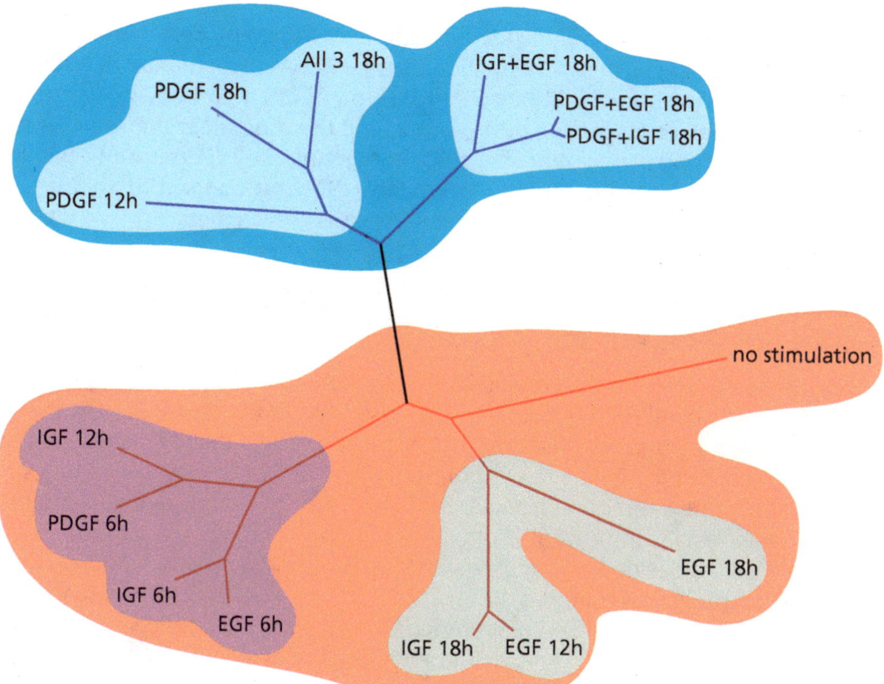


**Figure 15.16**
**A tree obtained by clustering of protein expression data obtained under different conditions of cell stimulation.** The data are clearly divided into two groups. One group (outlined in red) clusters along with the unstimulated sample. The other branch (blue) clusters together those samples that were stimulated for a longer time and those that were stimulated by more than one growth factor and PDGF (these two subclusters are outlined in light blue). For definitions of growth factors, see Figure 15.15.

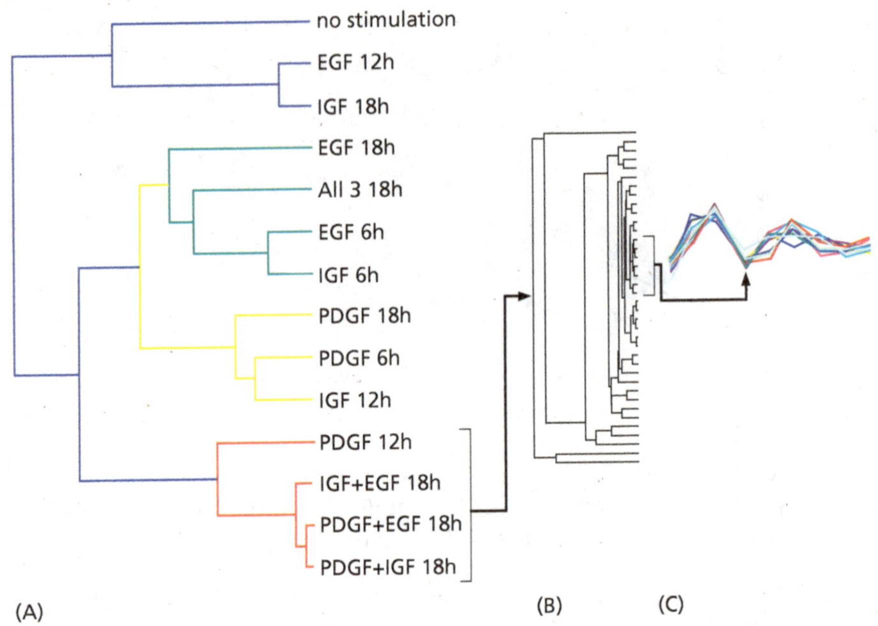


**Figure 15.17**
**A subset of spots picked out can be reclustered and reanalyzed.** Reclustering of a similar group such as the group with two growth factors and PDGF at 12h stimulation (A), using the individual protein expressions can pick out more detailed expression patterns as shown in (B). (C) These can then be plotted to show the actual expression profile. The expression patterns are similar to those obtained by SOM or SOTA techniques.

## Principal component analysis (PCA) is an alternative to clustering for analyzing microarray and 2D gel data

The purpose of most analysis of expression data, whether for genes or proteins, is to recognize classes or groups (of genes, proteins, or samples) that can provide some biologically significant information. Such groups consist of a set of objects described by a number of properties. In protein expression data, an object could be a spot, and a property could be the level of expression of that spot under a particular condition; that is, in a particular gel. These properties contain the signals that are used as the basis for defining the classes, but often none of the properties has a strong signal. A PCA combines this set of properties into an alternative set in such a way that most of the signal is present in just a few of them. The data are then plotted using just the two or three new combined properties with the strongest signals, in two (or three) dimensions. (PCA is described in detail in Section 16.1.) This reduction to fewer properties often makes it possible for class membership to be determined by visual inspection. Figure 15.18 shows a two-dimensional PCA representation of a subset of data from the growth-factor-induced cell line mentioned earlier. Two groups are apparent: one clusters at the top of the $y$-axis, and the other near the value 1 on the $x$-axis. This grouping can be reproduced by subjecting the PCA output to the other clustering methods (see Figure 15.18). Similarly PCA can be used to analyze DNA microarray data (see Figure 16.7) and any other quantitative data.

## The changes in a set of protein spots can be tracked over a number of different samples

Given the steady stream of technical improvements that have increased the information that can be obtained from a 2D gel, user-friendly tools to identify features that are interesting for identification by MS are essential. Most available programs calculate the changes in a set of spots between two gels; however, analysis of changes in protein expression is facilitated if this can be calculated between more than two gels. This is important if one wishes to track changes in protein expression over a set of defined parameters such as drug-induced activation or simply time. A set of programs that will do this are ChiClust and ChiMap, which use the parameters of spot volume and percentage volume to track changes in the expression of proteins detected by a particular spot in a set of gels. The programs calculate percentage differentials across a set of matched spots in many gels with respect

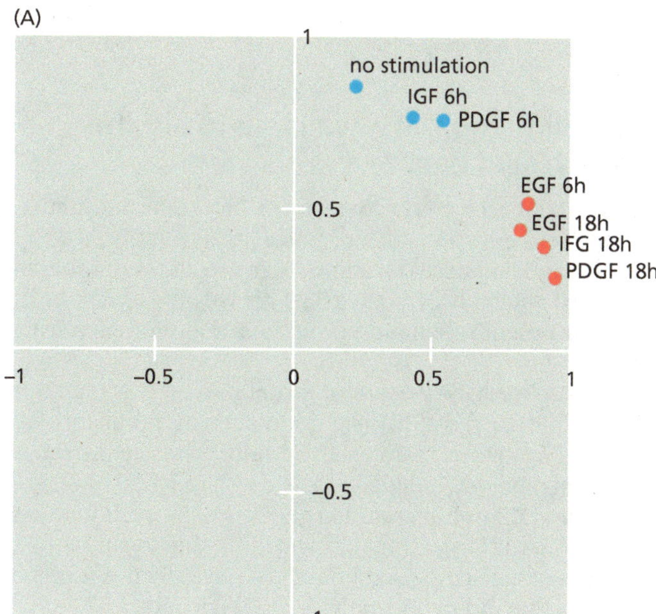


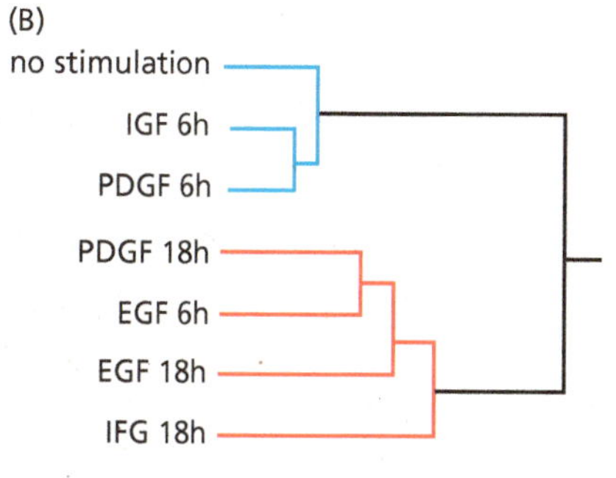


**Figure 15.18**
**The output obtained from a principal component analysis (PCA).** (A) Division into two groups is immediately apparent. This division is also reflected in the hierarchical tree (B). Again, samples (gels) stimulated for only a short time cluster close to the unstimulated sample, while the treatments over a longer period generally make a separate cluster.

to a reference gel. The differentials are given both as numbers and in a graphical format (see Figure 15.19). The graphical format allows easy identification of spots that have either increased expression (red boxes) or decreased expression (blue boxes) with respect to the reference gel. The output can be sorted according to the average increase or decrease in protein expression and the standard deviation. Spots above or below a chosen cut-off can be selected for further analysis by MS. Differential expression can be simply calculated by measuring the difference between the intensity of one spot with respect to another matched spot, using

(A)

-99 -50 0 50 99

| Basket | Group ID | Average | Stdev | %Vol/Ref | S006G1 | S005G1 | S004G1 | S003G1 | S002G1 |
|---|---|---|---|---|---|---|---|---|---|
| ☐ | 535 | 11.954 | 36.196 | 0.010 | -16.53 | -30.99 | 54.25 | -2.48 | 55.51 |
| ☐ | 833 | 12.682 | 63.058 | 0.006 | 9.90 | 9 | 77.97 | -1000 | 66.54 |
| ☐ | 117 | 14.164 | 38.15 | 0.015 | -21.31 | -39.34 | 29.89 | 61.39 | 40.2 |
| ☐ | 478 | -56.391 | 29.050 | 0.007 | -24.81 | -23.31 | -1000 | -68.42 | -65.41 |
| ☐ | 493 | 23.036 | 29.588 | 0.013 | 22.69 | 26.71 | -31.98 | 49.44 | 48.33 |
| ☐ | 497 | 41.842 | 10.409 | 0.022 | 26.60 | 38.93 | 42.47 | 59.18 | 42.02 |
| ☐ | 523 | 39.399 | 6.757 | 0.041 | 39.46 | 28.21 | 42.12 | 49.09 | 38.12 |
| ☐ | 533 | 23.986 | 16.344 | 0.546 | 7.75 | 21.56 | 5.07 | 41.78 | 43.78 |
| ☐ | 535 | 11.954 | 36.196 | 0.010 | -16.53 | -30.99 | 54.25 | -2.48 | 55.51 |
| ☐ | 704 | -31.832 | 13.383 | 0.037 | -17.28 | -25.13 | -20.94 | -46.6 | -49.21 |
| ☐ | 740 | 35.173 | 11.454 | 0.008 | 24.09 | 25.81 | 47.00 | 51.06 | 27.9 |
| ☐ | 808 | 17.893 | 26.725 | 0.014 | 18.31 | -16.14 | 43.54 | 51.23 | -7.47 |
| ☐ | 826 | 24.919 | 18.548 | 0.016 | 18.97 | -0.43 | 48.01 | 13.60 | 44.44 |
| ☐ | 877 | 35.43 | 10.067 | 0.032 | 39.14 | 22.87 | 24.05 | 46.22 | 44.88 |
| ☐ | 926 | -35.934 | 23.826 | 0.011 | -12.86 | -4.98 | -67.63 | -40.25 | -53.94 |
| ☐ | 978 | 32.108 | 28.821 | 0.013 | 7.29 | 10.98 | 70.61 | 7.71 | 63.94 |
| ☐ | 1143 | -34.226 | 8.423 | 0.013 | -18.45 | -33.56 | -42.83 | -37.57 | -38.72 |
| ☐ | 1189 | 29.478 | 19.415 | 0.011 | 1.32 | 18.18 | 52.73 | 25.5 | 49.66 |

(B)

| | | | | | | | | | |
|---|---|---|---|---|---|---|---|---|---|
| ☐ | 497 | 41.842 | 10.409 | 0.022 | 26.60 | 38.93 | 42.47 | 59.18 | 42.02 |
| ☐ | 1607 | 42.609 | 9.109 | 0.018 | 25.18 | 47.16 | 50.96 | 47.16 | 42.58 |
| ☐ | 1286 | 46.711 | 29.39 | 0.083 | 11.40 | 12.36 | 56.42 | 77.52 | 75.85 |

(C)

| | | | | | | | | | |
|---|---|---|---|---|---|---|---|---|---|
| ☐ | 2682 | -53.998 | 56.921 | 0.018 | 27.82 | 2.19 | -1000 | -1000 | -1000 |
| ☐ | 1447 | -45.764 | 15.149 | 0.013 | -35.45 | -29.97 | -38.04 | -53.75 | -71.61 |
| ☐ | 1880 | -39.430 | 11.236 | 0.056 | -19.74 | -34.77 | -45.87 | -51.57 | -45.19 |
| ☐ | 2079 | -38.949 | 6.691 | 0.013 | -28.44 | -35.88 | -40.24 | -41.43 | -48.76 |
| ☐ | 1474 | -38.830 | 7.31 | 0.022 | -29.25 | -31.48 | -40.39 | -47.08 | -45.96 |

**Figure 15.19**
**An example output from the program ChiMap.** ChiMap calculates the differential protein expression between a reference gel and any number of other matched gels. The differential expression is given in percentages. A 50% increase or decrease is equivalent to a two-fold change. Colors that are blue (light to dark blue) indicate a decrease in protein expression with respect to the reference gel, while colors in the red spectrum indicate an increase in expression. Green colors show no significant difference. White spaces with a high negative number indicate that no equivalent spot was found in that gel. The results can be sorted according to the average increase or decrease (B and C, respectively) and the standard deviation is also given.

distance measures such as the Euclidean distance or the Pearson correlation coefficient distance function (see Section 16.2).

## Databases and online tools are available to aid the interpretation of 2D gel data

Many sites maintain databases of the data from 2D gel experiments and their protein annotations to aid interpretation of future experiments (see Chapter 3 on databases and the Garland Science Web site). There are also a number of sites that offer online tools for 2D gel analysis, such as 2D gel viewing and comparison software, pI and molecular weight calculation programs, and sequence searches.

The ExPASy (Expert Protein Analysis System) proteomics server maintained by the Swiss Institute for Bioinformatics (SIB) is a good starting point for exploring the wealth of data obtained by 2D gel analyses. In addition, the site provides a free gel-viewing program that can be downloaded to the user's machine (Melanie-Viewer). It maintains an annotated 2D polyacrylamide gel electrophoresis database (SWISS-2D-PAGE) which provides 2D gel images showing the location of annotated proteins. As well as the exhaustive 2D gel database, ExPASy has a myriad of tools accessible over the Web that aid in the analysis of 2D gel data.

Flicker is a JAVA-based downloadable method for comparing images from different Internet sources as well as the user's own gels. Flicker is an image viewer that will read any two images from a gel database such as the SWISS-2D-PAGE and display them. The program then compares the gels by alternately displaying, in the same window, the two images being compared as well as showing them side by side (see Figure 15.20). The images are aligned by aligning similar morphologic features. The program also provides for limited gel analysis, such as quantification.

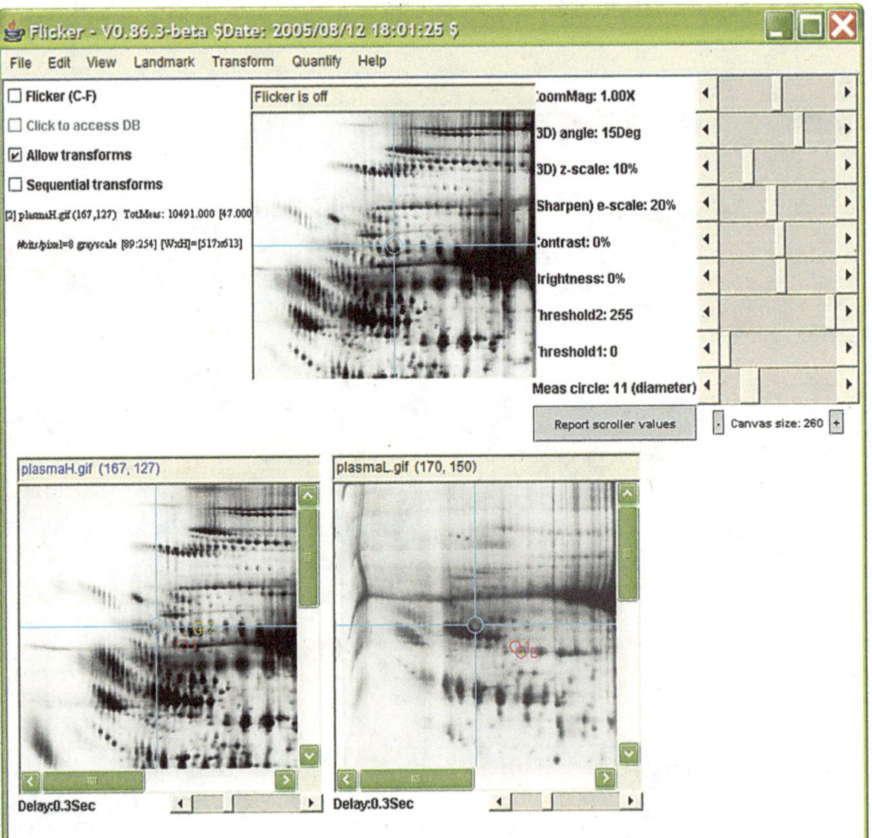


**Figure 15.20**
**Illustration of Flicker.** An example of two gels being compared in the Flicker program.

## Protein microarrays allow the simultaneous detection of the presence or activity of large numbers of different proteins

Protein microarrays on glass slides can measure the presence or function of thousands of proteins simultaneously and are the protein counterparts to DNA microarrays. To make a protein array or a protein chip, purified proteins are spotted onto chemically derivatized glass slides, to which they attach covalently. Once attached, the proteins retain their ability to interact specifically with other proteins, or even with small molecules, as if they were in solution. In protein-chip technology known proteins, such as antibodies against a particular range of proteins or the substrates for an enzyme family, are spotted on the slide and probed with the sample, which may contain many proteins, both known and unknown.

The array is designed so that specific binding, or the action of an enzyme on its substrate, can be detected. The interactions may be used to quantify protein abundance and analyze protein modifications in cells or tissues. For example, protein microarrays have been used to identify substrates for a set of known protein kinases. In this case, the substrates were spotted onto three slides, each slide was incubated with a different kinase, and phosphorylation of the substrate was monitored. Only the specific substrates for each enzyme were phosphorylated by the specific kinase. Similarly, small molecule arrays are being developed, which will help in drug discovery, and additionally whole cell arrays are also being developed, these will enable the study of cellular pathways and the effect drugs have on these pathways.

## Mass spectrometry can be used to identify the proteins separated and purified by 2D gel electrophoresis or other means

Having separated the proteins in a sample into spots on a gel, we want to know the identity of the protein in each spot, or at least the spots that change between samples. The current method for determining protein identity on a large scale is MS. Proteins are usually identified by ion trap or time-of-flight (TOF) mass spectrometers, and common methods of ionizing the sample to be injected into the mass spectrometer include electrospray ionization (ESI) or matrix-assisted laser desorption/ionization (MALDI), giving rise to acronyms for the techniques such as MALDI-TOF, ESI ion trap, or Q-TOF (quantitative TOF). Another common technique is the more sensitive tandem MS (MS/MS) where the ion fragments analyzed in a first round of MS are reanalyzed by a second round of fragmentation. For all types of MS, a small sample of a protein (say a spot from a gel) is digested with a proteolytic enzyme such as trypsin to give a mixture of peptide fragments. These fragments are then analyzed by MS. Figure 15.21 illustrates schematically the steps involved in processing a protein spot from a gel through MS to identification.

A mass spectrometer fragments each peptide into ions, measures the mass–charge ratio of each ion fragment, and produces a spectrum from which the mass of the peptide can be calculated. To identify the protein, the peptide mass fingerprint obtained from each protein spot is then searched against a database of protein sequences that have been theoretically digested by various enzymes and the resulting peptide masses calculated. The particular peptide mass fingerprint obtained by MS is matched against the peptide mass spectra predicted from individual proteins in the database. A good match means that the corresponding peptide is present in the protein under investigation. A protein is usually identified when a number of its component peptides match a number of theoretical peptides for a single database entry. A problem arises with the search algorithms because the protein masses in the databases are calculated for the protein as translated from its genomic sequence. But many proteins are posttranslationally modified, which means that their masses will be different from those in the database and they will be difficult to identify. A number of programs that attempt to deal with this problem have been written.

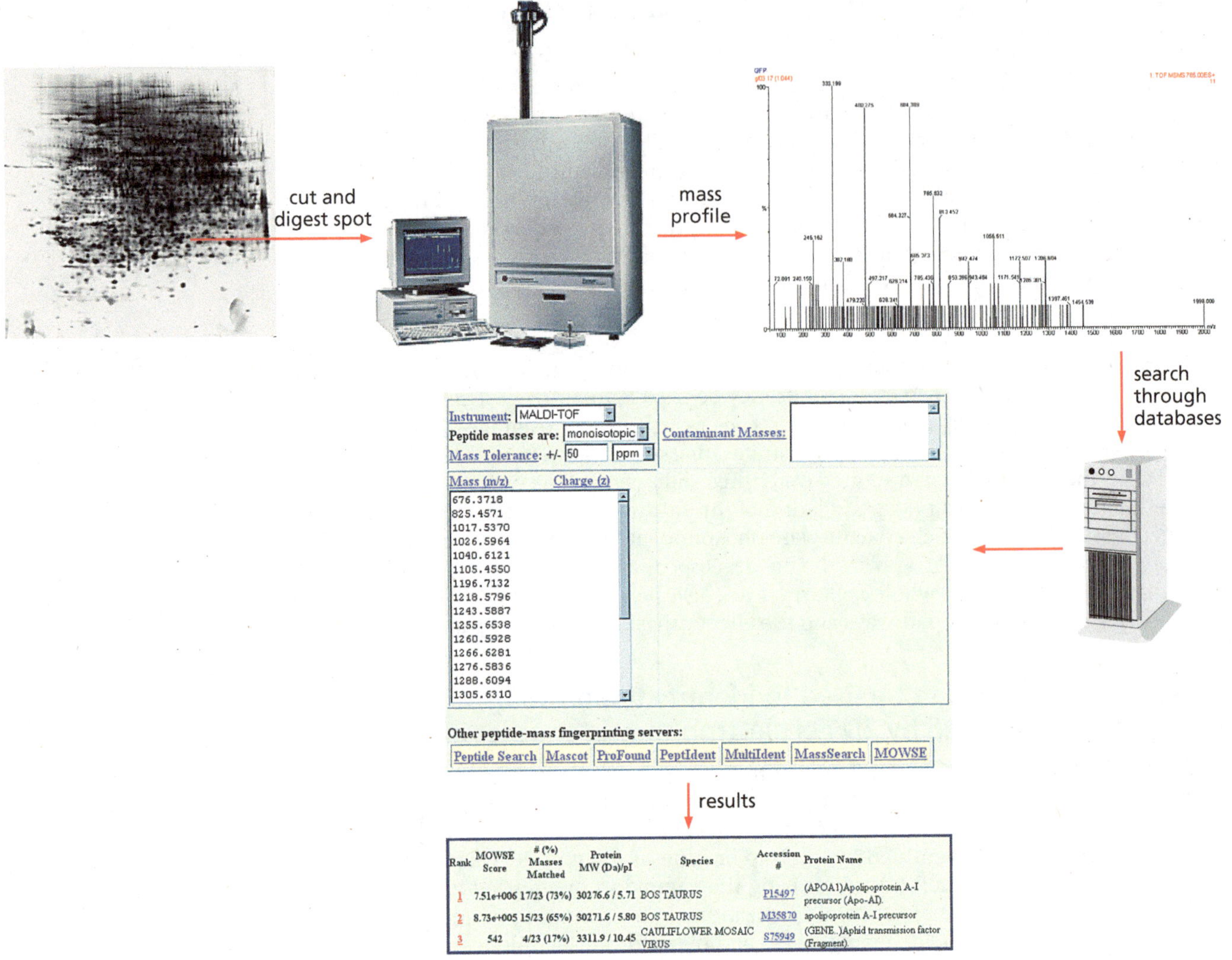


**Figure 15.21**
**Generalized steps in mass spectrometry analysis of a 2D gel.** First the spots of interest are cut out of the gel and digested and treated so that the mass spectrometer can analyze it to create a mass spectrum for each spot it is given. The data from MS are submitted for a search through specialized databases, which match fragment masses to those calculated for proteins. In this example the fragment masses matched, with high scores and high percentage of masses (73 and 65% matched), to bovine apolipoprotein precursor twice. The next hit in the results is a low-scoring match, with only 17% of the masses in the list matching.

One of these algorithms takes into account the mass differences between the peaks of the spectra being compared. If the peptides differ only due to a mutation or a small modification then the masses in the spectra will also differ by a constant number. The program identifies unexpected mass differences and uses the mass difference to determine the type of modification. Another method has focused on the alignment of the spectra to be matched. It uses conventional sequence alignment methods and applies them to the alignment of spectra. The peaks of the two spectra to be aligned are represented as elements in a two-dimensional matrix and the alignment algorithm searches for the most appropriate path through the matrix using the dynamic programming method (see Section 5.2). In this way, mismatches caused by mutations or other modifications are allowed for.

## Protein-identification programs for mass spectrometry are freely available on the Web

A number of MS search programs such as ProteinProspector and MASCOT are freely available through a Web site. They allow the user to search and fit a set of mass peaks, and search for peptides whose masses are identical or similar. The experimental mass values are compared with calculated peptide masses or fragment ion mass values obtained by applying theoretical digestions to the sequences in a sequence database. By using an appropriate scoring algorithm, the closest match or matches can be identified. If the unknown protein is present in the sequence database, then the aim is to identify that particular entry. If the sequence database does not contain the unknown protein, then the aim is to match those sequences that have the closest homology; these are often the same or similar proteins but from different species.

MASCOT is a powerful online search engine that uses the MS data to identify proteins from primary sequence databases. The program integrates a number of known methods of searching into one algorithm to search the sequence database with a peptide mass fingerprint, which provides a very specific signature. A search using a fingerprint is often all that is needed to identify the protein in a database, although when the protein sequence is not in the database identification may not be possible. The other method used by MASCOT is the sequence query, which combines one or more peptide molecular masses with sequence, composition, and fragment ion data. The sequence information is obtained by a partial interpretation of a tandem mass spectrometry (MS/MS) spectrum. Analysis of such a spectrum can often provide three to four residues of sequence data. The last method is similar to the first one but uses MS/MS ions and contains mass and intensity pairs. This search mode is often used to analyze a liquid chromatography/MS/MS run containing data from multiple peptides. Obtaining matches with MS/MS data to a number of peptides from a single protein increases the confidence level for the identification. The same search methods are available with the ProteinProspector package of searching programs.

### Mass spectrometry can be used to measure protein concentration

Liquid chromatography is now often used instead of 2D gel electrophoresis to separate proteins in preparation for MS. The mass spectrometer can be used to measure relative protein concentrations and with the use of liquid chromatography it will be possible to digest, separate, and measure protein concentration without the use of 2D gels. This technique uses isotopically labeled proteins from samples obtained under one particular condition and then mixes these proteins with a differentially labeled sample obtained under a different condition. The quantity of each protein can thus be measured relative to a reference state, and quantitative changes in protein expression can be detected. The same type of database searching and mass spectra fitting will have to take place as with MS intended for protein identification.

## Summary

This chapter presents an introduction to the fast-changing methods for measuring the expression of large numbers of genes or proteins simultaneously. DNA microarrays are at present the main method used for measuring simultaneous gene expression, and capture a snapshot of the transcriptome of a cell or tissue at a particular time or in a particular condition. A common use of microarrays is to compare the gene expression profiles of different tissues or conditions, including different types of tumors, with a view to developing better methods of diagnosis and treatment regimes. Such experiments produce a vast amount of data and specialized analysis methods have been developed both to convert the raw data into a measurement of gene or protein expression levels and to then analyze the differences in expression for their biological significance. The methods used for the analysis of microarray or protein expression results include hierarchical clustering, self-organizing maps, and principal component analysis. Other methods for measuring simultaneous gene expression and differences in expression between different samples include serial analysis of gene expression (SAGE) and the computational method known as digital differential display (DDD), which detects differences in gene representation between different EST libraries.

The set of proteins expressed by a particular cell or tissue (the proteome) can be detected by separation techniques such as 2D gel electrophoresis and liquid chromatography. In the case of 2D gel electrophoresis, the separated proteins can sometimes be identified simply from their position on the gel. Protein identification on a large scale is, however, now usually carried out by the powerful technique of mass spectrometry. The purified proteins are digested into peptides before submitting the protein to mass spectrometry. The spectra produced are compared with specialized databases of protein spectral information, from which the protein can be identified.

# Further Reading


## 15.1 Analysis of Large-scale Gene Expression

**Functional genomics**

Bork P & Eisenberg D (2000) Sequences and topology: genome and proteome informatics. An overview. *Curr. Opin. Struct. Biol.* 10, 341–342.

Diehn JJ, Diehn M, Marmor MF & Brown PO (2005) Differential gene expression in anatomical compartments of the human eye. *Genome Biol.* 6, R74.

Nau GJ, Richmond JFL, Schlesinger A et al. (2002) Human macrophage activation programs induced by bacterial pathogens. *Proc. Natl Acad. Sci. USA* 99, 1503–1508.

Newman JC & Weiner AM (2005) L2L: a simple tool for discovering the hidden significance in microarray expression data. *Genome Biol.* 6, R810.

Hunt S & Livesey F (eds) (2000) Functional Genomics: A Practical Approach, p 256. New York: Oxford University Press.

**Microarray quality control focus**

There is a variety of articles in *Nature Biotechnology* 2006; 24, 1115–1162.

**MIAME**

Brazma A, Hingamp P, Quackenbush J et al. (2001) Minimum information about a microarray experiment (MIAME)—toward standards for microarray data. *Nat. Genet.* 29, 365–371.

## 15.2 Analysis of Large-scale Protein Expression

**Proteomics**

Pennington SR & Dunn M (eds) (2000) Proteomics: From Protein Sequence to Function. Oxford: BIOS Scientific Publishers.

**Protein microarrays**

MacBeath G (2002) Protein microarrays and proteomics. *Nat. Genet.* 32, 526–532.

Tomlinson IM & Holt LJ (2001) Protein profiling comes of age. *Genome Biol.* 2, 1004.1–1004.3.

Zhu H, Klemic JF, Chang S et al. (2000) Analysis of yeast protein kinases using protein chips. *Nat. Genet.* 26, 283–288.

**MASCOT**

Perkins DN, Pappin DJ, Creasy DM & Cottrell JS (1999) Probability-based protein identification by searching sequence databases using mass spectrometry data. *Electrophoresis* 20, 3551–3567.

**ProteinProspector**

Clauser KR, Baker PR & Burlingame AL (1999) Role of accurate mass measurement (+/- 10 ppm) in protein identification strategies employing MS or MS/MS and database searching. *Anal. Chem.* 71, 2871– 2882.

**Databases and software**

Harris RA, Yang A, Stein RC et al. (2002) Cluster analysis of an extensive human breast cancer cell line protein expression map database. *Proteomics* 2, 212–223.

Ng A, Bursteinas B, Gao Q et al. (2006) pSTIING: a 'systems' approach towards integrating signalling pathways, interaction and transcriptional regulatory networks in inflammation and cancer. *Nucleic Acids Res.* 34, D527–D534.

# CLUSTERING METHODS AND STATISTICS

**When you have read Chapter 16, you should be able to:**

- Prepare expression level data for analysis.
- State the various ways of quantifying the difference between measurements.
- Identify similar expression level variation through cluster analysis.
- Describe the alternative clustering schemes produced by different methods of cluster analysis.
- Assess the value of clusters.
- Identify significant differences in expression levels using statistical methods.
- Use special techniques to control the rate of false positives when applying many statistical tests to a data set.
- Classify samples on the basis of a set of expression levels.

In Chapter 15 we discussed experiments for measuring protein and gene expression levels and introduced methods of analysis. In this chapter we look in more detail at the methods for analyzing these types of data and the principles behind them. Many of the analytical techniques rely on the assumption that the data show variation according to an idealized statistical distribution, usually the normal distribution. The data often show deviations from this assumption, and so we will first examine techniques available to normalize and transform the raw data so as to make them better approximate an idealized distribution. We then explore three key approaches to analyzing the data: cluster analysis is used to determine similar expression patterns and similar sample conditions; statistical tests are applied to determine the true significance of observed similarities and differences in the data; and finally, it is possible to define functions that can be used to classify an unknown sample.

This field is developing very rapidly, both in terms of the experimental methods and the techniques of analysis. It is beyond the scope of this book to give a comprehensive survey of the area, so emphasis is placed on general principles with some examples of the kinds of analytical methods available. One area in particular that will not be covered and yet can have a profound influence on the outcome is the careful design of experiments to reduce **systematic errors** and make data correction easier. More detailed information on the analytical techniques can be found in the references listed under Further Reading.

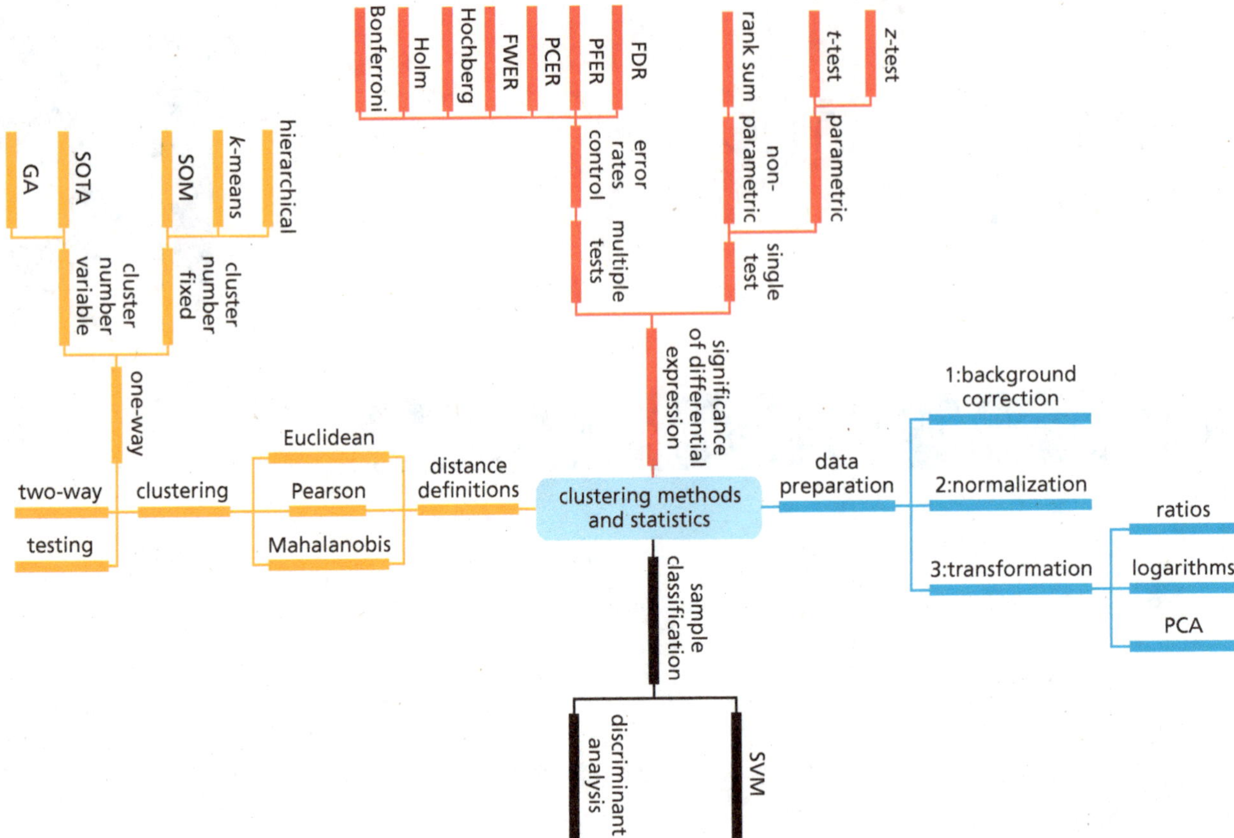


**Mind Map 16.1**
**Many of the methods described in the previous chapter use statistics and mathematical analysis.** This chapter describes these methods in detail and the mind map here is to help the reader remember all the steps needed for a careful analysis of expression data.

## 16.1 Expression Data Require Preparation Prior to Analysis

Experiments that measure expression levels result in a value for each of the set of genes or proteins in the experiment. In most cases these data will be analyzed to identify changes in the expression level of individual genes or proteins when compared at two or more states of the system. (The system here typically refers to the organism, tissue, or cell in which the expression occurs.) Two main quantitative analyses are performed: identifying significant changes in expression levels, and identifying groups of genes or proteins whose expression levels vary in similar ways when the system state is altered. As will be explained below, the directly measured expression levels are not appropriate for these analyses without prior modification. In this section we will look at some techniques of normalization and transformation that are used to prepare the data for analysis. Different stages will be considered here as shown in the scheme of Figure 16.1, but it should be noted that alternative methods may use a different scheme.

The aim of these manipulations is to remove errors and make the data have a distribution sufficiently close to a normal (Gaussian) distribution that parametric statistical tests can be used in the analysis with confidence. Normal distributions are symmetrical about the central mean value and have a characteristic decay of the distribution away from the mean. There are only two parameters to define—the mean and the variance—and sufficiently large subsets of data from the same distribution should have similar values for these parameters. These are the features that will be used to judge the suitability of the data for further analysis.

It should be noted that the methods described here relate most closely to the Cy5/Cy3 label gene expression experiments. Other methods such as Affymetrix arrays use different normalization procedures, but the general principles remain the same.

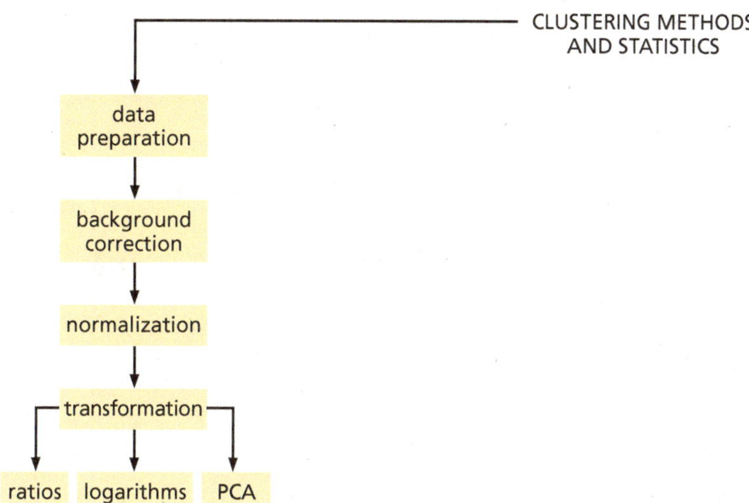


**Flow Diagram 16.1**
**In this section the key steps are described that might be necessary to prepare expression data for analysis.**

There are many different kinds of experiments that measure protein and gene expression. In most cases the level of expression of an individual gene or protein is obtained by measuring the intensity of a particular spot. Due to the heterogeneity of this type of experiment the spot value must be modified to account for the background intensity. Correcting for background intensity may involve measuring values near each spot, or the average intensity of the unexpressed genes. Sometimes, after the value has been subtracted from spot intensities, it may result in negative values. The data for such spots are occasionally discarded, but often retained, assigned a minimal positive value, and marked as absent or unobserved. It is not unusual in these kinds of experiments for this to affect a significant fraction of the spot intensities. This is because only a small proportion of all genes are expressed at a given time and under given conditions. It can also be due to some of the products having short half-lives. This treatment allows for future reanalysis of these spots. The observed absence of gene expression is often as significant a result as identifying high expression, so that such data are of great value.

## Data normalization is designed to remove systematic experimental errors

Normalization methods are mainly based on knowledge of the particular experimental methodology and potential sources of systematic error (such as using different quantities of different samples). Their application often involves simple manipulations such as taking averages, yet can achieve very good results. Many of the techniques, such as removing the background level, correct the mean but do not pay much attention to the variance. It has often been observed in general experimental work that the variance tends to increase as the signal increases. This is a **random error** (nonsystematic) associated with the general process of measurement. This effect can be seen in microarray data, and several normalization methods have been proposed that correct for this deviation from a normal distribution. Note that systematic errors can in the best of cases be completely removed from the data, whereas it is only possible to approximate the form of random error, and not to remove it entirely.

David Rocke and colleagues have proposed a model with two sources of random error. One source is an additive error $\varepsilon$, which is assumed to be normally distributed, with a mean of zero and standard deviation $\sigma_\varepsilon$. (An additive error is simply one that is added as a separate term to the true value to obtain the measured value.) The second source they propose is a multiplicative error term $e^\eta$ that is proportional to the level of expression, with mean zero and standard deviation $\sigma_\eta$. If the actual

Apply normalization techniques to the intensity values to remove as much systematic bias (error) as possible

↓

Assign floor value, or remove any data which do not exceed a threshold related to the background

↓

Convert the data to expression ratios and take logarithms

↓

Use further normalization techniques to improve the log ratio distribution

**Figure 16.1**
**The scheme described in this section for preparing data for further analysis.**

level of expression is written $\chi$, then according to this model the measured level of expression can be written

$$X = \chi e^{\eta} + \varepsilon \qquad \text{(EQ16.1)}$$

The measured intensity $I$ is linearly proportional to $X$, and can be written

$$I = \beta_1 + \beta_2 \chi e^{\eta} + \varepsilon \qquad \text{(EQ16.2)}$$

where $\beta_1$ and $\beta_2$ are constants and the additive error term $\varepsilon$ has been adjusted to account for the multiplier $\beta_2$. This model can be used to derive a method of normalizing the data to obtain a constant standard deviation over the range of intensities (see Further Reading for details).

## Expression levels are often analyzed as ratios and are usually transformed by taking logarithms

Figure 16.2 shows the frequency of occurrence of intensity values in a gene expression experiment. They do not follow a normal distribution and therefore need further modification before statistical analysis. In fact almost all experiments that measure gene and protein expression levels do not require the determination of the absolute values, such as 10 transcriptions of the gene per cell, or 150 protein molecules per cell. Instead the quantity of interest is usually the level of expression of a particular protein or gene relative to the level of the same protein or gene when the system is in a reference condition. In addition, the use of such a ratio has two benefits. Firstly, it removes any potential errors that would occur if different proteins or genes had differing abilities to produce a signal under the experimental conditions, for example due to different abilities to bind a dye. Secondly, there is always a large range of different absolute levels of expression across a library of genomic products, but the sensitivity of the system tends to be related to proportionate changes in expression such as double the reference level. Hence taking ratios will help to determine whether a change in expression should be recognized as significant.

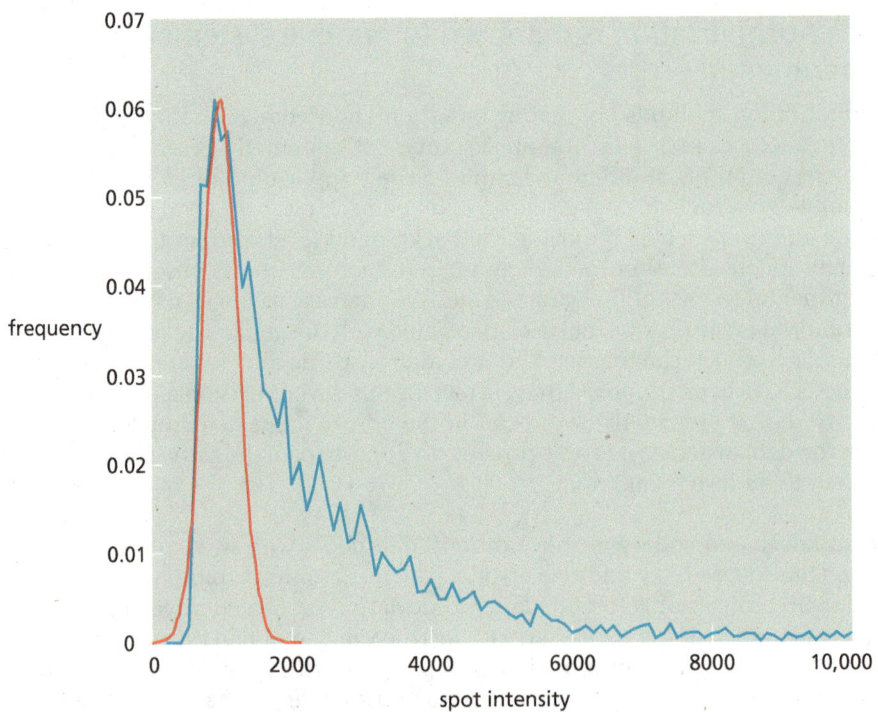


**Figure 16.2**
**Plot of (in blue) the frequency of intensity measurements, compared with (in red) an approximately equivalent normal distribution.** There are far more high intensity measurements than would be expected if sampling from a normal distribution. (The data are a small sample from Courcelle et al., Comparative gene expression profiles following UV exposure in wild-type and SOS-deficient *Escherichia coli*, *Genetics* 158:41–64, 2001.

Although expression level ratios have clear advantages over the absolute values, they are still not ideal quantities in further analysis. In most experiments both increases and decreases in expression levels may be important features of a biological response. However, the expression level ratio does not treat both of these situations equally. Increases in the expression level relative to the reference state are shown by values greater than 1, and can be very large. In contrast, decreases in expression level give values in the small range of 0 to 1. This situation is shown in Figure 16.3A, which also shows a typical situation for expression data, in that most of the genes or proteins do not show a change in expression level with respect to the reference state, i.e., have ratio values very close to 1. In this figure the ratios of the measured intensities that are proportional to the expression level are plotted against the square root of the product of the two intensities. This form of plot shows how the distribution of ratios varies with the level of expression. At all levels of expression the range of ratio values is noticeably skewed, with several points above 2, but none of course below 0, which would require negative intensities.

A simple solution to this is provided by taking logarithms of the ratios, typically to base 2. This is because logarithms have the property $\log(X) = -\log(1/X)$, so that for example $\log_2(2) = 1$ and $\log_2(1/2) = -1$, and the ranges are the same for both increases and decreases in expression levels. A ratio of 1 (no change) becomes 0 on taking logarithms. Figure 16.3B shows the same data as before after taking logarithms. The data are now centered about the value 0, and are much more equally distributed about that value. Note that taking logarithms of the measured expression intensity ratios is problematic if negative numbers are encountered, although usually any such data will not be included in later analysis.

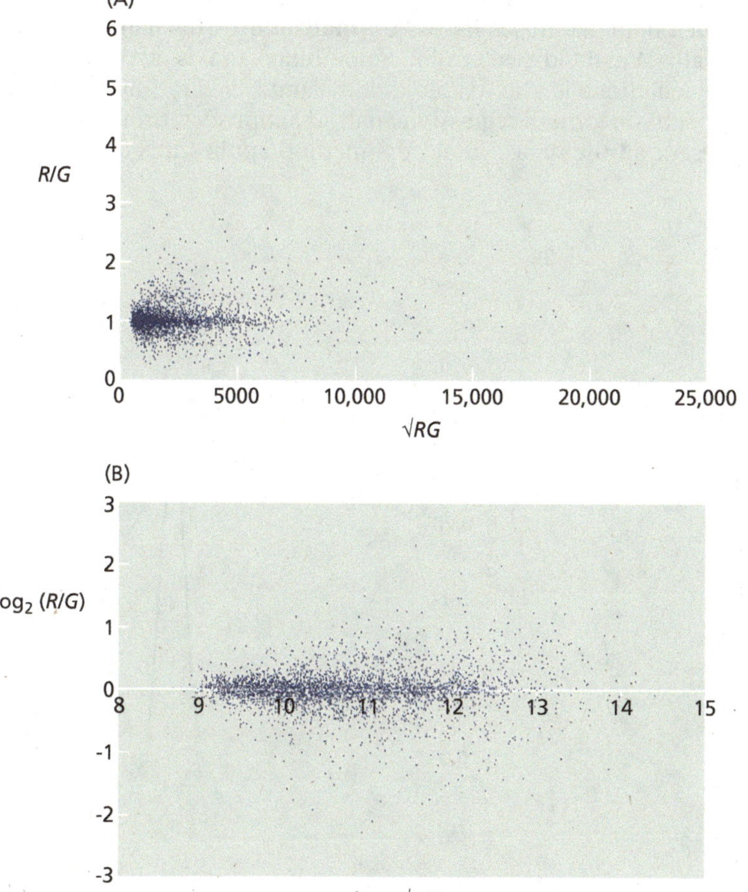


**Figure 16.3**
**Two alternative ways of viewing expression data.** The data shown were obtained in an experiment using two fluorescent labels (see Section 15.2). Cy3 is green, and the measured intensity will be written $G$, while Cy5 is red, and the measured intensity will be written $R$. Plots are of (A) the ratio $(R/G)$ and (B) $\log_2 R/G$, against $\sqrt{RG}$ and $\log_2\sqrt{RG}$, respectively. Plot B is often referred to as an R–I plot or an M–A plot. There are two commonly used functions plotted on the abscissa—$\log_2\sqrt{RG}$ as here or $\log_{10}(RG)$, but these differ only by a multiplicative factor. The data are the same as used in Figure 16.2.

Taking logarithms of the expression level ratios clearly makes the data behave better in that increases and decreases in expression are treated equally, and the data appear to be closer to the normal distribution. Figure 16.4 shows the frequency distributions of values for the ratios before and after taking logarithms, in both cases compared with similar normal distributions. The tails of these ratio distributions are much larger than in the normal distribution, which is to be expected as the tails will include those genes that have been up- or downregulated. However, close inspection of the tails reveals that the distribution of the ratios after taking logarithms is much more symmetrical.

Figure 16.3B also illustrates that the experimental data show deviations from the ideal distribution in that the extreme points are not evenly distributed, but show a dependence on the overall level of expression. There are no changes in expression level greater than two-fold amongst those genes or proteins that show very low levels of expression (on the left of the graph), whereas there are many at higher levels of expression. Also, at the highest expression levels (on the right of the graph) far fewer of the ratios are close to 0. These are signs of a correlation between the standard deviation and the level of expression. As mentioned previously, methods have been proposed to correct for this.

## Sometimes further normalization is useful after the data transformation

The log ratio data plotted in Figure 16.3B appear to be well behaved as regards the average value over different ranges of level of expression ($\log_2 \sqrt{RG}$), in that the center of the distribution is always close to 0. In Figure 16.5A another example is shown in which this is not the case, and furthermore the data are not simply shifted from 0, but show a pronounced curvature. (If there was just a simple shift these data could still have a normal distribution.) One of the most common methods for correcting such data features is **lowess normalization**. This name is an abbreviation of LOcally WEighted Scatterplot Smoothing, and is a type of regression analysis. The technique is also written as loess and called locally weighted regression. The most basic form of regression analysis simply determines the best single line to represent all the data. The lowess method applies this approach to small

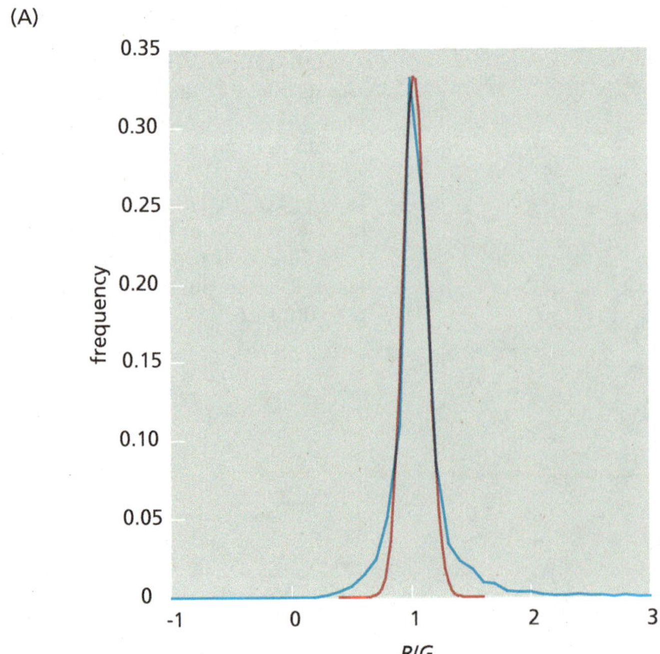


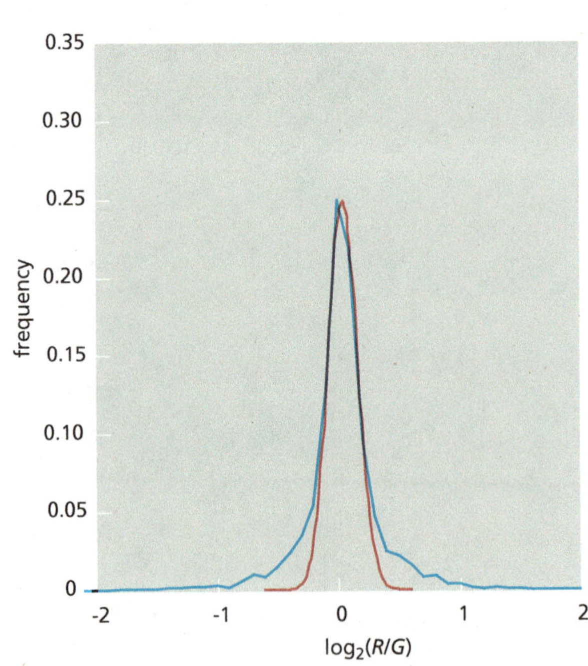


**Figure 16.4**
**Comparison of ratios and log ratios of expression data with normal distributions.** The blue lines show (A) plot of the distribution of ratios, and (B) plot of the log ratios. The data are centered about 1 and 0, respectively, and compared with (in red) approximately equivalent normal distributions. The log ratio distribution is much more symmetrical in the tails. The data are as in Figure 16.2.

regions of data, usually using either linear or quadratic fitting. This local regression analysis defines the mean values, which can then be used to shift the local data to the desired mean, in this case 0. The result of applying this technique to the data of Figure 16.5A is shown in Figure 16.5B.

There are equivalent procedures that can be applied on a local basis to correct for any dependence of the variance on expression levels. These usually take the form of simple scaling of local points following a lowess normalization. Following the application of all these manipulations the data should now have a distribution close enough to a normal distribution that parametric statistical tests can be applied with confidence.

## Principal component analysis is a method for combining the properties of an object

The different properties measured in an expression experiment may not be truly independent of each other. A simple instance is the measurement in two samples of the relative expression levels of a set of genes that are co-expressed. Another example of data dependency that is hard to avoid occurs when using tissue samples from many different individuals. Such samples are inherently heterogeneous, and this heterogeneity can cause random variation or unexpected dependencies. **Principal component analysis** (PCA) is a method that can identify these dependencies and simplify the data to minimize their effect. PCA is a technique that can also be used to identify patterns in numerical data. The method transforms the measurements into new variables—components—that are truly independent of each other. Each of the new variables is a linear combination of the actual measurements. In addition to this, the technique determines the fraction of the variation present in the data that comes from each new variable. New variables that vary most in a dataset and thus hold most of the variation of the experiment are the principal components.

PCA is often used to reduce the data to only two or three principal components holding most of the data variation and to discard the others, which will be more uniform across the data; of course, if the other components describe important variation in the data, that will now be missed. This small number of principal components can be used to plot out samples, and can be used in sample classification (see Figure 16.6).

The calculation of PCA is essentially the diagonalization of a matrix, $\boldsymbol{X}$, that contains the expression data. If an experiment consists of $N$ genes or proteins and $M$ different samples (or conditions), $\boldsymbol{X}$ will be an $N \times M$ matrix (see Figure 16.7). A row of this matrix corresponds to a different gene or protein and a column represents a specific

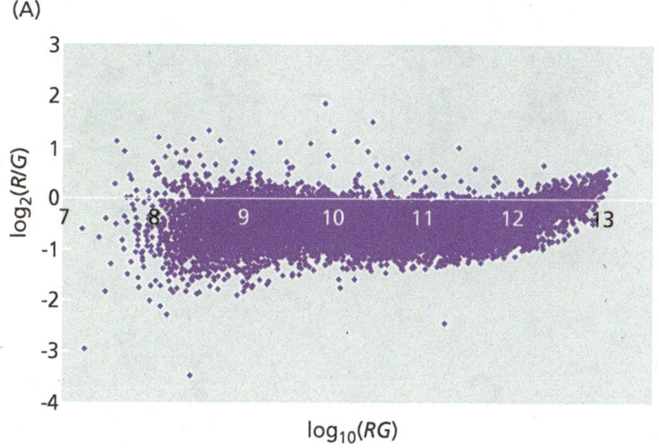


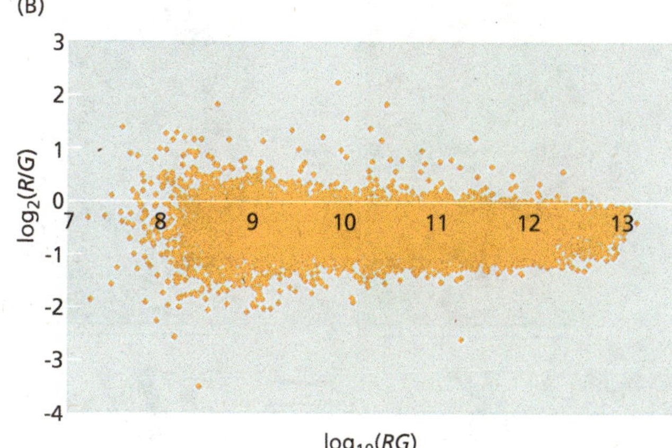


**Figure 16.5**
**A demonstration of the effect of lowess normalization on expression data.** (A) Plot of $\log_2(R/G)$ against $\log_{10}(RG)$, where $R$ and $G$ refer to Cy5/Cy3 measurements as described in Section 15.2. Note the different form of the abscissa compared with Figure 16.3B. This plot form is often referred to as an R–I plot. In this example there is a clear deviation of the mean log ratio value from 0, and also a distinct curvature can be seen. (B) The same data after applying a lowess normalization. (Reprinted by permission from Macmillan Publishers Ltd: *Nature Genetics* 32:496–501, J. Quackenbush, Microarray data normalization and transformation, 2002.)

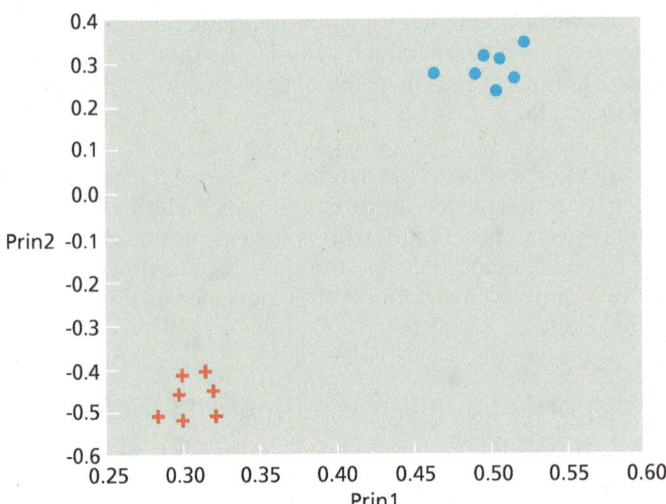


**Figure 16.6**
**An idealized PCA result in which the groups of data fall into well-separated sets.** Following PCA the coordinates for each data point in the first two principal components (Prin1 and Prin2) have been plotted. The two sets of data are shown in blue and red, colored according to their known group membership, and they appear in two clearly separated groups in this plot. In less-favorable cases the two clusters might not have been so distinct, or the data may have remained as a single group.

sample. Then for genes, a matrix element $X_{i,A}$ is the expression ratio for gene $i$ under condition $A$ with respect to the same gene measured under a different condition. For proteins, this would be the expression value for protein $i$ under condition $A$ with respect to the same protein measured under a different condition.

In the calculation, the matrix $\boldsymbol{X}$ will be re-expressed as a product of three new matrices: the first of these is a matrix, $\boldsymbol{U}$, that is also $N \times M$; the second, $\boldsymbol{\varepsilon}$, is a square matrix of dimensions $M \times M$ (assuming that $M < N$); and the third matrix, $\boldsymbol{V}^T$, is of dimensions $M \times M$ (see Figure 16.7).

$$\boldsymbol{X} = \boldsymbol{U} \boldsymbol{\varepsilon} \boldsymbol{V}^T \qquad \text{(EQ16.3)}$$

**Figure 16.7**
**Illustration of the main equation used in PCA.** The data used come from 14 cDNA arrays (samples) of 5981 genes that were used to measure genome-wide mRNA levels in a synchronized culture of *S. saccharomyces* at intervals over a complete cell cycle (the times in minutes are listed at the top of the plot), with the purpose of identifying genes linked to the cell cycle. The reference mRNA was taken from an unsynchronized culture. The normalized data is shown on the left side in the matrix *X*. Refer to the main text for a description of the three components on the right-hand side of the equation. (From O. Alter et al., Singular value decomposition for genome-wide expression data processing and modeling, *Proc. Natl Acad. Sci. USA* 97 (18):10101–10106, copyright (2000) National Academy of Sciences, USA.)

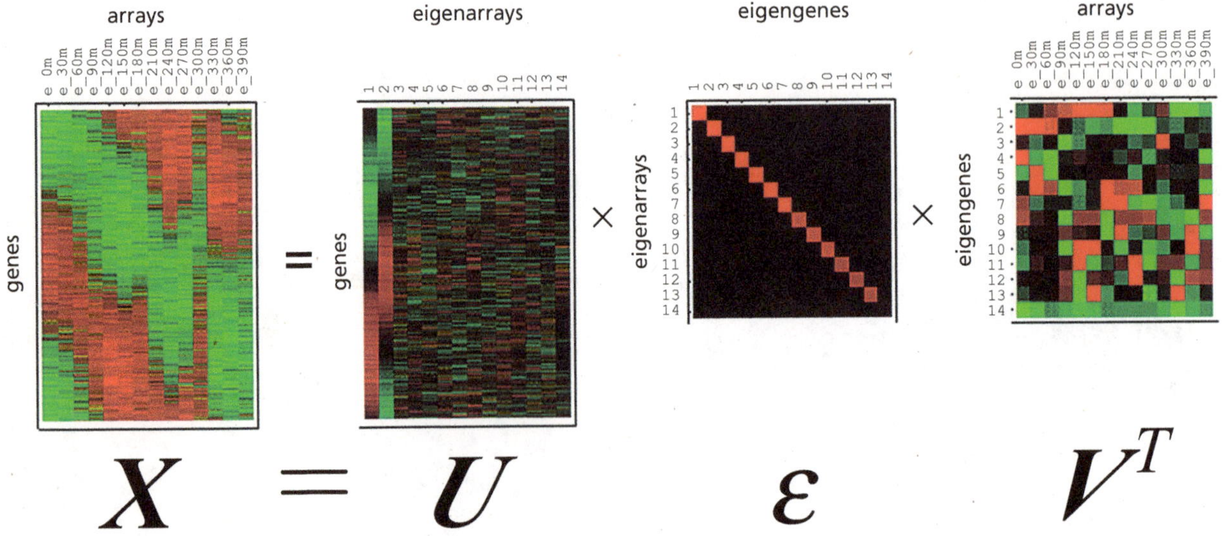

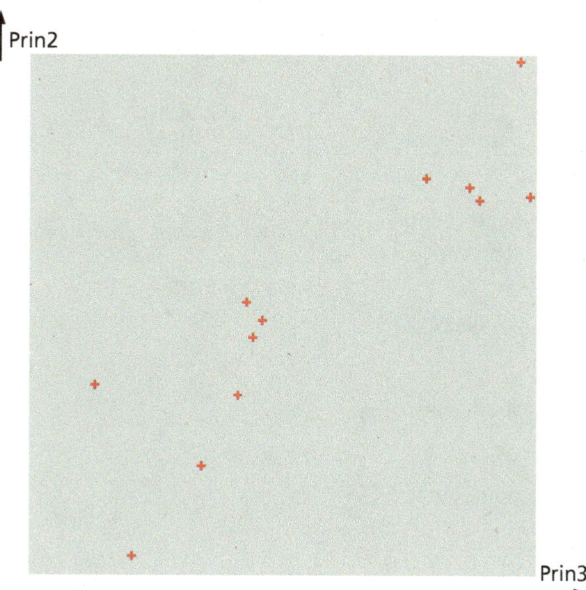


**Figure 16.8**
**A real example of PCA. In this case, using the second and third principal components (Prin2 and Prin3), the 12 samples appear to separate into two groups, with the five samples plotted at top right seeming separate from the others.** The data are not color-coded because this is for now the only evidence for this grouping. Further study of the samples might produce backup evidence for the groupings.

The outcome of the calculation is two sets of $M$ components—one set for the genes (sometimes referred to as eigengenes by analogy to the eigenvectors of matrix mathematics) or eigenproteins, and one set for the samples (called eigensamples or eigenarrays). For simplicity in the rest of this section we will refer to gene expression only. The expression level of every gene in an eigensample is given in matrix $\boldsymbol{U}$, and the expression level of every eigengene in each sample is given in matrix $\boldsymbol{V}^T$. The $A$th eigengene is only expressed in the $A$th eigensample, with the eigenexpression level $\varepsilon_A$ (analogous to an eigenvalue).

Each eigenvector defines a principal component. The expression data can then be plotted for each gene/protein $i$ along the axis defined by the $p$th principal component. Figure 16.8 shows a plot of data with two principal components. Figure 16.6 showed an idealized PCA result where the data fell into two well-separated sets. Figure 16.8 shows PCA on protein expression data from a series of 2D gels (samples) representing different conditions, where, although the sets are not as well defined, the PCA does separate the gels into two groups. In this case the samples have not been classified, so it remains to be seen if there is any useful correlation between the grouping and the samples. This situation could arise when the experiment is of a diagnostic nature, attempting to identify the sample condition rather than exploring the features of the condition.

## 16.2 Cluster Analysis Requires Distances to be Defined Between all Data Points

All clustering techniques identify clusters according to a distance between each pair of data points and therefore need a definition of this distance measure. As will be discussed later in the chapter, cluster analysis can be performed not only to identify genes whose expression levels change in similar ways, but also to identify samples that have similar expression patterns. These samples could for example be different organisms or different conditions, or a combination of the two. The distance must be defined as a number, and therefore each gene or sample in the experiment requires a set of quantitative parameters.

**Flow Diagram 16.2**
**The key concept introduced in this section is that there is a choice of definitions of the distance between two sets of measurements.**

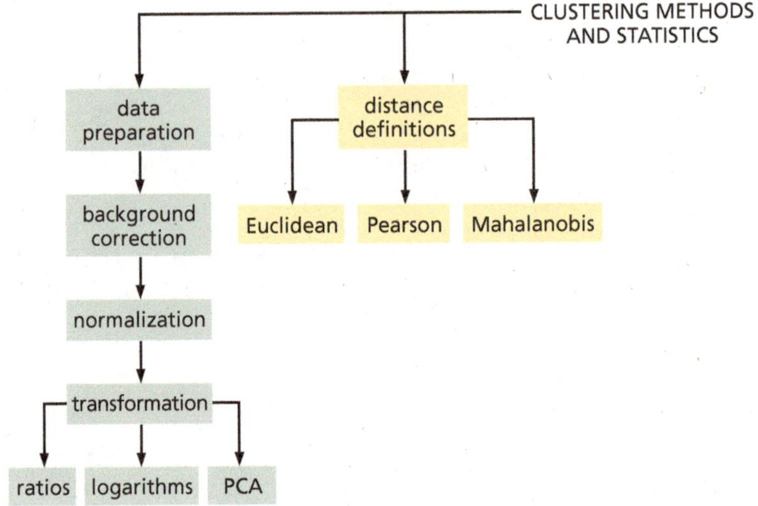


The parameters of the genes which are used to define distances are almost invariably the log ratios of expression described above. The normal definition of a distance between two gene expressions involves the log ratios of the two genes measured for every sample in the experiment. The sample parameters may well measure a more diverse set of properties. For example in many experiments measurements are taken at a set of time intervals after a particular stimulus was applied, so that time would be a sample parameter. Samples can have many parameters, as in the case of a study of changes in human gene expression following infection. Each (human) sample might be described in terms of a set of measurements whose aim is to define the stage and/or severity of the infection, such as body temperature, white blood cell count, or the time since onset of symptoms. In addition, other measures such as age or weight may be included. Any parameters that are not quantitative need to be suitably transformed; for example gender is not suitable, but assigning arbitrary values of –1 for males and +1 for females will resolve the difficulty.

The distance to be calculated is usually one of two alternatives. In most cases the variation in expression level of two genes or proteins is compared for each of the different samples/states in the experiment. An example of this is the comparison of patterns of expression level at different stages of cell division. The other common distance measured compares two samples or states, by comparing the expression levels of every gene or protein.

Having a set of quantitative parameters enables distances to be calculated, but there are several alternative distance measures. In this section we will look at some of the measures used, and discuss their advantages and disadvantages.

## Euclidean distance is the measure used in everyday life

If two points $A$ and $B$ in three-dimensional space have coordinates $(x_A, y_A, z_A)$ and $(x_B, y_B, z_B)$ then the distance between them is usually defined as

$$d_{AB} = \sqrt{\left(x_A - x_B\right)^2 + \left(y_A - y_B\right)^2 + \left(z_A - z_B\right)^2} \qquad \text{(EQ16.4)}$$

This formula is an example of the Euclidean definition of distance. Looking now at expression data, as discussed above there may be more than three different parameters, which are the coordinates in the example just given. Suppose there are $N$ different parameters (genes or protein expression levels) measured for each of the

$M$ samples so that the $A$th sample has a value $X_{i,A}$ for the $i$th gene. The **Euclidean distance**, between the $A$th and $B$th samples is defined as

$$d_{AB} = \sqrt{\sum_{i=1}^{N}\left(X_{i,A} - X_{i,B}\right)^2} \quad \text{(EQ16.5)}$$

with the sum over all genes or proteins. Note that the distance between genes or proteins can be defined by simply identifying them with $A$ and $B$ and the samples (now $M$ of them rather than $N$) with $i$ in the above equation. The key feature of this definition is that all parameters are treated in an identical way without any modification. This is not necessarily appropriate in the case of expression measurements. The distance measure in this case should ideally be proportional to the physiological relevance of the difference in expression, a concept that is not clearly defined and also is usually not known prior to completion of the analysis, as it is normally closely related to the intended outcome of the experiment. There is almost certainly a complex variation in sensitivity in the system, not just depending on the gene or protein, but probably also on the existing level of expression relative to a normal resting state and on the current state of the system. Why should doubling the expression of a kinase and of a cytochrome contribute equivalently to the distance? Assuming the sensitivity is proportional to the difference in expression ratios is convenient from the point of view of calculation, but may not be realistic. In addition, the distance will be different if log ratios are used as opposed to ratios (and one distance will not simply be the logarithm of the other), and it is not clear which value should be preferred.

This situation becomes perhaps even more acute when considering sample distances, in which case the different coordinates may represent entirely different properties. For example, if measuring weight in kilograms and temperature in kelvin, should a 1 kg difference in weight really contribute the same amount to the distance as a 1 K temperature difference?

The Euclidean distance measure is very commonly used because it is easy to evaluate, and obtaining a measure based on the true system response is generally totally unrealistic. However, the underlying problems could potentially result in significantly distorted conclusions from cluster analysis, which as will be seen must always be treated with a degree of scepticism until they are seen to correlate well with other experimental knowledge.

## The Pearson correlation coefficient measures distance in terms of the shape of the expression response

One of the problems with the Euclidean measure, as mentioned above, is that quantitative changes in expression ratios are treated equally for all genes. Genes whose transcription is coordinated will not necessarily produce such equivalent responses, and it may be more useful to give greater emphasis to the observation that two genes have correlated expression changes. In other words, two such genes' expression levels may not increase to the same degree, but their correlation is still an important and useful observation.

A frequently used measure of the correlation between two series of numbers is the Pearson correlation coefficient ($r_{AB}$). Consider as before the distance between two samples $A$ and $B$. The definition of $r_{AB}$ for two sets of expression levels $X_A = \{X_{1,A}, X_{2,A},\ldots, X_{N,A}\}$ and $X_B = \{X_{1,B}, X_{2,B},\ldots, X_{N,B}\}$ is given by:

$$r_{AB} = \frac{1}{(N-1)}\sum_{i=1}^{N}\left(\frac{X_{i,A} - \bar{X}_A}{s_A}\right)\left(\frac{X_{i,B} - \bar{X}_B}{s_B}\right) \quad \text{(EQ16.6)}$$

where $\bar{X}_A$ is the average of the values $X_A$, and $s_A$ is the standard deviation of these values; similarly for $\bar{X}_B$. Note that each of the terms in brackets refers to just one sample and is the deviation of the particular value from the mean value in units of standard deviations, a mathematical form that will be seen again in the context of statistical tests (see Section 16.4). For a typical expression experiment the sets $X_A$ and $X_B$ will be the expression log ratios of two genes or proteins for a series of different samples. The values given by this formula range from –1 for a completely negative correlation between $X_A$ and $X_B$, through 0 for no correlation to +1 for a perfect correlation.

The Pearson correlation coefficient measures distance in terms of the shape of the pattern, and not its absolute value, and it will therefore identify two protein features as similar if their expression pattern across samples is similar, even if their absolute expression values are different. This is not always an advantage, because we would probably want to assign greater significance when both genes are highly expressed than when they are both poorly expressed. Figure 16.9 shows the effect of using different distance calculations to analyze an experiment comparing protein expression patterns in healthy people and cancer patients. Using alternative distance measures does not always result in such differences in **dendrograms**, but when it does the conclusions can be significantly different.

## The Mahalanobis distance takes account of the variation and correlation of expression responses

When discussing the Euclidean distance measure a potentially serious problem was mentioned in that the different variables are treated equally even though some variables may show much greater absolute variation than others. One way to overcome this problem is to scale each variable by a measure of its variability such as the variance $s_i^2$ for the $i$th gene or protein. This is sometimes called the normalized Euclidean distance, and is defined by

$$d_{AB} = \sqrt{\sum_{i=1}^{N} \frac{\left(X_{i,A} - X_{i,B}\right)^2}{s_i^2}} \qquad \text{(EQ16.7)}$$

Note that these terms are not simply squares of the Pearson correlation coefficient terms, as they involve both samples $A$ and $B$, and the variance is of the expression level of a particular gene or protein. Again, an equivalent definition can be written for the distance between genes or proteins, in which case the summation is over the different samples/conditions for two particular genes or proteins.

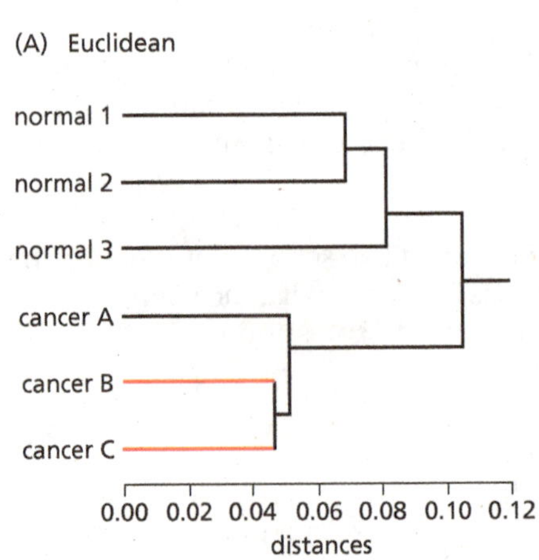


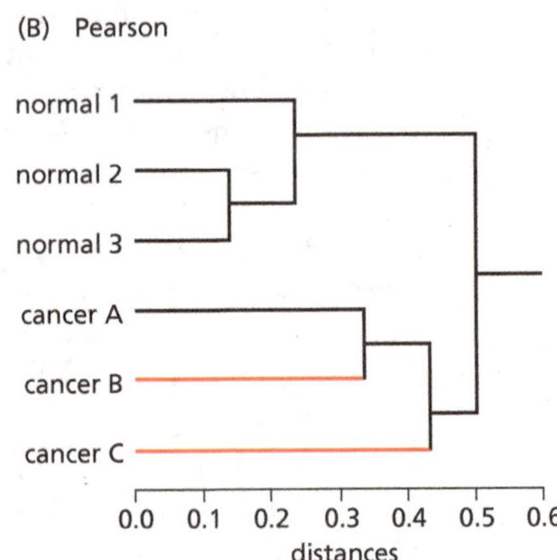


**Figure 16.9**
**An example showing that alternative distance measures can affect the conclusions of an analysis.** Six samples are analyzed from an experiment in which 2D gel electrophoresis was used to compare protein expression patterns in healthy individuals and in patients with breast cancer. The samples are labeled normal 1 to normal 3 and cancer A to cancer C. In this case all the protein expression levels are used to define the distance between pairs of samples and the distances are then used to define a dendrogram as described in Section 16.3. In (A) the Euclidean distance (Equation EQ16.5) is used, whereas in (B) the Pearson correlation coefficient (Equation EQ16.6) is used. Two key differences can be seen in these dendrograms. Firstly, in the dendrogram of (A) sample cancer A branches off before samples cancer B and cancer C whereas in (B) sample cancer C branches off first. The other major difference is that the normal samples branch at larger distance than the cancer samples in (A) but the reverse is true in (B).

If there is correlation between the expression of different genes or proteins, as might be expected due to frequent sharing of promoters and the existence of pathways, an alternative distance definition has been proposed. This is the **Mahalanobis distance**, which takes such correlations into account. This has a similar form to Equation EQ16.7 but also involves terms of the form $(X_{i,A}-X_{i,B})(X_{j,A}-X_{j,B})/\text{cov}_{i,j}$ where $\text{cov}_{i,j}$ is the covariance of measurements of genes or proteins $i$ and $j$.

## 16.3 Clustering Methods Identify Similar and Distinct Expression Patterns

In its simplest form, clustering can be defined as the task of classifying $N$ objects into $k$ groups (clusters) in such a way that the objects within a group are similar to each other but the groups are different from each other. The set of clusters and their membership is called a partition. The number of possible partitions is extremely large, and there are several methods that can be used to propose relevant and useful partitions. In this case a partition is considered relevant and useful if it identifies a relatively small number of clusters that have meaning in the context of the experiment, usually assessed by whether the constituent objects share important characteristics. Such characteristics might be the pattern of variation of expression level or the medical diagnosis of the patients from whom the samples were taken. Whichever clustering method is used, a distance measure must be defined as described in the previous section.

Before describing the clustering methods that are the topic of this section it is useful to look at the relationship between the main types of **data classification** methods (see Figure 16.10). All the classification methods described in this chapter involve assigning every object to a single group, given the general term **exclusive classification**. Examples of such classification are grouping animals according to age or height, since at a given time an animal can only have one age or height. There are situations where although there are several groups into which an object can be classified it might belong to more than one group. Examples of this situation are all

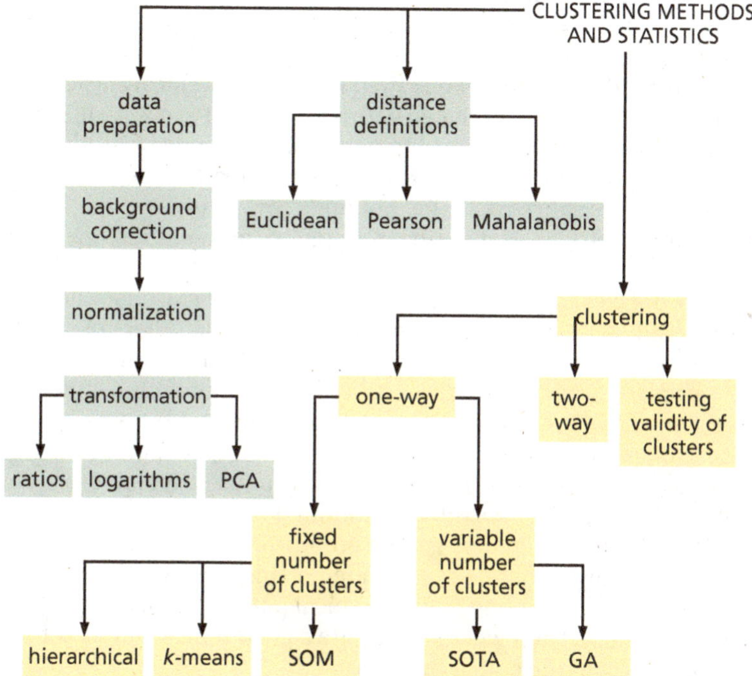


**Flow Diagram 16.3**
In this section the variety of methods that can be used to cluster expression data into clusters of similar expression are described.

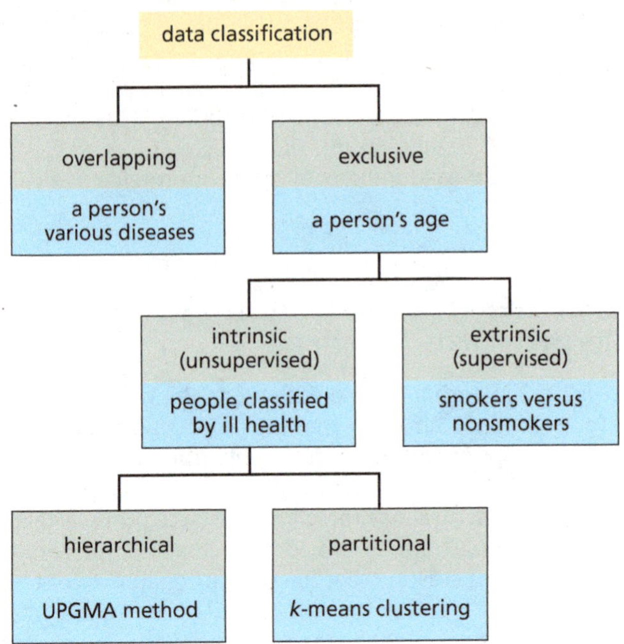


**Figure 16.10**
**The different kinds of data classification organized to show the relationships between them.** For all except the hierarchical and partitional classifications, examples are given of the kinds of data for which these classifications are appropriate.

the diseases carried by an individual or the drugs being administered to a patient. This is referred to as **overlapping classification**, but no methods described here fall into this category.

Exclusive classifications can be further divided into **intrinsic** and **extrinsic classifications**, which correspond to supervised and unsupervised learning, respectively. In the extrinsic case, objects are classified according to already known criteria, such as smokers or nonsmokers or gender, and the resulting groups further analyzed to try to identify discriminating features that can be used to predict the classification in unknown cases. One form of supervised learning was described in Section 12.4 in the case of training a neural network to predict protein secondary structures. Such methods are described further in Section 16.5. In this section we are solely concerned with intrinsic classification, commonly referred to as clustering.

Clustering can be divided into two main types. In hierarchical clustering a large set of alternative partitions are produced with a range from 1 to $N$ of different numbers of clusters. There is a clear relationship between all defined partitions and clusters, best expressed by a dendrogram. **Partitional classification** results in a single partition, although alternatives can be obtained by rerunning the methods with different parameters. We will describe examples of both of these methods below.

It is generally not clear at the outset how many different clusters will be appropriate for the data and some methods are more capable than others of determining an appropriate number of classes into which to put the objects. Therefore the choice of which algorithm to use is nontrivial, as it can have a profound effect on the interpretation of the results. Clusters of genes or proteins that have a similar pattern of expression can be said to be co-expressed. In addition, the patterns can fall into, or be similar to, an idealized set of predefined responses. Figure 16.11 shows eight characteristic patterns that are often encountered in time-series and dosage-regulated studies where proteins (or genes) are up- and downregulated in a specific pattern. In the analysis of such experiments some clusters would be expected whose members all exhibit one of these typical patterns. The difficulty of assessing the partitions obtained and whether they are a good representation of the data will be briefly discussed at the end of this section.

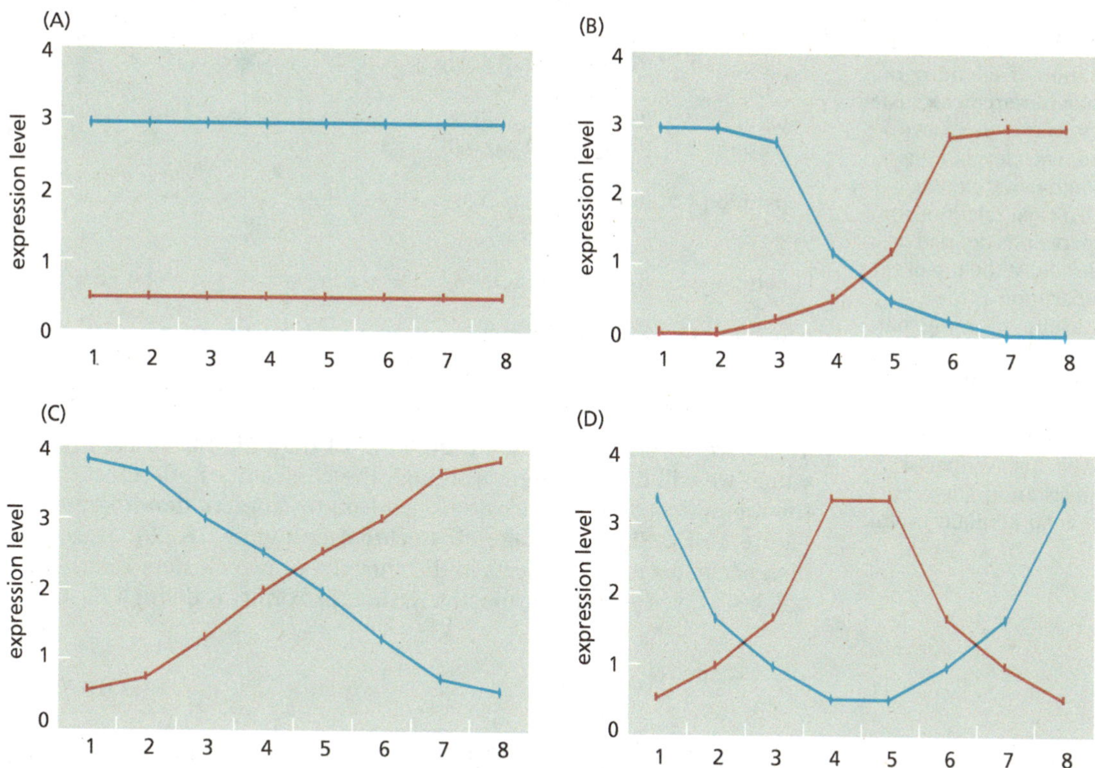


**Figure 16.11**
**Examples of the kinds of patterns that might occur in gene expression experiments when the samples are related to each other by time, amount of stimulation, or dosage.** The horizontal axis of these graphs is time, dosage, or whatever other quantitative variation was involved in the particular experiment. The patterns are shown in complementary pairs. In (A) the expression pattern shows no up- or downregulation; in other words, no effect of whatever stimulation or time. The two lines show consistent high or low expression. In (B) there is a sudden and rapid effect of either the upregulation or downregulation. In (C) the effect is a more gradual increase in either time or dosage, while in (D) the effect on expression is only transient and either lasts for a short time or increasing dosages negate the initial effect.

## Hierarchical clustering produces a related set of alternative partitions of the data

The hierarchical clustering methods produce a large number of alternative partitions for the data. At least two of these are trivial, namely the partitions where all the data are in one cluster and that where each of the *N* objects is in a cluster by itself. The other partitions have an intermediate number of clusters. There is a relationship defined between all the partitions allowing them to be represented as a single structure (see Figure 16.12). We will refer to this as a dendrogram to distinguish it from phylogenetic trees. The larger clusters (in terms of the number of members) are always formed by merging entire smaller clusters found in other partitions. In this way potentially important subgroups can be identified. A visual inspection of the dendrogram may allow the data to be split into several different groups. However, unless this split is based on additional knowledge (for example, which samples are normal and which are cancerous) it will be purely subjective.

In order to perform hierarchical clustering it is necessary to define the proximity of objects and clusters to each other. An initial proximity matrix for the whole dataset is calculated using an appropriate measure, often one of those described in the previous section. In addition the method for obtaining proximities involving clusters with several objects must be defined. As will be seen below, there are several options available, but the key steps in the method are the same in all cases.

The method starts with all *N* objects in separate clusters (the disjoint partition). The pair of clusters at the minimum proximity (least dissimilar) is selected. These two clusters are merged into a new cluster, defining a new partition with *N*–1 clusters. The parts of the proximity matrix that relate to the two merged clusters are replaced by the new cluster, whose proximity to all other clusters must be calculated. This procedure is repeated until all clusters are joined into a single cluster (the conjoint partition). The UPGMA method discussed in Section 8.2 in relation to phylogenetic tree construction is of this form (for example see Figure 8.3), and in fact UPGMA is a standard hierarchical clustering method applied in general data classification problems.

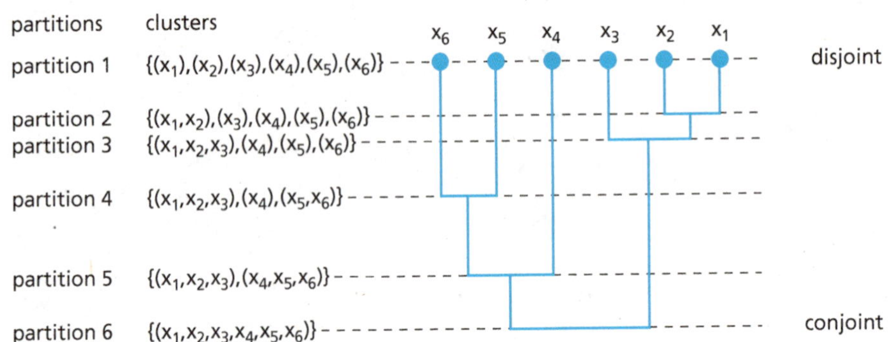


**Figure 16.12**
**The interpretation of a dendrogram resulting from a hierarchical cluster analysis.** Six objects $x_1$ to $x_6$ have been clustered. Each level of the dendrogram defines a different partition, with a clear relationship between the partitions defined on neighboring levels. At the top of the dendrogram (partition 1) the clustering is disjoint, meaning that each object is in its own cluster. At the bottom of the dendrogram (partition 6) the clustering is conjoint, meaning that all objects are in a single cluster. Neither of these partitions is informative, but the other four are potentially useful.

There are several commonly used ways of defining the proximity of clusters, of which we will describe four. Although the choice does not affect the operation of the method as described above, it can lead to different dendrograms, as shown in Figure 16.13. **Single linkage clustering** (see Figure 16.13A) uses the minimum distance between two objects in the cluster. For two clusters *Y* and *Z* with member objects labeled *A* and *B*, respectively, their proximity is defined as

$$\delta_{YZ} = \min\left\{d_{AB}\right\}_{A\in Y,\, B\in Z} \tag{EQ16.8}$$

while, in contrast, **complete linkage clustering** (see Figure 16.13B) defines the proximity as the maximum of the distances between all possible pairs of objects in each cluster:

$$\delta_{YZ} = \max\left\{d_{AB}\right\}_{A\in Y,\, B\in Z} \tag{EQ16.9}$$

In both of these formulae $d_{AB}$ is the distance measure between data points *A* and *B* as described in the previous section, for example the Euclidean distance. The maximum or minimum is defined as over all objects *A* and *B* in the two clusters. Another very common definition for cluster distances is the (average) centroid method (see Figure 16.13C), where the distances or similarities are calculated between the centroids of the clusters:

$$\delta_{YZ} = d_{\hat{Y}\hat{Z}} \tag{EQ16.10}$$

where $\hat{Y}$ is the centroid of cluster *Y* defined as the mean position of the $M_Y$ objects *A* within it and $\hat{Z}$ is the equivalent centroid of the other cluster *Z*. The *i*th component of the centroid of *Y* is defined as

$$\hat{Y}_i = \frac{1}{M_Y}\sum_{A\in Y} X_{i,A} \tag{EQ16.11}$$

The centroid method described here is called UPGMC, meaning unweighted pair group method using centroids. A fourth alternative not illustrated here is UPGMA, in which the distance between clusters is defined as the average of the distances between each object in one cluster and each object in the other. The different definitions of cluster proximity can produce different classifications for the same data, even if the same distance measure is used. No one method seems to give consistently better results than the others.

Because hierarchical clustering provides a series of alternative partitions, further analysis is necessary after the clustering to decide which of them are useful. It may

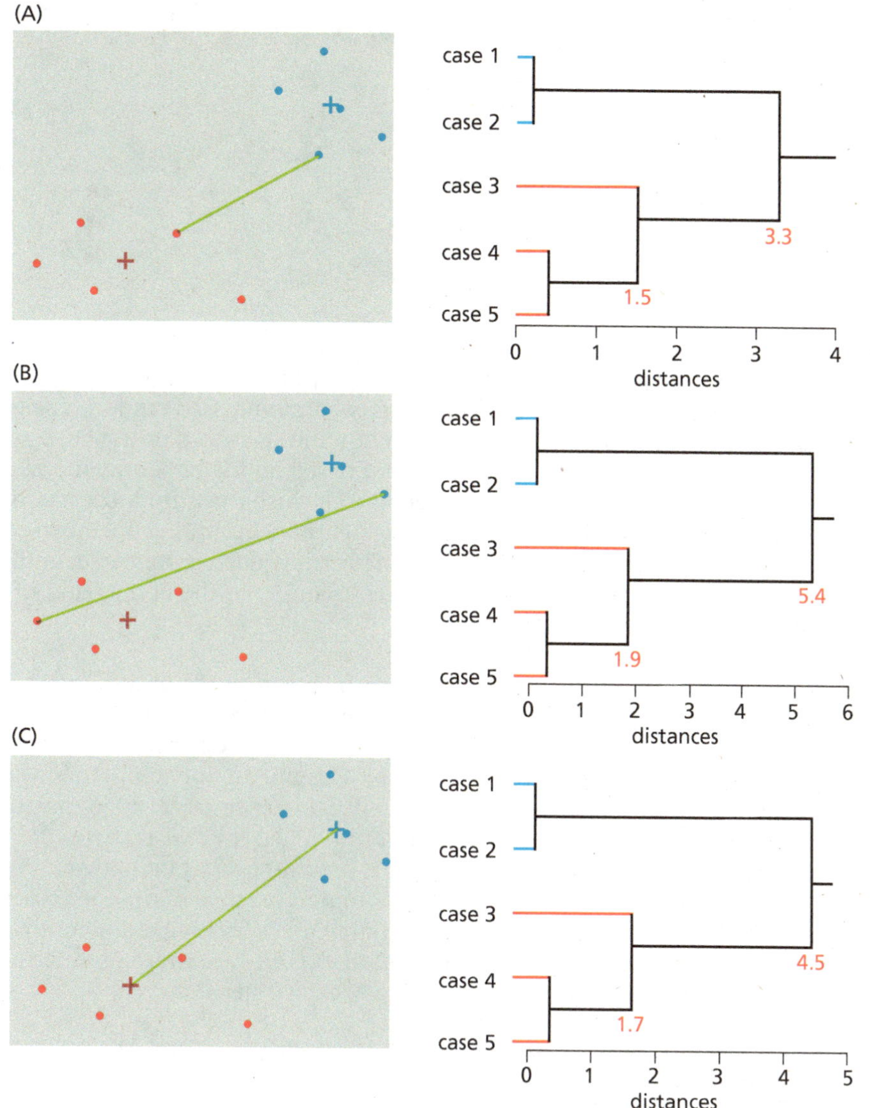


**Figure 16.13**
**The distance between any two clusters can be defined in several different ways, resulting in different results for hierarchical clustering.** The left side shows two clusters of different colors, with the centroids of the clusters drawn as crosses, and in green lines the alternative distances defined between the two clusters. (A) Single linkage clustering uses the minimum distance between two objects (see Equation EQ16.8). (B) Complete linkage clustering uses the maximum of the distances between pairs of objects (see Equation EQ16.9). (C) The centroid method finds the distance between the centroids of the clusters (see Equation EQ16.10). On the right side the effect on the dendrogram is shown, using as an example five measured expression levels: case 1 = 2.0, case 2 = 2.2, case 3 = 5.5, case 4 = 7.0, and case 5 = 7.4. The corresponding dendrograms are shown next to the three different cluster distance definitions. In this case all dendrograms have the same topology, but the clusters merge at different cluster distance values in each case. When typical expression datasets are used, these alternative distance definitions can produce different dendrogram topologies, and can potentially result in different conclusions being drawn from the data.

be that there are some clusters that seem more distinct than others, being present in several partitions. The choice of partition(s) to use in experimental interpretation is almost always subjective, but can be governed by extra information not available during the clustering process. For example the vast majority of a set of genes in a cluster may share a function, suggesting that the cluster has some biological meaning.

## *k*-means clustering groups data into several clusters but does not determine a relationship between clusters

The clustering methods described in this section use a fixed number of clusters. In this section we will discuss a commonly used partitional clustering method. One run of the *k*-means clustering method produces a single partition of the data into *k* clusters. Although there are ways of changing the value of *k* during a run, as will be discussed below, in general the number of clusters *k* is specified at the outset. The choice of *k* may be entirely subjective; for example, one could analyze a particular gene expression experiment using five clusters, but other techniques may suggest that the data require more clusters for adequate classification.

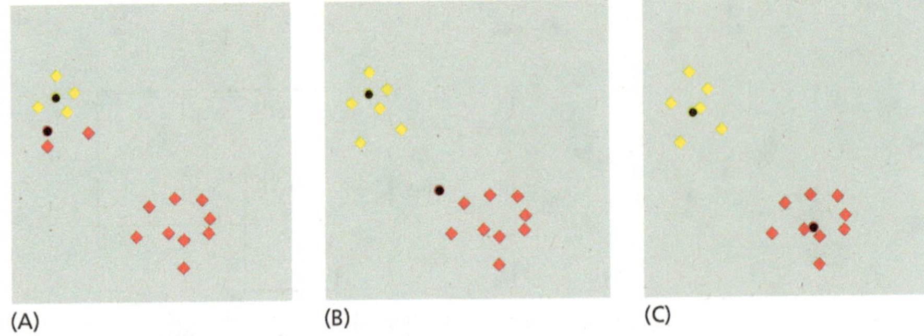


**Figure 16.14**
**A simple example of *k*-means clustering applied to data that occur in two distinct clusters.** In this example the method is run with two clusters, i.e., $k = 2$. Three stages are shown. In each case the locations of the *k*-means cluster centroids are shown by colored circles with a black dot, and the data are colored according to their assignment to a cluster based on these centroid positions. Thus the average of the data points of a given color gives the location of the centroids of the same color in the next diagram. (A) Two *k*-means cluster centroids are initially randomly located (without reference to any of the actual data), and happen to both occur in the same data cluster. The data are then assigned to clusters according to which centroid is the nearest. One data cluster is entirely assigned to the red *k*-means cluster, which is only marginally nearer than the yellow one. (B) After one cycle the *k*-means cluster centroids are already situated such that the data clusters are correctly partitioned. (C) In the final step the *k*-means cluster centroids are located at the centroids of the actual data clusters. The number of steps required to achieve this correct clustering depends on several factors, including the cluster separation relative to the radius of the clusters.

The algorithm initially generates $k$ data points, which are the centroids of the $k$ clusters. These data points are either generated randomly, or a randomly selected sample can be taken from the data. Subsequently, the algorithm proceeds to assign each data point to the nearest cluster centroid based on the proximity measure in use. After all data points have been assigned to a cluster centroid, the algorithm recalculates the centroids of the now redefined clusters by finding the mean of the cluster members as in Equation EQ16.11. This procedure is repeated until the membership of all clusters is stable. A simple example of the application of this method is given in Figure 16.14.

A different initial location of the cluster centroids can result in a different final partition, so that it is advisable to use several different starting points, generating several partitions. However, in spite of its apparent simplicity the method can be surprisingly robust, as shown by Figure 16.14. Despite an apparently poor initial configuration, in this example the method rapidly finds a good partition. Unfortunately, in most real cases this cannot be relied on and many alternative starting points should be tried. As a result there may be several alternative results to consider from *k*-means clustering. It is also common practice to run these classification methods several times with different parameters, including different values of $k$. However, in contrast to hierarchical methods, in this case alternative partitions and the clusters in them have no relationship to each other.

Although it may not be apparent from this description, *k*-means clustering is an optimization with an error function that is related to the sum of the squared distances from the cluster centroids. This squared error can be used as a crude measure of the degree to which the clusters represent the data. By calculating several partitions with different numbers of clusters and then plotting the squared error against the number of clusters $k$, often there is a value of $k_{opt}$ beyond which the error decreases little, if at all. In such cases there is reason to select the partition with $k_{opt}$ clusters as the most appropriate one.

It is possible to modify the *k*-means method to allow it to vary the number of clusters automatically. Once the set of $k$ clusters has been identified, the data can be examined to identify any data points that are relatively distant from all the centroids. In such cases, an extra cluster centroid can be added at that data point. Methods have also been proposed to identify and ignore outliers, but in the case of expression data analysis such outliers may be of great interest and are best left in the analysis.

As for hierarchical clustering, once the partitions have been obtained they must be examined to see if they make sense in the light of other information. In addition it is advisable to check that the set of data in a cluster really do have features in common, as it is possible that more clusters are needed to obtain partitions with clearly distinct patterns of expression. Manual inspection of the clusters is necessary to identify the general characteristics of each of them. In Box 16.1 the same dataset is analyzed using three different clustering methods for comparison: hierarchical and *k*-means clustering as described above, and the SOM method, which will now be presented.

## Box 16.1 An example of cluster analysis showing a comparison of the hierarchical, *k*-means, and SOM methods

*Streptomyces coelicolor* is often used to study the genetic control of antibiotic production. It produces quite a range of chemically diverse antibiotics, which include the red-pigmented tripyrrole undecylprodigiosin (Red), lipopeptide calcium-dependent antibiotic (CDA), and the deep blue pigmented polyketide actinorhodin (Act). Stanley Cohen and colleagues used DNA microarray analysis to study the antibiotic biosynthetic pathway. They collected samples at nine time points during the growth cycle for analysis using microarray experiments. The first six times were during rapid growth, but points 7, 8, and 9 were during the stationary phase. In addition the start of synthesis of different antibiotics was also monitored. The cluster analysis of these data illustrates hierarchical, SOM, and *k*-means clustering and allows some comparisons to be made. Using DNA microarrays the authors showed, among other things, that the antibiotic pathway genes are coordinately regulated at the level of transcription during *S. coelicolor* growth.

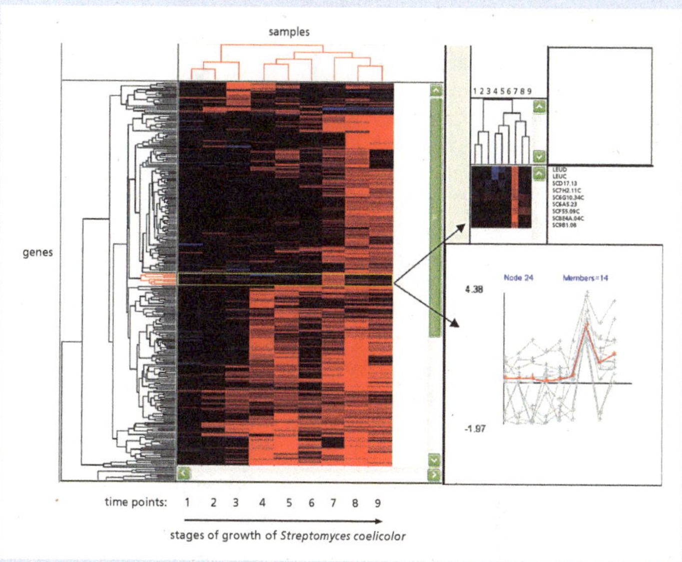


**Figure B16.1**
**A dendrogram and a map showing the clustering of genes with an expression profile of a few selected genes.**

Figure B16.1 shows on the left side a section of the map obtained by hierarchical clustering, using Pearson correlation and the average method to define inter-cluster distances. The red dendrogram on top of the map clusters the samples according to the time points, while the dendrogram on the side shows the clustering according to genes. We can see that the dendrogram divides the time points into three main groups. These correlate to some extent to the growth phase and antibiotic production. A small cluster of genes is selected (highlighted red on the side dendrogram) where the expression increases at time point 7 and then decreases again. The data for this small cluster are shown at higher magnification on the top right of the figure, and include the gene identities. The lower right part of Figure B16.1 shows a cluster produced by a SOM analysis of these data, containing 14 members, which includes all the nine genes found by the hierarchical cluster. Figure B16.2 shows on the left side a *k*-means clustering of genes resulting from analysis of these data, with the clusters identified by the vertical colored bar. A similar group of genes with increased expression at time point 7 is found, indicated by the pale magenta line on the left side of the cluster. This group contains 12 members including all 9 found in the hierarchical clustering case, and 10 that are also found in this SOM cluster. This example shows that there are cases where all three methods can reveal essentially the same result, but also illustrates the different ways in which the results are presented. The results obtained with these methods are frequently sufficiently different that it is hard to find regions that can easily be compared.

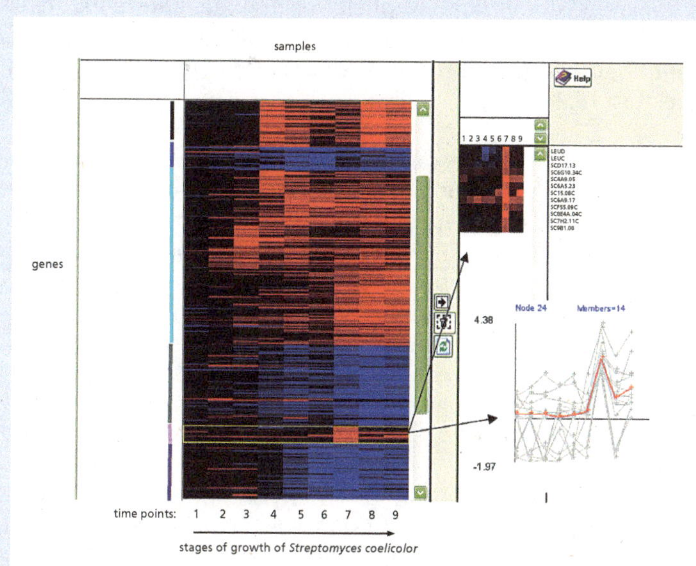


**Figure B16.2**
**A map obtained with the *k*-means method and an expression profile of the same set of genes as in Figure B16.1.**

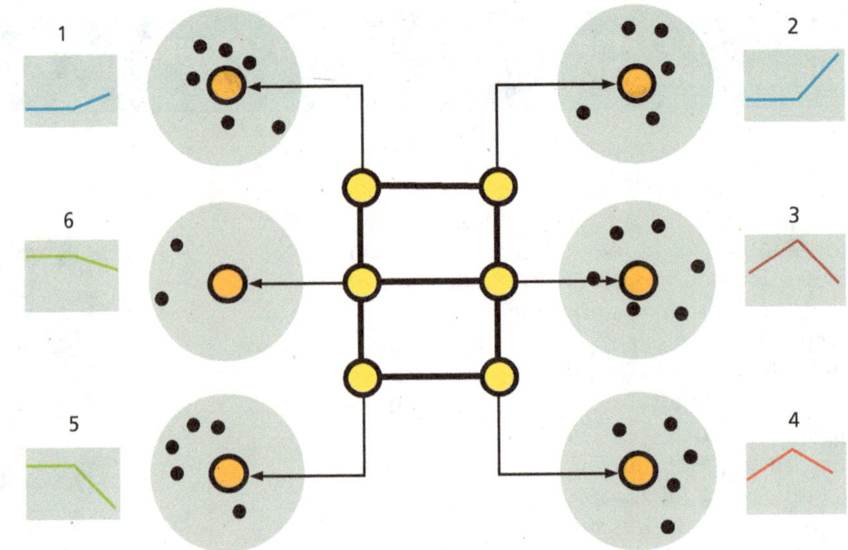


**Figure 16.15**
**A diagrammatic representation of a self-organizing map (SOM).** The initial geometry of the nodes in a 3 × 2 rectangular grid is given by solid lines connecting the nodes (yellow circles). The nodes migrate as they adapt to fit the data during successive iterations of the SOM training algorithm (arrows and orange circles). Individual data points are represented by black dots and the six clusters associated with the nodes in their final positions by large gray circles. Average patterns are shown for each of the nodes. Nodes such as 3 and 4, which are geometrically close, contain patterns that are similar. Adapted from Figure 1 of Tamayo et al., 1999.

## Self-organizing maps (SOMs) use neural network methods to cluster data into a predetermined number of clusters

Many researchers have investigated the possibility of using neural networks as an alternative way of clustering large datasets. The self-organizing map (SOM) (or **Kohonen network**, named after its inventor) is one such method, briefly introduced in Section 15.1. Unlike the neural networks described in Section 12.4 for secondary structure prediction, a SOM uses unsupervised learning. This means that the parameters of the network are determined by the data with no extra information supplied. (Recall that in the case of supervised learning by the back-propagation method the parameters were adjusted according to the output signal errors, requiring a dataset whose true secondary structure was known.)

The structure of the SOM is different from the neural networks described in other sections of this book, all of which involved lines of units that were called layers. A SOM typically has a grid-like topology between units (here called nodes) which have a two-dimensional connectivity (for example, a 2 × 3 grid of nodes as in Figure 16.15). Despite the apparent geometrical simplicity, each node represents a cluster of data points, and is defined within the same variable-space as the data. The variable-space is the coordinate system defined by the variables of the expression data, as summarized in Section 16.2.

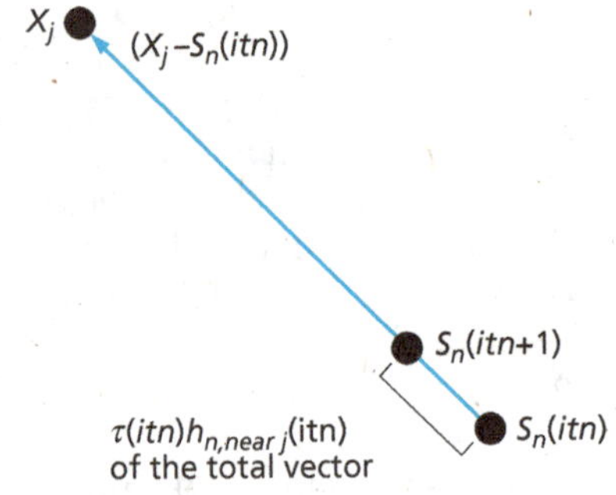


**Figure 16.16**
**A geometrical illustration of the formula for modifying the location of SOM nodes. (Equation EQ16.12.)** During the $itn^{th}$ training iteration, SOM node $S_n(itn)$ is close enough to data point $X_j$ for its location to be modified. The difference between the location of the node and data point is shown as a blue vector. The node is only moved a fraction $\tau(itn)$ $h_{n,near\,j}(itn)$ along this vector, a fraction which decreases with iteration number as well as with distance of the node $S_n(itn)$ from $S_{near\,j}(itn)$, the nearest node to $X_j$.

During the training of the SOM to the data, the individual nodes move toward nearby data points, and simultaneously nodes that are closely connected on the grid also move. This is in contrast to *k*-means clustering, where a data point only affects a single cluster during any step. The effect of each data point is considered separately, cycling through all the data for many iterations. Once training is complete the members of a cluster are defined as those data points for which a particular node is the nearest. Hence this procedure produces as many clusters as there are SOM nodes. Adjacent nodes on the SOM grid tend to have associated with them data clusters with closely related patterns (see Figure 16.15), as they will be in a close-by region of the variable-space.

The *j*th expression data is written $X_j$, and contains many components, of which the *A*th is written $X_{j,A}$, and corresponds to the expression level of the *j*th gene/protein in the *A*th sample/condition. The initial location of the *n*th SOM node is written $S_n(0)$ and is defined by a set of coordinates, one corresponding to each sample/condition. The position of the same node during iteration *itn* is written $S_n(itn)$. During the training procedure the nearest node $S_{near\,j}$ is identified for each data $X_j$, usually in terms of Euclidean distance. A neighborhood function $h_{n,near\,j}(itn)$ defines a coefficient for each node $S_n$, which is used to modify the effect that $X_j$ has on altering its position. This function has the value 1 for node $S_{near\,j}$ and decays to 0 for nodes at a distance away from $S_{near\,j}$, and is often a Gaussian. As well as the node separation, $h_{n,near\,j}(itn)$ depends on the iteration number $itn$ such that as $itn$ increases its range decreases, so that by the end of the training period only node $S_{near\,j}$ moves in response to $X_j$.

For each data $X_j$ during iteration $itn$ all the SOM nodes are moved toward $X_j$ according to the formula:

$$S_n(itn+1) = S_n(itn) + \tau(itn)h_{n,near\,j}(itn)\left(X_j - S_n(itn)\right) \quad \text{(EQ16.12)}$$

where the learning rate $\tau$ decreases with iteration number $itn$. This formula can be illustrated geometrically (see Figure 16.16) and shows that SOM nodes are moved toward nearby data points. The closest node $S_{near\,j}$ is moved the most, whereas other nodes are moved by smaller amounts according to their distance from $S_{near\,j}$ and the number of iterations already completed.

Often two phases of the training are distinguished. The first of these is when the major changes in node location occur, and self-organizing is very apparent. During this phase larger values of $\tau$ are used (often 0.1 at the beginning, decreasing to 0.01) and the neighborhood function extends over most of the SOM. The second stage involves smaller changes, and ideally should reach convergence. Although $\tau$ may stay at 0.01, the neighborhood function will at first extend over only a few nearest nodes, and ultimately not extend beyond $S_{near\,j}$.

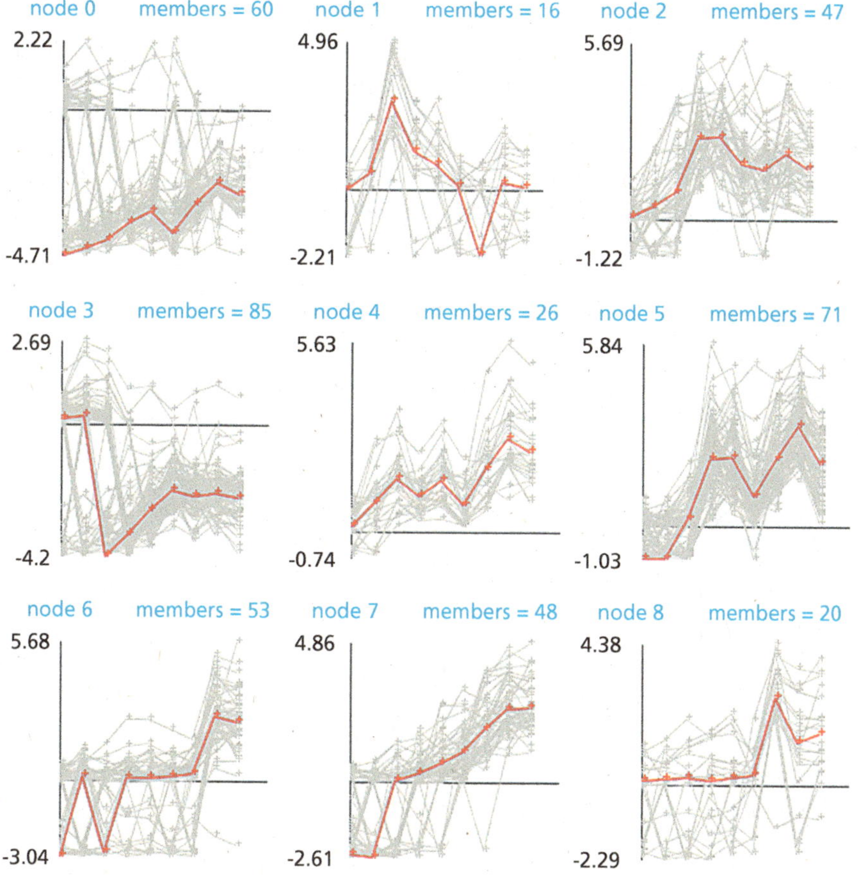


**Figure 16.17**
**An example of the results from a SOM analysis.** The SOM used is arranged as a 3 × 3 grid, labeled nodes 0 to 8. For each node the expression level patterns are shown in gray for all members of the node, with the average for the node shown in red. There is a considerable variation in the number of members of each node and the variability of the patterns in each node.

The properties of the data points associated with each SOM node can be investigated to identify their general characteristics and the variation present. An example is shown in Figure 16.17, which shows the data classified into nine clusters of a 3×3 SOM. If too few nodes were used, some or all the clusters will have a large variation in properties, indicating that another SOM should be trained with more nodes. The data associated with nodes 0 and 3 of Figure 16.17 show considerable variation that might be indicative of a need for more nodes. Eventually in ideal circumstances all the clusters should have well-defined characteristics, such as shown in node 5, indicating that the classification is complete. In this way the SOM methods can identify the required number of nodes for a particular dataset.

## Evolutionary clustering algorithms use selection, recombination, and mutation to find the best possible solution to a problem

Genetic algorithms use the principle of evolution to find a solution to a problem, and we have already seen one in operation in multiple sequence alignment (the SAGA algorithm in Section 6.5). The process mimics the factors found in biological evolution, such as natural selection, recombination, and mutation. Genetic algorithm methods use the concept of a population (a set of individuals) where each individual represents a possible solution to a problem. With each generation of the algorithm, the population is changed by selecting fit individuals, modifying and/or reproducing these, and then combining them with others to create a new population for the next generation (see Figure 16.18). The fitness measure gives an indication of the fitness of an individual; when clustering protein or gene expression data a fitness measure is an assessment of the quality of the clustering. Over time, the population gets fitter as solutions of better quality emerge. There are a number of ways in which genetic algorithms can be used in cluster analysis. We will briefly describe some aspects of one technique that analyzes $N$ objects into $k$ clusters.

The **greedy permutation encoding method** is fully defined by an ordered list of $N$ objects such as $(x_1\ x_2\ \ldots\ x_N)$. The first $k$ of these objects is used as the initial members of the $k$ clusters. Each subsequent object is taken in order and added to the cluster that gives the optimal clustering. In these methods optimal clustering is often defined as the sum of the squares of the distances of each object from the cluster centroid, which results in objects being added to the cluster with the nearest

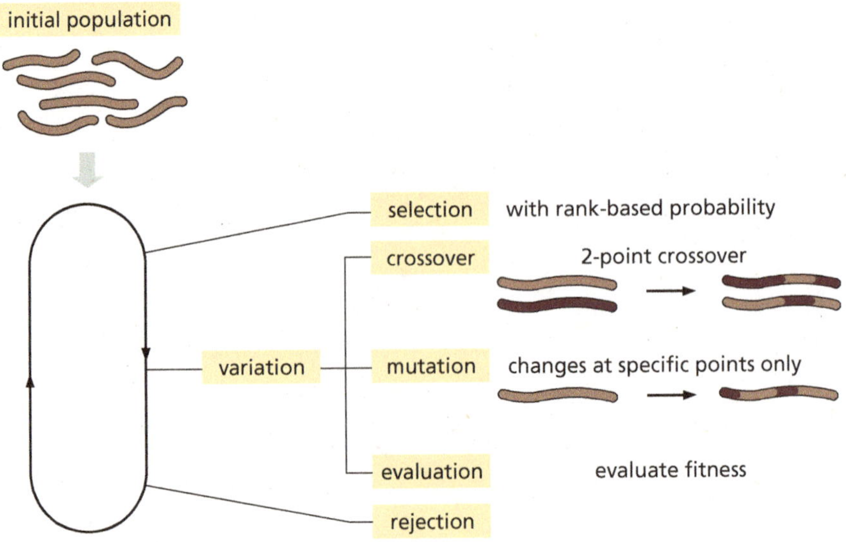


**Figure 16.18**
**The procedure used in the evolutionary clustering method.** The initial population is subjected to numerous cycles of modification to obtain the next generation, assessment of fitness, and further modification. Two types of modification of the existing clustering are available: crossover and mutation. These both involve modifications that have a random element, such as randomly choosing the clusters to change. Following modification, assessments must be made to determine whether the new clustering is an improvement on the previous one, so that changes that do not improve the fit to the data are rejected. Eventually the cycle will end, either because the desired level of fitness has been achieved, or more often because the required large number of cycles has been completed.

centroid. In this way the clustering is fully defined by the object order. The set of individuals that comprise a generation will have different object orders, and the new generation will be formed by modifying the object order of some or all of the existing individuals.

This modification process will be illustrated here by the order-based crossover operator, which creates two new individuals based on two existing ones. This operator uses two parents to produce two offspring by taking a sub-sequence of objects from one parent and adding missing objects while preserving the relative order in the other parent. In the following example the sub-sequence is defined as between the | symbols. Given the two parents:

$$P_1 = (1\,2\,3\,|\,4\,5\,6\,7\,|\,8\,9)$$

$$P_2 = (4\,5\,2\,|\,1\,8\,7\,6\,|\,9\,3) \qquad \text{(EQ16.13)}$$

this crossover operator produces offspring as follows. First the sub-sequences of each parent are copied into offspring:

$$O_1 = (x\,x\,x\,|\,4\,5\,6\,7\,|\,x\,x)$$

$$O_2 = (x\,x\,x\,|\,1\,8\,7\,6\,|\,x\,x) \qquad \text{(EQ16.14)}$$

Then, starting from the second "|" of one parent ($P_2$), the objects from the other parent are copied in the same order, omitting symbols already present. When the end of the string is reached, the transfer of objects continues from the first string position in the parent ($P_2$). The sequence of the objects in the second parent (from the second "|") is:

9 – 3 – 4 – 5 – 2 – 1 – 8 – 7 – 6

Objects that are already in the first offspring (4, 5, 6, and 7) are removed to leave the following objects:

9 – 3 – 2 – 1 – 8

The objects are transferred in this order into the first offspring (starting from the second | and then continuing from the start):

$$O_1 = (2\,1\,8\,|\,4\,5\,6\,7\,|\,9\,3) \qquad \text{(EQ16.15)}$$

In a similar fashion the second offspring is born:

$$O_2 = (3\,4\,5\,|\,1\,8\,7\,6\,|\,9\,2) \qquad \text{(EQ16.16)}$$

These define new clusterings, which can be evaluated by the measures as already described. Because this crossover operator preserves some of the object ordering, the clustering defined by the offspring should in most cases show relatively small changes from the parent clusterings. Other methods of modifying individuals to obtain the next generation can also be used, including small-scale changes equivalent to point mutations. Usually a random selection is made of which alternative method to use on any particular individual, using all possible methods during each cycle.

Each new generation is assessed for their fitness, and the methods by which successive generations are created are designed to tend to increase the fitness of the individuals. After a specific number of generations, or when the measured fitness exceeds a threshold, the process is stopped and the population used to obtain a clustering.

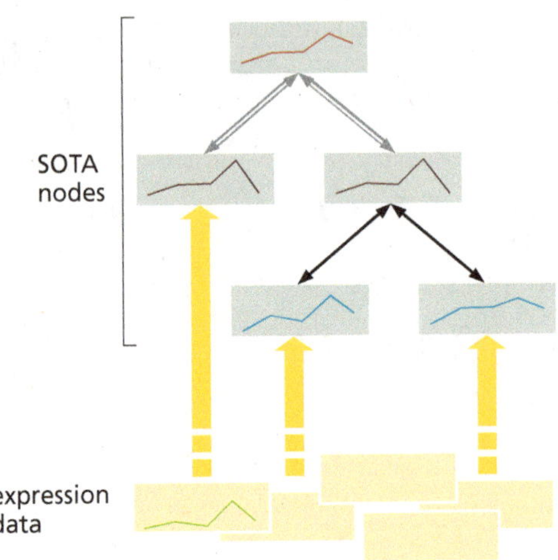


**Figure 16.19**
**A schematic illustration of the SOTA method.** Initially three nodes are defined whose vectors are the average of all the data. These nodes are then adapted using the data. The gene data become associated with their nearest cell. The results are the ancestral node with the red expression pattern and the two daughter nodes with brown expression patterns. As these are adapted, the two daughter nodes are the cells presented with data (shown on the bottom of the figure as yellow boxes), but all three nodes are adapted according to Equation EQ16.17, shown by the double arrows between them. After this initial adaptation, the right daughter cell has the greatest pattern variation (resource $R_n$, Equation EQ16.18) and so two daughter nodes (blue expression patterns) are added to it. The diagram shows the current adaptation phase, when data are presented to the three cells (external nodes) as shown by the yellow arrows. According to the rules, when the leftmost cell is adapted, its mother and sister nodes are not moved because the sister node is not a cell. The gray color of the double arrows indicates these interactions are not currently active. However, the blue pattern sister cells and their mother node do adapt together, as shown by the black double arrows. Adapted from Figure 1 of Herrero et al., 2001.

## The self-organizing tree algorithm (SOTA) determines the number of clusters required

All the clustering methods described above suffer from the limitation that the user has to arbitrarily fix the number of clusters at the start, leading to rather subjective data analysis. Methods such as *k*-means clustering and SOM permit the number of clusters to be chosen, so that to some extent the dependence of the results on this choice can be examined. The final decision on cluster number in these cases will, however, be less objective than the following method, which circumvents this problem by varying the number of clusters in a controlled manner.

The self-organizing tree algorithm (SOTA) is a combination of a Kohonen network (SOM) similar to that described previously, which allows network nodes to move (adapt) in response to the data, and a technique to selectively increase the number of nodes. This technique works equally well on gene and protein expression data. Each gene is represented by a vector of expression measurements and each network node is represented by an equivalent vector. The distance between the *j*th gene expression vector $X_j$ and the *n*th network node $S_n$ is defined by one of the measures discussed previously, and will be written $|X_j - S_n|$. The SOTA network differs in its organization from that used in SOM, in that it is hierarchical with each internal node being the ancestor of two daughter nodes (see Figure 16.19). A second significant difference is that only a limited set of nodes can adapt to the expression data during a stage in training the network. The external nodes are called cells, and only cells and their direct ancestors can be modified in the further training of the network.

The initial SOTA network consists of three nodes whose vectors are initialized to the average of all the data. One of these nodes is the ancestor of the other two sister nodes. These two descendants are external nodes and are called cells. During SOTA training the cells (external nodes) of the network are treated differently from the internal nodes. The algorithm consists of a repeating series of alternating steps, firstly moving the network nodes toward the data, and secondly creating new nodes.

Firstly, the cells are adapted in a similar manner to that applied to a SOM, in that all expression vectors are associated with the nearest cell, and that cell is then moved toward the data. The complete set of expression vectors is presented several times, each set being referred to as an epoch. In the *itr*th epoch the closest cell (winning cell) at current position $S_n(itr)$ is moved nearer to the expression vector $X_j$ according to the formula

$$S_n(itr+1) = S_n(itr) + \tau\left|X_j - S_n(itr)\right| \qquad \text{(EQ16.17)}$$

where $\tau$ is a small constant, typically 0.01. If the sister cell of $S_n$ is also an external node (see Figure 16.19), both it and the direct ancestor node are moved closer to $X_j$, but smaller values of $\tau$ are used so that the effect is less than for cell $X_j$. Typical values of $\tau$ for the sister and ancestor nodes are 0.001 and 0.005, respectively. If the sister cell of $S_n$ is an internal node only $S_n$ is adapted.

During this procedure convergence is monitored as a function of the average distance of associated expression vectors from each cell $S_n$. This is referred to as the resource $R_n$, defined by

$$R_n = \sum_{j \in S_n} \frac{|X_j - S_n|}{\|S_n\|} \quad \text{(EQ16.18)}$$

where the summation is over all expression vectors associated with cell $S_n$, and $\|S_n\|$ is the number of genes associated with this cell. This value is an indication of the heterogeneity of the expression vectors associated with the cell. The sum of the resource for all the cells in the network is calculated after each epoch, and is used with predefined thresholds to determine when the network has converged.

Once the network has converged it is examined to identify a suitable location for new nodes. The decision is based on the resource of each cell and a predefined threshold value. If any cell has a resource that exceeds the threshold, two new daughter cells will be generated from the cell with the highest resource. Once the daughter cells have been created the adaptation step is rerun, followed by reassessment of all cell resources. The result of applying SOTA to a gene expression dataset is a hierarchical dendrogram of clusters, with each cluster having a limited degree of heterogeneity as defined by the resource threshold. The vectors $S_n$ define the averages of the expression vectors associated with them. If the resource threshold chosen is zero, the network will continue to evolve until every cell contains just one expression profile, at which point the resource will be zero for all cells.

## Biclustering identifies a subset of similar expression level patterns occurring in a subset of the samples

There are many occasions when it is informative to cluster the samples together with the gene or protein expression patterns. One example of this arises in studies of tissue samples obtained from patients with a specific medical condition, where the samples are classified according to a medical diagnosis on the basis of clinical symptoms and tests. There is usually considerable heterogeneity within groups of patients (in age, habits, smoking or nonsmoking, living conditions, other disease conditions, and so on), which affects the sample and may potentially mislead analysis. Clustering the samples can confirm classifications (see Section 16.5); that is, different sample types will, hopefully, form separate clusters. It is often beneficial to do this before studying the expression level features, but even more so to combine the two cluster analyses. The technique of jointly defining clusters that involve a subset of similar expression level patterns and a subset of the samples is called **biclustering**.

In most systems only a relatively small set of all the genes or proteins in a system will have significantly different levels of expression under different conditions. If this set can be identified, they should provide a clear signal that can distinguish the conditions, as well as being informative about the key molecular processes and responses involved. The components that do not show significant variation will only serve to mask the signal with random noise. A major benefit of two-way clustering is to identify the genes or proteins that are highly correlated with specific sample conditions.

In recent years a large number of methods have been proposed for biclustering expression data. It is beyond the scope of this book to describe any of them in

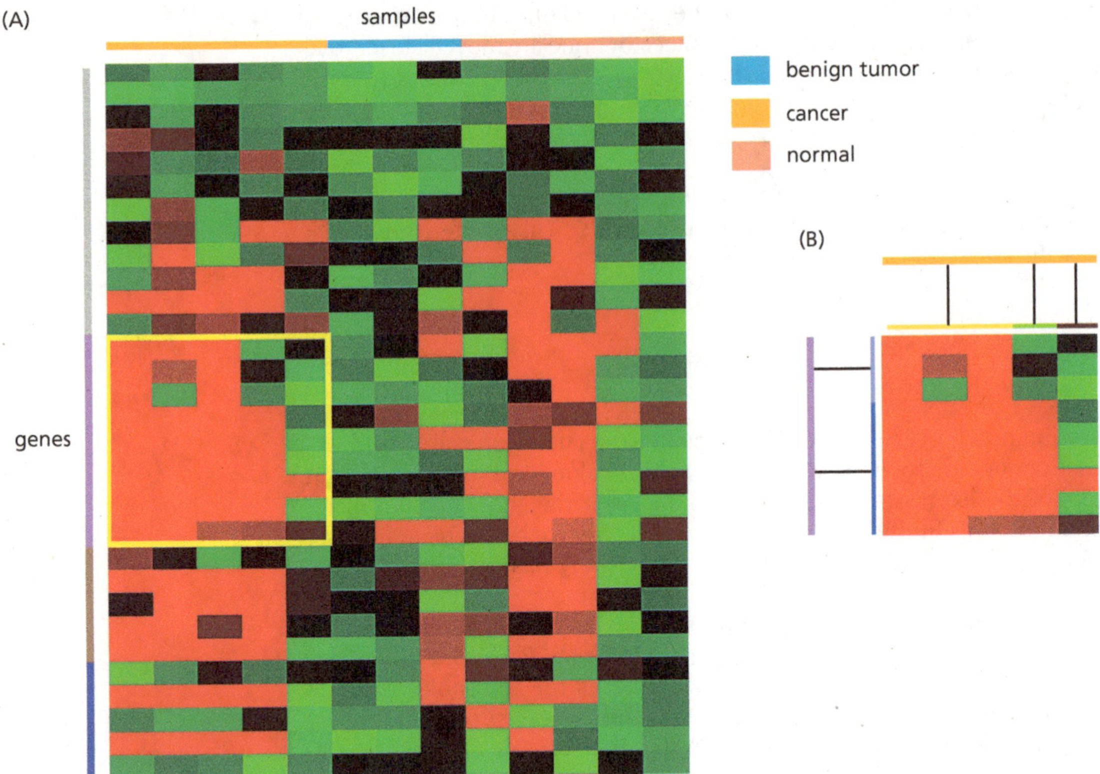


**Figure 16.20**
**An example of biclustering using a separate SOM in each dimension.** The example is for gene expression data that come from samples that have been identified as coming from normal tissue, cancerous tissue, or a benign tumor. (A) The first cycle of clustering results in three clusters for the samples, correlating with the known sample conditions as given by the color key. Four gene clusters are defined. The SOM clusters are shown by the colored bars alongside the grid of color-coded expression values. The yellow rectangle highlights the combined gene and sample cluster whose further analysis is shown in (B). (B) The further clustering of this rectangle of data shows three sample subclusters, and two gene subclusters. These steps could be continued if desired to define further subclusters.

detail. (See Further Reading for details.) The methods can be divided into two major groups: those that perform the complete analysis of both dimensions (samples and expression) in a unified algorithm, and those that use combinations of separate cluster analysis in each direction. In all cases the resulting partition consists of blocks of expression levels of a subset of genes or proteins as measured in a subset of the samples. Ideally these techniques should include clustering methods that automatically determine an appropriate number of clusters.

We will illustrate biclustering in the case of a method using cycles of combining SOM clustering in each dimension (see Figure 16.20). In the first cycle the expression levels and samples are clustered independently as described previously, to produce two separate sets of clusters for the complete sets of expression levels and samples (see Figure 16.20A). These clusters are then used to divide up the data for further analysis. For each pair of clusters—one expression level cluster and one sample cluster—the data relating to just the intersection of the two clusters is reanalyzed in the second cycle to obtain subclusters of expression levels and of samples (see Figure 16.20B). This procedure can be repeated further, as much as is found useful. This analysis can lead to new insights into the data that could be missed by the simpler clustering methods described previously.

## The validity of clusters is determined by independent methods

All the clustering methods described are exploratory data analysis tools and are mainly used as a means of inspecting the data to extract insights. All clustering results should be assessed to evaluate how well the clusters produced fit the data. It should be noted that clustering algorithms will report clusters even when the data are not truly clustered. Therefore the results need to be treated with some caution. In most cases the assessment of clusters relies on the ability to find the most meaningful clustering based on other knowledge. Meaningful clusters will commonly contain proteins or genes that act together in some biologically significant process, such as the same regulatory pathway. Therefore, once the proteins or genes that cluster together have been identified, they then have to be characterized further to establish whether the clustering has any biological meaning. A common approach is to use gene ontology (GO) terms to assess whether a cluster of genes or proteins has known shared functionality. For further information about GO see Section 3.3.

Some quantitative measures have been proposed that can be used to test the relative merits of one clustering solution over another. One example is the total resource in SOTA, equivalents of which are readily defined for other clustering methods. However, these measures usually only give a clear preference in the simplest of circumstances. For example, if the data form clusters that are cleanly separated from each other the average distance of data from their cluster center will be significantly less than the distances between cluster centers. In general for real data such assessment is not straightforward. For further details of some proposed methods to assess the validity of clusters and the results of biclustering see Further Reading.

## 16.4 Statistical Analysis can Quantify the Significance of Observed Differential Expression

It is often important to identify those genes or proteins that are expressed at different levels in different samples. However, all the measured expression levels can be expected to show differences, even if only on account of measurement errors. We need to identify the genes or proteins whose expression level is really different in the experimental samples. Another analysis that is frequently required compares two samples to determine if they are significantly different. This involves comparing the expression levels of a large number of genes or proteins for the two samples. The methods for determining which measured expression level differences are real and significant and for comparing samples are the subject of this section.

The data we have are measurements of the same property (expression of a particular gene or protein) under different conditions or in different samples. We want to know whether the measurements for a given gene/protein/condition/sample are truly different. The result of such an analysis will be a list of genes or proteins that

**Flow Diagram 16.4**
**This section describes some of the statistical tests which can be applied to expression data to determine if the differences in expression levels are significant.**

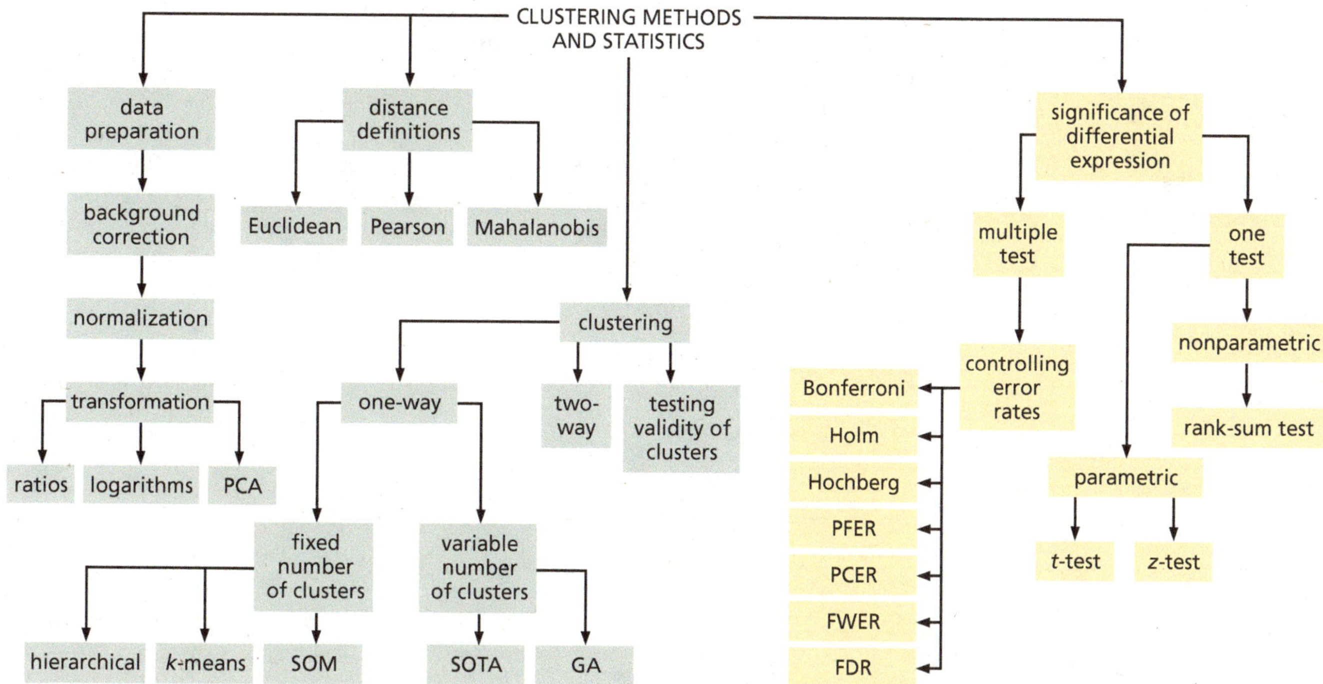

are identified as differentially expressed under the various conditions of the experiment. This list is just the starting point for further analysis, which may be computational or experimental. We want the most accurate list possible, so as to neither miss important expression changes due to false negatives nor to waste time on false positives.

The simplest technique for determining differential expression is to look at the ratios of expression in different samples. A gene or protein can be considered to be differentially expressed in different samples when the ratio exceeds a given threshold. The threshold is often set at 2 (indicating a two-fold increase or decrease in expression). If the threshold applied is large enough (say 10) then we can be more confident in this assignment, but many differentially expressed genes will be missed. However, if a lower threshold is applied more differential expression will be identified, but there will be a greater probability of false-positive assignments where the measured expression difference has a large contribution from random fluctuations.

We need to use statistical tests if true differences in expression are to be identified more effectively. The benefit of using such tests is that they provide a quantitative measure of the significance of the difference. The repeated measurements of a particular expression level are assumed to be random selections from a statistical distribution, so that the mean and variance of the measurements can be taken as estimates of those of the distribution. If in different samples or under different conditions the two expression levels are actually the same, the individual measurements will be random samples from the same distribution. The statistical tests we will describe estimate the probability that the two sets of expression level measurements come from the same distribution, so that there is no differential expression. If this probability is small enough, such as 0.01, this can be interpreted as evidence of significant differential expression, with the measurements being samples of two different distributions.

We need to determine whether the two mean values of expression level measured for the same gene or protein in different samples are really different. It is more likely that there is differential expression the less that the two distributions overlap, which will clearly depend not only on the difference in the means but also to some degree on the variances (see Figure 16.21). In fact the area of overlap is proportional to the difference in the measured means expressed in units of the standard deviation (square root of the variance) of the assumed distribution (see Figure 16.22). The tests described here involve several different forms of this quantity, often called the test statistic. Under the particular assumptions of a test (usually the type of statistical distribution from which the measurements are made and whether all variances are the same or not) this test statistic has a known distribution itself. The significance of the difference in mean measurements is assessed by calculating the area of the two tails of the distribution of the test statistic, as shown in Figure 16.23. This area is the probability that the two mean measurements are really different, as opposed to their difference being simply due to random effects. A smaller value of

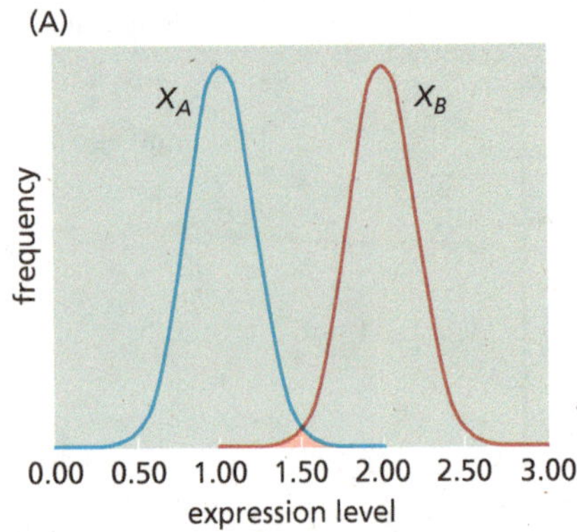


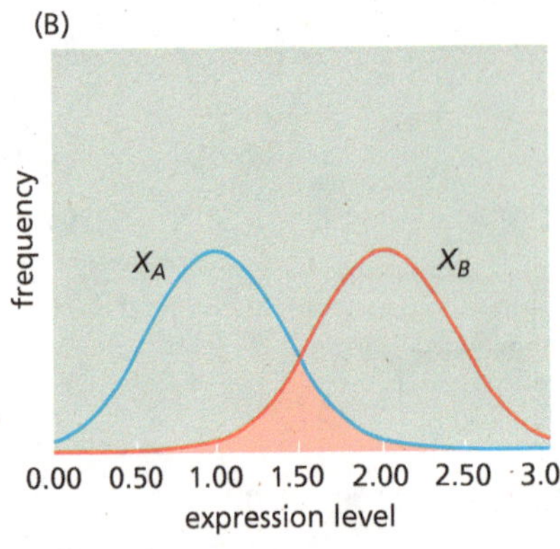


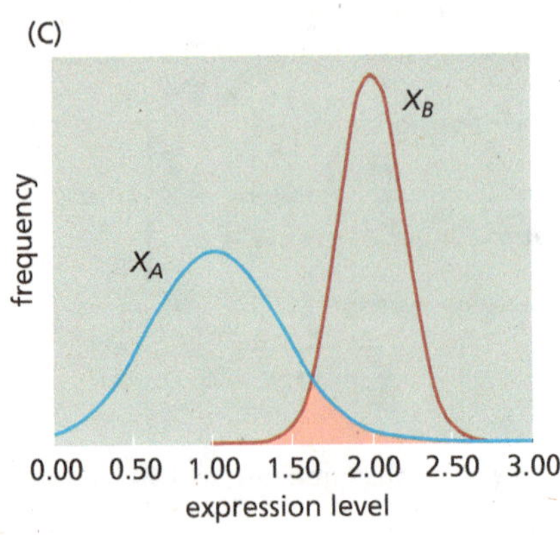


**Figure 16.21**
**The importance of variance in detecting significant differences.** A two-fold change in average expression level is only a significant difference if the distribution variances are sufficiently small. The significance is related to the area of overlap of the two distributions. In all three examples the two distributions have the same mean values ($\bar{X}_A = 1.0$, $\bar{X}_B = 2.0$). (A) Two distributions of measurements *A* and *B* are clearly different (variance $s^2 = 0.04$). (B) When the variance is larger the overlap increases ($s^2 = 0.16$). (C) Even if one measurement has a small variance, if the other has a large variance the overlap makes the difference in distributions less significant than in part (A).

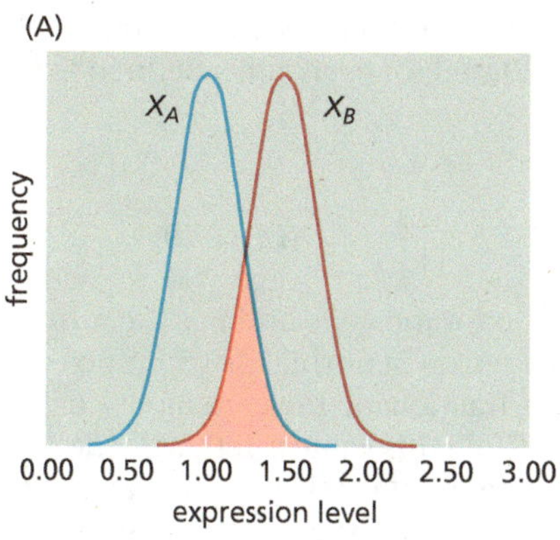


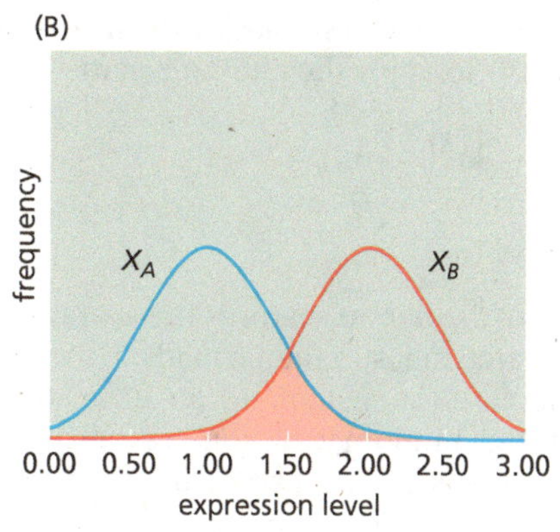


**Figure 16.22**
**The overlap of two distributions is proportional to the difference in means expressed in units of the standard deviations.** (A) Two distributions with means $\bar{X}_B = 1.0$, $\bar{X}_B = 1.5$, and variance $s^2 = 0.04$. (B) Two distributions with means $\bar{X}_A = 1.0$, $\bar{X}_B = 2.0$, and variance $s^2 = 0.16$. In both cases $(\bar{X}_A - \bar{X}_B)/s = -2.5$, and the same proportion of overlap occurs.

this area indicates a more significant measured difference. The area is compared to a preassigned value, typically 0.05, which equates to a 5% significance level, in order to define which tests indicate differential expression.

A distinction is made between testing whether the mean measurements are different, which is called a **two-tailed test**, and specifying that one mean measurement is, say, lower than the other, which is a **one-tailed test** and involves just one of the areas shown in Figure 16.23. All the applications described below use the two-tailed test, as we are interested in any differences, whether increases or decreases in expression level. The further the calculated test statistic is from zero, the more likely it is that the two mean measurements are statistically significantly different. A significance level is set to define the false-positive rate that can be tolerated. Here a **false positive** refers to an incorrect deduction that the two mean measurements are different when they are not, which is also called a **type I error**. The false-positive rate, often called $\alpha$, is typically set at 5%, which means that in 5 out of 100 such tests a statistically significant difference between the means is reported even if there is none.

If the expression data have been converted to log ratios and procedures such as lowess have been used to make the data have an approximately normal distribution as described in Section 16.1, the standard ***z*-test** can be applied to determine significant expression differences. In this case the *z*-test calculates the probability that a gene has the same expression level in state *A* as in state *B*, based on the observed log-ratio. The test assumes that the log ratios are normally distributed, and the distribution mean is taken to be 0. The variance $s^2$ is calculated over a set of genes

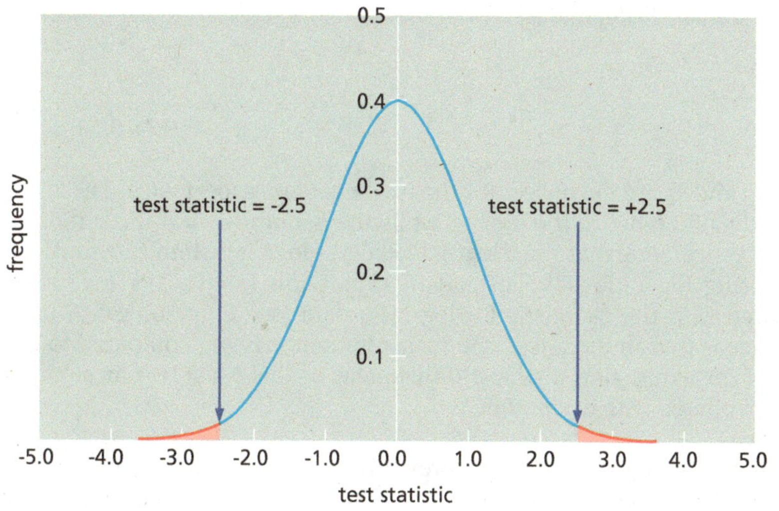


**Figure 16.23**
**Calculating the probability of a real measured difference in means.** A test statistic, in this case with the value –2.5, is plotted against the distribution of the test statistic itself. The area under the complete curve is exactly 1. The probability that there is no real difference in the means is given by the area of the two distribution tails defined by the positive and negative values of the measured test statistic. This is known as a two-tailed test.

within a range of similar expression level to that of the gene being tested (see Figure 16.5B). To apply the $z$-test a $z$-statistic is calculated for each gene, defined as

$$z = \frac{\log\left(x_A / x_B\right)}{s / \sqrt{n}} \qquad \text{(EQ16.19)}$$

where the denominator is the average log ratio for the gene and $n$ is the number of repeated measurements made. Using the properties of normal distributions we can calculate (often looking up in a table) the fraction of measurements expected through random variation to exceed in magnitude this value $z$. If this fraction is less than a predefined threshold $\alpha$ then the measurement is taken to indicate a significant difference, meaning that the expression levels of the gene in states $A$ and $B$ show a real difference.

This test relies on the log ratio data being normally distributed with a mean value of 0. In general this cannot be assumed, and it is better to use the tests described below, as they are likely to be more accurate in identifying differential expression.

## *t*-tests can be used to estimate the significance of the difference between two expression levels

A small number of independent measurements will have been made of the expression level of each gene or protein in each sample. An estimate of the actual value of the expression is given by the mean of these measurements. The degree of uncertainty in this value can be quantified by the variance in the measurements. The standard statistical test used to compare two such average measurements and estimate the possibility that they are not really different is the ***t*-test**. Note that there is an implicit assumption that the measurements are samples from a normal distribution.

There are several different versions of the $t$-test, depending on whether a single set of measurements is being compared to a known value, whether two sets are being compared with each other, and whether the variances of the two sets are different. The problems under consideration here all involve comparing two sets of measurements. To perform the test a value called the $t$-statistic must be computed. This is then compared with the theoretical distribution of this statistic (Student's $t$-distribution) in combination with another parameter called the **degrees of freedom (*df*)** to estimate the probability that the two measurements are in fact different. A threshold value of this probability is used to assess whether the difference is statistically significant.

When there are $n$ measurements in each set, and both sets have the same variance $s^2$, the $t$-statistic is defined by

$$t = \frac{\bar{x}_A - \bar{x}_B}{s / \sqrt{n}} \qquad \text{(EQ16.20)}$$

where $\bar{x}_A$ and $\bar{x}_B$ are the means for the two sets of measurements. The numerator is simply the difference in the means, while the denominator is the standard error of this difference. Note that since $\log(A/B) = \log A - \log B$ it is simple to apply these tests when using log ratio data instead of expression levels. This $t$-statistic can be compared with the $z$-statistic for log-ratio data (see Equation EQ16.19). The key difference is that in the $z$-test one measurement is being compared to the known mean of the expression distribution whereas in this $t$-test two measurements are being compared with each other.

It is generally recognized that in expression experiments the variances of the measurements cannot be assumed to be the same. When the experimental design has the same number $n$ of measurements for each expression level the $t$-statistic is modified to

$$t = \frac{\bar{x}_A - \bar{x}_B}{\sqrt{\dfrac{s_A^2 + s_B^2}{n}}} \qquad \text{(EQ16.21)}$$

In the case where the number of measurements differs, with $n_A$ and $n_B$ in the two sets, the $t$-statistic is defined by

$$t = \frac{\bar{x}_A - \bar{x}_B}{\sqrt{\dfrac{(n_A - 1)s_A^2 + (n_B - 1)s_B^2}{n_A + n_B - 2}\left(\dfrac{1}{n_A} + \dfrac{1}{n_B}\right)}} \qquad \text{(EQ16.22)}$$

In all these cases the number of degrees of freedom ($df$) is given by

$$df = n_A + n_B - 2 \qquad \text{(EQ16.23)}$$

which in the case of equal numbers of measurements becomes $2(n - 1)$. An alternative form of the test for unequal variances has been given, sometimes called Welsh's $t$-test, in which the $t$-statistic takes the form

$$t = \frac{\bar{x}_A - \bar{x}_B}{\sqrt{\dfrac{s_A^2}{n_A} + \dfrac{s_B^2}{n_B}}} \qquad \text{(EQ16.24)}$$

which requires an alternative definition of the degrees of freedom that can result in non-integer values:

$$df = \frac{\left(s_A^2/n_A + s_B^2/n_B\right)^2}{\left(s_A^2/n_A\right)^2/(n_A - 1) + \left(s_B^2/n_B\right)^2/(n_B - 1)} \qquad \text{(EQ16.25)}$$

Comparing Equations EQ16.22 and EQ16.24, the former has as denominator the correct standard error for the difference in means, but the latter still has a Student's $t$-distribution although requiring the modified value for $df$ to assess significance.

The following example shows how the $t$-test is applied to experimental data. In this case there are eight gels in two groups of four: one group is the control ($C_i$) and the other the treated samples ($T_i$). The measured spot volume for the same protein feature in each gel is given in Table 16.1. To see whether this protein feature changes significantly between the groups we have to analyze the data as shown in the table. The resulting $t$-statistic given by Equation EQ16.21 is

$$t = \frac{0.076175 - 0.1037}{\sqrt{\dfrac{0.019499^2 + 0.007775^2}{4}}} = -2.625 \qquad \text{(EQ16.26)}$$

The value of $-2.625$ is compared with a $t$-test table, using $df = 6$ and $\alpha = 0.05$. The table gives a critical value of 2.447 for this level of significance. As our calculated value is greater than this, it is significant according to the 5% level.

**Table 16.1**
Analysis of protein gel data using the *t*-test and the rank-sum test. The data for analysis come from eight gels. One group of four are the controls ($C_i$) and the other group are the treated samples ($T_i$). The data are the measured spot volume for the same protein feature in each gel.

| Sample | $C_1$ | $C_2$ | $C_3$ | $C_4$ | $T_1$ | $T_2$ | $T_3$ | $T_4$ |
|---|---|---|---|---|---|---|---|---|
| Expression level | 0.0766 | 0.0644 | 0.0602 | 0.1035 | 0.1138 | 0.0981 | 0.0971 | 0.1058 |
| Rank | 6 | 7 | 8 | 3 | 1 | 4 | 5 | 2 |
| *t*-test statistics | $\bar{x}_1 = \bar{C}_i = 0.076175$<br>$s_1 = 0.019499$ | | | | $\bar{x}_2 = \bar{T}_i = 0.1037$<br>$s_2 = 0.007775$ | | | |
| Rank-sum statistics | $R_1 = 24$ | | | | $R_2 = 12$ | | | |

## Nonparametric tests are used to avoid making assumptions about the data sampling

The statistical tests described above rely on the assumption that the measured expression levels are samples taken from a normal distribution. In many expression experiments the small numbers of measurements and large background errors prevent this assumption being properly tested, so that the validity of the tests is highly questionable. Another situation often encountered is the study of expression levels in two sets of patients distinguished by being either healthy or with a particular medical condition. (A common alternative situation involves the two sets of patients receiving and not receiving a given treatment.) The number of patients available is likely to be relatively small, and also to have additional unknown genetic and health variations. It would be unrealistic to assume a normal distribution for the expression of any particular gene or protein in such a set of patients. However, they might show differential expression of interest, and need tests that can cope with such data.

There are tests that do not rely on these assumptions about the data sampling. They are given the general name of nonparametric tests because they do not rely on assuming a distribution defined by parameters such as means and variances. Two classes of these tests can be distinguished: those that attempt to use the data to numerically reconstruct distributions, and those that use more general properties such as ranking.

Some nonparametric methods have been developed specifically with expression experiments in mind. In the parametric tests described earlier a test statistic is calculated and compared with the expected distribution of this statistic under the assumption that there is no difference in expression level. This distribution is often called the **null distribution** as it corresponds to the null hypothesis of the test, namely that the two measurements are the same. Several methods have been proposed that use the expression measurements to estimate the null distribution or related quantities. For this, techniques are used such as resampling the measurements (the significance analysis of microarrays, or SAM, method) and using Bayesian analysis (in this case nonparametric, unlike the method described in Box 16.2). For details of these methods see references in Further Reading.

The standard nonparametric statistical test that is the equivalent of the *t*-test for parametric data is the **rank-sum test**. There are several variations of this test, which is also known as the Mann–Whitney U or Wilcoxon test. We will consider its application to detect differential expression of a gene whose expression level has been measured in $n_A$ healthy patients and $n_B$ patients with a particular disease. The total

## Box 16.2 Bayesian techniques can be used to deal with a lack of replicates

In much of the discussion in this section it has been assumed that there are sufficient replicates to determine the values of the standard deviations of individual expression level measurements with sufficient accuracy. This is usually not the case, often due to a combination of expense and lack of material. Several methods have been proposed to obtain improved estimates of the standard deviations. As has been mentioned in several other contexts, Bayesian techniques are very useful for coping with such a lack of data and are explained in more detail in Appendix A. In this case, one can use priors based on a normal distribution with different parameters for each gene or protein under different conditions. The calculations of standard deviations are modified to a form such as

$$\sigma^2 = \frac{\upsilon_0\sigma_0^2 + (n-1)s^2}{\upsilon_0 + n - 1} \tag{BEQ16.1}$$

where $\sigma$ is the estimated standard deviation, $\sigma_0$ is the standard deviation in the prior, $n$ is the number of replicates in the experiment, and $s$ is the standard deviation from the experimental data alone. $\upsilon_0$ is a parameter that represents confidence in $s$, and should be larger for small $n$, to increase the weight of the prior in such cases. $\sigma_0$ is often taken as the average of $s$ for several (or even all) genes.

number of measurements is $n = n_A + n_B$. These measurements are assigned ranks from 1 for the highest measured expression level to $n$ for the lowest. The sum of the ranks is calculated for each set of patients, $R_A$ and $R_B$. If the expression level of the gene is the same in both patient sets these two rank sums would be expected to be equal within sampling errors. The test compares a measure of their difference (often written as $U$) with a table of expected values to obtain a probability of the gene being differentially expressed. (The appropriate version of the test to use depends on the number of measurements available, details of which can be found in standard works of statistics.) Once a significance level $\alpha$ has been chosen it can be used to identify differential expression. Applying the rank-sum test to the data of Table 16.1, no significant difference is found even at the 10% significance level, in contrast to the 5% significance result with the parametric *t*-test. This illustrates that nonparametric tests are less powerful, and therefore parametric tests should be used whenever possible.

## Multiple testing of differential expression requires special techniques to control error rates

The tests described above are intended for the analysis of individual measurements. In the case of expression experiments often many hundreds or even thousands of genes or proteins are involved. If each is tested as described above, at a 5% significance level ($\alpha = 0.05$) we can expect 50 false-positive results (incorrect assignment of expression levels as significantly different in the two states) for every 1000 tests. This level of error can be expected to cause serious problems in analysis of the experiment. A variety of modifications have been proposed to the basic *t*-test method to try to improve on this situation, and these will now be described. However, it is useful to start by defining some of the terms used to describe different kinds of errors and the rates at which they occur.

Consider an experiment involving the measurement of expression levels for $N$ different genes or proteins in each of two samples. A specified level of significance $\alpha$ is assumed to apply. Suppose that $N_0$ of these genes or proteins do not differ significantly in their levels of expression in the two samples, but that when statistical tests are performed $R$ of them are declared to have differential expression levels. A summary of the tests is shown in Table 16.2. Of the values shown in the table only $N$ and $R$ are known. Note that the different methods described below may have different appropriate values of $\alpha$.

There are a number of different definitions of rates of false-positive errors (type I errors). The **per-family error rate (PFER)** is defined as the expected number of false positives in all the tests, written $E(V)$. (The term family is used in this context for the complete set of $N$ tests.) The **per-comparison error rate (PCER)** is defined as $E(V)/N$, the fraction of all tests that were false positives. If there is concern to try to avoid even a single false-positive result, the **family-wise error rate (FWER)** can be used, defined as $P(V \geq 1)$, the probability of at least one false positive in all the tests. Finally, the **false discovery rate (FDR)** is defined as $E(V)/R$, the fraction of all the tests which result in a declaration of significant differential expression that are in fact false positives, given that some results were deemed significant ($R > 0$). If no results were found significant ($R = 0$) then the FDR is defined as 0.

The first modification we will describe that tries to control error rates in multiple testing is the **Bonferroni correction**. If there are $N$ $t$-tests to perform, the significance level $\alpha$ used for a single test is divided by $N$ so that the chance of a false-positive result remains at, say, 5%. Thus, the condition that must be satisfied by a probability $p$ given by one of the $N$ $t$-tests is

$$p \leq \frac{\alpha}{N} \qquad \text{(EQ16.27)}$$

If this condition is satisfied, the test has detected significant differential expression. This controls the FWER to be $\alpha$ or less. This approach is very cautious, and has the effect of severely reducing the chance of identifying significantly different measurements, because the difference in their mean values must now be very much larger for their difference to be deemed significant.

In fact the FWER can be controlled to be $\alpha$ or less in a slightly modified method that increases the chances of detecting true positives. Each of the $N$ tests produces a calculated probability of there being no differential expression. The **step-down Holm** method requires this calculated probability for each of the $N$ tests to be ordered from the smallest to the largest probability, written $p_i$ for the $i$th ordered test. Starting with the smallest $p_1$, all $p_i$ are compared in turn to see if

$$p_i > \frac{\alpha}{(N - i + 1)} \qquad \text{(EQ16.28)}$$

Note that if this condition is satisfied the test does not detect significant differential expression. Once the smallest $p_i$ has been identified for which this condition is true, all smaller $p_i$ (from $p_1$ to $p_{i-1}$) are declared to indicate significant differential expression, and all other tests are not significant differential expression. Note that the first comparison is equivalent to the Bonferroni correction, but thereafter the required condition for $p_i$ is less harsh. In statistical terminology this method has greater power than the Bonferroni correction.

**Table 16.2**
**A summary of the $N$ tests performed to determine differential expression levels, of which $N_0$ are not differentially expressed but the tests result in $R$ being declared significantly differentially expressed.** A level of significance is assumed to have been defined.

| | Declared not significantly differentially expressed | Declared significantly differentially expressed | Total |
|---|---|---|---|
| Actually not significantly differentially expressed | $U$ (*i.e., true negatives*) | $V$ (*i.e., false positives*) | $N_0$ |
| Actually significantly differentially expressed | $T$ (*i.e., false negatives*) | $S$ (*i.e., true positives*) | $N - N_0$ |
| | $N - R$ | $R$ | $N$ |

An alternative but similar method for controlling the FWER has been proposed called the **step-up Hochberg** method. In this case, the same ordering of $p_i$ is used, defining the index $i$, but this time they are compared in the order from the largest to the smallest. Starting with this largest value $p_N$, all $p_i$ are compared in turn to see if

$$p_i \leq \frac{\alpha}{(N - i + 1)} \tag{EQ16.29}$$

Once the largest $p_i$ has been identified for which this condition is true, that and all smaller $p_i$ (from $p_1$ to $p_i$) are declared to indicate significant differential expression, and all other tests are not significant differential expression.

In the case of expression experiments it is better to accept a few false positives than to miss many true positives, so that controlling the FWER as described above is often too cautious an approach. Yoav Benjamini and Yosef Hochberg have argued that it is more appropriate to be concerned with the FDR than the FWER. As in the step-up Hochberg method the tests proceed from $p_N$ to $p_1$, and all $p_i$ are compared in turn to see if

$$p_i \leq \frac{i}{N}\alpha \tag{EQ16.30}$$

Once the largest $p_i$ has been identified for which this condition is true, that and all smaller $p_i$ (from $p_1$ to $p_i$) are declared to indicate significant differential expression, and all other tests are not significant differential expression. This method controls the FDR to be $\alpha$ or less. If $\alpha$ is set at the value 0.05, this means that 5 of every set of 100 identified cases of significant differential expression would be expected to be false positives. Further progress in defining improved methods of significance detection based on measures related to the FDR has been reported by John Storey and co-workers, as referenced in Further Reading.

Many other statistical methods have been applied to the problem of distinguishing significant changes of expression. These have included different parametric tests, for example, analysis of variance (ANOVA), as well as analysis of covariance (ANCOVA). More complicated Bayesian models have been proposed that can determine the optimal number of expression classes, which is akin to automatically determining the number of clusters in methods such as SOTA. These techniques are beyond the scope of this book and the reader is referred to Further Reading for pointers into the vast literature on this topic.

## 16.5 Gene and Protein Expression Data Can be Used to Classify Samples

In the previous section techniques are described that help identify significant differences in expression levels in different samples. The results of such analyses can be applied in two distinct ways. The set of genes or proteins that are differentially expressed may shed light on the physiological response or differences in symptoms between the samples, leading to a better understanding of the biological processes involved. Alternatively the set of differentially expressed genes or proteins can be used to determine a method for automatically distinguishing between sample classes on the basis of expression levels, of particular interest in the field of medical diagnostics. It is important to appreciate that there is a clear difference between genes or proteins with significantly different expression levels and those genes or proteins that can discriminate between sample types (see Figure 16.24).

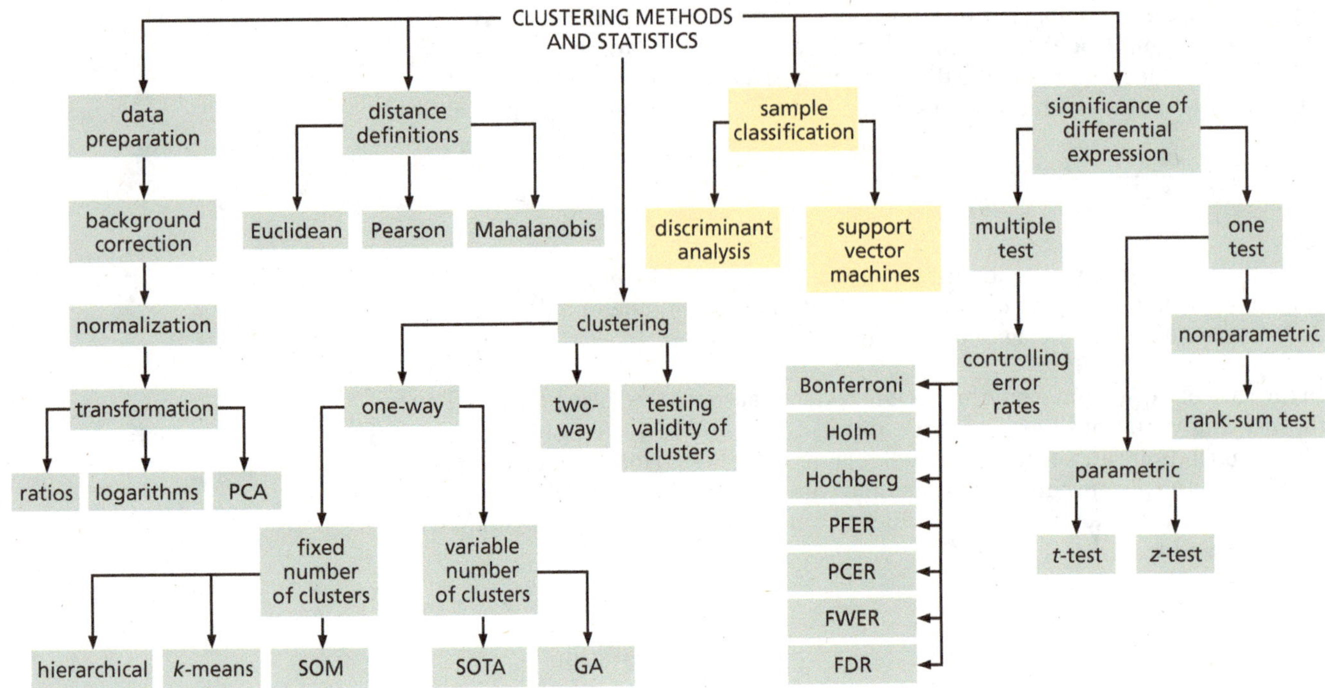


**Flow Diagram 16.5**
**In this section are described some of the range of methods that can be used to classify a sample into one of a limited number of categories on the basis of expression data.**

As an example application, the samples may have been taken from a set of patients with one of a limited number of medical conditions, and could be used to derive a method to classify samples taken from patients with unknown conditions. A particularly useful potential application can arise when patients show different responses to treatments despite having identical symptoms. Expression analysis could potentially identify signals that can then be used to indicate appropriate treatments. These signals can be used to design a **sample classifier**, which is an algebraic method of using expression levels to predict the nature of an unknown sample. It is often possible to obtain a slightly more accurate classifier by increasing the number of parameters involved. However, an optimal classifier will be accurate while only requiring a small number of measurements of the sample.

It should be noted that a potential problem with such classifiers is that they will predict the unknown sample to belong to one of the classes of sample used in deriving the classifier. Thus, for example, if the classifier was constructed using data from samples taken from patients who all have one of two different tumor types, unknown samples will always be predicted to have come from a patient with one of these tumors, even if the patient had no tumor. Care must be taken to include control samples wherever possible to minimize this potential problem.

## Many alternative methods have been proposed that can classify samples

We will now explore some of the ways in which classifiers can be derived. There are many techniques available, of which only two will be described. It should be noted that some of the methods described earlier in this chapter, such as PCA and biclustering, can also be useful in sample classification. Another method that can be used that has previously been described is the nearest-neighbor or *k*-nearest-neighbor method, which was described in the context of the prediction of amyloidogenic sequences in Section 12.3. In the nearest-neighbor method the expression levels of the unknown sample are compared with those of known samples to identify that sample which is most similar according to a given distance measure. The unknown sample is then assigned to the same group as this nearest-neighbor sample. The

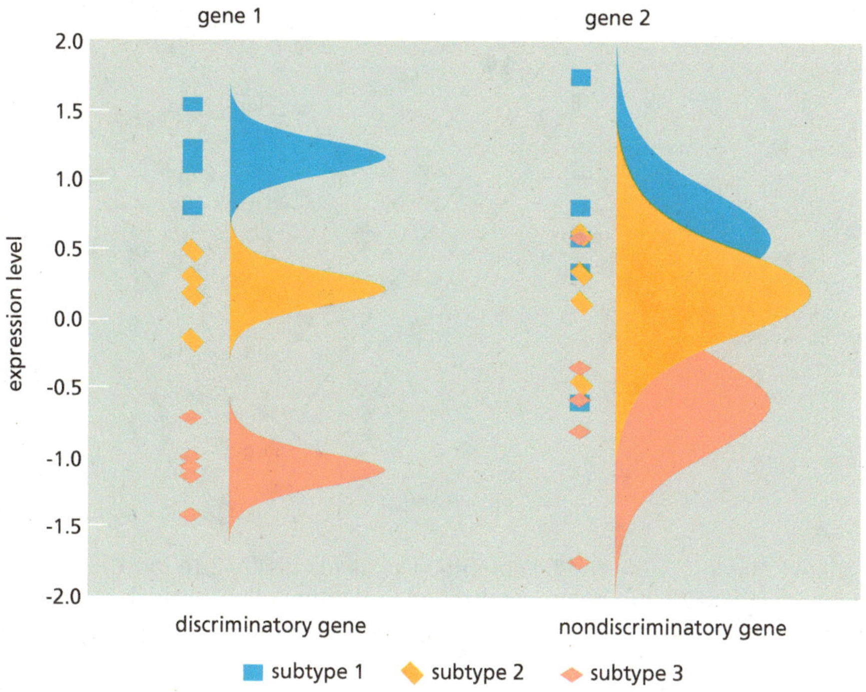


**Figure 16.24**
**The distinction between significant differential expression and discriminatory genes.** The expression levels of the two genes 1 and 2 have been measured for a number of different samples that have been classified into three distinct subtypes. Both genes show differential expression for the three subtypes, and these differences could be statistically significant if sufficient different sample measurements were available. (A large number of measurements would alter the *t*-statistic of Equation EQ16.20.) However, the large degree of overlap between gene 2 expression levels for the different subtypes makes this gene a poor discriminant of the subtypes, unlike gene 1. This is because the unknown sample will give only a single measurement, so that overlapping distributions are likely to give unclear predictions. (Adapted from D. Hwang et al., Determination of minimum sample size and discriminatory expression patterns in microarray data, *Bioinformatics* 18:1184–1193, 2002.)

*k*-nearest-neighbor method simply extends this to average the prediction over the *k* most similar samples.

Another method that has sometimes been employed to classify samples is based on decision trees, which were described in Section 10.1 and Figure 10.3 in the context of identifying tRNA genes. A series of tests is applied, usually in a specified order. In this case the tests relate to measured expression levels compared for two genes or proteins in the same sample. The genes selected are ones found to show clear expression differences in the different classes of sample, for example one gene being expressed at a higher level than the other in one sample class, and vice versa. Sometimes the result of one test determines which of several alternative tests is subsequently applied. At some point in the decision tree the sample is classified according to the last test result. Cases have been reported of successful classifiers involving only between 2 and 20 genes.

A commonly used classification method is **discriminant analysis**, which was briefly described in application to genome sequence analysis in Figure 10.16. The key concept of this method is that a transformation of the data is found which has a clear separation between the different sample groups. A simple measure of this separation can be defined as the ratio of the sum of squared distances between groups divided by the sum of squared distances within groups. The optimal transformation will maximize this ratio, and has been shown to be given by an algebraic manipulation involving covariance matrices of the groups. The end result is a formula—usually a linear or quadratic combination of the original data—whose value for the unknown sample indicates the appropriate sample group prediction. The real situation can be more complex than Figure 10.16 might suggest, as in reality multiple dimensions will be involved, but the algebraic form is still relatively simple. The line separating the two groups is called a **separating hyperplane**.

## Support vector machines are another form of supervised learning algorithm that can produce classifiers

As was presented in Section 12.6 in the context of secondary structure prediction, support vector machines (SVMs) are capable of classifying data into one of two

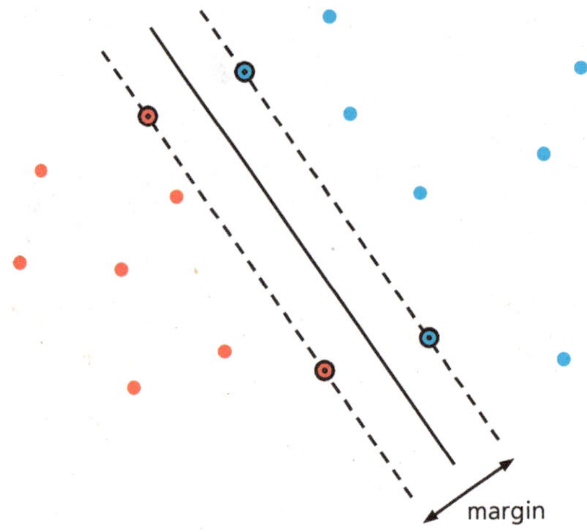


**Figure 16.25**
**The key concept behind the support vector machine.** In this simplified example all the data (represented by the circles) fall into two classes represented by their color, and the two classes are separable by a linear separating hyperplane shown as the solid black line. Two parallel hyperplanes shown as dashed lines indicate the closest that any of the data points approach the separating hyperplane. These closest data points are called the support vectors, and are indicated by double circles. The SVM method determines the separating hyperplane that maximizes the distance between these two hyperplanes, called the margin.

alternative classes. These can be combined so as to classify into more classes (see Figures 12.34 and 12.35), but generalizations exist that allow multiclass classification within a single SVM. Support vector machines are trained using a dataset that contains examples of the alternative classes, and can then be subsequently run with unclassified data to predict the class.

During training, the SVM algorithm identifies the separating hyperplane that has the greatest margin between the classes (see Figure 16.25). This involves an optimization, which turns out to be relatively tractable. The example shown has complete discrimination between the two classes, but SVM methods exist that allow for a controlled degree of classification errors, in which case some data will appear on the wrong side of the separating hyperplane. The hyperplanes do not need to be linear, and can be of complex form, improving the chances of accurate classification. For details of the algorithms see references in Further Reading. The training can be done using only those genes or proteins that have been identified as significant, but it can also use all the data. The formulae often involve coefficients that can be interpreted as weighting each gene or protein, and these can be used to select some data as having suitably small weights that they can be removed and a new SVM trained. In this way data that might have no real signal and merely contribute to the noise level can be removed to improve the performance of the classifier.

An example of the use of SVMs is shown in Figure 16.26, which also illustrates how the classification works in practice. In this example the samples come from patients with one of 14 different tumors. An SVM was trained for each tumor type to classify the sample as either having or not having that tumor. These are named classifier 1 to 14 according to the tumor type. When an unknown sample is presented, the distance from each classifier's separating hyperplane and the side on which it falls can be used to determine the predicted class and the strength of that prediction. In the example shown, the test sample gives a coherent set of classifier results indicating the sample is from a patient with breast cancer.

## Summary

Large-scale DNA microarrays or 2D gel electrophoresis followed by mass spectrometry are used to determine the simultaneous expression levels of large numbers of genes or proteins, respectively. These techniques can be used to detect genes or proteins with similar expression patterns across different samples and thus to track changes in gene expression at different times, or in different conditions, such as healthy and diseased tissue. There are a number of problems associated with this

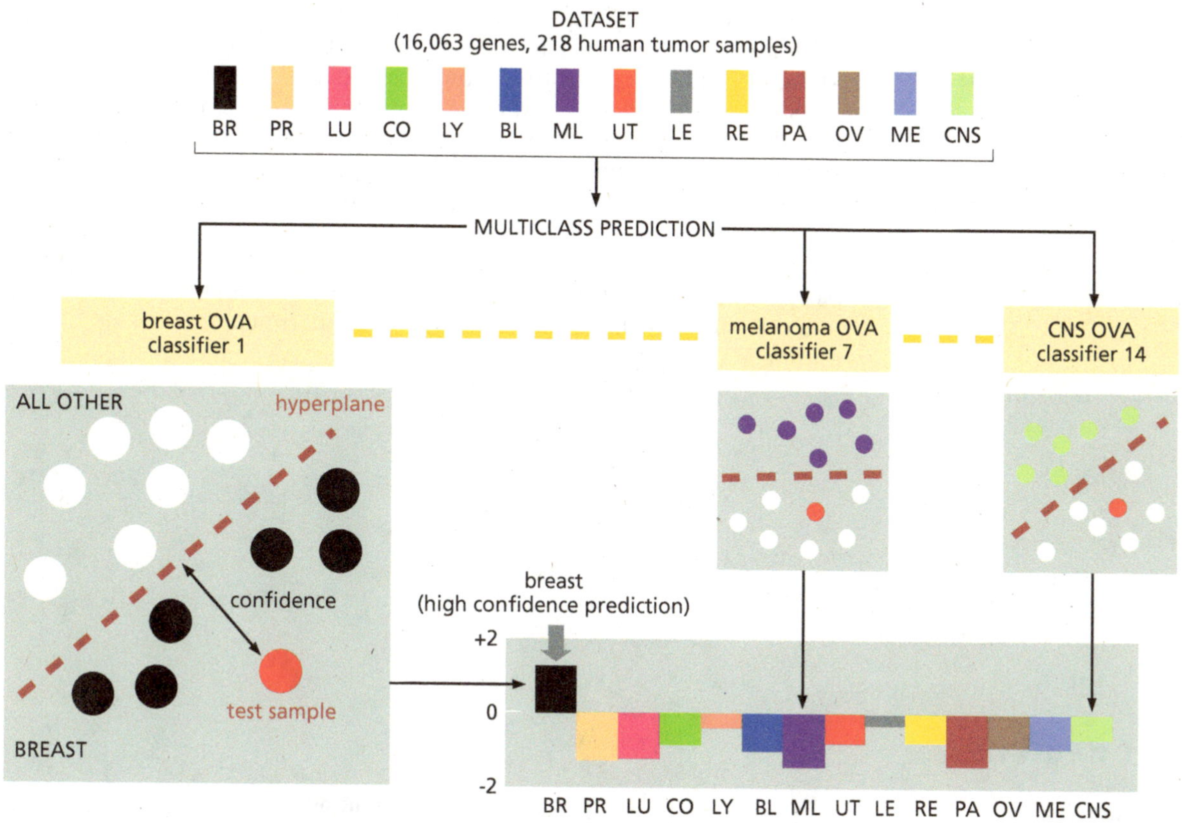


**Figure 16.26**
**The application of SVM classifiers to classify an unknown sample into one of 14 human tumors.** The SVMs were fitted to data from all 14 tumors, color-coded as in the top key. The classification of a test sample is shown. All of these SVM classifiers are of the one-versus-all (OVA) type that distinguish between X and not-X where X is one of the types of human tumor. Classifier 1, which predicts the likelihood that the sample comes from a patient with breast adenocarcinoma, gives a strong signal that it does. The other 13 classifiers give negative results, as shown by the bottom graph. In this case the result is a high confidence prediction that the test sample came from a patient with breast adenocarcinoma. The tumors are BR, breast adenocarcinoma; PR, prostate adenocarcinoma; LU, lung adenocarcinoma; CO, colorectal adenocarcinoma; LY, lymphoma; BL, bladder transitional cell carcinoma; ML, melanoma; UT, uterine adenocarcinoma; LE, leukemia; RE, renal cell carcinoma; PA, pancreatic adenocarcinoma; OV, ovarian adenocarcinoma; ME, pleural mesothelioma; CNS, central nervous system. (Adapted from S. Ramaswamy et al., Multiclass cancer diagnosis using tumor gene expression signatures, *Proc. Natl Acad. Sci. USA* 98:15149–15154, 2001.)

type of experiment, which mainly have their origins in statistical properties of the data, such as sources of error and variance. To some extent these difficulties can be overcome by suitable normalization and transformation of the data.

To analyze the transformed data further, they are then usually organized into clusters, or groups, of similar data. The hierarchical clustering method produces a single cluster within which the data (for example, the expression pattern of individual genes or proteins) are organized into a dendrogram that indicates their relative similarities to each other. Other clustering methods can be divided into different types depending on whether or not they are able to determine the appropriate number of clusters for the dataset, and whether or not they indicate the relationships between different clusters. From a theoretical viewpoint, it may be preferable to allow the method to determine the number of clusters, but in practice this is probably of minor benefit. In general, the data should be analyzed by using more than one clustering method and any conclusions drawn should, where possible, be supported by several independent analyses.

To study differential expression between a set of proteins or genes, statistical methods should be used to assess the significance of the measured difference. These methods can range from simple tests (such as expression ratios exceeding a threshold) to the widely used *t*-test, to complicated techniques that use Bayesian models. All these tests would be expected to identify the most extreme differential expression. More sophisticated analysis is required, however, to find all differential expressions without also identifying a large number of false positives.

One of the key applications of this area is in classifying unknown samples on the basis of the expression levels. This is likely to have a major impact on medical diagnosis in the near future, especially as the experimental methods become more routine and cost-effective.

## Further Reading


**Monograph**

Baldi P & Hatfield GW (2002) DNA Microarrays and Gene Expression. Cambridge: Cambridge University Press.

Causton H, Quackenbush J & Brazma A (2003) Microarray Gene Expression Data Analysis: A Beginner's Guide. Oxford: Blackwell Science.

Jain AK & Dubes RC (1988) Algorithms Clustering Data. New Jersey: Prentice Hall.


### 16.1 Expression Data Require Preparation Prior to Analysis


**Variance-stabilizing transformation**

Durbin BP, Hardin JS, Hawkins DM & Rocke DM (2002) A variance-stabilizing transformation for gene-expression microarray data. *Bioinformatics* 18, S105–S110.

Geller SC, Gregg JP, Hagerman P & Rocke DM (2003) Transformation and normalization of oligonucleotide microarray data. *Bioinformatics* 19, 1817–1823.

**Lowess normalization**

Cleveland WS & Devlin SJ (1988) Locally weighted regression: an approach to regression analysis by local fitting. *J. Am. Stat. Assoc.* 83, 596–610.

**General review of normalization and transformation of data**

Quackenbush J (2002) Microarray data normalization and transformation. *Nat. Genet.* 32, 496–501.

**Data used to generate Figures 16.2 to 16.4**

Courcelle J, Khodursky A, Peter B et al. (2001) Comparative gene expression profiles following UV exposure in wild-type and SOS-deficient *Escherichia coli*. *Genetics* 158, 41–64.

**Principal component analysis**

Alter O, Brown PO & Botstein D (2000) Singular value decomposition for genome-wide expression data processing and modeling. *Proc. Natl Acad. Sci. USA* 97, 10101–10106.


### 16.2 Cluster Analysis Requires Distances to be Defined Between all Data Points


**Distance definitions**

Quackenbush J (2001) Computational analysis of microarrays. *Nat. Rev. Genet.* 2, 418–427.


### 16.3 Clustering Methods Identify Similar and Distinct Expression Patterns


**Self-organizing maps (SOMs)**

Tamayo P, Slonim D, Mesirov J et al. (1999) Interpreting patterns of gene expression with self-organizing maps: Methods and application to hematopoietic differentiation. *Proc. Natl Acad. Sci. USA* 96, 2907–2912.

**Clustering using genetic algorithms**

Jones DR & Beltramo MA (1991) Solving partitioning problems with genetic algorithms, Proceedings 4th International Conference on Genetic Algorithms (KR Belew, LB Booker, eds), pp 442–449. San Francisco: Morgan Kaufman.

Lu Y, Lu S, Fotouhi F et al. (2004) Incremental genetic K-means algorithm and its application in gene expression data analysis. *BMC Bioinformatics* 5, 172.

Mitchell M (1999) An Introduction to Genetic Algorithms. Cambridge, MA: MIT Press.

**SOTA**

Herrero J, Valencia A & Dopazo J (2001) A hierarchical unsupervised growing neural network for clustering gene expression patterns. *Bioinformatics* 17, 126–136.

**Biclustering techniques**

Madeira SC & Oliveira AL (2004) Biclustering algorithms for biological data analysis: a survey. *IEEE/ACM Trans. Comput. Biol. Bioinform.* 1, 24–45. *And references therein.*

**Validation of partitions produced by clustering methods**

Handl J, Knowles J & Kell DB (2005) Computational cluster validation in post-genomic data analysis. *Bioinformatics* 21, 3201–3212.

Preliæ A, Bleuler S, Zimmermann P, Wille A et al. (2006) A systematic comparison and evaluation of biclustering methods for gene expression data. *Bioinformatics* 22, 1122–1129.

Suzuki R & Shimodaira H (2006) Pvclust: an R package for assessing the uncertainty in hierarchical clustering. *Bioinformatics* 22, 1540–1542.

## 16.4 Statistical Analysis can Quantify the Significance of Observed Differential Expression

**Bayesian estimation of variance**

Baldi P & Long AD (2001) A Bayesian framework for the analysis of microarray expression data: regularized t-test and statistical inferences of gene changes. *Bioinformatics* 17, 509–519.

**Nonparametric tests**

Efron B, Tibshirani R, Storey JD & Tusher V (2001) Empirical Bayes analysis of a microarray experiment. *J. Am. Stat. Assoc.* 96, 1151–1160.

Tusher VG, Tibshirani R & Chu G (2001) Significance analysis of microarrays applied to the ionizing radiation response. *Proc. Natl Acad. Sci. USA* 98, 5116–5121.

Zhao Y & Pan W (2003) Modified nonparametric approaches to detecting differentially expressed genes in replicated microarray experiments. *Bioinformatics* 19, 1046–1054.

**Multiple hypothesis testing**

Benjamini Y & Hochberg Y (1995) Controlling the false discovery rate: a practical and powerful approach to multiple testing. *J. R. Stat. Soc. B* 57, 289–300.

Dudoit S, Shaffer JP & Boldrick JC (2003) Multiple hypothesis testing in microarray experiments. *Stat. Sci.* 18, 71–103.

Storey JD (2003) The positive false discovery rate: A Bayesian interpretation and the q-value. *Ann. Stat.* 31, 2013–2035.

Storey JD, Dai JY & Leek JT (2005) The optimal discovery procedure for large-scale significance testing, with applications to comparative microarray experiments. *UW Biostatistics Working Papers* 260.

## 16.5 Gene and Protein Expression Data Can be Used to Classify Samples

Dudoit S, Fridlyand J & Speed TP (2002) Comparison of discrimination methods for the classification of tumours using gene expression data. *J. Am. Stat. Soc.* 97, 77–87.

Hwang D, Schmitt WA, Stephanopoulos G & Stephanopoulos G (2002) Determination of minimum sample size and discriminatory expression patterns in microarray data. *Bioinformatics* 18, 1184–1193.

Tan AC, Naiman DQ, Xu L et al. (2005) Simple decision rules for classifying human cancers from gene expression profiles. *Bioinformatics* 21, 3896–3904.

**Support vector machines**

Brown MSP, Grundy WN, Lin D et al. (2000) Knowledge-based analysis of microarray gene expression data by using support vector machines. *Proc. Natl Acad. Sci. USA* 97, 262–267.

Burges CJC (1998) A tutorial on support vector machines for pattern recognition. *Data Min. Knowl. Disc.* 2, 121–167.

Ramaswamy S, Tamayo P, Rifkin R et al. (2001) Multiclass cancer diagnosis using tumor gene expression signatures. *Proc. Natl Acad. Sci. USA* 98, 15149–15154.

# 17

# SYSTEMS BIOLOGY

**When you have read Chapter 17, you should be able to:**

Recount the origins of systems biology.

Summarize the uses of system biology.

Summarize the many types of system biological analyses.

Show how networks are starting points in many system biology studies.

Show how networks have to be extended with more information.

Discuss why systems need to be both robust and fragile for control.

Expalin why modularity is important in robustness.

Explain why more redundant systems tend to be more adaptable.

Over the past few years, systems biology has become a popular and much-talked-about concept. Unlike the rest of bioinformatics, however, its origins began in the first half of the twentieth century. As far back as 1934, the Austrian biologist Ludwig von Bertalanffy applied general systems theory to biology as well as to other fields, formulating his Organismic System Theory. He stated that it was old-fashioned science that attempted to explain phenomena by reducing them to interplay of individually investigable units but that contemporary science recognized the importance of the whole. He defined wholeness by characteristics such as the organization of dynamic interactions, which is manifested in the difference in the behavior of the individual parts when isolated or when they are combined together. In other words, systems cannot be fully understood by analysis of their components in isolation. This definition of systems biology is still valid.

To fully understand the functioning of cellular processes, whole cells, organs, and even organisms, it is not enough to simply assign functions to individual genes, proteins, and other cellular components. We need to analyze the organization and control of the system in an integrated way by looking at the dynamic networks of genes and proteins, and their interactions with each other. These interacting pathways are complex dynamic systems, and often behave in a nonlinear and adaptive way. **Nonlinearity** means, for example, that doubling a stimulus (the input) does not necessarily double the response, and may even cause a qualitatively different response. **Adaptive systems** can modify themselves to respond in a more appropriate way in the light of previous stimuli. The general goal of theoretical systems biology is to develop computer models that predict the properties of the large,

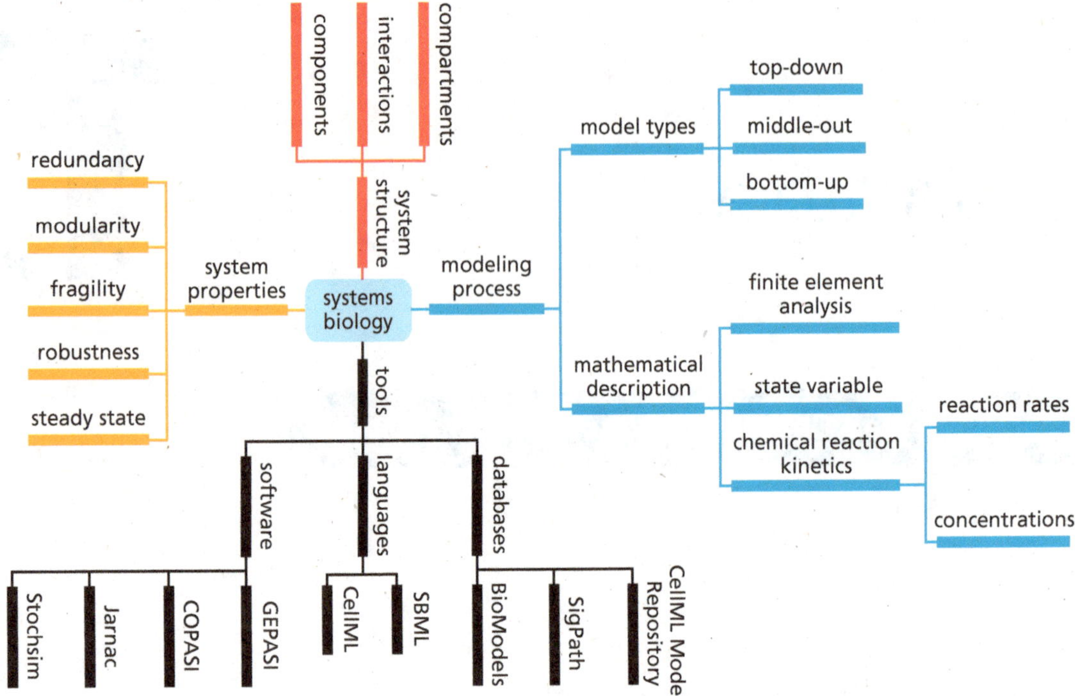


**Mind Map 17.1**
**Systems biology is a fast-growing part of computational research.** This mind map shows some of the topics introduced in this chapter and is a useful aid in memorizing the important aspects of this exciting field.

adaptive, interconnected networks that are found in living things. This type of modeling will allow us to investigate, for example, how extracellular signals are processed to produce functional cellular responses. Similar modeling can also be used to investigate larger-scale multicellular systems such as the development and physiology of whole organs or even whole organisms.

In systems biology, mathematical descriptions of the processes under study are used along with knowledge from engineering and physics and the power of modern computers. These techniques are used to obtain a detailed description of the parts (the components) and their interactions, and then to reassemble them into an interconnected whole. In other words, mathematical models are applied to biological processes to identify rules about molecular or cellular associations or dependencies. These are called causal dependencies. Genomics and proteomics have provided large datasets that can be used to describe the parts of a biological process at the gene and protein level. Models of cardiac cells and the heart, genomic regulatory networks, developmental genetic networks, and metabolic pathways are just a few of the biological systems that are currently being modeled using the techniques of systems biology.

International projects are under way to model all the molecular processes in a single cell, one such being the E-Cell Project. This aims to develop programs that will allow researchers to simulate the functioning of a complete cell starting from its molecular components. Another project is under way at the Pacific Northwest National Laboratory (PNNL) where several groups are collaborating, some focusing on the study of thousands of proteins and other cellular components, while the bioinformaticists work with the groups to integrate the data to enable the understanding of gene and protein networks that are part of cell signaling, cellular metabolic pathways, or intercellular communication. There are many other institutions bringing together various groups of scientists with computational methodologies to study complex biological questions.

In this chapter we will give a general overview of systems biology, what it can provide, and why it is important. It is based on complex mathematical modeling, using advanced techniques that are outside the scope of this book. Those who wish

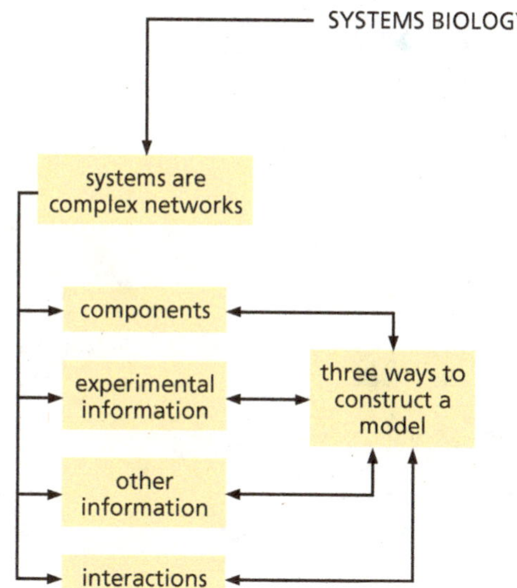


**Flow Diagram 17.1**
**In this section the concept of a system is introduced and some types and ways of constructing system models are described.**

to know more should look at the Further Reading at the end of the chapter. Here we will discuss some examples of biological processes currently being modeled to illustrate the scope of the field.

## 17.1 What is a System?

There are good reasons for the systems biology approach. First, biological systems tend to be so complex that it is difficult, without modeling, to know how they behave and to understand the actions of their control mechanisms. Second, such systems can have higher-order properties that are their main biological function, but are not apparent from the properties of the separate components; for example we cannot understand a book from just looking at the collection of words in a random way.

The higher-order properties and functions that arise from the interaction of the parts of a system are called **emergent properties**. For example, the human brain, capable of thought, depends on practically all the cells in the brain and their interconnections. But a single brain cell is incapable of the property of thought; therefore thought is an emergent property of all the cells in the brain. However, as a part of a whole made up of many interconnecting cells, a single cell is an important component of a complex system.

### A system is more than the sum of its parts

Although very useful information can be obtained from the analysis of individual parts of a complex system, the ultimate aim is to understand how all the parts act together in real time and how the functioning of the system is controlled (see Figure 17.1). This in turn will shed light on how each individual component contributes to the whole system. Any working system, whether a cell or a car engine, is not just a static assembly of its parts. It has a specific structure (the way the parts relate to each other) and dynamics (the ways in which it changes over time). A description of a fully functional system must take into account the spatial organization of elements, their interactions, and their response to external stimuli, including those processes that control and stabilize the system.

Let us examine an underground railway system like the London Underground or the New York subway. Our hypothetical subway consists of components (the

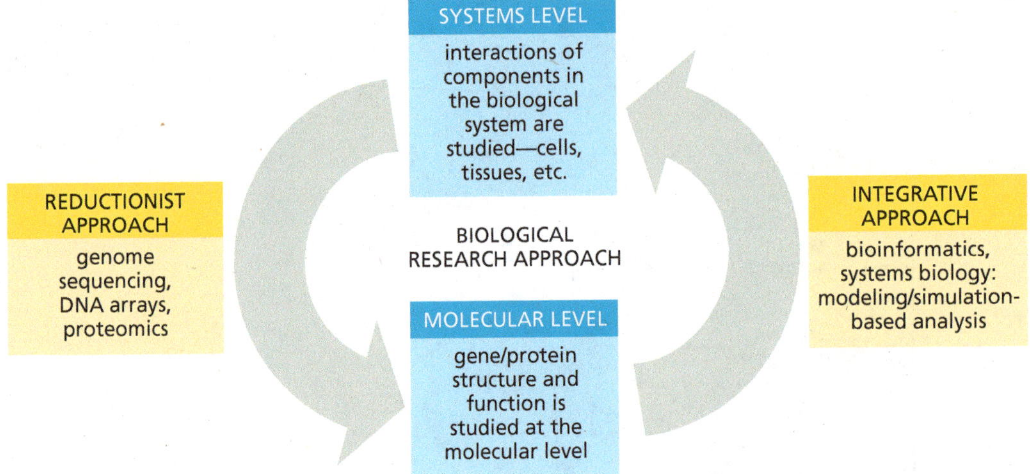


**Figure 17.1**
**A schematic of the reductionist approach and the integrative approach to research that is part of systems biology.**

stations) linked by interactions (the rail tracks) (see Figure 17.2). These define the structure of an interconnecting network but we need far more information to understand the way the railway works. We need to know what type of signals travel along the interactions (the passengers), what the signal carriers are (the trains), how fast they travel, and what controls their movement and prevents two trains being on the same track at the same time. In addition, it is important to know how perturbations will affect them, for example what effect a faulty power supply or a workers' strike will have on the functioning of the subway system. Different networks can also interact with each other; for example a city subway will interact with the overground railway system and the bus system. External influences can have a highly significant influence on the network. For example, people's working hours will determine the periods of high and low activity. The system must be studied at a quantitative level, because we need to know if there are sufficient trains for all passengers, and so on.

## A biological system is a living network

Biological networks—for example a network of intracellular signaling pathways or a set of interconnecting metabolic pathways—have many similarities with the transport systems just mentioned. They are assemblies of components, in this case the individual proteins and small molecules in the pathways, which are connected by interactions along which signals pass and have an effect on the function of the whole network. These networks too have inbuilt controls that activate or terminate particular signals. The biological system can also be perturbed by outside interference, ranging from the food we eat to therapeutic drugs. Therefore, to properly understand the function and the effect of individual proteins or genes and the consequence of their interactions and products we need to study the whole system in which these proteins (or genes) play a part. Once we have a quantitative mathematical model that can be manipulated, and understand the normal functioning of the system, we can see what the effects of various perturbations are *in silico*. This type of study is useful, for example, in the pharmaceutical industry to study the potential effects of new drugs, for identifying feedback mechanisms (controls) that might offset the effects of the drugs, and for predicting side effects.

Building simple networks of molecular interactions is the first stage in modeling a complex entity such as a cell. Biological systems consist of hierarchies—subsystems that can be studied at different levels—such as the molecular pathway, the organelle, the individual cell, and so on. Networks of interacting proteins are the

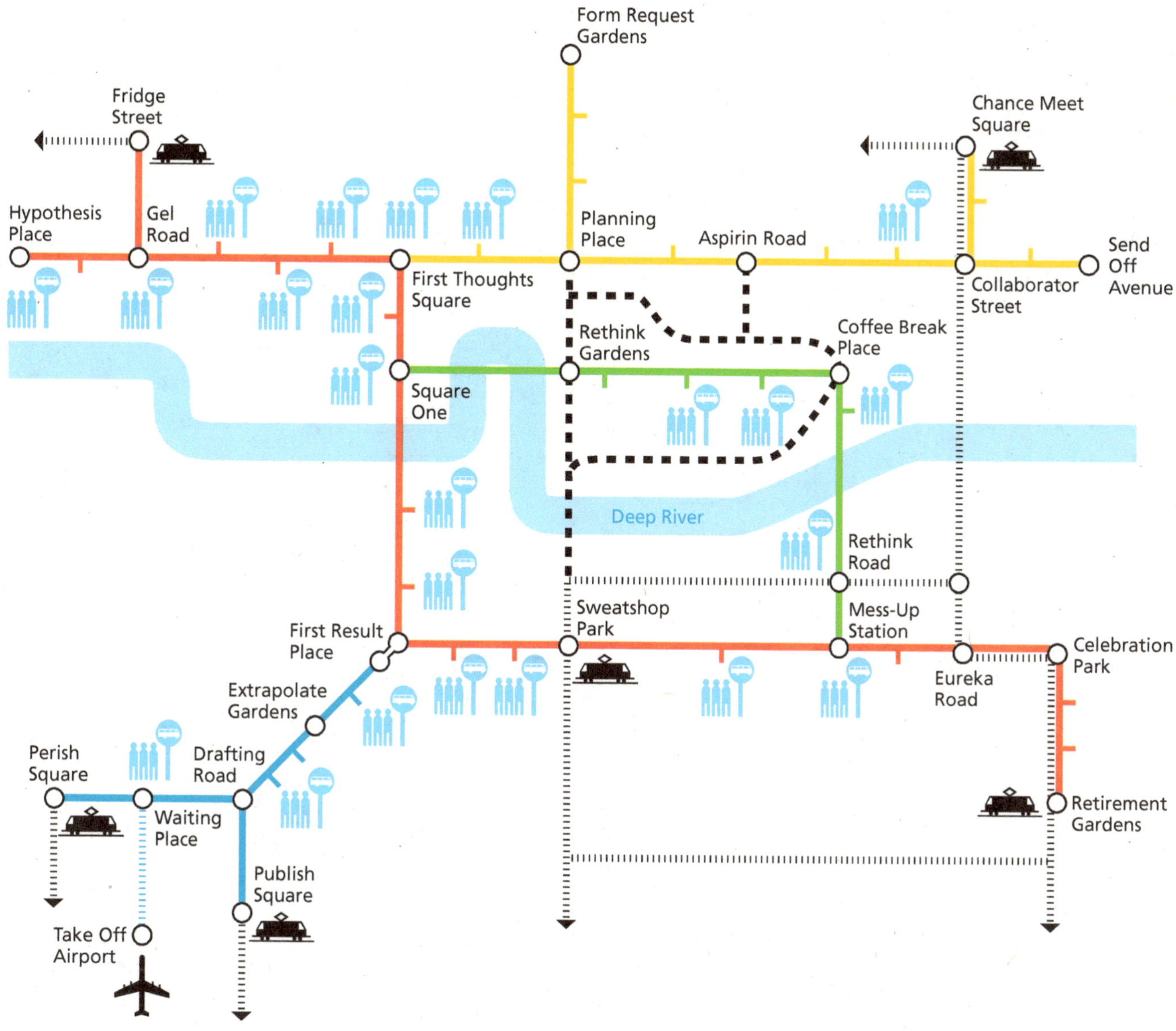


starting point in understanding cellular mechanisms. To make a dynamic model of such a network, we first need to build the basic network structure. Various databases exist that can be useful in creating an initial structure.

**Figure 17.2**
**A fabricated transport system of Research City.** The lines represent an underground railway system, the dotted thin lines are rail tracks. The circles represent main stations with links to the trains, the small lines are substations. Main train stations are illustrated by a train. The line of people shows the bus station and there is one airport to the south of our Research City.

## Databases are useful starting points in constructing a network

To construct a network you need several different types of information. A number of Web sites make available information about all or most of the interacting proteins in a particular pathway, for example the Kyoto Encyclopedia of Genes and Genomes (KEGG). The gene catalog in KEGG organizes the genes into functional hierarchies according to the classification of their protein products into biochemical pathways. KEGG includes information about the functional and/or structural protein type, RNA, and other small molecule ligands. For example Figure 17.3 shows the pathways of glycolysis. EcoCyc is an example of a similar catalog for a particular species, in this case the bacterium *E. coli.* It describes both the genome and the biochemical machinery of *E. coli*. Figure 17.4 shows the glycolytic pathway of *E. coli* as defined in EcoCyc with the same compounds highlighted in red as those given in Figure 17.3. Other Web sites provide interactive protein-interaction diagrams; one example is the Biomolecular Interaction Network Database (BIND) (see Figure 17.5). pSTIING (Protein, Signaling, Transcriptional Interactions & Inflammation Networks Gateway)

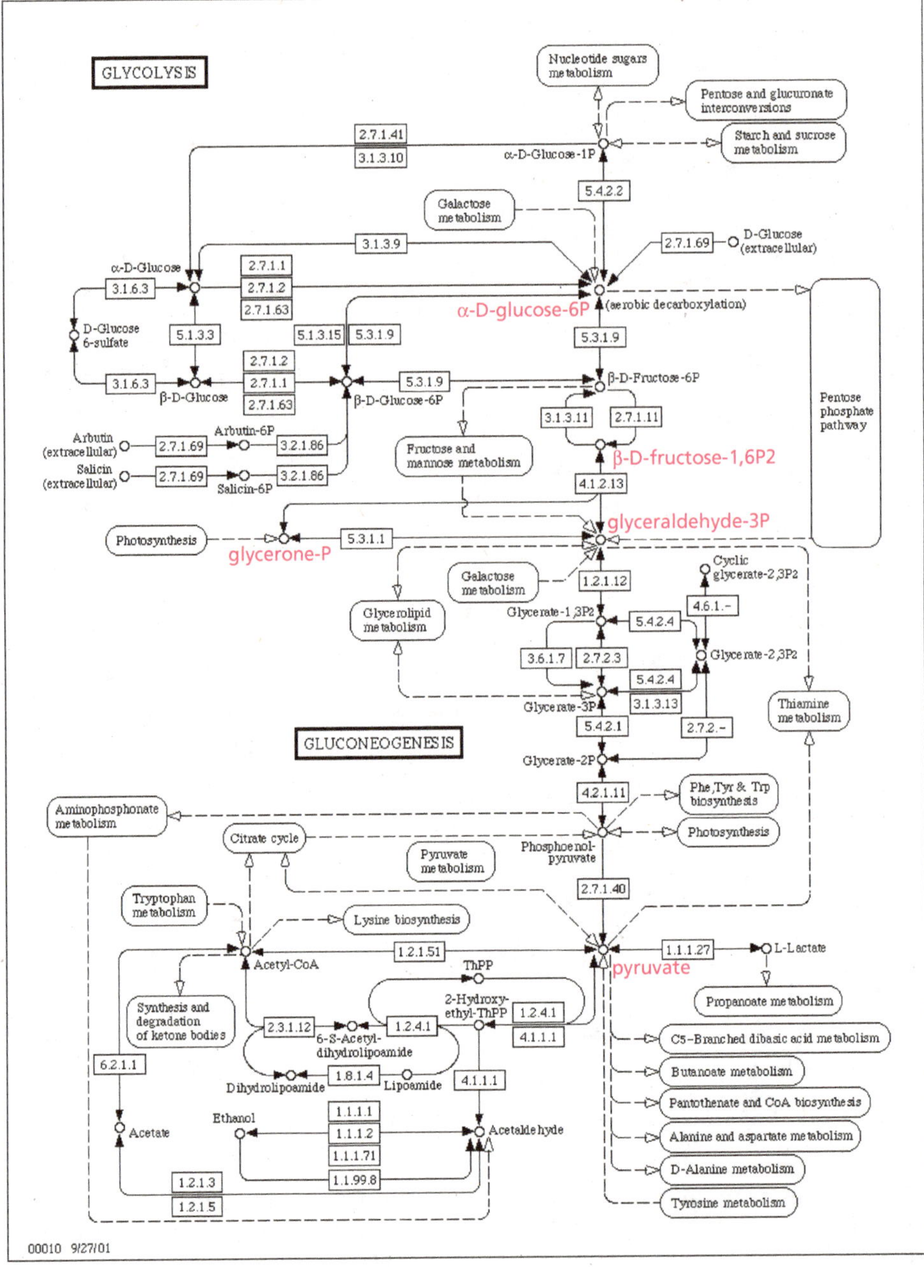


**Figure 17.3**
**The components and interactions in the glycolytic pathway as obtained from the KEGG database.** The components are the protein enzyme numbers (boxed), the products, and the type of reaction performed by the enzymes. The arrows indicate the direction of the reaction. Note that labels in round boxes (e.g. glycerolipid metabolism) represent other pathways.

is another publicly available database, which includes protein–protein, protein–lipid, protein–small molecule, and ligand–receptor interactions. In addition it also incorporates transcriptional regulatory associations with protein interaction information (see Figure 17.6) and allows for the integration of protein interaction information with experimental results, such as microarray data. Table 17.1 gives a selection of some of the commonly used databases in constructing pathways.

## To construct a model more information is needed than a network

A network structure can be defined using these databases. However, to obtain a dynamic network for modeling, whether using kinetic or statistical data (such as whether a gene is on or off), the information from the databases has to be augmented by other information from published experimental work. To complete the network definition the interactions between components must be defined. To

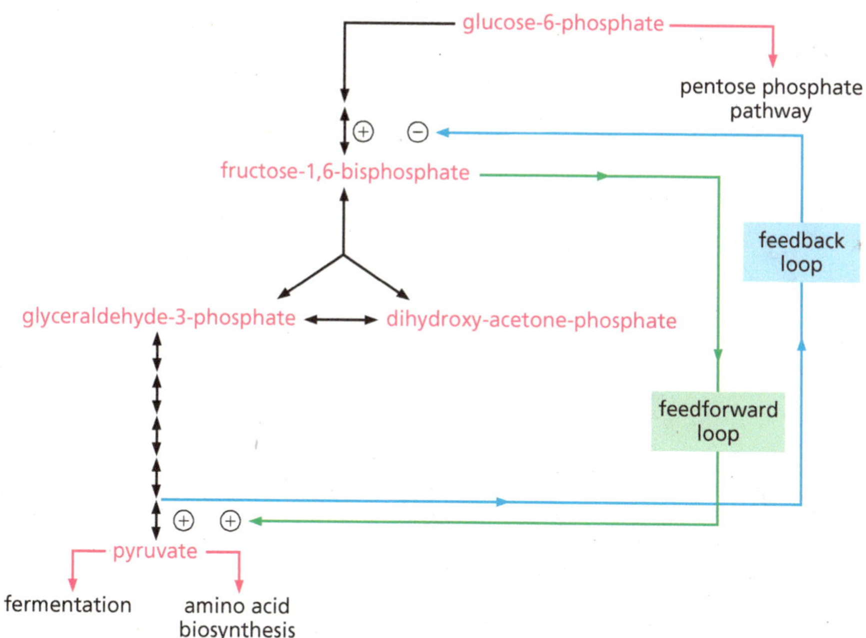


**Figure 17.4**
**EcoCyc is a database that has data on the *E. coli* genome and its biochemical system.** The figure illustrates one of the views (least detailed) of the glycolysis pathway. The names of the most important molecules involved are given as well as the black arrows indicating the direction of the reaction, while the green and blue arrows show control aspects in the pathway. The red arrows show the interaction with other pathways.

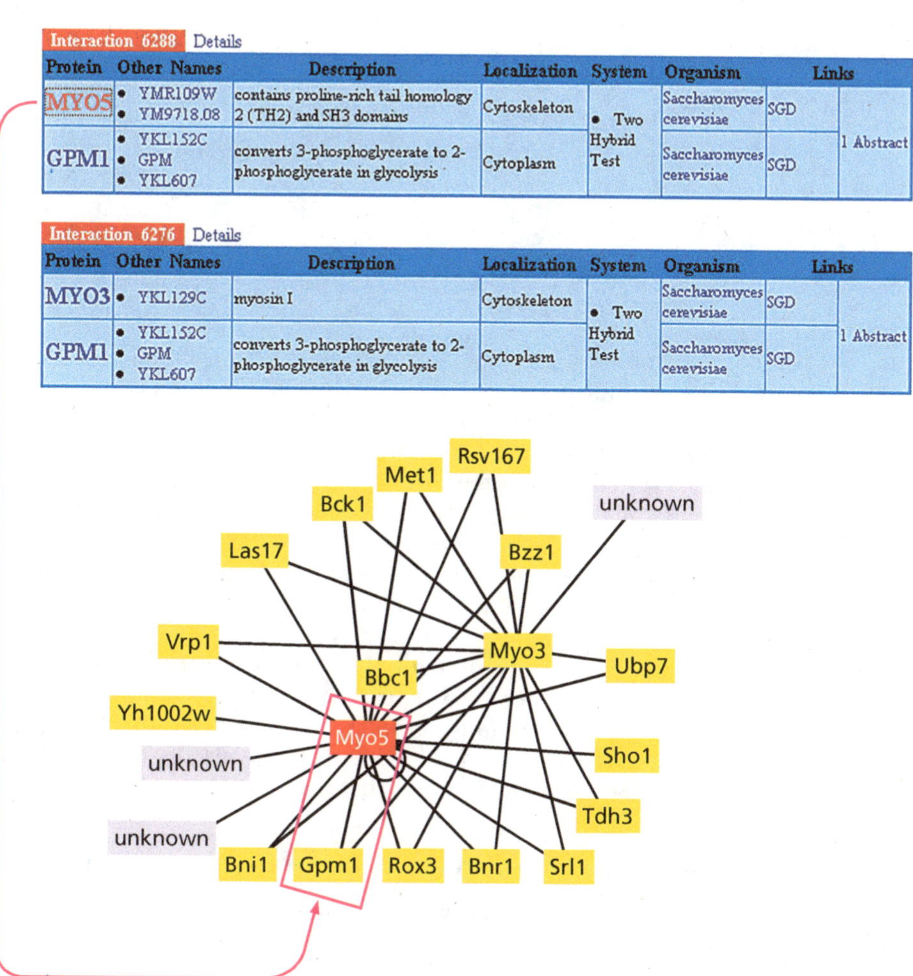

Interaction 6288 Details

| Protein | Other Names | Description | Localization | System | Organism | | Links |
|---|---|---|---|---|---|---|---|
| MYO5 | • YMR109W<br>• YM9718.08 | contains proline-rich tail homology 2 (TH2) and SH3 domains | Cytoskeleton | • Two Hybrid Test | Saccharomyces cerevisiae | SGD | 1 Abstract |
| GPM1 | • YKL152C<br>• GPM<br>• YKL607 | converts 3-phosphoglycerate to 2-phosphoglycerate in glycolysis | Cytoplasm | | Saccharomyces cerevisiae | SGD | |

Interaction 6276 Details

| Protein | Other Names | Description | Localization | System | Organism | | Links |
|---|---|---|---|---|---|---|---|
| MYO3 | • YKL129C | myosin I | Cytoskeleton | • Two Hybrid Test | Saccharomyces cerevisiae | SGD | 1 Abstract |
| GPM1 | • YKL152C<br>• GPM<br>• YKL607 | converts 3-phosphoglycerate to 2-phosphoglycerate in glycolysis | Cytoplasm | | Saccharomyces cerevisiae | SGD | |



**Figure 17.5**
**BIND is a database of components and interactions, where each interaction includes information on cellular location, experimental conditions, conserved sequence, molecular location of interaction, and so on.** In this figure the interaction summary is shown between proteins MYO5 and GPM1, which make part of the glycolytic pathway. The interaction can be visualized and expanded to include other interactions with MYO5 and interactions between any other protein present, such as GPM1 and its interaction partners (e.g., MYO3).

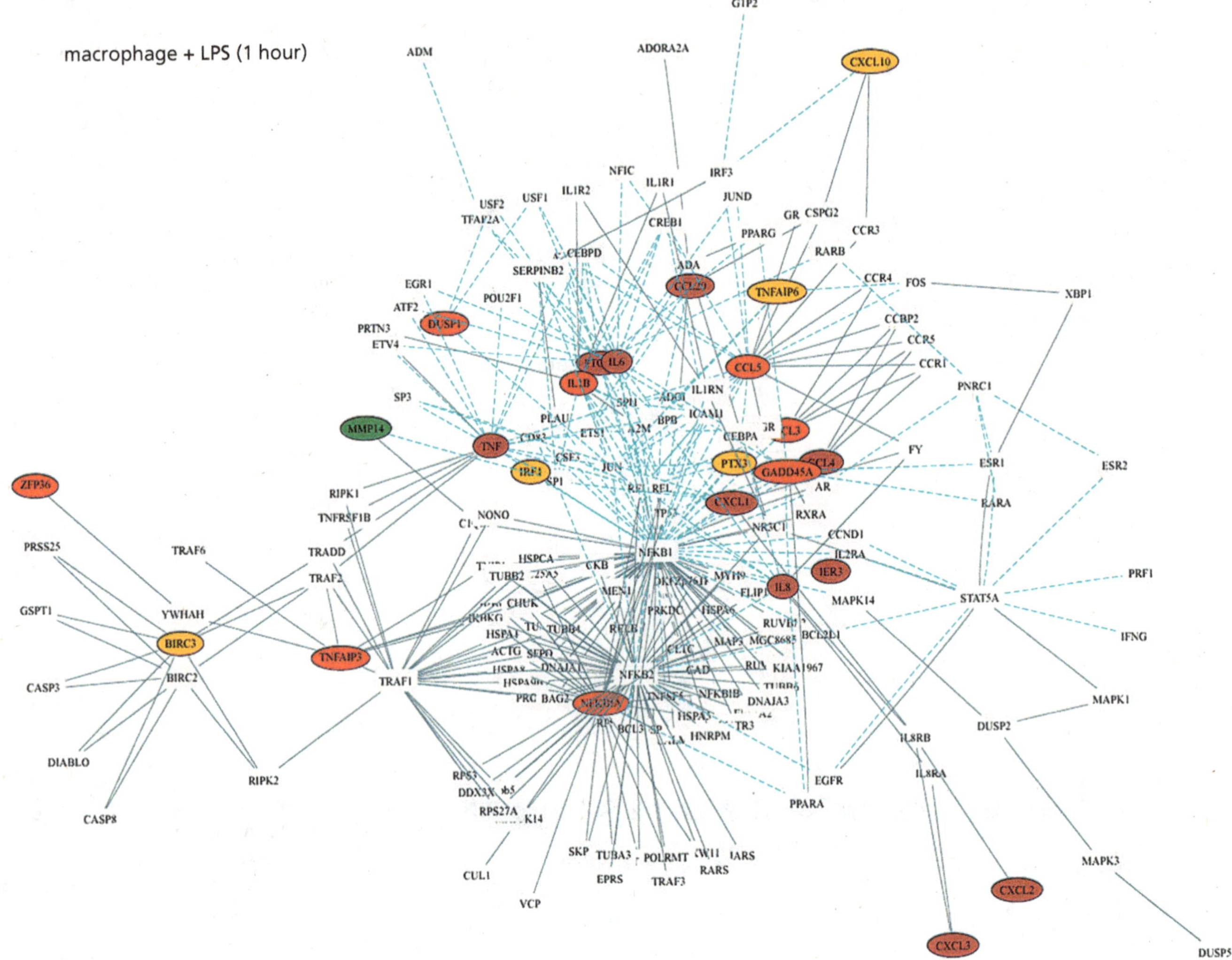


**Figure 17.6**
**pSTIING is a database that holds more than protein–protein interactions, shown by solid lines.** It also has information about which proteins/genes have transcriptional influence on other proteins/genes. This is shown by dashed lines. Experimental results can also be mapped onto the interaction diagram in pSTIING. In this figure the genes/proteins are colored according to gene expression results, where red indicates more than six times upregulation, dark orange more than four times, and light orange more than two times upregulated products. Green indicates downregulated products with respect to a control set of genes.

do this, you need to know which components interact with each other, and also must have quantitative information about the kinetics of the interaction, such as whether a reaction occurs rapidly or slowly and the starting concentrations of components. These kinetic parameters are hard to obtain, as their experimental determination is not yet commonplace, and sometimes they must be estimated from a related system that has been better characterized. Thus the construction of dynamic kinetic models relies on the in-depth knowledge of all enzymatic rate equations and their parameter values. The parameter values also depend on factors such as tissue type or physiological (and experimental) conditions. Therefore to obtain an accurate model all parameter values should be obtained using the same conditions and the same tissue type. An additional problem is that many enzyme kinetic rates are obtained from *in vitro* studies, and it is not certain if these rate functions will have the same values *in vivo*. Finally, the external signals that influence the network must be defined and independent knowledge of the actual responses to some of these are useful to check that the model reproduces reality.

## There are three possible approaches to constructing a model

There are a number of different ways that a mathematical model of a dynamic network can be constructed. The three main approaches are the bottom-up, top-down, and middle-out methods (see Figure 17.7). The bottom-up approach

**Table 17.1**
A small selection of Web-based tools and databases for systems biology.

| Web link | Name |
|---|---|
| http://pstiing.icr.ac.uk | pSTIING |
| http://wwwmgs.bionet.nsc.ru/mgs/gnw/genenet/ | GeneNet |
| http://string.embl.de/ | STRING |
| http://www.ebi.ac.uk/intact/index.jsp | IntAct |
| http://mint.bio.uniroma2.it/mint/ | MINT |
| http://www.bind.ca/Action | BIND |
| http://dip.doe-mbi.ucla.edu/ | DIP |
| http://www.hprd.org/ | HPRD |
| http://mips.gsf.de/proj/ppi/ | MIPS / MPPI |
| http://www.biocyc.org/ | BioCyc |
| http://metacyc.org/ | MetaCyc |
| http://www.pantherdb.org/ | PANTHER |
| http://www.genome.jp/kegg/ | KEGG |
| http://biozon.org/ | Biozon |
| http://www.biocarta.com/genes/index.asp | BioCarta |
| http://www.genmapp.org/ | GenMAPP |
| http://stke.sciencemag.org/ | STKE |
| http://www.signaling-gateway.org/ | AfCS |
| http://www.grt.kyushu-u.ac.jp/spad/ | SPAD |
| http://biodata.mshri.on.ca/osprey/servlet/Index | Osprey |
| http://www.cytoscape.org/ | Cytoscape |
| http://strc.herts.ac.uk/bio/maria/NetBuilder/index.html | NetBuilder |
| http://vlado.fmf.uni-lj.si/pub/networks/pajek/default.htm | Pajek |
| http://visant.bu.edu/ | VisANT |
| http://sbw.kgi.edu/software/jdesigner.htm | Jdesigner/SBW |
| http://projects.villa-bosch.de/bcb/software/software/Ulla/SimWiz/ | SimWiz |
| http://pavesy.mpimp-golm.mpg.de/PaVESy.htm | PaVESy |
| http://www.genomicobject.net/member3/index.html | Genomic Object Net |
| http://www.celldesigner.org/ | CellDesigner |
| http://www.ncbs.res.in/~bhalla/kkit/index.html | Genesis/Kinetikit |
| http://www.e-cell.org/ | E-Cell |
| http://page.mi.fu-berlin.de/~trieglaf/PNK2e/index.html | PNK 2e |
| http://www.bii.a-star.edu.sg/research/sbg/cellware/index.asp | Cellware |
| http://www.biouml.org | BioUML |
| http://www.nrcam.uchc.edu | Virtual Cell |
| http://icb.med.cornell.edu/services/sp-prod/sigpath/mainMenu.action | SigPath |
| http://www.cellml.org/ | CellML |
| http://sbml.org/index.psp | SBML |
| http://www.ebi.ac.uk/biomodels/ | BioModels |
| http://jjj.biochem.sun.ac.za/index.html | JWS |
| http://jigcell.biol.vt.edu/index.html | JigCell |
| http://www.biospice.org | Bio-SPICE |
| http://www.mcell.psc.edu | MCell |
| http://sbw.kgi.edu/software/jarnac.htm | Jarnac |
| https://biodynamics.indiana.edu/CellModeling/AboutCellX.html | CellX |
| http://sbml.org/software/sbmltoolbox/ | SBMLToolbox |
| http://sbml.org/software/mathsbml/index.html | MathSBML |
| http://wishart.biology.ualberta.ca/SimCell/ | SimCell |

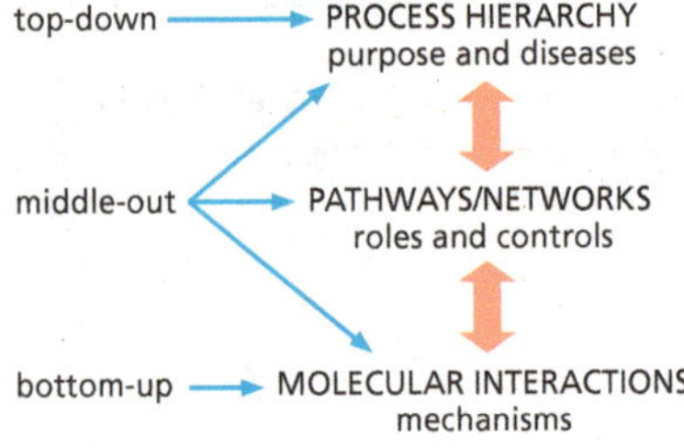


**Figure 17.7**
**A schematic showing the relationship between the top-down, bottom-up, and middle-out approaches to modeling a system.**

constructs networks and aims to predict their behavior starting with a collection of experimental data. All the information available about all the components in the system is gathered together first and is then combined to form a complete picture. For example, a study modeling the behavior of an engineered $O_RO_{lac}$ *E. coli* promoter under different conditions was carried out using a bottom-up approach. This promoter (see Figure 17.8) is repressed by LacI and activated by the protein CI. A green fluorescent protein (GFP) gene was placed under the control of the $O_RO_{lac}$ promoter to enable the measurement of activation and repression. A mathematical model was used to look at the behavior of this modular system. Figure 17.8 shows an example prediction from this study, which shows that this approach can be used as a predictive tool to provide insights into gene regulation. In the graph the red line represents experimental results, while the blue line shows the predicted behavior of the promoter. It is obvious that in this case the mathematical modeling accurately predicts the real behavior of the system.

The top-down approach starts from observed behavior and then fills in the components and interactions required to generate these observations by iterative experimental results and simulation. This increases and fine-tunes the mechanistic detail of the model. Many models start with the top-down approach, such as protein–protein interactions, enzymatic pathways, and signaling pathways. The architecture of such networks is an important feature. The architecture—the way the interacting nodes connect—will formulate how a network may behave. The same nodes in a network can be connected in different ways. For example, the nodes can interact by only connecting to their nearest nodes (see Figure 17.9A), by connecting to any other node randomly selected (see Figure 17.9B), or where some nodes have many connections, while others have few, often called a scale-free network (see Figure 17.9C). The scale-free network is of considerable interest, as it seems that proteins in a network that have many connections may act as important hubs central to the activity of the network.

An example of the top-down model is the mathematical modeling of tumor invasion in cancer (see Figure 17.10). The observed behavior—tumor invasion—has been subjected to quantitative simulation with a model including dependence on

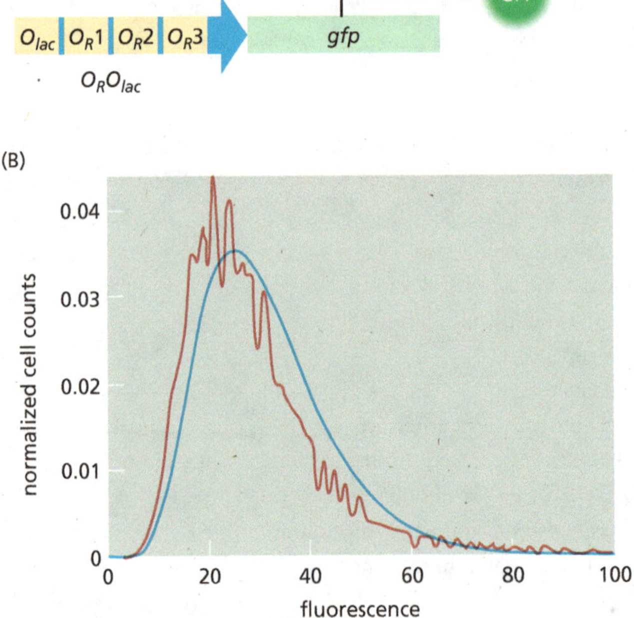


**Figure 17.8**
**The engineered $O_RO_{lac}$ promoter of *E. coli.*** (A) The promoter can be observed because it controls the expression of *gfp*, which is detected by fluorescence. The proteins CI and LacI bind to the promoter. (B) This shows curves of experimentally detected cell counts in red and cell counts from mathematical modeling in blue. (Adapted from N.J. Guido et al., A bottom-up approach to gene regulation, *Nature* 439:856–860, 2006.)

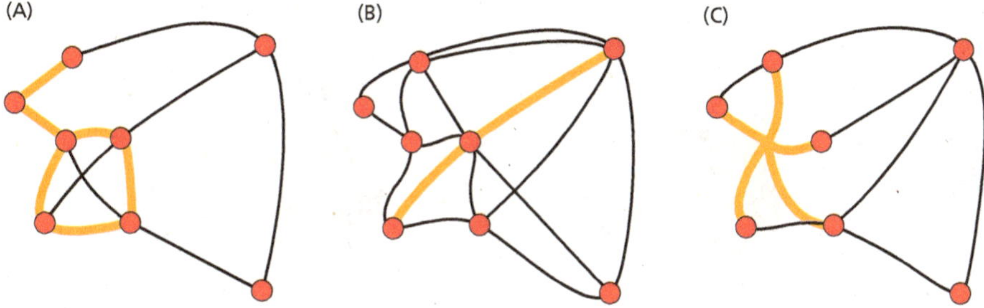


**Figure 17.9**
**The various architectures that networks can adopt.** (A) Only close-by nodes connect, (B) the connections are random, and (C) certain nodes adopt more connections than others.

the extracellular matrix (ECM) composition, metalloproteinases (MMP) that can digest the ECM, and the tumor cells. The small tumor at the start of the simulation grows symmetrically, which leads to further invasion. This type of simulation mimics metastatic behavior.

The middle-out approach is a more flexible method of investigation than the above methods. The data used to construct these mathematical models are collected from both molecular experiments and system-level observations. This model can start at any point for which data are available, as long as it is supported by a hypothesis. It can then expand either up or down in terms of both resolution and coverage. With the currently available biological data and techniques of experimental investigation, the middle-out approach is the one most often used.

The virtual heart project (see Further Reading) is an excellent example of this approach. Mathematical models of how heart muscle functions to keep the heart beating are now extremely sophisticated and include models of all the main types of cardiac muscle cells. This model is a compilation of a large number of mathematical equations that describe the protein, cells, and various tissues of the heart. The model can, for example, represent the variations in expression levels of large numbers of genes within a particular part of the heart, for example the atrium, and the consequences of their perturbation on heart function. These gene variations are important in interpreting electrocardiograms and have been helpful in understanding the various types of heart diseases. Figure 17.11 shows a model of mechanically induced sustained arrhythmia in a part of the heart. This type of modeling helps the understanding of, for example, pacemaker activity.

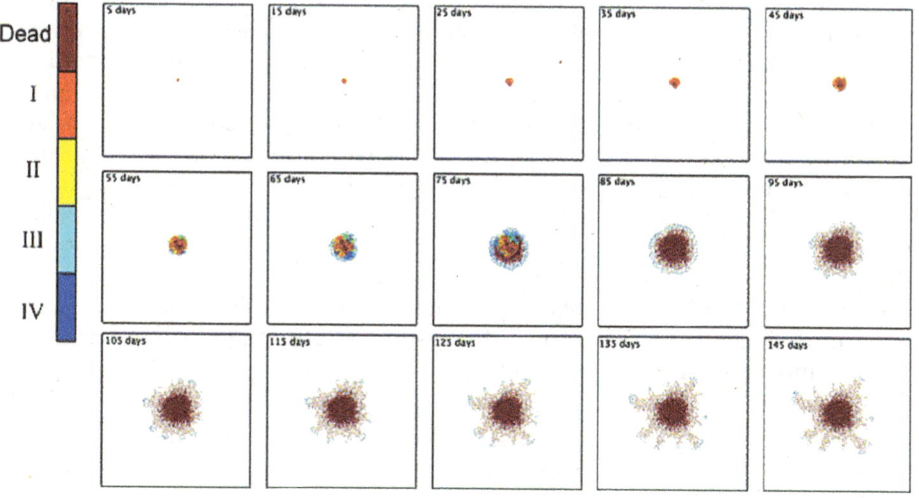


**Figure 17.10**
**Computer simulation of the way a tumor can spread based on simulation of the extracellular matrix.** At first the tumor is small and forms a neat circle. As time goes on the cells begin to proliferate and some cells die (brown). Further on some cells become aggressive (blue). From day 85 the tumor consists of dead and aggressive cells leading to more rapid invasion. (Reprinted from *Clinica Chimica Acta*, 357, V. Quaranta et al., Mathematical modeling of cancer: the future of prognosis and treatment, 173–179, 2005, with permission from Elsevier.)

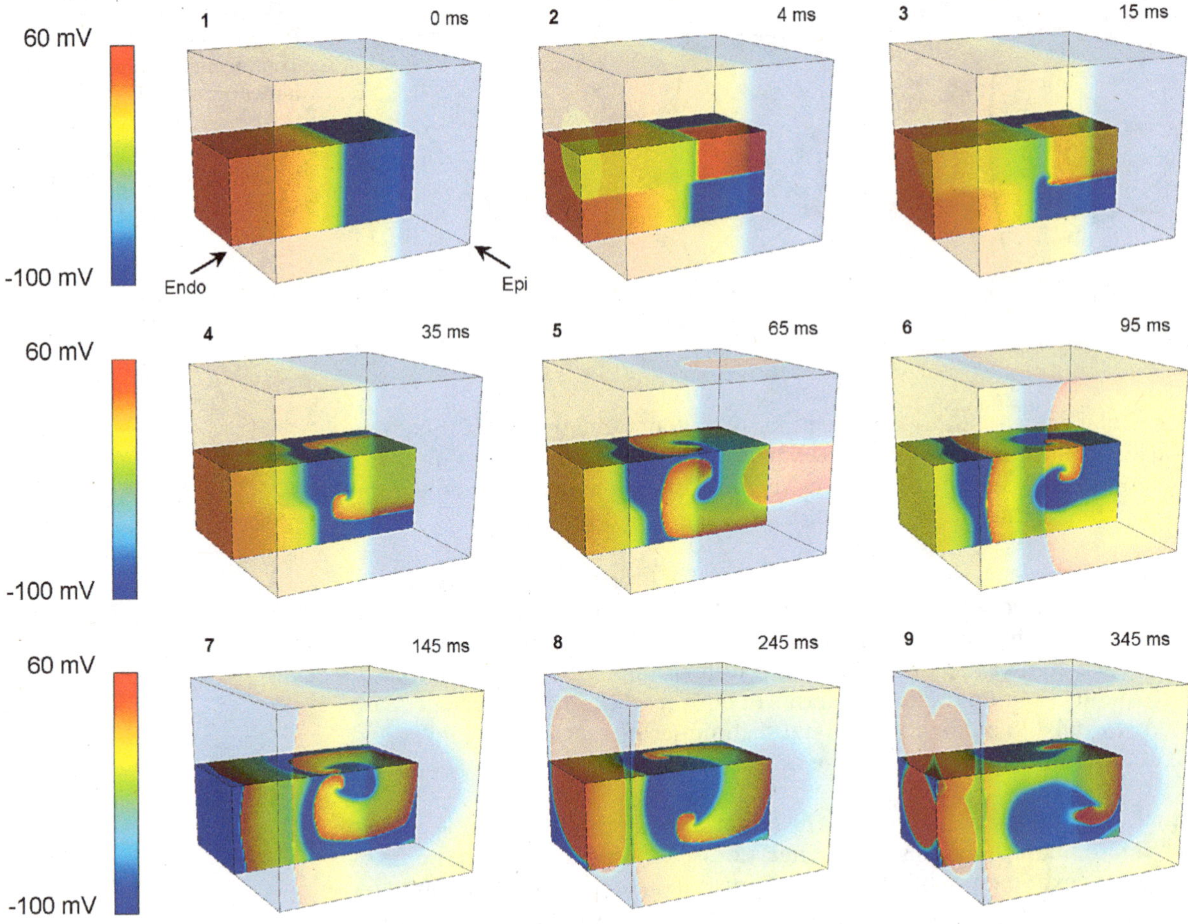


**Figure 17.11**
**A mathematical model of mechanically induced sustained arrhythmia in a cube representation of the ventricular wall of the heart.** The model is of the transmembrane potentials, where blue is –100 mV and red shows potentials of 60 mV. (Reprinted from *Prog. Biophys. Mol. Biol.*, 87, A. Garny et al., Dimensionality in cardiac modeling, 47–66, 2005, with permission from Elsevier.)

## Kinetic models are not the only way in systems biology

The rest of the discussion in this chapter applies mainly to kinetic models but they are not the only way of analyzing large-scale networks. Data-driven constraints-based models (topological models) have also been successfully applied. As all biological processes are dependent on physicochemical constraints (such as osmotic pressure, thermodynamics, and so forth) topological modeling is based on applying these physicochemical constraints on the function of computer-based recreated genome-scale networks to deduce their possible functions. The topological approach is especially useful when there are not enough kinetic data to enable the construction of a kinetic model. The methods used to analyze topological models include Flux Balance Analysis (FBA) and Extreme Pathways. The former uses a linear approach to optimizing the modeling **parameters** while the latter looks at all possible distributions. It is not possible to describe in detail these methods here and the interested reader should refer to Further Reading. The FBA topological approach has been successful for example in modeling the cellular metabolism of *E. coli*.

One other aspect of models that one should keep in mind is that there are metabolic models that are mainly based on steady-state kinetics, while signaling network models are time-dependent.

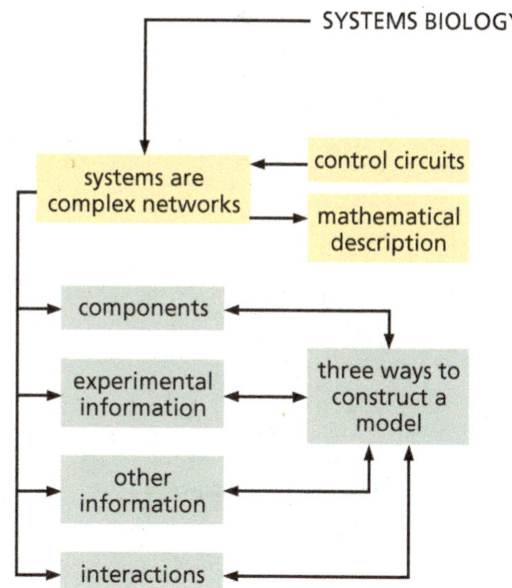


**Flow Diagram 17.2**
**In this section some types of system model are described in more detail.**

## 17.2 Structure of the Model

In this part of the chapter we will discuss the definition of a system structure and some of the typical substructures that are found within biological systems. After a brief discussion of the components and parameters that are the building blocks of the structure, we will discuss some very simple molecular systems in detail. System structure is delineated in terms of components and the interactions between them.

As we saw above, the first step in modeling a system is the description of its structure in terms of the constituent components and their interactions. The components can be proteins, genes, metabolites, or other molecules, or larger-scale objects such as cell membranes, whole cells, or even whole organs or organisms. The interactions can be the regulatory interactions of genes with each other, the formation of protein complexes, the relationship between enzymes and their substrates, or any other interrelationship between components (see for example Figure 17.3).

The system interactions are defined in terms of parameters. These describe factors such as the concentrations of the components and the reaction rates of individual steps in the system (see Figure 17.12). Some of these parameters will be treated as constants, such as those describing reaction rates, whereas others can vary in value during the system's lifetime (for example, temperature, or concentrations of components). In some cases, a parameter such as a reaction rate will be constant within a particular species but vary across species. Special attention may be paid to parameters that represent properties of the system that are under external control, such as temperature, concentration of particular chemical species, or activation levels of particular genes or gene products. Such parameters (sometimes called **state variables**) are often varied during the study to explore how the system responds.

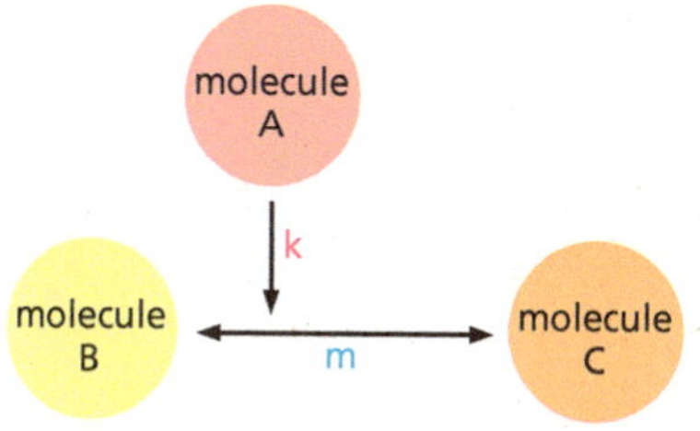


**Figure 17.12**
**The interaction of molecule B on molecule C, which has a parameter m, while k is the parameter that is associated with modifying the interaction of molecules B and C.** Rate of production of C $\propto$ k[B] and k can be either negative or positive. If k is negative then A is inhibitory; if k is positive then A is stimulatory.

Dependence of the function of the system on state variables such as temperature may be explicitly defined using specific terms. For example the standard theory of chemical reactions includes the dependence of reaction rate on the system temperature. Incorporating this into the model will thus include temperature effects for

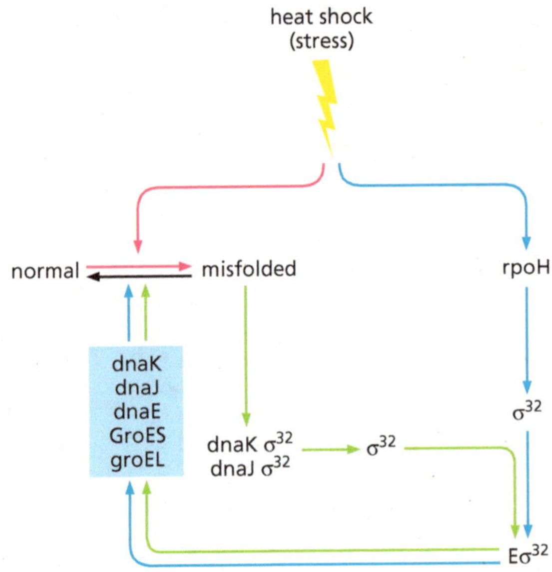


**Figure 17.13**
**A very simplified diagram of heat shock response in *E. coli*.** Upon heat shock a rapid translational modulation facilitates the production of $\sigma^{32}$ by acting on rpoH mRNA, which encodes $\sigma^{32}$. Then $\sigma^{32}$ RNAP holoenzyme ($E\sigma^{32}$) is formed, which activates hsp proteins (in blue box). These repair the misfolded proteins. This process is a feedforward control and is illustrated by the blue arrows in the diagram. Concurrently a second mechanism (green arrows) is in action, which is dependent upon the detection of the misfolded proteins. dnaK and dnaJ detect misfolded proteins and release their bound $\sigma^{32}$. The free $\sigma^{32}$ then activates transcription of the hsps. This mechanism is an example of feedback control because the level of misfolded proteins is monitored and that level controls the activity of $\sigma^{32}$.

each reaction. However, temperature effects for other components of the system will have a specific formula to model them.

In addition, the system structure may consist of aspects such as the physical structure of the cell and other components. For example if some of the components of a system are associated with the cell membrane this should be taken into account.

## Control circuits are an essential part of any biological system

Most real biological systems make a particular response (the **output**) to a particular stimulus (the input). This response is strictly controlled to give optimal function. In order to exert control there must be the possibility of some type of interaction between later components and those earlier in the pathway. This will not necessarily be a direct interaction and may involve a separate group of components that are not in the direct line of the pathway. The existence of regulation implies that circuits will be an essential part of the system structure.

There are two major control mechanisms in biological systems: feedforward and feedback (see Figure 17.13). In **feedforward control** a set of reactions is triggered by a specific action or input. **Feedback control** is more sophisticated and more precise; it detects the difference between the desired output and the actual output of a system and compensates for the difference. There are two types of feedback control: positive (stimulatory) and negative (inhibitory). Systems can use both feedforward and feedback control in the same pathway (see Figure 17.13). An example of a more complex system is given in Figure 17.14. Note that this complex pathway is built up of a number of smaller closed circuits, for example the p53 **module** that is activated by DNA damage or cellular stress.

## The interactions in networks can be represented as simple differential equations

All the interactions between components in a model need to be represented mathematically. These will define the relationships between the components and state variables. The nonmolecular structural components of systems may be modeled using finite element analysis techniques borrowed from engineering (see Further Reading). The molecular components of the models are also represented by differential equations as will now be described.

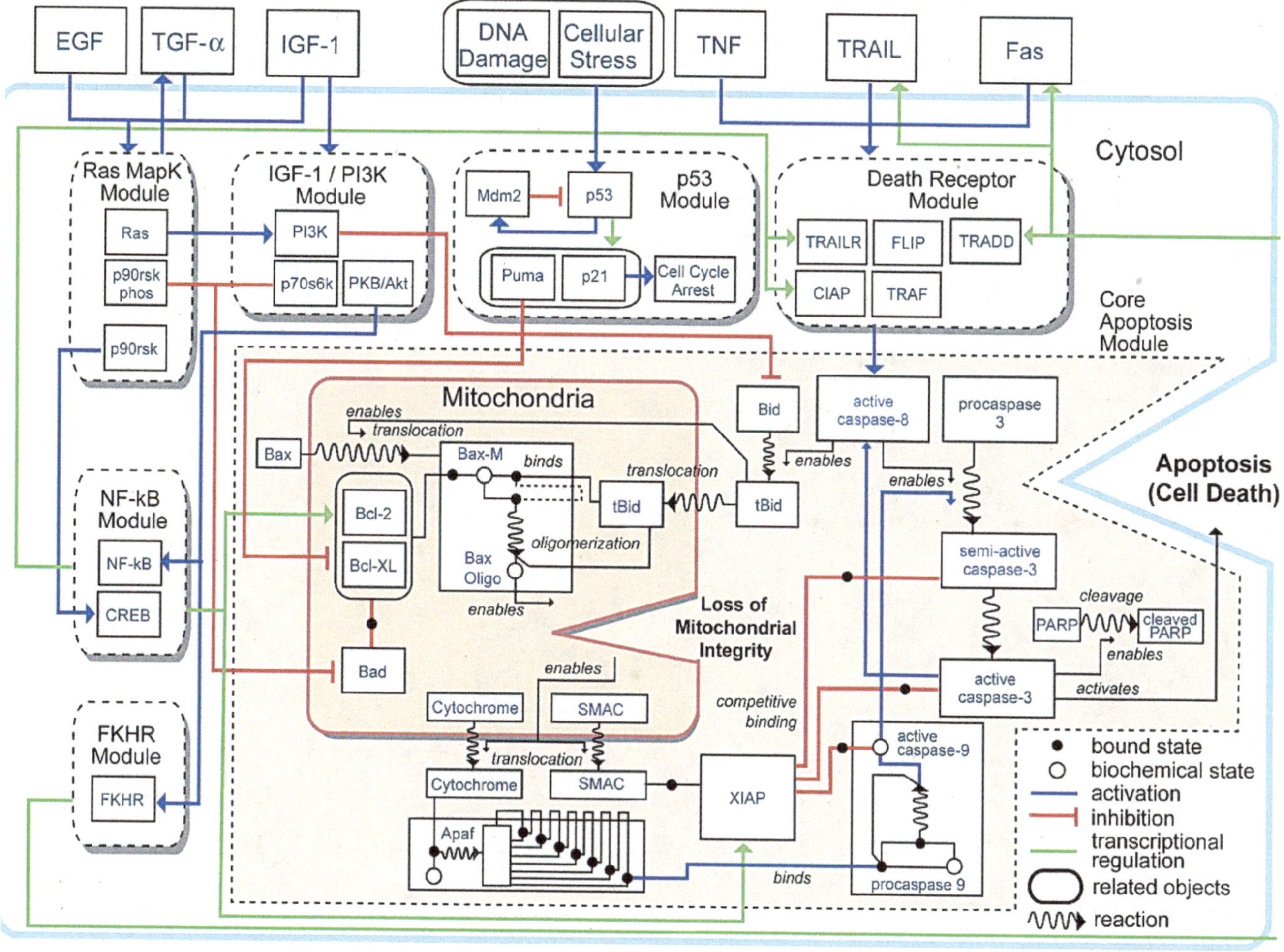


The epidermal growth factor receptor (EGFR) pathway (see Figure 17.15A) will be used as an example of a system with molecular components. Networks of metabolic and signaling pathways control many biological processes and their disruption is associated with many diseases. Studying the dynamics of these networks can provide new insights into their workings, give clues to their involvement in diseases such as cancer and heart disease, and help in the search for new treatments.

The main interactions in both metabolic and signaling pathways are between different proteins and between proteins and small molecules (such as metabolites) (see Figure 17.15A and B). The equations for a single interaction are taken from basic reaction theory as applied in chemistry and biochemistry. Suppose two components $A$ and $B$ interact to produce components $C$ and $D$, with forward rate constant $k_1$ and reverse rate constant $k_{-1}$, that is:

$$A + B \underset{k_{-1}}{\overset{k_1}{\rightleftharpoons}} C + D \qquad \text{(EQ17.1)}$$

The rate of accumulation of component $C$ with time $t$ is given by the first derivative

$$\frac{d[C]}{dt} = k_1[A][B] - k_{-1}[C][D] \qquad \text{(EQ17.2)}$$

where $[A]$, $[B]$, and so on, represent component concentrations. The first term corresponds to the forward reaction, and the second to the reverse reaction. Similarly, we

**Figure 17.14**
**The apoptopic pathway represented as a circuit board.** This type of diagram will help in understanding the quantitative way in which all the components interact within a cell to start apoptosis (cell death) or not. Eventually when most of the components and their variables are known, mathematical modeling will be able to predict how the pathway will respond to various extracellular and intracellular perturbations.

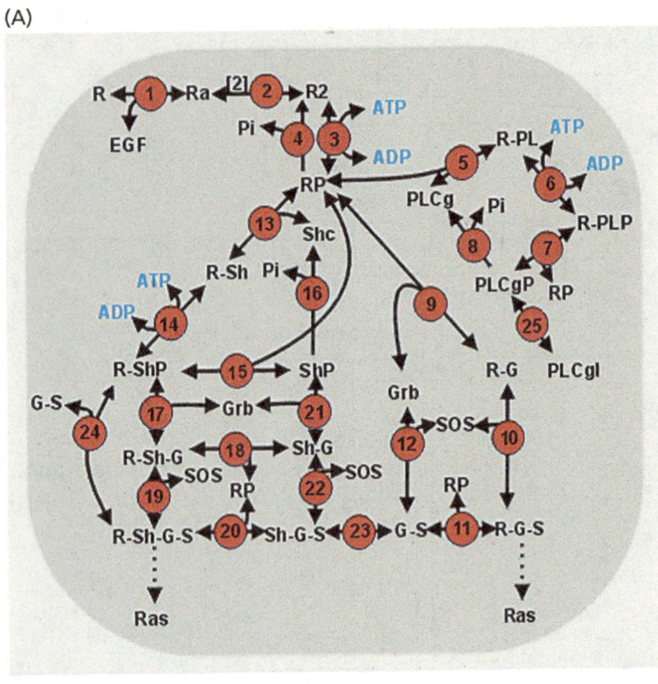


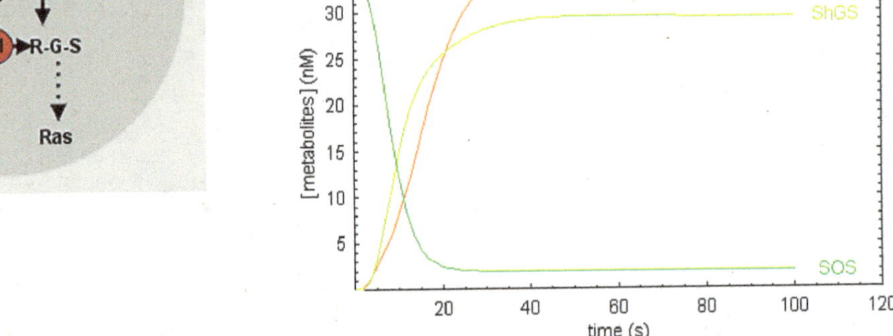

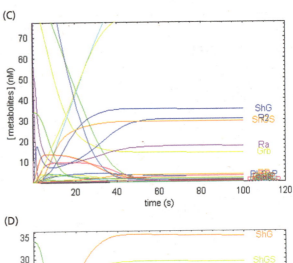

**Figure 17.15**
**The EGFR pathway is very important in carcinogenesis.** Phosphorylation of tyrosine residues on EGFR leads to the activation of many downstream proteins and enzymes. This in turn starts a signal transmission through a number of interacting cascades. To predict the signaling dynamics, testable computational models are of utmost importance. (A) The first mechanistic model of the EGFR network was published in 1999 and explained some of the dynamics of signaling responses in liver cells stimulated with EGF. The components of the pathway are named, while the associated reactions and equations are given by circled numbers. (B) This shows equations 4 and 13 of the pathway. (C) and (D) The graphs show some of the simulation results, where some components decrease as they are used up and others increase. The figure and graphs were generated with the program JWSAPPLET at http://www.jjj.bio.vu.nl

can obtain

$$\frac{d[A]}{dt} = -k_1[A][B] + k_{-1}[C][D] \qquad \text{(EQ17.3)}$$

and equivalent expressions for [B] and [D]. Given the initial concentrations of the four components and values for $k_1$ and $k_{-1}$, we can calculate the variation in component concentrations with time.

If component $C$ is subsequently involved in another interaction, for example dissociation into two constituent subunits:

$$C \underset{k_{-2}}{\overset{k_2}{\rightleftharpoons}} E + F \qquad \text{(EQ17.4)}$$

we can obtain from this

$$\frac{d[C]}{dt} = -k_2[C] + k_{-2}[E][F] \qquad \text{(EQ17.5)}$$

The two interactions are independent of each other, so that we can sum Equations EQ17.2 and EQ17.5 to obtain the variation in [C] over time as

$$\frac{d[C]}{dt} = k_1[A][B] + k_{-2}[E][F] - k_{-1}[C][D] - k_2[C] \qquad \text{(EQ17.6)}$$

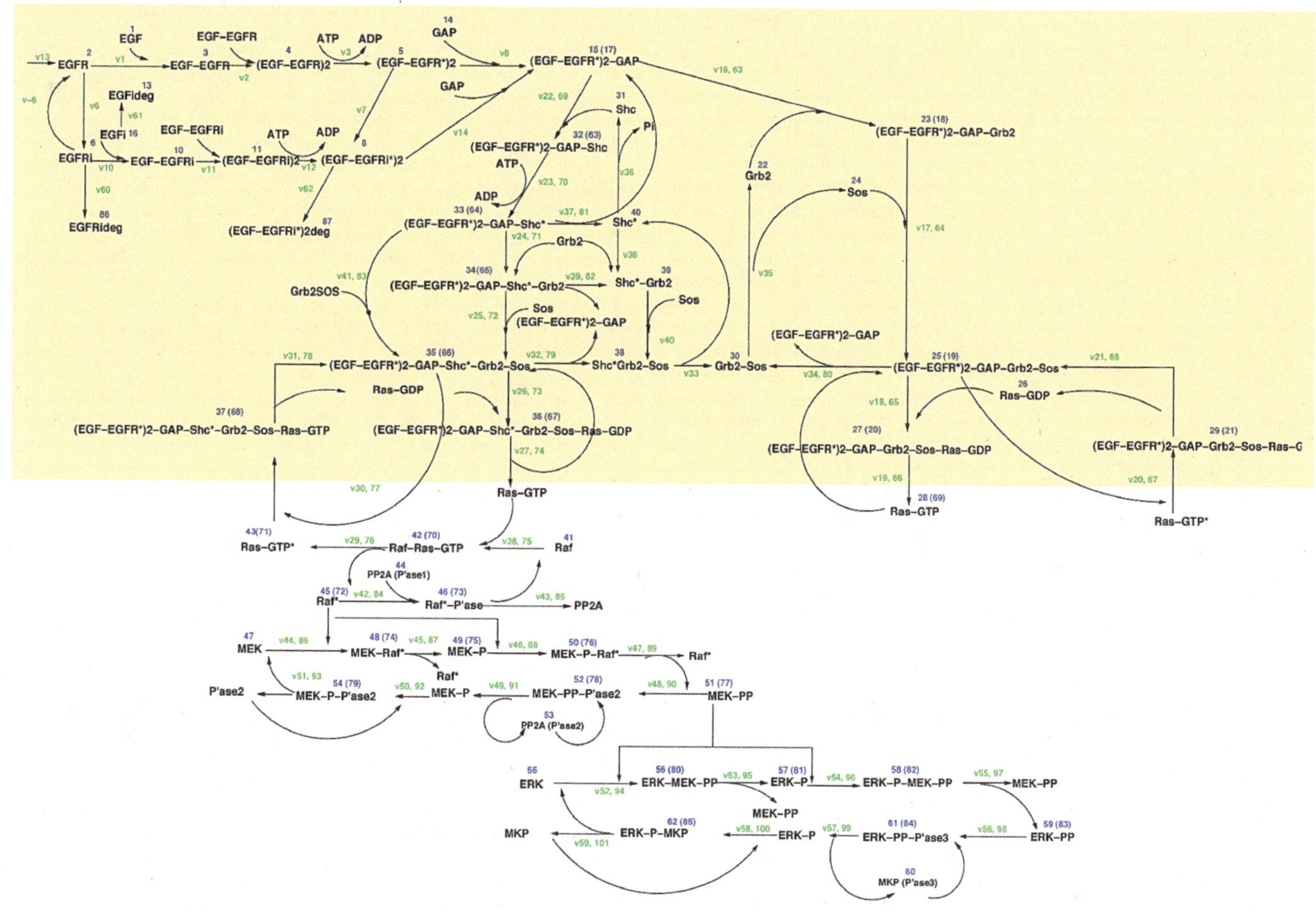


This procedure can be applied even to systems with a very large number of components and interactions. This set of **ordinary differential equations** (**ODEs**) defines the system. Figure 17.15B shows a representative example of two such equations, which describe the system in Figure 17.15A. All the equations must be solved together to determine the system properties. The example in Figure 17.15 is part of a larger and more complete network that has been modeled recently and is shown in Figure 17.16.

These conceptually simple equations can be used to represent even highly complex molecular systems, such as the interactions of transmembrane transporter proteins with their cargoes. They also form the basis of modeling higher-level cellular functions such as secretion and electrical activity, and multicellular functions, such as the cardiac conducting pathways (see Box 17.1).

**Figure 17.16**
**Structure of the EGF receptor-induced mitogen-activated protein (MAP) kinase system.** The blue numbers identify the different components (94 in total). The green numbers represent reaction rates (125 in total). (Reprinted by permission from Macmillan Publishers Ltd: *Nature Biotechnology* 20:370–375, B. Schoeberl et al., Computational modeling of the dynamics of the MAP kinase cascade activated by surface and internalized EGF receptors, 2002.)

## 17.3 Robustness of Biological Systems

A feature of any biological system, whether an individual cell or a developing embryo, is its **robustness**. In simple terms, robustness is the property that allows a system to maintain certain functions despite variability in components or in the environment. The robustness of biological systems is the reason that embryonic development, brain function, and life in general usually continue in the face of the inevitable internal variations in concentrations of components and rates of intracellular reactions, as well as external influences such as variations in nutrients and changes in temperature.

**Flow Diagram 17.3**
**In this section a number of principal properties of system models are described, especially robustness.**

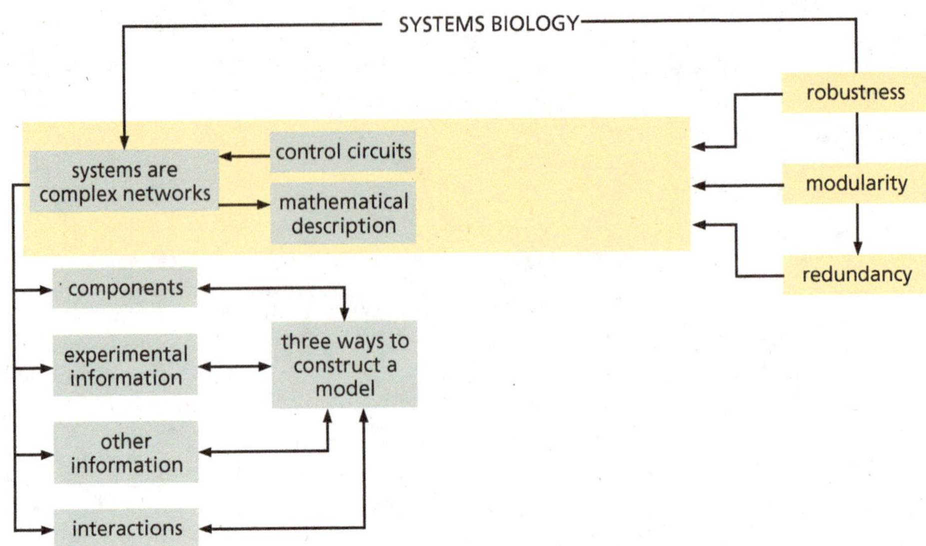


## Robustness is a distinct feature of complexity in biology

Understanding robustness and the mechanisms that underlie it is particularly important, because it provides in-depth understanding of how a system maintains its functional properties against disturbance. Systems biology should help us to understand how cells and organisms respond both to changes in their environment (such as a lack of nutrients, exposure to chemical agents, temperature changes, and so on) and to internal sources of malfunction (such as DNA damage, genetic defects in metabolic pathways, and so on). It is crucial to understand the intrinsic functions of the system if we are eventually to find cures for many diseases. Robustness is not always a desirable characteristic, especially when we are trying to disrupt or kill a cell; a robust cancer cell is not a good thing for the patient. As in engineered systems, robustness and stability can by achieved in biological systems by some distinct elements.

Some cellular life, such as the bacterium *Mycoplasma* (an obligate intracellular parasite of mammalian cells), gets by with only a few hundred genes. These organisms can live under specific environmental conditions but are very sensitive to any fluctuations, and are therefore not very robust. Even slightly more complex life forms have many more genes. For example, *E. coli* has some 4000 genes of which only about 300 are classified as essential (by the relatively crude test of whether they are essential for survival in standard laboratory conditions). The additional genes are thought to contribute to the backup functions and complex regulatory networks that confer the robustness *E. coli* requires to accommodate the various stresses it will encounter. In contrast to *Mycoplasma, E. coli* copes with the fluctuating environment in its normal habitat, the mammalian gut.

It is argued that robustness comes hand in hand with the complexity of a system as well as its opposite fragility. The relationship between the complexity and the fragility of life is well illustrated by the ways in which some living organisms have adapted to use molecular oxygen ($O_2$) as the electron acceptor in their energy-generating systems. Oxygen is potentially toxic to life as it is readily converted into highly active free radicals and compounds that can react with and damage biological molecules. Organisms that use $O_2$ as an electron acceptor have evolved complex feedback control mechanisms to ensure that their cells receive sufficient $O_2$ to live but that its concentration does not reach toxic levels. A complex set of networks provides the required robustness by maintaining almost invariant $O_2$ levels in the body, even if external concentrations vary. However, the dependence

## Box 17.1 Cellular modeling of the heart

To model multicellular systems accurately we need to be able to represent more than the simple biochemical or binding interactions described above. Multicellular systems, such as the heart, are composed of mechanical and electrophysiological interactions that occur on a variety of timescales and spatial scales, involving both intermolecular and intercellular interactions. To model the working heart, for example, at least two levels of data simulation are required: data from the cellular level and data at the organ level. At the cellular level, data on components such as the transported ions are used to model the behavior of cardiac muscle, while at the organ level, data such as the fiber structure of the muscles are used to build detailed three-dimensional anatomical models. The two levels are then linked to produce a useful predictive model of the heart and its action.

Cellular-level modeling has, for example, been used to explain the behavior of a pacemaker, and to improve understanding of what an electrocardiogram (the recording of the electrical activity of the heart) is telling us. In turn, this has shed light on some heart diseases such as the arrhythmias, where the electrical activity of the heart becomes uncontrolled, leading to a highly irregular heartbeat that can, in the most severe cases, prove fatal (see Figure 17.11).

Linking the two levels—cells and organ—enables researchers to model the spread of electrical excitation through the heart, which has allowed the accurate reconstruction of some dangerous arrhythmias in terms of the underlying molecular and cellular events and processes. Figure B17.1 shows images of a modeled heart.

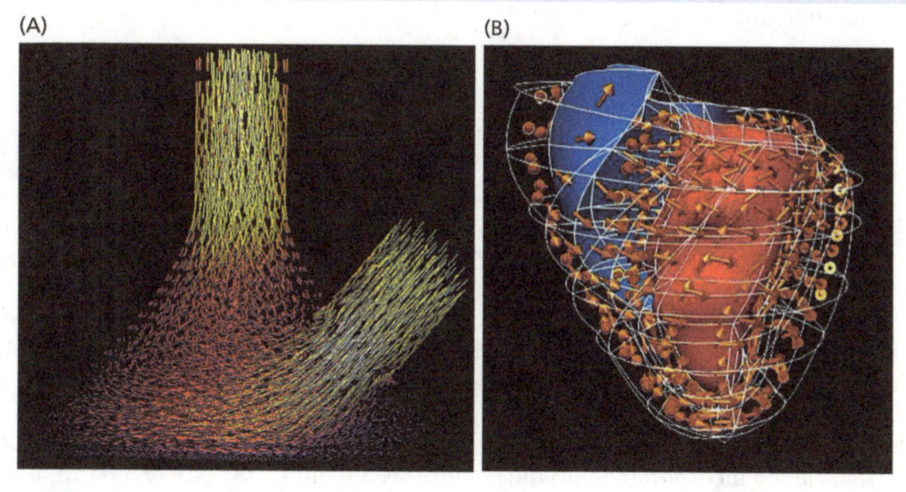


**Figure B17.1**
**Modeling of the heart at the organ level.** (A) The patterns of the flow of the fluid (blood) in a ventricular model. (B) How heart muscle fibers align where the muscle fibers are shown with arrows in the volume of the heart mesh. (Courtesy of Andrew Pullan.)

on such complex regulatory systems makes any failure of these systems lethal. Thus fragility is introduced because of the complexity of the regulatory system.

## Modularity plays an important part in robustness

Modules are individual components or subsystems of the whole system, maintain at least partial identity when isolated or rearranged, but also derive additional properties from the rest of the system. An example of a module in a molecular system is the Krebs cycle or tricarboxylic acid (TCA) cycle (see Figure 17.17) in the cellular respiratory pathway, which can be summarized in the form of a single reaction step:

$$\begin{matrix} C_6H_{12}O_6 + 10\,NAD^+ + 4ADP \\ + 4P_i + 2FAD + 2H_2O \end{matrix} \longrightarrow \begin{matrix} 6CO_2 + 10NADH + 10H^+ \\ + 4ATP + 2FADH_2 \end{matrix}$$

The interactions of modules with other components of the system can be complex, as the individual parts of the module can all potentially make interactions with

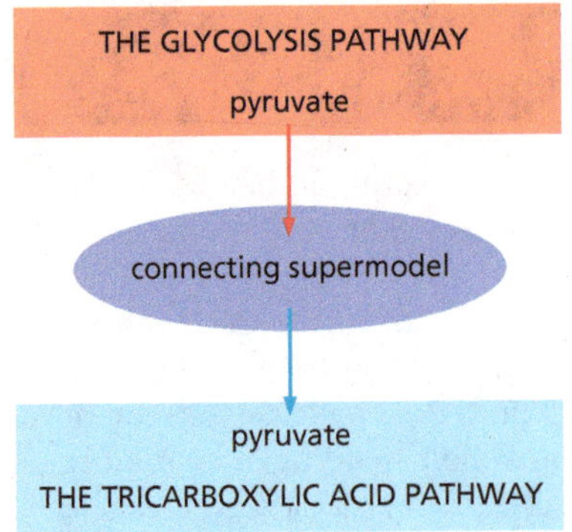


**Figure 17.17**
**A diagram of the three modules that make up the complete TCA cycle model.** The supermodel imports both the glycolysis pathway and the TCA cycle modules, and it allows the exchange of the variable pyruvate. (From CellML, www.cellml.org)

components in the rest of the system. The TCA cycle, for example, not only generates energy-rich products that are used in the next respiratory module, but in the process also generates metabolites that are used by other modules to synthesize amino acids, fatty acids, and other small molecules (see Figure 17.18).

In systems biology, interactions between different modules are sometimes referred to as **protocols**. There are advantages to having the interactions between modules as simple as possible. Modifications within modules may have little or no effect on the overall system properties as long as the inter-module protocols are maintained. The simpler the protocols, the easier it is to maintain them. In this way, modularity plays an important part in the robustness.

The LEGO® system of toy building bricks has been used to illustrate the concept of modularity and simple protocols and the flexibility they confer on a system (see Further Reading). The LEGO® bricks are the basic modules, which interact with each other via a reversible snap-connection protocol (reversibility is often important in biological systems too; many interactions between components are theoretically reversible). Many different shapes and sizes of LEGO®-brick modules exist, all of which interact by the same protocol. This allows many varied architectures to be built. Individual parts can be reused in new combinations, lost or damaged parts are easily replaced, and new modules are constantly being designed. The system can therefore evolve.

But despite creating a robust system, the snap-connection protocol has an inherent fragility. Small amounts of damage to the snap interfaces can cause the entire construction to break apart, whereas large amounts of damage at noninteraction faces will not cause any loss of robustness. If we compare the snap-connection protocol to something more robust, say a protocol that glues components together, we would indeed find that bricks glued to each other are better able to withstand trauma (throwing the toy across the room, for example) but the parts cannot be reused to make a new toy, and the possibility of evolution is decreased, if not eliminated.

For the LEGO® system to be made more versatile it can be made mobile by snap-protocol-compatible axles and wheels. Increased complexity comes with motorization, which requires new protocols for motor and battery interconnection and gearing. All these protocols can be combined to make a complex motorized module that can be incorporated into the system (see Figure 17.19), where its complexity is largely hidden from the user. Added complexity, especially in the control systems, also adds greater fragility, however.

Biological systems too are complex systems based on protocols and regulatory feedback loops that give the system robustness and enable it to evolve. In a large multicellular organism the death of one or even many cells will not kill the organism, but some types of malfunction in the control systems of a single cell can lead to fatal diseases such as cancer, thus illustrating the inherent fragility of this complex system. In the biological world, common protocols include gene regulation, covalent modifications to proteins (such as phosphorylation), the generation of action potentials (which are used as the single protocol by neurons to convey a vast number of different messages), and so on. Some associated modules would be the control of the expression of a gene, composed of amongst other things transcriptional repressor and activator proteins.

## Redundancy in the system can provide robustness

Most living systems have a high degree of **redundancy** built in, and this contributes to their robustness. Redundancy means that a function can be accomplished by several different pathways. This means that if one component is damaged, the

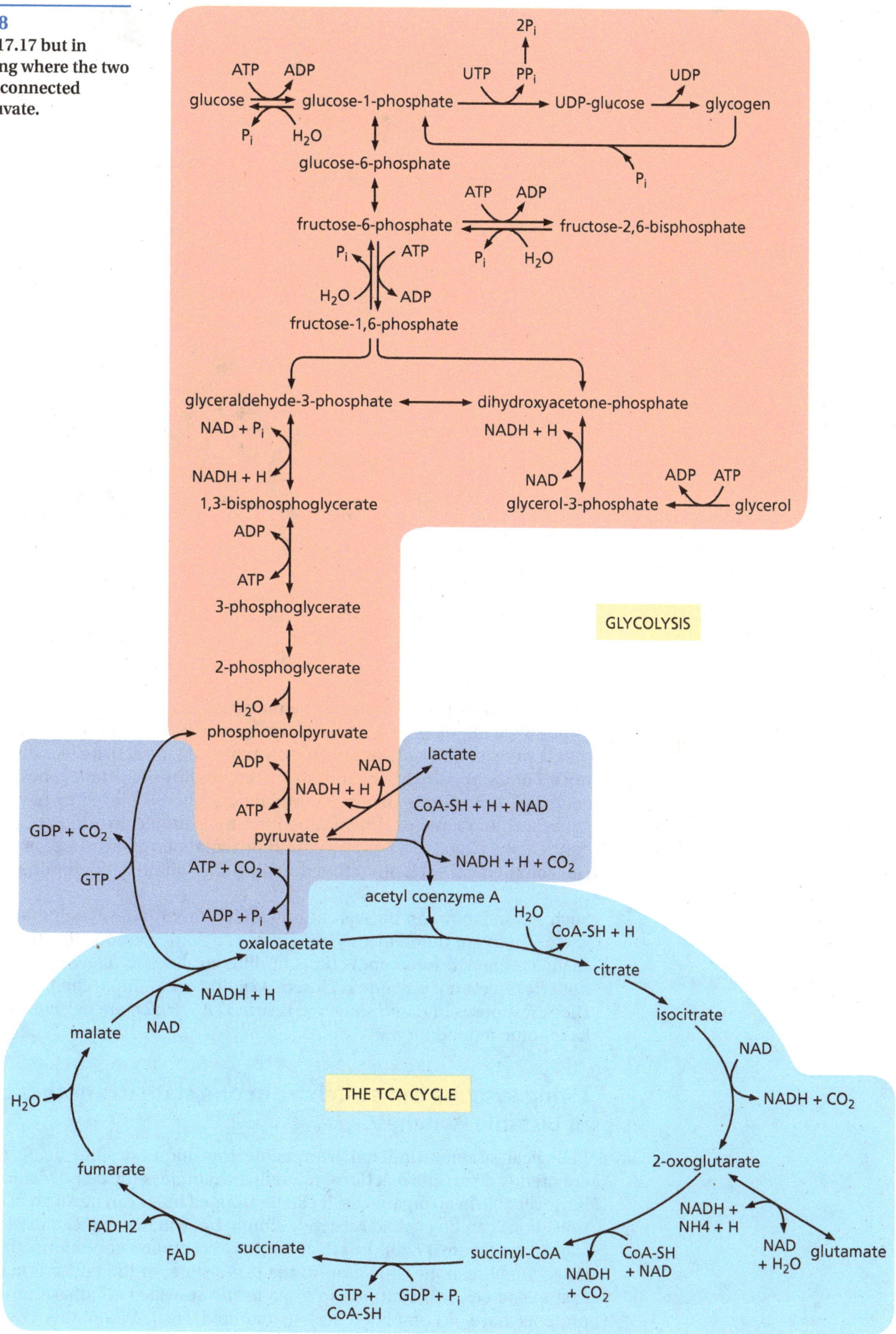

**Figure 17.18**
**As in Figure 17.17 but in detail, showing where the two modules are connected through pyruvate.**

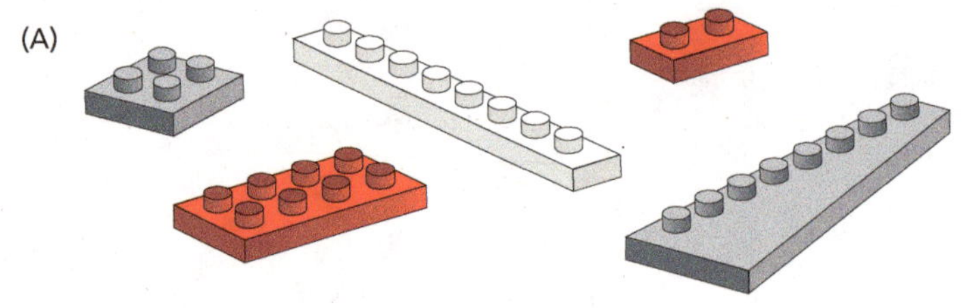

**Figure 17.19**
**(A) The building block used by the LEGO® system.** (B) The complex system that can be built using the LEGO® components and protocols.

function can still be carried out by pathways not involving this component. The importance of this in biological systems became apparent when knockout experiments were made involving specifically removing the activity of individual genes in mice. For a surprisingly large number of apparently important genes, the loss of the gene had quite minor effects. Redundancy is often achieved by having duplicated genes with the same or similar functions, for example. In addition, there is redundancy at the circuit level, such as multiple metabolic or transduction pathways that can complement each other functionally under different conditions.

Such redundancy can be seen in our underground railway map (see Figure 17.2) where there are different ways of getting to some stations. In the central region there are enough interconnections of different lines to allow most journeys to be undertaken even if one line is closed. A biological example can be seen in the heat shock responses in *E. coli* shown in Figure 17.13, which are partly independent and have some redundancy.

## Living systems can switch from one state to another by means of bistable switches

Biological systems undergo irreversible transitions despite being composed of apparently reversible reactions. A familiar example is the cell-division cycle, which is irreversible in all organisms. It can be stopped but it can never go backward. Such transitions can be created relatively simply by switches that can exist in two stable states (see Figure 17.20). In certain strictly controlled conditions, the system will make an all-or-none transition to the other state. In the eukaryotic cell cycle, for example, a cell will continue to grow in the so-called G1 phase until certain key proteins have accumulated to the required level. When this concentration is reached, the cell then switches irreversibly to the cell-division phases of the cycle.

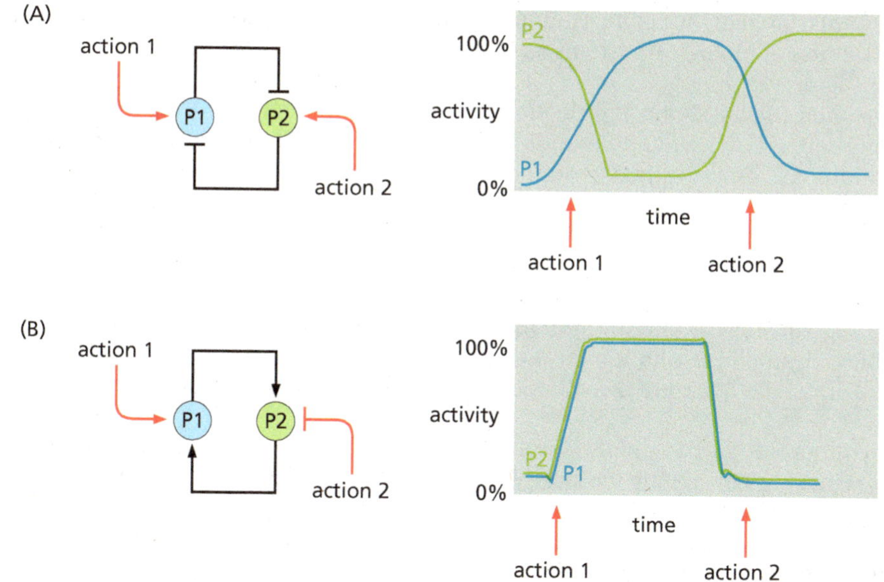


**Figure 17.20**
**An illustration of two bistable circuits as adapted from Ferrell.** (A) The first circuit shows a double-negative feedback loop where protein 1 (P1) inhibits protein 2 (P2), and vice versa (inhibition is illustrated by —|). A steady state could exist with one of the proteins switched off but not both. (B) A positive feedback loop where P1 activates P2, and P2 activates P1. Therefore in a stable state both P1 and P2 can be off or similarly both can be on. A stable state cannot exist with either P1 or P2 on and the other off. (Adapted from J.E. Ferrell Jr, Self-perpetuating states in signal transduction: positive feedback, double-negative feedback and bistability, *Curr. Opin. Chem. Biol.* 6:140–148, 2002.)

It replicates its DNA and then undergoes mitosis (nuclear division and chromosome segregation) and finally divides to give two daughter cells. Once the transition from growth mode to cell-division mode has been made, the cell cannot go back, nor can it skip a stage. The cell cycle may be stopped at these later stages if internal or external conditions become unfavorable, but if the block cannot be overcome the cell usually dies. What it definitely cannot do is return to the G1 phase without completing DNA replication and mitosis. The cell cycle is just one example of how a system can have higher-level emergent properties that could not be predicted just by looking at the properties of its isolated components.

## 17.4 Storing and Running System Models

**Flow Diagram 17.4**
**This section describes some of the practical aspects of storing and running system models.**

To model any biological system requires quantitative data as well as modeling and simulation tools. Simulation of how networks behave is one of the main aspects of

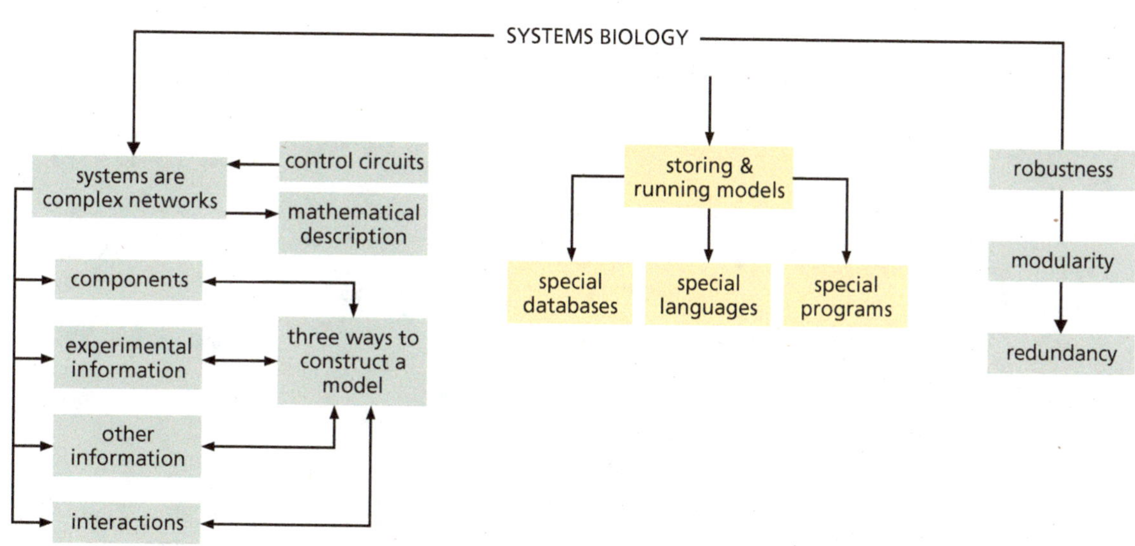

systems biology. Accurate simulation models are necessary to understand the dynamics of biological systems as well as for the design process.

The molecular systems described in Section 17.2 (see Figures 17.15A and 17.16) are fully defined by a set of differential equations, values of rate constants, and starting conditions. The starting conditions will be the initial concentrations of each component and the initial values of some state variables such as temperature. We need to calculate the variation of these concentrations with time, possibly including the variation with time of some other variables, such as those that represent time-dependent external stimuli. The techniques required are standard in numerical analysis and the systems often involve thousands of equations and variables. Example results for the EGFR pathway are shown in Figures 17.15C and 17.15D which illustrate the variation of concentration of components with time.

An important property of the system to characterize is the robustness, which can be examined by calculating the steady state of the system. The steady state is defined as that in which concentrations do not change with time. In this case the left-hand side of the set of equations such as Equation EQ17.6 are all set to 0. Solving the equations under these conditions defines the steady state. This enables the sensitivity (robustness) of the system to be examined for individual components.

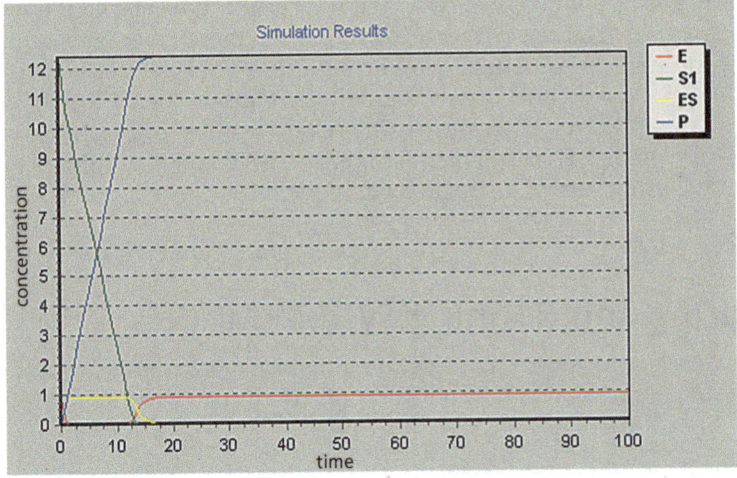


**Figure 17.21**
**A simple simulation of enzyme kinetics is submitted to the Jarnac program:**

$$E + S1 \underset{k2}{\overset{k1}{\rightleftharpoons}} ES \xrightarrow{k3} E + P$$

The code to calculate the graph is depicted in the boxed area.

```
DefaultModel Simple
    var S1, ES, P, E;

    E + S1 -> ES; k1*E*S1 - k2*ES;
    ES -> E + P; k3*ES;

end;

println "Model exhibiting two conserved cycles", nl;
println "E + S1 -> ES";
println "ES -> E + P";

k1 = 130.0;
k2 = 1.0;
k3 = 1.0;

E = 0.9; ES = 0.0;
S1 = 12.4241;   P = 0;

m = sim.eval (0, 100, 100, [<Time>, <E>, <S1>, <ES>, <P>]);
graph (m);
```

## Specialized programs make simulating systems easier

General-purpose simulators, such as GEPASI or Jarnac for example, predict the behavior of metabolic pathways by constructing differential equations from user-defined chemical reactions, which are then solved using numerical integration. Jarnac runs on a text-based language that can be used to describe a pathway model (see Figure 17.21). GEPASI simulates the steady-state and time-course behavior of reactions. Both programs need information on the structure of the pathway, the kinetics of each reaction, and initial concentrations of the chemical members involved. GEPASI builds the differential equations, which control the behavior of the system, and solves them (see Figure 17.22), while Jarnac is in addition an actual language that is used for modeling integrated cellular systems including multicellular

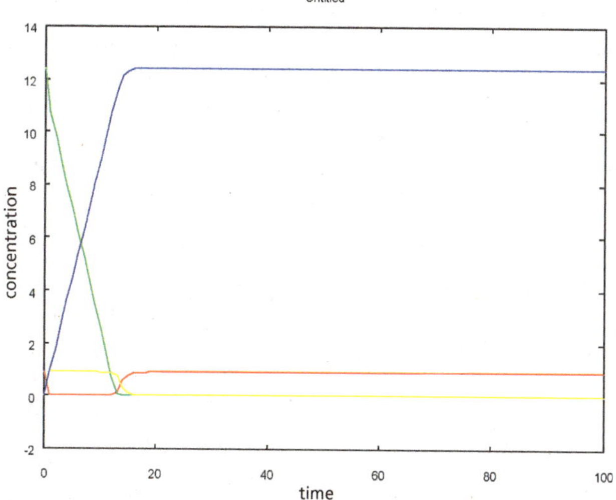


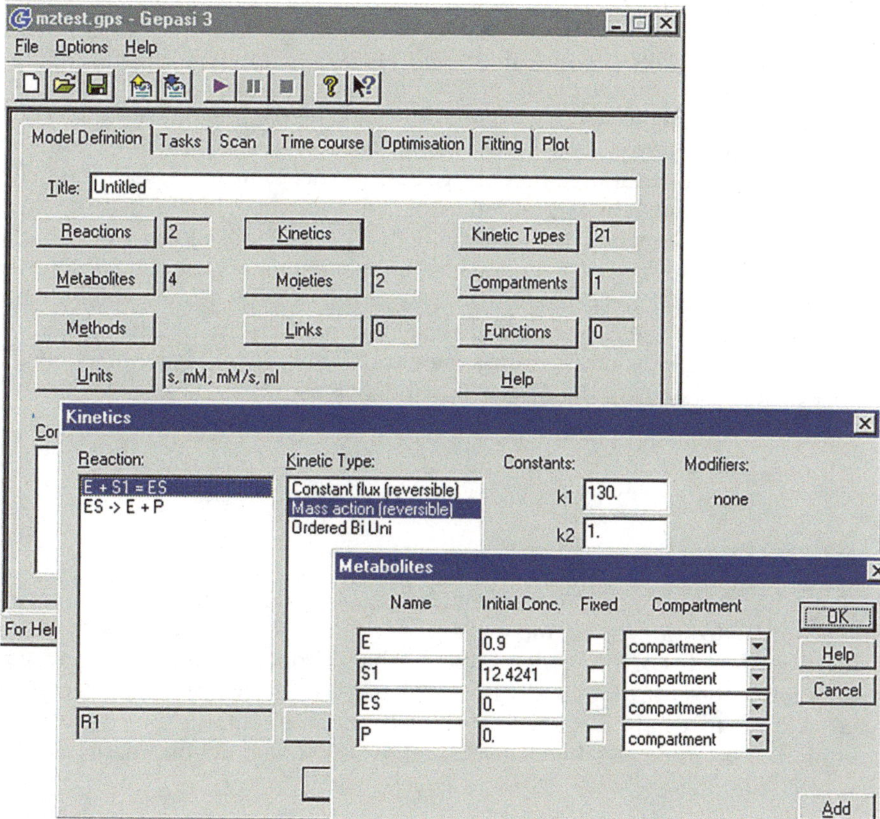


**Figure 17.22**
**An example of system simulation using GEPASI.** The same enzyme reactions are simulated as in Figure 17.21. The graph shows the variation of concentration with time of E (red line), S1 (green line), ES (yellow line), and P (blue line). In this program the components, initial volumes, equations, and other details are input through a set of windows (shown below the graph).

systems, creating a very powerful environment for system modeling. Figure 17.21 and Figure 17.22 illustrate the same simple simulation showing how the different methods can be used to obtain the same result.

There are also other tools that provide more than one simulation technique. For example, GEPASI has been superseded by COPASI, which employs differential equations and stochastic methods. In addition, programs are available to facilitate mechanistic organ modeling. For example, CMISS (Continuum Mechanics, Image analysis, Signal processing and System Identification) is a mathematical modeling environment developed to aid modeling of a variety of complex bioengineering problems; similarly EMAP, which is an Electrocardiac Mapping system tool.

## Standardized systems descriptions aid their storage and reuse

To obtain an accurate definition and comparison of systems, a standard way of describing models and their data is required. Such a description should contain sufficient additional information so that the model can be reproduced or used for further analysis. The development of well-structured controlled vocabularies, the adoption of standardized data exchange formats, and the use of model representation languages have made flexible data integration and model exchange possible.

There are two main descriptor (markup) languages available. The Systems Biology Markup Language (SBML) is a machine-readable language for describing qualitative and quantitative models of biological networks. The main aim of SBML is to enable the exchange of models between different programs. SBML is widely adopted with almost 100 software systems and databases based on it. The other widely used language is Cell Markup Language (CellML). This is mainly used for describing and exchanging models of cellular and subcellular processes. CellML describes the structure and underlying mathematics of cellular models.

MIRIAM (Minimum Information Requested In the Annotation of biochemical Models) is a standard for curating quantitative models of biological systems. To pass the MIRIAM rule a model must be encoded in a standard markup language, and values must be given for all initial conditions and parameters. In addition there is a strict set of rules for the model annotation that specifies the documentation of the model. For example, a PubMed identifier for the complete description of the model must be provided. Several databases have been designed for storing and making available quantitative models. CellML Model Repository uses the CellML format and is currently the largest repository with 188 models representing several types of cellular processes including models of electrophysiology, metabolism, signal transduction, and mechanics. Other model databases use the SBMLformat, for example the BioModels Database and SigPath. The latter also provides an interactive interface for the user to build their own pathway model using their own components, reactions, and quantitative data. Some model repositories like JWS Online Cellular System Modeling also provide tools for simulation, while others like CellML Model Repository enable the user to visualize the model.

# Summary

Biological systems are composed of a large number of components that have many interactions. They are too complex to understand fully on the basis of a visual inspection of the list of interactions, and can have properties that are only revealed by simulating the entire system. This is particularly true because biological systems have evolved to carefully control their cell environments despite the external conditions, and must also be robust to survive the harsh environments in which they often find themselves.

We have seen how systems can be modeled not only at the molecular level, but also at the cellular or organism level. Although the latter are still relatively rare, recent genomic experiments have greatly eased the problem of defining molecular systems, at least to the point of listing interactions. As a result they have become more commonly studied by these methods. This has focused more attention on the experimental determination of the parameters required to simulate the models. The numerical techniques themselves are often borrowed from other scientific fields, which has the advantage that they are well established and practical issues are well understood.

Quantitative models will have an increasingly important role in the future and provide new insights into diseases and how to treat them. For example systems biology is starting to be used in synthetic biology, where the aim is to redesign the cell and its constituent parts to provide novel abilities. Systems biology can be used to study first the existing system and then the redesigned system. Systems biology is not restricted to the purely biological field; for example, systems biology is being used for research in helping the environment by focusing on biological energy systems such as bio-based fuels and bioenergy.

## Further Reading


### 17.1 What is a System?

von Bertalanffy L (1976) General System Theory: Foundations, Development, Applications. New York: George Braziller Inc.

Kholodenko BN (2006) Cell-signalling dynamics in time and space. *Nat. Rev. Mol. Cell. Biol.* 7, 165–176.

Takahashi K, Ishikawa N, Sadamoto Y et al. (2003) E-Cell2: Multi-platform E-Cell simulation system. *Bioinformatics* 19, 1727–1729.

Genomics: GTL. Systems Biology for Energy and Environment. US Department of Energy. http://genomicsgtl.energy.gov/

*Science* (2002) 295, 1661–1682 *Includes papers by Csete and Doyle and Denis Noble on heart modeling. Also see:* http://www2.auckland.ac.nz/esc/Images/Heart/TOP.html *and* http://paterson.physiol.ox.ac.uk/research/CardiacMapping/

**Databases**

Bader GD, Donaldson I, Wolting C et al. (2001) BIND—The Biomolecular Interaction Network Database. *Nucleic Acids Res.* 29, 242–245.

Kanehisa M (1997) A database for post-genome analysis. *Trends Genet.* 13, 375–376.

Kanehisa M & Goto S (2000) KEGG: Kyoto Encyclopedia of Genes and Genomes. *Nucleic Acids Res.* 28, 27–30.

Kanehisa M, Goto S, Hattori M et al. (2006) From genomics to chemical genomics: new developments in KEGG. *Nucleic Acids Res.* 34, D354–D357.

Ng A, Bursteinas B, Gao Q et al. (2006) pSTIING: a 'systems' approach towards integrating signalling pathways, interaction and transcriptional regulatory networks in inflammation and cancer. *Nucleic Acids Res.* 34, D527–D534.

EcoCyc. Encyclopedia of *Escherichia coli* K-12 Genes and Metabolism.: http://ecocyc.org/

### 17.2 Structure of the Model

Bray D (2003) Molecular networks: The top-down view. *Science* 301, 1864–1865.

**Topological modeling**

Edwards JS & Palsson BØ (2000) Metabolic flux balance analysis and the *in silico* analysis of *Escherichia coli* K-12 gene deletions. *BMC Bioinformatics* 1, 1.

Price ND, Famili I, Beard DA & Palsson BO (2002) Extreme pathways and Kirchhoff's second law. *Biophys. J.* 83, 2879–2882.

### 17.3 Robustness of Biological Systems

Brandman O, Ferrell JE Jr, Li R & Meyer T (2005) Interlinked fast and slow positive feedback loops drive reliable cell decisions. *Science* 310, 496–498.

Chen B, Wang Y, Wu W & Li W (2005) A new measure of the robustness of biochemical networks. *Bioinformatics* 21, 2698–2705.

Cseste ME & Doyle JC (2002) Reverse engineering of biological complexity. *Science* 295, 1664–1669.

**Molecular systems modeling**

Barrett CL, Kim TY, Kim HU et al. (2006) Systems biology as a foundation for genome-scale synthetic biology. *Curr. Opin. Biotechnol.* 17, 488–492.

Feist AM, Scholten JCM, Palsson BØ et al. (2006) Modeling methanogenesis with a genome-scale metabolic reconstruction of *Methanosarcina barkeri. Mol. Syst. Biol.* 2, 2006.0004.

Ferrell JE Jr (2002) Self-perpetuating states in signal transduction: positive feedback, double-negative feedback and bistability. *Curr. Opin. Chem. Biol.* 6, 140–148.

Garny A, Noble D & Kohl P (2005) Dimensionality in cardiac modelling. *Prog. Biophys. Mol. Biol.* 87, 47–66.

Guido NJ, Wang X, Adalsteinsson D et al. (2006) A bottom-up approach to gene regulation. *Nature* 439, 856–860.

Helm PA, Younes L, Beg MF et al. (2006) Evidence of structural remodeling in the dyssynchronous failing heart. *Circ. Res.* 98, 125–132.

Hunter P & Nielsen P (2005) A strategy for integrative computational physiology. *Physiology* 20, 316–325.

Le Novere N, Bornstein B, Broicher A et al. (2006) BioModels Database: A free, centralized database of curated, published, quantitative kinetic models of biochemical and cellular systems. *Nucleic Acids Res.* 34, D689–D691.

Quaranta V, Weaver AM, Cummings PT & Anderson ARA (2005) Mathematical modeling of cancer: The future of prognosis and treatment. *Clin. Chim. Acta* 357, 173–179.

Schoeberl B, Eichler-Jonsson C, Gilles ED & Muller G (2002) Computational modeling of the dynamics of the MAP kinase cascade activated by surface and internalized EGF receptors. *Nat. Biotechnol.* 20, 370–375.

Winslow RL, Greenstein JL, Tomaselli GF & O'Rourke B (2001) Computational models of the failing myocyte: Relating altered gene expression to cellular function. *Philos. Trans. R. Soc. Lond. A* 359, 1187–1200.

# APPENDIX A: PROBABILITY, INFORMATION, AND BAYESIAN ANALYSIS

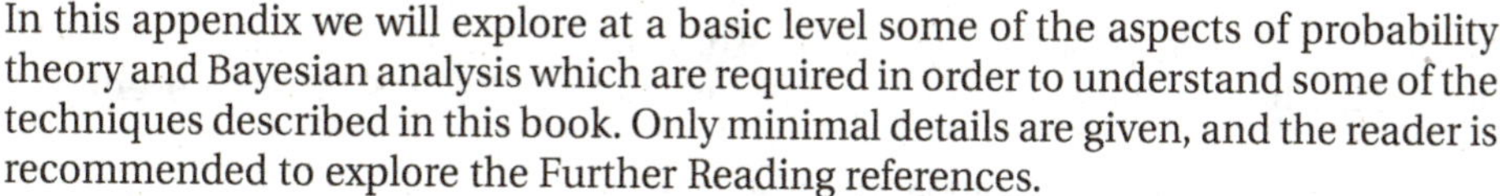

In this appendix we will explore at a basic level some of the aspects of probability theory and Bayesian analysis which are required in order to understand some of the techniques described in this book. Only minimal details are given, and the reader is recommended to explore the Further Reading references.

## Probability Theory, Entropy, and Information

The results given below are required to understand some of the presentation of sequence logos in Section 6.1 and aid appreciation of several other methods described. In addition, the notation is used in Bayesian analysis as presented in the following section.

### Mutually exclusive events

Suppose that there are $n$ alternative events of which just one can occur, often referred to as mutually exclusive events. If each of the possible events is labeled $x_i$ and has probability that is written $P(x_i)$, then

$$\sum_i P(x_i) = 1 \qquad \text{(EQA.1)}$$

where the summation is over all $n$ possible events. This situation arises, for example, when considering the amino acid that occurs at a particular position in a protein sequence, when the possible events are the 20 amino acids.

The uncertainty about which of the events will occur depends on the probabilities $P(x_i)$. If one of the possible events has a probability of 1 (i.e., is certain to occur) all the others will have zero probability, and the outcome will always be known. Alternatively, all the possible events might have an equal probability of $1/n$, in which case the outcome would be very uncertain. A measure of the uncertainty has been proposed, often called the Shannon entropy, defined by

$$H(X) = -\sum_i P(x_i) \log_2 P(x_i) \qquad \text{(EQA.2)}$$

where $X$ refers to the set of all possible events. If $P(x_i)$ is zero, the term $P(x_i) \log_2 P(x_i)$ is also taken to be zero. Hence when one event is certain $H(X)$ is zero, indicating that there is no uncertainty in the outcome. When all events are equal, the uncertainty takes its maximum value $H_{max}(X)$ of $\log_2 n$. The uncertainty as defined in Equation EQA.2 is measured in units of bits, as the logarithm base used is 2, but alternative definitions using other logarithm bases are equally valid. The maximum uncertainty $H_{max}(X)$ of a position in a sequence is 2 bits for nucleotide and approximately 4.32 bits for protein sequences. Note that the word entropy, although commonly used in this context, is not to be confused with the same term used in discussions of thermodynamics and free energy.

Information can be seen as a loss of uncertainty, so that at a given sequence position the amount of information can be equated to the difference between $H_{max}(X)$ and $H(X)$. The sequence logos described in Section 6.1 and illustrated in Figures 6.5 and 10.4 are plots of $\{H_{max}(X) - H(X)\}$ at each sequence position. (As mentioned in Chapter 6 the formula is modified to account for a lack of data.)

## Occurrence of two events

We will now consider the probabilities of two events that are not mutually exclusive. In what follows we will write the probability of event X occurring as P(*X*), and the probability of event *Y* occurring as P(*Y*). These are also called marginal probabilities, being the probability of one event occurring regardless of the other event.

If the events *X* and *Y* are independent of each other the joint probability of both occurring, written P(*X*,*Y*), is given by the product P(*X*)P(*Y*). However, the events *X* and *Y* might not be independent of each other. The probability of event *Y* occurring, given that event *X* has already occurred, is written $P(Y \mid X)$, and is the conditional probability of *Y* given *X*. Similarly, the probability of event *X* occurring given that event *Y* has already occurred is written $P(X \mid Y)$. If the two events are independent, then $P(Y \mid X) = P(Y)$ and $P(X \mid Y) = P(X)$. Otherwise, the marginal probability of event *X* can be written

$$P(X) = \sum_{Y} P(X \mid Y)\, P(Y) \qquad \text{(EQA.3)}$$

The joint probability P(*X*,*Y*) that events *X* and *Y* both happen is given by

$$P(X,Y) = P(Y \mid X)P(X) \qquad \text{(EQA.4)}$$

if *X* occurs first, or

$$P(X,Y) = P(X \mid Y)P(Y) \qquad \text{(EQA.5)}$$

if *Y* occurs first. These two products are equal, giving

$$P(Y \mid X)\, P(X) = P(X \mid Y)\, P(Y) \qquad \text{(EQA.6)}$$

from which

$$\frac{P(Y \mid X)}{P(Y)} = \frac{P(X \mid Y)}{P(X)} \qquad \text{(EQA.7)}$$

## Occurrence of two random variables

Suppose that the events *X* and *Y* each have a distribution of outcomes *x* and *y*, respectively. In the context of this book, either or both of these could correspond to many different things, for example a protein sequence, the secondary structural assignments of a protein sequence, or an expression level. In some cases the distribution will be a finite set of mutually exclusive alternatives such as the 20 amino acids, but in other cases the distribution will be a random sampling of a continuous statistical distribution.

A quantity called the mutual information, I($X$,$Y$), can be defined which is a measure of the degree of independence of distributions $X$ and $Y$. This is defined as

$$I(X,Y) = \sum_{x \in X} \sum_{y \in Y} P(x,y) \log \frac{P(x,y)}{P(x)\,P(y)} \tag{EQA.8}$$

When $X$ and $Y$ are independent, P($x$,$y$) = P($x$)P($y$), the log term becomes 0, and the mutual information is 0. This term represents the reduction in the uncertainty of one variable once the other has been observed, and is zero or positive. If the term P($x$,$y$) is substituted by either side of Equation EQA.6, the log term simplifies to the equivalent side of Equation EQA.7. This log term is identical to that employed in deriving the GOR protein secondary structure prediction method (see Equation EQ12.6) which is referred to by the same term. Furthermore, application of Equation EQA.7 to this form of the log term shows that I($X$,$Y$) is identical to I($Y$,$X$).

Another term, the relative entropy or Kullback–Leibler distance, is a measure of the distance between the two distributions $X$ and $Y$. When the outcomes from the two distributions can be paired as $x_i$ with $y_i$, it is defined as

$$H(X,Y) = \sum_i P(x_i) \log \frac{P(x_i)}{P(y_i)} \tag{EQA.9}$$

This measure is also zero or positive, but H($X$,$Y$) is not the same as H($Y$,$X$). This term arises in log-odds substitution scoring matrices, where it is the expected score for a matrix (see Equation EQ5.12 and Figure 5.7).

## Bayesian Analysis

There are many occasions in bioinformatics where a set of data has to be fitted to a model that has uncertain parameters. Additionally, in many cases there are insufficient data available to fit the large number of model parameters. The technique of Bayesian analysis is frequently used to solve these problems, and a brief explanation of the technique is presented here. This area is complex and of increasing importance. References are provided which will guide the reader to a fuller understanding of the technique and its applications.

### Bayes' theorem

Consider a situation in which there are a number of systems or models which differ slightly in their properties in a way that can be defined by the value of a parameter $y$ in each system. There will be a distribution of values of $y$ as found in the set of systems, such that the probability of randomly selecting a system with a particular value $Y$ can be written P($Y$). Usually there will be a hypothesis about the distribution $y$. These systems produce observed data $X$ which depend in some way on the value of the parameter $y$. Typically, we want to use the observed data to deduce the value $Y$ for the system. (Or in more complex instances we want to obtain the distribution $y$.)

We can regard the occurrence of the system with parameter value $Y$ and of the data $X$ as two separate events. It will be useful to calculate the term P($Y$|$X$) which is the probability of the parameter $y$ having the value $Y$ given that the system produced observed data $X$. Equation EQA.4 can be applied here, and rearranged into the form usually called Bayes' theorem:

$$P(Y|X) = \frac{P(X|Y)P(Y)}{P(X)} \tag{EQA.10}$$

The term $P(Y|X)$ is called the posterior probability of the hypothesis that the system has parameter $Y$, which is to be compared with the prior probability $P(Y)$. The description prior is used as this is the probability before any data have been observed from the system. In contrast the observed data are involved in calculating the posterior probability. The term $P(X|Y)$, the probability of obtaining the data $X$ given that the system parameter has value $Y$, is called the likelihood of the hypothesis that the system has parameter $Y$.

Usually, the prior probability is assigned as a standard probability distribution according to the model used for the system. There are many instances where there is a lack of knowledge of the true distribution, but so long as there are enough data the equation can be quite robust, so that even a uniform probability distribution can on occasion be used for the prior. However, sometimes specific prior distributions are to be preferred which depend on the form of the probability distribution for the likelihood (distribution of the data given the model). Because the prior and likelihood are multiplied together to obtain the posterior distributions certain combinations of distributions will result in the posterior distribution being a standard form. Those combinations of distributions which result in the posterior distribution having the same form as the prior are called conjugate priors. In this case the prior distribution has the same form as the observed data, so that the prior has the appearance of extra data. In this way, the conjugate prior links to the simple pseudocount ideas described in Section 6.1 for dealing with a lack of data.

Three examples of such conjugate distributions are the normal (or Gaussian) distribution which is conjugate to itself; the binomial and beta distributions and the multinomial and Dirichlet distributions. Thus if the likelihood distribution is multinomial, as occurs with the random sequence models discussed in Section 6.1, it is sensible to use a Dirichlet prior. Objections have been raised to the Bayesian approach because the prior distribution contains the assumptions (i.e., bias) about the model, which may well be incorrect. It is important to be aware that the conclusions from the analysis can depend on the prior used.

## Inference of parameter values

Frequently, the theory described above is used to estimate the value of model parameters by inference from the data and prior distribution. The standard method of statistical inference uses the largest value (i.e., mode) of the likelihood to predict the value of the model parameter $Y$, a method known as maximum likelihood or ML. However, the likelihood obtained is a statistical distribution, and it can also be used in other ways to estimate parameters. For example the mean likelihood could be used or the value averaged over the distribution. A comparative discussion of these methods is beyond the scope of this book.

## Further Reading

This appendix only shows how some of the basic concepts used in this book are connected, and does not do full justice to the field. There are a number of textbooks which will repay close study, but unfortunately they all proceed rapidly to an advanced level. For presentations of Bayesian theory see for example:

Gelman AR, Carlin JB, Stern HS & Rubin DB (2004) Bayesian Data Analysis, 2nd ed. Boca Raton, FL: Chapman & Hall/CRC Press.

There are also useful introductions to aspects of Bayesian and statistical analysis within the books:

Durbin R, Eddy S, Krogh A & Mitchison G (1998) Biological Sequence Analysis. Cambridge, UK: Cambridge University Press.

Baldi P & Brunak S (2001) Bioinformatics: The Machine Learning Approach 2nd ed. Cambridge, MA: MIT Press.

Finally, there are some useful articles on Bayesian methods:

Eddy SR & MacKay DJC (1996) Is the Pope the Pope? *Nature* 382, 490.

Liu JS & Logvinenko T (2003) Bayesian methods in biological sequence analysis. In Handbook of Statistical Genetics, 2nd ed, chap. 3 (DJ Balding, M Bishop, C Cannings eds), pp 66–93. Chichester, UK: John Wiley & Sons.

Eddy SR (2004) What is Bayesian statistics? *Nat. Biotechnol.* 22, 1177–1178.

# APPENDIX B: MOLECULAR ENERGY FUNCTIONS

In this appendix we will look at the form of some of the algebraic functions that are used to represent the energetics of molecular systems, in particular proteins. In principle, quantum mechanics calculations of molecular orbitals will provide the most accurate energies of molecular systems. However, these are too computationally demanding to be practical for the problems described in this book. Therefore there is a need to use relatively simple functions that approximate the exact molecular energies.

Quantum mechanics calculations obtain the energy of formation of a molecule from its constituent parts (usually the bare nuclei and free electrons). However, in the context of the topics in this book we are not interested in the energy of formation. Instead we are interested in the relative energy, which is the difference in energy of two states of the system. These two states might be two alternative conformations of the same molecule or a bound complex of several molecules and the same molecules at infinite separation. The latter two states are used to determine the interaction energy between the molecules. The focus on energy differences rather than absolute energies means that the zero can be arbitrarily assigned at any point on the energy scale. This is very important for the development of simple energy functions. For example, instead of having to know with accuracy the energy of formation of a covalent bond, the energy of formation of the bond with ideal geometry can be defined as the energy zero. It only remains to define the energy of distorting the bond from ideal geometry, which as will be described below can be done with simple yet quite accurate formulae.

A further consequence of only requiring relative system energies is that any interactions which remain constant for all the system states of interest can be ignored. A typical example of interactions assumed constant is the average solvent interaction with the system. Care is required when making this assumption, as there are many cases where it does not hold, but it is often used to justify omitting the solvent energetics from calculations. If the solvent interactions can be ignored, or at least greatly simplified, it can very significantly reduce computational requirements. This is because molecules interact over relatively long ranges—at least to 8 Å separation, albeit very weakly at that distance—so that many hundreds if not thousands of water molecules are required for an accurate model of the aqueous environment. There are several active areas of research, for example the modeling of molecular function, where solvent must be explicitly and accurately represented, but calculations in the areas of modeling described in this book generally ignore or greatly approximate the solvent energetics. For this reason we will only describe the few terms of relevance here, and those interested in learning about accurate solvent representation and calculation should consult the references in Further Reading.

Two distinct approaches have been used to derive molecular energy functions. In the first approach an attempt is made to reproduce in detail the bonding and nonbonding interaction energies of the molecular conformation(s) of interest. The alternative approach tries to represent the averaged energetics as seen in a collection of systems. The first approach mostly results in terms with algebraic forms that have a basis in the detailed physics of specific interactions, and the collection of these terms used to represent the system is often referred to as a force field. The second approach, most commonly used in threading, tends to produce a more empirical algebraic form that is obtained by fitting to observed data. We will describe the terms obtained by each approach in turn.

## Force Fields for Calculating Intra- and Intermolecular Interaction Energies

In this section we will discuss in detail some functions that have been proposed to represent specific inter-atomic interactions. This approach to molecular modeling was pioneered in studies of small molecules, especially of their vibrational spectroscopy, and subsequently the applications were extended to include macromolecules. Even for small molecules their conformational and interaction energetics can have a complicated form, but it was noticed that this could often be separated into simpler terms that when added together described the whole system well. Initially these were intramolecular energy functions, and all related to the covalent bonding structure of individual molecules, often referred to as bonding terms. For example, terms were proposed to model bond length stretching and bond angle bending. As this research area developed further terms were added that described interactions which did not involve covalent bonds, often called nonbonding terms.

All of these terms have certain features in common. Insights from molecular spectroscopy suggest that individual chemical groups have quasi-independent energetics. (These chemical groups are frequently very small, such as -OH, $-CH_3$, or $-CO_2H$.) This means that the total molecular energy can be divided into a number of terms, each relating to one of these groups. Because the energetics of the different chemical groups within a molecule are not totally independent of each other, for greater accuracy we should also include energy terms involving two or more groups, but often in the protein work described here this is not done. The bonding terms used tend to each involve two, three, or four atoms, but are limited to atoms bonded to each other. The nonbonding terms used involve only two atoms, although often there is a term for every pair of atoms in the system, which for proteins can be a large number. The complete force field for a protein may be composed of several thousand force field terms, but each of them will be relatively simple and only involve a small number of atoms.

One of the features of this approach is that the total system energy is represented by a large set of terms that are added together, often referred to as an additive force field. To a large extent this is intuitive, as for example the hydrogen bonding energy in a system will to a fair degree of accuracy be the sum of the energies of the individual hydrogen bonds of the system. However, this is an approximation, as in many cases there is an element of cooperativity, meaning for example that the energy of a hydrogen bond is influenced by the presence of others around it. The only way to accurately represent this is to use a nonadditive force field, but this usually results in much longer calculations. As the effect of cooperativity is often 10% or less of the total energy frequently only the additive terms are used.

Force fields have been parameterized several times and for many different types of molecules. Commonly used force fields for protein calculations are CHARMM and AMBER, details of which will be found in the references of Further Reading. In all cases the force fields are defined not just by the formulae used but also by the sets of parameters, which are usually updated on a regular basis.

In what follows we will discuss the major terms used in protein energy calculations of the kind described in this book, but will not present their theoretical background. Readers interested in understanding the origin of these terms and gaining a better appreciation of their strengths and weaknesses should explore the references given in Further Reading.

## Bonding terms

Intramolecular energies can usefully be separated into small independent components that correspond to elements of the molecular geometry such as variations in bond length, bond angle, and torsion angle. As mentioned above, if the molecular energy is required to very high accuracy, cross-terms need to be included such as terms involving the lengths of the two covalent bonds associated with a particular bond angle, but usually the individual components are sufficient for the applications covered in this book. Some workers classify force fields according to the presence of these cross-terms. Force fields without such cross-terms are called Class I, and those that include them are referred to as Class II.

There are several simple expressions that can represent the energetics of bond length variation. To some extent the choice of form to use depends on the processes that need to be modeled. When the bond length $l$ is close to its reference value $l_0$ the bond energy $E_l$ can be represented by the form

$$E_l = \frac{k_l}{2}\left(l - l_0\right)^2 \qquad \text{(EQB.1)}$$

where $k_l$ is a force constant. The force constant determines the range of length variation, and is a parameter often obtained from vibrational spectroscopy. This is known as the harmonic approximation and the energy minimum is zero at the reference bond length and positive for deviations (see Figure B.1). Different types of bonds will have different values for the parameters $k_l$ and $l_0$. A given force field will define the bond types as well as provide these parameters.

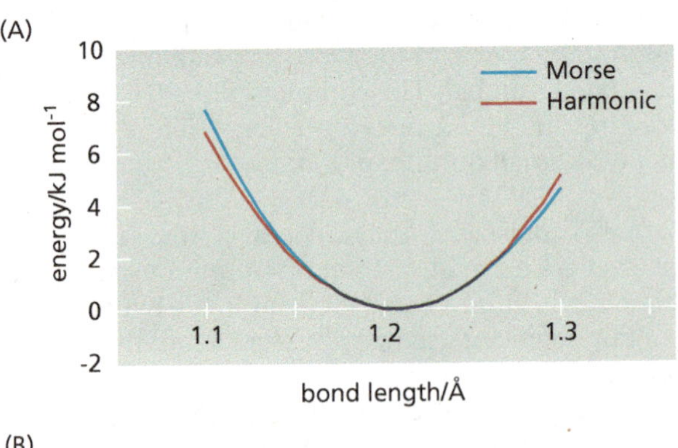


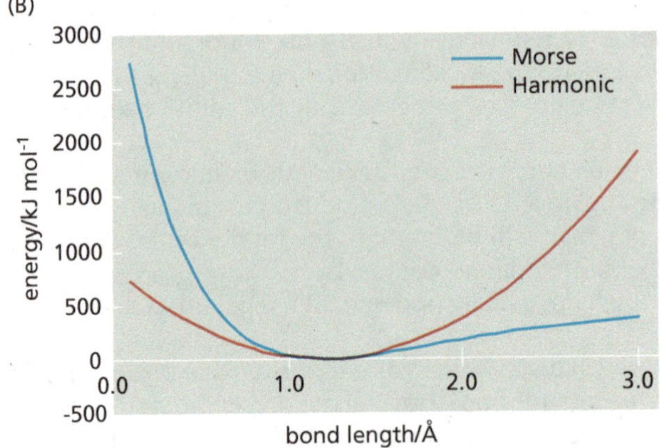


**Figure B.1**
**Two alternative terms used to represent the variation of energy with bond length.** The Morse potential can model bond dissociation as well as equilibrium dynamics. The harmonic approximation is much easier to use, but has limited accuracy. The example shown represents molecular oxygen ($O_2$) and has parameters $D_e = 493.50$ kJ mol$^{-1}$, $l_0 = 1.21$ Å, and $k_l = 1176.8$ N m$^{-1}$. The exponential coefficient $a$ of Equation EQB.2 is given by $\sqrt{k_l/2D_e}$. (A) The two functions have the same energy minimum and are very similar in this local region. (B) There is considerable difference between the functions far from the optimal bond length.

Usually for the applications of interest here, Equation EQB.1 provides sufficient accuracy. However, large distortions will not be represented well by the harmonic approximation. These can be represented by adding more terms of the form $(l - l_0)^n$ to the expression, which will increase the region of the energy well that is accurately reproduced. However, it is important to note that Equation EQB.1 will not allow bond dissociation to be modeled. The Morse potential gives a more accurate representation for large distortions and bond breaking, and has the form

$$E_l = D_e\left\{1 - exp\left[-a\left(l - l_0\right)\right]\right\}^2 \qquad \text{(EQB.2)}$$

where $D_e$ and $a$ are parameters that are supplied in the force field. ($D_e$ is in fact the energy minimum of this term.) As can be seen from Figure B.1 these two functions have very different forms away from the energy minimum.

Bond angles are usually represented by the harmonic approximation, now using a reference bond angle $\theta_0$. The bond angle energy $E_b$ is given by

$$E_b = \frac{k_b}{2}\left(\theta - \theta_0\right)^2 \qquad \text{(EQB.3)}$$

with parameters $k_b$ and $\theta_0$ being defined for different types of bond angle in a given force field.

The term torsion angle is used outside molecular mechanics to refer almost exclusively to the angle defined by four bonded atoms, as shown in Figure 2.6A. The general form for a torsion angle potential $E_\omega$ is given as

$$E_\omega = \sum_{n=0}^{N} \frac{V_n}{2}\left\{1 + \cos\left(n\omega - \alpha\right)\right\} \qquad \text{(EQB.4)}$$

for a torsion angle $\omega$, where there may be three terms in the sum ($N = 2$) each with its own $V_n$ and $\alpha$ defined for each type of torsion angle. An example of such a term is shown in Figure B.2. In molecular mechanics, however, it is often useful to define pseudo-torsion angles (also called improper torsion angles) using four atoms that are not bonded. An example of such use is for a planar ring, where these terms are often used to keep the ring planar.

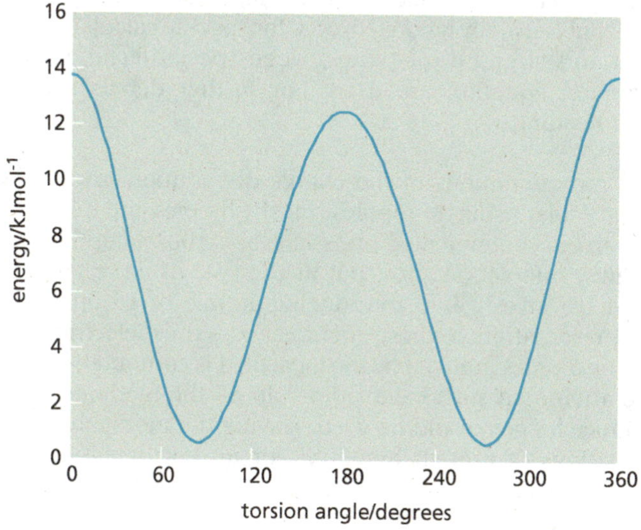


**Figure B.2**
**An example of a torsion angle potential.** This term describes the energetics of the torsion angle N-$C_\delta$-$C_\gamma$-$O_\delta$ in hydroxyproline, and uses the formula 0.625{1+cos3$\omega$} +{6.23{1+cos$\omega$} kJ mol$^{-1}$. Note that the energy maximum at 180° has a different value from that at 0°. (Formula taken from S. Park et al., A new set of molecular mechanics parameters for hydroxyproline and its use in molecular dynamics simulations of collagen-like peptides, *J. Comput. Chem.* 26:1612–1616, 2005.)

The cross-terms mentioned above will be combinations of these terms, usually formed by simply multiplying terms together. However, some of the parameters may well be unrelated to those of the simpler terms. Several different cross-terms are thought useful, including ones combining two similar terms, e.g., bond stretching terms for two bonds with a common atom. In general the combinations are only made using terms that share atoms, so that the set of these extra terms is relatively limited and does not greatly increase the size of the calculation.

## Nonbonding terms

The bonding terms presented above must be augmented by extra nonbonding terms when dealing with more than one molecule or a molecule which has two regions that can approach each other in space, i.e., interact but not by covalent bonds. Unlike the bonding terms, these must be represented by terms between every pair of atoms in the system. For an $N$ atom system, there will in general be approximately $N$ bonds, bond angles, and torsion angles but approximately $N^2$ nonbonding terms. These therefore tend to dominate calculations in terms of computational resources. However, beyond 8–10 Å the magnitude of the interaction is very small, so that in most cases the interactions between atoms more distant than this are ignored. For large systems such as proteins this can have a considerable impact on the size of calculations.

The electrostatic interactions form one of the major components of the nonbonding terms. There are many ways of representing the charge distribution of a molecule, but here we will only present the most commonly used: atomic charges. Each atom has a charge assigned to it, usually but not necessarily in the range –1 to +1. Usually every atom has a non-zero charge, even if only very small. In protein force fields the backbone is usually represented identically in all residues, and all instances of a given residue have the same atomic charges. The charges on atoms in a residue add up to –1, 0, or +1 according to the ionization state of the residue.

The formula used to calculate the electrostatic energy $E_q$ is given by

$$E_q = \sum_{i=1}^{N} \sum_{j>i} \frac{q_i q_j}{4\pi\varepsilon r_{ij}} \qquad \text{(EQB.5)}$$

where $r_{ij}$ is the distance between atoms $i$ and $j$ with charges $q_i$ and $q_j$, respectively, and $\varepsilon$ is the dielectric constant of the medium. The dielectric constant is a measure of the shielding ability of the medium between the charges, which depends on the polarizability of the medium. The value used for proteins is often 1, 2, or 4; polar solvents have a much larger dielectric constant. The disparity between $\varepsilon$ in the protein and solvent is usually ignored except for specific calculations of the electrostatic potential and of group pK values, when special techniques based on the Poisson–Boltzmann equation are used. For further details the reader should consult Further Reading.

More accurate representations of the charge distribution in proteins have been proposed, in particular using multipoles (mostly dipoles and quadrupoles) usually not on each atom but centered on groups such as carboxylates. This has the benefit of requiring fewer interaction sites, but at the cost of more complex functional forms. Others have investigated introducing atomic or group polarizabilities to improve the representation. In these circumstances the dielectric constant is set at 1. Calculating the dipoles induced by the local field is computationally intensive, so very few have attempted such calculations. In all the methods discussed in this book only atomic charges would be used, leaving the more advanced representations for simulations of protein function. For further details the reader should consult Further Reading.

The other nonbonding terms required are often referred to as van der Waals terms. They also occur between every pair of atoms in the system, and consist of two components: one repulsive (i.e., positive energy) and the other attractive (i.e., negative energy). The repulsion modeled is that between the two highly charged nuclei, and occurs at very close distances. The attractive energy is due to an effect called dispersion that in its most simple form can be explained as due to instantaneous polarizing of the electrons of each atom by the other. This force, which for ground state atoms is always attractive, is more long-range than the repulsion, and the strongest component varies as $r^{-6}$.

There are a few common variants, depending on the representation of the repulsive component. The most commonly used are the Lennard–Jones terms that are written as either

$$E_{LJ} = \sum_{i=1}^{N} \sum_{j>i} \left[ \frac{A_{ij}}{r_{ij}^{12}} - \frac{B_{ij}}{r_{ij}^{6}} \right] \quad \text{(EQB.6)}$$

or

$$E_{LJ} = \sum_{i=1}^{N} \sum_{j>i} \varepsilon \left[ \left( \frac{\overline{r}_{ij}}{r_{ij}} \right)^{12} - 2\left( \frac{\overline{r}_{ij}}{r_{ij}} \right)^{6} \right] \quad \text{(EQB.7)}$$

where $A_{ij}$ and $B_{ij}$ are constants that depend on the type of the atoms involved, $\varepsilon$ is the value of the energy minimum, and $\overline{r}_{ij}$ is the atom separation for this minimum. (Yet more forms exist involving the distance $\sigma$ at which the energy is zero.) The power 12 actually has as much to do with computational convenience as theory. An example of the Lennard–Jones potential is shown in Figure B.3. Some force fields use other powers such as 9 for the repulsive component. An alternative form that has more theoretical justification involves an exponential function for the repulsion.

More complex forms have been used to represent van der Waals forces, including extra terms for higher order dispersion forces and terms relating to the dispersive energies of more than two atoms, but none of these are used in calculations on macromolecules as yet.

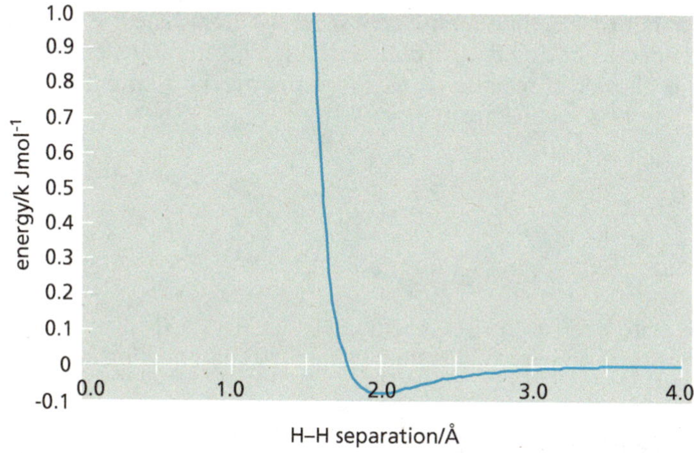


**Figure B.3**
**An example of a Lennard–Jones potential.** The parameters are for two approaching hydrogen atoms, and use parameters $\varepsilon = 0.084$ kJ mol$^{-1}$ and $\overline{r}_{ij} = 2.0$ Å (see Equation EQB.7).

## Potentials used in Threading

The force fields described above can provide reasonably accurate molecular energies, but because they involve all (or most) atoms in the system the calculation can be very demanding. Moreover, the terms used calculate the enthalpic component of the system free energy directly, but not the entropic component. The calculation of the entropic component using these energy terms is extremely demanding and rarely done. This might be an acceptable approach when studying proteins in their native folded state, but when trying to predict a protein fold the entropic contribution cannot be ignored.

Prediction of the protein fold adopted by a sequence using threading algorithms as described in Section 13.2 requires a more computationally efficient energy function to make the method practical. This increased efficiency has only been achieved with an accompanying reduction in the accuracy of the function, by using radially averaged functions as described below. In addition similarly crude expressions are often used to estimate the entropic contributions. All the functions used are derived from observed protein structures, usually by analysis of a carefully curated database of structures selected to have minimal sequence similarity to try to prevent any bias.

### Potentials of mean force

Manfred Sippl proposed treating the observed protein structures as a representative set of equilibrium states, such that the interactions between any two types of residue would occur in proportion to their Boltzmann distributions. According to statistical thermodynamics the frequency of occurrence of a state $j$ of energy $E_j$ is proportional to the quantity

$$e^{-\frac{E_j}{kT}} \qquad \text{(EQB.8)}$$

where $k$ is Boltzmann's constant, and $T$ is the absolute temperature of the system. This quantity is called a Boltzmann factor. The constant of proportionality is called the partition function $Z$ and is given by the sum of all these factors for all possible states of the system, i.e.,

$$Z = \sum_j e^{-\frac{E_j}{kT}} \qquad \text{(EQB.9)}$$

In the case of a continuous set of states, the summation is replaced by an integral.

The states considered are defined by the separation distance, originally of the $C_\alpha$ atoms of a pair of specified amino acid residues. To reduce the number of states a state is defined as having a $C_\alpha$ distance within a given range, resulting in about 20 states for each pair of residue types. From the previous equations the fraction of observed interactions of state $r$ can be written

$$f(r) = \frac{e^{-\frac{E(r)}{kT}}}{Z} \qquad \text{(EQB.10)}$$

where the state $r$ refers to two specific residues whose $C_\alpha$ distance is in a defined range around $r$, the energy of this state being $E(r)$. This equation can be rearranged as

$$E(r) = -kT\ln[f(r)] - kT\ln[Z] \qquad \text{(EQB.11)}$$

from which in principle we can obtain the energies by observing the frequencies of occurrence of the states.

These functions were determined for pairs of residues that are separated by a specified number $k$ of other residues in the protein sequence. The residue types were distinguished, and functions $\Delta E_k^{ab}(r)$ were examined, defined as the distance-dependent difference between the energy of pairs of residue types $a$ and $b$ and the average energy for all residue types. This has the form

$$\begin{aligned}\Delta E_k^{ab}(r) &= E_k^{ab}(r) - E_k(r) \\ &= -kT\ln\left[\frac{f_k^{ab}(r)}{f_k(r)}\right] - kT\ln\left[\frac{Z_k^{ab}}{Z_k}\right]\end{aligned} \quad \text{(EQB.12)}$$

Calculating partition functions is very difficult, but an argument can be made for ignoring the last term in Equation EQB.12 in the case of threading problems. Firstly, note that this term does not depend on the separation of the residues, i.e., is conformation-independent, and is therefore a constant for a given $a$, $b$, and $k$. Secondly, for a given sequence the same residue pairs occur at a given separation, so the same set of terms always appears regardless of the particular protein fold then under consideration. Thus, these terms will add to the same constant value for a given sequence for all folds onto which it is threaded.

There is a lack of available data to determine these functions. To overcome this, the observed frequency of specific residue pairs in a given separation range $r$ is modified by an application of the pseudocount method discussed in Section 6.1.

Typically such threading potentials are obtained for pairs of atoms other than $C_\alpha$, often the backbone N and O atoms and the side chain $C_\beta$, although usually only for a limited set of the possible atom pairs. In some implementations the position of a $C_\beta$ atom is constructed for glycine residues. The THREADER program, for example, uses seven pairs, namely $C_\beta$ to $C_\beta$, $C_\beta$ to N, $C_\beta$ to O, N to $C_\beta$, N to O, O to $C_\beta$, and O to N. (The first atom of the pair refers to the residue nearer the N-terminal.) In addition, often functions are obtained using data for a range of sequence separations $k$, THREADER using three ranges: short-range ($k \le 10$), medium-range ($11 \le k \le 30$), and long-range ($k > 30$). This can significantly reduce the problem of a lack of data, although at the expense of losing some of the fine detail. Sample potentials are shown in Figure 13.6.

## Potential terms relating to solvent effects

Protein structures are clearly strongly affected by the effects of solvation and hydrophobicity that are largely entropy-driven. Many workers include specific energy terms for these, often in addition to the potentials of mean force discussed above. We will describe two such terms here, one explicitly using the surface area accessible to the solvent, the other based on the number of residue contacts.

If a residue prefers to be in the protein interior it is likely to have a larger average number of residue contacts than a residue that prefers to be on the protein surface. Typically residues are defined to be in contact when their $C_\beta$ atoms are within a specified distance such as 7 Å. Usually all contacting residues are counted regardless of type. The energy can be assigned by observing the proportion of residues found to have a particular number of such contacts. One such example, called a contact capacity potential, is calculated from a structure database according to the formula

$$E_{cc}(i,n) = -\log\left[\frac{f_{i,n}}{f_i N_n}\right] \quad \text{(EQB.13)}$$

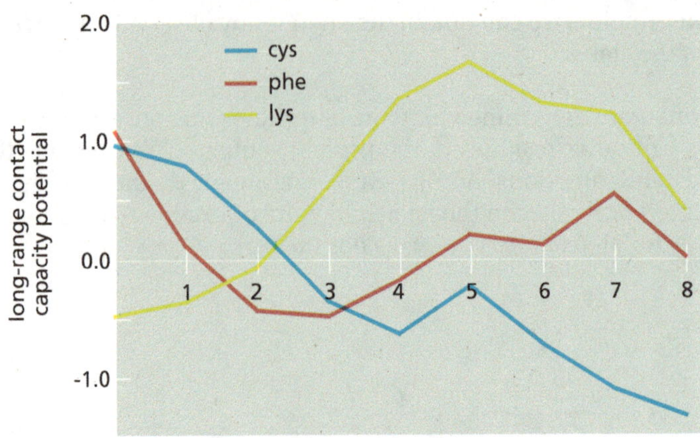


**Figure B.4**
Examples of long-range contact capacity potentials. The residues will prefer to have numbers of long-range contacts which have low potentials. Thus lysine residues will prefer to have 0 or 1 contacts, i.e., not be in the protein core, as would be expected due to their very hydrophilic nature. Conversely, cysteine residues prefer to be within the core. Phenylalanine residues are often found in the core, but also on the protein surface, as reflected in the potential. (Values taken from N.N. Alexandrov et al., Biocomputing: Proceedings of the 1996 Pacific Symposium (L. Hunter and T. Klein, eds), Singapore: World Scientific Publishing Co., 1996.)

where $f_i$ is the fraction of residues of type $i$, $f_{i,n}$ is the fraction of those residues that make $n$ contacts with any other residue, and $N_n$ is the fraction of all residues of any type which have exactly n contacts. Further detail can be obtained by distinguishing between contacting residues close in sequence (i.e., local) and those separated by many other residues, as well as by the secondary structure of the type $i$ residue. The potentials obtained by only including contacts between residues separated by at least five other residues are called long-range contact capacity potentials. In general, the hydrophobic residues will have lower energies for greater contacts, and conversely for the polar residues (see Figure B.4). When attention is paid to the types of residues involved in the contact, a modified form of the potential can be defined called the pairwise contact potential. These have also been employed in threading calculations.

## Further Reading

Molecular force fields are described in:

Field MJ (1999) A Practical Introduction to the Simulation of Molecular Systems. Cambridge, UK: Cambridge University Press.

Leach AR (2001) Molecular Modelling, 2nd ed. Upper Saddle River, NJ: Prentice Hall.

The force field parameters shown in Figure B.2 were taken from:

Park S, Radmer RJ, Klein TE & Pande VS (2005) A new set of molecular mechanics parameters for hydroxyproline and its use in molecular dynamics simulations of collagen-like peptides. *J. Comput. Chem.* 26, 1612–1616.

Threading potential references:

Alexandrov NN, Nussinov R & Zimmer RN (1996) Fast protein fold recognition via sequence to structure alignment and contact capacity potentials. In Biocomputing: Proceedings of the 1996 Pacific Symposium (L Hunter L and T Klein eds) pp 53–72, Singapore: World Scientific Publishing Co.

Sippl MJ (1993) Boltzmann's principle, knowledge-based mean fields and protein folding. An approach to the computational determination of protein structures. *J. Comput. Aided Mol. Des.* 7, 473–501.

# APPENDIX C: FUNCTION OPTIMIZATION

There are numerous occasions in bioinformatics when one wants to find the maximum or minimum of a function. Examples of such functions include the alignment score in sequence alignment (see Chapters 5 and 6), the likelihood of occurrence of a tree in the maximum-likelihood method of generating phylogenetic trees (see Chapter 8), and various energy-related functions used in comparative modeling and threading (see Chapter 13 and Appendix B). These problems have many common features that can usefully be considered together, which is the purpose of this appendix. It should be noted that optimization is also often required during the development of new techniques to obtain the best parameterization.

One other optimization method has been described in this book: the genetic algorithm method. This has been described and discussed in the context of multiple alignments in Section 6.5 and for data clustering in Section 16.3. It will not be discussed further in this appendix, except to say that it is closest to the Monte Carlo methods, in that random changes are made during each step, only some of which are accepted according to certain criteria.

The functions of interest are often very complex, involving many variables. The numerical optimization of a general function is not a trivial matter except for very simple functions that only involve a few variables. (Algebraic solutions are only feasible in the most trivial of cases.) A general function can have (very) many local minima and local maxima in the function space defined by the variables (see Figure C.1). Usually only one or a small subset of these will be the global extremum. We will discuss the methods used to determine a local optimum separately from those used for global optimization.

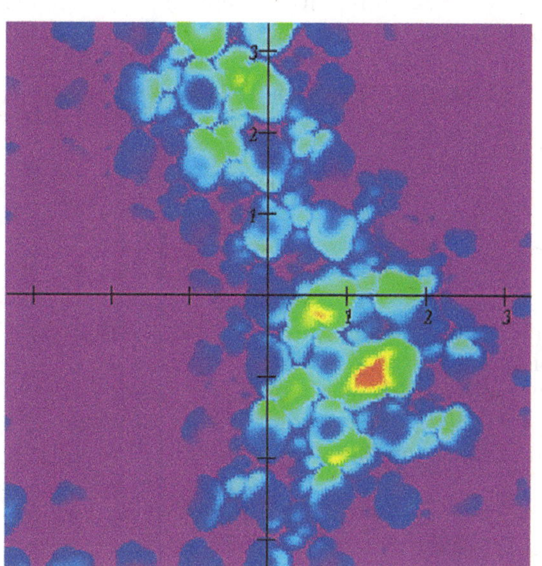


**Figure C.1**
**An example of a complex function showing many local optima.** This is a section through a multidimensional space, colored according to the function value, with the global optimum in this section being red. The function shown is not related to any systems described in this book, but illustrates well the complexity that can occur. The real global optimum may well not be in this section, but somewhere else in the multidimensional space.

In the first section we will examine the techniques used to locate points in the function space where the function is locally optimal. These methods can be divided into those that require calculation of the function gradient and those that do not. The latter include the dynamic programming method of sequence alignment described in Chapter 5. Whereas we can usually be confident of identifying a local optimum, locating the global optimum is much more difficult, and usually cannot be guaranteed. In the second section we will present the common techniques used in the field of molecular mechanics for determining global optima.

The methods will only be presented here in their most basic form. There are many modifications that have been proposed which can result in substantially more useful performance. For details of these, as well as practical advice on function optimization, refer to the references in Further Reading.

## Full Search Methods

One way to identify the global extrema of functions, in principle at least, is to calculate the value of the function at all points on the function surface, saving the highest and/or lowest found. While this is clearly not practical for functions such as atomic coordinates with continuous variables, it can be achieved in limited cases for functions with discrete variables, such as sequence alignment scores. The focus of this appendix is on methods for finding the extrema of functions with continuous variables, but since some of these full search methods have been presented in the book, we will briefly mention them here for completeness.

### Dynamic programming and branch-and-bound

The technique of dynamic programming, as used in pairwise sequence alignment and described in Section 5.2, is an optimization method that efficiently considers all the possible alignments, and is therefore a full search method. The score of a pairwise alignment can be obtained by summing the scores for separate segments of the alignment. Using this property, once a nonoptimal alignment segment has been identified no alignments need be considered which contain this nonoptimal segment. In this way the dynamic programming method reduces the number of alignments to consider to a level where a full search is practical.

Similarly, the branch-and-bound method of phylogenetic tree reconstruction, presented in Chapter 8, can effectively make a full search of the tree space to identify the optimal tree. As was shown to be the case for pairwise sequence alignments, the score of a tree can often be separated into additive parts, allowing the method to safely ignore those trees with nonoptimal topology. These methods will not be discussed further here, and the reader should see Chapters 7 and 8 for more details.

## Local Optimization

There are many different ways to find a local optimum, all involving moving in discrete steps over the surface of the function toward the nearest optimum. They differ in their computational requirements and the number of steps required for convergence to a solution. Some methods are regarded as more robust, in that they can successfully locate the optimum when starting at a greater distance from the solution, often referred to as having a larger radius of convergence. Unfortunately the success of these methods can be quite unpredictable for complex functions.

The approach to the optimum can be monitored at each step by observing the value of the function and function gradient at that point. Although the gradient is exactly zero at the optimum, in practice one must normally accept a small nonzero value as indicative of convergence, often with an additional stipulation that no component of the gradient should exceed a given threshold. For problems involving macromolecular structures this is because of the precision used in the calculations, and the properties of most molecular energy functions. Usually the variable values that define the optimum are obtained to high accuracy even allowing this small residual gradient.

In the discussions that follow we will present each method as a function minimization problem. This is for convenience only, but any maximization problem can be simply restated as a minimization problem by using $\hat{f} = -f$.

## The downhill simplex method

The downhill simplex method involves a structure called the simplex, defined for a function of $n$ variables by $n+1$ distinct points in the function space, the straight lines connecting all pairs of points, and the faces which are thus defined. For a two-variable function this is a triangle; for a three-variable function, it is a tetrahedron; etc. The first simplex defined will probably be at a random location in the function space, but the method defines steps by which it is moved to the location of the function optimum.

Given a simplex, the value of the function is calculated at all the points and those that have the highest, second highest, and lowest values are identified. The highest-valued and lowest-valued points are labeled $X_h$ and $X_l$, respectively. A new simplex is derived from this one by one of three processes illustrated in Figure C.2. The highest-valued point ($X_h$) is replaced by another point ($X_{new}$) along a vector defined from $X_h$ to the centroid ($X_0$) of the other points. (The centroid is defined as the point with variable values given by the arithmetic average of those of the other points.) Initially $X_r$ is located by reflection (i.e., $X_h - X_0 = X_0 - X_{new}$) (see Figure C.2B). If the value of the function at $X_{new}$ is less than that at $X_l$ it is possible that the minimum lies further along this direction, and an expansion is tried (i.e., $k_e\{X_h - X_0\} = X_0 - X_{new}$ with $k_e > 1$) (Figure C.2C) to see if this reduces the value further. If the value of the function at $X_{new}$ is greater than the second-highest value point, it is possible that the minimum lies within the existing simplex, and a contraction is tried (i.e., $k_c\{X_h - X_0\} = X_{new} - X_0$ with $0 < k_c < 1$) (Figure C.2D). The values of $k_e$ and $k_c$ are usually fixed at the start of the calculation. If this contraction does not improve the value at $X_{new}$, the simplex is contracted from all points towards the lowest, $X_l$ (Figure C.2E). The whole procedure is repeated many times, terminating when all the points have sufficiently similar values or when the simplex remains stationary within some defined tolerance.

To start the scheme, a set of $n+1$ distinct points must be chosen. Usually this is done by randomly choosing one point in the function space and then defining each of the other $n$ points as being a distance $\lambda$ along one of the $n$ variable axes. The value chosen for $\lambda$ (possibly different for each variable) will define the initial size of the simplex, and should reflect initial ideas about the scale of the features in the function space.

This scheme is inefficient when compared to the methods described below, in that many steps are usually required to identify the function optimum, but has the advantage of being easy to apply and moderately robust. In particular it has the advantage of not requiring calculation of gradients, but at the cost of a loss in efficiency, in that it requires many separate calculations of the function value at specific points, as well as many steps.

## The steepest descent method

This technique is based on the concept that to find the local minimum we simply move down the gradient of the function space from the current position. The next

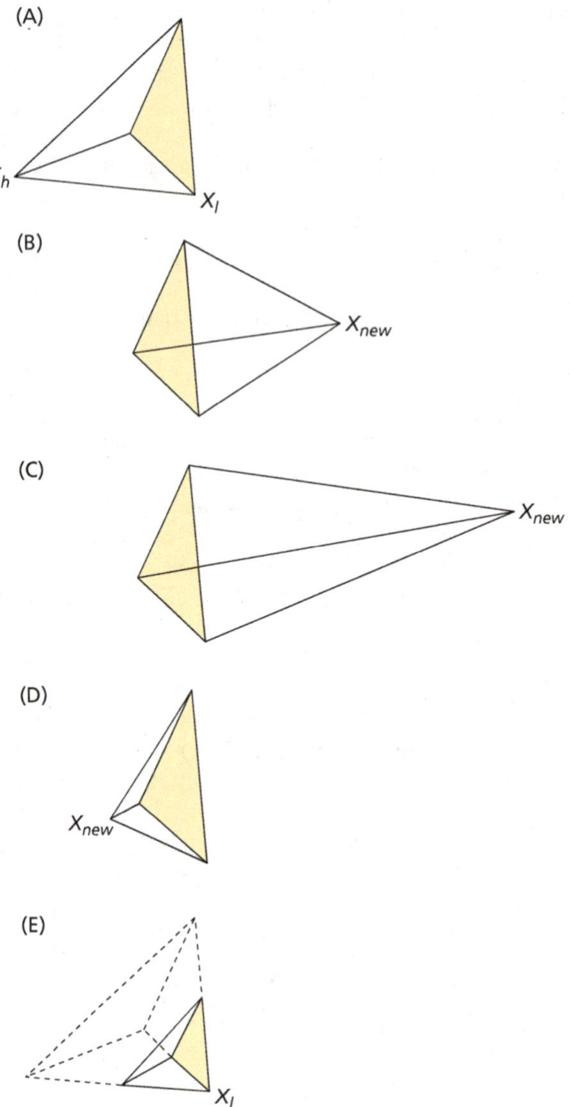


**Figure C.2**
**Illustration of the possible steps used in the downhill simplex method.** (A) The initial simplex at the start of the step, in which the highest and lowest function value points, $X_h$ and $X_l$, respectively, have been identified. The new position $X_{new}$ can be defined using (B) the reflection step; (C) the reflection and expansion step; or (D) the contraction step. (E) Alternatively, the multiple contraction step generates the next simplex. (From WH Press et al., Numerical Recipes in C: The Art of Scientific Computing, 2nd ed. Cambridge: Cambridge University Press, 1992.)

position is located along this gradient direction at the minimum along the line. Each iteration produces a new position as a better estimate of the minimum.

Suppose we have a function $f(X)$ where $X$ is the set of $n$ variables $x_1$ *to* $x_n$, which in the case of protein energy minimization could number several thousand. If the $j$th estimate of the coordinates of the local minimum is labeled $X_j$, then the value of the function at this point is $f(X_j)$. The gradient of the function at this point is given by

$$\nabla f\big|_{X_j} = \left\{ \begin{matrix} \partial f / \partial x_1 \\ \partial f / \partial x_2 \\ \vdots \end{matrix} \right\}\Bigg|_{X_j} \qquad \text{(EQC.1)}$$

where the notation indicates that the gradients are to be calculated using the values of the variables at $X_j$. This gradient defines a vector that points in the direction of fastest increasing values of the function.

For function minimization we want to move in the opposite direction from this gradient, and to search along this direction for the point at which the function is a minimum. The vector of this search direction $S$ is given simply as

$$S = -\nabla f\Big|_{X_j} \qquad \text{(EQC.2)}$$

If the minimum occurs at $\lambda S$ along this search vector from $X_j$, $\lambda$ being a number determined by searching along the line for the minimum, then the coordinates of the new estimate of the local minimum are given by

$$X_{j+1} = X_j + \lambda S \qquad \text{(EQC.3)}$$

The function and gradient are calculated at this new point, and if the gradient is not sufficiently small, the whole procedure is repeated to obtain a succession of points, each closer to the minimum. The process is stopped when the gradient is below a predetermined threshold value, when the current point is taken to be the location of the minimum.

For the potential functions described in Appendix B one can readily write the analytical form of the gradient, so that the gradient can be calculated with relative ease. However, obtaining the minimum along the search line is more difficult. One approximate method calculates the function at two other points along the line, making three in all. A quadratic form can then be fitted to this and used to estimate the minimum. The value of the function at this estimated minimum can then be used together with the next two lowest-value points in a further quadratic estimation. In fact robust methods of estimating the minimum along the search direction are quite complicated and beyond the scope of this book. See references in Further Reading for details. Often in practice the step size along the search line is restricted to a maximum value so as to remain within the same region of coordinate space.

An example of steepest descents minimization is given in Figure C.3 for a cubic function of two variables. In this example the minimum well extends horizontally, and many steps are required to walk down the valley to the actual minimum.

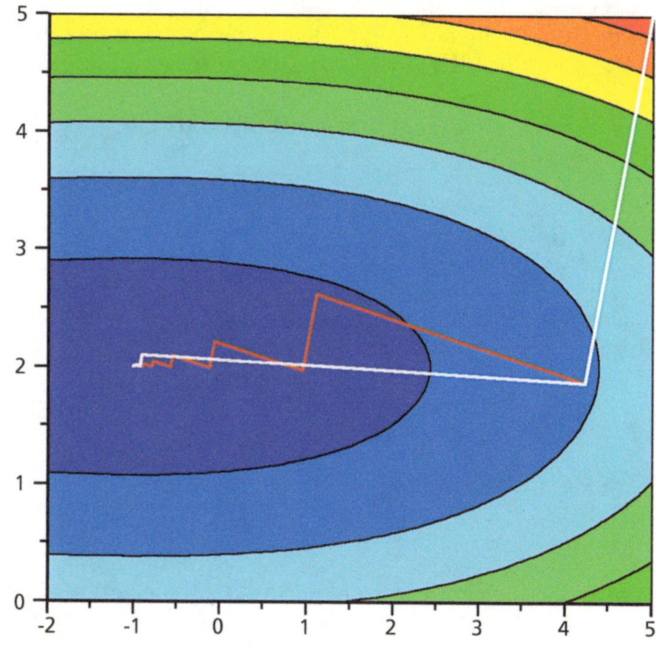


**Figure C.3**
**Practical example of the steepest descent and conjugate gradient methods compared.** The steepest descent path is shown by the red lines, each straight section representing one step. The conjugate gradients path is shown by the white lines. The first step is the same in both methods.

## The conjugate gradient method

One of the features of the steepest descents method is that alternate search vectors are orthogonal, leading to a zig-zag path that can be very inefficient at locating the minimum, and taking many iterations to converge. The conjugate gradient method attempts to resolve this by a modification to the definition of the search vectors.

The search vector is now usually given by the Fletcher–Reeves formula

$$S_{j+1} = -\nabla f_{j+1} + \frac{\left|\nabla f_{j+1}\right|^2}{\left|\nabla f_j\right|^2} S_j \qquad \text{(EQC.4)}$$

where the subscripts here relate to iterations $j$+1 and $j$, not individual variables. Alternative but related formulae have been proposed, details of which can be found in the references of Further Reading. The first search vector $S_1$ cannot be obtained with this formula without assuming $S_0$ to be zero, which makes the first step identical to that of the steepest descent method. There is a need to locate the minimum along each search direction, a problem solved in the same way as for the steepest descent method.

If the calculations were made exactly, this method would converge to the minimum for an $n$-dimensional problem in at most $n$ steps. However, effects such as rounding errors in computer-based calculations make this method less efficient than that, and may even prevent convergence. Also, when working with large macromolecular structure systems $n$ could be several tens of thousands, making a calculation of this many steps impractical. Thus, usually the system is regarded as converged when the gradient and all its components are below predetermined thresholds. In most cases this method produces a search path that locates the minimum in fewer steps than the steepest descent method, as shown in Figure C.3.

## Methods using second derivatives

Both steepest descent and conjugate gradient methods only require calculation of the first derivative of the function. By using the second derivatives as well, minimization methods have been proposed that are more stable, as the first derivative methods can encounter problems when there are large differences in the magnitudes of different components of the gradient. However, the cost of this improvement is that these methods involve considerably more computational resources.

There are many variants on second derivative methods, but they are all based in some way on the formula

$$X_{j+1} = X_j - H_j^{-1} \nabla f_j \qquad \text{(EQC.5)}$$

where the subscripts refer to steps $j$ and $j$+1, and $H_j$ is the matrix of second derivatives (called the Hessian) defined by

$$H_j = \begin{bmatrix} \partial^2 f/\partial x_1^2 & \partial^2 f/\partial x_1 \partial x_2 & \cdots \\ \partial^2 f/\partial x_1 \partial x_2 & \partial^2 f/\partial x_2^2 & \cdots \\ \vdots & \vdots & \ddots \end{bmatrix} \qquad \text{(EQC.6)}$$

Note that this formula does not involve determining the position of the minimum along a line.

The calculation of the Hessian itself can be time consuming, and the calculation of the inverse of this matrix, especially in the case of a protein force field, can be a considerable undertaking. Most optimization methods used in practice apply techniques that approximate this inverse Hessian. The details of practical second derivative methods are beyond the scope of this book (see Further Reading). They have the advantage of being more robust.

## Thermodynamic Simulation and Global Optimization

Probably all the functions of interest in this book have several local minima and/or maxima of which often only one is the global minimum or maximum. In the majority of cases we want to locate the global extremum. As in the previous section, for convenience we will assume that it is the minima that are of interest.

The function space can be divided up into regions, each of which is identified with a particular local minimum. For a simple one-dimensional function this is illustrated in Figure C.4, which has three such regions associated with the minima A, B, and C. At least one path exists for every point in a region connecting it with the local minimum such that as one travels toward this minimum the value of the function never increases. The lines that define the boundaries of the regions join local maxima via the tops of ridges.

The local optimization methods described in the previous section identify in a series of steps one path to the local minimum, as they do not allow the function value to increase at any stage. Note that some practical applications of these methods have modifications which may occasionally result in such increases. Thus, starting at point $P_1$ in Figure C.4 we will arrive at A, whereas starting at $P_2$ we will arrive at B. To find the global minimum A from $P_2$ we need to go uphill over the local maximum before dropping into the well of A.

One way in which these local optimization methods can be modified to locate the global optimum is to run them from many random starting points. Hopefully one of these points will lie in the region of the global optimum, which will be easily identified from its function value. For problems involving few variables this might be a reliable way of identifying the global optimum, but the problems we are interested in will almost always be too complex for such a scheme to be very effective.

The essence of the solution has already been hinted at; the method must allow for steps that go uphill. All the methods discussed below permit such moves, but do so in a restricted way. The application of Monte Carlo and molecular dynamics methods to molecular and atomic systems was originally proposed for thermodynamic simulation in which the properties of a system are calculated for particular

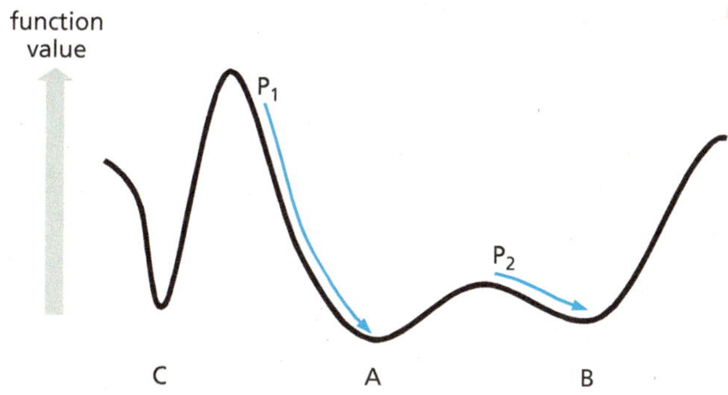


**Figure C.4**
**Diagram illustrating a one-dimensional minima problem.** For the function shown, of the three minima A, B, and C, A is the global minimum. The minimum found by an optimization method depends on the starting point and topology of the surface. If an optimization method is started at point $P_1$, it will reach A. However, if it starts at $P_2$, just to the right of the maximum (barrier) between A and B, the calculation will find minimum B. This assumes that the methods always move to lower function values, which is the case in most normal circumstances.

values of thermodynamic state functions such as temperature. However, this does not preclude their use in other situations, and the modifications of simulated annealing can convert both methods to perform global minimization.

## Monte Carlo and genetic algorithms

Starting at a given point in coordinate space, we would like to know in which direction to move, and how far, in order to approach the minimum. As we saw above, some of the local optimization methods use the function gradient to define the appropriate direction and line searches to determine the step size. The steepest descent and conjugate gradient methods uniquely define the direction. The Monte Carlo method and genetic algorithms take a different approach, allowing moves in different directions, and specifying a probability for each of them. Note that the standard Monte Carlo simulation method does not locate minima, but instead samples states in their vicinity according to a well-defined probability distribution.

We will examine the Monte Carlo method first, which is mostly used with energy functions. It is based on the same concepts from statistical thermodynamics that were discussed in Appendix B for potentials of mean force, namely the probability of occurrence of states of the system. A state 1 is defined by the positions of all the atoms in the system, and can be assigned energy $E_1$ by using a suitable force field. When the system is in equilibrium, the relative probability of a given state 1 occurring is given by the Boltzmann weighting $E^{-E_1/kT}$ where $k$ is Boltzmann's constant, and $T$ the absolute temperature.

The exact probability of state 1 is given by dividing this by the partition function $Z$ (see Equation EQB.9). It is impractical to calculate $Z$, but the Monte Carlo method neatly sidesteps this problem by looking at the ratio of probabilities for two states. If we consider a second state 2 with energy $E_2$, the ratio of probabilities is given by the term

$$\frac{e^{-E_2/kT}/Z}{e^{-E_1/kT}/Z} = e^{-\frac{E_2 - E_1}{kT}} = e^{-\frac{\Delta E_{21}}{kT}} \qquad \text{(EQC.7)}$$

Thus starting from a state 1 we can readily determine if a new state 2 is more likely to occur at equilibrium. If $\Delta E_{21}$ is negative (i.e., state 2 has the lower energy) the above term has a value greater than one (i.e., state 2 is more likely) and the move to state 2 is accepted. If state 2 has a higher energy than state 1 (i.e., the move is uphill on the energy surface) the above term has a value between 0 and 1. Instead of just

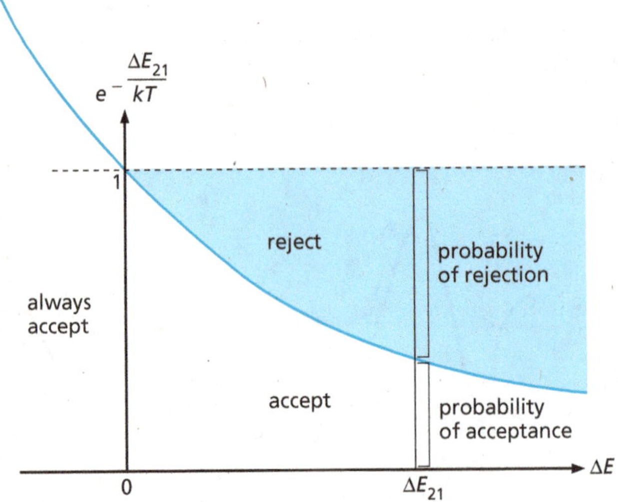


**Figure C.5**
**Illustration of the Monte Carlo acceptance/rejection criteria.**

rejecting this move as unfavorable, we now select a random number from a uniform distribution in the interval 0 to 1. If this number happens to be less than the above ratio term the move is accepted, otherwise it is rejected (see Figure C.5).

By choosing moves in this way, the Monte Carlo method will (under suitable conditions) locate the region of the global energy minimum, which will be the state with the highest probability at equilibrium. The suitable conditions required are that the moves are of an appropriate magnitude to allow efficient coverage of the available states (discussed further below) and that all the low energy states are accessible from each other without crossing unduly high barriers. Systems that obey the latter condition are known as ergodic systems. In theory nonergodic systems can be treated by starting a Monte Carlo procedure in each separate region of low-energy states, but in practice we rarely know the number and location of these regions.

In the Monte Carlo method, the direction of movement from the current state, and the distance to move are chosen randomly. In atomic systems, an atom or whole molecule is chosen at random. The distance moved and/or rotation about an axis is also chosen at random within specified limits. Usually a new state is defined by the move of a single atom, although this is not a requirement. The limits of movement are usually adjusted to obtain a 50% acceptance rate for new states, as this is thought optimal for exploring the state space efficiently. Often many hundreds or thousands of moves per atom are required to locate the well of the energy minimum.

This minimum is determined by monitoring the energy of the system and seeing when it levels off, at which point it is assumed that the system has found the minimum. (By running the method for much longer it is quite possible that this will prove to be a local minimum, and another set of lower states will be found. As with all these methods, there is no real proof that the global optimum has been found.)

Monte Carlo methods are usually used in molecular simulations of energetic and structural properties, not to locate minima, but to measure properties that are usually averages over states. Statistical thermodynamics tells us that molecular systems at non-zero temperatures have many low-energy states that contribute to these properties, and the Monte Carlo method gives them their correct relative weights. In calculating such properties if a proposed move is rejected the property value of the current state is counted again, before attempting another move.

Genetic algorithms, whilst not having such close links with thermodynamics as the Monte Carlo method, have many features in common. In this case the target function to be optimized may be quite unrelated to energy (for example the sequence alignment scores, as in the SAGA method, described in Section 6.5). Several alternative moves are available, some of which may be uphill as measured by the target function, but again all will be weighted and all will have some probability of being accepted.

The key difference between the two methods is that whereas the Monte Carlo method involves a succession of individual system states, genetic algorithms involve a succession of generations of states, where each generation may contain one hundred or more states. The origin of genetic algorithms is by analogy to real genetics, in that two (or more) states in one generation are often combined in some way to generate the members of the next generation. Some particularly favorable states are usually passed through to the next generation unchanged. Random combinations of random members of the preceding generation generate the remaining new states.

The methods of combination are usually highly specific for the problem in hand, as demonstrated for sequence alignment by the crossover operator used in the SAGA method (see Figure 6.22). Similarly, there are many ways of selecting the new generation such as to allow unfavorable moves, and these are also usually designed for the particular problem. Many generations are usually required, and the process is monitored for improvements in the target function. In the final generation, the state with optimal target function is taken as the result, but again one cannot be absolutely certain that the true optimum has been located.

## Molecular dynamics

If a molecular system is in a state which is not an energy minimum it will be subject to forces that are related to the energy gradients. These forces will result in atomic motions, leading to a trajectory through various states of the system. Any given state is defined by the positions and velocities of the component atoms, in contrast to the states in a Monte Carlo simulation, which do not include velocities. In a molecular dynamics simulation a trajectory is calculated step by step, using a very short time increment to ensure accuracy.

A thermodynamic state is defined by more than just the atomic coordinates and velocities. Some other state properties must also be given that are constants throughout the simulation. In most Monte Carlo simulations these fixed properties are the number of atoms or molecules (N), the volume of the system (V), and the temperature (T), called the canonical or NVT ensemble. In the simplest form of molecular dynamics the constants are microcanonical or NVE ensembles, E being the total energy of the system, which is the sum of the potential and kinetic energy. The potential energy is given by the sum of the force field terms, and is the function whose surface was drawn in Figure C.4. This is the function we wish in general to optimize, and which defines the global minimum. The kinetic energy is proportional to the square of the velocities of the system, with a basic formula $\frac{1}{2}mv^2$ for a single object of mass $m$ and velocity $v$.

According to classical mechanics, a system can only change in total energy if it interacts with some external force. In the absence of external interactions the total energy must remain constant, but the relative amounts of potential and kinetic energy can change. In principle a trajectory can cross suitably low energy barriers by converting kinetic energy into potential energy. As the trajectory approaches the barrier top the velocities decrease proportionately, and then increase again on descending the other side.

Note that such an NVE system is fully determined, so that it is similar to the gradient methods of optimization; there is only one way for a trajectory to go. In this sense it is very different from the Monte Carlo method. In fact the NVE simulation is rarely used as all systems of interest have some interaction with the outside world.

For most molecular dynamics simulations of interest we want to specify the temperature rather than the total energy (NVT). The temperature is in fact defined from the kinetic energy, and this equates to simulating with a constant kinetic energy. This can be done in a number of ways whose common feature is to add an external heat bath which can exchange energy with the system. The methods differ in how often the kinetic energy is adjusted, and whether all parts of the system are equally affected or not. Note that if kinetic energy is kept approximately constant, and if the system is at a sufficiently low temperature, it will not be able to explore areas of high potential energy, keeping the system in the vicinity of minima.

Each step involves calculating new positions, velocities, and accelerations using simple algebraic formulae. These formulae are only accurate over very short time periods, and typically a step size of one femtosecond ($10^{-15}$ s) is used, so that many thousands of steps are involved in a simulation. There are many other details that need to be addressed in order to have a full understanding of the practicalities of these calculations. The issues of ergodic systems and calculating properties are similar here to those mentioned for Monte Carlo methods. Here ergodicity means that the trajectory would ultimately pass through all the possible states below a certain energy cut-off. For further details the reader is referred to the references given in Further Reading.

### Simulated annealing

Neither Monte Carlo or molecular dynamics are designed for function optimization, being intended for calculating properties averaged over states. However, in both cases the temperature of the system determines the size of the energy barriers that can potentially be crossed. If the temperature is very low, both methods cannot stray far from the energy minimum they find.

Simulated annealing methods exist for both Monte Carlo and constant temperature molecular dynamics, and are simple modifications of these methods that transform them into global optimizers. At the beginning of the simulation the system is given a very high temperature (perhaps 1000 K). This allows the system to jump over reasonably high energy barriers. The system temperature is gradually lowered, ultimately causing the system to be confined to a single energy well. As these methods will spend more time in lower energy states, there is a good chance that the lowest energy state will be found. There is no absolute guarantee of this, however, and preferably the run would be repeated a few times from different starting points to see if a lower minimum could be found. The rate of temperature drop and the starting temperature are variables that can have a strong influence on the success, but can vary according to the system.

## Summary

Knowledge of the techniques of function optimization is crucial to many aspects of bioinformatics. The emphasis in this appendix has been on the key concepts behind the methods, rather than their detailed algorithms.

It is relatively easy to perform local optimization, and far harder to find global optima. There are cases where a local optimum will often suffice, but this is inadequate for many problems. The techniques can be divided into those involving explicit (first or second) gradient calculation and those relying on function evaluations alone. The first of these tend to be more powerful at the expense of requiring substantially more computer resources. However, the techniques for global optimization and simulation require orders of magnitude more computer time than any local optimization method

The importance of the subject has resulted in many good specialist texts dealing in great detail with both the theoretical and practical aspects. For this reason the treatment given here is rather brief. All of the techniques mentioned are prone to failure, sometimes in ways that are not immediately apparent. Any readers intending to write their own codes are strongly urged to consult the references listed below. A similar warning applies to those considering performing thermodynamic simulations, which have received only the briefest treatment here.

## Further Reading

Good references for many aspects of molecular force fields and thermodynamic simulation techniques are:

Allen MP & Tildesley DJ (1987) Computer Simulation of Liquids. Oxford: Oxford University Press.

Field MJ (1999) A Practical Introduction to the Simulation of Molecular Systems. Cambridge: Cambridge University Press.

Leach AR (2001) Molecular Modelling, 2nd ed. Harlow: Prentice Hall.

The general optimization techniques are well covered including practical aspects by:

Rao SS (1984) Optimization Theory and Applications, 2nd ed. New Delhi: Wiley Eastern Ltd.

Press WH, Teukolsky SA, Vetterling WT & Flannery BP (1992) Numerical Recipes in C: The Art of Scientific Computing, 2nd ed. Cambridge: Cambridge University Press.

The downhill simplex method was first presented in:

Nelder JA & Mead R (1965) A simplex method for function minimization. *Comp. J.* 7, 308–313.

An excellent presentation of the steepest descent and conjugate gradient methods is given in:

Shewchuk JR (1994) An Introduction to the Conjugate Gradient Method Without the Agonizing Pain. Available at:
http://www.cs.cmu.edu/~quake-papers/painless-conjugate-gradient.pdf

# LIST OF SYMBOLS

These symbols are in addition to the standard one-letter and three-letter codes for amino acids and nucleic acids. The following list gives each symbol used in this book, its definition, the significant occurrences in equations (EQ) and figures in this book, as well as the chapter(s) where it is mentioned.

## Notation concepts

- Sequences will be called $\boldsymbol{x}$, $\boldsymbol{y}$ and will have residues $x_i$, with indices $i$, $j$
- Alignments will have columns/positions $u$, $v$
- Residues will be of types $a$, $b$ and compositions will be $p_a$
- Scoring matrices will be $s_{a,b}$ and gaps $g$ giving total alignment scores $S$
- Expectation values will be written $\mathrm{E}(x)$ , probability $\mathrm{P}(x)$, information content $\mathrm{I}(x)$ and relative entropy $H$

| *Symbol* | *Definition* | *Chapter number* |
|---|---|---|
| $a$ | residue type | 5,10 |
| $a$ | Gamma distribution parameter; Gamma correction (EQ8.4, EQ8.13) | 8 |
| $\boldsymbol{a}$ | index array used in chaining | 5 |
| $\boldsymbol{A}$ | accepted point mutation matrix; PAM/MDM matrix theory (Figure 5.1C) | 5 |
| $[A]$ | Concentration of species $A$ (EQ17.1) | 17 |
| $A_{a,b}$ | element of the accepted point mutation matrix $\boldsymbol{A}$ for residue types $a$ and $b$; PAM/MDM matrix theory (EQ5.4) | 5 |
| $A(\boldsymbol{x},\boldsymbol{y})$ | pairwise alignment of sequences $\boldsymbol{x}$ and $\boldsymbol{y}$ extracted from a multiple alignment; COFFEE scoring function (EQ6.38) | 6 |
| $\lvert A(\boldsymbol{x},\boldsymbol{y})\rvert$ | the number of alignment positions in $A(\boldsymbol{x},\boldsymbol{y})$ that contain at least one residue from either of sequences $\boldsymbol{x}$ and $\boldsymbol{y}$; COFFEE scoring function (EQ6.38) | 6 |
| $AC$ | approximate correlation coefficient for gene prediction (BEQ10.4) | 10 |
| $ACP$ | average conditional probability (BEQ10.4, BEQ10.5) | 10 |
| $AE$ | actual exons; exon level gene prediction accuracy (BEQ10.6) | 10 |
| $b$ | residue type | 5 |
| $\boldsymbol{b}$ | index array used in chaining | 5 |
| $\boldsymbol{b}$ | column vector of branch lengths; least-squares method (EQ8.36) | 8 |
| $B$ | Beta function used in Dirichlet mixture density component PSSMs (EQ6.17, EQ6.18) | 6 |
| $B$ | back part of BRNN which uses information from the C-terminal part of the sequence (Figure 12.29) | 12 |
| $b_1b_2b_3b_4b_5$ | a specific nucleotide pentamer; GeneMark and CorePromoter methods (EQ10.1, EQ10.9) | 10 |
| $b_1b_2b_3b_4b_5b_6$ | a specific nucleotide hexamer; PromFind method (EQ10.8) | 10 |

| *Symbol* | *Definition* | *Chapter number* |
|---|---|---|
| $b_i$ | length of branch *i* of phylogenetic tree (EQ8.17, EQ8.18, Figure 8.4) | 8 |
| $b_{ij}$ | length of phylogenetic tree branch connecting nodes *i* and *j*; neighbor-joining method (EQ8.19, EQ8.21) | 8 |
| $b_u(x_i)$ | total probability of all possible HMM paths starting at state *u* having emitted all residues of sequence $\boldsymbol{x}$ up to $x_i$ and emitting the remainder of the sequence during progression to the HMM final state; backward algorithm (EQ6.33) | 6 |
| $c$ | number of occurrences of a component (nucleotide or amino acid) in a sequence; SEG method (BEQ5.3, BEQ5.6, BEQ5.7) | 5 |
| $C$ | percentage identity threshold used to cluster aligned sequences; BLOSUM matrix derivation (Figure 5.4, Figure 5.7) | 5 |
| $c_i$ | numerical representation of a k-tuple in hashing (EQ5.28) | 5 |
| $C_n$ | Zvelebil conservation number | 11 |
| $C_{helix}$ | Matthews correlation coefficient measure of the accuracy of prediction of helix secondary structure (EQ12.1) | 12 |
| $c_u$ | adjustment constant in deriving weight matrices (EQ10.7) | 10 |
| ${}^nC_r$ | number of combinations of *r* distinct objects chosen from amongst *n* distinct objects (EQ6.21) | 6 |
| $CE$ | correct exons; exon level gene prediction accuracy (BEQ10.6, BEQ10.7) | 10 |
| $\text{cov}_{i,j}$ | covariance of the expression measurements for genes/proteins *i* and *j*; Mahalanobis distance | 16 |
| $d$ | index of antidiagonal in the dynamic programming matrix; X-drop method (Figure 5.18) | 5 |
| $d$ | evolutionary distance measured in terms of the average number of mutations that have occurred per sequence site | 8 |
| $\mathbf{d}$ | column vector of data distances; least-squares method (EQ8.39) | 8 |
| $D$ | gapless pairwise local alignment (diagonal); DIALIGN method | 6 |
| $D$ | number of positions in an alignment of two sequences with two different residues aligned (EQ8.1) | 8 |
| $D$ | probability of a branch of length $3\alpha t$ between nodes with different bases; maximum-likelihood method (EQ8.55) | 8 |
| $d_{JC}$ | Jukes–Cantor corrected evolutionary distance; Jukes–Cantor model (EQ8.9) | 8 |
| $d_{JC+\Gamma}$ | Jukes–Cantor + Gamma corrected evolutionary distance; JC+$\Gamma$ model (EQ8.13) | 8 |
| $d_{JCprot}$ | Jukes–Cantor-type protein model corrected evolutionary distance; Jukes–Cantor-type protein model (EQ8.14) | 8 |
| $d_{K2P}$ | Kimura-2-parameter corrected evolutionary distance; K2P model (EQ8.11) | 8 |
| $d_N$ | proportion of nonsynonymous mutations (BEQ7.2) | 7 |
| $d_P$ | Poisson correction to the evolutionary distance measure *d* (EQ8.3) | 8 |
| $d_S$ | proportion of synonymous mutations (BEQ7.1) | 7 |
| $d_{sum}$ | sum of all the intersequence evolutionary distances; neighbor-joining method (EQ8.26) | 8 |
| $d_\Gamma$ | Gamma correction to the evolutionary distance measure *d* (EQ8.4) | 8 |
| $D_1(b_1b_2b_3b_4b_5b_6)$ | differential hexamer measure comparing promoter and noncoding regions; PromFind method (EQ10.8) | 10 |
| $D_2(b_1b_2b_3b_4b_5b_6)$ | differential hexamer measure comparing promoter and coding regions; PromFind method (EQ10.8) | 10 |
| $d_{AB}$ | Euclidean distance (EQ16.4, EQ16.5) or normalized Euclidean distance (EQ16.7) between samples *A* and *B* | 16 |

| Symbol | Definition | Chapter number |
|---|---|---|
| $d_{ij}$ | distance between two samples *i* and *j* (EQ16.8–EQ16.10) | 16 |
| $d_{ij}$ | evolutionary distance between sequences *i* and *j* | 8 |
| $d_{XY}$ | evolutionary distance between clusters X and Y; UPGMA method (EQ8.15) | 8 |
| $d_{j-i}$ | the $(j-i)$th diagonal in the dynamic programming matrix; FASTA method (Figure 5.21) | 5 |
| $D_{CF}$ | difference in amino acid composition between query and database sequences using Chou–Fasman propensities; NNSSP method (EQ12.23) | 12 |
| $D_{comp}$ | difference in amino acid composition between query and database sequences; NNSSP method (EQ12.22) | 12 |
| $D_u$ | delete state of the *u*th alignment position of a profile HMM (Figures 6.6–6.8) | 6 |
| *df* | degrees of freedom; *t*-test (EQ16.23, EQ16.25) | 16 |
| *E* | linear gap parameter (EQ5.13, EB5.1) and gap extension penalty (GEP) (EQ5.14, EQ5.15, EQ5.22, EQ5.24, EQ5.25, EQ6.8, Figure 6.16) | 5,6 |
| *E* | probability of a branch of length $3\alpha t$ between nodes with identical bases; maximum-likelihood method (EQ8.54) | 8 |
| *E* | error in the output of a neural network, using in training (BEQ12.1–BEQ12.3, BEQ12.6–BEQ12.9) | 12 |
| *E* | expected number of false-positive results in a series of tests | 16 |
| $\mathrm{E}(s_{a,b})$ | expected score for aligning residue types *a* and *b* when using substitution matrix ***s*** (EQ5.3) | 5 |
| $e_u(a)$ | probability of emission of residue *a* when in the state *u* of an HMM (EQ6.25–EQ6.27) | 6 |
| $e_u(a)$ | expected number of occurrences of base *a* at position *u* in an alignment of a signal sequence (EQ10.7) | 10 |
| $e_{a,b}$ | estimated probability of residues of type *a* and *b* aligning by chance; BLOSUM matrix derivation (EQ5.11) | 5 |
| *f* | response to transfer function of a neural network (Figure 12.26, EQ12.27) | 12 |
| *F* | forward part of BRNN which uses information from the N-terminal part of the sequence (Figure 12.29) | 12 |
| *f′* | first derivative of the response function *f* (BEQ12.6, Figure B12.8) | 12 |
| $f_a$ | normalized total exposure of residue type *a* (approximately the compositional frequency); PAM/MDM matrix theory (EQ5.6, EQ5.7) | 5 |
| $f_{av}(b_1b_2b_3b_4b_5)$ | average of the frequency of occurrence of pentamer $b_1b_2b_3b_4b_5$ in the windows $w_{i;1}$ and $w_{i+1}$; CorePromoter method (EQ10.9) | 10 |
| $f_i(b_1b_2b_3b_4b_5)$ | frequency of occurrence of pentamer $b_1b_2b_3b_4b_5$ in the window $w_i$; CorePromoter method (EQ10.9) | 10 |
| $f_{\text{non-coding}}(b_1b_2b_3b_4b_5b_6)$ | observed frequency of hexamer $b_1b_2b_3b_4b_5b_6$ in the noncoding region; PromFind method (EQ10.8) | 10 |
| $f_{\text{promoter}}(b_1b_2b_3b_4b_5b_6)$ | observed frequency of hexamer $b_1b_2b_3b_4b_5b_6$ in the promoter region; PromFind method (EQ10.8) | 10 |
| $F(i)$ | the set of alternative bases at an internal node *i*; parsimony method (Figure 8.14) | 8 |
| $f_{a,b}$ | weighted frequency of occurrence of aligned residue types *a* and *b*; BLOSUM matrix derivation (EQ5.8, EQ5.9) | 5 |
| $f_{u,b}$ | fraction of the residues in PSSM alignment column *u* that are of type *b* (EQ6.1) | 6 |
| $f'_{u,b}$ | modified form of $f_{u,b}$ used in deriving the PSSM scores $m_{c,a}$ (EQ6.3) | 6 |
| $f''_{u,b}$ | sequence-weighted form of $f_{u,b}$ (EQ6.9) | 6 |
| $f^A_{u,a}$ | fraction of the residues in intermediate alignment *A* column *u* that are of type *a* (EQ6.39, EQ6.40) | 6 |
| $f_u(x_i)$ | total probability of all possible HMM paths ending at state *u* having emitted all residues of sequence ***x*** up to $x_i$; forward algorithm (EQ6.31) | 6 |

| *Symbol* | *Definition* | *Chapter number* |
|---|---|---|
| $F_u(x_i)$ | total log-odds score of all possible HMM paths ending at state $u$ having emitted all residues of sequence $\boldsymbol{x}$ up to $x_i$; forward algorithm (EQ6.32) | 6 |
| $f(x_i x_{i+1} x_{i+2})$ | frequency of occurrence of each codon $x_i x_{i+1} x_{i+2}$ in the ORF set; ORPHEUS method (EQ10.4, EQ10.5) | 10 |
| $FN$ | number of false-negative residue predictions (of gene prediction EQ10.1, EQ10.5) (of secondary structure, EQ12.1) | 10,12 |
| $FP$ | number of false-positive residue predictions (of gene prediction EQ10.2, EQ10.3, EQ10.5) (of secondary structure, EQ12.1) | 10,12 |
| $g$ | gap penalty used in scoring a sequence alignment (EQ5.16, EQ5.17) | 5 |
| $g(n_{gap})$ | gap penalty for a gap of length $n_{gap}$ used in sequence alignment score (EQ5.13–EQ5.15, EQ5.18–EQ5.23, EQ5.27, EQ5.29) | 5 |
| $g'_u$ | position-specific multiplier of the gap penalty for alignment column $u$ in a profile (EQ6.8) | 6 |
| $g_{u,a}$ | estimated appropriate number of pseudocounts to apply for residue type $a$ in alignment column $u$ of a PSSM (EQ6.14, EQ6.15) | 6 |
| G | gap state in HMM alignment; HHsearch method (Figure 6.12) | 6 |
| $H$ | relative entropy (EQ5.12, Figure 5.7) | 5 |
| $h(i)$ | value of the Eisenberg hydrophobicity scale for residue $i$ | 11 |
| $H_u$ | entropy associated with the residue distribution at column $u$ of a multiple alignment; sequence logo (EQ6.19, EQ6.20) | 6 |
| $H(i)$ | the set of secondary alternative bases at an internal node $i$; parsimony method (Figure 8.14) | 8 |
| $h_{n,near\,j}(itn)$ | neighborhood function for node $S_n$ given that node $S_{near\,j}$ is the nearest node to data $X_j$; SOM method (Figure 16.16, EQ16.12) | 16 |
| $i_L$ | lower limit of antidiagonal; X-drop method (Figure 5.18) | 5 |
| $i_U$ | upper limit of antidiagonal; X-drop method (Figure 5.18) | 5 |
| $I$ | gap-opening penalty (GOP) (EQ5.14, EQ5.15, EQ5.24, EQ5.25, EQ6.8) | 5,6 |
| $I$ | measured intensity corresponding to the expression level of a gene/protein in expression measurement analysis (EQ16.2) | 16 |
| $\mathrm{I}(S_j = s \mid \hat{\boldsymbol{x}})$ | Fano mutual information which sequence $\hat{\boldsymbol{x}}$ contains about the structural state of the $j$th residue being $s$ (EQ12.6, EQ12.9, EQ12.10) | 12 |
| $\mathrm{I}(S_j = (s:\bar{s}); \hat{\boldsymbol{x}})$ | the preference (in information content) of $\hat{\boldsymbol{x}}$ for $s$ over $\bar{s}$ (EQ12.10, EQ12.13–EQ12.19, EQ12.21) | 12 |
| $I_u$ | insert state between the $M_u$ and $M_{u+1}$ match states of a profile HMM (Figures 6.6–6.8) | 6 |
| $I_u^{(\lambda)}$ | total input signal of the $u$th unit of neural network layer $\lambda$; (EQ12.25–EQ12.27, BEQ12.3–1BEQ2.8) | 12 |
| $I_c$ | information content at column $c$ of a multiple alignment; logo (EQ6.20) | 6 |
| $I_{SEG}$ | a measure of the information required per sequence position to specify a particular sequence, given the composition; SEG method (BEQ5.5) | 5 |
| $I'_{SEG}$ | an approximation to the information measure $I_{SEG}$ used in the first pass; SEG method (BEQ5.6) | 5 |
| $\mathrm{int}[x]$ | truncation of the value of $x$ to an integer (EQ12.2, EQ12.3) | 12 |
| $K$ | a constant involved in assessing alignment score significance that depends on the scoring matrix used and the sequence composition (EQ5.30, EQ5.34) | 5 |
| $k$ | length of a short sequence called a word, k-tuple, or k-mer; FASTA and BLAST methods | 4,5,6 |
| $k$ | number of clusters; $k$-means clustering method | 16 |

| Symbol | Definition | Chapter number |
|---|---|---|
| $k_i$ | forward reaction rate constant for reaction $i$ (EQ17.1) | 17 |
| $k_{-i}$ | reverse reaction rate constant for reaction $i$ (EQ17.1) | 17 |
| $k_{opt}$ | most appropriate number of clusters; $k$-means clustering method | 16 |
| $l$ | length of the gap between k-tuples on the same diagonal; FASTA method (EQ5.29) | 5 |
| $l$ | length of diagonal $D$; DIALIGN method | 6 |
| $\text{len}(s_{obs})$ | length in residues of any segment of secondary structure $s_{obs}$; Sov measure (EQ12.2, EQ12.3) | 12 |
| $L$ | length of a sequence | 2,5 |
| $L$ | length of sequence in discussion of sequence complexity; SEG method (BEQ5.2–BEQ5.7) | 5 |
| $L$ | length of sequence emitted by a profile HMM (EQ6.21, EQ6.22) | 6 |
| $L$ | length of an alignment of two sequences, excluding positions with gaps (EQ8.1) | 8 |
| $L$ | total likelihood of a tree topology given the sequence data; maximum-likelihood method (EQ8.53) | 8 |
| $L_{aln}$ | length of a sequence alignment | 5 |
| $L_i$ | likelihood of the observed bases at alignment position $i$ given a phylogenetic tree topology $T$ (EQ8.52) | 8 |
| $L_{\boldsymbol{x}}$ | length of sequence $\boldsymbol{x}$; MEME method | 6 |
| $m$ | length of a sequence aligned with another sequence or entire database of length $n$ (EQ5.30, EQ5.34) | 5 |
| $m$ | power of difference between data and patristic distances; least-squares method (EQ8.34) | 8 |
| $m_a$ | relative mutability of residue type $a$; PAM/MDM matrix derivation (EQ5.4–EQ5.6) | 5 |
| $m_{u,a}$ | PSSM score for residue type $a$ (row $a$) in alignment column $u$ (EQ6.2, EQ6.3, E6.4, EQ6.7) | 6 |
| $m^A_{u,b}$ | the score for aligning a residue type $b$ with column $u$ of intermediate alignment $A$ (EQ6.37, E6.39, E6.40) | 6 |
| $\bar{m}^A_{u,b}$ | mean over all residue types $b$ of $m^A_{u,b}$ in column $u$ of intermediate alignment $A$; LAMA method (EQ6.37) | 6 |
| $m_{u,v}$ | number of transitions observed in the training data between HMM states $u$ and $v$ (EQ6.24) | 6 |
| $\boldsymbol{M}$ | mutation probability matrix; PAM/MDM matrix derivation | 5 |
| $M$ | number of samples/conditions whose gene/protein expression levels have been measured; PCA method (EQ16.3) | 16 |
| $M_u$ | match state of the $u$th alignment position of a profile HMM (Figures 6.6–6.8) | 6 |
| $M_Y$ | number of objects in cluster $Y$; clustering methods (EQ16.11) | 16 |
| $M_{ins}$ | maximum allowed difference between the number of insertions in two aligned sequences; calculation restricted to band of matrix | 5 |
| $M_{a,b}$ | element of the mutation probability matrix $\boldsymbol{M}$ for residue types $a$ and $b$; PAM/MDM matrix derivation (EQ5.4–EQ5.7) | 5 |
| $ME$ | missing exons; exon level gene prediction accuracy (BEQ10.6) | 10 |
| $MM$ | aligned match states in HMM alignment; HHsearch method (Figure 6.12) | 6 |
| $\text{maxov}(s_{obs}, s_{pred})$ | total extent in residues of an overlapping pair of secondary structure segments from the set $S_o$ for which at least one residue is that state; Sov measure (EQ12.2, EQ12.3) | 12 |
| $\text{minov}(s_{obs}, s_{pred})$ | length of actually overlapping residues of an overlapping pair of secondary structure segments from the set $S_o$; Sov measure (EQ12.2, EQ12.3) | 12 |
| $N$ | number of match states in HMM (EQ6.22) | 6 |
| $N$ | total number of sequences in data used to reconstruct phylogenetic tree | 8 |

| *Symbol* | *Definition* | *Chapter number* |
|---|---|---|
| $N$ | number of genes/proteins whose expression level has been measured; PCA method and distance definitions (EQ16.3, EQ16.5–EQ16.7) | 16 |
| $N$ | number of objects to be clustered; *k*-means clustering method | 16 |
| $N$ | number of *t*-tests performed (EQ16.26–EQ16.29) | 16 |
| $N'$ | number of different residue types observed in a PSSM alignment column, including gaps; PSI–BLAST | 6 |
| $N_X$ | number of sequences in cluster X; UPGMA method (EQ8.15) | 8 |
| $N_{database}$ | number of residues in the database sequences; NNSSP method (EQ12.23) | 12 |
| $N_{exp}$ | expected number of appearances of a pattern in the *N* sequences; MEME method | 6 |
| $N_{query}$ | number of residues in the query sequence; NNSSP method (EQ12.23) | 12 |
| $N_{type}$ | number of different components in a sequence (20 for proteins, 4 for DNA) | 5 |
| $N_{seq}$ | number of sequences in a multiple alignment/PSSM or pattern search (EQ6.1, EQ6.3, EQ6.10, EQ6.11, EQ6.16–EQ6.18, EQ6.38, EQ6.41) | 6 |
| $N_{Sov}$ | number of observed residues in all overlap pairs of set $S_o$ plus the number of residues in the set $S_n$; Sov measure (EQ12.2) | 12 |
| $N_\lambda$ | number of units in feed-forward neural network layer $\lambda$ (EQ12.25) | 12 |
| $n$ | length of a sequence or the total summed length of all sequences in a database that is being compared with a query sequence of length *m* (EQ5.30, EQ5.34, EQ6.11) | 2,5 |
| $n$ | number of mutations at a given site in a sequence; Poisson correction | 8 |
| $n$ | number of codons used to define the normalized coding potential $R(x_m \ldots x_{m+3n;1})$; ORPHEUS method (EQ10.6) | 10 |
| $n$ | number of repeat measurements of $x_A$ (BEQ16.1, EQ16.19–EQ16.21) | 16 |
| $n$ | index of the SOTA network node $S_n$ (EQ16.17, EQ16.18) | 16 |
| $n_A$ | number of repeat measurements of $x_A$ (EQ16.22–EQ16.25) | 16 |
| $n_A$ | number of sequences in intermediate alignment A (EQ6.40) | 6 |
| $n_{gap}$ | length of a gap in an alignment gap penalty score (EQ5.13–EQ5.15, EQ5.18-23, EQ5.27, EQ6.8) | 5,6 |
| $n_a$ | number of times the *a*th component occurs in a sequence; SEG method (BEQ5.2, BEQ5.4–BEQ5.7) | 5 |
| $n_i(k)$ | the number of occurrences of base *k* in the F(*j*) of daughter nodes *j* of internal node *i*; parsimony method (Figure 8.14) | 8 |
| $n_{ident}(R(\boldsymbol{x},\boldsymbol{y}), A(\boldsymbol{x},\boldsymbol{y}))$ | the number of identical alignment columns in pairwise alignments $R(\boldsymbol{x},\boldsymbol{y})$ and $A(\boldsymbol{x},\boldsymbol{y})$; COFFEE scoring function (EQ6.38) | 6 |
| $\vec{n}_l$ | the vector of all $n_{l,b}$ occurrences of residue type *b* in PSSM alignment column *l* (EQ6.16–EQ6.18) | 6 |
| $n_{b_1b_2b_3b_4b_5a}$ | the number of times the pentamer $b_1b_2b_3b_4b_5$ followed by nucleotide *a* occurs in the training data; GeneMark method (EQ10.1) | 10 |
| $n_u(a)$ | number of occurrences of base *a* at position *u* in an alignment of a signal sequence (EQ10.7) | 10 |
| $n_{u,a}$ | number of residues in PSSM alignment column *u* that are of type *a* (EQ6.1, EQ6.10, EQ6.11, EQ6.16–6.18) | 6 |
| $n_{u,a}$ | number of emissions from HMM state *u* of residue type *a* observed in the training data (EQ6.25, E6.27) | 6 |
| $n_{u,a}$ | number of times residue type *a* is found at position *u* in the pattern located in the other ($N_{seq}$; 1) sequences; GIBBS method (EQ6.41) | 6 |
| $O_1$ | offspring 1 from a generation of clustering; greedy permutation encoding method (EQ16.14, EQ16.15) | 16 |
| $P$ | fraction of aligned sites whose two bases are related by a transition mutation; K2P model (EQ8.11) | 8 |

| *Symbol* | *Definition* | *Chapter number* |
|---|---|---|
| $P_{SEG}$ | probability of occurrence of a sequence based on its complexity; SEG method (BEQ5.7) | 5 |
| $P_{splice}$ | splicing efficiency; SplicePredictor method (EQ10.10) | 10 |
| $P(x)$ | probability of observing the entire sequence $x$ from an HMM (EQ6.34–EQ6.36) | 6 |
| $P(\alpha \mid low)_{NN}$ | fractions of the low-TC nearest neighbors with a central residue in the α-helical state; NN method (Figure 12.19) | 12 |
| $P_{nc}(x_1x_2x_3x_4x_5)$ | the probability of finding the pentamer $x_1x_2x_3x_4x_5$ of sequence $\boldsymbol{x}$ in a noncoding region; GeneMark method | 10 |
| $P(a \mid b_1b_2b_3b_4b_5)$ | the probability of the sixth base of a dicodon being $a$ given that the previous five bases in the sequence were $b_1b_2b_3b_4b_5$; GeneMark method (EQ10.1) | 10 |
| $P_1(a \mid x_1x_2x_3x_4x_5)$ | $P(a \mid b_1b_2b_3b_4b_5)$ for coding reading frame 1 with $a$ immediately following sequence $\boldsymbol{x}$ pentamer $x_1x_2x_3x_4x_5$; GeneMark method (EQ10.2) | 10 |
| $P_{nc}(i \mid x_1x_2x_3x_4x_5)$ | $P(i \mid x_1x_2x_3x_4x_5)$ for a noncoding region of sequence; GeneMark method | 10 |
| P(nc) | the *a priori* probability of the noncoding model; GeneMark method (EQ10.3) | 10 |
| P(3) | the *a priori* probability of the coding frame 3 model; GeneMark method (EQ10.3) | 10 |
| $P(3 \mid \boldsymbol{x})$ | the probability that model 3 applies given that we have sequence $\boldsymbol{x}$; GeneMark method (EQ10.3) | 10 |
| $P(\boldsymbol{x} \mid 3)$ | the probability of obtaining sequence $\boldsymbol{x}$ given that model 3 applies; GeneMark method (EQ10.3) | 10 |
| $P(S_i = s)$ | probability of the secondary structural state of the $i$th residue being $s$ (EQ12.6, EQ12.7, EQ12.10, EQ12.11, EQ12.13) | 12 |
| $P(S_j = s \mid \hat{\boldsymbol{x}})$ | probability of the secondary structural state of the $i^{th}$ residue being $s$ given that the sequence contains residues $\hat{\boldsymbol{x}}$ (EQ12.6, EQ12.8, EQ12.10, EQ12.12, EQ12.13, EQ12.20) | 12 |
| $P_1$ | parent 1 of a generation of clustering; greedy permutation encoding method (EQ16.13) | 16 |
| $P_{s,a}$ | propensity of residues of type $a$ to occur in proteins in structural type $s$ (EQ12.4, E12.23, E12.24) | 12 |
| $P'_{s,a}$ | log-likelihood form of the propensity of residues of type $a$ to occur in proteins in structural type $s$ (EQ12.5) | 12 |
| $p$ | $p$-distance, the proportion of alignment positions with different residues aligned (EQ8.1) | 8 |
| $p$ | modified significance level used in multiple testing with the Bonferroni correction (EQ16.26) | 16 |
| $p_a$ | proportion of amino acid type $a$ in a sequence/database (also known as background residue frequencies) (EQ5.1–EQ5.3, EQ5.10, EQ5.12, EQ5.38, EQ6.4–EQ6.7, E6.11–EQ6.14, EQ6.28–EQ6.32) (EQ12.4, E12.5, E12.20, E12.21) | 5,6,12 |
| $p_a$ | background residue frequency of residue type $a$ when not in the pattern; GIBBS method (EQ6.41) | 6 |
| $p(t)$ | fraction of different sites between two related sequences after a time $t$; Jukes–Cantor model (EQ8.6) | 8 |
| $p_a^{database}$ | frequency of occurrence of residue type $a$ in the database sequence; NNSSP method (EQ12.22) | 12 |
| $p_a^{query}$ | frequency of occurrence of residue type $a$ in the query sequence; NNSSP method (EQ12.22) | 12 |
| $p_i$ | modified significance level for use in multiple testing for the $i$th test (EQ16.27–EQ16.29) | 16 |
| $p_N$ | proportion of nonsynonymous mutations observed in a sequence alignment (BEQ7.2) | 7 |
| $p_S$ | proportion of synonymous mutations observed in a sequence alignment (BEQ7.1) | 7 |
| $p_{s,a}$ | proportion of residues in database that are in secondary structure state $s$ and are amino acid type $a$ (EQ12.4, E12.5, E12.20, E12.21) | 12 |
| $P_{y_i}$ | probability of getting observed pattern starting at $y_i$ in the $\boldsymbol{y}$th sequence by chance given the $\rho_a$; GIBBS method | 6 |
| $P_{k,m}$ | probability of obtaining the pattern starting at position $m$ in sequence $k$ given the $q_{i,j}$; MEME method | 6 |

| Symbol | Definition | Chapter number |
|---|---|---|
| $PE$ | predicted exons; exon level gene prediction accuracy (BEQ10.7) | 10 |
| $Q$ | fraction of aligned sites whose two bases are related by a transversion mutation; K2P model (EQ8.11) | 8 |
| $Q_3$ | measure of the accuracy of secondary structure prediction | 11,12 |
| $Q_{y_i}$ | probability of getting observed pattern starting at $y_i$ in the $\mathbf{y}$th sequence given the $q_{u,a}$; GIBBS method | 6 |
| $Q(x_m \ldots x_{m+3n;1})$ | coding potential of an $n$ codon region of sequence starting at base $x_m$; ORPHEUS method (EQ10.6) | 10 |
| $q(t)$ | fraction of identical sites between two related sequences after a time $t$; Jukes–Cantor model (EQ8.6, EQ8.8) | 8 |
| $q_a$ | base composition of the whole genome for nucleotide type $a$; ORPHEUS method (EQ10.4, E10.5) | 10 |
| $q_{i,j}$ | probability in nonrandom model of two residues of types $i$ and $j$ being aligned (EQ5.1, EQ5.2, EQ5.8, EQ5.10–EQ5.12) | 5 |
| $q_{u,a}$ | probability of residue of type $a$ being at position $u$ in the pattern; GIBBS method (EQ6.41) | 6 |
| $q_{u,a}$ | probability of residues of type $a$ occurring in a PSSM at alignment column $u$ (EQ6.4–EQ6.7, EQ6.10–EQ6.16, EQ6.14, EQ6.15) | 6 |
| $q'_{u,a}$ | unnormalized set of $q_{u,a}$ for residue type $a$ in PSSM alignment column $u$ as derived using Dirichlet mixture density components (EQ6.17) | 6 |
| $R$ | number of differential expression tests which identify differential expression | 16 |
| $R$ | transition–transversion ratio | 7,8 |
| $r$ | uniform mutation rate per sequence site per time unit; distance corrections | 8 |
| $r_c$ | number of components which occur $c$ times in a given sequence; SEG method (BEQ5.3, BEQ5.4, BEQ5.7) | 5 |
| $r_{AB}$ | Pearson correlation coefficient between two samples $A$ and $B$ (EQ16.6) | 16 |
| $R_A$ | rank of measurements relating to condition/sample $A$; rank-sum test | 16 |
| $R_n$ | resource of $n$th SOTA node $S_n$; SOTA method (EQ16.18) | 16 |
| $R(x_m \ldots x_{m+3n;1})$ | normalized coding potential of an $n$ codon region of sequence starting at base $x_m$; ORPHEUS method (EQ10.6) | 10 |
| $r_{A_u, B_v}$ | Pearson correlation coefficient for two multiple alignment columns $u$ of alignment $A$ and $v$ of alignment $B$; LAMA method (EQ6.37) | 6 |
| $R_{1,j}$ | $j$th element of the row vector used in the space-saving dynamic programming method (BEQ5.1) | 5 |
| $R(\mathbf{x},\mathbf{y})$ | pairwise alignment of sequences $\mathbf{x}$ and $\mathbf{y}$ in a reference alignment library; COFFEE scoring function (EQ6.38) | 6 |
| $s$ | score for aligning identical k-tuples; FASTA method (EQ5.29) | 5 |
| $s$ | residue secondary structure state (EQ12.4–EQ12.21, EQ12.23, EQ12.24) | 12 |
| $s^2$ | variance of a set of data (BEQ16.1, EQ16.19, EQ16.20) | 16 |
| $\bar{s}$ | not residue secondary structure state $s$ (EQ12.10–EQ12.19, EQ12.21) | 12 |
| $\mathbf{s}$ | substitution score matrix (Figure 5.3, Figure 5.5, Figure 5.6) | 5 |
| $S$ | overall score of a pairwise sequence alignment (EQ5.2, EQ5.32–EQ5.34) | 4,5 |
| $S$ | score of a multiple sequence alignment; COFFEE scoring function (EQ6.38) | 6 |
| $S$ | sum of the branch lengths of a phylogenetic tree; neighbor-joining method | 8 |
| $S$ | function based on tree structure whose optimal value indicates the optimal phylogenetic tree | 8 |

| *Symbol* | *Definition* | *Chapter number* |
|---|---|---|
| $\mathbf{S}$ | dynamic programming matrix of optimal scores of subalignments | 5 |
| $S_g$ | minimum HSP score to avoid hit being discarded; BLAST method | 5 |
| $S_i$ | the secondary structural state of the *i*th residue of the sequence (EQ12.6–EQ12.21) | 12 |
| $S_i(b_1b_2b_3b_4b_5)$ | score of pentamer $b_1b_2b_3b_4b_5$ in the window $w_i$; CorePromoter method (EQ10.9) | 10 |
| $S(i)$ | the score at an internal node *i*; parsimony method (Figure 8.14, EQ8.48) | 8 |
| $S_m$ | current optimal value of *S;* branch-and-bound method | 8 |
| $S_n$ | the set of all segments $s_{obs}$ that are not overlapped by a predicted segment of the same state; Sov measure | 12 |
| $\|\|S_n\|\|$ | number of genes associated with the nth SOTA node; SOTA method (EQ16.18) | 16 |
| $S_{near\,j}$ | the nearest SOM node to the set of expression ratios for gene/protein *j;* SOM method (EQ16.12) | 16 |
| $S_n(itn)$ | location of the nth SOM/SOTA node in iteration *itn;* SOM/SOTA method (Figure 16.16, EQ16.12, EQ16.17) | 16 |
| $S_o$ | the set of all overlapping pairs of $s_{obs}$ and $s_{pred}$ where the segments are the same state; Sov measure (EQ12.2) | 12 |
| $S_{\text{acceptor}}(j)$ | score of an acceptor splice site at sequence position *j*; GeneSplicer (EQ10.12) | 10 |
| $S_{\text{code}}(j)$ | scores for a coding region starting at base *j*; GeneSplicer (EQ10.11, E10.12) | 10 |
| $S_{\text{donor}}(j)$ | score of a donor splice site at sequence position *j*; GeneSplicer (EQ10.11) | 10 |
| $S_{\text{noncode}}(j)$ | scores for a noncoding region starting at base *j*; GeneSplicer (EQ10.11, E10.12) | 10 |
| $S_{\text{a}}(i)$ | minimal cost at the ancestral node when the base *i* is assigned to it; weighted parsimony (EQ8.49) | 8 |
| $S_{\text{d1}}(i)$ | minimal cost at daughter node d1 when the base *i* is assigned to it; weighted parsimony (EQ8.49) | 8 |
| $S_{12}$ | sum of the branch lengths of the phylogenetic tree of Figure 8.7B (EQ8.20, EQ8.27) | 8 |
| $s_A$ | standard deviation of the values $X_{i,A}$ for all *i* (EQ16.6, EQ16.21–EQ16.25) | 16 |
| $s_i$ | standard deviation of the values $X_{i,A}$ for all *A* (EQ16.7) | 16 |
| $s_{j-i}$ | score of the $d_{j-i}$th diagonal; FASTA method (EQ5.29) | 5 |
| $s_{obs}$ | the set of observed segments of secondary structure; Sov measure (EQ12.2, EQ12.3) | 12 |
| $s_{pred}$ | the set of predicted segments of secondary structure; Sov measure (EQ12.2, EQ12.3) | 12 |
| $S_{A_u,B_v}$ | score for aligning column *u* from intermediate alignment *A* and column *v* from intermediate alignment *B* (EQ6.40) | 6 |
| $s(a_i,b_j)$ | element of the substitution score matrix **s** for residue types $a_i$ and $b_j$ (EQ5.16–EQ5.18, EQ5.26, EQ5.27, EQB5.1) | 5 |
| $s_{a,b}$ | element of the substitution score matrix **s** for residue types *a* and *b* (EQ5.2, EQ5.3, EQ5.7, EQ5.11, EQ5.12, EQ5.31, EQ6.2, EQ6.3, EQ6.5, EQ6.6, EQ6.13) | 5,6 |
| $S_{i,j}$ | element of the dynamic programming matrix of optimal scores of subalignments (Figures 5.16–5.27) | 5 |
| *Sn* | sensitivity of gene prediction—the fraction of the bases in real genes—that are correctly predicted to be in genes (BEQ10.1) | 10 |
| Sov | a measure of the accuracy of secondary structure prediction at the structural segment level (EQ12.2) | 11,12 |
| *Sp* | specificity of gene prediction; the fraction of those bases which are predicted to be in genes that actually are (BEQ10.2) | 10 |
| $T$ | current best score; X-drop method | 5 |

| Symbol | Definition | Chapter number |
|---|---|---|
| $T$ | threshold minimum value of k-mer score; BLAST method | 5 |
| $T$ | tree topology; maximum-likelihood method (EQ8.51) | 8 |
| $t$ | probability of transition between HMM states (EQ6.22, Figure 6.9) | 6 |
| $t$ | evolutionary time period; Poisson correction | 8 |
| $t$ | time since ancestral sequence diverged into two related sequences; Jukes–Cantor model (EQ8.6, EQ8.8) | 8 |
| $t$ | test statistic of *t*-tests (EQ16.20–EQ16.22, EQ16.24) | 16 |
| $\mathbf{T}$ | tree topology matrix; least-squares method (EQ8.36, EQ8.37) | 8 |
| $T_N$ | number of distinct fully resolved unrooted tree topologies with *N* leaves (EQ8.33) | 8 |
| $t(u,v)$ | transition probability from state *u* to state *v* in HMM (EQ6.23, EQ6.24, EQ6.28–EQ6.35) | 6 |
| $TN$ | number of true-negative residue predictions (of gene prediction BEQ10.3, BEQ10.5) (of secondary structure, EQ12.1) | 10,12 |
| $TP$ | number of true-positive residue predictions (of gene prediction BEQ10.1–BEQ10.3, BEQ10.5) (of secondary structure EQ12.1) | 10,12 |
| $u$ | position in an alignment (EQ5.1, EQ5.2) | 5 |
| $u$ | state in HMM (EQ6.23, EQ6.24, EQ6.28–EQ6.35) | 6 |
| $U$ | score at the maximum of the extreme value distribution of ungapped local alignment scores (EQ5.30, EQ5.32, EQ5.33) | 5 |
| $\mathbf{U}$ | matrix; principal component analysis (EQ16.3, Figure 16.7) | 16 |
| $U_i$ | sum of the evolutionary distances from sequence *i* to all others; neighbor-joining method (EQ8.24, EQ8.28) | 8 |
| $\mathbf{V}$ | matrix; principal components analysis (EQ16.3, Figure 16.7) | 16 |
| $v_0$ | measure of confidence in the accuracy of the standard deviation of the data *s* (BEQ16.1) | 16 |
| $v_u(x_i)$ | probability of most probable HMM path ending at state *u* having emitted all residues of sequence *x* up to $x_i$; Viterbi algorithm (EQ6.28, E6.29) | 6 |
| $V_u(x_i)$ | log-odds score of most probable HMM path ending at state *u* having emitted all residues of sequence *x* up to $x_i$; Viterbi algorithm (EQ6.30) | 6 |
| $V_{i,j}$ | element of the matrix used for efficient affine gap sequence alignment (EQ5.19–EQ5.24, EQ5.26) | 5 |
| $\boldsymbol{w}$ | weight matrix; least-squares method (EQ8.38, EQ8.40) | 8 |
| $w_i$ | sequence window in CorePromoter method | 10 |
| $w_{y_i}$ | probability used for selecting new start position $y_i$ of the pattern in the sequence $\boldsymbol{y}$; GIBBS method | 6 |
| $w_u(a)$ | weight matrix term for base *a* at position *u* in a signal sequence (EQ10.7) | 10 |
| $w_u^{(\lambda)}$ | bias of the *u*th unit of neural network layer λ (EQ12.25) | 12 |
| $w'_{x_i}$ | $N_{exp}$ × probability of pattern starting at position $x_i$ in sequence $\boldsymbol{x}$; MEME method | 6 |
| $w_{R(\boldsymbol{x},\boldsymbol{y})}$ | weight assigned to pairwise sequence alignment $R(\boldsymbol{x},\boldsymbol{y})$ in a reference library; COFFEE scoring function (EQ6.38) | 6 |
| $w_u^x$ | weight assigned to sequence *x* at alignment column *u* in a PSSM (EQ6.9) | 6 |
| $w_{ij}$ | weights applied to distances measured between sequences *i* and *j*; least-squares method (EQ8.34) | 8 |
| $w_{ij}$ | weights applied to mutations; parsimony method (Figure 8.49) | 8 |
| $w_{uv}^{(\lambda)}$ | weight applied to the signal arriving at the *v*th unit of neural network layer λ from the *u*th unit of the previous layer (BEQ12.1–BEQ12.4, BEQ12.6, BEQ12.8) | 12 |

| *Symbol* | *Definition* | *Chapter number* |
|---|---|---|
| $W$ | length of sequence pattern searched for; GIBBS method | 6 |
| $W_{\text{signal}}$ | measure of the three upstream and four downstream bases surrounding the splice site; SplicePredictor method (EQ10.10) | 10 |
| $W_{i,j}$ | element of the matrix used for efficient affine gap sequence alignment (EQ5.25, EQ5.26) | 5 |
| $WE$ | wrong exons; exon level gene prediction accuracy (BEQ10.7) | 10 |
| $x$ | score of an ungapped local alignment (EQ5.32–EQ5.34) | 5 |
| $\boldsymbol{x}$ | a sequence | 6,10 |
| $\hat{\boldsymbol{x}}$ | region of sequence $\boldsymbol{x}$ (EQ12.6, EQ12.10, EQ12.12, EQ12.13) | 12 |
| $\hat{\boldsymbol{x}}_1$ | part of the regionof $\hat{\boldsymbol{x}}$ sequence $\boldsymbol{x}$ (EQ12.7–EQ12.9, EQ12.14) | 12 |
| $X$ | arbitrary constant added or multiplying an alignment score | 5 |
| $X$ | maximum allowed drop in score below current optimal value along a diagonal; X-drop method | 5 |
| $X$ | measured expression level of a gene/protein in expression measurement analysis (EQ16.1) | 16 |
| $\boldsymbol{X}$ | matrix of measured expression level ratios of a gene/protein in different samples/conditions (EQ16.3, Figure 16.7) | 16 |
| $x_A$ | measurement of expression of gene in condition/sample $A$ (EQ16.19 ) | 16 |
| $\overline{x}_A$ | mean of the set of $n$ measurements $x_A$ (EQ16.20–EQ16.22, EQ16.24) | 16 |
| $x_i$ | the residue at the $i$th position in sequence $\boldsymbol{x}$ | 5,6,10,12 |
| $x_i$ | the $i$th object of $N$; greedy permutation encoding method | 16 |
| $x_s$ | residue indicating the start of the pattern in sequence $\boldsymbol{x}$; GIBBS method | 6 |
| $x_u^A$ | residue in the $u$th column of the $\boldsymbol{x}$th sequence of intermediate alignment A (EQ6.40) | 6 |
| $X_A$ | set of expression ratios for all genes/proteins $i$ in sample/condition $A$ with respect to a control condition (EQ16.6) | 16 |
| $\overline{X}_A$ | average of the values $X_{i,A}$ over all $i$; Pearson correlation coefficient (EQ16.6) | 16 |
| $X_g$ | maximum allowed score drop below current optimal value during gapped hit extension; BLAST method | 5 |
| $X_j$ | set of expression ratios for gene/protein $j$ in all samples/conditions $A$ with respect to a control condition; SOM/SOTA method (EQ16.17, EQ16.18, Figure 16.12) | 16 |
| $X_u$ | maximum allowed score drop below current optimal value during ungapped hit extension; BLAST method | 5 |
| $X_{i,A}$ | expression ratio for gene $i$ under condition $A$ with respect to a control condition, an element of matrix $\boldsymbol{X}$ (EQ16.5–EQ16.7, EQ16.11) | 16 |
| $\boldsymbol{y}$ | a sequence | 6 |
| $y_i$ | the residue at the $i$th position in sequence $\boldsymbol{y}$ of an alignment | 5 |
| $\hat{Y}$ | centroid of cluster $Y$ with components $\hat{Y}$; clustering (EQ16.11) | 16 |
| $z$ | test statistic for the $z$-test (EQ16.19) | 16 |
| $\alpha$ | weighting parameter for the observed data when including pseudocounts (EQ6.12, EQ6.15) | 6 |
| $\alpha$ | rate of nucleotide substitution per unit time; Jukes–Cantor model (EQ8.5, EQ8.8) | 8 |
| $\alpha$ | rate of nucleotide transition substitutions per unit time; K2P model (EQ8.10) | 8 |
| $\alpha$ | false-positive rate, also called significance level (EQ16.26–EQ16.29) | 16 |
| $\beta$ | rate of nucleotide transversion substitutions per unit time; K2P model (EQ8.10) | 8 |

| Symbol | Definition | Chapter number |
|---|---|---|
| $\beta$ | scaling parameter for total number of pseudocounts in an alignment column of a PSSM (EQ6.11, EQ6.12, EQ6.15) | 6 |
| $\beta$ | weighting of pseudocounts; GIBBS method (EQ6.41) | 6 |
| $\beta_1$ | constant in expression measurement analysis (EQ16.2) | 16 |
| $\beta_2$ | constant in expression measurement analysis (EQ16.2) | 16 |
| $\beta_i$ | sum of all $\beta_{i,a}$ for all residue types $a$ in Dirichlet mixture density component $i$ (EQ6.16–EQ6.18, Figure 6.2) | 6 |
| $\vec{\beta}_i$ | the vector formed with all the $\beta_i$ for all Dirichlet mixture density components $i$ (EQ6.16–EQ6.18) | 6 |
| $\beta_{i,a}$ | amino acid composition for residue type $a$ in Dirichlet mixture density component $i$ (EQ6.16–EQ6.18, Figure 6.2) | 6 |
| $\Gamma$ | Gamma functions used to define Beta functions $B$ used in Dirichlet mixture density component PSSMs (EQ6.18) | 6 |
| $\Delta$ | column vector of patristic distances; least-squares method (EQ8.36) | 8 |
| $\Delta G_t$ | free energy of transfer of residue from hydrophilic to hydrophobic environment | 11 |
| $\Delta_{GC}$ | difference in fractional GC content between the 50 bases upstream and the 50 bases downstream of the splice site; SplicePredictor method (EQ10.10) | 10 |
| $\Delta_U$ | difference in fractional U content between the 50 bases upstream and the 50 bases downstream of the splice site; SplicePredictor method (EQ10.10) | 10 |
| $\delta(s_{obs}, s_{pred})$ | term which accounts for flexibility in the definition of secondary structure segment boundaries; Sov measure (EQ12.2, EQ12.3) | 12 |
| $\delta_{ab}$ | difference in amino acid composition between query and database sequences using Chou–Fasman propensities; NNSSP method (EQ12.24) | 12 |
| $\delta_{ij}$ | modified distances between sequences; neighbor-joining method (EQ8.28) | 8 |
| $\delta_{YZ}$ | proximity of clusters $Y$ and $Z$; clustering methods (EQ16.8–EQ16.10) | 16 |
| $\delta^x_{u,b}$ | has value 1 if sequence $\boldsymbol{x}$ has residue type $b$ at alignment column $u$ of a PSSM, and value 0 otherwise (EQ6.9) | 6 |
| $\Delta_{ij}$ | patristic distance between sequences $i$ and $j$; those calculated directly from a phylogenetic tree; least-squares method (Figure 8.5, EQ8.35) | 8 |
| $\Delta w^{(\lambda)}_{uv}$ | correction applied to $w^{(\lambda)}_{uv}$ during the training of a neural network (BEQ12.1, BEQ12.2) | 12 |
| $\varepsilon$ | momentum term in training a neural network (BEQ12.2) | 12 |
| $\varepsilon$ | additive random error term in expression measurement analysis (EQ16.1, EQ16.2) | 16 |
| $\boldsymbol{\varepsilon}$ | square matrix; principal component analysis (EQ16.3, Figure 16.7) | 16 |
| $\lambda$ | parameter of the extreme-value distribution of ungapped local alignment scores (EQ5.30–EQ5.34, EQ6.6, EQ6.7, EQ6.13) | 5,6 |
| $\lambda$ | index of a layer of a feed-forward neural network (EQ12.25, EQ12.26, BEQ12.1–BEQ12.8) | 12 |
| $\Lambda$ | constant used in the PAM/MDM matrix derivation (EQ5.4–EQ5.6) | 5,6 |
| $\mu$ | average value of $\log[f(x_i x_{i+1} x_{i+2})]$; ORPHEUS method (EQ10.5, EQ10.6) | 10 |
| $\pi_a$ | fraction of nucleotide type $a$ in sequence data; HKY85 model (EQ8.12) | 8 |
| $\rho_a$ | probability of residue of type $a$ when not in the pattern; GIBBS method in the whole dataset; GIBBS method | 6 |
| $\sigma$ | estimated standard deviation of data (BEQ16.1) | 16 |

| *Symbol* | *Definition* | *Chapter number* |
|---|---|---|
| $\sigma_0$ | standard deviation of prior distribution (BEQ16.1) | 16 |
| $\sigma_\varepsilon$ | standard deviation of additive random error term $\varepsilon$ in expression measurement analysis (EQ16.1) | 16 |
| $\sigma_\eta$ | standard deviation of additive random error term $\eta$ in expression measurement analysis (EQ16.1) | 16 |
| $\sigma^2$ | variance of $\log[f(x_i x_{i+1} x_{i+2})]$; ORPHEUS method (EQ10.5, EQ10.6) | 10 |
| $\sigma_\upsilon^{(\lambda-1)}$ | signal sent from the $\upsilon$th unit of the $\lambda$–1 layer to all units of the $\lambda$ layer (EQ12.25, EQ12.26, BEQ12.3–BEQ12.9) | 12 |
| $\eta$ | learning rate for neural network parameters (BEQ12.1, BEQ12.2) | 12 |
| $\eta$ | multiplicative random error term in expression measurement analysis (EQ16.1, EQ16.2) | 16 |
| $\theta$ | the parameters of the Dirichlet mixture density components (EQ6.16) | 6 |
| $\phi$ | protein chain torsion angle (Figure 2.6C) | 2 |
| $\psi$ | protein chain torsion angle (Figure 2.6C) | 2 |
| $\tau$ | learning rate for SOM/SOTA nodes; SOM/SOTA method (EQ16.17, Figure 16.12) | 16 |
| $\omega$ | dipeptide bond torsion angle (Figure 2.6B) | 2 |
| $\chi$ | actual expression level of a gene/protein in expression measurement analysis (EQ16.1, EQ16.2) | 16 |
| $\Omega(x_m \ldots x_{m+3n;1})$ | coding quality of an $n$ codon region of sequence starting at base $x_m$; ORPHEUS method | 10 |

# GLOSSARY

**3′ end** The end of a nucleotide sequence according to convention.

**5′ end** The beginning of a nucleotide sequence according to convention.

**α-helix** One of the regular (secondary) structures found in proteins. The structure forms a helix, with one turn of the helix formed by 3.6 residues. The helical structure is stabilized by backbone hydrogen bonds between the carbonyl oxygen of residue $i$ and the backbone -NH of residue $i$+4.

***ab initio* approach** Computational prediction of protein structure based entirely on energy calculations without using templates of generic protein folds.

**accepted mutations** Mutations of the genomic nucleotide sequence that are preserved and passed on to subsequent generations.

**acceptor splice sites** The 3′ intron side and the 5′ side of an exon where the spliceosome binds. Generally this is the beginning of an exon. See also **donor splice sites**.

**activators** DNA-binding proteins that increase the rate of transcription initiation.

**adaptive systems** Systems that modify themselves in response to a stimulus in a beneficial way to the system.

**additive tree** A tree where the evolutionary distance between any two nodes is equal to the sum of the branch lengths that connect them.

**affine gap penalty** The affine gap penalizes insertions or deletions using a length-independent and a length-dependent term. The normal gap penalty assigns a fixed gap penalty for each gap. The affine gap penalty promotes gap extension rather than the introduction of new gaps.

**Akaike information criterion (AIC)** A measure of the goodness of fit to an estimated statistical model, based on entropy.

**alternative splicing** Occurs in pre-mRNA and is a post-translational process. Pre-mRNA contains introns as well as exons. The removal of certain bits of pre-mRNA is called RNA splicing. During this process different exons can be kept or removed to give different products that can have different functional implications. This is called alternative splicing.

**amino acid(s)** (residue) A molecule containing one or more amino groups (-$NH_2$) and one or more carboxyl groups (-COOH). There are 20 standard amino acids found in proteins. The different amino acids have the same main chain and different side chains.

**amino acid sequence** The order of amino acid residues along a protein chain, defined as starting at the N-terminal and proceeding to the C-terminal..

**amino terminus** (or **N terminus**) The N-terminal end of a polypeptide chain. The start of the protein chain, usually on the left side when the sequence is written out.

**amphipathic helix** A helix that has both a hydrophilic and a hydrophobic side. This type of helix will lie close to or form part of the protein's surface. Due to the periodicity of 3.6 residues per helical turn, a residue pattern having hydrophobic residues at positions $i$, $i$+4, $i$+7, and so on with hydrophilic residues in between.

**analogous enzymes** Enzymes that are similar in function but are not evolutionarily related.

**ancestral state** The property (which could simply be the base at a specific sequence site) that is postulated to have occurred in the common ancestor of a group of organisms or proteins.

**anchor points** One or more residues in the conserved region of the protein that is being modeled, which are used as anchors when searching for insertions.

**annotation** In the context of database entries and genome sequences this term is used to refer to additional information, often based on the comparison of the key data with other similar data, or the results of applying prediction methods to the key data.

**annotation** (**sequence** or **genome**) Additional information about the sequence or genome, such as may have been obtained from experiment or by similarity analysis.

**anticoding strand** See **noncoding strand**.

**anticodon** The three nucleotides in a tRNA molecule that bind (by base pairs) to a codon (formed by three nucleotides) in an mRNA molecule, enabling the mRNA to code for the specific amino acids via the tRNA.

**antigen** An object that causes the production of an antibody.

**antigen-binding site** The parts of an antibody that bind to antigens, usually specific for certain antigens.

**antisense strand** See **noncoding strand**.

**β-strand** One of the commonly occurring regular structures in a protein. The backbone of a β-strand is almost fully extended (in contrast to the α-helix structure in which the backbone is coiled to form a helical structure).

**β-turn** A regular protein secondary structure found between β-strands or α-helices.

**backbone** Amino acids connect together to form the protein chain. The atoms that compose the uninterrupted chain are referred to as the backbone atoms. Each amino acid contributes its -N-CαC(O)- atoms to the backbone.

**back-propagation method** A technique for determining suitable parameters of a feed-forward neural network by analysis of the origin of the output errors obtained during training.

**backward algorithm** A method for obtaining the total log-odds score for all possible hidden Markov model paths that emit a particular sequence. Also used in **Baum–Welch expectation maximization.**

**balanced training** When methods are parameterized by training against a selected set of data, this term refers to using a set that is not biased; so, for example, the balanced training of a globular cytoplasmic protein secondary structure prediction method would involve a set of data whose secondary structure content mirrors that found in globular cytoplasmic proteins.

**balloting probabilities** (neural network) A method of estimating the confidence of the prediction based on the difference between the two highest output signals.

**base** One of the building blocks of nucleic acids. DNA contains four bases: adenine, thymine, cytosine, and guanine. In RNA, thymine is replaced by uracil.

**base sequence** The order of bases present in a long chain linear polymer of DNA or RNA.

**Baum–Welch expectation maximization** A method used to estimate the parameters of a hidden Markov model using data from unaligned sequences.

**Bayesian information criterion** (**BIC**) A method, based on Bayesian statistics, for comparing models of evolution to see which is the most appropriate for a given dataset.

**Bayesian methods** A group of methods of statistical inference based on the Bayesian approach in which observed data are combined with prior information or supposition about the system, resulting in the posterior probability distribution.

**biased mutation pressure** The tendency observed in some DNA sequences, due to a variety of evolutionary factors, to accept point mutations preferentially for certain bases, revealed as a substitution rate bias when comparing homologous sequences.

**biclustering** Simultaneous clustering of data on the basis of two separate characteristics, for example in gene expression data clustering by both gene and sample conditions.

**bifurcating** A topological property of phylogenetic trees: which results from each ancestor producing two offspring, so that each internal node has three branches, one connecting to the ancestor of the node and the other two to the two offspring.

**BLOSUM matrix** An amino acid substitution matrix for protein sequence analysis whose derivation and properties are described in Section 5.1.

**Bonferroni correction** This is a correction applied to the significance level used when simultaneously carrying out multiple statistical tests, in an overzealous attempt to reduce the number of false positives, usually resulting in many false negatives.

**bootstrap analysis** A method of estimating the support present in the sequence data for the topological features of a phylogenetic tree, in which many randomized selections of the data are examined to determine their support for each split.

**bootstrap interior branch test** A technique that is used to obtain an estimate of the uncertainty in the branch lengths of a phylogenetic tree obtained from a set of data.

**branch** Within a phylogenetic tree each branch connects two nodes, which represent a descendant and its immediate ancestor.

**branch-and-bound** A computer method to find the best solution to a problem by keeping the best solution found at any particular time. At any point where different ways to obtain a solution can be sampled the best one is kept and the others not explored.

**branch swapping** The general term for a group of methods that are used to explore alternative phylogenetic tree topologies by exchanging regions of a tree.

**$C_\alpha$ models/chain** Representation of protein structure by linking the $C_\alpha$ atoms of the backbone. This gives a simple but good overview of the folds of the proteins.

**C-terminal** The end of a protein chain which has a free carboxylate group.

**carboxy terminus** (or **C terminus**) The C-terminal end of an amino acid chain. Usually, when writing out a peptide sequence the C-terminal end is located on the right.

**cDNA** Complementary DNA, which is the DNA that comes from the mRNA template.

**chaining** A method to obtain more data from available data by rules that construct inferences.

**chloroplasts** Organelles, found generally in plants, which are involved in photosynthesis.

**chromatography** A method to separate and analyze complex mixtures of components, such as proteins.

**cis conformation** The conformation of a molecule in which two specified chemical groups are on the same side of a reference line, which may be a covalent bond. There are frequently one or more alternative conformations, often energetically favored. See **trans conformation**.

**clade** A term used in cladistics that refers to the group of all descendants that originate from a single common ancestor. Examples are the major animal groups of reptiles, mammals, etc.

**cladogram** A form of phylogenetic tree that represents the evolutionary relationships present in the data in a purely qualitative manner.

**cluster analysis** The process by which observed data are organized, or clustered, into meaningful groups, on the basis of common properties.

**clustering method** A method that builds up a tree by starting from a small number of sequences (or any other objects) then adds another sequence (or object) at each consecutive step.

**coding** The process of encoding the amino acids from the nucleotide codons.

**coding strand** The DNA strand that has the same nucleotide sequence as the transcribed mRNA (apart from T being U). Also known as a sense strand.

**codons** A codon is formed by three nucleotides. There are 64 codons that code for the amino acids.

**codon-pair** See **dicodon**.

**co-expressed** Genes transcribed at the same time and under similar conditions.

**comparative, homology,** or **knowledge-based modeling** The prediction of a three-dimensional protein structure based on the sequence alignment to a protein with an experimentally solved three-dimensional structure.

**complementary** Used in the context of two strands of nucleic acid when their base sequence is such that a full set of **Watson–Crick base pairs** can be formed and the two strands can interact to form the standard DNA double-helical structure.

**complete linkage clustering** Opposite to single linkage clustering. The distance between groups is calculated as the distance between the most outlying pair of objects in the groups.

**condensation reaction** The formation of a single product from two or more reactants with the accompanying formation of water.

**condensed tree** A phylogenetic tree whose branch structure illustrates the degree of support present in a set of trees obtained by bootstrap analysis of a set of data, in that branches with relatively less support are removed.

**confidence index** A numerical measure of the degree of confidence that can be placed in a prediction, often at the level of individual bases or amino acids.

**conformation** A three-dimensional arrangement of atoms forming a particular molecule.

**conformational flexible docking** A method of docking two molecules where at least one molecule can change its conformation. Usually, in protein–ligand docking it is the smaller ligand that can be flexible.

**conformers** Corresponding structures that differ only in the values of their torsion angles.

**conjugate gradient method** An iterative method used in energy minimization for finding the nearest local minimum of a function, based on first derivatives.

**consensus features** Features, especially but not exclusively of a sequence, that are conserved fully or with high frequency.

**consensus sequence** A sequence deduced from a sequence alignment that highlights the conserved nucleotides or amino acids.

**consensus tree** A phylogenetic tree whose branch structure illustrates the degree of support present in a set of alternative trees obtained from the data, in that branches with relatively less support are removed.

**convergent evolution** Independent evolution of similar characteristics in organisms that do not have to be closely related. For example, wings in the different animals that fly, ranging from insects to bats. Structures resulting from convergent evolution are termed analogous.

**core promoter** A promoter is a region of DNA sequence that allows for the start of gene transcription. A core promoter is the minimal part of the promoter needed for transcription.

**data** A collection of measurements, observations, variables, and other such components.

**data classification** Grouping of data according to a particular descriptor.

**data warehouse** A special type of database allowing for easy access from a wide community.

**decision tree** A method for deciding between several alternative final outcomes in which the overall decision is separated into a number of linked simpler tests. Each of these tests will have a limited set of mutually exclusive possible results, some of which may lead directly to final outcomes, while others will lead to further tests. The connections between the set of tests can have the appearance of a tree, leading to the name used for this general technique. For any given data there will be a single path leading from the initial test through the decision tree to the final outcome.

**degenerate** Used in the context of the genetic code to describe the existence of alternative codons that are used to represent many of the amino acids, so that a given protein sequence can be represented by many different DNA sequences.

**degrees of freedom** A measure of the number of freely varying parameters that a system has.

**denatured** The state of a protein after breaking of non-covalent interactions by chemical or physical intervention.

**dendrogram** A tree-like diagram showing the relationship between groups of components.

**deoxyribonucleic acid (DNA)** Long polymer molecules made up from a large number of deoxyribonucleotides. DNA usually exists as a double-stranded molecule held together by base-pairing between the nucleotides on opposing strands.

**deoxyribonucleotides** The nucleotide building blocks of DNA.

**deterministic finite-state automaton** This is a model of a system involving a limited set of states and a limited set of transitions between states. At any instant the system will exist in one of the set of states, and may then subsequently move to another state connected to the first by one of the transitions. The transition used will be selected in a nonrandom way that is related to the system data.

**diagonals (dynamic programming pairwise sequence alignment)** Two sequence segments that are aligned without gaps are represented during traceback of a pairwise sequence alignment dynamic programming matrix along a diagonal.

**dichotomous** See **bifurcating**.

**dicodon** A six-nucleotide segment of a sequence used in gene detection methods, which could represent two sequential translated codons, but may not be in the coding frame and may even be in a noncoding sequence region.

**dihedral angle** In proteins, dihedral angles refer to the backbone and side-chain atoms. There are three backbone dihedral angles: $\phi$, ($C_{i-1}$-$N_i$-$C_{\alpha,i}$-$C_i$), $\psi$ ($N_i$-$C_{\alpha,i}$-$C_i$-$N_{i+1}$), and $\omega$ ($C_{\alpha,i}$-$C_i$-$N_{i+1}$-$C_{\alpha,i+1}$). The $\phi$ and $\psi$ dihedral angles establish the secondary structure of the protein chain. Also called **the torsion angle**.

**directed acyclic graph (DAG)** A form of **decision tree** in that cyclic topology is not allowed, so that any test is encountered at most once.

**directional information** The information about the protein secondary structure at residue *j* carried by a residue at another position, usually defined as independent of the type of residue found at *j*.

**Dirichlet distribution density** A statistical distribution that is the conjugate prior of the multinomial distribution, and is therefore useful in representing missing data in models of biological sequences.

**Dirichlet mixture** A linear combination of Dirichlet distribution densities sometimes used in obtaining parameters of profile HMMs when the models include different amino acid preferences at different sequence positions.

**discriminant analysis** A technique for classifying an observation into one of a set of predefined classes.

**distance** A measure or score of difference. (See **evolutionary distance** for a specific example related to sequence data.)

**distance correction** A term used to describe any of the modified formulae used to correct the *p*-distance to a more accurate estimate of the evolutionary distance between two aligned sequences.

**distributed database** A database controlled by a central database management system but the storage place can be distributed between multiple computers located in the same location, or dispersed over a network of connected computers.

**divergent evolution** The process by which a single ancestor or ancestral gene is modified over time into two or more descendants that have an increasing degree of dissimilarity as the time since they diverged increases. Also called adaptive evolution.

**divide-and-conquer method** A specific method that can be used in sequence alignment algorithms. The problem to be solved is split into smaller, independently solvable entities. The solutions are then combined to give the overall result.

**DNA microarray technology** A set of known oligonucleotides complementary to protein-coding sequences is attached to a solid surface and used to probe different samples for the presence of these sequences. This technique enables the measurement of gene expression under different conditions or in different tissues. Also referred to as DNA chip or gene array technology.

**DNA repeats** DNA sequences that occur multiple times throughout the genome. Some repeats occur more often than others. Their function is often poorly understood, although some are implicated in various diseases.

**DNA replication** The process of synthesizing a duplicate genomic DNA molecule from an existing one; in cellular life forms this must occur prior to cell division.

**docking** The computational method of identifying suitable structural complexes in which two or more molecules, often a ligand and a protein, are bound together.

**domain** In the three-dimensional structure of globular proteins this term refers to a part of the structure that appears to form a relatively self-contained unit. It often (but not always) consists of a contiguous segment of peptide chain, in which case the domain may be able to exist in a stable structure independent of other parts of the protein.

**domain** In the context of the tree of life or classification of life, see **superkingdom**.

**donor splice sites** Sites in the nucleotide sequence that mark the boundary between exons and introns. 5′ splice sites—those at the 5′ end of the intron—are called donor sites. See also **acceptor splice sites**.

**dot-plot** A method for comparing two sequences, represented along the sides of a two-dimensional matrix, in which the values of the matrix elements are related to the similarity score of the two residues. Often only identical residues give a non-zero score, which is represented by a dot, so that stretches of identical sequence appear in the plot as diagonal lines of dots.

**downstream sequences** This term refers to anything after the 3′ or C-terminal end of a particular sequence, often relating to specific sequence signals that have some functional connection to the particular sequence. See also **upstream sequences**.

**dynamic programming algorithms** A set of algorithms effective for certain general optimization problems. In bioinformatics, the technique is almost exclusively used to calculate the best (highest scoring) alignment between protein or nucleotide sequences using a two-dimensional matrix.

**E-value** See **expectation value**.

**edge** See **branch**.

**emergent properties** Functions and properties, usually at the level of the complete system, that result from the interaction of parts of the system.

**emission** In the context of hidden Markov models this is the output which results from some of the states. In a protein profile HMM these emissions produce a protein sequence.

**end state** The non-emitting state in a hidden Markov model, which must be the final state visited by all valid paths and terminates the path.

**energy gradient** The first derivative of the energy with respect to system coordinates.

**energy minimization** Method for locating the coordinates at which the system is in an energy well.

**epitope** The part of a molecule or system that is recognized by an antibody.

**Euclidean distance** The distance between two points in Euclidean space, defined by the square root of the sum of squared differences for each of the system coordinates.

**eukarya** See **eukaryotes**.

**eukaryotes** Single-cell or multi-cell organisms with cells that have a membrane-bound and distinct nucleus. The word means "true kernel."

**evolutionary distance** An estimate of the amount of evolution that has occurred since the last common ancestor of two species or sequences, for the latter often measured in units of numbers of mutations per sequence site or, less often, in units of time.

**exclusive classification** A classification of data into groups that are nonoverlapping, so that each item of the data belongs to just one classification group.

**exon** A segment of the gene sequence that once transcribed remains part of the mRNA molecule that is ultimately translated into protein. See also **intron**.

**expectation value** A statistical measure for estimating the significance of alignments between a pair of sequences when one sequence (the query sequence) has been submitted for a homology search against a sequence database. The smaller the *E*-value, the better the chance that the sequence that matched the query sequence is truly related.

**expected score** See **expectation value**.

**expressed** The adjective applied to a gene or protein indicating that it has been transcribed or synthesized, respectively.

**expressed sequence tag (EST)** A fragment of a gene around 300 nucleotides long that has been sequenced from a cDNA.

**eXtensible Markup Language (XML)** A general and extremely flexible format for structured data or documents.

**external nodes or leaves** These occur at the end of a branch in a phylogenetic tree, and usually correspond to an organism or nucleotide/protein sequence that is in the observed data.

**extrinsic classification** A classification of data in which each item of data is labeled with the group it belongs to, and this information is used to teach the system to distinguish between the groups. See also **intrinsic classification**.

**extrinsic methods** A term applied to describe a gene detection method which includes information based on other known sequences than just the sequence under analysis, typically using sequence similarity. See **intrinsic methods**.

**false discovery rate (FDR)** The fraction of predictions in a set of multiple pairwise tests that are false positives, given that some of the tests indicate significant effects.

**false negative** A prediction that an object does not have a particular property or does not belong to a particular group when in fact it does.

**false negative** (sequence analysis) Missed homologous sequences in a database search because the match criteria are set too high or the homologous sequence is very distantly related.

**false positive** A prediction that an object has a particular property or belongs to a particular group when in fact it does not.

**false positive** (sequence analysis) Sequences that have been reported as similar to the query sequence during a database search but in fact are unrelated.

**family-wise error rate (FWER)** The probability of one or more false positives in a set of multiple pairwise tests (or a family of tests).

**feedback control** A mechanism to maintain equilibrium in a system. A signal is sent back to a system component to control the output of the system. For example, if the amount of an enzyme increases to high levels, a signal will be sent back to stop the production of this enzyme at the genetic level.

**feedforward control** A mechanism to maintain equilibrium in a system by reacting to the changes of the system in a predefined manner.

**Felsenstein zone** In a phylogenetic tree reconstruction, when there are some long branches, certain reconstruction methods are susceptible to supporting an incorrect topology in which these long branches occur next to each other. The region of tree space where this occurs is called the Felsenstein zone.

**field** (databases) One unit of data that is part of a database record.

**finite-state automata (FSA)** In the context of this book, a model system that accepts input and responds to it by moving in a deterministic way from the current state to another state chosen from a limited set of alternatives.

**fold** See **protein fold**.

**fold recognition method** Methods that predict the three-dimensional structure of a protein from its amino acid sequence by selecting the best matching fold from a set of unique folds.

**force field** A set of algebraic terms that represent the enthalpic energy of a molecular system as a function of atomic coordinates.

**forward algorithm** A method for obtaining the total log-odds score for all possible hidden Markov model paths that emit a particular sequence. Also used in **Baum–Welch expectation maximization**.

**fractional alignment difference** The uncorrected distance measured from a sequence alignment as the fraction of sites that have a difference in base or residue.

**frameshift** A change in the reading frame of mRNA, often a consequence of the insertion or deletion of a single nucleotide, resulting in a change to the protein product downstream from the mutation.

**free energy** The total energy of a system that can be converted to work.

**free insertion modules** (**FIMs**) A special component of hidden Markov models used by the SAM package, which allows the sequences that flank the region that aligns to the profile to be modeled without affecting the score.

**fully resolved tree** Refers to a phylogenetic tree in which every internal node is linked to three other nodes: one ancestor and two descendants in the case of ultrametric trees. See also **partially resolved tree**.

**gamma distance** An evolutionary distance between two sequences, which includes a correction for the variation of mutation rate at different sequence sites.

**gamma distribution** A standard probability distribution based on the gamma function, used in some models of evolution to model the variation of mutation rate at different sequence sites.

**gap** Position in one sequence of a sequence alignment where no residue is shown, and where with respect to the other sequence(s) the alignment proposes that there has either been a deletion or an insertion in another sequence.

**gap extension penalty** A penalty when adding gaps to an existing gap, so that more than one residues are not aligned or are aligned to gaps.

**gap opening penalty** Penalty for starting a gap in a sequence during alignment.

**gap penalty** A numerical penalty (a score) for introducing a gap between the residues of one sequence with respect to the second sequence in order to align similar residues.

**gene** A region of a DNA (or RNA in some viruses) that codes for one or more molecular products.

**gene duplication** The process when a gene is duplicated and inserted into another part of the genomic DNA. This need not necessarily involve only a single gene, and can occur at all scales up to complete genome duplication.

**gene expression** The production of a gene's molecular products such as mRNA and encoded proteins.

**gene loss** During evolution some genes may lose their function and be lost from genomes by deletion of the gene or by mutations that prevent transcription.

**general time-reversible model** (known as **GTR** or **REV**) A model that is used in modeling nucleotide sequence evolution and allows for different rates for every type of base substitution and a specific base composition.

**general transcription initiation factors** Auxiliary factors (proteins) that are involved in the transcriptional initiation process along with RNA polymerase.

**generation** (genetic algorithms) In a method using a genetic algorithm, this is a set of objects that represents possible solutions to a problem, for example multiple alignments, and which is used together with a set of rules to obtain a subsequent generation in such a way that the new generation hopefully contains better solutions to the problem.

**genetic algorithms** Computational methods that attempt to solve a problem by a random process that involves simulating the processes of evolution and survival of the fittest.

**genetic code** The relationship between the three-nucleotide codon and the corresponding amino acid or stop signal, resulting from the interaction of tRNA with the mRNA in a ribosome during protein synthesis. This code allows us to readily translate a protein-coding nucleotide sequence into the corresponding protein sequence.

**genetic distance** (sequences) A measure of dissimilarity between sequences arising from evolution.

**genome** The total hereditary blueprint of an organism encoded in its DNA (or RNA in some viruses).

**genomic imprinting** The process by which the gene activity of the parents can be transferred to the offspring, especially by the pattern of methylation of genomic DNA.

**global alignment** Sequence alignment spanning all the nucleotides or amino acids in the sequences that have been submitted for alignment. See **local alignment**.

**greedy permutation encoding method** This is a clustering method in which the first $k$ data in an ordered list initialize the $k$ clusters, and clusters are further refined by the ordered addition of subsequent data.

**guide tree** Guide trees are used in some multiple sequence alignment methods to determine the order of adding sequences to construct the multiple alignment. The guide tree is obtained from the distances between sequences calculated from pairwise alignments.

**harmonic approximation** A standard method of approximating the energetic cost of the displacement of a system from an optimum situation. The energy is assumed to be proportional to the square of the displacement.

**hashing** Producing values (hashes) for accessing data. The hash is a number generated from a string of text, and in many cases enables faster analysis of the string.

**hidden layer** (neural network) In neural networks hidden layers are additional layers between the input and output layers. The user does not see the signals of these hidden layers, but they transform the input layer signal and feed into the output layer, having a significant effect on the network output.

**hidden Markov model (HMM)** Probabilistic methods that can be used to analyze biological sequences and other sequential data, shown in this book applied to problems such as profile alignment and secondary structure prediction.

**hidden neural network (HNN)** A model that is a combination of features of a neural network and a hidden Markov model.

**hierarchical clustering** A classification technique in which a series of data partitions are defined, classifying data into successively smaller and more similar groups. See **partitional classification**.

**hierarchical likelihood ratio test (hLRT)** A method, based on likelihoods, for comparing models of evolution to see which is the most appropriate for a given dataset.

**high-scoring segment pair (HSP)** A term used in BLAST sequence database searching to describe a short gapless segment of a pairwise sequence alignment that scores sufficiently highly to be identified as a potential nucleus of a significant alignment of two sequences.

**homologous (homology)** Protein or DNA sequences are said to be homologous if they have been derived from a common ancestor. Homology implies an evolutionary relationship and is distinct from similarity.

**homology modeling** See **comparative modeling**.

**homoplasy** Observed similarity that is not due to common ancestry, as opposed to the case of **homology**.

**horizontal gene transfer (HGT)** The transfer of a gene (probably as part of a larger segment of the genome) from one species to another.

**hybridize** The formation of base pairs between two complementary regions of DNAs, used in many experimental techniques such as gene expression arrays to identify identical or extremely similar segments of nucleic acids.

**hybridization** The formation of base pairs between two complementary regions of DNAs, used in many experimental techniques such as gene expression arrays to identify identical or extremely similar segments of nucleic acids.

**hydropathic profile** A graph representing the hydropathic character of a protein sequence, often used in identifying possible transmembrane structures.

**hydrophilic** By contrast to the term **hydrophobic** this refers to polar or charged amino acids that like to be surrounded by other polar or water atoms.

**hydrophobic** In a molecular context, hydrophobic molecules tend not to occur in aqueous environments, preferring nonpolar environments. In particular, certain amino acids prefer to be surrounded by other hydrophobic residues rather than water or other polar groups. See **hydrophilic**.

**hydrophobic moment** A vector quantity for each amino acid based on hydrophobic scales.

**hydrophobic scales** Numerical scales that classify the hydrophobicity of amino acids.

**identity** (sequence alignment) In a sequence alignment this refers to when two aligned residues are identical.

**indels** An acronym representing insertions and deletions in one sequence or structure with respect to another sequence or structure.

**indexing techniques** A set of methods that generate an index of the locations of certain features within a data sequence. In many instances using such an index results in significantly faster analysis of the data.

**informative sites** In the parsimony method of phylogenetic tree reconstruction, this refers to alignment positions where the combination of observed bases will show a preference amongst the alternative tree topologies under consideration. Such alignment positions are used to identify the tree topology supported by the data.

**input** (neural network) In neural network methods the data/information that is given to the input layer units.

**input layer** (neural network) The first layer of units in a neural network, which takes **input** from the data and sends output signals to subsequent layers for further processing.

**integral proteins** Also known as integral membrane proteins. These are proteins that are largely embedded in a membrane.

**intermediate alignment** An alignment of a subset of all the sequences, which is created during the process of constructing a multiple alignment using progressive methods.

**intermediate sequences** Protein sequences that link two related but evolutionarily very diverged sequences together. If two homologous sequences are so diverged that direct comparison does not show their homology, an intermediate sequence that is homologous to both can be used to detect the homology between the first two sequences.

**internal node** In a phylogenetic tree an internal node has more than one branch, and corresponds to an ancestor of one or more leaves.

**intrinsic classification** A classification of data that depends only on the data itself, i.e., is unsupervised as compared with **extrinsic classification**, in which the data have group labels.

**intrinsic methods** A term applied to describe a gene detection method that only includes information based on the sequence under analysis. See **extrinsic methods**.

**intron** A segment of the gene sequence that, following transcription into mRNA, is spliced out before the final mRNA molecule is translated into protein. See also **exon**.

**invariable sites** In phylogenetic tree methods applied to sequence data, this refers to those alignment positions which have identical bases/residues in all sequences. In multiple alignments these are often referred to as fully conserved sites.

**isochores** A large (>300 kb) DNA segment in a genome that is homogeneous in base composition. For example, regions of DNA that are rich in A and T bases.

**iterated sequence search (ISS)** Searching the sequence database for homologs using regular expressions from usually an initial alignment and then using these to search for more distantly related sequences.

**iterative alignment** Stepwise alignment of multiple sequences, usually in the first instance the most homologous sequences are aligned first, followed by the alignment of the next most homologous sequence to the preceding two, and so on until all of the sequences are aligned.

**jury decision** When combining the results of several prediction methods this refers to taking the majority prediction from a set of independent predictions, by analogy to the way jury voting works in a court of law. In the case of the PHDsec method, however, this refers to assigning as the prediction the state with the highest average signal over several networks.

***k*-means clustering** A method of grouping data where the user specifies the number of clusters (*k*) into which the data should be grouped.

**k-tuple** A word (short sequence segment) that is used to make a hashing in the program FASTA. k refers to the number of amino acids/nucleotides in the word.

**key** A method of indexing database data. A primary key is one that will refer uniquely to each record in a table of a database.

**knowledge-based scoring** Scoring based on knowledge-based potentials, which are derived from statistical analysis of molecular structures.

**Kohonen network** A neural network method for unsupervised object clustering. It uses a self-organizing network of artificial neurons made up of an array of weights corresponding to the inputs.

**lateral gene transfer (LGT)** See **horizontal gene transfer.**

**learning rate** (neural networks) A parameter that is used to control the modification of neural network parameters during training, affecting the speed and efficiency of the training.

**least-squares method** (phylogenetic trees) A technique which constructs a tree by minimizing the sum of squares of the differences between the evolutionary distances as measured on the tree and as determined directly from the alignments.

**leucine zipper** A protein helical motif consisting of a set of consecutive leucines repeated every seven amino acids.

**linear gap penalty** A score applied in sequence alignment problems to account for insertions and deletions. The score is directly proportional to the length of the alignment gap.

**links** Pointers in a database entry to relevant entries in the same or other databases. These may simply be the identifiers of the database entries or may be hyperlinks that can retrieve the material in one click.

**local alignment** An alignment of the most similar regions of a nucleotide or amino acid sequence ignoring other segments of the sequences. See **global alignment**.

**log-odds ratio** The logarithm of the ratio of calculated likelihoods of two alternatives, often alternative hypotheses, on the basis of observed data. Used as a measure of the relative preference of one alternative to the other.

**logo** A visual representation of a set of aligned sequences that indicates the positional preferences as given by information theory.

**long-branch attraction** See **Felsenstein zone**.

**loop** α-helices and β-strands are connected to each other by structural elements called loops. These are also referred to as coils, but loops often have some definite structure while coil regions are unstructured.

**low-complexity regions** Low-complexity regions are sequence segments that have a relatively simple structure, often composed of only a few different types of base or amino acid. They are often removed from protein sequences before a database search as they can result in misleading hits.

**lowess normalization** A method often used to smooth the data and remove intensity-dependent variation in data from a two fluorescent dye gene expression experiment.

**machine-learning methods** A general term describing computation techniques that include methods for determining optimal parameterization.

**Mahalanobis distance** A definition of distance between two sets of measurements that takes account of the statistical distributions of the measurements.

**main chain** See **backbone.**

**majority-rule consensus trees** A form of consensus tree in which all branches are shown that are supported by a majority (i.e., >50%) of the trees.

**Markov chain** The series of discrete steps in a **Markov model** as the system changes its state.

**Markov chain Monte Carlo (MCMC)** A numerical technique often used to calculate complex integrals, in which the function is sampled at random points, with each usually being a small move from the previous point. The points are obtained using a Markov process, and form a Markov chain.

**Markov model** A Markov model is a probabilistic model of a system in which it can occur in many (possibly an infinite number of) states, and at each step can move to another state according to probabilities that only depend on a limited number of previous moves. In the simplest Markov models (first order) the probability of the transitions only depends on the immediately previous state.

**mass spectrometry** An experimental technique that can be used to identify molecules based on their mass and charge.

**Matthews correlation coefficient** A number between –1 and +1 that measures the degree of correlation between two sets of data.

**maximum likelihood** (**ML**) The name for a set of procedures used to generate phylogenetic trees, but also a general statistical procedure for determining the best value of a parameter as that with the greatest probability of resulting in the observed data.

**maximum parsimony** A technique for phylogenetic tree reconstruction based on determining the tree that requires the least number of substitutions to explain the sequence data.

**maximal segment pair** (**MSP**) A term used in the BLAST sequence alignment algorithm to describe a short gapless alignment of two sequences that has a high score, and may be used as the starting point of a complete alignment.

**messenger RNA** (**mRNA**) The RNA that is produced by transcription of a gene and which will be translated into a protein.

**microarray** A solid surface array of known DNA probes (or other molecular probes) that is used to identify DNA in samples or to measure their expression.

**midnight zone** The level of protein sequence identity where sequence comparison fails to detect structural similarity.

**minimum evolution** The concept that in the absence of any other evidence it is preferable to choose the model that requires the least number of evolutionary events (such as mutations).

**minimum evolution method** A method of phylogenetic tree reconstruction that is based on the principle of **minimum evolution**.

**mitochondria** Eukaryotic organelles involved in energy production, which contain their own genomic DNA.

**model surgery** A term used to describe modifying the structure of a hidden Markov model on the basis of preliminary results, to improve its accuracy when used for a particular set of data. A typical model change involves altering the number of sets of match, delete, and insert states for a profile HMM.

**modules** A term used in systems biology. Modules are components of subsystems of a whole system, which retain some identity when isolated.

**molecular clock** A measure of time for nucleotide substitutions per year.

**molecular configuration** The connectivity between the constituent atoms of a molecule, defined by the covalent bonds of the molecule.

**molecular dynamics** A method of simulating the motion of molecular systems, which involves repeatedly solving the equations that approximate the motion of the atoms in the system over a very short time period.

**molecular mechanics** A general term used to describe the set of techniques that use energy calculations of molecular systems to estimate their properties.

**monophyletic** Entities that share a common ancestor, whether it be a sequence or an organism.

**Monte Carlo method** (molecular mechanics) A method in which the current system conformation is slightly modified in a random way to produce a new conformation, which may be accepted or rejected according to the difference in energy of the old and new conformations. Often, the system properties are obtained by suitable averaging over the conformations.

**motifs** A structurally or a sequentially conserved element in a protein, which can correlate with a particular function.

**multifurcating tree** A phylogenetic tree with one or more internal nodes having more than three branches.

**multiple alignment** An alignment of more than two nucleotide or amino acid sequences.

**mutation** A change in the genomic nucleotide sequence.

**mutation probability matrix** A scoring matrix used in alignments based on the probability of a nucleotide/amino acids mutating in a given period of time.

**n-terminus** The end of a protein chain which has a free amino group.

**native structure** or **native state** The natural, usually low energy, structure of a protein.

**nearest-neighbor interchange** (**NNI**) **method** A phylogenetic tree reconstruction search method for generating a new tree topology from an existing one.

**nearest-neighbor methods** A classification algorithm based on assigning a known character of a nearby example to the item with unknown characters. When used in secondary structure prediction this means that the structure of residue *i* is assigned according to the structural state of the closest homolog with known structure.

**negative selection** A term used in the theory of evolution to describe the situation when a mutation occurs that has a deleterious effect for the organism, reducing its reproductive success, ultimately resulting in the mutation disappearing from the population.

**neighbor-joining** (**NJ**) A method for reconstructing phylogenetic trees in a stepwise fashion based on identifying neighboring sequences and on the principle of minimal evolution.

**nested genes** Genes that are situated in the introns of other genes.

**neural network method** A machine-learning method that was inspired by the structure of neurons in the brain, and consists of nodes (neurons) which receive signals, process them, and transmit signals.

**Newick** or **New Hampshire format** A specific representation of phylogenetic trees that can be read by a computer.

**Newton–Raphson method** A minimization method that uses both first and second derivatives of the energy function.

**node** (neural network) See **unit**.

**noncoding RNA** (**ncRNA**) **genes** RNA that does not code for a protein.

**noncoding strand** This is the complementary strand to the coding DNA strand, and in transcription is used as the template in synthesizing the mRNA. Also known as the antisense strand or anticoding strand.

**nonlinearity** A property of a system in which its response does not simply depend on the sum of all input stimuli, but may be qualitatively different in different stimulus ranges.

**nonredundant database** A database that has no duplicate entries.

**nonsynonymous change** or **mutation** A substitution or other mutation of protein-coding nucleotide sequence that alters the amino acid sequence of the resulting protein.

**nucleotide** The chemical building block of DNA or RNA.

**nucleotide sequence** The order of a linear polymer of nucleotides in a molecule of DNA or RNA.

**null distribution** The distribution of values of the test statistic expected if the null hypothesis is true.

**null model** In the context of profile HMMs with log-odds scoring this is the alternative model to the protein sequence family, and represents sequences which are not related to the profile. The choice of null model has a major impact on the performance of the profile HMM.

**object-oriented database** A database where the information is represented as objects used in object-oriented programming.

**odds ratio** The ratio of odds of an event occurring in one model with respect to the odds of the same event occurring in another model, giving an easily interpretable measure of the preference for different models.

**one-tailed test** A statistical test that is explicitly concerned with determining if the test statistic is significantly greater or significantly less than the null hypothesis. See **two-tailed test.**

**ontologies** A set of field-specific descriptors enabling the sharing of same concepts and definitions for specific terms. One of the most common ontologies is the gene ontology, which provides a controlled vocabulary for genes.

**open reading frames** (**ORFs**) A sequence segment that consists solely of a set of codons encoding for a protein.

**operational taxonomic units** (**OTU**) One of the taxa (organisms, sequence, and so on) whose data are used to reconstruct a phylogenetic tree.

**operon** A segment of a genome which consists of several consecutive genes whose expression is controlled as a single unit. These are transcribed into a single mRNA molecule that codes for several proteins.

**optimal alignment** The optimal alignment of two sequences is the highest-scoring alignment out of many possible alignments sampled by the alignment program. The score is calculated using one of the scoring systems based on mutation matrices and gap penalties.

**ordinary differential equations** (**ODEs**) A set of equations involving one independent variable and derivatives (but not partial derivatives) with respect to it.

**orthogonal encoding** A structure of certain neural networks that take sequence information as their input signal. Each unit in the input layer corresponds to exactly one type of base or amino acid at a certain position in the sequence window. In this method every window position must be represented by at least as many units as there could be different types of base/residue in the sequence.

**orthologous sequences** (**orthologs**) Similar sequences or genes in different species that arose through speciation and mutation and not from gene duplication.

**outgroup** A small set of sequences or other data is known to be relatively evolutionarily distant from the main data. When the complete set of data is used to reconstruct a phylogenetic tree, the root will be known to lie on the internal branch connecting the outgroup to the rest of the tree.

**output** A term in systems biology for a particular response of a system after a defined simulation of the system.

**output expansion** (neural network) A structure used in some secondary structure prediction neural networks where the output from the **sequence-to-structure network** corresponds to more than just the central residue of the window. This output subsequently becomes the input to the **structure-to-structure network**.

**output layer** In neural networks the layer of the network that gives the output.

**overall alignment score** A measure of accuracy for the whole alignment.

**overlapping classification** In contrast to many forms of classification, in overlapping classifications an object can belong to more than one group at the same time, so that the defined groups are not distinct.

***p*-distance** A measure of the dissimilarity of two aligned sequences which is the fraction of aligned residues which differ between the sequences.

**paired-site tests** A statistical test in which each value in one set of data is to be compared with a specific value in the other set. Such tests are applicable when the pairs of measurements are correlated in some way, such as because they were taken simultaneously.

**pair information** Used in some protein secondary structure prediction methods such as GOR, this refers to the secondary structural preferences for a given amino acid type that arise as a consequence of a specific amino acid type occurring at a particular sequential separation.

**pairwise alignment** An alignment between two nucleotide or protein sequences.

**PAM matrix** An amino acid substitution matrix for protein sequence analysis whose derivation and properties are described in Section 5.1.

**paralogous genes** A set of genes in the same genome, which has similar sequence (and has arisen usually by gene duplication) but different function due to mutation and sequence changes after duplication.

**paralogs** Related genes (or proteins) in the same genome. The related genes have arisen by gene duplication. Usually the gene duplication event is followed by independent mutations and sequence changes in the duplicated genes, resulting in a set of paralogous genes.

**parameters** (systems biology) In systems biology a term to describe variables associated with systems interactions such as the concentration and reaction rates of individual components.

**partially resolved tree** Refers to a phylogenetic tree in which one or more internal nodes are linked to four or more other nodes. See also **fully resolved tree**.

**partition** (classification) A term used to describe the definition of a set of groups and group membership, which is obtained from a classification of data.

**partitional classification** An intrinsic classification method in which a single set of exclusive groups is defined and the data are assigned to individual groups. See **hierarchical clustering**.

**path** In the context of hidden Markov models the path is the ordered series of model states from the start to the end state and any emissions from these states. In a protein profile HMM these emissions produce a protein sequence.

**patristic distance** A distance that is the sum of branch lengths of a path between a pair of species or sequences (taxa).

**Pearson correlation coefficient** A measure of the linear dependency between two variables.

**percentage/percent identity** The percentage of residues that are identical between aligned sequences.

**percent similarity** The percent of residues that have similar physiological properties between aligned sequences.

**perceptron** The most simple form of a neural network, consisting of only an input and an output layer with no hidden layers.

**per-comparison error rate** (**PCER**) The fraction of false positives in a set of multiple pairwise tests.

**per-family error rate** (**PFER**) The expected number of false positives in a set of multiple pairwise tests given a threshold level of significance.

**phosphatidylinositol-OH-3 kinases** (**PI3 kinases**) A family of enzymes that phosphorylate the 3 position hydroxyl group on the inositol ring of a phosphatidylinositol. They are associated with many diseases.

**phosphorylation** A chemical reaction that results in the addition of a phosphate to a molecule.

**phylogenetic tree** A diagram that represents the evolutionary relationships of sequences and/or species.

**phylogenomics** An area of research that combines phylogenetic tree analyses for a large number of sequences from a set of related genomes in order to gain an understanding of evolutionary relationships that might not be apparent from individual trees.

**plasmids** A self-replicating circular DNA molecule separate from the bacterial genome. It is not essential for cell survival under normal conditions.

**Poisson corrected distance** An evolutionary distance measure between two sequences, which takes account of multiple substitutions at the same sequence site.

**polyadenylation** A posttranslational modification of a nucleotide sequence that leads to the addition of a number of adenosine nucleotides to the 3′ end of mRNA.

**polytomous** See **multifurcating tree**.

**positive selection** A term used in the theory of evolution to describe the situation when a mutation occurs that has a beneficial effect for the organism that improves its reproductive success, ultimately resulting in the mutation being preserved in the population

**position-specific scoring matrix** (**PSSM**) A scoring scheme for sequence alignment, which uses specific scores at individual positions along the sequence. Typically used for aligning protein sequences to a representation of a family of domains. See Section 6.1 for details.

**positive-inside rule** An observation used in the prediction of transmembrane helices. The rule states that there are more positive charges in the cytoplasmic part of the protein than the extracytoplasmic segments. This aids the prediction and localization of the N and C termini of the membrane protein.

**post-order traversal** In phylogenetic tree reconstruction by parsimony a technique whereby all nodes of the tree are processed by initially processing all subtrees recursively and then processing the root of the tree.

**potential energy** (molecular mechanics) The energy as calculated by a force field that is associated with the conformation of the system, and does not include the kinetic energy associated with atomic movement or entropy.

**potential energy function** See **force field**.

**potential energy surface** A surface representation of potential energy of a system.

**prediction confidence level** (**PCL**) A measure of confidence in a prediction.

**prediction filtering** The stage after a prediction has been made for each residue or base in which the predictions are modified, typically to impose minimal lengths for certain features. For example, when predicting $\alpha$-helices often a minimal length of four residues is imposed.

**primary structure** See **amino acid sequence**.

**principal components analysis** (**PCA**) A mathematical technique that in some situations, such as the analysis of gene expression experiments, can be used to identify major features of variation in the data.

**prior distribution** A term used in Bayesian analysis, it describes an assumed probability distribution for the system.

**profile** See **position-specific scoring matrix (PSSM)**.

**profile HMM** A hidden Markov model that represents the common features of a set of (usually protein) sequences and is used to align further sequences.

**progressive alignment** A multiple alignment approach in which the sequences are initially clustered and then added one by one in decreasing similarity to the alignment.

**prokaryotes** Organisms with cells that do not have a membrane-bound distinct nucleus.

**promoter** A sequence motif close to the start site of transcription (TSS) and therefore the place where initiation of transcription starts.

**protein backbone** See **backbone**.

**protein families** Protein families are a group of proteins that are recognized as having the same ancestral protein and which have retained similarities in their sequences and functions through divergent evolution.

**protein fold** The fold refers to the topology a three-dimensional domain of a protein sequence adopts as it folds.

**protein fold recognition** See **threading**.

**protein kinases** Enzymes that phosphorylate proteins.

**proteome** The complete set of proteins in a cell, tissue, or organism.

**protocols** (systems biology) The interactions between separate modules.

**pseudocounts** A term for nonexistent data, which are added to real data in an attempt to overcome problems due to a lack of real data. Such data are added at a risk, as they can unduly bias the results unless great care is taken.

**pseudo-energy functions** Functions that are derived based on statistical analysis of proteins, but do not correspond with individual energy terms in the way that force field terms do.

**pseudogene** A genomic DNA sequence that looks like a gene sequence coding for a functional protein but contains sequence changes that prevent transcription or translation.

**purifying selection** see **negative selection.**

**$Q_3$** A measure of prediction accuracy used in secondary structure prediction, which for the example of $\alpha$-helices is the fraction of correctly predicted residues out of all actual $\alpha$-helical residues.

**quaternary conformation** The structural arrangement of two or more interacting macromolecules.

**random error** Error arising from random effects, as opposed to **systematic error**.

**rank-sum test** See the **Mann–Whitney *U* test**.

**reading frames** A nucleotide sequence is read in sets of three nucleotides (codons) to produce specific amino acids. The reading frame specifies the sequence in which the codons will be read. Different reading frames will give different amino acid products.

**reconciled trees** A form of phylogenetic trees that represents the differences between gene family trees and the phylogeny of the species that the genes are taken from.

**record** A term used in databases, where a record corresponds to a row.

**redundancy** In systems biology, the presence of more than one mechanism that can take over a function if one mechanism fails.

**redundant** A term used in the context of databases to refer to a database having more than one identical item, such as the same sequence present in two or more records.

**regulatory element** Elements that regulate transcription, and include promoters, response elements, enhancers, and silencers.

**relational database** A database where the data and relations between these are organized in tables forming a database.

**relative entropy (*H*)** In information theory, this is a measure of the distance between two probability distributions.

**relative mutability** Is the number of times each amino acid has changed divided by the number of occurrences and the number of times it has been subjected to mutation.

**reliability** See **confidence index.**

**repressors** Proteins that bind to particular sites on the DNA to prevent specific gene transcription.

**response function** (neural network) The algebraic function that is used by a node in a neural network to convert the received signals into an output signal.

**ribonucleic acid (RNA)** One of the two major forms of nucleic acid molecules, the other being DNA. The main difference between a DNA and an RNA is the sugar moiety of the nucleic acids; in DNA it is the deoxyribose and in RNA it is a ribose.

**ribonucleotides** The nucleotide building blocks of RNA, which consist of a base (purine or pyrimidine) linked to a ribose sugar.

**ribosomal RNA (rRNA)** Specialized RNAs that are core molecules of the ribosome.

**RNA capping** The addition of a 5′ cap structure to the growing RNA transcript.

**RNA secondary structure** The RNA molecule does not exist as a linear peptide but folds into a structural unit called the secondary structure.

**RNA splicing** The removal of introns from nascent translated RNA to produce the mature mRNA.

**Robinson-Foulds difference** see **symmetric difference.**

**robustness** A system-level phenomenon. Robustness indicates adaptability of a system to change.

**root** or **last common ancestor** (phylogenetic trees) The most recent common ancestor of all the species (taxa) that comprise the tree.

**rooted tree** A phylogenetic tree in which the last common ancestor is identified and the direction of time is defined for all branches.

**root mean square deviation (RMSD)** Defined as the square root of the sum of squared deviations of the observed values from the mean value, divided by the number of observations.

**rotamer** Short for rotational isomer. It describes a single protein side-chain conformation as a set of values, one for each dihedral-angle degree of freedom.

**sample classifier** A method that classifies data into groups, and is used to identify the group to which an unknown sample belongs. A practical application of this might be to identify whether a sample comes from a cancerous patient.

**Sankoff algorithm** One of several dynamic programming algorithms proposed by David Sankoff, of which two are described in this book: one for predicting RNA structures and another for phylogenetic tree reconstruction using parsimony.

**scatterplot** A diagram that can show graphically associations between two variables.

**score** A measure obtained in different ways by different methods, and which is intended to correlate in some way with the predictive value, confidence, and/or significance. There are many different types of scores.

**scoring schemes/matrices** (sequence alignment) A table of values representing the scores to be applied when aligning two nucleotides or amino acid residues. These scores are used to calculate alignment scores in sequence comparison methods.

**secondary structure** Secondary structure refers to the local conformation of a protein. Secondary structures can be regular or irregular. The regular conformations are formed by hydrogen bonds between main-chain atoms, the most common ones being the α-helix and the β-strand.

**self-information** In the context of the GOR protein secondary structure prediction method, this is the information a residue carries about its own structural state.

**self-organizing maps (SOMs)** Unsupervised machine-learning methods designed to learn the association between groups of inputs. Often used in clustering techniques.

**self-organizing tree algorithm (SOTA)** A classification method that monitors the divergence of the data in each partition group, and subdivides any group whose degree of divergence exceeds a specified threshold. One of the few classification methods that can automatically determine an appropriate number of groups to use based on the data.

**semiglobal alignment** An optimally scoring alignment of two sequences that involves the complete length of one sequence but only part of the other. The unaligned end(s) of the other sequence are not included in the score.

**semi-Markov model, HMM with duration,** or **explicit state duration HMM** A hidden Markov model in which the duration of the system in a particular state (or model section) is determined by a separate method. This is typically used to overcome the length dependence that is built in to certain HMMs, in order to give models that produce suitable distributions of exon lengths, for example.

**sense strand** The DNA strand that has the same nucleotide sequence as the transcribed mRNA (apart from T being U).

**sensitivity** (As applied to predictive methods.) This is the fraction of real positives that are predicted to be positives.

**separating hyperplane** A term that relates to support vector machines, and describes the boundary between two distinct classification groups of data once the data have been transformed.

**sequence-to-structure network** The initial section of some secondary structure prediction neural networks, which takes the protein sequence as the input and produces an output that is fed as input to a **structure-to-structure network**, which produces the final prediction.

**serial analysis of gene expression (SAGE)** An experimental method for the measurement of cellular transcipts.

**Shine–Dalgarno sequence** The sequence upstream of a prokaryotic gene where a ribosome binds.

**shuffle test** A test that estimates the accuracy of matches in fold recognition (threading) techniques. In this method the score for a match is compared to that of a randomly shuffled sequence.

**silent states** A term used in hidden Markov models to describe those states that do not emit a residue.

**similarity** (sequence alignment) A measure of sequence relationship from an alignment. Similarity is measured based on matched positions of identical or related amino acids.

**simple sequences** See **low-complexity regions**. Regions in protein sequences that contain a string of simple amino acid composition with overrepresentation of a few residues.

**simulated annealing** A technique in molecular mechanics that is designed to improve the chance of detecting the global energy minimum by initially using high temperatures, such as 1000 K, to generate new conformations and then reducing the temperature to trap the system in a minimum.

**single linkage clustering** A measure of the distance between clusters, which is defined as the shortest distance between items in a group.

**singleton sites** (phylogenetic trees) A term used in the parsimony method to describe an alignment position at which only one sequence has a different base/residue from all the others. Such positions cannot distinguish between alternative tree topologies, but will have an effect on the estimation of branch lengths.

**singular value decomposition (SVD)** A method based on matrix algebra, which can be used to transform data to identify independent components that can be separately analyzed.

**Sippl test** A test to assess the accuracy of potentials used in threading algorithms. The Sippl test, named after Manfred Sippl, calculates a score by aligning the target sequence to all the possible folds in a database without insertion of gaps, including the fold of the target sequence itself. The scores for all the alignments are evaluated and the native score is expressed in terms of standard deviations.

**softmax** (neural networks) A function that is sometimes applied to the signals from the output layer of a neural network to make them sum to 1, enabling interpretation of the signals as probabilities.

**Sov** A measure for assessment of secondary structure prediction methods. This measure uses secondary structure segments rather than individual residues as in $\mathbf{Q_3}$.

**spacer unit** In neural network prediction programs, a spacer unit in the input layer is used to represent the absence of a residue (i.e., a gap) at that position in a sequence.

**speciation duplication inference** (**SDI**) A method of determining the root of a phylogenetic tree based on the identification of gene duplication and speciation events.

**speciation event** A term used for the point at which the population of an existing species divides into two separate groups that subsequently diverge into separate species.

**species tree** A phylogenetic tree representing the evolutionary relationship between species.

**specificity** (As applied to predictive methods.) Usually this is the fraction of real negatives that are predicted to be negatives. However, in the case of some gene prediction methods this measure has been redefined as the fraction of all correct predictions that are true positives.

**splice variants** Alternative mature mRNA resulting from different combinations of exons, which result from cutting out different sets of introns, sometimes also cutting out some exons.

**split** A division of a phylogenetic tree into two parts (**subtrees**) by cutting one of the tree branches. There will be as many different splits of a given tree as there are branches.

**star decomposition** A method of reconstructing phylogenetic trees, which begins with a single internal node connected to all leaves, and by successive steps removes pairs of leaves or a leaf and internal node to create another internal node, until the tree consists only of bifurcated nodes.

**star tree** (phylogenetic tree) A tree in which all the taxa are connected to a single central internal node, such as is sometimes the starting point for tree reconstruction methods such as neighbor-joining.

**start state** The non-emitting state in a hidden Markov model, which must be the initial state visited by all valid paths and begins the path.

**state** (hidden Markov model) A component of a hidden Markov model, which in sequence models might correspond to aligning two residues at a particular alignment position (a match state). States are connected by a set of transitions and may have a set of possible emissions.

**state variables** Parameters in a system biology simulation that are under external control, such as temperature.

**steepest descent method** A technique of function minimization in which the function gradient is calculated and used to identify the direction to move over the function surface.

**step-down Holm** A method for modifying the significance level used in a set of statistical tests, which aims to reduce the incidence of false positives.

**step-up Hochberg** A method for modifying the significance level used in a set of statistical tests, which aims to reduce the incidence of false positives.

**stepwise addition** A technique of generating a phylogenetic tree by adding the data one at a time to generate progressively larger trees with increasing numbers of external nodes until all the data are present.

**steric hindrance** Repulsive interactions between atoms of the same molecule caused by the two atoms being too close together.

**strict consensus tree** A form of **consensus tree** in which branches are only shown which are supported by all of the trees.

**Structured Query Language** (**SQL**) A relational database management computer language.

**structure-to-structure network** The final section of some secondary structure prediction neural networks, which takes the output of the **sequence-to-structure network** as the input and produces the final secondary structure prediction. (In some methods this output is further processed to a final reported prediction.)

**suboptimal alignment** An alignment that has a score that is very near to the optimal alignment. In implementations allowing overlap with the optimal alignment this will indicate regions of the alignment that have several possible alignments. Implementations that prevent overlap with the optimal or other suboptimal alignments will identify alternative regions of the sequences that are similar.

**substitution groups** A set of residues that are found to often be substituted for each other in an alignment. Such groups are used in some sequence pattern identification methods.

**substitution matrix** A matrix containing the information on the frequency of mutation of one residue to another. Widely used substitution matrices are the PAM and the BLOSUM matrices obtained by statistical analysis of observed mutations.

**subtree** A part of a phylogenetic tree obtained by cutting one of the branches. See **split**.

**subtree pruning and regrafting** (**SPR**) A heuristic search algorithm used in reconstruction of phylogenetic trees. It works by breaking off part of the tree and connecting it to another part of the tree. If it finds a better tree, then this new tree is used as a starting tree for further cycles of SPR.

**suffix** The final section of a sequence or data string, including the end.

**suffix tree** A representation of a sequence (or in general any data string) which can be used to efficiently search for substrings.

**sum-of-pairs** (or **SP**) A specific scoring scheme for multiple alignments, which assigns a score to each possible aligned pair and then adds all the scores together.

**superfamily** Frequently occurring domain structures found in unrelated proteins. The best example of this is the TIM domain.

**superkingdom** (or **domain)** A major classification division of life, such as the Archaea, Bacteria, and Eukaryota.

**supervised learning** A type of parameterization of prediction or classification techniques in which a **training dataset** is used. The error in the output can then be determined and used to modify the parameters to reduce the error.

**support vector machine** (**SVM**) A machine-learning method that is used to classify data into one of a few alternative classes, often just two.

**symmetric difference** A measure of the topological difference between two trees, defined as the number of splits present in only one of the two trees.

**synonymous mutation** A substitution or other mutation of protein-coding nucleotide sequence that does not alter the amino acid sequence of the resulting protein.

**syntenic region** A region of a genome which contains a series of genes in a similar order to that found in a region of the genome of a different species, implying that those regions have a common evolutionary ancestry. (Note that other definitions are in use in other fields.)

**systematic error** Error that results from the experimental method and relates to all measurements made, such as due to an instrument being incorrectly calibrated.

***t*-test** A statistical test based on an assumption that measurements are samples of a normal distribution, which is used to determine if the means of two populations are statistically different.

**taxa** (phylogenetic trees) The group of organisms whose data are being analyzed and whose evolutionary history is being reconstructed.

**target protein** A term used for the protein of unknown structure that will be modeled based on a homologous protein of known structure, referred to as the template.

**TATA box** The TATA box is a short nucleotide sequence that binds the TATA-binding protein (TBP), which is an important general transcription factor that is required to initiate **transcription**.

**template protein** The term used for the homologous protein of known structure that is used in protein structure modeling as the template for the protein of unknown structure (the target).

**terminator signal** A nucleotide sequence signal that causes the termination of **transcription**.

**tertiary structure** The three-dimensional structure of a protein.

**test dataset** A dataset that is not used as a training dataset to parameterize a predictive algorithm, but is used to measure the performance accuracy.

**threading** (**fold recognition**) A protein structure modeling procedure that involves substituting the side chains of the query protein at positions along a backbone structure of each of a set of representative protein folds. Each structure so generated is scored, the highest-scoring alignment and fold being predicted to be a good starting model of the query sequence structure.

**torsion angle** Also known as the **dihedral angle** is the angle between the A-B bonds and the C-D bonds of four atoms connected in the order A-B-C-D or between two planes defined as A-B-C and B-C-D. In protein structures torsion angles are defined to be in the range of –180° to +180°. The most commonly referred to torsion angles are those along the backbone between the peptide bonds, written $\psi$ and $\phi$.

**topological families** A group of proteins whose three-dimensional structures have related topologies of secondary structural elements.

**traceback** The second part of a **dynamic programming algorithm**. In this step the best alignment is determined by following the traceback arrows from the matrix element that defines the end of the alignment. The traceback arrows indicate which of the three alternative matrix elements were used in determining the scores in each matrix element.

**training dataset** A dataset that is used to parameterize a predictive algorithm, and for which the predicted property is known.

**trans conformation** Molecular conformation where two atomic groups are on the opposite sides of a reference line, which may be a covalent bond, but in other cases may correspond to a more complex rigid structure. See **cis conformation**.

**transcription** Transcription is the process of producing an RNA molecule from genomic DNA.

**transcription start site** (**TSS**) The location where transcription starts at the 5′ end of a gene sequence.

**transcriptome** The total set of transcripts produced by a specific genome.

**transfer function** See **response function**.

**transfer RNA** (**tRNA**) An RNA molecule involved in the synthesis of peptide chains that is the key component which embodies the genetic code, in that it binds to an RNA codon at one end and to the related amino acid at the other.

**transition** (**1**) A nucleotide mutation in which a purine is substituted by a purine, or a pyrimidine is substituted by a pyrimidine. See also **transversion**.

**transition** (**2**) In the context of hidden Markov models a transition refers to the directed connection between two states. There may be several transitions from a given state with defined probabilities for each transition.

**transition/transversion ratio** The ratio of the number of **transitions** to the number of **transversions**, either as observed in a sequence alignment or as proposed in a model of evolution.

**translation** The process of decoding the mRNA codons and using these to synthesize a protein, catalyzed by the ribosome.

**transmembrane proteins** See **integral proteins**.

**transposons** Segments of DNA sequence that are able to move from one site in a genome to another site within the same genome.

**transversion** A nucleotide mutation in which a purine is substituted by a pyrimidine, or vice versa. See also **transition**.

**tree bisection and reconnection (TBR)** An algorithm to search through the space of possible tree topologies in that a phylogenetic tree is separated into two parts which are then reconnected at all possible branches. Each new tree is scored to identify those that are an improved fit to the data.

**tree topology** That part of the definition of the structure of a tree that is only concerned with which nodes are connected and ignores the branch lengths.

**twilight zone** When the percentage identity in a protein sequence alignment falls below about 25%, unless the sequences are very long the alignment cannot be taken as clear evidence that the sequences are related. This region is often referred to as the twilight zone.

**two-dimensional gel electrophoresis** An experimental method that is used to separate proteins in two dimensions according to their molecular weight and charge.

**two-tailed test** A statistical test that is concerned with determining if the test statistic is significantly different from the null hypothesis. See **one-tailed test.**

**type I error** See **false positive**.

**ultrametric tree** This is a rooted additive phylogenetic tree where the terminal nodes are all equally distant from the root, implying a constant rate of mutation.

**unit** (neural network) A component of a neural network, which receives input signals from one or more sources, and uses the input to generate an output signal that is sent to one or more units or is an output signal of the network.

**unrooted tree** A phylogenetic tree in which the last common ancestor has not been identified, so that the direction of time is uncertain along the branches.

**unweighted parsimony** A method of reconstructing phylogenetic trees in which the score used for alternative trees is the number of mutations that must be proposed to generate the data from the tree.

**UPGMA** An abbreviation for unweighted pair group method using arithmetic averaging, this is a common method of clustering used in both phylogenetic tree reconstruction and as a general data classification technique.

**upstream sequences** This term refers to anything before the 5′ or N-terminal end of a particular sequence, often relating to specific sequence signals that have some functional connection to the particular sequence. See also **downstream sequences**.

**URL** An abbreviation for uniform resource locator, a technical term used in the World Wide Web, which gives the location of a specific page, or other resources.

**variable region** Structurally or sequentially variable segment, based on a comparison of homologous proteins.

**virus** An organism that can only reproduce within the cell of a host and is the cause of many diseases.

**Viterbi algorithm** A method for obtaining the most probable path (that with the highest score) for a hidden Markov model that emits a particular sequence.

**Watson–Crick base pair** The structure that occurs in the standard double-helical DNA structure, in which two bases—either cytosine and guanine, or thymine and adenine—interact via hydrogen bonds and are coplanar. In the case of RNA, adenine pairs with uracil.

**weight sharing** (neural networks) A technique used in neural networks to reduce the number of parameters by using the same weight parameter for a group of units.

**window** A sliding segment of a sequence that is analyzed in a single stage by various programs. Sometimes this is in an attempt to include the modifying effects of sequentially close residues on properties of the central residue. It may also be used to determine average properties across the window.

**winner takes all** (neural networks) The method by which the final prediction of a neural network with several output layer units is determined to correspond to that unit which has the largest signal.

**wobble base-pairing** The ability of a single tRNA to bind to more than one codon due to different ways the third base in an **anticodon** can form hydrogen bonds.

***X*-drop method** A technique in dynamic programming sequence alignment, which results in faster calculations by not calculating matrix elements beyond any that have a score $X$ less than the current best score.

**xenologous** Xenologous applies to homologous genes obtained by horizontal transfer of genetic material between different species.

***z*-statistic** A test statistic that measures the deviation of an observation from the expected normal distribution mean in units of the number of standard deviations of the distribution. Also known as a $z$-score.

***z*-test** A statistical test based on an assumption of measurements being samples from a normal distribution, which calculates the $z$-statistic and uses it to assess the significance of the deviation of a measurement from the expected distribution.

**Zvelebil conservation number ($C_n$)** A score that measures sequence conservation in an alignment based on physicochemical properties of the amino acids.

# INDEX

Note: Entries which are simply page numbers refer to the main text. Other entries have the following abbreviations immediately afer the page number: B, box; F, figure; FD, flow diagram; MM, mind map; T, table.

## D

## E

## L



## M

## O



## P

## T

## U



## V



## W



## X



## Y



## Z